The GALE ENCYCLOPEDIA of SCIENCE

THIRD EDITION

The GALE ENCYCLOPEDIA of SCIENCE

THIRD EDITION

VOLUME 2
Charge-coupled device – Eye

K. Lee Lerner and
Brenda Wilmoth Lerner,
Editors

GALE®

THOMSON
★
GALE

Detroit • New York • San Diego • San Francisco • Cleveland • New Haven, Conn. • Waterville, Maine • London • Munich

Gale Encyclopedia of Science, Third Edition

K. Lee Lerner and Brenda Wilmoth Lerner, Editors

Project Editor
Kimberley A. McGrath

Editorial
Deirdre S. Blanchfield, Chris Jeryan, Jacqueline Longe, Mark Springer

Editorial Support Services
Andrea Lopeman

Indexing Services
Synapse

Permissions
Shalice Shah-Caldwell

Imaging and Multimedia
Leitha Etheridge-Sims, Lezlie Light, Dave Oblender, Christine O'Brien, Robyn V. Young

Product Design
Michelle DiMercurio

Manufacturing
Wendy Blurton, Evi Seoud

LIBRARY OF CONGRESS CATALOGING-IN-PUBLICATION DATA

Gale encyclopedia of science / K. Lee Lerner & Brenda Wilmoth Lerner, editors.— 3rd ed.
  p. cm.
    Includes index.
    ISBN 0-7876-7554-7 (set) — ISBN 0-7876-7555-5 (v. 1) — ISBN 0-7876-7556-3 (v. 2) — ISBN 0-7876-7557-1 (v. 3) — ISBN 0-7876-7558-X (v. 4) — ISBN 0-7876-7559-8 (v. 5) — ISBN 0-7876-7560-1 (v. 6)
    1. Science—Encyclopedias. I. Lerner, K. Lee. II. Lerner, Brenda Wilmoth.
Q121.G37 2004
503—dc22
                                                    2003015731

This title is also available as an e-book.
ISBN: 0-7876-7776-0 (set)
Contact your Gale sales representative for ordering information.

Printed in Canada
10 9 8 7 6 5 4 3 2 1

CONTENTS

TOPIC LIST

A

Aardvark
Abacus
Abrasives
Abscess
Absolute zero
Abyssal plain
Acceleration
Accelerators
Accretion disk
Accuracy
Acetic acid
Acetone
Acetylcholine
Acetylsalicylic acid
Acid rain
Acids and bases
Acne
Acorn worm
Acoustics
Actinides
Action potential
Activated complex
Active galactic nuclei
Acupressure
Acupuncture
ADA (adenosine deaminase)
   deficiency
Adaptation
Addiction
Addison's disease
Addition
Adenosine diphosphate
Adenosine triphosphate
Adhesive

Adrenals
Aerobic
Aerodynamics
Aerosols
Africa
Age of the universe
Agent Orange
Aging and death
Agouti
Agricultural machines
Agrochemicals
Agronomy
AIDS
AIDS therapies and vaccines
Air masses and fronts
Air pollution
Aircraft
Airship
Albatrosses
Albedo
Albinism
Alchemy
Alcohol
Alcoholism
Aldehydes
Algae
Algebra
Algorithm
Alkali metals
Alkaline earth metals
Alkaloid
Alkyl group
Alleles
Allergy
Allotrope
Alloy

Alluvial systems
Alpha particle
Alternative energy sources
Alternative medicine
Altruism
Aluminum
Aluminum hydroxide
Alzheimer disease
Amaranth family (Amaranthaceae)
Amaryllis family (Amaryllidaceae)
American Standard Code for
   Information Interchange
Ames test
Amicable numbers
Amides
Amino acid
Ammonia
Ammonification
Amnesia
Amniocentesis
Amoeba
Amphetamines
Amphibians
Amplifier
Amputation
Anabolism
Anaerobic
Analemma
Analgesia
Analog signals and digital signals
Analytic geometry
Anaphylaxis
Anatomy
Anatomy, comparative
Anchovy
Anemia

Anesthesia
Aneurism
Angelfish
Angiography
Angiosperm
Angle
Anglerfish
Animal
Animal breeding
Animal cancer tests
Anion
Anode
Anoles
Ant-pipits
Antarctica
Antbirds and gnat-eaters
Anteaters
Antelopes and gazelles
Antenna
Anthrax
Anthropocentrism
Anti-inflammatory agents
Antibiotics
Antibody and antigen
Anticoagulants
Anticonvulsants
Antidepressant drugs
Antihelmintics
Antihistamines
Antimatter
Antimetabolites
Antioxidants
Antiparticle
Antipsychotic drugs
Antisepsis
Antlions
Ants
Anxiety
Apes
Apgar score
Aphasia
Aphids
Approximation
Apraxia
Aqueduct
Aquifer
Arachnids
Arapaima

Arc
ARC LAMP
Archaebacteria
Archaeoastronomy
Archaeogenetics
Archaeology
Archaeometallurgy
Archaeometry
Archeological mapping
Archeological sites
Arithmetic
Armadillos
Arrow worms
Arrowgrass
Arrowroot
Arteries
Arteriosclerosis
Arthritis
Arthropods
Arthroscopic surgery
Artifacts and artifact classification
Artificial fibers
Artificial heart and heart valve
Artificial intelligence
Artificial vision
Arum family (Araceae)
Asbestos
Asexual reproduction
Asia
Assembly line
Asses
Associative property
Asteroid 2002AA29
Asthenosphere
Asthma
Astrobiology
Astroblemes
Astrolabe
Astrometry
Astronomical unit
Astronomy
Astrophysics
Atmosphere, composition and
   structure
Atmosphere observation
Atmospheric circulation
Atmospheric optical phenomena
Atmospheric pressure

Atmospheric temperature
Atomic clock
Atomic models
Atomic number
Atomic spectroscopy
Atomic theory
Atomic weight
Atoms
Attention-deficit/Hyperactivity
   disorder (ADHD)
Auks
Australia
Autism
Autoimmune disorders
Automatic pilot
Automation
Automobile
Autotroph
Avogadro's number
Aye-ayes

B

Babblers
Baboons
Bacteria
Bacteriophage
Badgers
Ball bearing
Ballistic missiles
Ballistics
Balloon
Banana
Bandicoots
Bar code
Barberry
Barbets
Barbiturates
Bariatrics
Barium
Barium sulfate
Bark
Barley
Barnacles
Barometer
Barracuda

C

Calibration

Caliper

Calorie

Calorimetry

Camels

Canal

Cancel

Cancer

Canines

Cantilever

Capacitance

Capacitor

Capillaries

Capillary action

Caprimulgids

Captive breeding and reintroduction

Capuchins

Capybaras

Carbohydrate

Carbon

Carbon cycle

Carbon dioxide

Carbon monoxide

Carbon tetrachloride

Carbonyl group

Carboxyl group

Carboxylic acids

Carcinogen

Cardiac cycle

Cardinal number

Cardinals and grosbeaks

Caribou

Carnivore

Carnivorous plants

Carp

Carpal tunnel syndrome

Carrier (genetics)

Carrot family (Apiaceae)

Carrying capacity

Cartesian coordinate plane

Cartilaginous fish

Cartography

Cashew family (Anacardiaceae)

Cassini Spacecraft

Catabolism

Catalyst and catalysis

Catastrophism

Catfish

Catheters

Cathode

Cathode ray tube

Cation

Cats

Cattails

Cattle family (Bovidae)

Cauterization

Cave

Cave fish

Celestial coordinates

Celestial mechanics

Celestial sphere: The apparent motions of the Sun, Moon, planets, and stars

Cell

Cell death

Cell division

Cell, electrochemical

Cell membrane transport

Cell staining

Cellular respiration

Cellular telephone

Cellulose

Centipedes

Centrifuge

Ceramics

Cerenkov effect

Cetaceans

Chachalacas

Chameleons

Chaos

Charge-coupled device

Chelate

Chemical bond

Chemical evolution

Chemical oxygen demand

Chemical reactions

Chemical warfare

Chemistry

Chemoreception

Chestnut

Chi-square test

Chickenpox

Childhood diseases

Chimaeras

Chimpanzees

Chinchilla

Chipmunks

Chitons

Chlordane

Chlorinated hydrocarbons

Chlorination

Chlorine

Chlorofluorocarbons (CFCs)

Chloroform

Chlorophyll

Chloroplast

Cholera

Cholesterol

Chordates

Chorionic villus sampling (CVS)

Chromatin

Chromatography

Chromosomal abnormalities

Chromosome

Chromosome mapping

Cicadas

Cigarette smoke

Circle

Circulatory system

Circumscribed and inscribed

Cirrhosis

Citric acid

Citrus trees

Civets

Climax (ecological)

Clingfish

Clone and cloning

Closed curves

Closure property

Clouds

Club mosses

Coal

Coast and beach

Coatis

Coca

Cocaine

Cockatoos

Cockroaches

Codeine

Codfishes

Codons

Coefficient

Coelacanth

Coffee plant
Cogeneration
Cognition
Cold, common
Collagen
Colloid
Colobus monkeys
Color
Color blindness
Colugos
Coma
Combinatorics
Combustion
Comet Hale-Bopp
Comets
Commensalism
Community ecology
Commutative property
Compact disc
Competition
Complementary DNA
Complex
Complex numbers
Composite family
Composite materials
Composting
Compound, chemical
Compton effect
Compulsion
Computer, analog
Computer, digital
Computer languages
Computer memory, physical and virtual memory
Computer software
Computer virus
Computerized axial tomography
Concentration
Concrete
Conditioning
Condors
Congenital
Congruence (triangle)
Conic sections
Conifer
Connective tissue
Conservation
Conservation laws

Constellation
Constructions
Contaminated soil
Contamination
Continent
Continental drift
Continental margin
Continental shelf
Continuity
Contour plowing
Contraception
Convection
Coordination compound
Copepods
Copper
Coral and coral reef
Coriolis effect
Cork
Corm
Cormorants
Corn (maize)
Coronal ejections and magnetic storms
Correlation (geology)
Correlation (mathematics)
Corrosion
Cosmic background radiation
Cosmic ray
Cosmology
Cotingas
Cotton
Coulomb
Countable
Coursers and pratincoles
Courtship
Coypu
Crabs
Crane
Cranes
Crayfish
Crestfish
Creutzfeldt-Jakob disease
Crickets
Critical habitat
Crocodiles
Crop rotation
Crops
Cross multiply

Cross section
Crows and jays
Crustacea
Cryobiology
Cryogenics
Cryptography, encryption, and number theory
Crystal
Cubic equations
Cuckoos
Curare
Curlews
Currents
Curve
Cushing syndrome
Cuttlefish
Cybernetics
Cycads
Cyclamate
Cyclone and anticyclone
Cyclosporine
Cyclotron
Cystic fibrosis
Cytochrome
Cytology

D

Dams
Damselflies
Dark matter
Dating techniques
DDT (Dichlorodiphenyl-trichloroacetic acid)
Deafness and inherited hearing loss
Decimal fraction
Decomposition
Deer
Deer mouse
Deforestation
Degree
Dehydroepiandrosterone (DHEA)
Delta
Dementia
Dengue fever
Denitrification

Density
Dentistry
Deoxyribonucleic acid (DNA)
Deposit
Depression
Depth perception
Derivative
Desalination
Desert
Desertification
Determinants
Deuterium
Developmental processes
Dew point
Diabetes mellitus
Diagnosis
Dialysis
Diamond
Diatoms
Dielectric materials
Diesel engine
Diethylstilbestrol (DES)
Diffraction
Diffraction grating
Diffusion
Digestive system
Digital Recording
Digitalis
Dik-diks
Dinosaur
Diode
Dioxin
Diphtheria
Dipole
Direct variation
Disease
Dissociation
Distance
Distillation
Distributive property
Disturbance, ecological
Diurnal cycles
Division
DNA fingerprinting
DNA replication
DNA synthesis
DNA technology
DNA vaccine

Dobsonflies
Dogwood tree
Domain
Donkeys
Dopamine
Doppler effect
Dories
Dormouse
Double-blind study
Double helix
Down syndrome
Dragonflies
Drift net
Drongos
Drosophila melanogaster
Drought
Ducks
Duckweed
Duikers
Dune
Duplication of the cube
Dust devil
DVD
Dwarf antelopes
Dyes and pigments
Dysentery
Dyslexia
Dysplasia
Dystrophinopathies

E

e (number)
Eagles
Ear
Earth
Earth science
Earth's interior
Earth's magnetic field
Earth's rotation
Earthquake
Earwigs
Eating disorders
Ebola virus
Ebony
Echiuroid worms

Echolocation
Eclipses
Ecological economics
Ecological integrity
Ecological monitoring
Ecological productivity
Ecological pyramids
Ecology
Ecosystem
Ecotone
Ecotourism
Edema
Eel grass
El Niño and La Niña
Eland
Elapid snakes
Elasticity
Electric arc
Electric charge
Electric circuit
Electric conductor
Electric current
Electric motor
Electric vehicles
Electrical conductivity
Electrical power supply
Electrical resistance
Electricity
Electrocardiogram (ECG)
Electroencephalogram (EEG)
Electrolysis
Electrolyte
Electromagnetic field
Electromagnetic induction
Electromagnetic spectrum
Electromagnetism
Electromotive force
Electron
Electron cloud
Electronics
Electrophoresis
Electrostatic devices
Element, chemical
Element, families of
Element, transuranium
Elements, formation of
Elephant
Elephant shrews

Formula, structural
Fossa
Fossil and fossilization
Fossil fuels
Fractal
Fraction, common
Fraunhofer lines
Freeway
Frequency
Freshwater
Friction
Frigate birds
Frog's-bit family
Frogs
Frostbite
Fruits
Fuel cells
Function
Fundamental theorems
Fungi
Fungicide

G

Gaia hypothesis
Galaxy
Game theory
Gamete
Gametogenesis
Gamma-ray astronomy
Gamma ray burst
Gangrene
Garpike
Gases, liquefaction of
Gases, properties of
Gazelles
Gears
Geckos
Geese
Gelatin
Gene
Gene chips and microarrays
Gene mutation
Gene splicing
Gene therapy
Generator

Genetic disorders
Genetic engineering
Genetic identification of
   microorganisms
Genetic testing
Genetically modified foods and
   organisms
Genetics
Genets
Genome
Genomics (comparative)
Genotype and phenotype
Geocentric theory
Geochemical analysis
Geochemistry
Geode
Geodesic
Geodesic dome
Geographic and magnetic poles
Geologic map
Geologic time
Geology
Geometry
Geomicrobiology
Geophysics
Geotropism
Gerbils
Germ cells and the germ cell line
Germ theory
Germination
Gerontology
Gesnerias
Geyser
Gibbons and siamangs
Gila monster
Ginger
Ginkgo
Ginseng
Giraffes and okapi
GIS
Glaciers
Glands
Glass
Global climate
Global Positioning System
Global warming
Glycerol
Glycol

Glycolysis
Goats
Goatsuckers
Gobies
Goldenseal
Gophers
Gorillas
Gourd family (Cucurbitaceae)
Graft
Grand unified theory
Grapes
Graphs and graphing
Grasses
Grasshoppers
Grasslands
Gravitational lens
Gravity and gravitation
Great Barrier Reef
Greatest common factor
Grebes
Greenhouse effect
Groundhog
Groundwater
Group
Grouse
Growth and decay
Growth hormones
Guenons
Guillain-Barre syndrome
Guinea fowl
Guinea pigs and cavies
Gulls
Guppy
Gutenberg discontinuity
Gutta percha
Gymnosperm
Gynecology
Gyroscope

H

Habitat
Hagfish
Half-life
Halide, organic
Hall effect

Halley's comet
Hallucinogens
Halogenated hydrocarbons
Halogens
Halosaurs
Hamsters
Hand tools
Hantavirus infections
Hard water
Harmonics
Hartebeests
Hawks
Hazardous wastes
Hazel
Hearing
Heart
Heart diseases
Heart, embryonic development and changes at birth
Heart-lung machine
Heat
Heat capacity
Heat index
Heat transfer
Heath family (Ericaceae)
Hedgehogs
Heisenberg uncertainty principle
Heliocentric theory
Hematology
Hemophilia
Hemorrhagic fevers and diseases
Hemp
Henna
Hepatitis
Herb
Herbal medicine
Herbicides
Herbivore
Hermaphrodite
Hernia
Herons
Herpetology
Herrings
Hertzsprung-Russell diagram
Heterotroph
Hibernation
Himalayas, geology of
Hippopotamuses

Histamine
Historical geology
Hoatzin
Hodgkin's disease
Holly family (Aquifoliaceae)
Hologram and holography
Homeostasis
Honeycreepers
Honeyeaters
Hoopoe
Horizon
Hormones
Hornbills
Horse chestnut
Horsehair worms
Horses
Horseshoe crabs
Horsetails
Horticulture
Hot spot
Hovercraft
Hubble Space Telescope
Human artificial chromosomes
Human chorionic gonadotropin
Human cloning
Human ecology
Human evolution
Human Genome Project
Humidity
Hummingbirds
Humus
Huntington disease
Hybrid
Hydra
Hydrocarbon
Hydrocephalus
Hydrochlorofluorocarbons
Hydrofoil
Hydrogen
Hydrogen chloride
Hydrogen peroxide
Hydrogenation
Hydrologic cycle
Hydrology
Hydrolysis
Hydroponics
Hydrosphere
Hydrothermal vents

Hydrozoa
Hyena
Hyperbola
Hypertension
Hypothermia
Hyraxes

I

Ibises
Ice
Ice age refuges
Ice ages
Icebergs
Iceman
Identity element
Identity property
Igneous rocks
Iguanas
Imaginary number
Immune system
Immunology
Impact crater
Imprinting
In vitro fertilization (IVF)
In vitro and in vivo
Incandescent light
Incineration
Indicator, acid-base
Indicator species
Individual
Indoor air quality
Industrial minerals
Industrial Revolution
Inequality
Inertial guidance
Infection
Infertility
Infinity
Inflammation
Inflection point
Influenza
Infrared astronomy
Inherited disorders
Insecticides
Insectivore

Insects
Insomnia
Instinct
Insulin
Integers
Integral
Integrated circuit
Integrated pest management
Integumentary system
Interference
Interferometry
Interferons
Internal combustion engine
International Space Station
International Ultraviolet Explorer
Internet file transfer and tracking
Internet and the World Wide Web
Interstellar matter
Interval
Introduced species
Invariant
Invasive species
Invertebrates
Ion and ionization
Ion exchange
Ionizing radiation
Iris family
Iron
Irrational number
Irrigation
Island
Isobars
Isomer
Isostasy
Isotope
Isthmus
Iteration

J

Jacanas
Jacks
Jaundice
Jellyfish
Jerboas
Jet engine

Jet stream
Juniper
Jupiter

K

K-T event (Cretaceous-Tertiary event)
Kangaroo rats
Kangaroos and wallabies
Karst topography
Karyotype and karyotype analysis
Kelp forests
Kepler's laws
Keystone species
Killifish
Kingfishers
Kinglets
Koalas
Kola
Korsakoff's syndrome
Krebs cycle
Kuiper belt objects
Kuru

L

Lacewings
Lactic acid
Lagomorphs
Lake
Lamarckism
Lampreys and hagfishes
Land and sea breezes
Land use
Landfill
Landform
Langurs and leaf monkeys
Lantern fish
Lanthanides
Larks
Laryngitis
Laser
Laser surgery
Latitude and longitude

Laurel family (Lauraceae)
Laws of motion
LCD
Leaching
Lead
Leaf
Leafhoppers
Learning
Least common denominator
Lecithin
LED
Legionnaires' disease
Legumes
Lemmings
Lemurs
Lens
Leprosy
Leukemia
Lewis structure
Lice
Lichens
Life history
Ligand
Light
Light-year
Lightning
Lilac
Lily family (Liliaceae)
Limit
Limiting factor
Limpets
Line, equations of
Linear algebra
Lipid
Liquid crystals
Lithium
Lithography
Lithosphere
Lithotripsy
Liverwort
Livestock
Lobsters
Lock
Lock and key
Locus
Logarithms
Loons
LORAN

Lorises
Luminescence
Lungfish
Lycophytes
Lyme disease
Lymphatic system
Lyrebirds

M

Macaques
Mach number
Machine tools
Machine vision
Machines, simple
Mackerel
Magic square
Magma
Magnesium
Magnesium sulfate
Magnetic levitation
Magnetic recording/audiocassette
Magnetic resonance imaging (MRI)
Magnetism
Magnetosphere
Magnolia
Mahogany
Maidenhair fern
Malaria
Malnutrition
Mammals
Manakins
Mangrove tree
Mania
Manic depression
Map
Maples
Marfan syndrome
Marijuana
Marlins
Marmosets and tamarins
Marmots
Mars
Mars Pathfinder
Marsupial cats
Marsupial rats and mice

Marsupials
Marten, sable, and fisher
Maser
Mass
Mass extinction
Mass number
Mass production
Mass spectrometry
Mass transportation
Mass wasting
Mathematics
Matrix
Matter
Maunder minimum
Maxima and minima
Mayflies
Mean
Median
Medical genetics
Meiosis
Membrane
Memory
Mendelian genetics
Meningitis
Menopause
Menstrual cycle
Mercurous chloride
Mercury (element)
Mercury (planet)
Mesoscopic systems
Mesozoa
Metabolic disorders
Metabolism
Metal
Metal fatigue
Metal production
Metallurgy
Metamorphic grade
Metamorphic rock
Metamorphism
Metamorphosis
Meteorology
Meteors and meteorites
Methyl group
Metric system
Mice
Michelson-Morley experiment
Microbial genetics

Microclimate
Microorganisms
Microscope
Microscopy
Microtechnology
Microwave communication
Migraine headache
Migration
Mildew
Milkweeds
Milky Way
Miller-Urey Experiment
Millipedes
Mimicry
Mineralogy
Minerals
Mining
Mink
Minnows
Minor planets
Mint family
Mir Space Station
Mirrors
Miscibility
Mistletoe
Mites
Mitosis
Mixture, chemical
Möbius strip
Mockingbirds and thrashers
Mode
Modular arithmetic
Mohs' scale
Mold
Mole
Mole-rats
Molecular biology
Molecular formula
Molecular geometry
Molecular weight
Molecule
Moles
Mollusks
Momentum
Monarch flycatchers
Mongooses
Monitor lizards
Monkeys

Organic farming
Organism
Organogenesis
Organs and organ systems
Origin of life
Orioles
Ornithology
Orthopedics
Oryx
Oscillating reactions
Oscillations
Oscilloscope
Osmosis
Osmosis (cellular)
Ossification
Osteoporosis
Otter shrews
Otters
Outcrop
Ovarian cycle and hormonal
    regulation
Ovenbirds
Oviparous
Ovoviviparous
Owls
Oxalic acid
Oxidation-reduction reaction
Oxidation state
Oxygen
Oystercatchers
Ozone
Ozone layer depletion

P

Pacemaker
Pain
Paleobotany
Paleoclimate
Paleoecology
Paleomagnetism
Paleontology
Paleopathology
Palindrome
Palms
Palynology

Pandas
Pangolins
Papaya
Paper
Parabola
Parallax
Parallel
Parallelogram
Parasites
Parity
Parkinson disease
Parrots
Parthenogenesis
Particle detectors
Partridges
Pascal's triangle
Passion flower
Paternity and parentage testing
Pathogens
Pathology
PCR
Peafowl
Peanut worms
Peccaries
Pedigree analysis
Pelicans
Penguins
Peninsula
Pentyl group
Peony
Pepper
Peptide linkage
Percent
Perception
Perch
Peregrine falcon
Perfect numbers
Periodic functions
Periodic table
Permafrost
Perpendicular
Pesticides
Pests
Petrels and shearwaters
Petroglyphs and pictographs
Petroleum
pH
Phalangers

Pharmacogenetics
Pheasants
Phenyl group
Phenylketonuria
Pheromones
Phlox
Phobias
Phonograph
Phoronids
Phosphoric acid
Phosphorus
Phosphorus cycle
Phosphorus removal
Photic zone
Photochemistry
Photocopying
Photoelectric cell
Photoelectric effect
Photography
Photography, electronic
Photon
Photosynthesis
Phototropism
Photovoltaic cell
Phylogeny
Physical therapy
Physics
Physiology
Physiology, comparative
Phytoplankton
Pi
Pigeons and doves
Pigs
Pike
Piltdown hoax
Pinecone fish
Pines
Pipefish
Placebo
Planck's constant
Plane
Plane family
Planet
Planet X
Planetary atmospheres
Planetary geology
Planetary nebulae
Planetary ring systems

Plankton
Plant
Plant breeding
Plant diseases
Plant pigment
Plasma
Plastic surgery
Plastics
Plate tectonics
Platonic solids
Platypus
Plovers
Pluto
Pneumonia
Podiatry
Point
Point source
Poisons and toxins
Polar coordinates
Polar ice caps
Poliomyelitis
Pollen analysis
Pollination
Pollution
Pollution control
Polybrominated biphenyls (PBBs)
Polychlorinated biphenyls (PCBs)
Polycyclic aromatic hydrocarbons
Polygons
Polyhedron
Polymer
Polynomials
Poppies
Population growth and control
   (human)
Population, human
Porcupines
Positive number
Positron emission tomography
   (PET)
Postulate
Potassium aluminum sulfate
Potassium hydrogen tartrate
Potassium nitrate
Potato
Pottery analysis
Prairie
Prairie chicken

Prairie dog
Prairie falcon
Praying mantis
Precession of the equinoxes
Precious metals
Precipitation
Predator
Prenatal surgery
Prescribed burn
Pressure
Prey
Primates
Prime numbers
Primroses
Printing
Prions
Prism
Probability theory
Proboscis monkey
Projective geometry
Prokaryote
Pronghorn
Proof
Propyl group
Prosimians
Prosthetics
Proteas
Protected area
Proteins
Proteomics
Protista
Proton
Protozoa
Psychiatry
Psychoanalysis
Psychology
Psychometry
Psychosis
Psychosurgery
Puberty
Puffbirds
Puffer fish
Pulsar
Punctuated equilibrium
Pyramid
Pythagorean theorem
Pythons

Q

Quadrilateral
Quail
Qualitative analysis
Quantitative analysis
Quantum computing
Quantum electrodynamics (QED)
Quantum mechanics
Quantum number
Quarks
Quasar
Quetzal
Quinine

R

Rabies
Raccoons
Radar
Radial keratotomy
Radiation
Radiation detectors
Radiation exposure
Radical (atomic)
Radical (math)
Radio
Radio astronomy
Radio waves
Radioactive dating
Radioactive decay
Radioactive fallout
Radioactive pollution
Radioactive tracers
Radioactive waste
Radioisotopes in medicine
Radiology
Radon
Rails
Rainbows
Rainforest
Random
Rangeland
Raptors
Rare gases
Rare genotype advantage

Rate
Ratio
Rational number
Rationalization
Rats
Rayleigh scattering
Rays
Real numbers
Reciprocal
Recombinant DNA
Rectangle
Recycling
Red giant star
Red tide
Redshift
Reflections
Reflex
Refrigerated trucks and railway cars
Rehabilitation
Reinforcement, positive and negative
Relation
Relativity, general
Relativity, special
Remote sensing
Reproductive system
Reproductive toxicant
Reptiles
Resins
Resonance
Resources, natural
Respiration
Respiration, cellular
Respirator
Respiratory diseases
Respiratory system
Restoration ecology
Retrograde motion
Retrovirus
Reye's syndrome
Rh factor
Rhesus monkeys
Rheumatic fever
Rhinoceros
Rhizome
Rhubarb
Ribbon worms
Ribonuclease

Ribonucleic acid (RNA)
Ribosomes
Rice
Ricin
Rickettsia
Rivers
RNA function
RNA splicing
Robins
Robotics
Rockets and missiles
Rocks
Rodents
Rollers
Root system
Rose family (Rosaceae)
Rotation
Roundworms
Rumination
Rushes
Rusts and smuts

S

Saiga antelope
Salamanders
Salmon
Salmonella
Salt
Saltwater
Sample
Sand
Sand dollars
Sandfish
Sandpipers
Sapodilla tree
Sardines
Sarin gas
Satellite
Saturn
Savanna
Savant
Sawfish
Saxifrage family
Scalar
Scale insects

Scanners, digital
Scarlet fever
Scavenger
Schizophrenia
Scientific method
Scorpion flies
Scorpionfish
Screamers
Screwpines
Sculpins
Sea anemones
Sea cucumbers
Sea horses
Sea level
Sea lily
Sea lions
Sea moths
Sea spiders
Sea squirts and salps
Sea urchins
Seals
Seamounts
Seasonal winds
Seasons
Secondary pollutants
Secretary bird
Sedges
Sediment and sedimentation
Sedimentary environment
Sedimentary rock
Seed ferns
Seeds
Segmented worms
Seismograph
Selection
Sequences
Sequencing
Sequoia
Servomechanisms
Sesame
Set theory
SETI
Severe acute respiratory syndrome (SARS)
Sewage treatment
Sewing machine
Sex change
Sextant

Sexual reproduction
Sexually transmitted diseases
Sharks
Sheep
Shell midden analysis
Shingles
Shore birds
Shoreline protection
Shotgun cloning
Shrews
Shrikes
Shrimp
Sickle cell anemia
Sieve of Eratosthenes
Silicon
Silk cotton family (Bombacaceae)
Sinkholes
Skates
Skeletal system
Skinks
Skuas
Skunks
Slash-and-burn agriculture
Sleep
Sleep disorders
Sleeping sickness
Slime molds
Sloths
Slugs
Smallpox
Smallpox vaccine
Smell
Smog
Snails
Snakeflies
Snakes
Snapdragon family
Soap
Sociobiology
Sodium
Sodium benzoate
Sodium bicarbonate
Sodium carbonate
Sodium chloride
Sodium hydroxide
Sodium hypochlorite
Soil
Soil conservation

Solar activity cycle
Solar flare
Solar illumination: Seasonal and diurnal patterns
Solar prominence
Solar system
Solar wind
Solder and soldering iron
Solstice
Solubility
Solution
Solution of equation
Sonar
Song birds
Sonoluminescence
Sorghum
Sound waves
South America
Soybean
Space
Space probe
Space shuttle
Spacecraft, manned
Sparrows and buntings
Species
Spectral classification of stars
Spectral lines
Spectroscope
Spectroscopy
Spectrum
Speech
Sphere
Spider monkeys
Spiderwort family
Spin of subatomic particles
Spina bifida
Spinach
Spiny anteaters
Spiny eels
Spiny-headed worms
Spiral
Spirometer
Split-brain functioning
Sponges
Spontaneous generation
Spore
Springtails
Spruce

Spurge family
Square
Square root
Squid
Squirrel fish
Squirrels
Stalactites and stalagmites
Standard model
Star
Star cluster
Star formation
Starburst galaxy
Starfish
Starlings
States of matter
Statistical mechanics
Statistics
Steady-state theory
Steam engine
Steam pressure sterilizer
Stearic acid
Steel
Stellar evolution
Stellar magnetic fields
Stellar magnitudes
Stellar populations
Stellar structure
Stellar wind
Stem cells
Stereochemistry
Sticklebacks
Stilts and avocets
Stimulus
Stone and masonry
Stoneflies
Storks
Storm
Storm surge
Strata
Stratigraphy
Stratigraphy (archeology)
Stream capacity and competence
Stream valleys, channels, and floodplains
Strepsiptera
Stress
Stress, ecological
String theory

Stroke
Stromatolite
Sturgeons
Subatomic particles
Submarine
Subsidence
Subsurface detection
Subtraction
Succession
Suckers
Sudden infant death syndrome (SIDS)
Sugar beet
Sugarcane
Sulfur
Sulfur cycle
Sulfur dioxide
Sulfuric acid
Sun
Sunbirds
Sunspots
Superclusters
Superconductor
Supernova
Surface tension
Surgery
Surveying instruments
Survival of the fittest
Sustainable development
Swallows and martins
Swamp cypress family (Taxodiaceae)
Swamp eels
Swans
Sweet gale family (Myricaceae)
Sweet potato
Swifts
Swordfish
Symbiosis
Symbol, chemical
Symbolic logic
Symmetry
Synapse
Syndrome
Synthesis, chemical
Synthesizer, music
Synthesizer, voice
Systems of equations

T

T cells
Tanagers
Taphonomy
Tapirs
Tarpons
Tarsiers
Tartaric acid
Tasmanian devil
Taste
Taxonomy
Tay-Sachs disease
Tea plant
Tectonics
Telegraph
Telemetry
Telephone
Telescope
Television
Temperature
Temperature regulation
Tenrecs
Teratogen
Term
Termites
Terns
Terracing
Territoriality
Tetanus
Tetrahedron
Textiles
Thalidomide
Theorem
Thermal expansion
Thermochemistry
Thermocouple
Thermodynamics
Thermometer
Thermostat
Thistle
Thoracic surgery
Thrips
Thrombosis
Thrushes
Thunderstorm
Tides

Time
Tinamous
Tissue
Tit family
Titanium
Toadfish
Toads
Tomato family
Tongue worms
Tonsillitis
Topology
Tornado
Torque
Torus
Total solar irradiance
Toucans
Touch
Towers of Hanoi
Toxic shock syndrome
Toxicology
Trace elements
Tragopans
Trains and railroads
Tranquilizers
Transcendental numbers
Transducer
Transformer
Transgenics
Transistor
Transitive
Translations
Transpiration
Transplant, surgical
Trapezoid
Tree
Tree shrews
Trichinosis
Triggerfish
Triglycerides
Trigonometry
Tritium
Trogons
Trophic levels
Tropic birds
Tropical cyclone
Tropical diseases
Trout-perch
True bugs

True eels
True flies
Trumpetfish
Tsunami
Tuatara lizard
Tuber
Tuberculosis
Tumbleweed
Tumor
Tuna
Tundra
Tunneling
Turacos
Turbine
Turbulence
Turkeys
Turner syndrome
Turtles
Typhoid fever
Typhus
Tyrannosaurus rex
Tyrant flycatchers

U

Ulcers
Ultracentrifuge
Ultrasonics
Ultraviolet astronomy
Unconformity
Underwater exploration
Ungulates
Uniformitarianism
Units and standards
Uplift
Upwelling
Uranium
Uranus
Urea
Urology

V

Vaccine

Vacuum
Vacuum tube
Valence
Van Allen belts
Van der Waals forces
Vapor pressure
Variable
Variable stars
Variance
Varicella zoster virus
Variola virus
Vegetables
Veins
Velocity
Venus
Verbena family (Verbenaceae)
Vertebrates
Video recording
Violet family (Violaceae)
Vipers
Viral genetics
Vireos
Virtual particles
Virtual reality
Virus
Viscosity
Vision
Vision disorders
Vitamin
Viviparity
Vivisection
Volatility
Volcano
Voles
Volume
Voyager spacecraft
Vulcanization
Vultures
VX agent

W

Wagtails and pipits
Walkingsticks
Walnut family
Walruses

Warblers
Wasps
Waste management
Waste, toxic
Water
Water bears
Water conservation
Water lilies
Water microbiology
Water pollution
Water treatment
Waterbuck
Watershed
Waterwheel
Wave motion
Waxbills
Waxwings
Weasels
Weather
Weather forecasting
Weather mapping
Weather modification
Weathering
Weaver finches
Weevils
Welding
West Nile virus
Wetlands
Wheat
Whisk fern
White dwarf
White-eyes
Whooping cough
Wild type
Wildfire
Wildlife
Wildlife trade (illegal)
Willow family (Salicaceae)
Wind
Wind chill
Wind shear
Wintergreen
Wolverine
Wombats
Wood
Woodpeckers
Woolly mammoth
Work

Wren-warblers
Wrens
Wrynecks

X

X-ray astronomy
X-ray crystallography
X rays
Xenogamy

Y

Y2K
Yak
Yam
Yeast
Yellow fever
Yew
Yttrium

Z

Zebras
Zero
Zodiacal light
Zoonoses
Zooplankton

ORGANIZATION OF THE ENCYCLOPEDIA

The *Gale Encyclopedia of Science, Third Edition* has been designed with ease of use and ready reference in mind.

- Entries are alphabetically arranged across six volumes, in a single sequence, rather than by scientific field

- Length of entries varies from short definitions of one or two paragraphs, to longer, more detailed entries on more complex subjects.

- Longer entries are arranged so that an overview of the subject appears first, followed by a detailed discussion conveniently arranged under subheadings.

- A list of key terms is provided where appropriate to define unfamiliar terms or concepts.

- Bold-faced terms direct the reader to related articles.

- Longer entries conclude with a "Resources" section, which points readers to other helpful materials (including books, periodicals, and Web sites).

- The author's name appears at the end of longer entries. His or her affiliation can be found in the "Contributors" section at the front of each volume.

- "See also" references appear at the end of entries to point readers to related entries.

- Cross references placed throughout the encyclopedia direct readers to where information on subjects without their own entries can be found.

- A comprehensive, two-level General Index guides readers to all topics, illustrations, tables, and persons mentioned in the book.

AVAILABLE IN ELECTRONIC FORMATS

Licensing. *The Gale Encyclopedia of Science, Third Edition* is available for licensing. The complete database is provided in a fielded format and is deliverable on such media as disk or CD-ROM. For more information, contact Gale's Business Development Group at 1-800-877-GALE, or visit our website at www.gale.com/bizdev.

ADVISORY BOARD

A number of experts in the scientific and libary communities provided invaluable assistance in the formulation of this encyclopedia. Our advisory board performed a myriad of duties, from defining the scope of coverage to reviewing individual entries for accuracy and accessibility, and in many cases, writing entries. We would therefore like to express our appreciation to them:

ACADEMIC ADVISORS

Marcelo Amar, M.D.
Senior Fellow, Molecular Disease Branch
National Institutes of Health (NIH)
Bethesda, Maryland

Robert G. Best, Ph.D.
Director
Divison of Genetics, Department of Obstetrics and
  Gynecology
University of South Carolina School of Medicine
Columbia, South Carolina

Bryan Bunch
Adjunct Instructor
Department of Mathematics
Pace University
New York, New York

Cynthia V. Burek, Ph.D.
Environment Research Group, Biology Department
Chester College
England, UK

David Campbell
Head
Department of Physics
University of Illinois at Urbana Champaign
Urbana, Illinois

Morris Chafetz
Health Education Foundation
Washington, DC

Brian Cobb, Ph.D.
Institute for Molecular and Human Genetics
Georgetown University
Washington, DC

Neil Cumberlidge
Professor
Department of Biology

Northern Michigan University
Marquette, Michigan

Nicholas Dittert, Ph.D.
Institut Universitaire Européen de la Mer
University of Western Brittany
France

William J. Engle. P.E.
Exxon-Mobil Oil Corporation (Rt.)
New Orleans, Louisiana

Bill Freedman
Professor
Department of Biology and School for Resource and
  Environmental Studies
Dalhousie University
Halifax, Nova Scotia, Canada

Antonio Farina, M.D., Ph.D.
Department of Embryology, Obstetrics, and
  Gynecology
University of Bologna
Bologna, Italy

G. Thomas Farmer, Ph.D., R.G.
Earth & Environmental Sciences Division
Los Alamos National Laboratory
Los Alamos, New Mexico

Jeffrey C. Hall
Lowell Observatory
Flagstaff, Arizona

Clayton Harris
Associate Professor
Department of Geography and Geology
Middle Tennessee State University
Murfreesboro, Tennesses

Lyal Harris, Ph.D.
Tectonics Special Research Centre
Department of Geology & Geophysics

CONTRIBUTORS

Nasrine Adibe
Professor Emeritus
Department of Education
Long Island University
Westbury, New York

Mary D. Albanese
Department of English
University of Alaska
Juneau, Alaska

Margaret Alic
Science Writer
Eastsound, Washington

James L. Anderson
Soil Science Department
University of Minnesota
St. Paul, Minnesota

Monica Anderson
Science Writer
Hoffman Estates, Illinois

Susan Andrew
Teaching Assistant
University of Maryland
Washington, DC

John Appel
Director
Fundación Museo de Ciencia y
    Tecnología
Popayán, Colombia

David Ball
Assistant Professor
Department of Chemistry
Cleveland State University
Cleveland, Ohio

Dana M. Barry
Editor and Technical Writer
Center for Advanced Materials
    Processing
Clarkston University
Potsdam, New York

Puja Batra
Department of Zoology
Michigan State University
East Lansing, Michigan

Donald Beaty
Professor Emeritus
College of San Mateo
San Mateo, California

Eugene C. Beckham
Department of Mathematics and
    Science
Northwood Institute
Midland, Michigan

Martin Beech
Research Associate
Department of Astronomy
University of Western Ontario
London, Ontario, Canada

**Julie Berwald, Ph.D. (Ocean
    Sciences)**
Austin, Texas

Massimo D. Bezoari
Associate Professor
Department of Chemistry
Huntingdon College
Montgomery, Alabama

John M. Bishop III
Translator
New York, New York

T. Parker Bishop
Professor
Middle Grades and Secondary
    Education
Georgia Southern University
Statesboro, Georgia

Carolyn Black
Professor
Incarnate Word College
San Antonio, Texas

Larry Blaser
Science Writer
Lebanon, Tennessee

Jean F. Blashfield
Science Writer
Walworth, Wisconsin

Richard L. Branham Jr.
Director
Centro Rigional de
    Investigaciones Científicas y
    Tecnológicas
Mendoza, Argentina

Patricia Braus
Editor
American Demographics
Rochester, New York

David L. Brock
Biology Instructor
St. Louis, Missouri

Leona B. Bronstein
Chemistry Teacher (retired)
East Lansing High School
Okemos, Michigan

Brandon R. Brown
Graduate Research Assistant
Oregon State University
Corvallis, Oregon

Lenonard C. Bruno
Senior Science Specialist
Library of Congress
Chevy Chase, Maryland

Janet Buchanan, Ph.D.
Microbiologist
Independent Scholar
Toronto, Ontario, Canada.

Scott Christian Cahall
Researcher
World Precision Instruments, Inc.
Bradenton, Florida

G. Lynn Carlson
Senior Lecturer
School of Science and
    Technology
University of Wisconsin—
    Parkside
Kenosha, Wisconsin

James J. Carroll
Center for Quantum Mechanics
The University of Texas at Dallas
Dallas, Texas

Steven B. Carroll
Assistant Professor
Division of Biology
Northeast Missouri State
    University
Kirksville, Missouri

Rosalyn Carson-DeWitt
Physician and Medical Writer
Durham, North Carolina

Yvonne Carts-Powell
Editor
Laser Focus World
Belmont, Massachustts

Chris Cavette
Technical Writer
Fremont, California

Lata Cherath
Science Writer
Franklin Park, New York

Kenneth B. Chiacchia
Medical Editor
University of Pittsburgh Medical
    Center
Pittsburgh, Pennsylvania

M. L. Cohen
Science Writer
Chicago, Illinois

Robert Cohen
Reporter
KPFA Radio News
Berkeley, California

Sally Cole-Misch
Assistant Director
International Joint Commission
Detroit, Michigan

George W. Collins II
Professor Emeritus
Case Western Reserve
Chesterland, Ohio

Jeffrey R. Corney
Science Writer
Thermopolis, Wyoming

Tom Crawford
Assistant Director
Division of Publication and
    Development
University of Pittsburgh Medical
    Center
Pittsburgh, Pennsylvania

Pamela Crowe
Medical and Science Writer
Oxon, England

Clinton Crowley
On-site Geologist
Selman and Associates
Fort Worth, Texas

Edward Cruetz
Physicist
Rancho Santa Fe, California

Frederick Culp
Chairman
Department of Physics
Tennessee Technical
Cookeville, Tennessee

Neil Cumberlidge
Professor
Department of Biology
Northern Michigan University
Marquette, Michigan

Mary Ann Cunningham
Environmental Writer
St. Paul, Minnesota

Les C. Cwynar
Associate Professor
Department of Biology
University of New Brunswick
Fredericton, New Brunswick

Paul Cypher
Provisional Interpreter
Lake Erie Metropark
Trenton, Michigan

Stanley J. Czyzak
Professor Emeritus
Ohio State University
Columbus, Ohio

Rosi Dagit
Conservation Biologist
Topanga-Las Virgenes Resource
    Conservation District
Topanga, California

David Dalby
President
Bruce Tool Company, Inc.
Taylors, South Carolina

Lou D'Amore
Chemistry Teacher
Father Redmund High School
Toronto, Ontario, Canada

Douglas Darnowski
Postdoctoral Fellow
Department of Plant Biology
Cornell University
Ithaca, New York

Sreela Datta
Associate Writer
Aztec Publications
Northville, Michigan

Sarah K. Dean
Science Writer
Philadelphia, Pennsylvania

Sarah de Forest
Research Assistant
Theoretical Physical Chemistry
   Lab
University of Pittsburgh
Pittsburgh, Pennsylvania

Louise Dickerson
Medical and Science Writer
Greenbelt, Maryland

Marie Doorey
Editorial Assistant
Illinois Masonic Medical Center
Chicago, Illinois

Herndon G. Dowling
Professor Emeritus
Department of Biology
New York University
New York, New York

Marion Dresner
Natural Resources Educator
Berkeley, California

John Henry Dreyfuss
Science Writer
Brooklyn, New York

Roy Dubisch
Professor Emeritus
Department of Mathematics
New York University
New York, New York

Russel Dubisch
Department of Physics
Sienna College
Loudonville, New York

Carolyn Duckworth
Science Writer
Missoula, Montana

**Laurie Duncan, Ph.D.
   (Geology)**
Geologist
Austin, Texas

Peter A. Ensminger
Research Associate
Cornell University
Syracuse, New York

Bernice Essenfeld
Biology Writer
Warren, New Jersey

Mary Eubanks
Instructor of Biology
The North Carolina School of
   Science and Mathematics
Durham, North Carolina

Kathryn M. C. Evans
Science Writer
Madison, Wisconsin

William G. Fastie
Department of Astronomy and
   Physics
Bloomberg Center
Baltimore, Maryland

Barbara Finkelstein
Science Writer
Riverdale, New York

Mary Finley
Supervisor of Science Curriculum
   (retired)
Pittsburgh Secondary Schools
Clairton, Pennsylvania

Gaston Fischer
Institut de Géologie
Université de Neuchâtel
Peseux, Switzerland

Sara G. B. Fishman
Professor
Quinsigamond Community
   College
Worcester, Massachusetts

David Fontes
Senior Instructor
Lloyd Center for Environmental
   Studies
Westport, Maryland

Barry Wayne Fox
Extension Specialist,
   Marine/Aquatic Education
Virginia State University
Petersburg, Virginia

Ed Fox
Charlotte Latin School
Charlotte, North Carolina

Kenneth L. Frazier
Science Teacher (retired)
North Olmstead High School
North Olmstead, Ohio

Bill Freedman
Professor
Department of Biology and
   School for Resource and
   Environmental Studies
Dalhousie University
Halifax, Nova Scotia

T. A. Freeman
Consulting Archaeologist
Quail Valley, California

Elaine Friebele
Science Writer
Cheverly, Maryland

Randall Frost
Documentation Engineering
Pleasanton, California

Agnes Galambosi, M.S.
Climatologist
Eotvos Lorand University
Budapest, Hungary

Robert Gardner
Science Education Consultant
North Eastham, Massachusetts

Gretchen M. Gillis
Senior Geologist
Maxus Exploration
Dallas, Texas

**Larry Gilman, Ph.D. (Electrical
   Engineering)**
Engineer
Sharon, Vermont

Kathryn Glynn
Audiologist
Portland, Oregon

David Goings, Ph.D. (Geology)
Geologist
Las Vegas, Nevada

Natalie Goldstein
Educational Environmental
   Writing
Phoenicia, New York

David Gorish
TARDEC
U.S. Army
Warren, Michigan

Louis Gotlib
South Granville High School
Durham, North Carolina

Hans G. Graetzer
Professor
Department of Physics
South Dakota State University
Brookings, South Dakota

Jim Guinn
Assistant Professor
Department of Physics
Berea College
Berea, Kentucky

Steve Gutterman
Psychology Research Assistant
University of Michigan
Ann Arbor, Michigan

Johanna Haaxma-Jurek
Educator
Nataki Tabibah Schoolhouse of
  Detroit
Detroit, Michigan

Monica H. Halka
Research Associate
Department of Physics and
  Astronomy
University of Tennessee
Knoxville, Tennessee

Brooke Hall, Ph.D.
Professor
Department of Biology
California State University at
  Sacramento
Sacramento, California

Jeffrey C. Hall
Astronomer
Lowell Observatory
Flagstaff, Arizona

C. S. Hammen
Professor Emeritus
Department of Zoology
University of Rhode Island

Lawrence Hammar, Ph.D.
Senior Research Fellow
Institute of Medical Research
Papua, New Guinea

William Haneberg, Ph.D.
  (Geology)
Geologist
Portland, Oregon

Beth Hanson
Editor
The Amicus Journal
Brooklyn, New York

Clay Harris
Associate Professor
Department of Geography and
  Geology
Middle Tennessee State
  University
Murfreesboro, Tennessee

Clinton W. Hatchett
Director Science and Space
  Theater
Pensacola Junior College
Pensacola, Florida

Catherine Hinga Haustein
Associate Professor
Department of Chemistry
Central College
Pella, Iowa

Dean Allen Haycock
Science Writer
Salem, New York

Paul A. Heckert
Professor
Department of Chemistry and
  Physics
Western Carolina University
Cullowhee, North Carolina

Darrel B. Hoff
Department of Physics
Luther College
Calmar, Iowa

Dennis Holley
Science Educator
Shelton, Nebraska

Leonard Darr Holmes
Department of Physical Science
Pembroke State University
Pembroke, North Carolina

Rita Hoots
Instructor of Biology, Anatomy,
  Chemistry
Yuba College
Woodland, California

Selma Hughes
Department of Psychology and
  Special Education
East Texas State University
Mesquite, Texas

Mara W. Cohen Ioannides
Science Writer
Springfield, Missouri

Zafer Iqbal
Allied Signal Inc.
Morristown, New Jersey

Sophie Jakowska
Pathobiologist, Environmental
  Educator
Santo Domingo, Dominican
  Republic

Richard A. Jeryan
Senior Technical Specialist
Ford Motor Company
Dearborn, Michigan

Stephen R. Johnson
Biology Writer
Richmond, Virginia

Kathleen A. Jones
School of Medicine
Southern Illinois University
Carbondale, Illinois

Harold M. Kaplan
Professor
School of Medicine
Southern Illinois University
Carbondale, Illinois

Anthony Kelly
Science Writer
Pittsburgh, Pennsylvania

Amy Kenyon-Campbell
Ecology, Evolution and
    Organismal Biology Program
University of Michigan
Ann Arbor, Michigan

Judson Knight
Science Writer
Knight Agency
Atlanta, Georgia

Eileen M. Korenic
Institute of Optics
University of Rochester
Rochester, New York

Jennifer Kramer
Science Writer
Kearny, New Jersey

Pang-Jen Kung
Los Alamos National Laboratory
Los Alamos, New Mexico

Marc Kusinitz
Assistant Director Media
    Relations
John Hopkins Medical Institution
Towsen, Maryland

Arthur M. Last
Head
Department of Chemistry
University College of the Fraser
    Valley
Abbotsford, British Columbia

Nathan Lavenda
Zoologist
Skokie, Illinios

Jennifer LeBlanc
Environmental Consultant
London, Ontario, Canada

Nicole LeBrasseur, Ph.D.
Associate News Editor
Journal of Cell Biology
New York, New York

Benedict A. Leerburger
Science Writer
Scarsdale, New York

Betsy A. Leonard
Education Facilitator

Reuben H. Fleet Space Theater
    and Science Center
San Diego, California

Adrienne Wilmoth Lerner
Graduate School of Arts &
    Science
Vanderbilt University
Nashville, Tennessee

Lee Wilmoth Lerner
Science Writer
NASA
Kennedy Space Center, Florida

Scott Lewis
Science Writer
Chicago, Illinois

Frank Lewotsky
Aerospace Engineer (retired)
Nipomo, California

Karen Lewotsky
Director of Water Programs
Oregon Environmental Council
Portland, Oregon

Kristin Lewotsky
Editor
Laser Focus World
Nashua, New Hamphire

Stephen K. Lewotsky
Architect
Grants Pass, Oregon

Agnieszka Lichanska, Ph.D.
Department of Microbiology &
    Parasitology
University of Queensland
Brisbane, Australia

Sarah Lee Lippincott
Professor Emeritus
Swarthmore College
Swarthmore, Pennsylvania

Jill Liske, M.Ed.
Wilmington, North Carolina

David Lunney
Research Scientist
Centre de Spectrométrie
    Nucléaire et de Spectrométrie
    de Masse
Orsay, France

Steven MacKenzie
Ecologist
Spring Lake, Michigan

J. R. Maddocks
Consulting Scientist
DeSoto, Texas

Gail B. C. Marsella
Technical Writer
Allentown, Pennsylvania

Karen Marshall
Research Associate
Council of State Governments
    and Centers for Environment
    and Safety
Lexington, Kentucky

Liz Marshall
Science Writer
Columbus, Ohio

James Marti
Research Scientist
Department of Mechanical
    Engineering
University of Minnesota
Minneapolis, Minnesota

Elaine L. Martin
Science Writer
Pensacola, Florida

Lilyan Mastrolla
Professor Emeritus
San Juan Unified School
Sacramento, California

Iain A. McIntyre
Manager
Electro-optic Department
Energy Compression Research
    Corporation
Vista, California

Jennifer L. McGrath
Chemistry Teacher
Northwood High School
Nappanee, Indiana

Margaret Meyers, M.D.
Physician, Medical Writer
Fairhope, Alabama

G. H. Miller
Director
Studies on Smoking
Edinboro, Pennsylvania

J. Gordon Miller
Botanist
Corvallis, Oregon

Kelli Miller
Science Writer
NewScience
Atlanta, Georgia

Christine Miner Minderovic
Nuclear Medicine Technologist
Franklin Medical Consulters
Ann Arbor, Michigan

David Mintzer
Professor Emeritus
Department of Mechanical
    Engineering
Northwestern University
Evanston, Illinois

Christine Molinari
Science Editor
University of Chicago Press
Chicago, Illinois

Frank Mooney
Professor Emeritus
Fingerlake Community College
Canandaigua, New York

Partick Moore
Department of English
University of Arkansas at Little
    Rock
Little Rock, Arkansas

Robbin Moran
Department of Systematic Botany
Institute of Biological Sciences
University of Aarhus
Risskou, Denmark

J. Paul Moulton
Department of Mathematics
Episcopal Academy
Glenside, Pennsylvania

Otto H. Muller
Geology Department

Alfred University
Alfred, New York

Angie Mullig
Publication and Development
University of Pittsburgh Medical
    Center
Trafford, Pennsylvania

David R. Murray
Senior Associate
Sydney University
Sydney, New South Wales,
    Australia

Sutharchana Murugan
Scientist
Three Boehringer Mannheim
    Corp.
Indianapolis, Indiana

Muthena Naseri
Moorpark College
Moorpark, California

David Newton
Science Writer and Educator
Ashland, Oregon

F. C. Nicholson
Science Writer
Lynn, Massachusetts

James O'Connell
Department of Physical Sciences
Frederick Community College
Gaithersburg, Maryland

Dúnal P. O'Mathúna
Associate Professor
Mount Carmel College of
    Nursing
Columbus, Ohio

Marjorie Pannell
Managing Editor, Scientific
    Publications
Field Museum of Natural History
Chicago, Illinois

Gordon A. Parker
Lecturer
Department of Natural Sciences
University of Michigan-Dearborn
Dearborn, Michigan

David Petechuk
Science Writer
Ben Avon, Pennsylvania

Borut Peterlin, M.D.
Consultant Clinical Geneticist,
    Neurologist, Head Division of
    Medical Genetics
Department of Obstetrics and
    Gynecology
University Medical Centre
    Ljubljana
Ljubljana, Slovenia

John R. Phillips
Department of Chemistry
Purdue University, Calumet
Hammond, Indiana

Kay Marie Porterfield
Science Writer
Englewood, Colorado

Paul Poskozim
Chair
Department of Chemistry, Earth
    Science and Physics
Northeastern Illinois University
Chicago, Illinois

Andrew Poss
Senior Research Chemist
Allied Signal Inc.
Buffalo, New York

Satyam Priyadarshy
Department of Chemistry
University of Pittsburgh
Pittsburgh, Pennsylvania

Patricia V. Racenis
Science Writer
Livonia, Michigan

Cynthia Twohy Ragni
Atmospheric Scientist
National Center for Atmospheric
    Research
Westminster, Colorado

Jordan P. Richman
Science Writer
Phoenix, Arizona

Kitty Richman
Science Writer
Phoenix, Arizona

Vita Richman
Science Writer
Phoenix, Arizona

Michael G. Roepel
Researcher
Department of Chemistry
University of Pittsburgh
Pittsburgh, Pennsylvania

Perry Romanowski
Science Writer
Chicago, Illinois

Nancy Ross-Flanigan
Science Writer
Belleville, Michigan

Belinda Rowland
Science Writer
Voorheesville, New York

Gordon Rutter
Royal Botanic Gardens
Edinburgh, Great Britain

Elena V. Ryzhov
Polytechnic Institute
Troy, New York

David Sahnow
Associate Research Scientist
John Hopkins University
Baltimore, Maryland

Peter Salmansohn
Educational Consultant
New York State Parks
Cold Spring, New York

Peter K. Schoch
Instructor
Department of Physics and
    Computer Science
Sussex County Community
    College
Augusta, New Jersey

Patricia G. Schroeder
Instructor
Science, Healthcare, and Math
    Division
Johnson County Community
    College
Overland Park, Kansas

Randy Schueller
Science Writer
Chicago, Illinois

Kathleen Scogna
Science Writer
Baltimore, Maryland

William Shapbell Jr.
Launch and Flight Systems
    Manager
Kennedy Space Center
KSC, Florida

Kenneth Shepherd
Science Writer
Wyandotte, Michigan

Anwar Yuna Shiekh
International Centre for
    Theoretical Physics
Trieste, Italy

Raul A. Simon
Chile Departmento de Física
Universidad de Tarapacá
Arica, Chile

Michael G. Slaughter
Science Specialist
Ingham ISD
East Lansing, Michigan

Billy W. Sloope
Professor Emeritus
Department of Physics
Virginia Commonwealth
    University
Richmond, Virginia

Douglas Smith
Science Writer
Milton, Massachusetts

Lesley L. Smith
Department of Physics and
    Astronomy
University of Kansas
Lawrence, Kansas

Kathryn D. Snavely
Policy Analyst, Air Quality Issues
U.S. General Accounting Office
Raleigh, North Carolina

Charles H. Southwick
Professor
Environmental, Population, and
    Organismic Biology
University of Colorado at Boulder
Boulder, Colorado

John Spizzirri
Science Writer
Chicago, Illinois

Frieda A. Stahl
Professor Emeritus
Department of Physics
California State University, Los
    Angeles
Los Angeles, California

Robert L. Stearns
Department of Physics
Vassar College
Poughkeepsie, New York

Ilana Steinhorn
Science Writer
Boalsburg, Pennsylvania

David Stone
Conservation Advisory Services
Gai Soleil
Chemin Des Clyettes
Le Muids, Switzerland

Eric R. Swanson
Associate Professor
Department of Earth and Physical
    Sciences
University of Texas
San Antonio, Texas

Cheryl Taylor
Science Educator
Kailua, Hawaii

Nicholas C. Thomas
Department of Physical Sciences
Auburn University at
    Montgomery
Montgomery, Alabama

W. A. Thomasson
Science and Medical Writer
Oak Park, Illinois

Marie L. Thompson
Science Writer
Ben Avon, Pennsylvania

Laurie Toupin
Science Writer
Pepperell, Massachusetts

Melvin Tracy
Science Educator
Appleton, Wisconsin

Karen Trentelman
Research Associate
Archaeometric Laboratory
University of Toronto
Toronto, Ontario, Canada

Robert K. Tyson
Senior Scientist
W. J. Schafer Assoc.
Jupiter, Florida

James Van Allen
Professor Emeritus
Department of Physics and
    Astronomy
University of Iowa
Iowa City, Iowa

Julia M. Van Denack
Biology Instructor
Silver Lake College
Manitowoc, Wisconsin

Kurt Vandervoort
Department of Chemistry and
    Physics
West Carolina University
Cullowhee, North Carolina

Chester Vander Zee
Naturalist, Science Educator
Volga, South Dakota

Rashmi Venkateswaran
Undergraduate Lab Coordinator
Department of Chemistry
University of Ottawa
Ottawa, Ontario, Canada

R. A. Virkar
Chair
Department of Biological
    Sciences
Kean College
Iselin, New Jersey

Kurt C. Wagner
Instructor
South Carolina Governor's
    School for Science and
    Technology
Hartsville, South Carolina

Cynthia Washam
Science Writer
Jensen Beach, Florida

Terry Watkins
Science Writer
Indianapolis, Indiana

Joseph D. Wassersug
Physician
Boca Raton, Florida

Tom Watson
Environmental Writer
Seattle, Washington

Jeffrey Weld
Instructor, Science Department
    Chair
Pella High School

Pella, Iowa

Frederick R. West
Astronomer
Hanover, Pennsylvania

Glenn Whiteside
Science Writer
Wichita, Kansas

John C. Whitmer
Professor
Department of Chemistry
Western Washington University
Bellingham, Washington

Donald H. Williams
Department of Chemistry
Hope College
Holland, Michigan

Robert L. Wolke
Professor Emeritus
Department of Chemistry
University of Pittsburgh
Pittsburgh, Pennsylvania

Xiaomei Zhu, Ph.D.
Postdoctoral research associate
Immunology Department
Chicago Children's Memorial
    Hospital, Northwestern
    University Medical School
Chicago, Illinois

Jim Zurasky
Optical Physicist
Nichols Research Corporation
Huntsville, Alabama

Charge-coupled device

Charge-coupled devices (CCDs) have made possible a revolution in image processing. They consist of a series of light-sensitive elements, called pixels, arranged in a **square** or rectangular array. When CCDs are exposed to **light**, an image of the object being observed is formed; this image can be extracted from the CCD and stored on a computer for later analysis. CCDs are used in a variety of modern instruments, ranging from scanners and photocopiers to video cameras and digital still cameras. They have transformed the way scientists measure and chart the universe. Because CCDs are available in a wide price range, they are accessible to amateurs as well as professionals, and enable both to make significant contributions to modern **astronomy**.

How the devices work

All CCDs work on the same principle. The CCD surface is a grid of pixels (pixel is a contraction for "picture element"). Small CCDs may have a grid of 256 x 256 pixels, while large CCDs may have 4,096 x 4,096 pixel grids. Although many CCD pixel grids are square, this is not always the case; scanners and photocopiers, for example, have a single line of pixels that passes over the picture or page of text being imaged. The pixels are tiny; some CCDs have pixels only 9 microns across, while others may have 27-micron pixels. The scale and resolution of the image a camera is able to form on the CCD depends both on the pixel size and the grid size. Regardless of the pixel or grid size, however, each pixel on the CCD has the ability to convert the light striking it into an electric signal. The voltage accumulated by each pixel during an exposure is directly proportional to the amount of light striking it. When the CCD is exposed to light for a length of time, an image of whatever is being observed—whether a distant **galaxy** or cars in a parking lot—forms on the CCD as an array of differing electric voltages.

After an image has been recorded on the CCD, the device can be "read out," meaning that the voltages are extracted from the CCD for storage on a computer. The analogy that is almost universally used to describe this process is the "bucket brigade" analogy. Picture each pixel on the CCD as a bucket with a certain amount of **water** in it. When the CCD is read out, the water in each row of buckets is emptied into the adjacent row. The water in the first row goes into a special row of storage buckets, the water in each bucket in the second row goes into its neighbor bucket in the first row, and so on across the whole CCD. Then, the amount of water in each of these buckets is emptied, measured, and stored in a computer's memory. This process is repeated until all of the

rows have been shifted into the storage buckets, emptied, and measured. If you now replace the water with electric voltages, and replace the measurement of water with the digital measurement of the analog electric signal, you have the basic process by which an image is extracted from the CCD. The actual process of reading out the CCD is performed by fairly complicated and exquisitely synchronized **electronics** that move all the electric charges between the "buckets," convert the analog voltages into digital numbers, and make the data available for storage on a computer.

Once the pixel outputs have been measured and stored on a computer, they can be used in a variety of ways. For simple line drawings, the image processing software may render the data from the CCD in black and white. For pictures, a 256-level grayscale may be appropriate. In either case, a grid of numbers, corresponding to the original light intensity, is present and can be analyzed in any way the person studying the image desires.

From the description above, it may seem that CCDs cannot be used for **color** imaging, since they respond only to light intensity. In fact, color CCDs are available, although they are used in video equipment such as camcorders and almost never in astronomy. If an astronomer wanted to create a color image using a CCD, the old practice of taking three images through three different color filters is still the usual way to go. True color CCDs have pixels with built-in filters, alternating red, green, and blue. They can produce real-time color images, but they are undesirable for scientific work because they introduce significant difficulties into the data analysis process, as well as reducing the effective resolution of the CCD by a factor of three.

Applications in astronomy

Astronomers began using charge-coupled devices in their work in the early 1980s, when the increasing power and clock speed of semiconductors, and the computers needed to drive the hardware and analyze the data became both fast and affordable. Almost every field of astronomy was directly impacted by CCDs: for observations of asteroids, galaxies, stars, and planets, whether by direct imaging or the recording of spectra, the CCD rapidly became the detector of choice.

CCDs are also useful to astronomers because an average, CCDs are about ten times more light-sensitive than film. Astronomers are notorious for finding desperately faint objects to observe, so the CCD gave them the ability not only to see fainter objects than they could before, but to reduce the amount of time spent tracking and observing a given object. A CCD camera can record in a 15 minute exposure the same information that would take a standard camera loaded with film two hours or more. While film typically records only 2–3% of the light that strikes it, charge-coupled device cameras can record between 50–80% of the light they detect. Furthermore, CCDs can capture light outside the visible **spectrum**, which film cannot do. The devices operate without darkrooms or chemicals, and the results can be reconstructed as soon as the information is loaded into an image processing program.

However, CCD cameras do have some drawbacks. The small size of the most affordable arrays results in a much smaller field of view. Large celestial bodies such as the **moon**, which are easily photographed with a 35mm camera, become very difficult to reproduce as a single image with a CCD camera. Although larger arrays are coming to the market, they remain pricy and beyond the resources of the amateur astronomer. They require complicated systems to operate, any many of them have to be cooled to typical temperatures of -112°F (-80°C) to reduce their background electronic noise to an acceptable level. Finally, color images for astronomical CCD cameras (unlike commercially-available video and digital still cameras) require three separate exposures for each filter used. The final image has to be created by combining the data from each exposure within the computer.

CCDs, professionals, and amateurs

With web-based **star** catalogues and other Internet and electronic resources, such as the Hubble Guide Star Catalog and the Lowell Observatory Asteroid Database, professional and amateur astronomers have begun sharing resources and comparing data in hopes of creating a more accurate and complete picture of the heavens. Organizations such as the Amateur Sky Survey help individuals coordinate and share data with others. Thanks to CCDs, amateurs have often contributed as significantly to these projects as professional astronomers have. Paul Comba, an amateur based in Arizona, discovered and registered some 300 previously unknown asteroids in 1996–97, after adding a digital camera to his **telescope**. In 1998, **astrophysics** student Gianluca Masi recorded the existence of an unknown variable star, discovered with the use of his Kodak KAF-0400 CCD, mounted in a Santa Barbara Instrument Group ST-7 camera. CCDs help level the playing field in the science of **astrometry**, drastically reducing the equipment barrier between the amateur and the professional.

Resources

Periodicals

di Cicco, Dennis. "Measuring the Sky with CCDs." *Sky & Telescope* 94 (December 1997): 115-18.

Gombert, Glenn, and Tom Droege. "Boldness: The Amateur Sky Survey." *Sky & Telescope* 95 (February 1998): 42- 45.

Hannon, James. "Warming Up to Digital Imaging." *Sky & Telescope* 97 (March 1999): 129.

Masi, Gianluca. "CCDs, Small Scopes, and the Urban Amateur." *Sky & Telescope* 95 (February 1998): 109-12.

Terrance, Gregory. "Capture the Sky on a CCD: Digital Imaging with a CCD Camera Is Revolutionizing the Way Amateur Astronomers Record Planets and Galaxies." *Astronomy* 28 (February 2000): 72.

Kenneth R. Shepherd

Charles's law *see* **Gases, properties of**

Cheetah *see* **Cats**

Chelate

A chelate is a type of **coordination compound** in which a single metallic ion is attached by coordinate covalent bonds to a **molecule** or an ion called a **ligand**. The term chelate comes from the Greek word *chela*, meaning "crab's claw." The term clearly describes the appearance of many kinds of chelates, in which the ligand surrounds the central atom in a way that can be compared to the grasping of food by a crab's claw.

Bonding in a chelate occurs because the ligand has at least two pairs of unshared electrons. These unshared pairs of electrons are regions of **negative** electrical charge to which are attracted cations such as the copper(I) and copper(II), silver, nickel, platinum, and **aluminum** ions. A ligand with only two pairs of unshared electrons is known as a bidentate ("two-toothed") ligand; one with three pairs of unshared electrons, a tridentate ("three-toothed") ligand, and so on.

The geometric shape of a chelate depends on the number of ligands involved. Those with bidentate ligands form linear molecules, those with four ligands form planar or tetrahedral molecules, and those with six ligands form octahedral molecules.

One of the most familiar examples of a chelate is hemoglobin, the molecule that transports **oxygen** through the **blood**. The "working part" of a hemoglobin molecule is heme, a complex molecule at whose core is an iron(II) ion bonded to four **nitrogen atoms** with coordinate covalent bonds.

Among the most common applications of chelates is in **water** softening and treatment of poisoning. In the former instance, a compound such as **sodium** tripolyphosphate is added to water. That compound forms chelates with **calcium** and **magnesium** ions, ions responsible for the hardness in water. Because of their ability to "tie up" **metal** ions in chelates, compounds like sodium tripolyphosphate are sometimes referred to as sequestering agents.

A typical sequestering agent used to treat poison victims is ethylenediaminetetraacetic acid, commonly known as EDTA. Suppose that a person has swallowed a significant amount of **lead** and begins to display the symptoms of lead poisoning. Giving the person EDTA allows that molecule to form chelates with lead ions, removing that toxic material from the bloodstream.

Chemical bond

A chemical bond is any **force** of attraction that holds two **atoms** or ions together. In most cases, that force of attraction is between one or more electrons held by one of the atoms and the positively charged nucleus of the second atom. Chemical bonds vary widely in their stability, ranging from relatively strong covalent bonds to very weak **hydrogen** bonds.

History

The concept of bonding as a force that holds two particles together is as old as the concept of ultimate particles of **matter** itself. As early as 100 B.C., for example, Asklepiades of Prusa speculated about the existence of "clusters of atoms," a concept that implies the existence of some force of attraction holding the particles together. At about the same **time**, the Roman poet Lucretius in his monumental work *De Rerum Natura* ("On the nature of things") pictured atoms as tiny spheres to which were attached fishhook-like appendages. Atoms combined with each other, according to Lucretius, when the appendages from two adjacent atoms became entangled with each other.

Relatively little progress could occur in the field of bonding theory, of course, until the concept of an atom itself was clarified. When John Dalton proposed the modern **atomic theory** in 1803, he specifically hypothesized that atoms would combine with each other to form "compound atoms." Dalton's concept of bonding was essentially non-existent, however, and he imagined that atoms simply sit adjacent to each other in their compound form.

The real impetus to further speculation about bonding was provided by the evolution of the concept of a **molecule**, originally proposed by Amedeo Avogadro in 1811 and later refined by Stanislao Cannizzaro more than four decades later.

The origin of bond symbolism

Some of the most vigorous speculation about chemical bonding took place in the young field of organic **chemistry**. In trying to understand the structure of organic compounds, for example, Friedrich Kekulé suggested that the **carbon** atom is tetravalent; that is, it can bond to four other atoms. He also hypothesized that carbon atoms could bond with each other almost endlessly in long chains.

Kekulé had no very clear notion as to how atoms bond to each other, but he did develop an elaborate system for showing how those bonds might be arranged in **space**. That system was too cumbersome for everyday use by chemists, however, and it was quickly replaced by another system suggested earlier by the Scottish chemist Archibald Scott Couper. Couper proposed that the bond between two atoms (what the real physical nature of that bond might be) be represented by a short dashed line. Thus, a molecule of **water** could be represented by the structural formula: H-O-H.

That system is still in existence today. The arrangement of atoms in a molecule is represented by the symbols of the elements present joined by dashed lines that show how the atoms of those elements are bonded to each other. Thus, the term chemical bond refers not only to the force of attraction between two particles, but also to the dashed line used in the structural formula for that substance.

Development of the modern theory of bonding

The discovery of the **electron** by J. J. Thomson in 1897 was, in the long run, the key needed to solve the problem of bonding. In the short run, however, it was a serious hindrance to resolving that issue. The question that troubled many chemists at first was how two particles with the same electrical charge (as atoms then seemed to be) could combine with each other.

An answer to that dilemma slowly began to evolve, beginning with the work of the young German chemist Richard Abegg. In the early 1900s, Abegg came to the conclusion that inert gases are stable elements because their outermost shell of electrons always contain eight electrons. Perhaps atoms combine with each other, Abegg said, when they exchange electrons in such as way that they all end up with eight electrons in their outer **orbit**. In a simplistic way, Abegg had laid out the principle of ionic bonding. Ionic bonds are formed when one atom completely gives up one or more electrons, and a second atom takes on those electrons.

Since Abegg was killed in 1910 at the age of 41 in a **balloon** accident, he was prevented from improving upon his original hypothesis. That work was taken up in the 1910s, however, by a number of other scientists, most prominently the German chemist Walther Kossel and the American chemists Irving Langmuir and Gilbert Newton Lewis.

Working independently, these researchers came up with a second method by which atoms might bond to each other. Rather than completely losing or gaining electrons, they hypothesized, perhaps atoms can share electrons with each other. One might imagine, for example, that in a molecule of methane (CH_4), each of the four **valence** electrons in carbon is shared with the single electron available from each of the four hydrogen atoms. Such an arrangement could provide carbon with a full outer shell of eight electrons and each hydrogen atom with a full outer shell of two. Chemical bonds in which two atoms share pairs of electrons with each other are known as covalent bonds.

In trying to illustrate this concept, Lewis developed another system for representing chemical bonds. In the Lewis system (also known as the electron-dot system), each atom is represented by its chemical symbol with the number of electrons in its outermost orbit, its bonding or valence electrons. The formula of a compound, then, is to be represented by showing how two or more atoms share electrons with each other.

Bond types

Credit for the development of the modern theory of chemical bonding belongs largely to the great American chemist Linus Pauling. Early in his career, Pauling learned about the revolution in **physics** that was taking place largely in **Europe** during the 1920s. That revolution had come about with the discovery of the relativity theory, **quantum mechanics**, the uncertainty principle, the duality of matter and **energy**, and other new and strikingly different concepts in physics.

Most physicists recognized the need to reformulate the fundamental principles of physics because of these discoveries. Relatively few chemists, however, saw the relevance of the revolution in physics for their own subject. Pauling was the major exception. By the late 1920s, he had already begun to ask how the new science of quantum mechanics could be used to understand the nature of the chemical bond.

In effect, the task Pauling undertook was to determine the way in which any two atoms might react with each other in such a way as to put them in the lowest possible energy state. Among the many discoveries he made was that, for most cases, atoms form neither a purely ionic nor purely covalent bond. That is, atoms typically do not completely lose, gain, or share equally the electrons that form the bond between them. Instead,

the atoms tend to form hybrid bonds in which a pair of shared electrons spend more time with one atom and less time with the second atom.

Electronegativity

The term that Pauling developed for this concept is electronegativity. Electronegativity is, in a general sense, the tendency of an atom to attract the electrons in a covalent bond. The numerical values for the electronegativities of the elements range from a maximum of 4.0 for fluorine to a minimum of about 0.7 for cesium. A bond formed between fluorine and cesium would tend to be ionic because fluorine has a much stronger attraction for electrons than does cesium. On the other hand, a bond formed between cobalt (electronegativity = 1.9) and silicon (electronegativity = 1.9) would be a nearly pure covalent bond since both atoms have an equal attraction for electrons.

The modern concept of chemical bonding, then, is that bond types are not best distinguished as purely ionic or purely covalent. Instead, they can be envisioned as lying somewhere along a continuum between those two extremes. The position of any particular bond can be predicted by calculating the difference between the two electronegativities of the atoms involved. The greater that difference, the more ionic the bond; the smaller the difference, the more covalent.

Bond polarity

The preceding discussion suggests that most chemical bonds are polar; that is, one end of the bond is more positive than the other end. In the bond formed between hydrogen (electronegativity = 2.2) and **sulfur** (electronegativity = 2.6), for example, neither atom has the ability to take electrons completely from the other. Neither is equal sharing of electrons likely to occur. Instead, the electrons forming the hydrogen-sulfur bond will spend somewhat more time with the sulfur atom and somewhat less time with the hydrogen atom. Thus, the sulfur end of the hydrogen-sulfur bond is somewhat more **negative** (represented as δ-) and the hydrogen end, somewhat more positive (δ+).

Coordination compounds

Some chemical bonds are unique in that both electrons forming the bond come from a single atom. The two atoms are held together, then, by the attraction between the pair of electrons from one atom and the positively charged nucleus of the second atom. Such bonds have been called coordinate covalent bonds.

An example of this kind of bonding is found in the reaction between copper(II) ion and **ammonia**. The **nitrogen** atom in ammonia has an unshared pair of electrons that is often used to bond with other atoms. The copper(II) ion is an example of such an **anion**. It is positively charged and tends to surround itself with four ammonia molecules to form the cupric ammonium ion, $Cu(NH_3)_4^{2+}$. The bonding in this ion consists of coordinate covalent bonds with all bonding electrons supplied by the nitrogen atom.

Multiple bonds

The bonds described thus far can all be classified as single bonds. That is, they all consist of a single pair of electrons. Not uncommonly, two atoms will combine with each other by sharing two pairs of electrons. For example, when **lead** and sulfur combine to form a compound, the molecules formed might consist of two pairs of electrons, one electron from lead and one electron from sulfur in each of the pairs. The standard shorthand for a double bond such as this one is a double dashed line (=). For example, the formula for a common double-bonded compound, ethylene, is: $H_2C=CH_2$.

Compounds can also be formed by the sharing of three pairs of electrons between two atoms. The formula for one such compound, acetylene, shows how a triple bond of this kind is represented: HC-CH.

Other types of bonds

Other types of chemical bonds also exist. The atoms that make up a **metal**, for example, are held together by a metallic bond. A metallic bond is one in which all of the metal atoms share with each other a cloud of electrons. The electrons that make up that cloud originate from the outermost energy levels of the atoms.

A hydrogen bond is a weak force of attraction that exists between two atoms or ions with opposite charges. For example, the hydrogen-oxygen bonds in water are polar bonds. The hydrogen end of these bonds are slightly positive and the **oxygen** ends, slightly negative. Two molecules of water placed next to each other will feel a force of attraction because the oxygen end of one molecule feels an electrical force of attraction to the hydrogen end of the other molecule. Hydrogen bonds are very common and extremely important in biological systems. They are strong enough to hold substances together, but weak enough to break apart and allow chemical changes to take place within the system.

Van der Waals forces are yet another type of chemical bond. Such forces exist between particles that appear to be electrically neutral. The rapid shifting of electrons

that takes place within such molecules means that some parts of the molecule are momentarily charged, either positively or negatively. For this reason, very weak, transient forces of attraction can develop between particles that are actually neutral.

KEY TERMS

. . . .

Coordinate covalent bond—A type of covalent bond in which all shared electrons are donated by only one of two atoms.

Covalent bond—A chemical bond formed when two atoms share a pair of electrons with each other.

Double bond—A covalent bond consisting of two pairs of shared electrons that hold the two atoms together.

Electronegativity—A quantitative method for indicating the relative tendency of an atom to attract the electrons that make up a covalent bond.

Ionic bond—A chemical bond formed when one atom gains and a second atom loses electrons.

Lewis symbol—A method for designating the structure of atoms and molecules in which the chemical symbol for an element is surrounded by dots indicating the number of valence electrons in the atom of that element.

Molecule—A collection of atoms held together by some force of attraction.

Multiple bond—A double or triple bond.

Polar bond—A covalent bond in which one end of the bond is more positive than the other end.

Structural formula—The chemical representation of a molecule that shows how the atoms are arranged within the molecule.

Triple bond—A triple bond is formed when three pairs of electrons are shared between two atoms..

Valence electrons—The electrons in the outermost shell of an atom that determine an element's chemical properties.

Resources

Books

Bynum, W.F., E.J. Browne, and Roy Porter. *Dictionary of the History of Science.* Princeton, NJ: Princeton University Press, 1981, pp. 433-435.

Kotz, John C., and Paul Treichel. *Chemistry and Cehmical Reactivity.* Pacific Grove, CA: Brooks/Cole, 1998.

Lide, D.R., ed. *CRC Handbook of Chemistry and Physics.* Boca Raton: CRC Press, 2001.

Oxtoby, David W., et al. *The Principles of Modern Chemistry.* 5th ed. Pacific Grove, CA: Brooks/Cole, 2002.

Pauling, Linus. *The Nature of the Chemical Bond and the Structure of Molecules and Crystals: An Introduction to Modern Structural Chemistry.* 3rd edition. Ithaca, NY: Cornell University Press, 1960.

Chemical compound *see* **Compound, chemical**

Chemical element *see* **Element, chemical**

Chemical equilibrium *see* **Equilibrium, chemical**

Chemical evolution

Chemical evolution describes chemical changes on the primitive **Earth** that gave rise to the first forms of life. The first living things on Earth were prokaryotes with a type of **cell** similar to present-day **bacteria**. **Prokaryote** fossils have been found in 3.4-million-year-old rock in the southern part of **Africa**, and in even older **rocks** in **Australia**, including some that appear to be photosynthetic. All forms of life are theorized to have evolved from the original prokaryotes, probably 3.5-4.0 billion years ago.

The primitive Earth

The chemical and physical conditions of the primitive Earth are invoked to explain the **origin of life**, which was preceded by chemical evolution of organic chemicals. Astronomers believe that 20-30 billion years ago, all **matter** was concentrated in a single **mass**, and that it blew apart with a "big bang." In time, a disk-shaped cloud of dust condensed and formed the **Sun**, and the peripheral matter formed its planets. **Heat** produced by compaction, **radiation**, and impacting meteorites melted Earth. Then, as the **planet** cooled, Earth's layers formed. The first atmosphere was made up of hot **hydrogen** gas, too light to be held by Earth's gravity. **Water** vapor, **carbon monoxide**, **carbon dioxide**, **nitrogen**, and methane replaced the hydrogen atmosphere. As Earth cooled, water vapor condensed and torrential rains filled up its basins, thereby forming the seas. Also present were **lightning**, volcanic activity, and ultraviolet radiation. It was in this setting that life began.

According to one theory, chemical evolution occurred in four stages.

In the first stage of chemical evolution, molecules in the primitive environment formed simple organic substances, such as amino acids. This concept was first proposed in 1936 in a book entitled, "The Origin of Life on Earth," written by the Russian scientist, Aleksandr Ivanovich Oparin. He considered hydrogen, **ammonia**, water vapor, and methane to be components in the early atmosphere. **Oxygen** was lacking in this chemically- reducing environment. He stated that ultraviolet radiation from the Sun provided the **energy** for the transformation of these substances into organic molecules. Scientists today state that such spontaneous synthesis occurred only in the primitive environment. Abiogenesis became impossible when photosynthetic cells added oxygen to the atmosphere. The oxygen in the atmosphere gave rise to the **ozone** layer which then shielded Earth from ultraviolet radiation. Newer versions of this hypothesis contend that the primitive atmosphere also contained **carbon** monoxide, carbon dioxide, nitrogen, hydrogen sulfide, and hydrogen. Present-day volcanoes emit these substances.

In 1957, Stanley Miller and Harold Urey provided laboratory evidence that chemical evolution as described by Oparin could have occurred. Miller and Urey created an apparatus that simulated the primitive environment. They used a warmed flask of water for the **ocean**, and an atmosphere of water, hydrogen, ammonia and methane. Sparks discharged into the artificial atmosphere represented lightning. A condenser cooled the atmosphere, causing rain that returned water and dissolved compounds back to the simulated sea. When Miller and Urey analyzed the components of the **solution** after a week, they found various organic compounds had formed. These included some of the amino acids that compose the **proteins** of living things. Their results gave credence to the idea that simple substances in the warm primordial seas gave rise to the chemical building blocks of organisms.

In the second stage of chemical evolution, the simple organic molecules (such as amino acids) that formed and accumulated joined together into larger structures (such as proteins). The units linked to each other by the process of dehydration synthesis to form polymers. The problem is that the abiotic synthesis of polymers had to occur without the assistance of enzymes. In addition, these reactions give off water and would, therefore, not occur spontaneously in a watery environment. Sydney Fox of the University of Miami suggested that waves or rain in the primitive environment splashed organic monomers on fresh lava or hot rocks, which would have allowed polymers to form abiotically. When he tried to do this in his laboratory, Fox produced proteinoids—abiotically synthesized polypeptides.

The next step in chemical evolution suggests that polymers interacted with each other and organized into

KEY TERMS
. . . .

Abiogenesis—Origin of living organisms from nonliving material.

Autotroph—This refers to organisms that can synthesize their biochemical constituents using inorganic precursors and an external source of energy.

Heterotroph—Organism that requires food from the environment since it is unable to synthesize nutrients from inorganic raw materials.

Prokaryote—Type of cell that lacks a membraneenclosed nucleus. Found solely in bacteria.

aggregates, known as protobionts. Protobionts are not capable of reproducing, but had other properties of living things. Scientists have successfully produced protobionts from organic molecules in the laboratory. In one study, proteinoids mixed with cool water assembled into droplets or microspheres that developed membranes on their surfaces. These are protobionts, with semipermeable and excitable membranes, similar to those found in cells.

In the final step of chemical evolution, protobionts developed the ability to reproduce and pass genetic information from one generation to the next. Some scientists theorize RNA to be the original hereditary **molecule**. Short polymers of RNA have been synthesized abiotically in the laboratory. In the 1980s, Thomas Cech and his associates at the University of Colorado at Boulder discovered that RNA molecules can function as enzymes in cells. This implies that RNA molecules could have replicated in prebiotic cells without the use of protein enzymes. Variations of RNA molecules could have been produced by mutations and by errors during replication. Natural **selection**, operating on the different RNAs would have brought about subsequent evolutionary development. This would have fostered the survival of RNA sequences best suited to environmental parameters, such as **temperature** and **salt concentration**. As the protobionts grew and split, their RNA was passed on to offspring. In time, a diversity of prokaryote cells came into existence. Under the influence of natural selection, the prokaryotes could have given rise to the vast variety of life on Earth.

See also Amino acid.

Resources

Books

Keeton, William T., and James L. Gould. *Biological Science.* New York: W.W. Norton and Co., 1993.

Periodicals

Franklin, Carl. "Did Life Have a Simple Start?" *New Scientist* (October 2, 1993).

Radetsky, Peter. "How Did Life Start?" *Discover* (November 1992).

Bernice Essenfeld

Chemical oxygen demand

Chemical **oxygen** demand (COD) is a measure of the capacity of **water** to consume oxygen during the **decomposition** of organic **matter** and the oxidation of inorganic chemicals such as **ammonia** and nitrite. COD measurements are commonly made on samples of waste waters or of natural waters contaminated by domestic or industrial wastes. Chemical oxygen demand is measured as a standardized laboratory assay in which a closed water **sample** is incubated with a strong chemical oxidant under specific conditions of **temperature** and for a particular period of **time**. A commonly used oxidant in COD assays is potassium dichromate ($K_2Cr_2O_7$) which is used in combination with boiling **sulfuric acid** (H_2SO_4). Because this chemical oxidant is not specific to oxygen-consuming chemicals that are organic or inorganic, both of these sources of oxygen demand are measured in a COD assay.

Chemical oxygen demand is related to **biochemical oxygen demand** (BOD), another standard test for assaying the oxygen-demanding strength of waste waters. However, biochemicaloxygen demand only measures the amount of oxygen consumed by microbial oxidation and is most relevant to waters rich in organic matter. It is important to understand that COD and BOD do not necessarily measure the same types of oxygen consumption. For example, COD does not measure the oxygen-consuming potential associated with certain dissolved organic compounds such as acetate. However, acetate can be metabolized by **microorganisms** and would therefore be detected in an assay of BOD. In contrast, the oxygen-consuming potential of **cellulose** is not measured during a short-term BOD assay, but it is measured during a COD test.

Chemical reactions

Chemical reactions describe the changes between reactants (the initial substances that enter into the reaction) and products (the final substances that are present at the end of the reaction). Describing interactions among chemical species, chemical reactions involve a rearrangement of the **atoms** in reactants to form products with new structures in such a way as to conserve atoms. Chemical equations are notations that are used to concisely summarize and convey information regarding chemical reactions.

In a balanced chemical reaction all of the **matter** (i.e., atoms or molecules) that enter into a reaction must be accounted for in the products of a reaction. Accordingly, associated with the symbols for the reactants and products are numbers (stoichiometry coefficients) that represent the number of molecules, formula units, or moles of a particular reactant or product. Reactants and products are separated by **addition** symbols (addition signs). The addition signs represent the interaction of the reactants and are used to separate and list the products formed. The chemical equations for some reactions may have a lone reactant or a single product. The subscript numbers associated with the chemical formula designating individual reactants and products represent the number of atoms of each element that are in each **molecule** (for covalently bonded substances) or formula unit (for ironically associated substances) of reactants or products.

For a chemical reaction to be balanced, all of the atoms present in molecules or formula units or moles of reactants to the left of the equation arrow must be present in the molecules, formula units and moles of product to the right of the equation arrow. The combinations of the atoms may change (indeed, this is what chemical reactions do) but the number of atoms present in reactants must equal the number of atoms present in products.

Charge is also conserved in balanced chemical reactions and therefore there is a conservation of electrical charge between reactants and products.

Although chemical equations are usually concerned only with reactants and products chemical reactions may proceed through multiple intermediate steps. In such multi-step reactions the products of one reaction become the reactants (intermediary products) for the next step in the reaction sequence.

Reaction catalysts are chemical species that alter the **energy** requirements of reactions and thereby alter the speed at which reactions run (i.e., control the **rate** of formation of products).

Combustion reactions are those where **oxygen** combines with another compound to form **water** and **carbon dioxide**. The equations for these reactions usually designate that the reaction is exothermic (**heat** producing). Synthesis reactions occur when two or more simple compounds combine to form a more complicated compound. **Decomposition** reactions reflect the reversal of synthesis reactions (e.g., reactions where complex molecules are broken down into simpler molecules). The **electrolysis** of water to make oxygen and **hydrogen** is an excellent example of a decomposition reaction.

Equations for single displacement reactions, double displacement, and acid-base reactions reflect the appropriate reallocation of atoms in the products.

In accord with the laws of **thermodynamics**, all chemical reactions change the energy state of the reactants. The change in energy results from changes in the in the number and strengths of chemical bonds as the reaction proceeds. The heat of reaction is defined as the quantity of heat evolved or absorbed during a chemical reaction. A reaction is called exothermic if heat is released or given off during a chemical transformation. Alternatively, in an **endothermic** reaction, heat is absorbed in transforming reactants to products. In endothermic reactions, heat energy must be supplied to the system in order for a reaction to occur and the heat content of the products is larger than that of the reactants. For example, if a mixture of gaseous hydrogen and oxygen is ignited, water is formed and heat energy is given off. The chemical reaction is an exothermic reaction and the heat content of the product(s) is lower than that for the reactants. The study of energy utilization in chemical reactions is called chemical kinetics and is important in understanding chemical transformations.

A chemical reaction takes place in a vessel which can be treated as a system. If the heat "flows" into the vessel during reaction, the reaction is said to be "endothermic" (e.g., a decomposition process) and the amount of heat, say, q, provided to the system is taken as a positive quantity. On the other hand, when the system has lost heat to the outside world, the reaction is "exothermic" (e.g., a combustion process) and q is viewed as a **negative** number. Normally the heat change involved in a reaction can be measured in an adiabatic bomb calorimeter. The reaction is initiated inside a constant-volume container. The observed change in **temperature** and the information on the total **heat capacity** of the colorimeter are employed to calculate q. If the heat of reaction is obtained for both the products and reactants at the same temperature after reaction and also in their standard states, it is then defined as the "standard heat of reaction," denoted by ΔH°.

Both chemical kinetics and thermodynamics are crucial issues in studying chemical reactions. Chemical kinetics help us search for the factors that influence the rate of reaction. It provides us with the information about how fast the chemical reaction will take place and about what the sequence of individual chemical events is to produce observed reactions. Very often, a single reaction like A → B may take several steps to complete. In other words, a chain reaction mechanism is actually involved which can include initiation, propagation, and termination stages, and their individual reaction rates may be very different. With a search for actual reaction mechanisms, the expression for overall reaction rate can be given correctly. As to determining the maximum extent to which a chemical reaction can proceed and how much heat will be absorbed or liberated, we need to estimate from thermodynamics data. Therefore, kinetic and thermodynamic information is extremely important for reactor design.

As an example of a chemical reaction, Hydrogen (H_2) and oxygen (O_2) gases under certain conditions can react to form water (H_2O). Water then exists as solid (**ice**), liquid, or vapor (steam); they all have the same composition, H_2O, but exhibit a difference in how H_2O molecules are brought together due to variations in temperature and **pressure**.

Chemical reactions can take place in one phase alone and are termed "homogeneous." They can also proceed in the presence of at least two phases, such as reduction of **iron ore** to iron and **steel**, which are normally described as "heterogeneous" reactions. Quite frequently, the rate of chemical reaction is altered by foreign materials, so-called catalysts, that are neither reactants nor products. Although usually used to accelerate reactions, reaction catalysts can either accelerate or hinder the reaction process. Typical examples are found in Pt as the catalyst for oxidation of **sulfur dioxide** (SO_2) and iron promoted with Al_2O_3 and K as the catalyst for **ammonia** (NH_3) synthesis.

Chemical reactions can characterized as irreversible, reversible, or oscillating. In the former case, the equilibrium for the reaction highly favors formation of the products, and only a very small amount of reactants remains in the system at equilibrium. In contrast to this, a reversible reaction allows for appreciable quantities of all reactants and products co-existing at equilibrium. $H_2O + 3NO_2 \rightleftarrows 2HNO_3 + NO$ is an example of a reversible reaction. In an oscillating chemical reaction, the concentrations of the reactants and products change with **time** in a periodic or quasi-periodic manner. Chemical oscillators exhibit chaotic behavior, in which concentrations of products and the course of a reaction depend on the initial conditions of the reaction.

Chemical reactions may proceed as a single reaction A → B, series reactions A → B → C, side-by-side **parallel** reactions A → B and C → D, two competitive parallel reactions → B and A → C, or mixed parallel and series reactions A + B → C and C + B → D. In order for chemical reactions to occur, reactive species

KEY TERMS

Chemical kinetics—The study of the reaction mechanism and rate by which one chemical species is converted to another.

Equilibrium—The conditions under which a system shows no tendency for a change in its state. At equilibrium the net rate of reaction becomes zero.

Phase—A homogeneous region of matter.

Standard state—The state defined in reaction thermodynamics for calculation purposes in which the pure gas in the ideal-gas state at 1 atm and pure liquid or solid at 1 atm are taken for gas and liquid or solid, respectively.

Thermodynamics—Thermodynamics is the study of energy in the form of heat and work, and the relationship between the two.

have to first encounter each other so that they can exchange atoms or groups of atoms. In gas phases, this step relies on collision, whereas in liquid and solid phases, **diffusion** process (**mass** transfer) plays a key role. However, even reactive species do encounter each other, and certain energy state changes are required to surmount the energy barrier for the reaction. Normally, this minimum energy requirement (e.g., used to break old chemical bonds and to form new ones) is varied with temperature, pressure, the use of catalysts, etc. In other words, the rate of chemical reaction depends heavily on encounter rates or frequencies and energy availability, and it can vary from a value approaching **infinity** to essentially **zero**.

See also Catalyst and catalysis; Chemical bond; Chemistry; Conservation laws; Entropy; Enzyme; Equation, chemical; Equilibrium, chemical; Molecular formula; Moles; Stereochemistry.

Resources

Books

Housecroft, Catherine E., et al. *Inorganic Chemistry*. Prentice Hall, 2001.

Incropera, Frank P., and David P. DeWitt. *Fundamentals of Heat and Mass Transfer*. 5th ed. John Wiley & Sons, 2001.

Moran, Michael J., and Howard N. Shapiro. *Fundamentals of Engineering Thermodynamics*. 4th ed. John Wiley & Sons, 2000.

K. Lee Lerner
Pang-Jen Kung

Chemical warfare

Chemical warfare involves the use of natural or synthetic substances to incapacitate or kill an enemy or to deny them the use of resources such as agricultural products or screening foliage. The effects of the chemicals may last only a short time, or they may result in permanent damage and death. Most of the chemicals used are known to be toxic to humans or **plant** life. Other normally benign (mild) chemicals have also been intentionally misused in more broadly destructive anti-environmental actions, called ecocide, and as a crude method of causing mayhem and damaging an enemy's economic system. The deliberate dumping of large quantities of crude oil on the land or in the **ocean** is an example.

Chemical warfare dates back to the earliest use of weapons. Poisoned arrows and darts used for hunting were also used as weapons in intertribal conflicts (and primitive peoples still use them for these purposes today). In 431 B.C., the Spartans used burning **sulfur** and pitch to produce **clouds** of suffocating **sulfur dioxide** in their sieges against Athenian cities. When the Romans defeated the Carthaginians of North **Africa** in 146 B.C. during the last of a series of Punic Wars, they levelled the city of Carthage and treated the surrounding fields with **salt** to destroy the agricultural capability of the land, thereby preventing the rebuilding of the city.

The attraction of chemicals as agents of warfare was their ability to inflict **mass** casualties or damage to an enemy with only limited risk to the forces using the chemicals. Poisoning a town's **water** supply, for example, posed almost no threat to an attacking army, yet resulted in the death of thousands of the town's defenders. In many cases, the chemicals were also not detectable by the enemy until it was too late to take action.

Chemical agents can be classified into several general categories. Of those that attack humans, some, like tear gas, cause only temporary incapacitation. Other agents cause violent skin irritation and blistering, and may result in death. Some agents are poisonous and are absorbed into the bloodstream through the lungs or skin to kill the victim. Nerve agents attack the **nervous system** and kill by causing the body's vital functions to cease. Still others cause psychological reactions including disorientation and hallucinations. Chemical agents which attack vegetation include defoliants that kill plant leaves, **herbicides** that kill the entire plant, and **soil** sterilants that prevent the growth of new vegetation.

Antipersonnel agents—chemicals used against people

The first large-scale use of poisonous chemicals in warfare occurred during World War I. More than 100,000 tons (90,700 metric tons) of lethal chemicals were used by both sides during several battles in an effort to break the stalemate of endless trench warfare. The most commonly used chemicals were four lung-destroying poisons: chlorine, chloropicrin, phosgene, and trichloromethyl chloroformate, along with a skin-blistering agent known as **mustard gas**, or bis (2-chloroethyl) sulfide. These poisons caused about 100,000 deaths and another 1.2 million injuries, almost all of which involved military personnel.

Despite the agreements of the Geneva Protocol of 1925 to ban the use of most chemical weapons, the United States, Britain, Japan, Germany, Russia and other countries all continued development of these weapons during the period between World War I and World War II. This development included experimentation on animals and humans. Although there was only limited use of chemical weapons during World War II, the opposing sides had large stockpiles ready to deploy against military and civilian targets.

During the war in Vietnam, the United States military used a nonlethal "harassing agent" during many operations. About 9,000 tons (8,167 tonnes) of tear gas, known as CS or o-chlorobenzolmalononitrile, were sprayed over 2.5 million acres (1.0 million ha) of South Vietnam, rendering the areas uninhabitable for 15-45 days. Although CS is classified as nonlethal, several hundred deaths have been reported in cases when CS has been used in heavy concentrations in confined spaces such as underground bunkers and bomb shelters.

Poisonous chemicals were also used during the Iran-Iraq War of 1981-1987, especially by Iraqi forces. During that war, both soldiers and civilians were targets of chemical weapons. Perhaps the most famous incident was the gassing of Halabja, a town in northern Iraq that had been overrun by Iranian-supported Kurds. The Iraqi military attacked Halabja with two rapidly acting neurotoxins, known as sabin and tabun, which cause rapid death by interfering with the transmission of nerve impulses. About 5,000 people, mostly civilians, were killed in this incident.

Use of herbicides during the Vietnam War

During the Vietnam War, the U.S. military used large quantities of herbicides to deny their enemies agricultural food production and forest cover. Between 1961 and 1971, about 3.2 million acres (1.3 million ha) of forest and 247,000 acres (100,000 ha) of croplands were sprayed at least once. This is an area equivalent to about one-seventh of South Vietnam.

The most commonly used herbicide was called **agent orange**, a one-to-one blend of two phenoxy herbi-

Soldiers at Assaf Harofe Hospital washing "victims" in simulated chemical attack. *Photograph by Jeffrey L. Rotman. Corbis. Reproduced by permission.*

cides, 2,4-D and 2,4,5-T. Picloram and cacodylic acid were also used, but in much smaller amounts. In total, this military action used about 25,000 tons of 2,4-D; 21,000 tons of 2,4,5-T; and 1,500 tons of picloram. Agent orange was sprayed at a **rate** of about 22.3 lb/acre (25 kg/ha), equivalent to about ten times the rate at which those same chemicals were used for plant control in **forestry**. The spray rate was much more intense during warfare, because the intention was to destroy the ecosystems through ecocide, rather than to manage them towards a more positive purpose.

The ecological damages caused by the military use of herbicides in Vietnam were not studied in detail; however, cursory surveys were made by some visiting ecologists. These scientists observed that coastal mangrove **forests** were especially sensitive to herbicides. About 36% of the mangrove **ecosystem** of South Vietnam was subjected to herbicides, amounting to 272,000 acres (110,000 ha). Almost all of the plant **species** of mangrove forests proved to be highly vulnerable to herbicides, including the dominant species, the red mangrove. Consequently, mangrove forests were devastated over large areas, and extensive coastal barrens were created.

There were also severe ecological effects of herbicide spraying in the extremely biodiverse upland forests of Vietnam, especially rain forests. Mature tropical forests in this region have many species of hardwood trees. Because this forested ecosystem has such a dense and complexly layered canopy, a single spraying of herbicide killed only about 10% of the larger trees. Re-sprays of upland forests were often made, however, to achieve a greater and longer-lasting defoliation. To achieve this effect, about 34% of the area of Vietnam that was subjected to herbicides was treated more than once.

The effects on animals of the herbicide spraying in Vietnam were not well documented; however, there are many accounts of sparse populations of **birds, mammals, reptiles,** and other animals in the herbicide-treated mangrove forests and of large decreases in the yield of near-shore fisheries, for which an intact mangrove ecosystem provides important spawning and nursery **habitat.** More than a decade after the war, Vietnamese ecologists examined an inland valley that had been converted by herbicide spraying from a rich upland tropical forest into a degraded ecosystem dominated by **grasses** and shrubs. The secondary, degraded landscape only supported 24 species of birds and five species of mammals, compared with 145-170 birds and 30-55 mammals in nearby unsprayed forests.

The effects on wild animals were probably caused mostly by habitat changes resulting from herbicide spraying. There were also numerous reports of domesticated agricultural animals becoming ill or dying. Because of the constraints of warfare, the specific causes of these illnesses and deaths were never studied properly by veterinary scientists; however, these ailments were commonly attributed to toxic effects of exposure to herbicides, mostly ingested with their food.

Use of petroleum as a weapon during the Gulf War

Large quantities of **petroleum** are often spilled at sea during warfare, mostly through the shelling of tankers or facilities such as offshore production platforms. During the Iran-Iraq War of the 1980s and the brief Gulf War of 1991-1992, **oil spills** were deliberately used to gain tactical advantage, as well as inflicting economic damages on the postwar economy.

The world's all-time largest oceanic spill of petroleum occurred during the Gulf War, when the Iraqi military deliberately released almost 1.0 million tons (907,441 tonnes) of crude oil into the Persian Gulf from several tankers and an offshore facility for loading tankers. In part, the oil was spilled to establish a defensive barrier against an amphibious counter-invasion of Kuwait by coalition forces. Presumably, if the immense quantities of spilled petroleum could have been ignited, the floating inferno might have provided an effective barrier to a seaborne invasion. The spilled oil might also have created some military advantage by contaminating the seawater intakes of Saudi Arabian **desalination** plants, which supply most of that nation's fresh water and, therefore, have great strategic value.

Another view is that this immense spillage of petroleum into the ocean was simply intended to wreak economic and ecological havoc. Certainly, there was no other reason for the even larger spillages that were deliberately caused when Iraqi forces sabotaged and ignited the wellheads of 788 Kuwaiti oil wells on land. This act caused enormous releases of petroleum and **combustion** residues to the land and air for the following year. Although the wells were capped, there will be lingering **pollution** of the land for many decades.

Controls over the use of chemical weapons

The first treaty to control the use of chemical weapons was negotiated in 1925 and subsequently signed by the representatives of 132 nations. This Geneva Protocol was stimulated by the horrific uses of chemical weapons during World War I, and it banned the use of asphyxiating, poisonous, or other gases, as well as bacteriological methods of warfare. In spite of their having signed this treaty, it is well known that all major nations subsequently engaged in research towards the development of new, more effective chemical and bacteriological weapons.

In 1993, negotiators for various nations finalized the Chemical Weapons Convention, which would require the destruction of all chemical weapons within 10–15 years of the ratification of the treaty. This treaty has been signed by 147 nations but is not yet being enforced. The Chemical Weapons Convention is an actual pact to achieve a disarmament of chemical weapons; however, its effectiveness depends on its ratification by all countries having significant stockpiles of chemical weapons, their subsequent good faith actions in executing the provisions of the treaty, and the effectiveness of the associated international monitoring program to detect non-compliance.

It is important to understand that the destruction of existing chemical weapons will not be an inexpensive activity. It has been estimated that it could cost $16–20 billion just to safely destroy the chemical weapons of the United States and Russia.

Terrorism and chemical weapons

Chemicals in the hands of terrorists hold a grave potential for disaster, according to experts in "weapons of mass destruction," which include chemical, biological, and nuclear (or radiological) devices. In 1995, the Japanese sect called Aum Shinrikyo unleashed **sarin gas** in a Tokyo subway. Twelve people died, and 5,000 got sick from the attack; and experts claim that the toll should have been higher but for the terrorists' minor errors. Counterterrorism forces have been established by various Federal agencies, but the substances and handbooks for synthesizing chemicals are available on the Internet, through the mail, and at survivalist shows. Dangers of dispersing chemicals (including effects of the **weather**) may dissuade some; but the unknown quantities of who might have a grudge, the knowledge to choose chemicals as weapons, and the location and speed of a chemical attack make **nuclear weapons** seem better controlled and less dangerous by comparison.

See also Poisons and toxins.

Resources

Books

Fleming, D. O., and D. L. Hunt. *Biological Safety: Principles and Practices.* 3rd ed. Washington: American Society for Microbiology, 2000.

KEY TERMS

. . . .

Defoliant—A chemical that kills the leaves of plants and causes them to fall off.

Ecocide—The deliberate carrying out of antienvironmental actions over a large area as a tactical element of a military strategy.

Harassing agent—A chemical which causes temporary incapacitation of animals, including humans.

Herbicide—A chemical that kills entire plants, often selectively.

Nerve agent—A chemical which kills animals, including humans, by attacking the nervous system and causing vital functions, such as respiration and heartbeat, to cease.

Franz, David R., and Nancy K. Jaax. "Ricin Toxin." *Medical Aspects of Chemical and Biological Warfare.* Washington, DC: Borden Institute, Walter Reed Army Medical Center, 1997. p. 632.

Freedman, B. *Environmental Ecology.* 2nd ed. Academic Press, 1994.

Harris, Robert, and Jeremy Paxman. *A Higher Form of Killing.* Hill and Wang, 1982.

Proliferation: Threat and Response. Washington, DC: Department of Defense, 2001. p.15.

Sivard, R. L. *World Military and Social Expenditures, 1993.* World Priorities, 1993.

Tucker, Jonathan B., and Jason Pate. "The Minnesota Patriots Council." *Toxic Terror: Assessing Terrorist Use of Chemical and Biological Weapons.* Cambridge, MA: MIT Press, 2000. pp. 159-183.

Periodicals

"Better Killing Through Chemistry." *Scientific American* (December 2001).

Byrne, W. Russell, et al."Clinical Recognition and Management of Patients Exposed to Biological Warfare Agents," *Laboratory Aspects of Biowarfare (Clinics in Laboratory Medicine)* 21, no. 3, (September 2001): 459.

Dire, D. J., and T. W. McGovern. "CBRNE—Biological Warfare Agents." *eMedicine Journal* no. 4 (April 2002): 1-39.

Dutton, Gail. "Biotechnology Counters Bioterrorism." *Genetic Engineering News* no. 21 (December 2000): 1-22ff.

Dyer, G. "Environmental Warfare in the Gulf." *Ecodecision* (1991): 21-31.

Greenfield, Ronald A. "Microbiological, Biological, and Chemical Weapons of Warfare And Terrorism." *American Journal of The Medical Sciences* 323 no. 6 (2002): 326-340.

Reutter, S. "Hazards of Chemical Weapons Release During War: New Perspectives." *Environmental Health Perspectives* 107, no. 12 (1999): 985-990.

Other

"US Knew of Bioterror Tests in Iraq." BBC News. August 20, 2002.

Bill Freedman
Chris Cavette

Chemistry

Chemistry is the science that studies why materials have their characteristic properties, how these particular qualities relate to their simplest structure, and how these properties can be modified or changed. The term chemistry is derived from the word alchemist, which finds its roots in the Arabic name for Egypt *al-Kimia.* The Egyptians are credited with being the first to study chemistry. They developed an understanding of the materials around them and became very skillful at making different types of metals, manufacturing colored **glass**, dying cloth, and extracting oils from plants. Today, chemistry is divided into four traditional areas: organic, inorganic, analytical, and physical. Each discipline investigates a different aspect of the properties and reactions of the substances in our universe. The different areas of chemistry have the common goal of understanding and manipulating **matter**.

Organic chemistry is the study of the chemistry of materials and compounds that contain **carbon atoms**. Carbon atoms are one of the few elements that bond to each other. This allows vast variation in the length of carbon atom chains and an immense number of different combinations of carbon atoms, which form the basic structural framework for millions of molecules.

The word organic is used because most natural compounds contain carbon atoms and are isolated from either plants or animals. Rubber, vitamins, cloth, and **paper** represent organic materials we come in contact with on a daily basis. Organic chemistry explores how to change and connect compounds based on carbon atoms in order to synthesize new substances with new properties. Organic chemistry is the backbone in the development and manufacture of many products produced commercially, such as drugs, food preservatives, perfumes, food flavorings, dyes, etc. For example, scientists recently discovered that chlorofluorocarbon containing compounds, or CFCs, are depleting the **ozone** layer around the **earth**. One of these CFCs is used in refrigerators to keep food cold. Organic chemistry was used to make new carbon atom containing compounds that offer the same physical capabilities as the chlorofluorocarbons in maintaining a cold environment, but do not deplete the ozone layer. These compounds are called hydrofluoro-carbons or HFCs and are not as destructive to the earth's protective layer.

Inorganic chemistry studies the chemistry of all the elements in the **periodic table** and their compounds, except for carbon-hydrogen compounds. Inorganic chemistry is a very diverse field because it investigates the properties of many different elements. Some materials are solids and must be heated to extremely high temperatures to react with other substances. For example, the powder responsible for the **light** and **color** of **fluorescent light** bulbs is manufactured by heating a mixture of various solids to very high temperatures in a poisonous atmosphere. An inorganic compound may alternatively be very unreactive and require special techniques to change its chemical composition. Electronic components such as transistors, diodes, computer chips, and various **metal** compounds are all constructed using inorganic chemistry. In order to make a new gas for refrigerators that does not deplete the ozone layer, inorganic chemistry was used to make a metal catalyst that facilitated the large scale production of HFCs for use throughout the world.

Physical chemistry is the branch of chemistry that investigates the physical properties of materials and relates these properties to the structure of the substance. Physical chemistry studies both organic and inorganic compounds and measures such variables as the **temperature** needed to liquefy a solid, the **energy** of the light absorbed by a substance, and the **heat** required to accomplish a chemical transformation. Computers may be used to calculate the properties of a material and compare these assumptions to laboratory measurements. Physical chemistry is responsible for the theories and understanding of the physical phenomenon utilized in organic and inorganic chemistry. In the development of the new refrigerator gas, physical chemistry was used to measure the physical properties of the new compounds and determine which one would best serve its purpose.

Analytical chemistry is the area of chemistry that develops methods to identify substances by analyzing and quantifying the exact composition of a mixture. A material may be identified by measurement of its physical properties. Examples of physical properties include the **boiling point** (the temperature at which the physical change of state from a liquid to a gas occurs) and the refractive index (the **angle** at which light is bent as it shines though a **sample**). Materials may also be identified by their reactivity with various known substances. These characteristics that distinguish one compound from another are also used to separate a mixture of materials into their component parts. If a liquid contains two materials with different boiling points, then the liquid can be separated into its components by heating the mixture until one of the mate-

KEY TERMS

Analytical chemistry—That area of chemistry that develops ways to identify substances and to separate and measure the components in a mixture.

Inorganic chemistry—The study of the chemistry of all the elements in the periodic table and their compounds except for carbon-hydrogen compounds.

Organic chemistry—The study of the chemistry of materials and compounds that contain carbon atoms.

Physical chemistry—The branch of chemistry that investigates the properties of materials and relates these properties to the structure of the substance.

rials boils out and the other remains. By measuring the amount of the remaining liquid, the component parts of the original mixture can be calculated. Analytical chemistry can be used to develop instruments and chemical methods to characterize, separate, and measure materials. In the development of HFCs for refrigerators, analytical chemistry was used to determine the structure and purity of the new compounds tested.

Chemists are scientists who work in the university, the government, or the industrial laboratories investigating the properties and reactions of materials. These people research new theories and **chemical reactions** as well as synthesize or manufacture drugs, **plastics**, and chemicals. Today's chemists also explore the boundaries of chemistry and its connection with the other sciences, such as **biology**, **physics**, **geology**, environmental science, and **mathematics**.

Applications of new theories and reactions are important in the field of chemical technology. Many of the newest developments are on the atomic and molecular level. One example is the development of "smart molecules" such as a **polymer** chain that could replace a fiber optic cable. The chemist of today may have many so-called non-traditional occupations such as a pharmaceutical salesperson, a technical writer, a science librarian, an investment broker, or a patent lawyer, since discoveries by a traditional chemist may expand and diversify into a variety of fields which encompass our whole society.

Further Reading

Castellan, G.W. *Physical Chemistry.* Addison-Wesley, 1983.
Hargis, L. *Analytical Chemistry: Principles & Techniques.* Prentice-Hall, 1988.
Huheey, J. *Inorganic Chemistry.* New York: Harper & Row, 1983.

McMurry, J. *Organic Chemistry*. Pacific Grove, CA: Brooks/ Cole Publishing Co., 1992.

Segal, B. *Chemistry, Experiment and Theory*. New York: John Wiley & Sons, 1989.

Chemoreception

Chemoreception is the biological recognition of chemical stimuli, by which living organisms collect information about the **chemistry** of their internal and external environments. Chemoreception has three sequential stages: detection, amplification, and signaling.

In detection, a **molecule** typically binds to a chemoreceptor protein on the surface of a **cell**, changing the shape of the chemoreceptor. All chemoreceptors therefore have some degree of specificity, in that they only bind to specific molecules or specific classes of molecules.

In amplification, the cell uses **energy** to transform the shape change of the chemoreceptor into biochemical or electrical signals within the cell. In many cases, amplification is mediated by formation of cAMP (cyclic adenosine monophosphate), which increases the cell's permeability to **sodium** ions and alters the electrical potential of the cell **membrane**.

In signaling, the amplified signal is transformed into a physiological or behavioral response. In higher animals the **nervous system** does the signaling, while in single-celled organisms signaling is intracellular, which may be manifested as chemotaxis, a directional movement in response to a chemical **stimulus**.

Detection, amplification, and signaling are often connected by feedback pathways. Feedback pathways allow adjustment of the sensitivity of the chemoreceptive system to different **concentration** ranges of the elicitor molecule. Thus, the sensitivity decreases as the background concentration of the molecule increases; the sensitivity increases as the background concentration of the molecule decreases.

Chemoreceptive systems detect chemical changes within an **organism** (interoreception) or outside an organism (exteroreception). The most familiar examples of exteroreception in humans are the senses of **taste** and **smell**.

Humans have chemoreceptor cells for taste in taste buds, most of which are on the upper surfaces of the tongue. Each human has about 10,000 taste buds and each taste bud consists of about 50 cells. An individual taste bud is specialized for detection of a sweet, sour, salty, or bitter taste. The sense of smell is important in discriminating among more subtle differences in taste.

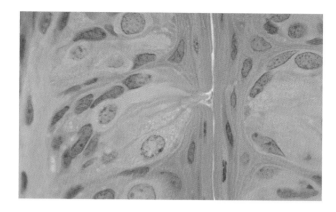

Taste buds on the tongue. *Carolina Biological Supply/Phototake NYC. Reproduced by permission.*

Human chemoreceptors in the nasal cavity can discriminate thousands of different odors. One theory of odor **perception** in humans proposes that each chemoreceptive cell is connected to a single **neuron** and that an odorant molecule binds to many different chemoreceptors with different affinities. Thus, the neural signals from many different neurons can be integrated in many different ways to yield a rich panoply of odor sensations.

Many chemoreception systems also collect information about the internal environment of multicellular organisms. For example, the carotid body in the carotid artery of humans has chemoreceptive cells which respond to changes in the **pH** and **oxygen** levels in the **blood**. As the amount of dissolved oxygen in the blood decreases, chemoreceptive cells in the carotid body emit an electrical discharge, which stimulates specific neurons in the hind **brain** respiratory centers to increase the **rate** of breathing. The hypothalamus in the human brain has chemoreceptive cells which respond to changes in blood glucose levels. When blood glucose levels fall, the chemoreceptive system causes a person to feel hungry; when blood glucose levels rise, this chemoreceptive system causes a person to feel satiated. The endocrine and nervous systems also have many other chemoreceptive cells which signal different organs within the body to change their activity.

Cherry *see* **Rose family (Rosaceae)**

Chestnut

Chestnuts are **species** of trees in the genus *Castanea*, family Fagaceae. They are species of temperate hardwood (angiosperm-dominated) **forests** found in the Northern Hemisphere and are indigenous to eastern **North America** and Eurasia. Species in the genus *Castanea* can grow to be 100 ft (30 m) tall. They have simple leaves with a broadly toothed margin and sweet-smelling, yellowish, insect-pollinated, early-summer flowers aggregated on a long flowering axis. Fertilized flowers develop into prickly, tough-coated fruit, containing two to three large, rich-brown colored, edible **seeds** (or nuts). True chestnut seeds should not be confused with horse chestnuts, or buckeyes, genus *Aescellus*, which have somewhat poisonous seeds.

The **wood** of all chestnut species can be manufactured into an open-grained, decay-resistant, lumber. This wood has a rich brown **color** and can be worked easily to manufacture fine furniture and musical instruments. Chestnut is also used for its durability in construction timber, railway ties, pit props, and shingles.

The sweet chestnut

The sweet chestnut (*Castanea sativa*) is a cultivated species originally native to southern **Europe** and **Asia** Minor. There are extensive plantations in parts of southern France, Italy, and some other countries. This **tree** grows 30-100 ft tall (9-30 m), with wide, spreading branches. The nuts of the sweet chestnut are highly nutritious, containing about 80% starch and 4% oil. Chestnuts are eaten roasted or boiled, or sometimes ground into a flour and used to make cakes. In 1999, the global crop of sweet chestnut was harvested from about 630,000 acres (255,000 ha) and had a production of 573,000 tons (521,000 tonnes).

The American chestnut

The American chestnut tree, *Castanea dentata*, is native to the rich hardwood forests of the northeastern United States and southeastern Canada. It was once a dominant species in forests of this region, occurring over an area of approximately 9 million acres (3.6 million ha), and particularly abundant in southern Appalachia. The American chestnut was an economically important tree. At one time, chestnuts contributed about one-tenth of the sawlog production in the United States. Its nuts were gathered as food for humans, were eaten by **livestock**, and were a staple for wild species such as **turkeys**, passenger pigeons, and forest **rodents**.

The American chestnut was nearly wiped out by the introduction of chestnut blight fungus (*Endothia parasitica*). This fungus is a wind- and animal-dispersed pathogen that was inadvertently introduced with horticultural planting stock of an Asian species of chestnut (*Castanea crenata*) imported to New York. The first

symptoms of chestnut blight in the American chestnut were seen in 1902, and within 25 years it had been virtually eliminated as a canopy species in the deciduous forest of eastern North America, being replaced by other shade-tolerant species of trees. Many individuals of the American chestnut survived the blight, as their **root system** was not killed. They regenerated by growing stump-sprouts, but once these trees grew tall, they were again attacked by the fungus and knocked back. Efforts have been made to breed a light-resistant variety by crossing the American chestnut with Asian species. This has been somewhat successful and the **hybrid** chestnuts are now available as shade trees. Unfortunately, this is not likely to help the American chestnut, a native tree species, become prominent in the forests of eastern North America once again. It is possible, however, that the introduced blight pathogen will evolve to be less deadly to the American chestnut.

Other chestnuts

The Japanese chestnut (*Castanea crenata*) and Chinese chestnut (*Castanea mollissima*) are species of eastern Asia. They are resistant to chestnut blight and have been introduced to North America as shade trees.

Bill Freedman

Chi-square test

The chi-square test ($KHGR^2$) is the most commonly used method for comparing frequencies or proportions. It is a statistical test used to determine if observed data deviate from those expected under a particular hypothesis. The chi-square test is also referred to as a test of a measure of fit or "goodness of fit" between data. Typically, the hypothesis tested is whether or not two samples are different enough in a particular characteristic to be considered members of different populations. Chi-square analysis belongs to the family of univariate analysis, i.e., those tests that evaluate the possible effect of one **variable** (often called the independent variable) upon an outcome (often called the dependent variable).

The chi-square analysis is used to test the null hypothesis (H_0), which is the hypothesis that states there is no significant difference between expected and observed data. Investigators either accept or reject H_0, after comparing the value of chi-square to a probability distribution. Chi-square values with low probability lead to the rejection of H_0 and it is assumed that a factor other than chance creates a large deviation between expected and observed results. As with all non-parametric tests (that do not require normal distribution curves), chi-square tests only evaluate a single variable, thus they do not take into account the interaction among more than one variable upon the outcome.

A chi-square analysis is best illustrated using an example in which data from a population is categorized with respect to two qualitative variables. Table 1 shows a **sample** of patients categorized with respect to two qualitative variables, namely, **congenital heart** defect (CHD; present or absent) and karyotype (trisomy 21, also called **Down syndrome**, or trisomy 13, also called Patau **syndrome**). The classification table used in a chi-square analysis is called a contingency table and this is its simplest form (2 x 2). The data in a contingency table are often defined as row (r) and column (c) variables.

In general, a chi-square analysis evaluates whether or not variables within a contingency table are independent, or that there is no association between them. In this example, independence would mean that the proportion of individuals affected by CHD is not dependent on karyotype; thus, the proportion of patients with CHD would be similar for both Down and Patau syndrome patients. Dependence, or association, would mean that the proportion of individuals affected by CHD is dependent on kayotype, so that CHD would be more commonly found in patients with one of the two karyotypes examined.

Table 1 shows a 2 x 2 contingency table for a chi-square test—CHD (congenital heart defects) found in patients with Down and Patau syndromes

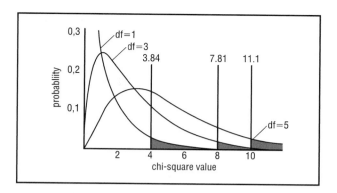

Figure 1. Chi-square distributions for 1, 3, and 5 degrees of freedom. The shaded region in each of the distributions indicates the upper 5% of the distribution. *Illustration by Argosy. The Gale Group.*

Chi-square is the sum of the squared difference between observed and expected data, divided by the expected data in all possible categories:

$$\chi^2 = (O_{11} - E_{11})^2 / E_{11} + (O_{12} - E_{12})^2 / E_{12} + (O_{21} - E_{21})^2 / E_{21} + (O_{22} - E_{22})^2 / E_{22},$$

where O_{11} represents the observed number of subjects in column 1, row 1, and so on. A summary is shown in Table 2.

The observed frequency is simply the actual number of observations in a cell. In other words, O_{11} for CHD in the Down-syndrome-affected individuals is 24. Likewise, the observed frequency of CHD in the Patau-syndrome-affected patients is 20 (O_{12}). Because the null hypothesis assumes that the two variables are indepen-

TABLE 1.				
		Karyotype		
		Down syndrome	**Patau syndrome**	**Total**
Congenital Heart Defects	CHD present	24	20	44
	CHD absent	36	5	41
	Total	60	25	85

TABLE 2.				
		Karyotype		
		Down syndrome	**Patau syndrome**	**Total**
Congenital Heart Defects	CHD present	O_{11}	O_{12}	r1
	CHD absent	O_{21}	O_{22}	r2
	Total	c1	c2	N

TABLE 3.

Observed (o)	Expected (e)	o − e	(o − e)²	(o − e)²/e
24	31.1	−7.1	50.41	1.62
20	12.9	7.1	50.41	3.91
36	28.9	7.1	50.41	1.74
5	12.1	−7.1	50.41	4.17
85	**85.0**			$\chi^2 = 11.44$

dent of each other, expected frequencies are calculated using the **multiplication** rule of probability. The multiplication rule says that the probability of the occurrence of two independent events X and Y is the product of the individual probabilities of X and Y. In this case, the expected probability that a patient has both Down syndrome and CHD is the product of the probability that a patient has Down syndrome (60/85=0.706) and the probability that a patient has CHD (44/85 = 0.518), or 0.706 x 0.518 = 0.366. The expected frequency of patients with both Down syndrome and CHD is the product of the expected probability and the total population studied, or 0.366 x 85 = 31.1.

Table 3 presents observed and expected frequencies and χ^2 for data in Table 1.

Before the chi-square value can be evaluated, the degrees of freedom for the data set must be determined. Degrees of freedom are the number of independent variables in the data set. In a contingency table, the degrees of freedom are calculated as the product of the number of rows minus 1 and the number of columns minus 1, or $(r-1)(c-1)$. In this example, $(2-1)(2-1) = 1$; thus, there is just one **degree** of freedom.

Once the degrees of freedom are determined, the value of χ^2 is compared with the appropriate chi-square distribution, which can be found in tables in most statistical analyses texts. A relative standard serves as the basis for accepting or rejecting the hypothesis. In biological research, the relative standard is usually $p = 0.05$, where p is the probability that the deviation of the observed frequencies from the expected frequencies is due to chance alone. If p is less than or equal to 0.05, then the null hypothesis is rejected and the data are not independent of each other. For one degree of freedom, the critical value associated with $p = 0.05$ for χ^2 is 3.84. Chi-square values higher than this critical value are associated with a statistically low probability that H_0 is true. Because the chi-square value is 11.44, much greater than 3.84, the hypothesis that the proportion of trisomy-13-affected patients with CHD does not differ significantly from the corresponding proportion for trisomy-21-affect-

ed patients is rejected. Instead, it is very likely that there is a dependence of CHD on karyotype.

Figure 1 shows chi-square distributions for 1, 3, and 5 degrees of freedom. The shaded region in each of the distributions indicates the upper 5% of the distribution. The critical value associated with $p = 0.05$ is indicated. Notice that as the degrees of freedom increases, the chi-square value required to reject the null hypothesis increases.

Because a chi-square test is a univariate test; it does not consider relationships among multiple variables at the same time. Therefore, dependencies detected by chi-square analyses may be unrealistic or non-causal. There may be other unseen factors that make the variables appear to be associated. However, if properly used, the test is a very useful tool for the evaluation of associations and can be used as a preliminary analysis of more complex statistical evaluations.

See also Bioinformatics and computational biology; Statistics.

Resources

Books

Grant, Gregory R., and Warren J. Ewens. *Statistical Methods in Bioinformatics*. New York: Springer Verlag, 2001.
Nikulin, Mikhail S., and Priscilla E. Greenwood. *A Guide to Chi-Square Testing*. New York: Wiley-Interscience, 1996.

Other

Rice University. *Rice Virtual Lab in Statistics*. "Chi Square Test of Deviations." (cited November 20, 2002). <http;://www.ruf.rice.edu/~lane/stat_sim/chisq_theor/>.

Antonio Farina

Chickenpox

Chickenpox, a **disease** characterized by skin lesions and low-grade fever, is common in the United States and other countries located in areas with temperate climates. The incidence of chickenpox is extremely high-almost everyone living in the United States contracts chickenpox, usually during childhood, but sometimes in adulthood. In the United States, about 3.9 million people a year contract chickenpox. A highly contagious disease, chickenpox is caused by Varicella-Zoster **virus** (VZV), the same virus that causes the skin disease **shingles**. For most cases of chickenpox, no treatment besides **pain** relief and management of itching is necessary. In some cases, however, chickenpox may evolve into more serious conditions, such as bacterial **infection** of the skin lesions or **pneumonia**. These complications tend to occur in persons with weakened immune systems, such as children receiving chemotherapy for **cancer** or people with Acquired Immune Deficiency Syndrome (**AIDS**). A **vaccine** for chickenpox is now receiving widespread use.

Despite its name, chickenpox has nothing to do with chickens. Its name has two possible origins. Some think that "chicken" comes from the French word *chiche* (chick-pea) because at one stage of the disease, the lesions do indeed resemble chick-peas. Others think that "chicken" may have evolved from the Old English word *gigan* (to itch). Interestingly, the term "varicella" is a diminutive form of the term "variola," the Latin term for **smallpox**. Although both chickenpox and smallpox are viral diseases that cause skin lesions, smallpox is more deadly and its lesions cause severe scarring.

Symptoms of chickenpox

Chickenpox is spread by breathing in respiratory droplets spread through the air by a cough or sneeze of an infected individual. Contact with the fluid from skin lesions can also spread the virus. The incubation period—or the **time** from exposure to VZV to the onset of the disease—is about 14-15 days. The most contagious period is just prior to the appearance of the rash, and early in the illness when fresh pox are still appearing. The first sign of chickenpox in children is often the appearance of the chickenpox rash. Adults and some children may have a prodrome, or series of warning symptoms. This prodrome is typical of the flu, and includes headache, fatigue, backache, and a fever. The onset of the rash is quite rapid. First, small red "dots" appear on the skin. Soon, a vesicle containing clear fluid appears in the center of the dot. This small, reddish bump with a central clear fluid is sometimes referred to as "dewdrop on a rose petal" appearance. The vesicle rapidly dries, forming a crust. This cycle, from the appearance of the dot to the formation of the crust, can take place within eight to 12 hours. As the crust dries, it falls off, leaving a slight depression that eventually recedes. Scarring from chickenpox is rare.

Over the course of a case of chickenpox, an individual may develop between 250 and 500 skin lesions. The lesions occur in waves, with the first set of lesions drying up just as successive waves appear. The waves appear over two to four days. The entire disease runs its course in about a week, but the lesions continue to heal for about two to three weeks. The lesions first appear on the scalp and trunk. Most of the lesions in chickenpox are found at the center of the body; few lesions form on the soles and palms. Lesions are also found on the mucous membranes, such as the respiratory tract, the gastrointestinal tract, and the urogenital tract. Researchers think that the lesions on the respiratory tract may help transmit the disease. If a person with respiratory lesions coughs, they may spray some of the vesicle fluid into the atmosphere, to be breathed by other susceptible persons.

Although the lesions look serious, chickenpox in children is usually a mild disease with few complications and a low fever. Occasionally, if the rash is severe, the fever may be higher. Chickenpox is more serious in adults, who usually have a higher fever and general malaise. The most common complaint about chickenpox from both children and adults is the itching caused by the lesions. It is important not to scratch the lesions, as scratching may cause scarring.

Treatment

Because chickenpox is usually a mild disease, no drug treatment is prescribed. Pain relief, in the form of acetaminophen (i.e., Tylenol) is recommended rather than salicylate, or aspirin. Salicylate may cause **Reye's syndrome**, a very serious neurological condition that is especially associated with aspirin intake and chickenpox; in fact, 20-30% of the total cases of Reye's syndrome occur in children with chickenpox. It is therefore important to control pain in children with chickenpox (or any other respiratory illness) with acetaminophen, not aspirin. Adults should also take acetaminophen if they have chickenpox.

The itching of the lesions can sometimes be controlled with calamine lotion or special preparations that are put into bath **water**. **Antihistamines** may also help relieve itching. The itching becomes less severe as the lesions heal; however, because children are more likely to scratch the lesions, the potential for scarring is greater in children than in adults.

Chickenpox, although not deadly for most people, can be quite serious in those who have weakened immune systems, and drug therapy is recommended for these cases. Antiviral drugs (such as acyclovir) have been shown to lessen the severity and duration of the disease, although some of the side effects, such as gastrointestinal upset, can be problematic.

Complications

If the lesions are severe and the person has scratched them, bacterial infection of the lesions can result. This complication is managed with antibiotic treatment. A more serious complication is pneumonia. Pneumonia is rare in otherwise healthy children and is more often seen in older patients or in children who already have a serious disease, such as cancer. Pneumonia is also treated with **antibiotics**. Another complication of chickenpox is shingles. Shingles are painful outbreaks of skin lesions that occur some years after a bout with chickenpox. Shingles are caused by VZV left behind in the body which then becomes reactivated. Shingles causes skin lesions and burning pain along the region served by a specific nerve. It is not clear why VZV is reactivated in some people and not in others, but many people with compromised immune systems can develop severe, even life-threatening cases of shingles.

Pregnant women are more susceptible to chickenpox, which also poses a threat to both prenatal and newborn children. If a woman contracts chickenpox in the first trimester (first three months) of pregnancy, the fetus may be at increased risk for **birth defects** such as scarring and **eye** damage. A newborn may contract chickenpox in the uterus if the mother has chickenpox five days before **birth**. Newborns can also contract chickenpox if the mother has the disease up to two days after birth. Chickenpox can be a deadly disease for newborns-the fatality **rate** from chickenpox in newborns up to five days old is about 30%. For this reason, women contemplating pregnancy may opt to be vaccinated with the new VZV vaccine prior to conception if they have never had the disease. If this has not been done, and a pregnant woman contracts chickenpox, an injection of varicella-zoster immunoglobulin can lessen the chance of complications to the fetus.

Chickenpox and environmental factors

Researchers have long noted the seasonality of chickenpox. According to their research, chickenpox cases occur at their lowest rate during September. Numbers of cases increase throughout the autumn, peak in March and April, and then fall sharply once summer begins. This cycle corresponds to the typical school year in the United States. When children go back to school in the fall, they begin to spread the disease; when summer comes and school ends, cases of chickenpox diminish. A typical "mini-epidemic" within a school occurs when one child contracts chickenpox. This child rapidly infects other susceptible children. Soon, all the children who had not had chickenpox contract the disease within two or three cycles of transmission. It is not uncommon for high numbers of children to be infected during one

"mini-epidemic;" one school with 69 children reported that the disease struck 67 of these students.

KEY TERMS

Immunocompromised—A condition in which the immune system is weakened, as during chemotherapy for cancer or AIDS.

Reye's syndrome—A neurological condition that usually occurs in children; associated with a respiratory illness and aspirin intake.

Immunity and the new vaccine

Contrary to popular belief, it is possible to get chickenpox a second time. If a person had a mild case during childhood, his or her immunity to the virus may be weaker than that of someone who had a severe childhood case. In order to prevent chickenpox, especially in already-ill children and immunocompromised patients, researchers have devised a VZV vaccine, consisting of live, attenuated (modified) VZV. Immunization recommendations of the American Academy of Pediatrics state that children between 12 and 18 months of age who have not yet had chickenpox should receive the vaccine. Immunization can be accomplished with a single dose. Children up to the age of 13 who have had neither chickenpox nor the immunization, should also receive a single dose of the vaccine. Children older than age 13 who have never had either chickenpox or the vaccine should be immunized with two separate doses, given about a month apart. The vaccine provokes strong immunity against the virus. Although some side effects have been noted, including a mild rash and the reactivation of shingles, the vaccine is considered safe and effective.

See also Childhood diseases.

Resources

Periodicals

Gorman, Christine. "Chicken Pox Conundrum." *Time* 142 (July 19, 1993): 53.

Kolata, Gina. "Curtailing Chicken Pox." *Reader's Digest* 137 (August 1990).

Kump, Theresa. "Chicken Pox Survival Guide." *Parents' Magazine* 69 (May 1994): 29.

Malhotra, R. "Nasolacrimal Duct Obstruction Following Chicken Pox." *Eye* 16, no. 1 (2002): 88–89.

Ortolon, Ken. "Short On Shots." *Texas Medicine* 98, no. 5 (2002): 30–33.

Plotkin, Stanley A. "Vaccines for Chicken Pox and Cytomegalovirus: Recent Progress." *Science* 265 (September 2, 1994): 1383.

"Vaccine Found to Be Effective on Chicken Pox; FDA Panel Still Has Questions to Resolve." *New York Times,* January 29, 1994.

"Varicella Seroprevalence In A Random Sample Of The Turkish Population." *Vaccine* 20, Issue: no. 10 (2002): 1425-1428.

Kathleen Scogna

Chickens *see* **Livestock**

Chicory *see* **Composite family**

Childhood diseases

Diseases that are more common among children than among adults are referred to as childhood diseases. That is not to say that adults cannot or will not contract these illnesses; but usually children contract these diseases and form the immunity against them that will protect them as adults. In fact, some of these diseases may be quite uncomplicated in children, but may be life-threatening when contracted by an adult who never had the **disease** as a child. Vaccines provide immunization against some of these diseases; others, however, can neither be prevented nor cured.

Although the first vaccination for any disease was administered in the late eighteenth century, the development of **immunology** proceeded slowly for the next hundred years. Dr. Edward Jenner, an English physician, noticed that milkmaids who developed cowpox from contact with cows were immune to the plague (**smallpox**). He correctly hypothesized that exposure to the cowpox somehow conferred protection on the milkmaids. Jenner withdrew some of the material from the skin pustules of the milkmaids and injected it under the skin of his own children, and they never developed smallpox. His methodology, however, was not accepted by the mainstream medical world of the time. Also, the means to develop, test, and produce vaccines were unknown. Physicians had not yet connected disease to the existence of microscopic organisms (**bacteria**).

Some childhood diseases are brought on by a bacterium or **virus** (**chickenpox** or measles, for example), others are inherited (**Tay-Sachs disease** or **sickle cell anemia**), and still others are caused by heavy use of **alcohol** or drugs by the mother while she is pregnant.

Many of these diseases are contagious—that is, can be passed on from one person to another by transmission of the bacterium or virus. Children are brought together in school buses and classrooms, and these close quarters are ideal for the transmission of the etiologic agents that cause diseases. When one child contracts measles, usually several in the same bus or class will also get the disease before steps can be taken to slow the **rate** of transmission, because such infections are often most contagious before any outward symptoms appear. A child with mumps can infect a number of other children before he actually shows signs of being ill. This is particularly true of the common cold. The virus that causes the cold is especially numerous in the early stages of the cold, before the patient actually starts to sneeze and develop a fever.

Much research is directed toward developing vaccines and remedies for diseases which are presently incurable (e.g., the common cold or **AIDS**), but virus-borne diseases are much more difficult to cure or prevent than the bacterial diseases. A virus cannot be seen under the normal light **microscope** used in laboratories. A special **electron** microscope must be used to see a virus. Viruses can also change or mutate to fend off any natural immunity that may develop against them. Some diseases, such as the common cold, are caused by more than one virus. A cold can be brought on by any one of some 200 viruses. A **vaccine**, if developed, would be effective against only one virus; many scientists feel such a vaccine would hardly be worth the trouble to develop. Even diseases that are caused by a single virus (such as AIDS) can defy the development of an effective vaccine.

Contagious diseases

The etiologic agents of contagious diseases can be passed from one person to another in any number of ways. They are present in droplets of saliva and mucus sprayed by sneezing and coughing. They can be conveyed by passing an object from the infected person to someone else. With the close proximity of children in classrooms, the agent can be passed quickly through the class.

Chickenpox

Chickenpox is one of the most easily transmitted childhood diseases; it is second only to measles. It is caused by the varicella-zoster virus, which may reactivate to cause **shingles** in individuals who have already had chickenpox. After exposure to the virus, a 7-21 day period of incubation occurs before any symptoms appear. Chickenpox begins with a low fever and general feeling of tiredness. Soon a rash develops on the abdomen and chest, which may or may not spread to the extremities, but usually affects the scalp. The rash appears in successive stages; as a result, some of the bumps are mature while others are just appearing. The rash progresses from the initial red bumps through a vesicle stage, in which they are filled with liquid, to a mature, crusty stage. Itching is intense, but scratching can cause localized **infection** of the broken vesicles and may require **antibiotics**. Within a week after the appearance of the rash, the patient is no longer infectious.

There is no treatment for chickenpox, but the Food and Drug Administration approved a varicella-zoster vaccine in March 1995. The live-virus vaccine, developed from a strain of the virus isolated in Japan in 1981, is recommended for children between the ages of 12-18 months. A person also develops immunity to the virus if he has experienced an outbreak of chickenpox. The outbreak is usually harmless to children and passes within a week without any noticeable permanent effect. In adults the disease is much more serious; it can cause damage to the eyes and, in males, the testes. An adult with chickenpox also requires a longer period of recuperation than does a child. Pregnant women who contract chickenpox may pass the infection on to their unborn child, with serious consequences. Varicella-zoster immunoglobulin may be given to such pregnant women, in an effort to decrease the complications to the fetus. Children who have weakened immune systems (due to chemotherapy treatments for **cancer**, for example) may have severe, life-threatening complications from chickenpox. Anti-viral medications (such as acyclovir) may be given in an attempt to decrease the severity and shorten the duration of the illness.

Even when the infection has disappeared, however, the virus remains in the person's body and can cause shingles later in life. The chickenpox virus lies dormant in some nerve cells and can become active in an **individual** after the age of 50 years. In this case, the nerve root becomes inflamed, and the area of the body served by that nerve is affected. Again an eruption occurs, but in this case it is very painful. A rash may appear on the abdomen or any area of the arms or legs. The outbreak lasts for five to six days, unless the patient has an underlying cancer, in which case the rash may persist for two weeks or longer.

It is not possible to predict who will have shingles after they have had chickenpox as a child. There is no treatment for shingles, but usually an individual will have it only once and then be immune to any further outbreaks. If the virus infects certain facial nerves, care must be taken to prevent damage to the eyes. Unlike chickenpox, shingles is not contagious. The virus is confined to the nerve fiber and is not released into the air.

Measles

Measles generally refers to nine-day measles, also called rubeola, a highly contagious disease spread by a virus. A person who has the measles virus can pass it to others before he shows signs of the disease. Once exposed to the virus, it will be 7-18 days before the typical measles rash develops. The patient is infectious, however, for the two to four days immediately before the rash appears; thus he spreads the disease unknowingly. Present in mucus and saliva droplets from the nose and mouth, the virus is spread by coughing or sneezing.

The initial symptoms of measles include headaches, a low fever, tiredness, and itchy eyes. Spots appearing on the roof of the mouth look like white grains of **sand** surrounded by an inflamed area. These are called Koplik's spots. A sore throat may also develop. The rash appears three to five days later: a bright red outbreak usually begins on the side of the head in front of the ears and spreads over the body within the next day or two. The **temperature** may climb to 104°F (40°C). **Inflammation** of the eyes may cause painful sensitivity to light.

The disease is short lived; the rash fades within three to five days and the body temperature returns to normal. The disease, while active, renders the patient much more susceptible to bacterial infections and may worsen diseases such as **tuberculosis**, if present. **Pneumonia** and **ear** infections are common complications of measles, especially in infants and very young children. Also, the virus can penetrate the central **nervous system** and cause **encephalitis** (inflammation of the **brain tissue**), which can lead to convulsions, **coma**, and even death. A person with measles should have bed rest during the active stage of the disease and be protected from exposure to any bacterial infections.

Fortunately, a vaccine has been developed against measles. The vaccine is a suspension of a live, attenuated (weakened) virus which is given to children at the age of approximately 15 months. The vaccine causes the formation of antibodies against the measles virus that will protect the child from future infections.

Another form of measles, known as three-day measles, German measles, or rubella, is also caused by a virus. Contagion is high because the infected person can transmit the virus to others for a week before showing any symptoms, and remains infectious for up to a week after the measles rash disappears.

Rubella is less infectious than the nine-day measles, and some infections may be so mild that the patient's case of rubella goes undetected. After exposure to the virus, an incubation period of 14-21 days passes before any symptoms appear. Usually the symptoms afflict only young children; teenagers and adults will not develop the typical rash.

The rubella rash is similar to that of nine-day measles but is less extensive; it appears on the face and neck and may spread to other areas. The rash lasts about three days before fading. No other symptoms, such as a sore throat, accompany the rash.

The most serious complication of three-day measles is its effect on a woman in the early stages of pregnancy. The virus can cause loss of the fetus or stillbirth, or it may result in **congenital (birth)** defects. These **birth defects** can include **heart** defects, **eye** defects (including glaucoma and cataracts), deafness, bleeding problems, mental retardation, and an increased risk of **diabetes mellitus**, thyroid problems, and future encephalitis (brain inflammation). A woman in the first three months of pregnancy should be protected from exposure to individuals who have measles. This form of measles can also be prevented by vaccination.

Mumps

Mumps, also called **epidemic** parotitis, is a viral infection of the salivary **glands**, especially the parotid glands. The mumps virus is spread in droplets of saliva sprayed during sneezing or coughing, and can be passed along on any object that has the infected saliva on it. The virus is present in the saliva of the infected person for up to six days before symptoms appear. Late winter and early spring are the peak periods of mumps epidemics, and children aged 5-15 years are most commonly infected. The disease is not as infectious as chickenpox or measles, and it is rare in children under two years of age.

The first symptom of mumps is **pain** during chewing or swallowing. The pain is worsened by acidic foods such as vinegar or lemon juice. The parotid gland, located in the area at the angle of the jaw, becomes sensitive to pressure. Body temperature increases to 103–104°F (40°C) once the inflammation of the parotid gland begins. The infected parotid gland becomes inflamed and swollen; the swelling may extend beyond the gland to the ear and the lower area of the jaw. The swelling reaches its maximum within two days and then recedes.

The mumps virus can also penetrate the central nervous system and cause abnormally high numbers of cells to accumulate in the spinal fluid. Usually this form of encephalitis has no residual effects, although rarely some facial paralysis or deafness due to auditory nerve damage may result. Mumps afflicting an adult male can cause atrophy of the testes and, in some cases, subsequent sterility. Patients should remain in bed until the fever accompanying the disease has subsided.

There is no treatment for mumps. Usually it is benign and will leave no residual effects other than a natural immunity against catching it again. During the illness, patients may take aspirin to ease the pain in the jaw and lower the fever. Eating soft food also helps to ease jaw pain. Anyone who has been in contact with a mumps patient should be watched closely for up to four weeks to see whether he or she will also develop the disease. A live-virus mumps vaccine is available for administration to children who are 15 months of age.

Other infectious childhood diseases

As recently as the early decades of the twentieth century, childhood was fraught with diseases that often entailed suffering and premature death. Many of those diseases were highly contagious; a child who contracted one of them was immediately isolated at home and a "quarantine" sign was posted conspicuously on the door to warn others. These once-perilous diseases included **diphtheria**, **whooping cough** (pertussis), and **tetanus** (lockjaw), which have been effectively controlled by vaccines; **scarlet fever**, another such disease, is now easily treated with antibiotics.

Diphtheria is caused by a toxin-producing bacterium, *Corynebacterium diphtheriae*, which infects the nervous tissue, kidneys, and other organs. The disease is spread by contact with the secretions of an infected person or objects that have the bacterium on them. Diphtheria develops rapidly after a short incubation period of one to four days. The bacterium usually lodges in the tonsils, where it multiplies and begins to produce a toxin. The toxic exudate, or secretion, is lethal to the cells around the infected area and can be carried to distant organs by the **blood**. Areas of infection and damage can be found in the kidneys, heart muscle, and respiratory tissues as well as in the brain. A **membrane** that is characteristic of the infection forms over the area affected by the toxin.

If left untreated, diphtheria can cause serious heart damage that can result in death, nerve damage resulting in a palsy, or kidney damage, which is usually reversible. A penicillin treatment and a diphtheria antitoxin are used to neutralize the bacterial secretions. One of the earliest vaccines now given to children is a combined vaccine for diphtheria, pertussis, and tetanus; as a result, diphtheria is rare now.

Pertussis, or whooping cough, is another highly infectious bacterial disease so named because of the characteristic high-pitched crowing sound of the breath between coughs. The etiologic agent is the bacterium*Bordetella pertussis*.

Pertussis is known throughout the world. It is transmitted in the saliva of coughing patients who have the bacterium, usually in the early stages of the disease. Patients are no longer infectious after eight weeks. The bacterium invades the nose, pharynx (back of the throat), trachea (windpipe), and bronchi. Symptoms appear after an incubation period of about one to two weeks. The earliest stage of the disease consists of sneezing, fatigue, loss of appetite, and a bothersome nighttime cough. This stage lasts for about two weeks, after which the coughs become rapid (paroxysmal) and are followed by the characteristic whoop, a few normal breaths, and another paroxysm of coughing. The coughing spells expel copious amounts of a thick mucus, which may cause gagging and vomiting. This stage of the disease can last up to four weeks, after which the patient begins a recovery; the coughing lessens and the mucus decreases.

Pertussis may be fatal in very young children; it is rarely serious in older children and adults. Fatalities in young children are usually caused by a subsequent bout of pneumonia. Infected individuals should be isolated, but do not necessarily need bed rest. Very young children should be hospitalized so that mucus may be suctioned from the throat area. A pertussis vaccine is available and is part of the early inoculation program in children. It is given with the vaccines for diphtheria and tetanus. Newer "acellular" pertussis vaccines cause fewer side effects than the old whole **cell** vaccines.

Poliomyelitis

Poliomyelitis, also called polio or infantile paralysis, is caused by a virus and once appeared in epidemic proportions. It occurs mostly in young children and appears primarily in the summer or fall. The poliovirus, the causative agent, is found in three forms-types I, II, and III. Type I is the most likely to cause paralysis.

Most people who host the poliovirus do not develop any symptoms but can still spread the virus. Because it is present in the throat of infected individuals, the virus is spread by saliva. Polio is known worldwide, and cases of it occur year round in tropical areas. Fortunately, only one of every 100 people who have the virus actually exhibits the symptoms of polio.

At one time polio was so widespread that young children developed an immunity to polio very early in life, because they would acquire the virus without fail. With the onset of hygienic sanitation, however, the disease began to appear as epidemics in developed countries. Since children no longer develop a natural immunity, as they did prior to the installation of modern sewage facilities, an outbreak of polio can quickly sweep through the younger population.

The onset of polio is divided into two phases: a minor illness and a major illness. The minor illness, experienced by about 90% of those who contract the virus, consists of vague symptoms such as headaches, nausea, fatigue, and a mild fever. These symptoms pass within 72 hours, and for most victims the minor illness is the extent of the disease. Those who acquire the major illness, however, combat a much more drastic form of the disease. It begins with severe headaches during the seven to 35 days following exposure to the virus. A fever develops, and

stiffness and pain in certain muscles appear. The affected muscles become weak and the nerve reflexes to those muscles are lost. This is the beginning of the paralysis, which results because the virus infects certain areas of the nervous system, preventing control of muscle groups.

A vaccine is available for the prevention of polio. The first polio vaccine was given by injection, but a later version was given orally. The latest guidelines for polio immunization state that IPV (inactivated poliovirus vaccine, supplied via injection) should be given at two and four months of age; subsequent immunizations at 6-18 months and 4-6 years may be given either as injection or orally. These newer guidelines were designed to decrease the rare complication of vaccine-induced polio infections which increased when all doses were given orally (which is a live virus vaccine).

Noncontagious childhood diseases

Noncontagious childhood diseases are acquired by heredity—that is, passed from parents to offspring. In fact, neither of the parents may have any physical signs of the disease, but if they are carriers—people who have the recessive **gene** for the disease—they can pass it on to their children.

Some of these conditions are serious or even fatal; there is no cure for the person who has the disease. Some effective preventive measures can be taken to keep the disease in an inactive state, but even these measures are sometimes not effective.

Sickle cell anemia

Sickle cell **anemia** is named for the shape assumed by some of the red blood cells in persons who have this disease. It primarily affects people of African descent, but it can also be present in people of Mediterranean descent, such as Arabs and Greeks.

Some people carry the gene for sickle cell anemia without having any active symptoms. For those in whom the disease is active, however, a "sickle cell crisis" can be a painful and debilitating experience.

When the red blood cell undergoes changes that alter its shape from a disk to a sickle, the cells can no longer pass easily through the tiniest blood vessels, the **capillaries**. The sickle cells stick in these vessels and prevent the passage of normal cells; as a result, the **organ** or muscle dependent on blood flow through the affected capillaries is no longer getting **oxygen**. This causes a very painful crisis that may require the hospitalization of the patient.

No treatment exists for sickle cell anemia, so the person who has the active disease must avoid infections and maintain a healthy lifestyle. Any activity that is strenuous enough to cause shortness of breath can also bring on a crisis.

Tay-Sachs disease

Tay-Sachs disease is an inherited, invariably fatal condition in which a missing **enzyme** allows certain toxic substances to accumulate in the brain. Under ordinary circumstances the enzyme, hexosaminidase A, breaks down these toxins, but without its presence the toxins accumulate.

The condition causes the development of red spots in the eye, retarded development, blindness, and paralysis. The child usually dies by the age of three or four. Tay-Sachs disease primarily affects Jews from eastern **Europe**.

Parents who carry the gene for Tay-Sachs can be counseled about having children. Statistically for parents who are both carriers of the gene, one in four children will have the active disease, two of the four will be unaffected carriers of the gene, and one of four will have neither the gene nor the disease.

Congenital diseases

Some conditions are passed from mother to child not as a result of an infection or a genetic malfunction, but because the mother has failed to provide an optimal prebirth condition for the developing baby. The placenta, which lines the womb and serves to nourish the developing infant, can be penetrated by substances such as alcohol, **nicotine**, **cocaine**, and heroin. Also, a mother who has AIDS can pass the virus to the child during gestation.

The mother who smokes, drinks alcohol, or uses drugs while pregnant can cause developmental problems for the child. The fetus is especially susceptible to these influences during the first three months (first trimester) of pregnancy. The organs are formed and the **anatomy** and **physiology** of the infant are established during the first trimester.

Fetal alcohol syndrome (FAS) is a well-recognized affliction brought about by the mother's heavy consumption of alcohol during pregnancy. Alcohol consumed very early in the pregnancy can cause brain damage by interfering with the fetus' brain development. Other features of a child with FAS are wide-set eyes, flattened bridge of the nose, and slowed growth and development.

A child born of a mother addicted to drugs will also be addicted. Often the baby will exhibit signs of withdrawal, such as shaking, vomiting, and crying with pain. These children usually have a low birth weight and are slow to develop. In time, once the drug is out of his system, the child will assume a normal life pattern.

Resources

Books

Ziegleman, David. *The Pocket Pediatrician.* New York: Doubleday Publishing, 1995.

Periodicals

Ortolon, Ken. "Short On Shots." *Texas Medicine* 98, no. 5 (2002): 30–33.
Schultz, D. "That Spring Fever May Be Chickenpox." *FDA Consumer* 27 (March 1993): 14–17.
Stix, G. "Immuno-Logistics." *Scientific American* 270 (June 1994): 102–103.
Tsukahara, Hirokazu. "Xidant And Antioxidant Activities In Childhood Meningitis." *Life Sciences* 71, no. 23 (2002): 2797.
"Varicella Seroprevalence In A Random Sample Of The Turkish Population." *Vaccine* 20, no. 10 (2002): 1425–1428.

Larry Blaser

Chimaeras

The chimaeras (order Chimaerae, class Bradyodonti) are a most peculiar looking group of **fish** that live near the sea bed off continental shelves and in deep offshore waters at a depth of 985-1,640 mi (300-500 m). Collectively these **species** form a small, cohesive group of about 25 species. They are all exclusively marine species. Closely related to **sharks**, **rays**, and dogfish, chimaeras are characterized by their cartilaginous skeletons-in contrast to the bony skeletons of most fish. One feature that distinguishes them from rays and dogfish is the fact that the upper jaw is firmly attached to the cranium. They also have flattened teeth (two on the upper jaw and one on the lower) that are modified for crushing and grinding their food. There is one gill opening on either side of the head, each of which is covered with a fleshy flaplike cover.

Also known as rabbit fish, rat fish, or elephant fish, these species have large heads, a tapering body and a long, rat-like tail. Relatively large fish, they range from 2 to 6 ft (0.61 to 2 m) in length. Unlike their larger relatives—the sharks, for example—they are all weak swimmers. The fins are highly modified: one striking feature is the presence of a strong, sharp spine at the front of the first dorsal fin. In some species this may be venomous. When the fin is relaxed, for example when the fish is resting, this spine folds into a special groove in the animal's back, but it may be quickly erected if disturbed. Chimaeras have a very reduced, elongated tail fin. The large pectoral fins, unlike those of sharks, play a major role in swimming, while additional propulsion is gained through lateral body movements.

Chimaeras have a smooth skin and one gill opening on either side—a morphological change that lies between that of a shark and the **bony fish**. A series of mucus-secreting canals occur on the head. The males of some species such as *Callorhinchus* have a moveable club-shaped growth on the head, the function of which is not known.

Little is known about the **ecology** of chimaeras. Most are thought to feed on a wide range of items, including seaweed, worms, **crabs**, **shrimp**, brittle stars, molluscs and small fish. Most rat fish are thought to be nocturnal—one explanation for their peculiarly large eyes. Male chimaeras have small appendages known as claspers, used to retain their hold on females during copulation; this act is also observed in sharks and rays. Unlike the fertilized eggs of bony fishes, those of chimaeras are enclosed in toughened capsules. In some species these capsules may measure 6 in (15 cm) in length and are pointed to stick into the soft substrate, possibly to prevent the eggs from drifting too far.

Chimpanzees

Chimpanzees belong to the order **Primates**, which includes **monkeys, apes**, and humans. Chimpanzees are assigned to the family Pongidae, which includes all of the other apes: **gorillas**, orang-utans, and gibbons. Compared with monkeys, apes are larger, have no tail, and have longer arms and a broader chest. When apes stand upright, their long arms reach below their knees. They also have a great deal of upper body strength, needed for a life spent mostly in the forest canopy.

Chimpanzee species and habitat

There are two **species** of chimpanzees: the common chimpanzee (*Pan troglodytes*) and the bonobo or pygmy chimpanzee (*Pan paniscus*). The common chimpanzee occurs in forested West and Central **Africa**, from Senegal to Tanzania. Their **habitat** includes rain forest and deciduous woodland, from **sea level** to above 6,000 ft (1,830 m). Common chimpanzees are rarely found in open habitat, except temporarily when there is ready access to fruit-bearing trees. The pygmy chimpanzee is found in Central Africa, and is confined to Zaire between the Kasai and Zaire **rivers**. Its habitat is restricted to closed-canopy tropical **rainforest** below 5,000 ft (1,525 m).

Chimpanzees live within the borders of some 15 African countries. Both species are endangered by habitat destruction, a low **rate** of reproduction, and by hunting by humans as meat and for the live-animal trade.

Physical characteristics

The height of chimpanzees varies from about 39 in (1 m) in males to about 36 in (90 cm) in females. An adult wild male weighs about 132 lb (60 kg), but in captivity can grow up to 220 lb (100 kg). The weight of females range from about 66 lb (30 kg) in the wild to as much as 190 lb (87 kg) in captivity.

Chimpanzee pelage is primarily black, turning gray on the back after about 20 years of age. Both sexes have a short white beard, and baldness frequently develops in later years. The skin on the hands and feet is black, and that of the face ranges from pink to brown or black. The ears are large and the nostrils are small. Chimpanzees have heavy eyebrows, a flattened forehead, large protruding ears, and a short neck. The jaw is heavy and protruding, and the canine teeth are large. Male chimpanzees have larger canines than females, and these are used in battle with other males and during predation. Although chimpanzees have long powerful fingers, they have small weak thumbs. Their big toes function like thumbs, and their feet as well as their hands are used for grasping.

The genitalia of both sexes are prominent. Areas of the female's genital skin become pink during estrus, a period that lasts two to three weeks and occurs every four to six weeks. The characteristic gait on the ground is the "knuckle-walk," which involves the use of the knuckles for support. Chimps spend much of their time climbing in trees, and they **sleep** alone every night in nests that they make with branches and leaves. A mother sleeps with her baby until her next infant is born. Chimpanzees live up to 40-45 years.

Behavior

Life in the wild presents many challenges for chimpanzees. They have complex social systems, and daily use their considerable mental skills for their survival. They are presented with a multitude of choices in their natural habitat, and **exercise** highly developed social skills. For instance, males aspire to attain a high position of dominance within the hierarchy of chimpanzee society, and consequently low-ranking individuals must learn the art of deception, doing things in secret to satisfy their own needs.

Researchers have discovered that chimpanzees experience a full range of emotions, from joy to grief, fear, anger, and curiosity. Chimps also have a relatively good ability to learn and understand concepts and the elements of language.

Chimpanzees (*Pan troglodytes*) in Gombe National Park, Tanzania. *Photograph by Kennan Ward. Stock Market. Reproduced by permission.*

Chimpanzees have a sophisticated social organization. They band together in groups, which vary in size and the age of its members. Between 15 and 120 individuals will form a community, with generally twice as many adult females as adult males in the group. The range and territory of a particular group depends on the number of sexually mature males.

Chimps generally do not travel as an entire social unit. Instead, they move around in smaller groups of three to six individuals. While travelling about, individuals may separate and join other chimpanzees. Temporary bonds with other chimps are created by an abundant food source, or by a female in estrus. The strongest social bond is between a mother and her young. Offspring that are under eight years of age are always found with their mother.

Parenting

A female's estrus cycle averages 38 days, which includes 2-4 days of menstruation. When a female begins her estrus cycle, her genital area swells for approximately ten days, and this is when she is sexually attractive and receptive. The last three or four days of estrus is when the likelihood of conception is highest. Mating is seemingly **random** and varied, and receptive females are often mounted by most of the mature males in the community. A high-ranking male may, however, claim "possession" and prevent other males from mating with a female. Regardless of rank and social status in the community, all males and females have a chance to pass on their genes.

On average, female chimpanzees give **birth** every five to six years. Gestation is 230-240 days. Newborn chimps have only a weak grasping **reflex**, and initially require full support from their mother as she moves about. After a few days, however, the infant is able to cling securely to its mother's underside. About the age of five to seven months, the youngster is able to ride on its mother's back. At the age of four years, a young chimp can travel well by walking. Weaning occurs before its third year, but the youngster will stay with its mother until it is five to seven years old.

When an infant chimp is born, its older sibling will start to become more independent of its mother. It will look for food and will build its own sleeping nest, but a close relationship remains with the mother and develops between the siblings. Young males stay close to their family unit until about the age of nine. At this time, they find an adult male to follow and watch his **behavior**. Thus begins the long process by which the male develops his place in the community.

Young females stay with their mothers until about ten years of age. After her first estrus, a young female typically withdraws from her natal group and moves to a neighboring one, mating with its males. At this time a female may transfer out of her initial group to form a family of her own. This exchange helps to prevent inbreeding and maintains the diversity of the **gene** pool.

Eating habits

Chimpanzees are omnivorous, eating both meat and **plant** material. Their diet includes **fruits**, leaves, buds, **seeds**, pith, **bark**, **insects**, bird eggs, and smaller **mammals**. Chimpanzees have been observed to kill **baboons**, other monkeys, and young bush **pigs**, and they sometimes practice cannibalism. Chimps eat up to 200-300 species of plants, depending on local availability.

Chimpanzees seem to know the medicinal value of certain plants. In the Gombe National Forest in Tanzania, chimps have been seen to eat the plant *Apilia mossambicensis* to help rid themselves of **parasites** in their **digestive system**. A branch of science, zoopharmacognosy, has recently developed to study the medicinal use of plants by wild animals.

Fruit is the main component of the chimpanzee diet, and they spend at least four hours a day finding and eating varieties of this food. In the afternoon chimps also spend another hour or two feeding on young leaves. They also eat quantities of insects that they collect by hand, or in the case of **termites**, using simple tools. Chimpanzees break open the hard shells of nuts with sticks or smash them between two **rocks**. **Animal** prey is eaten less regularly than fruits and leaves. Chimpanzees (usually males) will regularly kill and eat young pigs, monkeys, and antelopes.

Chimpanzees are able to devise simple tools to assist in finding food and for other activities. They use stones to smash open nuts, sticks for catching termites, and they peel leaves from bamboo shoots for use as wash cloths to wipe off dirt or **blood**, and to collect rainwater from tree-cavities. The use of tools by chimpanzees varies from region to region, which indicates that it is a learned behavior. Young chimps have been observed to imitate their elders in the use of tools, and to fumble with the activity until they eventually become proficient.

Communication

Chimpanzees use a multitude of calls to communicate. After being separated, chimpanzees often embrace, kiss, **touch**, stroke, or hold hands with each other. When fighting, the opponents may strike with a flat hand, kick, bite, or stomp, or drag the other along the ground. Scratching and hair pulling are favorite tactics of females. When the fighting is over, the loser will approach the winner and weep, crouch humbly, or hold out its hand. The victor usually responds by gently touching, stroking, embracing, or grooming the defeated chimp.

Body contact is of utmost importance in maintaining social harmony in a chimpanzee community. Chimpanzees will often groom each other for hours. Grooming a way to maintain calmness and tranquility, while preserving close relationships.

Chimpanzees also communicate through a combination of posture, gesture, and noise. While avoiding direct contact, a male chimpanzee will charge over the ground and through the trees, swinging and pulling down branches. He will drag branches, throw sticks and stones, and stomp on the ground. By doing this, he gives the impression that he is a dangerous and large opponent. The more impressive this display, the better the position achieved in the male ranking order.

Confrontations between members of different communities can, however, be extremely violent. Fighting is ferocious and conducted without restraint, often resulting in serious injury and sometimes death. These encounters usually occur in places where communities overlap. Chimpanzees behave cautiously in such places, often climbing trees to survey the area for members of the neighboring community.

When two community groups of balanced strength meet, they may show aggression by performing wild dances, throwing rocks, beating **tree** trunks, and making fierce noises. This display is usually followed by retreat into their territory. However, when only one or several strangers, whether male or female, is met by a larger group it is in danger of being viciously attacked. Chimpanzees have been seen to twist the limbs, tear the flesh, and drink the blood of strangers they have murdered in such aggressive encounters.

This hostile activity often occurs when male chimpanzees are routinely involved in "border patrols." Males may patrol for several hours, keenly watching and listening for signs of nearby activity. It is not known if the purpose of the patrols is to protect the local food source of the community, or if the males are engaged in **competition** for females, or even engaging in predatory cannibalism.

Jane Goodall's observations

In 1960, Jane Goodall, a young Englishwoman, first set up camp in Gombe, Tanzania, to conduct a long-term study of chimpanzees. Louis Leakey, a famous anthropologist, helped Goodall by providing the initial funding for her research and by serving as her mentor. Leakey is best known for his discovery of hominid fossils in eastern Africa, and his contributions to the understanding of **human evolution.**

Goodall was not initially a trained scientist, but Leaky felt that this could be an advantage in her work, because she would not bring pre-conceived scientific bias into her research. One of the most difficult hurdles that Goodall had to overcome when presenting the results of her work to the scientific community was to avoid making references to the fact that chimps have feelings. Projecting human emotions onto animals is thought to signal anthropomorphic bias, and is often regarded as a scientific flaw. However, as Goodall demonstrated, chimps do experience a wide range of emotions, and they perceive themselves to be individuals. These are some of the compelling similarities that they share with humans.

Goodall made several particularly significant discoveries early in her research. Her first chimpanzee friend was a male individual that she named David Greybeard. One day she was observing him when he walked over to a termite mound, picked up a stiff blade of grass, carefully trimmed it, and poked it into a hole in the mound. When he pulled the grass out of the mound, termites were clinging to it, and he ate them. This remarkable discovery showed that chimps are toolmakers.

Goodall's second discovery also involved David Greybeard; she observed him eating the carcass of an infant bushpig (a medium-sized forest mammal). David Greybeard shared the meat with some companions, although he kept the best parts for himself. The use of tools and the hunting of meat had never before been observed in apes. Numerous other observations have since been made of chimps making and using simple tools, and engaging in sometimes well-organized group hunts of monkeys and other prey.

The bonobo (pygmy chimpanzee)

Several projects begun in the early 1970s were the first to study bonobos in the wild. Their alternative name, pygmy chimp, is inaccurate because these animals are only slightly smaller than common chimpanzees. The reference to "pygmy" has more to do with so-called pedomorphic traits of bonobos, meaning they exhibit certain aspects of adolescence in early adulthood, such as a rounded shape of their head.

Another characteristic of the bonobo that differs from the common chimp is the joining of two digits of their foot. Additionally, the bonobo's body frame is thinner, its head is smaller, its shoulders narrower, and legs are longer and stretch while it is walking. Furthermore, the eyebrow ridges of the bonobo are slimmer, its lips are reddish with a black edge, its ears are smaller, and its nostrils are widely spaced. Bonobos also have a flatter and broader face with a higher forehead than do common chimpanzees, and their hair is blacker and finer.

Bonobos also have a somewhat more complex social structure. Like common chimpanzees, bonobos belong to large communities and form smaller groups of six to 15 that will travel and forage together. Groups of bonobos have an equal sex **ratio**, unlike those of the common chimpanzee. Among bonobos the strongest bonds are created between adult females and between the sexes. Bonds between adult males are weak (whereas in common chimps they may be strong). Bonobo females take a more central position in the social hierarchy of the group.

Sex is an important pastime among bonobo chimpanzees. Female bonobos are almost always receptive and are willing to mate during most of their monthly cycle. The ongoing sexual exchanges within their communities help to maintain peace and ease **friction.** Bonobos try to avoid social conflict, especially when it relates to food.

Bonobos are extremely acrobatic and enjoy spending time in the trees. They do not fear **water** and have been observed to become playful on rainy days, unlike common chimpanzees who hate the rain. It is believed that there are fewer than 100,000 bonobos in the wild; they are threatened by hunting as food and for sale in foreign trade, and by the destruction of their natural forest habitat.

Language

Sign language has been used successfully for communication between human beings and captive chimpanzees. Sign language research has shown that some chimpanzees are able to create their own symbols for communication when none has been given for a specific object. Other studies of the use of language by chimps suggest that they understand the syntax of language, that is, the relationship of words to action and to the actor. Chimpanzees also have pre-mathematical skills, and are able to differentiate and categorize. For example, they can learn the difference between fruits and **vegetables,** and to divide sundry things into piles of similar objects.

Use in research

The chimpanzee is the closest living relative of humans. In fact, the DNA of humans and chimpanzees differs by less than 1%. Because of this genetic closeness, and anatomical and biochemical similarities, chimps

Periodicals

Gouzoules, Harold. "Primate Communication by Nature Honest, or by Experience Wise." *International Journal of Primatology* 23, no. 4 (2002): 821-848.

Matsumoto, Oda. "Behavioral Seasonality in Mahale Chimpanzees." *Primates* 43, no. 2 (2002): 103-117.

Savage-Rumbaugh, Sue, and Roger Lewin. "Ape at the Brink." *Discover* (September 1994): 91-98.

Sheeran, L. K. "Tree of Origin: What Primate Behavior Can Tell Us About Human Society." *American Journal off Human Biology* 14, no. 1 (2002): 82-83.

Small, Meredith F. "What's Love Got to Do with It?" *Discover* (June 1992): 48-51.

Kitty Richman

KEY TERMS

Anthropomorphic—Ascribing human feelings or traits to other species of animals.

Bipedal—The ability to walk on two legs.

Border patrol—Routine visits that common chimpanzees make to the edges of their communal areas to observe neighboring territories.

Estrus—A condition marking ovulation and sexual receptiveness in female mammals.

Pedomorphic—Having juvenile traits in adulthood.

Zoopharmacognosy—A field of research that studies the medicinal values of plants that animals eat.

---

have been widely used for testing new vaccines and drugs in biomedical research.

Chimpanzees can also become infected by certain diseases that humans are susceptible to, such as colds, flu, **AIDS**, and **hepatitis** B. Gorillas, gibbons, and orangutans are the only other animals that show a similar susceptibility to these diseases. Consequently, these species are used in biomedical research seeking cures for these ailments, including work that would be considered ethically wrong if undertaken on human subjects. However, many people are beginning to object to using chimpanzees and other apes in certain kinds of invasive biomedical research. This is because of the recent understanding that chimpanzees, other apes, and humans are so closely related, and that all are capable of experiencing complex emotions, including **pain** and suffering. Some people are even demanding that apes should be given legal rights and protection from irresponsible use in research.

Resources

Books

Crewe, S. *The Chimpanzee.* Raintree/Steck Vaughn, 1997.

De Waal, Frans. *Chimpanzee Politics: Power and Sex Among Apes.* Johns Hopkins University Press, 1998.

Goodall, Jane. *Peacemaking Among Primates.* Cambridge, MA: Harvard University Press, 1989.

Goodall, Jane. *Through a Window: My Thirty Years with the Chimpanzees of Gombe.* Boston: Houghton Mifflin, 1990.

Goodall, Jane, and Dale Peterson. *Visions of Caliban: On Chimpanzees and People.* Boston: Houghton Mifflin, 1993.

MacDonald, David, and Sasha Norris, eds. *Encyclopedia of Mammals.* New York: Facts on File, 2001.

Montgomery, Sy. *Walking with the Great Apes.* Boston: Houghton Mifflin, 1991.

Chinchilla

Chinchillas and viscachas are seven **species** of small, South American **rodents** in the family Chinchillidae. Chinchillas have a large head, broad snout, large eyes, rounded ears, and an extremely fine and dense fur. Their forelimbs are short and the paws small, while the hindlegs and feet are larger and relatively powerful, and are used for a leaping style of locomotion, as well as for running and creeping.

Two species of true chinchillas are recognized. The short-tailed chinchilla (*Chinchilla brevicaudata*) is native to Andean **mountains** of Argentina, Bolivia, and Peru, while the long-tailed chinchilla (*C. laniger*) is found in the mountains of Chile. Chinchillas are alpine animals, living in colonies in rock piles and scree, basking at dawn and dusk, and feeding at night on vegetation and occasional **arthropods**.

Chinchillas have an extremely thick, warm, and soft pelage, considered to be perhaps the finest of any fur. When the commercial implications of this fact were recognized in the late nineteenth century, a relentless exploitation of the wild populations of both species of chinchillas ensued. The over-harvesting of these animals brought both species to the brink of **extinction** by the early twentieth century.

Fortunately, methods have been developed for breeding and growing chinchillas in captivity, and large numbers are now raised on fur ranches. This development made it possible to stop most of the unregulated exploitation of wild chinchillas. Unfortunately, this happened rather late, and both species are widely extirpated from their original native habitats. The short-tailed chinchilla may, in fact, no longer be found in the wild, and this species is not commonly ranched on fur farms. The long-tailed chinchilla has fared much better, and al-

A chinchilla. *Photograph by Janet Stone/National Audubon Society/Photo Researchers, Inc. Reproduced by permission.*

though it remains rare in the wild, it is common in captivity, and is often kept as a pet. Attempts are being made to re-stock wild populations of chinchillas, but it is too soon to tell whether these efforts will be successful.

The plains viscacha (*Lagostomus maximus*) is another species in the Chinchillidae, found in the dry, lowland pampas of Argentina. These animals are much larger than the true chinchillas, and can reach a body length of 24 in (60 cm). Viscachas live in colonies of about 20-50 individuals, which inhabit complexes of underground burrows. The diggings from the burrows are piled in large heaps around the entrances, and the viscachas have a habit of collecting odd materials and placing them on those mounds. These can include natural objects such as bones and vegetation, but also things scavenged from people, such as watches, boots, and other unlikely items.

Viscachas are sometimes hunted as a source of wild meat. More importantly, viscachas have been widely exterminated because their diggings are considered to be a hazard to **livestock**, which can fall and break a leg if they break through an underground tunnel.

Four species of mountain viscachas (*Lagidium* spp.) occur in rocky habitats in the Andean **tundra**. Mountain viscachas live in colonies located in protective crevices, from which they forage during the day. Mountain viscachas are eaten by local people, and their fur is used in clothing. Sometimes the hair is removed from the skin of trapped animals, and used to weave an indigenous Andean cloth.

Chipmunks

Chipmunks are small **mammals** in the order Rodentia, the **rodents**. Specifically, they are classified with the squirrel-like rodents, the Sciuridae. Chipmunks are divided into two genera: *Gallos* and *Tamias*.

North America is home to 17 **species** of chipmunk, 16 in the West and only one, *Tamias striatus*, the eastern chipmunk, in the East. The eastern chipmunk is about 5-6 in (12.7-15 cm) long, and the tail adds another 4 in (10 cm) or so to the animal's length. They have stripes on their face and along the length of their body. In all species the tail is bushy, although not quite as much as the **tree squirrels**. Their eyes are large, their **vision** is excellent, and their sensitive whiskers give them a well-developed sense of **touch**.

Like other squirrels, chipmunks are opportunists, making a comfortable home where other animals would not dare. They are generally unafraid of human beings, and are frequent visitors to campgrounds. They are burrowers, digging holes among **rocks**, under logs, and within scrub, to make a burrow as long as 15 ft (4.6 m) and extend downward about 3 ft (0.9 m).

Like squirrels, chipmunks are active during the day. They emerge from their burrows in the morning to forage on **mushrooms**, **fruits**, **seeds**, berries, and acorns. The chipmunk will store food, particularly items with a long "shelf-life," in its cheek pouches for transport back to the burrow. The Siberian chipmunk can carry more than a quarter of an ounce of seed for half a mile. This food is stored in an underground larder, which can contain between 4.5 and 13 lb (2-5.9 kg) of food. They do not hibernate like **bears**; instead they become more lethargic than normal during the winter months.

In the spring, the female bears a litter of pups, numbering up to eight, which are born naked, toothless, and with closed eyes. The young grow quickly and are weaned at five weeks, but stay with the female for several months.

Chipmunks can live as long as five years, providing they avoid predators such as **weasels**, **owls**, **hawks**, bobcats, pine martens, and coyotes. Many chipmunks die after eating rodenticide set out for **rats**; these poisons have effectively eliminated chipmunks in some locations.

Chitons

Chitons are small **mollusks**, oval in outline, with a broad foot, and a mantle that secretes, and sometimes extends over, the shell. They live on rocky seashores in much the same life-style as **limpets**. They are easily distinguishable from limpets, however, by their shell made of eight plates (or valves) with transverse sutures. Also, unlike limpets and other **snails**, the chitons have no tentacles or eyes in the head region, just a mouth and a radula. The shell is so different from those of other mollusks that one might think chitons are segmented (or metameric), but contrary to the general rule, this is incorrect. Internally, there is no evidence of segmentation, and the eight valves are actually derived from a single embryonic shell.

Except for the **color**, the uniformity of external appearance tempts one to regard chitons as races of a single **species**, but the small variations are very important to other chitons and to chiton specialists, who like to count the notches and slits along the valve edges. The word chiton is a Greek word meaning a gown or tunic, usually worn next to the skin, and covered with a cloak on going outdoors. The chiton was worn by both men and women, just as the eight plates are worn by both male and female chitons. There are about 600 species of chitons in all, about 75 of them are on the U.S. Pacific Coast. Among the most common species are *Chaetopleura apiculata* of New England and *Mopalia muscosa* of California.

Chitons are classified as subclass Polyplacophora in the class Amphineura, one of the six classes of mollusks. The other subclass contains the Aplacophora, a group of wormlike mollusks lacking a shell, but possessing in some genera calcareous spicules embedded in the mantle. Amphineura means nerves on both sides, and Polyplacophora means bearing many plates; chitons have two pairs of **parallel** nerve cords running the length of the body. The **nervous system** is simple and straight, not twisted as in prosobranch snails. In spite of their anatomical simplicity, there is no reason to suppose that chitons represent a form ancestral to all the mollusks. Rather the opposite, the fossil record suggests that they followed the gastropods and **bivalves** in **evolution**, and lost some structures or traits as they became adapted to a restricted **niche**. The lack of tentacles and eyes, for example, means that chitons cannot function as predators. The shell is obviously defensive. When pried loose from their preferred spot, chitons roll up in a ball, much like certain isopod crustaceans, called pill bugs, and like the armadillo, an armored mammal.

Most chitons are 0.8-1.2 in (2-4 cm) long, but there is a giant Pacific coast species, *Cryptochiton stelleri*, up to 11.8 in (30 cm) long. This species is unusual also for the mantle or girdle that completely covers the shell (crypto = hidden). Other surprises include a species of *Callochiton septemvalvis* (seven valves). Eggs are laid singly or in a jelly string, and are fertilized by sperm released into the sea **water**. The larvae of a few species develop within the female, but most larvae are planktonic.

The giant chiton *Cryptochiton stelleri* was included in a classic study of nucleotide sequences in RNA of a great variety of animals, in which the goal was to establish relations of the phyla. On the resultant phylogenetic tree, the chiton appeared at the end of a branch close to a polychaete worm and a brachiopod, and not far from two clams. Another analysis of the same data put *Cryptochiton* on a branch next to a nudibranch *Anisodoris nobilis* and ancestral to the two clams. There is reason to suspect that living chitons are highly evolved creatures, and not good subjects for deductions about the initial metazoan radiation in the pre-Cambrian. The reasoning is as follows: Many marine mollusks have oxidative enzymes that use an **amino acid** to produce products such as octopine, alanopine, etc. while serving to reoxidize coenzyme, and keep **anaerobic metabolism** going. These opine enzymes are most varied in archaeogastropods, which are regarded as primitive on numerous grounds. The trend in evolution has been to lose some or all of the opine enzymes, and come to depend entirely on **lactic acid** production for their function. This is what has happened in a few bivalves and polychaete worms, and in fishes and other **vertebrates**. It is also the case with the chitons *Chaetopleura apiculata* and *Mopalia muscosa*, which have only a lactate oxidase and no opine enzymes. The earliest chitons may have had a great variety of genes that modern species no longer possess, but this is something that would be difficult to investigate.

Resources

Books

Abbott, R.T. *Seashells of the Northern Hemisphere.* New York: Gallery Books, 1991.

Periodicals

Field, K.G., G.J. Olsen, D.J. Lane, S.J. Giovannoni, M.T. Ghiselin, E.C. Raff, N.R. Pace, and R.A. Raff. "Molecular Phylogeny of the Animal Kingdom." *Science* 239 (1988) 748-753.

Hammen, C.S., and R.C. Bullock. "Opine Oxidoreductases in Brachiopods, Bryozoans, Phoronids, and Molluscs." *Biochemical Systems and Ecology* 19 (1991): 263-269.

Chlordane

Chlordane is an organochlorine insecticide, more specifically a chlorinated cyclic **hydrocarbon** within the cyclodiene group. The proper scientific name for chlordane is 1,2,4,5,6,7,8,8-Octachloro-3a,4,7,7a-tetrahydro-4,7-methanoindan. However, the actual technical product is a mixture of various **chlorinated hydrocarbons**, including isomers of chlordane and other closely related compounds.

The first usage of chlordane as an insecticide was in 1945. Its use was widespread up until the 1970s and included applications inside of homes to control **insects** of stored food and clothing, as well as usage to control **termites**, carpenter **ants**, and wood-boring **beetles**. An especially intensive use was to kill earthworms in golf-course putting greens and in prize lawns, for which more than 9 kg/ha might be applied. The major use of chlordane in agriculture was for the control of insect **pests** in **soil** and on plants. In 1971 about 25 million lb (11.4 million kg) of chlordane was manufactured in the United States, of which about 8% was used in agriculture, and most of the rest in and around homes. Today the use of chlordane is highly restricted, and limited to the control of fire ants.

Like other chlorinated hydrocarbon **insecticides** such as DDT, chlordane is virtually insoluble in **water** (5 ppm), but highly soluble in organic solvents and oils. This property, coupled with the persistence of chlordane in the environment, gives it a propensity to accumulate in organisms (i.e., to bioaccumulate), especially in animals at the top of food webs. Because of its insolubility in water, chlordane is relatively immobile in soil, and tends not to leach into surface or ground water.

The acute toxicity of chlordane to humans is considered to be high to medium by oral ingestion, and hazardous by inhalation. Chlordane causes damage to many organs, including the liver, testicles, **blood**, and the neural system. It also affects hormone levels and is a suspected **mutagen** and **carcinogen**. Chlordane is very toxic to **arthropods** and to some **fish**, **birds**, and **mammals**.

See also Bioaccumulation; Pesticides.

Chlorinated hydrocarbons

A very large and diverse group of organic molecules are chlorinated hydrocarbons. Hydrocarbons are molecules composed entirely of **hydrogen** and **carbon atoms**, often derived from carbon-based **fossil fuels** like **petroleum** oils and **coal**. Chlorinated hydrocarbons are specific **hydrocarbon** molecules that also have atoms of the element **chlorine** chemically bonded to them. The number of chlorine atoms bonded to a specific chlorinated hydrocarbon determines, in part, the properties of the **molecule**. The number of carbon atoms and how they are arranged in three-dimensions also determines the chemical and physical properties of chlorinated hydrocarbons. Because there is such an immense number of possible forms of chlorinated hydrocarbons, this class of useful compounds has a wide set of applications that are of great economic and practical importance. For example, chlorinated hydrocarbons produced from the refinement of crude oil comprise such things as synthetic rubbers

used in **automobile** tires and tennis shoes. They also create **plastics** used in packaging, and products like fluid pipes, furniture, home siding, credit cards, fences, and toys, to name just a few. Chlorinated hydrocarbons can also be used as anesthetics, industrial solvents, and as precursors in the production of non-stick coatings like Teflon. Chlorinated hydrocarbons are some of the most potent and environmentally persistent **insecticides**, and when combined with the element fluorine, they function as refrigerants called chlorofluorocarbons, or CFCs. Because of their wide array of uses, chlorinated hydrocarbons are among the most important industrial organic compounds. Since they are derived from distillates of petroleum fossil fuels, however, the depletion of global oil and coal reserves looms as a concern for the future.

Organic chemistry and chlorinated hydrocarbons

Chemistry, the study of **matter** and its interactions, can be divided broadly into two groups: inorganic chemistry and organic chemistry. Inorganic chemistry is concerned with atoms and molecules that, by and large, do not contain the element carbon. For example, table **salt**, or **sodium chloride** (NaCl) is an inorganic compound. The production of table salt and **water** from the reaction of **sodium hydroxide** (NaOH) and hydrochloric acid (HCl) is an example of a reaction in inorganic chemistry, since none of the elements within the compounds are carbon. Exceptions to the no-carbon rule are oxides like **carbon dioxide**, which we exhale when we breathe, and carbonates, like **calcium carbonate** (blackboard chalk). Although these substances contain carbon, they are considered to be inorganic in nature.

Organic chemistry, then, is the branch of chemistry dealing with most carbon-containing compounds. Carbon, the sixth element listed in the **periodic table** of elements, is a very versatile element. Atoms of carbon have the capacity to form chemical bonds with other carbon atoms in many configurations. This great variety makes carbon containing, or organic, molecules very important. Most biological molecules involved in the very chemical processes of life and in most of the cellular structures of living things are organic molecules. Approximately 98% of all living things are composed of organic molecules containing the three elements carbon, hydrogen, and **oxygen**.

Organic molecules vary both in the number of carbon atoms they contain and in the spatial arrangement of the member carbon atoms. Examples of organic molecules containing only one carbon atom are methane (**natural gas**), **chloroform** (a general anesthetic), and **carbon tetrachloride** (an industrial solvent). But most organic molecules contain more than one carbon atom.

Like people holding hands, carbon atoms can form molecules that look like chains. The carbon atoms are chemically linked with each other, like people linked together hand-in-hand in a chain.

Some organic molecules are very short chains of three or four carbon atoms. Other organic molecules are very long chains, containing many carbon atoms linked together. Also, just as people in a chain can form a ring when the first person and the last person in the chain join hands, the carbon atoms in an organic molecule can form ring structures, called aromatic rings. The most common rings are five- and six-member rings, containing five or six atoms of carbon respectively.

Hydrocarbons, then, are specific organic molecules that contain only carbon and hydrogen chemically bound together in chains or in rings. Many hydrocarbons are actually combinations of chains and rings, or multiple rings linked together. Also, some hydrocarbons can be branched chains. These carbon chains have portions branching from a main chain, like limbs from the trunk of a tree. The number of carbon atoms involved, and the pattern of chain formation or aromatic ring formation determines the unique chemical and physical properties of particular organic hydrocarbons (like rubber, or plastic, or volatile liquid). Chlorinated hydrocarbons are organic hydrocarbon chains and/or aromatic rings that also contain chlorine atoms chemically linked within the molecule.

Many organic molecules, and many chlorinated hydrocarbons, are actually polymers. Organic polymers are large molecules made of many smaller repeating units joined together. The smaller subunits of polymers are called monomers. Just as a locomotive train is made of many train cars linked together, polymers are many smaller monomers linked together in a line. For example, DNA (deoxyribonucleic acid) in the chromosomes of cells is a **polymer** of nucleotide monomers. Many repeating nucleotide subunit molecules are joined together to form a large molecule of DNA. Similarly, polystyrene plastic that is used to make foam cups, toys, and insulation, is a hydrocarbon polymer consisting of carbon chains and aromatic rings. To illustrate their importance, of all the organic petrochemicals (hydrocarbons and their derivatives) produced industrially, over three-fourths are involved in the production of polymers. Some of the most important polymers are chlorinated hydrocarbons.

Chloroform and carbon tetrachloride: simple chlorinated hydrocarbons

Chloroform is the name given to the chlorinated hydrocarbon compound trichloromethane. Trichloromethane, as its name implies (tri-, meaning three) contains three chlorine atoms. Carbon tetrachloride (tetra-, four),

in contrast, has four atoms of chlorine bonded to a single atom of carbon. Both molecules are essentially a methane molecule with chlorine atoms substituted in place of hydrogen atoms. Since both contain only one atom of carbon, they are among the simplest chlorinated hydrocarbon molecules.

Chloroform

Chloroform is a colorless liquid at room **temperature** and is very volatile. Characterized as having a heavy sweet odor somewhat like **ether**, chloroform is actually sweeter than cane sugar. Chloroform cannot mix well with water. Like salad oil in water, chloroform separates into a layer. However, it does mix well with other hydrocarbons, so one of its uses is as a solvent or cleaner to dissolve other organic substances like gums, waxes, **resins**, and fats. Also, chloroform is used in the industrial synthesis of the non-stick coating called Teflon (polytetrafluoroethylene), which is an organic polymer. However, in the past, the primary use for chloroform was as a general anesthetic.

General anesthetics are drugs that cause the loss of consciousness in order to avoid sensations of extreme **pain**, such as those encountered during **surgery**. First synthesized in the laboratory in 1831, chloroform was used as a general anesthetic for the first time in 1847 by British physician Sir James Simpson during an experimental surgical procedure. Before the discovery of chloroform's utility as a general anesthetic, drugs such as opium, **alcohol**, and **marijuana** were used to dull the pain of medical procedures. However, none were effective enough to allow pain-free surgery within the body. Other substances, like ether and nitrous oxide, were also used as general anesthetics around that same time. Because it was used for Queen Victoria of England's labor pain during childbirth in 1853, chloroform became very popular. It was soon discovered that chloroform can cause fatal cardiac paralysis in about one out of every 3,000 cases, and therefore is seldom used as an anesthetic today.

Carbon tetrachloride

Carbon tetrachloride, like chloroform, is a clear, organic, heavy liquid. Consisting of one carbon atom and four chlorine atoms, carbon tetrachloride has a sweet odor and evaporates very easily and so is most often encountered as a gas. The compound does not occur naturally. Rather, it is manufactured industrially in large amounts for use as a solvent to dissolve other organic materials, or as a raw material in the production of **chlorofluorocarbons (CFCs)** used as aerosol propellants and refrigeration fluids. For many years, carbon tetrachloride was used as a cleaning agent to remove greasy stains from carpeting, draperies, furniture upholstery, and clothing. Also, prior to 1960, carbon tetrachloride was used in fire extinguishers since it is inflammable. Because it is an effective and inexpensive pesticide, before 1986 carbon tetrachloride was used to fumigate grain. These applications, however, have been discontinued since the discovery that the compound is probably carcinogenic, or **cancer** causing. Given its potential to cause cancer in humans, carbon tetrachloride is especially dangerous since it does not break down in the environment very easily. It can take up to 200 years for carbon tetrachloride to degrade fully in **contaminated soil**. Fortunately, the carcinogenic effects seen in laboratory experiments were due to very high levels of exposure that are not characteristic of the levels encountered by most people. Currently, it is not known what long-term low levels of exposure might have on human health.

Important complex chlorinated hydrocarbons

The chlorinated hydrocarbons discussed above are considered to be simple because they contain only one carbon atom in their molecules. Many chlorinated hydrocarbon substances, however, are much larger than this. Having molecules consisting of numerous carbon atoms, some of the most important examples of complex chlorinated hydrocarbons are polymers and biologically active compounds that act as poisons.

Chlorinated hydrocarbon polymers

Organic polymer materials are prevalent in our modern society. The common term plastic really refers to synthetic organic polymer materials consisting of long carbon chains. One of the best known polymers, home plastic wraps used to cover food, is polyethylene. For use as plastic food wrap, polyethylene is prepared as a thin sheet. A greater thickness of polyethylene is used to mold plastic houseware products, like plastic buckets, carbonated drink bottles, or brush handles. Another common and closely related polymer is polypropylene. In addition to also being used for similar products, polypropylene is used to make clothing, the fibers of which are woven synthetic polymer strands. Both polypropylene and polyethylene are used extensively. As chemicals, though, they are classified as hydrocarbons. The chemical addition of chlorine atoms into the molecular structure of hydrocarbon polymers gives the polymers different useful properties. Organic polymers containing chlorine are called chlorinated hydrocarbon polymers.

Perhaps the best known chlorinated hydrocarbon polymer is Polyvinyl Chloride, or PVC. Because PVC also contains atoms of chlorine incorporated into the polymer molecule structure, it has different uses than polyethylene or polypropylene. PVC polymer molecules are created by chemically linking monomers of vinyl chloride molecules into very long chains. Vinyl chloride is a two carbon molecule unit, also containing chlorine. The polymer structure is very similar to polyethylene. Vinyl chloride is made from the addition of chlorine to ethylene with hydrochloric acid as a byproduct. In the United States, about 15% of all ethylene is used in PVC production.

PVC was first discovered as a polymer in 1872 when sealed tubes containing vinyl chloride were exposed to sunlight. The **solution** inside the tubes polymerized into PVC. In the United States, the first patents for the industrial production of PVC were submitted in 1912, making PVC one of the earliest plastics in use. The prevalence of PVC and its importance to our everyday lives is immense. To name just a few products made, PVC is found in pipes for household plumbing and **waste management**, **phonograph** records, soles and heels of shoes, electrical wire insulation, coated fabrics like Naugahyde, plastic films like Saran Wrap, patio furniture, vinyl floor tiles, novelty toys, yard fences, home siding, and credit cards. Its properties make it very useful in making many of the products that we take for granted each day. PVC is inexpensive to synthesize relative to other polymers, making it an attractive material to use.

Because the polymer molecules of PVC are able to fit closely together, they prevent the seepage of fluids through the plastic. Therefore, PVC has important advantages over other organic polymers in clean water transport, preventing food **contamination**, and securing sterile products. For instance, PVC **blood** bags allow blood products to be stored longer than do **glass** containers while allowing for flexibility. PVC packaging protects fresh food from deterioration, and PVC pipes and liners provide safe drinking water supplies from reservoirs, preventing contamination during transport.

PVC is also fire retardant, making it a very safe chlorinated hydrocarbon polymer. Because they are derived from petroleum products, organic polymers are often very flammable. PVC, however, is difficult to ignite. When PVC is burned, it releases less **heat** than other materials. PVC is used to insulate cables that can build up heat because of its heat resistant property. Additional safety characteristics of PVC include its durability and shatterproof qualities. Therefore, PVC is used to make protective eyewear, shatterproof bottles, life jackets and inflatable personal flotation devices. The durability and **corrosion** resistance of PVC makes it useful in auto underbody sealing, gutters, window frames, shutters, and cladding of homes.

In addition to the previously listed uses, PVC is an important polymer because it requires less **energy** to manufacture than other plastics, and can be recycled in to new products after first use. Other closely related chlorinated hydrocarbon polymers include polychloroethylene, and trichloroethylene.

A very important example of a chlorinated hydrocarbon polymer that is a synthetic rubber is polychloroprene. Polychloroprene is an example of an elastomer, or polymer that has the elastic properties of rubber. Along with butadiene, isoprene, and styrene, polychloroprene accounts for 90% of all worldwide synthetic rubbers produced. Closely related in chemical structure to natural rubber extracted from rubber-tree plants, polychloroprene is used to make hoses, belts, shoe heels, and fabrics because it is resistant to corrosive chemicals.

Chlorinated hydrocarbon insecticides

In addition to making very useful polymers, rubbers, plastics, solvents, and cleaners, chlorinated hydrocarbons also are potent pesticide substances. Perhaps the best known chlorinated hydrocarbon insecticide is DichloroDiphenylTrichloroethane, or DDT. First synthesized in the 1800s, the insecticidal properties of DDT were not discovered until 1939. Paul Muller, while working for the Swiss company Geigy, first uncovered the effectiveness of DDT against **insects**. After demonstrations of its effectiveness and relative safety to humans was established, the use of DDT exploded around the globe in the war against disease-carrying and agricultural insect **pests**.

The first major use of DDT was during World War II to control **lice** infestation in Allied troops. Its success led to the large-scale use of DDT to control the blood-sucking insects spreading **yellow fever**, **malaria**, **typhus**, and plague. Its initial use met with exceptional results. For example, by the 1960s, malaria cases in India fell from tens of millions of infections, to fewer than 200,000. Fatal cases of malaria in India dropped from near one million, to just two thousand per year.

However, the success of DDT application in the fight against insect transmitted **disease** led to massive misuse of the chemical. Widespread overuse quickly led to the development of resistant insects, upon which the poison had no effect. At the same time, evidence was accumulating that toxic levels of DDT were accumulating in the fatty tissues of animals higher on the food chain, including **fish**, **mammals**, **birds**, and humans. Like other chlorinated hydrocarbons, the persistence of DDT in the environment allows for biological magnification in na-

ture, a process where minute quantities in run-off water is concentrated into toxic levels as it travels upward in the food chain. Because of its harmful effects on **vertebrates** as well as insects, its creation of resistant insect **species**, its environmental persistence, and its biological magnification, the use of DDT has been banned in many countries despite its general effectiveness.

Apart from DDT, there are other chlorinated hydrocarbon **pesticides** that have been developed. These include **Chlordane**, Aldrin, Mirex, and Toxaphene. Because other, less persistent, insecticide alternatives have been developed, the use of chlorinated hydrocarbon insecticides in general has fallen by the wayside in most places.

Closely related compounds

A close cousin to chlorinated hydrocarbons like chloroform or carbon tetrachloride are chlorofluorocarbons, or CFCs. Chlorofluorocarbons are single carbon atoms with both the elements chlorine and fluorine chemically bonded to them. The compounds trichlorofluoromethane (Freon-11) and dichlorodifluoromethane (Freon-12) are widely used CFCs. They are odorless, nonflammable, very stable compounds used as refrigerants in commercial refrigerators and air conditioners. CFCs are also used as aerosol propellants, which launch products like hairspray and spray paint outward from cans. Very useful compounds, over 1,500 million lb (700 million kg) of CFCs were made worldwide in 1985. However, because they destroy the **ozone** layer, they are being phased-out as propellants and refrigerants.

The future

Chlorinated hydrocarbons are incredibly useful in an astounding variety of products, making them an important part of modern life. Because they are environmentally persistent, new ways of cleaning up areas contaminated with chlorinated hydrocarbons are being developed. The term **bioremediation** refers to the use of living organisms to clean up chemically contaminated habitats. Currently, scientists are using **genetic engineering** to develop **microorganisms** that can degrade chlorinated hydrocarbons, and plants that can absorb them from contaminated **soil**. In this way, **pollution** with chlorinated hydrocarbons, and hydrocarbons in general, can be efficiently remedied.

The limited supply of fossil fuels is a threat to future industry because all of the uses of chlorinated hydrocarbons, including the wide array of polymers, depends on building-block **monomer** molecules extracted from crude oil and other carbon-based fossil fuels. Global oil and gas reserves are dwindling at the same time demand for their use is skyrocketing. One possibility to meet de-

KEY TERMS

. . . .

Chlorinated hydrocarbon—An important class of organic molecules composed of carbon, hydrogen, and chlorine atoms chemically bonded to one another.

Chloroform—A simple chlorinated hydrocarbon compound consisting of one carbon atom, one hydrogen atom, and three chlorine atoms. An important chemical, chloroform was widely used as a general anesthetic in the past.

DDT—The chlorinated hydrocarbon, Dichloro-diphenyltrichloroethane. Once widely used an a potent insecticide, DDT use has been reduced due to its persistence in the environment over long periods of time.

Elastomer—An organic polymer that has rubber-like, elastic qualities.

Monomer—The repeating chemical unit of a polymer. Like cars in a train, monomers are chemically linked together to form long chains within a polymer.

Organic chemistry—Chemistry dealing with carbon-containing molecules. Most molecules containing carbon are considered to be organic.

Polymer—A large organic molecule consisting of a chain of smaller chemical units. Many plastic and rubber compounds are polymers.

PVC—PVC, or polyvinylchloride, is an important chlorinated hydrocarbon polymer plastic used in thousands of everyday household and industrial products. PVC is, perhaps, best known for its use in water pipes.

mand for chlorinated hydrocarbon products in the future might be the synthesis of organic molecules from coal, which is much more abundant than oil.

Resources

Books

Boyd, Richard H., and Paul J. Phillips, *The Science of Polymer Molecules.* Cambridge University Press, 1996.

Interrante, Leonard V. *Chemistry of Advanced Materials: An Overview.* Vch Publishing, 1997.

Johnson, Rebecca L. *Investigating the Ozone Hole.* Minneapolis, MN: Lerner Publishing, 1993.

Matthews, George. *PVC: Production, Properties, and Uses.* Institute of Materials, 1997.

Terry Watkins

Chlorination

Chlorination is the process by which the element **chlorine** reacts with some other substance. Chlorination is a very important chemical reaction both in pure research and in the preparation of commercially important chemical products. For example, the reaction between chlorine and methane gas produces one or more chlorinated derivatives, the best known of which are trichloromethane (**chloroform**) and tetra-chloromethane (**carbon tetrachloride**). The **chlorinated hydrocarbons** constitute one of the most commercially useful chemical families, albeit a family surrounded by a myriad of social, political, economic, and ethical issues. One member of that family, as an example, is dichlorodiphenyltrichloroethane (DDT). Although one of the most valuable **pesticides** ever developed, DDT is now banned in most parts of the world because of its deleterious effects on the environment.

The term chlorination is perhaps best known among laypersons in connection with its use in the purification of **water** supplies. Chlorine is widely popular for this application because of its ability to kill **bacteria** and other disease-causing organisms at relatively low concentrations and with little risk to humans. In many facilities, chlorine gas is pumped directly into water until it reaches a **concentration** of about one ppm (part per million). The exact concentration depends on the original purity of the water supply. In other facilities, chlorine is added to water in the form of a solid compound such as **calcium** or **sodium hypochlorite**. Both of these compounds react with water releasing free chlorine. Both methods of chlorination are so inexpensive that nearly every public water purification system in the world has adopted one or the other as its primary means of destroying disease-causing organisms.

Chlorine

Chlorine is the non-metallic chemical element of **atomic number** 17, symbol Cl, **atomic weight** 35.45, melting point -149.8°F (-101°C), and **boiling point** -29.02°F (-33.9°C). It consists of two stable isotopes, of **mass** numbers 35 and 37. Ordinary chlorine is a mixture of 75.77% chlorine-35 **atoms** and 24.23% chlorine-37 atoms.

Chlorine is a highly poisonous, greenish yellow gas, about two and a half times as dense as air, and with a strong, sharp, choking odor. It was, in fact, one of the first poisonous gases used in warfare—in 1915 during

World War I. In spite of its disagreeable nature, there are so many everyday products that contain chlorine or are manufactured through the use of chlorine that it is among the top ten chemicals produced in the United States each year. In 1994, more than 24 billion lb (11 billion kg) of chlorine were produced.

In nature, chlorine is widely distributed over the **earth** in the form of the **salt** (**sodium chloride**) in sea **water**. At an average **concentration** of 0.67 oz (19 g) of chlorine in each liter of sea water, it is estimated that there are some 10^{16} tons of chlorine in the world's oceans. Other compounds of chlorine occur as **minerals** in the earth's crust, including huge underground deposits of solid **sodium** chloride.

Along with fluorine, bromine, iodine and astatine, chlorine is a member of the halogen family of elements in group 17 of the periodic table-the most non-metallic (least metallic) and most highly reactive group of elements. Chlorine reacts directly with nearly all other elements; with metals, it forms salts called chlorides. In fact, the name *halogen*, meaning salt producer, was originally defined for chlorine (in 1811 by J. S. C. Schweigger), and it was later applied to the rest of the elements in this family.

History of chlorine

The most common compound of chlorine, sodium chloride, has been known since ancient times; archaeologists have found evidence that rock salt was used as early as 3000 B.C. The first compound of chlorine ever made by humans was probably hydrochloric acid (**hydrogen chloride** gas dissolved in water), which was prepared by the Arabian alchemist Rhazes around A.D.900. Around A.D.1200, aqua regia (a mixture of nitric and hydrochloric acids) began to be used to dissolve gold; it is still the only liquid that will dissolve gold. When gold dissolves in aqua regia, chlorine is released along with other nauseating and irritating gases, but probably nobody in the thirteenth century paid much attention to them except to get as far away from them as possible.

The credit for first preparing and studying gaseous chlorine went to Carl W. Scheele (1742-1786) in 1774. Scheele was a Swedish chemist who discovered several other important elements and compounds, including **barium**, manganese, **oxygen**, **ammonia,** and glycerin. Scheele thought that chlorine was a compound, which he called dephlogisticated marine acid air. All gases were called airs at that time, and what we now know as hydrochloric acid was called marine acid because it was made from sea salt. The word dephlogisticated came from a completely false theory that slowed the progress of **chemistry** for decades and which is best left unexplained.

It was not until 1811 that Sir Humphry Davy (1778-1829) announced to the Royal Society of London that chlorine gas was an element. He suggested the name chlorine because it is the same pale, yellowish green **color** that sick plants sometimes develop, a color that is known as *chloros* in Greek. (The sick plants are said to have chlorosis.)

Properties and uses of chlorine

Because it is so reactive, chlorine is never found alone-chemically uncombined—in nature. It is prepared commercially by passing **electricity** through a water **solution** of sodium chloride or through molten sodium chloride.

When released as the free element, chlorine gas consists of diatomic (two-atom) molecules, as expressed by the formula Cl_2. The gas is very irritating to the mucous membranes of the nose, mouth and lungs. It can be smelled in the air at a concentration of only 3 parts per million (ppm); it causes throat irritation at 15 ppm, coughing at 30 ppm, and is very likely to be fatal after a few deep breaths at 1,000 ppm.

Chlorine gas dissolves readily in water, reacting chemically with it to produce a mixture of hydrochloric acid (HCl) and hypochlorous acid (HOCl), plus some unreacted Cl_2. This solution, called chlorine water, is a strong oxidizing agent that can be used to kill germs or to **bleach** paper and fabrics. It is used to obtain bromine (another member of its halogen family) from sea water by oxidizing bromide ions to elemental bromine.

In organic chemistry, chlorine is widely used, not only as an oxidizing agent, but as a way of making many useful compounds. For example, chlorine atoms can easily replace **hydrogen** atoms in organic molecules. The new molecules, with their chlorine atoms sticking out, are much more reactive and can react with various chemicals to produce a wide variety of other compounds. Among the products that are manufactured by the use of chlorine somewhere along the way are antiseptics, dyes, **explosives**, foods, **insecticides**, medicines, metals, paints, **paper**, **plastics**, refrigerants, solvents, and **textiles**.

Probably the most important use of chlorine is as a water purifier. Every water supply in the United States and in much of the rest of the world is rendered safe for drinking by the addition of chlorine. Several chlorine-releasing compounds are also used as general disinfectants.

Bleaching is another very practical use of chlorine. Until it was put to use as a bleach around 1785, bright sunlight was the only way people could bleach out stains and undesired colors in textiles and paper. Today, in the form of a variety of compounds, chlorine is used almost exclusively. Here's how it works: many compounds are colored because their molecules contain loose electrons that can absorb specific colors of **light**, leaving the other colors unabsorbed and therefore visible. An oxidizing agent such as chlorine water or a compound containing the hypochlorite ion OCl⁻ removes those electrons (an oxidizing agent is an **electron** remover), which effectively removes the substance's light-absorbing power and therefore its color. Ordinary laundry bleach is a 5.25% solution of **sodium hypochlorite** in water.

Among the important organic compounds containing chlorine are the chlorinated hydrocarbons-hydrocarbons that have had some of their hydrogen atoms replaced by chlorine atoms. A variety of **chlorinated hydrocarbons** have been used as insecticides. One of the earliest to be used was DDT, dichlorodiphenyltrichloroethane. Because it caused serious environmental problems, its use has largely been banned in the United States. Other chlorinated hydrocarbons that are used as **pesticides** include dieldrin, aldrin, endrin, lindane, **chlordane**, and heptachlor. Because all of these compounds are very stable and do not degrade easily, they also have serious environmental drawbacks.

Compounds of chlorine

Following are a few of the important compounds of chlorine.

Calcium hypochlorite, CaOCl: A white powder known as bleaching powder and used for bleaching and as a swimming pool disinfectant. Both its bleaching and its disinfectant qualities come from its chemical instability: it decomposes to release chlorine gas.

Chlorates: Chlorates are compounds of metals with the **anion** ClO_3^-. An example is potassium chlorate, $KClO_3$. Chlorates can cause explosions when mixed with flammable materials, because the chlorate ion decomposes under **heat** to release oxygen, and the oxygen speeds up the **combustion** process to explosive levels. Potassium chlorate is used in fireworks.

Chlorides: Chlorides are the salts of hydrochloric acid, HCl. They are compounds of a **metal** with chlorine and nothing else. Some common examples are sodium chloride (NaCl), ammonium chloride (NH_4Cl), calcium chloride ($CaCl_2$), and **magnesium** chloride ($MgCl_2$). When dissolved in water, these salts produce chloride ions, Cl⁻. Polyvinyl chloride, the widely used plastic known as PVC, is a **polymer** of the organic chloride, vinyl chloride.

Freons are hydrocarbons with fluorine and chlorine atoms substituted for some of the hydrogen atoms in their molecules. They have been widely used as the liquids in refrigerating machines and as propellants in aerosol spray cans. They have been implicated in destroying the **ozone** layer in the upper atmosphere, however, and their use is now severely restricted.

See also Chlorination; Chloroform; Dioxin; Halogenated hydrocarbons; Halogens; Hydrochlorofluorocarbons; Mercurous chloride.

Resources

Books

Emsley, John. *Nature's Building Blocks: An A-Z Guide to the Elements.* Oxford: Oxford University Press, 2002.
Greenwood, N.N., and A. Earnshaw. *Chemistry of the Elements.* Oxford: Butterworth-Heinneman Press, 1997.
Sconce, J.S. *Chlorine, Its Manufacture, Properties and Uses.* New York: Reinhold, 1962.

Periodicals

"Chlorine Industry Running Flat Out Despite Persistent Health Fears." *Chemical & Engineering News* (November 21, 1994).

Robert L. Wolke

Chlorofluorocarbons (CFCs)

Chlorofluorocarbons (CFCs) are man-made chemical compounds used as refrigerants, cleaning solvents, aerosol propellants, and blowing agents for foam packaging in many commercial applications. CFCs do not spontaneously occur in nature. They were developed by industrial chemists searching for a safer alternative to refrigerants used until the late 1920s. CFCs are non-toxic, chemically non-reactive, inflammable, and extremely stable near Earth's surface. Their apparent safety and commercial effectiveness led to widespread use, and to steadily rising concentrations of CFCs in the atmosphere, throughout the twentieth century.

CFCs are generally non-reactive in the troposphere, the lowest layer of the atmosphere, but intense ultraviolet **radiation** in the outer layer of the atmosphere, called the stratosphere, decomposes CFCs into component molecules and **atoms** of **chlorine**. These subcomponents initiate a chain of **chemical reactions** that quickly breaks down molecules of radiation-shielding **ozone** (O_3) in the lower stratosphere. The stratospheric ozone layer absorbs ultraviolet radiation and protects Earth's surface from destructive biological effects of intense solar radiation, including cancers and cataracts in humans.

CFCs and ozone destruction

Laboratory chemists first recognized CFCs as catalysts for ozone destruction in the 1970s, and atmospheric scientists observed that CFCs and their subcomponents had migrated into the lower stratosphere. When scientists discovered a zone of depleted stratospheric ozone over **Antarctica**, CFCs were identified as the culprit. Announcement of accelerated loss of stratospheric ozone in 1985 spurred research into the exact chemical and atmospheric processes responsible for the depletion, extensive mapping of the Antarctic and Arctic "ozone holes," and confirmation of overall thinning of the ozone layer. These discoveries also precipitated an international regulatory effort to reduce CFC emissions, and to replace them with less destructive compounds. The Montreal Protocol of 1987, and its 1990 and 1992 amendments, led to a near-complete ban on CFCs and other long-lived chemicals responsible for stratospheric ozone depletion. Atmospheric scientists predict that the phase-out of CFCs and other ozone-destroying chemicals should result in disappearance of the Antarctic ozone hole by about 2050.

CFCs are **halogens**, a group of synthetic compounds containing atoms of the elements fluorine, chlorine, bromine and iodine. CFC-11 ($CFCl_3$), CFC-12 (CF_2Cl_2), CFC-113 ($CF_2ClCFCl_2$), and CFC-114 (CF_2ClCF_2Cl) are the most common forms. Materials scientists first recognized the utility of CFC-12 in 1928 as a replacement for the extremely toxic **sulfur dioxide**, methylchloride, and ammonia-based refrigerants used in turn-of-the-century appliances. CFC-12 and other CFCs were then rapidly developed for other industrial applications and widely distributed as commercial products. Because of their stability, low toxicity, low **surface tension**, ease of liquification, thermodynamic properties, and non-flammability, CFCs were used as refrigerants in **heat** pumps, refrigerators, freezers, and air conditioners; as propellants in **aerosols**; as blowing agents in the manufacture of plastic foam products and insulation, such as expanded polystyrene and polyurethane; as cleaning and de-greasing agents for metals and electronic equipment and components, especially circuit boards; as carrier gases for chemicals used in the sterilization of medical instruments; and as dry-cleaning fluids.

CFC-11 and CFC-12 were the most widely-used CFCs. Large industrial air-conditioning equipment and centrifugal systems typically used CFC-11. Residential, automotive and commercial refrigeration and air-conditioning equipment generally contained CFC-12. Some commercial air-conditioning equipment also contained CFC-113 and CFC-114.

Emissions of CFCs to the atmosphere peaked in 1988, when air conditioners, refrigerators and factories released 690 million lb (315 million kg) of CFC-11, and 860 million lb (392 million kg) of CFC-12. At that time, about 45% of global CFC use was in refrigeration, 38% in the manufacture of foams, 12% in solvents, and 5% in aerosols and other uses. Depending on its size, a typical domestic refrigerator sold in 1988 contained about 0.4–0.6 lb (0.2–0.3 kg) of CFCs, a freezer 0.6–1.1 lb (0.3–0.5 kg), and a central air-conditioning unit about 65.5 lb (13.5 kg). About 90% of new automobiles sold in the United States and 60% of those in Canada have air conditioning units, and each contained 3–4 lb (1.4–2.0 kg) of CFCs. CFC production and use in the United States in 1988 involved about 5,000 companies in 375,000 locations, employing 700,000 people, and generating $28 billion worth of goods and services.

Chemical activity of CFCs

CFCs are highly stable, essentially inert chemicals in the troposphere, with correspondingly long residence times. For example, CFC-11 has an atmospheric lifetime of 60 years, CFC-12 120 years, CFC-113 90 years, and CFC-114 200 years. The atmospheric **concentration** of total CFCs in the early 1990s was about 0.7 ppb (parts per billion), and was increasing about 5–6% per year. Because of continued releases from CFC-containing equipment and products already in use, CFC emissions to the lower atmosphere have continued since their manufacture was banned in 1990. However, CFC concentrations in the troposphere declined in 2000 for the first time since the compounds were introduced. Model calculations show that it will take 20–40 years to return to pre-1980 levels.

Because of their long life spans and resistance to chemical activity, CFCs slowly wend their way into the stratosphere, 5–11 mi (8–17 km) above the earth's surface, where they are exposed to intense ultraviolet and other short-wave radiation. CFCs degrade in the stratosphere by photolytic breakdown, releasing highly reactive atoms of chlorine and fluorine, which then form simple compounds such as chlorine monoxide (ClO). These secondary products of stratospheric CFC **decomposition** react with ozone (O_3), and result in a net consumption of this radiation-shielding gas.

Ozone is naturally present in relatively large concentrations in the stratosphere. Stratospheric O_3 concentrations typically average 0.2–0.3 ppm, compared with less than 0.02–0.03 ppm in the troposphere. (Ozone, ironically, is toxic to humans, and tropospheric O_3 is a component of the photochemical **smog** that pollutes the air in urban areas.) Stratospheric O_3 is naturally formed and destroyed during a sequence of photochemical reactions called the Chapman reactions. Ultraviolet radiation decomposes O_2 molecules into single **oxygen** atoms, which then combine with O_2 to form O_3. Ultraviolet **light** then breaks the O_3 molecules back into O_2 and oxygen atoms by photodissociation. Rates of natural ozone creation and destruction were essentially equal, and the concentration of stratospheric ozone was nearly constant, prior to introduction of ozone-depleting compounds by human activity. Unlike the Chapman reactions, reactions with trace chemicals like ions or simple molecules of chlorine, bromine, and fluorine, results in rapid one-way depletion of ozone. CFCs account for at least 80% of the total stratospheric ozone depletion. Other man-made chemical compounds, including halogens containing bromide and **nitrogen** oxides, are responsible for most of the remaining 20%.

The stratospheric O_3 layer absorbs incoming solar ultraviolet (UV) radiation, thereby serving as a UV shield that protects organisms on Earth's surface from some of the deleterious effects of this high-energy radiation. If the ultraviolet radiation is not intercepted, it disrupts the genetic material, DNA, which is itself an efficient absorber of UV. Damage to human and **animal** DNA can result in greater incidences of skin cancers, including often-fatal melanomas; cataracts and other **eye** damage such as snow blindness; and **immune system** disorders. Potential ecological consequences of excessive UV radiation include inhibition of **plant** productivity in regions where UV light has damaged pigments, including **chlorophyll**.

Ozone "hole" and other CFC environmental effects

British Antarctic Survey scientists first observed a large region of depleted stratospheric O_3 over Antarctica during the late 1970s, and dubbed it an "ozone hole." Springtime decreases in stratospheric ozone averaged 30–40% during the 1980s. By 1987, 50% of the ozone over the Antarctic **continent** was destroyed during the austral spring (September to December), and 90% was depleted in a 9–12 mi (15–20 km) wide zone over Queen Maud Land. Atmospheric scientists also observed an Arctic ozone hole, but it is smaller, more variable, and less depleted than its southern hemisphere counterpart. Rapid ozone depletion and the development of seasonal ozone holes is most pronounced at high latitudes where intensely cold, dark conditions and isolating **atmospheric circulation** promote ozone-destroying chemical reactions and inhibit synthesis of new ozone molecules during the darkest spring months. Stratospheric O_3 destruction by the secondary compounds of CFCs also occurs at lower latitudes, but the depletion is much slower because constant light and atmospheric mixing foster ozone regeneration. A seasonal thinning of the stratospheric ozone layer also occurs at lower latitudes when O_3-depleted air disperses from the poles in late spring.

In addition to accelerating the loss of stratospheric ozone, CFCs may also contribute to an intensification of the so-called "greenhouse effect," and to long-term **global climate** change. The **greenhouse effect** is a phenomenon by which an envelope of atmospheric gases and vapors like **carbon dioxide** and **water** vapor maintains the earth's average surface **temperature** at about 77°F (25°C) by trapping a portion of the heat emitted by the planet's surface. Insulation by the greenhouse gases keeps Earth's temperature approximately 33 degrees warmer than would be possible if the atmosphere was transparent to long-wave infrared **energy**. The greenhouse effect also permits existence of the large reservoirs of liquid water that sustain biological life on the **planet**. The pre-industrial concentration of greenhouse gases was chemically balanced to allow global cooling at a **rate** that maintained temperatures within an acceptable temperature range. However, environmental scientists are concerned that natural and synthetic radioactive gases emitted by human activities will slow the earth's cooling rate, and lead to **global warming**. CFCs are greenhouse gases; they are very efficient absorbers of infrared energy. On a per-molecule basis, CFC-11 is 3,000–12,000 times as efficient as **carbon** dioxide as a greenhouse gas, and CFC-12 is 7,000–15,000 times as effective. Atmospheric concentrations of total CFCs increased from almost **zero** several decades ago to about 0.5 ppb in 1992.

CFC reduction efforts

Concerns about the environmental effects of CFCs led to partial restrictions on their use in the early 1980s, when they were prohibited as propellants in aerosol cans. Industrial chemists also began a search for chemical compounds to replace CFCs in refrigerators, air conditioners, manufacturing processes and aerosol generators. **Hydrochlorofluorocarbons** (HCFCs) and hydrofluorocarbons (HFCs) can replace CFCs for many purposes, and **ammonia** functions well as a refrigerant in modern cooling units. HCFCs and HFCs are much less durable than CFCs in the lower atmosphere because they contain **hydrogen** atoms in their molecular structure, are thus less likely to persist and carry their ozone-destroying chlorine and fluorine atoms into the stratosphere.

The United Nations Environment Program (UNEP) convened the Montreal Protocol on Substances that Deplete the Ozone Layer to regulate production, use, and **emission** of CFCs in 1987. The Montreal Protocol was a comprehensive, international agreement designed to slow and eventually reverse stratospheric ozone depletion. The 1987 protocol called for a 50% reduction of CFC emis-

Periodicals

"For the Ozone Layer, a New Look." *New York Times* (October 8, 2002): D1.

Manzer, L. E. "The CFC-Ozone Issue: Progress on the Development of Alternatives." *Science* (1990).

Other

CEISN Thematic Guides. "Chlorofluorocarbons and Ozone Depletion." [cited October 19, 2002]. <http://www.ciesin.org/TG/OZ/cfcozn.html>.

University of Cambridge Centre for Atmospheric Science. "The Ozone Hole Tour." [cited October 19, 2002]. <http://www.atm.ch.cam.ac.uk/tour/index.html>.

National Oceanographic and Atmospheric Administration Climate Monitoring and Diagnostics Laboratory (NOAA). "Chlorofluorocarbons (CFCs)." [cited October 19, 2002]. <http://www.cmdl.noaa.gov/noah/publictn/elkins/cfcs.html>.

Bill Freedman
Laurie Duncan

KEY TERMS

Greenhouse effect—The physical process that allows Earth's surface to be maintained at an average of 77°F (25°C), about 33° warmer than would otherwise be possible if certain gases and vapors, especially carbon dioxide and water, did not interfere with the rate of dissipation of absorbed solar radiation.

Ozone holes—Decreased concentrations of stratospheric ozone, occurring at high latitudes during the early springtime. Ozone holes are most apparent over Antarctica, where they develop under intensely cold conditions during September and November, allowing a greater penetration of deleterious solar ultraviolet radiation to Earth's surface.

Stratosphere—A layer of the upper atmosphere above an altitude of 5–11 mi (8–17 km) and extending to about 31 mi (50 km), depending on season and latitude. Within the stratosphere, air temperature changes little with altitude, and there are few convective air currents.

Troposphere—The layer of air up to 15 mi (24 km) above the surface of the earth, also known as the lower atmosphere.

---

sions by 2000. Scientific advances prompted amendments to the protocol in 1990 and 1992, the most recent of which required signatories to cease production of the main CFCs by 1995. (Exceptions were allowed for limited essential uses, including medical sprays.) Some major industrial users of CFCs committed to earlier phaseouts, and the European community agreed to an even stricter set of regulations that requires even tighter restrictions on CFCs and other ozone-depleting chemicals. As a result of these regulatory measures, CFC concentrations declined in lower atmosphere, and remained constant in the upper atmosphere in 2000. Computational models predict that the Antarctic ozone hole will disappear by about 2050. Because of the unusually rapid and effective international response to the problem of stratospheric ozone depletion caused by emissions of CFCs, the Montreal Protocol and subsequent agreements on CFCs have been described as an environmental success story.

See also Air pollution; Halogenated hydrocarbons; Ozone layer depletion.

Resources

Books

Freedman, B. *Environmental Ecology.* 2nd ed. San Diego: Academic Press, 1994.

Chloroform

Chloroform is the common name of the organic compound whose chemical formula is $HCCl_3$. The **molecule** of trichloromethane, as it is also called, consists of a central **carbon** atom bonded to a **hydrogen** atom and three **chlorine atoms**. Chloroform is a nonflammable colorless liquid (**boiling point** 141.8°F [61°C]) that has a heavy sweet odor and **taste**. The compound was first prepared in 1831 simultaneously by Justus von Liebig (1803-1873) in Germany and by Eugene Soubeiran (1797-1858) in France using different procedures. Samuel Guthrie (1782-1848), in the United States, also discovered chloroform in that same year.

Chloroform was originally used to calm people suffering from **asthma**. In 1847, James Y. Simpson, a Professor of Midwifery at the University of Edinburgh, began using chloroform as an anesthetic to reduce **pain** during childbirth. From this initial experiment, chloroform began to be used as general **anesthesia** in medical procedures throughout the world. The use of chloroform in this application was eventually abandoned because of its harmful side effects on the **heart** and liver.

Chloroform has commonly been used as a solvent in the manufacture of **pesticides**, dyes, and drugs. The use of the chemical in this manner was important in the preparation of penicillin during World War II. Chloroform was used as a sweetener in various cough syrups and to add flavor "bursts" to toothpaste and mouthwash. Its pain relieving properties were incorporated into various liniments and toothache medicines. The chemical was also used in photographic processing and dry cleaning. All of these applications for chloroform were stopped by the Federal Drug Administration (FDA) in 1976, when the compound was discovered to cause **cancer** in laboratory **mice**. Today, chloroform is a key starting material for the production of chemicals used in refrigerators and air conditioners.

Chlorophyll

Chlorophyll is a green pigment contained in the foliage of plants, giving them their notable coloration. This pigment is responsible for absorbing sunlight required for the production of sugar molecules, and ultimately of all biochemicals, in the **plant**.

Chlorophyll is found in the thylakoid sacs of the **chloroplast**. The chloroplast is a specialized part of the **cell** that functions as an organelle. Once the appropriate wavelengths of **light** are absorbed by the chlorophyll into the thylakoid sacs, the important process of **photosynthesis** is able to begin. In photosynthesis, the chloroplast absorbs light **energy**, and converts it into the chemical energy of simple sugars.

Vascular plants, which can absorb and conduct moisture and **nutrients** through specialized systems, have two different types of chlorophyll. The two types of chlorophyll, designated as chlorophyll *a* and *b*, differ slightly in chemical makeup and in **color**. These chlorophyll molecules are associated with specialized **proteins** that are able to penetrate into or span the **membrane** of the thylakoid sac.

When a chlorophyll **molecule** absorbs light energy, it becomes an excited state, which allows the initial chain reaction of photosynthesis to occur. The pigment molecules cluster together in what is called a photosynthetic unit. Several hundred chlorophyll *a* and chlorophyll *b* molecules are found in one photosynthetic unit.

A photosynthetic unit absorbs light energy. Red and blue wavelengths of light are absorbed. Green light cannot be absorbed by the chlorophyll and the light is reflected, making the plant appear green. Once the light energy penetrates these pigment molecules, the energy is passed to one chlorophyll molecule, called the reaction center chlorophyll. When this molecule becomes excited, the light reactions of photosynthesis can proceed. With **carbon dioxide**, **water**, and the help of specialized enzymes, the light energy absorbed creates chemical energy in a form the cell can use to carry on its processes.

In addition to chlorophyll, there are other pigments known as accessory pigments that are able to absorb light where the chlorophyll is unable to. Carotenoids, like B-carotenoid, are also located in the thylakoid membrane. Carotenoids give carrots and some autumn leaves their color. Several different pigments are found in the chloroplasts of **algae**, **bacteria**, and **diatoms**, coloring them varying shades of red, orange, blue, and violet.

See also Plant pigment.

Chloroplast

Chloroplasts are organelles—specialized parts of a **cell** that function in an organ—like fashion. They are found in vascular plants, mosses, liverworts, and **algae**. Chloroplast organelles are responsible for **photosynthesis**, the process by which sunlight is absorbed and converted into fixed chemical **energy** in the form of simple sugars synthesized from **carbon dioxide** and **water**.

Chloroplasts are located in the mesophyll, a green **tissue** area in **plant** leaves. Four layers or zones define the structure of a chloroplast. The chloroplast is a small lens-shaped organelle which is enclosed by two membranes with a narrow intermembrane space, known as the chloroplast envelope. Raw material and products for photosynthesis enter in and pass out through this double **membrane**, the first layer of the structure.

Inside the chloroplast envelope is the second layer, which is an area filled with a fluid called stroma. A series of **chemical reactions** involving enzymes and the incorporation of carbon dioxide into organic compounds occur in this region.

The third layer is a membrane-like structure of thylakoid sacs. Stacked like poker chips, the thylakoid sacs form a grana. These grana stacks are connected by membranous structures. Thylakoid sacs contain a green pigment called **chlorophyll**. In this region the thylakoid sacs, or grana, absorb **light** energy using this pigment. Chlorophyll absorbs light between the red and blue spectrums and reflects green light, making leaves appear green. Once the light energy is absorbed into the final layer, the intrathylakoid sac, the important process of photosynthesis can begin.

Scientists have attempted to discover how chloroplasts convert light energy to the chemical energy stored in organic molecules for a long time. It has only been since the beginning of this century that scientists have begun to understand this process. The following equation is a simple formula for photosynthesis:

$$6CO_2 + 6H_2O \rightarrow C_6H_{12}O_6 + 6O_2$$

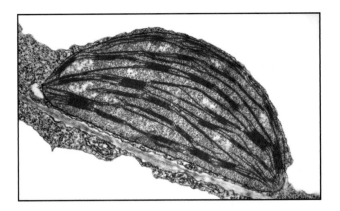

A transmission electron micrograph (TEM) of a chloroplast from a tobacco leaf (*Nicotiana tabacum.*). The stacks of flattened membranes that can be seen within the chloroplast are grana. The membranes that run between the stacks are stroma. The faint white patches within the chloroplast are nucleoids, where chloroplast DNA is stored. *© Dr. Jeremy Burgess/Science Photo Library, National Audubon Society Collection/ Photo Researchers, Inc. Reproduced with permission.*

Carbon dioxide plus water produce a **carbohydrate** plus **oxygen**. Simply, this means that the chloroplast is able to split water into **hydrogen** and oxygen.

Many questions still remain unanswered about the complete process and role of the chloroplast. Researchers continue to study the chloroplast and its **evolution**. Based on studies of the evolution of early complex cells, scientist have devised the serial endosymbiosis theory. It is suspected that primitive microbes were able to evolve into more complex ones by incorporating other photosynthetic microbes into their cellular structures and allowing them to continue functioning as organelles. As **molecular biology** becomes more sophisticated, the origin and genetic makeup of the chloroplast will be more clearly understood.

See also Leaf.

Cholera

Cholera is one of the most devastating of all human diseases. Although **endemic** in some areas of the world, cholera is usually associated with massive migrations of people, such as those occurring during war or famine. Cholera is also common in developing countries, where suboptimal sanitation practices are responsible for its spread. If not treated, cholera has a fatality **rate** of over 60%. Death results from dehydration, a consequence of the severe diarrhea and vomiting that characterize this **disease**. In the last 15 years, treatment

strategies have been devised that have cut the fatality rate of cholera to 1%. Preventive measures have also reduced the incidence of cholera outbreaks. These measures, however, require swift intervention, which is not always possible during the social upheavals that can lead to cholera epidemics.

The cause of cholera

Cholera is caused by a **bacteria** called *Vibrio cholerae*, which secretes a toxin, or poison, that binds to the cells of the small intestine. One of the functions of the small intestine in humans is to regulate the amount of fluid that is absorbed by cells. Normally, small intestine cells absorb most of the fluid that is ingested; only a small amount of fluid is excreted in the feces. Under abnormal conditions, such as in response to a pathogen, cells do not absorb fluid, and as a result a lot of fluid enters the small intestine and is excreted in the feces. These frequent, watery stools are called diarrhea. Diarrhea can actually be helpful, as the rapid movement of fluid flushes the gastrointestinal tract of harmful bacteria and other **pathogens**. But if diarrhea is severe or long lasting, such as occurs in cholera, too much fluid is lost and the body becomes dehydrated. If fluids are not replaced, death can result.

Along with causing fluid loss, the binding of cholera toxin to small intestine cells also results in loss of electrolytes. Electrolytes are chemicals that the body needs to function properly, such as potassium chloride, **sodium chloride (salt)**, and bicarbonate. Electrolytes are crucial in the control of **blood** pressure, excretion of metabolic wastes, and maintenance of blood sugar levels. If the amount of electrolytes in the body deviates even slightly, these crucial body functions are imperiled. Cholera toxin prompts the small intestine cells to secrete large amounts of electrolytes into the small intestine. These electrolytes are then excreted in the watery diarrhea.

The cholera toxin consists of two subunits, the A subunit and the B subunit. The B subunit is a ring, and the A subunit is suspended within it. By itself, the B subunit is nontoxic; the A subunit is the poisonous part of the toxin. The B subunit binds to the small intestine **cell** and creates a channel within the cell **membrane** through which the A subunit enters. Once inside the small intestine cell, the A subunit disrupts the cascade of reactions that regulates the cell's fluid and **electrolyte** balance. Fluid and electrolytes leave the cell and enter the small intestine. The resultant diarrhea may cause a fluid loss that exceeds 1 qt (1 liter) per hour.

V. cholerae lives in aquatic environments, and especially favors salty or **brackish** waters. *V. cholerae* frequently colonize shellfish; in fact, cholera cases in the

United States are almost always traced to eating raw or undercooked shellfish. Interestingly, *V. cholerae* can also cause skin and other soft **tissue** infections. Cases of such **infection** with these bacteria have been found in persons who have sustained injuries in marine environments; apparently, *V. cholerae* in **water** can penetrate broken skin and cause infection.

Transmission of cholera

Cholera is endemic in several areas of the world, including parts of India and Bangladesh. From these areas, cholera has been disseminated throughout the world during several pandemics, or worldwide outbreaks. In the United States, a cholera pandemic that lasted from 1832 to 1849 killed 150,000 people; in 1866, another cholera pandemic killed 50,000 U.S. citizens. The most recent pandemic, which began in the 1960s and lasted until the early 1980s, involved **Africa**, Western **Europe**, the Philippines, and Southeast **Asia**. Smaller outbreaks, such as the Rwanda **epidemic** of 1994, are characteristic of wartime and famine conditions, in which large numbers of people concentrate in one place where sanitary conditions are poor to nonexistent.

Because of the nature of *V. cholerae* infection, past epidemics can lead to future epidemics. People recovering from cholera continue to shed the **organism** in their feces for weeks to months after the initial infection. These people are called convalescent carriers. Another kind of carrier, called a chronic carrier, continues to shed the bacteria for years after recovery. In both carrier types, no symptoms are present. With the ease of worldwide transportation, carriers can travel throughout the world, spreading *V. cholerae* wherever they go. If a carrier visits an area with less-than-ideal sanitary conditions or does not wash his or her hands after using the bathroom, the deadly *V. cholerae* bacteria can be easily transmitted.

Symptoms and treatment of cholera

Cholera is characterized by sudden onset. Within several hours or days after ingesting *V. cholerae*, severe diarrhea and vomiting occur. Fluid losses can be up to 4-5 gal (15-20 liters) per day. As a consequence of this severe fluid loss, the eyes and cheeks can appear sunken, and the skin loses its pliancy.

Treatment of cholera involves the rapid replacement of fluid and electrolytes. Most patients can be treated using special rehydration formulations which utilize **rice** or grain as a base, and which are supplemented with appropriately balanced electrolytes. If a patient is too severely ill to drink even small, frequent sips, infusions of

electrolytes and fluids may be necessary. Once fluid and electrolyte balance is restored, rapid reversal of symptoms occurs. Treatment with **antibiotics**, typically tetracycline, neutralizes the *V. cholerae*, and decreases the number of bacteria passed into the stool.

Prevention

In the United States, **sewage treatment** and water purification plants are ubiquitous, and consequently, the incidence of cholera is low. Almost all cases of cholera in the U.S. are caused by improperly cooked shellfish. Experts recommend that all shellfish be boiled for 10 minutes; steaming does not kill *V. cholerae*. Raw shellfish should be avoided.

Another way to prevent cholera is to identify and treat cholera carriers in areas where cholera is endemic. Treating carriers would eliminate a major route of cholera transmission.

Currently, several cholera vaccines are being developed, but only one is likely to be effective. Injectable vaccines are impractical in many areas. Oral vaccines are more easily delivered to the population, but are not nearly as effective. A genetically engineered **vaccine** that consists of an altered *V. cholerae* organism appears to stimulate an immune response in a small number of volunteers. Larger vaccine trials in endemic populations are necessary, however, to determine the efficacy of this vaccine. Currently available vaccines confer only partial, short-term immunity, and therefore are not being recommended for most travelers. Instead, preventive measures are advised. For cholera endemic areas, suggestions include drinking only boiled or chlorine- or iodine-treated water; avoiding **ice**; avoiding **fruits** and **vegetables** unless cooked thoroughly or peeled; eating only very thoroughly cooked seafood.

Resources

Books

Delaporte, François. *Disease and Civilization: The Cholera in Paris, 1832*. Cambridge: MIT Press, 1986.

Hayhurst, Chris. *Cholera*. New York: Rosen Publishing Group, 2001.

Van Heyningen, Willian Edward, and John R. Seal. *Cholera: The American Scientific Experience, 1947-1980*. Boulder, CO: Westview Press, 1983.

Periodicals

Besser, R.E., D.R. Feiken, and P.N. Griffin. "Diagnosis and Treatment of Cholera in the United States: Are We Prepared?" *Journal of the American Medical Association* 272 (October 19, 1993): 1203.

Royal, Louis, and Iain McCoubrey. "International Spread of Disease by Air Travel." *American Family Physician* 40 (November 1, 1989): 129.

Spangler, Brenda D. "Structure and Function of Cholera Toxin and the Related *Escherichia coli* Heat-Labil Enterotoxin." *Microbial Reviews* 56 (December 1, 1992): 622.

Cholesterol

Cholesterol is a complex organic compound with the **molecular formula** $C_{27}H_{46}O$. It is a member of the biochemical family of compounds known as the lipids. Other lipids, such as the waxes, fats, and oils, share not a structural similarity (as is the case with most families of compounds), but a physical property-they are all insoluble in **water**, but are soluble in organic liquids.

Cholesterol belongs more specifically to a class of compounds known as the steroids. Most steroids are naturally occurring compounds that play critical roles in **plant** and **animal physiology** and **biochemistry**. Other steroids include sex **hormones**, certain vitamins, and adrenocorticoid hormones. All steroids share a common structural unit, a four-ring structure known as the perhydrocyclopentanophenanthrene ring system or, more simply, the steroid nucleus.

History

Although cholesterol had been isolated as early as 1770, productive research on its structure did not begin until the twentieth century. Then, in about 1903, a young German chemist by the name of Adolf Windaus decided to concentrate on finding the molecular composition of the compound. Windaus, sometimes referred to as the Father of Steroid Chemistry, eventually worked out a detailed structure for cholesterol, an accomplishment that was partially responsible for his earning the 1928 Nobel Prize in chemistry.

Late research showed that the structure proposed by Windaus was in error. By the early 1930s, however, addi-tional evidence from x-ray analysis allowed Windaus' long-time colleague Heinrich Wieland (among others) to determine the correct structure for the cholesterol **molecule**.

The next step in understanding cholesterol, synthesizing the compound, was not completed for another two decades. In 1951, the American chemist Robert B. Woodward completed that line of research when he synthesized cholesterol starting with simple compounds. For this accomplishment and his other work in synthesizing large molecule compounds, Woodward was awarded the 1965 Nobel Prize in chemistry.

Properties and occurrence

Cholesterol crystallizes from an alcoholic **solution** as pearly white or pale yellow granules or plates. It is waxy in appearance and has a melting point of 299.3°F (148.5°C) and a **boiling point** of 680°F (360°C) (with some **decomposition**). It has a specific gravity of 1.067. Cholesterol is insoluble in water, but slightly soluble in **alcohol** and somewhat more soluble in **ether** and **chloroform**.

Cholesterol occurs in almost all living organisms with the primary exception of **microorganisms**. Of the cholesterol found in the human body, about 93% occurs in cells and the remaining 7% in the **circulatory system**. The **brain** and spinal cord are particularly rich in the compound. About 10% of the former's dry weight is due to cholesterol. An important commercial source of the compound is spinal fluid taken from cattle. Cholesterol is also found in myelin, the material that surrounds nerve strands. Gallstones are nearly pure cholesterol.

The **concentration** of cholesterol in human **blood** varies rather widely, from a low of less than 200 mg/dL (milligrams per deciliter) to a high of more than 300 mg/dL. It is also found in bile, a source from which, in fact, it gets its name: chole (Greek for bile) + stereos (Greek for solid).

Cholesterol in the human body

Cholesterol is a critically important compound in the human body. It is synthesized in the liver and then used in the manufacture of bile, hormones, and nerve **tissue**.

But cholesterol is also a part of the human diet. A single egg yolk for example, contains about 250 mg of cholesterol. **Organ** meats are particularly rich in the compound. A 3 oz (85 g) serving of beef liver, for example, contains about 372 mg of cholesterol and a similar-size serving of calves' brain, about 2,700 mg of the compound. Because diets differ from culture to culture, the amount of cholesterol an **individual** consumes differs widely around the world. The average European diet includes about 500 mg of cholesterol a day, but the average Japanese diet, only about 130 mg a day. The latter fact reflects a diet in which **fish** rather than meat tends to predominate.

The human body contains a feedback mechanism that keeps the serum concentration of cholesterol approximately constant. The liver itself manufactures about 600 mg of cholesterol a day, but that output changes depending on the intake of cholesterol in the daily diet. As a person consumes more cholesterol, the liver reduces it production of the compound. If one's intake of cholesterol greatly exceeds the body's needs, excess cholesterol may then precipitate out of blood and be deposited on arterial linings.

Cholesterol and health

Some of the earliest clues about possible ill effects of cholesterol on human health came from the research of Russian biologist Nikolai Anitschow in the 1910s. Anitschow fed rabbits a diet high in cholesterol and found that the animals became particularly susceptible to circulatory disorders. Post-mortem studies of the animals found the presence of plaques (clumps) of cholesterol on their arterial walls.

Since Anitschow's original research, debate has raged over the relationship between cholesterol intake and circulatory **disease**, particular atherosclerosis (the blockage of coronary **arteries** with deposits of fatty material). Over time, it has become increasingly obvious that high serum cholesterol levels do have some association with such diseases. A particularly powerful study in forming this conclusion has been the on-going Framingham Study, conducted since 1948 by the National Heart Institute in the Massachusetts town that has given its name to the research. Among the recommendations evolving out of that study has been that a reduced intake of cholesterol in one's daily diet is one factor in reducing the risk of **heart** disease.

The cholesterol-heart disease puzzle is not completely solved. One of the remaining issues concerns the role of lipoproteins in the equation. Since cholesterol is not soluble in water, it is transported through the blood stream bound to molecules containing both **fat** and protein components, lipoproteins. These lipoproteins are of two kinds, high **density** lipoproteins (HDLs) and low density lipoproteins (LDLs). For some time, researchers have thought that LDL is particularly rich in cholesterol and, therefore, "bad," while HDL is low in cholesterol and, therefore, "good." While this analysis may be another step in the right direction, it still does not provide the final word on the role of cholesterol in the development of circulatory diseases.

See also Lipid; Nervous system.

Resources

Books

Byrne, Kevin P. *Understanding and Managing Cholesterol: A Guide for Wellness Professionals.* Champaign, IL: Human Kinetics Books, 1991.

Other

National Cholesterol Education Program. *Second Report of the Expert Panel on Detection, Evaluation, and Treatment of High Blood Cholesterol in Adults (Adult Treatment Panel II).* Bethesda, MD: National Institute of Health, National Heart, Lung, and Blood Institute, 1993.

David E. Newton

Chordates

Chordates are a diverse group of animals that comprise the phylum Chordata. There are approximately 44,000 **species** of chordates, ranging in size from several millimeters to 105 ft (32 m) long. The simplest and earliest chordates are pre-vertebrate animals such as ascidians, tunicates, and *Amphioxus*. The major group of chordates are the sub-phylum Vertebrata, the **vertebrates**. Listed more-or-less in the order of their first appearance in the fossil record, vertebrates include **sharks**, lampreys, bony fishes, **amphibians**, **reptiles**, **birds**, and **mammals**.

Chordates exhibit bilateral **symmetry**, and they have a body cavity (the coelom), which is enclosed within a **membrane** (the peritoneum), and which develops from the middle **tissue** layer known as the mesoderm. A defining feature of chordates is a structure known as the notochord. This is a rod-like, flexible structure that runs along the upper, mid-line of chordates, and a notochord is present for at least some part of the life of all chordates. In the earliest chordates, the notochord stiffens the body against the pull of muscles. This function is less important in the more advanced vertebrate chordates, whose bodies are supported by the cartilaginous and bony elements of the skeleton. In vertebrates, the notochord is only present during the embryonic, developmental stages.

Other defining features of chordates are the presence of pharyngeal gill slits (which are precursors of the gill arches in **fish** and amphibians), a hollow nerve cord on the upper surface of the **animal** (that eventually develops into the spinal cord in vertebrates), and a tail extending beyond the anal opening. As with the notochord, these features may only occur during a part of the life of the animal, especially in the more recently evolved chordates, such as the vertebrates.

Chordate animals have a closed **circulatory system**, in which **blood** is transported around the body inside **veins** and **arteries** of various sizes. The blood is circulated by the pumping action of the **heart**; the respiratory gases in the blood diffuse across the thin walls of the smallest vessels (**capillaries**) in the tissues. The most recently evolved vertebrates have a four-chambered heart, and a double circulation of the blood, which involves a separate circulation for the heart and the lungs, and for the heart and the rest of the body (systemic circulation).

Most chordates have two sexes, and the male and female individuals tend to be different in many aspects of their form and function (dimorphic). **Fertilization** is external in the earlier-evolved groups of chordates (fish and amphibians), and internal in later groups (reptiles, birds, and mammals). Many chordates lay eggs, or are **oviparous**, while others give **birth** to live young (viviparous).

Chordates utilize a wide range of habitats. The earliest evolved chordates and some of the more recent groups are aquatic, while others are primarily terrestrial.

See also Sea squirts and salps.

Chorionic villus sampling (CVS)

Chorionic villus sampling (CVS) is a relatively new form of a prenatal testing, completed earlier in gestation than the more traditional testing method, **amniocentesis**.

Through CVS, small samples of the trophoblast are obtained for **chromosome** or DNA analysis. The use of

CVS as a tool for fetal karyotyping at 10 weeks' gestation was introduced in 1969 but it became accepted with the introduction of an ultrasound guided technique for aspiration of material. A larger amount of DNA is derived from CVS than from cells obtained at amniocentesis, thereby allowing more reliable DNA and biochemical analysis in a shorter period of **time**. A transvaginal sonographic examination usually precedes the procedure to determine the number of embryos and chorionicity (number and type of placentas), and to determine fetal viability.

CVS can be performed in two ways. In transcervical testing, after a speculum insertion into the vagina, a catheter is gently passed through the cervix into the uterus. Guided by ultrasound, the catheter is inserted **parallel** to the placenta. The stylet is then removed and a syringe attached in order to have a negative **pressure** and to obtain adequate **tissue sample** called chorionic villi from within the placenta, which contain the same genetic material as does the fetus.

These same fibers can be extracted by transabdominal CVS, when a thin needle guided by ultrasound real time monitoring is inserted through the abdomen into the uterus. Again, ultrasound assists in identifying the exact location of the placenta.

The advantage of the CVS procedure is that it can be completed as early as 10 weeks in the pregnancy, but preferably closer to 12 or 13 weeks. Villi are processed for cytogenetic analysis in two ways. The cytogenetic results are usually available within 48 hours in the so called "direct preparation" that is performed on the trophoblastic cells. Results of "culture method" are usually available within one week and are performed on the mesenchymal cells. Several laboratories wait in order to report a single result at the same time.

In multiple gestations, uncertain results requiring further investigation are more frequently encountered in CVS than by amniocentesis. Mesenchymal cells, are more related to the embryo than trophoblastic cells because they are derived from the more recently separated lineage of a tissue that envelops the embryo at this stage of pregnancy, the so called extraembryonic mesoderm. In some cases, CVS results in the confined placental mosaicism—defined as a discrepancy between the chromosomes of placental and fetal tissues. An abnormal karyotype in the mesenchymal cells, instead, is more likely to be associated to an abnormal fetal karyotype.

CVS was initially considered riskier because of an increased chance of miscarriage. With advances in procedures and training since its introduction in 1983, however, CVS is now considered as safe as amniocentesis and the preferred method for those women who need to obtain early information about the fetus. An excess **rate** of fetal losses of 0.6–0.8% for the CVS over amniocentesis has been reported.

Maternal complications of CVS, even if not frequent, can occur although such correlations with CVS are controversial.

See also Chromosomal abnormalities; Genetic testing.

Resources

Periodicals

Wijnberger, L.D., Y.T. van der Schouw, and G.C. Christiaens. "Learning in Medicine: Chorionic Villus Sampling." *Prenat Diagn* 20 (2000): 241.

Chromatin

Chromatin is the masses of fine fibers comprising the chromosomes in the nucleus of a eukaryotic **cell** in a nondividing state. During **cell division** (**mitosis** or **meiosis**) the chromatin fibers pull together into thick shortened bodies which are then called chromosomes. Chromatin is present only in cells with a nuclear **membrane**; it is not found in prokaryotic cells (e.g., **bacteria**) that lack a nucleus.

Chromatin earned its name from early biologists who examined cells using **light** microscopes. These scientists found that in cells stained with a basic dye, the granular material in the nucleus turned a bright **color**. They named this material "chromatin," using the Greek word *chroma*, which means color. When the chromatin condensed during cell division, the researchers called the resulting structures chromosomes, which means "colored bodies." Chromatin granules which form the chromosomes are known as chromomeres, and these may correspond to genes.

Chemically, chromatin fibers consist of DNA (deoxyribonucleic acid) and two types of **proteins** found in the cell nucleus (nucleoproteins): histones and nonhistones. The histones are simple proteins found in chromosomes bound to the nucleic acids. Histones may be important in switching off **gene** action. While all cells of the body contain the same DNA instructions for every type of body cell, the specialized cells do not use all of these instructions. The unneeded DNA is put into storage, by binding with proteins, forming a complex called a nucleosome. Histones link the nucleosomes, forming large pieces of chromatin. DNA contains the genetic material that determines heredity. That chromatin contains DNA is to be expected, since chromosomes are made of chromatin. The compact structure of chromatin chromomeres, where DNA is wrapped around protein balls, is an efficient means of storing long stretches of DNA.

The discovery in 1949 of a condensed X **chromosome** of sex (termed the Barr body) which was visible at interphase (nondividing) in body cells of female **mammals**, provided physicians with a new means of determining the genetic sex of hermaphrodites. The sex chromatin test has now given way to direct M chromosomal analysis.

The Barr (X chromosome) is the inactive partner of the two X sex chromosomes in female mammals, including humans. Its dense, compact form led researcher M. F. Lyon to hypothesize in 1962 that it was inactive. In effect, then, both sexes have only one active X chromosome. Males (XY) have the sex chromosomes X and Y, while females (XX) have one functional almost inactive X chromosome. The influence of the inactive X chromosome expression in offspring is known as "lyonization," and is responsible for female tortoiseshell **cats**.

See also Eukaryotae; Nucleus, cellular.

Chromatography

Chromatography is a family of laboratory techniques for separating mixtures of chemicals into their individual compounds. The basic principle of chromatography is that different compounds will stick to a solid surface, or dissolve in a film of liquid, to different degrees.

To understand chromatography, suppose that all the runners in a race have sticky shoe soles, and that some runners have stickier soles than others. The runners with the stickier shoes will not be able to run as fast. All other things being equal, the runners will cross the finish line in the exact order of their shoe stickiness—the least sticky first and the stickiest last. Even before the race is over, they will spread out along the track in order of their stickiness.

Similarly, different chemical compounds will stick to a solid or liquid surface to varying degrees. When a gas or liquid containing a mixture of different compounds is made to flow over such a surface, the molecules of the various compounds will tend to stick to the surface. If the stickiness is not too strong, a given **molecule** will become stuck and unstuck hundreds or thousands of times as it is swept along the surface. This repetition exaggerates even tiny differences in the various molecules' stickiness, and they become spread out along the "track," because the stickier compounds move more slowly than the less-sticky ones do. After a given time, the different compounds will have reached different places along the surface and will be physically separated from one another. Or, they can all be allowed to reach the far end of the surface—the "finish line"—and be detected or measured one at a time as they emerge.

Using variations of this basic phenomenon, chromatographic methods have become an extremely powerful and versatile tool for separating and analyzing a vast variety of chemical compounds in quantities from picograms (10^{-12} gram) to tons.

Chromatographic methods all share certain characteristics, although they differ in size, shape, and configuration. Typically, a stream of liquid or gas (the mobile phase) flows constantly through a tube (the column) packed with a porous solid material (the stationary phase). A **sample** of the chemical mixture is injected into the mobile phase at one end of the column, and the compounds separate as they move along. The individual separated compounds can be removed one at a time as they exit (or "elute from") the column.

Because it usually does not alter the molecular structure of the compounds, chromatography can provide a non-destructive way to obtain pure chemicals from various sources. It works well on very large and very small scales; chromatographic processes are used both by scientists studying micrograms of a substance in the laboratory, and by industrial chemists separating tons of material.

The technology of chromatography has advanced rapidly in the past few decades. It is now possible to obtain separation of mixtures in which the components are so similar they only differ in the way their **atoms** are oriented in **space**, in other words, they are isomers of the same compounds. It is also possible to obtain separation of a few parts per million of a contaminant from a mixture of much more concentrated materials.

The development of chromatography

The first **paper** on the subject appeared in 1903, written by Mikhail Semyonovich Tsvet (1872-1919), a Russian-Italian biochemist, who also coined the word chromatography. Tsvet had managed to separate a mixture of **plant** pigments, including **chlorophyll**, on a column packed with finely ground **calcium carbonate**, using **petroleum ether** as the mobile phase. As the colored mixture passed down the column, it separated into individual colored bands (the term chromatography comes from the Greek words *chroma*, meaning **color**, and *graphein*, meaning writing, or drawing). Although occasionally used by biochemists, chromatography as a science lagged until 1942, when A. J. P. Martin (1910-2002) and R. L. M. Synge (1914-1994) developed the first theoretical explanations for the chromatographic separation process. Although they eventually received the Nobel Prize in **chemistry** for this work, chromatography did not come into wide use until 1952, when Martin, this time working with A. T. James, described a way of using a gas instead of a liquid as the mobile phase, and a highly viscous liquid coated on solid particles as the stationary phase.

Gas-liquid chromatography (now called gas chromatography) was an enormous advance. Eventually, the stationary phase could be chemically bonded to the solid support, which improved the **temperature** stability of the column's packing. Gas chromatographs could then be operated at high temperatures, so even large molecules could be vaporized and would progress through the column without the stationary phase vaporizing and bleeding off. Additionally, since the mobile phase was a gas, the separated compounds were very pure; there was no liquid solvent to remove. Subsequent research on the technique produced many new applications.

The shapes of the columns themselves began to change, too. Originally vertical tubes an inch or so in diameter, columns began to get longer and thinner when it was found that this increased the efficiency of separation. Eventually, chemists were using coiled **glass** or fused silica capillary tubes less than a millimeter in diameter and many yards long. **Capillaries** cannot be packed, but they are so narrow that the stationary phase can simply be a thin coat on the inside of the column.

A somewhat different approach is the set of techniques known as "planar" or "thin layer" chromatography (TLC), in which no column is used at all. The stationary phase is thinly coated on a glass or plastic plate. A spot of sample is placed on the plate, and the mobile phase migrates through the stationary phase by **capillary action**.

In the mid-1970s, interest in liquid mobile phases for column chromatography resurfaced when it was discovered that the efficiency of separation could be vastly improved by pumping the liquid through a short packed column under **pressure**, rather than allowing it to flow slowly down a vertical column by gravity alone. High-pressure liquid chromatography, also called high performance liquid chromatography (HPLC), is now widely used in industry. A variation on HPLC is Supercritical Fluid Chromatography (SFC). Certain gases (**carbon dioxide**, for example), when highly pressurized above a certain temperature, become a state of **matter** intermediate between gas and liquid. These "supercritical fluids" have unusual **solubility** properties, some of the advantages of both gases and liquids, and appear very promising for chromatographic use.

Most chemical compounds are not highly colored, as were the ones Tsvet used. A chromatographic separation of a colorless mixture would be fruitless if there were no way to tell exactly when each pure compound eluted from the column. All chromatographs thus must have a device attached, and some kind of recorder to capture the output of the detector—usually a chart recorder or its computerized equivalent. In gas chromatography, several kinds of detectors have been developed; the most common are the thermal conductivity detector, the flame ionization detector, and the **electron** capture detector. For HPLC, the UV detector is standardized to the **concentration** of the separated compound. The sensitivity of the detector is of special importance, and research has continually concentrated on increasing this sensitivity, because chemists often need to detect and quantify exceedingly small amounts of a material.

Within the last few decades, chromatographic instruments have been attached to other types of analytical instrumentation so that the mixture's components can be identified as well as separated (this takes the concept of the "detector" to its logical extreme). Most commonly, this second instrument has been a mass spectrometer, which allows identification of compounds based on the masses of molecular fragments that appear when the molecules of a compound are broken up. Currently, chromatography as both science and practical tool is intensively studied, and several scientific journals are devoted exclusively to chromatographic research.

Types of chromatographic attraction

Absorption chromatography (the original type of chromatography) depends on physical forces such as **dipole** attraction to hold the molecules onto the surface of the solid packing. In gas chromatography and HPLC, however, the solubility of the mixture's molecules in the stationary phase coating determines which ones progress through the column more slowly. Polarity can have an influence here as well. In gel **filtration** (also called size-exclusion or gel permeation) chromatography, the relative sizes of the molecules in the mixture determine which ones exit the column first. Large molecules flow right through; smaller ones are slowed down because they spend time trapped in the pores of the gel. **Ion exchange** chromatography depends on the relative strength with which ions are held to an ionic resin. Ions that are less strongly attached to the resin are displaced by more strongly attached ions. Hence the name ion exchange: one kind of ion is exchanged for another. This is the same principle upon which home **water** softeners operate. Affinity chromatography uses a stationary phase composed of materials that have been chemically altered. In this type of chromatography, the stationary phase is attached to a compound with a specific affinity for the desired molecules in the mobile phase. This process is similar to that of ion exchange chromatography, and is used mainly for the recovery of biological compounds. Hydrophobic Interaction Chromatography is used for amino acids that do not carry a positive or **negative** charge. In this type of chromatography, the hydrophobic amino acids are attracted to the solid phase, which is composed of materials containing hydrophobic groups.

Chemists choose the mobile and stationary phases carefully because it is the relative interaction of the mixture's compounds with those two phases that determines how efficient the separation can be. If the compounds have no attraction for the stationary phase at all, they will flow right through the column without separating. If the compounds are too strongly attracted to the stationary phase, they may stick permanently inside the column.

Industrial applications of chromatography

Chromatography of many kinds is widely used throughout the chemical industry. Environmental testing laboratories look for trace quantities of contaminants such as PCBs in waste oil, and **pesticides** such as DDT in **groundwater**. The Environmental Protection Agency uses chromatography to test drinking water and to monitor air quality. Pharmaceutical companies use chromatography both to prepare large quantities of extremely pure materials, and also to analyze the purified compounds for trace contaminants.

A growing use of chromatography in the pharmaceutical industry is for the separation of chiral compounds. These compounds have molecules that differ slightly in the way their atoms are oriented in space. Although identical in almost every other way, including **molecular weight**, element composition, and physical properties, the two different forms—called optical isomers, or enantiomers—can have enormous differences in their biological activity. The compound **thalidomide**, for example, has two optical isomers. One causes **birth defects** when women take it early in pregnancy; the other **isomer** does not. Because this compound looks promising for the treatment of certain drug-resistant illnesses, it is important that the benign form be separated completely from the dangerous isomer.

Chromatography is used for quality control in the food industry, by separating and analyzing additives, vitamins, preservatives, **proteins**, and amino acids. It can also separate and detect contaminants such as aflatoxin, a cancer-causing chemical produced by a **mold** on peanuts. Chromatography can be used for purposes as varied as finding drug compounds in urine or other body fluids, to looking for traces of flammable chemicals in burned material from possible arson sites.

See also Compound, chemical; Mixture, chemical.

Resources

Books

Ebbing, Darrell. *General Chemistry*. 3d ed. Boston: Houghton Mifflin, 1990.

Periodicals

Poole, F., and S.A. Schuette. *Contemporary Practice of Chromatography* Amsterdam: Elsevier, 1984.

Gail B. C. Marsella

Chromium *see* **Element, chemical**

Chromosomal abnormalities

Chromosome abnormalities describe alterations in the normal number of chromosomes or structural problems within the chromosomes themselves. Both kinds of chromosome abnormalities may result from an egg (ovum) or sperm **cell** with the incorrect number of chromosomes, or with a structurally faulty chromosome uniting with a normal egg or sperm during conception.

Some chromosome abnormalities may occur shortly after conception. In this case, the zygote, the cell formed during conception that eventually develops into an embryo, divides incorrectly. Other abnormalities may lead to the death of the embryo. Zygotes that receive a full extra set of chromosomes, a condition called polyploidy, usually do not survive inside the uterus, and are spontaneously aborted (a process sometimes called a miscarriage).

Chromosomal abnormalities can cause serious mental or physical disabilities. **Down syndrome**, for instance, is caused by an extra chromosome 21 (trisomy 21). People with Down syndrome are usually mentally retarded and have a host of physical defects, including **heart** disorders. Other individuals, called Down syndrome *mosaics,* have a mixture of normal cells and cells with three copies of chromosome 21, resulting in a mild form of the disorder.

Normal number and structure of human chromosomes

A chromosome consists of the body's genetic material, the deoxyribonucleic acid, or DNA, along with many kinds of protein. Within the chromosomes, the DNA is tightly coiled around these **proteins** (called histones) allowing huge DNA molecules to occupy a small space within the nucleus of the cell. When a cell is not dividing, the chromosomes are invisible within the cell's nucleus. Just prior to **cell division**, the chromosomes uncoil and begin to replicate. As they uncoil, the individual chromosomes look somewhat like a fuzzy "X." Chromosomes contain the genes, or segments of DNA that encode for proteins, of an individual. When a chromosome is structurally faulty, or if a cell contains an abnormal number of chromosomes, the types and amounts of the proteins encoded by the genes is changed. When proteins are altered in the human body, the result can be serious mental and physical defects and **disease**.

Humans have 22 pairs of autosomal chromosomes and one pair of sex chromosomes, for a total of 46 chromosomes. These chromosomes can be studied by constructing a karyotype, or organized depiction, of the chromosomes. To construct a karyotype, a technician stops cell division just after the chromosomes have replicated using a chemical such as colchicine; the chromosomes are visible within the nucleus at this point. The chromosomes are photographed, and the technician cuts up the photograph and matches the chromosome pairs according to size, shape, and characteristic stripe patterns (called banding).

Normal cell division

In most animals, two types of cell division exist. In **mitosis**, cells divide to produce two identical daughter cells. Each daughter cell has exactly the same number of chromosomes. This preservation of chromosome number is accomplished through the replication of the entire set of chromosomes just prior to mitosis.

Sex cells, such as eggs and sperm, undergo a different type of cell division called **meiosis**. Because sex cells each contribute half of a zygote's genetic material, sex cells must carry only half the full complement of chromosomes. This reduction in the number of chromosomes within sex cells is accomplished during two rounds of cell division, called meiosis I and meiosis II. Prior to meiosis I, the chromosomes replicate, and chromosome pairs are distributed to daughter cells. During meiosis II, however, these daughter cells divide without a prior replication of chromosomes. It is easy to see that mistakes can occur during either meiosis I and meiosis II. Chromosome pairs can be separated during meiosis I, for instance, or fail to separate during meiosis II.

Meiosis produces four daughter cells, each with half of the normal number of chromosomes. These sex cells are called haploid cells (haploid means "half the number"). Non-sex cells in humans are called diploid (meaning "double the number") since they contain the full number of normal chromosomes.

Alterations in chromosome number

Two kinds of chromosome number defects can occur in humans: aneuploidy, an abnormal number of chromosomes, and polyploidy, more than two complete sets of chromosomes.

Aneuploidy

Most alterations in chromosome number occur during meiosis. During normal meiosis, chromosomes are distributed evenly among the four daughter cells. Sometimes, however, an uneven number of chromosomes are distributed to the daughter cells. As noted in the previous section, chromosome pairs may not move apart in meiosis I, or the chromosomes may not separate in meiosis II. The result of both kinds of mistakes (called nondisjunction of the chromosomes) in that one daughter cell receives an extra chromosome, and another daughter cell does not receive any chromosome.

When an egg or sperm that has undergone faulty meiosis and has an abnormal number of chromosomes unites with a normal egg or sperm during conception, the zygote formed will have an abnormal number of chromosomes. This condition is called aneuploidy. There are several types of aneuploidy. If the zygote has an extra chromosome, the condition is called trisomy. If the zygote is missing a chromosome, the condition is called monosomy.

If the zygote survives and develops into a fetus, the chromosomal abnormality is transmitted to all of its cells. The child that is born will have symptoms related to the presence of an extra chromosome or absence of a chromosome.

Examples of aneuploidy include trisomy 21, also known as Down syndrome that occurs approximately 1 in 700 newborns, Trisomy 18 also called Edward syndrome (1:3500) and trisomy 13, also called Patau syndrome (1:5000). Trisomy 18 and 13, are more severe than of Down syndrome. Prenatal ultasound anomalies of fetuses affected by Down syndrome include nuchal thickening, duodenal stenosis, short femur and many other. Children with trisomy 21 have the characteristic face with a flat nasal bridge, epicanthic folds, protruding tongue and small ears. Among possible malforamations cleft palates, hare lips, and cardiac malformations (atrial and ventricular septal, and atriventricular canal defects). Mental retardation is always present at various degree. The life span once the individual has survived ranges between 50 and 60 years. Trisomy 18, known as Edwards' syndrome, results in severe multi-system defects. Most trisomies 18 results in spontaneous abortion. Affected infants have a small facies, small ears, overlapping fingers and rocker-bottom heels. Cardiac and other internal malformations are very frequent. Newborns with trisomy 13 have midline anomalies as well as scalp cutis aplasia, **brain** malformations, cleft lip and or palate, omphalocele and many others. Polydactyly is also frequent. Children with trisomy 13 and trisomy 18 usually survive less than a year after **birth** and are more likely to be females (Figure 1).

Aneuploidy of sex chromosomes

Sometimes, nondisjunction occurs in the sex chromosomes. Humans have one set of sex chromosomes.

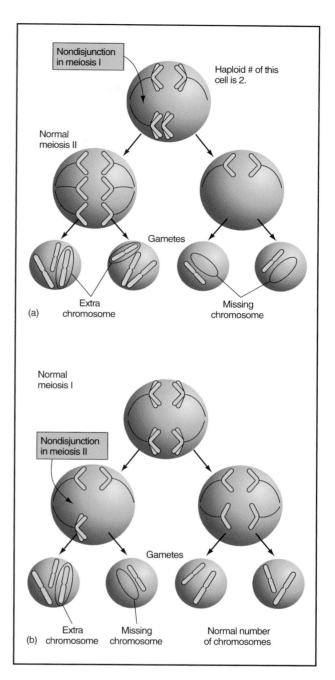

Figure 1. Karyotype from a child with trisomy 13 (a) or trisomy 18 (b). *Illustration by Hans & Cassidy. Courtesy of Gale Group.*

These sex chromosomes are called "X" and "Y" after their approximate shapes in a karyotype. Males have both an X and a Y chromosome, while females have two X chromosomes. Remarkably, abnormal numbers of sex chromosomes usually result in less severe defects than those that result from abnormal numbers of the other 22 pairs of chromosomes. The lessened severity may be due

Klinefelter's syndrome	XXY
Extra Y	XYY
Metafemale	XXX
Turner's syndrome	XO

Figure 2. Aneuploidy of sex chromosomes. *Illustration by Hans & Cassidy. Courtesy of Gale Group.*

to the fact that the Y chromosome carries few genes, and any extra X chromosomes become inactivated shortly after conception. Nevertheless, aneuploidy in sex chromosomes causes changes in physical appearance and in fertility (Figure 2).

In Klinefelter's syndrome, for instance, a male has two X chromosomes (XXY). This condition occurs in 1 out of every 2,000 births (1:800 males about). Men with Klinefelter's syndrome have small testes and are usually sterile. They also have female sex characteristics, such as enlarged breasts (gynecomastia). Males who are XXY are of normal intelligence or affected by a mild delay or behavioural immaturity. However, males with more than two X chromosomes, such as XXXY, XXXXY, or XXXXXY are mentally retarded.

Males with an extra Y chromosome (XYY) have no physical defects, although they may be taller than average. XYY males occur in 1 out of every 2,000 births (1:800 males about). It is not associated with increased aggressive and criminal behaviour as initially thought.

Females with an extra X chromosome (XXX) are called metafemales. This defect occurs in 1 out of every 800 females Metafemales have lowered fertility (oligomenorrhea and premature **menopause**), but their physical appearance is normal.

Females with only one X chromosome (XO) have **Turner syndrome**. Turner syndrome is also called monosomy X and occurs in 1 out of every 5,000 births (1:2,500 females). People with Turner syndrome have sex organs that do not mature at **puberty** and are usually sterile. They are of short stature and have no mental deficiencies.

Uniparental disomy is the presence in a diploid cell line of both chromosomes of a given pair from only one of the two parents. Possible consequence of uniparental disomy include **imprinting** of single genes (abnormal levels of **gene** product), homozygosity for mutant **alleles**. Microdeletion syndromes associated to a disomy of the chromosome 15 include Prader-Willi (maternal disomy) and Angelman Syndrome (patermal disomy).

Polyploidy

Polyploidy is lethal in humans. Normally, humans have two complete sets of 23 chromosomes. Normal human cells, other than sex cells, are thus described as diploid. Two polyploid conditions occur in humans. Triploidy (three set of chromosomes that results in 69 chromosomes with XXX, XXY or XYY sex chromosome) and tetrapolidy (92 chromosome and either XXXX or XXYY sex chromosome). Triploidy could result from the **fertilization** of an abnormal diploid sex cell with a normal sex cell. Tetraploidy could result from the failure of the zygote to divide after it replicates its chromosomes. Human zygotes with either of these conditions usually die before birth, or soon after. Interestingly, polyploidy is common in plants and is essential for the proper development of certain stages of the **plant** life cycle. Also, some kinds of cancerous cells have been shown to exhibit polyploidy. Rather than die, the polyploid cells have the abnormally accelerated cell division and growth characteristic of **cancer**.

Alterations in chromosome structure

Another kind of chromosomal abnormality is alteration of chromosome structure. Structural defects arise during replication of the chromosomes just prior to a meiotic cell division. Meiosis is a complex process that often involves the chromosomes exchanging segments with each other in a process called crossing-over. If the process is faulty, the structure of the chromosomes changes. Sometimes these structural changes are harmless to the zygote; other structural changes, however, can be lethal.

Five types of general structural alterations occur during replication of chromosomes (Figure 3). All four types begin with the breakage of a chromosome during replication. In a deletion, the broken segment of the chromosome is "lost." Deletion can be terminal (arises from one break) or interstitial (arises from 2 breaks). A particular case is the ring chromosome where two points of breakdown are present and then fused. Segment distal to the breaks are lost and sucha alost involves both arms of the chromosome. In isochromosomes one of the chromosome arms (p or q) is duplicated, and all material from the other arm is lost. The arm of one side of the centromere is a mirror image of the other. In a duplication, the segment joins to the other chromosome of the pair. In an inversion, the segment attaches to the original chromosome, but in a reverse position. It can be pericentric (break and rearrangements of both sides of the centromere) or paracentric (break and rearrangment on the same of centromere) In a translocation, the segment attaches to an entirely different chromosome. Translocaton can be Robertsonian (involves chromosomes 13–15, 21, and 22) where the whole arms of these chromosomes named acrocentric are fused and reciprocal. Reciprocal translocation results from breakage and exchange of segments between chromosomes.

Because chromosomal alterations in structure cause the loss or misplacement of genes, the effects of these defects can be quite severe. Deletions are usually fatal to a zygote. Duplications, inversions, and translocations can cause serious defects, as the expression of the gene changes due to its changed position on the chromosomes.

Examples of structural chromosomal abnormalities include *cri du chat* syndrome. *Cri du chat* means "cat cry" in French. Children with this syndrome have an abnormally developed larynx that makes their cry sound like the mewing of a cat in distress. They also have a small head, misshapen ears, and a rounded face, as well as other systemic defects. These children usually die in infancy. *Cri du chat* is caused by a deletion of a segment of DNA in chromosome 5.

A structural abnormality in chromosome 21 occurs in about 4% of people with Down syndrome. In this abnormality, a translocation, a piece of chromosome 21 breaks off during meiosis of the egg or sperm cell and attaches to chromosome 13, 14, or 22.

Some structural chromosomal abnormalities have been implicated in certain cancers. For instance, myelogenous **leukemia** is a cancer of the white **blood** cells. Researchers have found that the cancerous cells contain a translocation of chromosome 22, in which a broken segment switches places with the tip of chromosome 9.

Some unusual chromosomal abnormalities include Prader-Willi syndrome, Angelman's syndrome, and Fragile X syndrome. These structural defects are unusual because the severity or type of symptoms associated with the defect depend on whether the child receives the defect from the mother or the father.

Both Prader-Willi syndrome and Angelman's syndrome are caused by a deletion in chromosome 5. Prader-Willi syndrome is characterized by mental retardation, **obesity**, short stature, and small hands and feet. Angelman's syndrome is characterized by jerky movements and neurological symptoms. People with this syndrome also have an inability to control laughter, and may laugh inappropriately at odd moments. If a child inherits the defective chromosome from its father, the result is Prader-Willi syndrome. But if the child inherits the defective chromosome from its mother, the child will have Angelman's syndrome. Researchers believe that genes from the deleted region

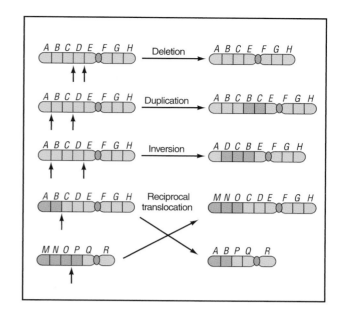

Figure 3. The four types of chromosome structure alterations. *Illustration by Hans & Cassidy. Courtesy of Gale Group.*

function differently in offspring depending on whether the genes come from the mother or father, but they are not sure about the exact nature of these differences.

Another condition that depends on whether the defect is inherited from the mother or father is fragile X syndrome. In fragile X, an extra X chromosome hangs off the normal X by a thin "thread" of genetic material. The syndrome occurs in 1 out of 1,000 male births and 1 out of 2,000 female births. Males are affected more severely than females, and thus the syndrome is more pronounced if the child inherits the defect from its mother. To understand why this is so, remember that a male is XY and a female is XX. A male child receives a Y chromosome from the father and an X chromosome from the mother. A female child, however, can receive an X from either the mother or the father. Again, researchers believe this difference in fragile X symptoms between boys and girls stems from a difference in gene function that depends on whether they come from the mother or father.

Genetic counseling

Currently, no cures exist for any of the syndromes caused by chromosomal abnormalities. For many of these conditions, the age of the mother carries an increased risk for giving birth to a child with a chromosomal abnormality. The risk for Down syndrome, for instance, jumps from 1 in 1,000 when the mother is age 15-30 to 1 in 400 at age 35, increasing risk with increasing maternal age. One theory postulates that this is due to the build-up of toxins

KEY TERMS

Amniocentesis—A method of detecting genetic abnormalities in a fetus; in this procedure, amniotic fluid is sampled through a needle placed in the uterus; fetal cells in the amniotic fluid are then analyzed for genetic defects.

Aneuploidy—An abnormal number of chromosomes.

Angelman's syndrome—A syndrome caused by a deletion in chromosome 5 inherited from the mother.

Chorionic villi sampling—A procedure in which hair-like projections from the chorion, a fetal structure present early in pregnancy, are suctioned off with a catheter inserted into the uterus. These fetal cells are studied for the presence of certain genetic defects.

Chromosomes—he structures that carry genetic information in the form of DNA. Chromosomes are located within every cell and are responsible for directing the development and functioning of all the cells in the body.

Cri du chat syndrome—A syndrome caused by a deletion in chromosome 5; characterized by a strange cry that sounds like the mewing of a cat.

Deletion—Deletion of a segment of DNA from a chromosome.

Deoxyribonucleic acid (DNA)—The genetic material in a cell.

Diploid—Means "double number;" describes the normal number of chromosomes for all cells of the human body, except for the sex cells.

Down syndrome—A syndrome caused by trisomy 13; characterized by distinct facial characteristics, mental retardation, and several physical disorders, including heart defects.

Duplication—A type of chromosomal defect in which a broken segment of a chromosome attaches to the chromosome pair.

Edwards' syndrome—A syndrome caused by trisomy 18; characterized by multi-system defects; is usually lethal by age 1.

Fragile X syndrome—A condition in which an extra X chromosome hangs from the X chromosome by a "thread" of genetic material.

Gene—A discrete unit of inheritance, represented by a portion of DNA located on a chromosome. The gene is a code for the production of a specific

over **time** within the ovaries, damaging the egg cells that are present in females since early childhood. By the time they are ovulated after age 35, the chances of fertilization of a damaged egg are greater.

People at high risk for these abnormalities may opt to know whether the fetus they have conceived has one of these abnormalities. **Amniocentesis** is a procedure in which some of the amniotic fluid that surrounds and cushions the fetus in the uterus is sampled with a needle placed in the uterus. The amniotic fluid contains some of the fetus's skin cells, which can be tested for chromosomally-based conditions. Another test, called chorionic villi sampling, involves taking a piece of **tissue** from a part of the placenta. If a chromosomal defect is found, the parents can be advised of the existence of the abnormality. Some parents opt to abort the pregnancy; others can prepare before birth for a child with special needs.

In addition to amniocentesis and chorionic villi sampling, researchers are working on devising easier tests to detect certain abnormalities. A new test for Down syndrome, for instance, measures levels of certain **hormones** in the mother's blood. Abnormal levels of these hormones indicate an increased risk that the fetus has Down syn-

drome. These **enzyme** tests are safer and less expensive than the sampling tests and may be able to diagnose chromosomally-based conditions in more women. Most recently, scientists have devised a procedure called *in situ* hybridization which uses molecular tags to locate defective portions of chromosomes collected from amniocentesis. The process uses fluorescent molecules that seek out and adhere to specific faulty portions of chromosomes. Chromosomes having these faulty regions then "glow" (or fluoresce) under special lighting in regions where the tags bind to, or hybridize with, the chromosome. The procedure makes identification of defects easier and more reliable.

See also Birth defects; Embryo and embryonic development; Genetic disorders.

Resources

Books

Harper, Peter S. *Practical Genetic Counseling.* Boston: Butterworth-Heineman, 1993.

Nussbaum, Robert L., Roderick R. McInnes, Huntington F. Willard. *Genetics in Medicine.* Philadelphia: Saunders, 2001.

Rimoin, David L. *Emery and Rimoin's Principles and Practice of Medical Genetics.* London; New York: Churchill Livingstone, 2002.

kind of protein or RNA molecule, and therefore for a specific inherited characteristic.

Haploid—Nucleus or cell containing one copy of each chromosome; the number of chromosomes in a sex cell.

Inversion—A type of chromosomal defect in which a broken segment of a chromosome attaches to the same chromosome, but in reverse position.

Klinefelter's syndrome—A syndrome that occurs in XXY males; characterized by sterility, small testes, and female sex characteristics.

Meiosis—Cell division that results in four haploid sex cells.

Metafemale—An XXX female.

Mitosis—Cell division that results in two diploid cells.

Monosomy—A form of aneuploidy in which a person receives only one chromosome of a particular chromosome pair, not the normal two.

Patau's syndrome—A syndrome caused by trisomy 13; characterized by a hare lip, cleft palate, and many other physical defects; usually lethal by age 1.

Polyploidy—A condition in which a cell receives more than two complete sets of chromosomes.

Prader-Willi syndrome—A syndrome caused by a deletion in chromosome 5 inherited from the father.

Tetraploidy—A form of polyploidy; four sets of chromosomes.

Translocation—A genetic term referring to a situation during cell division in which a piece of one chromosome breaks off and sticks to another chromosome.

Triploidy—A form of aneuploidy; three sets of chromosomes.

Trisomy—A form of aneuploidy in which a person receives an extra chromosome of a particular chromosome pair, not the normal two.

Turner syndrome—A syndrome that occurs in X0 females; characterized by sterility, short stature, small testes, and immature sex organs.

Zygote—The cell resulting from the fusion of male sperm and the female egg. Normally the zygote has double the chromosome number of either gamete, and gives rise to a new embryo.

Periodicals

Bos, A.P., et al. "Avoidance of Emergency Surgery in Newborn Infants with Trisomy 18." *The Lancet* 339, no. 8798 (April 11, 1992): 913-6.

D'Alton, Mary E., et al. "Prenatal Diagnosis." *New England Journal of Medicine* 328, no. 2 (January 14, 1995): 114-21,

Day, Stephen. "Why Genes have a Gender." *New Scientist* 138, no.1874 (May 22, 1993): 34-39.

Hoffman, Michelle. "Unraveling the Genetics of Fragile X Syndrome." *Science* 252, no. 5010, (May 24, 1991): 1070.

Money, John. "Specific Neurological Impairments associated with Turner and Kline-Felter Syndromes: A Review." *Social Biology* 40, no. 1-2, (Spring-Summer 1993): 147-152.

Nicolaides, K.H., et al. "Ultrasonographically Detectable Markers of Fetal Chromosomal Abnormalities." *The Lancet* 340, no. 8821, (September 19, 1992): 704-8.

Solomon, Ellen, et al. "Chromosome Aberrations and Cancer." *Science* 254, no. 5035, (November 22, 1991): 1153-61.

Antonio Farina
Brenda Wilmoth Lerner
Kathleen Scogna

Chromosome

A chromosome is a threadlike structure found in the nucleus of most cells. that carries the genetic material in the form of a linear sequence of **deoxyribonucleic acid (DNA)**. In prokaryotes, or cells without a nucleus, the chromosome represents circular DNA containing the entire **genome**. In eukaryotes, or cells with a distinct nucleus, chromosomes are much more complex in structure. The function of chromosomes is to package the extremely long DNA sequence. A single chromosome (uncoiled) could be as long as 3 in (7.6 cm) and therefore visible to the naked **eye**. If DNA were not coiled within chromosomes, the total DNA in a typical eukaryotic **cell** would extend thousands of times the length of the cell nucleus.

DNA and protein synthesis

DNA is the genetic material of all cells and contains information necessary for the synthesis of **proteins**. DNA is composed of two strands of nucleic acids arranged in a **double helix**. The **nucleic acid** strands are composed of a sequence of nucleotides. The nucleotides

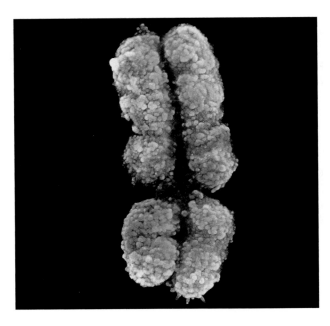

A scanning electron micrograph (SEM) of a human X-chromosome. © Biophoto Associates/Photo Researchers, Inc. Reproduced by permission.

in DNA have four kinds of nitrogen-containing bases: adenine, guanine, cytosine, and thymine. Within DNA, each strand of nucleic acid is partnered with the other strand by bonds that form between these nucleotides. Complementary base pairing dictates that adenine pairs only with thymine, and guanine pairs only with cytosine (and vice versa). Thus, by knowing the sequence of bases in one strand of the DNA helix, you can determine the sequence on the other strand. For instance, if the sequence in one strand of DNA were ATTCG, the other strand's sequence would be TAAGC.

DNA functions in the cell by providing a template by which another nucleic acid, called **ribonucleic acid (RNA)**, is formed. Like DNA, RNA is also composed of nucleotides. Unlike DNA, RNA is single stranded and does not form a helix. In addition, the RNA bases are the same as in DNA, except that uracil replaces thymine. RNA is transcribed from DNA in the nucleus of the cell. **Gene** are expressed when the chromosome uncoils with the help of enzymes called helicases and specific DNA binding proteins. DNA is transcribed into RNA.

Newly transcribed RNA is called messenger RNA (mRNA). Messenger RNA leaves the nucleus through the nuclear pore and enters into the cytoplasm. There, the mRNA **molecule** binds to a ribosome (also composed of RNA) and initiates protein synthesis. Each block of three nucleotides called **codons** in the mRNA sequence encodes for a specific **amino acid**, the building blocks of a protein.

Genes

Genes are part of the DNA sequence called coding DNA. Noncoding DNA represents sequences that do not have genes and only recently have found to have many new important functions. Out of the 3 billion base pairs that exist in the human DNA, there are only about 40,000 genes. The noncoding sections of DNA within a gene are called introns, while the coding sections of DNA are called exons. After transcription of DNA to RNA, the RNA is processed. Introns from the mRNA are excised out of the newly formed mRNA molecule before it leaves the nucleus.

Chromosome numbers

The human genome (which represents the total amount of DNA in a typical human cell) has approximately 3×10^9 base pairs. If these nucleotide pairs were letters, the genome book would number over a million pages. There are 23 pairs of chromosomes, for a total number of 46 chromosomes in a dipoid cell, or a cell having all the genetic material. In a haploid cell, there is only half the genetic material. For example, sex cells (the sperm or the egg) are haploid, while many other cells in the body are diploid. One of the chromosomes in the set of 23 are X or Y (sex chromosomes), while the rest are assigned numbers 1 through 22. In a diplod cell, males have both an X and a Y chromosome, while females have two X chromosomes. During **fertilization**, the sex cell of the father combines with the sex cell of the mother to form a new cell, the zygote, which eventually develops into an embryo. If the one of the sex cells has the full complement of chromosomes (diploidy), then the zygote would have an extra set of chromosomes. This is called triploidy and represents an anomaly that usually results in a miscarriage. Sex cells are formed in a special kind of **cell division** called **meiosis**. During meiosis, two rounds of cell division ensure that the sex cells receive the haploid number of chromosomes.

Other **species** have different numbers of chromosomes in their nuclei. Mosquitos, for instance, have 6 chromosomes. Lilies have 24 chromosomes, earthworms have 36 chromosomes, chimps have 48 chromosomes, and **horses** have 64 chromosomes. The largest number of chromosomes are found in the Adders tongue fern, which has more than 1,000 chromosomes. Most species have, on average, 10–50 chromosomes.

Chromosome shape

Chromosomes can be visible using a **microscope** just prior to cell division, when the DNA within the nucleus uncoils as it replicates. By visualizing a cell during metaphase, a stage of cell division or **mitosis**, researchers can take pictures of the duplicated chromosome and match the pairs of chromosomes using the characteristic patterns of bands that appear on the chromosomes when they are stained with a dye called giemsa. The resulting arrangement is called a karyotype. The ends of the chromosome are referred to as telomeres, which are required to maintain stablility and recently have been associated with aging. An **enzyme** called telomerase maintains the length of the telomere. Older cells tend to have shorter telomeres. The telomere has a repeated sequence (TTAGGG) and intact telomeres are important for proper **DNA replication** processes.

Karyotypes are useful in diagnosing some genetic conditions, because the karyotype can reveal an aberration in chromosome number or large alterations in structure. For example, **Down syndrome** is can be caused by an extra chromosome 21 called trisomy 21. A karyotype of a child with Down syndrome would reveal this extra chromosome.

A chromosome usually appears to be a long, slender rod of DNA. Pairs of chromosomes are called homologues. Each separate chromosome within the duplicate is called a sister chromatid. The sister chromatids are attached to each other by a structure called the centromere. Chromosomes appear to be in the shape of an X after the material is duplicated. The bottom, longer portion of the X is called the long arm of the chromosome (q-arm), and the top, shorter portion is called the short arm of the chromosome (p-arm).

The role of proteins in packaging DNA

Several kinds of proteins are important for maintaining chromosomes in terms of its organization and gene expression. Some proteins initiate DNA replication when the cell prepares to divide. Other proteins control gene transcription in the preliminary stages of protein synthesis. Structural proteins help the DNA fold into the intricate configurations within the packaged chromosome.

DNA in chromosomes is associated with proteins and this complex is called **chromatin**. Euchromatin refers to parts of the chromosome that have coding regions or genes, while heterchromatin refers to regions that are devoid of genes or regions where gene transcription is turned off. DNA binding proteins can attach to specific regions of chromatin. These proteins mediate DNA replication, gene expression, or represent structural proteins important in packaging the chromosomes. Histones are structural proteins of chromatin and are the most abundant protein in the nucleus. In fact, the **mass** of histones in a chromosome is almost equal to that of DNA. Chromosomes contain five types of these small

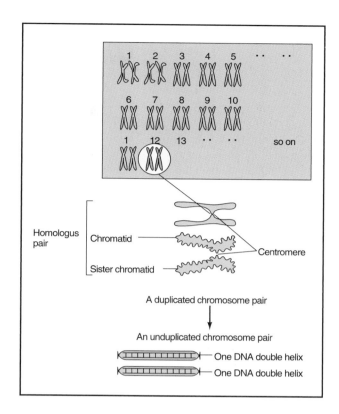

Figure 1. A human karyotype. *Illustration by Hans & Cassidy. Courtesy of Gale Group.*

proteins: H1, H2A, H2B, H3, and H4. There are two of each of latter four histones that form a structure called the octomeric histone core. The H1 histone is larger than the other histones, and performs a structural role separate from the octomeric histone core in organizing DNA within the chromosome.

The octomeric histone core functions as a spool from which DNA is wound two times. Each histone-DNA spool is called a nucleosome. Nucleosomes occur at intervals of every 200 bases pairs of the DNA helix. In photographs taken with the help of powerful microscopes, DNA wrapped around nucleosomes resembles beads (the nucleosome) threaded on a string (the DNA molecule). The DNA that exists between nucleosomes is called linker DNA. Chromosomes can contain some very long stretches of linker DNA. Often, these long linker DNA sequences are the regulatory portions of genes. These regulatory portions switch genes on when certain molecules bind to them.

Nucleosomes are only the most fundamental organizing structure in the chromosome. They are packaged into structures that are 30 nanometers in size and called the chromatin fiber (compared to the 2 nm DNA double helix, and 11 nm histone core). The 30 nanometer fibers are then

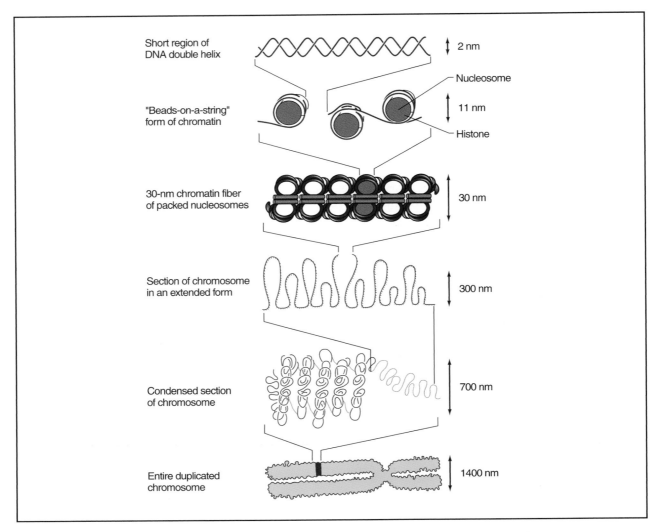

Short region of
DNA double helix — 2 nm

Nucleosome

"Beads-on-a-string"
form of chromatin — 11 nm

Histone

30-nm chromatin fiber
of packed nucleosomes — 30 nm

Section of chromosome
in an extended form — 300 nm

Condensed section
of chromosome — 700 nm

Entire duplicated
chromosome — 1400 nm

Figure 2. How DNA is packaged within chromosomes. *Illustration by Hans & Cassidy. Courtesy of Gale Group.*

further folded into a larger chromatin fiber sometimes that is approximately 300 nanometers thick and represent on of the arms of the chromsome. The chromatin fibers are formed into loops by another structural protein. Each loop contains 20,000–30,000 nucleotide pairs. These loops are then arranged within the chromosomes, held in place by more structural proteins. Metaphase chromosomes are approximately 1400 nm wide.

Chromosomes and mitosis

Chromosomes in eukaryotes perform a useful function during mitosis, the process in which cells replicate their genetic material and then divide into two new cells (also called daughter cells). Because the DNA is packaged within chromosomes, the distribution of the correct amount of genetic material to the daughter cells is maintained during the complex process of cell division.

Before a cell divides, the chromosomes are replicated within the nucleus. In a human cell, the nucleus just prior to cell division contains 46 pairs of chromosomes. When the cell divides, the sister chromatids from each duplicated chromosome separate. Each daughter cell ends up with 23 pairs of chromosomes and after DNA replication, the daughter cells will have a diploid number of chromosomes.

In meiosis, the type of cell division that leads to the production of sex cells, the division process is more complicated. Two rounds of cell division occur in meiosis. Before meiosis, the chromosomes replicate, and the nucleus has 46 pairs of chromosomes. In the first round of meiotic cell division, the homologous chromosomes pairs separate as in mitosis (a stage called meiosis I). In the second round of cell division (meiosis II), the sister chromatids of each chromosome separate at the cen-

KEY TERMS

Chromatin—The material that comprises chromosomes; consists of DNA and proteins.

Chromatin fiber—The fiber that is formed by the gathering of nucleosomes by H1 histones.

Chromosome puffs—The regions of active DNA that are transcribing RNA; appear as puffed regions in a chromosome.

Deoxyribonucleic acid (DNA)—The genetic material of cells that are packed into chromosomes.

Eukaryote—A cell whose genetic material is carried on chromosomes inside a nucleus encased in a membrane. Eukaryotic cells also have organelles that perform specific metabolic tasks and are supported by a cytoskeleton which runs through the cytoplasm, giving the cell form and shape.

Exons—The regions of DNA that code for a protein or form tRNA or mRNA.

Genome—The complete set of genes an organism carries.

Histone—A structural protein that functions in packaging DNA in chromosomes.

Homologue—The partner of a chromosome in a chromosome pair.

Introns—The sections of DNA that do not code for proteins or RNAs.

Karyotype—An arrangement of chromosomes according to number.

Linker DNA—The sections of DNA between nucleosomes.

Meiosis—The process of sex cell division; results in four haploid daughter cells.

Messenger RNA—The RNA that is transcribed from DNA in the nucleus; functions in protein synthesis.

Mitosis—The process of body cell division; results in two diploid daughter cells.

Nitrogen-containing base—Part of a nucleotide; in DNA, the bases are adenine, guanine, thymine, and cytosine; in RNA, the bases are adenine, guanine, uracil, and cytosine.

Nucleic acid—The chemical component of DNA and RNA.

Nucleosome—DNA wrapped around a histone core.

Nucleotide—The building blocks of nucleic acids.

Octomeric histone core—The "spool" in a nucleosome; consists of four small histones.

Ribonucleic acid—RNA; the molecule translated from DNA in the nucleus that directs protein synthesis in the cytoplasm; it is also the genetic material of many viruses.

Ribosomal RNA—A type of RNA that functions in protein synthesis.

Sister chromatids—Two copies of the same chromosome produced by DNA replication.

Transcription—The process of synthesizing RNA from DNA.

tromere, so that each of the four daughter cells receives the haploid number of chromosomes.

Protein synthesis and chromosomes

DNA is bound up within chromatids, which serve as storage unit for the DNA. In order for an mRNA molecule to be transcribed from a DNA template, the DNA needs to be freed from its tightly bound and condensed conformation so that the RNA molecule can form on its exposed strands during transcription. Some evidence exists that transcription can take place through histones. However, most often the genes on the DNA that have been activated after DNA binding protein unwind the chromatid structure. This loosened, transcriptionally active regions of DNA is microscopically resembles puffs on the chromosomes. When RNA transcription con-cludes, the puffs receed and the chromosome is thought to resume its original conformation.

See also Genetics; Nucleus, cellular; Prokaryote.

Resources

Books

Nussbaum, Robert L., Roderick R. McInnes, Huntington F. Willard. *Genetics in Medicine.* Philadelphia: Saunders, 2001.

Rimoin, David L. *Emery and Rimoin's Principles and Practice of Medical Genetics.* London; New York: Churchill Livingstone, 2002.

Other

United States Department of Energy Office of Science. "Human Genome Project Information." (October 28, 2002). <http://www.ornl.gov/Tech Resources/Human_Genome/home.html>.

Kathleen Scogna

Chromosome mapping

Chromosome mapping is the assignment of genes to specific locations on a chromosome. A **gene** map serves many important functions and is much like understanding the basic human **anatomy** to allow doctors to diagnose patients with **disease**. A doctor requires knowledge of where each **organ** is located as well as the function of this organ to understand disease. A map of the human **genome** will allow scientist to understand where genes are located so that its function within the human genome can be elucidated. A detailed chromosome map also provides methods to study how genes are segregated and how genetic heterogeneity (variation between a particular gene maternally inherited and the same gene with a slightly different sequence that is paternally inherited) can help identify disease genes. Gene mapping can provide clinicians with useful information regarding genes that are linked, or segregate closely together.

Scientists use several methods to map genes to the appropriate locations. These methods include family studies, somatic **cell** genetic methods, cytogenetic techniques, and gene dosage studies. Family studies are used to determine whether two different genes are linked close together on a chromosome. If these genes are linked, it means they are close together on the same chromosome. Additionally, the frequency with which the genes are linked is determined by recombination events (crossing over of the chromosomes during **meiosis**) between known locations or markers, and determines the linear order or genetic **distance**. In somatic cell genetic methods, chromosomes are lost from a special type of cell and the remaining chromosome that has one gene, but not a different gene, would suggest that they are located on different chromosomes. This method allows scientists to identify which chromosome contains the gene, and represents one of the first mapping methods used by scientists.

Cytogenetic techniques refer to utilization of karyotype preparations, a technique that allows scientists to visualize of chromosomes, using **fluorescence** so that a fluorescently-labeled gene will reveal where the gene is found on the chromosome. Gene dosage studies uses, for example, numerical abnormalities to determine indirectly the location of the gene on a chromosome. In **Down syndrome**, there can be three chromosome number 21 (Trisomy 21), resulting in three copies of the gene and therefore, three times as much protein. In this case, a gene can be localized to chromosome 21 if there is three times as much protein in a cell with three 21 chromosomes. In this method, the amount of **deoxyribonucleic acid (DNA)** is assumed to be directly proportional to the amount of protein.

KEY TERMS

Genetic linkage map—Genetic maps constructed by using data on the frequency of occurrence of genetic markers. Specifically how such markers are coinherited.

Physical genetic map—A genetic map are based upon actual distances between genes on a chromosome. Contigs are physical maps are based on collections of overlapping DNA fragments.

Using these methods, various maps of chromosomes can be developed. These maps are called cytogenetic maps, linkage maps, physical maps, or a DNA sequence map. A cytogenetic map uses bands produced by a dye that stains chromosomes in a karyotpe and assigns genes to these bands. A linkage map, also referred to as a genetic map, orders genes along the DNA strand based on recombination frequency. Linkage mapping involves using two characteristics (and hence their responsible genes), both of which are present in one parent, combined with the frequency in which they occur together in the offspring to construct the map. For example, the Moravian-born Augustinian monk and science teacher Gregor Johann Mendel (1823–1884) studied the **flower color** and **plant** height of peas. He found that various heights were observed just as frequently with white flowers as with other colored flowers and similarly, dwarf plants occurred just as frequently with the two flower types. Mendel concluded that the forms of the two genes were transmitted from parent to offspring independently of each other. This later became known as the Law of Independent Assortment, a concept that enhanced chromosome mapping techniques. A physical map orders genes or markers along the DNA strand of a chromosome. Finally, a DNA sequence, strung together, is the most precise type of map in that it contains both coding (gene-containing) and noncoding DNA. It is felt that obtaining the complete DNA sequence from the genome of many different organisms will provide scientists with vital information that will unlock many biological mysteries.

See also Chromosomal abnormalities; DNA technology; Human Genome Project.

Resources

Books

Friedman, J., F. Dill, M. Hayden, and B. McGillivray. *Genetics.* Maryland: Williams & Wilkins, 1996.
Wilson, G.N. *Clinical Genetics: A Short Course.* New York: Wiley-Liss, Inc., 2000.

Other

The National Health Museum. <http://www.accessexcellence.
org/AB/BC/Gregor_Mendel.html>. (October 28, 2002).

Brian Cobb

Cicadas

Cicadas are **insects** in the order Homoptera, family
Cicadidae. Male cicadas make a well-known, loud, stri-
dent, buzzing sound during the summer, so these unusual
insects are often heard, but not necessarily seen. **Species**
of cicadas are most diverse in closed and open **forests** of
the temperate and tropical zones.

Biology of cicadas

Cicadas are large dark-bodied insects, with a body
length of 2 in (5 cm), membranous wings folded tent-
like over the back, and large eyes.

Male cicadas have a pair of small drum-like organs
(tymbals), located at the base of their abdomen. These
structures have an elastic, supporting ring, with a **membrane** extending across it (the tymbal membrane). The
familiar very loud, buzzing noises of cicadas are made
by using powerful muscles to move the tymbal membrane rapidly back and forth, as quickly as several hundred times per second. The actual sound is made in a
manner similar to that by which a clicking noise is made
by moving the center of the lid of a **metal** can back and
forth. The loudness of the cicada song is amplified using
resonance chambers, known as opercula. Each species
of cicada makes a characteristic sound.

Cicadas are herbivorous insects, feeding on the sap
of the roots of various types of perennial plants, most
commonly woody species. Cicadas feed by inserting
their specialized mouth parts, in the form of a hollow
tube, into a **plant** root, and then sucking the sap.

Life cycle of cicadas

Cicadas have prolonged nymphal stages, which are
spent within the ground, sucking juices from the roots of
plants, especially woody species. Most cicada species
have overlapping generations, so that each year some of
the population of subterranean nymphs emerges from the
ground and transforms into a fairly uniform abundance
of adults, as is the case of the dog-day or annual cicada
(*Tibicen pruinosa*).

Other species of cicadas have non-overlapping generations, so there are periodic events of the great abun-

An adult periodical cicada. *Photograph by Alan & Linda
Detrick. The National Audubon Society Collection/Photo
Researchers, Inc. Reproduced by permission.*

dance of adults and their noisy summer renditions, interspersed with much longer periods during which the adult
animals are not found in the region. The irruptive adult
phase occurs at intervals as long as 17 years, in the case
of northern populations of the periodical cicada (*Magici-
cada septendecum*), which has the longest generation
time of any plant-sucking insect. Southern populations
of the periodical cicada gave generation times as short as
13 years, and are usually treated as a different species.

The periodical cicada spends most of its life in the
ground, in its developmental nymph stages. During years
of irruption or peak emergence, the ground in late spring
and early summer can be abundantly pock-marked with
the emergence holes of the mature nymphs of this
species, at a density greater than one thousand per square
meter. The stout-bodied nymphs emerge from the ground
and then climb up upon some elevated object, where they
metamorphose into the adult form, which lives for about
one month. During years when periodical cicadas are
abundant, the strange-looking, cast exoskeletons of their
mature nymphs can be found in all manner of places.

The adult periodic cicada is black or dark brown,
with large, membranous wings folded over its back, and
large, red eyes. The females have a strong, chisel-like
ovipositor, which is used to make incisions in small
branches and twigs, into which her eggs are deposited.
The incisions severely injure the affected twigs, which
generally die from the point of the incision to the tip.
Soon after hatching, the small nymphs drop to the
ground and burrow in, ready for a relatively long life of
17 years. The subterranean nymph excavates a chamber
beside the root of a woody plant, into which the cicada
inserts its beak and feeds on sap.

Cicadas and people

When they are breeding in abundance, some species of cicadas cause economic damage by the injuries that result when the females lay their eggs in the branches and twigs of commercially important species of trees. The periodical cicada is the most important species in this respect in **North America**. This species can cause a great deal of damage in hardwood-dominated forests in parts of eastern North America. The damage is not very serious in mature forests, but can be important in younger forests and nurseries.

Although cicadas are not often seen, their loud buzzing noises are a familiar noise of hot, sunny days in many regions. As such, cicadas are appreciated as an enjoyable aspect of the outdoors.

Bill Freedman

Cigarette smoke

The World Health Organization (WHO) has named tobacco one of the greatest public health threats of the twenty-first century. As of 2001, more than 1.2 billion people worldwide smoke, and 3.5 million people are expected to die from causes directly related to tobacco use. This death **rate** is expected to rise to 10 million by the year 2030. Seventy **percent** of these deaths will occur in developing countries where the proportion of smokers is growing, particularly among women. Calling tobacco "a global threat," WHO says these figures do not include the enormous physical, emotional, and economic costs associated with **disease** and disability caused by tobacco use.

In the United States alone, 25.2 million men, 23.2 million women, and 4.1 million teens between 12 and 17 years of age smoke. Every day, more than three million youths under the age of 18 begin smoking. The gruesome **statistics** show that more than five million children alive today will die prematurely because, as adolescents, they decided to use tobacco. Nationally, one in five of all deaths is related to tobacco use. It kills more than 430,000 people every year—more than **AIDS**, **alcohol**, drug abuse, **automobile** accidents, murders, suicides, and fires combined. Five million years of potential life is lost every year due to premature death caused by tobacco use. Medical costs total more than $50 billion annually, and indirect cost another $50 billion.

Components of cigarette smoke

Of the 4,000 or more different chemicals present in cigarette smoke, 60 are known to cause **cancer** and others to cause cellular genetic mutations that can lead to cancer. Cigarette smoke contains **nicotine** (a highly addictive chemical), tars, nitrosamines, and polycyclic hydrocarbons, all of which are carcinogenic. It also contains **carbon monoxide** which, when inhaled, interferes with transportation and utilization of **oxygen** throughout the body.

Environmental tobacco smoke

Cigarette smoke is called mainstream smoke when inhaled directly from a cigarette. Sidestream smoke is smoke emitted from the burning cigarette and exhaled by the smoker. Sidestream smoke is also called environmental tobacco smoke (ETS) or secondhand smoke. Inhalation of ETS is known as passive smoking. In 1993, the Environmental Protection Agency (EPA) classified ETS as a Group A (known human) carcinogen—the grouping reserved for the most dangerous carcinogens. By 1996, the Department of Health and Human Services' Centers for Disease Control and Prevention (CDC) found that nine out of 10 non-smoking Americans are regularly exposed to ETS. A study by the American Heart Association reported in 1997 that women regularly exposed to ETS have a 91% greater risk of heart attack and those exposed occasionally a 58% greater risk—rates which are believed to apply equally to men. The EPA estimates that, annually, ETS is responsible for more than 3,000 lung cancer deaths, 35,000-62,000 deaths from heart attacks, and lower respiratory tract infections (such as **bronchitis** [300,000 cases annually] and **asthma** [400,000 existing cases]), and middle **ear** infections in children.

ETS may be more carcinogenic than mainstream smoke as it contains higher amounts of carcinogenic materials with smaller particles. These smaller particles are more likely to lodge in the lungs than the larger particles in mainstream smoke. Researchers found that no safe threshold exists for exposure to ETS. With this information, many municipal governments and workplaces have banned cigarette smoking altogether.

The health consequences of tobacco use

Scientific evidence has proven that smoking can cause cancer of the lung, larynx, esophagus, mouth, and

A normal lung (left) and the lung of a cigarette smoker (right). *Photograph by A. Glauberman. National Audubon Society Collection/Photo Researchers, Inc. Reproduced by permission.*

bladder; cardiovascular disease; chronic lung ailments; coronary heart disease; and **stroke**. Smokeless tobacco has equally deadly consequences. When cigarette smoke is inhaled, the large surface area of the lung tissues and alveoli quickly absorb the chemical components and nicotine. Within one minute of inhaling, the chemicals in the smoke are distributed by the bloodstream to the **brain**, **heart**, kidneys, liver, lungs, gastrointestinal tract, muscle, and **fat tissue**. In pregnant women, cigarette smoke crosses the placenta and may effect fetal growth.

Cardiovascular disease

Cardiovascular disease, or diseases of the **blood** vessels and heart, includes stroke, heart attack, peripheral vascular disease, and aortic aneurysm. In 1990 in the United States, one fifth of all deaths due to cardiovascular disease were linked to smoking. Specifically, 179,820

deaths from general cardiovascular disease, 134,235 deaths from heart disease, and 23,281 deaths from cerebrovascular disease (stroke) were directly linked to smoking. In addition, researchers have noted a strong dose-response relationship between the duration and extent of smoking and the death rate from heart disease in men under 65. The more one smokes, the more one is likely to develop heart disease. Researchers have also seen a similar trend in women.

Cigarette smoking leads to cardiovascular disease in a number of ways. Smoking damages the inside of the blood vessels, initiating changes that lead to atherosclerosis, a disease characterized by blood vessel blockage. It also causes the coronary **arteries** (that supply the heart muscle with oxygen) to constrict, increasing vulnerability of the heart to heart attack (when heart muscle dies as a result of lack of oxygen) and cardiac arrest (when the heart stops beating). Smoking also raises the levels of low-density lipoproteins (the so-called "bad" **cholesterol**) in the blood, and lowers the levels of high-density lipoproteins (the so-called "good" cholesterol), a situation that has been linked to atherosclerosis. Finally, smoking increases the risk of stroke by 1.5 to 3 times the risk for nonsmokers.

Cancer

Smoking causes 85% of all lung cancers, and 14% of all cancers—among them cancers of the mouth, pharynx (throat), larynx (voice-box), esophagus, stomach, pancreas, cervix, kidney, ureter, and bladder. More than 171,500 new diagnoses were expected in 1998. Other environmental factors add to the carcinogenic qualities of tobacco. For example, alcohol consumption combined with smoking accounts for three-quarters of all oral and pharyngeal cancers. Also, persons predisposed genetically to certain cancers may develop cancer more quickly if they smoke. Only 14% of lung cancer patients survive five years after **diagnosis**.

Lung disease

Smoking is the leading cause of lung disease in the United States. Among the direct causes of death are **pneumonia**, **influenza**, bronchitis, **emphysema**, and chronic airway obstruction. Smoking increases mucus production in the airways and deadens the respiratory cilia, the tiny hairs that sweep debris out from the lungs. Without the action of the cilia, **bacteria** and inhaled particles from cigarette smoke are free to damage the lungs.

In the smaller airways of the lungs—the tiny bronchioles that branch off from the larger bronchi—chronic **inflammation** is present in smokers which causes airway to constrict causing cough, mucus production, and shortness of breath. Eventually, this inflammation can lead to chronic obstructive pulmonary disease (COPD), a condition in which oxygen absorption by the lungs is greatly reduced, severely limiting the amount of oxygen transported to body tissues.

Other health problems

For the 40 years prior to 1987, breast cancer was the leading cause of cancer death among women in the United States. In 1987, lung cancer took the lead. As well as increased risk of cancer and cardiovascular disease, women smokers are at increased risk of **osteoporosis** (a disease in which bones become brittle and vulnerable to breakage), cervical cancer, and decreased fertility. Pregnant women have increased risk for spontaneous abortion, premature separation of the placenta from the uter-

ine wall (a life-threatening complication for mother and fetus), placenta previa (in which the placenta implants much lower in the uterus than normal, which may lead to hemorrhage), bleeding during pregnancy, and premature rupture of the placental membranes (which can lead to **infection**). Infants born to women who smoke during pregnancy are at increased risk for low **birth** weight (18,600 cases annually), and other developmental problems. In men, smoking lowers testosterone levels, and appears to increase male **infertility**.

Numerous other health problems are caused by smoking such as poor circulation in the extremities due to constricted blood vessels. This not only leads to constantly cold hands and feet, it often requires **amputation** of the lower extremities. Smoking also deadens the **taste** buds and the receptors in the nasal epithelium, interfering with the senses of taste and **smell**, and may also contribute to periodontal disease.

Nicotine—addiction or habit?

In 1992, the Surgeon General of the United States declared nicotine to be as addictive as **cocaine**. An article published in the December 17, 1997 issue of the Journal of the National Cancer Institute stated nicotine **addiction** rates are higher than for alcohol or cocaine—that of all people trying only one cigarette, 33-50% will ultimately become addicted. The article concluded that simply knowing the harmful effects of tobacco is insufficient to help people kick the addiction and that behavioral intervention and support methods similar to those applied in alcohol and drug addictions appear to be most helpful.

The physical effects of cigarette smoke include several neurological responses which, in turn, stimulate emotional responses. When serotonin, a **neurotransmitter** (substances in the brain used by cells to transmit nerve impulses) is released, a person feels more alert. Nicotine stimulates serotonin release. Soon, however, serotonin release becomes sluggish without the boost from nicotine and the smoker becomes dependent on nicotine to prompt the release of serotonin. Other neurotransmitters released in response to nicotine include **dopamine**, opioids (naturally-occurring pain-killing substances), and various **hormones**, all of which have powerful effects on the brain where addiction occurs.

Genes and nicotine addiction

In 1998, scientists found a defective **gene** which makes the **metabolism** of nicotine difficult. The normal gene produces a liver **enzyme** needed to break down nicotine. The defective gene, found in about 20% of non-smokers, may lessen the likelihood of nicotine addiction.

In 1999, researchers discovered a version of a gene which increases the levels of dopamine in the brain. Because nicotine stimulates the release of dopamine, researchers believe the new-found gene may reduce the individual's desire to "pump up" dopamine production with nicotine.

The effects of quitting

Quitting smoking significantly lowers the risk of cancer and cardiovascular disease. In fact, the risk of lung cancer decreases from 18.83 at one to four years after quitting, to 7.73 at five to nine years, to below 5 at 10-19 years, to 2.1 at 20-plus years. The risk of lung cancer for nonsmokers is 1.

Weight gain is a common side effect of quitting, since smoking interferes with pancreatic function and **carbohydrate** metabolism, leading to a lower body weight in some people. However, not all people experience this lowered body weight from smoking, thus, not all people who quit gain weight. Taste buds and smell are reactivated in nonsmokers, which may lead to increased food intake.

Methods of treatment

About 80% of people who quit relapse within the first two weeks. Less than 3% of smokers become nonsmokers annually. Nicotine gum and patches, which maintain a steady level of nicotine in the blood, have met with some success but are more successful when combined with other support programs. Researchers now believe that smoking may be linked to **depression**, the withdrawal symptom causing most people who quit to begin again. In 1997, the FDA approved the antidepressant medication bupropion to help treat nicotine dependence.

Offense is the best defense

In 1998, a $206 billion settlement from tobacco companies to 46 states included a ban on all outdoor advertising of tobacco products. In 1999, the CDC appropriated more than $80 million to curtail tobacco use among young people. Coordinated education and prevention programs through schools have lowered the onset of smoking by 37% in seventh-grade students alone. By educating today's youth to the dangers of tobacco use, adults of tomorrow will have a longer, healthier, more productive life.

See also Respiratory system.

Resources

Periodicals

Bertrecchi, Carl E., et al. "The Human Costs of Tobacco Use, Part I" *New England Journal of Medicine* 330 (March 1994): 907.

Boyle, Peter. "The Hazards of Passive-and Active-Smoking." *New England Journal of Medicine* 328 (June 1993): 1708.
Brownlee, Shannon. "The Smoke Next Door." *U.S. News and World Report* 116 (June 1994): 66.
Hurt, Richard D., et al. "Nicotine Patch Therapy for Smoking Cessation Combined with Physician Advice and Nurse Follow-Up: One Year Outcome and Percentage of Nicotine Replacement." *Journal of the American Medical Association* 271 (February 1994): 595.
MacKenzie, Thomas D., et al. "The Human Costs of Tobacco Use, Part II" *New England Journal of Medicine* 330 (April 1994): 975.
Rogge, Wolfgang F. "Cigarette Smoke in the Urban Atmosphere." *Environmental Science and Technology* 28 (July 1994): 1375.
Sekhon, Harmanjatinder S., et al. "Cigarette Smoke Causes Rapid Cell Proliferation in Small Airways and Associated Pulmonary Arteries." *American Journal of Physiology* 267 (November 1994): L557.

Kathleen Scogna

Cinchona *see* **Quinine**
Cinematography *see* **Motion pictures**
Cinnamon *see* **Laurel family (Lauraceae)**
Circadian rhythm *see* **Biological rhythms**

Circle

In the language of **geometry**, a circle is the **locus** of points in a **plane** that are all at an equal **distance** from a single **point**, called the center of the circle. The fixed distance is called the radius of the circle. A line segment with each of its endpoints on the circle, that passes through the center of the circle, is called a diameter of the circle. The length of a diameter is twice the radius. The distance around a circle, called its circumference, is the length of the line segment that would result if the circle were broken at a point and straightened out. This length is given by $2\pi r$, where r is the radius of the circle and π (the Greek letter π, pronounced "pie") is a constant equal to approximately 3.14159. Points lying outside the circle are those points whose distance from the center is greater than the radius of the circle, and points lying in the circle are those points whose distance from the center is less than the radius of the circle. The area covered by a circle, including all the points within it, is called the area of the circle. The area of a circle is also related to its radius by the formula $A = \pi r^2$, where A is the area, r is the radius, and π is the same constant as that in the formula for the circumference. In the language of **algebra**, a circle corresponds to the set of ordered pairs (x,y) such that $(x - a)^2 + (y - b)^2 = r^2$, where the point corresponding to the ordered pair (a,b) is the center of the circle and the radius is equal to r.

Circuit *see* **Electric circuit**

Circulatory system

Living things require a circulatory system to deliver food, **oxygen**, and other needed substances to all cells, and to take away waste products. Materials are transferred between individual cells and their internal environment through the **cell membrane** by **diffusion**, **osmosis**, and active transport. During diffusion and osmosis, molecules move from a higher **concentration** to a lower concentration. During active transport, carrier molecules push or pull substances across the cell membrane, using **adenosine triphosphate** (ATP) for **energy**. Unicellular organisms depend on passive and active transport to exchange materials with their watery environment. More complex multicellular forms of life rely on transport systems that move material-containing liquids throughout the body in specialized tubes. In vascular plants, tubes transport food and **water**. Some **invertebrates** rely on a closed system of tubes, while others have an open system. Humans and other higher **vertebrates** have a closed system of circulation.

Circulation in vascular plants

Water and dissolved **minerals** enter a plant's roots from the **soil** by means of diffusion and osmosis. These substances then travel upward in the **plant** in xylem vessels. The **transpiration** theory ascribes this ascending flow to a pull from above, caused by transpiration, the **evaporation** of water from leaves. The long water column stays intact due to the strong cohesion between water molecules. Carbohydrates, produced in leaves by **photosynthesis**, travel downward in plants in specialized **tissue**, phloem. This involves active transport of sugars into phloem cells and water **pressure** to **force** substances from cell to cell.

Circulation in invertebrates

Animal circulation depends on the contraction of a pump—usually a **heart** that pumps **blood** in one direction through vessels along a circulatory path. In a closed path, the network of vessels is continuous. Alternately, an open path has vessels that empty into open spaces in the body. The closed system in the earthworm uses five pairs of muscular hearts (the aortic arches), to pump blood. Located near the anterior or head end of the animal, the aortic arches contract and force blood into the ventral blood vessel that runs from head to tail. Blood then returns back to the hearts in the dorsal blood vessel. Small ring vessels in each segment connect dorsal and ventral blood vessels. As blood circulates throughout the body, it delivers **nutrients** and oxygen to cells and picks up **carbon dioxide** and other wastes.

Most **arthropods** and some advanced molluscs such as **squid** and octopuses have an open circulatory system. In the grasshopper, a large blood vessel runs along the top of the body, and enlarges at the posterior or tail end to form a tubelike heart. Openings in the heart (ostia) have valves that permit only the entry of blood into the heart. The heart contracts, forcing blood forward in the blood vessel and out into the head region. Outside the heart, the blood goes into spaces that surround the insect's internal organs. The blood delivers food and other materials to cells and picks up wastes. Animals with open circulatory systems depend on the **respiratory system** to transport oxygen and **carbon** dioxide. The blood moves slowly from the head to the tail end of the animal. At the posterior, the blood re-enters the heart through the openings. Contraction of muscles helps speed up the blood flow.

Human circulatory system

The human circulatory system is termed the cardiovascular system, from the Greek word *kardia*, meaning heart, and the Latin *vasculum*, meaning small vessel. The basic components of the cardiovascular system are the heart, the blood vessels, and the blood. The work done by the cardiovascular system is astounding. Each year, the heart pumps more than 1,848 gal (7,000 l) of blood through a closed system of about 62,100 mi (100,000 km) of blood vessels. This is more than twice the **distance** around the equator of the **earth**. As blood circulates around the body, it picks up oxygen from the lungs, nutrients from the small intestine, and **hormones** from the endocrine **glands**, and delivers these to the cells. Blood then picks up carbon dioxide and cellular wastes from cells and delivers these to the lungs and kidneys, where they are excreted. Substances pass out of blood vessels to the cells through the interstitial or tissue fluid which surrounds cells.

The human heart

The adult heart is a hollow cone-shaped muscular **organ** located in the center of the chest cavity. The lower tip of the heart tilts toward the left. The heart is about the size of a clenched fist and weighs approximately 10.5 oz (300 g). Remarkably, the heart beats more than 100,000 times a day and close to 2.5 billion times in the average lifetime. A triple-layered sac, the pericardium, surrounds, protects, and anchors the heart. A liquid pericardial fluid located in the space between two of the layers, reduces **friction** when the heart moves.

The heart is divided into four chambers. A partition or septum divides it into a left and right side. Each side is further divided into an upper and lower chamber. The upper chambers, atria (singular atrium), are thin-walled. They receive blood entering the heart, and pump it to the ventricles, the lower heart chambers. The walls of the ventricles are thicker and contain more cardiac muscle than the walls of the atria, enabling the ventricles to pump blood out to the lungs and the rest of the body. The left and right sides of the heart function as two separate pumps. The right atrium receives oxygen-poor blood from the body from a major vein, the vena cava, and delivers it to the right ventricle. The right ventricle, in turn, pumps the blood to the lungs via the pulmonary artery. The left atrium receives the oxygen-rich blood from the lungs from the pulmonary **veins**, and delivers it to the left ventricle. The left ventricle then pumps it into the aorta, a major artery that leads to all parts of the body. The wall of the left ventricle is thicker than the wall of the right ventricle, making it a more powerful pump able to push blood through its longer trip around the body.

One-way valves in the heart keep blood flowing in the right direction and prevent backflow. The valves open and close in response to pressure changes in the heart. Atrioventricular (AV) valves are located between the atria and ventricles. Semilunar (SL) valves lie between the ventricles and the major **arteries** into which they pump blood. The "lub-dup" sounds that the physician hears through the stethoscope occur when the heart valves close. The AV valves produce the "lub" sound upon closing, while the SL valves cause the "dup" sound. People with a heart murmur have a defective heart valve that allows the backflow of blood.

The **rate** and rhythm of the heartbeat are carefully regulated. We know that the heart continues to beat even when disconnected from the **nervous system**. This is evident during heart transplants when donor hearts keep beating outside the body. The explanation lies in a small **mass** of contractile cells, the sino-atrial (SA) node or **pacemaker**, located in the wall of the right atrium. The SA node sends out electrical impulses that set up a wave of contraction that spreads across the atria. The wave reaches the atrio-ventricular (AV) node, another small mass of contractile cells. The AV node is located in the septum between the left and right ventricle. The AV node, in turn, transmits impulses to all parts of the ventricles. The bundle of His, specialized fibers, conducts the impulses from the AV node to the ventricles. The im-

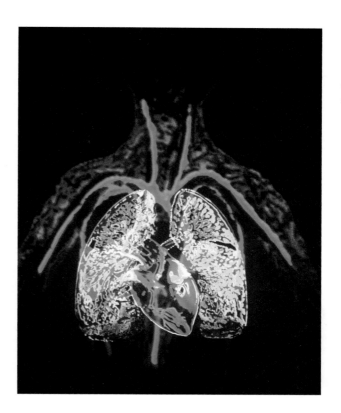

Colorized image of the main components of the human circulatory system. The heart (placed between the lungs) delivers blood to lungs to pick up oxygen and circulates it throughout the body by means of a system of blood vessels (red). *Photograph by Howard Sochurek. The Stock Market. Reproduced with permission.*

pulses stimulate the ventricles to contract. An electrocardiogram, ECG or EKG, is a record of the electric impulses from the pacemaker that direct each heartbeat. The SA node and conduction system provide the primary heart controls. In patients with disorganized electrical activity in the heart, surgeons implant an artificial pacemaker that serves to regulate the heart rhythm. In addition to self-regulation by the heart, the autonomic nervous system and hormones also affect its rate.

The heart cycle refers to the events associated with a single heartbeat. The cycle involves systole, the contraction phase, and diastole, the relaxation phase. In the heart, the two atria contract while the two ventricles relax. Then, the two ventricles contract while the two atria relax. The heart cycle consists of a systole and diastole of both the atria and ventricles. At the end of a heartbeat all four chambers rest. The rate of heartbeat averages about 75 beats per minute, and each **cardiac cycle** takes about 0.8 seconds.

Heart **disease** is the number one cause of death among people living in the industrial world. In coronary heartdisease (CHD), a clot or stoppage occurs in a blood vessel of the heart. Deprived of oxygen, the surrounding tissue becomes damaged. Education about prevention of CHD helps to reduce its occurrence. We have learned to prevent heart attacks by eating less **fat**, preventing **obesity**, exercising regularly, and by not smoking. Medications, medical devices and techniques also help patients with heartdisease. One of these, the **heart-lung machine**, is used during open-heart and bypass **surgery**. This device pumps the patient's blood out of the body, and returns it after having added oxygen and removed carbon dioxide. For patients with CHD, physicians sometimes use coronary artery bypass grafting (CABG). This is a surgical technique in which a blood vessel from another part of the body is grafted into the heart. The relocated vessel provides a new route for blood to travel as it bypasses the clogged coronary artery. In addition, cardiologists can also help CHD with angioplasty. Here, the surgeon inflates a balloon inside the aorta. This opens the vessel and improves the blood flow. For diagnosing heartdisease, the echocardiogram is used in conjunction with the ECG. This device uses high **frequency sound waves** to take pictures of the heart.

Blood vessels

The blood vessels of the body make up a closed system of tubes that carry blood from the heart to tissues all over the body and then back to the heart. Arteries carry blood away from the heart, while veins carry blood toward the heart. **Capillaries** connect small arteries (arterioles) and small veins (venules). Large arteries leave the heart and branch into smaller ones that reach out to various parts of the body. These divide still further into smaller vessels called arterioles that penetrate the body tissues. Within the tissues, the arterioles branch into a network of microscopic capillaries. Substances move in and out of the capillary walls as the blood exchanges materials with the cells. Before leaving the tissues, capillaries unite into venules, which are small veins. The venules merge to form larger and larger veins that eventually return blood to the heart. The two main circulation routes in the body are the pulmonary circulation, to and from the lungs, and the systemic circulation, to and from all parts of the body. Subdivisions of the systemic system include the coronary circulation, for the heart, the cerebral circulation, for the **brain**, and the renal circulation, for the kidneys. In addition, the hepatic portal circulation passes blood directly from the digestive tract to the liver.

The walls of arteries, veins, and capillaries differ in structure. In all three, the vessel wall surrounds a hollow center through which the blood flows. The walls of both arteries and veins are composed of three coats. The inner coat is lined with a simple squamous endothelium, a single flat layer of cells. The thick middle coat is composed of smooth muscle that can change the size of the vessel when it contracts or relaxes, and of stretchable fibers that provide **elasticity**. The outer coat is composed of elastic fibers and **collagen**. The difference between veins and arteries lies in the thickness of the wall of the vessel. The inner and middle coats of veins are very thin compared to arteries. The thick walls of arteries make them elastic and capable of contracting. The repeated expansion and recoil of arteries when the heart beats creates the pulse. We can feel the pulse in arteries near the body surface, such as the radial artery in the wrist. The walls of veins are more flexible than artery walls and they change shape when muscles press against them. Blood returning to the heart in veins is under low pressure often flowing against gravity. One-way valves in the walks of veins keep blood flowing in one direction. Skeletal muscles also help blood return to the heart by squeezing the veins as they contract. Varicose veins develop when veins lose their elasticity and become stretched. Faulty valves allow blood to sink back thereby pushing the vein wall outward. The walls of capillaries are only one cell thick. Of all the blood vessels, only capillaries have walls thin enough to allow the exchange of materials between cells and the blood. Their extensive branching provides a sufficient surface area to pick up and deliver substances to all cells in the body.

Blood pressure is the pressure of blood against the wall of a blood vessel. Blood pressure originates when the ventricles contract during the heartbeat. In a healthy young adult male, blood pressure in the aorta during systole is about 120 mm Hg, and approximately 80 mm Hg during diastole. The sphygmomanometer is an instrument that measures blood pressure. A combination of nervous carbon and hormones help regulate blood pressure around a normal range in the body. In addition, there are local controls that direct blood to tissues according to their need. For example, during **exercise**, reduced oxygen and increased carbon dioxide stimulate blood flow to the muscles.

Two disorders that involve blood vessels are **hypertension** and atherosclerosis. Hypertension, or high blood pressure, is the most common circulatory disease. For about 90% of hypertension sufferers, the blood pressure stays high without any known physical cause. Limiting **salt** and **alcohol** intake, stopping smoking, losing weight, increasing exercise, and managing **stress** help reduce blood pressure. Medications also help control hypertension. In atherosclerosis, the walls of arteries thicken and lose their elasticity. Fatty material such as **cholesterol** accumulates on the artery wall forming plaque that obstructs blood flow. The plaque can form a clot that breaks

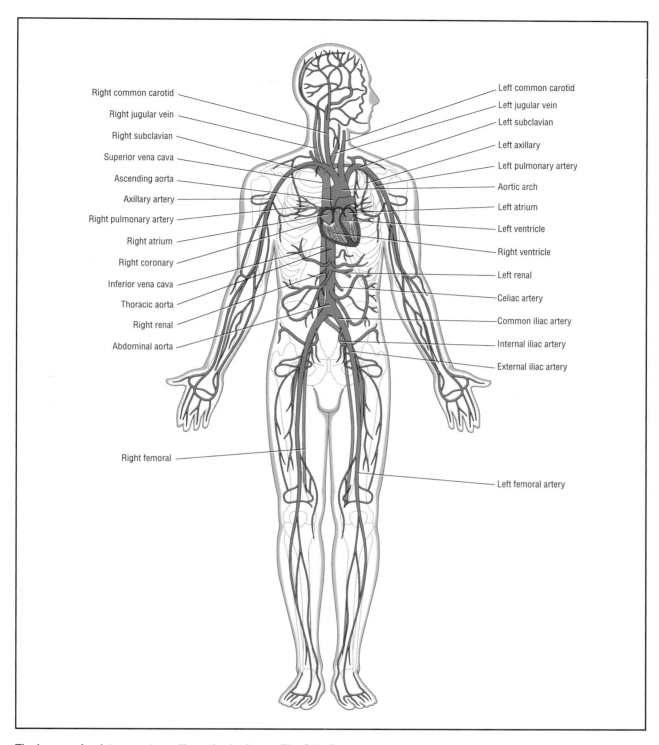

Right common carotid
Right jugular vein
Right subclavian
Superior vena cava
Ascending aorta
Axillary artery
Right pulmonary artery
Right atrium
Right coronary
Inferior vena cava
Thoracic aorta
Right renal
Abdominal aorta

Right femoral

Left common carotid
Left jugular vein
Left subclavian
Left axillary
Left pulmonary artery
Aortic arch
Left atrium
Left ventricle
Right ventricle
Left renal
Celiac artery
Common iliac artery
Internal iliac artery
External iliac artery

Left femoral artery

The human circulatory system. *Illustration by Argosy. The Gale Group.*

off, travels in the blood, and can block a smaller vessel. For example, a **stroke** occurs when a clot obstructs an artery or capillary in the brain. Treatment for atherosclerosis includes medication, surgery, a low-fat, high-fiber diet, and exercise. The type of cholesterol carried in the blood indicates the risk of atherosclerosis. Low **density** lipoproteins (LDLs) deposit cholesterol on arteries, while high density lipoproteins (HDLs) remove it.

Blood

Blood is liquid **connective tissue**. It transports oxygen from the lungs and delivers it to cells. It picks up carbon dioxide from the cells and brings it to the lungs. It carries nutrients from the **digestive system** and hormones from the endocrine glands to the cells. It takes **heat** and waste products away from cells. The blood helps regulate the body's base-acid balance (**pH**), **temperature**, and water content. It protects the body by clotting and by fighting disease through the **immune system**.

When we study the structure of blood, we find that it is heavier and stickier than water, has a temperature in the body of about 100.4°F (38°C), and a pH of about 7.4. Blood makes up approximately 8% of the total body weight. A male of average weight has about 1.5 gal (5-6 l) of blood in his body, while a female has about 1.2 gal (4-5 l). Blood is composed of a liquid portion (the **plasma**), and blood cells.

Plasma is composed of about 91.5% water which acts as a solvent, heat conductor, and suspending medium for the blood cells. The rest of the plasma includes plasma **proteins** produced by the liver, such as albumins, that help maintain water balance, globulins, that help fight disease, and fibrinogen, that aids in blood clotting. The plasma carries nutrients, hormones, enzymes, cellular waste products, some oxygen, and carbon dioxide. Inorganic salts, also carried in the plasma, help maintain osmotic pressure. Plasma leaks out of the capillaries to form the interstitial fluid (tissue fluid) that surrounds the body cells and keeps them moist, and supplied with nutrients.

The cells in the blood are erythrocytes (red blood cells), leukocytes (white blood cells), and thrombocytes (platelets). More than 99% of all the blood cells are erythrocytes, or red blood cells. Red blood cells look like flexible biconcave discs about 8 nm in diameter that are capable of squeezing through narrow capillaries. Erythrocytes lack a nucleus and therefore are unable to reproduce. Antigens, specialized proteins on the surface of erythrocytes, determine the ABO and Rh blood types. Erythrocytes contain hemoglobin, a red pigment that carries oxygen, and each red cell has about 280 million hemoglobin molecules. An **iron** ion in hemoglobin combines reversibly with one oxygen **molecule**, enabling it to pick up, carry and drop off oxygen. Erythrocytes are formed in red bone marrow, and live about 120 days. When they are worn out, the liver and spleen destroy them and recycle their breakdown products. **Anemia** is a blood disorder characterized by too few red blood cells.

Leukocytes are white blood cells. They are larger than red blood cells, contain a nucleus, and do not have hemoglobin. Leukocytes fight disease organisms by destroying them or by producing antibodies. Lymphocytes

are a type of leukocyte that bring about immune reactions involving antibodies. Monocytes are large leukocytes that ingest **bacteria** and get rid of dead **matter**. Most leukocytes are able to squeeze through the capillary walls and migrate to an infected part of the body. Formed in the white/yellow bone marrow, a leukocyte's life ranges from hours to years depending on how it functions during an **infection**. In **leukemia**, a malignancy of bone marrow tissue, abnormal leukocytes are produced in an uncontrolled manner. They crowd out the bone marrow cells, interrupt normal blood cell production, and cause internal bleeding. Treatment for acute leukemia includes blood transfusions, anticancer drugs, and, in some cases, **radiation**.

Thrombocytes or platelets bring about clotting of the blood. Clotting stops the bleeding when the circulatory system is damaged. When tissues are injured, platelets disintegrate and release the substance thromboplastin. Working with **calcium** ions and two plasma proteins, fibrinogen and prothrombin, thromboplastin converts prothrombin to thrombin. Thrombin then changes soluble fibrinogen into insoluble fibrin. Finally, fibrin forms a clot. **Hemophilia**, a hereditary blood disease in which the patient lacks a clotting factor, occurs mainly in males. Hemophiliacs hemorrhage continuously after injury. They can be treated by transfusion of either fresh plasma or a concentrate of the deficient clotting factor.

The lymphatic system and the circulatory system

The **lymphatic system** is an open transport system that works in conjunction with the circulatory system. Lymphatic vessels collect intercellular fluid (tissue fluid), kill foreign organisms, and return it to the circulatory system. The lymphatic system also prevents tissue fluid from accumulating in the tissue spaces. Lymph capillaries pick up the intercellular fluid, now called lymph, and carry it into larger and larger lymph vessels. Inside the lymph vessels, lymph passes through lymph nodes, where lymphocytes attack viruses and bacteria. The lymphatic system transports lymph to the large brachiocephalic veins below the collarbone where it is re-enters the circulatory system. Lymph moves through the lymphatic system by the squeezing action of nearby muscles, for there is no pump in this system. Lymph vessels are equipped with one-way valves that prevent backflow. The spleen, an organ of the lymphatic system, removes old blood cells, bacteria, and foreign particles from the blood.

Resources

Books

Berne, R.M., and M.N. Levy. *Cardiovascular Physiology*. St. Louis: C.V. Mosby, 1992.

Guyton & Hall. *Textbook of Medical Physiology*. 10th ed. New York: W. B. Saunders Company, 2000.

Kapit, Wynn, and Lawrence M. Elson. *The Anatomy Coloring Book*. New York: Harper & Row, 1995.

Periodicals

Acierno, L.J., and T. Worrell. "Profiles in Cardiology: James Bryan Herrick." *Clinical Cardiology* no. 23 (2000): 230-232.

Fackelmann, K. A. "Immune Cell Triggers Attack on Plaque." *Science News* (October 22, 1994).

Vogel, Steven. "Nature's Pumps." *American Scientist* (September/October 1994).

Other

Two Hearts That Beat as One. Films for Humanities and Science, 1995. Videocassette.

Bernice Essenfeld

Circumscribed and inscribed

The terms circumscribed and inscribed refer, respectively, to geometric figures that have been drawn around the outside of or within some other geometric figure. For example, imagine that a **circle** is drawn around a triangle so that the circle passes through all three vertices of the triangle. Then the circle is said to be circumscribed around the triangle, and the triangle is said to be inscribed within the circle.

Many combinations of figures could be substituted for the triangle and circle described above. For example, a circle circumscribed about any kind of polygon is one that passes through all of the vertices of the polygon. Then the polygon itself is said to be inscribed within the circle. Conversely, a polygon can be circumscribed around a circle if all of the sides of the polygon are tangent to the circle. Then the circle is inscribed within the polygon.

Three-dimensional figures can be circumscribed around and inscribed within each other also. For example, a cone can be circumscribed around a **pyramid** if the vertices of the cone and pyramid coincide with each other, and the base of the cone circumscribes the base of the pyramid. In such a case, the pyramid is inscribed within the cone. As another example, a **sphere** can be inscribed within a cylinder if all parts of the cylinder are tangent to the sphere's surface. Then the cylinder is circumscribed around the sphere.

Cirrhosis

Cirrhosis is a degenerative liver **disease** in which the lobes of the liver become infiltrated with **fat** and fibrous **tissue** (fibrous tissue is a type of **connective tissue** composed of protein fibers called **collagen**). The word "cirrhosis" is derived from the Greek words *kirrhos*, meaning "yellowish orange" and *osis*, meaning "condition," and connotates the appearance of the liver of a patient with cirrhosis. The infiltration of these substances within the liver disrupts liver functions, including the conversion of the storage **carbohydrate** glycogen into glucose; detoxification of drugs and other substances; **vitamin** absorption; gastrointestinal functions; and hormone **metabolism**. Because the blood vessels of the liver are affected by fibrous tissue formation, **blood** flow through the liver is impaired, leading to increased blood pressure within the liver. Impaired blood flow in the liver also causes a condition called esophageal varices, in which blood vessels in the esophagus bleed due to increased blood pressure. In addition, patients with cirrhosis are often weakened by hypoglycemia, or low blood sugar.

Complications of cirrhosis include **coma**, gastrointestinal hemorrhage, and kidney failure. No definitive treatment for cirrhosis is available besides management of symptoms and complications. The mortality **rate** of cirrhosis ranges from 40-63%, and it is the ninth leading cause of death in the United States.

Causes of cirrhosis

Cirrhosis is most often caused by excessive ingestion of **alcohol**. In the United States, 90% of all cases of cirrhosis are related to alcohol overconsumption. Although researchers still do not have a clear understanding of how alcohol damages the liver, it is thought that consuming more than 1 pint of alcohol (86 proof) daily for 10 years increases the risk of cirrhosis by 10%. The risk of cirrhosis increases to 50% if this consumption lasts 25 years. Women may be more susceptible to alcoholic cirrhosis than men. Women have lower levels of the **enzyme** alcohol dehydrogenase, which breaks down alcohol in the stomach. Lower levels of this enzyme lead to higher blood alcohol levels.

Researchers believe that the main cause of alcohol-induced cirrhosis is acetaldehyde, the first product created when alcohol is broken down in the stomach. Acetaldehyde combines with **proteins** in the body and damages the liver cells, leading to fat accumulation and fibrous tissue formation. Further damage to liver cells may be caused by a secondary effect of alcohol on the liver: the acetaldehyde may make the liver cells vulnerable to the potentially damaging effects of substances such as acetaminophen (a common over-the-counter **pain** reliever), industrial solvents, and certain anesthetics on the liver cells.

Although excessive alcohol intake is considered the leading cause of cirrhosis, there are numerous other causes of the disease. These include several types of viral **hepatitis**, nutritional factors, genetic conditions, and others. In addition, these are often contributing factors in alcoholic cirrhosis. Chronic **infection** with hepatitis B **virus** (HBV) can lead to cirrhosis and is also related to liver **cancer**.

Progression of cirrhosis

Cirrhosis is a progressive disease. In cases of alcoholic cirrhosis, it begins with a condition called alcoholic fatty liver. In this condition, fat accumulates in the liver. The liver enlarges, sometimes to as much as 10 lb (5000 g), and appears yellow and greasy. Patients with this condition often have no symptoms beyond low blood sugar or some digestive upset. Some patients have more severe symptoms, such as **jaundice** (a condition caused by the accumulation of a yellowish bile pigment in the blood; patients with jaundice have yellow-tinged skin) and weight loss. Complete recovery is possible at this stage of liver disease if the patient abstains from alcohol.

The next stage in the progression of cirrhosis is often a condition called hepatitis. Hepatitis is a general term meaning **inflammation** of the liver. Hepatitis may

be caused by alcohol, a virus, or other factors. In acute alcoholic hepatitis, the inflammation is caused by alcohol. Fibrous tissue is deposited in the liver, the liver cells degenerate, and a type of connective tissue called hyaline infiltrates the liver cells. Regardless of the cause, patients with hepatitis have serious symptoms, including general debilitation, loss of muscle mass, jaundice, fever, and abdominal pain. The liver is firm, tender, and enlarged. Vascular "spiders," or varicose **veins** of the liver, are present. Again, no definitive treatment is available, although some patients respond well to corticosteroids. General treatment for this condition includes treatment of symptoms and complications.

Acute alcoholic hepatitis often progresses to cirrhosis. In cirrhosis, the fibrous tissue and fat accumulation is prominent. Collagen is deposited around the veins of the liver, leading to impairment of blood flow. The liver cells degenerate further and die, leading to formation of nodules in the liver, and the liver atrophies (shrinks). Symptoms of cirrhosis include nausea, weight loss, jaundice, and esophageal varices. Gastrointestinal symptoms, such as diarrhea and gastric distention, are also features of cirrhosis. The mortality rate of cirrhosis is high. Five year survival of patients with cirrhosis is 64% for patients who stop drinking, and 40% for patients who continue to drink. Death results from kidney failure, coma, **malnutrition**, and cardiac arrest.

Treatment of cirrhosis depends on the type and cause. For patients with alcoholic cirrhosis, it includes general support, treatment of complications, a nutritious diet, and abstention from alcohol consumption. In selected patients, liver transplant may be indicated. Although the liver has a remarkable ability to regenerate, the damage that cirrhosis inflicts on the liver may be so severe that recovery is not possible. In cirrhosis caused by viral hepatitis, the use of experimental drugs has had some success.

Resources

Periodicals

Cohen, Carl, et. al. "Alcoholics and Liver Transplantation." *Journal of the American Medical Association* 265 (March 1991): 1299.

Hegarty, Mary. "The Good News (Preventing Cirrhosis of the Liver)." *Health* 5 (March-April 1991): 11.

Mann, Robert E.G. and Reginald G. Smart. "Alcohol and the Epidemic of Liver Cirrhosis." *Alcoholic Health and Research World* 16 (Summer 1992): 217.

Parrish, Kiyoko M., et al. "Average Daily Alcohol Consumption During Adult Life among Descendants with and without Cirrhosis: The 1986 National Mortality Followback Survey." *Journal of Studies on Alcohol* 54 (July 1993): 450.

Poynard, Thierry, et al. "Evaluation of Efficacy of Liver Transplantation in Alcoholic Cirrhosis by a Case Control Study and Simulated Controls." *The Lancet* 344 (August 1994): 502.

Kathleen Scogna

Citric acid

Citric acid is an organic (**carbon** based) acid found in nearly all citrus **fruits**, particularly lemons, limes, and grapefruits. It is widely used as a flavoring agent, preservative, and cleaning agent. The structure of citric acid is shown below. The COOH group is a carboxylic acid group, so citric acid is a tricarboxylic acid, possessing three of these groups.

Citric acid is produced commercially by the **fermentation** of sugar by several **species** of **mold**. As a flavoring agent, it can help produce both a tartness [caused by the production of **hydrogen** ions (H^+)] and sweetness (the result of the manner in which citric acid molecules "fit" into "sweet" receptors on our tongues). Receptors are protein molecules that recognize specific other molecules.

Citric acid helps to provide the "fizz" in remedies such as Alka-Seltzer trademark. The fizz comes from the production of **carbon dioxide** gas which is created when **sodium bicarbonate** (baking soda) reacts with acids. The source of the acid in this case is citric acid, which also helps to provide a more pleasant **taste**.

Citric acid is also used in the production of hair rinses and low **pH** (highly acidic) or slightly acidic shampoos and toothpaste. As a preservative, citric acid helps to bind (or sequester) **metal** ions that may get into food

via machinery used in processing. Many metals ions speed up the degradation of fats. Citric acid prevents the metal ions from being involved in a reaction with fats in foods and allows other preservatives to function much more effectively. Citric acid is also an intermediate in metabolic processes in all mammalian cells. One of the most important of these metabolic pathways is called the citric acid cycle (it is also called the **Krebs cycle**, after the man who first determined the role of this series of reactions). Some variants of citric acid containing fluorine have been used as rodent poisons.

See also Metabolism.

Citron *see* **Gourd family (Cucurbitaceae)**

Citrus trees

Citrus trees are various **species** of trees in the genus *Citrus*, in the rue family, or Rutaceae. There are 60 species in the genus *Citrus,* of which about 10 are used in agriculture. The center of origin of most species of *Citrus* is southern and southeastern **Asia**. Citrus trees are widely cultivated for their edible **fruits** in sub-tropical and tropical countries around the world. The sweet orange (*Citrus sinensis*) is the most common species of citrus in cultivation, and is one of the most successful fruits in agriculture.

The rue family consists of about 1,500 species and 150 genera. Most species in this family are trees or shrubs. The greatest richness of species occurs in the tropics and subtropics, especially in South **Africa** and **Australia**. However, a few species occur in the temperate zone. Several species native to **North America** are the shrubs known as prickly ash (*Zanthoxylum americanum*), southern prickly ash (*Z. clava-herculis*), and three-leaved hop **tree** (*Ptelea trifoliata*).

Biology of citrus

Citrus trees are species of sub-tropical and tropical climates. They are intolerant of freezing, and their foliage and fruits will be damaged by even a relatively short exposure to freezing temperatures for just a few hours. Colder temperatures can kill the entire tree.

Species of citrus trees range in size from shrubs to trees. Most species have thorny twigs. The leaves are alternately arranged on the twigs, and the foliage is dark green, shiny, aromatic, leathery, and evergreen. The roots of citrus trees do not develop root hairs, and as a result citrus trees are highly dependent for their mineral **nutri-**

tion on a mutualistic **symbiosis** with **soil** fungi called mycorrhizae.

Citrus trees have small white or purplish flowers, which are strongly scented and produce **nectar**. Both scent and nectar are adaptations for attracting **insects**, which are the pollinators of the flowers of citrus trees. Some species in the genus *Citrus* will easily hybridize with each other. This biological trait can make it easier for **plant** breeders to develop profitable agricultural varieties using controlled hybridization experiments to incorporate desirable traits from one species into another. However, the occurrence of **hybrid** *Citrus* plants makes it difficult for plant taxonomists to designate true species. As a result, there is some controversy over the validity of some *Citrus* species that have been named.

The ripe fruit of citrus trees is properly classified as a hesperidium, which is a type of berry, or a fleshy, multi-seeded fruit. The fruits of citrus trees have a relatively leathery outer shell, with a more spongy rind on the inside. The rind of citrus fruits is very rich in **glands** containing aromatic oils, which can be clearly detected by **smell** when these fruits are being peeled. The interior of the fruit is divided into discrete segments that contain the **seeds** surrounded by a large-celled, juicy pulp. The seeds of citrus trees are sometimes called "pips." Edible fruits like those of citrus trees are an **adaptation** to achieve dispersal of their seeds. The attractive and nutritious fruits are sought out by many species of animals, who eat the pulp and seeds. However, the citrus seeds generally survive the passage through the gut of the **animal**, and are excreted with the feces. In the meantime, the animal has likely moved somewhere, and the seeds have been dispersed far from the parent tree, ready to germinate, and hopefully, develop into a new citrus plant.

Cultivation and economic products of citrus trees

The fruits of citrus trees contain large concentrations of sour-tasting **citric acid**. Nevertheless, the fruits of some species can be quite sweet because they contain large concentrations of fruit sugar. Plant breeders have developed various sorts of cultivated varieties, or cultivars, from the wild progenitors of various species of citrus trees. This has resulted in the selective breeding of varieties with especially sweet fruits, others which peel relatively easily, and yet others that are seedless and therefore easier to eat or process into juice.

Once plant breeders discover a desirable cultivar of a species of citrus tree, it is thereafter propagated by rooting stem cuttings or by grafting. The latter procedure involves taking a stem of the cultivar, and attaching it to the rootstock of some other, usually relatively hardy, variety.

The graft is carefully wrapped until a protective callus **tissue** is formed. Because the attributes of the cultivar are genetically based, these methods of propagation avoid the loss of desirable genetic attributes of the new variety that would inevitably occur through sexual cross-breeding.

Citrus trees can also be propagated using relatively new techniques by which small quantities of cells can be grown and induced to develop into fully-formed plants through specific hormone treatments. These relatively new techniques, known as micro propagation or tissue culture, allow for the rapid and inexpensive production of large numbers of trees with identical genetic qualities.

The most important economic products of cultivated citrus trees are, of course, their fruits. In agriculture, the fruits of oranges and grapefruits are commonly picked when they are ripe or nearly so, while those of lemons and limes are usually picked while they are still unripened, or green.

The fruits of the sweet orange can be eaten directly after peeling, or they may be processed into a juice, which can be drunk fresh. It may also be concentrated by evaporating about three-quarters of its **water** content, and then frozen for transport to far-away markets. This is a more economical way of moving orange juice around, because removal of much of the water means that much less weight must be transported. In addition, juice concentrates can also be used to manufacture flavorings for various types of drinks.

The juice of citrus fruits is relatively rich in ascorbic acid, or **vitamin** C. For example, a typical orange contains about 40 mg of vitamin C, compared with only 5 mg in an apple. Vitamin C is an essential nutrient for proper nutrition of animals. However, animals cannot synthesize their own vitamin C and must obtain the micronutrient from their diet. In the absence of a sufficient dietary supply of vitamin C, a debilitating and eventually lethal **disease** known as scurvy develops. In past centuries scurvy often afflicted mariners on long oceanic voyages, during which foods rich in vitamin C or its biochemical precursors could not be readily obtained. Because they stored relatively well, lemons were an important means by which sailors could avoid scurvy, at least while the supply of those fruits lasted.

At one time, citric acid was commercially extracted from the fruits of citrus trees, mostly for use in flavoring drinks. Today, citric acid is used in enormous quantities to flavor carbonated soft drinks and other beverages. However, most of this industrial citric acid is synthesized by fungi in huge **fermentation** vats.

In addition, the shredded peel and juice of citrus fruits can be sweetened and jelled for use in such sweet spreads as marmalade.

Grapefruit ready to pick. © 1980 Ken Brate. National Audubon Society Collection/Photo Researchers, Inc. Reproduced by permission.

The major economic value of oranges lies in their fruits, but several fragrant oils can also be extracted from their flowers, or, more commonly, their peel as a by-product of the orange-juice industry. These essences can be used to manufacture so-called Neroli and Portugal oils. These fragrances were originally used in the manufacturing of perfumes and to scent potpourri, and they are still used for these purposes. In addition, many household products, such as liquid detergents, shampoos, and soaps, are pleasantly scented using the aromatic oils extracted from citrus trees.

Pomanders, which are oranges studded with cloves, are an archaic use of the fruit. Originating in Spain, pomanders were worn around the neck for several purposes-as perfumery, to ward off infections, or to attract a person of the opposite sex. Today pomanders are more commonly used to pleasantly scent closets and drawers.

In regions with a warm climate, citrus trees are sometimes grown as ornamental shrubs and trees. The citron was reputedly grown in the ancient Hanging Gardens of Babylon in what is now Iraq. During those times the citron was used in scenting toilet water and in making an aromatic ointment known as pomade.

The sweet orange

The sweet orange (*Citrus sinensis*) is a 16-46 ft (5-14 m) tall tree with evergreen foliage, white flowers, and spherical fruits. This species is originally from southern China or perhaps Southeast Asia. However, the original range is somewhat uncertain, because wild plants in natural habitats are not known. The sweet orange has been cultivated in China and elsewhere in southern Asia for thousands of years, being mentioned in dated Chinese scripts from 2200 B.C. The sweet orange reached **Europe** as a cultivated species sometime before the fourteenth century. Sweet oranges are now grown around the world wherever the climate is suitably subtropical or tropical.

The sweet orange tends to **flower** and fruit during periods of relatively abundant rainfall, and becomes dormant if a pronounced drier period occurs during the summer. The sweet orange is commonly cultivated in plantations or groves. These are widely established in subtropical parts of the southern United States, particularly in southern Florida and California. Oranges are also widely grown in Mexico, Spain, the Middle East, North Africa, and many other countries, for both local use and export.

The global production of sweet oranges is more than 38.5 million tons (35 million metric tonnes) per year.

Orange fruits are very tasty and nutritious, containing 5-10% sugar, 1-2% citric acid, along with vitamin C and beneficial fiber and pulp. Most sweet oranges have an orange-colored rind when they are ripe as well as an orange interior and juice. However, some cultivated varieties of sweet oranges have a yellow or green rind, while still others have a deep-red interior and juice. Some varieties have been bred to be seedless, including navel, Jaffa, and Malta oranges.

Oranges were a scarce and expensive fruit in past centuries, and many children were delighted to find a precious orange in their Christmas stocking. Today, however, oranges are grown in enormous quantities and are readily available as an inexpensive fruit at any time of the year.

The tangerine or mandarin orange

The tangerine and mandarin (*Citrus reticulata*) are a species of small tree native to southern China. The fruits of this species are similar to those of the sweet orange, but they are generally smaller, their rind is much easier to separate from the interior pulp, and the segments separate more readily.

Compared with the sweet orange, the tangerine and mandarin do not store very well. As a result, these citrus fruits tend to be sold fairly close to where they are grown, and less of a long-distance export market exists for these **crops**. However, in some countries a tradition has developed of eating mandarin oranges at certain festive times of year. For example, North Americans and many western Europeans often eat mandarins around Christmas time. Because a premium price can be obtained for these fruits during the Christmas season, there is a well organized market that keys on about a one-or-two month export market for mandarins during that festive season.

The grapefruit

The grapefruit or pomelo (*Citrus paradisi*) is a variety of cultivated citrus tree whose geographic origin is not known, but is likely native to Southeast Asia. The fruit of the grapefruit has a yellowish rind, and is relatively large, as much as 1 lb (0.5 kg) in weight. The pulp and juice of the grapefruit are rather bitter and acidic and are often sweetened with cane sugar before being eaten.

The lemon

The lemon (*C. limon*) is an evergreen tree native to Indochina and cultivated there for thousands of years. The lemon was later imported to the **basin** of the Mediterranean Sea, where it has been cultivated for at least 2,000 years. Lemon trees are very attractive, especially when their fragrant white or yellow flowers are in bloom. However, the fruits of lemons are quite tart and bitter, containing about 5% citric acid, but only 0.5% sugar.

The fruits of lemons are picked when they are not yet ripe and their rinds are still green. This is done because lemon fruits deteriorate quickly if they are allowed to ripen on the tree. Commercial lemons develop their more familiar, yellow-colored rinds some time after they are harvested while they are being stored or transported to markets.

Although few people have the fortitude to eat raw lemons, the processed juice of this species can be used to flavor a wide range of sweetened drinks, including lemonade. Lemon flavoring is also used to manufacture many types of carbonated beverages, often in combination with the flavoring of lime. A bleaching agent and stain remover can also be made from lemon juice.

The lime

The lime is native to Southeast Asia and is very susceptible to frost. More sour than the lemon, the lime (*C. aurantifolia*) cannot be eaten raw. However, the lime can be used to make a sweetened beverage known as limeade, and an extract of its juice is widely used to flavor commercially prepared soft drinks.

Other citrus trees

The Seville, sour or bitter orange (*C. media*), is derived from a wild progenitor that grows in the foothills of the Himalayan Mountains of south Asia. The flowers of this species are exceedingly fragrant and have been used to produce aromatic oils for perfumery. The large orange-red fruits of the sour orange are rather bitter and acidic. These are not often eaten, but are used to make flavorings, marmalades, candied peels, aromatic oils, and to flavor a liquor known as curacao.

The citron (*C. medica*) is another species native to the southern Himalayas, probably in northern India. This may be the oldest of the cultivated citrus trees, perhaps going back as far as 6,000 years. The fruit of this species is very large in comparison to those of other citrus trees, weighing as much as 6.5 lb (3 kg). The rind of the citron is thick and has a lumpy surface, and the pulpy interior is bitter. The peel of the citron is soaked in salty water, which removes much of the bitter **taste**. It is then candied with sugar and used to flavor cakes, pastries, and candies.

The shaddock or pomelo (*C. maxima*) is probably native to Southeast Asia. This species is mostly used to

KEY TERMS

Cultivar—A distinct variety of a plant that has been bred for particular agricultural or culinary attributes. Cultivars are not sufficiently distinct in the genetic sense to be considered a subspecies.

Cutting—A section of a stem of a plant, which can be induced to root and can thereby be used to propagate a new plant that is genetically identical to the parent.

Grafting—A method of propagation of woody plants whereby a shoot, known as a scion, is taken from one plant and inserted into a rootstock of another plant. The desired traits of the scion for horticultural or agricultural purposes are genetically based. Through grafting, large numbers of plants with these characteristics can be readily and quickly developed.

Scurvy—A disease of humans that is caused by an insufficient supply of ascorbic acid, or vitamin C, in the diet. The symptoms of scurvy include spongy, bleeding gums, loosening and loss of teeth, and subcutaneous bleeding. It can ultimately lead to death.

Tissue culture—This is a relatively recently developed method of growing large numbers of genetically identical plants. In tissue culture, small quantities of undifferentiated cells are grown on an artificial growth medium, and are then caused to develop into small plantlets by subjecting them to specific treatments with growth-regulating hormones.

manufacture candied rind. The pomelo develops a large, spherical, thick-rinded fruit, weighing as much as 13 lb (6 kg), and having a diameter of up to 6 in (16 cm). The name shaddock comes from the name of a sea captain who first introduced this species to the West Indies.

Other relatively minor species of citrus trees include the Panama orange or calamondin (*C. mitis*) and the bergamot (*C. bergamia*).

See also Mycorrhiza.

Resources

Books

Hvass, E. *Plants That Serve and Feed Us.* New York: Hippocrene Books, 1975.
Judd, Walter S., Christopher Campbell, Elizabeth A. Kellogg, Michael J. Donoghue, and Peter Stevens. *Plant Systematics: A Phylogenetic Approach.* 2nd ed. with CD-ROM. Suderland, MD: Sinauer, 2002.
Klein, R. M. *The Green World. An Introduction to Plants and People.* New York: Harper and Row, 1987.

Bill Freedman

Civets

Small to medium-sized carnivores, civets are in the Viverridae family which includes **genets**, linsangs, and **mongooses**. There are 35 **species** of civets and genets in 20 genera. Their natural distribution is restricted to the warmer regions of the Old World, and they occupy a **niche** similarly filled by **weasels** and their relatives found in temperate deciduous **forests**. Civets vary in size and form, but most present a catlike appearance with long noses, slender bodies, pointed ears, short legs and generally a long furry tail.

Civets with a spotted or striped coat have five toes on each foot. There is webbing between the toes, and the claws are totally or semi-retractile. The pointed ears extend above the profile of the head. The **ear** flaps have pockets or bursae on the outside margins, similar to domestic **cats**. Their teeth are specialized for an omnivorous diet, including shearing carnassial teeth and flat-crowned molars in both upper and lower jaws. Teeth number from 38 to 40, depending on the species.

Primarily nocturnal foragers with semiarboreal and arboreal habits, civets typically ambush their **prey**. During the day, civets usually rest in a hollow **tree**, rock crevice or empty, shallow burrow. They are solitary animals maintaining a wide home range (250 acres [101 ha]) by scent marking trees on the borders of their territory. The term "civet" is derived from an Arabic word describing the oily fluid and its odor secreted by the perineal **glands**. Scent marking is important in civet communication, but the method differs among species from passively passing the scent when moving about causing the gland to rub vegetation, to squatting and then wiping or rubbing the gland on the ground or some prominent object.

Civet oil has been used in the perfume industry for centuries and has been recorded as being imported from **Africa** by King Solomon in the tenth century B.C. Once refined, civet oil is prized for its odor and long lasting properties. Civet oil is also valued for its medicinal uses which include the reduction of perspiration, a cure for some skin disorders and claims of aphrodisiac powers. Although the development of sensitive chemical substitutes has decreased the value of civet oil, it is still a part of some East African and Oriental economies.

The Viverridae family can be broken into six subfamilies. There are seven southern Asian and one African species of palm civets (subfamily Paradoxurinae). The African palm civet, also known as the two-spotted palm civet, spends most of its time in the forest canopy where it feeds primarily on fruit, occasionally supplemented with small **mammals**, **birds**, **insects**, and lizards. It is distinguished by its semi-retractile claws and perineal gland covered by a single fold of skin. All other species of palm civets live in the forests of **Asia**. Their semiarboreal life style is supported by sharp, curved, retractile claws, hairless soles, and partially fused third and fourth toes which add to a more sure-footed grasp. Although skillful climbers, they spend considerable time foraging on the ground for animals and fallen **fruits**. The broad-faced binturong, or bear cat, has a strong, muscular long-haired tail that is prehensile at the tip. This characteristic is unique among viverrids. The body hair is long and coarse. The ears have long black tufts of hair with white margins. Nearly four feet long, it is the largest member of the civet family. Despite its mostly vegetarian diet, the binturong has been reported to swim in **rivers** and prey on **fish**. The celebes, giant, or brown palm civet may be fairly common in certain limited areas (known by its tracks and feces, rather than by actual sightings). It is quite adept at climbing and has a web of thin skin between the toes.

There are five species of banded palm civets and otter civets (subfamily Hemigalinae), all living in the forests of Southeast Asia. Perhaps best known is the banded palm civet, named for the dark brown markings on its coat. The general coat **color** ranges from pale yellow to grayish buff. The face is distinctly marked by several dark longitudinal stripes. The coloration of the body is broken by about five transverse bands stretching midway down the flank. The tail is dark on its lower half, with two broad dark rings at the base. Foraging at night on the ground and in trees, the banded palm civet searches for **rats**, lizards, **frogs**, **snails**, **crabs**, earthworms and **ants**.

Otter civets could be mistaken for long-nosed **otters**. Similar in habit and appearance, otter civets are excellent swimmers and capable of climbing trees. Their toes are partially webbed, but their dense **water** repellant fur, thick whiskers, and valve-like nostrils are effective adaptations for living in water and preying on fish. There are two species which show differences in their coat coloration and number of teeth. Otter civets have smaller ears, blunter muzzles, shorter tails and more compact bodies than most banded palm civets.

There are 19 species of true civets and genets classified in the subfamily Viverrinae. One of the best known is the African civet. A rather large, heavily built, long-bodied and long-legged **carnivore**, it is the most doglike viverrid. Preferring to be near water, it lives in a variety of habitats ranging from moist tropical forest to dry scrub savannah. It is considered terrestrial, climbing

KEY TERMS

Aphrodisiac—Stimulating or intensifying sexual desire.

Arboreal—Living in trees.

Bursae—Pockets; a saclike body cavity.

Carnassial teeth—Specialized teeth of mammals in the order Carnivora, which are longer and sharper than other teeth and useful in tearing meat.

Niche—The area within a habitat occupied by an organism.

Omnivorous—Eating all kinds of food.

Perineal—The region between the scrotum and the anus in males and between the posterior vulva junction and the anus in females.

Prehensile—Adapted for seizing or holding, especially by wrapping around an object.

Retractile—Capable of being drawn back or in.

Terrestrial—Of or pertaining to the earth and its inhabitants.

trees only in an emergency such as when hunted. The African civet hunts exclusively on the ground at night, resting in thickets or burrows during the day. An opportunistic and omnivorous **predator**, it will eat carrion, but prefers small mammals. Birds, eggs, **amphibians**, **reptiles**, **invertebrates**, fruit, berries, and vegetation round out its diet. African civets deposit their droppings in one place creating middens or "civetries." Although typically found at territorial boundaries, they also mark their territory using their perineal gland.

Betsy A. Leonard

Clay *see* **Sediment and sedimentation**

Clays *see* **Minerals**

Climax (ecological)

Climax is a theoretical, ecological notion intended to describe a relatively stable community that is in equilibrium with environmental conditions, and occurring as the terminal, end-point of **succession**.

One of the early proponents of the concept of climax was the American ecologist, Frederic Clements. In an important publication in 1916, he theorized that there was only one true climax community for any given climatic region. This so-called climatic climax would be the eventual end-point of all successions, whether they started after fire, deglaciation, or other disturbances, or even from a pond or **lake** filling in, and regardless of **soil** type. This monoclimax theory was criticized as too simple, and was challenged by other ecologists. A.G. Tansley proposed a more realistic polyclimax theory that accommodated the important successional influences of local soil type, topography, and disturbance history. In the early 1950s, R.H. Whittaker suggested that there were gradually varying climax types on the landscape, associated with continuous gradients of environmental variables. According to Whittaker, ecological communities vary continuously, and climax communities cannot be objectively divided into discrete types.

In a practical sense, it is not possible to identify the occurrence of a climax community. The climax condition may be suggested by relatively slow rates of change in the structure and function of old-growth communities, compared with earlier, more dynamic stages of succession. However, change in ecological communities is a universal phenomenon, so the climax state cannot be regarded as static. For example, even in old-growth communities, microsuccession is always occurring, associated perhaps with the death of individual trees. Moreover, if the frequency of return of stand-level disturbance events is relatively short, the old-growth or climax condition will not be reached.

Clingfish

Clingfish are about 100 **species** of small, ray-finned **bony fish** found primarily in tropical marine waters. They belong to the family Gobiesocidae in the order Gobiesociformes. Clingfish are shaped like tadpoles with a wide, flattened head; they have no scales and are covered with a thick coating of slime that makes them very slippery. Clingfish are characterized by a large suction disc formed by the union of the pelvic fins and adjacent folds of flesh. This disc allows clingfish to attach themselves to the bottom and, in this way, they are able to withstand strong **currents**. Clingfish have a single dorsal fin and no spines. Most species of clingfish are small, about 4 in (10 cm) or less in length, but the rocksucker (*Chorisochismus dentex*) clingfish of South **Africa** may grow as large as 12 in (30 cm) long.

Clingfish are most commonly found in the intertidal zone of oceans worldwide. A number of species inhabit Caribbean waters, and about 20 species are found along the Pacific coast of **North America**. Among these North American species is the northern clingfish (*Gobiesox meandricus*) which is found from California to Alaska; this species averages 6 in (15 cm) in length. Six other species of *Gobiesox* inhabit the North American coastal waters of the Atlantic **Ocean**. The most common of these species is the skilletfish (*G. strumosus*), a drab, dusky **fish** measuring 4 in (10 cm) in length.

Clone and cloning

A clone is a **molecule** (DNA), **cell**, or **organism** that is genetically identical to its parental molecule, cell, or organism. There are three types of cloning. One method, **gene** cloning, utilizes copying fragments of DNA for easier manipulation and study. Another cloning method involves producing genetically identical animals through a process called twinning. The final cloning method involves producing an organism through a nuclear transfer of genetic material from adult cell into an egg. Although before the **time** of **genetic engineering**, people cloned plants by grafts and stem cuttings, cloning involving complex laboratory techniques is a relatively recent scientific advance that is at the forefront of modern **biology**. Cloning has many promising applications in medicine, industry, **conservation**, and basic research.

History of cloning

Humans have manipulated **plant asexual reproduction** through methods like grafting and stem cuttings for more than 2,000 years. The modern era of laboratory cloning began in 1958 when F. C. Steward cloned carrot plants from mature single cells placed in a nutrient culture containing **hormones**. The first cloning of **animal** cells took place in 1964 when John B. Gurdon took the nuclei from intestinal cells of toad tadpoles and injected them into unfertilized eggs whose nuclei containing the original parents' genetic information had been destroyed with ultraviolet **light**. When the eggs were incubated, Gurdon found that 1-2% of the eggs developed into fertile, adult **toads**.

The first successful cloning of **mammals** was achieved nearly 20 years later when scientists in both Switzerland and the United States successfully cloned **mice** using a method similar to Gurdon's approach; but their method required one extra step. After the nuclei were taken from the embryos of one type of mouse, they were transferred into the embryos of another type of mouse who served as a surrogate mother that went

These calves were cloned in Japan from cells found in cow's milk. *AP/Wide World Photos. Reproduced by permission.*

through the birthing process to create the cloned mice. The cloning of cattle **livestock** was achieved in 1988 when nuclei from embryos of prize cows were transplanted to unfertilized cow eggs whose own nuclei had been removed.

In 1997, Scottish scientists cloned a **sheep** named Dolly, using cells from the mammary **glands** of an adult sheep and an egg cell from which the nucleus had been removed. This was the first time adult cells, rather than embryonic cells, had been used to clone a mammal. Since then, mice, cattle, **goats**, and other mammals have been cloned by similar methods. Some of these clones have been genetically altered so that the animals can produce drugs used in human medicine. Scientists are trying to clone organs for human transplant and may soon be able to clone human beings.

In 2001, scientists from Advanced Cell Technology cloned the first endangered animal, a bull gaur (a wild ox from **Asia**). The newborn died after two days due to **infection**. Currently, the same group is trying to clone another **species** of wild cattle, bantengs. In meantime, a group led by P. Loi cloned an endangered mouflon using somatic cells from post-mortem samples.

The cloning process

The cloning of specific genes can provide large numbers of copies of the gene for use in **genetics**, medical research, and systematics. Gene cloning begins by separating a specific length of DNA that contains the target gene. This fragment is then placed into another DNA molecule called the vector, which is then called a **recombinant DNA** molecule. The recombinant DNA molecule is used to transport the gene into a host cell, such as a bacterium or **yeast**, where the vector DNA replicates independently of the nuclear DNA to produce many copies, or clones, of the target gene. Recombinant DNA can also be introduced into plant or animal cells, but if the cells are to produce a particular protein (e.g., hormone or **enzyme**) for a long time, the introduced DNA molecule has to integrate into the nuclear DNA.

In nature, the simple organisms such as **bacteria**, yeast, and some other small organisms use cloning (asexual reproduction) to multiply, mainly by budding.

The cloning of animal cells in laboratories has been achieved through two main methods, twinning and nuclear transfer. Twinning occurs when an early stage embryo is divided in vitro and inserted into surrogate mothers to develop to term. Nuclear transfer relies on the transfer of the nucleus into a fertilised egg from which the nucleus was removed. The progeny from the first procedure is identical to each other while being different from their parents. In contrast progeny from the second procedure share only the nuclear DNA with the donor, but not mitochondrial DNA and in fact it is not identical to the donor.

In 1993, the first human embryos were cloned using a technique that placed individual embryonic cells (blastomeres) in a nutrient culture where the cells then divided into 48 new embryos. These experiments were conducted as part of some studies on in vitro (out of the body) **fertilization** aimed at developing fertilized eggs in test tubes, which could then be implanted into the wombs of women having difficulty becoming pregnant. However, these fertilized eggs did not develop to a stage that was suitable for transplantation into a human uterus.

Cloning cells intially held promise to produce many benefits in farming, medicine, and basic research. In agriculture, the goal is to clone plants containing specific traits that make them superior to naturally occurring plants. For example, in 1985, field tests were conducted using clones of plants whose genes had been altered in the laboratory (by genetic **engineering**) to produce resistance to **insects**, viruses, and bacteria. New strains of plants resulting from the cloning of specific traits could also lead to **fruits** and **vegetables** with improved nutritional qualities and longer shelf lives, or new strains of plants that can grow in poor **soil** or even under **water**. A cloning technique known as twinning could induce livestock to give **birth** to twins or even triplets, thus reducing the amount of feed needed to produce meat.

In medicine, gene cloning has been used to produce vaccines and hormones, for example: **insulin** for treating diabetes and of **growth hormones** for children who do not produce enough hormones for normal growth. The use of monoclonal antibodies in **disease** treatment and research involves combining two different kinds of cells (such as mouse and human **cancer** cells), to produce large quantities of specific antibodies, which are produced by the **immune system** to fight off disease.

Biopysical problems associated with cloning

Recent years revealed that some cloned animals suffer from age-related diseases and die prematurely.

Although, clones from other species still appear healthy, mice cloned using somatic cells have a higher than expected death **rate** from infections and hepatic failure.

Plagued with a chronic and progressive lung disease, veterinarians were forced to humanly euthanize Dolly in February 2003. Dolly lived approximately 60% of the normal lifespan of sheep (normally 11 to 12 years) and developed other conditions (e.g., chronic **arthritis**) much earlier than expected in a sheep's normal lifespan. Dolly's seemingly fragile health, along with more generalized fears of premature aging in cloned animals, renewed fears first raised by a study in 1999 that Dolly's telomeres were shorter than normal.

Telomeres are the physical ends eukaryotic chromosomes and play a role in the replication and stabilization of the chromosomes upon which they reside. Telomeres are synthesized by the enzyme telomerase. Telomerase is one of the factors believed to control the length of telomeres that may act as a biological clock for the cell.

According to the telomere theory of aging, during **DNA synthesis**, DNA polymerase fails to replicate all of the nucleic acids resulting in shortened telomeres—and hence shortened chromosomes—with each successive generation of **cell division**. Eventually, the cell will no longer divide and after enough critical regions have been deleted, the cell cycle will arrest and the cell will die.

Because telomerase is not active all the time, nor is it found in every cell of the body, the genetic regulation of telomerase is under intense study. Researchers have discovered that if the action of telomerase is interrupted, the telomere will abnormally shorten and thus accelerate the general aging process of the cell.

At a minimum, cloning eliminates genetic variation and thus, can be detrimental in the long term, leading to inbreeding and increased susceptibility to diseases. Although cloning also holds promise for saving certain rare breeds of animals from **extinction**, for some of them, finding the surrogate mothers can be a challenge best illustrated by failed trials with cloning of **pandas**.

KEY TERMS

Blastomeres—Individual embryonic cells.

Cell cycle—A cycle of growth and cellular reproduction that includes nuclear division (mitosis) and cell division (cytokinesis).

Chromosomes—The structures that carry genetic information in the form of DNA. Chromosomes are located within every cell and are responsible for directing the development and functioning of all the cells in the body.

DNA—Deoxyribonucleic acid; the genetic material in a cell.

Embryo—The earliest stage of animal development in the uterus before the animal is considered a fetus (which is usually the point at which the embryo takes on the basic physical form of its species).

Gene—A discrete unit of inheritance, represented by a portion of DNA located on a chromosome. The gene is a code for the production of a specific kind of protein or RNA molecule, and therefore for a specific inherited characteristic.

Genetic engineering—The manipulation of genetic material to produce specific results in an organism.

Genetics—The study of hereditary traits passed on through the genes.

Heredity—Characteristics passed on from parents to offspring.

Hybrid—The offspring resulting from combination of two different varieties of plants.

Nucleus (plural nuclei)—The part of the cell that contains most of its genetic material, including chromosomes and DNA.

The ethics of cloning

Despite the benefits of cloning and its many promising avenues of research, certain ethical questions concerning the possible abuse of cloning have been raised. At the heart of these questions is the idea of humans tampering with life in a way that could harm society, either morally, or in a real physical sense. Despite these concerns, there is little doubt that cloning will continue to be used. Cloning of particular genes for research and production of medicines is usually not opposed, **gene therapy** is much more controversial, while cloning of human beings is nearly uniformly opposed. **Human cloning** was banned in most countries and even the use of human embryonic **stem cells** is being reviewed in many countries. On the other hand, cloning plants or animals will probably continue.

See also Transgenics.

Resources

Books

Schaefer, Brian C. *Gene Cloning and Analysis: Current Innovations.* Norfolk: Horizon Press, 1997.

Wilmut, I., Kieth Campbell, and Colin Tudge. *The Second Creation: Dolly and the Age of Biological Control.* New York: Farrar Straus & Giroux, 2000.

Pence, Gregory C. *Who's Afraid of Human Cloning?* Landham, MD: Rowman and Littlefield, 1998..

Periodicals

Loi, Pasqualino, Grazyna Ptak, Barbara Barboni, Josef Fulka Jr., Pietro Cappai, and Michael Clinton, "Genetic Rescue of an Endangered Mammal by Cross-species Nuclear Transfer Using Post-mortem Somatic Cells." *Nature Biotechnology* (October 2001):962–964

Ogonuki, Narumi, et al. "Early Death of Mice Cloned from Somatic Cells." *Nature Genetics* (February 11, 2002):253-254

David Petechuk

Cloning, shotgun method *see*
Shotgun cloning

Closed curves

A closed **curve** is one which can be drawn without lifting the pencil from the **paper** and which ends at the **point** where it began. In Figure 1, A, B, and C are closed curves; D, E, and F are not.

Curve A is a **circle**. Although the starting point is not indicated, any of its points can be chosen to serve that purpose. Curve B crosses itself and is therefore not a "simple" closed curve, but it is a closed curve. Curve C

Figure 2. *Illustration by Hans & Cassidy. Courtesy of Gale Group.*

has a straight portion, but "curve" as used in **mathematics** includes straight lines as well as those that bend. Curve D, in which the tiny circle indicates a single missing point, fails for either of two reasons. If the starting point is chosen somewhere along the curve, then the one-point gap is a discontinuity. One cannot draw it without lifting the pencil. If one tries instead to start the curve at the point next to the gap, there is no point "next to" the gap. Between any two distinct points on a continuous curve there is always another point. Whatever point one chooses, there will always be an undrawn point between it and the gap-in fact an infinitude of such points. Curve E has a closed portion, but the tails keep the curve as a whole from being closed. Curve F simply fails to end where it began.

Curves can be described statically as sets of points. For example, the sets {P: PC = r, where *C* is a fixed point and *r* is a positive constant} and {(x,y): $x^2 + y^2 = r^2$} describe circles. The descriptions are static in the way a pile of bricks is static. Points either belong to the sets or they do not; except for set membership, there is no obvious connection between the points.

Although such descriptions are very useful, they have their limitations. For one thing, it is not obvious that they describe curves as opposed, say, to surfaces (in fact, the first describes a **sphere** or a circle, depending on the space in which one is working). For another, they omit any sense of continuity or movement. A planet's path around the **sun** is an **ellipse**, but it does

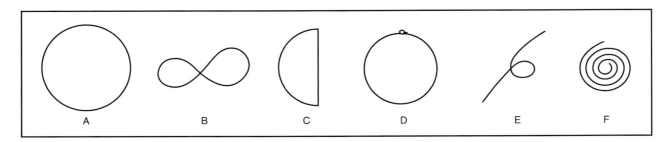

Figure 1. *Illustration by Hans & Cassidy. Courtesy of Gale Group.*

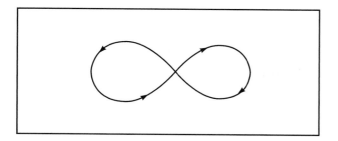

Figure 3. *Illustration by Hans & Cassidy. Courtesy of Gale Group.*

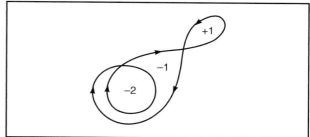

Figure 4. *Illustration by Hans & Cassidy. Courtesy of Gale Group.*

not occupy every point on the ellipse simultaneously, and it does not hop from point to point. It moves continuously along the curve as a **function** of **time**. A dynamic description of its path is more useful than a static one. Curves are therefore often described as paths, as the position of a point, P(t), which varies continuously as a function of time (or some analogous **variable**), which increases from some value a to value b. If the variable t does not represent time, then it represents a variable which increases continuously as time does, such as the **angle** formed by the **earth**, the sun, and a reference **star**.

However it is described, a curve is one-dimensional. It may be drawn in three-dimensional space, or on a two-dimensional surface such as a **plane**, but the curve itself is one-dimensional. It has one degree of freedom.

A person on a roller coaster experiences this. The car in which he is riding loops, dives, and twists, but it does not leave the track. As time passes, it makes its way from the starting point, along the track, and finally back to the starting point. Its position is a function of a single variable, called a "parameter," time.

A good way to describe a circle as a path is to use trigonometric functions: P(t) = (r cos t, r sin t), where P(t) is a point on the coordinate plane. This describes a circle of radius r with its center at the origin. If t is measured in degrees, varying over the closed **interval** [90,450], the curve will start at (0,1) and, since the sine and cosine of 90° are equal to the sine and cosine of 450°, will end there. If the interval over which t varies is the open interval (90,450), it will be the circle with one point missing shown in Figure 1, and will not be closed.

Not all curves can be represented as neatly as circles, but from a topological point of view, that doesn't matter. There are properties which certain closed curves share regardless of the way in which P(t) is represented.

One of the properties is that of being a "simple" closed curve. In Figure 1, curve B was not a simple curve; it crossed itself. There were two values of t, call them t_1 and t_2, for which $P(t_1) = P(t_2)$. If there are no such

values—if different values of t always give different points, then the curve is simple.

Another property is a sense of direction along the curve. On the roller coaster, as time passes, the car gets closer and closer to the end of the ride (to the relief of some of its passengers). On the circle described above, P(t) moves counterclockwise around it. If, in Figure 2, a toy train were started so that it traversed the left loop in a counterclockwise direction, it would traverse the other loop in a clockwise direction. This can be indicated with arrowheads.

Associated with each closed part of a closed curve is a "winding number." In the case of the circle above, the winding number would be +1. A person standing inside the circle watching P(t) move around the circle would have to make one complete revolution to keep the point in view. Since such a person would have to rotate counterclockwise, the winding number is arbitrarily considered positive.

In the same way, a person standing inside the left loop of Figure 2 would rotate once counterclockwise. Watching the point traverse the right loop would necessitate some turning first to the left and then to the right, but the partial revolutions would cancel out. The winding number for the left loop is therefore +1. The winding number for the right loop, based on the number of revolutions someone standing inside that loop would have to make, would be -1.

In Figure 3 the winding numbers are as shown.

The reader can check this. In Figure 4, although the curve is quite contorted, the winding number is +1.

This illustrates a fundamental fact about simple closed curves: their winding numbers are always +1 or -1.

Being or not being closed is a propery of a curve that can survive a variety of geometrical transformations. One can rotate a figure, stretch it in one direction or two, shrink it, shear it, or reflect it without breaking it open or closing it. One transformation that is a notable exception to this is a projection. When one projects a circle, as with a slide projector, one can, by adjusting the angle of the screen, turn that circle into an open **parabola** or **hyperbola**.

Resources

Books

Chinn, W.G., and N.E. Steenrod. *First Concepts of Topology.* Washington, DC: The Mathematical Association of America, 1966.

J. Paul Moulton

Closed interval *see* **Interval**

Closure property

"Closure" is a property which a set either has or lacks with respect to a given operation. A set is closed with respect to that operation if the operation can always be completed with elements in the set.

For example, the set of even **natural numbers**, 2, 4, 6, 8,..., is closed with respect to addition because the sum of any two of them is another even natural number. It is not closed with respect to **division** because the quotients 6/2 and 4/8, for instance, cannot be computed without using odd numbers or fractions.

Knowing the operations for which a given set is closed helps one understand the nature of the set. Thus one knows that the set of natural numbers is less versatile than the set of **integers** because the latter is closed with respect to **subtraction**, but the former is not. Similarly one knows that the set of **polynomials** is much like the set of integers because both sets are closed under addition, **multiplication**, negation, and subtraction, but are not closed under division.

Particularly interesting examples of closure are the positive and **negative** numbers. In mathematical structure these two sets are indistinguishable except for one property, closure with respect to multiplication. Once one decides that the product of two positive numbers is positive, the other rules for multiplying and dividing various combinations of positive and negative numbers follow. Then, for example, the product of two negative numbers must be positive, and so on.

The lack of closure is one reason for enlarging a set. For example, without augmenting the set of rational numbers with the irrationals, one cannot solve an equation such as $x^2 = 2$, which can arise from the use of the **pythagorean theorem**. Without extending the set of **real numbers** to include imaginary numbers, one cannot solve an equation such as $x^2 + 1 = 0$, contrary to the fundamental **theorem** of **algebra**.

Closure can be associated with operations on single numbers as well as operations between two numbers. When the Pythagoreans discovered that the **square root** of 2 was not rational, they had discovered that the rationals were not closed with respect to taking roots.

Although closure is usually thought of as a property of sets of ordinary numbers, the concept can be applied to other kinds of mathematical elements. It can be applied to sets of rigid motions in the **plane**, to vectors, to matrices, and to other things. For example, one can say that the set of three-by-three matrices is closed with respect to addition.

Closure, or the lack of it, can be of practical concern, too. Inexpensive, four-function calculators rarely allow one to use negative numbers as inputs. Nevertheless, if one subtracts a larger number from a smaller number, the **calculator** will complete the operation and display the negative number which results. On the other hand, if one divides 1 by 3, the calculator will display 0.333333, which is close, but not exact. If an operation takes a calculator beyond the numbers it can use, the answer it displays will be wrong, perhaps significantly so.

Cloud chambers *see* **Particle detectors**

Clouds

All clouds are a form of **water**. Clouds are condensed atmospheric moisture in the form of minute water droplets or **ice** crystals. The creation of a cloud begins at ground level. The **sun** heats the earth's surface, the warm ground heats the air, which rises. The air contains variable amounts of water, as vapor, that has evaporated from bodies of water and plants. Air at ground level is denser than air higher up, and as the warm air rises, it expands and becomes less dense. Expansion cools the air and as the air cools, the water vapor that is present in the air, condenses into tiny microscopic droplets. Cloud formation depends on how much water is in the atmos-

Cirrus clouds (top) and cumulus clouds (bottom). © *John Deeks, National Audubon Society Collection/Photo Researchers, Inc. Reproduced with permission.*

phere, the **temperature**, the air current, and topography. If there is no water, no clouds can form. If condensation occurs below the freezing point, the cloud is made of ice crystals. Warm and cold air fronts, as well as topography can control how air rises. Clouds that form during vigorous uplift of air have a tall, stacked appearance and clouds formed by gentle uplift of air currents have a flat or stratified appearance. One can make short-term forecasts by observing clouds, as any change in the way a cloud looks indicates a change in the **weather**.

Classification

A couple of hundred years ago clouds were not identified by name. Luke Howard, an English pharmacist and amateur naturalist, developed a system of classification (from Latin) for clouds in 1803. Howard categorized clouds into three major groups: cumulus (accumulate or piled up heaps and puffs), cirrus (fibrous and curly), and stratus (stretched out and layered). To further describe clouds, he combined those terms and used other descriptive words such as, alto (high), and nimbus (rain). Today, the International Cloud Classification is based on Howard's system.

There are three basic forms of clouds: cirrus, cumulus, and stratus. All clouds are either purely these forms, or a combination or modification of the basic forms. Since there is more water vapor at lower elevations, lower clouds appear denser than higher, thin clouds.

Cloud categories

Today, there are 10 characteristic forms or genera of clouds recognized by the International Cloud Classification, and there are three height categories with an established altitude range for each category. Low-level clouds range from the surface to 6,500 ft (2,000 m), mid-level from 6,500-23,000 ft (2,000-7,000 m), and high-level, generally above 20,000 ft (6,000 m). Below is a brief description of each category and their genera.

Nimbus category

There are two genera of rain clouds, cumulonimbus and nimbostratus. Nimbostratus clouds are usually mid level clouds, thick, dark, gray, and sometimes seen with "virga" or skirts of rain trailing down. These clouds are made of water droplets that produce either rain or snow.

Cirrocumulus clouds fill the evening sky. *Photograph by Robert J. Huffman. Field Mark Publications. Reproduced by permission.*

Cumulonimbus clouds are **thunderstorm** clouds, and arise from cumulus clouds that have reached a great height. When the cloud normally reaches the height of the tropopause above, it flattens out, resembling an anvil. All phases of water, gas, liquid, and solid, are contained in these clouds. There are powerful updrafts and downdrafts that can create violent storms.

High clouds

The altitude range for these clouds is 16,500-45,000 ft (5,032-13,725 m) but they usually form between 20,000-25,000 ft (6,000-7,500 m). There are three genera of high level clouds and they are all labeled with the term cirrus. Cirrus clouds are the highest clouds, forming around 30,000 ft (9,150 m). They are totally made of ice crystals (or needles of ice) because they form where freezing temperatures prevail. Pure cirrus clouds look wispy, with a slight curl, and very white. Because of their appearance, they are often called mares' tails. Cir-

rocumulus clouds, the least common cloud, are small, white or pale gray, with a rippled appearance. Sometimes they appear like a sky full of **fish** scales; this effect is called a mackerel sky. These clouds usually cover a large area. They form around 20,000-25,000 ft (6,000-7,500 m) and are made of either supercooled water droplets or ice crystals. Cirrostratus also form at 20-25,000 ft, but are made completely of ice crystals. They usually cover the sky as a thin veil or sheet of white. These clouds are responsible for the halos that occur around the sun or **moon**. The term "on cloud nine" (feeling of euphoria) is derived from the fact that the highest clouds are labeled category nine.

Middle level clouds

The mid level clouds 6,500-23,000 ft (2,000-7,000 m) typically have the prefix "alto" added to the two genera in this category. Altostratus clouds appear as a uniform bluish or gray sheet covering all, or large areas of the sky. The sun or moon may be totally covered or shine through very weakly. These clouds are complex as they are usually layered, with ice crystals at the higher, top layers, ice and snow in the middle, and water droplets in the lower layers. Altostratus clouds often yield **precipitation**. Altocumulus are elliptical, dense, fluffy balls. They are seen as singular units or as closely bunched groups in a clear sky. When the sun or moon shines through these clouds, one can sometimes see the "sun's rays" or corona.

Low level clouds

There are three genera in the low level (surface to 6,500 ft [2,000 m]). Stratus clouds are usually the lowest of the three genera. Stratus clouds blanket the sky and usually appear gray. They form when a large mass of air rises slowly and the water vapor condenses as the air becomes cooler, or when cool air moves in over an area close to ground level. These clouds often produce mist or drizzle. **Fog** is a stratus cloud at ground level. Cumulus clouds have flat bases, are vertically thick, and appear puffy. Inside a cumulus cloud are updrafts that create the cloud's appearance. They form when a column of warm air rises, expands, cools, and condenses. Cumulus clouds occur primarily in warm weather. They consist of water droplets and appear white because the sunlight reflects off the droplets. Thick clouds appear darker at the bottom because the sunlight is partially blocked. Cumulus clouds can develop into cumulonimbus clouds. Stratocumulus clouds are large, grayish masses, spread out in a puffy layer. Sometimes they appear as rolls. These clouds appear darker and heavier than the altocumulus cloud. They can transform into nimbostratus clouds.

Unusual clouds

Beside the basic cloud types, there are subgroups and some unusual cloud formations. Terms such as humulus (fair weather), and congestus (rainshower) are used to further describe the cumulus genus. Fractus (jagged), castellanus (castle shaped), and uncinus (hook shaped) are other descriptive terms used together with some basic cloud types. In mountainous regions, lenticular clouds are a common sight. They form only over mountain peaks and resemble a stack of different layers of cloud matter. Noctilucent clouds form only between sunset and sunrise, and are only seen in high latitude countries. Contrails (condensation trails) are artificial clouds formed from the engine exhaust of high altitude **aircraft**.

See also Weather forecasting; Weather modification.

Resources

Books

Day, John A., and Vincent J. Schaefer. *Peterson First Guide to Clouds and Weather.* Boston, Houghton Mifflin Co., 1998.
Roth, Charles E. *The Sky Observer's Guidebook.* New York: Prentice Hall Press, 1986.
Rubin Sr., Louis D. and Jim Duncan. *The Weather Wizard's Cloud Book.* Chapel Hill: Algonquin Books of Chapel Hill, 1984.
Shafer, Vincent J., and John A. Day. *A Field Guide to the Atmosphere.* Boston: Houghton Mifflin Co., 1981.

Christine Minderovic

Clover *see* **Legumes**
Cloves *see* **Myrtle family (Myrtaceae)**

Club mosses

Club mosses, also called **lycophytes**, are flowerless and seedless plants in the family Lycopodiaceae, that belong to an ancient group of plants of the division Lycophyta. The lycophytes were one of the dominant plants during the Coal age (360-286 million years ago) and many were shrubs or large trees. By 250 million years ago, most of the woody **species** had died out. Between 10 and 15 living genera have been recognized, consisting of about 400 species. Lycopodiaceae are cosmopolitan, occurring in arctic to tropical regions. Nowhere do they dominate **plant** communities today as they did in the past. In arctic and temperate regions, club mosses are terrestrial; whereas in the tropics, they are mostly epiphytes near the tops of trees and seldom seen. The classification of club mosses has changed radically in recent years. Most temperate species were grouped within the genus *Lycopodium*, from the Greek *lycos*, meaning wolf, and *pous* meaning foot, in an imaginative reference to the resemblance in some species of the densely-leaved branch tips to a wolf's foot. However, it is now clear that fundamental differences exist among the club mosses with respect to a variety of important characters. Seven genera and 27 species have been recognized in the **flora** of **North America**. Four of the common genera, formerly all within the genus *Lycopodium*, are *Lycopodium*, the **tree** club mosses (6 species), *Diphasiastrum*, the club mosses (5 species), *Huperzia*, the fir mosses (7 species), and *Lycopodiella*, the bog club mosses (6 species); all are terrestrial. The sole epiphytic member of the club **moss** family in North America is the hanging fir moss (*Phlegmariurus dichotomus*), which is common in subtropical and tropical Central and **South America**. In North America it is known only from Big Cypress Swamp, Florida.

Unlike some of the other ancient plants, such as liverworts, the sporophytes of club mosses are clearly differentiated into root, stem, and leaves. All living club mosses are perennial herbs that typically possess underground stems that branch and give rise to shoots that rarely exceed 7.9 in (20 cm) in height. Although the photosynthetic organs of club mosses are commonly called leaves, technically speaking they are microphylls and differ from true leaves in that they contain only one unbranched strand of conducting **tissue**. The "micro" in the name does not necessarily mean that these photosynthetic organs are small, in fact some microphylls of extinct tree lycophytes were 3.3 ft (1 m) long. Micro refers to the **evolution** of the structure from an initially very small flap of tissue that grew along the stem of primitive leafless plants, and that eventually, through evolution, grew larger and had a strand of conducting tissue enter it to produce the modern type of microphyll. Microphylls are generally needle-like, spear-shaped, or ovate and arranged spirally along the stem, but occasionally appear opposite or whorled. The habit of evergreen leaves on stems that in some species run along the ground has given rise to the common name of ground or running **pines**. Stems have a primitive vascular tissue composed of a solid, central column.

Spores, all of one type, are produced in sporangia that occur either singly on fertile leaves (sporophylls) that look much like non-fertile leaves or on modified leaves that are tightly appressed on the tip of a branch producing a cone or club-like structure, hence the name club moss. The cones may or may not be stalked. The spores germinate to produce bisexual gametophytes that are either green and photosynthetic on the **soil** surface or are underground and non-photosynthetic, in the latter

KEY TERMS

Gametophyte—Individual plant containing only one set of chromosomes per cell that produces gametes, i.e., reproductive cells that must fuse with other reproductive cells to produce a new individual.

Sporophyte—The diploid, spore-producing generation in a plant's life cycle.

case deriving some of their nourishment from mycorrhyzae. The maturation of a gametophyte may require six to 15 years. Biflagellated sperm are produced in an antheridium (male reproductive **organ**) and an egg is produced in a flask-shaped archegonium (female reproductive organ). **Water** is required for the sperm to swim to another gametophyte and down the neck of an archegonium to reach the egg at the bottom. The young sporophyte produced after **fertilization** may remain attached for many years, and in some species the gametophyte may continue to grow and produce a succession of young sporophytes.

Club mosses are ecologically minor components of all the ecosystems in which they occur. Their economic importance is also slight. Many club mosses produce masses of sulpher-colored spores that are highly inflammable and were therefore once used as a constituent of flash powder in early **photography** and in fireworks. The spores were also formerly used by pharmacists to coat pills. In parts of eastern North America, local cottage industries have sprung up to collect club mosses in order to make the most elegant of Christmas wreaths. Spores of common club moss (*Lycopodium clavatum*) are used by paleoecologists to calibrate the number of fossil pollen grains in samples of **lake** mud. Some Druid sects considered club mosses to be sacred plants and had elaborate rituals to collect club mosses and display them on their alters for good luck.

See also Liverwort; Spore.

Resources

Books

Flora of North America Editorial Committee, eds. *Pteridophytes and Gymnosperms*. Vol. 2 of *Flora of North America*. New York: Oxford University Press, 1993.

Raven, Peter, R.F. Evert, and Susan Eichhorn. *Biology of Plants*. 6th ed. New York: Worth Publishers Inc., 1998.

Les C. Cwynar

Coal

Coal is a naturally occurring combustible material consisting primarily of the element **carbon**, but with low percentages of solid, liquid, and gaseous hydrocarbons and other materials, such as compounds of **nitrogen** and **sulfur**. Coal is usually classified into the sub-groups known as anthracite, bituminous, lignite, and peat. The physical, chemical, and other properties of coal vary considerably from **sample** to sample.

Origins of coal

Coal forms primarily from ancient **plant** material that accumulated in surface environments where the complete decay of organic **matter** was prevented. For example, a plant that died in a swampy area would quickly be covered with **water**, silt, **sand**, and other sediments. These materials prevented the plant debris from reacting with **oxygen** and decomposing to **carbon dioxide** and water, as would occur under normal circumstances. Instead, **anaerobic** bacteria (**bacteria** that do not require oxygen to live) attacked the plant debris and converted it to simpler forms: primarily pure carbon and simple compounds of carbon and **hydrogen** (hydrocarbons). Because of the way it is formed, coal (along with **petroleum** and **natural gas**) is often referred to as a fossil fuel.

The initial stage of the decay of a dead plant is a soft, woody material known as peat. In some parts of the world, peat is still collected from boggy areas and used as a fuel. It is not a good fuel, however, as it burns poorly and with a great deal of smoke.

If peat is allowed to remain in the ground for long periods of time, it eventually becomes compacted as layers of sediment, known as overburden, collect above it. The additional **pressure** and **heat** of the overburden gradually converts peat into another form of coal known as lignite or brown coal. Continued compaction by overburden then converts lignite into bituminous (or soft) coal and finally, anthracite (or hard) coal. Coal has been formed at many times in the past, but most abundantly during the Carboniferous Age (about 300 million years ago) and again during the Upper Cretaceous Age (about 100 million years ago).

Today, coal formed by these processes is often found in layers between layers of **sedimentary rock**. In some cases, the coal layers may lie at or very near the earth's surface. In other cases, they may be buried thousands of feet or meters under ground. Coal seams range from no more than 3-197 ft (1-60 m) or more in thickness. The location and configuration of a coal seam determines the method by which the coal will be mined.

Composition of coal

Coal is classified according to its heating value and according to its relative content of elemental carbon. For example, anthracite contains the highest proportion of pure carbon (about 86%-98%) and has the highest heat value—13,500–15,600 Btu/lb (British thermal units per pound)—of all forms of coal. Bituminous coal generally has lower concentrations of pure carbon (from 46% to 86%) and lower heat values (8,300–15,600 Btu/lb). Bituminous coals are often sub-divided on the basis of their heat value, being classified as low, medium, and high volatile bituminous and sub-bituminous. Lignite, the poorest of the true coals in terms of heat value (5,500-8,300 Btu/lb) generally contains about 46%-60% pure carbon. All forms of coal also contain other elements present in living organisms, such as sulfur and nitrogen, that are very low in absolute numbers, but that have important environmental consequences when coals are used as fuels.

Properties and reactions

By far the most important property of coal is that it combusts. When the pure carbon and hydrocarbons found in coal burn completely only two products are formed, carbon dioxide and water. During this chemical reaction, a relatively large amount of **energy** is released. The release of heat when coal is burned explains the fact that the material has long been used by humans as a source of energy, for the heating of homes and other buildings, to run ships and trains, and in many industrial processes.

Environmental problems associated with the burning of coal

The complete **combustion** of carbon and hydrocarbons described above rarely occurs in nature. If the **temperature** is not high enough or sufficient oxygen is not provided to the fuel, combustion of these materials is usually incomplete. During the incomplete combustion of carbon and hydrocarbons, other products besides carbon dioxide and water are formed, primarily **carbon monoxide**, hydrogen, and other forms of pure carbon, such as soot.

During the combustion of coal, minor constituents are also oxidized. Sulfur is converted to **sulfur dioxide** and sulfur trioxide, and nitrogen compounds are converted to nitrogen oxides. The incomplete combustion of coal and the combustion of these minor constituents results in a number of environmental problems. For example, soot formed during incomplete combustion may settle out of the air and deposit an unattractive coating on homes, cars, buildings, and other structures. Carbon monoxide formed during incomplete combustion is a

A coal seam in northwest Colorado. *JLM Visuals. Reproduced by permission.*

toxic gas and may cause illness or death in humans and other animals. Oxides of sulfur and nitrogen react with water vapor in the atmosphere and then are precipitated out as **acid rain**. Acid rain is thought to be responsible for the destruction of certain forms of plant and **animal** (especially **fish**) life.

In **addition** to these compounds, coal often contains a few **percent** of mineral matter: quartz, calcite, or perhaps clay **minerals**. These do not readily combust and so become part of the ash. The ash then either escapes into the atmosphere or is left in the combustion vessel and must be discarded. Sometimes coal ash also contains significant amounts of **lead**, **barium**, arsenic, or other compounds. Whether air borne or in bulk, coal ash can therefore be a serious environmental hazard.

Coal mining

Coal is extracted from the **earth** using one of two major techniques, sub-surface or surface (strip) **mining**. The former method is used when seams of coal are located at significant depths below the earth's surface. The first step in sub-surface mining is to dig vertical tunnels into the earth until the coal seam is reached. Horizontal tunnels are then constructed laterally off the vertical tunnel. In many cases, the preferred method of mining coal by this method is called room-and-pillar mining. In this method, vertical columns of coal (the pillars) are left in place as coal around them is removed. The pillars hold up the ceiling of the seam preventing it from collapsing on miners working around them. After the mine has been abandoned, however, those pillars may often collapse, bringing down the ceiling of the seam and causing **subsidence** in land above the old mine.

Surface mining can be used when a coal seam is close enough to the earth's surface to allow the overburden to be removed economically. In such a case, the first step is to strip off all of the overburden in order to reach the coal itself. The coal is then scraped out by huge power shovels, some capable of removing up to 100 cubic meters at a time. Strip mining is a far safer form of coal mining, but it presents a number of environmental problems. In most instances, an area that has been strip mined is terribly scarred, and restoring the area to its original state is a long and expensive procedure. In addition, any water that comes in contact with the exposed coal or overburden may become polluted and require treatment.

Resources

Coal is regarded as a non-renewable resource, meaning that it was formed at times during the earth's history, but significant amounts are no longer forming. Therefore, the amount of coal that now exists below the earth's surface is, for all practical purposes, all the coal that humans have available to them for the foreseeable future. When this supply of coal is used up, humans will find it necessary to find some other substitute to meet their energy needs.

Large supplies of coal are known to exist (proven reserves) or thought to be available (estimated resources) in **North America**, the former Soviet Union, and parts of **Asia**, especially China and India. According to the most recent data available, China produces the largest amount of coal each year, about 22% of the world's total, with the United States 19%, the former members of the Soviet Union 16%, Germany 10%, and Poland 5% following. China is also thought to have the world's largest estimated resources of coal, as much as 46% of all that exists. In the United States, the largest coal-producing states are Montana, North Dakota, Wyoming, Alaska, Illinois, and Colorado.

Uses

For many centuries, coal was burned in small stoves to produce heat in homes and factories. Today, the most important use of coal, both directly and indirectly, is still as a fuel. The largest single consumer of coal as a fuel is the electrical power industry. The combustion of coal in power generating plants is used to make steam which, in turn, operates turbines and generators. For a period of more than 40 years, beginning in 1940, the amount of coal used in the United States for this purpose doubled in every decade. Coal is no longer widely used to heat homes and buildings, as was the case a half century ago, but it is still used in industries such as **paper** production, cement and ceramic manufacture, **iron** and **steel** production, and chemical manufacture for heating and for steam generation.

KEY TERMS

Anthracite—Hard coal; a form of coal with high heat content and high concentration of pure carbon.

Bituminous—Soft coal; a form of coal with less heat content and pure carbon content than anthracite, but more than lignite.

British thermal unit (Btu)—A unit for measuring heat content in the British measuring system.

Coke—A synthetic fuel formed by the heating of soft coal in the absence of air.

Combustion—A form of oxidation that occurs so rapidly that noticeable heat and light are produced.

Gasification—Any process by which solid coal is converted to a gaseous fuel.

Lignite—Brown coal; a form of coal with less heat content and pure carbon content than either anthracite or bituminous coal.

Liquefaction—Any process by which solid coal is converted to a liquid fuel.

Peat—A primitive form of coal with less heat content and pure carbon content than any form of coal.

Strip mining—A method for removing coal from seams that are close to the earth's surface.

Another use for coal is in the manufacture of coke. Coke is nearly pure carbon produced when soft coal is heated in the absence of air. In most cases, one ton of coal will produce 0.7 ton of coke in this process. Coke is of value in industry because it has a heat value higher than any form of natural coal. It is widely used in steel making and in certain chemical processes.

Conversion of coal

A number of processes have been developed by which solid coal can be converted to a liquid or gaseous form for use as a fuel. Conversion has a number of advantages. In a liquid or gaseous form, the fuel may be easier to transport, and the conversion process removes a number of impurities from the original coal (such as sulfur) that have environmental disadvantages.

One of the conversion methods is known as gasification. In gasification, crushed coal is reacted with steam and either air or pure oxygen. The coal is converted into a complex mixture of gaseous hydrocarbons with heat values ranging from 100 Btu to 1000 Btu. One suggestion has been to construct gasification systems within a coal mine, making it much easier to remove the coal (in a gaseous form) from its original seam.

In the process of liquefaction, solid coal is converted to a petroleum-like liquid that can be used as a fuel for motor vehicles and other applications. On the one hand, both liquefaction and gasification are attractive technologies in the United States because of our very large coal resources. On the other hand, the wide availability of raw coal means that new technologies have been unable to compete economically with the natural product.

During the last century, coal oil and coal gas were important sources of fuel for heating and lighting homes. However, with the advent of natural gas, coal distillates quickly became unpopular, since they were somewhat smoky and foul smelling.

See also Air pollution; Hydrocarbon.

Resources

Books

Gorbaty, Martin L., John W. Larsen, and Irving Wender, eds. *Coal Science.* New York: Academic Press, 1982.

Periodicals

Jia, Renhe. "Chemical Reagents For Enhanced Coal Flotation." *Coal Preparation* 22, no. 3 (2002): 123-149.
Majee, S.R. "Sources Of Air Pollution Due To Coal Mining And Their Impacts In The Jaharia Coal Field." *Environment International* 26, no. 1-2 (2001): 81-85.
Ryan III, T.W. "Coal-Fueled Diesel Development: A Technical Review." *Journal of Engineering for Gas Turbines and Power* (1994).

David E. Newton

Coast and beach

The coast and beach, where the continents meet the sea, are dynamic environments where agents of **erosion** vie with processes of deposition to produce a set of features reflecting their complex interplay and the influences of changes in **sea level**, climate, sediment supply, etc. "Coast" usually refers to the larger region of a **continent** or **island** that is significantly affected by its proximity to the sea, whereas "beach" refers to a much smaller region, usually just the areas directly affected by wave action.

Observing erosion and deposition

Earth is constantly changing. **Mountains** are built up by tectonic forces, weathered, and eroded away. The erosional debris is deposited in the sea. In most places

The coast of Long Island, New York. *Photograph by Claudia Parks. Stock Market. Reproduced by permission.*

these changes occur so slowly they are barely noticeable, but at the beach, they are often observable.

Most features of the beach environment are temporary, steady state features. To illustrate this, consider an excavation in **soil**, where **groundwater** is flowing in, and being pumped out by mechanical pumps. The level of the **water** in the hole is maintained because it is being pumped out just as fast as it is coming in. It is in a steady state, but changing either **rate** will promptly change the level of the water. A casual observer may fail to notice the pumps, and erroneously conclude that the water in the hole is stationary. Similarly, a casual observer may think that the **sand** on the beach is stationary, instead of in a steady state. The size and shape of a spit, which is a body of sand stretching out from a point, **parallel** to the shore, is similar to the level of the water in this example. To stay the same, the rate at which sand is being added to the spit must be exactly balanced by the rate at which it is being removed. Failure to recognize this has often led to serious degradation of the coastal environment.

Sea level is the point from which elevation is measured. A minor change in elevation high on a mountain is undetectable without sophisticated surveying equipment. The environment at 4,320 ft (1,316.7 m) above sea level is not much different from that at 4,310 ft (1,313.6 m).

The same 10-ft (3-m) change in the elevation of a beach would expose formerly submerged land, or inundate formerly exposed land, making it easy to notice. Not only is the environment different, but also the dominant geologic processes are different: Erosion occurs above sea level, deposition occurs below sea level. As a result, coasts where the land is rising relative to sea level (emergent coasts) are usually very different from those where the land is sinking relative to sea level (submergent coasts).

Emergent coasts

If the coast rises, or sea level goes down, areas that were once covered by the sea will emerge and form part of the landscape. The erosive action of the waves will attack surfaces that previously lay safely below them. This wave attack occurs at sea level, but its effects extend beyond sea level. Waves may undercut a cliff, and eventually the cliff will fail and fall into the sea, removing material from higher elevations. In this manner, the cliff retreats, while the beach profile is extended at its base. The rate at which this process continues depends on the material of the cliff and the profile of the beach. As the process continues, the gradual slope of the bottom extends farther and farther until most waves break far from shore and the rate of cliff retreat slows, resulting in a stable profile that may persist for long periods of time. Eventually another episode of **uplift** is likely to occur, and the process repeats.

Emergent coasts, such as the coast along much of California, often exhibit a series of terraces, each consisting of a former beach and wave cut cliff. This provides evidence of both the total uplift of the coast, and its incremental nature.

Softer **rocks** erode more easily, leaving resistant rock that forms points of land called headlands jutting out into the sea. Subsurface depth contours mimic that of the shoreline, resulting in wave refraction when the change in depth causes the waves to change the direction of their approach. This refraction concentrates wave **energy** on the headlands, and spreads it out across the areas in between. The "pocket beaches" separated by jagged headlands, which characterize much of the scenic coastline of Oregon and northern California were formed in this way. Wave refraction explains the fact that waves on both sides of a headland may approach it from nearly opposite directions, producing some spectacular displays when they break.

Submergent coasts

If sea level rises, or the elevation of the coast falls, formerly exposed topography will be inundated. Valleys carved out by **rivers** will become estuaries like Chesa-peake Bay. Hilly terrains will become collections of islands, such as those off the coast of Maine.

The ability of rivers to transport sediment depends on their velocities. When rivers flow into a deep body of water, they slow down and **deposit** their sediment in what will eventually become a **delta**. Thus, the **flooding** of estuaries causes deposition further inland. As the estuary fills in with sediment, the depth of the water will decrease, and the **velocity** of the water flowing across the top of the delta will increase. This permits further sediment transport. The delta builds out toward, and eventually into, the sea. The additional load of all the sediment may cause the crust of the earth to deform, submerging the coast further.

The sand budget

Wave action moves incredible amounts of sand. As waves approach shallow water, they slow down because of **friction** with the bottom, then get steeper, and finally break. It is during this slowing and breaking that sand is transported. When waves reach the shore, the approach is almost straight on, so that the wave front is nearly parallel to the shore as it breaks.

When a breaking wave washes up onto the beach at a slight angle it moves sand on the beach with it. This movement is mostly towards shore, but also slightly down the beach. When the water sloshes back, it goes directly down the slope, without any oblique component. As a result, sand moves in a zigzag path with a net **motion** parallel to the beach. This is called "longshore drift." Although most easily observed and understood in the swash zone, the area of the beach that gets alternately wet and dry with each passing wave, longshore drift is active in any water shallow enough to slow waves down.

Many features of sandy coasts are the result of longshore drift. Spits build out from projecting land masses, sometimes becoming hooked at their end, as sand moves parallel to the shore. At Cape Cod, Massachusetts, glacial debris deposited thousands of years ago is still being eroded and redistributed by wave action.

An artificial jetty or "groin" can trap sand on one side of it, broadening the beach there. On the other side, however, wave action will transport sand away. Because of the jetty it will not be replenished, and erosion of the beach will result.

The magnitude and direction of transport of longshore drift depends on the strength and direction of approach of waves, and these may vary with the season. A beach with a very gentle slope, covered with fine sand every July may be a steep pebble beach in February.

A rocky coastline at Boardman State Park, Oregon. *Photograph by George Ranalli. Photo Researchers, Inc. Reproduced by permission.*

Barrier islands

Long, linear islands parallel to the shore are common along the Atlantic coast. Attractive sites for resorts and real estate developments, these **barrier islands** are in flux. A hurricane can drive **storm** waves over low spots, cutting islands in two. Conversely, migration of sand can extend a spit across the channel between two islands, merging them into one.

Interruptions in sand supply can result in erosion. This has happened off the coast of Maryland, where Assateague Island has become thinner and moved shoreward since jetties were installed at Ocean City, just to the north.

Society and the beach environment

Coastal areas of the continental United States comprise only 17% of the land area of the country, but house over one-half of the population. Currently, the population of this zone is above 139 million and expected to rise to 165 million by the year 2015, a rate of growth greater than that of the country as a whole. The coast is attractive for a wide variety of reasons and economic growth of the zone typically follows the growth of the population. Unfortunately, the impacts on the coastal environment are not so positive. Environmental degradation accompanies shoreline development in a variety of forms. Furthermore, development within the coastal zone is increasingly located within high-risk areas of natural or man-made shore degradation.

In many cases, the public or property developer in the coastal region has an incomplete understanding of the coastal environment. Often, the dynamic nature of the beach environment is not properly respected. At higher elevations, where rates of erosion and deposition are much slower, man can construct huge hills to support interstate highways, level other hills to make parking lots, etc., expecting the results to persist for centuries, or at least decades. In a beach environment, however, modifications are ephemeral. Maintaining a parking lot where winds would produce a **dune** requires removal of tons of sand every year. Even more significantly, because the flow of sediment is so great, modifications intended to have only a local, beneficial effect may influence erosion and deposition far down the beach, in ways which are not beneficial.

Coastal retreat is a significant issue along many areas of coastline. Efforts are ongoing to quantify the rate of retreat along the coast, to designate areas of particular risk, and to match appropriate uses with the location. Even in areas with minimal retreat, the movement of sediment along the shore can impact the property owner significantly. Utilization of engineered **shoreline protection** can affect nearby properties. Sediment budgets for a shoreline can be impacted by the damming of rivers upstream. Even artificial means of **beach nourishment** can have unintended environmental impacts. One might be able to protect the beach in front of a beach house by installing a **concrete** barrier, but this might result in eroding the supports to the highway giving access to the beach house.

The costs of shoreline protection are high, yet they are rarely considered when the development of a coastal property is contemplated. Furthermore, these costs are often borne by the taxpayer rather than the property owner. The long-range outlook for all such costs is that they will ultimately exceed the value of the property and are likely to be exacerbated by rising sea levels associated with **global climate** alterations.

Many scientists encourage that the migratory nature of the coast should be recognized, and claim it is unwise to assume that structures built upon a moving coastline are immovable.

See also Beach nourishment; Ocean; Shoreline protection; Tides.

Resources

Books

Bird, E. C. F. *Submerging Coasts: The Effects of a Rising Sea Level on Coastal Environments.* Chichester, New York: John Wiley & Sons, 1993.

Carter, R. W. G. *Coastal Environments: An Introduction to the Physical, Ecological, and Cultural Systems of Coastlines.* London; San Diego: Academic Press, 1988.

Carter, R. W. G., and C. D. Woodroffe. *Coastal Evolution.* Cambridge: Cambridge University Press, 1994.

Griggs, Gary, and Lauret Savoy, eds. *Living with the California Coast.* Durham, NC: Duke University Press, 1985.

KEY TERMS

. . . .

Emergent coast—A coast rising relative to sea level, characterized by exposed terraces consisting of older wave cut cliffs and formerly submerged beaches.

Longshore drift—Movement of sand parallel to the shore, caused by waves approaching the shore obliquely, slowing and breaking.

Refraction—The bending of light that occurs when traveling from one medium to another, such as air to glass or air to water.

Submergent coast—A coast sinking relative to sea level, characterized by drowned river valleys.

Other

Geological Society of America (GSA). *Beach Nourishment: The Wave of the Future for Erosion Control (Part A).* Southeast Section Annual Meeting Abstracts. 2001 [cited October 19, 2002]. <gsa.confex.com/gsa/2001SE/finalprogram/session_77.htm>.

National Oceanic and Atmospheric Administration (NOAA). *Managing Coastal Resources.* NOAA's State of the Coast Report, 1998 [cited October 19, 2002]. <http://state-of-coast.noaa.gov/bulletins/html/crm_13/crm.html>.

National Oceanic and Atmospheric Administration (NOAA). *Monitoring the Coastal Environment.* NOAA's State of the Coast Report. 1998 [cited October 19, 2002]. <http://state-of-coast.noaa.gov/bulletins/html/mcwq_12/mcwq.html> (October 18, 2002).

National Oceanic and Atmospheric Administration (NOAA). *Population at Risk from Natural Hazards.* NOAA's State of the Coast Report. 1998 [cited October 19, 2002]. <http://state-of-coast.noaa.gov/bulletins/html/par_02/par.html>.

Otto H. Muller

Coatis

Coatis are raccoon-like **mammals** in the family Procyonidae, which have a long ringed tail, typically held perpendicular to the body, and a flexible, upturned, elongated snout. Coatis are also distinctive socially. The females live together in highly organized groups called bands, composed of 5-12 allied individuals, while adult males are solitary. The difference in the social patterning of the sexes initially confused biologists who described the males as a separate **species**. The use of the name

A coati. *Photograph by Renee Lynn. Photo Researchers, Inc. Reproduced by permission.*

"coatimundi" for this species, meaning "lone coati" in Guarani, reflects this error.

There are four species of coatis in two genera, all fairly similar in appearance. Coatis are versatile animals found in a variety of vegetation ranging from thorn scrub, moist tropical forest, and grassland areas stretching from southwestern through Central and **South America** to northern Argentina. The ringtailed coati (*Nasua nasua*) is a common species. Its coat is tawny red in **color** with a black face. Patches of white accentuate the coat above and below each **eye** and one on each cheek. The throat and belly are also white, while the feet and the rings on the tail are black, and the ears are short and heavily furred. The head-to-tail length ranges from 32 to 51 in (80 to 130 cm) with a little more than half the length being tail. Other species of coati vary slightly in size, coat color, and markings.

Coatis have strong forelimbs that are shorter than the hind legs. Despite a plantigrade foot posture the animal is arboreal as well as terrestrial. The long, strong claws are nonretractile, and the long tail is used for balance. Coatis can reverse the direction of their ankles, facilitating a head-first descent from the trees. The upper mandible is longer than the lower contributing to the flexible use of the snout. The coati has 38-40 teeth with long, slender and sharp canines.

Contrary to other members of the family Procyonidae, coatis are primarily diurnal searching for their omnivorous diet by probing their sensitive snout in the **leaf** litter and rotting logs on the forest floor. Predominately insectivorous, eating such things as **beetles**, grubs, **ants**, and **termites**, coatis also eat a considerable amount of fruit when it is in season, foraging both on the ground and high in trees. Additionally, they are opportunistic predators on **vertebrates**, occasionally catching **frogs**, lizards, and **mice**. Coatis have been known to unearth and eat turtle and lizard eggs. Foraging alone, male coatis typically catch more lizards and

KEY TERMS

....

Arboreal—Living in trees.

Chitter—To twitter or chatter.

Diurnal—Refers to animals that are mainly active in the daylight hours.

Insectivorous—Feeding on insects.

Mandible—A jaw.

Omnivorous—Eating both animal and vegetable substances; eating all kinds of food.

Plantigrade—Walking with the entire lower surface of the foot on the ground, as humans and bears do.

Retractile—Capable of being drawn back or in.

Terrestrial—Of or pertaining to the earth and its inhabitants.

rodents than the females and young, who forage in small groups.

Most of the day is spent foraging for food. During rest periods, coatis will groom each other. They curl up and **sleep** in trees at night. Coatis are highly vocal and use a variety of communication calls. They make a wide diversity of grunts, whines and shrieks, including a sharp whistle. When enraged, coatis will produce a chittering sound.

In the larger ringtailed and white-nosed coatis, females mature in their second year; males mature in their third year. During most of the year, females chase males away from the band because they often will kill juveniles. However, around February and March females become more tolerant and allow a dominant male to interact with the band. The male wins favor with the band females by submissively grooming them. Actual mating occurs in trees. Soon after the male has bred with all the females, he is expelled from the group by the once again aggressive females.

About four weeks before **birth**, females leave their bands to build stick platform nests in trees. After a 77 day gestation period, females bear a litter of three to five poorly developed young, weighing only 3.5-6.4 oz (100-180 g). For the next five to six weeks, females care for the young in the nest. Once the females and youngsters return to the band, young coatis will join their mothers in search of food, but they also play much of the time, wrestling and chasing each other among the trees.

Home ranges of bands of coatis cover about 0.6 mi (1 km), though there is considerable overlapping of range territory of neighboring bands. Despite friendly re-

lations among bands, coatis maintain stable and distinct band membership.

Coatis have little interaction with humans in the wild, however, they are hunted for their fur and meat. Coatis have been kept in captivity, both as pets and as exhibit animals. They typically live about seven years, but have been known to survive fourteen years in captivity.

Resources

Books

Burton, Maurice, ed. *The New Larousse Encyclopedia of Animal Life.* New York: Bonanza Books, 1984.

MacDonald, David, and Sasha Norris, eds. *Encyclopedia of Mammals.* New York: Facts on File, 2001.

National Geographic Society, ed. *Book of Mammals, Volume One.* Washington, DC: National Geographic Society, 1981.

Redford, Kent H., and John F. Eisenberg. *Mammals of the Neotropics: The Southern Cone, Volume 2.* Chicago: University of Chicago Press, 1989.

Vaughan, Terry A. *Mammalogy.* New York: Saunders College Publishing, 1986.

Betsy A. Leonard

Cobalt *see* **Element, chemical**

Cobras *see* **Elapid snakes**

Coca

The coca **plant**, genus *Erythroxylum*, family Erythroxylaceae, order Linales, is native to the Andean slopes of **South America**. The genus *Erythroxylum* comprises approximately 250 **species**, of which the most cultivated species are *Erythroxylum coca* (southern Peru and Bolivia) and *Erythroxylum novogranatense* (Colombia and northern coastal Peru). The coca plant is a shrub, growing to about 15 ft (5 m). Cultivated plants are pruned to about 6 ft (2 m). The leaves are oval, smooth-edged, dark green, and 1.6-3.1 in (4-8 cm) long, 1-1.6 in (2.5-4 cm) wide. Unlike other short term **crops** such as maize and **rice**, or other mountain grown commodities such as coffee, coca plants require little care. Coca plants can thrive in poor **soil**, have few **pests** or predators—an ideal crop for the bleak growing conditions in the Andes. After planting, leaves can be harvested by 6-12 months. Coca plants can yield 4-5 crops per year for 30-40 years.

The coca plant is the source of **cocaine**, one of about 14 alkaloids obtained from the leaves. The **concentration** of cocaine in the leaves varies from about 23% to 85%, depending on the species and growing conditions. The cocaine **alkaloid** was first extracted from the leaves in the 1840s. It soon became a popular addition to powders, medicines, drinks, and potions. The popular American soft drink Coca-cola, introduced in 1885 by John Pemberton, uses coca leaves in its preparation. Since 1906, when the Pure Food and Drug Law was passed, Coca-Cola has been made with decocainized coca leaves. Today, some species of *Erythroxylum* are grown in other regions of the world, such as India and Indonesia, where the climate is similar to the Andean tropics, and cultivated primarily for cocaine extraction. Before cocaine became a popular street drug, coca was grown mainly for traditional consumption among the Andean peoples, and for legal industrial medicinal use.

The indigenous people of the Andean mountain range have been chewing the leaves of the coca plant for thousands of years. Archeological evidence indicates that Peruvians were chewing coca as early as 1800 B.C. Ancient sculptures show the heads of warriors with the characteristic "bulge" in the cheek, depicting coca chewing. The coca plant was one of the first cultivated and domesticated plants in the New World. During the reign of the Incas, coca was regarded as sacred and it was used only by chieftains, priests, or privileged classes. Coca was the link between man and the supernatural. It was used for various social and religious rituals, ceremonies, and fortune telling. Leaves of the coca plant were buried with the dead to help with the journey to the afterworld. Coca leaves were used in traditional medical practices, aiding in **diagnosis** and treatment. When the leaves are chewed with an alkaline substance such as lime, or plant ash, the active ingredients that stimulate the central **nervous system** are released. Stamina was increased, hunger depressed, **pain** eased—a feeling of well being and strength was achieved.

After the Spanish conquered the Incas in the sixteenth century, coca was given to the peasants, or working classes. The Spanish realized that coca enabled the peasants to work harder, longer, and that they needed less food. What was once exclusive to the ruling class was made available to the common people. Thus chewing coca leaves became a way of life for an entire peasant nation. The Indians, then and now, chew the leaves with other substances and never ingest the cocaine alkaloid alone, and apparently do not experience the addictive mind altering effects associated with cocaine. Coca **leaf** chewing is still identified with religious practices, social rituals, traditional medicine, and work situations. The leaves are used for bartering or, as a form of currency, to obtain other goods such as food items. In the past few decades however, growing coca has become associated with obtaining material goods and becoming rich. An entirely new economy, mostly illegal or underground, has developed around coca. Many plantation

The leaves and fruit of a coca plant (*Erythroxylum coca*) in Costa Rica, the plant from which cocaine is extracted. © *Gregory G. Dimijian 1990, National Audubon Society Collection/ Photo Researchers, Inc. Reproduced with permission.*

owners have changed their focus from leaf production to the extraction of the cocaine alkaloid in paste form. Coca production is now the most lucrative industry in Peru and Bolivia, the world's leading producers. The coca industry is heavily scrutinized by several international governmental groups. Realizing the cultural significance of coca chewing among certain sectors of people living in the Andes, the Peruvian government developed a separate agency to protect and supervise legal trade. Most of the annual production of coca however, goes to the black market.

Christine Miner Minderovic

Cocaine

Cocaine is a colorless or white **narcotic** crystalline **alkaloid** derived from the leaves of the South American **coca** plant—*Erythroxylum coca*. Aside from its use as a local anesthetic, which has largely been supplanted by safer drugs, its medical applications failed to live up to the hopes of physicians and chemists of the late nineteenth century. They administered cocaine to themselves and others in the hope that it would be a cure-all wonder drug. After about two decades of wide use in prescription and patented medicine, the harmful effects of cocaine became manifest, and its use as a drug in medical practice was eventually banned.

It subsequently became an illegal drug used for its mood-altering effects, which include euphoria and bursts of short-lived physical **energy**. The "high" produced by cocaine lasts for a short time. The "crash" that follows leaves the user in need of another "fix" to get back to the former high. But each encounter produces diminished highs, so that increasing doses are required to recapture the initial experience. The physical and social consequences of cocaine **addiction** are devastating both to the individual and society. It leads to impoverishment and the destruction of the individual's health. When young people begin to use cocaine, communities begin to feel the effects of increased crime, violence, and social decay.

In the late 1970s cocaine was "snorted," or sniffed through the nose, in its crystalline form, then known as "snow." Because of its high cost, the number of users was limited. In order to get a faster and stronger high, cocaine was also taken by injection with a hypodermic needle. By the 1980s a cheaper version of pure cocaine made its appearance on the illegal market in the form of "crack," which is smoked, primarily in the "crack houses" where it is produced. In the form of crack, cocaine has reached a larger population, making it one of the chief drug problems of the present.

History

Coca plants, which are the source for cocaine, are indigenous to Central and **South America**. The name of the **plant** is derived from the Inca word *Kuka*. Archaeological evidence points to the use of coca plants in South America as early as seven thousand years ago. They were used for many centuries by the Incas as part of their religious ceremonies. To help the dead in the afterworld, mounds of stored coca leaves were left at burial sites in the area of modern Peru. These sites are estimated to be about 4,500 years old. The Incas may also have been using liquid coca **leaf** compounds to perform **brain surgery** 3,500 years ago. Inca records dating from the thirteenth through the sixteenth century indicate that coca was revered as a sacred object with magical powers. The magic plant of the Incas was chewed by priests to help induce trances that led them into the spirit world to determine the wishes of their gods. Artifacts dating back thousands of years to the earliest Incan periods show the cheeks of their high priests distended with what in all probability were the leaves of the coca plant.

Even before the Spanish conquest, Indians working in silver mines of the northern Andes chewed the coca leaf to help overcome **pain**, fatigue, and the respiratory problems common at high altitudes. Early European explorers in the fifteenth century compared the common sight of Indians they saw chewing the coca leaves to cattle chewing cud. After the Spanish conquest the Church sought to ban the practice of chewing the coca leaf, mainly because of its association with Incan religious ceremonies. When the ban failed, the Spanish allowed the Incan survivors to continue their ancient practice of coca leaf chewing in order to maintain **mining** production. South American farmers, who are descendants of the Incas, continue the practice to the present day.

Introduction to the West

The main alkaloid in the leaves of the coca plant was extracted in 1859 by Albert Niemann, a German scientist, who gave it the name cocaine. Reports soon followed of therapeutic benefits of cocaine in the treatment of a number of physical and mental disorders. These reports also praised cocaine for being a highly effective stimulant, able to conquer the most severe cases of fatigue.

Sigmund Freud, two decades after Niemann's work, began experimenting with cocaine, thinking it could be used to treat "nervous fatigue," an ailment many upper- and middle-class Viennese were diagnosed with. He gave his fiancée cocaine and also administered it to himself. Then he wrote a **paper** praising the curative powers of cocaine in the treatment of such problems as **alcohol** and **morphine** addiction, gastrointestinal disorders, **anxiety**, **depression**, and respiratory problems. In this paper, Freud completely dismissed the powerful addictive properties of the drug, insisting that the user would develop an aversion to, rather than a craving for, its continued use.

Freud shortly afterwards became aware of his mistake when he attempted to cure a friend's morphine addiction with the use of cocaine. At first the treatment seemed to work, but he soon saw that the friend developed an addiction to cocaine instead. Soon afterwards Freud's friend suffered a complete nervous breakdown.

Unfortunately, there were other physicians and chemists who misjudged the properties of cocaine in the same way Freud had done. For example, a neurologist and former surgeon general of the United States, William Hammond, also praised the healing powers of cocaine and pronounced it no more addictive than coffee or tea.

Coca-Cola

In the 1880s, John Pemberton, a pharmacist from Atlanta, concocted a drink called Coca-Cola from a prescription syrup that had been used to treat headache, hysteria, and depression. Pemberton's elixir drink contained coca leaves, **kola** nuts, and a small amount of cocaine in a sugary syrup. His secret formula was picked up by Asa Chandler, who formed the Coca-Cola company. The drink was praised by the *New York Times* as the new wonder drug. At about the same time, cocaine was sold in a cigarette produced by the Parke-Davis pharmaceutical company. The cigarettes were marketed as a cure for infections of the throat.

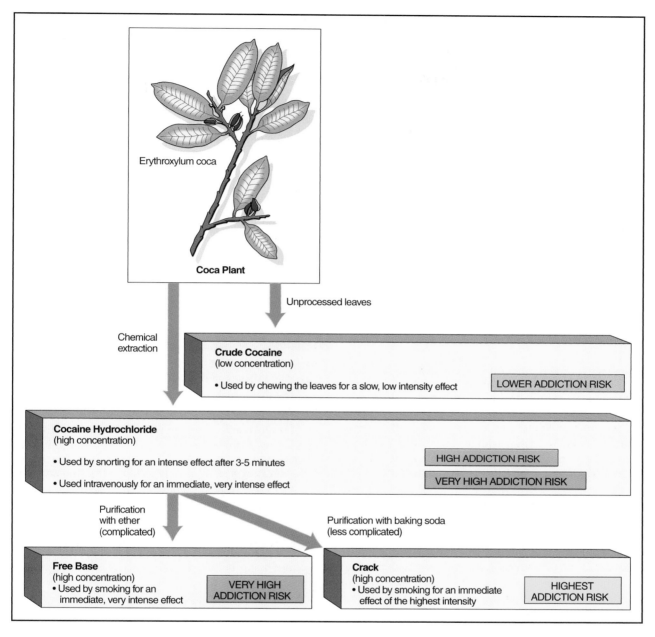

Erythroxylum coca

Coca Plant

Unprocessed leaves

Chemical extraction

Crude Cocaine
(low concentration)

• Used by chewing the leaves for a slow, low intensity effect   LOWER ADDICTION RISK

Cocaine Hydrochloride
(high concentration)

• Used by snorting for an intense effect after 3-5 minutes   HIGH ADDICTION RISK

• Used intravenously for an immediate, very intense effect   VERY HIGH ADDICTION RISK

Purification with ether (complicated)

Purification with baking soda (less complicated)

Free Base
(high concentration)
• Used by smoking for an immediate, very intense effect   VERY HIGH ADDICTION RISK

Crack
(high concentration)
• Used by smoking for an immediate effect of the highest intensity   HIGHEST ADDICTION RISK

Various forms of cocaine and the addiction risks associated with them. *Illustration by Hans & Cassidy. Courtesy of Gale Group.*

Early drug laws

From all the nineteenth-century hopes for the possible medical uses of cocaine, the only practical application that held up was its use as a local anesthetic. All the other efforts to prove that cocaine was a wonder drug were dismal failures in light of the powerful addictive effects of the drug. By the early twentieth century, it had become clear that cocaine posed a serious hazard to any user.

In 1904 the cocaine was removed from the Coca-Cola syrup. In 1906 the Pure Food and Drug Act was en-acted to stop the sale of patent medicines containing substances such as cocaine. Before that date, manufacturers were not required to list the ingredients of their patent medicines. The 1906 act made truthful labeling of patent medicines sold across state borders mandatory, but it did not stop the sale of cocaine products. New York State tried to curtail cocaine sales by passing a law in 1907 that limited the right to distribute cocaine to physicians. That law merely paved the way for the illicit street traffic. Dealers obtained cocaine from physicians and then sold it on the street.

The cocaine drug problem continued to rise until 1914, when the Harrison Act was passed. This legislation used the federal Treasury Department to levy taxes on all phases of cocaine trafficking and imposed further strict measures on the sale and distribution of cocaine. From that time to the 1960s, cocaine use dwindled, in part because of the rising popularity of **amphetamines** on the illegal market. The medical use of cocaine as a topical anesthetic and as an ingredient in cough medicines continued, while its illegal use was largely confined to the very rich.

After the 1960s

By the 1970s, when illegal drug use became more widespread in the general population, middle- and upper-class groups began to use cocaine in its white crystalline form. A mythology of its effectiveness as an aphrodisiac (a substance supposed to enhance the sex drive), a mental energizer, and a self-esteem booster began to develop. Along with benefits that active and ambitious middle-class people hoped for in their drug of choice, came reports of the relative safety of cocaine use in comparison to other drugs. The harsh lessons learned around the turn of the century were all but forgotten.

Crack

By the late 1970s, cocaine addiction in the United States had reached **epidemic** proportions. In the mid-1980s people started smoking cocaine after "freebasing" it, that is, dissolving the cocaine alkaloid from its white powder base to create a smokable form of pure cocaine. **Ether** is used to remove the hydrochloride base, which does not burn. The smoked cocaine goes straight into the bloodstream and gives a quicker and stronger high. Freebasing with ether can be dangerous because if any ether remains in the freebase cocaine, it can quickly ignite into flames when smoked. The comedian Richard Pryor, to mention one of the more famous cases, was severely burned when he freebased cocaine.

Besides freebase cocaine, there is another form of smokable cocaine, called "crack," which also gives a fast and potent high. Crack is safer and easier to obtain than freebase cocaine because baking soda is used instead of ether to remove the hydrochloride. The baking soda produces pure forms of cocaine in small pellets that can be smoked in a pipe (where it makes the crackling sound that gave this form of cocaine its name). The cost of crack is so low that anybody, even a child, can afford it, and the drug soon began to wreak its devastations on the working classes. The widespread use of crack cocaine has led most visibly to rising crime rates, with gang wars erupting over control of territory and with users resorting to theft, prostitution, and other crimes to support their habits. Other consequences have been impaired workplace performance, new public health problems including such phenomena as crack babies, and a host of other social and economic evils.

Biochemistry

Used as a local anesthetic, cocaine constricts the **blood** vessels, thereby slowing down blood circulation. It also reduces the sensitivity of nerve endings, especially in the skin, eyes, and areas of the mouth. Because cocaine is a stimulant, it increases the **heart** and pulse **rate** and raises blood **pressure**, causing alertness, **insomnia**, loss of appetite, and dilated pupils.

Several theories have been proposed to explain the addictive effects of cocaine, which differs from other stimulants in its ability to trap the user in a cycle of continued use. Experiments using animals who are able to self-administer cocaine show that, once the need for cocaine is established, an **animal** will neglect its hunger and sex drives in order to satisfy the craving for the drug. **Rats** took cocaine until they died, while **monkeys** indulged until they exhibited such behaviors as paranoia, hyperactivity, convulsions, and heart failure.

Cocaine, like the opioids morphine and heroin, causes addiction by arousing an intense sense of pleasure. Certain parts of the brain induce pleasurable sensations when stimulated. Unlike the opioids, though, cocaine appears to have a greater access to those parts of the brain known as the limbic system, which controls the emotions. Cocaine stimulates the release of the **neurotransmitter dopamine**, which is responsible for the stimulation of the limbic system. The drug is therefore more potent than other drugs in being more psychologically rewarding. According to a recent theory, most of the mood and **behavior** changes brought about by cocaine use is due to the release of excess amounts of dopamine in the reward centers of the brain. The ensuing depression and craving for the drug are caused by dopamine depletion after the effects of the drug wear off.

Treatment and prevention

Breaking a cocaine dependency is difficult, and treatment is costly and prolonged, involving treatment centers and support groups. Since addiction is a chronic disorder, the detoxification process is just the first step, and there is no final cure. Remissions can be expected, and the goal of treatment may have to be the control and reduction of use and dependency.

Prevention efforts in the United States have for a long time been focused primarily on stopping cocaine imports from South America, mainly Peru and Colombia,

KEY TERMS

Alkaloid—A nitrogen-based chemical, usually of plant origin, also containing oxygen, hydrogen, and carbon. Many are very bitter and may be active if ingested. Common alkaloids include nicotine, caffeine, and morphine.

Amphetamines—Stimulant drugs discovered in the 1930s that were widely prescribed as diet pills and became a staple in the illegal drug traffic.

Aphrodisiac—A drug that is supposed to stimulate sexual impulses.

Coca leaves—Leaves of the coca plant that were chewed by the Incas and are still used by farmers of certain regions in South America.

Crack—A smokable and inexpensive form of pure cocaine sold in the form of small pellets, or "rocks."

Dopamine—The neurotransmitter believed to be responsible for the cocaine high.

Euphoria—Feelings of elation and well being produced by drugs such as cocaine.

Freebasing—Processes used to free drugs such as cocaine from their hydrochloride base.

Local anesthetic—A pain killer that acts on a particular site of the body without affecting other sites or causing unconsciousness.

Snow—The white powder of cocaine hydrochloride that is inhaled through the nostrils. This way of taking cocaine is called "snorting" and was popular in the 1970s before the advent of crack cocaine.

and these efforts have had some success in breaking up the powerful and wealthy cartels that control the cultivation and trade of the coca leaf. However, these producers are still sending coca to the United States and continue to seek other markets worldwide for their deadly crop.

Studies have shown that a recent decline of cocaine usage in the United States is directly correlated to educational programs targeting young people and aiming to enhance their understanding of the dangers of cocaine use. Such educational programs are more likely to lead to results than interdiction efforts and provide the best hope of curtailing the current epidemic of cocaine abuse and preventing similar epidemics in the future.

Resources

Books

Flynn, John C. *Cocaine.* New York: Carol Publishing, 1991.
Gold, Mark S. *Cocaine.* New York: Plenum Publishing, 1993.
Rice-Licare, Jennifer, and Katharine Delaney-McLaughlin. *Cocaine Solutions.* Binghamton: Haworth Press, 1990.
Washton, Arnold M., and Mark S. Gold. *Cocaine: A Clinician's Handbook.* New York: Guilford Press, 1987.

Jordan P. Richman

Cockatoos

Cockatoos are **species** of **birds** in the family Cacatuidae, in the order Psittaciformes, which also contains the typical **parrots** (family Psittacidae).

Parrots and cockatoos all have powerful, curved bills, short legs, and strong, dexterous feet with two toes pointing forward and two backward. These birds also have specialized feathers known as powder down, which disintegrates into a powder that is used for dressing the feathers during preening. Parrots and cockatoos are colorful, intelligent birds. They mostly eat **fruits** and **seeds**.

Cockatoos are relatively simply colored birds, with a crest on the top of the head that can be erected at will. Species in this family mostly occur in **Australia**, New Guinea, and nearby islands, with species occurring as far north as the Philippines. Cockatoos usually nest in holes in trees. They feed on a wide range of fruits and seeds, as well as flowers, roots, rhizomes, palm shoots, and beetle and moth larvae.

The best-known species is the sulfur-crested cockatoo (*Cacatua galerita*), which occurs widely in eastern Australia and New Guinea. This large bird has a pure-white body, with a yellow-colored crest, and a black beak and feet. The galah (*Cacatua roseicapilla*) has a rosy breast and face, and a gray back and wings. This species occurs widely in woodlands, savannas, **grasslands**, and parks throughout most of Australia. The pink cockatoo (*Cacatua leadbeateri*) has white wings, a salmon-colored belly and head, and occurs in woodlands of western and central Australia. The little corella (*C. sanguinea*) is a smaller species, with a white body, dark eye-patch, and yellow under the wings. This species occurs widely in Australia and southern New Guinea.

The palm cockatoo (*Probosciger aterrimus*) is a large, dark-grey bird, with orange-pink facial skin, and a long and erectile, black crest. This cockatoo occurs in tropical rainforests and eucalyptus woodlands of northeastern Australia and New Guinea. The red-tailed black cockatoo (*Calyptorhynchus magnificus*) is a sooty black bird with red streaks on the tail, occurring in woodlands of northern Australia, and in scattered places elsewhere on that **continent**.

Cockatoos. *Photograph by David Woods. Stock Market. Reproduced by permission.*

Cockatoos are kept as entertaining, sometimes "talking" pets. The sulfur-crested cockatoo is the species most commonly kept in this way. Some species of cockatoos are hunted as food by aboriginal peoples in Australia and New Guinea. Most species of cockatoos are endangered, particularly those that are **endemic** to Pacific islands.

Cockroaches

Cockroaches are **insects** in the order Blattaria. They are somewhat flat, oval shaped, leathery in texture, and are usually brown or black in **color**. Cockroaches range in body size from 0.1 to 2.3 in (2.5 to 60 mm), and are rampant pest insects in human inhabited areas, as well as common outdoor insects in most warm areas of the world.

These insects were formerly classified in the order Orthoptera, which consists of the **grasshoppers** and katydids. Now they are often classified along with the mantises in an order referred to as Dictyoptera. The separate order Blattaria, however, is the more common classification for them, and this order is placed in the **phylogeny**, or evolutionary history, of the class Insecta between the orders Mantodea, the mantids, and Isoptera, the **termites**.

The primitive wood-boring cockroaches in the family Cryptocercidae, a family in which there is only a single **species** in the United States, *Cryptocercus punctulatus*, are thought to have shared a common ancestor with termites. The evidence for this is a close phylogenetic relationship between the two groups' obligate intestinal symbionts, single-celled organisms called protozoans which break down the **wood** that the insects eat into a form that is useful to the insect, and in turn receive **nutrition** from the **matter** which is not nutritive to the insect. There are also behavioral similarities between these wood-boring cockroaches and termites, including the fact that they both live gregariously in family groups, a characteristic not shared by any of the other types of cockroaches. Finally, the relationship between the two orders is evidenced by the resemblance between *Cryptocercus* nymphs, and adult termites.

Interesting morphological characteristics of these insects are their chewing mouthparts and large compound **eye**. The pronotum, or segment of the thorax that is closest to the head, conceals the head, and in most species, both male and female are winged, although they rarely fly. They exhibit many fascinating behaviors, such as the ability to stridulate, that is, produce sound by rubbing a comb-like structure with a scraper. Other species, however, communicate by producing sound by drumming the tip of their abdomen on a substrate. An important developmental feature of these insects is their paurometabolous life-history. Paurometabolism is a type of simple **metamorphosis** in which there are definite egg, immature or nymph, and adults stages, but no larval or pupal stages and in which the nymphs resemble the adults except in size, development of wings, and body proportions. **Habitat** requirements of nymphs and adults do not vary in paurometabolous insects, a fact which helps cockroaches to thrive under a rather generalized set of environmental conditions at all stages of their lives.

Cockroaches are saprophagous insects, or scavengers, feeding on a great variety of dead and decaying **plant** and **animal** matter such as **leaf** litter, rotting wood, and carrion, as well as live material. They are, thus, very flexible in their diet, and this flexibility allows them to exist under a wide range of conditions. In fact, the habitat types of this order span such areas as wood rat nests, grain storage silos, forest leaf litter, and nests of leaf cutter **ants**. They thrive in areas with moisture and warmth.

Cockroaches produce oothecae, sacs in which the eggs are held and protected by secretions produced by the female. These egg sacs may be carried by the female externally until the time of hatching, or internally until the female gives **birth** to live young, or they may simply be deposited on a suitable substrate and left to hatch without any care from the female.

Besides being rather fast runners, many cockroaches have other adaptations that allow them to escape from predation. One such defensive **adaptation** is the ability to produce an offensive odor which they emit when disturbed. Other species, such as the Madagascaran cockroach, *Gromphadorhina laevigata*, force air out through their spiracles, thus producing an intimidating hissing sound.

Cockroaches are worldwide in distribution, although most of this order's approximately 4,000 species occur in the tropics. In the United States and Canada, there are some 29 different genera, and about 50 species. Most of these species occur in the southern United States.

Due to their preference for moist, warm places, flexible diet, and nocturnal activity, cockroaches are very successful at living uncontrolled in human areas. Although annoying and often feared by people, they are not known to be significant **disease** carriers, crop **pests**, or agents of other large-scale damage to human areas. There are four species in **North America** which are common as household insects. They are the German cockroach (*Blattella germanica*), the American cockroach (*Periplaneta americana*), the brown-banded cockroach (*Supella longipalpa*), and the oriental cockroach (*Blatta orientalis*). The oothecae of these indoor species are often deposited on common household items such as cardboard boxes, and in this way may be transported from place to place before they actually hatch, thereby spreading to new areas.

In contrast to its image as a fearsome, although relatively harmless pest, the cockroach has for many years actually benefitted humans—a benefit that has resulted from its abundance and often large size. The cockroach has contributed greatly to our general understanding of **physiology** due to its use as a model study **organism** in biological research investigating nutrition, neurophysiology, and endocrinology. Medical knowledge has expanded as a result of the data gained from such studies of the cockroach.

Resources

Books

Arnett, Ross H. *American Insects.* New York: CRC Publishing, 2000.

KEY TERMS

Obligate intestinal symbiont—An organism that lives in the intestinal tract of another organism, and whose presence is necessary for the survival of the host. Intestinal symbionts are usually bacteria or protozoans.

Oothecae—Egg sacs produced by some insects including cockroaches.

Paurometabolism—A type of simple metamorphosis in which the nymph, or immature, stage of the insect resembles the adult except in size, proportion, and wing length, and whose habitat requirements are the same as those of the adult.

Phylogeny—A hypothesized shared evolutionary history between members of a group of organisms based on shared traits of the organisms; also known as "evolutionary trees."

Saprophagous—Refers to decomposer organisms that eat dead and decaying plant and animal matter.

Simple metamorphosis—A developmental series in insects having three life-history stages: egg, nymph, and adult.

Borror, D. J., C A. Triplehorn, and N. F. Johnson. *An Introduction to the Study of Insects.* 6th ed. Orlando, FL: Harcourt Brace College Publishers, 1989.

Carde, Ring, and Vincent H. Resh, eds. *Encyclopedia of Insects.* San Diego, CA: Academic Press, 2003.

Cornell, P.B. *The Cockroach.* London: Hutchinson, 1968.

Elzinga, R. J. *Fundamentals of Entomology.* 3rd ed. Englewood Cliffs, NJ: Prentice-Hall, 1987.

Periodicals

Roth, L. M. "Evolution and Taxonomic Significance of Reproduction in Blattaria." *Annual Review of Entomology* 15 (1970): 75-96.

Puja Batra

Coconuts *see* **Palms**

Codeine

Codeine is a type of medication belonging to a class of drugs known as opioid analgesics, which are derived from the *Papaver somniferum,* a type of poppy **flower**, or are manufactured to chemically resemble the products of that poppy. In Latin, *Papaver* refers to any flower of the poppy variety, while *somniferum* translates to mean "maker of sleep." The **plant** has been used for over 6,000 years, beginning with the ancient cultures of Egypt, Greece, Rome, and China, to cause **sleep**. Analgesics are drugs which provide relief from **pain**. Codeine, an opioid analgesic, decreases pain while causing the user to feel sleepy. At lower doses, codeine is also helpful for stopping a cough.

Although codeine is present in nature within the sticky substance latex which oozes out of the opium poppy's seed pod, it is present in only small concentrations (less than 0.5). However, **morphine**, another opioid analgesic, is present in greater concentrations (10) within the opium poppy's latex, and codeine can be made from morphine via a process known as methylation, which is the primary way that codeine is prepared.

Codeine is a centrally acting drug, meaning that it goes to specific areas of the central **nervous system** (in the **brain** and spinal cord) to interfere with the transmission of pain, and to change your **perception** of the pain. For example, if you have your wisdom teeth removed and your mouth is feeling very painful, codeine will not go to the hole in your gum where the tooth was pulled and which is now throbbing with pain, but rather will act with the central nervous system to change the way you are perceiving the pain. In fact, if you were given codeine after your tooth was pulled, you might explain its effect by saying that the pain was still there, but it just was not bothering you anymore.

Codeine's anti-tussive (cough stopping) effects are also due to its central actions on the brain. It is believed that codeine inhibits an area of the brain known as the medullary cough center.

Codeine is not as potent a drug as is morphine, so it tends to be used for only mild-to-moderate pain, while morphine is useful for more severe pain. An advantage to using codeine is that a significant degree of pain relief can be obtained with oral medication (taken by mouth), rather than by injection. Codeine is sometimes preferred over other opioid analgesics because it has a somewhat lower potential for **addiction** and abuse than do other drugs in that class.

Scientists are currently trying to learn more about how codeine and other opioid analgesics affect the brain. It is interesting to note that there are certain chemicals (endorphins, enkephalins, and dynorphins) which are made within the brains of **mammals**, including humans, and which closely resemble opioid analgesics. In fact, the mammalian brain itself actually produces tiny amounts of morphine and codeine! Some of these chemicals are produced in the human brain in response to certain behaviors, including **exercise**. Exploring how and when human brains produce these chemicals could help scientists understand more about ways to control pain

with fewer side effects, as well as helping to increase the understanding of addictive substances and behaviors.

See also Narcotic.

Resources

Books

Berkow, Robert, and Andrew J. Fletcher. *The Merck Manual of Diagnosis and Therapy.* Rahway, NJ: Merck Research Laboratories, 1992.

Katzung, Bertram G. *Basic & Clinical Pharmacology.* Norwalk, CT: Appleton & Lange, 1992.

Marieb, Elaine Nicpon. *Human Anatomy & Physiology.* 5th ed. San Francisco: Benjamin/Cummings, 2000.

Rosalyn Carson-DeWitt

Codfishes

Codfish (family Gadidae) are a family of bottom-feeding **fish** that live in cool or cold seas, mostly in the Northern Hemisphere. There are about 21 genera and 55 **species** of codfishes. The most commonly utilized marine habitats are inshore waters and continental shelves, generally in depths of less than about 300 ft (100 m), but sometimes considerably deeper. Codfishes are voracious predators of smaller species of fish and **invertebrates**. Some species of codfish are of great economic importance, supporting very large fisheries.

The Gadidae family is divided into two subfamilies. The Gadinae includes cod (e.g., Atlantic cod, *Gadus morhua*) and haddock (*Melanogrammus aeglefinus*), while the Lotinae includes hake (e.g., silver hake, *Merluccius bilinearis*), rockling (e.g., silver rockling, *Gaidropsarus argentatus*), and burbot (e.g., the **freshwater** American burbot, *Lota lota*).

Atlantic cod and its fishery

The economically most important species of codfish is the Atlantic cod, which has supported one of the world's largest fisheries. This species is common on both sides of the Atlantic Ocean. The Atlantic cod extends from Novaya Zemlya and Spitzbergen in the northeastern Atlantic, to Baffin Island and central Greenland in the northwestern Atlantic. The Atlantic cod is found as far south as the Bay of Biscay in western **Europe**, and coastal North Carolina in **North America**. In the western Atlantic, the Atlantic cod is most abundant on the Grand Banks, a large region of open-ocean shelf east of Newfoundland.

Atlantic cod are ravenous feeders on a wide variety of **prey** found on the sea bottom or in the **water** column.

A black cod. *Photograph by Harvey Lloyd. Stock Market. Reproduced by permission.*

Fry and juvenile cod eat smaller invertebrates and fish larvae. The most common prey of adult Atlantic cod is small species of fish, but cannibalistic feeding on smaller size classes of its own species is known to occur.

The Atlantic cod has long been the target of European fishers, and this species was one of the first natural resources to be heavily exploited by European settlers in the Americas. The extraordinarily large cod populations of the Grand Banks and the northern Gulf of Saint Lawrence were noted during the exploratory voyages of the Cabot brothers, who sailed on behalf of England during the late 1490s. By 1505, many Portuguese and Basque fishers were exploiting the bountiful cod resources of the New World. By 1550, hundreds of ships departed every year from European ports for the northwest Atlantic in search of cod, which were taken in large quantities, preserved by salting or drying, and transported to the kitchens of western Europe. By 1620, there were more than 1,000 fishing vessels in the waters off Newfoundland, and about 1,600 in 1812.

During this early stage of the cod fishery in the northwest Atlantic, fish were typically very large, commonly 3-6 ft (1-2 m) long and weighing more than 220 lb (100 kg). However, because of the long history of heavy exploitation of Atlantic cod, such large fish are very rare today.

During the eighteenth and nineteenth centuries the cod fishery on the banks off Newfoundland was an unregulated, open-access enterprise, involving large numbers of ships sailing from Europe, Newfoundland, Canada, and New England. In addition, flotillas of smaller, local boats were exploiting near-shore populations of cod. Most of the fishing during these times was done using hand lines and long lines, which are not very efficient methods. Still, the substantial fishing effort caused noticeable depletions of many of the near-shore cod stocks.

During the twentieth century, especially its second half, the new technologies allowed a much more intensive exploitation of the stocks of Atlantic cod. A variety of highly efficient trawls, seines, and gill nets have been developed, and their effective deployment is aided by fish-finding devices based on sonar technology. Moreover, storage and processing capacities of ships have significantly increased, which permits large vessels to stay at sea for long periods of time.

The 1960s saw the largest harvests of cod in the northwest Atlantic, as a result of efficient technology and open-access and unregulated fishery. The total catch in this region in 1968 was more than two million tons. These huge catches were not sustainable by the cod population, and the stocks of Atlantic cod began to collapse by the 1970s.

In 1977, the Canadian government began to manage fisheries within a 200 mi (320 km) wide zone around its coast. This was mostly accomplished by controlling access and allocating quotas of fish, especially cod, the most important species in the fishery. These **conservation** efforts led to small increases in cod stocks and catches. However, during the late 1980s and 1990s, the cod stocks suffered a more serious collapse. Because the populations of mature cod capable of reproducing the species are small, the stocks will probably recover quite slowly, despite the huge reductions in fishing beginning in 1991, and a ban on cod fishing in 1992. The fishing moratorium recognizes the sad fact that one of the world's greatest renewable resources, the stocks of cod in the northwest Atlantic, had been over fished to commercial **extinction**. This damage has been long lasting—there was still not much recovery by the year 2000.

Undoubtedly, the collapse of cod stocks was mostly caused by over fishing, that is, exploitation at a **rate** exceeding the productivity of the cod population. The over fishing occurred because of economic greed, faulty fish-population models that predicted excessively large quotas, and because politicians set fishing quotas exceeding those recommended by their fishery scientists. Moreover, this over fishing occurred along with other environmental changes that may have exacerbated the effects of the excessive harvesting. In particular, several years of unusually cold waters off Newfoundland and Labrador may have reduced spawning success, so the heavily fished population was not being replenished. Other factors have also been suggested, including a rapidly increasing population of **seals**. However, this particular seal species does not consume much cod and is therefore not considered a significant factor in the collapse of cod stocks. Unregulated fishing is clearly the major factor.

Fortunately, Atlantic cod populations, while low, are not threatened with extinction. In 1994, spawning schools of fish were once again observed in the waters off Newfoundland. If the regulating authorities maintain the moratorium on cod fishing, and if they subsequently regulate and monitor the fishery, then there is hope that this great natural resource will again provide food for humans, but this time in a sustainable fashion.

Resources

Books

Freedman, B. *Environmental Ecology.* 2nd ed. San Diego: Academic Press, 1995.

Whiteman, Kate. *World Encyclopedia of Fish & Shellfish.* New York: Lorenz Books, 2000.

Bill Freedman

Codons

Information for the genetic code is stored in a sequence of three nucleotide bases of DNA called base triplets, which act as a template for which messenger RNA (mRNA) is transcribed. A sequence of three successive nucleotide bases in the transcript mRNA is called a codon.

Codons are complimentary to base triplets in the DNA. For example, if the base triplet in the DNA sequence is GCT, the corresponding codon on the mRNA strand will be CGA.

When interpreted during protein syntheis, codons direct the insertion of a specific **amino acid** into the protein chain. Codons may also direct the termination of protein synthesis.

During the process of translation, the codon is able to code for an amino acid that is incorporated into a polypeptide chain. For example, the codon GCA, designates the amino acid arginine. Each codon is nonoverlapping so that each nucleotide base specifies only one amino acid or termination sequence. A codon codes for an amino acid by binding to a complimentary sequence of RNA nucleotides called an anticodon located on a **molecule** of tRNA. The tRNA binds to and transports the amino acid that is specific to the complementary mRNA codon. For example, the codon, GCA on the mRNA strand will bind to CGU on a tRNA molecule that carries the amino acid arginine.

Because there are four possible nucleotide bases to be incorporated into a three base sequence codon, there are 64 possible codons ($4^3 = 64$). Sixty-one of the 64 codons signify the 20 known amino acids in **proteins**. These codons are ambiguous codons, meaning that more than one codon can specify the same amino acid. For example, in addition to GCA, five additional codons specify the amino acid arginine. Because the RNA/DNA sequence cannot be predicted from the protein, and more than possible sequence may be derived from the same sequence of amino acids in a protein, the genetic code is said to be degenerate.

The remaining three codons are known as stop codons and signal one of three termination sequences that do not specify an amino acid, but rather stop the synthesis of the polypeptide chain.

Research began on deciphering the genetic code in several laboratories during the 1950s. By the early 1960s, an *in vitro* system was able to produce proteins through the use of synthetic mRNAs in order to determine the base composition of codons. The **enzyme**, polynucleotide phosphorylase, was used to catalyze the formation of synthetic mRNA without using a template to establish the nucleotide sequence. In 1961, English molecular biologist Francis Crick's research on the molecular structure of DNA provided evidence that three nucleotide bases on an mRNA molecule (a codon) designate a particular amino acid in a polypeptide chain. Crick's work then helped to establish which codons specify each of the 20 amino acids found in a protein. During that same year, Marshall W. Nirenberg and H. Matthei, using synthetic mRNA, were the first to identify the codon for phenylalanine. Despite the presence of all 20 amino acids in the reaction mixture, synthetic RNA polyuridylic acid (poly U) only promoted the synthesis of polyphenylalanine. Soon after Nirenberg and Matthei correctly determined that UUU codes for phenylalanine, they discovered that AAA codes for lysine and CCC codes for proline.

Eventually, synthetic mRNAs consisting of different nucleotide bases were developed and used to determine the codons for specific amino acids.

In 1968, American biochemists Marshall W. Nirenberg, Robert W. Holley, and Har G. Khorana won the Nobel Prize in Physiology or Medicine for discovering that a three nucleotide base sequence of mRNA defines a codon able to direct the insertion of amino acids during protein sythesis (translation).

See also Alleles; Chromosome mapping; Genetic engineering; Molecular biology.

Coefficient

A coefficient is a constant multiplier of variables and any part of an algebraic **term**, Thus, in the expression

$$3\,xy^2 \frac{x}{2} + \frac{4x}{3y}$$

the possible coefficients for the term $3xy^2$ would include 3, which is the coefficient of xy^2, and x, which is the coefficient of $3y^2$:

$$\frac{4x}{3y}$$

has 4 as a coefficient of

$$\frac{x}{3y}$$

Most commonly, however, the word coefficient refers to what is, strictly speaking, the numerical coefficient. Thus, the numerical coefficients of the expression $5xy^2 - 3x + 2y - 4\text{-}X$ are considered to be 5, -3, +2, and -4.

In many formulas, especially in **statistics**, certain numbers are considered coefficients, such as correlation coefficients in statistics or the coefficient of expansion in **physics**.

Coelacanth

The coelacanth (*Latimeria chalumnae*) is the only living representative of an ancient order of fishes, until recently thought to have become extinct 70 million years ago, at about the same time as the dinosaurs. In 1938, however, scientists were astonished when living coelacanths were discovered (this is described later).

The coelacanth is a sarcoptergian, or lobe-finned **fish**, distantly related to the **lungfish**. However, unlike other bony fishes, the pectoral and pelvic fins of the coelacanth are muscular, and even leg-like in appearance. (The evolutionary ancestor of **amphibians**, and thus of all land-dwelling animals, had fins similar to those of the coelacanth.) The fins are able to move over 180°, allowing the fish to swim forwards, backwards, and even upside down. While swimming, the coelacanth moves its fins like a quadrupedal land **animal** moves its legs while walking: the front left and right rear fins move in unison, and the front right and left rear do the same.

The bluish body of the coelacanth is covered with thick scales that are unique to the order. Its jaws are strong. A few specimens have been found with lanternfish in their stomach, indicating that the coelacanth is predatory in its feeding habits. The retina of its eyes has a reflective layer (similar to that of **cats**) that helps it see in dimly lit waters.

Most of what we know about coelacanths has come from the study of dead specimens. Some of the first females to be dissected were found to contain several baseball-sized eggs that lacked a shell or hard case. Although it had been hypothesized that fossil coelacanths bore their young alive, this was not conclusively demonstrated until 1975, when a female specimen at the American Museum of Natural History was dissected and five perfectly shaped fetuses were discovered. Each was about 14 in (35

A coelacanth (*Latimeria chalumnae*) preserved specimen. © *Tom McHugh/Steinhart Aquarium/Photo Researchers, Inc.*

cm) long, and had a yolk sac attached to its stomach. This demonstrated that the coelacanth is **ovoviviparous**, meaning the female keeps the eggs inside her body to protect them as they develop to the hatchling stage.

Little is known about the **ecology** of the coelacanth. In 1987, marine biologist Hans Fricke managed to film coelacanths in their deep-water environment off the Comoros Islands in the Indian **Ocean**. Not only did his team observe the unusual swimming gait described above, but they also saw coelacanths doing "headstands" when the researchers triggered an electronic lure. The coelacanth has an **organ** between its nostrils called the rostral organ, which is believed to detect the electrical field of **prey**, as the ampullae of Lorenzini do in **sharks**. Fricke's research strengthened the evidence that the rostral organ of coelacanths is an electrical sensor.

The first, astonishing find

A living coelacanth first came to the attention of science in 1938. At the time, a young biologist named Majorie Courteney-Latimer was the curator of a small natural history museum in South **Africa**. Because the museum was new, she had been given freedom to decide the direction its collections would take, and she chose to focus on marine life. Local fishermen often brought her unusual fish from their catches. One day, Capt. Hendrik Goosen contacted Courteney-Latimer, saying he had several fish she might be interested in. When she got to the dock, there was a pile of sharks waiting for her. But buried in the pile was a blue fish unlike any she'd ever seen. This one she took back with her to the museum.

The specimen was large, nearly 5 ft (1.5 m) in length, and far too big to fit into the museum's freezer. Desperate to preserve the fish against the hot South African weather, Courteney-Latimer asked a local hospital if she could keep it in their morgue, but was refused. Finally, she managed to get it to a taxidermist.

Knowing that the strange fish was unique, Courteney-Latimer wrote to L.B.J. Smith, an expert on South African fish. Smith recognized the fish as a coelacanth from a sketch provided by Courteney-Latimer. He was able to learn a great deal from the mounted specimen (he dissected one side of it). He then published a notice in the prestigious journal *Nature*, which announced to the world that coelacanths still existed.

This notice also gave Smith the right to name the new **species**. He chose to name the genus *Latimeria* in honor of Courteney-Latimer, and the species *chalumnae* after the river near which Goosen had caught the fish.

Fourteen years passed before another coelacanth turned up, despite Smith's peppering the eastern coast of Africa with posters describing the coelacanth and offering a reward. Eventually, in 1952, a second coelacanth was caught near the Comoro Islands in the Indian Ocean, many miles from where the first specimen had been caught off eastern South Africa. Smith thought that the second specimen represented another species, which he named *Malania anjouanae*. However, later scientists deduced that *Malania* was actually another specimen of *Latimeria*.

Since that second specimen, more than 200 additional coelacanths have been caught. Many of these are now in research collections, particularly in France, which had ruled the Comoros as a colonial power. Most coelacanths have been caught by local fishermen, using a handline from a dugout canoe. Unfortunately, the hooked fish quickly die, because of the great **pressure** difference between its deepwater **habitat** and the surface.

Some people have thought that catching a live coelacanth to display in an aquarium would be a good idea. Others, including Hans Fricke, have objected to this possibility, arguing that the numbers of coelacanths already taken from its small population have adversely affected the species. Part of the problem is that these animals only bear a few live young at a time, so their fecundity is highly limited and cannot sustain much mortality. Because of these objections, the American and British aquariums that had been considering an expedition to capture a breeding pair of coelacanths changed their minds. An Asian aquarium that did mount such an expedition was unable to capture any live fish.

The coelacanth is protected under the Convention on International Trade in Endangered Species (CITES). However, interest in this unusual fish continues and appears to have created a black market in coelacanth specimens. Unless the government of the Comoro Islands decides to vigorously preserve the endangered coelacanth, instead of allowing continued exploitation, the future of this "living fossil" is not encouraging.

As was just described, living coelacanths were only known from cold, dark waters 1,300-2,000 ft (400-600 m) deep off the Comoro Islands, near the northern tip of Madagascar. Remarkably, however, in 1998 another population of coelacanths was discovered in deep **water** off the Indonesian **island** of Sulawesi. Incredibly, the extremely rare fish was incidentally noticed in a local village market by an American zoologist, Mark Erdmann, who was on a honeymoon vacation. The two

known populations of coelacanths are located 7,000 mi (11,200 km) apart, and are likely different species. Essentially nothing is known yet about the **behavior**, ecology, or abundance of the newly discovered population of coelacanths.

F. C. Nicholson

Coffee plant

The coffee **tree**, genus *Coffea*, family Rubiaceae (Madder family), is native to Ethiopia. The name coffee also refers to the fruit (beans) of the tree and to the beverage brewed from the beans. Coffee is one of the world's most valuable agricultural **crops**.

There are about 30 **species** of *Coffea*, but only two species provide most of the world market for coffee. *Coffea arabica* is indigenous to Ethiopia and was the first cultivated species of coffee tree. *C. arabica* provides 75% of the world's supply of coffee. *Coffea robusta*, also known as *Coffea canephora*, was first discovered growing wild in what is now Zaire. This species was not domesticated and cultivated until the turn of the twentieth century, and now supplies about 23% of the world's coffee. *Coffea liberica* is also an important source of coffee beans, but is mostly consumed locally and does not enter the world market in great quantity. *C. robusta* and *C. liberica* were developed because of their resistance to **insects** and diseases.

Cultivation and harvesting

The coffee tree or shrub grows to 15-30 ft (3-9 m). The tree has shiny, dark green, simple, ovate leaves that grow opposite each other in an alternate fashion, and reach 3 in (7.5 cm) in length. Fragrant, white flowers that bloom for only a few days grow where the leaves join the branches. Clusters of fruit, called cherries, follow the flowers. The cherries are green while developing and growing. The green berries change to yellow, and then to red when the cherries are mature, and deep crimson when ripe and ready for picking. The cherries do not all ripen at once and trees that grow in lower, hotter regions often hold multicolored berries, flowers, and leaves all at once. Each cherry has two chambers or locules that hold two beans. The beans are oval and flat on one side with a lengthwise groove. They are covered by papery skin that must be removed before roasting. A soft, fleshy pulp surrounds the beans. Cherries with one bean, usually round, are called peaberries. Coffee trees raised from **seeds** gen-

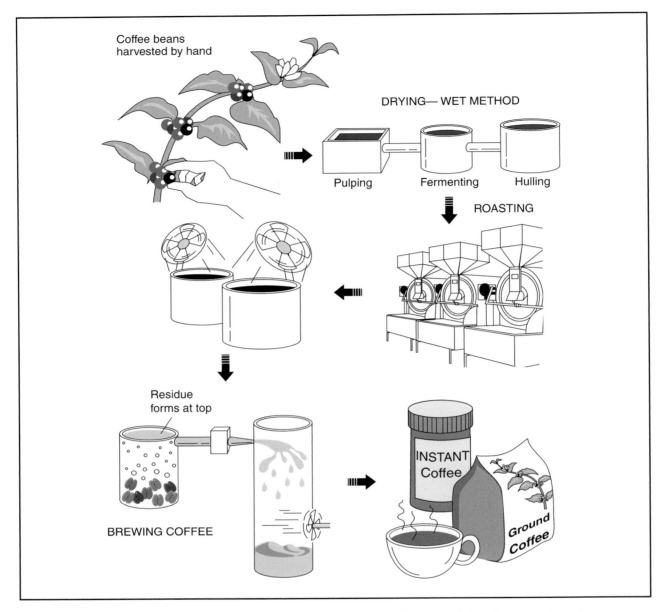

Coffee beans harvested by hand

DRYING— WET METHOD

Pulping   Fermenting   Hulling

ROASTING

Residue forms at top

BREWING COFFEE

INSTANT Coffee

Ground Coffee

The "wet method" of manufacturing coffee. First, the ripe coffee beans are picked and their hulls removed. The beans are then soaked in water during the pulping process, softening and removing most of the skin left on the beans. The remaining residue is removed by fermenting the beans, drying them, and putting them into a hulling machine. The beans are then roasted, cooled, and packaged. *Courtesy of Gale Research.*

erally **flower** the third or fourth year, and produce a good crop at five years. The trees can produce crops for about 15-20 years. Coffee trees can yield from about 1-8 lb (0.5-3.6 kg) in a year, with 1.5-2 lb (0.7-0.9 kg) being the average. It takes 5 lb (2.3 kg) of cherries to produce 1 lb (0.5 kg) of beans.

Coffee grows best in regions located between the Tropic of Cancer and the Tropic of Capricorn (25° north and south of the equator), also called the "coffee belt."

Coffee trees do not produce well in extremely hot weather, nor can they tolerate frost. Ideally, the annual **mean temperature** should be around 70°F (21.1°C). There should be adequate rainfall 70 in (178 cm) per year especially when the fruit is developing. *C arabica* grows best at higher altitudes 2,000-6,000 ft (610-1,830 m) and because the fruit of this species takes about 6-7 months to ripen after flowering, only one crop is harvested per year. *C. robusta* grows best at lower altitudes around 3,000 ft (915 m), and depending on the climate and **soil**, the fruit can be harvested two or three times per year. Coffee trees grow best in rich, well drained, organic soil, particularly in regions with disintegrated volcanic ash. The dangers for growing coffee trees are frost, the coffee bean borer, coffee **leaf** miner, and the fungus *Hemileia vastatrix*.

There are two methods of harvesting and processing the cherries. The wet method, used only where **water** is abundant, involves picking only the ripe cherries. The cherries are soaked in water to soften the skin and the skin and pulp are removed, leaving a sticky film. The cherries are put into tanks to ferment for about 24 hours and then washed to remove the sticky covering on the bean. The beans are spread out to dry, and put into hulling machines that remove the papery skin on the bean. Coffee beans processed by the wet method tend to be more expensive. They are considered to have a better flavor, probably because only ripe cherries are picked. The dry method involves stripping all the cherries from the branches. The cherries are thoroughly dried and put into machines that remove the dry outer covering, pulp, and papery skin. The dry method is the oldest type of processing and is currently used for about two-thirds of the world's coffee. Both processes result in a dried, green coffee bean. Dried, processed beans are then sorted, and graded for quality, type, and size. The beans are packed for transport into bags of 132 lb (60 kg) each. Coffee is exported all over the world and is usually roasted after it reaches its destination.

History

The first cultivated coffee, *C. arabica*, is native to Ethiopia. In **Africa**, coffee beans were consumed as food and later made into wine. The coffee **plant** made its way to neighboring Arabia around A.D. 1000 where it was made into and consumed as a beverage. Coffee beans were introduced to **Europe** during the spice trade (fifteenth century). The first coffee tree was brought to Europe by Jussieu and planted in the Jardin des Plantes, Paris in 1714. This tree was to become the source of all Latin American coffees. This same tree was stolen and later replanted (after a treacherous sea voyage) in Martinique. This species spread to the West Indies and later, Brazil. Until the late part of the seventeenth century, all coffee came from Arabia. The West Indian colonies of Spain and France became major world suppliers. Later, the Dutch successfully cultivated the coffee tree in Indonesia, and eventually became the leading coffee producer. The fungus, *Hemileia vastatrix*, wiped out most of the coffee trees in **Asia**, allowing the West Indian and Brazilian industry to gain dominance. By the late nineteenth century, Brazil had vast coffee plantations and was the leading coffee producer. This status fluctuated

A coffee plant in Costa Rica. © *Alan D. Carey, National Audubon Society Collection/Photo Researchers, Inc. Reproduced with permission.*

with the emancipation of its slaves, incoming European immigrant workers, the start of many small farms, and overproduction. Today, Brazil and Colombia are the world's leading producers of coffee beans.

Resources

Books

Clarke, R.C., and R. Macrae, eds. *Coffee*. 5 vol. New York: Elsevier, 1988.

Lewington, Anna. *Plants for People*. New York: Oxford University Press, 1990.

Christine Miner Minderovic

Cogeneration

Cogeneration is the simultaneous generation of two forms of **energy**, usually **heat** and **electricity**, from one energy source. Traditional energy generating systems produce only heat or electricity by burning a fuel source. In both cases, burning the fuel generates a lot of heat and the exhaust gases can be hotter than 932°F (500°C). Traditionally, this "waste heat" would be vented into the environment for disposal. Cogeneration facilities capture some of that waste heat and use it to produce steam or more electricity. Both systems produce the same amount of energy but cogeneration uses about 35% less fuel because it is designed to be a highly efficient process.

Cogeneration is widely used in some European countries, such as Denmark and Italy, where fuel costs are particularly large. In the United States, where fuel costs are relatively small, cogeneration produces about 5% of the energy supply. Some researchers estimate that if all large U. S. industrial plants used cogeneration technology, there would be enough energy-generating capacity to last until 2020 without building any new power plants.

Why cogenerate?

There are several reasons why cogeneration is a beneficial technology. Cogeneration is an excellent method of improving **energy efficiency**, which has positive environmental and economic results. It also buys time to find new energy sources, and is a reliable, well-understood process.

The most important environmental reason to cogenerate is that vast amounts of precious, non-renewable resources are being wasted by inefficient uses. For example, in the United States, only 16% of the energy used for industrial processes creates useful energy or products. About 41% of the waste is unavoidable because some energy is always lost whenever energy is transformed. However, 43% of the wasted energy could potentially be used in a more energy-efficient process. Cogeneration is an excellent way to increase energy efficiency, which reduces both environmental impacts and operating costs.

Another benefit of cogeneration is that it is an off-the-shelf technology. It has been used in some forms for over a century and therefore most technical problems have been solved. Because cogeneration is a reliable, proven technology, there are fewer installation and operating problems compared with new, untested technology.

History of cogeneration

At the beginning of the twentieth century, steam was the main source of mechanical power. However, as electricity became more controllable, many small "power houses" that produced steam realized they could also produce and use electricity, and they adapted their systems to cogenerate both steam and electricity. Then from 1940 to 1970, the concept developed of a centralized electric utility that delivered power to the surrounding area. Large utility companies quickly became reliable, relatively inexpensive sources of electricity, so the small power houses stopped cogenerating and bought their electricity from the utilities.

During the late 1960s and early 1970s, interest in cogeneration began to revive, and by the late 1970s the need to conserve energy resources became clear. In the United States, legislation was passed to encourage the development of cogeneration facilities. Specifically, the Public Utilities Regulatory Policies Act (PURPA) of 1978 encouraged this technology by allowing cogenerators to connect with the utility network to purchase and sell electricity. PURPA allowed cogenerators to buy electricity from utility companies at fair prices, in times of shortfall, while also allowing them to sell their electricity based on the cost the utility would have paid to produce that power, the so-called "avoided cost." These conditions have encouraged a rapid increase in cogeneration capacity in the United States.

In **Europe**, there has been little government support because cogeneration is not seen as new technology and therefore is not covered under "Thermie," the European Community's (EC) energy program. Under Thermie, 40% of the cost for capital projects is covered by the EC government. However, some individual European countries, like Denmark and Italy, have adopted separate energy policies. In Denmark, 27.5% of their electricity is produced by cogeneration, and all future energy projects must involve cogeneration or some form of alternative energy. In Italy, low-interest loans are provided to cover up to 30% of the cost of building new cogeneration facilities.

Barriers to cogeneration

There are several barriers to the large-scale implementation of cogeneration. Although the operating costs of cogeneration facilities are relatively small, the initial costs of equipment and installation are large. Also, multinational oil companies and central utility companies have substantial political influence in many countries. These companies emphasize their own short-term profits over the long-term environmental costs of inefficient use of non-renewable resources. Other barriers to cogeneration are the falsely low costs of **fossil fuels**, relative to their true, longer-term costs and future scarcity. In a world of plentiful, seemingly inexpensive energy, there is little incentive to use fuel wisely. In addition, national energy policies can have a tremendous effect, like the EC's Thermie policy which does not support cogeneration, and the recent cutbacks in the U. S. energy conservation policies and research, the effects of which remain to be seen.

In the United States, much of the energy research dollar is devoted to developing new energy sources, despite the fact that most of the country's current energy sources are wasted due to inefficient uses. In fact, energy efficiency has not increased much since 1985. As the world's largest user and waster of energy, the United States has a substantial impact on many forms of worldwide **pollution**, and therefore has a special responsibility to use its resources efficiently.

Current research

Current cogeneration research is examining ways of improving the old technology. One improvement in-

KEY TERMS

Avoided cost—Under PURPA, this is the price that the utility company must pay to buy electricity from a cogenerating company. It is calculated as the amount the utility would have paid if the utility company had generated the electricity itself.

Public Utilities Regulatory Policies Act (PURPA)—This is U. S. federal legislation that is designed to encourage the development of cogenerating plants.

Waste heat—This is heat that is released as fuels are burned but is not used.

volves steam-injected gas turbines, which would increase the electric output capacity of the turbines, and thereby increase the energy efficiency of cogeneration. Other improvements are making cogeneration more feasible for smaller plants. Currently, this technology is feasible only in larger facilities. Smaller cogeneration units would allow a more widespread application of this energy efficient technology.

See also Electrical power supply.

Resources

Books

Miller, Jr., G.T., *Environmental Science: Sustaining the Earth.* Belmont, CA: Wadsworth Publishing Company, 1991.
Orlando, J.A. *Cogeneration Planner's Handbook.* Lilburn, GA: Fairmont Press, 1991.
Payne, F.W., ed. *Cogeneration Sourcebook.* Atlanta, Fairmont Press, 1985.

Periodicals

Ganapathy, V. "Recovering Heat When Generating Power." *Chemical Engineering* (February 1993): 94–98.
Shelley, S., and K. Fouhy. "All Fired Up About Cogeneration." *Chemical Engineering* (January 1992): 39-45.

Jennifer LeBlanc

Cognition

Cognition is a complex mental process whereby an individual gains knowledge and understanding of the world. While cognition cannot be neatly dissected into constitutive processes, psychologists point out that it reveals the interplay of such critical psychological mechanisms as **perception**, attention, **memory**, imagery, verbal function, judgment, problem-solving, decision-making, with the admixture of other factors, including physical health, educational background, socio-economic status, and cultural identity. A dynamic process, since both the world and the individual are subject to change, cognition is a vital function which enables an individual to exist in the world as an independent and active participant.

Historical background

Before **psychology** existed as a scientific discipline, the study of cognition was the domain of philosophy. There are two fundamental scientific paradigms-with many variations-regarding cognition in Western philosophy: idealism and empiricism. According to idealistic view, represented by such thinkers as Plato (c. 427-347 B.C.) and René Descartes (1596-1650), innate ideas are the crucial component in cognition; in other words, knowledge is determined by what has been in an individual's mind since-or before-birth. The opposing, empiricist view, is succinctly expressed by John Locke's (1632-1704) dictum that, without sense-perceptions, the mind is an empty slate, a *tabula rasa*. While certain psychologists struggled to determine which of the two paradigms was dominant, the celebrated Swiss cognitive psychologist Jean Piaget (1896-1980) developed a theoretical model of cognition which recognized the importance of both the innate/genetic and the empirical components of cognition. It could be said that cognition depends on both components, just as the successful operation of a computer program requires both the hardware and the software.

How cognition works

Cognition starts with perception. Perception, which occurs in **space** and **time**, provides the general framework for cognition; perception is also the process of becoming aware of a **stimulus**, which can be external or internal. The next step is conceptualization: after realizing the existence of something, we try to figure out what it is: the percept becomes a concept. For example, cognition happens at the instance when the perception "something coming our way" crystallizes as the concept "dog." If the dog is unfriendly, we will use judgment to evaluate our newly acquired knowledge of the situation in an effort to avoid injury. Fortunately, while problem-solving is a key application of the power of judgment in everyday life, not all problems are unpleasant. Working on a mathematical problem, for instance, can be a pleasant, one could say esthetic, experience; the same could be said for any problems requiring creativity and ingenuity, abilities of a higher order than simpler methods, such as the trial-and-error approach. In the realm of the scientific imagination, cognition can, in rare moments, occur as an unexpected flash of illumination. The problem appears to solve itself. One such extraordinary experience is an often quoted mathematical discovery by the French mathematician and philosopher Henri Poincaré (1854-1912). Unable to fall asleep one night, Poincaré thought about a tough problem that he had been grappling with: "Ideas rose in crowds; I felt them collide until pairs interlocked, so to speak, making a stable combination. By the next morning I had established the existence of a new class of Fuchian functions."

Varieties of cognition

Poincaré's experience shows that cognition, while originally stemming from less complex psychological mechanisms, such as perception, is not literally tied to the world of sense-perception. Without contradicting the statement about perception providing the spatio-temporal context of cognition, we can say that cognition also operates in the seemingly unlimited expanses of imaginary space (as in art and **mathematics**) and inner space (as in introspection). In addition, while cognition is traditionally defined as rational and conceptual, it can contain such non-intellectual components as feelings, intuitions, and physical acts. The process of **learning** to play a musical instrument, for example, although a rationally structured endeavor, contains many repetitive, mechanical operations that could be defined as elements of unconscious learning. When the source of new knowledge is unknown, when we do not know why we know something, we are probably dealing with the hidden, silent, non-conceptual dimensions of cognition. A new skill, insight, ability, or perspective suddenly appears "out of nowhere." But this "nowhere" is not really outside the realm of cognition. As in the case of perception, conceptual thinking provides a framework but does not limit cognition. While cognition is certainly limited by human **biology**, it has no limits of its own. Finally, cognition is also never complete; despite repeated attempts, throughout the history of thought, to create closed intellectual systems postulating absolute knowledge as a theoretical goal, the human mind-as evidenced, for example, by the tremendous development of science since the Scientific Revolution-inevitably finds a way to widen the horizons of knowledge. That cognition is an open-ended process is also demonstrated by the seemingly unlimited human capacity for learning, introspection, change, and **adaptation** to a changing world.

See also Brain.

Resources

Books

Matlin, K.M. *Cognition.* 3d ed. San Diego: Harcourt Brace Jovanovich, 1994.

Morris, Charles G. *Psychology: An Introduction.* 7th ed. Englewood Cliffs, NJ: Prentice Hall, 1990.

KEY TERMS

. . . .

Concept—A mental construct, based on experience, used to identify and separate classes of phenomena. The perceived distinctions between cats and dogs allow us to formulate the concepts "cat" and "dog."

Creativity—The ability to find solutions to problems and answers to questions without relying on established methods.

Empiricism—A general philosophical position holding that all knowledge comes from experience, and that humans are not born with any ideas or concepts independent of personal experience.

Idealism—Scientific thinking based on the view that ultimate reality is immaterial.

Imagination—The ability to create alternate worlds without losing contact with reality.

Judgment—The ability to evaluate events and statements.

Percept—The mental representation of a single perceived event or object.

Scientific paradigm—A general view shared by groups of scientists.

Piaget, Jean. *Psychology and Epistemology: Towards a Theory of Knowledge.* New York: Viking, 1971.

Polanyi, Michael. *The Tacit Dimension.* Magnolia, MA: Peter Smith, 1983.

Sheldrake, Rupert. *The Presence of the Past: Morphic Resonance and the Habits of Nature.* New York: Random House, 1988.

Zoran Minderovic

Cold, common

The common cold, also often referred to as an upper respiratory **infection**, is caused by some 200 different viruses, and has defied both cure and **vaccine** for centuries. The United States alone will have about a half a billion colds a year, or two for each man, woman, and child.

Dedicated researchers have searched for a cure or even an effective treatment for years. The pharmaceutical company that discovers the antiviral agent that will kill the cold viruses will reap a great return. Discovering or constructing the agent that will be universally lethal to all the cold-causing viruses has been fruitless. A drug that will kill only one or two of the viruses would be of little use since the patient would not know which of the viruses was the one that brought on his cold. So at present, as the saying goes, if you treat a cold you can get rid of it in about a week. Left untreated it will hang around for about seven days.

The common cold differs in several ways from **influenza** or the flu. Cold symptoms develop gradually and are relatively mild. The flu has a sudden onset and has more serious symptoms the usually put the sufferer to bed, and the flu lasts about twice as long as the cold. Also influenza can be fatal, especially to elderly persons, though the number of influenza viruses is more limited than the number of cold viruses, and vaccines are available against certain types of flu.

Rhinoviruses, adenoviruses, influenza viruses, parainfluenza viruses, syncytial viruses, echoviruses, and coxsackie viruses all have been implicated as the agents that cause the runny nose, cough, sore throat, and sneezing that advertise that you have a cold. More than 200 viruses, each with its own favored method of being passed from one person to another, its own gestation period, each different from the others, wait patiently to invade the mucous membranes that line the nose of the next cold victim.

Passing the cold-causing **virus** from one person to the next can be done by sneezing onto the person, by shaking hands, or by an object handled by the infected person and picked up by the next victim. Oddly, direct contact with the infected person, as in kissing, is not an efficient way for the virus to spread. Only in about 10% of such contacts does the uninfected person get the virus. Walking around in a cold rain will not cause a cold. Viruses like warm, moist surroundings, so they thrive indoors in winter. Colds are easily passed in the winter, because people spend more time indoors then than they do outdoors. However, being outdoors in cold weather can dehydrate the mucous membranes in the nose and make them more susceptible to infection by a rhinovirus.

In addition, the viruses mutate with regularity. Each time it is passed from one person to the next the virus changes slightly, so it is not the virus the first person had. Viruses are tiny creatures considered to be alive, though they hover on the brink of life and lifelessness. They are obligate **parasites**, meaning that they can carry out their functions only when they invade another living thing, **plant** or **animal**.

The virus is a tough envelope surrounding its **nucleic acid**, the genetic structure for any living thing. Once it invades the body the virus waits to be placed in the location in which it can function best. Once there it attaches to a **cell** by means of receptor areas on its envelope and

KEY TERMS

Acute rhinitis—The medical term given to the common cold. No one knows where the name "cold" came from since a cold can be caught during warm as well as cold weather. Rhinitis means inflammation of the nose.

Mucous membrane—the moist lining of the respiratory and digestive systems. Cells that produce mucus maintain the condition of these membranes.

Vaccine—A substance given to ward off an infection. Usually made of attenuated (weakened) or killed viruses or bacteria, the vaccine causes the body to produce antibodies against the disease.

on the cell **membrane**. The viral nucleicacid then is inserted into the cell nucleus and it takes over the functions of the nucleus, telling it to reproduce viruses.

Taking regular doses of **vitamin** C will not ward off a cold. However, high doses of vitamin C once a person has a cold may help to alleviate symptoms and reduce discomfort. Over-the-counter drugs to treat colds treat only the symptoms. True, they may dry up the patient's runny nose, but after a few days the nose will compensate and overcome the effects of the medication and begin to drip again. The runny nose is from the loss of **plasma** from the **blood** vessels in the nose. Some researchers believe the nose drip is a defensive mechanism to prevent the invasion of other viruses. **Antibiotics** such as penicillin are useless against the cold because they do not affect viruses.

Scientists agree that the old wives' remedy of chicken soup can help the cold victim, but so can any other hot liquid. The steam and **heat** produced by soup or tea helps to liquify the mucus in the sinus cavities, allowing them to drain, reducing the pressure and making the patient feel better. The remedy is temporary and has no effect on the virus.

Ridding the body of the viral invaders and therefore easing the symptoms of the cold are the functions of the body's **immune system**. An assortment of white blood cells, each with a different function, gathers at the site of invasion and heaviest viral population and wages a life and death struggle against the invaders. It will take about a week, but in most cases the body's defenses will prevail.

Resources

Periodicals

Huntington, D. "How to Stop the Family Cold Before it Stops You." *Parents* 69 (February, 1994): 26-28.

Poppy, J. "How toMake Colds Less Common." *Men's Health* 9 (January/February, 1994): 30–31.

Larry Blaser

Collagen

Collagen is a protein found abundantly throughout the bodies of animals, including humans. In fact, collagen makes up about one-third of the total body weight. Collagen is an important component of the body's connective tissues, which perform a variety of functions in the body. These tissues provide the framework, or internal scaffolding, for various organs such as the kidneys and lymph nodes. Connective tissues also impart great support and strength to structures such as the bones and tendons. **Blood**, an important type of **connective tissue**, transports **oxygen** and **nutrients** throughout the body.

Connective **tissue** is composed of a nonliving, gel-like material called a matrix, in which living cells are embedded. The matrix is composed of different kinds of protein fibers, the most common of which is collagen.

Structure of collagen

Collagen is a fibrous protein; that is, it is composed of many fibers. Each fiber consists of three microscopic ropes of protein wrapped around each other. The fibers in collagen are arranged **parallel** to each other, and are often grouped together in bundles. The bundling of collagen fibers gives the fibers greater strength than if they occurred individually. Collagen fibers are extremely tough and can resist a pulling **force**, but because they are not taut, they allow some flexibility.

Locations and functions of collagen

Collagen is a primary component of the connective tissue located in the dermis, the tough inner layer of the skin. This kind of connective tissue is also found in mucous membranes, nerves, blood vessels, and organs. Collagen in these structures imparts strength, support, and a certain amount of **elasticity**. As the skin ages, it loses some of its elasticity, resulting in wrinkles. Recently, injections of **animal** collagen given under the surface of the skin have been used to "plump up" the skin and remove wrinkles. However, this treatment is controversial. Many people develop allergic reactions to the collagen, and the procedure must be performed by a qualified physician.

Collagen is also a component of a kind of connective tissue that surrounds organs. This connective tissue encases and protects delicate organs like the kidneys and spleen.

Other locations where collagen fibers are prominent are in the tendons and ligaments. Tendons are straps of tough tissue that attach muscles to bones, allowing for movement. Ligaments are structures that hold the many bones of a joint, such as the knee joint, in proper position. Tendons and ligaments differ slightly in structure. In ligaments, the collagen fibers are less tightly packed than in tendons; in some ligaments, the fibers are not parallel.

Collagen adds strength to tendons and ligaments, and it imparts some stretch to these structures by allowing for some flexibility. However, collagen is not extremely elastic. If tendons and ligaments are stretched too far, these structures will tear, which may lead to problems in movement and bone position. Many athletes tear tendons and ligaments. When tearing occurs, the joint or bone in which the structures occur must be immobilized to allow for proper healing.

Cartilage is a connective tissue found in various places throughout the body, including the tip of the nose, the outside of the ears, the knees, and parts of the larynx and trachea. Cartilage consists of collagen fibers and cartilage cells. At these locations, collagen provides flexibility, support, and movement. Cartilage soaks up **water** like a sponge and is therefore somewhat "springy" and flexible. If the tip of the nose is pushed in and let go, it springs immediately back into place.

Examples of collagen in the animal kingdom

In **vertebrates**, which include all animals with a backbone, connective tissues are highly organized and developed. In **invertebrates**, which include the animals without backbones, connective tissues are not as well organized. However, in nematodes, also known as **roundworms** (an invertebrate animal), collagen plays a role in movement. The outer covering of the nematode, called the cuticle, consists primarily of collagen. The collagen helps the nematode move and also imparts some longitudinal elasticity. Because the collagen fibers crisscross each other and are not parallel in the nematode cuticle, nematodes are limited in side-to-side movement.

See also Muscular system.

Resources

Books

Ayad, Shirley, et al. *The Extracellular Matrix Factsbook.* San Diego: Academic Press, 1994.
Hay, Elizabeth D., ed. *The Cell Biology of Extracellular Matrix.* New York: Plenum Press, 1991.
Kucharz, Eugene. *The Collagens: Biochemistry and Pathophysiology.* New York: Springer-Verlag, 1992.

Periodicals

Fackelmann, Kathy A. "Chicken Collagen Soothes Aching Joints." *Science News* 144 (September 25, 1993): 198.

KEY TERMS

. . . .

Cartilage—A connective tissue found in the knees, tip of the nose, and outside of the ears; it provides flexibility and resilience to these structures.

Ligaments—Structures that hold the bones of joints in the proper position.

Matrix—The nonliving, gel-like component of connective tissue.

Tendon—A strap of tissue that connects muscle to bone, providing for movement.

Johnstone, Iain L. "The Cuticle of the Nematode *Caenorhabditis elegans* " *BioEssays* 16 (March 1993): 171.

Young, Crain M., et al. "Smart Collagen in Sea Lilies." *Nature* 366 (December 9, 1993): 519.

Kathleen Scogna

Colloid

A colloid is a type of particle intermediate in size between a **molecule** and the type of particles we normally think of, which are visible to the naked **eye**. Colloidal particles are usually from 1 to 1,000 nanometers in diameter. When a colloid is placed in **water**, it forms a mixture which is similar in some ways to a **solution**, and similar in some ways to a suspension. Like a solution, the particles never settle to the bottom of the container. Like a suspension, the dispersion is cloudy.

The size of colloidal particles accounts for the cloudiness of a colloidal dispersion. A true solution, such as you might obtain by dissolving table **salt** in water, is transparent, and **light** will go through it with no trouble, even if the solution is colored. A colloidal dispersion, on the other hand, is cloudy. If it is held up to the light, at least some of the light scatters as it goes through the dispersion. This is because the light rays bounce off the larger particles in the colloid, and bounce away from your eye.

A colloidal dispersion does not ever settle to the bottom of the container. In this way, it is like a solution. The particles of the dispersion, though relatively large, are not large and heavy enough to sink. The solvent molecules support them for an indefinite time.

Everyone has seen what looks like dust particles moving about in a beam of sunlight. What you see is light reflected from colloid-sized particles, in **motion** because of tiny changes in air currents surrounding the sus-

pended particles. This type of motion, called *Brownian motion*, is typical of colloids, even those in suspension in solution, where the motion is actually caused by bombardment of the colloidal particles by the molecules of the liquid. This constant motion helps to stabilize the suspension, so the particles do not settle.

Another commonly visible property of a colloidal dispersion is the *Tyndall effect*. If you shine a strong light through a translucent colloidal dispersion that will let at least some of the light through, the light beam becomes visible, like a column of light. This is because the large particles of the dispersed colloid scatter the light, and only the most direct beams make it through the medium.

Milk is the best known colloidal dispersion, and it shows all these properties. They can be seen by adding several drops of milk to a **glass** of water. Most of its cloudiness is due to **fat** particles that are colloidal in size, but there is also a significant amount of protein in it, and some of these are also in the colloidal size range.

See also Brownian motion.

Colobus monkeys

Colobus **monkeys** and the closely related **langurs and leaf monkeys** are Old World monkeys in the subfamily Colobinae of the family Cercopithecidae. The **primates** in this subfamily share a common trait—they lack thumbs or have only small, useless thumbs. (The name *colobus* comes from a Greek word meaning mutilated.) However, lack of a thumb does not stop them from nimbly moving among the branches. They just grasp a branch between the palm of the hand and the other fingers. The Colobinae are distinguished from the other subfamily of Old World monkeys, the Cercopithecinae, by their lack of cheek pouches, their slender build, and their large salivary **glands**. They live throughout equatorial **Africa**, India, and Southeast **Asia**.

Unlike most monkeys, which are fairly omnivorous, eating both **plant** and **animal** foods, many colobus monkeys eat primarily leaves and shoots. In the tropical and subtropical regions in which these leaf-eaters live, they have an advantage over other monkeys in that they have a year-round supply of food. Leaf-eaters also have an advantage in that they can live in a wider variety of habitats than fruit-eaters, even in open **savanna**. Fruit-eaters, on the other hand, must keep on the move in **forests** to find a fruit supply in season. Most colobus monkeys also eat **fruits**, flowers, and **seeds** when they can find them.

Like all Old World monkeys, colobus monkeys have tough hairless pads, called ischial callosities, on their

A black and white colobus monkey in Lake Nakuru National Park, Kenya. © *Tim Davis, National Audubon Society Collection/ Photo Researchers, Inc. Reproduced with permission.*

rear ends. There are no nerves in these pads, a fact that allows them to sit for long periods of time on **tree** branches without their legs "going to sleep." Some colobus monkeys have loud, raucous calls that they use particularly at dawn. Apparently these signals tell neighboring troops of monkeys where a particular group is going to be feeding that day.

Leaves do not have much nourishment, so colobus monkeys must eat almost continuously, up to a quarter of their body weight each day. Their leafy diet requires them to have stomachs specially adapted for handling hard-to-digest materials. The upper portion of the stomach contains **anaerobic bacteria** that break down the **cellulose** in the foliage. These bacteria are also capable of digesting chemicals in leaves that would be poisonous to other primates. Colobus monkeys have large salivary glands which send large amounts of saliva into the fermenting food to help its passage through the **digestive system**.

Black and white colobus monkeys

The black and white colobus monkeys of central Africa (genus *Colobus)* have the least visible thumb of any genus in the subfamily Colobinae, although what little thumb remains has a nail on it. These monkeys have slender bodies, bare faces, and long tails with a puff of long fur on the end. Head and body length is 17.7-28.3 in (45-72 cm), tail length is 20.5-39.3 in (52-100 cm), and weight is 11.9-31.9 lb (5.4-14.5 kg).

The five **species** in this genus are distinguished by the amount of white markings and by the placement of long silky strands of fur in different locations. The black colobus (*C. satanus*) has a completely black, glossy coat. *Colobus polykomos* has a white chest and whiskers and a white, tuftless tail; *C. vellerosus* has white thigh patches, a white mane framing its face, and a white, tuftless tail; *C. guereza* has a flat black cap over a white beard and "hairline," a long white mantle extending from the shoulders to the lower back, and a large white tuft on its tail; *C. angolensis* has long white hairs around the face and on the shoulders, and a white tuft on the end of its tail.

Black and white colobus monkeys are typically highly arboreal (tree-dwelling) inhabitants of deep forests, but some species feed and travel on the ground where the trees are more widely spaced. When they sit still in the trees, these monkeys are well camouflaged because their black and white **color** blends with the patches of sunlight and shadow. When moving in the trees, colobus monkeys tend to walk along branches, either upright or on all fours, instead of swinging beneath them, and they often make amazingly long leaps from tree to tree. Their long hair apparently acts as a parachute to slow them down.

Black and white colobus monkeys often live in small social groups, consisting of both males and females. For example, *C. guereza* lives in groups of 3-15 individuals; most groups have a single adult male and several adult females with their young. The female membership in these groups seems stable, but adult males are sometimes ousted by younger males. Relations among members of the same group are generally friendly and are reinforced by mutual grooming.

Black and white colobus monkeys can apparently breed at any time of the year. Females become sexually mature at about four years of age, while males reach sexual maturity at about six years of age. Each pregnancy results in a single offspring. Infants are born with all white fur, which is shed before the regular coloring comes in. Child rearing seems to be shared among the females in the group.

All species of black and white colobus monkeys have declined over the last 100 years due to hunting for meat and the fur trade, the rapid expansion of human populations, and **habitat** destruction by logging or agriculture. The skins of black-and-white colobus monkeys were often used for clothing in **Europe** during the nineteenth century. They were still available as rugs in the early 1970s. The pelts of as many as 50 animals might have been used to make a single rug. The black colobus *(C. satanus)* is classified as vulnerable by IUCN—The World Conservation Union. Its continued survival is threatened by hunting and habitat disturbance and destruction.

Red colobus monkeys

Red colobus monkeys (genus *Procolobus* or *Piliocolobus*) live along the equator in Africa. They come in many different colors in addition to the reddish black that gives them their name. They often have whitish or grayish faces and chests, with the deep red color appearing only on their back, crown of the head, paws, and tip of the tail. This color variety has made these monkeys difficult to classify, and there is considerable disagreement in their grouping. Red colobus monkeys have a head and body length of 17.7-26.4 in (45-67 cm), a tail length of 20.5-31.5 in (52-80 cm), and weigh 11.2-24.9 lb (5.1-11.3 kg). These monkeys have no thumb at all, lacking even the small vestigial thumb seen in black and white colobus monkeys.

Red colobus monkeys are also arboreal. Most populations are found in rain forests, but they also inhabit savanna woodland, mangrove swamps, and floodplains. Red colobus monkeys also form stable groups, but the groups are much larger than those formed by black and white colobus monkeys-ranging in size from 12 to 82 with an average size of 50. These groups usually include several adult males and 1.5-3 times as many adult females. There is a dominance hierarchy within the group maintained by aggressive **behavior**, but rarely by physical fighting. Higher ranking individuals have priority access to food, space, and grooming.

Red colobus monkeys also seem to breed throughout the year. A single offspring is born after a gestation period of 4.5-5.5 months. The infant is cared for by the mother alone until it reaches 1-3.5 months old.

Most red colobus species are coming under increased pressure from timber harvesting. This activity not only destroys their **rainforest** habitat, but also makes them more accessible to hunters. At least one authority considers red colobus monkeys to be the easiest African monkeys to hunt. Several species and subspecies are considered endangered, vulnerable, or rare by interna-

> ## KEY TERMS
> ··
> **Ischial callosity**—A hard hairless pad of skin, or callus, located on the lower part of the buttocks, or ischium.

tional conservation organizations. For example, the Zanzibar red colobus (*Procolobus kirkii*) is seriously endangered—in 1981 less than 1,500 animals were estimated to survive.

Grouped with the red colobus monkeys because of its four-chambered stomach is the olive colobus, *Procolobus verus*, of Sierra Leone and central Nigeria. This monkey is actually more gray than olive or red. Its head and body length is 16.9–35.4 in (43–90 cm) and it weighs 6.4–9.7 lb (2.9–4.4 kg). This species is also arboreal and is restricted to rainforests. It forms small groups of 10–15 individuals usually with more than one adult male in each group. In a practice unique among monkeys and **apes**, mothers of this species carry newborns in their mouths for the first several weeks of the infant's life. This species is threatened by intensive hunting and habitat destruction; it is considered vulnerable by IUCN—The World Conservation Union.

Resources

Books

Kerrod, Robin. *Mammals: Primates, Insect-Eaters and Baleen Whales.* New York: Facts on File, 1988.

Napier, J. R., and P. H. Napier. *The Natural History of the Primates.* Cambridge, MA: The MIT Press, 1985.

Nowak, Ronald M., ed. *Walker's Mammals of the World.* 5th ed. Baltimore: Johns Hopkins University Press, 1991.

Peterson, Dale. *The Deluge and the Ark: A Journey Into Primate Worlds.* Boston: Houghton Mifflin, 1989.

Preston-Mafham, Rod, and Ken Preston-Mafham. *Primates of the World.* New York: Facts on File, 1992.

Jean F. Blashfield

Color

Color is a complex and fascinating subject. Several fields of science are involved in explaining the phenomenon of color. The **physics** of **light**, the **chemistry** of colorants, the **psychology** and **physiology** of human emotion are all related to color. Since the beginning of history, people of all cultures have tried to explain why there is light and why we see colors. Some people have regarded color with the same mixture of fear, reverence, and curiosity with which they viewed other natural phenomena. In recent years scientists, artists, and other scholars have offered interpretations of the **sun**, light, and color differences.

Color **perception** plays an important role in our lives and enhances the quality of life. This influences what we eat and wear. Colors help us to understand and appreciate the beauty of sunrise and sunset, the artistry of paintings, and the beauty of a bird's plumage. Because the sun rises above the horizon in the east, it gives light and color to our world. As the sun sets to the west, it gradually allows darkness to set in. Without the sun we would not be able to distinguish colors. The **energy** of sunlight warms the **earth**, making life possible; otherwise, there would be no plants to provide food. We are fortunate to have light as an essential part of our **planet**.

Light and color

Colors are dependent on light, the primary source of which is sunlight. It is difficult to know what light really is, but we can observe its effects. An object appears colored because of the way it interacts with light. A thin line of light is called a ray; a beam is made up of many rays of light. Light is a form of energy that travels in waves. Light travels silently over long distances at a speed of 190,000 mi (300,000 km) a second. It takes about eight minutes for light to travel from the sun to the earth. This great speed explains why light from shorter distances seems to reach us immediately.

When we talk about light, we usually mean white light. When white light passes through a **prism** (a triangular transparent object) something very exciting happens. The colors that make up white light disperse into seven bands of color. These bands of color are called a **spectrum** (from the Latin word for image). When a second prism is placed in just the right position in front of the bands of this spectrum, they merge to form invisible white light again. Isaac Newton (1642-1727) was a well known scientist who conducted research on the sun, light, and color. Through his experiments with prisms, he was the first to demonstrate that white light is composed of the colors of the spectrum.

Seven colors constitute white light: red, orange, yellow, green, blue, indigo, and violet. Students in school often memorize acronyms like ROY G BIV, to remember the seven colors of the spectrum and their order. Sometimes blue and indigo are treated as one color. In any spectrum the bands of color are always organized in this order from left to right. There are also wavelengths outside the visible spectrum, such as ultraviolet.

Rainbows

A rainbow is nature's way of producing a spectrum. One can usually see **rainbows** after summer showers, early in the morning or late in the afternoon, when the sun is low. Rain drops act as tiny prisms and disperse the white sunlight into the form of a large beautiful arch composed of visible colors. To see a rainbow one must be located between the sun and raindrops forming an **arc** in the sky. When sunlight enters the raindrops at the proper angle, it is refracted by the raindrops, then reflected back at an angle. This creates a rainbow. Artificial rainbows can be produced by spiraling small droplets of **water** through a garden hose, with one's back to the sun. Or, indoors, diamond-shaped **glass** objects, **mirrors**, or other transparent items can be used.

Refraction: the bending of light

Refraction is the bending of a light ray as it passes at an angle from one transparent medium to another. As a beam of light enters glass at an angle, it is refracted or bent. The part of the light beam that strikes the glass is slowed down, causing the entire beam to bend. The more sharply the beam bends, the more it is slowed down.

Each color has a different wavelength, and it bends differently from all other colors. Short wavelengths are slowed more sharply upon entering glass from air than are long wavelengths. Red light has the longest wavelength and is bent the least. Violet light has the shortest wavelength and is bent the most. Thus violet light travels more slowly through glass than does any other color.

Like all other wave phenomena, the speed of light depends on the medium through which it travels. As an analogy, think of a wagon that is rolled off a sidewalk onto a lawn at an oblique angle. When the first wheel hits the lawn, it slows down, pulling the wagon toward the grass. The wagon changes direction when one of its wheels rolls off the pavement onto the grass. Similarly, when light passes from a substance of high **density** into one of low density, its speed increases, and it bends away from its original path. In another example, one finds that the speed and direction of a car will change when it comes upon an uneven surface like a bridge.

Sometimes while driving on a hot sunny day, we see pools of water on the road ahead of us, which vanish mysteriously as we come closer. As the car moves towards it the pool appears to move further away. This is a mirage, an optical illusion. Light travels faster through hot air than it does through cold air. As light travels from one transparent material to another it bends with a different refraction. The road's hot surface both warms the air directly above it and interacts with the light waves reaching it to form a mirage. In a mirage, reflections of trees and buildings may appear upside down. The bending or refraction of light as it travels through layers of air of different temperatures creates a mirage.

Diffraction and interference

Similar colors can be seen in a thin film of oil, in broken glass and on the vivid wings of **butterflies** and other **insects**. Scientists explain this process by the terms, **diffraction** and **interference**. Diffraction and refraction both refer to the bending of light. Diffraction is the slight bending of light away from its straight line of travel when it encounters the edge of an object in its path. This bending is so slight that it is scarcely noticeable. The effects of diffraction become noticeable only when light passes through a narrow slit. When light waves pass through a small opening or around a small object, they are bent. They merge from the opening as almost circular, and they bend around the small object and continue as if the object were not there at all. Diffraction is the sorting out of bands of different wavelengths of a beam of light.

When a beam of light passes through a small slit or pin hole, it spreads out to produce an image larger than the size of the hole. The longer waves spread out more than the shorter waves. The rays break up into dark and light bands or into colors of the spectrum. When a ray is diffracted at the edge of an opaque object, or passes through a narrow slit, it can also create interference of one part of a beam with another.

Interference occurs when two light waves from the same source interact with each other. Interference is the reciprocal action of light waves. When two light waves meet, they may reinforce or cancel each other. The phenomenon called diffraction is basically an interference effect. There is no essential difference between the phenomena of interference and diffraction.

Light is a mixture of all colors. One cannot look across a light beam and see light waves, but when light waves are projected on a white screen, one can see light. The idea that different colors interfere at different angles implies that the wavelength of light is associated with its colors. A spectrum can often be seen on the edges of an aquarium, glass, mirrors, chandeliers or other glass ornaments. These colored edges suggest that different colors are deflected at different angles in the interference pattern.

The color effects of interference also occur when two or more beams originating from the same source interact with each other. When the light waves are in phase, color intensities are reinforced; when they are out of phase, color intensities are reduced.

When light waves passing through two slits are in phase there is constructive interference, and bright light will result. If the waves arrive at a point on the screen out of phase, the interference will be destructive, and a dark line will result. This explains why bubbles of a nearly colorless **soap** solution develop brilliant colors before they break. When seen in white light, a soap bubble presents the entire visible range of light, from red to violet. Since the wavelengths differ, the film of soap cannot cancel or reinforce all the colors at once. The colors are reinforced, and they remain visible as the soap film becomes thinner. A rainbow, a drop of oil on water, and soap bubbles are phenomena of light caused by diffraction, refraction, and interference. Colors found in **birds** such as the blue jay are formed by small air bubbles in its feathers. Bundles of white rays are scattered by suspended particles into their components colors. Interference colors seen in soap bubbles and oil on water are visible in the peacock feathers. Colors of the mallard duck are interference colors and are iridescent changing in hue when seen from different angles. Beetles, **dragonflies**, and butterflies are as varied as the rainbow and are produced in a number of ways, which are both physical and chemical. Here, the spectrum of colors are separated by thin films and flash and change when seen from different angles.

Light diffraction has the most lustrous colors of mother of pearl. Light is scattered for the blue of the sky which breaks up blue rays of light more readily than red rays.

Transparent, translucent, and opaque

Materials like air, water, and clear glass are called transparent. When light encounters transparent materials, almost all of it passes directly through them. Glass, for example, is transparent to all visible light. The color of a transparent object depends on the color of light it transmits. If green light passes through a transparent object, the emerging light is green; similarly if red light passes through a transparent object, the emerging light is red.

Materials like frosted glass and some **plastics** are called translucent. When light strikes translucent materials, only some of the light passes through them. The light does not pass directly through the materials. It changes direction many times and is scattered as it passes through. Therefore, we cannot see clearly through them; objects on the other side of a translucent object appear fuzzy and unclear. Because translucent objects are semitransparent, some ultraviolet rays can go through them. This is why a person behind a translucent object can get a sunburn on a sunny day.

Most materials are opaque. When light strikes an opaque object none of it passes through. Most of the light is either reflected by the object or absorbed and converted to **heat**. Materials such as **wood**, stone, and metals are opaque to visible light.

Mixing colors

We do not actually see colors. What we see as color is the effect of light shining on an object. When white light shines on an object it may be reflected, absorbed, or transmitted. Glass transmits most of the light that comes into contact with it, thus it appears colorless. Snow reflects all of the light and appears white. A black cloth absorbs all light, and so appears black. A red piece of **paper** reflects red light better than it reflects other colors. Most objects appear colored because their chemical structure absorbs certain wavelengths of light and reflects others.

The sensation of white light is produced through a mixture of all visible colored light. While the entire spectrum is present, the **eye** deceives us into believing that only white light is present. White light results from the combination of only red, green, and blue. When equal brightnesses of these are combined and projected on a screen, we see white. The screen appears yellow when red and green light alone overlap. The combination of red and blue light produces the bluish red color of magenta. Green and blue produce the greenish blue color called cyan. Almost any color can be made by overlapping light in three colors and adjusting the brightness of each color.

Color vision

Scientists today are not sure how we understand and see color. What we call color depends on the effects of light waves on receptors in the eye's retina. The currently accepted scientific theory is that there are three types of cones in the eye. One of these is sensitive to the short blue light waves; it responds to blue light more than to light of any other color. A second type of cone responds to light from the green part of the spectrum; it is sensitive to medium wavelengths. The third type of light sensitive cone responds to the longer red light waves. If all three types of cone are stimulated equally our **brain** interprets the light as white. If blue and red wavelengths enter the eye simultaneously we see magenta. Recent scientific research indicates that the brain is capable of comparing the long wavelengths it receives with the shorter wavelengths. The brain interprets electric signals that it receives from the eyes like a computer.

Nearly 1,000 years ago, Alhazen, an Arab scholar recognized that **vision** is caused by the reflection of light from objects into our eyes. He stated that this reflected light forms optical images in the eyes. Alhazen believed that the colors we see in objects depend on both the light striking these objects and on some property of the objects themselves.

Color blindness

Some people are unable to see some colors. This is due to an inherited condition known as **color blindness**. John Dalton (1766-1844), a British chemist and physicist, was the first to discover color blindness in 1794. He was color blind and could not distinguish red from green. Many color blind people do not realize that they do not distinguish colors accurately. This is potentially dangerous, particularly if they cannot distinguish between the colors of traffic lights or other safety signals. Those people who perceive red as green and green as red are known as red-green color blind. Others are completely color blind; they only see black, gray, and white. It is estimated that 7% of men and 1% of women on Earth are born color blind.

Color effects in nature

We often wonder why the sky is blue, the water in the sea or swimming pools is blue or green, and why the sun in the twilight sky looks red. When light advances in a straight line from the sun to the earth, the light is refracted, and its colors are dispersed. The light of the dispersed colors depends on their wavelengths. Generally the sky looks blue because the short blue waves are scattered more than the longer waves of red light. The short waves of violet light (the shortest of all the light waves) disperse more than those of blue light. Yet the eye is less sensitive to violet than to blue. The sky looks red near the horizon because of the specific angle at which the long red wavelengths travel through the atmosphere. Impurities in the air may also make a difference in the colors that we see.

Characteristics of color

There are three main characteristics for understanding variations in color. These are hue, saturation, and intensity or brightness. Hue represents the observable visual difference between two wavelengths of color. Saturation refers to the richness or strength of color. When a beam of red light is projected from the spectrum onto a white screen, the color is seen as saturated. All of the light that comes to the eye from the screen is capable of exciting the sensation of red. If a beam of white light is then projected onto the same spot as the red, the red looks diluted. By varying the intensities of the white and red beams, one can achieve any degree of saturation. In handling pigments, adding white or gray to a hue is equivalent to adding white light. The result is a decrease in saturation.

A brightly colored object is one that reflects or transmits a large portion of the light falling on it, so that it appears brilliant or luminous. The brightness of the resulting color will vary according to the reflecting quality of the object. The greatest amount of light is reflected on a white screen, while a black screen would not reflect any light.

Mixing colorants, pigments, dyes, and printing

Color fills our world with beauty. We delight in the golden yellow leaves of Autumn and the beauty of Spring flowers. Color can serve as a means of communication, to indicate different teams in sports, or, as in traffic lights, to instruct drivers when to stop and go. Manufacturers, artists, and painters use different methods to produce colors in various objects and materials. The process of mixing of colorants, paints, pigments and dyes is entirely different from the mixing of colored light.

Colorants are chemical substances that give color to such materials as ink, paint, crayons, and chalk. Most colorants consist of fine powders that are mixed with liquids, wax, or other substances that facilitate their application to objects. Dyes dissolve in water. Pigments do not dissolve in water, but they spread through liquids. They are made up of tiny, solid particles, and they do not absorb or reflect specific parts of the spectrum. Pigments reflect a mixture of colors.

When two different colorants are mixed, a third color is produced. When paint with a blue pigment is mixed with paint that has yellow pigments the resulting paint appears green. When light strikes the surface of this paint, it penetrates the paint layer and hits pigment particles. The blue pigment absorbs most of the light. The same color looks different against different background colors. Each pigment subtracts different wavelengths.

Additive and subtractive

All color is derived from two types of light mixture, an additive and a subtractive process. Both additive and subtractive mixtures are equally important to color design and perception. The additive and subtractive elements are related but different. In a subtractive process, blended colors subtract from white light the colors that they cannot reflect. Subtractive light mixtures occur when there is a mixture of colored light caused by the transmittance of white light. The additive light mixture appears more than the white light.

In a subtractive light mixture, a great many colors can be created by mixing a variety of colors. Colors apply only to additive light mixtures. Mixing colored light produces new colors different from the way colorants are mixed. Mixing colorants results in new colors because each colorant subtracts wavelengths of light. But mixing colored lights produces new colors by adding light of different wavelengths.

Both additive and subtractive mixtures of hues adjacent to each other in the spectrum produce intermediate hues. The additive mixture is slightly saturated or mixed with light, the subtractive mixture is slightly darkened. The complimentary pairs mixed subtractively do not give white pigments with additive mixtures.

Additive light mixtures can be used in a number of slide projectors and color filters in order to place different colored light beams on a white screen. In this case, the colored areas of light are added to one another. Subtractive color mixing takes place when a beam of light passes through a colored filter. The filter can be a piece of colored glass or plastic or a liquid that is colored by a dye. The filter absorbs and changes part of the light. Filters and paints absorb certain colors and reflect others.

Subtractive color mixing is the basis of color **printing**. The color printer applies the colors one at a time. Usually the printer uses three colors: blue, yellow, and red, in addition to black for shading and emphasis. It is quite difficult for a painter to match colors. To produce the resulting colors, trial and error, as well as experimenting with the colors, is essential. It is difficult to know in advance what the resulting color will be. It is interesting to watch a painter trying to match colors.

People who use conventional methods to print books and magazines make colored illustrations by mixing together printing inks. The color is made of a large number of tiny dots of several different colors. The dots are so tiny that the human eye does not see them individually. It sees only the combined effects of all of them taken together. Thus, if half of the tiny dots are blue and half are yellow, the resulting color will appear green.

Dying fabrics is a prehistoric craft. In the past, most dyes were provided solely from **plant** and **animal** sources. In antiquity, the color purple or indigo, derived from plants, was a symbol of aristocracy. In ancient Rome, only the emperor was privileged to wear a purple robe.

Today, many **dyes and pigments** are made of synthetic material. Thousands of dyes and pigments have since been created from natural **coal** tar, from natural compounds, and from artificially produced organic chemicals. Modern chemistry is now able to combine various arrangements and thus produce a large variety of color.

The color of a dye is caused by the absorption of light of specific wavelengths. Dyes are made of either natural or synthetic material. Chemical dyes tend to be brighter than natural dyes. In 1991 a metamorphic color system was created in a new line of clothes. The color of these clothes changes with the wearer's body **temperature** and environment. This color system could be used in a number of fabrics and designs.

Sometimes white clothes turn yellow from repeated washing. The clothes look yellow because they reflect more blue light than they did before. To improve the color of the faded clothes a blue dye is added to the wash water, thus helping them reflect all colors of the spectrum more evenly in order to appear white.

Most paints and crayons use a mixture of several colors. When two colors are mixed, the mixture will reflect the colors that both had in common. For example, yellow came from the sap of a **tree**, from the **bark** of the birch tree, and onion skins. There were a variety of colors obtained from leaves, **fruits**, and flowers.

Primary, secondary, and complimentary

Three colorants that can be mixed in different combinations to produce several other colors are the primary colorants. In mixing red, green, and blue paint the result will be a muddy dark brown. Red and green paint do not combine to form yellow as do red and green light. The mixing of paints and dyes is entirely different from the mixing of colored light.

By 1730, a German engraver named J. C. LeBlon discovered the primary colors red, yellow, and blue are primary in the mixture of pigments. Their combinations produce orange, green, and violet. Many different three colored combinations can produce the sensation of white light when they are superimposed. When two primary colors such as red and green are combined, they produce a secondary color. A color wheel is used to show the relationship between primary and secondary colors. The colors in this primary and secondary pair are called complimentary. Each primary color on the wheel is opposite the secondary color formed by the mixture of the two primary colors. And each secondary color produced by mixing two primary colors lies half-way between them on a color wheel. The complimentary colors produce white light when they are combined.

Colors are everywhere

Color influences many of our daily decisions, consciously or unconsciously from what we eat and what we wear. Color enhances the quality of our lives, it helps us to fully appreciate the beauty of colors. Colors are also an important function of the psychology and physiology of human sensation. Even before the ancient civilizations in prehistoric times, color symbolism was already in use.

Different colors have different meanings which are universal. Colors can express blue moods. On the other hand, these could be moods of tranquility or moods of conflict, sorrow or pleasure, warm and cold, boring or stimulating. In several parts of the world, people have

---

KEY TERMS

Beams—Many rays of light.

Colorant—A chemical substance that gives color to such materials as ink, paint, crayons and chalk.

Diffraction—The bending of light.

Hue—The observable visual different between two wavelengths of color.

Light—A form of energy that travels in waves.

Mirage—An optical illusion.

Pigment—A substance which becomes paint or ink when it is mixed with a liquid.

Ray—A thin line of light.

Reflection—The change in direction of light when it strikes a substance but does not pass through it.

Refraction—The bending of light that occurs when traveling from one medium to another, such as air to glass or air to water.

Spectrum—A display of the intensity of radiation versus wavelength.

---

specific meanings for different colors. An example is how Eskimos indicate the different numbers of snow conditions. They have seventeen separate words for white snow. In the west the bride wears white, in China and in the Middle East area white is worn for mourning. Of all colors, that most conspicuous and universal is red.

The color red can be violent, aggressive, and exciting. The expression "seeing red" indicates one's anger in most parts of the world. Red appears in more national and international colors and red cars are more often used than any other color. Homes can be decorated to suit the personalities of the people living in them. Warm shades are often used in living rooms because it suggests sociability. Cool shades have a quieting effect, suitable for study areas. Hospitals use appropriate colors depending on those that appeal to patients in recovery, in **surgery**, or very sick patients. Children in schools are provided bright colored rooms. Safety and certain color codes are essential. The color red is for fire protection, green is for first aid, and red and green colors are for traffic lights.

Human beings owe their survival to plants. The function of color in the flowering plants is to attract **bees** and other insects—to promote **pollination**. The color of fruits attract birds and other animals which help the distribution of **seeds**. The utmost relationship between humans and animals and plants are the chlorophylls. The green coloring substance of leaves and the yellowish green **chloro-**

phyll is associated with the production of carbohydrates by **photosynthesis** in plants. Life and the quality of the earth's atmosphere depends on photosynthesis.

Resources

Books

Birren, Faber. *Color—A Survey in Words and Pictures.* New Hyde Park, NY: University Books, Inc, 1963.

Birren, Faber. *History of Color in Painting.* New York: Reinhold Publishing Corporation, 1965.

Birren, Faber. *Principles of Color.* New York: Van Nostrand Reinhold Co., 1969.

Hewitt, Paul. *Conceptual Physics.* Englewood Cliffs, NJ: Prentice Hall, 2001.

Meadows, Jack. *The Great Scientists.* New York: Oxford University Press, 1992.

Verity, Enid. *Color Observed.* New York: Van Nostrand Reinhold Co., 1990.

Periodicals

Cunningham, James, and Norman Herr. "Waves." "Light." *Hands on Physics Activities with Real Life Applications* West Nyack, NY: Center for Applied Research Education, 1994.

Kosvanec, Jim. "Mixing Grayed Color." *American Artist* 1994.

Suding, Heather, and Jeanne Bucegross. "A Simple Lab Activity to Teach Subtractive Color and Beer's Law," *Journal of Chemical Education,* 1994.

Nasrine Adibe

Color blindness

The condition known as **color** blindness is a defect in **vision** that causes problems in distinguishing between certain colors. The condition is usually passed on genetically, and is more common in men than in women. About 6% of all men and about 0.6% of women inherit the condition. Individuals can also acquire the condition through various **eye** diseases. There is no treatment for color blindness.

Reds and greens

The first study of color blindness was published in 1794 by physicist John Dalton, who was color-deficient himself. The condition Dalton described is not actually any sort of blindness. Color blindness does not affect the overall visual acuity of individuals with the condition. A small number of people can not distinguish between any color and see all things in shades of gray.

People who are color blind often are not aware they have a problem until they are asked to distinguish between reds and greens. This is the most common problem among individuals who are color blind. Some people

who are color blind also have trouble telling the difference between green and yellow.

Color blindness stems from a problem in the cone cells of the retina. **Light** rays enter the eye in some combination of red, green, or blue. Normal cone cells contain light-sensitive molecules sensitive to one of the color spectrum's band of colors. Short-wave cone cells absorb blue, middle-wave cone cells absorb green, and long-wave cone cells absorb red.

Individuals with a color defect do not have a normal complement of these substances, and may be missing one or more of them. Some people who are color blind have trouble distinguishing between reds and greens when the light is dim, but are capable of seeing the difference between the two colors in good light. A less common type of color blindness makes distinguishing between reds and greens difficult regardless of the light quality.

A simple test for color blindness involves the use of cards with dots in different colors. Individuals who are color blind see different numbers or words than those who have a complete range of color vision.

Inherited or acquired defect

Most individuals who are color blind inherit the trait. Men are more likely to be color blind because of the way color blindness is inherited. The **gene** for the trait is located on the X **chromosome**. Men have one X chromosome and women have two. If a man inherits the gene for the trait, he will have a color vision defect. If a woman inherits a single gene for the trait, she will not, because the normal gene on her other X chromosome will dominate over the defective gene. Women must inherit the defective trait from both parents to be color blind.

Color blindness is a so-called sex-linked characteristic. This means it is a gene that occurs only on the X chromosome, which is passed to the child by the mother. The Y chromosome, which is passed to the child by the father, does not carry the defective gene. This means that children inherit color blindness only from their mothers. Children can inherit color blindness from a mother who is color blind or from a mother who is a carrier of the gene but is not color blind herself. Daughters of men who are color blind will carry the trait, but sons will not.

A more unusual way to become color blind is through **disease**. Cataracts are the most common cause of acquired color deficiency. In one of the most common eye diseases, cataracts, a cloudy layer in the **lens** or eye capsule develops. The condition can cause vision to worsen in bright sunlight. Other conditions that may cause acquired color deficiency are retinal and optic nerve diseases.

Resources

Books

Donn, Anthony. "The Eye and How it Works." *The Columbia University College of Physicians and Surgeons Complete Home Medical Guide*. 2nd Ed., 1989.

Kunz, Jeffrey R.M., and Asher J. Finkel. "Color Blindness." *The American Medical Association Family Medical Guide*. New York: Random House, 1987.

Periodicals

Chandler, David G. "Diabetes Problems Can Affect Color Vision." *Diabetes in the News* (May-June 1993): 42.

"Color Blindness Misconceptions." *USA Today* February 1992., 16.

Mollon, John. "Worlds of Difference." *Nature* Vol. 356. (April 2, 1992): 378-379.

"Not Seeing Red." *The University of California, Berkeley Wellness Letter* (August 1993): 2.

Patricia Braus

KEY TERMS

Cataract—Eye disease characterized by the development of a cloudy layer in the lens of the eye.

Chromosomes—The structures that carry genetic information in the form of DNA. Chromosomes are located within every cell and are responsible for directing the development and functioning of all the cells in the body.

Molecule—A chemical combination of atoms, and the smallest amount of a chemical substance.

Retina—An extremely light-sensitive layer of cells at the back part of the eyeball. The image formed by the lens on the retina is carried to the brain by the optic nerve.

Medications such as **digitalis**, a common medication for **heart** disease, and **quinine**, medicine for **malaria**, can also make color **perception** change. **Alcohol** has also been known to change the way people see color.

Adapting to a different world

Color blindness generally does not cause a great deal of hardship. However, there is evidence that individuals who are color blind may face higher risks on the road. A German study found that men who were color blind were twice as likely to have rear-end collisions as were men who had normal vision. About seven million North American drivers can not distinguish easily between red and green lights.

Designers of traffic signals are working to make driving easier for color-deficient motorists. Traffic lights are generally made in a standard format today, with red on top, amber in the middle and green at the bottom. One improvement would be changing the shape of each of the different signals, so that color-deficient drivers could more easily distinguish between stop and go. Another possible change would involve altering the color of brake lights. Experts bemoan the fact that people who are color-deficient can not see the red in brake lights clearly.

There is no cure or treatment for color blindness. However, there is an abundant amount of research concerning the nature of vision in people with normal and limited color discrimination. As researchers become more knowledgeable about the process of sight, correction of color blindness may become a possibility.

See also Eye; Vision disorders.

Colugos

A colugo is a furry mammal with a thin neck, a slender body, and large eyes. It is about the size of an average house cat, measuring between 15-16.5 in (38-42 cm) long with a tail adding another 8-10 in (20-25 cm). Also known as a flying lemur, the colugo neither truly flies nor is it a lemur. A gliding mammal, it is able to give the appearance of flight with the help of a **membrane** that stretches completely around its body starting behind its ears, going down its neck to its wrists and ankles, and ending at the tip of its tail. Colugos have characteristics of **lemurs**, **bats**, **prosimians**, and insectivores; recent studies suggest that their closest relatives are **primates**.

Because of its varying characteristics, colugos have been classified in their own order, the order Dermoptera. Belonging to the Cynocephalidae family, the only two **species** of colugo are the Malayan or Temminck colugo (*Cynocephalus temminckii*) and the Philippine colugo (*Cynocephalus volans*). Colugos inhabit the rainforests and rubber plantations in Southeast **Asia**, Thailand, Malaysia, Java, Borneo, Vietnam, Kampuchea, and the Philippines.

Characteristics

Colugos differ greatly in terms of their basic coloring. Their backs are usually shades of brown, brownish gray, or gray sprinkled with light yellow spots. Their undersides can be bright orange, brownish red, or yellow. In general, they have very thick, soft fur on their slender bodies. Their necks are long and their heads can be described as "doglike." Colugos females are slightly bigger than males. The Malayan colugo is the larger of the two species, weighing approximately 4 lb (1.8 kg) and measuring about 25 in (63.5 cm) from head to tail. Colugos have large eyes and very good eyesight, traits essential in judging the distances between trees. They have interesting teeth as well; for reasons unknown, their lower incisors are made up of ten thin prongs each, similar to the structure of a comb.

Flightskin

Many animals inhabiting rainforests evolve special mechanisms to enable them to move easily among the trees; thus, they avoid exposing themselves to predators living on the ground. In the colugo's case, this has been accomplished by the development of a membrane surrounding almost all of its body. On each colugo, there are three separate sections of this "parachute-skin." In the first section (called the propatagium), the skin comes out of both sides of the colugo's neck, down its shoulders to completely attached to the entire span of its arm all the way to its fingertips. The second section (called the plagiopatagium) begins on their underside of the colugos' arms and spans its entire body, thus connecting the animal's front and rear legs. The final section of this skin (uropatagium) connects the hind legs to the tail on both sides all the way to its tip.

Interestingly, the colugo is the only mammal that has developed the rearmost uropatagium section of its flightskin. The tails of all of the other **mammals** with the capacity to glide—such as honey gliders and flying squirrels—are free of the membrane; thus, they use their tails as steering mechanisms.

Covered in hair on both sides, this "parachute-skin" enables these animals to glide long distances—reportedly up to 426 ft (130 m)—while losing very little altitude. Although colugos have strong claws with which to grip branches, they are not skilled climbers. In fact, colugos are almost helpless on the ground.

Behavior

Colugos have never lived more than a few months in captivity; thus, there is only limited knowledge of their **behavior**. One fact that is known is that they are strict vegetarians. Specifically, they feed on shoots, fruit, buds, young seed pods, and flowers from multiple kinds of forest trees. They pull the vegetation out of trees using their very powerful tongues. Colugos get their **water** by licking it from leaves and **tree** hollows.

Colugos are nocturnal; when the **sun** goes down, they move quickly, skirting the underside of branches. To get the height they need to glide from tree to tree, they scramble up tree trunks with few powerful jumping mo-

---

KEY TERMS

Patagium—The entire flightskin of the colugo.

Plagiopatagium—The second section of the flightskin. It originates on the underside of the colugos' arms and spans its entire body, thus connecting the animal's front and rear legs.

Propatagium—The front portion of the colugo's flightskin, coming out of both sides of the colugo's neck, down its shoulders. It attaches the neck to the arm all the way to the fingertips.

Uropatagium—The final section of the flightskin, unique to colugo's. It connects the hind legs to the tail all the way to its tip.

---

tions. According to observation, colugos often reuse their routes when traveling from location to location. In the Philippines, the inhabitants seize on the easy hunting that this behavior presents; Philippine locals often wait on known colugo paths with weapons or traps ready.

During the daytime, colugos rest, hanging from the undersides of branches, in tree hollows, or in the shade of palm stalks using all four of their feet. Hanging in such a way makes colugos hard to see. Their brownish coloring helps to camouflage them further in the shady rain forest.

Reproduction

Colugos of the Philippines generally mate in February, although the mating behavior of colugos throughout Southeast Asia can occur from January to March. After a two month pregnancy, the female gives **birth** to a single offspring. (Although, on rare occasions, colugo females have twins.) Interestingly, because females cannot nurse more than one young at a time, they have the ability to give birth in rapid succession to stabilize the population. Thus, the female is able to become pregnant again before her young are weaned.

When born, the baby colugo measures about 10 in (25 cm) long and is fairly undeveloped; in fact, some authorities describe these young as "semi-fetal." After a baby's birth, the female carries it in a pouch which she creates by folding the flightskin under her tail. She holds her young tightly against her as she feeds; as she travels, the baby fastens itself to one of its mother's nipples. The female generally carries the baby around everywhere she goes, reducing the baby's vulnerability to predators. This relationship continues, until the young is too large and heavy for the mother to carry.

Threats to colugos

The main natural **predator** of the colugo is the Philippine monkey-eating eagle, which eats colugos almost to the exclusion of all other types of food. Humans also pose a significant threat to these animals. People operating rubber and coconut plantations often shoot colugos because they view them as **pests**. Furthermore, colugos are hunted for their meat. Most importantly, colugo habitats are continually shrinking due to **deforestation**. While neither species is endangered, their numbers will shrink as their habitats disappear.

Resources

Books

Grzimek, H.C. Bernard, Dr., ed. *Grzimek's Animal Life Encyclopedia.* New York: Van Nostrand Reinhold Company, 1993.

Gunderson, Harvey L. *Mammalogy.* New York: McGraw Hill, 1976.

Pearl, Mary Corliss, Ph.D. Consultant. *The Illustrated Encyclopedia of Wildlife.* London: Grey Castle Press, 1991.

Kathryn Snavely

Coma

Coma, from the Greek word *koma*, meaning deep **sleep**, is a state of extreme unresponsiveness in which an individual exhibits no voluntary movement or **behavior**. In a deep coma, stimuli, even painful stimuli, are unable to effect any response. Normal reflexes may be lost.

Coma lies on a **spectrum** with other alterations in consciousness. The level of consciousness which you, the reader, are currently enjoying is at one end of the spectrum, while complete **brain** death is at the other end of the spectrum. In between are such states as obtundation, drowsiness, and stupor, which all allow the individual to respond to stimuli, though such response may be brief and require a **stimulus** of greater than normal intensity.

Consciousness

In order to understand the loss of function suffered by a comatose individual, consider the important characteristics of the conscious state. Consciousness is defined by two fundamental elements: awareness and arousal.

Awareness allows us to receive and process information communicated by the five senses and thereby relate to ourselves and the rest of the world. Awareness has psychological and physiological components. The psychological component is governed by an individual's mind and its mental processes. The physiological component refers to the functioning—the physical and chemical condition—of an individual's brain. Awareness is regulated by areas within the cerebral hemispheres, the outermost layer of the brain, which separates humans from other animals because it allows greater intellectual functioning.

Arousal is regulated solely by physiological functioning. Its primitive responsiveness to the world is demonstrated by predictable **reflex** (involuntary) responses to stimuli. Arousal is maintained by the reticular activating system (RAS). This is not an anatomical area of the brain but rather a network of structures (including the brainstem, the medulla, and the thalamus) and nerve pathways which function together to produce and maintain arousal.

Causes of coma

Coma is the result of something which interferes with the functioning of the cerebral cortex and/or the functioning of the structures which make up the RAS. The number of conditions which could result in coma is mind-boggling. A good way of categorizing these conditions is to consider the anatomic and the metabolic causes of coma. Anatomic causes of coma are those conditions which disrupt the normal physical architecture of the brain structures responsible for consciousness. Metabolic causes of coma consist of those conditions which change the chemical environment of the brain and thereby adversely affecting function.

Anatomic causes of coma include brain tumors, infections, and head injuries. All three types of condition can affect the brain's functioning by actually destroying brain **tissue**. They may also affect the brain's functioning by taking up too much space within the skull. The skull is a very hard, bony structure which is unable to expand in size. If something within the skull begins to require more space (for example an expanding **tumor** or an injured/infected area of the brain which is swelling) other areas of the brain are compressed against the hard surface of the skull, which results in damage to these areas.

There are many metabolic causes of coma, including the following: (1) A decrease in the delivery of substances necessary for appropriate brain functioning, such as **oxygen**, glucose, and **sodium**. (2) The presence of certain substances disrupting the functioning of neurons. Drugs or **alcohol** in toxic quantities can result in neuronal dysfunction, as can some substances normally found in the body, but which accumulate at toxic levels due to some **disease** state. Accumulated substances which might cause coma include **ammonia** due to liver disease, ketones due to uncontrolled diabetes, or **carbon dioxide** due to a severe **asthma** attack. (3) The changes in chemical levels in the brain due to the electrical derangements caused by seizures.

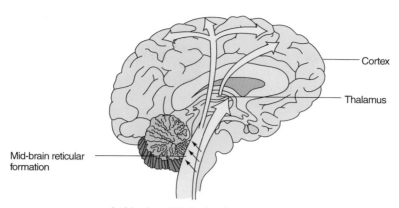

Cortex

Thalamus

Mid-brain reticular formation

A side-view of the brain, showing movement of the reticular activating substance (RAS) essential to consciousness

Diffuse and bilateral damage to the cerebral cortex (relative preservation of brain-stem reflexes)

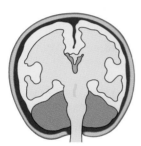

Possible causes
- Damage due to lack of oxygen or restricted blood flow, perhaps resulting from cardiac arrest, an anaesthetic accident, or shock
- Damage incurred from metabolic processes associated with kidney or liver failure, or with hypoglycemia
- Trauma damage
- Damage due to a bout with meningitis, encephalomyelitis, or a severe systemic infection

Mass lesions in this region resulting in compression of the brain-stem and damage to the reticular activating substance (RAS)

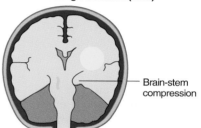

Brain-stem compression

Structural lesions within this region also resulting in compression of the brain-stem and damage to the reticular activating substance (RAS)

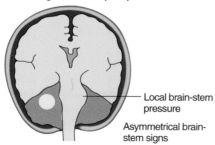

Local brain-stem pressure

Asymmetrical brain-stem signs

Possible causes • Cerebellar tumors, abscesses, or hemorrhages

Lesions within the brain-stem directly suppressing the reticular activating substance (RAS)

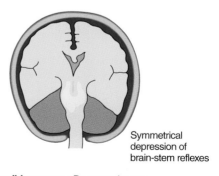

Symmetrical depression of brain-stem reflexes

Possible causes • Drug overdosage

The four brain conditions that result in coma. *Illustration by Hans & Cassidy. Courtesy of Gale Group.*

Outcome

It is extremely important for a physician to quickly determine the cause of a coma, so that potentially reversible conditions are treated immediately. For example, an **infection** may be treated with **antibiotics**, a brain tumor may be removed, brain swelling from an injury can be reduced with certain medications. Furthermore, various **metabolic disorders** can be addressed by supplying the individual with the correct amount of oxygen, glucose, or sodium, by treating the underlying disease in liver disease, asthma, or diabetes, and by halting seizures with medication.

Some conditions which cause coma can be completely reversed, restoring the individual to his or her original level of functioning. However, if areas of the brain have been sufficiently damaged because of the severity or duration of the condition which led to the coma, the individual may recover from the coma with permanent disabilities, or may never regain consciousness. Take the situation of someone whose coma was caused by brain injury in a car accident. Such an injury can result in one of three outcomes. In the event of a less severe brain injury, with minimal swelling, an individual may indeed recover consciousness and regain all of his or her original abilities. In the event of a more severe brain injury, with swelling which results in further pressure on areas of the brain, an individual may regain consciousness, but with some degree of impairment. The impairment may be physical, such as paralysis of a leg, or result in a change in the individual's intellectual functioning and/or personality. The most severe types of brain injury result in states in which the individual loses all ability to function and remains deeply unresponsive. An individual who has suffered such a brain injury may remain in a coma indefinitely.

Outcome from a coma depends on its cause and duration. In drug poisonings, extremely high rates of recovery can be expected, following prompt medical attention. Patients who have suffered head injuries tend to do better than patients whose coma was caused by other types of medical illnesses. Excluding drug-poisoning induced comas, only about 15% of patients who remain in a coma for more than a few hours make a good recovery. Adult patients who remain in a coma for more than four weeks have almost no chance of regaining their previous level of functioning. However, children and young adults have regained functioning after two months in a coma.

Glasgow Coma Scale

The Glasgow Coma Scale, a system of examining a comatose patient, can be helpful for evaluating the depth of the coma, tracking the patient's progress, and possibly predicting ultimate outcome of the coma. The Glasgow Coma Scale assigns a different number of points for

KEY TERMS

Anatomic—Related to the physical structure of an organ or organism.

Ethical—Referring to a system of values that provides the criteria for right behavior.

Metabolic—Related to the chemical processes of an organ or organism.

Neuron—The cells within the body that make up the nervous system.

Physiological—Pertaining to the functioning of an organ, as governed by its physical and chemical condition.

Psychological—Pertaining to the mind and its mental processes.

Stimulus—An action performed on an individual that predictably provokes a reaction.

exam results in three different categories: opening the eyes, verbal response (using words or voice to respond), and motor response (moving a part of the body). Fifteen indicates the highest level of functioning. An individual who spontaneously opens his or her eyes, gives appropriate answers to questions about his or her situation, and can follow a command (such as "move your leg," "nod your head") has the highest level of functioning. Three is the smallest possible number of points, and would be given to a patient who is unresponsive to a painful stimulus. In the middle are those patients who may be able to respond, but who require an intense or painful stimulus, and whose response may demonstrate some degree of brain malfunctioning. When performed as part of the admission examination, a Glasgow score of three to five points suggests that the patient likely has suffered fatal brain damage, while eight or more points indicates that the patient's chances for recovery are good.

The ethical dilemma presented by persistent coma

When a patient has not suffered brain death (the complete absence of any electrical activity within the brain) but has been in a deep coma for some time, a change in condition may occur. This condition is called a persistent vegetative state. The patient may open his or her eyes and move his or her limbs in a primitive fashion, demonstrating some degree of arousal. However, the patient lacks any element of awareness and is unable to have any measurably meaningful interactions with the surrounding world. This condition may last for years. The care of these patients has sparked some of the most heated debates within the field of medical ethics. The discovery of medical advances that allow various disease states to be arrested, without restoration of lost brain function, and the fact that medical resources are limited, have led to debates regarding when medical help should be withdrawn from an individual who has no hope of recovery.

See also Nervous system; Neuron; Psychology; Stimulus.

Resources

Books

Guberman, Alan. *An Introduction to Clinical Neurology.* Boston: Little, Brown, 1994.
Isselbacher, Kurt J., et al. *Harrison's Principles of Internal Medicine.* New York: McGraw-Hill, 1994.
Liebman, Michael. *Neuroanatomy Made Easy and Understandable.* Baltimore: University Park Press, 1991.

Rosalyn Carson-DeWitt

Combinations *see* **Combinatorics**

Combinatorics

Combinatorics is the study of combining objects by various rules to create new arrangements of objects. The objects can be anything from points and numbers to apples and oranges. Combinatorics, like **algebra**, numerical analysis and **topology**, is a important branch of **mathematics**. Examples of combinatorial questions are whether we can make a certain arrangement, how many arrangements can be made, and what the best arrangement for a set of objects is.

Combinatorics has grown rapidly in the last two decades making critical contributions to computer science, operations research, finite **probability theory** and cryptology. Computers and computer networks operate with finite data structures and algorithms which makes them perfect for enumeration and graph theory applications. Leading edge research in areas like neural networking rely on the contribution made by combinatorics.

Combinatorics can be grouped into two categories. Enumeration, which is the study of counting and arranging objects, and graph theory, or the study of graphs.

History of combinatorics

Leonhard Euler (1701-1783) was a Swiss mathematician who spent most of his life in Russia. He was responsible for making a number of the initial contributions to combinatorics both in graph theory and enumeration. One of these contributions was a **paper** he published in 1736. The people of an old town in Prussia called Königsberg (now Kaliningrad in Russia) brought to Euler's attention a stirring question about moving along **bridges**. Euler wrote a paper answering the question called "The Seven Bridges of Königsberg." The town was on an **island** in the Pregel river and had seven bridges. A frequently asked question there at the time was "Is it possible to take a walk through town, starting and ending at the same place, and cross each bridge exactly once?" Euler generalized the problem to points and lines where the island was represented by one point and the bridges were represented by lines. By abstracting the problem, Euler was able to answer the question. It was impossible to return to the same place by only crossing each bridge exactly once. The abstract picture he drew of lines and points was a graph, and the beginnings of graph theory. The study of molecules of hydrocarbons, a compound of **hydrogen** and **carbon atoms**, also spurred the development of graph theory.

Enumeration

To enumerate is to count. In combinatorics, it is the study of counting objects in different arrangements. The objects are counted and arranged by a set of rules called equivalence relations.

One way to count a set of objects is to ask, "how many different ways can the objects be arranged?" Each change in the original arrangement is called a permutation. For example, changing the order of the songs to be played on a **compact disc** (CD) player would be a permutation of the regular order of the songs. If there were only two songs on the CD, there would be only two orders, playing the songs in the original order or in reverse order, song two and then song one. With three songs on the CD, there are more than just two ways to play the music. There is the original order, or songs one, two, and three (123) and in reverse order, 321. There are two orders found by flipping the first two songs or the last two songs to get 213 or 132 respectively. There are another two orders, 312 and 231, found by rotating the songs to the right or left. This gives a total of six ways to order the music on a CD with three songs. By just trying different orders, it was intuitively seen how many combinations there were. If the CD had twelve or more songs on it, then this intuitive approach would not be very effective. Trying different arrangements would take a long time, and knowing if all arrangements were found would not be easy. Combinatorics formalizes the way arrangements are found by coming up with general formulas and methods that work for generic cases.

The power of combinatorics, as with all mathematics, is this ability to abstract to a point where complex problems can be solved which could not be solved intuitively. Combinatorics abstracts a problem of this nature in a recursive way. Take the CD example, with three songs. Instead of writing out all the arrangements to find out how many there are, think of the end arrangement and ask, "for the first song in the new arrangement, how many choices are there?" The answer is any three of the songs. There are then two choices for the second song because one of the songs has already been selected. There is only one choice for the last song. So three choices for the first song times two songs for the second choice gives six possibilities for a new arrangement of songs. Continuing in this way, the number of permutations for any size set of objects can be found.

Another example of a permutation is shuffling a deck of playing cards. There are 52 cards in a deck. How many ways can the cards be shuffled? After tearing off the plastic on a brand new deck of cards, the original order of the cards is seen. All the hearts, spades, clubs, and diamonds together and, in each suit, the cards are arranged in numerical order. To find out how many ways the cards can be shuffled, start by moving the first card to any of the 52 places in the deck. Of course, leaving the card in the first place is not moving it at all, which is always an option. This gives 52 shuffles only by moving the first card. Now consider the second card. It can go in 51 places because it can not go in the location of the first card. Again, the option of not moving the card at all is included. That gives a total of 52×51 shuffles which equals 2,652 already. The third card can be placed in 50 places and the fourth in 49 places. Continuing this way find to the last card gives a total of $52 \times 51 \times 50.... \times 4 \times 3 \times 2 \times 1$ possible shuffles which equals about 81 with sixty-six zeros behind it. A huge number of permutations. Once 51 cards were placed, the last card had only one place it could go, so the last number multiplied was one. Multiplying all the numbers from 52 down to one together is called 52 **factorial** and is written 52!.

Binomial coefficients

The importance of binomial coefficients comes from another question that arises. How many subsets are contained in a set of objects? A set is just a collection of objects like the three songs on the CD. Subsets are the set itself, the empty set, or the set of nothing, and any smaller groupings of the set. So the first two songs alone would be a subset of the set of three songs. Intuitively, eight subsets could be found on a three song CD by writing out all possible subsets, including the set itself.

Unfortunately, the number of subsets also gets large quickly. The general way to find subsets is not as easily

seen as finding the total number of permutations of a deck of cards. It has been found that the number of subsets of a set can be found by taking the number of elements of the set, and raising the number two to that power. So for a CD with three songs, the number of subsets is just two to the third power, $2 \times 2 \times 2$, or 8.

For a deck of 52 cards, the number of subsets comes to about 45 with fourteen zeros behind it. It would take a long time to write all of those subsets down. Binomial coefficients represent the number of subsets of a given size. Binomial coefficients are written C(r;c) and represent "the number of combinations of r things taken c at a time." Binomial coefficients can be calculated using factorials or with **Pascal's triangle** as seen below (only the first six rows are shown.) Each new row in Pascal's triangle is solved by taking the top two numbers and adding them together to get the number below.

row 0						1					
row 1					1		1				
row 2				1		2		1			
row 3			1		3		3		1		
row 4		1		4		6		4		1	
row 5	1		5		10		10		5		1

Pascal's Triangle

The triangle always starts with one and has ones on the outside.

So for our three song CD, to find the number of two song subsets we want to find C(3,2) which is the third row and second column, or three. The subsets being songs one and two, two and three, and one and three. Binomial coefficients come up in many places in algebra and combinatorics and are very important when working with **polynomials**. The other formula for calculating C(r;c) is r! divided by c! $\times$ (r-c)!.

Equivalence relations

Equivalence relations is a very important concept in many branches of mathematics. An equivalence **relation** is a way to partition sets into subsets and equate elements with each other. The only requirements of an equivalence relation are that it must abide by the reflexive, symmetric and **transitive** laws.

Relating cards by suits in the deck of cards is one equivalence relation. Two cards are equivalent if they have the same suit. Card **color**, red or black, is another equivalence relation. In algebra, "equals," "greater than" and "less than" signs are examples of equivalence relations on numbers. These relations become important when we ask questions about subsets of a set of objects.

Recurrence relations

A powerful application of enumeration to computers and algorithms is the recurrence relation. A sequence of numbers can be generated using the previous numbers in a sequence by using a recurrence relation. This recurrence relation either adds to, or multiplies one or more previous elements of the sequence to generate the next sequence number. The factorial, n!, is solved using a recurrence relation since n! equals n $\times$ (n-1)! and (n-1)! equals (n-1) $\times$ (n-2)! and so on. Eventually one factorial is reached, which is just one. Pascal's triangle is also a recurrence relation. Computers, being based on algorithms, are designed to calculate and count numbers in this way.

Graph theory

Graphs are sets of objects which are studied based on their interconnectivity with each other. Graph theory began when people were seeking answers to questions about whether it was possible to travel from one point to another, or what the shortest **distance** between two points was.

A graph is composed of two sets, one of points or vertices, and the other of edges. The set of edges represents the vertices that are connected to each other. Combinatorally, graphs are just a set of objects (the vertex set) and a set of equivalence relations (the edge set) regarding the arrangement of the objects. For example, a triangle is a graph with three vertices and three edges. So the vertex set may be (x,y,z) and the edge set (xy,yz,zx). The actual labeling of the points is not as important as fundamental concepts which differentiate graphs.

Sometimes graphs are not easy to tell apart because there are a number of ways we can draw a graph. The graph (x,y,z) with edges (xy,yz,zx) can be drawn as a **circle** with three points on the circumference. The lines do not have to be straight. The vertex and edge sets are the only information defining the graph. So a circle with three distinct points on the circumference, and a triangle, are the same graph. Graphs with hundreds of vertices and edges are hard to tell apart. Are they the same?

One of a number of ways to tell graphs apart is to look at their connectivity and cycles, inherent properties of graphs.

A graph is connected if every vertex can be reached to every other vertex by traveling along an edge. The triangle is connected. A **square**, thought of as a graph with four vertices, (x,y,z,w) but with only two edges (xy,zw) is not connected. There is no way to travel from vertex x to vertex z. A graph has a cycle if there is a path from a vertex back to itself where no edge is passed over twice. The triangle has one cycle. The square, as defined above, has

no cycles. Graphs can have many cycles and still not be connected. Ten disconnected triangles can be thought of as a graph with 10 cycles. The two properties, connectivity and cycles, do not always allow for the differentiation of two graphs. Two graphs can be both connected and have the same number of cycles but still be different.

Another four properties for determining if two graphs are different is explained in a very nice introduction to the subject, *Introduction to Graph Theory* by Richard Trudeau.

Computer networks are excellent examples of a type of graph that demonstrates how important graphs are to the computer field. Networks are a type of graph that has directions and weights assigned to each edge. An example of a network problem is how to find the best way to send information over a national computer network. Should the information go from Washington, D.C. through Pittsburgh, a high traffic point, and then to Detroit, or should the information be sent through Philadelphia and then through Toledo to Detroit? Is it faster to go through just one city even if there is more traffic through that city?

A similar issue involving networks is whether to have a **plane** make a stop at a city on the way to Los Angeles from Detroit, or should the trip be non-stop. Adding factors like cost, travel time, number of passengers, etc. along with the number of ways to travel to Los Angeles leads to an interesting network theory problem.

A traditional problem for the gasoline companies has been how to best determine their truck routes for refilling their gas stations. The gasoline trucks typically drive around an area, from gas station to gas station, refilling the tanks based on some route list, a graph. Driving to the nearest neighboring gas stations is often not the best route to drive. Choosing the cheapest path from vertex to vertex is known as the greedy **algorithm**. Choosing the shortest route based on distance between locations is often not the most cost effective route. Some tanks need refilling sooner than others because some street corners are much busier than others. Plus, like the Königsberg Bridge problem, traveling to the next closest station may lead to a dead end and a trucker may have to back track. The list of examples seems endless. Graph theory has applications to all professions.

Trees

Trees are yet another type of graph. Trees have all the properties of graphs except they must be connected with no cycles. A computer's hard drive directory structure is set up as a tree, with subdirectories branching out from a single root directory. Typically trees have a vertex labeled as the root vertex from which every other vertex can be reached from a unique path along the edges. Not

KEY TERMS

Binomial coefficients—Numbers that stand for the number of subsets of equal size within a larger set.

Combinatorics—The branch of mathematics concerned with the study of combining objects (arranging) by various rules to create new arrangements of objects.

Cycles—A graph has a cycle if there is a path from a vertex back to itself where no edge is passed over twice. The triangle has one cycle.

Enumeration—The methods of counting objects and arrangements.

Equivalence relations—A way to relate two objects which must abide by the reflexive, symmetric and transitive laws.

Factorial—An operation represented by the symbol!. The term n! is equal to multiplying n by all of the positive whole number integers that are less than it.

Graph—A graph is a finite set of vertices or points and a set of finite edges.

Königsberg Bridge Problem—A common question in the Königsberg town on how to travel through the city and over all the bridges without crossing one twice.

Network—A term used in graph theory to mean a graph with directions and weights assigned to each edge.

Permutations—Changing the order of objects in a particular arrangement.

Recurrence relation—A means of generating a sequence of numbers by using one or more previous numbers of the sequence and multiplying or adding terms in a repetitive way. Recurrence relations are especially important for computer algorithms.

Trees—Graphs which have no cycles.

all vertices can be a root vertex. Trees come into importance for devising searching algorithms.

Resources

Books

Berman, Gerald, and K.D. Fryer. *Introduction to Combinatorics.* Academic Press, 1972.

Bogard, Kenneth P. *Introductory Combinatorics.* Harcourt Brace Jovanovic Incorporated, 1990.

Bose R.C., and B. Manvel. *Introduction to Combinatorial Theory.* John Wiley & Sons, 1984.

Jackson, Bradley, and Dmitri Thoro. *Applied Combinatorics with Problem Solving.* Addison-Wesley, 1990.

Trudeau, Richard J. *Introduction to Graph Theory.* Dover, 1993.

David Gorsich

Combustion

Combustion is the chemical term for a process known more commonly as burning. It is certainly one of the earliest chemical changes noted by humans, at least partly because of the dramatic effects it has on materials. Today, the mechanism by which combustion takes place is well understood and is more correctly defined as a form of oxidation that occurs so rapidly that noticeable **heat** and **light** are produced.

History

Probably the earliest reasonably scientific attempt to explain combustion was that of Johannes (or Jan) Baptista van Helmont, a Flemish physician and alchemist who lived from 1580 to 1644. Van Helmont observed the relationship among a burning material, smoke and flame and said that combustion involved the escape of a "wild spirit" (*spiritus silvestre*) from the burning material. This explanation was later incorporated into a theory of combustion—the phlogiston theory—that dominated alchemical thinking for the better part of two centuries.

According to the phlogiston theory, combustible materials contain a substance—phlogiston—that is emitted by the material as it burns. A non-combustible material, such as ashes, will not burn, according to this theory, because all phlogiston contained in the original material (such as **wood**) had been driven out. The phlogiston theory was developed primarily by the German alchemist Johann Becher and his student Georg Ernst Stahl at the end of the seventeenth century.

Although scoffed at today, the phlogiston theory satisfactorily explained most combustion phenomena known at the time of Becher and Stahl. One serious problem was a quantitative issue. Many objects weigh more after being burned than before. How this could happen when phlogiston escaped from the burning material? One possible explanation was that phlogiston had negative weight, an idea that many early chemists thought absurd, while others were willing to consider. In any case, precise measurements had not yet become an important feature of chemical studies, so loss of weight was not an insurmountable barrier to the phlogiston concept.

Modern theory

As with so many other instances in science, the phlogiston theory fell into disrepute only when someone appeared on the scene who could reject traditional thinking almost entirely and propose a radically new view of the phenomenon. That person was the great French chemist Antoine Laurent Lavoisier (1743-1794). Having knowledge of some recent critical discoveries in **chemistry**, especially the discovery of **oxygen** by Karl Wilhelm Scheele (1742-1786) in 1771 and Joseph Priestley (1733-1804) in 1774, Lavoisier framed a new definition of combustion. Combustion, he said, is the process by which some material combines with oxygen. By making the best use of precise quantitative experiments, Lavoisier provided such a sound basis for his new theory that it was widely accepted in a relatively short period of time.

Lavoisier initiated another important line of research related to combustion, one involving the amount of heat generated during oxidation. His earliest experiments involved the study of heat lost by a guinea pig during **respiration**, which Lavoisier called "a combustion." In this work, he was assisted by a second famous French scientist, Pierre Simon Laplace (1749-1827). As a result of their research, Lavoisier and Laplace laid down one of the fundamental principles of **thermochemistry**, namely that the amount of heat needed to decompose a compound is the same as the amount of heat liberated during its formation from its elements. This line of research was later developed by the Swiss-Russian chemist Henri Hess (1802-1850) in the 1830s. Hess' development and extension of the work of Lavoisier and Laplace has earned him the title of father of thermochemistry.

Combustion mechanics

From a chemical standpoint, combustion is a process in which chemical bonds are broken and new chemical bonds formed. The net result of these changes is a release of **energy**, the heat of combustion. For example, suppose that a gram of **coal** is burned in pure oxygen with the formation of **carbon dioxide** as the only product. In this reaction, the first step is the destruction of bonds between **carbon atoms** and between oxygen atoms. In order for this step to occur, energy must be added to the coal/oxygen mixture. For example, a lighted match must be touched to the coal.

Once the carbon-carbon and oxygen-oxygen bonds have been broken, new bonds between carbon atoms and oxygen atoms can be formed. These bonds contain less energy than did the original carbon-carbon and oxygen-oxygen bonds. That energy is released in the form of heat, the heat of combustion. The heat of combustion of one **mole** of carbon, for example, is about 94 kcal.

Applications

Humans have been making practical use of combustion for millennia. Cooking food and heating homes have long been two major applications of the combustion reaction. With the development of the **steam engine** by Denis Papin, Thomas Savery, Thomas Newcomen, and others at the beginning of the eighteenth century, however, a new use for combustion was found: performing work. Those first engines employed the combustion of some material, usually coal, to produce heat that was used to boil **water**. The steam produced was then able to move pistons and drive machinery. That concept is essentially the same one used today to operate fossil-fueled electrical power plants.

Before long, inventors found ways to use steam engines in transportation, especially in railroad engines and steam ships. However, it was not until the discovery of a new type of fuel—gasoline and its chemical relatives—and a new type of engine—the internal combustion engine—that the modern face of transportation was achieved. Today, most forms of transportation depend on the combustion of a **hydrocarbon** fuel such as gasoline, kerosene, or diesel oil to produce the energy that drives pistons and moves the vehicles on which modern society depends.

When considering how fuels are burned during the combustion process, "stationary" and "explosive" flames are treated as two distinct types of combustion. In stationary combustion, as generally seen in gas or oil burners, the mixture of fuel and oxidizer flows toward the flame at a proper speed to maintain the position of the flame. The fuel can be either premixed with air or introduced separately into the combustion region. An explosive flame, on the other hand, occurs in a homogeneous mixture of fuel and air in which the flame moves rapidly through the combustible mixture. Burning in the cylinder of a gasoline engine belongs to this category. Overall, both chemical and physical processes are combined in combustion, and the dominant process depends on very diverse burning conditions.

Environmental issues

The use of combustion as a power source has had such a dramatic influence on human society that the period after 1750 has sometimes been called the Fossil Fuel Age. Still, the widespread use of combustion for human applications has always had its disadvantages. Pictorial representations of England during the **Industrial Revolution**, for example, usually include huge **clouds** of smoke emitted by the combustion of wood and coal in steam engines.

Today, modern societies continue to face environmental problems created by the prodigious combustion

KEY TERMS

Chemical bond—The force or "glue" that holds atoms together in chemical compounds.

Fossil fuel—A fuel that is derived from the decay of plant or animal life; coal, oil, and natural gas are the fossil fuels.

Industrial Revolution—That period, beginning about the middle of the eighteenth century, during which humans began to use steam engines as a major source of power.

Internal combustion engine—An engine in which the chemical reaction that supplies energy to the engine takes place within the walls of the engine (usually a cylinder) itself.

Thermochemistry—The science that deals with the quantity and nature of heat changes that take place during chemical reactions and/or changes of state.

of carbon-based fuels. For example, one product of any combustion reaction in the real world is **carbon monoxide**, a toxic gas that is often detected at dangerous levels in urban areas around the world. Oxides of **sulfur**, produced by the combustion of impurities in fuels, and oxides of **nitrogen**, produced at high **temperature**, also have deleterious effects, often in the form of **acid rain** and **smog**. Even carbon dioxide itself, the primary product of combustion, is suspected of causing **global climate** changes because of the enormous concentrations it has reached in the atmosphere.

See also Air pollution; Chemical bond; Internal combustion engine; Oxidation-reduction reaction.

Resources

Books

Gilpin, Alan. *Dictionary of Fuel Technology*. New York: Philosophical Library, 1969.

Joesten, Melvin D., et al. *World of Chemistry*. Philadelphia: Saunders, 1991.

Olah, George A., ed. *Chemistry of Energetic Materials*. San Diego: Academic Press, 1991.

Snyder, C.H. *The Extraordinary Chemistry of Ordinary Things*. 4th ed. New York: John Wiley and Sons, 2002.

Periodicals

Rutland, Christopher. "Probability Density Function Combustion Modeling of Diesel Engine." *Combustion Science and Technology* 174, no. 10 (2002): 19-54.

Other

"Combustion Modelling For Direct Injection Diesel Engines." *Proceedings Of The Institution Of Mechanical Engineers* 215, no. 5 (2001): 651–663.

David E. Newton

Comet Hale-Bopp

In the spring of 1997, the night sky provided a spectacular **light** show as Comet Hale-Bopp, one of the brightest **comets** of the century, traversed the heavens. The comet was more closely studied than any before it, and scientists continue to make discoveries based on the data gathered by terrestrial and orbital telescopes and instrumentation.

Originating primarily from the Oort Cloud, a belt of stellar debris ringing our **solar system**, comets are balls of **water ice** interspersed with other elements and compounds. Occasionally, a passing **star** or one of the larger planets perturbs the **orbit** of one of the balls of debris, tipping it into the solar system toward the **Sun** on an elliptical orbit. As the comet approaches the Sun, the nucleus of the comet heats. Gas and dust begins to boil off, creating a bright **coma** around the head and trailing tails of dust and ionized gases, stretching for hundreds of thousands of kilometers behind the nucleus. Depending on the size of the orbit and whether it originates in the solar system or in the Oort cloud, a comet can return in a handful of years or in several thousand years.

Streaking across the heavens

On July 22, 1995, two men peered into their telescopes to view an obscure **star cluster** called M70. One was **concrete** company worker and amateur astronomer Thomas Bopp, who had taken a group of friends out into the Arizona **desert** for a star party. The other was professional astronomer Alan Hale, head of the Southwest Institute for Space Research, who was conducting his own observation session hundreds of mi (80-90 million km) away in New Mexico. Within minutes of the same time, it was later determined, each of the men noticed small fuzzy spot near **galaxy** M70, a spot where none had appeared previously. Both Hale and Bopp reported their findings by e-mail to the International Astronomical Union, which verified the discovery of Comet 1995-01. Shortly thereafter, the object was renamed Comet Hale-Bopp.

Hale-Bopp was one of the most spectacular comets to appear this century, remaining visible to the naked **eye** for more than 19 months. At its peak, the comet was as bright as the most brilliant stars in the sky, with a tail that stretched 50-60 million mi (80–90 million km). Orbital analysis shows that the comet last passed through the inner solar system in 2213 B.C.; it will next visit in the year A.D. 4300. During its recent visit, it passed within 122 million mi (196 million km) of **Earth** and a mere 85 million mi (137 million km) from the Sun.

Boasting a cometary nucleus approximately 25 mi (40.25 km) in diameter, Hale-Bopp was more than four times as large as **Halley's comet**; later analysis showed that the primary nucleus may have a lesser companion of approximately half the size. As the comet streaked through the sky at about 98,000 MPH (157,789 km/h) at perihelion, the nucleus was rotating with a period of about 11.34 hours.

Most comets feature two tails. One, a streak of dust and debris emitted by the nucleus, trails behind the comet. The other, a stream of ionized gas stripped off by the **solar wind**, faces away from the sun. Hale-Bopp, however, boasted a third tail consisting of electrically neutral **sodium atoms**.

Astronomers from the Isaac Newton Telescope on La Palma, Italy, observed the tail while imaging the comet using a special filter that rejected all wavelengths of light but that emitted by sodium atoms. In most comets, sodium emission is observed only around the nucleus. Instead of observing the expected small halo of light around the nucleus of Hale-Bopp, however, the group discovered a tail 372,000 mi (600,000 km) wide and stretching out 31 million mi (50 million km) behind the comet. Though scientists do not fully understand the phenomenon, they believe that the tail was formed as sodium atoms released by the coma interacted with the solar wind and fluoresced.

From a scientific point of view, comet Hale-Bopp generated a tremendous amount of data and insight in the astronomical and astrophysical community. Soon after its discovery, scientists began to study the comet; by the time it made its pass around the sun, plans for observation and data gathering were well in place. To obtain ultraviolet data that is normally blocked by the earth's atmosphere, for example, NASA sent three instrument packages up on suborbital rockets to gather data in five-minute intervals.

Astronomers assert that comets were formed of the debris left behind during the early stages of the formation of the solar system. A study of **hydrogen** cyanide molecules ejected by Hale-Bopp provided astronomers with evidence linking icy nuclei of comets to the ices found in interstellar gas **clouds**, which are believed to condense into stars and planets. By studying the **chemistry** of the comet with techniques such as **spectroscopy**, astronomers hope to better understand the conditions under which the Sun and planets formed.

Using **radio** observations, scientists discovered eight molecules that had never before been seen in a comet, including **sulfur** monoxide, and other unique organic molecules and isotopes.

As the comet hurtles toward the outer solar system, astronomers continue to observe it, studying the differences in gas emission patterns, monitoring the evolution of the cooling nucleus, as well as the shrinking tail. New discoveries about the nature of comets and of our solar system will undoubtedly result from the analysis of this data and that collected during the spectacular transit of Hale-Bopp through our skies.

Kristin Lewotsky

Comets

A comet is an object with a dark, solid core (the nucleus) some miles in diameter. The core is composed mostly of **water ice** and frozen gas and is surrounded—whenever the comet is close enough to the **Sun** for part of the core to vaporize—by a cloud of glowing vapor (the coma). Together, the core and coma comprise the comet's head, which appears as a bright, well-defined cloud. As a comet nears the Sun, charged particles streaming outward from the Sun—the "solar wind"—sweep the coma out into a long, luminous tail which may be visible from the **earth**. Comets spend most of their time far from the Sun, beyond the **orbit** of **Pluto**, where hundreds of billions of comets (too dark to observe directly from Earth) orbit the Sun in a mass called the Oort cloud; only a few ever approach the Sun closely enough for to be observed. A comet that does approach the Sun follows either an elliptical or parabolic orbit. An elliptical orbit is oval in shape, with the Sun inside the oval near one end; a parabolic orbit is an open **curve** like the cross-section of a valley, with the Sun inside the curve near the bottom. A comet following an elliptical orbit will eventually return to the Sun, perhaps after tens, hundreds, or thousands of years; a comet following a parabolic orbit never returns to the Sun.

Perhaps among the most primitive bodies in the **solar system**, comets are probably debris from the formation of the Sun and its planets some 4.5 billion years ago. Astronomers believe that the Oort cloud is a dense shell of debris surrounding the solar system. Occasionally, disruptive gravitational forces (perturbations) destabilize one or more comets, causing a piece of debris from the cloud to fall into the gravitational pull of one of the large planets, such as **Jupiter**. Ultimately, such an object may take up an elliptical or parabolic orbit around the Sun. Comets on elliptical (returning) orbits are either short-period, with orbits of less than 200 years, or long-period, with enormous, nearly parabolic orbits with periods of more than 200 years. Of the 710 individual comets recorded from 1680 to mid 1992, 121 were short-period and 589 long-period.

Age-old fascination

Evidence of a human fascination with the night sky goes back as far as recorded history. Records on Babylonian clay tablets unearthed in the Middle East, dating to at least 3000 B.C. and rock carvings found in prehistoric sites in Scotland, dating to 2000 B.C., record astronomical phenomena that may have been comets. Until the Arabic astronomers of the eleventh century, the Chinese were by far the world's most astute sky-watchers. By 400 B.C., their intricate cometary classification system included sketches of 29 comet forms, each associated with a past event and predicting a future one. Comet type 9, for example, was named Pu-Hui, meaning "calamity in the state, many deaths." Any comet appearing in that form supposedly foretold such a calamity. In fact, from Babylonian civilization right up until the seventeenth century and across cultures, comets have been viewed as omens portending catastrophe.

Of all the Greek and Roman theories on comets— "comet" is a Greek word meaning "long-haired one"— the most influential, though entirely incorrect, was that of Greek philosopher, Aristotle (384–322 B.C.). His view of the solar system put Earth at the center circled by the **Moon**, Mercury, **Venus**, the Sun, **Mars**, Jupiter, and **Saturn** (in order of increasing **distance** from the earth). Stars were stationary, and temporary bodies like comets traveled in straight lines. Aristotle believed that comets were fires in the dry, sublunar "fiery sphere," which he supposed to be a combustible atmosphere "exhaled" from Earth which accumulated between Earth and the Moon. Comets were therefore considered technically terrestrial—originating from Earth—rather than celestial heavenly bodies.

Aristotle's model left many unexplained questions about the movement of bodies through the solar system. His theory came in the Middle Ages to be so strongly supported by the Christian Church, however, that those who challenged it were often called heretics. His theory remained standard for over 2,000 years, bringing European investigations into comets virtually to a halt. Fortunately, prolific and accurate cometary records were kept by the Chinese during this period.

Sporadic European scientific investigations into comets also had some influence, however. Although the Church held his theory in check, Polish astronomer Nicolaus Copernicus (1473–1543) suggested that a heliocentric (Sun-centered) solar system would help explain the motions of the planets and other celestial bodies. Through acute observation of the "Great Comet" of 1577, Danish astronomer Tycho Brahe (1546–1601) calculated that it must be four times further away from Earth than the Moon, refuting Aristotle's sublunar theory of comets. Also, Brahe found that the comet's tail pointed away from the Sun and that its orbit might be oval.

Study of the Great Comet by Brahe and his contemporaries was a turning point for astronomical science. Throughout the seventeenth and eighteenth centuries, mathematicians and astronomers proposed conflicting ideas on the origin, formation, orbits, and meaning of comets. In the early 1600s, English physicist and mathematician Isaac Newton (1642–1727) built on theories from the likes of German astronomer Johannes Kepler (1571–1630), who developed the three laws of planetary **motion**; Polish astronomer Johannes Hevelius (1611–1687), who suggested comets move on parabolas around the Sun; and English physicist Robert Hooke (1635–1703), who introduced the possibility of a universal gravitational influence. Newton developed an mathematical model for the parabolic motion of comets, published in 1687 in his book *Principia*, one of the most important scientific works ever written.

By this time, comets were viewed as celestial rather than terrestrial, and the focus turned from superstition to science. They were, however, still viewed as singular rather than periodic occurrences. In 1687, English astronomer Edmond Halley (1656–1742) suggested to Newton that comets may be periodic, following elliptical paths. Newton did not agree. Using Newton's own mathematical model, Halley argued that the comets of 1531, 1607, and 1682—the latter observed by both he and Newton—were actually one and the same, and predicted that this comet should return late in 1758. It did, and was subsequently named **Halley's comet**. Halley's comet has continued to return on a regular schedule.

By the end of the eighteenth century, comets were believed to be permanent celestial bodies composed of solid material, the movement of which could be calculated using Newton's laws of planetary motion. The return of two more comets in 1822 and 1832 was accurately predicted. The first, comet Enke, did not follow Newton's law of planetary motion, as its orbital period of re-

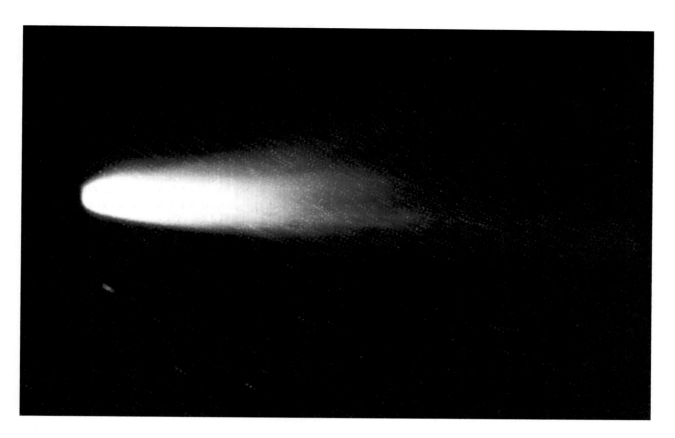

Optical image of Halley's comet. *Royal Observatory, Edinburgh/Science Photo Library/Photo Researchers, Inc. Reproduced by permission.*

currence was decreasing. In 1835, German astronomer Friedreich Bessel (1784–1846) accurately suggested that this was because gases given off by the comet as it passed near the Sun acted like a rocket, thrusting the comet either closer to or further away from the Sun and so affecting its orbital length.

The second predicted periodic comet, comet Biela, with a periodic orbit of 6.75 years, upset Newton's idea of comets'when it split in two in 1846. The now-twinned comet reappeared in 1852 for the last time.

Earlier in the nineteenth century, scientists had speculated that meteor showers may be flying debris from disintegrating comets. In November 1872, when Biela should have returned, the meteor shower predicted by some astronomers did indeed appear, strengthening the connection between meteors and dying comets.

Stargazing and discovering comets

The first observation of a comet through a **telescope** was made in 1618. Previously, comets were discovered with the naked **eye**. Today, most new comet discoveries are made from telescopic photographs and electronic de-

tectors; many comets are discovered by amateur astronomers, and are named after their discoverers.

The long-focal-length, refracting telescope—the primary astronomical observation tool of the 1800s—worked well for direct viewing although, with the relatively insensitive photographic emulsions of the period, it did not collect sufficient **light** to allow astronomical **photography**. In 1858, an English artist, William Usherwood, used a short focal-length **lens** to produce the first photograph of a comet. In 1864, by using a **spectroscope**, an instrument which separates the wave lengths of light into spectral bands, Italian astronomer Giovanni Donati (1826–1873) first identified a chemical component in a comet's atmosphere. The first cometary spectrogram (photographed **spectrum** of light from a comet) was taken by English amateur astronomer William Huggins (1824–1910) of London in 1881.

The early twentieth century saw the development of short-focal-length spectrographs which, by the 1950s, allowed identification of several different chemical components in a comet's tail. Infrared spectrography was introduced in the 1960s and, in 1983, the **Infrared Astronomy Satellite** (IRAS) began gathering information on cometary dust particles that was unobtainable by ground-based technology. Today, observations are also made by **radio astronomy** and ultraviolet spectrography.

Composition, origin, and extinction

The questions of the **birth**, composition, and death of comets still defy definitive answers. Increasing knowledge and advancing twentieth-century technology have brought many and, as usual, conflicting theories.

Composition of the nucleus

Two major theories on the composition of the nucleus have developed over time. The "flying sandbank" model, first proposed by Richard Proctor in the mid 1800s and again in the mid 1900s by Raymond Lyttleton, conjectured swarms of tiny solid particles bound together by mutual gravitational attraction. In 1950, U.S. astronomer Fred Whipple (1906–) introduced the "icy-conglomerate" or "dirty-snowball" model, which describes a comet's core as a solid nucleus of meteoric rock particles and dust frozen in ice. Observations of Halley's comet by spacecraft in 1986 strongly support this model.

No one knows the exact composition of the core, but it is believed that **rocks** and dust are held together by ices of water, methane, **ammonia**, and **carbon monoxide** that are contaminated by **carbon** and **sulfur**. The 1986 spacecraft encounter showed Halley's nucleus as peanut-shaped or potato-shaped, 9 mi (15 km) long, and 5.5 mi (8 km) wide.

The nuclei of comets are too small for observation through telescopes. As they approach the Sun, however, they produce one of the largest, most spectacular sights in the solar system—a magnificent, glowing tail often visible even to the naked eye. Cometary nuclei have been seen to produce sudden, bright flares and some to split into as many as five pieces.

Development of the coma

As the nucleus of a comet nearing the Sun approaches the distance of the asteroid belt (outside the orbit of Mars), its ices begin to sublimate (turn to gas), releasing **hydrogen**, carbon, **oxygen**, **nitrogen**, and other substances in the form of vapors and particles. Carried away from the nucleus by the **solar wind** at several hundred meters per second, they create an enormous coma and tail hundreds of thousands of kilometers long, hiding the nucleus. The Sun's ultraviolet light excites the gaseous molecules, causing them to fluoresce (shine). Microscopic mineral particles in the dust reflect and scatter the Sun's light. In 1970, during the first space-based observation of a comet, a gigantic hydrogen cloud was discovered surrounding the coma. Depending on the size of a cometary nucleus and its proximity to the Sun, this cloud can be larger than the Sun itself.

Tail configuration

As the comet swings around the Sun on its elliptical orbit, the gas and dust particles streaming from the coma form two types of tails: a gaseous ion tail (Type I) or a dust tail (Type II). In a Type I tail, ionized gases form a thin, usually straight tail, sometimes millions of miles long. (The tail of the Great Comet of 1843 stretched out more than 136 million mi [220 million km].) The ion tail, glowing brightly, does not trail behind the core along its path of motion but is blown away from the core along a line pointing directly away from the sun. The head collides with the solar wind, which wraps around the nucleus, pulling the ionized particles of the coma with it. Depending on its position relative to the Sun, a comet's tail may even be traveling *ahead* of the nucleus. A Type II tail is usually shorter and wider, and curves slightly because its heavier particles are carried away from the nucleus at a slower **rate**.

Comet Hale-Bopp, which streamed across the skies in the spring of 1997, boasted a feature hitherto unseen in comets: a third type of tail composed of electrically neutral **sodium atoms**. Observers using instruments with spectral filters that eliminated all but the yellow light emitted by fluorescing sodium atoms found that the tail was 373,000 mi (600,000 km) wide and 31,000 million mi (50 million km) long, streaming in a direction slightly different from that of the ion tail. The exact mechanism producing this type of tail is still not understood.

Origins

As the solar system moves slowly through the center of the **galaxy**, it encounters interstellar molecular gas **clouds** which, under certain circumstances, strip comets from the Oort cloud. How, then, is the Oort cloud replenished? One theory proposes "capture" of comets from interstellar **space**. The popularity of the interstellar theory waxes and wanes, and new hypotheses are again being proposed. One suggests the presence of comets in high-density clouds within the galaxy's inner **spiral** arms. The Sun may capture comets while passing through one of these arms, which happens once every 100 million years. Also, comets may be captured from the very same molecular gas clouds which, under other circumstances, so severely deplete the Oort cloud population. Mathematical calculations and known chemical composition of comets and stars indicate the possibility of interstellar origins.

Death of a comet

Although vaporization during passages by the Sun diminishes the nucleus, it is not believed to be enough to cause a comet's **extinction**. There are two commonly accepted reasons for a comet's death: splitting, which may result in deterioration and ultimately a meteor shower; and flaring, bright explosions visible in the coma. Another theory postulates that asteroids may be extinct comets.

Comets and Earth

The paths of comets and asteroids cross the orbital path of the planets and are believed to be the cause of some impact craters on Earth and the Moon. In 1979, United States Air Force satellite P78–1 took the first photograph of a comet colliding with the Sun. Late in 1994, comet Shoemaker-Levy collided with Jupiter. Asteroidal impacts on Earth may have caused the extinction of many **species**, including the dinosaurs, while making the development of new species—including ourselves—possible.

For millennia, humans have predicted the "end of the world" from the impact of a giant comet. Now, however, some scientists argue that molecules released by comets' vaporized gases may have supplied important molecules in Earth's early atmosphere. When exposed to the Sun's **radiation**, these molecules undergo the formation of organic compounds. During the recent passage of Hale-Bopp, for example, scientists discovered a variety of complex organic chemicals in the comet.

The theory gained evidence from data gathered by the Polar spacecraft, launched by NASA in 1996. According to observations by the probe, cometlike objects 30–40 ft (9.1–12.1 m) in diameter are hitting the atmosphere at the rate of 43,000 per day. These cosmic slushballs are too small to vaporize and provide the standard show we associate with comets; most disintegrate in the upper atmosphere, entering the **weather** cycle and eventually reaching the terrestrial surface as **precipitation**. According to estimates by scientists associated with the study, this cosmic rain has added one inch of water to the earth's surface each 10,000–20,000 years, supplying a large quantity of water over **geologic time**.

Bright objects keep us in the dark

In a pair of space missions planned for the early part of the twenty-first century, space probes will rendezvous with a pair of short-period comets, hopefully to help scientists reach a better understanding of the **physics** of comets. NASA's Stardust mission, launched in 1999, is

on its way to capture dust from the tail of Comet Wild (pronounced "vilt") 2 in 2004, returning the samples to Earth in 2006 for analysis. In February 2003, the European Space Agency's Rosetta mission—originally scheduled to rendezvous with Comet Wirtanen on its trip around the Sun—was postponed due to launch failures suffered by Europe's Ariane 5 rocket. In March 2003, ESA scientists retasked the Rosetta mission spacecraft to rendezvous with 67P/Churyumov-Gerasimenko. With a launch planned as early as January 2004, Rosetta will orbit the comet and send a probe to the surface. An early 2004 launch date will permit a rendezvous in 2014. The larger size of 67P/Churyumov-Gerasimenko—and thus a stronger gravitational field—poses some problems for the lander that will require recalculation of the landing impact stress on the lander legs.

At present, however, despite spaceships probing the outer limits of our solar system; gigantic telescopes in deserts, atop **mountains**, and floating in space; and satellites designed specifically to capture meteor dust hurtling through Earth's atmosphere from interstellar space, significant questions about the origin, nature, and fate of comets remains unsolved.

See also Meteors and meteorites; Space probe.

Resources

Books

Bailey, M.E., S.V.M. Clube, and W.M. Napier. *The Origin of Comets.* Oxford: Pergamon Press, 1990.

Gibilisco, Stan. *Comets, Meteors & Asteroids: How They Affect Earth.* Blue Ridge Summit, PA: Tab Books, 1985.

Levy, David H. *The Quest for Comets: An Explosive Trail of Beauty and Danger.* New York: Plenum Press, 1994.

Yeomans, Donald K. *Comets, A Chronological History of Observation, Science, Myth, and Folklore.* New York: John Wiley & Sons, 1991.

Other

National Aeronautics and Space Administration. "STARDUST Mission." 2000. (cited October 19, 2002). <http://stardust.jpl.nasa.gov/mission/>.

Marie L. Thompson

Commensalism

Commensalism is a type of **symbiosis**, specifically, a biological relationship in which one **species** benefits from an interaction, while the host species is neither positively or negatively affected to any tangible degree.

For example, epiphytic plants (which grow on other plants but are not parasitic) gain an enormous ecological benefit from living on larger plants, because they gain access to a substrate upon which to grow relatively high in the canopy. The host trees, however, are not affected in any significant way by this relationship, even in cases when they are supporting what appears to be a large population of epiphytes. Some plants are specialized as epiphytes, for example, many species of air-plants or bromeliads (family Bromeliaceae), orchids (Orchidaceae), and **ferns** (Pterophyta). Many **lichens**, mosses, and liverworts are also epiphytes on trees. There are also **animal** analogues of this relationship. Sometimes **sea anemones** (order Actiniaria) will gain a benefit in terms of food availability by growing on the upper carapace of a hermit crab (crustacean infraorder Anomura) which is apparently unaffected by the presence of the epiphyte.

Another commensal relationship, known as phoresy, is a type of biological hitch-hiking in which one **organism** benefits through access to a mode of transportation while the animal providing this service is not significantly affected by its role. For example, many plants produce **fruits** that adhere to fur and are thereby dispersed by the movement of **mammals**. Some North American examples of such animal-dispersed plants are the burdock (*Arctium lappa*), beggar-tick or stick-tight (*Bidens frondosa*), and tick-trefoil (*Desmodium canadense*). The fruits of these plants have special anatomical adaptations for adhering to fur-in fact, those of the burdock are the botanical model from which the idea for the very useful fastening material known as velcro was developed. In a few cases, **individual** animals may become heavily loaded with these sorts of sticky fruits causing their fur to mat excessively, perhaps resulting in a significant detriment. This is not common, however, and usually this biological relationship is truly commensal.

Common denominator *see* **Least common denominator**

Community ecology

Within the science of **ecology**, a *community* is a set of organisms coexisting within a defined area. Community ecology, then, is the study of the interactions that occur among groups of **species** coexisting within a region. For example, a community ecologist might consider the ways in which plants and animals within a forest affect one another's growth. Contrary to popular usage, the term ecology itself does not refer to **recycling (conservation)**. Rather, it refers to the study of the distribution and abundance of living organisms and their interchange with each other and with the non-living (abiotic) world. Community ecology is concerned with the distribution, abundance, and interactions of *mixtures* of many different kinds of living things in a defined area.

By definition, a community is composed of two different species. Therefore, a set of **birds** consisting of two species of **finches** would constitute a very simple community. Most communities are much more complex, however, containing many coexisting **plant**, **animal**, bacterial, fungal, and protozoal (single-celled) individuals. Interestingly, a community is really defined by the person who is considering the community; the geographical boundaries of a community are arbitrary. The concept of a "community," depends on how it is defined.

Community ecology seeks to understand how species interact by studying many different kinds of relationships between organisms. Animal-animal interactions, animal-plant interactions, and plant-plant interactions are examples of community relationships considered. A plant-plant interaction might be the ways in which weeds affect growth of tomatoes in a garden, or how a tall **tree** blocks sunlight from smaller plants. An example of an animal-animal interaction is **competition** between birds for limited **seeds** for food, or how **snakes prey** upon **mice** in meadows. Plant-animal interactions include herbivory (eating of plants by animals) or **Venus** fly trap plants capturing and digesting **insects**.

Commuity ecology also looks to understand such concepts as **niche**, diversity, biogeography, diversity, species assembly, predation, competition, and **symbiosis** (the beneficial and parasitic coexistence of organisms). The major theme within community ecology is competition as the driving force for change in environments. This specialized field of **biology** is important because it helps scientists understand how communities are structured and how they change over **time**. Also, an understanding of community structure is vital to predict the effects of decline in, or **extinction** of, species (loss of diversity). Therefore, in order to fully understand the effect that humankind has upon our environment, knowledge of community ecology is needed.

Terry Watkins

Commutative property

"Commutativity" is a property an operation between two numbers (or other mathematical elements) may or may not have. The operation is commutative if it does not matter which element is named first.

For example, because **addition** is commutative, 5 + 7 has the same value as 7 + 5. **Subtraction**, on the other hand, is not commutative, and the difference 5 - 7 does not have the same value as 7-5.

Commutativity can be described more formally. If * stands for an operation and if A and B are elements from a given set, then * is commutative if, for all such elements A * B = B * A.

In ordinary **arithmetic** and **algebra**, the commutative operations are **multiplication** and addition. The non-commutative operations are subtraction, **division**, and exponentiation. For example, x + 3 is equal to 3 + x; xy is equal to yx; and (x + 7)(x - 2) is equal to (x - 2)(x + 7). On the other hand, 4 - 3x is not equal to 3x - 4; 16/4 is not equal to 4/16; and 5^2 is not equal to 2^5.

The commutative property can be associated with other mathematical elements and operations as well. For instance, one can think of a translation of axes in the coordinate **plane** as an "element," and following one translation by another as a "product." Then, if T_1 and T_2 are two such **translations**, T_1T_2 and T_2T_1 are equal. This operation is commutative. If the set of transformations includes both translations and rotations, however, then the operation loses its commutativity. A **rotation** of axes followed by a translation does not have the same effect on the ultimate position of the axes as the same translation followed by the same rotation.

When an operation is both commutative and associative (an operation is associative if for all A, B, and C, (A * B) * C = A * (B * C), the operation on a finite number of elements can be done in any order. This is particularly useful in simplifying an expression such as $x^2 + 5x + 8 + 2x^2 + x + 9$. One can combine the squared terms, the linear terms, and the constants without tediously and repeatedly using the associative and commutative properties to bring like terms together. In fact, because the terms of a sum can be combined in any order, the order need not be specified, and the expression can be written without parentheses. Because ordinary multiplication is both associative and commutative, this is true of products as well. The expression $5x^2y^3z$, with its seven factors, requires no parentheses.

Compact disc

In 1978, Philips and Sony together launched an effort to produce an audio compact disc (CD) as a method of delivering digital sound and music to consumers. The two companies continued to cooperate through the 1980s and eventually worked out standards for using the CD technology to store computer data. These recommendations evolved into the CD-ROM technology of today.

The CD-ROM (compact disc-read only memory) is a read-only optical storage medium capable of holding 600 megabytes of data (approximately 500,000 pages of text), 70 minutes of high fidelity audio, or some combination of the two. Legend has it that the size of the compact disc was chosen so that it could contain a slow-tempo rendition of Beethoven's Ninth Symphony. As can be seen from Table 1, compact discs offer a high volume of data storage at a lower cost than other media.

The first users of CD-ROMs were owners of large databases: library catalogs, reference systems, and parts lists. Typical applications of CD-ROMs as storage media now include storage of such information as the following:

- every listing from all of the Yellow Pages in the United States
- maps of every street in the country
- facsimile numbers for all publicly held businesses and government institutions
- a 21-volume encyclopedia

The extremely thin metal layer of the CD, usually pure aluminum, reflects light from a tiny infrared laser as the disc spins in the CD player. The reflections are transformed into electrical signals and then further converted into meaningful data for use in digital equipment. © *Kelly A. Quin. Reproduced by permission.*

CD-ROMs are expected to achieve significant impact in storage of the following kinds of documents:

• business reference materials
• interactive educational materials for schools
• scholarly publications
• government archives
• home-based reference materials

Manufacture of a compact disc

A compact disc is a thin wafer of clear polycarbonate plastic and **metal** measuring 4.75 in (120 mm) in diameter with a small hole in its center. The metal layer is usually pure **aluminum** that is spread on the polycarbonate surface in a layer that is only a few molecules thick.

The metal reflects **light** from a tiny infrared **laser** as the disc spins in the CD player. The reflections are transformed into electrical signals and then further converted to meaningful data for use in digital equipment.

Information (either audio on a music CD or data of many kinds on a CD-ROM) is stored in pits on the CD that are 1-3 microns long, about 0.5-micron wide, and 0.1-micron deep. There may be more than 3 mi (4.8 km) of these pits wound around the center hole on the disc. The CD is coated with a layer of lacquer that protects the surface. By convention, a label is usually silkscreened on the backside.

Compact discs are made in a multistep process. First a **glass** master is made using photolithographic techniques. An optically ground glass disc is coated with a layer of photoresist material 0.1-micron thick. A pattern is produced on the disc using a laser; then the exposed areas on the disc are washed away, and the disc is silvered to produce the actual pits. The master disc is next coated with single-molecule-thick layers of nickel, one layer at a time, until the desired thickness has been achieved. The nickel layer is next separated from the glass disc and used as a metal **negative**.

For low production runs, the metal negative is used to make the actual discs. Most projects require that several positives be produced by plating the surface of the metal negative. Molds or stampers are then made from the positives and used in injection molding machines.

Plastic pellets are heated and injected into the molds, where they form the disc with pits in it. The plastic disc is coated with a thin aluminum layer for reflectance and with a protective lacquer layer. The disc is then given its silkscreened label and packaged for delivery. Most of these operations take place in a cleanroom because a single particle of dust larger than a pit can destroy data. Mastering alone takes about 12 hours of work.

Retrieving information from a disc

The primary unit of data storage on a compact disc is a sector, which is 1/75-second long. Each sector on a CD contains 2352 bytes (processable units) of data, and each

TABLE 1. COSTS OF INFORMATION STORAGE		
Medium	**Capacity**	**Cost per megabyte**
Hard disk	100 megabytes	~$7.00
Paper	2 kilobytes per page	~$5.00
Magnetic tape	60 megabytes	<$1.00
Floppy disk	1.44 megabytes	<$0.50
CD-ROM	650 megabytes	~$0.01

TABLE 2. COMPACT DISC AND DRIVE FORMATS

Format name	Description	Notes
CD-ROM ISO 9660	Read-only memory	Applies to MS-DOS and Macintosh files. This standard evolved from the Yellow Book specifications of Philips and Sony. Defined the Volume Table of Contents that tells the CD reader where and how the data are laid out on the disc.
CD-ROM High Sierra	Read-only memory	Based on a standard worked out in 1985 to resolve differences in leading manufacturers' implementations of ISO 9660.
CD-DA	Digital audio	Data drives that can read data and audio are called CD-DA.
PhotoCD	Compressed images	KODAK multisession XA system. Customers present an exposed 35 mm roll of color film for wet processing, and purchase a Photo CD for an additional charge. The negatives are then processed by a technician who scans each image at an imaging workstation. The images are written onto the Photo CD write-once media, color thumbnails of all of the images are printed, and the Photo CD is returned to the consumer in a CD jewel case with the index sheet inserted as a cover. The customer may return the same Photo CD to have more images written onto it, with the result that a multisession disc is produced. A Photo CD can hold 125 or more high resolution images. Photo CDS may be viewed using a Kodak Photo CD player connected to a television at the customer's home. Photo CD images can also be viewed using a CD-ROM/XA player attached to a computer. Photo CD images can be converted to other formats for incorporation into multimedia applications.

sector is followed by 882 bytes of data for detecting errors, correcting information, and controlling timing. Thus, a CD actually requires 3234 bytes to store 2352 bytes of data.

The disc spins at a constant linear **velocity**, which means that the rotational speed of the disc may vary from about 200 rpm when the data being read are near the outer of the disc to about 530 rpm when the data are located near the center of the disc. (This is because rotational speed = linear velocity × radius of sector.) The CD is read at a sustained **rate** of 150K (150,000) bytes per second, which is sufficient for good audio but very

TABLE 2. COMPACT DISC AND DRIVE FORMATS (cnt'd)

Format name	Description	Notes
CD-ROM/XA	Read-only memory	Extended architecture. Data are read off a disc in alternating pieces, and synchronized at play-back. The result is a simultaneous presentation of graphics and audio. CD-ROM/XA defined a new sector format to allow computer data, compressed audio data, and video/image information to be read and played back apparently simultaneously. Special enabling hardware is required on CD-ROM/XA players because the audio must be separated from the interleaved data, decompressed, and sent to speakers, at the same time the computer data are being sent to the computer.
CD-R (CD-WO or CD-WORM)	Write-once	May use multiple sessions to fill disc. Instead of burning pits into a substrate, the CD-R uses differences in reflectivity to fool the reader into believing that a pit actually exists. This format allows you to write your own CDS.
CD-ROM HFS	Read-only memory	The Macintosh Hierarchical File System (HFS) is Apple's method for managing files and folders on the Macintosh desktop. The HFS driver provides Macintosh users with the expected and familiar Apple desktop. This is the preferred format for delivery to Macintosh platforms, even though it does not conform to the ISO 9660 standard.
CD-I or CD-RTOS	Interactive	Philips Interactive motion video. CD-I discs are designed to work with Philips CD-I players, but the CD-I system also hooks up to the customer's TV or stereo. It can play audio CDS, and can also read Kodak PhotoCD discs. The CD-I is marketed for education, home, and business use.
CD-I Ready	Interactive/Ready	CD-Audio with features for CD-I player.
CD-Bridge	Bridge	Allows XA track to play on CD-I player.

TABLE 2. COMPACT DISC AND DRIVE FORMATS (cnt'd)		
Format name	**Description**	**Notes**
CD-MO	Magneto-optical	Premastered area readable on any CD player.
CD+G	Mixed mode	CD+G stands for CD audio plus graphics. This format allows the customer to play CD audio along with titles, still pictures, and song lyrics synchronized to the music. This CD may be best suited to karaoke-style discs, i.e., music playing along with on-screen lyrics.
CDTV	ISO 9660 variant	Commodore proprietary system.

slow for large image files, **motion** video, and other multimedia resources. Newer drives spin at twice or even three to six times this rate. Still, CD access speeds and transfer rates are much slower than those from a hard disc in a computer. This is expected to change as discs are made to spin faster and different types of lasers are perfected for use in computers.

The surface of the CD is essentially transparent. It must allow a finely focused beam of laser light to pass through it twice, first to the metallic layer beneath the plastic where the data reside, and then back to the receptors. Dirt, scratches, fingerprints, and other imperfections interfere with retrieval of the stored data.

CD-ROM drives

Each CD-ROM drive for a personal computer (PC) may be characterized according to the following:

- drive specifications
- formats the drive can read
- interface the drive requires to connect the computer

Drive specifications

The drive specifications tell the drive's performance capabilities. These specifications commonly include the following:

- The data transfer rate, which specifies how much data the drive can read from a data CD and transfer to the host computer when reading one large, sequential chunk of data.

- The access time, which is the delay between the drive receiving the command to read and its actual first reading of the data.

- The buffer size, which is the size of the cache of memory stored in the drive. Not all drives are equipped with memory buffers.

Drive formats

The data on compact discs need to be organized if the CD-ROM drive and computer are to make sense of the data. The data are therefore encoded to conform to certain standards. Although advanced CD-ROM standards are still evolving, most drives today comply with earlier CD-ROM formats.

Interfaces

The CD-ROM interface is the physical connection of the drive to the PC's expansion bus. The three typical interfaces are SCSI Standard, SCSI-2, and ASPI, and nonstandard SCSI.

SCSI standard interfaces

Small Computer System Interface (SCSI) refers to a group of adapter cards that conform to a set of common commands. These adapter cards allow a chain of devices to be strung from a single adapter. Consequently, SCSI interfaces are preferred for connecting a CD-ROM drive to a personal computer.

SCSI-2 and ASPI interfaces

SCSI-2 and Advanced SCSI Programming Interfaces (ASPI) take into account rapid enhancements of computer interface technology. SCSI-2 incorporates several enhancements, including greater data throughput and improved read and write technologies. ASPI provides a standard software interface to the host adapter hardware.

TABLE 3. MULTIMEDIA STORAGE REQUIREMENTS	
One Minute Of...	**Storage Space Required**
audio, mono	700 kilobytes
audio, stereo	more than 1.5 megabytes
animation	2.5 to 5.5 megabytes
video	20 to 30 megabytes, compressed

Nonstandard SCSI interfaces

Nonstandard SCSI interfaces may not accept installation of multiple SCSI devices; in cases where this is not a problem, they may prove acceptable.

Care of CD-ROMs

Audio CDs tend to be more forgiving than CDs that will be read by computers. The audio CD player can fill in any missing data on the disc because audio data are easily interpolated, with the result that scratches do not have much effect on the quality of the sound produced when the CD is played.

Computer data are less predictable than audio data. Consequently, it is not as easy to interpolate when data are missing. Because computer data are digital (either 0s or 1s), the computer cannot represent missing data by an "average" value lying somewhere between 0 and 1. Because small scratches are inevitable, the CD-ROM incorporates a scheme that makes it possible to reconstruct any bit from the surrounding data. This scheme is called Error Correction Code (ECC). ECC permits the CD-ROM to undergo some surface damage and still remain usable, but it does not replace the need to **exercise** care when handling a disc.

Multimedia

Multimedia is a computer application that employs more than one medium to convey information. Examples of multimedia include:

- text with graphics
- text with photos
- text with sound
- text with animation
- text with video
- graphics with sound
- photos with sound
- animation with sound
- video with sound

As indicated in Table 3, some multimedia combinations require very large amounts of disc storage space.

To date, most multimedia applications have been text-based with multimedia features added. Many multimedia applications, however, make heavy use of memory-intensive features such as video and sound. Although the CD-ROM is not a requirement for multimedia on the personal computer, its impressive storage capacity makes it a logical choice for delivering multimedia documentation.

Compact discs of the near future

Recordable and erasable CDs will give the compact disc greater versatility. Compact disc recorders went on the market in 1997 and allow the user to record audio from various sources on CDs. The recorders require some finesse in use because the recording procedure used depends on the type, quality, and input device of the source material. If the source is a CD that can be played on a machine with digital optical output, it can be connected directly to the CD recorder as input and be dubbed much like audio tapes. The recorder evaluates the sonic range of the original and digitally synchronizes it; if tracks are recorded from several CDs, the recorder must resynchronize with each track. Also, the CD recorder does not erase, so care is needed during recording to copy the desired tracks only.

Erasable CDs became common by 2001. Erasable CDs or CD-RWs allow flexible data storage because they can be overwritten when the data on them becomes obsolete. CD-RWs are important as a publishing medium because they can be used to display multi-media presentations. Consequently, today's desktop-published newsletter may become text with audio and video displays. High-density CD-Rs and CD-RWs are also being developed.

Improvements are also coming to audio CDs as manufacturers seek new features that will improve sales. Enhanced audio CDs now include music videos, lyrics, scores that the home musician can play, and interviews with the musicians. Enhanced audio CDs can be played on a CD-ROM drive and viewed on a monitor or connected **television** set. High Definition Compatible Digitals, or HDCDs, are also being marketed. They produce more realistic sound but require a CD player with a built-in decoder.

Developments in technical and scientific uses of CDs are also being explored. One of the most promising is a portable medical laboratory called the LabCD. A drop of **blood** is placed on the CD near the center hole. As the CD spins, it acts like a **centrifuge** and separates the cells in the blood. They slip into receptacles in the CD that contain testing chemicals, and sensors read the results of a range of blood tests including DNA tests. This technology opens the possibility for ambulances to carry LabCDs and the CD-sensing machine and to perform on-the-spot analyses for drug and **alcohol** use or DNA tests at crime scenes.

See also Computer, digital.

Resources

Books

Bosak, S,. J. Sloman, and D. Gibbons, *The CD-ROM Book.* Indianapolis, IN: Que Corporation, 1994.

Vaughan, T., *Multimedia: Making It Happen.* Berkeley, CA: Osborne-McGraw Hill, 1994.

Randall S. Frost

Competition

Competition is a biological interaction among organisms of the same or different **species** associated with the need for a common resource that occurs in a limited supply relative to demand. In other words, competition occurs when the capability of the environment to supply resources is smaller than the potential biological requirement so that organisms interfere with each other. Plants, for example, often compete for access to a limited supply of **nutrients**, **water**, sunlight, and **space**.

Intraspecific competition occurs when individuals of the same species vie for access to essential resources, while interspecific competition occurs between different species. Stresses associated with competition are said to be symmetric if they involve organisms of similar size and/or abilities to utilize resources. Competition is asymmetric when there are substantial differences in these abilities, as occurs in the case of large trees interacting with plants of a forest understory.

Competition as an ecological and evolutionary factor

Individuals of the same species have virtually identical resource requirements. Therefore, whenever populations of a species are crowded, intraspecific competition is intense. Intraspecific competition in dense populations results in a process known as self-thinning, which is characterized by mortality of less-capable individuals and relative success by more-competitive individuals. In such situations, intraspecific competition is an important regulator of population size. Moreover, because **individual** organisms vary in their reproductive success, intraspecific competition can be a selective factor in **evolution**.

Interspecific competition can also be intense if individuals of the various species are crowded and have similar requirements of resources. One ecological theory, known as the competitive exclusion principle, states that species with ecologically identical life styles and resource needs cannot coexist over the longer term; the competitively less-fit species will be displaced by the better fit species. Although it is debatable that different species could have identical ecological requirements, it is not difficult to comprehend that intense competition must occur among similar species living in the same, resource-limited **habitat**. In such situations, interspecific competition must be important in structuring ecological communities and as an agent of natural **selection**.

The term competitive release refers to a situation in which an **organism** or species is relieved of the stresses associated with competition allowing it to become more successful and dominant in its habitat. For example, by the early 1950s the American **chestnut** (*Castanea dentata*) had been eliminated as a dominant canopy species in deciduous **forests** of eastern **North America** by the accidental introduction of a fungal pathogen known as chestnut blight (*Endothia parasitica*). Other **tree** species took advantage of their sudden release from competition with the chestnut by opportunistically filling in the canopy gaps that were left by the demise of mature chestnut trees. Similarly, competitively suppressed plants may be released when a mature forest is disturbed, for example, by **wildfire**, a windstorm, or harvesting by humans. If the disturbance kills many of the trees that formed the forest canopy but previously suppressed plants survive, then these understory plants will gain access to an abundance of environmental resources such as **light**, moisture, and nutrients, and they will be able to grow relatively freely.

Competitive displacement is said to occur when a more competitive species causes another to utilize a distinctly sub-optimal habitat. A number of interesting cases of competitive displacement have been described by ecologists, many involving interactions of **plant** species. In eastern North America, for example, the natural habitat utilized by the silver maple tree (*Acer saccharinum*) is almost entirely restricted to forested **wetlands**, or swamps. However, the silver maple is more productive of **biomass** and **fruits** if it grows on well-drained, upland sites, and for this reason it is commonly cultivated in cities and towns. In spite of this habitat preference, the silver maple does not occur in the natural forest community of well drained sites. It appears that the silver maple is not sufficiently competitive to co-occur in well-drained sites with more vigorous tree species such as the sugar maple (*Acer saccharum*), **basswood** (*Tilia americana*), or the red oak (*Quercus rubra*). Consequently, the silver maple is displaced to swamps, a distinctly sub-optimal habitat in which there is frequent physiological stress associated with **flooding**.

Over long periods of **time**, competitive displacement may lead to evolutionary changes. This happens as species displaced to marginal environments evolve to become better adapted to those conditions, and they may eventually become new species. Competitive displacement is believed to be the primary force leading to the evolution of species swarms on isolated islands such as those of fruit flies (*Drosophila* spp.) and **honeycreepers** (Drepaniidae) on the Hawaiian Islands and Darwin's **finches** (Geospizinae) on the Galapagos Islands.

In the cases of the honeycreepers and Darwin's finches, the islands are believed to have been colonized by a few individuals of a species of finch. These founders then developed a large population which saturated the **carrying capacity** of the common habitats so that intraspecific competition became intense. Some individuals that were less competitive in the usual means of habitat exploitation were relegated to marginal habitats or to unusual means of exploiting resources within a common habitat. Natural selection would have favored genetically based adaptations that allowed a more effi-

cient exploitation of the marginal habitats or lifestyles of the populations of displaced **birds**, leading to evolutionary changes. Eventually, a condition of reproductive isolation would have developed, and a new species would have evolved from the founder population. Competitive displacements among species of finches could then have further elaborated the species swarms. The various species of Darwin's finches and Hawaiian honeycreepers are mostly distinguished on the basis of differences in the size and shape of their bills and on behavioral differences associated with feeding styles.

It must be understood that not all environments are resource limited, and in such situations competition is not a very important process. There are two generic types of non-competitive environments—recently disturbed and environmentally stressed. In habitats that have recently been subjected to a catastrophic disturbance, the populations and biomass of organisms is relatively small, and the biological demand for resources is correspondingly not very intense. Species that are specialized to take advantage of the resource-rich and competition-free conditions of recent disturbances are known as ruderals. These species are adapted to rapidly colonizing disturbed sites where they can grow freely and are highly fecund. However, within several years the ruderals are usually reduced in abundance or eliminated from the community by slower growing, but more competitive species that eventually take over the site and its resources and dominate later successional stages.

Some habitats are subject to intense environmental stress such as physical stress associated with climate or toxic stress associated with nutrient deficiency or **pollution**. Because of the severe intensity of environmental stress in such habitats, the productivity of organisms is highly constrained, and there is little competition for resources. The arctic **tundra**, for example, is an **ecosystem** that is highly stressed by climate. If the density of individual plants of the tundra is experimentally decreased by thinning, the residual plants do not grow better because their productivity was not constrained by competition. However, the intensity of environmental stress can be experimentally alleviated by enclosing an area of tundra in a greenhouse and by fertilizing with nutrients. In such a situation, competition among arctic plants can become a significant ecological interaction, and this change can be experimentally demonstrated.

Because the effects of competition can be profound and are nearly always measurable in at least some parameter, the processes surrounding and affecting competition, as well as the environmental forces affected or shaped by competition, are an active area for research by ecologists. Competition is believed to have a strong result on, for example, the process of speciation. Speciation is the formation of two distinct species from a single one over time. Therefore, ecologists might compare the divergence of genetic characteristics between organisms in an area with high levels of intraspecific competition for a limiting resource versus those that are not.

Similarly, competition as a major force that structures communities of organisms within ecosystems is a major area or research. The relative abundances of different organisms in a community, for example, is determined in part on the levels of competition for resources found in their habitat. Diversity, another very popular topic of active research in **ecology**, also deals with competition. Competitive interactions are believed to increase the amount of diversity in an environment. In other words, the number of species present in a given ecosystem increases in areas with increased competition. The current global **biodiversity** project, which is attempting to catalog all of the species found on **Earth**, has helped to establish the link between diversity and competition.

See also Stress, ecological.

Resources

Books

Begon, M., J. L. Harper, and C. R. Townsend. *Ecology: Individuals, Populations and Communities.* 2nd ed. London: Blackwell Sci. Pub., 1990.

Ricklefs, R. E. *Ecology.* New York: W. H. Freeman, 1990.

Bill Freedman

Complementary DNA

Complementary **deoxyribonucleic acid (DNA)** is DNA in which the sequence of the constituent molecules on one strand of the double stranded structure chemically matches the sequence on the other strand.

A useful analog is to picture a key and a lock. While there are many different types of keys, only one design matches the contours of the lock and so will fit into the lock. The different chemical molecules that make up DNA also do not pair up nonspecifically. A "lock in key" fit operates at the molecular level.

The chemical molecules that make up DNA are known as nucleotide bases. There are four common types of bases: adenine (A), cytosine (C), guanine (G), and thymine (T). In the chemical "lock and key" fit, an A on one strand always pairs with a T on the other strand. As well, a C on one strand always pairs with a G on the other strand. The two strands are described as complementary to one another.

Complementary DNA (cDNA) is a copy of a region of a strand of DNA. For example, if the original DNA stand had a sequence of ATT, the complementary sequence will be TAA. The cDNA will bind to the complementary site on the DNA strand.

Complementary DNA is important naturally, in the manufacture of new copies of DNA, and has become an important experimental tool. In **DNA replication**, the two strands are unwound from one another. A **molecule** called DNA polymerase runs the length of each strand, making a complementary copy of each strand. In other words, each strand acts as a blueprint to produce a complementary strand. The two new strands are complementary to one another, and so can join together in a process called annealing. The old strands also anneal. The result is two complete copies of DNA.

Complementary DNA has been exploited to develop research techniques and to produce genetically altered commercial products. A classic example of cDNA is the technique of polymerase chain reaction (**PCR**). PCR mimics the process of DNA manufacture in a test tube. In a series of reactions, a target stretch of DNA is copied, and the copies themselves serve as templates for more copies. The original DNA sequence is amplified to make a billion copies within minutes.

Because **ribonucleic acid (RNA)** is made using DNA as the blueprint, the phenomenon of complementary strands also extends to RNA. RNA is made of four bases; adenine (A), cytosine (C), guanine (G), and uracil (U; instead of the thymine found in DNA). In the lock in key scenario, an A pairs with the U) on the other strand, and a C always pairs with a G. Complementary RNA (cRNA) is a copy of a strand of RNA that will bind to the appropriate region of the original molecule. If the original RNA stand had a base sequence of AUU, for example, the sequence of the cRNA strand would be UAA.

The association of a DNA or RNA strand to its complement is one of the basic research tools of the molecular biologist. Binding of a compliment can identify target regions of DNA or RNA, and can be used to disrupt the process of DNA manufacture. If the complementary DNA is labeled with a compound that fluoresces, then the binding of the fluorescent probe can actually be visualized using a **microscope**. This permits the "real time" examination of **DNA synthesis**.

See also DNA technology; Microorganisms.

Resources

Books

Synder, L., and W. Champness. *Molecular Genetics of Bacteria.* 2nd ed. Washington, DC: American Society for Microbiology Press, 2002.

Periodicals

Aizaki, H., Y. Aoki, T. Harada, et al., "Full-length Complementary DNA of Hepatitis C Virus From an Infectious Blood Sample." *Hepatology* 27 (February 1998): 621–627.

Chien, Y.-C., Y.-J. Zhu, and C.-M. Chuen. "Complementary DNA Cloning and Analysis of Gene Structure of Pyruvate Kinase From *Drosophila melanogaster*." *Zoological Studies* 38 (February 1999): 322–332.

Volchkov, V.E., V.A. Volchkova, E. Mühlberger, et al. "Recovery of Infectious Ebola Virus From Complementary DNA: RNA Editing of the GP Gene and Viral Cytotoxicity." *Science* 229 (March 2001): 1965–1969.

Complex

A complex is a species in which the central atom is surrounded by a group of Lewis bases that have covalent bonds to the central atom. The Lewis bases that surround the central atom are generally referred to as ligands. Complexes are so named because when they were first studied, they seemed unusual and difficult to understand. Primarily, transition metals form complexes and their most observable property is their vivid **color**. The color of transition **metal** complexes is dependent on the identity and **oxidation state** of the central atom and the identity of the **ligand**.

Acids and bases were originally defined by Arrhenius as substances that donated protons (positively charged **hydrogen** ions) and hydroxide ions (consisting of one hydrogen atom bonded to one **oxygen** atom and having an overall charge of minus one) respectively. Subsequently, Bronsted and Lowry redefined acids and bases as being **proton** donors and proton acceptors, respectively. This broadened definition made it possible to include substances that were known to behave as bases but did not contain the hydroxide ion. Much later, Lewis defined acids as substances that could accept an **electron** pair and bases as substances that could donate an electron pair, and this is currently the broadest definition of acids and bases as it includes substances that have neither protons nor hydroxide ions. Thus, the Lewis bases, or ligands, in complexes have electron pairs they can share with the central atom. Covalent bonds are bonds in which a pair of electrons is shared between two **atoms**; as opposed to ionic bonds in which one atom more or less appropriates the electron(s), acquiring a **negative** charge, while the other atom loses the electrons, resulting in a positive charge.

Transition metals are the elements that appear in the central block of the **periodic table** (atomic numbers 21–30 and the columns below them). The transition metals are capable of having different oxidation states (a measure of the number of electrons arranged around the central atom). A complex having the same ligands and the same central atom but with different oxidation states will have different colors. A complex with the same ligands but different central atoms with the same oxidation state will have different colors. Similarly, a complex with the same central atom and oxidation state, but having different ligands will have different colors.

A number of biologically important molecules are dependent on the presence of transition metals and the biological role of the transition elements usually depends on the formation of complexes. For example, hemoglobin is an **iron** complex important in the transport of oxygen in the body. Chromium is a part of the glucose tolerance factor that, along with **insulin**, controls the removal of glucose from the **blood**. More than 300 enzymes contain zinc, one of them being the digestive **enzyme** that hydrolyzes protein. In addition, many synthetic **dyes and pigments** are transition metal complexes, such as Prussian blue. Transition metal complexes are also used as catalysts in many important industrial processes, such as the formation of **aldehydes** from alkenes, the extraction of gold from **ore** and the purification of nickel.

Rashmi Venkateswaran

Complex numbers

Complex numbers are numbers which can be put into the form a + bi, where a and b are real numbers and $i^2 = -1$.

Typical complex numbers are 3 - i, 1/2 + 7i, and -6 - 2i. If one writes the real number 17 as 17 + 0i and the **imaginary number** -2.5i as 0 - 2.5i, they too can be considered complex numbers.

Complex numbers are so called because they are made up of two parts which cannot be combined. Even though the parts are joined by a plus sign, the **addition** cannot be performed. The expression must be left as an indicated sum.

Complex numbers are occasionally represented with ordered pairs, (z,b). Doing so shows the two-component nature of complex numbers but renders the operations with them somewhat obscure and hides the kind of numbers they are.

Inklings of the need for complex numbers were felt as early as the sixteenth century. Cardan, in about 1545, recognized that his method of solving **cubic equations** often led to solutions with the **square root** of **negative** numbers in them. It was not until the seventeenth and early eighteenth centuries that de Moivre, the Bernoullis, Euler, and others gave formal recognition to imaginary and complex numbers as legitimate numbers.

Arithmetic

Complex numbers can be thought of as an extension of the set of real numbers to include the imaginary numbers. These numbers must obey the laws, such as the distributive law, which are already in place. This they do with two exceptions, the fact that the "sum" a + bi must be left uncombined, and the law $i^2 = -1$, which runs counter to the rule that the product of two numbers of like sign is positive.

Arithmetic with complex numbers is much like the "arithmetic" of binomials such as 5x + 7 with an important exception. When such a binomial is squared, the **term** $25x^2$ appears, and it doesn't go away. When a + bi is squared, the i^2 in the term b^2i^2 does go away. It becomes $-b^2$. These are the rules:

Equality: To be equal two complex numbers must have equal real parts and equal imaginary parts. That is a + bi = c + di if and only if a = c and b = d.

Addition: To add two complex numbers, add the real parts and the imaginary parts separately. The sum of a + bi and c + di is (a + c) + (b + d)i. The sum (3 + 5i) + (8 - 7i) is 11 - 2i.

Subtraction: To subtract a complex number, subtract the real part from the real part and the imaginary part from the imaginary part. The difference (a + bi) - (c + di) is (a - c) + (b - d)i; (6 + 4i) - (3 - 2i) is 3 + 6i.

Zero: To equal zero, a complex number must have both its real part and its imaginary part equal to zero: a + bi = 0 if and only if a = 0 and b = 0.

Opposites: To form the opposite of a complex number, take the opposite of each part: -(a + bi) = -a + (-b)i. The opposite of 6 - 2i is -6 + 2i.

Multiplication: To form the product of two complex numbers multiply each part of one number by each part of the other: (a + bi)(c + di) = ac + adi + bci + bdi², or (ac - bd) + (ad + bc)i. The product (5 - 2i)(4 - 3i) is 14 - 23i.

Conjugates: Two numbers whose imaginary parts are opposites are called "complex conjugates." These complex numbers a + bi and a - bi are conjugates. Pairs of complex conjugates have many applications because the product of two complex conjugates is real: (6 - 12i)(6 + 12i) = 36 - 144i², or 180.

Division: Division of complex numbers is an example. Except for division by zero, the set of complex numbers is closed with respect to division: If a + bi is not zero, then (c + di)/(a + bi) is a complex number. To di-

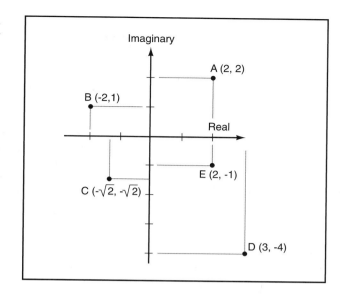

Figure 1. *Illustration by Hans & Cassidy. Courtesy of Gale Group.*

vide c + di by a + bi, multiply them both by the conjugate a - bi, which eliminates the need to divide by a complex number. For example

$$\frac{5 - i}{6 + 2i} = \frac{(5 - i)(6 - 2i)}{(6 + 2i)(6 - 2i)}$$

$$= \frac{28 - 16i}{36 + 4}$$

$$= 7/10 - 2/5i$$

While the foregoing rules suffice for ordinary complex-number arithmetic, they must often be coupled with ingenuity for non-routine problems. An example of this can be seen in the problem of computing a square root of 3 - 4i.

One starts by assuming that the square root is a complex number a + bi. Then 3 - 4i is the square of a + bi, or $a^2 - b^2 + 2abi$.

For two complex numbers to be equal, their real and imaginary parts must be equal

$$a^2 - b^2 = 3$$
$$2ab = -4$$

Solving these equations for a yields four roots, namely 2, -2, i, and -i. Discarding the imaginary roots and solving for b gives 2 - i or -2 + i as the square roots of 3 - 4i. These roots seem strange, but their squares are in fact 3 - 4i.

Graphical representation

The mathematicians Wessel, Argand, and Gauss, separately devised a graphical method of representing

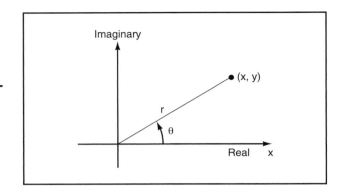

Figure 2. *Illustration by Hans & Cassidy. Courtesy of Gale Group.*

complex numbers. Real numbers have only a real component and can be represented as points on a one-dimensional number line. Complex numbers, on the other hand, with their real and imaginary components, require a two-dimensional complex-number **plane**.

The complex number plane can also by represented with **polar coordinates**. The **relation** between these coordinates and rectangular coordinates are given by the equations

$$x = r \cos \theta$$
$$y = 4 \sin \theta$$
$$r = \sqrt{x^2 + y^2}$$
$$\theta = \arc \tan y/x$$

Thus the complex number $x + iy$ can also be written in polar form $r \cos \theta + ir \sin \theta$ or $r(\cos \theta + i \sin \theta)$, abbreviated r cis θ. When written in polar form, r is called the "modulus" or "absolute value" of the number. When the **point** is plotted on a Gauss-Argand diagram, r represents the **distance** from the point to the origin.

The **angle** θ is called the "argument" or the "amplitude" of the complex number, and represents the angle shown in Figure 2. Because adding or subtracting 360° to θ will not change the position of the ray drawn to the point, r cis θ, r cis $(\theta + 360°)$, r cis $(\theta - 360°)$, and others all represent the same complex number. When θ is measured in radians, adding or subtracting a multiple of 2π will change the representation of the number, but not the number itself.

In polar form the five points shown in Figure 1 are A: $\sqrt{\theta}$ (cos 45° + I sin 45°) or $\sqrt{\theta}$ cis 45°. B: $\sqrt{+5}$ cis 153.4°. C: 2 cis 225°. D: 5 cis 306.9°. E: $\sqrt{+5}$ cis 333.4°. Except for A and C, the polar forms seem more awkward. Often, however, it is the other way about. 1 cis 72°, which represents the fifth root of 1 is simple in its polar form, but considerably less so in its rectangular form: .3090 +.9511 i, and even this is only an **approximation**.

The polar form of a complex number has two remarkable features which the rectangular form lacks. For one, multiplication is very easy. The product of r_1 cis θ_1 and r_2 cis θ_2 is simply $r_1 r_2$ cis $(\theta_1 + \theta_2)$. For example, (3 cis 30°) (6 cis 60°) is 18 cis 90°. The other feature is known as de Moivre's **theorem**: $(r$ cis $\theta)^n = r^n$ cis $n\theta$, where n is any real number (actually any complex number). This is a powerful theorem. For example, if one wants to compute $(1 + i)^n$, multiplying it out can take a lot of time. If one converts it to polar form, $\neq 12$ cis 45°, however, $(\neq 2$ cis 45°$)^5$ $\neq 32$ cis 225° or -4 -4i.

One can use de Moivre's theorem to compute roots. Since the nth root of a real or complex number z is $z^{1/n}$, the nth root of r cis θ is $r^{1/n}$ cis θ/n.

It is interesting to apply this to the cube root of 1. Writing 1 as a complex number in polar form one has 1 cis 0°. Its cube root is $1^{1/3}$ cis 0/3°, or simply 1. But 1 cis 0° is the same number as 1 cis 360° and 1 cis 720°. Applying de Moivre's theorem to these alternate forms yields 1 cis 120° and 1 cis 240°, which are not simply 1. In fact they are $-1/2 + \sqrt{+2}$ /2 i and $-1/2 - \sqrt{43/2}$i in rectangular form.

Uses of complex numbers

Complex numbers are needed for the fundamental theorem of **algebra**: Every polynomial equation, $P(x) = 0$, with complex coefficients has a complex root. For example, the polynomial equation $x^2 + 1 = 0$ has no real roots. If one allows complex numbers, however, it has two: 0 + i and - i.

Complex numbers are also used in the branch of **mathematics** known as "functions of complex variables." Such functions can often describe physical situations, as in electrical **engineering** and **fluid dynamics**, which real-valued functions cannot.

Resources

Books

Ball, W.W. Rouse. *A Short Account of the History of Mathematics*. London: Sterling Publications, 2002.

Bittinger, Marvin L., and Davic Ellenbogen. *Intermediate Algebra: Concepts and Applications*. 6th ed. Reading, MA: Addison-Wesley Publishing, 2001.

Jourdain, Philip E. B. "The Nature of Mathematics." *The World of Mathematics*. Newman, James, ed. New York: Simon and Schuster, 1956.

Stein, Sherman K. *Mathematics, the Man-made Universe*. San Francisco: W. H. Freeman and Co., 1969.

J. Paul Moulton

Composite family (Compositaceae)

The composite or aster family (Asteraceae) is one of the largest families of plants, containing about 20,000 **species**, distributed among more than 1,000 genera, and occurring widely on all continents, except **Antarctica**. This family is commonly regarded by modern botanists as the most advanced of the **plant** families, because of the complex, highly evolved structure of its multi-flowered, composite reproductive structures.

The members of the composite family display a remarkable range of growth forms, ranging from tiny, herbaceous annual plants, to vine-like lianas, and tall, tree-like perennials. For example, some species in the genus *Senecio* are small, annual plants, such as the widespread common groundsel (*Senecio vulgaris*). In contrast, the giant senecio (*S. adnivalis*) species found on a mountain in Uganda, is a perennial plant that grows as tall as 26 ft (8 m).

The most species-rich genera in the aster family are *Senecio* (about 1,500 species), *Vernonia* (900 species), *Hieracium* (800 species), and *Eupatorium* (600 species). Various members of the aster family are familiar species in natural habitats, while others are cultivated plants in gardens, and some are grown as foods. Some species in the aster family are considered to have negative values as weeds of agriculture or lawns.

Characteristics of the Asteraceae

Members of the Asteraceae are most readily characterized by their unique floral structure. The flowers of members of this family are aggregated within a composite grouping known as an inflorescence, which in this family is known as a head. In the head, the small, individual flowers, called florets, are attached to a basal structure known as a receptacle. The latter is surrounded by one or more rows of bracts, that make up the involucre.

Artichokes in Salinas, California. *© 1983 Lawrence Midgale, National Audubon Society Collection/Photo Researchers, Inc. Reproduced with permission.*

The heads may be present singly, or they may occur in various sorts of aggregated groupings. Typically, each head gives the visual impression of being a single, large **flower**, even though the structure is actually a composite of several to many, small florets. This visual display is best developed in insect-pollinated species of the Asteraceae, and is ultimately designed to attract pollinators.

In many cases, the individual flowers may occur as disk florets that have functional stamens and pistils but lack petals, or as ray florets that have an elongate, strap-shaped petal known as a ligule or ray. In some species, the head is entirely composed of disk florets, and is known as a discoid head. Discoid heads occur, for example, in the tansies (*Tanacetum* spp.). In other species the head is entirely made up of ray florets, and is referred to as a ligulate head, for example, in the dandelions (*Taraxacum* spp.).

In other species, the disk florets occur in the center of the head, and ray florets on the periphery, giving a particularly striking resemblance to a single, large flower. The ray florets of these latter, relatively complex inflorescences are commonly sterile, and are only designed to aid in attracting pollinating **insects**. One familiar example of this head structure is the ox-eye daisy (*Chrysanthemum leucanthemum*), which has a central packing of bright-yellow disc florets, and a white fringe of long, ray florets. The ox-eye daisy is the species of wildflower that love-struck young people use to tell whether their adoration is returned by their sweetheart-the petals are picked off one by one, to determine whether "he/she love me, or loves me not." The **seeds** of plants in the aster family are borne in dry **fruits** known as achenes. In many cases, the achenes of these plants are small, and have a fine, filamentous attachment known as pappus, which serves to give the fruits aerodynamic qualities that favor their dispersal by the **wind**.

This seed form and dispersal method can be illustrated by the familiar dandelion, whose fruiting heads develop as whitish puffs of pappus-bearing seeds, which eventually disintegrate and blow about on the wind.

In some other cases, such as the sunflowers (*Helianthus* spp.), the seeds are encased in a relatively hard coat, and are only dispersed locally. The seeds of some other plants in the composite family are specialized to stick to the fur of **mammals**, and are dispersed in this way. Two common examples of this hitch-hiking strategy are the beggar ticks (*Bidens* spp.) and the burdock (*Arctium* spp.).

Horticultural species

Many species in the aster family have very attractive inflorescences, and some of these are commonly grown as ornamentals in parks and gardens.

Many of the ornamental species in the aster family are annuals, and are used as bedding plants, in annual gardens, and in self-seeding gardens. Some common examples include the cosmos (*Cosmos bipinnatus*), sunflower (*Helianthus annuus*), summer chrysanthemum (*Chrysanthemum coronarium*), blanket flower (*Gaillardia pulchella*), strawflower (*Helichrysum bracteatum*), iceplant or living-stone daisy (*Mesembryanthemum criniflorum*), marigolds (*Tagetes patula*, *T. erecta*, and *T. tenuifolia*), and zinnia (*Zinnia elegans*).

A few horticultural species are biennials, or species that can complete their life cycle in two years. Two examples are the daisy (*Bellis perennis*) and ox-eye daisy (*Chrysanthemum leucanthemum*).

Many other horticultural species in the composite family are longer-lived, herbaceous perennials, and can be used in perennial gardens. Some common examples include various species of asters (*Aster* spp., such as New England aster, *A. novae-angliae*), black-eyed Susan (*Rudbeckia hirta*), shasta daisy (*Chrysanthemum maximum*), **hemp** agrimony (*Eupatorium purpureum*), Scotch **thistle** (*Onopordum acanthium* and *O. arabicum*), yarrow (*Achillea* spp., such as *A. filipendulina*), yellow chamomile (*Anthemis tinctoria*), knapweed (*Centaurea montana*), blanket flower (*Gaillardia aristata*), and goldenrods (*Solidago* spp., such as Canada goldenrod, *S. canadensis*).

Wormwoods (*Artemisia* spp.) have rather unattractive, greenish inflorescences, but are commonly cultivated for their attractive foliage.

Agricultural species of composites

A few species of composites have been domesticated for agricultural purposes. The cultivated sunflower (*Helianthus annuus*) is an annual plant native to Mexico and **South America** that is now widely grown for its seeds, which are eaten roasted or raw. Sunflower seeds contain about 40-50% oil, which can be extracted as a fine edible oil, the remaining cake being used as **animal** fodder. This sunflower grows as tall as 11.5 ft (3.5 m), and can have enormous flowering heads, up to 14 in (35 cm) in diameter. The long ray florets make this sunflower very attractive, and it is often grown as an ornamental. Safflower (*Carthamus tinctorius*) is another species that is sometimes grown as a source of edible oil.

The Jerusalem artichoke (*Helianthus tuberosus*) is a perennial sunflower, native to the prairies of **North America**. The Jerusalem artichoke has underground rhizomes, on which grow starchy and nutritious tubers. The tubers can be eaten as a vegetable, or may be processed into **alcohol**.

The globe artichoke (*Cynara scolymus*) is another perennial composite, originally from the Mediterranean region. The pre-flowering heads of this species are cut off before they begin to expand. These are boiled or steamed, and the thick, fleshy material of the involucral leaves of the receptacle (that is, the base of the flowering structure) are eaten, commonly by peeling them between the teeth. The fleshy interior of the receptacle itself, known as the artichoke heart, is also eaten.

The lettuce, or cabbage lettuce (*Lactuca sativa*), originated in southern **Europe**, and is grown for its greens, which are mostly used in salads, or as a green decoration for other foods.

Chicory (*Cichorium intybus*) is grown for its roots, which can be roasted and used as a substitute for coffee. A leafy variety of chicory is used as a salad green. Endive (*Cichorium endivia*) is a related species, also used as a salad green. These plants are originally from western Europe. Species of dandelions are also used as salad greens, for example, the common dandelion (*Taraxacum officinalis*).

Other useful species of composites

Several species of composites have minor uses in medicine. Chamomile (*Anthemis nobilis*) is an annual European species that is collected and dried, and brewed into an aromatic tea that has a calming effect. The dried leaves and flowers of common wormwood (*Artemisia absinthium*) of Europe are used to make a tonic known as bitters, while the flower buds are used to flavor a liquor known as vermouth. The seeds of the wormwoods *Artemisia cina* and *A. maritima*, species native to the steppes of central **Asia**, are given as a treatment against several types of intestinal **parasites**.

Some other species in the aster family have been erroneously ascribed medicinal qualities. This occurred as a result of a theory of medicine that was developed during the Middle Ages, known as the "Doctrine of Signatures." According to this ideology, the potential medicinal usefulness of plants was revealed through some sort of sign, such as a similarity between their shape, and that of a part of the human **anatomy**. In the case of the herbaceous plant known as boneset (*Eupatorium perfoliatum*), the leaves are arranged opposite each other on the stem, and they lack a petiole, and are fully joined to each other by a band of leafy **tissue** that broadly clasps the stem. This unusual growth form, or signature, was interpreted by herbalists to suggest that boneset must have therapeutic properties in helping broken bones to heal. As a result, boneset was spread as a moist poultice over a broken bone, which was then encased within a bandage, plaster, or splint.

Several species of chrysanthemums are used to manufacture an organic insecticide known as pyrethrum. *Chrysanthemum roseum*, *C. coccinium*, and *C. cinerariaefolium* of southern regions of Asia have been widely cultivated for the production of these chemicals. Sometimes, living chrysanthemums are inter-cultivated with other plants in gardens, in order to deter some types of herbivorous insects.

The latex of the rubber dandelion (*Taraxacum bicorne*) contain 8-10% rubber latex, and is potentially useful for the commercial production of rubber.

Composites as weeds

Some members of the aster family have become regarded as important weeds. In many cases, these are aesthetic weeds, because they occur abundantly in places where people, for whatever reason, do not want to see these plants. For example, the common dandelion (*Taraxacum officinale*), originally from Europe but now widely distributed in North America and elsewhere, is often regarded to be a weed of lawns and landscapes. This is largely because many people only want to see certain species of **grasses** in their lawns, so that any dicotyledonous plants, such as dandelions, are considered to be weeds. As a result, many people put a great deal of time and effort into manually digging dandelions out of their lawns, or they may use a herbicide such as 2,4-D to rid themselves of these perceived weeds.

Interestingly, many other people consider the spectacular, yellow displays that dandelion flowers can develop in lawns and pastures in the springtime to be very pleasing. Dandelions are also favored by some people as a food, especially the fresh leaves that are collected in the early springtime. Clearly, the judgement of a plant as

Late goldenrod (*Solidago gigantea*). © *John Dudak/ Phototake NYC. Reproduced with permission.*

an aesthetic weed is substantially a matter of perspective and context.

However, a few species in the aster family are weeds for somewhat more important reasons. Some species are weeds because they are poisonous to **livestock**. For example, the ragwort or stinking-Willie (*Senecio jacobea*) has alkaloids in its foliage that are toxic to liver of cattle. The natural range of the ragweed is Eurasia, but it has become an important weed in pastures in parts of North America and elsewhere, possibly having been introduced as an ornamental plant. Recently, several insect species that are herbivores of ragweed in its native habitats have been introduced to some of its invasive range, and these are showing promise as agents of biological control of this important pest.

Some other species in the aster family are important weeds of pastures because they are very spiny, and livestock cannot eat them. These inedible plants can become abundant in pastures, displacing valuable forage species. Some examples of these sorts of weeds in North America include various thistles introduced from Europe, such as bull thistle (*Cirsium vulgare*), field thistle (*C. arvense*), nodding thistle (*Carduus nutans*), and Scotch thistle (*Onopordum acanthium*).

Some species in the aster family have anatomical mechanisms of attaching their seeds to the fur of mammals, for the purposes of dispersal. Animals with large numbers of these seeds in their fur can become very irritated by the matting, and they may scratch themselves so much that wounds develop, with a risk of **infection**. Examples of weeds that stick to animals, and to the clothing of humans, include the beggar-ticks (for example, *Bidens frondosa*), and several **introduced species** known as burdock (for example, the greater burdock, *Arctium lappa*). Interestingly, the finely hooked bristles of the globular fruits of burdock were the inspiration for the development of the well-known fastening material known as velcro.

KEY TERMS

Achene—A dry, indehiscent, one-seeded fruit, with the outer layer fused to the seed.

Bract—A small, scale-like, modified leaf that is associated with a flower or inflorescence.

Floret—This is a small flower, often with some reduced or missing parts. Florets are generally arranged within a dense cluster.

Head—A dense cluster of flowers attached to a common receptacle. This is the characteristic arrangement of the flowers of members of the aster family.

Inflorescence—A grouping or arrangement of florets or flowers into a composite structure.

Pappus—A distinctive tissue of members of the aster family, attaching to the top of the achene, and resembling small scales or fine hairs, sometimes intricately branched to achieve aerodynamic buoyancy. The pappus is derived from modified tissues of the calyx.

Receptacle—The enlarged tip of a peduncle where the parts of a flower are attached. Four distinct whorls of modified leaves, the sepals, petals, stamens, and carpels make up the parts of the flower.

Weed—Any plant that is growing abundantly in a place where humans do not want it to be.

The several species that are known as ragweed (*Ambrosia artemesiifolia* and *A. trifida*) are the major causes of hay-fever during the summer and early autumn. The ragweeds are wind pollinated, and to achieve this function they shed large quantities of tiny, spiny-surfaced pollen grains to the wind. Many people have an **allergy** to ragweed pollen, and they may suffer greatly from hay-fever caused by ragweeds.

Interestingly, at about the same time that ragweeds are shedding their abundant pollen to the air, some other, more conspicuous species in the aster family are also flowering prolifically. For example, pastures, fields, and other habitats may develop spectacular shows of yellow goldenrods (*Solidago* spp.) and white, blue, or purple asters (*Aster* spp.) at that time of year. Because people notice these brightly colored plants, but not the relatively small and drab ragweeds, the asters and goldenrods are commonly blamed for hay-fever. For this reason, fields of these attractive plants may be mowed or herbicided to deal with this perceived weed-management problem. However, the asters and goldenrods are insect-pollinated,

and they do not shed their pollen to the wind. Therefore, these plants are not the cause of hay-fever—they are merely implicated by their association in time with the guilty but inconspicuous ragweed. Indeed, even people who suffer badly from hay-fever, often do not recognize the rather plain-green, unobtrusive-looking ragweeds as the cause of their allergy.

Resources

Books

Hvass, E. *Plants That Serve and Feed Us.* New York: Hippocrene Books, 1975.

Judd, Walter S., Christopher Campbell, Elizabeth A. Kellogg, Michael J. Donoghue, and Peter Stevens. *Plant Systematics: A Phylogenetic Approach.* 2nd ed. with CD-ROM. Suderland, MD: Sinauer, 2002.

Klein, R. M. *The Green World. An Introduction to Plants and People.* New York: Harper and Row, 1987.

Bill Freedman

Composite materials

A composite material is a microscopic or macroscopic combination of two or more distinct materials with a recognizable interface between them. For structural applications, the definition can be restricted to include those materials that consist of a reinforcing phase such as fibers or particles supported by a binder or matrix phase. Other features of composites include the following: (1) The distribution of materials in the composite is controlled by mechanical means; (2) The term composite is usually reserved for materials in which distinct phases are separated on a scale larger than atomic, and in which the composite's mechanical properties are significantly altered from those of the constituent components; (3) The composite can be regarded as a combination of two or more materials that are used in combination to rectify a weakness in one material by a strength in another. (4) A recently developed concept of composites is that the composite should not only be a combination of two materials, but the combination should have its own distinctive properties. In terms of strength, **heat** resistance, or some other desired characteristic, the composite must be better than either component alone.

Composites were developed because no single, homogeneous structural material could be found that had all of the desired characteristics for a given application. Fiber-reinforced composites were first developed to replace **aluminum** alloys, which provide high strength and fairly high stiffness at low weight but are subject to **corrosion** and fatigue.

An example of a composite material is a glass-reinforced plastic fishing rod in which **glass** fibers are placed in an epoxy matrix. Fine individual glass fibers are characterized by their high tensile stiffnesses and a very high tensile strengths, but because of their small diameters, have very small bending stiffnesses. If the rod were made only of epoxy plastic, it would have good bending stiffness, but poor tensile properties. When the fibers are placed in the epoxy plastic, however, the resultant structure has high tensile stiffness, high tensile strength, and high bending stiffness.

The discontinuous filler phase in a composite is usually stiffer or stronger than the binder phase. There must be a substantial **volume** fraction of the reinforcing phase (~10%) present to provide reinforcement. Examples do exist, however, of composites where the discontinuous phase is more compliant and ductile than the matrix.

Natural composites include **wood** and bone. Wood is a composite of **cellulose** and lignin. Cellulose fibers are strong in tension and are flexible. Lignin cements these fibers together to make them stiff. Bone is a composite of strong but soft **collagen** (a protein) and hard but brittle apatite (a mineral).

Particle-reinforced composites

A particle has no long dimension. Particle composites consist of particles of one material dispersed in a matrix of a second material. Particles may have any shape or size, but are generally spherical, ellipsoidal, polyhedral, or irregular in shape. They may be added to a liquid matrix that later solidifies; grown in place by a reaction such as age-hardening; or they may be pressed together and then inter-diffused via a powder process. The particles may be treated to be made compatible with the matrix, or they may be incorporated without such treatment. Particles are most often used to extend the strength or other properties of inexpensive materials by the addition of other materials.

Fiber-reinforced composites

A fiber has one long dimension. Fiber-reinforced materials are typified by fiberglass in which there are three components: glass filaments (for mechanical strength), a **polymer** matrix (to encapsulate the filaments); and a bonding agent (to bind the glass to the polymer). Other fibers include **metal**, **ceramics**, and polymers. The fibers can be used as continuous lengths, in staple-fiber form, or as whiskers (short, fine, perfect, or nearly perfect single crystals). Fiber-reinforcement depends as much on fabrication procedure as on materials.

Laminar composites

Platelets or lamina have two long dimensions. Laminar composites include plywood, which is a laminated composite of thin layers of wood in which successive layers have different grain or fiber orientations. The result is a more-or-less isotropic composite sheet that is weaker in any direction than it would be if the fibers were all aligned in one direction. The stainless **steel** in a cooking vessel with a copper-clad bottom provides corrosion resistance while the **copper** provides better heat distribution over the base of the vessel.

Mechanical properties

The mechanical properties of composite materials usually depend on structure. Thus these properties typically depend on the shape of inhomogenities, the volume fraction occupied by inhomogenities, and the interfaces between the components. The strength of composites depends on such factors as the brittleness or ductility of the inclusions and matrix.

For example, failure mechanisms in fiber-filled composites include fracture of the fibers; shear failure of the matrix along the fibers; fracture of the matrix in tension normal to the fibers or failure of the fiber-matrix interface. The mechanism responsible for failure depends on the **angle** between the fibers and the specimen's axis.

If a mechanical property depends on the composite material's orientation, the property is said to be anisotropic. Anisotropic composites provide greater strength and stiffness than do isotropic materials. But the material properties in one direction are gained at the expense of the properties in other directions. For example, silica fibers in a pure aluminum matrix produce a composite with a tensile strength of about 110,000 psi along the fiber direction, but a tensile strength of only about 14,000 psi at right angles to the fiber axis. It therefore only makes sense to use anisotropic materials if the direction that they will be stressed is known in advance.

Isotropic material are materials properties independent of orientation. Stiff platelet inclusions are the most effective in creating a stiff composite, followed by fibers, and then by spherical particles.

High performance composites

High performance composites are composites that have better performance than conventional structural materials such as steel and aluminum alloys. They are almost all continuous fiber-reinforced composites, with organic (resin) matrices.

Fibers for high performance composites

In a high-performance, continuous fiber-reinforced composite, fibers provide virtually all of the load-carrying characteristics of the composite, i.e., strength and stiffness. The fibers in such a composite form bundles, or filaments. Consequently, even if several fibers break, the load is redistributed to other fibers, which avoids a catastrophic failure.

Glass fibers are used for nonstructural, low-performance applications such as panels in **aircraft** and appliances to high-performance applications such as rocket-motor cases and **pressure** vessels. But the sensitivity of the glass fiber to attack by moisture poses problems for other applications. The most commonly used glass fiber is a **calcium** aluminoborosilicate glass (E-glass). High silica and quartz fibers are also used for specialized applications.

Carbon fibers are the best known and most widely used reinforcing fiber in advanced composites.The earliest carbon fibers were produced by thermal **decomposition** of rayon precursor materials. The starting material is now polyacrylonitrile.

Aramid fibers are aromatic polyamide fibers. The aramid fiber is technically a thermoplastic polymer like nylon, but it decomposes when heated before it reaches its projected melting point. When polymerized, it forms rigid, rod-like molecules that cannot be spun from a melt. Instead they have to be spun from a liquid crystalline **solution**. Early applications of aramid fibers included filament-wound motor cases, and gas pressure vessels. Aramid fibers have lower compressive strengths than do carbon fibers, but their high specific strengths, low densities, and toughness keep them in demand.

Boron fibers were the first high-performance reinforcement available for use in advance composites. They are, however, more expensive and less attractive for their mechanical properties than carbon fibers. Boron filaments are made by the decomposition of boron halides on a hot tungsten wire. Composites can also be made from whiskers dispersed in an appropriate matrix.

Continuous silicon carbide fibers are used for large-diameter monofilaments and fine multifilament yarns. Silicon carbide fibers are inherently more economical than boron fibers, and the properties of silicon carbide fibers are generally as good or better than those of boron.

Aluminum oxide (alumina) fibers are produced by dry spinning from various solutions. They are coated with silica to improve their contact properties with molten metal.

There is usually a size effect associated with strong filaments. Their strengths decrease as their diameter increases. It turns out that very high strength materials have diameters of about 1 micrometer. They are consequently not easy to handle.

Matrices for high performance composites

The matrix binds fibers together by virtue of its cohesive and **adhesive** characteristics. Its purpose is to transfer load to and between fibers, and to protect the fibers from hostile environments and handling. The matrix is the weak link in the composite, so when the composite experiences loading, the matrix may crack, debond from the fiber surface, or break down under far lower strains than are usually desired. But matrices keep the reinforcing fibers in their proper orientation and position so that they can carry loads, distribute loads evenly among fibers, and provide resistance to crack propagation and damage. Limitations in the matrix generally determine the overall service **temperature** limitations of the composite.

Polyester and vinyl esterresins are the most widely used matrix materials in high performance continuous-fiber composites. They are used for chemically resistant piping and reactors, truck cabs and bodies, appliances, bathtubs and showers, **automobile** hoods, decks, and doors. These matrices are usually reinforced with glass fibers, as it has been difficult to adhere the matrix suitably to carbon and aramid fibers. Epoxies and other **resins**, though more expensive, find applications as replacements for polyester and vinyl **ester** resins in high performance sporting goods, piping for chemical processing plants, and printed circuit boards.

Epoxy resins are used more than all other matrices in advanced composite materials for structural aerospace applications. Epoxies are generally superior to polyesters in their resistance to moisture and other environmental influences.

Bismaleimide resins, like epoxies, are fairly easy to handle, relatively easily processed, and have excellent composite properties. They are able to withstand greater fluctuations in hot/wet conditions than are epoxies, but they have worse failure characteristics.

Polyimide resins release volatiles during curing, which produces voids in the resulting composite. However, these resins do withstand even greater hot/wet temperature extremes than bismaleimide matrices, and work has been underway to minimize the void problem.

The thermoplastic resins used as composite matrices such as polyether etherketone, polyphenylene sulfide, and polyetherimide are very different from the commodity thermoplastics such as polyethylene and polyvinyl chloride. Although used in limited quantities, they are attractive for applications requiring improved hot/wet properties and impact resistance.

Other composites

In addition to the examples already given, examples of composites materials also include: (1) Reinforced and prestressed **concrete**, which is a composite of steel and concrete. Concrete is itself a composite of **rocks** (coarse aggregate), **sand** (fine aggregate), hydrated Portland cement, and usually, voids. (2) Cutters for machining made of fine particles of tungsten carbide, which is extremely hard, are mixed with about 6% cobalt powder and sintered at high temperatures. (3) Ordinary grinding wheels, which are composites of an abrasive with a binder that may be plastic or metallic. (4) Walls for housing, which have been made of thin aluminum sheets epoxied to polyurethane foam. The foam provides excellent thermal insulation. This composite has a higher structural rigidity than aluminum sheets or polyurethane foam alone. The polyurethane foam is itself a composite of air and polyurethane. (5) Underground electrical cables composed of **sodium** metal enclosed in polyethylene. (6) Superconducting ribbons made of Nb_3Sn deposited on copper. (7) Synthetic hard superconductors made by forcing liquid **lead** under pressure into porous glass fibers. (8) Microelectronic circuits made from silicon, which are oxidized to form an insulating layer of SiO_2. This insulating layer is etched away with hydrofluoric acid, and phosphorous is diffused into the silicon to make a junction. Aluminum or another metal can be introduced as a microconductor between points. The microelectronic circuit is thus a tailored composite. (9) Ceramic fiber composites including graphite or pyrolytic carbon reinforced with graphite fibers; and borosilicate glass **lithium** aluminum silicate glass ceramics reinforced with silicon carbide fibers. It was possible to drive a tungsten carbide spike through such a composition without secondary cracking in much the same way that a nail can be driven through wood.

Resources

Books

Reinhart, Theodore J. "Introduction to Composites." In *Engineered Materials Handbook* Vol. 1. Metals Park, OH: ASM International, 1987.
Smith, Charles O. *The Science of Engineering Materials.* Englewood Cliffs, NJ: Prentice-Hall, Inc. 1969.
Sperling, L.H. *Introduction to Physical Polymer Science.* New York, John Wiley & Sons, Inc. 1992.

Randall Frost

Composting

Composting is the process of arranging and manipulating organic wastes so that they are gradually broken down, or decomposed, by **soil microorganisms** and animals. The resulting product is a black, earthy-smelling, nutritious, crumbly mixture called compost or **humus**. Compost is usually mixed with other soil to improve the soil's structural quality and to add **nutrients** for **plant** growth. Composting and the use of compost in gardening are important activities of gardeners who prefer not to use synthetic **fertilizers**.

Nature itself composts materials by continually recycling nutrients from dead organic **matter**. Living things take in inorganic nutrients to grow. They give off waste, die, and decompose. The nutrients contained in the plant or **animal** body become available in soil for plants to take up again. Composting takes advantage of this natural process of **decomposition**, usually speeding up the process, by the creation of a special pile of organic materials called a compost heap.

The major benefit of compost is its organic content. Humus added to soil changes its structure, its ability to hold **oxygen** and **water**, and its capacity to adsorb certain nutrient ions. It improves soils that are too sandy to hold water or contain too much clay to allow oxygen to penetrate. Compost also adds some mineral nutrients to the soil. Depending on the organic material of the compost and microorganisms present, it can also balance the **pH** of an acidic or alkaline soil.

History

Prehistoric farming people discovered that if they mixed manure from their domesticated animals with straw and other organic waste, such as crop residues, the mixture would gradually change into a fertile soil-like material that was good for **crops**. Composting remained a basic activity of farming until the twentieth century, when various synthetic fertilizers were found to provide many of the nutrients occurring naturally in compost.

The steadily increasing population of the world has come to require large supplies of food. In order to increase productivity, farmers have come to depend on synthetic fertilizers made in factories from nonrenewable resources. However, regular use of these fertilizers does not improve the structure of the soil, and can, in fact, gradually harm the soil. Also, synthetic fertilizers are expensive, an important consideration to farmers in less developed countries.

It was in an underdeveloped country—India—that modern composting got its big start. Sir Albert Howard, a government agronomist, developed the so-called Indore method, named after a city in southern India. His method calls for three parts garden clippings to one part manure or kitchen waste arranged in layers and mixed periodically. Howard published his ideas on organic gardening in the 1940 book *An Agricultural Testament*.

The first articulate advocate of Howard's method in the United States was J.I. Rodale (1898-1971), founder of *Organic Gardening* magazine. These two men made composting popular with gardeners who prefer not to use synthetic fertilizers.

People are attracted to composting for a variety of reasons. Many wish to improve their soil or help the environment. Compost mixed with soil makes it darker, allowing it to warm up faster in the spring. Compost adds numerous naturally occurring nutrients to the soil. It improves soil quality by making the structure granular, so that oxygen is retained between the granules. In addition, compost holds moisture. This is good for plants, of course, but it is also good for the environment because it produces a soil into which rain easily soaks. When water cannot soak directly into soil, it runs across the surface, carrying away soil granules and thus eroding the soil. In addition, the use of compost limits the use of **natural gas**, petrochemicals, and other nonrenewable resources that are used in making synthetic fertilizers. Composting also recycles organic materials that might otherwise be sent to landfills.

Despite its many benefits, making and using compost does have its disadvantages. Composting releases methane, a greenhouse gas which traps solar **heat** in the earth's atmosphere and may contribute to **global warming**. The kitchen wastes and warmth of compost heaps may attract **pests** such **mice**, **rats**, and **raccoons**.

Composting on any scale

Composting can be done by anyone. A homeowner can use a small composting bin or a hole where kitchen wastes (minus meats and fats) are mixed with grass clippings, small branches, shredded newspapers, or other coarse, organic debris.

Communities may have large composting facilities to which residents bring grass, leaves, and branches to be composted. Such communities often have laws against burning garden waste and use composting as an alternative to disposal in a **landfill**. Sometimes sewage sludge, the semisolid material from **sewage treatment** plants, is added. The heat generated in the heap kills any disease-causing **bacteria** in the sludge. The materials are usually arranged in long rows, called windrows, which may be covered by roofs. The resulting humus is used to condition soil on golf courses, parks, and other municipal grounds.

The largest scale of composting is done commercially by companies that collect organic materials, including **paper**, from companies and private citizens. Commercial composting is usually mechanized, using large machines called composters. Raw solid waste is loaded onto a slow-moving belt, then is dumped into a device which turns the waste, and compost comes out the other end within a few days or weeks. This in-vessel or container process allows careful control of moisture and air. Some communities are looking toward such mechanized digesters as a way of helping to solve the municipal solid waste problem as more and more landfills close in the future.

Materials to compost

Most organic materials can be used in a compost heap—shredded paper, hair clippings, food scraps from restaurants (omitting meats), coffee grounds, eggshells, fireplace ashes, chopped-up Christmas trees, seaweed, anything that originally came from a living thing. Meat is omitted because it can putrefy, giving off bad odors. It can also attract rats and other pests. Soil or finished humus is added to supply the microorganisms needed to make the heap work. To work most efficiently, the materials are layered, with woody materials, **grasses**, kitchen waste, and soil alternating. Farmyard or zoo manure mixed with straw makes an excellent addition to compost. However, feces from household pets may carry diseases.

A **ratio** of approximately 25 parts **carbon** to 1 part **nitrogen** should be available in the compost heap. If the ratio is quite different, **ammonia** smells can be given off, or the process may not work efficiently. Chopped-up **tree** branches, fallen leaves, and sawdust are good sources of carbon. Alfalfa is a good nitrogen source.

How it works

A compost heap needs to have both water and oxygen to work efficiently. In dry weather it may need to be

Backyard composting in Livonia, Michigan. This system includes a 16 cubic foot (0.5 cubic meter) composting bin made from chicken wire and plywood, a soil screen made from 1/2 inch galvanized mesh wire and 1x6 boards, a wheelbarrow, and a digging fork. The system produces about 10 cubic feet (0.3 cubic meters) of compost per year. *Photograph by Robert J. Huffman. Field Mark Publications. Reproduced by permission.*

watered. More importantly, however, the compost heap must be turned regularly. The more often it is turned, the more the compost materials are exposed to oxygen, which raises their **temperature** and increases the efficiency of the process.

A compost heap needs to be at least 3 ft (0.9 m) in diameter and about 3 ft (0.9 m) high to work properly. The heap can be just piled on the ground or layered within a shallow hole. It can also be placed inside a small fenced enclosure or even in a large plastic or **metal** tub with holes cut into it. A smaller compost heap will probably be unable to achieve the high internal temperature-about 120–140°F (49–60°C)-necessary to work efficiently.

The chemical process

The processes that occur within a compost heap are microbiological, chemical, and physical. Microorganisms break down the carbon bonds of organic materials in the presence of oxygen and moisture, giving off heat in the process.

High temperatures can be achieved most easily in a compost heap that is built all at once and tended regularly. Enclosed bins, often used by city gardeners, may produce humus within a month. An open heap, such as in the back corner of a garden, will probably achieve lower temperatures, but it will still eventually decompose. However, a year or more may be required to produce humus, and it will contain some undecomposed materials. In northern winters, the composting process will slow down and almost stop except at the core of a large, well-arranged heap.

The byproducts of composting can also be used. Some composters run water pipes through their compost heaps and utilize the heat generated to warm greenhouses and even houses. The methane given off can also be collected and used as a fuel called biogas for cooking.

The organisms

The most heat is given off at the beginning of the composting process, when readily oxidized material is decomposing. Digestion of the materials by bacteria is strongest at that time. Later, the temperature within the pile decreases, and the bacterial activity slows down, though it continues until all the waste is digested. Other microorganisms take over as the heap cools.

Microorganisms, such as bacteria, **protozoa**, **fungi**, and actinomycetes (the latter resemble both bacteria and fungi), work to change the **chemistry** of the compost. They produce enzymes that digest the organic material. Bacteria are most important initially and fungi later. If the pile is not turned regularly, the decomposition will be **anaerobic** and produce foul-smelling odors. By turning the pile, a gardener creates conditions for **aerobic** decomposition, which does not produce odors.

Some organisms work on the compost pile physically instead of chemically. They tend to arrive only after the pile has cooled to normal air temperature. These organisms include **mites**, **millipedes**, sowbugs and pillbugs (isopods), **snails** and **slugs**, **springtails**, and **beetles**. Finally, the worms—nematodes, **flatworms**, and earthworms—do their part. These animals eat and digest the organic materials, adding their nutrient-filled excrement to the humus. In addition, they give off substances that bind the material in granules or clumps. The movement of these animals, especially earthworms, through the material helps to aerate it.

The nutrients

During the composting process, the material oxidizes, breaking down into **proteins** and carbohydrates. The proteins break down into peptides and amino acids, then into ammonium compounds. These compounds are changed by certain bacteria into nitrates, a form of nitrogen which can be used by plants to make **chlorophyll** and essential proteins. The carbohydrates break down into simple sugars, organic acids, and **carbon dioxide**.

Nutrients in humus enter plant tissues by a process called base exchange. In this process, **hydrogen** ions in the fine root hairs of plants are exchanged for the nutrient ions in the soil moisture. The nutrients are then free to move up into the plant.

Composting with worms

Some composters use a somewhat different form of composting, especially during winter. Called vermicomposting, it consists of maintaining worms (preferably redworms, or *Eisenia foetida*) in a container filled with a plant-based material (such as shredded corrugated paper, manure, or peat **moss**) that they gradually consume.

Kitchen waste is pushed into the soil and digested by the worms. Their excrement, called castings, along with partially decomposed waste, can be "harvested" in

about four months and used as a nutrient-laden addition to soil. Worm castings are even more nutrient-filled than garden compost.

See also Waste management.

Resources

Books

Appelhof, Mary. *Worms Eat My Garbage*. Kalamazoo, MI: Flower Press, 1982.

Blashfield, Jean F., and Wallace B. Black. *Recycling*. Saving Planet Earth series. Chicago: Childrens Press, 1991.

Campbell, Stu. *Let It Rot! The Gardener's Guide to Composting*. Rev. ed. Pownal, VT: Storey Communications, 1990.

Culen, Gerald, William Bluhm, Preethi Mony, Janice Easton, and Larry Schnell. *Organics: A Wasted Resource? an Extended Case Study for the Investigation and Evaluation of Composting and Organic Waste Management Issues*. Champaign, IL: Stipes Publishing, 2001.

Martin, Deborah L., and Grace Gershuny, eds. *The Rodale Book of Composting*. Rev. ed. Emmaus, PA: Rodale Press, 1992.

Whitehead, Bert. *Don't Waste Your Wastes-Compost 'Em: The Homeowner's Guide to Recycling Yard Wastes*. Sunnyvale, TX: Sunnyvale Press, 1991.

Jean F. Blashfield

Compound, chemical

A compound is a substance composed of two or more elements chemically combined with each other. Historical-

---

KEY TERMS

Aerobic—Requiring or in the presence of oxygen.

Anaerobic—Describes biological processes that take place in the absence of oxygen.

Decomposition—The breakdown of the complex molecules composing dead organisms into simple nutrients that can be reutilized by living organisms.

Microorganism—Any living thing that can only be seen through a microscope.

Nutrient—Any substance required by a plant or animal for energy and growth.

Organic—Made of or requiring the materials of living things. In pure chemistry, organic refers to compounds that include carbon.

Vermicomposting—Using the digestive processes of worms to compost organic materials.

---

ly, the distinction between compounds and mixtures was often unclear. Today, however, the two can be distinguished from each other on the basis of three primary criteria. First, compounds have constant and definite compositions, while mixtures may exist in virtually any proportion. A **sample** of **water** always consists of 88.9% **oxygen** and 11.1% **hydrogen** by weight. However, a mixture of hydrogen and oxygen gases can have any composition whatsoever.

Second, the elements that make up a compound lose their characteristic elemental properties when they become part of the compound, while the elements that make up a mixture retain those properties. In a mixture of **iron** and **sulfur**, for example, black iron granules and yellow sulfur crystals can often be recognized. Also, the iron can be extracted from the mixture by means of a magnet, or the sulfur can be dissolved out with **carbon** disulfide. One part of the compound is called iron(II) sulfide, however, both iron and sulfur lose these properties.

Third, the formation of a compound is typically accompanied by the evolution of **light** and **heat**, while no observable change is detectable in the making of a mixture. A mixture of iron and sulfur can be made simply by stirring the two elements together. But the compound iron(II) sulfide is produced only when the two elements are heated. Then, as they combine with each other, they give off a glow.

Non-chemical definitions

The term compound is often used in fields of science other than **chemistry**, either as an adjective or a verb. For example, medical workers may talk about a compound fracture in referring to a broken bone that has cut through the flesh. Biologists use a compound **microscope**, one that has more than one **lens**. Pharmacologists may speak of compounding a drug, that is, putting together the components of which that medication consists. In the case of the last example, a compounded drug is often one that is covered by a patent.

History

Prior to the 1800s, the term compound had relatively little precise meaning. When used, it was often unclear as to whether one was referring to what scientists now call a mixture or to what they now know as a compound. During the nineteenth century, the debate as to the meaning of the word intensified, and it became one of the key questions in the young science of chemistry.

A critical aspect of this debate focused on the issue of constant composition. The issue was whether all compounds always had the same composition, or whether their composition could vary. The primary spokesman

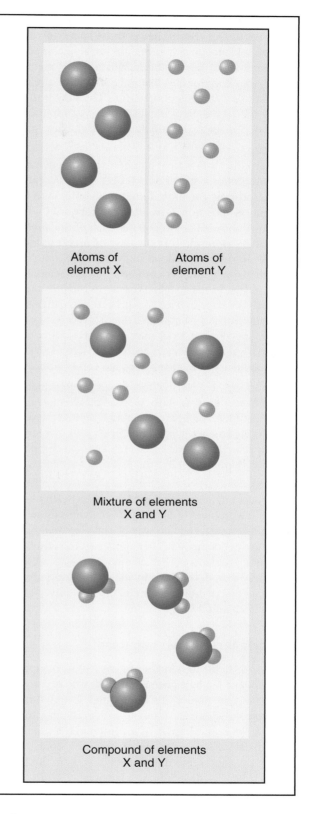

Atoms of element X   Atoms of element Y

Mixture of elements X and Y

Compound of elements X and Y

A mixture versus a compound. *Illustration by Argosy. The Gale Group.*

for the latter position was the French chemist Claude Louis Berthollet. Berthollet pointed to a considerable body of evidence that suggested a variable composition for compounds. For example, when some metals are heated, they form oxides that appear to have a regularly changing percentage composition. The longer they are heated, the higher the percentage of oxygen found in the oxide. Berthollet also mentioned alloys and amalgams as examples of substances with varying composition.

Berthollet's principal antagonist in this debate was his countryman Joseph Louis Proust. Proust argued that Dalton's **atomic theory** required that compounds have a constant composition, a position put forward by Dalton himself. Proust set out to counter each of the arguments set forth by Berthollet. In the case of **metal** oxides, for example, Proust was able to show that metals often form more than one oxide. As **copper** metal is heated, for example, it first forms copper(I) or cuprous oxide and then, copper(II) or cupric oxide. At any one time, then, an experimenter would be able to detect some mixture of the two oxides varying from pure copper(I) oxide to pure copper(II) oxide. However, each of the two oxides itself, Proust argued, has a set and constant composition.

Working in Proust's favor was an argument that nearly everyone was willing to acknowledge, namely that quantitative techniques had not yet been developed very highly in chemistry. Thus, it could be argued that what appeared to be variations in chemical composition were really nothing other than natural variability in results coming about as a result of imprecise techniques.

Proust remained puzzled by some of Berthollet's evidence, the problem of alloys and amalgams as an example. At the time, he had no way of knowing that such materials are not compounds but are in fact mixtures. These remaining problems notwithstanding, Proust's arguments eventually won the day and by the end of the century, the constant composition of compounds was universally accepted in chemistry.

Early theories of compounds

It is difficult for a reader in the 1990s to appreciate the challenge facing a chemist in 1850 who was trying to understand the nature of a compound. Today it is clear that **atoms** of elements combine with each other to form, in many cases, molecules of a compound. Even the beginning chemistry student can express this concept with facility by using symbols and formulas, as in the formation of iron(II) sulfide from its elements:

$$Fe + S \rightarrow FeS.$$

The chemist of 1850 was just barely comfortable with the idea of an atom and had not yet heard of the concept of a **molecule**. Moreover, the connection between ultimate particles (such as atoms and molecules) and materials encountered in the everyday work of a laboratory was not at all clear. As a result, early theories about the nature of compounds were based on empirical data (information collected from experiments), not from theoretical speculation about the behavior of atoms.

One of the earliest theories of compounds was that of the Swedish chemist Jons Jacob Berzelius. Berzelius argued that all compounds consist of two parts, one charged positively and one, negatively. The theory was at least partially based on Berzelius' own studies of **electrolysis**, studies in which compounds would often be broken apart into two pieces by the passage of an electrical current. Thus, he pictured salts as being composed of a positively charged metal oxide and a negatively charged non-metallic oxide. According this theory, sodium sulfate, Na_2SO_4 could be represented as $Na_2O \cdot SO_3$.

Other theories followed, many of them developed in an effort to explain the rapidly growing number of organic compounds being discovered and studied. According to the radical theory, for example, compounds were viewed as consisting of two parts, one of which was one of a few standard radicals, or groups of atoms. Organic compounds were explained as being derived from the methyl, ethyl, benzyl, cyanogen, or some other radical.

The type theory, proposed by Charles Gerhardt in the 1840s, said that compounds could be understood as derivatives of one or more basic types, such as water or **ammonia**. According to this theory, bases such as **sodium hydroxide** (NaOH) were thought to be derivatives of water (HOH) in which one hydrogen atom is replaced by a metal.

Modern theory of compounds

The most fundamental change that has taken place in chemistry since the nineteenth century is that atomic theory now permits an understanding of chemical compounds from the particle level rather than from purely empirical data. That is, as our knowledge of atomic structure has grown and developed, our understanding of the reasons that atoms (elements) combine with each other has improved. For example, the question of how and why iron and sulfur combine with each other to form a compound is now approached in terms of how and why an iron atom combines with a sulfur atom to form a molecule of iron(II) sulfide.

A key to the solution of that puzzle was suggested by the German chemist Albrecht Kossel in 1916. In considering the unreactivity of the inert gases, Kossel came to the conclusion that the presence of eight electrons in the outermost **energy** level of an atom (as is the case with all inert gases) conferred a certain stability on a substance. Perhaps, Kossel said, the tendency of atoms to exchange electrons in such a way as to achieve a full octet (eight) of electrons could explain **chemical reactions** in which elements combine to form compounds.

Although Kossel had hit on a key concept, he did not fully develop this theory. That work was left to the American chemist Gilbert Newton Lewis. At about the same time that Kossel was proposing his octet theory, Lewis was developing a comprehensive explanation showing how atoms can gain a complete octet either by the gain and loss or by the sharing of pairs of electrons with other atoms. Although Lewis' theory has undergone many transformations, improvements, and extensions (especially in the work of Linus Pauling), his explanation of compound formation still constitutes the **heart** of such theory today.

Types of compounds

Most of the ten million or so chemical compounds that are known today can be classified into a relatively small number of subgroups or families. More than 90% of these compounds are, in the first place, designated as organic compounds because they contain the element carbon. In turn, organic compounds can be further subdivided into a few dozen major families such as the alkanes, alkenes, alkynes, alcohols, **aldehydes**, ketones, **carboxylic acids**, and amines. Each of these families can be recognized by the presence of a characteristic functional group that strongly determines the physical and chemical properties of the compounds that make up that family. For example, the functional group of the alcohols is the hydroxyl group (-OH) and that of the carboxylicacids, the **carboxyl group** (-COOH).

An important subset of organic compounds are those that occur in living organisms, the biochemical compounds. Biochemical compounds can largely be classified into four major families: the carbohydrates, **proteins**, nucleic acids, and lipids. Members of the first three families are grouped together because of common structural features and similar physical and chemical properties. Members of the **lipid** family are so classified on the basis of their **solubility**. They tend not to be soluble in water, but soluble in organic liquids.

Inorganic compounds are typically classified into one of five major groups: acids, bases, salts, oxides, and others. Acids are defined as compounds which ionize or dissociate in water solution to yield hydrogen ions. Bases are compounds that ionize or dissociate in water solution to yield hydroxide ions. Oxides are compounds whose only **negative** part is oxygen. Salts are compounds whose cations are any ion but hydrogen and

---

KEY TERMS

. . . .

Alloy—A mixture of two or more metals with properties distinct from the metals of which it is made.

Amalgam—An alloy that contains the metal mercury.

Coordination compounds—Compounds formed when metallic ions or atoms are joined to other atoms, ions, or molecules by means of coordinate covalent bonds.

Empirical—Evidence that is obtained from some type of experimentation.

Family—A group of chemical compounds with similar structure and properties.

Functional group—A group of atoms that give a molecule certain distinctive chemical properties.

Mixture—A combination of two or more substances that are not chemically combined with each other and that can exist in any proportion.

Molecule—A particle made by the chemical combination of two or more atoms; the smallest particle of which a compound is made.

Octet rule—An hypothesis that atoms that have eight electrons in their outermost energy level tend to be stable and chemically unreactive.

Oxide—An inorganic compound whose only negative part is the element oxygen.

Radical—A group of atoms that behaves as if it were a single atom.

---

whose anions are any ion but the hydroxide ion. Salts are often described as the compounds formed (other than water) when an acid and a base react with each other.

This system of classification is useful in grouping compounds that have many similar properties. For example, all acids have a sour **taste**, impart a pink **color** to litmus **paper**, and react with bases to form salts. One drawback of the system, however, is that it may not give a sense of the enormous diversity of compounds that exist within a particular family. For example, the element **chlorine** forms at least five common acids, known as hydrochloric, hypochlorous, chlorous, chloric, and perchloric acids. For all their similarities, these five acids also have important distinctive properties.

The "others" category of compound classification includes all those compounds that don't fit into one of the other four categories. Perhaps the most important group of compounds contained in this "others" category is the coordination compounds. Coordination compounds are different from acids, bases, salts, and oxides primarily because of their method of bonding. Members of the last four groups are formed when atoms give or take electrons to form ionic bonds, share pairs of electrons to form covalent bonds, or exchange electrons in some fashion intermediary between these cases to form polar covalent bonds. Coordination compounds, on the other hand, are formed when one or more ions or molecules contributes both electrons in a bonding pair to a metallic atom or ion. The contributing species in such a compound is (or are) known as ligands and the compound as a whole is often called a metal **complex**.

See also Element, chemical; Mixture, chemical.

Resources

Books

Masterson, William L., Emil J. Slowinski, and Conrad L. Stanitski. *Chemical Principles.* Philadelphia: Saunders, 1983, Chapter 3.

Moore, John, and Nicholas D. Spencer. *Encyclopedia of Chemical Physics and Physical Chemistry.* Washington, DC: Institute of Physics, 2001.

Williams, Arthur L., Harland D. Embree, and Harold J. DeBey. *Introduction to Chemistry.* 3rd edition. Reading, MA: Addison-Wesley Publishing Company, 1986.

David E. Newton

Compound microscope *see* **Microscopy**

Compton effect

The Compton effect (sometimes called Compton scattering) occurs when an x ray collides with an **electron**. In 1923, Arthur H. Compton did experiments bouncing **x rays** off the electrons in graphite **atoms**. Compton found the x rays that scattered off the electrons had a lower **frequency** (and longer wavelength) than they had before striking the electrons. The amount the frequency changes depends on the scattering **angle**, the angle that the x ray is deflected from its original path.

Imagine playing pool. Only the cue ball and 8 ball are left on the table. When the cue ball strikes the 8 ball, which was initially at rest, the cue ball is scattered at some angle. It also loses some of its **momentum** and kinetic **energy** to the 8 ball as the 8 ball begins to move. The x-ray **photon** scattering off an electron behaves sim-

ilarly. The x ray loses energy and momentum to the electron as the electron begins to move. The energy and frequency of **light** and other electromagnetic **radiation** are related so that a lower frequency x-ray photon has a lower energy. The frequency of the x ray decreases as it loses energy to the electron.

In 1905, Albert Einstein explained the **photoelectric effect**, the effect that causes solar cells to produce **electricity**, by assuming light can occur in discrete particles, photons. This photon model for light still needed further experimental confirmation. Compton's x ray scattering experiments provided additional confirmation that light can exhibit particle-like behavior. Compton received the 1927 Nobel Prize in **physics** for his work. Additional experiments show that light can also exhibit wave-like behavior and has a wave particle duality.

Compulsion

The main concern of psychiatrists and therapists who treat people with compulsions is the role they play in a mental illness called obsessive-compulsive disorder (OCD). Compulsions need to be distinguished from obsessions in order to understand how they interconnect with compulsive **behavior** and reinforce this debilitating illness.

In psychiatric literature, compulsions are defined as repetitive behavior, such as hand washing, counting, touching, and checking and rechecking an action (like turning the **light** off and on again and again to be sure it is off or on). Performing the specific act relieves the tension of the **obsession** that the light may not be on or off. The person feels no pleasure from the action. On the contrary, the compulsive behavior and the obsession cause a great deal of distress for the person.

Compulsive behavior also needs to be distinguished from excessive or addictive behaviors where the person feels pleasure from the activity, such as in compulsive eating or compulsive gambling.

Obsessive-compulsive disorder (OCD)

Obsessive-compulsive disorder is classified as an **anxiety** disorder. Other anxiety disorders are panic attacks, agoraphobia (the fear of public places), **phobias** (fear of specific objects or situations), and certain **stress** disorders. This illness becomes increasingly more difficult to the patient and family because it tends to consume more and more of the individual's time and **energy**. While a person who is suffering from an obsessive-compulsive disorder is aware of how irrational or senseless the fear is, he or she is overwhelmed by the need to carry out compulsive behavior in order to avoid anxiety that is created if the behavior is not carried out.

Therapists categorize compulsions into motor compulsions and ideational or mental compulsions. A motor compulsion involves the need to use physical action. Touching or hand washing would be classified as motor compulsions. An ideational compulsion involves thinking processes. Examples of ideation or mental compulsions are counting, following a ritualized thought pattern, such as picturing specific things, or repeating what someone is saying in the mind.

Obsessive-compulsive personality disorder

People with personality traits, like being a perfectionist or rigidly controlling, may not have OCD, but may have obsessive-compulsive personality disorder. In this illness, the patient may spend excessive amounts of energy on details and lose perspective about the overall goals of a task or job. They become compulsively involved with performing the details but disregard larger goals.

Like obsessive-compulsive disorder, obsessive-compulsive personality disorder can be time-consuming. The compulsive personality may be able to function successfully in a work environment but may make everyone else miserable by demanding excessive standards of perfection.

Treatments for obsessive-compulsive illnesses

The problem for treatment of obsessive-compulsive illnesses must follow careful **diagnosis** of the specific nature of the disorder.

Methods used to treat these illnesses include a careful physical and psychological diagnosis, medications, and therapies. Besides the compulsive behavior symptoms a person with OCD exhibits, he or she may also have physical symptoms, such as tremors, dry mouth, stammering, dizziness, cramps, nausea, headaches, sweating, or **butterflies** in the stomach. Since these and the major symptoms are found in other illnesses, a careful diagnosis is important before treatment is prescribed.

In behavior therapy, the patient is encouraged to control behavior, which the therapist feels can be accomplished with direction. The patient is also made to understand that thoughts cannot be controlled, but that when compulsive behavior is changed gradually through modified behavior, obsessive thoughts diminish. In this therapy, patients are exposed to the fears that produce anxiety in them, called **flooding**, and gradually learn to deal with their fears.

KEY TERMS

. . . .

Anxiety disorder—An illness in which anxiety plays a role.

Behavior therapy—A therapeutic program that emphasizes changing behavior.

Cognitive therapy—A therapeutic program that emphasizes changing a patient's thinking.

Compulsive behavior—Behavior that is driven by an obsession.

Diagnosis—A careful evaluation by a medical professional or therapist to determine the nature of an illness or disorder.

Flooding—Exposing a person with an obsession to his or her fears as a way of helping him or her face and overcome them.

Ideational or mental compulsions—Compulsions of a mental nature, such as counting or repeating words.

Motor compulsions—Compulsions where a specific, ritualized act is carried out.

Obsessive-compulsive disorder—A mental illness in which a person is driven to compulsive behavior to relieve the anxiety of an obsession.

Obsessive-compulsive personality disorder—The preoccupation with minor details to the exclusion of larger issues; exhibiting overcontrolling and perfectionistic attitudes.

Prompting and shaping—A therapeutic technique that involves using a helper to work with a person suffering from compulsive slowness.

Cognitive therapists feel it is important for OCD patients to learn to think differently in order to improve their condition. Because OCD patients are rational, this type of therapy can sometimes be useful. Most professionals who treat obsessive-compulsive illnesses feel that a combination of therapy and medication is helpful. Some antidepressants, like Anafranil (clomipramine) and Prozac (fluoxetine), are prescribed to help alleviate the condition.

When patients exhibit compulsive slowness, prompting and shaping techniques are used. Persons who are compulsively slow work with a helper who prompts them along gradually until they can perform actions in a more reasonable time frame, such as reducing a two-hour morning grooming period to half an hour. The shaping aspect is the reduction of time.

Resources

Books

Amchin, Jess. *Psychiatric Diagnosis: A Biopsychosocial Approach Using DSM-III-R.* Washington, DC: Psychiatric Press, 1991.

Baer, Lee. *Getting Control.* Boston: Little, Brown, 1991.

Green, Stephen A. Green. *Feel Good Again.* Mt. Vernon, NY: Consumers Union, 1990.

Jamison, Kay Redfield. *Touched with Fire.* New York: Free Press, 1993.

Neziroglu, Fugen, and Jose A. Yaryura-Tobias. *Over and Over Again.* Lexington, Mass: D.C. Heath, 1991.

Vita Richman

Computer-aided design *see* **CAD/CAM/CIM**

Computer-aided manufacture *see* **CAD/CAM/CIM**

Computer, analog

A digital computer employs physical device states as symbols; an analog computer employs them as models. An analog computer models the behaviors of smoothly varying mathematical variables—usually representing physical phenomena such as temperatures, pressures, or velocities—by translating these variables into (usually) voltages or gear movements. It then manipulates these physical quantities so as to solve the equations describing the original phenomena. Thermometers and scales can be viewed as rudimentary analog computers: they translate an unwieldy physical phenomenon (i.e., a patient's **temperature** or weight) into a manageable physical model or analog (i.e., the **volume** of a fixed quantity of colored **alcohol** or the displacement of a spring) that has been designed to vary linearly with the phenomenon to be measured; the device then derives a numerical result from the model (i.e., by aligning alcohol level or a pointer with a printed scale). Another example: pouring three equal units of **water** into an empty vertical tube and measuring the height of the result would be an analog method of computing that $x + x + x = 3x$.

The earliest known analog computer is the **astrolabe**. First built in Greece during the first century B.C., this device used pointers and scales on its face and a complex arrangement of bronze **gears** to predict the motions of the **Sun**, planets, and stars.

Other early measuring devices were also analog computers. Sundials, for example, traced a shadow's path to show the **time** of day. Springweight scales, which have been used for centuries, convert the pull on a stretched spring to units of weight. The slide rule was invented about 1620 and was used until superseded by the electronic **calculator** in the late twentieth century.

In 1905, Rollin Harris (1863–1918) and E. G. Fisher (1852–1939) of the United States Coast and Geodetic Survey started work on a calculating device that would forecast **tides**. It was not the first such device, but was the most complex to be built. Dubbed the Great Brass **Brain**, it was 11 ft (3.35 m) long and 7 ft (2.1 m) high, weighed 2,500 lb (1135 kg), and contained a maze of cams, gears, and rotating shafts. Completed in 1910, the machine worked as follows: an operator set 37 dials (each representing a particular geological or astronomical variable) and turned a crank. The computer then drew up tidal charts for as far into the future as the operator wished. It made accurate predictions and was used for 56 years before being retired in 1966.

Vannevar Bush (1890–1974), an electrical engineer at the Massachusetts Institute of Technology, created what is considered to be the first modern analog computer in the 1930s. Bush, with a team from MIT's electrical **engineering** staff, discouraged by the time-consuming mathematical computations of differential equations that were required to solve certain engineering problems, began work on a device to solve these equations automatically. The first version of their device, dubbed the differential analyzer, was unveiled in 1930; the second in 1935. The latter weighed 100 tons, contained 150 motors, and hundreds of miles of wires connecting relays and **vacuum** tubes; instructions could be fed to the machine using hole-punched **paper** tape. Three copies of the machine were built for military and research use. Over the next 15 years, MIT built several new versions of the computer. By present standards the machine was slow, only about 100 times faster than a human operator using a desk calculator. Like most analog computers since, the MIT machines modeled phenomena using voltages, and contained a number of standard voltage-manipulating modules—integrators, differentiators, adders, inverters, and so forth—whose connections could be reconfigured to model, within limits, any desired equation.

In the 1950s, RCA produced the first reliable design for a fully electronic analog computer, but by this time, many of the most complex functions of analog computers were being assumed by faster and more accurate digital computers. Analog computers are still used today for specialized applications in scientific calculation, engineering design, industrial process control, and spacecraft navigation. Neural networks are an active research subfield in analog computing. Recent research suggests that, in theory, an ideal analog computer might be able to solve certain problems beyond the reach even of an ideal digital computer with unlimited processing power.

Resources

Other

Bains, Sunny. "Analog Computer Trumps Turing Model." EE Times. November 3, 1998 [cited January 6, 2003]. <http://www.eetimes.com/ story/OEG19981103S0017>.

Computer, digital

A digital computer is a programmable device that processes information by manipulating symbols according to logical rules. Digital computers come in a wide variety of types, ranging from tiny, special-purpose devices embedded in cars and other devices to the familiar desktop computer, the minicomputer, the mainframe, and the supercomputer. The fastest supercomputer, as of early 2003, can execute up to 36 trillion instructions (elementary computational operations) per second; this record is certain to be broken. The impact of the digital computer on society has been tremendous. It is used to run everything from spacecraft to factories, healthcare systems to telecommunications, banks to household budgets. Since its invention during World War II, the electronic digital computer has become essential to the economies of the developed world.

The story of how the digital computer evolved goes back to the beyond the calculating machines of the 1600s to the pebbles (in Latin, *calculi*) that the merchants of imperial Rome used for counting, to the **abacus** of the fifth century B.C. Although the earliest devices could not perform calculations automatically, they were useful in a world where mathematical calculations, laboriously performed by human beings in their heads or on **paper**, tended to be riddled with errors. Like writing itself, mechanical helps to calculation such as the abacus may have first developed to make business easier and more profitable to transact.

By the early 1800s, with the **Industrial Revolution** well under way, errors in mathematical data had assumed new importance; faulty navigational tables, for example, were the cause of frequent shipwrecks. Such errors were a source of irritation to Charles Babbage (1792–1871), a young English mathematician. Convinced that a machine could do mathematical calculations faster and more accurately than humans, Babbage, in 1822, produced a small working model of what he called his "difference engine." The difference engine's **arithmetic** was limited, but it could compile and print mathematical tables with no more human intervention than a hand to turn the handles at the top of the device. Although the British government was impressed enough to invest £17,000 in the construction of a full-scale difference engine—a sum equivalent to millions of dollars in today's money—it was never built. The project came to a halt in 1833 in a dispute over payments between Babbage and his workmen.

By that time, Babbage had already started to work on an improved version—the analytical engine, a programmable machine that could perform all types of arithmetic functions. The analytical engine had all the essential parts of the modern computer: a means of entering a program of instructions, a memory, a central processing unit, and a means of outputting results. For input and programming, Babbage used punched cards, an idea borrowed from French inventor Joseph Jacquard (1757–1834), who had used them in his revolutionary weaving loom in 1801.

Although the analytical engine has gone down in history as the prototype of the modern computer, a full-scale version was never built. Among the obstacles were lack of funding and manufacturing methods that lagged well behind Babbage's **vision**.

Less than 20 years after Babbage's death, an American by the name of Herman Hollerith (1860–1929) was able to make use of a new technology, **electricity**, when he submitted to the United States government a plan for a machine that could compute census data. Hollerith's electromechanical device tabulated the results of the 1890 U.S. census in less than six weeks, a dramatic improvement over the seven years it had taken to tabulate the results of the 1880 census. Hollerith went on to found the company that ultimately emerged as International Business Machines, Inc. (IBM).

World War II was the driving force behind the next significant stage in the evolution of the digital computer: greater complexity, greater programmability, and greater speed through the replacement of moving parts by electronic devices. These advances were made in designing the Colossus, a special-purpose electronic computer built by the British to decipher German codes; the Mark I, a gigantic electromechanical device constructed at Harvard University under the direction of U.S. mathematician Howard Aiken (1903–1973); and the ENIAC, a large, fully electronic machine that was faster than the Mark I. Built at the University of Pennsylvania under the direction of U.S. engineers John Mauchly (1907–1980) and J. Presper Eckert (1919–1995), the ENIAC employed some 18,000 **vacuum** tubes.

The ENIAC was general-purpose in principle, but to switch from one program to another meant that a part of the machine had to be disassembled and rewired. To avoid this tedious process, John von Neumann (1903–1957), a Hungarian-born American mathematician, proposed the concept of the stored program—that is, the technique of coding the program in the same way

A possible future direction for computer technology is the optical computer. The Bit-Serial Optical Computer (BSOC) shown here is the first computer that both stores and manipulates data and instructions as pulses of light. To enable this, the designers developed bit-serial architecture. Each binary digit is represented by a pulse of infrared laser light 13 ft (4 m) long. The pulses circulate sequentially through a tightly wound, 2.5-mile-long (4-km-long) loop of optical fiber some 50,000 times per second. Other laser beams operate lithium niobate optical switches which perform the data processing. This computer was developed by Harry Jordan and Vincent Heuring at the University of Colorado and was unveiled on January 12, 1993. *Photograph by David Parker. National Audubon Society Collection/Photo Researchers, Inc. Reproduced by permission.*

as the stored data and keeping it in the computer's own memory for as long as needed. The computer could then be instructed to change programs, and programs could even be written to interact with each other. For coding, von Neumann proposed using the binary numbering system, which uses only 0 and 1, as opposed to the decimal system, which uses the ten digits 0 through 9. Because 0 and 1 can readily be symbolized by the "on" or "off" states of a switched **electric current**, computer design was greatly simplified.

Von Neumann's concepts were incorporated in the first generation of large computers that followed in the late 1940s and 1950s. All these machines were dinosaurs by today's standards, but in them all the essential design

principles on which today's billions of digital devices operate were worked out.

The digital computer is termed "digital" to distinguish it from the analog computer. Digital computers manipulate symbols—not necessarily digits, despite the name—while analog computers manipulate electronic signals or other physical phenomena that act as models or analogs of various other phenomena (or mathematical variables). Today, the word "computer" has come to be effectively synonymous with "digital computer," due to the rarity of analog computation.

Although all practical computer development to date has obeyed the principles of binary logic laid down by von Neumann and the other pioneers, and these principles are sure to remain standard in digital devices for the near future, much research has focused in recent years on quantum computers. Such devices will exploit properties of **matter** that differ fundamentally from the on-off, yes-no logic of conventional digital computers.

See also Analog signals and digital signals; Computer, analog; Computer software.

Resources

Books

Lee, Sunggu. *Design of Computers and Other Complex Digital Devices.* Upper Saddle River, NJ: Prentice Hall, 2000.

White, Ron, and Timothy Downs. *How Computers Work.* 6th ed. Indianapolis, IN: Que Publishers, 2001.

Other

Associated Press. "Study: Japan Has Fastest Supercomputer." December, 2002. (cited January 5, 2003). <http://www.govtech.net/news/news. phtml?docid=2002.11.15-30715>.

Computer languages

A computer language is the means by which instructions and data are transmitted to computers. Put another way, computer languages are the interface between a computer and a human being. There are various computer languages, each with differing complexities. For example, the information that is understandable to a computer is expressed as zeros and ones (i.e., binary language). However, binary language is incomprehensible to humans. Computer scientists find it far more efficient to communicate with computers in a higher level language.

First-generation language

First-generation language is the lowest level computer language. Information is conveyed to the computer by the programmer as binary instructions. Binary instructions are the equivalent of the on/off signals used by computers to carry out operations. The language consists of zeros and ones. In the 1940s and 1950s, computers were programmed by scientists sitting before control panels equipped with toggle switches so that they could input instructions as strings of zeros and ones.

Second-generation language

Assembly or assembler language was the second generation of computer language. By the late 1950s, this language had become popular. Assembly language consists of letters of the alphabet. This makes programming much easier than trying to program a series of zeros and

ones. As an added programming assist, assembly language makes use of mnemonics, or memory aids, which are easier for the human programmer to recall than are numerical codes.

Second-generation language arose because of the programming efforts of Grace Hopper, an American computer scientist and Naval officer. Hopper developed FLOW-MATIC, a language that made programming easier for the naval researchers using the ENIAC computer in the 1940s. FLOW-MATIC used an English-based language, rather than the on-off switch language the computer understood. FLOW-MATIC was one of the first "high-level" computer languages. A high-level computer language is one that is easier for humans to use but which can still be translated by another program (called a compiler) into language a computer can interpret and act on.

Third-generation language

The introduction of the compiler in 1952 spurred the development of third-generation computer languages. These languages enable a programmer to create program files using commands that are similar to spoken English. Third-level computer languages have become the major means of communication between the digital computer and its user.

By 1957, the International Business Machine Corporation (IBM) had created a language called FORTRAN (FORmula TRANslater). This language was designed for scientific work involving complicated mathematical formulas. It became the first high-level programming language (or "source code") to be used by many computer users.

Within the next few years, refinements gave rise to ALGOL (ALGOrithmic Language) and COBOL (COmmon Business Oriented Language). COBOL is noteworthy because it improved the record keeping and data management ability of businesses, which stimulated business expansion.

In the early 1960s, scientists at Dartmouth College in New Hampshire developed BASIC (Beginner's All-purpose Symbolic Instruction Code). This was the first widespread computer language designed for and used by nonprofessional programmers. BASIC enjoyed widespread popularity during the 1970s and 1980s, particularly as personal computers grew in use.

Since the 1960s, hundreds of programming languages have been developed. A few noteworthy examples include PASCAL, first developed as a teaching tool; LISP, a language used by computer scientists interested in writing programs they hoped would give computers some abilities usually associated with intelligence in humans; and the C series of programs (i.e., C, C+, C++). The latter are object-oriented languages, where the object (data) is used by what are known as routines. The C series of programs first allowed a computer to use higher-level language programs like store-bought software.

The actual program written in the third-generation language is called the source program. This is the material that the programmer puts into the computer to obtain results. The source program can usually be translated into an object program (the language of zeros and ones that is interpretable by the computer).

Information in a source program is converted into the object program by an intermediate program called an interpreter or compiler. An interpreter is a program that converts (or executes, in programming jargon) a source program, usually on a step-by-step, line-by-line, or unit-by-unit basis. The price for this convenience is that the programs written in third-generation languages require more memory and run more slowly than those written in lower level languages.

A compiler is a program that translates a source program written in a particular programming language to an object program that a particular computer can run. The compiler is a very specific interpreter, which is both language- and machine-dependent.

Block-structured language

Block-structured language grew out of research leading to the development of structured programming. Structured programming is based on the idea that any computer program can be written using only three arrangements of the information. The arrangements are called sequential, selection, and iteration. In a sequential arrangement, each programming instruction (statement) is executed one after the other. This order is vital. The execution of the second statement is dependent on the prior execution of the first statement. There is more flexibility built into the selection arrangement, where choices are typically made with an IF...THEN...ELSE structure. Iteration is also known as loop structure. Loop structures specify how many times a loop will be executed. In other words, a command can be executed a number of times until the task is completed.

PASCAL, ALGOL, and MODULA-2 are examples of block-structured languages. Examples of non-block structured languages are BASIC, FORTRAN, and LISP. Refinements of BASIC and FORTRAN produced more structured languages.

Block-structured languages rely on modular construction. A module is a related set of commands. Each module in a block-structured language typically begins with a "BEGIN" statement and ends with an "END" statement.

Fourth-generation language

Fourth-generation languages attempt to make communicating with computers as much like the processes of thinking and talking to other people as possible. The problem is that the computer still only understands zeros and ones, so a compiler and interpreter must still convert the source code into the machine code that the computer can understand. Fourth-generation languages typically consist of English-like words and phrases. When they are implemented on microcomputers, some of these languages include graphic devices such as icons and on-screen push buttons for use during programming and when running the resulting application.

Many fourth-generation languages use Structured Query Language (SQL) as the basis for operations. SQL was developed at IBM to develop information stored in relational databases. Eventually, it was adopted by the American National Standards Institute (ANSI) and later by the International Standards Organization (ISO) as a means of managing structured, factual data. Many database companies offer an SQL-type database because purchasers of such databases seek to optimize their investments by buying open databases, i.e., those offering the greatest compatibility with other systems. This means that the information systems are relatively independent of vendor, operating system, and computer platform.

Examples of fourth-generation languages include PROLOG, an **artificial intelligence** language that applies rules to data to arrive at solutions; and OCCAM and PARLOG, both parallel-processing languages. Newer languages may combine SQL and other high-level languages. IBM's Sonnet is being modified to use sound rather than visual images as a computer interface.

In 1991, development began on a refinement of C++ that would be adaptable to the Internet. The result, in 1995, was Java. The program formed the basis of the Netscape Internet browser. Java enables files to be acquired from the Internet in order to run programs or subprograms. This adaptability has made Java a very popular language.

See also Modular arithmetic; Virtual reality.

Resources

Books

Block, Joshua. *Effective Java Programming Language Guide.* New York: Addison Wesley Professional, 2001.

Cockburn, Alistar *Agile Software Development.* Boston: Addison-Wesley, 2001

Other

Bangladesh University of Engineering & Technology. "History of Linux." Department of Computer Science & Engineering. July 24, 2002 [cited January 16, 2002]. <http://ragib. hypermart.net/linux/>.

Randall S. Frost

Computer memory, physical and virtual memory

Physical and virtual memory are forms of memory (internal storage of data). Physical memory exists on chips (RAM memory) and on storage devices such as hard disks. Before a process can be executed, it must first load into RAM physical memory (also termed main memory).

Virtual memory is a process whereby data (e.g., programming code,) can be rapidly exchanged between physical memory storage locations and RAM memory. The rapid interchanges of data are seamless and transparent to the user. The use of virtual memory allows the use of larger programs and enables those programs to run faster.

In modern operating systems, data can be constantly exchanged between the hard disk and RAM memory via virtual memory. A process termed swapping is used to exchange data via virtual memory. The use of virtual memory makes it appear that a computer has a greater RAM capacity because virtual memory allows the emulation of the transfer of whole blocks of data, enabling programs to run smoothly and efficiently. Instead of trying to put data into often-limited volatile RAM memory, data is actually written onto the hard disk. Accordingly, the size of virtual memory is limited only by the size of the hard disk, or the space allocated to virtual memory on the hard disk. When information is needed in RAM, the exchanges system rapidly swaps blocks of memory (also often termed pages of memory) between RAM and the hard disk.

Modern virtual-memory systems replace earlier forms of physical file swapping and fragmentation of programs.

In a sense, virtual memory is a specialized secondary type of data storage, and a portion of the hard drive is dedicated to the storage of specialized virtual-memory files (also termed pages). The area of the hard drive dedicated to storing blocks of data to be swapped via virtual memory interface is termed the page file. In most operating systems, there is a preset size for the page file area of the hard disk, and page files can exist on multiple disk drives. Users of most modern operating systems can, however, vary the size of the page file to meet specific performance requirements. As with the page file size, although the actual size of the pages is preset, modern operating systems usually allow the user to vary the size of the page. Virtual memory pages range in size from a thousand bites to many megabytes.

The use of virtual memory allows an entire block of data or programming (e.g., an application process) to reside in virtual memory, while only the part of the code being executed is in physical memory. Accordingly, the use of virtual memory allows operating systems to run many programs and thus, increase the degree of multiprogramming within an operating system.

Virtual memory integration is accomplished through either a process termed demand-segmentation or through another process termed demand-paging. Demand-paging is more common because it is simpler in design. Demand-paging virtual-memory processes do not transfer data from disk to RAM until the program calls for the page. There are also anticipatory paging processes utilized by operating systems that attempt to read ahead and execute the transfer of data before the data is actually required to be in RAM. After data is paged, paging processes track memory usage and constantly call data back and forth between RAM and the hard disk. Page states (valid or invalid, available or unavailable to the CPU) are registered in the virtual page table. When applications attempt to access invalid pages, a virtual-memory manager that initiates memory swapping intercepts the page fault message. Rapid translation of virtual addresses into a real physical address is via a process termed mapping. Mapping is a critical concept to the virtual-memory process. Virtual-memory mapping works by linking real hardware addresses (a physical storage address) for a block or page of stored data to a virtual address maintained by the virtual-memory process. The registry of virtual address allows for the selective and randomized translation of data from otherwise serial reading drives. In essence, virtual-memory processes supply alternate memory addresses for data, and programs can rapidly utilize data by using these virtual addresses instead of the physical address of the data page.

Virtual memory is a part of many operating systems, including Windows, but is not a feature of DOS. In addition to increasing the speed of execution and operational size of programs (lines of code), the use of virtual memory systems provide a valuable economic benefit. Hard-disk memory is currently far less expensive than RAM memory. Accordingly, the use of virtual memory allows the design of high-capacity computing systems at a relatively low cost.

Although swaps of pages of data (specific lengths of data or clocks of data) via virtual-memory swaps between the hard drive and RAM memory are very fast, an over-reliance upon virtual-memory swaps can slow overall system performance. If the amount of the hard drive dedicated to storing page files is insufficient to meet the demands of a system that relies heavily on the exchange of data via virtual memory, it is possible for users to receive "OUT OF MEMORY" messages and faults, even though they have large amounts of unused hard disk space.

By early 2003, personal computers with RAM capacities of 1024 MB (1 mega byte = 1,000,000 bytes) were widely available in the United States and many brands of personal computers boasted hard disk capacities of 60 GB (1 giga byte equals 1 billion bytes). The relative limits of both hard disk capacity and RAM memory capacity improve steadily with advances in microchip technology.

See also Computer languages; Computer memory, physical and virtual memory; Computer software; Computer, analog; Computer, digital.

Computer software

Computers are built of electronic components encased in a sturdy **metal** container and attached to the outside of the container. Examples of hardware include electrical connections, circuit boards, hard and disk drives, viewing monitor, and printer. The components and the container are referred to as hardware. On their own, computer hardware is functionally useless. For the electronic circuitry to be of use, commands must be supplied. Every task that a computer performs—from mathematical calculations, to the composition and manipulation of text, to modeling—requires instructions. These instructions are referred to as computer software.

The software instructions tells a computer what to do and how to do it. It is the "brain" that tells the hardware or "body" of a computer what to do. The operation of the hardware depends upon the procedures in a software program. The software instructions can be under the commands of the person using the computer (i.e., spreadsheet or word processing programs). Or, software

may run automatically in the background, without user intervention (i.e., **virus** monitoring programs).

Computers must receive instructions to complete every task they perform. Some microcomputers, such as hand-held calculators, wristwatches, and **automobile** engines contain built-in operating instructions. These devices can be called "dedicated" computers. Personal computers found in businesses, homes, and schools, however, are general-purpose machines because they can be programmed to do many different types of jobs. They contain enough built-in programming to enable them to interpret and use commands provided by a wide range of external software packages which can be loaded into their memory.

Computer software are a convenient collection of highly detailed and specific commands written by computer programmers and recorded on disks or tapes in a defined order. Once a software program has been loaded into the memory of a computer, the instructions remain until they are deliberately deleted or become corrupt in some accidental or malicious way. The instructions do not have to be entered by the user every time the computer is used.

Origin of computer software

English mathematician Charles Babbage conceived the ancestor of modern computers and computer software in 1856. Babbage dubbed the sophisticated calculating machine the analytical engine. While the analytical engine never became fully operational, Babbage's design contained all the crucial parts of modern computers. It included an input device for entering information, a processing device (that operated without **electricity**) for calculating answers, a system of memory for storing answers, and even an automated printer for reporting results.

The analytical engine also included a software program devised by the daughter of poet Lord Byron, Ada Augusta. Her "programs" were the coordinated set of steps designed to turn the **gears** and cranks of the machine. The instructions were recorded as patterns of holes on punch cards—a system that had been used since the 1750s by operators of weaving looms to produce woven cloth having specific and desired patterns. Depending on the pattern of holes in a card, and on the sequence of different cards, the computer would translate the instructions into physical movements of the machines' calculating mechanical parts.

The design of this software utilized features that are still used by software programmers today. One example is the subroutine; a series of instructions that could be used repeatedly for different purposes. Subroutines simplify the writing of the software program, as one set of instructions can be applied to more than one task. Another example is called the conditional jump, which is an instruction for the machine to jump to different cards if certain criteria were met. A final example is called looping. Looping is the ability to repeat the instructions on a set of cards as often as needed to satisfy the demands of a task.

Modern day computer software

The first modern computers were developed by the United States military during World War II. Their purpose was to calculate the paths of artillery shells and bombs. By present day standards, these machines were primitive. They used bulky **vacuum** tubes that were the predecessors of today's electronic circuitry. As a result, the machines were massive, occupying large rooms. Additionally, the myriad of settings were controlled by on-off switches, which had to be reset by hand for each operation. This was time-consuming and difficult for the programmers.

John von Neumann, a Princeton mathematician, suggested that computers would be more efficient if they stored their programs in memory. This would eliminate the need to manually setting every single instruction each time a problem was to be solved. This and other suggestions by von Neumann transformed the computer from a fancy adding machine into a machine capable of simulating many real-life, complex problems.

The language of software

The instructions computers receive from software are written in a computer language, a set of symbols that convey information. Like spoken languages used by humans, **computer languages** come in many different forms.

Computers use a very basic language to perform their jobs. The language ultimately can be reduced to a pattern of "on-or-off" responses, called binary digital information or Machine Language. Computers work using nothing but electronic "switches" that are either on or off, as represented by "1" and "0."

Human beings have trouble writing complex instructions using binary, "1 or 0" language. A simple command to a computer might look something like this: 00010010 10010111001 010101000110. Because such code is tedious and time consuming to write, programmers invented "assembly language." It allows programmers to assign a separate code to different machine language commands. Another, special program called a compiler translates the codes back into 1s and 0s for the computer.

Assembly language was a problem in that it only worked with computers that had the same type of "computer chip" or microprocessor (an electrical component that controls the main operating parts of the computer).

The development of what are referred to as high-level languages helped to make computers common objects in work places and homes. They allowed instructions to be written in languages that many people and their computers could recognize and interpret. Recognizable commands like READ, LIST, and PRINT could be used when writing instructions for computers. Each word may represent hundreds of instructions in the 1s and 0s language of the machine.

Because the electronic circuitry in a computer responds to commands expressed in terms of 1s and 0s, the high-level language command must be translated back into machine language. The various types of software needed to translate high-level language back into machine language are called translator programs.

Operating system software is vital for the performance of a computer. Before a computer can use application software, such as a word processing or a game-playing package, the computer must run the instructions through the operating system software. The operating system software contains many built-in instructions, so that each piece of application software does not have to repeat simple instructions (i.e., **printing** a document).

Disk Operating System or DOS is a popular operating system software program for many personal computers in use in the late 1990s. The Microsoft corporation has modified this software. MS-DOS (or Windows) has become the most popular computer operating system software. IBM-compatible machines can all run the same type of software, but Macintosh computers cannot. Indeed, the requirement for MS-DOS before other software programs could be run prompted charges that Microsoft has monopolized the software industry. In April 2000, the United States district court ruled that Microsoft has violated antitrust laws, and that the company was to be broken up, in order to foster competition. However, upon appeal, the breakup order was reversed in 2002.

Another operating software that is gaining in popularity is called Linux. Linux was initially written by Linus Torvalds in 1991. Linux is an **open-source software**. This means that anyone can modify the software. In contrast, the key codes that permit Windows to be modified are the property of Microsoft, and are not made available to the consumer.

A Linux-based operating system called Lindows has been devised. Commercially available as of January 2003, Lindows allows a user to run the Windows program on a Linux-based operating system. This development allows users to run Microsoft programs without having to purchase the Microsoft operating system. Not surprisingly, this concept is facing legal challenges from Microsoft.

Application software

Software provides the information and instructions that allow computers to create documents, solve simple or complex calculations, operate games, create images, maintain and sort files, and complete hundreds of other tasks.

Word-processing software, for example, makes writing, rewriting, editing, correcting, arranging, and rearranging words convenient.

Database software enables computer users to organize and retrieve lists, facts and inventories, each of which may include thousands of items.

Spreadsheet software allows users to keep complicated and related figures straight, to graph the results and to see how a change in one entry affects others. It is useful for financial and mathematical calculations. Each entry can be connected, if necessary, to other entries by a mathematical formula. If a spreadsheet keeps track of a computer user's earnings and taxes, for example, when more earnings are added to the part of the spreadsheet that keeps track of them (called a" register"), the spreadsheet can be set up, or programmed, to automatically adjust the amount of taxes. After the spreadsheet is programmed, entering earnings will automatically cause the spreadsheet to recalculate taxes.

Graphics software lets the user draw and create images. Desktop publishing software allow publishers to arrange photos, pictures, and words on a page before any printing is done. With desktop publishing and word processing software, there is no need for cutting and pasting layouts. Today, thanks to this type of computer software, anyone with a computer, a good quality printer, and the right software package can create professional looking documents at home. Entire books can be written and formatted by the author. The printed copy or even just a computer disk with the file can be delivered to a traditional printer without the need to reenter all the words on a typesetting machine.

Sophisticated software called CAD, for computer-aided design, helps architects, engineers, and other professionals develop complex designs. The software uses high-speed calculations and high-resolution graphics to let designers try out different ideas for a project. Each change is translated into the overall plan, which is modified almost instantly. This system helps designers create structures and machines such as buildings, airplanes, scientific equipment, and even other computers.

Software for games can turn a computer into a space ship, a battlefield, or an ancient city. As computers get more powerful, computer games get more realistic and sophisticated.

KEY TERMS

. . . .

Computer hardware—The physical equipment used in a computer system.

Computer program—Another name for computer software, a series of commands or instructions that a computer can interpret and execute.

Communications software allows people to send and receive computer files and faxes over phone lines. Education and reference software makes tasks such as **learning** spoken languages and finding information easier. Dictionaries, encyclopedias, and other reference books can all be searched quickly and easily with the correct software.

Utility programs keep computers more efficient by helping users to search for information and inspecting computer disks for flaws.

A development of the Internet era has been the creation of file sharing. This allows the sharing of computer files across the electronic network of the Internet. A host computer equipped with the necessary software can download other programs from the Internet. A prominent example of this concept—currently defunct because of copyright infringement implications—is Napster, which is used to download music files (MP3 files) to a personal computer. Via file sharing, a user can freely acquire many files that would otherwise have to be purchased.

See also LED; Optical data storage; Virtual reality.

Resources

Books

Auletta, Ken. *World War 3.0: Microsoft and Its Enemies.* New York: Random House, 2001.

Negus, Christopher. *Red Hat Linux 8 Bible.* New York: John Wiley & Sons, 2002.

Patterson, Daniel, and John Hennessy. *Computer Organization and Design: The Hardware/Software Interface.* 2nd. ed. New York: Elsevier Science, 1997.

Dean Allen Haycock

Computer virus

A computer **virus** is a program or segment of executable computer code that is designed to reproduce itself in computer memory and, sometimes, to damage data. Viruses are generally short programs; they may either stand-alone or be embedded in larger bodies of code. The term "virus" is applied to such code by analogy to biological viruses, **microorganisms** that force larger cells to manufacture new virus particles by inserting copies of their own genetic code into the larger cell's DNA. Because DNA can be viewed as a data-storage mechanism, the parallel between biological and computer viruses is remarkably exact.

Many viruses exploit computer networks to spread from computer to computer to computer, sending themselves either as e-mail messages over the Internet or directly over high-speed data links. Programs that spread copies of themselves over network connections of any kind are termed "worms," to distinguish them from programs that actively copy themselves only within the memory resources of a single computer. Some experts have sought to restrict the term "virus" to self-replicating code structures that embed themselves in larger programs and are executed only when a user runs the host program, and to restrict the term "worm" to stand-alone code that exploits network connections to spread (as opposed to, say, floppy disks or CD ROMs, which might spread a virus). However, virus terminology has shifted over the last decade, as computers that do not communicate over networks have become rare. So many worm/virus hybrids have appeared that any distinction between them is rapidly disappearing. In practice, any software that replicates itself may be termed a "virus," and most viruses are designed to spread themselves over the Internet and are therefore "worms."

A program that appears to perform a legitimate or harmless function, but is in fact designed to propagate a virus is often termed a Trojan Horse, after the hollow, apparently-harmless, giant wooden horse supposedly used by the ancient Greeks to sneak in inside the walls of Troy and overthrow that city from within. Another interesting subclass of viruses consists of chain letters that purport to warn the recipient of a frightening computer virus currently attacking the world. The letter urges its recipient to make copies and send them to friends and colleagues. Such hoax letters do not contain executable code, but do exploit computerized communications and legitimate concern over real, executable-code viruses to achieve self-replication, spread fear, and waste time. Chain letters have also been used as carriers for executable viruses, which are attached to the chain letter as a supposedly entertaining or harmless program (e.g., one that will draw a Christmas card on the screen).

The first "wild" computer viruses, that is, viruses not designed as computer-science experiments but spreading through computers in the real world, appeared in the early 1980s and were designed to afflict Apple II personal computers. In 1984, the science fiction book *Necromancer*, by William Gibson, appeared; this book romanticized the hacking of giant corporate computers by brilliant freelance rebels, and is thought by some experts to have increased interest among young programmers in writing real-world viruses. The first IBM PC computer viruses appeared in 1986, and by 1988 virus infestations on a global scale had become a regular event. An anti-virus infrastructure began to appear at that time, and anti-virus experts has carried on a sort of running battle with virus writers ever since. As anti-virus software increases in sophistication, however, so do viruses, which thrive on loopholes in software of ever-increasing complexity. As recently as January 28, 2003, a virus dubbed "SQL Slammer" (SQL Server 2000, targeted by the virus, is a large software package run by many businesses and governments) made headlines by suspending or drastically slowing Internet service for millions of users worldwide. In the United States alone, some 13,000 automatic teller machines were shut down for most of a day.

All viruses cause some degree of harm by wasting resources, that is, filling a computer's memory or, like SQL Slammer, clogging networks with copies of itself. These effects may cause data to be lost, but some viruses are designed specifically to delete files or issuing physically harmful series of instructions to hard drives. Such viruses are termed *destructive*. The number of destructive viruses has been rising for over a decade; in 1993 only about 10% of viruses were destructive, but by 2000 this number had risen to 35%.

Because even nonmalicious or nondestructive viruses may clog networks, shut down businesses or websites, and cause other computational harm (with possible real-world consequences, in some cases), both the private sector and governments are increasingly dedicating resources to the prevention, detection, and defeat of viruses. Twenty to 30 new viruses are identified every day, and over 50,000 viruses have been detected and named since the early 1980s, when computers first became integrated with the world economy in large numbers. Most viruses are written merely as egotistical pranks, but a successful virus can cause serious losses. The ILOVEYOU virus that afflicted computers globally in May of 2000 is a dramatic recent case that illustrates many of the properties of viruses and worms.

The ILOVEYOU virus was so named because in its most common form (among some 14 variants) it spread by looking up address-book files on each computer it infected and sending an e-mail to all the addresses it found, including a copy of itself as an attachment named LOVE-LETTER-FOR-YOU.TXT.VBS. ("VBS" stands for Visual Basic Script, a type of file readable by World Wide

Web browsers.) If a recipient of the e-mail opened the attachment, the ILOVEYOU virus code would run on their computer, raiding the recipient's address book and sending out a fresh wave of e-mails to still other computers.

ILOVEYOU first appeared in **Asia** on May 4, 2000. Designed to run on PC-type desktop computers, it rapidly spread all over the world, infecting computers belonging to large corporations, media outlets, governments, banks, schools, and other groups. Many organizations were forced to take their networks off line, losing business or suspending services. The United States General Accounting Office later estimated that the losses inflicted by the ILOVEYOU virus may have totaled $10 billion worldwide. Monetary losses occurred because of lost productivity, diversion of staff to virus containment, lost business opportunities, loss of data, and loss of consumer confidence (with subsequent loss of business).

National security may also be threatened by computer viruses and similar software objects. Creating or sending a computer virus is often a crime. Because of the interstate nature of the Internet, computer virus crimes are investigated by special units and divisions of the Federal Bureau of Investigation.

See also Computer software.

Resources

Books

Fites, Philip, Peter Johnston, and Martin Kratz. *The Computer Virus Crisis.* New York: Van Nostrand Reinhold, 1992.

Periodicals

"Virus Hits A.T.M.s and Computers Across Globe." *New York Times,* January 28, 2003.

Other

Brock, Jack L. "'ILOVEYOU' Computer Virus Highlights Need for Improved Alert and Coordination Capabilities." United States General Accounting Office. Testimony before the Subcommittee on Financial Institutions, Committee on Banking, Housing and Urban Affairs, U.S. Senate. May 18, 2000. [cited January 28, 2003]. <nsi.org/library/virus/ai00181t.pdf>

Larry Gilman

Computerized axial tomography

Computerized axial tomography (CAT) is a diagnostic procedure that employs **x rays** in a unique manner. The CAT scan machine is computer controlled to assure accuracy in placement of the x-ray beam. Axial refers to the fact that the x-ray tubes are arranged in an **arc** about an axis. Tomography is a combination of *tomo*, from the Greek meaning "to cut," and graph, "to draw," a reference to the fact that the CAT scan image reveals a cross-section of the body or body part.

The tomograph was developed in England in 1972. After a number of years of fine tuning the apparatus, it became a part of clinical medicine that is widely relied on now. Prior to the development of the CAT, x rays were done on the familiar table by a single x-ray tube that passed the rays through a given part of the body and exposed a plate of x-ray film. That film had to be developed and then viewed by a physician. This form of x ray was displayed on a film plate that offered a one-dimensional view of the body part under the x-ray tube. If a different angle was needed, the patient had to be turned over. The CAT offers a number of improvements over the old method.

The CAT scan machine, often referred to as the CT machine, consists of a horizontal pad on which the patient lies. Sandbags are placed around him to insure that he lies motionless. At one end of the pad is a circular structure that contains an array of x-ray tubes. The patient lies on the pad which is advanced into the **circle** until the desired area of the body is under the x-ray tubes. The x-ray tubes are focused to provide a very narrow angle of exposure, approximately 0.4 in (1 cm). The first x rays are made after which the array of tubes rotates and another exposure is made, the tubes rotate again, and so on until x rays have been made from all angles around the body.

Each x-ray tube is connected to the controlling computer. As the x rays pass through the patient's body they fall upon a sensitive window. The image from each tube is fed into the computer, and this is repeated whenever the x-ray tube fires, which is from a different angle each time.

In this way, the x-ray image is projected into the computer from different angles. The computer constructs a cross-sectional image of the body each time the array of x-ray tubes has completed a revolution around the patient.

Following each x ray exposure the patient is advanced another centimeter into the machine and the process repeats. X rays are made and the patient is advanced until exposures have been made in 0.4-in (1-cm) increments for the length of the **organ** being examined.

The images from each x-ray tube are fed through a computer, giving numerical values to the **density** of **tissue** through which the beam passed. The computer uses the numerical values to reconstruct an image of the cross-section of the body at the level the x rays passed through. The image is printed onto a screen for the physician to see and on a panel of x-ray film.

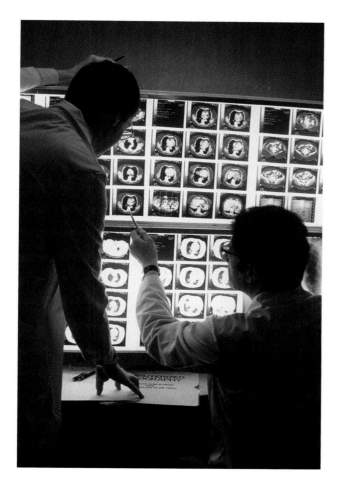

Doctors examining CAT scan x rays. *Photograph by Joseph Nettis. National Audubon Society Collection/Photo Researchers, Inc. Reproduced by permission.*

The differences in tissue density give the CT scan its definition. The liver is more dense than the pancreas, bone is more dense than liver, and so forth. The structures appear in different shades of gray on the screen and the film. The film is printed as a series of cross sectional images showing, for example, the liver from top to bottom. Any of the images can be called up on the computer screen for closer evaluation if the physician needs to do so.

A CT scan is described is a noninvasive procedure; that is, nothing is inserted into the body. At times the physician may want more contrast or definition to a given organ and may inject a contrast medium to accomplish this. A contrast medium is a substance that is visible on x rays. The medium, injected into the **blood**, will concentrate in an organ and will outline the organ or a cavity within it. In this way, the size of a kidney **tumor** may be determined, for example, as may other forms of **pathology**.

KEY TERMS

Contrast medium—A liquid that can be injected into a given part of the body to give shape to an organ. The contrast medium resists the passage of x rays, so it shows on the film as a dense area shaped like the cavity in which it has settled. It can show imperfections or blockages not seen on the unaided x ray.

Radiologist—A physician who specializes in imaging techniques such as x rays, CAT scans, MRI scans, and certain scans using radioactive isotopes.

Roentgenogram—A technical term for the x-ray image, named after the discoverer of x rays.

X ray—Electromagnetic radiation of very short wavelength, and very high energy.

X-ray tube—Evacuated tube in which electrons moving at high velocities are made to hit a metal target producing x rays.

Obviously, the CAT scan is a specialized form of **diagnosis** and is not practical for such cases as bone fractures. The procedure requires more time to complete than does the ordinary, one-dimensional x ray and is not cost-effective for simpler procedures.

For diagnosis of soft-tissue tumors, which are difficult to print on an ordinary x ray, the CAT scan is superior.

All x rays rely on differences in tissue density to form the x-ray image. Bone resists the passage of the x-ray beam more than muscle, which resists more than a softer tissue, such as liver. Thus, the x-ray image from a single x-ray beam is a plate somewhat like a film negative showing various tones of gray. Small differences in tissue density, as would be seen with a tumor in the liver, where both tissues are nearly the same density, would not be seen as two separate structures. The liver would appear as a uniformly dense organ.

The CAT scan, however, takes x rays from different angles and the machine is capable after several exposures to determine the slight difference in densities of nearly similar tissues. The liver will appear as an organ of a certain shade of gray, and a tumor within it will be discernable as a spot of slightly lighter gray because of the minute variation in density. Also, by finding the panel on which the tumor first appears and following it through to the panel on which it disappears, the radiologist can determine the size of the tumor.

See also Radioactive tracers.

Resources

Books

Cukier, Daniel, and Virginia E. McCullough. *Coping With Radiation Therapy*. Los Angeles: Lowell House, 1993.

Periodicals

"Finding the Brain's Autopilot." *USA Today* 122 (April 1994): 13.

Nadis, S.J. "Kid's Brainpower: Use It or Lose It." *Technology Review* 96 (November-December 1993): 19-20.

Larry Blaser

Concentration

Concentration is a **ratio** of how much of one ingredient is present in a mixture, compared to the whole mixture or compared to the main ingredient, often the solvent. The amounts of each substance can be expressed in **mass** or **volume** units, and many different units can be used. The components of the mixture can be gases, liquids, or solids.

Earth's atmosphere, for example, is a mixture of gases, and 78% of the total volume is **nitrogen** gas. (Percentages are the number of parts of a certain substance per hundred parts of the mixture.) Different types of **steel** are mixtures of **iron** with other elements. For example, stainless steel has close to 20%, by weight, of chromium. Sometimes a combination of mass and volume measurements are used; vinegar can be said to be a 5% **solution** of **acetic acid**, meaning 5 g of aceticacid per 100 mL of solution. Because it is not usually practical to analyze the whole substance in question, (Earth's atmosphere, for example) only samples are taken. Getting a true representation of the whole is crucial, or the concentration determined will not be accurate. For example, the concentration of nitrogen in the atmosphere changes slightly with altitude.

Many commonly used mixtures are liquid solutions. For **chemical reactions**, molarity is a useful unit of concentration. It is the number of moles of solute per liter of solution. Concentrated hydrochloric acid is 12 M, meaning that there are 12 moles of **hydrogen chloride** per liter of **water** solution. Other useful units of concentrations are molality, or the number of moles of solute per kilogram of solvent; **mole** fraction, which is the ratio of the numbers of moles of solute and solution; and normality, which is the number of chemical equivalents per liter of solution.

When very small amounts of a substance are present, parts per million or parts per billion may be used. A **sample** of tap water may contain 35 parts per million of dissolved solids. The concentration of a radioactive gas such as **radon** in air can be reported in picocuries per liter, or the amount of radioactivity per unit volume of air. For homes that are tested for the presence of radon, the safe limit is about 4 picocuries per liter. Exposure levels of dust or vapors in air may be given in units of mass of substance per volume of air. The maximum acceptable level for human exposure to **ammonia** vapor, for example, is 27 mg per cubic meter of air for short term exposure, that is, during a 15-minute period. Because modern analytical instruments require only small samples, results are often reported in milligrams per milliliter or nanograms per milliliter. Clinical laboratory reports of substances in **blood** or urine may be reported as milligrams per deciliter.

Concentration also refers to the process of removing solvent from a solution to increase the proportion of solute. The Dead Sea becomes more concentrated in salts as water evaporates from the surface. Ores are produced by the concentration of valuable **minerals**, such as those containing gold or silver, in small region of Earth's crust.

Concrete

Concrete, from the Latin word *concretus* meaning "having grown together," generally consists of Portland cement, **water**, and a relatively unreactive filler called aggregate. The filler is usually a conglomerate of gravel, **sand**, and blast-furnace stony **matter** known as slag.

Portland cement consists of finely pulverized matter produced by burning mixtures of lime, silica, alumina, and **iron** oxide at about 2,642°F (1,450°C). Chemically, Portland cement is a mixture of **calcium aluminum** silicates, typically including tricalcium silicate ($3CaO\ SiO_2$), dicalcium silicate ($2CaO\ SiO_2$), and tricalcium aluminate ($3CaO\ Al_2O_3$); it may also contain tetracalcium aluminoferrate ($4CaO\ Al_2O_3\ Fe_2O_3$). Small amounts of **sulfur**, potassium, **sodium**, and magnesia may also be present. The properties of the Portland cement may be varied by changing the relative proportions of the ingredients, and by grinding the cement to different degrees of fineness.

When Portland cement is mixed with water, the various ingredients begin to react chemically with the water. For a short time, the resultant mix can be poured or formed, but as the **chemical reactions** continue, the mix begins to stiffen, or set. Even after the mix has finished setting, it continues to combine chemically with the water, acquiring rigidity and strength. This process is called hardening.

Ordinarily, an excess of water is added to the concrete mix to decrease its **viscosity** so that it can be

Concrete being used to pave a road. *Photograph by Chris Jones. Stock Market. Reproduced by permission.*

poured and shaped. When the chemical reactions have more or less finished taking place, the excess water, which is only held by secondary chemical bonds, either escapes, leaving behind voids, or remains trapped in tiny **capillaries**.

It is important to recognize that setting and hardening result from chemical reactions between the Portland cement and the water. They do not occur as the result of the mixture drying out. In the absence of water, the reactions stop. Likewise, hardening does not require air to take place, and will take place even under water.

The strength of concrete is determined by the extent to which the chemical reactions have taken place by the filler size and distribution, the void **volume**, and the amount of water used.

Concrete is usually much stronger in compression than in tension.

Concrete may be modified with plastic (polymeric) materials to improve its properties. Reinforced concrete is made by placing **steel** mesh or bars into the form or mold and pouring concrete around them. The steel adds strength.

Concrete is used in buildings, **bridges**, **dams**, aqueducts, and road construction. Concrete is also used as **radiation** shielding.

See also Bond energy.

Conditioning

Conditioning is a term used in **psychology** to refer to two specific types of associative **learning** as well as to the operant and classical conditioning procedures which produce that learning. Very generally, operant conditioning involves administering or withholding reinforcements based on the performance of a targeted response, and classical conditioning involves pairing a **stimulus** that naturally elicits a response with one that does not until the second stimulus elicits a response like the first. Both of these procedures enabled the scientific study of associative learning, or the forming of connections between two or more stimuli. The goal of conditioning research is to discover basic laws of learning and **memory** in animals and humans.

Historical roots

Theories of conditioning and learning have a number of historical roots within the philosophical doctrine of associationism. Associationism holds that simple associations between ideas are the basis of human thought and knowledge, and that complex ideas are combinations of these simple associations. Associationism can be traced as far back as Aristotle (384-322 B.C.), who proposed three factors—contrast, similarity, and contiguity, or nearness in **space** or **time** of occurrence—that determine if elements, things, or ideas will be associated together.

British associationist-empiricist philosophers of the 1700s and 1800s such as Locke, Hume, and Mills, held that the two most fundamental mental operations are association and sensation. As empiricists, they believed all knowledge is based on sensory experience, and complex mental processes such as language, or ideas such as truth, are combinations of directly experienced ideas. This school of thought differs from nativist views which generally **stress** inherited genetic influences on **behavior** and thought. According to these views, we are born with certain abilities or predispositions that actively shape or limit incoming sensory experience. For example, Plato (c. 427–347 B.C.) believed we are born with certain pre-formed ideas as did Rene Descartes (1596-1650). Many contemporary psychologists believe we are born with certain skill-based potentials and capacities such as those involved in language. In the 1880s the German psychologist Hermann Ebbinghaus brought this philosophical doctrine within the realm of scientific study by creating experimental methods for testing learning and memory that were based on associationistic theory. Associationist ideas are also at the root of behaviorism, a highly influential school of thought in psychology that was begun by John B. Watson in the 1910s. And conditioning experiments enabling the standardized investigation of associations formed, not between ideas, but between varying stimuli, and stimuli and responses, are also based on associationism.

Classical and operant conditioning

The systematic study of conditioning began with the Russian physiologist Ivan P. Pavlov. Working in the late 1800s, Pavlov developed the general procedures and terminology for studying classical conditioning wherein he could reliably and objectively study the conditioning of reflexes to various environmental stimuli.

Pavlov initially used a procedure wherein every few minutes a hungry dog was given dry meat powder that was consistently paired with a bell tone. The meat powder always elicited salivation, and after a few experimental trials the bell tone alone was able to elicit salivation.

In Pavlov's terminology, the meat powder is an unconditional stimulus because it reliably or unconditionally led to salivation. The salivation caused by the meat powder is an unconditional response because it did not have to be trained or conditioned. The bell tone is a conditional stimulus because it was unable to elicit salivation until it had been conditioned to do so through repeated pairings with the unconditional stimulus. The salivation that eventually occurred to the conditional stimulus alone (the bell tone) is now called a conditional response. Conditional responses are distinctly different from unconditional responses even though they are superficially the same behavior. Conditioning is said to have occurred when the conditional stimulus will reliably elicit the conditional response, or when reflexive behaviors have come under the control of a novel stimulus.

In line with his physiological orientation, Pavlov interpreted his findings according to his hypotheses about **brain** functioning. He believed that **organism** responses are determined by the interaction of excitatory and inhibitory processes in the brain's cerebral hemispheres.

There are a number of different classical conditioning experimental designs. Besides varying the nature of the unconditional stimulus, many involve varying the timing of the presentation of the stimuli. Another type of experiment involves training a subject to respond to one conditional stimulus and not to any other stimuli. When this occurs it is called discrimination.

American psychologist Edward L. Thorndike developed the general procedures for studying operant conditioning (also referred to as instrumental conditioning) in the late 1800s. Thorndike's experimental procedure typically involved placing **cats** inside specially designed boxes from which they could escape and obtain food located outside only by performing a specific behavior such as pulling on a string. Thorndike timed how long it took individual cats to gain release from the box over a number of experimental trials and observed that the cats behaved aimlessly at first until they seemed to discover the correct response as if by accident. Over repeated trials the cats began to quickly and economically execute the correct response within just seconds. It seemed the initially **random** behaviors leading to release had become strengthened or reinforced by their positive consequences. It was also found that responses decreased and might eventually cease altogether when the food reward or reinforcement was no longer given. This is called **extinction**.

In the 1930s and 1940s, the American psychologist Burrhus F. Skinner modified Thorndike's procedures by, for instance, altering the box so that food could be delivered automatically. In this way the probability and **rate** of responding could be measured over long periods of time without needing to handle the **animal**. Initially, Skinner worked with **rats** but he eventually altered the box for use with pigeons.

In these procedures the response being conditioned, pressing the lever, is called the operant because it operates on the environment. The food reward or any consequence that strengthens a behavior is termed a reinforcer of conditioning. In operant conditioning theory, behaviors cease or are maintained by their consequences for the organism (Thorndike's "Law of Effect").

In most operant conditioning experiments, a small number of subjects are observed over a long period of time, and the dependent variable is the response rate in a given period of time. In traditional operant conditioning theory, physiological or biological factors are not used to explain behavior as they are in traditional classical conditioning theory.

Variations in operant conditioning experimental designs involve the nature of the reinforcement and the timing or scheduling of the reinforcers with respect to the targeted response. Reinforcement is a term used to refer to the procedure of removing or presenting negative or positive reinforcers to maintain or increase the likelihood of a response. Negative reinforcers are stimuli whose removal, when made contingent upon a response, will increase the likelihood of that response. Negative reinforcers then are unpleasant in some way, and they can range from uncomfortable physical sensations or interpersonal situations, to severe physical distress. Turning off one's alarm clock can be seen as a negative reinforcer for getting out of bed, assuming one finds the alarm unpleasant. Positive reinforcers are stimuli that increase the likelihood of a response when its presentation is made contingent upon that response. Giving someone pizza for achieving good grades is using pizza as a positive reinforcer for the desired behavior of achieving good grades (assuming the individual likes pizza). Punishment involves using aversive stimuli to decrease the occurrence of a response.

Reinforcement schedules are the timing and patterning of reinforcement presentation with respect to the response. Reinforcement may be scheduled in numerous ways, and because the schedule can affect the behavior as much as the reinforcement itself, much research has looked at how various schedules affect targeted behaviors. **Ratio** and **interval** schedules are two types of schedules that have been studied extensively. In ratio schedules, reinforcers are presented based on the number of responses made. In interval schedules, reinforcements are presented based on the length of time between reinforcements. Thus the first response to occur after a given time interval from the last reinforcement will be reinforced.

Conditioning and theory thrived from approximately the 1940s through the 1960s, and many psychologists viewed the learning theories based upon conditioning as one of psychology's most important contributions to the understanding of behavior. Psychologists created numerous variations on the basic experimental designs and adapted them for use with humans as well.

Comparison

Operant and classical conditioning have many similarities but there are important differences in the nature of the response and of the reinforcement. In operant conditioning, the reinforcer's presentation or withdrawal depends on performance of the targeted response, whereas in classical conditioning the reinforcement (the unconditional stimulus) occurs regardless of the organism's response. Moreover, whereas the reinforcement in classical conditioning strengthens the association between the conditional and unconditional stimulus, the reinforcement in operant conditioning strengthens the response it was made contingent upon. In terms of the responses studied, classical conditioning almost exclusively focuses on reflexive types of behavior that the organism does not have much control over, whereas operant conditioning focuses on non-reflexive behaviors that the organism does have control over.

Whether the theoretical underlying conditioning processes are the same is still an open question that may ultimately be unresolvable. Some experimental evidence supports an important distinction in how associations are formed in the two types of conditioning. Two-process learning theories are those that see classical and operant conditioning processes as fundamentally different.

Current research/future developments

How findings from conditioning studies relate to learning is an important question. But first we must define learning. Psychologists use the term learning in a slightly different way than it is used in everyday language. For most psychologists, learning at its most general is evidenced by changes in behavior due to experience. In traditional theories of conditioning learning is seen in the strengthening of a conditional **reflex**, and the creation of a new association between a stimulus and a response. Yet more recent and complex conditioning experiments indicate that conditioning involves more than the strengthening of stimulus-response connections or new reflexes. It seems conditioning may be more accurately described as a process through which the relationship between events or stimuli and the environment are learned about and behavior is then adjusted.

In addition, research comparing normal and retarded children, and older children and adults, suggests that people have language- or rule-based learning forms that are

KEY TERMS

. . . .

Associationism—A philosophical doctrine which holds that simple associations between ideas are the basis of all human thought and knowledge, and complex ideas are built upon combinations of the simple.

Behaviorism—A highly influential school of thought in psychology, it holds that observable behaviors are the only appropriate subject matter for psychological research.

Classical conditioning—A procedure involving pairing a stimulus that naturally elicits a response with one that does not until the second stimulus elicits a response like the first.

Conditional—Term used in classical conditioning to describe responses that have been conditioned to elicit certain responses. It also describes the stimuli that elicit such responses

Empiricism—A general philosophical position holding that all knowledge comes from experience, and that humans are not born with any ideas or concepts independent of personal experience.

Operant conditioning—A procedure involving administering or withholding reinforcements based on the performance, or partial performance, of a targeted response.

Unconditional—Term used in classical conditioning to describe responses that are naturally or unconditionally elicited, they do not need to be conditioned. It also describes the stimuli that elicit such responses.

more efficient than associative learning, and these types of learning can easily override the conditioning process. In sum, conditioning and associative learning seem to explain only certain aspects of human learning, and are now seen as simply another type of learning task. So, while conditioning had a central place in American experimental psychology from approximately the 1940s through the 1960s, its theoretical importance for learning has diminished. On the other hand, practical applications of conditioning procedures and findings continue to grow.

See also Reinforcement, positive and negative.

Resources

Books

Hearst, E. "Fundamentals of Learning and Conditioning." *Stevens' Handbook of Experimental Psychology.* 2nd ed. Edited by R.C. Atkinson, R.J. Herrnstein, G. Lindzey, and R. D. Luce. New York: John Wiley & Sons, 1988.

Mackintosh, N.J. "Classical and Operant Conditioning." In *Companion Encyclopedia of Psychology,* ed. A. W. Colman. New York: Routledge, 1994.

Schwartz, B. *Psychology of Learning and Behavior.* 3rd ed. New York: W.W. Norton & Co., Inc., 1988.

Marie Doorey

Condors

Condors are New World **vultures** that are among the largest of flying **birds**. There are only two **species**, the Andean condor (*Vultur gryphus*) and the critically endangered California condor (*Gymnogyps californianus*). They are related to the smaller vultures of the Americas, including the king vulture (*Sarcoramphus papa*) and turkey vulture (*Cathartes aura*), which also belong to family Cathartidae. In the same family, but extinct for about 10,000 years was the largest flying bird that ever lived. This was *Teratornis incredibilis*, a vulture found in the southwestern United States that had a wingspan of at least 16 ft (4.9 m).

The combined head and body length of the living condors is about 50 in (127 cm), and they weigh 20–25 lb (9–11 kg). They have black or dark brown plumage with white patches on the underside of the wings. The California condor has a wing span of 9 ft (2.7 m), while that of the Andean condor is 10 ft (3.1 m). Both species have a ruff of feathers around the neck, colored black on the California condor and white on the Andean. Both condors have a bald head and a short, sharply hooked beak. The Andean condor's naked skin is red, while that of the California condor is pinkish orange. The Andean male has an extra fleshy growth on top of its head, rather like a rooster's comb. The California condor does not have this growth.

The range of the Andean condor extends throughout the high Andean Mountains, and much of this **habitat** remains wild. It can fly over the highest peaks, but may land on lower-lying fields to scavenge dead animals. Although rare, this species still exists in relatively large numbers.

The California condor, however, is one of the most critically endangered animals on **Earth**. Historically, its range extended over much of **North America**, when it once foraged for carcasses of large ice age **mammals**. However, the condors began to decline at about the same time that many of these large mammals became extinct, around 10–12 thousand years ago. By the time

California condor. *U.S. Fish & Wildlife Service.*

of the European settlement, the range of the California condor was restricted to the western coast and **mountains**. As human populations in the region grew, the condor population declined further. By the 1950s its range was restricted to a small area of central California surrounding the southern end of the San Joaquin valley. In recent decades, condor habitat has been further disrupted by **petroleum** drilling, planting of citrus groves, and residential developments. In addition, **rangeland** where dead cattle might have been scavenged was extensively converted to cultivated fields of alfalfa and other **crops**. California condors have also suffered **lead** poisoning after ingesting lead shot or bullets in carrion. They have also been affected by DDT and other **insecticides**.

California condors lay only a single egg. After hatching, it takes the young condor 18 months to develop its wings sufficiently for flight. During that time, the chick is vulnerable to cold or hunger, especially when its parents fly far to forage for food, leaving the chick exposed for an extended time. It takes six years for a California condor to attain sexual maturity. Because of its low fecundity, its population cannot sustain much mortality.

Return to the wild

It was obvious by the 1950s that California condors were in danger of **extinction**. In 1978, there were only about 30 birds left in the wild, and seven years later only nine. At that time, all wild California condors were captured by the Fish and Wildlife Service and taken to the San Diego Wild Animal Park and the Los Angeles Zoo, where several other condors were already in residence. The zoos began a captive breeding program for the California condor, and by 1996, 103 individuals were alive. The population recovery has been sufficient to allow some birds to be introduced back into the wild.

To test ideas about how best to return condors to the wild, several Andean condors were brought to the United States and released in a national forest. It quickly became apparent that there were too many human activities and influences in the area for the condors to be reintroduced successfully. They had to be released farther from civilization. To this end, the Sespe Condor Sanctuary was acquired by the Fish and Wildlife Service as a wilderness habitat for these endangered birds. First two, then several additional California condors were released to this area. When several birds were poisoned by bullets in carrion they ate, it became clear that the reintroduced birds would have to be provided with safe food until their numbers increased. In late 1998, 22 condors were in the wild in southern California and 14 in Arizona. The ultimate goal is to establish at least two separate populations of more than 150 birds each.

The California condor has received a reprieve from extinction, but its survival depends on the continuation of intensive management efforts and the **conservation** of sufficient habitat to sustain a viable breeding population.

Resources

Books

Caras, Roger A. *Source of the Thunder: The Biography of a California Condor.* Lincoln, NE: University of Nebraska Press, 1991.

Peters, Westberg. *Condor.* New York: Crestwood House, 1990.

Silverstein, A., V. Silverstein, and L. Nunn. *The California Condor.* Millbrook Press, 1998.

Jean F. Blashfield

Congenital

The term congenital is used to describe a condition or defect that exists at **birth**. Congenital disorders are inborn. They are present in the developing fetus. Sickle cell disease, **Down syndrome**, and congenital rubella syndrome are three examples of congenital conditions in humans. Congenital disorders result from abnormalities in the fetus's genetic inheritance, conditions in the fetal environment, or a combination of the two. Two to three **percent** of babies in the United States are born with a major congenital defect. Prenatal testing can detect some congenital conditions.

Many congenital conditions are caused by chromosomal disorders. A human fetus inherits 23 chromosomes from its mother and 23 chromosomes from its father, making a total of 23 pairs. An extra **chromosome** or a missing chromosome creates havoc in fetal development. Down syndrome, marked by mental retardation and a distinctive physical appearance, is caused by an extra chromosome.

Each chromosome carries many genes. Like chromosomes, genes are present in pairs. Genes are responsible for many inherited traits, including eye color and **blood** type. **Genetic disorders** are caused by abnormal genes. Sickle cell disease, a blood disorder, occurs when a fetus inherits an abnormal **gene** from each parent. Polydactyly, the presence of extra fingers or toes, occurs when an abnormal gene is inherited from one parent.

Some congenital disorders are caused by environmental factors. It may be that certain genetic combinations leave some fetuses more vulnerable to the absence or presence of certain **nutrients** or chemicals. **Spina bifida**, also known as "open spine," occurs when embryo development goes awry and part of the neural tube fails to close. Adequate amounts of folic acid, a **vitamin**, help prevent spina bifida. Cleft palate, a hole in the roof of the mouth, is another congenital defect that seems to be caused by multiple factors. Congenital rubella syndrome, marked by mental retardation and deafness, is present in newborns whose mothers contracted rubella (German measles) during pregnancy.

Prenatal testing can diagnose certain congenital disorders. Ultrasound, which uses **sound waves** to produce an image of the fetus, can discover some defects of the **heart** and other organs. **Amniocentesis** and chorionic villi sampling are procedures that remove fetal cells from the pregnant uterus for **genetic testing**. These tests can determine the presence of Down syndrome, **sickle cell anemia**, **cystic fibrosis**, and other genetic diseases. Couples may choose to terminate the pregnancy if serious abnormalities are discovered.

See also Birth defects; Embryo and embryonic development.

Resources

Books

Davis, Joel. *Mapping the Code.* New York: John Wiley & Sons, 1990.

Edelson, Edward. *Birth Defects.* New York: Chelsea House Publishers, 1992.

Marshall, Liz. *The Human Genome Project: Cracking the Code Within Us.* Brookfield, CT: Millbrook Press, 1995.

Planning for Pregnancy, Birth, and Beyond. Washington, DC: American College of Obstetricians and Gynecologists, 1990.

Wills, Christopher. *Exons, Introns, and Talking Genes.* New York: Basic Books, 1991.

Congruence (triangle)

Two triangles are congruent if they are alike in every geometric respect except, perhaps, one. That one possible exception is in the triangle's "handedness." There are only six parts of a triangle that can be seen and measured: the three angles and the three sides. The six features of a triangle are all involved with congruence. If triangle ABC is congruent to triangle DEF, then

$$\angle A \cong \angle D \qquad AB \cong DE$$
$$\angle B \cong \angle E \qquad BC \cong EF$$
$$\angle C \cong \angle F \qquad CA \cong FD$$

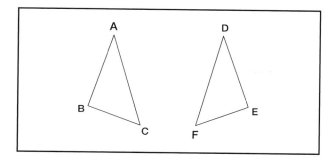

Figure 1. *Illustration by Hans & Cassidy. Courtesy of Gale Group.*

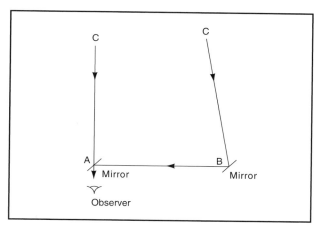

Figure 2. *Illustration by Hans & Cassidy. Courtesy of Gale Group.*

(The symbol for "is congruent to" is ≅.) Thus triangles ABC and DEF in Figure 1 are (or appear to be) congruent.

The failure of congruence to include handedness usually does not matter. If triangle DEF were a mirror, however, it would not fit into a frame in the shape of triangle ABC.

The term *congruent* comes from the Latin word *congruere*, meaning "to come together." It therefore carries with it the idea of superposition, the idea that one of two congruent figures can be picked up and placed on top of the other with all parts coinciding. In the case of congruent triangles the parts would be the three sides and the three angles.

Some authors prefer the word "equal" instead of "congruent." Congruence is usually thought of as a **relation** between two geometric figures. In most practical applications, however, it is not the congruence of two triangles that matters, but the congruence of the triangle with itself at two different times.

There is a remarkably simple **proof**, for instance, that the base angles of an isosceles triangle are equal, but it depends on setting up a correspondence of a triangle ABC with the triangle ACB, which is, of course, the same triangle.

Two triangles are congruent if two sides and the included **angle** of one are congruent to two sides and the included angle of the other. This can be proven by superimposing one triangle on the other. They have to match. Therefore the third side and the two other angles have to match. Modern authors typically make no attempt to prove it, taking it as a **postulate** instead.

Whatever its status in the logical structure, side-angle-side congruence (abbreviated S.A.S) is a very useful geometric property. The compass with which one draws circles works because it has legs of a fixed length and a tight joint between them. In each of the positions the compass takes, the spacing between the legs-the third side of the triangles-is unchanging. Common shelf brackets support the shelf with two stiff legs and a reinforced corner joining them. A builder frames a door with measured lengths of **wood** and a carpenter's **square** to make the included angle a right angle. In all these instances, the third side of the triangle is missing, but that does not matter. The information is sufficient to guarantee the length of the missing side and the proper shape of the entire triangle.

Two triangles are congruent if two angles and the included side of one are congruent respectively to two angles and the included side of the other. This is known as angle-side-angle (A.S.A.) congruence, and is usually proved as a consequence of S.A.S. congruence.

This is also a very useful property. The range finders, for example, which photographers, golfers, artillery observers, and others use are based on A.S.A. congruence. In Figure 2, the user sights the target C along line AC, and simultaneously adjusts angle B so that C comes into view along CB (small **mirrors** at B and at A direct the ray CB along BA and thence into the observer's **eye**). The angle CAB is fixed; the **distance** AB is fixed; and the angle CAB, although adjustable, is measurable. By A.S.A. congruence there is enough information to determine the shape of the triangle.

Although A.S.A. congruence calls for an included side, the side can be an adjacent side as well. Since the sum of the angles of a triangle is always a straight angle, if any two of the angles are given, then the third angle is determined. The correspondence has to be kept straight, however. The equal sides cannot be an included side in one triangle and an adjacent side in the other or adjacent sides of angles which are not the equal angles.

Two triangles are congruent if three sides of one triangle are equal, respectively, to the three sides of the other triangle. This is known as side-side-side (S.S.S.) congruence.

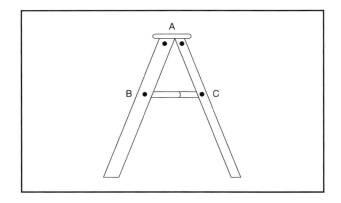

Figure 3. *Illustration by Hans & Cassidy. Courtesy of Gale Group.*

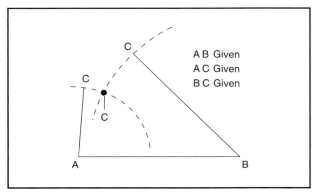

Figure 4. *Illustration by Hans & Cassidy. Courtesy of Gale Group.*

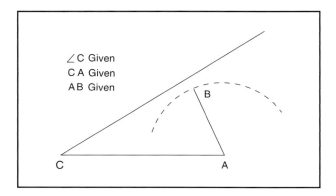

Figure 5. *Illustration by Hans & Cassidy. Courtesy of Gale Group.*

Probably the most widely exploited case of congruence is a folding ladder that is kept steady by a locking brace. It is only the lengths of the three sides (see Figure 3), that are fixed. The angles at A, B, and C are free to vary, and do so when the brace is unlocked and the ladder folded.

The draftsman can replicate a triangle easily and accurately with only a T-square, compass, and scale. Figure 4 shows the technique which he or she would use. A contractor, with nothing more than a tape, some pegs, and some string, can lay out a rectangular foundation.

One set of criteria for congruence that can be used with caution is side-side-angle congruence. Figure 5 illustrates this. Here the lengths AB and AC are given. So is the size of angle C, which is not an included angle. With AB given, the vertex B can lie anywhere on a **circle** with center at A. If the second side of angle C misses the circle, no triangle meeting the specifications is possible. If it is tangent to the circle, one triangle is possible; if it cuts the circle, two are. This type of congruence comes into play when using the law of sines

$$\frac{\text{sine of angle B}}{\text{side opposite B}} \qquad \frac{\text{sine of angle C}}{\text{side opposite C}}$$

to find an unknown angle, say angle B, by solving for sin B. If sin B is greater than 1, no such angle exists. If it equals 1, then B is a right angle. If it is less than 1, then sin B = sin (180 - B) gives two solutions.

Resources

Books

Coxeter, H.S.M., and S.L. Greitzer. *Geometry Revisited.* Washington, DC: The Mathematical Association of America, 1967.

Euclid. *Elements.* Translated by Sir Thomas L. Heath, New York: Dover Publications, 1956.

Hahn, Liang-shin. *Complex Numbers and Geometry.* 2nd ed. The Mathematical Association of America, 1996.

Hilbert, D., and S. Cohn-Vossen. *Geometry and the Imagination.* New York: Chelsea Publishing Co. 1952.

Moise, Edwin E. *Elementary Geometry from an Advanced Standpoint.* Reading, Massachusetts: Addison-Wesley Publishing Co., 1963.

J. Paul Moulton

Conic sections

A conic section is the **plane** curve formed by the intersection of a plane and a right-circular, two-napped cone. Such a cone is shown in Figure 1.

The cone is the surface formed by all the lines passing through a **circle** and a **point.** The point must lie on a line, called the "axis," which is **perpendicular** to the plane of the circle at the circle's center. The point is called the "vertex," and each line on the cone is called a "generatrix." The two parts of the cone lying on either side of the vertex are called "nappes." When the intersecting plane is perpendicular to the axis, the conic section is a circle (Figure 2).

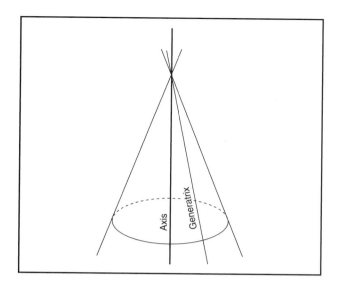

Figure 1. *Illustration by Hans & Cassidy. Courtesy of Gale Group.*

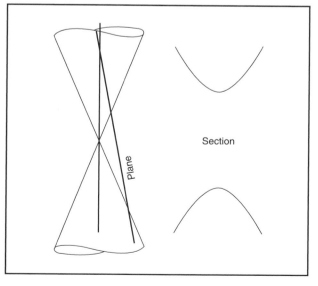

Figure 2. *Illustration by Hans & Cassidy. Courtesy of Gale Group.*

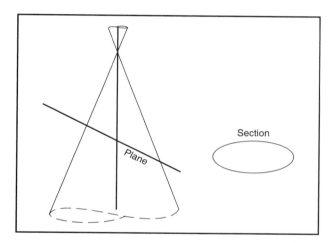

Figure 3. *Illustration by Hans & Cassidy. Courtesy of Gale Group.*

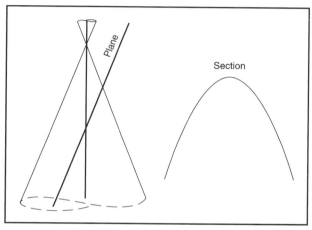

Figure 4. *Illustration by Hans & Cassidy. Courtesy of Gale Group.*

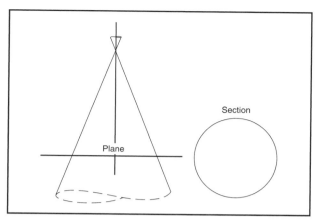

Figure 5. *Illustration by Hans & Cassidy. Courtesy of Gale Group.*

When the intersecting plane is tilted and cuts completely across one of the nappes, the section is an oval called an **ellipse** (Figure 3).

When the intersecting plane is **parallel** to one of the generatrices, it cuts only one nappe. The section is an open **curve** called a **parabola** (Figure 4).

When the intersecting plane cuts both nappes, the section is a **hyperbola**, a curve with two parts, called "branches" (Figure 5).

All these sections are curved. If the intersecting plane passes through the vertex, however, the section will be a single point, a single line, of a pair of crossed lines. Such sections are of minor importance and are known as "degenerate" conic sections.

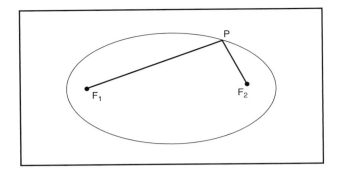

Figure 6. *Illustration by Hans & Cassidy. Courtesy of Gale Group.*

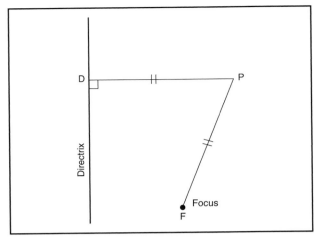

Figure 7. *Illustration by Hans & Cassidy. Courtesy of Gale Group.*

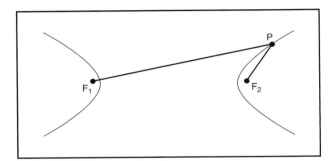

Figure 8. *Illustration by Hans & Cassidy. Courtesy of Gale Group.*

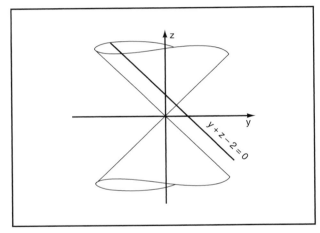

Figure 9. *Illustration by Hans & Cassidy. Courtesy of Gale Group.*

Since ancient times, mathematicians have known that conic sections can be defined in ways that have no obvious connection with conic sections. One set of ways is the following:

Ellipse: The set of points P such that $PF_1 + PF_2$ equals a constant and F_1 and F_2 are fixed points called the "foci" (Figure 6).

Parabola: The set of points P such that $PD = PF$, where F is a fixed point called the "focus" and D is the foot of the perpendicular from P to a fixed line called the "directrix" (Figure 7).

Hyperbola: The set of points P such that $PF_1 - PF_2$ equals a constant and F_1 and F_2 are fixed points called the "foci" (Figure 8).

If P, F, and D are shown as in Figure 7, then the set of points P satisfying the equation $PF/PD = e$ where e is a constant, is a conic section. If $0 < e < 1$, then the section is an ellipse. If $e = 1$, then the section is a parabola. If $e > 1$, then the section is a hyperbola. The constant e is called the "eccentricity" of the conic section.

Because the **ratio** PF/PD is not changed by a change in the scale used to measure PF and PD, all conic sections having the same eccentricity are geometrically similar.

Conic sections can also be defined analytically, that is, as points (x,y) which satisfy a suitable equation.

An interesting way to accomplish this is to start with a suitably placed cone in coordinate space. A cone with its vertex at the origin and with its axis coinciding with the z-axis has the equation $x^2 + y^2 - kz^2 = 0$. The equation of a plane in space is $ax + by + cz + d = 0$. If one uses substitution to eliminate z from these equations, and combines like terms, the result is an equation of the form $Ax^2 + Bxy + Cy^2 + Dx + Ey + F = 0$ where at least one of the coefficients A, B, and C will be different from **zero**.

For example if the cone $x^2 + y^2 - z^2 = 0$ is cut by the plane $y + z - 2 = 0$, the points common to both must satisfy the equation $x^2 + 4y - 4 = 0$, which can be simplified by a translation of axes to $x^2 + 4y = 0$. Because, in this example, the plane is parallel to one of the generatrices of the cone, the section is a parabola (Figure 9).

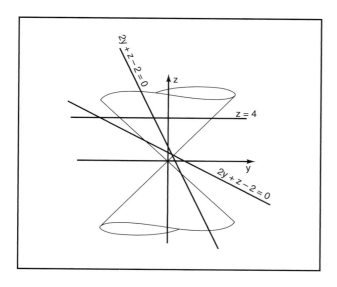

Figure 11. *Illustration by Hans & Cassidy. Courtesy of Gale Group.*

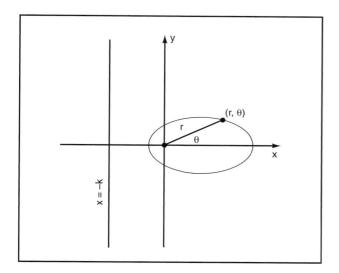

Figure 10. *Illustration by Hans & Cassidy. Courtesy of Gale Group.*

One can follow this procedure with other intersecting planes. The plane z - 5 = 0 produces the circle $x^2 + y^2 - 25 = 0$. The planes y + 2z - 2 = 0 and 2y + z - 2 = 0 produce the ellipse $12x^2 + 9y^2 - 16 = 0$ and the hyperbola $3x^2 - 9y^2 + 4 = 0$ respectively (after a simplifying translation of the axes). These planes, looking down the x-axis are shown in Figure 10.

As these examples illustrate, suitably placed conic sections have equations which can be put into the following forms:

Circle: $x^2 + y^2 = r^2$

Ellipse: $A^2x^2 + B^2y^2 = C^2$

Parabola: $y = Kx^2$

Hyperbola: $A^2x^2 - B^2y^2 = +C^2$

The equations above are "suitably placed." When the equation is not in one of the forms above, it can be hard to tell exactly what kind of conic section the equation represents. There is a simple test, however, which can do this. With the equation written $Ax^2 + Bxy + Cy^2 + Dx + Ey + F = 0$, the discriminant $B^2 - 4AC$ will identify which conic section it is. If the discriminant is positive, the section is a hyperbola; if it is **negative**, the section is an ellipse; if it is zero, the section is a parabola. The discriminant will not distinguish between a proper conic section and a degenerate one such as $x^2 - y^2 = 0$; it will not distinguish between an equation that has real roots and one, such as $x^2 + y^2 + 1 = 0$, that does not.

Students who are familiar with the quadratic formula

$$x = \frac{(-b \pm \sqrt{b^2-4ac})}{2a}$$

will recognize the discriminant, and with good reason. It has to do with finding the points where the conic section crosses the line at **infinity**. If the discriminant is negative, there will be no solution, which is consistent with the fact that both circles and ellipses lie entirely within the finite part of the plane. Parabolas lead to a single root and are tangent to the line at infinity. Hyperbolas lead to two roots and cross it in two places.

Conic sections can also be described with **polar coordinates**. To do this most easily, one uses the focus-directrix definitions, placing the focus at the origin and the directrix at x = -k (in rectangular coordinates). Then the polar equation is $r = Ke/(1 - e \cos \theta)$ where e is the eccentricity (Figure 11).

The eccentricity in this equation is numerically equal to the eccentricity given by another ratio: the ratio

CF/CV, where CF represents the **distance** from the geometric center of the conic section to the focus and CV the distance from the center to the vertex. In the case of a circle, the center and the foci are one and the same point; so CF and the eccentricity are both zero. In the case of the ellipse, the vertices are end points of the major axis, hence are farther from the center than the foci. CV is therefore bigger than CF, and the eccentricity is less than 1. In the case of the hyperbola, the vertices lie on the transverse axis, between the foci, hence the eccentricity is greater than 1. In the case of the parabola, the "center" is infinitely far from both the focus and the vertex; so (for those who have a good imagination) the ratio CF/CV is 1.

Resources

Books

Finney, Ross L., et al. *Calculus: Graphical, Numerical, Algebraic of a Single Variable.* Reading, MA: Addison Wesley Publishing Co., 1994.

Gullberg, Jan, and Peter Hilton. *Mathematics: From the Birth of Numbers.* W.W. Norton & Company, 1997.

J. Paul Moulton

Conies *see* **Lagomorphs**

Conifer

Conifer (common name for phylum Pinophyta) is a type of **tree** that thrives in temperate and boreal climates. Characterized by seed-bearing cones, conifers typically have narrow, needle-like leaves covered with a waxy cuticle, and straight trunks with horizontal branches. These trees are usually evergreen, meaning they do not shed their leaves all at once, and can photosynthesize continually. There are two orders of conifer, Pinales and Taxales.

There are two major seed-producing plants: gymnosperms (meaning naked seed) and angiosperms (meaning enclosed seed). These two groups get their names from their female reproductive characteristics: the gymnosperms have egg cells or **seeds** on the scales of the cone, while the **angiosperm** seeds are enclosed within ovaries, which, if fertilized, eventually turn into fruit. Conifers are one of the three groups of gymnosperms, which also include the **cycads** (tropical plants with palmlike leaves) and a group consisting of four plants having both **gymnosperm** and angiosperm features.

The female cones and male cones grow separately on the same tree. The female cones are larger and grow on the upper branches, while the male cones tend to grow on the lower branches. Both female and male cones have a central shaft with scales or leaflike projections called sporophylls that are specially shaped to bear sporangia (a reproductive unit). Each female sporophyll has an area where two ovules (each containing a group of fertile eggs) develop within a protective **tissue** called the nucellus. Male sporangia contain thousands of microspores, which divide (through **meiosis**) into more microspores, which eventually turn into grains of yellow pollen. The dispersal of pollen is dependent on air currents, and with dry, windy conditions, a grain of pollen can travel miles from where it was released. The pollen enters the female cone through an opening in the nucellus and sticks to the ovule. After **fertilization**, a little conifer seedling, complete with a root, develops within a seed coat. The seed is still attached to the scale of the cone, which, when caught by the **wind**, acts as a wing to carry the seed.

Some conifers can be shrublike while others grow very tall, like the giant **sequoia** (*Sequoia sempervirens*). Through fossils, it has been learned that conifers have existed since the Carboniferous Period, some 300 million years ago. Most **species** no longer exist. Currently there are approximately 550 known species. In **North America**, **firs** (*Abies*), larches (*Larix*), spruces (*Picea*), **pines** (*Pinus*), hemlocks (*Tsuga*), and junipers (*Juniperus*) are most common in mountain ranges of the Pacific Northwest and the Rocky Mountains. Conifers also extend through the northern regions of the United States and Canada, as well as into mountain ranges closer to the tropics. Some pine species grow in lowland areas of the southeastern United States.

Conifers are an important renewable resource; they provide the majority of **wood** for building as well as pulp for **paper**. **Resins**, oleoresins, and gums are important materials for the chemical industry for products such as soaps, hard resins, varnishes, and turpentine.

See also Juniper; Spruce; Yew.

Resources

Books

Wilkins, Malcolm. *Plantwatching.* New York: Facts On File, 1988.

Other

Chaw, S. M., et al. "Seed Plant Phylogeny Inferred From All Three Plant genomes: Monophyly of Extant Gymnosperms and Origin of Gnetales from Conifers." *Proceedings of the National Academy of Sciences of the United States of America* 97 (2000): 4086-4091.

Christine Miner Minderovic

Connective tissue

Connective tissue is found throughout the body and includes **fat**, cartilage, bone, and **blood**. The main functions of the different types of connective tissue include providing support, filling in spaces between organs, protecting organs, and aiding in the transport of materials around the body.

General structure of connective tissue

Connective tissue is composed of living cells and protein fibers suspended in a gel-like material called matrix. Depending on the type of connective tissue, the fibers are either **collagen** fibers, reticular fibers, or elastin fibers or a combination of two or more types. The type and arrangement of the fibers gives each type of connective tissue its particular properties.

Overview of connective tissue matrix

Of the three types of protein fibers in connective tissue collagen is by far the most abundant, and accounts for almost one third of the total body weight of humans. Under the **microscope**, collagen looks like a rope, with three individual protein fibers twined around each other. Collagen is extremely strong, but has little flexibility. Reticular fibers are composed of very small collagen fibers, but are shorter than collagen fibers, and they form a net-like supporting structure that gives shape to various organs. Elastin fibers have elastic properties and can stretch and be compressed, importing flexibility in the connective tissues where they are found.

Types of connective tissue

Two main types of fibrous connective tissue are found in the body: dense and loose. In dense connective tissue, almost all the space between the cells is filled by large numbers of protein fibers. In loose connective tissue, there are fewer fibers between the cells which imparts a more open, loose structure.

Dense connective tissue contains large numbers of collagen fibers, and so it is exceptionally tough. Dense regular connective tissue has **parallel** bundles of collagen fibers and forms tendons that attach muscles to bone and ligaments that bind bone to bone. Dense irregular connective tissue, with less orderly arranged collagen fibers, forms the tough lower layer of the skin known as the dermis, and encapsulates delicate organs such as the kidneys and the spleen.

Loose connective tissue has fewer collagen fibers than dense connective tissue, and therefore is not as tough. Loose connective tissue (also known as areolar connective tissue) is widely distributed throughout the body and provides the loose packing material between **glands**, muscles, and nerves.

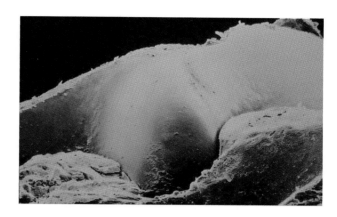

A scanning electron micrograph (SEM) of hyaline articular cartilage covering the end of a long bone. The smooth, slippery surface of the cartilage enables movement of the joint. *Photograph by Prof. P. Motta/Dept. of Anatomy/University "La Sapienza", Rome/Science Photo Library. National Audubon Society Collection/Photo Researchers, Inc. Reproduced by permission.*

Two other connective tissues with fibers are adipose tissue and reticular tissue. Adipose tissue is composed of specialized fat cells and has few fibers: this **tissue** functions as an insulator, a protector of delicate organs and as a site of **energy** storage. Reticular connective tissue is composed mostly of reticular fibers that form a net-like web, which forms the internal framework of organs like the liver, lymph nodes, and bone marrow.

Connective tissue composed of ground substance and protein fibers

Connective tissue composed of ground substance and protein fibers differs from fibrous connective tissue in that it contains more ground substance. Two main types of this kind of connective tissue are found in the body: cartilage and bone.

Cartilage is composed of cartilage cells, and collagen fibers or a combination of collagen and elastin fibers. An interesting characteristic of cartilage is that when it is compressed it immediately springs back into shape.

Hyaline cartilage is rigid yet flexible, due to evenly-spaced collagen fibers. Hyaline cartilage is found at the ends of the ribs, around the trachea (windpipe), and at the ends of long bones that form joints. Hyaline cartilage forms the entire skeleton of the embryo, which is gradually replaced by bone as the newborn grows.

Fibrocartilage contains densely-packed regularly arranged collagen fibers which impact great strength to this connective tissue. Fibrocartilage is found between the bones of the vertebrae as discs that act as a cushion.

Elastic cartilage contains elastin fibers and is thus more flexible that either hyaline cartilage or fibrocartilage. Elastic cartilage is found in the pinnas of the external **ear**.

Bone is composed of bone cells (osteocytes), suspended in a matrix consisting of collagen fibers and **minerals**. The mineral portion imparts great strength and rigidity to bone. Osteocytes are located in depressions called lacunae connected by canals called Haversian canals.

Two types of bone form the mammalian skeleton: cancellous bone and compact bone. Cancellous bone is more lattice-like than compact bone, and does not contain as many collagen fibers in its matrix. Cancellous bone is light-weight, yet strong, and is found in the skull, the sternum and ribs, the pelvis and the growing ends of the long bones. Compact bone is densely packed with fibers, and forms the outer shell of all bones and the shafts of the long bones of the arms and legs. Compact bone is heavier than cancellous bone, and provides great strength and support.

Mostly fluid connective tissue

Blood is a liquid connective tissue composed of a fluid matrix and blood cells. The blood cells include white blood cells, which function in the **immune system**, and red blood cells, which transport **oxygen** and **carbon dioxide**. The fluid part of the blood (the **plasma**) transports **hormones**, **nutrients**, and waste products, and plays a role in **temperature regulation**.

See also Skeletal system.

Resources

Periodicals

Brittberg, Mats, et al. "Treatment of Deep Cartilage Defects in the Knee with Autologous Chondrocyte Transplantation." *New England Journal of Medicine* 331 (October 1994).

Couzens, Gerald Seor, and Paula Derrow. "Weak in the Knees: New Ways to Protect—and Prevent—this Fragile Joint." *American Health* 12 (June 1993): 70.

Larkin, Marilynn. "Coping with Connective Tissue Diseases." *FDA Consumer* 26 (November 1992): 28.

Urry, Dan W. "Elastic Biomolecular Machines: Synthetic Chains of Amino Acids, Patterned After Those in Connective Tissue, Can Transform Heat and Chemical Energy Into Motion." *Scientific American* 272 (January 1995): 64.

Kathleen Scogna

Conservation

Conservation is the philosophy that natural resources should be used cautiously and rationally so that they will remain available for future generations of people.

American conservationist thought has evolved from its inception in the mid 1850s, when naturalists, businesspeople and statesmen alike foresaw environmental, economic and social peril in the unregulated use and abuse of North America's natural resources. Since those early attempts to balance the needs and desires of a growing, industrialized American public against the productivity and aesthetic beauty of the American wilderness, American environmental policy has experienced pendulum swings between no-holds-barred industrial exploitation, economically-tempered natural resource management, and preservationist movements that advocate protection of nature for nature's sake.

Government agencies instituted at the beginning of the twentieth century to guide the lawful, scientifically sound use of America's **forests**, **water** resources, agricultural lands, and **wetlands**, have had to address new environmental concerns such as air and **water pollution**, **waste management**, **wildfire** prevention, and **species extinction**. As the human population increased and technology advanced, American conservation policies and environmental strategies have had to reach beyond United States borders to confront issues like **global warming**, stratospheric **ozone** depletion, distribution of global **energy** and mineral resources, loss of **biodiversity**, and overuse of marine resources.

An organized, widespread conservation movement, dedicated to preventing uncontrolled and irresponsible exploitation of forests, land, **wildlife**, and water resources, first developed in the United States during the last decades of the nineteenth century. This was a time when accelerating settlement and resource depletion made conservationist policies appealing both to a large portion of the public and to government leaders. European settlement had reached across the entire North American **continent**, and the census of 1890 declared the American frontier closed. The era of North American exploration and the myth of an inexhaustible, virgin continent had come to an end. Furthermore, loggers, miners, settlers, and ranchers were laying waste to the nation's forests, prairies, **mountains**, and wetlands. Accelerating, wasteful commercial exploitation of natural resources went almost completely unchecked as political corruption and the economic power of lumber, **mining** and cattle barons made regulation impossible.

At the same time, American wildlife was disappearing. The legendary, immense flocks of passenger pigeons that migrated down the North American Atlantic coast disappeared entirely within a generation because of unrestrained hunting. Millions of **bison** were slaughtered by market hunters for their skins and meat, and by tourists shooting from passing trains. Logging, grazing, and hydropower development threatened America's most dramatic national landmarks. Niagara Falls, for example, nearly lost its untamed water flow. California's **sequoia** groves were considered for logging, and **sheep** grazed in Yosemite Valley.

Conservationist movement founded

Gifford Pinchot, the first head of the U.S. Forest Service, founded the conservation movement in the United States. He was a populist who fervently believed that the best use of nature was to improve the life of common citizens. Pinchot had extensive influence during the administration of President Theodore Roosevelt, himself an ardent conservationist, and helped to steer conservation policies from the turn of the century to the 1940s. Guided by the writing and thought of his conservationist predecessors, Pinchot brought science-based methods of resource management and a utilitarian philosophy to the Forest Service.

George Perkins Marsh, a Vermont forester and geographer, whose 1864 publication *Man and Nature* is a wellspring of American environmental thought, influenced Pinchot's ideas for American environmental policy. He was also inspired to action by John Wesley Powell, Clarence King, and other explorer-naturalists who assessed and cataloged the nation's physical and biological resources following the Civil War, as well as by his own observations of environmental destruction and social inequities precipitated by unregulated wilderness exploitation.

Conservation, as conceived by Pinchot, Powell, and Roosevelt, advocated thoughtful, rational use of natural resources, and not establishment of protected, unexploited wild areas. In their emphasis on wise resource use, the early conservationists were philosophically divided from the early preservationists. Preservationists, led by the eloquent writer and champion of Yosemite Valley, John Muir, bitterly opposed the idea that the best **vision** for the nation's forests was their conversion into agricultural land and timber tracts, developed to produce only species and products useful to humans. Muir, guided by the writing of the transcendentalist philosophers Emerson and Thoreau, argued vehemently that parts of the American wilderness should be preserved for their aesthetic value and for the survival of wildlife, and that all land should not be treated as a storehouse of useful commodities. Pinchot, however, insisted that: "The object of [conservationist] forest policy is not to preserve the forests because they are beautiful... or because they are refuges for the wild creatures of the wilderness... but the making of prosperous homes... Every other consideration is secondary." The motto of the U.S. National Forest Service, "The Land of Many Uses" reflects Pinchot's philosophy of land management.

Because of its more moderate and politically palatable stance, conservation became the more popular position by the turn of the century. By 1905, conservation had become a blanket term for nearly all defense of the environment. More Americans had come to live in cities, and to work in occupations not directly dependent upon resource exploitation. The urban population was sympathetic to the idea of preserving public land for recreational purposes, and provided much of the support for the conservation movement from the beginning. The earlier distinction from preservation was lost until it re-emerged in the 1960s as "environmentalists" once again raised vocal objections to conservation's anthropocentric (human-centered) emphasis. Late twentieth century naturalists like Rachel Carson, Edward Abbey, Aldo Leopold, as well as more radical environmental groups, including Greenpeace and Earth First!, owe much of their legacy to the turn of the century preservationists. More recently, deep ecologists and bioregionalists have likewise departed from mainstream conservation, arguing that other species have intrinsic rights to exist outside of the interests of humans.

As a scientific, humanistic, and progressive philosophy, conservation has led to a great variety of government and popular efforts to protect America's natural resources from exploitation by businesses and individuals at the expense of the American public. A professionally trained government forest service was developed to maintain national forests, and to limit the uncontrolled "timber mining" practiced by logging and railroad companies of the nineteenth century. Conservation-minded presidents and administrators set aside millions of acres of public land as national forests and parks for public use. A corps of scientifically trained **fish** and wildlife managers was established to regulate populations of gamebirds, sportfish, and hunted **mammals** for public use on federal lands.

Some of the initial conservation tactics seem strange by modern, ecological standards, and have had unintended consequences. For example, federal game conservation involved extensive programs of **predator** elimination leading to near extinction of some of America's most prized animals, including the timber wolf, the grizzly bear, the mountain lion, and the nation's symbol, the bald eagle. Decades of no-burn policies in national forests and parks, combined with encroachment by suburban neighborhoods, have led to destructive and dangerous forest fires in the American West. Extreme flood control measures have exposed a large population along the Mississippi river system to catastrophic **flooding**. However, early environmental policies were advised by the science of their time, and were unquestionably fairer and less destructive than the unchecked industrial development they replaced.

An important aspect of the growth of conservation has been the development of professional schools of **forestry**, game management, and wildlife management. When Gifford Pinchot began to study forestry, Yale University had only meager resources, and he gained the better part of his education at a school of forest management in Nancy, France. Several decades later, the Yale School of Forestry, initially financed largely by the wealthy Pinchot family, was able to produce such well-trained professionals as Aldo Leopold, who went on to develop the first professional school of game management in the United States at the University of Wisconsin. Today, most American universities offer courses in resource management and **ecology**, and many schools offer full-fledged programs in integrated ecological science and resource management.

During the administration of Franklin D. Roosevelt, conservation programs included such immense economic development projects as the Tennessee Valley Authority (TVA), which dammed the Tennessee River for flood control and **electricity** generation. The Bureau of Reclamation, formed in 1902 to manage surface water resources in 17 western states, constructed more than 600 **dams** in 1920s and 1930s, including the Hoover Dam and Glen Canyon dams across the Colorado River, and the Grand Coulee Dam on the Columbia River. The Civilian Conservation Corps developed roads, built structures, and worked on **erosion** control projects for the public good. The Soil Conservation Service was established to advise farmers in maintaining and developing their farmland.

Voluntary citizen conservation organizations have also done extensive work to develop and maintain natural resources. The Izaak Walton League, Ducks Unlimited, and local gun clubs and fishing groups have set up game sanctuaries, preserved wetlands, campaigned to control water **pollution**, and released young game **birds** and fish. Organizations with less directly utilitarian objectives have also worked and lobbied in defense of nature and wildlife, including the National Audubon Society, the Nature Conservancy, the Sierra Club, the Wilderness Society, and the World Wildlife Fund.

Global environmental efforts

From the beginning, American conservation ideas, informed by the science of ecology, and the practice of resource management on public lands, spread to other countries and regions. In recent decades, however, the rhetoric of conservation has taken a prominent role in international development and affairs, and the United States Government has taken a back-seat role in global environmental policy. United Nations Environment Program (UNEP), the Food and Agriculture Organization of the United Nations (FAO), the International Union for the Conservation of Nature and Natural Resources (IUCN), and the World Wildlife Fund (WWF) are some of today's most visible international conservation organizations.

The international community first convened in 1972 at the UN Conference on Earth and Environment in Stockholm to discuss global environmental concerns. UNEP was established at the Stockholm Convention. In 1980, the IUCN published a document entitled the *World Conservation Strategy*, dedicated to helping individual countries, including developing nations, plan for the maintenance and protection of their **soil**, water, forests, and wildlife. A continuation and update of this theme appeared in 1987 with the publication of the UN World Commission on Environment and Development's book, *Our Common Future*, also known as the Brundtland Report. The idea of **sustainable development**, with its vision of ecologically balanced, conservation-oriented economic development, was introduced in this 1987 **paper** and has gone on to become a dominant ideal in international development programs.

In 1992, world leaders gathered at the United Nations Conference on Environment and Development to discuss some of the issues set forth in the Brundtland Report. The Rio "Earth Summit" painted a grim picture of global environmental problems like **global climate** change, resource depletion, and pollution. The Rio summit inspired a number of ratified agreements designed to tackle some of these seemingly intractable issues, including stratospheric ozone depletion by man-made chemicals with the 1987 Montreal Protocol, and mitigation of possible global climate change caused by industrial emissions with the 2002 Kyoto Protocol.

The International community has tempered its philosophy of conservation since the 1992 Rio Summit. Sustainable development, a philosophy very similar to Pinchot's original conservation ideal, was the catch-phrase for the United Nation's 2002 Earth Summit in Johannesburg, South **Africa**. The 1992 Rio summit produced a laundry list of grim environmental problems: global warming, the ozone hole, biodiversity and **habitat** loss, **deforestation**, marine resource depletion, and suggested an "either-or" decision between economic development and environmental solutions. The 2002 Earth Summit, however, focused on international regulations that address environmental problems: water and air quality, accessibility of food and water, sanitation, agricultural productivity, and land management, that often accompany the human population's most pressing social issues: poverty, famine, **disease**, and war. Furthermore, new strategies for coping with environmental issues involve providing economic incentives to provide for the common good instead of punishing non-compliant governments and corporations.

See also Agrochemicals; Air pollution; Alternative energy sources; Animal breeding; Beach nourishment; Bioremediation; Blue revolution (aquaculture); Chlorofluorocarbons (CFCs); Crop rotation; Ecological economics; Ecological integrity; Ecological monitoring; Ecological productivity; Ecotourism; Environmental impact statement; Indicator species; Old-growth forests; Organic farming; Ozone layer depletion; Pollution control; Recycling; Restoration ecology; Slash-and-burn agriculture; Water conservation.

Resources

Books

Fox, S. *John Muir and His Legacy: The American Conservation Movement.* Boston: Little, Brown, 1981.

Marsh, G.P. *Man and Nature.* Cambridge: Harvard University Press, 1965 (originally 1864).

Meine, C. *Aldo Leopold: His Life and Work.* Madison, WI: University of Wisconsin Press, 1988.

Pinchot, G. *Breaking New Ground.* Washington, DC: Island Press, 1987 (originally 1947).

Periodicals

Kluger, Jefferey, and Andrea Dorfman. "The Challenges We Face." Special Report: "How to Preserve the Planet and Make this a Green Century." *Time Magazine* (August 26, 2002): A1-A60.

Other

United Nations World Summit on Sustainable Development. "Johannesburg Summit 2002." Johannesburg, South Africa. December 12, 2002 [cited January 7, 2003]. <http://www.johannesburgsummit.org/>.

United States Department of Agriculture Forest Service. "Caring for the Land and Serving People." January 6, 2003 [cited January 7, 2003]. <http://www.fs.fed.us/>.

Mary Ann Cunningham
Laurie Duncan

Conservation laws

Conservation laws refer to physical quantities that remain constant throughout the multitude of processes which occur in nature. If these physical quantities are carefully measured, and if all known sources are taken into account, they will always yield the same result. The validity of the conservation laws is tested through experiments. However, many of the conservation laws are suggested from theoretical considerations. The conservation laws include: the **conservation** of linear **momentum**, the conservation of angular momentum, the conservation

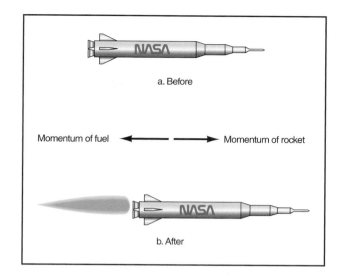

Figure 1. *Illustration by Hans & Cassidy. Courtesy of Gale Group.*

of **energy** and **mass**, and the conservation of **electric charge**. In addition, there are many conservation laws that deal with **subatomic particles**, that is, particles that are smaller than the atom.

Conservation of linear momentum

A rocket ship taking off, the recoil of a rifle, and a bank-shot in a pool are examples which demonstrate the conservation of linear momentum. Linear momentum is defined as the product of an object's mass and its **velocity**. For example, the linear momentum of a 220 lb (100 kg) football-linebacker traveling at a speed of 10 MPH (16 km/h) is exactly the same as the momentum of a 110 lb (50 kg) sprinter traveling at 20 MPH (32 km/h). Since the velocity is both the speed and direction of an object, the linear momentum is also specified by a certain direction.

The linear momentum of one or more objects is conserved when there are no external forces acting on those objects. For example, consider a rocket-ship in deep outer **space** where the **force** of gravity is negligible. Linear momentum will be conserved since the external force of gravity is absent. If the rocket-ship is initially at rest, its momentum is **zero** since its speed is zero (Figure 1a). If the rocket engines are suddenly fired, the rocket-ship will be propelled forward (Figure 1b). For linear momentum to be conserved, the final momentum must be equal to the initial momentum, which is zero. Linear momentum is conserved if one takes into account the burnt fuel that is ejected out the back of the rocket. The positive momentum of the rocket-ship going forward is equal to the **negative** momentum of the fuel going backward (note that the direction of **motion** is used to define positive and negative). Adding these two quantities yields

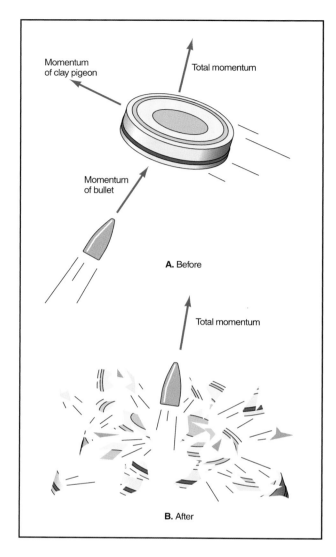

Momentum
of clay pigeon

Total momentum

Momentum
of bullet

A. Before

Total momentum

B. After

Figure 2. *Illustration by Hans & Cassidy. Courtesy of Gale Group.*

zero. It is important to realize that the rocket's propulsion is not achieved by the fuel pushing on anything. In outer space there is nothing to push on! Propulsion is achieved by the conservation of linear momentum. An easy way to demonstrate this type of propulsion is by propelling yourself on a frozen pond. Since there is little **friction** between your ice skates and the **ice**, linear momentum is conserved. Throwing an object in one direction will cause you to travel in the opposite direction.

Even in cases where the external forces are significant, the concept of conservation of linear momentum can be applied to a limited extent. An instance would be the momentum of objects that are affected by the external force of gravity. For example, a bullet is fired at a clay pigeon that has been launched into the air. The linear momentum of the bullet and clay pigeon at the in-

stant just before impact is equal to the linear momentum of the bullet and hundreds of shattered clay pieces at the instant just after impact. Linear momentum is conserved just before, during, and just after the collision (Figure 2). This is true is because the external force of gravity does not significantly affect the momentum of the objects within this narrow **time** period. Many seconds later, however, gravity will have had a significant influence and the total momentum of the objects will not be the same as just before the collision.

There are many examples that illustrate the conservation of linear momentum. When we walk down the road, our momentum traveling forward is equal to the momentum of the **earth** traveling backward. Of course, the mass of Earth is so large compared to us that its velocity will be negligible. (A simple calculation using the 220 lb [100 kg] linebacker shows that as he travels forward at 10 mph [16 kph]. Earth travels backward at a speed of 9 trillionths of an inch per century!) A better illustration is to walk forward in a row-boat and you will notice that the boat travels backward relative to the **water**. When a rifle is fired, the recoil you feel against your shoulder is due to the momentum of the rifle which is equal but in the opposite direction to the momentum of the bullet. Again, since the rifle is so much heavier than the bullet, its velocity will be correspondingly less than the bullet's. Conservation of linear momentum is the chief reason that heavier cars are safer than lighter cars. In a head-on collision with two cars traveling at the same speed, the motion of the two cars after the collision will be along the original direction of the larger car due to its larger momentum. Conservation of linear momentum is used to give space probes an extra boost when they pass planets. The momentum of the **planet** as it circles the **sun** in its **orbit** is given to the passing **space probe**, increasing its velocity on its way to the next planet. In all of the experiments ever attempted, there has been never been a violation of the law of conservation of linear momentum. This applies to all objects ranging in size from galaxies to subatomic particles.

Conservation of angular momentum

Just as there is the conservation of motion for objects traveling in straight lines, there is also a conservation of motion for objects traveling along curved paths. This conservation of rotational motion is known as the conservation of angular momentum. An object which is traveling at a constant speed in a **circle** (compare this to a race car on a circular track) is shown in Figure 3. The angular momentum for this object is defined as the product of the object's mass, its velocity, and the radius of the circle. For example, a 2,200 lb (1000 kg) car traveling at 30 MPH (50 km/h) on a 2–mi-radius (3-km) track, a

4,400 lb (2000 kg) truck traveling at 30 MPH (50 km/h) on a 1–mi-radius (1.6 km) track, and a 2,22200 lb (1000 kg) car traveling at 60 MPH (97 km/h) on a 1–mi-radius (1.6-km) track will all have the same value of angular momentum. In addition, objects which are spinning, such as a top or an ice skater, have angular momentum which is defined by their mass, their shape, and the velocity at which they spin.

In the absence of external forces that tend to change an object's **rotation**, the angular momentum will be conserved. Close to Earth, gravity is uniform and will not tend to alter an object's rotation. Consequently, many instances of angular momentum conservation can be seen every day. When an ice skater goes from a slow spin with her arms stretched into a fast spin with her arms at her sides, we are witnessing the conservation of angular momentum. With arms stretched, the radius of the rotation circle is large and the rotation speed is small. With arms at her side, the radius of the rotation circle is now small and the speed must increase to keep the angular momentum constant.

An additional consequence of the conservation of angular momentum is that the rotation axis of a spinning object will tend to keep a constant orientation. For example, a spinning Frisbee thrown horizontally will tend to keep its horizontal orientation even if tapped from below. To test this, try throwing a Frisbee without spin and see how unstable it is. A spinning top remains vertical as long as it keeps spinning fast enough. Earth itself maintains a constant orientation of its spin axis due to the conservation of angular momentum.

As is the case for linear momentum, there has never been a violation of the law of conservation of angular momentum. This applies to all objects, large and small. In accordance with the **Bohr model** of subatomic particles, the electrons that surround the nucleus of the atom are found to possess angular momentum of only certain discrete values. Intermediate values are not found. Even with these constraints, the angular momentum is always conserved.

Conservation of energy and mass

Energy is a state function that can be described in many forms. The most basic form of energy is kinetic energy, which is the energy of motion. A moving object has energy solely due to the fact that it is moving. However, many non-moving objects contain energy in the form of potential or stored energy. A boulder on the top of a cliff has potential energy. This implies that the boulder could convert this potential energy into kinetic energy if it were to fall off the cliff. A stretched bow and arrow have potential energy also. This implies that the stored energy in the bow could be converted into the kinetic energy of the arrow, after it is released. Stored energy may be more

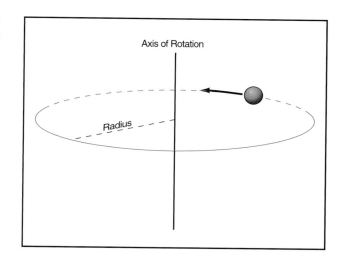

Figure 3. *Illustration by Hans & Cassidy. Courtesy of Gale Group.*

complicated than these mechanical examples, however, as in the stored electrical energy in a car **battery**. We know that the battery has stored energy because this energy can be converted into the kinetic energy of a cranking engine. There is stored chemical energy in many substances, for example gasoline. Again we know this because the energy of the gasoline can be converted into the kinetic energy of a car moving down the road. This stored chemical energy could alternately be converted into thermal energy by burning the gasoline and using the **heat** to increase the **temperature** of a bath of water. In all these instances, energy can be converted from one form to another, but it is always found that the total energy remains constant.

In certain instances, even mass can be converted into energy. For example, in a **nuclear reactor** the nucleus of the **uranium** atom is split into fragments. The sum of the masses of the fragments is always less than the original uranium nucleus. What happened to this original mass? This mass has been converted into thermal energy which heats the water to drive steam turbines which ultimately produces electrical energy. As first discovered by Albert Einstein (1879-1955), there is a precise relationship defining the amount of energy that is equivalent to a certain amount of mass. In instances where mass is converted into energy, or visa versa, this relationship must be taken into account.

In general, therefore, there is a universal law of conservation of energy and mass that applies to all of nature. The sum of all the forms of energy and mass in the universe is a certain amount which remains constant. As is the case for angular momentum, the energies of the electrons that surround the nucleus of the atom can possess only certain discrete values. And again, even with these constraints, the conservation of energy and mass is always obeyed.

Conservation of electric charge

Electric charge is the property that makes you experience a spark when you touch a **metal** door knob after shuffling your feet across a rug. It is also the property that produces **lightning**. Electric charge comes in two varieties, positive and negative. Like charges repel, that is, they tend to push one another apart, and unlike charges attract, that is, they tend to pull one another together. Therefore, two negative charges repel one another and, likewise, two positive charges repel one another. On the other hand, a positive charge will attract a negative charge. The net electric charge on an object is found by adding all the negative charge to all the positive charge residing on the object. Therefore, the net electric charge on an object with an equal amount of positive and negative charge is exactly zero. The more net electric charge an object has, the greater will be the force of attraction or repulsion for another object containing a net electric charge.

Electric charge is a property of the particles that make up an atom. The electrons that surround the nucleus of the atom have a negative electric charge. The protons which partly make up the nucleus have a positive electric charge. The neutrons which also make up the nucleus have no electric charge. The negative charge of the **electron** is exactly equal and opposite to the positive charge of the **proton**. For example, two electrons separated by a certain **distance** will repel one another with the same force as two protons separated by the same distance and, likewise, a proton and electron separated by this same distance will attract one another with the same force.

The amount of electric charge is only available in discrete units. These discrete units are exactly equal to the amount of electric charge that is found on the electron or the proton. It is impossible to find a naturally occurring amount of electric charge that is smaller than what is found on the proton or the electron. All objects contain an amount of electric charge which is made up of a combination of these discrete units. An analogy can be made to the winnings and losses in a penny ante game of poker. If you are ahead, you have a greater amount of winnings (positive charges) than losses (negative charges), and if you are in the hole you have a greater amount of losses than winnings. Note that the amount that you are ahead or in the hole can only be an exact amount of pennies or cents, as in 49 cents up or 78 cents down. You cannot be ahead by 32 and 1/4 cents. This is the analogy to electric charge. You can only be positive or negative by a discrete amount of charge.

If one were to add all the positive and negative electric units of charge in the universe together, one would arrive at a number that never changes. This would be analogous to remaining always with the same amount of money in poker. If you go down by five cents in a given hand, you have to simultaneously go up by five cents in the same hand. This is the statement of the law of conservation of electric charge. If a positive charge turns up in one place, a negative charge must turn up in the same place so that the net electric charge of the universe never changes. There are many other subatomic particles besides protons and electrons which have discrete units of electric charge. Even in interactions involving these particles, the law of conservation of electric charge is always obeyed.

Other conservation laws

In addition to the conservation laws already described, there are conservation laws that describe reactions between subatomic particles. Several hundred subatomic particles have been discovered since the discovery of the proton, electron, and the **neutron**. By observing which processes and reactions occur between these particles, physicists can determine new conservation laws governing these processes. For example, there exists a subatomic particle called the positron which is very much like the electron except that it carries a positive electric charge. The law of conservation of charge would allow a process whereby a proton could change into a positron. However, the fact that this process does not occur leads physicists to define a new conservation law restricting the allowable transformations between different types of subatomic particles.

Occasionally, a conservation law can be used to predict the existence of new particles. In the 1920s, it was discovered that a neutron could change into a proton and an electron. However, the energy and mass before the reaction was not equal to the energy and mass after the reaction. Although seemingly a violation of energy and mass conservation, it was instead proposed that the missing energy was carried by a new particle, unheard of at the time. In 1956, this new particle named the **neutrino** was discovered. As new subatomic particles are discovered and more processes are studied, the conservation laws will be an important asset to our understanding of the Universe.

Resources

Books

Feynman, Richard. *The Character of Physical Law.* Cambridge, MA: MIT Press, 1965.

Feynman, Richard. *Six Easy Pieces.* Reading, MA: Addison-Wesley, 1995.

Giancoli, Douglas. *Physics.* Englewood Cliffs, NJ: Prentice Hall, 1995.

Schwarz, Cindy. *A Tour of the Subatomic Zoo.* New York: American Institute of Physics, 1992.

Young, Hugh. *University Physics.* Reading, MA: Addison-Wesley, 1992.

Kurt Vandervoort

Constellation

A constellation is a group of stars that form a long-recognized pattern in the sky. The names of many constellations are Greek in origin and are related to ancient mythology. The stars that make up a constellation may be at very different distances from the **earth** and from one another. The pattern is one that we as humans choose to see and has no physical significance.

Novice stargazers are often taught that the pattern of stars in a constellation resembles an **animal** or a person engaged in some activity. For example, Sagittarius is supposed to be an archer, Ursa Major a large bear, and Ursa Minor a small bear. However, most people locate Sagittarius by looking for a group of stars that resemble an old-fashioned coffee pot. Ursa Major is more commonly seen as a Big Dipper and Ursa Minor as a Little Dipper. In fact, it is more likely that ancient stargazers named constellations to honor people, objects, or animals that were a part of their mythology and not because they thought the pattern resembled the honoree.

Today's modern stars divide the sky into 88 constellations that are used by astronomers to identify regions where stars and other objects are located in the celestial **sphere** (sky). Just as you might tell someone that Pike's Peak is near Colorado Springs, Colorado, so an astronomer refers to nebula (M 42) as the Orion Nebula, or speaks of **galaxy** M 31 in Andromeda and the globular cluster M 13 in Hercules.

The constellations that you see in the Northern Hemisphere's winter sky—Orion, Taurus, Canis Major, and others—gradually move westward with **time**, rising above the eastern horizon approximately four minutes earlier each evening. By late spring and early summer, the winter constellations are on the western horizon in the early evening and Leo, Bootes, Cygnus, and Sagittarius dominates the night sky. In the fall, Pegasus, Aquila, and Lyra brighten the heavens. A number of polar constellations (Cephus, Cassiopeia, and Ursa Minor in the north and Crux, Centaurus, and Pavo in the south) are visible all year as they rotate about points directly above the North and South Poles.

The westward movement of the constellations is the result of Earth's **motion** along its **orbit**. With each passing day and month, we see a different part of the celestial sphere at night. From our frame of reference on a **planet** with a tilted axis, the **sun**, **moon** and planets follow a path along the celestial sphere called ecliptic, which makes an **angle** of 23.5° with the celestial equator. As the sun moves along the ecliptic, it passes through 12 constellations, which ancient astronomers referred to as the Signs of the Zodiac—Aries, Taurus, Gemini, Cancer,

The constellation Orion, the Great Hunter. The three closely placed stars just left of center in this photo are Alnilam, Alnitak, and Mintaka, and they mark Orion's belt. The short line of less brilliant stars beneath the belt are his scabbard. In the upper left, at Orion's right shoulder, is the star Betelgeuse. His left shoulder is the star Bellatrix. His left foot is the star Rigel, and the bright stars to the right of top center are an animal skin that he carries as a shield. *U.S. National Aeronautics and Space Administration (NASA).*

Leo, Libra, Scorpius, Sagittarius, Capricorn, Aquarius, and Pisces. The planets also move along the ecliptic, but because they are much closer to us than the stars constellations along the zodiac their paths change with respect to the constellations. These wanderers, which is what the ancients called the planets, led to astrology—the belief that the motion of the sun, moon, and planets along the zodiac has some influence on human destiny. While there is no evidence to support such belief, the careful observations of early astronomers owes much to the pseudoscience of astrology.

See also Celestial coordinates; Milky Way; Star.

Constructions

Much of Euclidean **geometry** is based on two geometric constructions: the drawing of circles and the drawing of straight lines. To draw a **circle** with a compass, one needs to know the location of the center and some one **point** on the circle. To draw a line segment with a straightedge, one needs to know the location of its two end points. To extend a segment, one must know the location of it or a piece of it.

Three of the five postulates in Euclid's *Elements* say that these constructions are possible:

To draw a line from any point to any point.

To produce a finite straight line in a straight line.

To describe a circle with any center and **distance**.

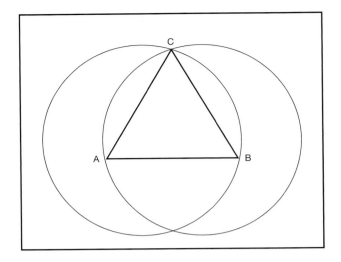

Figure 1. Construction of an equilateral triangle. *Illustration by Hans & Cassidy. Courtesy of Gale Group.*

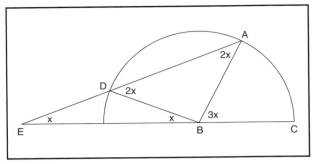

Figure 2. Illustration of Archimedes method for trisecting an arbitrary angle ABC. *Illustration by Hans & Cassidy. Courtesy of Gale Group.*

The constructions based on these postulates are called "straightedge and compass" constructions.

The *Elements* does not explain why these tools have been chosen, but one may guess that it was their utter simplicity which geometers found, and continue to find, appealing. These tools are certainly not the only ones which the Greeks employed, and they are not the only ones upon which modern draftsmen depend. They have triangles, french curves, ellipsographs, T-squares, scales, protractors, and other drawing aids which both speed the drawing and make it more precise.

These tools are not the only ones on which contemporary geometry courses are based. Such courses will often include a protractor **postulate** which allows one to measure angles and to draw angles of a given size. They may include a ruler-placement postulate which allows one to measure distances and to construct segments of any length. Such postulates turn problems which were once purely geometric into problems with an **arithmetic** component. Nevertheless, straightedge and compass constructions are still studied.

Euclid's first proposition is to show that, given a segment AB, one can construct an equilateral triangle ABC. (There has to be a segment. Without a segment, there will not be a triangle.) Using A as a center, he draws a circle through B. Using B as a center, he draws a circle through A. He calls either of the two points where the circles cross C. That gives him two points, so he can draw segment AC. He can draw BC. Then ABC is the required triangle (Figure 1).

Once Euclid has shown that an equilateral triangle can be constructed, the ability to do so is added to his tool bag. He now can draw circles, lines, and equilater-

al triangles. He goes on to add the ability to draw perpendiculars, to bisect angles, to draw a line through a given point **parallel** to a given line, to draw equal circles, to transfer a line segment to a new location, to divide a line segment into a specified number of equal parts, and so on.

There are three constructions which, with the given tools, neither Euclid nor any of his successors were able to do. One was to trisect an arbitrary **angle**. Another was to draw a **square** whose area was equal to that of a given circle. A third was to draw the edge of a cube whose **volume** was double that of a given cube. "Squaring the circle," as the second construction is called, is equivalent to drawing a segment whose length is π times that of a given segment. "Duplicating the cube" requires drawing a segment whose length is the cube root of 2 times that of the given segment.

In about 240 B.C., Archimedes devised a method of trisecting an arbitrary angle ABC. Figure 2 shows how he did it. Angle ABC is the given angle. ED is a movable line with ED = AB. It is placed so that E lies on BC extended; D lies on the circle; and the line passes through A. Then ED = DB = AB, so triangles EDB and ABD are isosceles. Because the base angles of an isosceles triangle are equal and because the exterior angle of a triangle is equal to the sum of the two non-adjacent interior angles, the sizes, in terms of x, of the various angles are as marked. Angle E is, therefore, one third the size of the given angle ABC; ABC has been trisected.

Why is this ingenious but simple construction not a solution to the problem of trisecting an angle? Line ED has to be movable. It requires a straightedge with marks on it. Simple as marking a straightedge might be, the Euclidean postulates don't make provision for doing so.

Archimedes' technique for trisecting an angle is by no means the only one which has been devised. Eves, in his *History of Mathematics,* describes several others, all ingenious. He also describes techniques for squaring the circle and duplicating the cube. All the constructions he describes, however, call for tools other than a compass and straightedge.

Actually doing these constructions is not just difficult with the tools allowed; it is impossible. This was proved using algebraic arguments in the nineteenth century. Nevertheless, because the goals of the constructions are so easily stated and understood, and because the tools are so simple, people continue to work at them, not knowing, or perhaps not really caring, that their task is a Sisyphean one.

The straightedge and compass are certainly simple tools, yet mathematicians have tried to get along with even simpler ones. In the tenth century Abul Wefa, a Persian mathematician, based his constructions on a straightedge and a rusty compass—one that could not be adjusted. Nine centuries later it was proved by mathematicians Poncelet and Steiner that, except for drawing circles of a particular size, a straightedge and rusty compass could do everything a straightedge and ordinary compass could do. They went even further, replacing the rusty compass with one already-drawn circle and its center.

In 1797, the Italian mathematician Mascheroni published a book in which he showed that a compass alone could be used to do anything that one could do with a compass and straightedge together. He could not draw straight lines, of course, but he could locate the two points that would determine the undrawn line; he could find where two undrawn lines would intersect; he could locate the vertices of a pentagon; and so on. Later, his work was found to have been anticipated more than 100 years earlier by the Danish mathematician Mohr. Compass-only constructions are now known as Mohr-Mascheroni constructions.

Resources

Books

Birkhoff, George David, and Ralph Beatley. *Basic Geometry.* New York: Chelsea Publishing Co., 1959.

Euclid. Sir Thomas L. Heath, trans. *Elements.* New York: Dover Publishing Co., 1956.

Gardner, Martin. *Mathematical Circus.* New York: Alfred A. Knopf, 1979.

Gullberg, Jan, and Peter Hilton. *Mathematics: From the Birth of Numbers.* W.W. Norton & Company, 1997.

Hahn, Liang-shin. *Complex Numbers and Geometry.* 2nd ed. The Mathematical Association of America, 1996.

J. Paul Moulton

Contaminated soil

The presence of toxic and radioactive chemicals in **soil** at concentrations above trace levels poses potential risks to human health and **ecological integrity**. Soil can be contaminated by many human actions, including discharge of solid or liquid materials to the soil surface, pesticide and fertilizer application, subsurface release from buried tanks, pipes, or landfills, and deposition of atmospheric contaminants such as dust and particles containing **lead**.

Contaminants can be introduced into the soil **horizon** at discrete locations called point sources, or across wide areas called non-point sources. **Point source** contamination typical of leaking tanks, pipes, and landfills is often concentrated and causes rapid, dramatic effects in a localized region near the original spill or leak. Soil contaminated by a plume emanating from a point source is, however, often easier to identify and remediate than the diffuse **pollution** caused by non-point sources like agriculture runoff or airfall from coal-burning **energy** plants. Governmental programs established to cleanup contaminated soil in the United States have made progress in cleaning up the nation's most polluted sites, but the technical difficulty and expense of remediation has made prevention the clear solution to soil **contamination** issues.

Frequently observed soil contaminants include volatile hydrocarbons such as **benzene**, toluene, ethylene, and xylene, and alkanes found in fuels. Heavy paraffins used in chemical processing, chlorinated organic compounds such as **polychlorinated biphenyls (PCBs)** that were used as coolants and lubricants in electrical equipment, **pesticides** and **wood** preservatives such as pentachlorophenol, and inorganic compounds of heavy metals like lead, cadmium, arsenic, and mercury, are all additional contaminants found in soil. Soil contaminated with **radioactive waste** has also been observed. Often, soil is tainted with a mixture of contaminants. The nature of the soil, the chemical and physical characteristics of the contaminant, environmental factors such as climate and **hydrology**, and proximity to human agricultural and municipal **water** sources interact to determine the accumulation, mobility, toxicity, and overall significance of the contamination in any specific instance.

Fate of soil contaminants

Contaminants in soil may be present in solid, liquid, or gaseous phases. When liquids are released, they move downward through the soil and may fill pore spaces, absorb onto mineral or organic surfaces, dissolve into soil water, or volatilize into the soil atmosphere. Most hydrocarbons exist in more than one phase in the soil horizon. Insoluble soil contaminants travel downward through the unsaturated, or vadose zone to reach the saturated zone, or water table, where voids between soil particles are filled with fluid. Their behavior in the saturated zone depends on their **density**. Light compounds float on the water table, while denser compounds may sink. Although many **hydrocarbon** compounds are not very soluble in water, even low levels of dissolved contaminants may produce unsafe or unacceptable **groundwater** quality. Other contaminants such as inorganic salts, nitrate **fertilizers** for example, are highly soluble and move rapidly through the soil environment. Metals like lead, mercury, and arsenic demonstrate a range of behaviors; some chemically bind to soil particles and are thus, immobile, while others dissolve in water and are transported widely.

Pore water containing dissolved contaminants, called leachate, is transported by groundwater flow, which moves both horizontally and vertically away from the contaminant source. Point source groundwater pollution often forms a three-dimensional plume that decreases in **concentration** with **distance** from the source and **time** since introduction of the contaminant. A portion of the contaminant, called the residual, is left behind as groundwater flow passes, resulting in a longer-term contamination of the soil after the contaminant plume has receded. If the groundwater **velocity** is fast, hundreds of feet per year, the zone of contamination may spread quickly, potentially affecting wells, surface water and plants that extract the contaminated water in a wide area.

Over years or decades, especially in sandy and other porous soils, groundwater contaminants and leachate may be transported over distances of miles, resulting in a situation that is extremely difficult and expensive to remedy. In such cases, immediate action is needed to contain and cleanup the contamination. However, if the soil is largely comprised of fine-grained silts and clays, contaminants will spread slowly. Comprehensive understanding of site-specific factors is important in evaluating the extent of soil contamination, and in selection of an effective cleanup strategy.

Superfund and other legislation

Prior to the 1970s, inappropriate waste disposal practices like dumping untreated liquids in lagoons and landfills were common and widespread. Some of these practices were undertaken with willful disregard for their environmental consequences, and occasionally for existing regulations, but many types of soil contamination occurred as a result of scientific ignorance. The toxicity of many contaminants, including PCBs and methyl mercury, was discovered long after the chemicals were spilled, dumped, or sprayed into soils. The ability of groundwater flow to transport contaminants was likewise unknown until decades after many contaminants had flowed away from unlined landfills, sewage outlets and chemical dumps.

The presence and potential impacts of soil contamination were finally brought to public attention after well-publicized disasters at Love Canal, New York, the Valley of the Drums, Kentucky, and Times Beach, Missouri. Congress responded by passing the Comprehensive Environmental Response, Compensation, and Liability Act in 1980 (commonly known as CERCLA, or Superfund) to provide funds with which to remediate contamination at the worst sites of point-source pollution. After five years of much litigation but little action, Congress updated CERCLA in 1986 with the Superfund Amendments and Reauthorization Act.

In 2003, a total of 1,300 sites across the United States were designated as National Priorities List (NPL) sites that were eligible under CERCLA for federal cleanup assistance. These Superfund sites are considered the nation's largest and most contaminated sites in terms of the possibility for adverse human and environmental impacts. They are also the most expensive sites to clean. While not as well recognized, numerous other sites have serious soil contamination problems.

The U.S. Office of Technology Assessment and the Environmental Protection Agency (EPA) have estimated that about 20,000 abandoned waste sites and 600,000 other sites of land contamination exist in the United States. These estimates exclude soils contaminated with lead paint in older urban areas, the accumulation of fertilizers and pesticides in agricultural land, salination of irrigated soils in arid regions, and other classes of potentially significant soil contamination. Federal and state programs address only some of these sites.

Currently, CERCLA is the EPA's largest program, with expenditures exceeding $3 billion during the 1990s. However, this is only a fraction of the cost to government and industry that will be needed to clean all hazardous and toxic waste sites. Mitigation of the worst 9,000 waste sites is estimated to cost at least $500 billion, and to take at least 50 years to complete. The problem of contaminated soil is significant not only in the United States, but in all industrialized countries.

U.S. laws such as CERCLA and the Resource Conservation and Recovery Act (RCRA) also attempt to prohibit practices that have led to extensive soil contamination in the past. These laws restrict disposal practices, and they mandate financial liability to recover cleanup, personal injury and property damage costs. Furthermore, the laws discourage polluters from willful misconduct or negligence by exposing non-compliant industries and individuals to criminal liability. Pollution prevention regulations also require record keeping to track waste, and provide incentives to reduce waste generation and improve **waste management**.

Soil cleanup

The cleanup or remediation of contaminated soil takes two major approaches: (1) source control and containment, and (2) soil and residual treatment and management. Typical containment strategies involve isolating potential contamination sources from surface and groundwater flow. Installation of a cover over the waste limits the infiltration of rain and snowmelt, and decreases the amount of potential leachate in the surrounding soil. Scrubbers and filters on energy **plant** smokestacks prevent contamination of rainwater by dissolved chemicals and particulate **matter**. Vertical slurry walls may control horizontal transport of pollutants in near-surface soil and groundwater. Clay, cement, or synthetic liners encapsulate soil contaminants. Groundwater pump-and-treat systems, sometimes coupled with injection of clean water, hydraulically isolate and manage contaminated water and leachate. Such containment systems reduce the mobility of the contaminants, but the barriers used to isolate the waste must be maintained indefinitely.

Soil and residual treatment strategies are after-the-fact remediation methods that reduce the toxicity and **volume** of soil contaminants. Treatment procedures are generally categorized as either extractive or *in situ* measures. Extractive options involve physical removal of the contaminated soil, off-site treatment of the contaminants by **incineration** or chemical **neutralization**, and disposal in a **landfill**. *In situ* processes treat the soil in place.

In situ options include thermal, biological, and separation/extraction technologies. Thermal technologies involve heating soils in place. The thermal desorption process breaks the chemical bonds between contaminants and soil particles. Vitrification, or glassification, involves melting the mineral component of the soil and encapsulating the contaminants in the resolidified glassy matrix. Biological treatment includes biodegradation by soil fungi and **bacteria**, ultimately rendering contaminants into **carbon dioxide**, other simple **minerals**, and water. This process is also called mineralization. Biodegradation may, however, produce long-lived toxic intermediate products and even contaminated organisms. Separation technologies attempt to isolate contaminants from pore fluids, and force them to the surface. Soil vapor extraction can successfully remove volatile organic compounds by enhancing volatilization with externally-forced subsurface air flow. Stabilization, or chemical fixation, uses additives that bind dissolved organic pollutants and heavy metals to eliminate contaminated leachate. Soil washing and flushing processes use dispersants and solvents to dissolve contaminants such as PCBs and enhance their removal. Groundwater pump-and-treat schemes purge the contaminated soil with clean water in a flushing action. A number of these *in situ* approaches are still experimental, notably soil vitrification and enhanced **bioremediation** using engineered **microorganisms**.

The number of potential remediation options is large and expanding due to an active research program driven by the need for more effective, less expensive solutions. The selection of an appropriate cleanup strategy for contaminated soil requires a thorough characterization of the site, and an analysis of the cost-effectiveness of suitable containment and treatment options. A site-specific analysis is essential because the **geology**, hydrology, waste properties, and source type determine the extent of the contamination and the most effective remediation strategies. Often, a demonstration of the effectiveness of an innovative or experimental approach may be required by governmental authorities prior to its full-scale implementation. In general, large sites use a combination of remediation options. Pump-and-treat and vapor extraction are the most popular technologies.

Cleanup costs and standards

The cleanup of contaminated soil can involve significant expense and environmental risk. In general, containment is cheaper and has fewer environmental consequences than soil treatment. The Superfund law establishes a preference for these supposedly permanents remedies, but many Superfund cleanups have occurred at sites that used both containment and treatment options. In cases where containment measures failed, or were never instituted, *in situ* treatment methods, such as groundwater pump-and-treat, are generally preferable to extractive approaches like soil incineration because they are often less expensive. Excavation and incineration of contaminated soil can cost $1,500 per ton, leading to total costs of many millions of dollars at large sites. (Superfund clean-ups have averaged about $26 million.) In contrast, small fuel spills at gasoline stations may be mitigated using vapor extraction at costs under $50,000. However, *in situ* options may not achieve cleanup goals.

Unlike air and water, which have specific federal laws and regulations detailing maximum allowable levels of contaminants, no levels have been set for contaminants in soil. Instead, the federal Environmental Protection Agency and state environmental agencies use subjective, case-specific criteria to set acceptable contaminant levels. For Superfund sites, cleanup standards must exceed applicable or relevant and appropriate requirements (ARARs) under federal environmental and public health laws. Cleanup standards are often determined by measuring background levels of the offending contaminant in similar, nearby, unpolluted soil. In some cases, soil contaminant levels may be acceptable if the soil does not produce leachate with concentration levels above drinking water standards. Such determinations are often based on a test called the Toxics Characteristic Leaching Procedure, which mildly acidifies and agitates the soil, followed by chemical analysis of the leachate. Contaminant levels in the leachate below the maximum contaminant levels (MCLs) in the federal Safe Drinking Water Act are considered acceptable. Finally, soil contaminant levels may be set in a determination of health risks based on typical or worst case exposures. Exposures can include the inhalation of soil as dust, ingestion (generally by small children), and direct skin contact. The flexible definition of acceptable toxicity levels reflects the complexity of contaminant mobility and toxicity in soil, and the difficulty of pinning down acceptable and safe levels.

Resources

Books

Fonnum, F., B. Paukstys, B. A. Zeeb, and K.J. Reimer. *Environmental Contamination and Remediation Practices at Former and Present Military Bases*. Kluwer Academic Publishers, 1998.

Jury, W. "Chemical Movement Through Soil." In *Vadose Modeling of Organic Pollutants*. Ed. S. C. Hern and S. M. Melancon. Chelsea, MI: Lewis Publishing, 1988.

Periodicals

Chen, C. T. "Understanding the Fate of Petroleum Hydrocarbons in the Subsurface Environment." *Journal of Chemical Education* 5 (1992): 357–59.

Other

Agency for Toxic Substances and Disease Registry. January 6, 2003 [Cited January 7, 2003]. <http://www.atsdr.cdc.gov/>.

United States Environmental Protection Agency. "Superfund." January 7, 2003 [Cited January 7, 2003]. <http://www.epa.gov/ superfund//>.

Stuart Batterman
Laurie Duncan

Contamination

Contamination generally refers to the occurrence of some substance in the environment. The contaminant may be present in a larger **concentration** than normally occurs in the ambient environment. However, contamination is only said to occur when the concentration is smaller than that at which measurable biological or ecological damage can be demonstrated. Contamination is different from **pollution**, which is judged to occur when a chemical is present in the environment at a concentration greater than that required to cause damage to organisms. Pollution results in toxicity and ecological change, but contamination does not cause these effects because it involves sub-toxic exposures.

Chemicals that are commonly involved in toxic pollution include the gases **sulfur dioxide** and **ozone**, elements such as arsenic, **copper**, mercury, and nickel, **pesticides** of many kinds, and some naturally occurring biochemicals. In addition, large concentrations of **nutrients** such as phosphate and nitrate can cause **eutrophication**, another type of pollution. All of these pollution-causing chemicals can occur in the environment in concentrations that are smaller than those required to cause toxicity or other ecological damages. Under these circumstances the chemicals would be regarded as contaminants.

Modern analytical **chemistry** has become extraordinarily sophisticated. As a result, trace contamination by potentially toxic chemicals can often be measured in amounts that are much smaller than the thresholds of exposure, or dose, that are required to demonstrate physiological or ecological damage.

Toxic chemicals

An important notion in **toxicology** is that any chemical can poison any **organism**, as long as a sufficiently large dose is experienced. In other words, all chemicals are potentially toxic, even **water**, **carbon dioxide**, sucrose (table sugar), **sodium chloride** (table **salt**), and other substances that are routinely encountered during the course of the day. However, exposures to these chemicals, or to much more toxic substances, do not necessarily result in a measurable poisonous response, if the dose is small enough. Toxicity is only caused if the exposure exceeds physiological thresholds of tolerance. According to this interpretation of toxicology, it is best to refer to "potentially toxic chemicals" in any context in which the actual environmental exposure to chemicals is unclear, or when the effects of small doses of particular chemicals are not known.

However, it is important to understand that there is scientific controversy about this topic. Some scientists believe that even exposures to single molecules of certain chemicals could be of toxicological significance, and that dose-response relationships can therefore be extrapolated in a linear fashion to a **zero** dosage. This might be especially relevant to some types of cancers, which could theoretically be induced by genetic damage occurring in a single **cell**, and potentially caused by a single **molecule** of a **carcinogen**. This is a very different view from that expressed above, which suggests that there are thresholds of physiological tolerance that must be exceeded if toxicity is to be caused.

The notion of thresholds of tolerance is supported by several lines of scientific evidence. It is known, for example, that cells have some capability of repairing damage caused to nuclear materials such as DNA (deoxyribonucleic acid), suggesting that minor damage caused by toxic chemicals might be tolerated because they could be repaired. However, major damage could overwhelm the physiological repair function, so that there would be a threshold of tolerance.

In addition, organisms have physiological mechanisms for detoxifying many types of poisonous chemicals. Mixed-function oxidases (MFOs), for example, are a class of enzymes that are especially abundant in the liver of vertebrate animals, and to a lesser degree in the bloodstream. Within limits, these enzymes can detoxify certain potentially toxic chemicals, such as chlorinated hydrocarbons, by rendering them into simpler, less toxic substances. Mixed-function oxidases are inducible enzymes, meaning that they are synthesized in relatively large quantities when there is an increased demand for their metabolic services, as would occur when an organism is exposed to a large concentration of toxic chemicals. However, the ability of the mixed-function oxidase system to deal with toxic chemicals can be overwhelmed if the exposure is too intense, a characteristic that would be represented as a toxicological threshold.

Organisms also have some ability to deal with limited exposures to potentially toxic chemicals by partitioning them within tissues that are not vulnerable to their poisonous influence. For example, chlorinated hydrocarbons such as the **insecticides** DDT and dieldrin, the industrial fluids known as **polychlorinated biphenyls** (**PCBs**), and the **dioxin** TCDD are all very soluble in fats, and therefore are mostly found in the fatty tissues of animals. Within limits, organisms can tolerate exposures to these chemicals by immobilizing them in fatty tissues. However, toxicity may still result if the exposure is too great, or if the **fat** reserves must be mobilized in order to deal with large metabolic demands, as might occur during **migration** or breeding. Similarly, plants have some ability to deal with limited exposures to toxic metals, by synthesizing certain **proteins**, organic acids, or other

biochemicals that bind with the ionic forms of metals, rendering them much less toxic.

Moreover, all of the chemicals required by organisms as essential nutrients are toxic at larger exposures. For example, the metals copper, **iron**, molybdenum, and zinc are required by plants and animals as micronutrients. However, exposures that exceed the therapeutic levels of these metals are poisonous to these same organisms. The smaller, sub-toxic exposures would represent a type of contamination.

Some chemicals are ubiquitous in the environment

An enormous variety of chemicals occurs naturally in the environment. For example, all of the natural elements are ubiquitous, occurring in all aqueous, **soil**, atmospheric, and biological samples in at least a trace concentration. If the methodology of analytical chemistry has sufficiently small detection limits, this ubiquitous presence of all of the elements will always be demonstrable. In other words, there is a universal contamination of the living and non-living environment with all of the natural elements. This includes all of the potentially toxic metals, most of which occur in trace concentrations.

Similarly, the organic environment is ubiquitously contaminated by a class of synthetic, persistent chemicals known as chlorinated hydrocarbons, including such chemicals as DDT, PCBs, and TCDD. These chemicals are virtually insoluble in water, but they are very soluble in fats. In the environment, almost all fats occur in the **biomass** of living or dead organisms, and as a result chlorinated hydrocarbons have a strong tendency to bioaccumulate in organisms, in strong preference to the non-organic environment. Because these chemicals are persistent and bioaccumulating, they have become very widespread in the **biosphere**. All organisms contain their residues, even in remote places such as **Antarctica**. The largest residues of chlorinated hydrocarbons occur in predators at the top of the ecological food web. Some of these top predators, such as raptorial **birds** and some marine **mammals**, have suffered toxicity as a result of their exposures to chlorinated hydrocarbons. However, toxicity has not been demonstrated for most other **species**, even though all are contaminated by various of the persistent chlorinated hydrocarbons.

One last example concerns some very toxic biochemicals that are synthesized by wild organisms, and are therefore naturally occurring substances. Saxitoxin, for example, is a biochemical produced by a few species of marine dinoflagellates, a group of unicellular **algae**. Saxitoxin is an extremely potent toxin of the vertebrate **nervous system**. When these dinoflagellates are abundant, filter-feeders such as **mollusks** can accumulate saxitoxin to a large concentration, and these can then be poisonous to birds and mammals, including humans, that eat the shellfish. This toxic **syndrome** is known as paralytic shellfish poisoning. Other species of dinoflagellates synthesize the biochemicals responsible for diarrhetic shellfish poisoning, while certain **diatoms** produce domoic acid, which causes amnesic shellfish poisoning. The poisons produced by these marine **phytoplankton** are commonly present in the marine environment as a trace contamination. However, when the algae are very abundant the toxins occur in large enough amounts to poison animals in their food web, causing a type of natural, toxic pollution.

See also Biomagnification; Poisons and toxins; Red tide.

Resources

Books

Freedman, B. *Environmental Ecology.* 2nd ed. San Diego: Academic Press, 1995.
Hemond, H.F., and E.J. Fechner. *Chemical Fate and Transport in the Environment.* San Diego: Academic Press, 1994.
Matthews, John A., E.M. Bridges, and Christopher J. Caseldine *The Encyclopaedic Dictionary of Environmental Change.* New York: Edward Arnold, 2001.

Bill Freedman

Continent

A continent is a large land **mass** and its surrounding shallow **continental shelf**. Both are composed of felsic crust. Continents, as by-products of plate tectonic activity, have grown to cover about one-third of Earth's surface over the last four billion years. Continents are unique to **Earth**, as **plate tectonics** does not occur on the other planets of our **solar system**.

Crusts compared

Earth's crust comes in two varieties, continental and oceanic. All crust consists primarily of silicate **minerals**. These contain silica (SiO_2), which consists of the elements silicon and **oxygen** bonded together, and a variety of other elements. Continental crust is silica-rich, or felsic. Oceanic crust is mafic, relatively rich in **iron** and **magnesium** and silica-poor. The mantle has silicate minerals with a greater abundance of iron and magnesium and even less silica than oceanic crust, so it is called ultramafic.

Continental crust is less dense and thicker (specific gravity = 2.7, thickness = 20-25 mi; 30-40 km) than oceanic crust (S.G. = 2.9, thickness = 2.5-3.75 mi; 6-7 km), and much less dense than the upper mantle (S.G. 3.3). Continents are therefore very unlikely to subduct at an oceanic trench, while oceanic crust subducts rather easily. Consequently, Earth's oldest oceanic crust is less than 200 million years old, while the oldest existing continental crust is 3.8 billion years old and most is more than two billion years old.

Continental margins

Continents consist of large blocks of continental crust, evidenced by dry land, bordered by continental shelves—the part of the continental crust that is below **sea level**. Every continent is also surrounded by either passive or active continental margins. At a passive margin, the continental shelf is typically a broad, nearly flat, sediment-covered submarine platform which ends at a **water** depth of about 600 ft (180 m) and tens or hundreds of miles (tens or hundreds of km) offshore. Marked by an abrupt increase in slope, known as the shelf break, the true edge of the continent is hidden below a thick layer of sediments on the adjacent continental slope. Seaward of the continental slope, a thick sediment wedge forms a much lower angle surface, called the continental rise. These sediments generally rest on oceanic crust, not continental crust. However, both the felsic crust of the **continental margin** and the mafic crust of the **ocean** floor are part of the same tectonic plate.

At active continental margins, interactions between two or more plates result in a very abrupt boundary between one plate's continental crust, and the crust of the neighboring plate. Typically, the continental shelf is narrow. For example, off the coast of Washington State, the Juan de Fuca plate subducts below the North American plate and the profile of the coast is very steep. There is no continental slope or rise because sediments move from the shelf down into the nearby ocean trench.

Structure of a continent

Horizontal crustal structure

Continent interiors consist of several structural segments. A rock foundation called the craton composes most of every continent. This consists of several large masses, called terranes, composed of ancient igneous and metamorphic **rocks** joined together into a rigid, stable unit. Each terrane may be quite different in structure, rock type and age from adjoining terranes. Where exposed at the earth's surface, cratons are called shields. These typically are ancient surfaces planed flat by **erosion**. In areas where younger sedimentary rocks cover the craton, it is called the stable platform. The craton below these sedimentary layers is usually called basement rock. Ancient mountain chains, or orogenic belts, occur within the craton where two smaller cratons became sutured together in the distant past. Some of the world's highest mountain ranges, such as the Himalayas, developed when two cratons (continents) collided, and erosion has not yet leveled them.

The margins of continents host younger orogenic belts than their interiors. These belts usually form due to plate convergence along active continental margins. Younger orogenic belts, with their steep slopes, tend to shed large volumes of sediment. In coastal regions, these sediments form a seaward-facing, shallowly-sloping land surface called a coastal plain. Within the continental interior, sediments eroded from **mountains** to form an area of interior lowlands, such as the United States' Great Plains region.

Divergence within a continent's interior leads to rifting. Initially, a steep-sided rift valley forms accompanied by small- to moderate-sized volcanic eruptions. Eventually this rift valley extends to the coast and the valley floor drops below sea level. A small inland sea develops, like the Red Sea between **Africa** and the Arabian subcontinent, with an oceanic ridge at its center. Given sufficient **time** and continued sea floor spreading, this sea becomes an ocean, similar to the Atlantic, with passive margins and wide continental shelves along both its shores.

An unusual continental feature develops when a continental rift fails. For whatever reason, rather than divergence producing an inland sea, rifting ends and the structure that remains is called an aulocogen. Depending on the rift's degree of development when failure occurs, the aulocogen may range from an insignificant crustal flaw to a major crustal weakness. Geologists attribute many powerful mid-plate earthquakes, such as the three 1811-1812 New Madrid, Missouri earthquakes, to **fault** movements associated with failed rifts.

Other common, large-scale continental structures include basins and domes. Basins are circular areas where the crust has subsided, usually under the load of accumulated sediments or due to crustal weakness such as an aulocogen. Domes occur when the crust is uplifted, perhaps due to compression of the continental interior during plate convergence at a nearby active margin.

Vertical crustal structure

Continental crust is heterogenous; however, general trends in structure, composition, and rock type are known. Our knowledge of the subsurface character of the crust comes from two main sources. The crustal interior is observed directly in areas where **uplift** and erosion expose the cores of ancient mountain belts and other structures. In addition, seismic waves produced during earthquakes change speed and character when moving through the crust. These changes allow geophysicists to infer crustal structure and **density**.

Continents reach their greatest thickness (up to 45 mi; 70 km) below mountain ranges and are thinnest (10-15 mi; 16-24 km) beneath rifts, aulocogens, shields, and continental margins. Density tends to increase downwards, in part due to an increase in mafic content. The upper crust has an average composition similar to granite, while the lower crust is a mixture of felsic and mafic rocks. Therefore, the average composition of the continents is slightly more mafic than granite. Granite contains an average of 70-75% silica; basalt about 50%. The continental crust is composed of 65% silica, the composition of the igneous rock granodiorite. The intensity of **metamorphism** and **volume** of **metamorphic rock** both increase downward in the crust as well.

Crustal origins

Whether directly or indirectly, the source of all Earth's crust is the mantle. **Radioactive decay** in **Earth's interior** produces **heat**, which warms regions of the mantle. This causes mantle rock, although solid, to convect upward where **pressure** is inevitably lower. Pressure and melting **temperature** are directly related, so decreasing pressure eventually causes the rock to begin melting, a process called pressure-relief melting. Every mineral, due to its composition and atomic structure, has its own melting temperature, so not all the minerals in the convecting rock melt. Instead, the first minerals to melt are the ones with the lowest melting temperatures. Generally, the higher the silica content of a mineral, the lower its melting temperature.

The mantle is composed of the ultramafic rock peridotite. Partial melting of peridotite produces molten rock, or **magma**, with a mafic, or basaltic, composition. This magma, now less dense due to melting, continues convecting upwards until it arrives below an oceanic ridge where it crystallizes to form new ocean crust. Over time, the crust slowly moves away from the oceanic ridge allowing more new crust to form, a process called sea floor spreading. For its first 100 million years or so, the older oceanic crust is, the cooler it becomes. This increases its density and, therefore, its likelihood of subducting. By the time subduction occurs, the crust is rarely more than 150-170 million years old.

Growing pains

Continents develop and grow in several ways as a byproduct of sea floor spreading and subduction. When subduction occurs where oceanic and continental crust con-

KEY TERMS

. . . .

Basalt—A dense, dark colored igneous rock, with a composition rich in iron and magnesium (a mafic composition).

Continental shelf—A relatively shallow, gently sloping, submarine area at the edges of continents and large islands, extending from the shoreline to the continental slope.

Felsic—A term applied to light-colored igneous rocks, such as rhyolite, that are rich in silica. Felsic rocks are rich in the minerals feldspar and quartz.

Granite—A felsic igneous rock that composes the bulk of the upper continental crust.

Mafic—Pertaining to igneous rocks composed of silicate minerals with low amounts of silicon and abundant iron and magnesium.

Peridotite—An ultramafic rock that composes the bulk of the mantle.

Silicate—Any mineral with a crystal structure consisting of silicon and oxygen, either with or without other elements.

Specific gravity—The weight of a substance relative to the weight of an equivalent volume of water; for example, basalt weighs 2.9 times as much as water, so basalt has a specific gravity of 2.9.

Ultramafic—Pertaining to igneous rocks composed of silicate minerals with very low amounts of silicon and abundant iron and magnesium.

verge, the plate margin carrying oceanic crust invariably subducts below the other. As the oceanic plate margin subducts into the upper mantle, volatiles (primarily water) escape the subducting crust. These volatiles lower the melting temperature of the overlying mantle, producing mafic magma due to partial melting. The magma moves upwards through the base of the felsic crust overhead, causing more partial melting. The resulting magma tends to have an intermediate to felsic composition. As crystallization begins within magma, the early crystals tend to be more mafic due to their higher crystallization temperature. Therefore, this fractional crystallization further refines the magma toward a felsic composition. When the magma crystallizes, the upper crust gains additional felsic rock.

Alternatively, accumulated sediments can be scraped off the subducted plate onto the overriding one, forming an accretionary wedge. A volcanic **island** arc may eventually arrive at the plate boundary as well. Subduction of such a large mass is not possible, however, so part of the

arc will also be welded, or accreted, to the continental margin. The accretionary wedge may be metamorphosed and welded to the crust too. Much of California formed in this way, one small piece—known as a microterrane—at a time. Finally, fragments of oceanic crust, called ophiolites, sometimes resist subduction and get shoved, or obducted, onto the continental margin as well. So continents grow by magma emplacement, accretion and obduction. In time, through intensive metamorphic and igneous activity, alteration of these microterranes will produce a more-or-less stable, homogenous, rock unit and a new terrane will become a part of the ancient craton.

Primeval continents

Over four billion years ago, before the first continents developed, the mantle was much hotter than today. Accordingly, the **rate** of plate **tectonics** was much faster. This meant that plates were smaller, thinner and much warmer when subducted. The associated crust was nearer to its melting point and so partially melted along with the mantle. Early subduction zone volcanism therefore produced magmas with higher silica content, since partially melting basalt produces an intermediate magma composition. The early Earth consequently developed volcanic island arcs with relatively low density, which resisted subduction. These formed Earth's earliest microcontinents and built the first cratons when sutured together at subduction zones. Throughout the 1990s, research focused on early continent development and the questions of when continents first appeared, how they originated and at what rate growth occurred.

Resources

Books

Trefil, James. *Meditations At 10,000 Feet: A Scientist In The Mountains.* New York: Macmillan Publishing Company. 1986.

Vogel, Shawna. *Naked Earth: The New Geophysics.* New York: Penguin Books. 1995.

Periodicals

Taylor, S. Ross, and Scott McLennan. "The Evolution of Continental Crust." *Scientific American* (January 1996).

Weiss, Peter. "Land Before Time." *Earth* (February 1998).

Clay Harris

Continental drift

The relative movement of the continents is explained by modern theories of **plate tectonics**. The relative movement of continents is explained by the move-

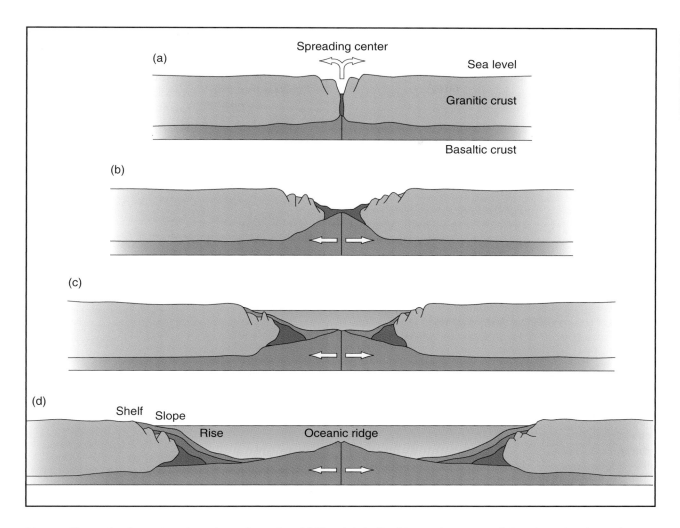

Diagram illustrating formation of continental margins. (a) Materials in Earth's mantle move up (often through volcanos) expanding the continental crust and causing rifting. (b) As it divides, the continental crust thins, and continues to separate. (c) New oceanic crust forms from the mantle materials that have surfaced. (d) The oceanic crust is further widened by sea floor spreading. *Illustration by Argosy. The Gale Group.*

ment of lithospheric plates—of which the visible continents are a part—over the athenosphere (the molten, ductile, upper portion of Earth's mantle). In a historical sense, the now discarded explanations of continental drift were rooted in antiquated concepts regarding Earth's structure.

Explanations of continental drift that persisted well into the twentieth century made the improbable geophysical assertion that the continents moved through and across an underlying oceanic crust much as **ice** floats and drifts through **water**. Eventually multiple lines of evidence allowed modern tectonic theory to replace continental drift theory.

In the 1920s, German geophysicist Alfred Wegener's writings advanced the hypothesis of continental drift

depicting the movement of continents through an underlying oceanic crust.

Wegner's hypothesis met with wide skepticism but found support and development in the work and writings of South African geologist Alexander Du Toit who discovered a similarity in the fossils found on the coasts of **Africa** and **South America** that were seemingly derived from a common source. Other scientists also attempted to explain orogeny (mountain building) as resulting from Wegner's continental drift.

Technological advances necessitated by the Second World War made possible the accumulation of significant evidence regarding Wegener's hypothesis, eventually refining and supplanting Wegner's theory of continental drift with modern plate tectonic theory. Although Wegener's theory accounted for much of the then existing geological evidence, Wegener's hypothesis was specifically unable to provide a verifiable or satisfying mechanism by which continents—with all of their bulk and drag—could move over an underlying mantle that was solid enough in composition to be able to reflect seismic S- waves.

History of Wegener's theory

At one time—estimated to be 200 to 300 million years ago—continents were united in one supercontinent or protocontinent named Pangaea (or Pangea, from the Greek pan, meaning all, and gaea, meaning world) that first split into two halves. The two halves of the protocontinent were the northern **continent** Laurasia and the southern continent named Gondwanaland or Gondwana. These two pieces were separated by the Tethys Sea. Laurasia later subdivided into **North America**, Eurasia (excluding India), and Greenland. Gondwana is believed to have included **Antarctica**, **Australia**, Africa, South America, and India. Two scientists, Edward Suess and Alexander Du Toit, named Pangaea, Gondwanaland, and Laurasia.

In Wegener's 1915 book, *The Origin of Continents and Oceans,* he cited the evidence that Pangaea had existed; most of the evidence came from Gondawana, the southern half of the supercontinent, and included the following: glacially gouged **rocks** in southern Africa, South America, and India; the fit of the coastlines and undersea shelves of the continents, especially eastern South America into western Africa and eastern North America into northwestern Africa; fossils in South America that match fossils in Africa and Australia; mountain ranges that start in Argentina and continue into South Africa and Australia; and other mountain ranges like the Appalachians that begin in North America and trend into **Europe**. He even measured Greenland's **distance** from Europe over many years to show that the two are drifting slowly apart.

Although Wegener's ideas are compelling today, scientists for decades dismissed the Continental Drift theory because Wegener could not satisfactorily explain how the continents moved. His assertion that continents plowed through oceanic rock riding **tides** in the **earth** like an icebreaker through sea ice brought derision from the world's geophysicists (scientists who study the physical properties of Earth including movements of its crust). Harold Jeffreys, a leading British geophysicist of the 1920s, calculated that, if continents did ride these Earth tides, mountain ranges would collapse and Earth would stop spinning within a year.

Wegener's fossil arguments were countered by a widely-held belief that defunct land bridges (now sunken below **sea level**) once connected current continents. These bridges had allowed the small fossil **reptiles** Ly-strosaurus and Mesosaurus (discovered on opposite sides of the Atlantic) to roam freely across what is now an **ocean** too wide to swim. The cooling and shrinking of Earth since its formation supposedly caused the **flooding** of the bridges. Furthermore, critics explained that Wegener's fossil **plant** evidence from both sides of the Atlantic resulted from wind-blown **seeds** and plant-dispersing ocean **currents**.

Measurements of Greenland's movements proved too imprecise for the equipment available to Wegener at that time. The fit of continents could be dismissed as coincidence or by a counter theory claiming that Earth is expanding. Like shapes drawn on an expanding balloon, the continents move farther from each other as Earth grows.

Evidence of the theory

Technological improvements after World War II supported many of Wegener's ideas about continental drift. New methods of dating and drilling for rock samples, especially from deep-sea drilling ships like the Glomar Challenger, have allowed more precise matching of Pangaea's rocks and fossils. Data from magnetometers (instruments that measure the **magnetism** of the **iron** in sea floor rocks) proved that the sea floors have spread since Pangaea's breakup. Even satellites have clocked continental movement.

Geologists assume that, for the 100 million years that Pangaea existed, the climatic zones were the same as those operating today: cold at the poles, temperate to desert-like at the mid-latitudes, and tropical at the equator. The rocks and fossils deposited in the early days of Pangaea show that the equator crossed a clockwise-tilted North America from roughly southern California through the mid-Atlantic United States and into Northwestern Africa. Geological and archaeological evidence from the Sahara **desert** indicate the remains of a tropical world beneath the sands. Rock layers in southern Utah point to a warm sea that gradually altered into a large sandy desert as the west coast of the North American section of Pangaea slid north into a more arid latitude. Global climates changed as Pangaea rotated and slowly broke up over those 100 million years.

Meanwhile, dinosaurs, **mammals**, and other organisms evolved as they mingled across the connected Earth for millions of years, responding to the changing climates created by the shifting landmass. Fossil dinosaurs unearthed in Antarctica proved that it was connected to the rest of Pangaea, and **dinosaur** discoveries in the Sahara desert indicate that the last connection between Laurasia (the northern half of Pangaea) and Gondwana was severed as recently as 130 million years ago.

All these discoveries also helped to develop the companion theory of plate **tectonics**, in which moving plates (sections of Earth's outer shell or crust) constantly smash into, split from, and grind past each other. Wegener's theory of Continental Drift predated the theory of plate tectonics. He dealt only with drifting continents, not knowing that the ocean floor drifts as well. Parts of his Continental Drift theory proved wrong, such as his argument that continental movement would cause the average height of land to rise, and parts proved correct. The Continental Drift theory and plate tectonics, although demonstrating many interrelated ideas, are not synonymous.

Formation of Pangaea

With improved technology, geologists have taken the Continental Drift theory back in time to 1,100 million years ago (Precambrian **geologic time**) when another supercontinent had existed long before Pangaea. This supercontinent named Rodinia split into the two half-continents that moved far apart to the north and south extremes of the **planet**. About 514 million years ago (in the late Cambrian), Laurasia (what is now Eurasia, Greenland, and North America) drifted from the north until 425 million years ago when it crashed into Gondwana (also called Gonwanaland and composed of South America, Africa, Australia, Antarctica, India, and New Zealand). By 356 million years ago (the Carboniferous period), Pangaea had formed. The C-shaped Pangaea, united along Mexico to Spain/Algeria, was separated in the middle by the Tethys Sea, an ancient sea to the east, whose remnants include the Mediterranean, Black, Caspian, and Aral Seas. Panthalassa, a superocean ("All Ocean"), covered the side of the globe opposite the one protocontinent.

Pangaea splits

During the formation of Pangaea, the collision of North America and northwestern Africa uplifted a mountain range 621 mi (1,000 km) long and as tall as the Himalayas, the much-eroded roots of which can still be traced from Louisiana to Scandinavia. The Appalachians are remnants of these **mountains**, the tallest of which centered over today's Atlantic Coastal Plain and over the North American **continental shelf**. Pangaea's crushing closure shortened the eastern margin of North America by at least 161 mi (260 km).

The internal crunching continued after the formation of Pangaea, but most of the colliding shifted from the east coast of North America to the western edge of the continent. Pangaea drifted northward before it began to break up, and it plowed into the Panthalassa ocean floor

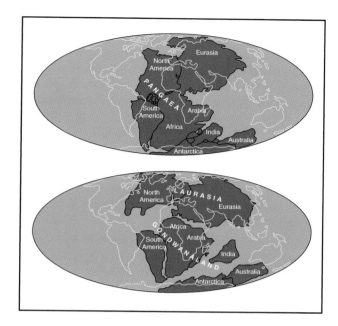

The Pangaea supercontinent (top) and its break-up into Laurasia and Gondwanaland (bottom). Contemporary continental outlines are shown in gray. *Illustration by Hans & Cassidy. Courtesy of Gale Group.*

and created some of today's western mountains. The rifting of Pangaea that began 200 million years ago (the end of the Triassic period) forced up most of the mountain ranges from Alaska to southern Chile as North and South America ground west into and over more ocean floor.

The tearing apart of Pangaea produced long valleys that ran roughly **parallel** to the east coasts of the Americas and the west coasts of Europe and Africa. The floors of these valleys dropped down tremendously in elevation. One of the valleys filled with seawater and became the Atlantic Ocean, which still grows larger every year. Other valleys gradually choked with sediment eroded off the ridges on either side. Today, many of these former low spots lie buried beneath thousands of feet of debris. The formation of the valleys did not occur along the same line as the collision 100 million years before; a chunk of Gondwana now sits below the eastern United States.

Pangaea began to break up around 200 million years ago with the separation of Laurasia from Gondwana, and the continents we know began to take shape in the late Jurassic about 152 million years ago. New oceans began to open up 94 million years ago (the Cretaceous period). India also began to separate from Antarctica, and Australia moved away from the still united South America and Africa. The Atlantic zippered open northward for the next few tens of millions of years until Greenland eventually tore from northern Eu-

KEY TERMS

. . . .

Continental Drift theory—Alfred Wegener's theory that all the continents once formed a giant continent called Pangaea and later shifted to their present positions.

Gondwana (Gondwanaland)—The southern half of Pangaea that included today's South America, Africa, India, Australia, and Antarctica.

Laurasia—The northern half of Pangaea that included today's North America, Greenland, Europe, and Asia.

Pangaea (Pangea)—The supercontinent from approximately 200–300 million years ago, which was composed of all today's landmasses.

Panthalassa—The ocean covering the opposite site of the globe from Pangaea. Panthalassa means "All Ocean."

Wegener, Alfred—German meteorologist (1880–1930), developer of the Continental Drift theory.

Resources

Books

Hancock, P.L. and B.J. Skinner, eds. *The Oxford Companion to the Earth.* New York: Oxford University Press, 2000.

Tarbuck, Edward. D., Frederick K. Lutgens, and Tasa Dennis. *Earth: An Introduction to Physical Geology.* 7th ed. Upper Saddle River, NJ: Prentice Hall, 2002.

Winchester, Simon. *The Map That Changed the World: William Smith and the Birth of Modern Geology.* New York: Harper Collins, 2001.

Periodicals

Buffett, Bruce A., "Earth's Core and the Geodynamo." *Science* (June 16, 2000): 2007–2012.

Hellfrich, George, and Wood, Bernard. "The Earth's Mantle." *Nature* (August 2, 2001): 501–507.

Other

United States Department of the Interior, U.S.Geological Survey. "This Dynamic Earth: The Story of Plate Tectonics." February 21, 2002 (cited February 5, 2003). <http://pubs.usgs.gov/publications/text/dynamic.html>.

Ed Fox
K. Lee Lerner

rope. By about 65 million years ago, all the present continents and oceans had formed and were sliding toward their current locations while India drifted north to smack the south side of **Asia**.

Current Pangaea research no longer focuses on whether or not it existed, but refines the matching of the continental margins. Studies also center on parts of Earth's crust that are most active to understand both past and future movements along plate boundaries (including earthquakes and volcanic activity) and continuing Continental Drift. For example, the East African Rift Valley is a plate boundary that is opening like a pair of scissors, and, between Africa and Antarctica, new ocean floor is being created, so the two African plates are shifting further from Antarctica. Eventually, 250 to 300 million years in the future, experts theorize that Pangaea will unite all the continents again. In another aspect of the study of Continental Drift, scientists are also trying to better understand the relatively sudden continental shift that occurred 500 million years ago that led to the formation of Pangaea. The distribution of the land mass over the spinning globe may have caused continents to relocate comparatively rapidly and may also have stimulated extraordinary evolutionary changes that produced new and diverse forms of life in the Cambrian period, also about 500 million years ago.

See also Earth's interior; Planetary geology.

Continental margin

The continental margin is that portion of the **ocean** that separates the continents from the deep ocean floor. For purposes of study, the continental margin is usually subdivided into three major sections: the **continental shelf**, the continental slope, and the continental rise. In addition to these sections, one of the most important features of the continental margin is the presence of very large submarine canyons that cut their way through the continental slope and, less commonly, the continental shelf.

Continental shelf

The continental shelf is a portion of the **continent** to which it is adjacent, and not actually part of the ocean floor. As a result of continual **earth** movement, the shelf is continuously exposed and covered by **water**. Even when covered by water, as it is today, it shows signs of once having been dry land. Fossil river beds, for example, are characteristic of some slopes. Remnants of glacial action can also be found in some regions of the continental shelf.

The continental shelf tends to be quite flat, with an average slope of less than 6.5 ft (2 m) for each mile (km) of **distance**. It varies in width from a few miles (kms) to more than 932 mi (1,500 km) with a world-wide average of about 43 mi (70 km). Some of the widest continental slopes are to be found along the northern coastline of Russia and along the western edge of the Pacific Ocean, from Alaska to **Australia**. Very narrow continental slopes are to be found along the western coastline of **South America** and the comparable coasts of west **Africa**. The average depth at which the continental shelf begins to fall off toward the ocean floor (the beginning of the continental slope) is about 440 ft (135 m).

Materials washed off the continents by **rivers** and streams gradually work their way across the continental shelf to the edge of the continental slope. In some instances, the flow of materials can be dramatically abrupt as, for example, following an **earthquake**. At the outer edge of the continental shelf, eroded materials are dumped, as it were, over the edge of the shelf onto the sea floor below.

The continental shelf is one of the best studied portions of the ocean bottom. One reason for this fact, of course, is that it is more accessible to researchers than are other parts of the sea floor. More than that, however, the waters above the continental shelf are the richest fishing grounds in the world. A number of nations have declared that their national sovereignty extends to the end of the continental shelf around their territory—often a distance of 120 mi (200 km)—to protect their marine resources.

Included among those resources are extensive mineral deposits. Many nations now have offshore wells with which they extract oil and **natural gas** from beneath the continental shelf.

The continental slope

At the seaward edge of the continental shelf, the ocean floor drops off abruptly along the continental slope. The break point between the shelf and slope is sometimes known as the continental shelf break. The continental slopes are the most dramatic cliffs on the face of the Earth. They may drop from a depth of 656 ft (200 m) to more than 9,840 ft (3,000 m) in a distance of about 62 mi (100 km). In the area of ocean trenches, the drop-off may be even more severe, from 656 ft (200 m) to more than 32,800 ft (10,000 m).

The average slope of sea floor along the continental slope is about 4°, although that value may range from as little as 1° to as much as 25°. In general, the steepest slopes tend to be found in the Pacific Ocean, and the least steep slopes in the Atlantic and Indian Oceans. Sedimentary materials carried to the continental slope from the continental shelf do not remain along the slope (be-

---

KEY TERMS

Continental rise—A region at the base of the continental slope in which eroded sediments are deposited.

Continental shelf—A relatively shallow, gently sloping, submarine area at the edges of continents and large islands, extending from the shoreline to the continental slope.

Continental shelf break—The outer edge of the continental shelf, at which the ocean floor drops off quite sharply in the continental slope.

Continental slope—A steeply-sloping stretch of the ocean that reaches from the outer edge of the continental shelf to the continental rise and deep ocean bottom.

Submarine canyon—A steep V-shaped feature cut out of the continental slope by underwater rivers known as turbidity currents.

Turbidity currents—Local, rapid-moving currents that result from water heavy with suspended sediment mixing with lighter, clearer water. Causes of turbidity currents are earthquakes or when too much sediment piles up on a steep underwater slope. They can move like avalanches.

---

cause of its steep sides), but flow downward into the next region, the continental rise.

Submarine canyons

The most distinctive features of the continental slopes are the submarine canyons. These are V-shaped features, often with tributaries, similar to canyons found on dry land. The deepest of the submarine canyons easily rivals similar landforms on the continents. The Monterrey Canyon off the coast of northern California, for example, drops from a water depth of 354 ft (108 m) below **sea level** near the coastline to 6,672 ft (2,034 m) below sea level. That vertical drop is half again as great as the depth of the Grand Canyon.

There seems little doubt that the submarine canyons, like their continental cousins, have been formed by **erosion**. But, for many years, oceanographers were puzzled as to the eroding **force** that might be responsible for formation of the submarine canyons. Today, scientists agree that canyons are produced by the flow of underwater rivers that travel across the continental slopes (and sometimes the continental shelf) carrying with them sediments that originated on the continents. These rivers are known as turbidity **currents**.

Evidence for the turbidity current theory of canyon formation was obtained in 1929 when an earthquake struck the Grand Banks region of the Atlantic Ocean off Newfoundland. An enormous turbidity current was set in **motion** that traveled at a speed ranging from 25 to 60 mph (40 to 100 km per hour), breaking a sequence of transatlantic **telegraph** cables along the way. The pattern of cable destruction was what made it possible, in fact, for scientists to track so precisely the movement of the giant turbidity current.

The continental rise

Sediments eroded off continental land, after being carried across the shelf and down the continental slope, are finally deposited at the base of the slope in a region of the ocean known as the continental rise. By some estimates, half of all the sediments laid down on the face of the **planet** are found in the continental rise.

In many regions, the continental rise looks very much like a river **delta** such as the one found at the mouth of the Mississippi River. In fact, these underwater deltas may also include a network of channels and natural levees similar to those found in the area of New Orleans. One of the most thoroughly studied sections of the continental rise is the Amazon Cone located northeast of the coast of Brazil. The Amazon Cone has a total width of about 30 mi (50 km) and a depth of about 1,000 ft (300 m). It is bisected by a primary channel that is 800 ft (250 m) deep and as much as 2 mi (3 km) wide.

See also Ocean zones.

Resources

Books

Duxbury, Alyn C., and Alison Duxbury. *An Introduction to the World's Oceans.* MA: Addison-Wesley Publishing Company, 1984.

Golden, Fred, Stephen Hart, Gina Maranto, and Bryce Walker. *How Things Work: Oceans.* Alexandria, VA: Time-Life Books, 1991.

Hancock, Paul L., Brian J. Skinner, and David L. Dineley, eds. *Oxford Companion to the Earth.* Oxford: Oxford University Press, 2001.

Skinner, Brian J., and Stephen C. Porter. *The Dynamic Earth: An Introduction to Physical Geology.* 4th ed. John Wiley & Sons, 2000.

Thurman, Harold V., and Alan P. Trujillo. *Essentials of Oceanography.* 7th ed. Englewood Cliffs, NJ: Prentice Hall, 2001.

Woodhead, James A. *Geology.* Boston: Salem Press, 1999.

David E. Newton

Continental rise *see* **Continental margin**

Continental shelf

The continental shelf is a gently sloping and relatively flat extension of a **continent** that is covered by the oceans. Seaward, the shelf ends abruptly at the shelf break, the boundary that separates the shelf from the continental slope.

The shelf occupies only 7% of the total **ocean** floor. The average slope of the shelf is about 10 ft per mi (1.9 m per km). That is, for every one kilometer of **distance**, the shelf drops 10 ft (1.9 m) in elevation until the shelf break is reached. The average depth of the shelf break is 440 ft (135 m). The greatest depth is found off **Antarctica** [1,150 ft (350 m)], where the great weight of the **ice** on the Antarctic continent pushes the crust downward. The average width of the shelf is 43 mi (70 km) and varies from tens of meters to approximately 800 mi (1,300 km) depending on location. The widest shelves are in the Arctic Ocean off the northern coasts of Siberia and **North America**. Some of the narrowest shelves are found off the tectonically active western coasts of North and **South America**.

The shelf's gentle slope and relatively flat terrain are the result of **erosion** and sediment deposition during the periodic fall and rise of the sea over the shelf in the last 1.6 million years. The changes in **sea level** were caused by the advance and retreat of **glaciers** on land over the same **time** period. During the last glacial period (approximately 18,000 years ago), sea level was 300-400 ft (90–120 m) lower than present and the shoreline was much farther offshore, exposing the shelf to the atmosphere. During lowered sea level, land plants and animals, including humans and their ancestors, lived on the shelf. Their remains are often found at the bottom of the ocean. For example, 12,000 year old bones of mastodons, extinct relatives of the **elephant**, have been recovered off the coast of the northeastern United States.

Continental shelves contain valuable resources, such as oil and gas and **minerals**. Oil and gas are formed from organic material that accumulates on the continental shelf. Over time the material is buried and transformed to oil and gas by **heat** and **pressure**. The oil and gas moves upward and is concentrated beneath geologic traps. Oil and gas is found on the continental shelf off the coasts of California and Louisiana, for example. Minerals come from **rocks** on land and are carried to the ocean by **rivers**. The minerals were deposited in river channels and beaches on the exposed continental shelf and sorted (concentrated) by waves and river **currents**, due to their different densities. Over time as the sea level rose, these minerals were again sorted by waves and ocean currents and finally deposited. The different colored bands of **sand** that one can see on a beach are an example of **density** sorting by waves. The concentrated minerals are often in sufficient enough quantities to be minable. Examples of important minerals on the shelf are diamonds, chromite (chromium **ore**), ilmenite (**titanium** ore), magnetite (**iron** ore), platinum, and gold.

Continental slope *see* **Continental margin**

Continuity

Continuity expresses the property of being uninterrupted. Intuitively, a continuous line or **function** is one that can be graphed without having to lift the pencil from the **paper**; there are no missing points, no skipped segments and no disconnections. This intuitive notion of continuity goes back to ancient Greece, where many mathematicians and philosophers believed that reality was a reflection of number. Thus, they thought, since numbers are infinitely divisible, **space** and **time** must also be infinitely divisible. In the fifth century B.C., however, the Greek mathematician Zeno pointed out that a number of logical inconsistencies arise when assuming that space is infinitely divisible, and stated his findings in the form of paradoxes. For example, in one paradox Zeno argued that the infinite divisibility of space actually meant that all **motion** was impossible. His argument went approximately as follows: before reaching any destination a traveler must first complete one-half of his journey, and before completing one-half he must complete one-fourth, and before completing one-fourth he must complete one-eighth, and so on indefinitely. Any trip requires an infinite number of steps, so ultimately, Zeno argued, no journey could ever begin, and all motion was impossible. Zeno's paradoxes had a disturbing effect on Greek mathematicians, and the ultimate resolution of his paradoxes did not occur until the intuitive notion of continuity was finally dealt with logically.

The continuity of space or time, considered by Zeno and others, is represented in **mathematics** by the continuity of points on a line. As late as the seventeenth century, mathematicians continued to believe, as the ancient Greeks had, that this continuity of points was a simple result of **density**, meaning that between any two points, no matter how close together, there is always another. This is true, for example, of the rational numbers. However, the rational numbers do not form a continuum, since irrational numbers like $\sqrt{2}$ are missing, leaving holes or discontinuities. The irrational numbers are required to complete the continuum. Together, the rational and irrational numbers do form a continuous set, the set

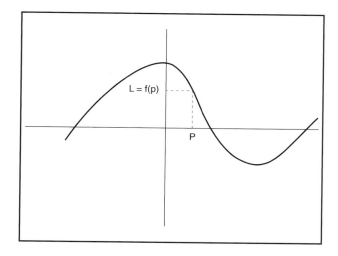

Figure 1. *Graph by Hans & Cassidy. Gale Group.*

of **real numbers**. Thus, the continuity of points on a line is ultimately linked to the continuity of the set of real numbers, by establishing a **one-to-one correspondence** between the two. This approach to continuity was first established in the 1820s, by Augustin-Louis Cauchy, who finally began to solve the problem of handling continuity logically. In Cauchy's view, any line corresponding to the graph of a function is continuous at a **point**, if the value of the function at x, denoted by f(x), gets arbitrarily close to f(p), when x gets close to a real number p. If f(x) is continuous for all real numbers x contained in a finite **interval**, then the function is continuous in that interval. If f(x) is continuous for every real number x, then the function is continuous everywhere.

Cauchy's definition of continuity is essentially the one we use today, though somewhat more refined versions were developed in the 1850s, and later in the nineteenth century. For example, the concept of continuity is often described in **relation** to limits. The condition for a function to be continuous, is equivalent to the requirement that the **limit** of the function at the point p be equal to f(p), that is:

$$\lim_{x \to p} f(x) = f(p).$$

In this version, there are two conditions that must be met for a function to be continuous at a point. First, the limit must exist at the point in question, and, second, it must be numerically equal to the value of the function at that point. For instance, polynomial functions are continuous everywhere, because the value of the function f(x) approaches f(p) smoothly, as x gets close to p, for all values of p.

However, a polynomial function with a single point redefined is not continuous at the point x = p if the limit

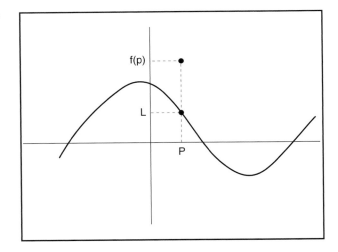

Figure 2. *Graph by Hans & Cassidy. Gale Group.*

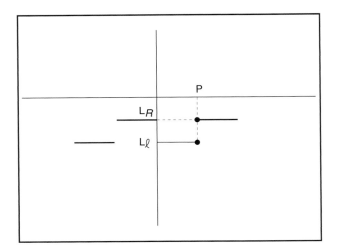

Figure 3. *Graph by Hans & Cassidy. Gale Group.*

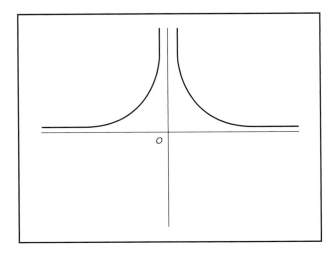

Figure 4. *Graph by Hans & Cassidy. Gale Group.*

KEY TERMS

. . . .

Function—A set of ordered pairs, defined by a rule, or mathematical statement, describing the relationship between the first element of each pair, called the independent variable, and the second element of each pair, called the dependent variable or value of the function.

Interval—An interval is a subset of the real numbers corresponding to a line segment of finite length, and including all the real numbers between its end points. An interval is closed if the endpoints are included and open if they are not.

Limit—The limit (L) of a function f(x) is defined for any point p to be the value that f(x) approaches when x gets infinitely close to p. If the value of the function becomes infinite, the limit does not exist.

of the function as x approaches p is L, and not f(p). This is a somewhat artificial example, but it makes the point that when the limit of f(x) as x approaches p is not f(p) then the function is not continuous at x = p. More realistic examples of discontinuous functions include the **square** wave, which illustrates the existence of right and left hand limits that differ; and functions with infinite discontinuities, that is, with limits that do not exist.

These examples serve to illustrate the close connection between the limiting value of a function at a point, and continuity at a point.

There are two important properties of continuous functions. First, if a function is continuous in a closed interval, then the function has a maximum value and a minimum value in that interval. Since continuity implies that f(x) cannot be infinite for any x in the interval, the function must have both a maximum and a minimum value, though the two values may be equal. Second, the fact that there can be no holes in a continuous **curve** implies that a function, continuous on a closed interval [a,b], takes on every value between f(a) and f(b) at least once. The concept of continuity is central to isolating points for which the **derivative** of a function does not exist. The derivative of a function is equal to the slope of the tangent to the graph of the function. For some functions it is not possible to draw a unique tangent at a particular point on the graph, such as any endpoint of a step function segment. When this is the case, it is not possible to determine the value of the derivative at that point. Today, the meaning of continuity is settled within the mathematics community, though it continues to present problems for philosophers and physicists.

Resources

Books

Allen, G.D., C. Chui, and B. Perry. *Elements of Calculus.* 2nd ed. Pacific Grove, CA: Brooks/Cole Publishing Co, 1989.

Boyer, Carl B. *A History of Mathematics.* 2nd ed. Revised by Uta C. Merzbach. New York: John Wiley and Sons, 1991.

Larson, Ron. *Calculus With Analytic Geometry.* Boston: Houghton Mifflin College, 2002.

Paulos, John Allen. *Beyond Numeracy, Ruminations of a Numbers Man.* New York: Alfred A. Knopf, 1991.

Silverman, Richard A. *Essential Calculus With Applications.* New York: Dover, 1989.

Periodicals

McLaughlin, William I. "Resolving Zeno's Paradoxes." *Scientific American* 271 (1994): 84-89.

J. R. Maddocks

Contour plowing

One of the earliest methods of **conservation** tillage came to be known as contour plowing, or "plowing on the contour." Tilling the **soil** along the gentle slopes of a piece of cropland, instead of up and down the gradient, prevents fertile topsoil from being carried downhill by flowing rainwater. This preventive measure is most important in areas which are prone to violent storms or heavy rains. Not only is the topsoil kept in place, **minerals** like **salt** or additives such as **fertilizers**, **insecticides** or weed control agents, as well as **bacteria** from **animal** waste are not swept away to pollute bodies of potable **water**.

In Thomas Jefferson's time, contour plowing was called more simply "horizontal plowing." Jefferson had won a coveted medal from the major agricultural society in France for his design of the moldboard plow, but he began to notice drawbacks to the heavy use of that instrument. One of his relatives, a politically active farmer named Thomas Mann Randolph, was inspired to develop a new plowing technique in order to salvage the hilly areas in Virginia. Instead of funneling water down, like shingles on the roof of a house, it caught the rain in little ridges of upturned **earth**. Jefferson commented on a noticeable improvement, specifying that the horizontal furrows retained surplus rainwater and allowed it to evaporate back into the soil.

Even after this successful experiment, later versions of the moldboard plow caused damage to the delicate top-

Contour rice farming in Arkansas. *Photograph by Paul Logsdon. Phototake NYC. Reproduced by permission.*

soil of the great plains and prairies of the Midwest United States. The most dramatic evidence of soil **erosion** took the form of huge dust storms and crop failures during the Great Depression. Since then, contour plowing and other forms of conservation tillage have been reinstituted.

Drawbacks to contour plowing have caused it to be less widely used than conventional tillage methods. Some farmers may not have been fully aware of erosion damage and prevention. Lack of access to equipment, funding, or training sometimes take their toll. One of the main limitations of contour plowing results from its contribution of pockets of untilled land. These untended spots eventually develop weeds, which require extra **herbicides**. Killing off the weeds sometimes destroys surrounding **grasses**, which in turn leaves another opportunity for rainwater runoff to arise. To combat this possibility, contour plowing is often applied in combination with other **soil conservation** techniques, such as **terracing**.

Contraception

Efforts to prevent pregnancy have been attempted since ancient times and in many cultures. Contraception methods ranged from the use of tampons treated with herbal spermicide by the Egyptians in 1550 B.C. to the use of **animal membrane** condoms in the eighteenth century. The introduction of the oral contraceptive pill in 1960 launched a new era, making contraception easier and more effective than earlier methods. However, sterilization remains the method used most frequently.

In the United States, about 64% of women between 15 and 44 years of age used contraception in 1995, a total of about 60 million. Worldwide, contraceptive use increased 10-fold from 1963 to 1993. However, contraception remains controversial, with some religious and political groups opposed to the distribution of contraceptives.

An ancient interest

A survey of early contraceptive methods reflects an odd combination of human knowledge and ignorance. Some methods sound absurd, such as the suggestion by the ancient Greek Dioscorides that wearing of cat testicles or asparagus would inhibit contraception. Yet some early methods used techniques still practiced today.

The Egyptian contraceptive tampon, described in the Ebers Papyrus of 1550 B.C., was made of lint and soaked in honey and tips from the acacia shrub. The acacia shrub contains gum arabic, the substance from which **lactic acid** is made, a spermicidal agent used in modern contraceptive jellies and creams.

Aristotle was one of many ancient Greeks to write about contraception. He advised women to use olive oil or cedar oil in the vagina, a method which helps inhibit contraception by slowing the movement of sperm. Other Greeks recommended the untrue contention that **obesity** was linked to reduced fertility.

Roman **birth** control practices varied from the use of woolen tampons to sterilization, which was typically performed on slaves. Another common ancient practice, still in use today, was the prolonged nursing of infants which makes conception less likely although still possible.

Ancient Asian cultures drew from a wide range of birth control methods. Women in China and Japan used bamboo **tissue paper** discs which had been oiled as barriers to the cervix. These were precursors of the modern diaphragm contraceptive device. The Chinese believed that **behavior** played a role in fertility, and that women who were passive during sex would not become pregnant. They suggested that women practice a total passivity beginning as early as 1100 B.C. In addition, they suggested that men practice intercourse without ejaculating.

The Chinese were not alone in promoting contraceptive methods based on men's ejaculation practices. The practice of withdrawal of the man's penis before ejaculation during intercourse, also known as coitus interruptus, has been called the most common contraceptive method in the world. While it was believed that coitus interruptus prevented conception, pre-ejaculatory fluid can contain sufficient sperm to cause pregnancy. A 1995 study found that 19% of women who depended on withdrawal became pregnant accidentally within their first year of using the method.

Magical potions were also used extensively throughout the world as contraceptives, including a wide range of herbal and vegetable preparations. Some may have actually caused abortions.

A controversial practice

Respectable physicians advocated contraceptive methods in ancient Greek and Roman society. However, by the Middle Ages, contraception had become controversial, in large part due to opposition by the Church. Early Christians were not outspoken about contraception. The first clear statement about sin and contraception was made in the fifth century by Saint Augustine who, with others, wrote that contraception was a mortal sin, a pronouncement that continues to resonate in modern culture. Since the fifth century, the Catholic Church has retained its opposition to all forms of birth control except abstinence and the so-called rhythm method, a calendar-based method involving timely abstinence.

As Christian influence took precedence during the Medieval period, contraceptive knowledge was suppressed. One measure of the primitive level of this knowledge was the writing of Albert the Great (1193–1280), a Dominican bishop, whose writing about sciences included contraceptive recipes. To avoid conception, people were advised to wear body parts of a dead fetus around the neck or to drink a man's urine.

Many scholars suggest that couples continued to practice contraception during the Middle Ages, in spite of the limited level of official contraceptive knowledge. Even when religious authorities condemned contraception, women passed their knowledge of such practices to one another.

In addition, other religious and cultural traditions maintained support for certain types of contraception during the Middle Ages. Most Jewish authorities supported women's use of contraceptive devices. Islamic physicians were not limited by the Christian opposition to birth control, and medical writings from the Middle Ages included a wide range of contraceptive information. The Koran, the holy book for Muslims, supported the use of prolonged nursing, and did not oppose other methods. While European Christians condemned contraception, the practice continued in other countries and cultures.

Evolution of the condom

Prior to the modern era, many of the most effective contraceptives evolved, rather than appearing suddenly as the result of an invention. The development and evolution of the condom is an example of a device that was present for hundreds of years, changing in function and manufacture to fit the times and needs of its users.

Contemporary condoms are used widely for contraception and to prevent the spread of sexually transmitted **disease**. Initially, condoms were developed for other reasons. Among the earliest wearers were the ancient Egyptians, who wore them as protection against *Schistosoma,* a type of parasite spread through **water**. Condoms were also worn as decoration or signs of rank in various cultures.

Condoms emerged as weapons against sexually transmitted disease in Renaissance **Europe** of the sixteenth century, when epidemics of a virulent form of syphilis swept through Europe. Gabriele Fallopio (1523-1562), an Italian anatomist who discovered the Fallopian tube, advised men to use a linen condom to protect against venereal disease.

By the eighteenth century, condoms were made of animal membrane. This made them waterproof and more effective as birth control devices. Condoms acquired a host of nicknames, including the English riding coat, instru-

ments of safety, and prophylactics. The great lover Casanova (1725–1798) described his use of condoms "to save the fair sex from anxiety." The **Industrial Revolution** transformed the condom once again. In 1837, condom manufacturers took advantage of the successful **vulcanization** of rubber, a process in which **sulfur** and raw latex were combined at a high **temperature**. This enabled manufacturers to make a cheaper yet durable product.

In the 1960s and 1970s, many women turned to modern medical contraceptives such as IUDs and birth control pills. However, by the 1980s, condoms experienced another resurgence due to the emergence of acquired immune deficiency syndrome (**AIDS**), and the discovery that condoms were most effective in preventing its transmission.

Currently, condoms—along with sterilization—are among the most common contraceptive used by Americans. In 1995, that number was 20.4%, or approximately 60 million, up from 13.2%, or 57.9 million, in 1988. A total of 12% of women who used condoms for contraception experienced accidental pregnancy in the first year of use in 1994.

Modern times

For centuries, limited knowledge of women's **physiology** slowed the development of effective contraceptives. There was no understanding of the accurate relationship between menstruation and ovulation until the early twentieth century. Yet contraceptive developers did make progress in the nineteenth century.

One major area was in updating the vaginal pessary. The rubber diaphragm and the rubber cervical cap, developed in the nineteenth century, are still in use today. The diaphragm is a disc-shaped object inserted into the vagina designed to prevent the passage of sperm while the cervical cap fits over the cervix.

Spermicides, substances developed to kill sperm, were mass produced by the late 1880s for use alone or for greater effectiveness with other devices such as the diaphragm. Vaginal sponges were also developed for contraceptive use in the late 1800s. Another popular nineteenth century method was douching, the use of a substance in the vagina following intercourse to remove sperm.

Contemporary use of these methods yields varying pregnancy rates. The diaphragm was used by 1.9% of American women who used contraceptives in 1995, down from 5% in 1988. In 1994, a total of 18% of women who used diaphragms with spermicide experienced accidental pregnancy in their first year of use. Women who depended on spermicides—1.3% of women using contraceptives—experienced a 21% accidental pregnancy **rate** in their first year of use. The cervical cap had an accidental pregnancy rate of 36% in 1995 among women who had previously given birth. The sponge, which had a first-year accidental pregnancy rate of 18% in 1990 among women who had never given birth, was take off the market in the mid-1990s.

While types of birth control increased in the nineteenth century, the topic of contraception was still considered sordid and unsuitable for public discourse. In the United States, the Comstock Law of 1873 declared all contraceptive devices obscene. The law prevented the mailing, interstate transportation, and importation of contraceptive devices. One effect of this was to eliminate contraceptive information from medical journals sent through the mail.

The social movement to make birth control legal and available challenged the Comstock Law and other restrictions against contraception. By the 1930s, the movement led by Margaret Sanger (1883-1966) had successfully challenged the Comstock Law, and the mailing and transportation of contraceptive devices was no longer illegal. Sanger was also instrumental in developing clinics to distribute birth control devices.

Advances in medical knowledge generated new contraceptive methods. The first intrauterine device (IUD), designed to be placed in the uterus, was described in 1909. The IUD was not used widely in the United States until the 1960s when new models were introduced. The **copper** IUD and the IUD with progesterone made the IUD more effective. A 1994 study found the typical accidental pregnancy rate in the first year of use was 3%.

The IUD works by causing a local inflammatory reaction within the uterus causing an increase in leukocytes, white **blood** cells, in the area. The product which results when the leukocytes break down is deadly to spermatozoa cells, greatly reducing the risk of pregnancy. The IUD devices available in the United States must be inserted by a health provider and typically must be replaced after one to four years. Possible dangers include bleeding, perforation of the uterus, and **infection**.

Use of the IUD fell from 2.0% in 1988 to 0.08% in 1995. One reason for this may be fear of lawsuits due to complications. In the United States, government officials pulled the Dalkon Shield IUD off the market in 1974, following reports of pelvic infections and other problems in women using the device. A second explanation may stem from the decision by two major IUD manufacturers to pull back from the U.S. market in the 1980s.

Another method which emerged in the early twentieth century was the rhythm method, which was approved for use by Catholics by Pope Pius XII in 1951. For centuries, various experts on birth control had speculated that certain periods during a woman's cycle were more fertile than others. But they often were wrong. For example, Soranos, a Greek who practiced medicine in second century Rome, believed a woman's fertile period occurred during her menstrual period.

As researchers learned more about female reproductive physiology in the early twentieth century, they learned that ovulation usually takes place about 14 days before a woman's next menstrual period. They also learned that an egg could only be fertilized within 24 hours of ovulation. The so-called calendar rhythm method calculates "safe" and "unsafe" days based on a woman's average **menstrual cycle**, and calls for abstinence during her fertile period. The method is limited by the difficulty of abstinence for many couples and the irregularity of menstrual cycles.

Several contemporary methods of natural contraception still used. Together, they are referred to as Periodic Abstinence and Fertility Awareness Methods, or natural family planning techniques. They include the rhythm method; the basal body temperature method, which requires the woman to take her temperature daily as temperature varies depending on time of ovulation; the cervical mucus method, which tracks the ovulation cycle based on the way a woman's cervical mucus looks; the symptothermal method, which combines all three; and the post-ovulation method, where abstinence or a barrier is used from the beginning of the period until the morning of the fourth day after predicted ovulation—approximately half the menstrual cycle. Accidental pregnancy rates for these methods were 20% in the first year of use. A total of 2.3% Americans used these methods in 1995.

As birth control became more acceptable in the twentieth century, major controversies grew about its social use. A series of mixed court decisions considered whether it is right to force an **individual** who is mentally deficient to be sterilized. In the 1970s, national controversy erupted over evidence that low-income women and girls had been sterilized under the federal Medicaid program. Federal regulations were added to prohibit the forced sterilization of women under the Medicaid program. Legal debates still continue on the issue of whether certain individuals, such as convicted child abusers, should be required to use contraceptives.

The pill and its offspring

The development of oral contraceptives has been credited with helping to launch the sexual revolution of the 1960s. Whether oral contraceptives, also known as "the pill," should take credit for broadening sexual activity or not, their development changed the contraceptive world dramatically. In 1988, oral contraceptives were the

Variety of contraceptive methods. *Photograph by Michael Keller. The Stock Market. Reproduced by permission.*

most popular reversible contraceptive in the United States, with 30.7% of all women who used birth control using them, second only to sterilization. That rate dropped to 26.9% by 1995. The accidental pregnancy rate among women using oral contraceptives is less than 3%.

The development of oral contraceptives incorporated great advances in basic scientific knowledge. These included the finding in 1919 that transplanted **hormones** made female animals infertile, and the isolation in 1923 of estrogen, the female sex hormones.

For years, the knowledge that hormones could make animals infertile could not be applied to humans because of the expense of obtaining naturally-occurring estrogen. Until chemist Russell Marker developed a technique for making estrogen from **plant** steroids in 1936, estrogen had to be obtained from animal ovaries. Scientists needed ovaries taken from 80,000 sows to manufacture a "fraction of a gram" of estrogen.

Once synthetic hormones were available, the creation of oral contraceptives was limited by a lack of interest in the development of new birth control devices among drug companies and other conventional funding sources. Gregory Pincus (1903–1967), who developed the oral contraceptive, obtained only limited funding from a drug company for his research. The bulk of his funding was from Katherine McCormick, a philanthropist, suffragist and Massachusetts Institute of Technology graduate who was committed to broadening birth control options.

The birth control pill, approved for use in the United States in 1960, uses steroids to alter the basic reproductive cycle in women. Pincus knew that steroids could interrupt the cyclic release of a woman's eggs during ovulation. Most pills use a combination of synthetic estrogen and progestin, although some only contain progestin. The steady levels of estrogen and progestin, obtained through daily oral contraceptive doses, prevent the release from the hypothalamus of gonadotrophin, a hormone which triggers ovulation. The pill also changes the cervical mucus so it is thicker and more difficult for sperm to penetrate.

Oral contraceptives can cause weight gain, nausea and headaches. In addition, women who smoke and are over 35 are advised not to take oral contraceptives due to risk of **stroke**. Oral contraceptives slightly increase the risk of cervical **cancer**, but they decrease the risk of endometrial and ovarian cancers.

Other contraceptives have drawn from oral contraceptive technology, including several which work for a long period of time and do not require daily doses of hormones. Norplant, approved for use in the United States in 1991, is a hormone-based contraceptive which is surgically implanted in the arm and which lasts approximately five years. It distributes a steady dose of the hormone progestin, which inhibits ovulation and alters cervical mucus to reduce movement of the sperm. The implant is highly effective. Of the 1.3% users in 1995, the accidental pregnancy rate was less than 0.95%. However, the side affects include excess bleeding and discomfort, which sometimes force removal of the device, and difficulty in removing the implants.

Several long-term contraceptives are injectable and also use progestin to inhibit ovulation. The most widely used is Depo-Medroxyprogesterone Acetate, also known as Depo-Provera or DMPA. The method is used in more than 90 countries but not widely in the United States. The drug, which is given every three months, is popular internationally, with as many as 3.5 million users worldwide. Fewer than 0.3% of women taking DMPA get pregnant accidentally. Most women who take DMPA for long periods of time stop having a regular menstrual cycle. The drug also causes temporary **infertility** after use.

Permanent contraception

Sterilization, the surgical alteration of a male or female to prevent them from bearing children, is a popular option. While sterilization can be reversed in some cases, it is not always possible, and should be considered permanent. In 1995, 38.6% of all contraceptive users aged 15 to 44 used sterilization, a slight decrease from the 1988 rate of 39%, which was an increase from 34% in 1982. Among married contraceptive users, 49% used sterilization in 1988, up from 42% in 1982 and 23.5% in 1973. Internationally, sterilization is also extremely common, with 70 million women and men sterilized in China alone, according to the Population Council.

Sterilization of women is more popular than sterilization of men, even though the male operation, called vasectomy, is simpler and takes less time than the female

operation, called tubal ligation. In 1995, a total of 27.7% U.S. women ages 15 to 44 used sterilization as their birth control method, compared to 10.9% of men.

Tubal ligation, which takes about an hour, calls for sealing the tubes that carry eggs to the uterus. The incision to reach the oviducts can either be made conventionally or by using laparoscopy, a technique which uses fiber optic **light** sources to enable surgeons to operate without making a large incision. Women continue to ovulate eggs following a tubal ligation, but the ovum is blocked from passage through the fallopian tube and sperm can not reach the ovum to fertilize it. The ovum eventually degenerates.

Vasectomy, which takes about 20 minutes, involves cutting the vas deferens to prevent sperm from reaching the semen. While semen is still produced, it no longer carries spermatozoa. Vasectomy and tubal ligation are considered to be safe procedures with few complications.

Challenges of contraception

Birth control policies and practices are controversial in the developed and the developing worlds. In developed countries, such as the United States, contraceptive methods fail frequently. Many of the types of contraceptives used commonly by Americans have well-documented rates of failure. One measure of the number of unwanted pregnancies is the rate of abortion, the surgical termination of pregnancy. Abortion and the controversial antigestation drug, RU 486 (Roussel-Uclaf), are not considered routine birth control methods in the United States. Although all individuals who receive abortions do not practice birth control, it is clear that many women do become pregnant when contraceptive methods fail.

Abortion rates typically are highest in countries where contraceptives are difficult to obtain. For example, in the Soviet Union in the early 1980s, when contraceptives were scarce, 181 abortions were performed annually for every 1,000 women aged 15 to 44; in 1990, 109 for every 1,000 women; and 1992, 98 per 1,000. In comparison, in the 1980s in selected western European countries, the rate did not exceed 20 per 1,000.

The abortion rate in the United States is typically higher than in many other developed countries. A 1995 survey showed the annual abortion rate was 20 per 1,000 women aged 15 to 44 (a total of 1,210,883 legal abortions), a decrease of 4.5% from 1994's rate of 28 per 1,000. The annual numbers have been decreasing since 1987, and 1995 was the lowest recorded since 1975. However, the rate in Great Britain in 1990 was less than half that of the U.S. at 13 per 1,000, and in the Netherlands, 5.6 per 1,000. A study of 20 western democracies found that countries with lower abortion rates tended to have contraceptive care accessible through primary care physicians.

Some experts believe that more access to contraceptive services would result in lower rates of accidental pregnancy and abortion. However, vigorous debate concerning programs to deliver contraceptives through school-based clinics and in other public settings have polarized the United States. Groups such as the Roman Catholic Church have opposed funding for greater accessibility of contraceptive services because they believe the use of any contraceptives is wrong.

Internationally, use of contraceptives has increased dramatically from the years 1960–1965, when 9% of married couples used contraceptives in the developing countries of Latin America, **Asia**, and **Africa**. By 1990, over 50% of couples in these countries used contraceptives.

China has taken an aggressive policy to limit population growth, which some experts have deemed coercive. Couples who agree to have one child and no more receive benefits ranging from increased income and better housing to better future employment and educational opportunities for the child. In addition, the Chinese must pay a fine to the government for each "extra" child.

Numerous problems exist which prevent the great demand for contraceptive services in developing countries from being met, due in part to the general difficulty of providing medical care to poor people. In addition, some services, such as sterilization, remain too expensive to offer to all those who could use them. Experts call for more money and creativity to be applied to the problem in order to avoid a massive population increase.

Future contraceptive methods

The high cost in time and money of developing new contraceptive methods in the United States creates a barrier to the creation of new methods. In the early 1990s, a new contraceptive device could take as long as 17 years and up to $70 million to develop. Yet new methods of contraception are being explored. One device in clinical trials is a biodegradable progestin implant which would last from 12 to 18 months. The device is similar to Norplant but dissolves on its own. A device being explored by a Dutch pharmaceutical company is a ring that rests against the uterus releasing low dosages of estrogen and progesterone. The ring would remain in place for an extended period. In May, 1998, a new oral contraceptive for women—the first to use a shortened "hormone-free interval" and lower daily doses of estrogen—was approved by the FDA.

Other research focuses on male contraceptive methods. In 1996, the World Health Organization hailed a contraceptive injection of testosterone that drastically reduces the sperm count and which is 99% effective. This contraceptive will not be available for five to 10 years. Two other studies in male birth control are ongoing—one focuses on preventing sperm from breaking the egg's gel-like protective coating; the other on blocking protein receptors on the sperm so it cannot "dock" with the egg.

Western controversy over contraception continues. There is still disagreement concerning how widely contraception should be made available and how much public money should be spent on birth control. The conclusion of a report from the Institute of Medicine released in May 1996 entitled *Contraceptive Research and Development: Looking to the Future*, reads, "despite the undeniable richness of the science that could be marshalled to give the women and men of the world a broader, safer, more effective array of options for implementing decisions about contraception, childbearing, and prevention of sexually transmitted disease, dilemmas remain. These dilemmas have to do with laws and regulations, politics and ideology, economics and individual behavior, all interacting in a very complex synergy that could lead to the conclusion that nothing can be done to resolve the dilemmas because everything needs to be done."

See also Fertilization; Sexual reproduction.

Resources

Books

Aubeny, E., H. Meden-Vrtovec, and C. Capdevila, eds. *Contraception in the Third Millennium: A (R)evolution in Reproductive and Sexual Health.* New York: Parthenon Publishers, 2001.

Connell, Elizabeth, and David A. Grimes. *The Contraception Sourcebook.* New York: McGraw-Hill/Contemporary Books, 2001.

Darney, Philip, et al. *A Clinical Guide for Contraception.* 3rd ed. Philadelphia: Lippincott, Williams & Wilkins Publishers, 2001.

Gordon, Linda. "Woman's Body, Woman's Right." *A Social History of Birth Control in America.* New York: Grossman Publishers, 1976.

Guillebaud, John. *Contraception Today: A Pocketbook for General Practitioners.* 4th ed. Dunitz Martin, 2000.

Kass-Annese, Barbara. *Natural Birth Control Made Simple.* 7th ed. Alameda, CA: Hunter House, 2003.

Sanger, Margaret. Esther Katz, Cathy Moran Hajo, and Peter C. Engelman, eds. *The Selected Papers of Margaret Sanger: The Woman Rebel, 1900-1928.* Urbana, IL: University of Illinois Press, 2001.

Periodicals

Drife, J.O. "Contraceptive Problems in the Developed World." *British Medical Bulletin* 49, no. 1 (1993): 17–26.

Lincoln, D.W. "Contraception for the Year 2020." *British Medical Journal* 49, no. 1 (1993): 222–236.

Schenker, Joseph G., and Vicki Rabenou. "Family Planning: Cultural and Religious Perspectives." *Human Reproduction* 8, no. 6 (1993): 969–976.

KEY TERMS

. . . .

Estrogen—A hormone present in both males and females. It is present in much larger quantities in females, however, and is responsible for many of those physical characteristics which appear during female sexual maturation. It is used in birth control pills, to reduce menopausal discomfort, and in osteoporosis.

Hormone—Chemical regulator of physiology, growth, or development which is typically synthesized in one region of the body and active in another and is typically active in low concentrations.

Ovary—Female sex gland in which ova, eggs used in reproduction, are generated.

Progestin—Synthetic form of progesterone, the hormone which prepares the uterus for development of the fertilized egg.

Sperm—Substance secreted by the testes during sexual intercourse. Sperm includes spermatozoon, the mature male cell which is propelled by a tail and has the ability to fertilize the female egg.

Steroids—A group of organic compounds that belong to the lipid family and that include many important biochemical compounds including the sex hormones, certain vitamins, and cholesterol.

Uterus—Organ in female mammals in which embryo and fetus grow to maturity.

Waites, G.M.H. "Male Fertility Regulation: The Challenges for the Year 2000." *British Medical Journal* 49, no. 1 (1993): 210-221.

Other

Baylor college of Medicine. "Contraception On-line." (cited April 2003). <http://www.contraceptiononline.org/>

Patricia Braus

Convection

Convection is the vertical transfer of **mass**, **heat**, or other properties in a fluid or substance that undergoes fluid-like dynamics. Convection takes place in the atmosphere, in the oceans, and in Earth's molten subcrustal **asthenosphere**. Convective currents of air in the atmosphere are referred to as updrafts and downdrafts.

In addition to **heat transfer**, convention can be driven by other properties (e.g., salinity, **density**, etc).

Thermal convection is one of the major forces in atmospheric dynamics and greatly contributes to, and directly influences, the development of **clouds** and **storm** systems. Convective air currents of rising warm and moist air allow a transfer of sensible and latent heat **energy** from the surface to the upper atmosphere.

One meteorological hypothesis, the convection theory of cyclones, asserts that convection resulting from very high levels of surface heating can be so strong that the current of air can attain cyclonic velocities and **rotation**.

The **temperature** differences in **water** cause **ocean** currents that vertically mix masses of water at different temperatures. In the atmosphere, convection drives the vertical transport of air both upward and downward. In both cases, convection acts toward equilibrium and the lowest energy state by allowing the properties of the differential air or water masses to mix.

Convection within Earth's mantle results from differential temperatures in mantle materials. In part, these differences can manifest as hot spots or convective currents where less dense and warmer mantle materials form slow moving vertical currents in the plastic (viscous or thick fluid-like) mantle. Phase change differences in materials also change their density and buoyancy. Convective currents in the mantle move slowly (at a maximum, inches per year), but may last millions of years.

Convection in Earth's mantle drives **motion** of the lithospheric plates. This convection is, in part, caused by temperature differences caused by the **radioactive decay** of the naturally radioactive elements, **uranium**, thorium, potassium.

See also Atmosphere, composition and structure; Atmospheric circulation; Earth's interior; Ocean zones; Solar illumination: Seasonal and diurnal patterns; Weather.

Coordinate covalent bond *see*
Chemical bond

Coordinate geometry *see* **Analytic geometry**

Coordination compound

A coordination compound is formed when groups of **atoms**, ions, or molecules chemically bond with each other by donating and accepting pairs of electrons. Groups donating **electron** pairs are called ligands. They are usually Lewis bases. Groups accepting electron pairs are often transition **metal** cations.

TABLE 1.
TYPICAL TRANSITION METAL CATIONS, M_T^{n+}

Iron(II)	Fe^{2+}	Iron(III)	Fe^{3+}
Cobalt(II)	Co^{2+}	Cobalt(III)	Co^{3+}
Nickel(II)	Ni^{2+}		
Copper(II)	Cu^{2+}		

In this context Iron(II) is read or spoken as 1, iron two. Cobalt(III) is referred to as "cobalt three," and so on.

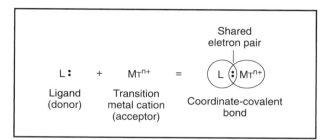

Figure 1. Formation and representations of a coordinate-covalent bond. *Illustration by Hans & Cassidy. Courtesy of Gale Group.*

TABLE 2. TYPICAL LIGANDS, L

	Name As Ligand	Ordinary Name
Electrically Neutral	:NH_3 ammine	ammonia
	:OH_2 aqua	water
	:C=O carbonyl	carbon monoxide
Negatively Charged	:Cl:- chloro	chloride
	:CN- cyano	cyanide
	:NCS:- thiocyanato	thiocyanate

TABLE 3: EXAMPLES OF HIGHLY COLORED COORDINATION COMPOUNDS

$[Co(NH_3)_6]^{3+}$	$[Fe^{II}(CN)_6]^{3-}$	$[Cu(NH_3)_4]^{2+}$	$[Fe(H_2O)_5(SCN)]^{2+}$
Yellow-Orange	Deep Blue	Deep Blue	Blood-Red
	Coordination unit found in blue print ink	Used to identify copper as Cu^{2+} ions	Used to identify iron as Fe^{3+} ions

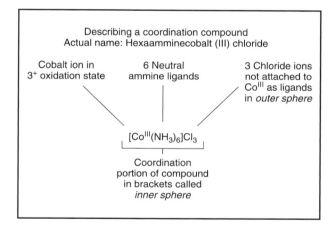

Figure 2. *Illustration by Hans & Cassidy. Courtesy of Gale Group.*

They are usually Lewis acids. Chemical bonds formed in this way are called coordinate-covalent, or dative bonds. As in any covalent bond, two electrons are shared between transition metal and **ligand**. But in a coordination compound, both electrons come from a pair found on the ligand (Figure 1).

The metal **cation** simply acts as the electron pair acceptor, itself donating no electrons to the bond. Because of the complicated nature of these arrangements, coordination compounds are often called coordination complexes or simply complexes.

It is most common to find six ligands coordinated to a single metal cation. The coordination number is then six. Think of ligands as **bees** swarming about and stinging a victim. The ligand-bee's stinger is its lone pair or non-bonding pair of electrons. These special ligand-bees attack their victim in groups of six, although ligand groups of four and other numbers do occur. Six coordination produces the shape of an eight-sided figure or octahedron. Four coordination produces one of two shapes, a flat **square** planar or a four-sided **tetrahedron**.

While nearly all cations can form coordination compounds with ligands, those listed in Table 1 are especially common.

Ligands come in all shapes and sizes, though they are usually non-metals from the right side of the **periodic table**. Those listed in Table 2 are typical.

The Swiss chemist Alfred Werner (1866-1919) is called the Father of Coordination Chemistry for his work in clarifying how coordination compounds are assembled. Figure 2 names and explains most of the parts.

Several theories help explain the nature of coordination compounds.

Effective atomic **number theory** matches the total number of electrons of the transition metal cation plus donated pairs of electrons from the ligands with the stable electron count of a noble gas atom. In $[Co^{III}(NH_3)_6]^{3+}$ above the central ion Co^{3+} contains 24 electrons. (A neutral cobalt atom has 27 electrons.) Each ammine ligand donates 2 electrons for a total of 12 electrons coming from the six ligands.

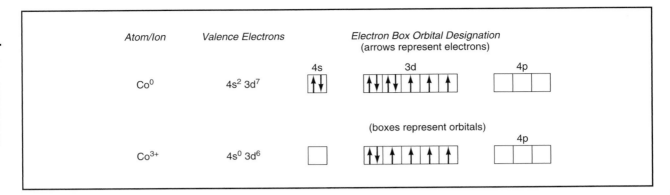

Figure 3. Electron boxes representing orbitals. *Illustrations by Hans & Cassidy. Courtesy of Gale Group.*

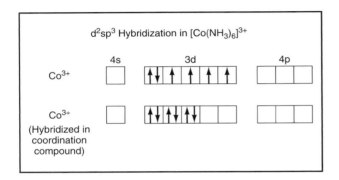

Figure 4. Six empty orbitals for six ligands. *Illustration by Hans & Cassidy. Courtesy of Gale Group.*

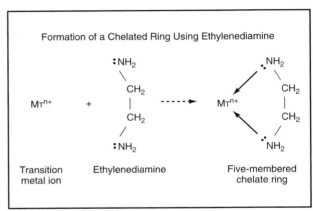

Figure 5. *Illustration by Hans & Cassidy. Courtesy of Gale Group.*

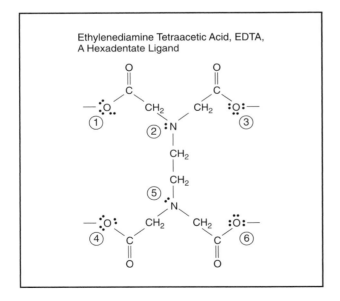

Figure 6. *Illustration by Hans & Cassidy. Courtesy of Gale Group.*

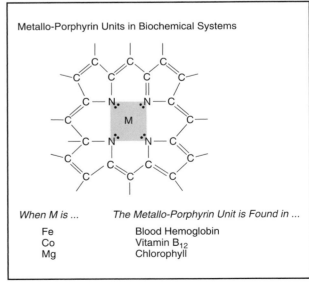

Figure 7. *Illustration by Hans & Cassidy. Courtesy of Gale Group.*

$$Co^{3+} \quad + \quad 6NH_3 \quad = \quad [Co\,(NH_3)_6]^{3+} \quad = \quad Krypton$$

24 + 6 x 2 = 36 = 36
electrons   electrons   electrons   electrons

If the total number of electrons associated with a co-ordination compound is the same as a noble gas, the compound will frequently have increased stability.

Valence Bond Theory locates and creates empty orbitals on the transition metal cation. The empty orbitals will accept electron pairs from the ligands. A neutral cobalt atom contains 27 electrons: $1s^2\,2s^2\,2p^6\,3s^2\,3p^6\,4s^2\,3d^7$. The 4s, 3d, and empty 4p orbitals are the valence orbitals containing the metal's valence electrons (Figure 3).

When forming cations, the 4s electrons are lost first, ahead of the 3d electrons.

When Co^{3+} forms a coordination compound with six electron pair donating ammine ligands, it must have six empty receiving orbitals. But Co^{3+} only shows four empty orbitals, the one 4s and the three 4p. However, if two electrons in the 3d orbitals of Co^{3+} move over and pair up, two additional empty 3d orbitals are created. Co^{3+} now has the required six empty orbitals, two 3ds, one 4s, and three 4ps (Figure 4).

This process is called hybridization, specifically $d^2s^1p^3$ or simply d^2sp^3 here.

An important magnetic phenomenon is associated with the Valence Bond Theory. Whenever at least one unpaired electron is present, that substance will be very weakly drawn towards a powerful magnetic. This is know as paramagnetism. Co^{3+} by itself has four unpaired electrons and is paramagnetic. But Co^{3+} in $[Co(NH_3)_6]^{3+}$ has no unpaired electrons. Most substances in nature do not have unpaired electrons and are said to be diamagnetic. These substances are very weakly repelled by a powerful magnetic. It is possible to measure paramagnetic and diamagnetic effects in a chemical laboratory.

Theories such as the Valence Bond Theory can thus be tested experimentally.

Crystal Field Theory is yet another approach to coordination compounds. It treats ligands as negatively charged anions, attracted to the positively charged transition metal cation.

This is what happens when a chloride **anion**, Cl^-, is attracted to the **sodium** ion, Na^+, in forming the ionic, crystalline compound, table **salt**, Na^+Cl^-. A crystal "field" is thus the electronic "field," produced in any ionic compound. Positive charges are attracted to **negative** charges and vice versa.

According to crystal field theory, when six negative ligands surround a transition metal cation, the energies of the metal's 3d electron orbitals are split up. Some 3d orbitals are stabilized and some are energized. The amount of stabilization **energy** predicted by the theory correlates well with that observed in laboratory experiments.

Many coordination compounds have vivid colors.

Crystal field theory predicts that these colors are due to metal 3d electrons jumping from stabilized to energized orbitals when visible **light** or daylight shines on them.

The bright colors of coordination compounds make them good candidates for dyes, paint pigments, and coloring agents of all sorts.

A very special application of coordination compounds occurs in what is called the **chelate** effect. Chelation takes place when ligands "bite" or "bee sting" the metal in more than one place at a time. Using dental "biting" terminology, if a ligand has two "teeth" to bite a transition metal, it is called a bidentate (two teeth) ligand. One example of this is called ethylenediamine, $NH_2\,CH_2$ $CH_2\,NH_2$. Figure 5 shows simultaneous donation of two electrons each from the two **nitrogen** atoms of ethylenediamine. A five membered ring is produced involving the metal, two nitrogen atoms, and two **carbon** atoms.

A hexadentate ligand such as ethylenediamine tetraacetic acid, EDTA, bites a metal cation in six places simultaneously by wrapping itself around the cation (Figure 6). This produces an extremely strong chelated compound. Using EDTA, such ions as **calcium**, Ca^{2+}, can be extracted from and tested in drinking **water**.

Chelated transition metal ions are also found in a wide variety of biochemical situations. A basic structural unit called metalloporphyrin is shown in Figure 7. It can be thought of as several ethylenediamine-like units fused and blended together into a single tetradentate (4 teeth) chelating ligand. Changing the central metal, M, changes the biochemical activity of the chelated coordination compound.

Research in this area is opening up a new field of chemistry called bio-inorganic chemistry.

See also Chemical bond.

Copepods

Copepods are pale or translucent crustaceans, measuring between 0.04 mm to several millimeters long. They have adapted to many different habitats; while they usually live in **salt water**, copepods can live in lakes and ponds as well. Furthermore, they have different modes of locomotion: some can swim purposefully but others are planktonic, floating with the current. Scientists generally

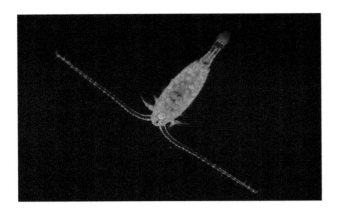

A copepod (*Diaptomus* sp.). © M.I. Walker/Science Photo Library, National Audubon Society Collection/Photo Researchers, Inc. Reproduced by permission.

distinguish between two basic forms of copepods, free-living and parasitic.

The phylum Arthropoda is the largest phylum in the **animal** kingdom, containing more than one million **species**. Within this phylum, the subphylum **Crustacea** contains some 35,000 species and can be broken down into eight classes. Copepods belong to the class Maxillopoda and the subclass Copepoda, containing seven orders and more than 7,500 species. Three of these orders—Calanoida, Cyclopoida, and Harpacticoida—are primarily free-living and are present in huge numbers. The other orders are: Misophrioida, Monstrilloida, Siphonostomatoida, and Poecilostomatoida.

Characteristics of free-living copepods

Given the incredible number of species, the physical structure of copepods varies greatly. However, the free-living forms of copepods have certain physical traits in common. For instance, the body is usually short and cylindrical, composed of a head, thorax, and abdomen. The lower part of the copepod's head is generally fused with its thorax; the front of its head often juts forward, like a tiny beak. Its thorax is divided into about six segments; each segment is connected to two appendages.

Generally, a free-living copepod has two pair of antennae and a single **eye**. The first pair of antennae is larger and has bristles. The male copepod can be distinguished from female because its antennae are slightly different from those of the female, modified for holding her during copulation. The free-living copepods' limbs are used for movement, sometimes with the help of the antennae. Its thin abdomen lacks limbs, except for the caudal furca—an appendage akin to a tail. Its tail has bristles similar to those found on its primary antennae.

Some tropical forms of copepods actually use their bristles to facilitate flotation.

The parasites

There are over 1,000 species of parasitic copepods. As larva—or nauplia—most look and act like typical copepods. It is only later, when the **parasites** reach various stages in their development, that they begin to attach themselves to a host creature and radically change in appearance. In fact, many of the adult parasitic copepods are incredibly deviant in physical structure from their free-living relatives. Indeed, it is nearly impossible to find a single trait that is common to all copepods, both free-living and parasitic. One of the only characteristics that they tend to share is that the females have egg sacs which secrete a sticky liquid when the eggs are laid, gluing them together.

Furthermore, parasitic copepods are vastly different in appearance from other crustaceans. In general, adult parasitic copepods are shapeless, having neither limbs nor antennae and sometimes no body segments. Because these creatures start their lives as free-living animals, scientists infer that their ancestors were free-living and that they only evolved parasitic **behavior** after their environments dictated it.

Parasitic copepods can inflict severe damage on their hosts. This damage is often worsened by the presence and infestation of accompanying **fungi**.

Place in the food chain

Free-living copepods form a crucial link in the food chain and are often assigned the role of "primary consumers." Although some large forms of copepods are predators, free-living copepods are generally herbivores, feeding only on **plant** plankton which they filter from the water. Specifically, they eat small plant **plankton** and are, in turn, eaten by larger animals, like herring or **mackerel**. More **fish** and other aquatic animals feed on copepods than any other kind of animal in existence. One of the dominant forms of animal plankton, some scientists estimate that there are more copepods on the **planet** than all other multicellular animals combined.

Order Calanoida

Calanoids are of major importance to the commercial fishing industry. Like other copepods, this species filters minute animal **algae** from the water and eats it in large clumps. In turn, these copepods are eaten in large numbers by fish, such as herring and **salmon**. Calanoids thrive close to the surface of large expanses of water, both seas and lakes. Anatomically, they are easy to recognize. The fused head and thorax of an individual calanoid is oval and clearly separated from its abdomen. The multisegmented first antennae are long, about the same length as its body. The abdomen is much slimmer than its thorax.

Like other animal plankton, calanoids allow themselves to float with the current, although they can use their first antennae to swim upward in the water. Unlike most other genera of copepods, calanoids move from the surface to the depths of the water each day. At dawn, they sink to several hundred feet in the water; at dusk they rise to the surface again. There are several theories explaining this activity. The most likely reason is that they are escaping the dangerous ultraviolet rays of the sun. Another theory is that they are avoiding predators. In any case, this activity facilitates their fairly even distribution in the sea, since **currents** at various depths often run in different directions.

Order Cyclopoida

All of the free-living cyclopoida are almost identical to each other in physical appearance. Their antennae are shorter than those of the calanoids, growing about half of the length of their bodies. Their bodies, relatively pair-shaped, have two clearly divided regions: the head and thorax in the front; and the last segment of the thorax fused with the abdomen in the rear. Their front portions narrow slowly into their abdomens. This order contains many marine forms and numerous **freshwater** representatives. They are rarely planktonic, rather they tend to swim near the bottom of the water, never migrating upwards. They thrive in small pools of water with large amounts of aquatic vegetation. Some of the larger species are carnivores, eating **insects** and other crustaceans.

There are also 12 or more families living in relation to other animals: either as hosts to parasites or as parasites themselves. Some freshwater species are important as temporary hosts to certain forms of worms that are parasitic to man. Other species are parasites on **mollusks**, **sea anemones**, or sea squirts. One specific group of parasitic cyclopoids live in the mouths or on the gills of certain fish, like frog-mouths. While the female can grow to 0.8 in (2 cm) long, the male never surpasses about 0.04 in (0.1 cm). The jaws of the female are shaped like sickles, enabling her to cling to her host and eat it.

Order Harpacticoida

Harpacticoid bodies, which rarely exceed 0.07 in (2 mm) in length, are not clearly divided into three distinct regions, and they vary dramatically in shape. Some harpacticoids are long and snake-like, while others are flat. Their antennae are very short and forked.

KEY TERMS

Caudal furca—An appendage on the free-living copepod, resembling a tail, that is attached to its abdomen.

Free-living copepod—The copepods that does not attach itself to a living host but, instead, feeds on algae or small forms of animal life.

Nauplia—Larva of either free-living or parasitic copepods; both kinds of larvae are similar in appearance.

Planktonic—Free-floating; not using limbs for locomotion.

Thorax—The area just below the head and neck; the chest.

Harpacticoids are planktonic, generally living in the muddy or sandy areas. Instead of swimming, these copepods hop, using the appendages on their thoraxes in combination with a rocking **motion** of their bodies. While most harpacticoids feed on organic waste or algae, some species are predators, swarming over small fish and immobilizing them by eating their fins.

Order Monstrilloida

Some of the most advanced species of parasitic copepods are found in this order. These copepods are worm parasites. Their nauplii appear quite typical, but have no stomach. When they find a suitable host, they shed their outer skeleton and all of their appendages and become a mass of cells. In this simple structure, they are able to reach the worms' body cavity. Once inside their host, these creatures form a thin outer skeleton; the copepods spend most of their lives without a mouth, intestines, or an anus. When they mature, they look like free-living copepods.

See also Zooplankton.

Resources

Books

George, David, and Jennifer George. *Marine Life: An Illustrated Encyclopedia of Invertebrates in the Sea.* New York: John Wiley and Sons, 1979.

Grzimek, H.C. Bernard, Dr., ed. *Grzimek's Animal Life Encyclopedia.* New York: Van Nostrand Reinhold Company, 1993.

The New Larousse Encyclopedia of Animal Life. New York: Bonanza Books, 1987.

Pearl, Mary Corliss, Ph.D. Consultant. *The Illustrated Encyclopedia of Wildlife.* London: Grey Castle Press, 1991.

Schmitt, Waldo L. *Crustaceans*. Ann Arbor: The University of Michigan Press, 1965.

Street, Philip. *The Crab and Its Relatives*. London: Faber and Faber Limited, 1966.

Trefil, James. *Encyclopedia of Science and Technology*. The Reference Works, Inc., 2001.

Kathryn Snavely

Copper

Copper is the metallic chemical element of **atomic number** 29, symbol Cu, **atomic weight** 63.55, specific gravity 8.96, melting point 1,985°F (1,085°C), and **boiling point** 4,645.4°F (2,563°C). It consists of two stable isotopes, of **mass** numbers 63 (69.1%) and 65 (30.9%).

Copper is one of only two metals that are colored, Copper is reddish brown, while gold is...gold—a unique **color** that is sometimes loosely described as yellow. All other metals are silvery, with various degrees of brightness or grayness. Almost everybody handles copper just about every day in the form of pennies. But because a piece of copper the size of a penny has become more valuable than one cent, today's pennies are made of zinc, with just a thin coating of copper.

Copper is in group 11 of the **periodic table**, along with silver and gold. This trio of metals is sometimes referred to as the coinage metals, because they are relatively valuable, corrosion-free and pretty, which makes them excellent for making coins. Strangely enough, the penny is the only American coin that is *not* made from a copper **alloy**. Nickels, dimes, quarters, and half dollars are all made from alloys of copper with other metals. In the case of nickels, the main **metal** is of course nickel.

Copper is one of the elements that are essential to life in tiny amounts, although larger amounts can be toxic. About 0.0004% of the weight of the human body is copper. It can be found in such foods as liver, shellfish, nuts, raisins and dried beans. Instead of the red hemoglobin in human **blood**, which has an **iron** atom in its **molecule**, **lobsters** and other large crustaceans have blue blood containing hemocyanin, which is similar to hemoglobin but contains a copper atom instead of iron.

History of copper

Copper gets its chemical symbol Cu from its Latin name, *cuprum*. It got that name from the **island** of Cyprus, the source of much of the ancient Mediterranean world's supply of copper.

But copper was used long before the Roman Empire. It is one of the earliest metals known to humans. One reason for this is that copper occurs not only as ores (compounds that must be converted to metal), but occasionally as native copper—actual metal found that way in the ground. In prehistoric times an early human could simply find a chunk of copper and hammer it into a tool with a rock. (Copper is very malleable, meaning that it can be hammered easily into various shapes, even without heating.)

Native copper was mined and used in the Tigris-Euphrates valley (modern Iraq) as long as 7,000 years ago. Copper ores have been mined for at least 5000 years because it is fairly easy to get the copper out of them. For example, if a copper oxide **ore** (CuO) is heated in a **wood** fire, the **carbon** in the charcoal can reduce the oxide to metal:

$$2CuO \ + \ C \ \rightarrow \ 2Cu \ + \ CO_2$$

| copper | charcoal | | copper | carbon |
| ore | | | metal | dioxide |

Making pure copper

Extremely pure copper (greater than 99.95%), called electrolytic copper, can be made by **electrolysis**. The high purity is needed because most copper is used to make electrical equipment, and small amounts of impurity metals in copper can seriously reduce its ability to conduct **electricity**. Even 0.05% of arsenic impurity in copper, for example, will reduce its conductivity by 15%. Electric wires must therefore be made of very pure copper, especially if the electricity is to be carried for many miles through high-voltage transmission lines.

To purify copper electrolytically, the impure copper metal is made the **anode** (the positive electrode) in an electrolytic cell. A thin sheet of previously purified copper is used as the **cathode** (the **negative** electrode). The **electrolyte** (the current-carrying liquid in between the electrodes) is a **solution** of copper sulfate and **sulfuric acid**. When current is passed through the cell, positively charged copper ions (Cu^{2+}) are pulled out of the anode into the liquid, and are attracted to the negative cathode, where they lose their positive charges and stick tightly as neutral **atoms** of pure copper metal. As the electrolysis goes on, the impure copper anode dissolves away and pure copper builds up as a thicker and thicker coating on the cathode. Positive ions of impurity metals such as iron, nickel, arsenic and zinc also leave the anode and go into the solution, but they remain in the liquid because the voltage is purposely kept too low to neutralize them at the cathode. Other impurities, such as platinum, silver and gold, are also released from the anode, but they are not soluble in the solution and simply fall to the bottom, where they are collected as a very valuable sludge. In fact, the silver and gold sludge is usually valuable enough to pay for the large amount of electricity that the electrolytic process uses.

Uses of copper

By far the most important use of copper is in electrical wiring; it is an excellent conductor of electricity (second only to silver), it can be made extremely pure, it corrodes very slowly, and it can be formed easily into thin wires—it is very ductile.

Copper is also an important ingredient of many useful alloys—combinations of metals, melted together. Brass is copper plus zinc. If it contains mostly copper, it is a golden yellow color; if it is mostly zinc, it is pale yellow or silvery. Brass is one of the most useful of all alloys; it can be cast or machined into everything from candle sticks to cheap, gold-imitating jewelry that turns your skin green. (When copper reacts with **salt** and acids in the skin, it produces green copper chloride and other compounds.) Several other copper alloys are common: bronze is mainly copper plus tin; German silver and sterling silver are silver plus copper; silver tooth fillings contain about 12% copper.

Probably the first alloy ever to be made and used by humans was bronze. Archaeologists broadly divide human history into three periods; the Bronze Age is the second one, after the Stone Age and before the Iron Age. During the Bronze Age, both bronze and pure copper were used for making tools and weapons.

Because it resists **corrosion** and conducts **heat** well, copper is widely used in plumbing and heating applications. Copper pipes and tubing are used to distribute hot and cold **water** through houses and other buildings.

Because copper is an extremely good conductor of heat, as well as of electricity (the two usually go together), it is used to make cooking utensils such as saute and fry pans. An even **temperature** across the pan bottom is important for cooking, so the food doesn't burn or stick to hot spots. The insides of the pans must be coated with tin, however, because too much copper in our food is toxic.

Copper corrodes only slowly in moist air—much more slowly than iron rusts. First it darkens in color because of a thin layer of black copper oxide, CuO. Then as the years goes by it forms a bluish green patina of basic copper carbonate, with a composition usually given as $Cu_2(OH)_2CO_3$. (The carbon comes from **carbon dioxide** in the air.) This is the cause of the green color of the Statue of Liberty, which is made of 300 thick copper plates bolted together. Without traveling to New York you can see this color on the copper roofs of old buildings such as churches and city halls.

Compounds of copper

In its compounds, copper can have a **valence** of either +1 (cuprous compounds) or +2 (cupric compounds). Cuprous compounds are not stable in water, and when dissolved they turn into a mixture of cupric ions and metallic copper.

Copper compounds and **minerals** are often green or blue. The most common minerals include malachite, a bright green carbonate, and azurite, a blue-green basic carbonate. Among the major copper ores are cuprite, CuO, chalcopyrite, $CuFeS_2$, and bornite, Cu_5FeS_4. Large deposits of copper ores are found in the United States, Canada, Chile, central **Africa**, and Russia.

Cupric sulfate, $CuSO_4 \cdot 5H_2O$, is also called blue vitriol. These poisonous blue crystals are used to kill **algae** in the purification of water, and as an agricultural dust or spray for getting rid of **insects** and **fungi**.

See also Electric conductor.

Resources

Books

Braungart, Michael, and William McDonough. *Cradle to Cradle: Remaking the Way We Make Things.* New York: North Point Press, 2002.

Emsley, John. *Nature's Building Blocks: An A-Z Guide to the Elements.* Oxford: Oxford University Press, 2002.

Greenwood, N. N., and A. Earnshaw. *Chemistry of the Elements.* Oxford: Butterworth-Heinneman Press, 1997.

Kirk-Othmer Encyclopedia of Chemical Technology. 4th ed. Suppl. New York: John Wiley & Sons, 1998.

Parker, Sybil P., ed. *McGraw-Hill Encyclopedia of Chemistry.* 2nd ed. New York: McGraw-Hill, 1993.

Robert L. Wolke

Coral and coral reef

Coral reefs are highly diverse ecosystems, supporting greater numbers of **fish** species and other organisms than any other marine **ecosystem**. Coral reefs are located in warm, shallow, tropical marine waters with enough **light** to stimulate the growth of the reef organisms. The primary reef-building organisms are invertebrate animals known as corals; corals secrete the bulk of the **calcium carbonate** (limestone) that makes up the inorganic reef structure, along with material deposited by coralline **algae**, **mollusks**, and **sponges**.

The builders: corals and coralline algae

Corals are small (0.06–0.5 in; 1.5–12 mm), colonial, marine **invertebrates**. They belong to the class Anthozoa, phylum Cnidaria (or Coelenterata). Corals are subdivided into (1) stony corals (reef-building or hermatypic)—order Scleractinia, subclass Hexacorallia—which have six tentacles, and (2) soft corals, sea fans, and sea whips—order Gorgonacea, subclass Octocorallia—with eight tentacles.

The limestone substrate of stony coral colonies develops because each individual **animal**, or polyp, secretes a hard, cup-like skeleton of **calcium** carbonate (limestone) around itself as a protection against predators and **storm** waves. These limestone skeletons, or corallites, make up the majority of the reef framework. Certain coral **species** produce distinctively shaped colonies, while others exhibit various shapes. Some species, such as staghorn coral, are intricately branched, and are sometimes called coral stands. **Brain** corals are almost spherical in outline and are often called coral heads; they often display surface convolutions reminiscent of those on a human brain.

Calcareous red, or coralline, algae also contribute to the framework of reefs by secreting their own encrusting skeleton that acts as cement, stabilizing loose sediment on the reef. Coralline algae often produce as much of a reef's limestone as do the stony corals. Other calcareous organisms that contribute reef sediments include sponges, bryozoans (another colonial animal), tube worms, clams, and **snails**.

Biology of corals

Adult corals are benthic (bottom-dwelling), sessile (attached) animals usually found in single-species colonies. These colonies may house hundreds or thousands of polyps. The polyps are joined to one another by a thin **tissue** layer called the coenosarc (pronounced SEE-na-sark). The coenosarc connects the entire coral colony and covers the underlying coral skeleton. Reproduction through an asexual budding process results in development of duplicate daughter polyps and allows for growth of the colony. A single polyp can develop into a massive coral head through multiple budding episodes. Corals also reproduce sexually, producing multitudes of planktonic larvae that **ocean currents** disperse widely. This allows colonization of suitable habitats, resulting in development of new colonies and new reefs.

Single-celled dinoflagellate algae known as zooxanthellae live symbiotically within coral polyps. Chemical exchanges occur between the coral polyps and zooxan-

thellae, and both thrive in a mutually beneficial relationship (**mutualism**). The zooxanthellae, which are essentially tiny green plants that can produce food from sunlight, **water**, and dissolved **minerals**, supply some coral species with more than 90% of their **nutrition** on sunny days. In exchange for **nutrients**, the coral polyps supply a **habitat** and essential minerals to the algae. Another result of this relationship is more rapid development of coral reefs. During **photosynthesis**, the zooxanthellae remove **carbon dioxide** from the water, which promotes calcium carbonate production, in turn allowing the coral to more easily secrete its home.

In addition to the food provided by their zooxanthellae, corals prey on tiny planktonic organisms. Some corals paralyze their **prey** using stinging cells, or nematocysts, located on their tentacles. Other corals feed by creating weak water currents with cilia to draw food into their mouth, or by producing sticky mucus with which to trap tiny planktonic animals. Most species feed at night; during the day, they retract into their corallites for protection. The members of the colony share nutrients by passing them to their neighbors through the coenosarc.

Coral reef distribution

Estimates of the total ocean floor area covered by coral reefs vary considerably for two reasons: difficulties in estimation due to their submarine location, and differences in deciding what constitutes a coral reef. A conservative estimate would be 235,000 sq mi (597,000 sq km)—only 0.1% of Earth's surface—for reef areas at depths less than 100 ft (30 m). Coral reefs occur in shallow, warm-water locations, primarily below 30° latitude in the western Atlantic and Indo-Pacific regions. Their distribution is strongly influenced by the environmental preferences of the coral animals.

Environmental setting and requirements

Although corals live in nearly all marine environments, hermatypic corals thrive in a rather narrow set of environmental conditions. These limitations also restrict the geographic distribution of well-developed coral reef tracts.

Coral reefs typically occur in water depths less than 190–230 ft (60–70 m) and maximum growth rates occur at depths less than 60 ft (18 m). This is because the corals, or rather their symbiotic zooxanthellae, depend on light for growth. The algae need access to light to accomplish their photosynthesis. Too much sediment in the water also causes problems, by limiting light penetration or suffocating organisms, and thereby slowing reef

Kayangel atoll, Belau. *Photograph by Douglas Faulkner. National Audubon Society Collection/Photo Researchers, Inc. Reproduced by permission.*

growth. Consequently, the amount of light and the clarity and depth of the water are important influences on the development of coral reefs.

Corals thrive in oligotrophic water; that is, water with low concentrations of nutrients such as phosphate, ammonium, and nitrate. Currents and wave activity help supply the continuous but low concentrations of nutrients that corals and algae require for survival, while also removing waste materials.

Water **temperature** is also an important environmental influence on the growth of stony corals. Typically, a water temperature of 74–78°F (23–26°C) is most conducive to coral growth, and temperatures must generally remain above 67°F (19°C) throughout the year. As reviewed below, **global warming** is having a disastrous impact on coral reefs worldwide by causing water temperatures to transgress these narrow bounds.

Stony corals also prefer marine waters with stable salinity. The **salt** concentration of the water must range between 35 and 38 parts per thousand, and the **concentration** of **oxygen** must remain high. Another important factor is the need for continuous submersion under water, although some corals can survive temporary exposure during low tide.

Coral reef development and zonation

Reefs tend to develop a definite depth profile and associated coral zonation under the influence of constant wave activity. This results from the decrease in wave **energy** with water depth. The reef crest is the shallowest reef area and subject to the highest wave energy; here coral and algae encrust the substrate to avoid being broken and swept away. The reef crest is located at the top of the seaward reef slope and may be exposed at low tide. Waves and **tides** cut channels across the reef crest. As tides rise and fall, water moves back and forth through these channels between the open sea and the lagoon.

Wave and storm energy are important controls on the character of the reef crest. Coralline algae tend to dominate reef crests if hurricanes are frequent and average daily wave conditions are rough. Grazing fish, which would normally consume the algae, are deterred by the consistent high wave energy. In areas with infrequent storms and calmer daily wave conditions, encrusting corals or robust branching corals tend to inhabit reef crests.

Moving down the seaward reef slope, the reef front lies just below the reef crest. Corals here are diverse and show the greatest range of forms. At the top of the slope, wave energy is high and coral forms are usually encrusting to massive, such as brain corals. Further down the slope, in deeper water, massive corals dominate, then give way to delicate branching corals as wave energy decreases with depth. Finally, at the base of the reef front, plate-like corals take advantage of the low wave energy. By orienting their flat, upper surface toward the **sun**, they attain maximum exposure to the diminished light of the deep reef. Further downslope, the fore reef consists of limestone boulders, coral branches and smaller sediments, all transported from above, as well as sponges, soft corals and algae thriving in place.

Shoreward of the reef crest lies the shallow reef flat. Reef rubble occurs here in high-energy reefs. In lower energy settings, carbonate **sand** may be present. These sediments are supplied by storm waves breaking on the reef crest. Even closer to shore is the back reef area, where fine-grained sediment inhibits reef growth; however, scattered stubby, branching or low, knobby corals usually develop in water depths of 3–4 ft (1–1.3 m).

Beyond the back reef, the water begins to deepen again—to as much as 100 ft (30 m) or more—within the lagoon (i.e., water between the reef and the shore, or completely surrounded by the reef). Here the sea floor is generally protected by the reef from significant wave agitation, so fine-grained sediments compose the lagoon floor. Hardy corals occur in scattered clusters, known as patch reefs.

Coral reef morphology and geology

There are three major kinds of coral reefs: fringing reefs, barrier reefs, and atolls. Fringing reefs are located in shallow water close to shore, either with or without a shallow lagoon. Barrier reefs are also located in shallow water, but with a deep lagoon separating the reef from the shoreline and a steep reef front. Both fringing and barrier reefs form on shallow continental shelves and along **island** shorelines. Atolls are ring-shaped coral reefs centered around a deep lagoon. They are typically found in the vicinity of volcanic **seamounts** and islands located in the deep ocean.

Major controls: crustal subsidence and sea level change

In addition to the environmental requirements for coral growth described above, other factors play a role in coral reef character over long **time** intervals, that is, during **geologic time** spans. The two most important controls are both related to water depth—sea level change and crustal movement.

World-wide fluctuations in **sea level** can be caused by **volume** changes of fresh water in global reservoirs (lakes, **groundwater**, or **glaciers**) and by changes in the volume of ocean basins. If sea level rises while environmental conditions remain favorable for reef growth, coral reefs may grow upward rapidly enough to keep pace with rising sea level. If conditions are unfavorable, upward growth will be slow and light levels on the reef will slowly decrease as water depth increases, causing the reef to "drown." If sea level drops, the crest of the reef may be exposed and eroded, while deeper zones will "back-step" down the reef slope as the water depth decreases.

Crustal movements—uplift or **subsidence** of the coast—result from tectonic events such as mountain building and continental rifting or from changes in crustal loading due to volcanism, **erosion**, or deposition. **Uplift** is analogous to a sea-level drop and results in coral reefs back-stepping from the coast. Subsidence has the same effects on coral as a sea level rise, and the coral reef must either grow to keep up or else drown. A "keep up" reef grows at a **rate** sufficient to keep up with the relative sea level rise and remains at or near sea level; a "give up" reef has a growth rate that falls behind the rate of relative sea level rise, and "drowns."

Coral reefs occur in two distinct settings: oceanic settings and continental shelves. Deep water surrounds oceanic coral reefs, which generally lie hundreds of miles from continental shelves. These may be fringing reefs, barrier reefs or atolls. Charles Darwin, who started his scientific career as a geologist, developed a theory in the mid-nineteenth century about the origins of and relationships between fringing reefs, barrier reefs, and atolls in oceanic settings. Darwin visited Pacific atolls and also observed barrier and fringing reefs in the south Atlantic. He hypothesized that the first stage of development of an atoll is the creation of a fringing reef around a volcanic island. The **volcano** subsides under its own weight over millions of years, but the reef's upward growth keeps pace with subsidence and so it remains in shallow water, developing into a barrier reef. If a volcanic island becomes completely submerged, the coral reef itself may be the only thing at sea level, forming an atoll.

Darwin was essentially correct, since the primary factor determining what types of reefs are present in oceanic settings is usually an island's rate of subsidence. However this model is less applicable to shelf reefs, which usually experience less significant rates of subsidence. Continental shelves are typically fairly stable, so

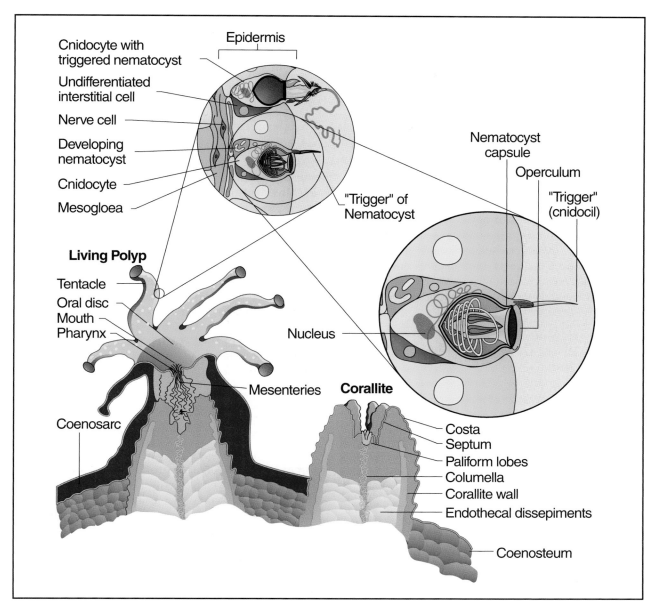

The anatomy of a hard coral (left) and the skeletal corallite that remains after the coral dies (right). *Illustration by Hans & Cassidy. Courtesy of Gale Group.*

sea-level changes tend to exert more control than subsidence on reef morphology (form). Shelf reefs occur on the margins of continents. A carbonate shelf is a broad, flat, shallow margin where conditions favor reef development. Most shelf reefs develop either on benches or banks or at the continental shelf's seaward margin.

Bench reefs form on the outer edge of a submarine erosional terrace, or bench, produced by near-shore erosion during times of lower sea level. This bench provides a substrate for reef development. Generally, bench reefs form at water depths of 35 ft (10 m) or less. Bank-barrier

reefs form on a shallow (less than 60 ft, 18 m) area of the shelf where, in the past, scattered corals trapped various coarse-grained skeletal material, forming a pile, or bank. Later, when water depths are suitable, corals use the bank as a substrate for reef development. Shelf-margin reefs are located at the outer edge of the shelf, beyond which water depths increase very rapidly.

When sea level rises or falls along a shelf, vast areas of land are flooded or exposed, respectively. If sea level drops on a shelf, bench or bank reefs cannot necessarily back-step—there may be no where to go, if the entire shelf becomes exposed. Shelf-margin reefs may back-step, however, sediments from erosion of the newly exposed shelf often overcome the reefs. Likewise, during rising sea level, water quality may be unfavorable to reefs due to coastal erosion of newly flooded land areas.

Ecology of coral reefs

Coral reefs around the world have similar plants and animals. This means that the same families and genera tend to be present, although the actual species may be different. They have the highest **biodiversity** and greatest ecological complexity of any marine ecosystem. Many coral-reef organisms have established balanced, mutualistic relationships that help sustain a great richness of species and a tremendous complexity of ecological interactions. Coral reefs develop under environmental conditions characterized by a restricted supply of nutrients, yet maintain a high rate of productivity because of their efficient use of nutrients.

Some ecologists believe that coral reefs maintain high biodiversity as a response to a stable but periodically disturbed environment that allows for the accumulation of species over time. The most competitive species are prevented from dominating the ecosystem, while small-scale disturbances maintain a shifting mosaic of relatively young habitats for colonization by less competitive species. Natural disturbances to which coral reefs must typically adapt include intense windstorms such as hurricanes, events of sediment deposition or volcanism, unusually low tides, short-term temperature extremes, and the population dynamics of reef species.

Coral-reef organisms

As an underwater environment, coral reefs offer a wide variety of habitats for plants and animals. **Phytoplankton**, benthic algae, and **bacteria** are at the base of the food web. They serve as food for the large variety of animals in the coral-reef ecosystem. If you ever visit a coral reef, you may find that greenery seems relatively scarce; however, six inconspicuous types of plants make coral reefs the marine community with the highest primary productivity. These are the calcareous and noncalcareous green algae, red algae, and brown algae; the hidden ("cryptic") zooxanthallae living in coral tissue and green filamentous algae living in the porous limestone substrate; and the microscopic phytoplankton.

In addition to the species of small **zooplankton** drifting over coral reefs, the reef water column supports **jellyfish** and other large invertebrates such as **cuttlefish**, **squid**, and **octopus**; many species of fish and marine **reptiles** (**snakes** and **turtles**); and the occasional mammal species (**seals**, porpoises, and manatees). As many as 500 species of fish may inhabit a given coral reef. Many of these fish species defend a territory, while others occur in large schools. Many benthic species occur in and on the reef as well. These include the corals themselves, **barnacles**, oysters, clams, lamp shells, and polychaete worms. All told, as many as 150,000 species may live in some coral reefs.

Coral reefs are subject to small-scale erosion caused by certain organisms that bore into the reef substrate, including bacteria, **fungi**, algae, sponges, worms, **sea urchins**, and predatory clams. Their activities create open spaces in the reef, increasing the diversity of habitats and allowing more species to inhabit the ecosystem.

Natural threats to coral reefs

Coral reefs are sometimes disturbed by natural forces, such as extreme rain events that dilute seawater, waves associated with hurricane-force winds, volcanism, earthquakes, and thermal stress from unusually warm water (such as El Niño events). These natural conditions rarely destroy entire reefs, and the ecosystem can recover over time.

The crown-of-thorns **starfish** (*Acanthaster planci*) has caused severe damage to coral reefs in the Pacific Ocean and elsewhere. These large, bottom-dwelling invertebrates feed on the corals, destroying them in the process. This damage has been well documented on the **Great Barrier Reef** of **Australia**, almost one-quarter of which was destroyed by a crown-of-thorns infestation in the 1980s. When corals are destroyed by starfish, algae and bacteria grow over the surface and inhibit the establishment of new coral.

Another threat to coral reefs, observed in early 1980's in the Caribbean and the Florida Keys, was a blight that decimated the population of spiny sea urchins. These invertebrates are important because they feed on benthic algae, preventing them from overgrowing the corals.

A phenomenon known as coral bleaching is caused when coral polyps expel their zooxanthellae so that the coral becomes pale or white. Without their algae corals become weak, and after several weeks may die. For some years marine scientists have been observing major bleaching events in many parts of the world; the bleaching event of 1998 was the worst ever observed, affecting most of the world's coral reefs simultaneously. In some cases, severe damage has been caused to the coral-reef ecosystem (e.g., 80% coral dieoff). Scientists believe that unusually warm water temperatures are responsible for these catastrophic bleaching episodes. The cause (or causes) of these unusually warm temperatures is not certainly known, but a majority of scientists believe that global warming caused by human alteration of the atmosphere (especially increases of atmospheric **carbon** dioxide [CO_2]) are responsible. Atmospheric CO_2—expected to increase to twice its natural level (i.e., its level just prior to the **Industrial Revolution**) by 2065—is also expected to harm coral reefs by changing the **chemistry** of seawater in such a way as to make calcium carbonate less available to reef-building organisms.

Marine biologists also suspect that other devastating infectious coral diseases are also becoming more common. These diseases have names such as "black band," "white plague," and "white pox." They are capable of wiping out much of the coral in an afflicted reef.

Events such as crown-of-thorns population explosions, spiny sea urchin population collapses, and coral diseases can all be considered natural events. However, in many cases, marine scientists suspect that human influences, such as **pollution**, **ozone** depletion, or global warming may ultimately be to blame.

Humans and coral reefs

Corals reefs provide extremely valuable environmental services for people, including the protection of shorelines from the full onslaught of storm-driven waves. Some of these services are of direct economic benefit to people, and they could be used on a sustained-yield basis, for example, through **ecotourism** or a controlled fishery.

Coral reefs serve as a natural laboratory where biologists are engaged in research that advances biological and ecological knowledge, while also being necessary for the wise **conservation** of coral reefs and their natural resources. Geologists study living coral reefs to better understand their ancient ancestors and because some of the world's largest oil and gas deposits are associated with **petroleum** reservoirs in ancient reefs. Geologists are interested in **learning** how these reefs entrap **fossil fuels**, and gain insight by studying modern reefs. Biochemical researchers have extracted more than 1,200 potentially useful compounds from coral-reef organisms, some of which are being used in the treatment of **AIDS**, multiple sclerosis, **cancer**, cardiovascular **disease**, **asthma**, **ulcers**, rheumatoid **arthritis**, and **pain**.

Coral reefs in decline

Regrettably, many human uses of coral reefs damage their physical and ecological structure. The most devastating direct damage is caused by **mining** of reefs to provide calcium-based material for the construction of buildings, including the manufacturing of cement.

Although fishing is a potentially sustainable activity, it is not often practiced as one. The most destructive fishing technique used in coral reef ecosystems involves the use of dynamite to stun or kill fish, which then float to the surface and are gathered. Dynamiting is extremely wasteful both of fish, many of which are not collected after they are killed, and of the coral reef ecosystem, which suffers serious physical damage from the explosions. Net fishing also physically damages reefs and depletes non-food species. Sometimes, poisons are used by divers to intoxicate fish so they can be collected by hand. This method is used both to catch fish as food for local people and for the international aquarium trade.

Coral reefs are also highly vulnerable to pollution of various kinds. Pollution by nutrients, or **eutrophication**, is most commonly associated with the discharge of sewage into the marine ecosystem as runoff from cities or coastal villages. Nutrients such as nitrate and phosphate can cause phytoplankton to become highly productive and abundant, and their **biomass** can prevent sunlight from reaching the corals in sufficient intensity to sustain their zooxanthellae. Nutrients can also cause free-living algae on the reef surface to become highly productive, and in some cases these algae smother the corals, causing further decline of the coral-reef ecosystem.

Many species of corals and their zooxanthellae are highly sensitive to toxic contaminants, such as **pesticides**, metals, and various petrochemicals. Coral reefs can easily be degraded by these chemical pollutants, for example, through agricultural runoff into the near-shore environment. Sometimes, coral reefs are severely damaged by **oil spills** from wrecked tankers, or from smaller but more frequent discharges from coastal refineries or urban runoff. Corals and their associated species can suffer damage from exposure to toxic hydrocarbons, and from the physical effects of smothering by dense, viscous residues of oil spills.

Sedimentation is a type of pollution that occurs when large quantities of fine **soil** particles erode from nearby land areas and settle out of water in near-shore marine waters, smothering coral reefs. Corals are tolerant of a certain amount of sedimentation. However, they may die if the rate of material deposition is greater than the corals can cope with through their natural cleansing mechanisms: ciliary action and outward growth of the colony. Sedimentation may also cause damage if its associated turbidity significantly reduces the amount of light available for the symbiotic zooxanthellae.

Even tourism can be extremely damaging to coral reefs. Physical damage may be caused by boat anchors, or by rarer events like ship groundings. Snorkelers swimming over or walking on shallow reefs may cause significant damage, because the less robust species of corals are easily broken. Also, some visitors like to collect pieces of coral as souvenirs, a practice that can contribute to the degradation of frequently visited reefs. Local residents may also harvest corals for sale to tourists or in the aquarium trade. Many coral reefs that support heavy recreational usage by tourists are in a relatively degraded condition, compared with more remote reefs that are less intensively visited.

The U.S. State Department estimated in 1999 that 58% of the world's reef area may be directly threatened by human activities such as coastal development and pollution. This figure does not include the possible effects of global warming and atmospheric CO_2 increases, which threaten all of the world's reefs.

Hope for the future: Coral reef management

Although coral reefs may be suffering a variety of ills, there is still hope. In the 1980s and 1990s, many countries began to realize the importance of coral reefs and to act accordingly. In response to requests from marine scientists for increased monitoring of reef condition, along with calls from environmental activists for enhanced reef conservation, several countries developed management plans for their reef areas.

In some cases, governments applied lessons learned elsewhere. For example, a small island north of Venezuela—Bonaire, in the Netherlands Antilles—established a marine park 1979. The park boundaries completely surround the island and ordinances provide some level of protection for all marine resources from the high water mark down to 190 ft (60 m). Bonaire's reefs are now some of the healthiest in the Caribbean, even though they too have been affected at times by reduced water quality, coral disease outbreaks, declining spiny sea urchin populations, and storm damage.

Recent establishment of similar management zones with enforced limitations and controls on marine resource exploitation has resulted in significant improvement in the health of some reef systems. However, it remains to be seen if more restrictive measures may be necessary. In particular, it may be imperative to improve effluent water quality in areas of high population density, such as the Florida Keys.

Global efforts to reduce CO_2 and other greenhouse-gas emissions, such as the Kyoto protocols (signed by most industrial nations, not including the United States), may provide eventual relief to coral reefs. However, because reefs are already suffering severely from global warming and atmospheric CO_2 is predicted to continue rising for some time, the outlook for continuing climatic damage to the world's coral reefs a serious concern to scientists.

KEY TERMS

Benthic—Dwelling in or on the ocean bottom.

Calcareous—Composed of calcite, or calcium carbonate ($CaCO_3$), a common mineral.

Mutualism—A mutually beneficial interaction, or symbiosis, between two different species.

Polyp—An individual animal in a coral colony.

Sessile—Unable to move about.

Zooxanthellae—The single-celled, dinoflagellate algae that live in a mutualistic symbiosis with corals.

Coral reefs are wonderfully diverse and complex ecosystems. They are also vulnerable to damage from a wide range of human influences. Stewardship of coral reefs must improve or these magnificent ecosystems and their many species will become rarer and more endangered than they already are—and many will disappear altogether.

See also Greenhouse effect; Oceanography; Pollution control.

Resources

Books

Clarke, Aurthur C. *Coast of Coral.* New York: iBooks, 2002.
Dubinsky, Z., ed. *Coral Reefs.* Ecosystems of the World, no. 25. New York: Elsevier, 1991.
Wolanski, Eric, ed. *Oceanographic Processes of Coral Reefs: Physical and Biological Links in the Great Barrier Reef.* Boca Raton, FL: CRC Press, 2000.

Periodicals

Kleypas, Joan A., et al. "Geochemical Consequences of Increased Atmospheric Carbon Dioxide on Coral Reefs." *Science* (April 2, 1999):118–120.
Pennisi, Elizabeth. "Survey Confirms Coral Reefs Are in Peril." *Science* (September 6, 2002): 1622–1623.

Other

Bureau of Oceans and International Environmental and Scientific Affairs, U.S. Department of State. "Coral Bleaching, Coral Mortality, and Global Climate Change: Report presented by Rafe Pomerance, Deputy Assistant Secretary of State for the Environment and Development, to the U.S. Coral Reef Task Force." March 5, 1999. (cited October 25, 2002). <http://www. state.gov/www/global/global_issues/ coral_reefs/990305_coralreef_rpt.html>

Clay Harris

Coral snakes *see* **Elapid snakes**

Coriander *see* **Carrot family (Apiaceae)**

Coriolis effect

The Coriolis effect is a mechanical principle demonstrating that, on a rotating solid body, an inertial **force** acts on the body at right angles to its direction of **motion**. The Coriolis effect (also called the Coriolis force) is based on the classic **laws of motion** introduced to the world by Sir Issac Newton (1642–1727). A rotating body not only moves according to Newtonian motion, but it is also acted on by an inertial force. If that body is rotating in a counterclockwise direction, the inertial force will deflect the body to its own right with respect to the observer. If the body is rotating in a clockwise motion, the inertial force acts to the left of the direction of motion.

Behavior of objects under the Coriolis effect

Within its rotating coordinate system, the object acted on by the Coriolis effect appears to deflect off of its path of motion. This deflection is not real. It only appears to happen because the coordinate system that establishes a frame of reference for the observer is also rotating. The Coriolis effect is, therefore, linked to the motion of the object, the motion of **Earth** (the rotating frame of reference), and the latitude on Earth at which the object is moving.

Several illustrations of the Coriolis effect are described below. First, imagine that a cannon on the equator is fired to the north. The cannon ball will land farther to the right than its target because the cannon ball moving on the equator moves faster to the east than its target, which started out farther to the north. If the cannon is fired from the North Pole at a target toward the equator, the cannon ball will again land to the right of its true path because the target area has moved farther to the east faster. In other words, in the Northern Hemisphere, the cannon ball will always land to the right of its target no matter where it is fired relative to the target. In the Southern Hemisphere, the effect is reversed and the cannon ball will always fall to the left of its target.

The second involves an experiment demonstrating the Coriolis effect. Imagine a **phonograph** record on a turntable. The center hole is the North Pole, and the rim of the record is the equator. As the record turns on the table, a chalk line drawn across the record from the hole to the rim toward the person drawing the line will **curve** to the right.

A third example uses a carousel or merry-go-round to illustrate the Coriolis effect. As the carousel goes around, a rider on the carousel stands at the center (the North Pole) and throws the ball to someone standing on the ground beyond the edge of the carousel. From the ground, the ball appears to travel in a straight line, but, to the rider, the ball seems to travel in a curve.

History

The Coriolis effect was first described by Gustave-Gaspard Coriolis, for whom the effect is named. Coriolis (1792–1843) was a French mathematician and engineer who graduated in highway **engineering**. He was a professor of mechanics at the École Centrale des Arts et Manufactures and later at the École des Ponts et Chaussées. Coriolis studied the motions of moving parts in machines relative to the fixed parts. His special gift was for interpreting and adapting theories to applied **mathematics** and mechanics. Coriolis was an assistant professor of mechanics and analysis and later director of studies at the École Polytechnique in Paris from 1816 to 1838 where he gave the terms "work" and "kinetic energy" their scientific and practical meanings, which are still used today.

Coriolis authored several books. His first, *Du calcul de l'effet des machines* (*On the calculation of mechanical action*) was published in 1829 and discussed applied mechanics. He also wrote *Théorie math,matique des effects du jeu de billiard* (*Mathematical theory of the game of billiards*) in 1835, and his understanding of the motions of billiard balls was directly related to his study of other solid bodies like the planets and especially **planet** Earth. In 1835, he published a **paper** called "Sur les équations du mouvement relatif des systemes des corps" ("On the equations of relative motion of systems of bodies"), which described the force later named the Coriolis effect. Publication of this paper changed the studies of **meteorology** (**weather**), **ballistics**, **oceanography**, and **astronomy** as well as mechanics. The Coriolis effect and other mechanical principles were also described in a book published in 1844 after Coriolis' death and titled *Traité de la méchanique des corps solides* (*Treatise on the mechanics of solid bodies*).

Significance of the Coriolis effect

The Coriolis effect is important to virtually all sciences that relate to Earth and planetary motions. It is critical to the dynamics of the atmosphere including the motions of winds and storms. In oceanography, it explains the motions of oceanic **currents**. Ballistics encompasses not only weapons but the motions of **aircraft** including launching and orbiting spacecraft. In the mechanics of machinery, rotating motors and other electrical devices generate instantaneous voltages (called Christoffel voltages) that must be calculated relative to the **rotation**. In astronomy, **astrophysics**, and studies of the dynamics of the stars, the Coriolis effect explains the rotation of **sunspots** and the true directions of **light** seen on Earth from the stars.

The Coriolis effect does not have any relationship to two other effects. For many years, geologists have used the Coriolis effect to suggest that right banks of **rivers** will tend to erode more rapidly than left banks in the Northern Hemisphere; this has been proven not to be true. Also, many people claim **water** in their sinks and toilet bowls drains away in counterclockwise or clockwise motion depending on whether the drain is located in the Northern or Southern Hemisphere. The Coriolis effect acts only on fluids over great distances or long lengths of **time**, so the motion of draining water is due to the shape of the drain not to the pseudoforce of the Coriolis effect.

Resources

Periodicals

Kearns, Graham. "The Great Coriolis Conspiracy." *Weatherwise* 5, no. 3 (June 1998): 63.

Other

National Oceanic and Atmospheric Administration (NOAA). *Coriolis Force.* [cited 2003]. <http://www.nws.noaa.gov./om/educ/activit/coriolis.htm>

Gillian S. Holmes

Cork

Cork is the outer, regenerative **bark** of the cork oak **tree**, *Quercus suber,* family Fagaceae. Unlike other oak **species**, the cork oak is an evergreen tree and dates from the Oligocene epoch of the Tertiary period. The oldest cork fossil, dating 10 million years old, was found in Portugal and is identical to modern cork. Today cork oak trees grow exclusively around the edge of the Mediterranean, primarily in Portugal, Spain, and Algeria, and to a lesser extent in Morocco, Tunisia, Italy, Sicily, and France.

A **cross section** of the tree trunk or large branch of *Q. suber* reveals three distinct layers: (1) the inner and largest area, the xylem (woody **tissue**), (2) a thin layer covering the xylem called the inner bark, and (3) the outer cork layer, also known as phellogen or cork-cambium. When the tree is about 10 years old, it is stripped of its outer bark (the cork) for the first time. Cork from the first stripping is called virgin cork. Strippers make vertical and horizontal cuts into the bark and lift the planks off. Care is taken not to damage the living, inner bark, which cannot replenish the cork if it is damaged. Strip-

ping takes place during spring or summer when new cork cells are developing near the inner bark, making the cork easier to strip off. Within three months after stripping, growth of the cork layer resumes. The cork layer stops growing when cold weather begins.

A cork oak is stripped about every 10 years. The second and subsequent strippings are called reproduction cork and are of better quality than virgin cork. A healthy tree can live for 150 years. An old tree with a large girth and branches can yield more than 1,000 lb (455 kg) of cork in a single harvest. Cork **oaks** aged between 35 and 45 years typically yield about 200 lb (91 kg) per year, and trees aged 50 or 60 years can yield 330 lb (150 kg) of cork.

The cork planks are seasoned for about six months in the open air. Exposure to rain, **wind**, and **sun** during seasoning cause chemical transformations in the cork, improving its quality. After the planks are seasoned, they are boiled to remove tannic acid and **resins**, to soften the cork, and to make the outermost, rough surface easier to remove. The cork planks are sorted according to quality. The highest quality cork is used for bottle stoppers. The rest of the cork is either cut into sections or granulated to make agglomerated cork. Agglomerated cork consists of small pieces (granules) of cork that are agglutinated (glued or adhered together) with either the cork oak's own resin, or with products such as rubber, asphalt, or synthetic resins, and then **heat** treated under **pressure** to solidify the composite material.

The cork consists of suberose tissue formed by the phellogen, a tissue between the inner bark and cork, from which cork develops. Suberin is a waxy, waterproof substance in the **cell** wall (suberose tissue), made of **fatty acids** and organic alcohols, and making cork impermeable to liquids and gases. Although cork is the bark of a living tree, it is actually a conglomeration of dead cells. Each cell is a multi-sided **polyhedron**, only 30–40 microns in diameter, and is filled with a gas almost identical to the normal atmosphere. One cubic centimeter of cork contains about 40 million cells. Cork is made up of more gas (90%) than solid material, making its **density** very low. Cork floats and does not rot. It is also fire resistant, compressible, and does not conduct heat or sound very well. These characteristics make cork a useful material for diverse purposes. Cork is used, for example, in the fishing industry, the electrical and building industries, **automobile** and aeronautic industries, the manufacture of sporting goods and home furnishings, shoes, and musical instruments. Some examples of cork products are floor and wall tiles, expansion or compression joints in **concrete** structures, insulation, safety helmets, several types of sporting good balls, heat shields, gaskets, shoe soles, and fishing tackle.

Today, Portugal is the leading producer of cork. In Portugal, the cork oak tree is protected by stringent laws regarding the growing, pruning, and stripping of the tree. Cork cultivators are given technical and financial assistance, and cork products are highly regulated to maintain high standards of quality.

Christine Miner Minderovic

Corm

A corm is a modified, upright, swollen, underground stem base of a herbaceous **plant**. Corms serve as a perennating **organ**, storing **energy** and producing new shoots and flowering stems from one or more buds located in the axils of the scale–like leaves of the previous year. Corms differ from superficially similar bulbs in that their leaves are thin rather than fleshy, and they are entirely composed of stem tissues.

Herbaceous plants are perennials, meaning that they have a lifespan of several to many years. However, after each growing season the above-ground parts of herbaceous plants die back to the ground, and new growth must issue from below ground to begin the following season. In the case of cultivated **species** such as gladiolus (*Gladiolus communis*), crocus (*Crocus sativus*), and water chestnut (*Eleocharis tuberosa*), the new herbaceous growth develops from underground corms. In fact, the corm of the water chestnut is eaten.

Horticulturalists usually propagate these species using corms, which develop small "cormels" from lateral buds on the sides of the parent corm. A relatively vigorous production of cormels can be stimulated by wounding the parent corm, for example, by making some sharp but shallow cuts on its base. To cultivate these plants, the cormels are split off and individually planted, right-side up, and a new plant will develop. The use of corms to propagate new plants in these ways does not involve any exchange of genetic information, and is referred to as vegetative propagation because the parent and progeny are genetically identical.

Cormorants

Cormorants or shags are long-necked, generally black or dark grey, aquatic **birds** in the family Phalacrocoracidae. These birds occur in most temperate and tropical marine coasts, and on many large lakes. There are 29 **species** of cormorants with fewer species occurring at higher latitudes.

The plumage of cormorants is not completely waterproof, since these birds lack an oil gland for preening, so their feathers get waterlogged when they swim under **water**. As a result, after swimming, cormorants spend time drying their feathers by standing with their wings spread to the **sun** and breeze.

The diet of cormorants is mostly small- to medium-sized species of **fish**. Cormorants dive for their **prey**, which they catch underwater in their bills. Cormorants power their swimming using their webbed feet, employing their wings and tails to assist with steering.

Cormorants are colonial breeders. They usually build their rather bulky nests of twigs and other debris in trees, and sometimes on artificial platforms such as old pilings. The young birds are initially without feathers and are fed by their parents by regurgitation. Cormorant colonies are loud, raucous places. These birds commonly kill the stand of trees that they nest in, mostly through the caustic influence of their copious defecations.

The most widespread species is the common or great cormorant (*Phalacrocorax carbo*), which occurs in **North America**, Eurasia, and **Australia**. This species and the double-crested cormorant (*P. auritus*) are the only cormorants on the coast of eastern North America. The double-crested cormorant also breeds abundantly on large, inland lakes.

The west coast of North America also has Brandt's cormorant (*P. penicillatus*) and the pelagic cormorant (*P. pelagicus*). The olivaceous cormorant (*P. olivaceous*) occurs on the southern coast of the Gulf of Mexico and off western Mexico, while the red-faced cormorant (*P. urile*) is a Eurasian species that occurs in the Aleutian islands of western Alaska.

The Peruvian cormorant (*P. bougainville*) breeds in enormous colonies on offshore islands of Chile and Peru, where its guano has long been mined as a source of phosphorus-rich fertilizer. This species is subject to occasional mass die-offs, caused by starvation resulting from periodic collapses of the stocks of its most important prey, the Peruvian **anchovy**. One unusual species, the Galápagos cormorant (*Nannopterum harrisi*), which lives on the remote and predator-free Galápagos Islands is flightless.

Captive cormorants in Japan and coastal China have been trained for fishing. When used for this purpose, the birds are tethered by tying a line to one of their feet, and their neck is constricted by a ring, so the cormorant can catch fish but not swallow them.

Cormorants, like this double-crested cormorant (*Phalacrocorax auritus*), are sometimes mistaken for geese in flight, but unlike geese, which flap steadily and honk as they fly, cormorants flap for a while and then glide, and they are silent in flight. *Photograph by Robert J. Huffman. Field Mark Publications. Reproduced by permission.*

Cormorants are considered to be a pest in many places, because they may eat species of fish that are also sought by human fishers. In some cases, cormorants are killed in large numbers for this reason. Sometimes, they are also considered to be **pests** because they kill vegetation in their nesting colonies.

Double-crested cormorants breeding on some of the Great Lakes of North America (eg., Lake Michigan) have rather large concentrations of **polychlorinated biphenyls (PCBs)** and other **chlorinated hydrocarbons** in their body **fat** and eggs. This effect of **pollution** has been blamed for apparent increases in the incidence of developmental deformities in some colonies of cormorants, especially the crossed-bill **syndrome**. However, in spite of this toxic stress, colonies of double-crested cormorants have been increasing rapidly on the Great Lakes during the past decade or so.

One species of cormorant, *P. perspicillatus,* bred on Bering Island in the Bering Sea, but was rendered extinct by humans.

Resources

Books

Brooke, M., and T. Birkhead, *The Cambridge Encyclopedia of Ornithology.* Cambridge, U.K.: Cambridge University Press, 1993.

Forshaw, Joseph. *Encyclopedia of Birds.* New York: Academic Press, 1998.

Harrison, Peter. *A Field Guide to Seabirds.* New York: Viking/Penguin, 1987.

Harrison, Peter. *Seabirds: An Identification Guide.* Beckenham, U.K.: Croom Helm, 1983.

Mackenzie, John P. *Seabirds.* Minocqua, WI: NorthWord, 1987.

Perkins, Simon. *Audubon Society Pocket Guide to North American Birds of Sea and Shore.* New York: Random House, 1994.

Bill Freedman

Corn (maize)

According to Native American legends of the American Southwest, the Indian people have occupied four (some say five) worlds since the creation of man. The present world is the fourth world and each of the three former worlds lies under the one succeeding it. Some legends say that maize, or corn as Europeans came to call it, was already present in the first world, at the time the first humans were created. The first people are said to have relied on maize for nourishment, and to have thought of the corn **plant** as a second mother (the first mother being the Earth Mother). Whenever a child was born, custom required that the Corn Mother—a perfect ear of corn with a tip ending in four full kernels—be placed beside the child, where it would remain for 20 days. The Corn Mother was also present at the naming of the child, and remained the child's spiritual mother for the rest of her or his life.

These myths attest to the profound importance of corn for many of the native peoples of the Americas, where corn originated—an importance comparable to that of **rice** for the peoples of Southeast **Asia** or of potatoes for the ancient peoples of the Andes Mountains in **South America**. Today corn is an important crop worldwide, but some rural Native American and Latin American farmers—especially in the highlands of Mexico and Guatemala—still have a special relationship to corn. Hundreds of genetically distinct varieties of corn are cultivated by small farmers in that region. The industrialized world relies on a few, genetically similar varieties of corn planted in very large, uniform fields, which renders its corn crop more vulnerable to **weather** changes and to blights (such as the **virus** that destroyed 15% of the U.S. corn crop in 1970); the genetic diversity preserved by the small, rural farmers of Central America is thus a unique asset, the "world's insurance policy," according to the International Maize and Wheat Improvement Center.

Science debates the origin of corn, **species** name *Zea mays*. Although some paleobotanists believe that corn developed from a grass known as teosinte in the vicinity of the Tehuacan Valley of Mexico more than 5,000 years ago, most believe that teosinte was itself an offshoot of a wild corn and another species of grass. Other discoveries suggest that corn did not develop in Mexico at all, arguing instead for a South American origin. The prevalent botanical theory today is that corn was domesticated from a wild, now-extinct, corn-like plant. Analyses of what appears to be wild-corn pollen found about 200 ft (60 m) below modern Mexico City have led some paleobotanists to conclude that the pollen came from an extinct wild corn that existed 25,000–80,000 years ago.

The corn plant is uniquely adapted for high productivity for two reasons: it has a very large **leaf** area, and it has a modified photosynthetic pathway that allows it to survive extended periods of **drought**. Known as the C4 **syndrome**, the modified pathway for **photosynthesis** provides a highly efficient way for the plant to exchange **water** vapor for atmospheric **carbon dioxide**.

At a biochemical level, carbon dioxide is converted into a **molecule** containing four carbon **atoms** in C4 photosynthesis, whereas in conventional photosynthesis **carbon** dioxide is converted into a molecule containing three carbon atoms (C3 photosynthesis). Thus C4 photosynthesis permits the corn plant to make more efficient use of carbon dioxide to build the carbon compounds needed to sustain plant growth than can be made by conventional plants. As a result, the corn plant can produce more dry **matter** (i.e., various combinations of carbohydrates, **proteins**, oils, and mineral **nutrients**) per unit of water transpired (released to the atmosphere) than can plants endowed with the conventional C3 photosynthetic pathway. The amount of grain produced by the plant depends upon the **rate** and length of time of this dry matter accumulation.

Successful cultivation of corn requires proper **fertilization** during the early stages of corn plant growth. The final size of the leaves, ear, and other plant parts depends largely upon maintaining an adequate supply of nutrients to the plant, especially during this time. Regionally adapted hybrids can be selected to accommodate local growing **seasons** and particular needs, and to make efficient use of specific types of land. Planting time should be carefully gauged. The highest yields are obtained only where envi-

Corn piled high outside of grain elevators, Magnolia, MN. *Photograph by Jim Mone. AP/Wide World Photos. Reproduced by permission.*

ronmental conditions are most favorable to growth. Weeds, diseases, and **insects** all reduce crop yield.

In recent years, genetically engineered (GE) varieties of corn have swiftly become a major fraction of the U.S. corn crop. In 2002, 34% of the corn grown in the United States (most of which was grown for the manufacture of corn syrup for human consumption) was genetically engineered. The most common variety of GE corn is Bt corn, a product of the Monsanto Corporation. Bt corn resists the corn borer, a common insect pest, by producing insecticide in its own tissues. Critics are concerned that Bt-corn-produced insecticide may linger in **soil** and sediments, threatening nontarget insect populations, and that genes from GE corn may contaminate non-GE varieties with unpredictable consequences. The discovery in 2001 that genes from GE corn (transgenes) had in only a few years' time managed to insinuate themselves into traditional maize varieties in Mexico has enhanced global concern about GE corn and other GE **crops**. The European Union and Zambia have both refused to accept the importation of GE corn from the United States because of concerns that the genetic changes made to the corn may produce unforeseen health effects. Many scientists argue that such concerns are unfounded, and that the benefits of **genetic engineering** outweigh the risks; others argue that the risks are too high, and that yields for genetically modified crops have not actually been greater than for traditional varieties.

Many historians believe that the course of American history was shaped more by corn than by any other plant. Their argument derives largely from the observation that for more than 250 years after the settlement of Jamestown, Virginia in 1607, corn remained the staple crop of about 90% of all European-American farmers. Not until 1979 did soybeans surpass corn as the United States' most important crop, and then only temporarily.

Currently, more bushels of corn are produced in the United States each year than of any other grain crop. (One bushel = 35.24 L.)

Corn is grown on every **continent** except **Antarctica**. This is possible because scientists have developed diverse **hybrid** varieties of corn that suit growing conditions and locations worldwide. In 1999, the United States led the world in corn production, with a crop valued at 9,437 million bushels. In second place was China with a crop value of 5,039 million bushels, followed by European countries with a crop valued at 1,457 million bushels.

Although U.S. citizens today eat much less corn than they did historically, large amounts of corn are consumed indirectly in the form of products derived from **livestock** fed a diet rich in corn, including beef, pork, poultry, eggs, milk, and many other food products. Eighty **percent** of the U.S. corn crop is fed to livestock. As a point of reference, one bushel of corn—56 lb (25 kg)—produces approximately 5.6 lb (2.5 kg) of beef.

Corn is still used for many traditional purposes (e.g., meal, corn on the cob, corn-cob pipes). However, new markets have developed for corn byproducts, including the manufacture of fiberboard panels, nylon products, furfural (used in **plastics** and solvents), and **ethanol** as a renewable vehicular fuel. In fact, more than 3,500 different uses have been found for corn products. Some of the more innovative uses include a basis for vitamins and amino acids; a substitute for phosphate for cleaning; a packing-peanut material; an ink base to replace **petroleum** ink bases; and an absorbent material for diapers and for **automobile** fuel filters.

Resources

Periodicals

Associated Press. "Zambia Bars Altered Corn From U.S." *New York Times* August 18, 2002.
Dalton, Rex. "Transgenic Corn Found Growing in Mexico." *Nature* 413 (September 27, 2001): 337.

Organizations

International Maize and Wheat Improvement Center (CIMMYT). Apartado #370, PO Box 60326, Houston, TX 77205. (650) 833 6655 (in U.S.). <http://www.cimmyt.org/> (cited May 25, 2003).

Randall Frost
Larry Gilman

Coronal ejections and magnetic storms

Coronal mass ejections (CMEs) are explosive and violent eruptions of charged, magnetic field-inducing particles and gas (plasma) from the Sun's outer coronal layer. The ejection from the Sun's corona can be massive (e.g., estimates of CME mass often range in the billions of tons. Ejections propel particles in specific directions, some directly crossing Earth's orbital position, at velocities up to 1,200 miles per second (1,931 km per second) or 4,320,000 miles per hour (6,952,366 km/h) in an ionized plasma (also known as the **solar wind**). Solar CMEs that are **Earth** directed disrupt and distort **Earth's magnetic field** and result in geomagnetic storms.

Although the solar wind is continuous, CMEs reflect large-scale increases in wind (i.e., particle) mass and **velocity** that are capable of generating geomagnetic storms.

Solar coronal ejections and magnetic storms interact with Earth's **magnetosphere** to produce spectacular auroral displays. Intense storms may interfere with communications and preclude data transfer from Earth orbiting satellites.

Solar coronal ejections and magnetic storms provide the charged particles that result in the northern and southern lights—aurora borealis and aurora australialis—electromagnetic phenomena that usually occur near Earth's polar regions. The auroras result from the interaction of Earth's magnetic field with ionic gas particles, protons, and electrons streaming outward in the solar wind.

The **rate** of solar coronal ejections is correlated to solar sunspot activity that cycles between maximum levels of activity (i.e., the solar maximum) approximately every 11 years. During solar maximums, it is not uncommon to observe multiple coronal ejections per day. At solar minimum, one solar coronal ejection per day is normal. The last peak of activity occurred in 2001.

Earth's core structure provides it with a relatively strong internal magnetic field (oriented about 10–12 degrees from the polar axis). Earth's magnetosphere protects the Earth from bombardment by Comes by deflecting and modifying the solar wind. At the interface of Earth's magentosphere and the solar wind there is a "bow wave" or magnetic shock wave to form a magnetosheath protecting the underlying magnetosphere that extends into Earth's ionosphere.

Coronal mass ejections not only interact with Earth's magnetic field, they also interact with each other. Stronger or faster ejections may subsume prior weaker ejections directed at the same region of space in a process known as CME cannibalization. Accordingly, the strength of magnetic storms on Earth may not directly correlate to observed coronal ejections. In addition, CME cannibalization can alter predicted arrival **time** of geomagnetic storms because the interacting CMEs can change the eruption velocity.

See also Atmospheric optical phenomena; Atomic theory; Bohr model; Element, chemical; Solar activity cycle; Solar flare; Solar prominence; Stellar evolution; Stellar magnetic fields; Stellar wind.

Correlation (geology)

In **geology**, the term correlation refers to the methods by which the age relationship between various **strata** of Earth's crust is established. Such relationships can be established, in general, in one of two ways: by comparing the physical characteristics of strata with each other (physical correlation); and by comparing the type of fossils found in various strata (fossil correlation).

Correlation is an important geological technique because it provides information with regard to changes that have taken place at various times in **Earth** history. It also provides clues as to the times at which such changes have occurred. One result of correlational studies has been the development of a **geologic time** scale that separates Earth history into a number of discrete **time** blocks known as eras, periods, and epochs.

The nature of sedimentary strata

Sedimentary **rocks** provide information about Earth history that is generally not available from igneous or metamorphic rocks. To understand why this is so, imagine a region in which sediments have been laid down for millions of years. For example, suppose that for many millions of years a river has emptied into an **ocean**, laying down, or depositing, sediments eroded from the land. During that period of time, layers of sediments would have collected one on top of the other at the mouth of the river.

These layers of sediments are likely to be very different from each other, depending on a number of factors, such as the course followed by the river, the climate of the area, the rock types exposed along the river course, and many other geological factors in the region. One of the most obvious differences in layers is thickness. Layers of **sedimentary rock** may range in thickness from less than an inch to many feet.

Sedimentary layers that are identifiably different from each other are called beds or strata. In many places on Earth's surface, dozens of strata are stacked one on top of each other. Strata are often separated from each other by relatively well-defined surfaces known as bedding planes.

In 1669, the Danish physician and theologian Nicolaus Steno made a seemingly obvious assertion about the nature of sedimentary strata. Steno stated that in any sequence of sedimentary rocks, any one layer (stratum) is older than the layer below it and younger than the layer above it. Steno's discovery is now known as the law of superposition.

The law of superposition applies only to sedimentary rocks that have not been overturned by geologic forces. **Igneous rocks**, by comparison, may form in any horizontal sequence whatsoever. A flow of **magma** may force itself, for example, underneath, in the middle or, or on top of an existing rock stratum. It is very difficult to look back millions of years later, then, and determine the age of the igneous rock compared to rock layers around it.

Physical correlation

Using sedimentary rock strata it should be possible, at least in theory, to write the geological history of the continents for the last billion or so years. Some important practical problems, however, prevent the full realization of this goal. For example, in many areas, **erosion** has removed much or most of the sedimentary rock that once existed there. In other places, strata are not clearly exposed to view but, instead, are buried hundreds or thousands of feet beneath the thin layer of **soil** that covers most of Earth's surface.

A few remarkable exceptions exist. A familiar example is the Grand Canyon, where the Colorado River has cut through dozens of strata, exposing them to view and making them available for study by geologists. Within the Grand Canyon, a geologist can follow a particular stratum for many miles, noting changes within the stratum and changes between that stratum and its neighbors above and below.

One of the characteristics observable in such a case is that a stratum often changes in thickness from one edge to another. At the edge where the thickness approaches **zero**, the stratum may merge into another stratum. This phenomenon is understandable when one considers the way the sediment in the rocks was laid down. At the mouth of a river, for example, the accumulation of sediments is likely to be greatest at the mouth itself, with decreasing thickness at greater distances into the **lake** or ocean. The principle of lateral continuity describes this phenomenon, namely that strata are three-dimensional features that extend outward in all directions, merging with adjacent deposits at their edges.

Human activity also exposes strata to view. When a highway is constructed through a mountainous (or hilly) area, for example, parts of a mountainside may be excavated, revealing various sedimentary rock strata. These strata can then be studied to discover the correlation among them and with strata in other areas.

Another problem is that strata are sometimes disrupted by earth movements. For example, an **earthquake** may lift one block of Earth's crust over an adjacent block or may shift it horizontally in comparison to the second block. The correlation between adjacent strata may then be difficult to determine.

Physical correlation is accomplished by using a number of criteria. For example, the **color**, grain size, and type of **minerals** contained within a stratum make it possible for geologists to classify a particular stratum quite specifically. This allows them to match up portions of that stratum in regions that are physically separated from each other. In the American West, for example, some strata have been found to cover large parts of two or more states although they are physically exposed in only a few specific regions.

Interpreting Earth history within a stratum

Imagine that geologists a million years from now began studying Earth's surface within a 3 mi (5 kilometer) radius of your home. What would they find? They would probably discover considerable variation in the sediment deposits that are accumulating in your region today. They might find the remains of a river bed, a swamp, a lake, and other features. Geologists living today who study strata laid down millions of years ago make similar discoveries. As they follow an individual stratum for many kilometers, they find that its characteristics change. The stratum tends to have one set of characteristics in one region, which gradually changes into another set of characteristics farther along in the stratum. Those characteristics also change, at some **distance** farther along, into yet another set of characteristics.

Rocks with a particular set of characteristics are called a facies. Facies changes, changes in the characteristics of a stratum or series of strata, are important clues to Earth history. Suppose that a geologist finds that the facies in a particular stratum change from a limestone to a shale to a sandstone over a distance of a few miles. The geologist knows that limestone is laid down on a sea bottom, shale is formed from compacted mud, and sandstone is formed when **sand** is compressed. The limestone to shale to sandstone facies pattern may allow an astute geologist to reconstruct what Earth's surface looked like when this particular stratum was formed. For example, knowing these rocks were laid down in adjacent environments, the geologist might consider that the limestone was deposited on a coral reef, the shale in a quiet lagoon or coastal swamp, and the sandstone in a nearby beach. So facies changes indicate differences in the environments in which adjacent facies were deposited.

KEY TERMS

Bedding plane—The top of a layer of rock.

Deposition—The accumulation of sediments after transport by wind, water, ice, or gravity.

Facies—A body of sedimentary rock with distinctive characteristics.

Fossil correlation—The matching of sedimentary strata based on fossils present in the strata.

Lateral continuity—The principle that sedimentary strata are three-dimensional features that extend horizontally in all directions and that eventually terminate against the margin of other strata.

Physical correlation—The matching of sedimentary strata based on the physical characteristics of rocks that make up the strata.

Radiometric dating—A process by which the age of a rock can be determined by studying the relative concentrations of a radioactive isotope and the products formed by its decay.

Superposition—The principle that a layer of rocks is older than any other layer that lies above it and younger than any other layer that lies below it.

Fossil correlation

One of the most important discoveries in the science of correlation was made by the English surveyor William Smith in the 1810s. One of Smith's jobs involved the excavation of land for canals being constructed outside of London. As sedimentary rocks were exposed during this work, Smith found that any given stratum always contained the same set of fossils. Even if the stratum were physically separated by a relatively great distance, the same fossils could always be found in all parts of the stratum.

In 1815, Smith published a **map** of England and Wales showing the geologic history of the region based on his discovery. The map was based on what Smith called his law of faunal **succession**. That law says simply that it is possible to identify the sequence in which strata are laid down by examining the fossils they contain. The simplest fossils are the oldest and, therefore, strata that contain simple fossils are older than strata that contain more complex fossils.

The remarkable feature of Smith's discovery is that it appears to be valid over very great distances. That is, suppose that a geologist discovers a stratum of rock in southwestern California that contains fossils A, B, and C. If another stratum of rock in eastern Texas is also discovered that contains the same fossils, the geologist can conclude that it is probably the same stratum—or at least of the same age—as the southwestern California stratum.

Absolute vs. relative ages of strata

The correlational studies described so far allow scientists to estimate the relative ages of strata. If stratum B lies above stratum A, B is the younger of the two. However determining the actual, or absolute, age of strata (for example, 3.5 million years old) is often difficult since the age of a fossil cannot be determined directly. The most useful tool in dating strata is radiometric dating of materials. A radioactive **isotope** such as uranium-238 decays at a very regular and well-known **rate**. That rate is known as its **half-life**, the time it takes for one-half of a **sample** of the isotope to decay. The half-life of uranium-238, for example, is 4.5 billion years. By measuring the **concentration** of uranium-238 in comparison with the products of its decay (especially lead-206), a scientist can estimate the age of the rock in which the **uranium** was found. This kind of **radioactive dating** has made it possible to place specific dates on the ages of strata that have been studied and correlated by other means.

See also Dating techniques; Deposit; Fossil and fossilization; Sediment and sedimentation.

David E. Newton

Correlation (mathematics)

Correlation refers to the **degree** of correspondence or relationship between two variables. Correlated variables tend to change together. If one **variable** gets larger, the other one systematically becomes either larger or smaller. For example, we would expect to find such a relationship between scores on an **arithmetic** test taken three months apart. We could expect high scores on the first test to predict high scores on the second test, and low scores on the first test to predict low scores on the second test.

In the above example the scores on the first test are known as the independent or predictor variable (designated as "X") while the scores on the second test are known as the dependent or response variable (designated as "Y"). The relationship between the two variables X and Y is a positive relationship or positive correlation when high measures of X correspond with high measures of Y and low measures of X with low measures of Y. It is also possible for the relationship between variables X and Y to be an inverse relationship or **negative** correlation. This occurs when high measures of variable

KEY TERMS

Correlation coefficient—The numerical index of a relationship between two variables.

Negative correlation—The changes in one variable are reflected by inverse changes in the second variable.

Positive correlation—The changes in one variable are reflected by similar changes in the second variable.

X are associated with low measures of variable Y and low measures on variable X are associated with high measures of variable Y. For example, if variable X is school attendance and variable Y is the score on an achievement test we could expect a negative correlation between X and Y. High measures of X (absence) would be associated with low measures of Y (achievement) and low measures of X with high measures of Y.

The correlation **coefficient** tells us that a relationship exists. The + or - sign indicates the direction of the relationship while the number indicates the magnitude of the relationship. This relationship should not be interpreted as a causal relationship. Variable X is related to variable Y, and may indeed be a good predictor of variable Y, but variable X does not cause variable Y although this is sometimes assumed. For example, there may be a positive correlation between head size and IQ or shoe size and IQ. Yet no one would say that the size of one's head or shoe size causes variations in intelligence. However, when two more likely variables show a positive or negative correlation, many interpret the change in the second variable to have been caused by the first.

Resources

Books

Gonick, Larry, and Woollcott Smith. *The Cartoon Guide to Statistics.* New York: Harper Row, 1993.
Moore, David, and George McCabe. *Introduction to the practice of Statistics.* New York: W. H. Freeman, 1989.
Walpole, Ronald, and Raymond Myers, et al. *Probability and Statistics for Engineers and Scientists.* Englewood Cliffs, NJ: Prentice Hall, 2002.

Selma Hughes

Corrosion

Corrosion is the deterioration of metals and other materials by chemical reaction. Corrosion of metals is the most common type of corrosion and is a process involving an exchange of electrons between two substances, one of them being the **metal**. In this process, the metal usually loses electrons, becoming oxidized, while the other substance gains electrons, becoming reduced. For this reason, corrosion is classified as an oxidation-reduction or redox reaction.

While many redox reactions are extremely important and beneficial to society (for example, those that are used to make batteries), the redox reactions involved in corrosion are destructive. In fact, close to $200 billion dollars (or 4% of the gross domestic product) is spent in the United States each year to prevent or repair the damage done by corrosion. Economically, one of the most important metals to corrode is **iron** and one of its alloys, **steel**. Almost 20% of the iron and steel produced in the United States each year is used to replace objects that have corroded.

Most metals react with **oxygen**, but not all metals become subject to corrosion. Many metals react, but their **rate** of reaction is so slow that no significant corrosion appears visible. In other cases, the metal reacts with oxygen to form a metal oxide which then coats the metal and prevents further reaction with oxygen. A few metals, such as gold and platinum, are very unreactive and they are used in applications where their tendency not to corrode is important.

In general, both oxygen and **water** are necessary for corrosion to occur. For this reason, corrosion occurs much more slowly in **desert** climates that are drier. The process of corrosion is accelerated in the presence of salts or acids. In the northern United States and in Canada, winter is especially harsh on cars because there is plenty of water in the form of snow and roads are salted to prevent the formation of **ice**. Warm temperatures also speed the process of corrosion.

There are many ways in which we can minimize, and sometimes even prevent the occurrence of corrosion. One simple way is to paint the metal surface and prevent oxygen and water from having access to the metal surface. If the paint chips, however, the exposed surface can cause rust to spread beneath the surface of the metal. Another way to protect the metal surface is to coat it with a layer of zinc metal. Zinc is more likely to oxidize but it then forms a coating that adheres firmly to the zinc and prevents further reaction, protecting both the zinc and the metal beneath the zinc. Metals protected by zinc are said to be galvanized. Other metals that form protective coatings can also be used to protect metals that corrode. For example, in stainless steel, chromium or nickel is added to the iron. It is also possible to deposit a thin film of one metal on the surface of another metal. This process is called electroplating. Iron cans are electroplat-

ed with tin to form tin cans. The tin is less easily oxidized than the iron. If the shape of the metal object is such that it is difficult to galvanize or to electroplate, a piece of **magnesium** or zinc can be attached to the object. This piece of attached metal is called a sacrificial **anode**. It will preferentially corrode, keeping the metal (such as underground tanks or buried pipeline) intact and free from breakage. It becomes necessary to replace the sacrificial anode from time to time, but that is less expensive than replacing the whole pipeline.

Rashmi Venkateswaran

Cosmetic surgery *see* **Plastic surgery**

Cosmic background radiation

In 1965, Arno Penzias and Robert Wilson announced the discovery of microwave **radiation** which uniformly filled the sky and had a blackbody **temperature** of about 3.5K. The pair had been testing a new **radio amplifier** that was supposed to be exceptionally quiet. What better way to do such a test than to tune the radio so that it should hear nothing at all? After many attempts to account for all extraneous sources of radio noise, they came to the conclusion that there was a general background of radiation at the radio **frequency** they were using. After discussions with a group led by Robert Dicke at nearby Princeton University it became clear that they had in fact detected remnant radiation from the origin of the universe.

Although neither Dicke's group or Penzias and Wilson realized it at the time they had confirmed a prediction made 17 years earlier by Alpher, Bethe, and Gamow. Although the temperature that characterized the detected radiation was somewhat different than predicted, the difference could be accounted for by changes to the accepted structure of the universe discovered between 1948 and 1965. The detection of this radiation and its subsequent verification at other frequencies was taken as confirmation of a central prediction of a **cosmology** known as the big bang.

The interpretation of the red-shifts of **spectral lines** in distant galaxies by Edwin Hubble 40 years earlier suggested a universe that was expanding. One interpretation of that expansion was that the universe had a specific origin in **space** and time. Such a universe would have a very different early structure from the present one.

It was George Gamow and colleagues who suggested that the early phases of the universe would have been hot and dense enough to sustain nuclear reactions. Following these initial phases, the expanding universe would eventually cool to the point at which the dominant material, **hydrogen**, would become relatively transparent to **light** and **radio waves**. We know that for hydrogen, this occurs when the gas reaches a temperature of between 5,000K–10,000K. From that point on in the evolution of the universe, the light and **matter** would go their separate ways.

As every point in the universe expands away from every other point, any observer in the universe sees all objects receding from him or her. The faster moving objects will appear at greater distances by virtue of their greater speed. Indeed, their speed will be directly proportional to their **distance** which is what one expects for material ejected from a particular point in space and time. However, this expansion results from the expansion of space itself and should not be viewed simply as galaxies rushing headlong away from one another through some absolute space. The space itself expands.

As it does, light traveling through it is stretched, becoming redder and appearing cooler. If one samples that radiation at a later date it will be characteristic of radiation from a much cooler source. From the **rate** of expansion of the universe it is possible to predict what that temperature ought to be. Current values of the expansion rate are completely consistent with the current measured temperature of about 2.7K. The very existence of this radiation is strong evidence supporting the expanding model of the universe championed by Gamow and colleagues and disparagingly named the "big bang" cosmology by Sir Fred Hoyle.

Fossil radiation

Since its discovery in 1965, the radiation has been carefully studied and found to be a perfect blackbody as expected from theory. Since, this radiation represents fossil radiation from the initial big bang, any additional **motion** of **Earth** around the **Sun**, the Sun around the galactic center, and the **galaxy** through space should be reflected in a slight asymmetry in the background radiation. The net motion of Earth in some specific direction should be reflected by a slight Doppler shift of the background radiation coming from that direction toward shorter wavelengths.

Doppler shift is the same effect that the police use to ascertain the **velocity** of approaching vehicles. Of course there will be a similar shift toward longer wavelengths for light coming from the direction from which we are receding. This effect has been observed indicating a combined peculiar motion of Earth, Sun, and galaxy on the order of 600 km/sec.

Finally, small fluctuations in the background radiation are predicted which eventually led to the formation

---

KEY TERMS

Blackbody—A blackbody (not to be confused with a black hole) is any object which absorbs all radiant energy which falls upon it and subsequently re-radiates that energy. The radiated energy can be characterized by a single dependent variable, the temperature. That temperature is known as the blackbody temperature.

Doppler shift—The change in frequency or wavelength resulting from the relative motion of the source of radiation and the observer. A motion of approach between the two will result in a compression of the waves as they pass the observer and a rise in "pitch" in the frequency of the wave and a shortening of the relative wavelength called a "blue shift." A relative motion of recession leads to a lowering of the "pitch" and a shift to longer "redder" wavelengths.

Microwave radiation—Electromagnetic radiation that occurs in the wavelength region of about 0.4 in to 3.3 f (1 cm to 1 m).

---

galaxies, clusters of galaxies. Such fluctuations have been found by the CO(smic) B(ackground) E(xplorer) **Satellite**, launched by NASA in 1989. COBE detected these fluctuations at about 1 part in 105 which was right near the detection limit of the satellite. The details of these fluctuations are crucial to deciding between more refined models of the expanding universe. COBE was decommissioned in 1993, but scientists are still unraveling the information contained in its data.

It is perhaps not too much of an exaggeration to suggest that cosmic background radiation has elevated cosmology from enlightened speculative metaphysics to an actual science. We may expect developments of this emerging science to lead to a definitive description of the evolutionary history of the universe in the near future.

George W. Collins, II

Cosmic ray

The term cosmic ray refers to tiny particles of **matter** that travel through **space**. Cosmic rays generally possess an electromagnetic charge and are highly energetic. Physicists divide cosmic rays into two categories: primary and secondary. Primary cosmic rays originate far outside Earth's atmosphere. Secondary cosmic rays are particles produced within Earth's atmosphere as a result of collisions between primary cosmic rays and molecules in the atmosphere.

Discovery of cosmic rays

The existence of cosmic **radiation** was first discovered in 1912, in experiments performed by the physicist Victor Hess. His experiments were sparked by a desire to better understand phenomena of **electric charge**. A common instrument of the day for demonstrating such phenomena was the electroscope. An electroscope contains thin **metal** leaves or wires that separate from one another when they become charged, due to the fact that like charges repel. Eventually the leaves (or wires) lose their charge and collapse back together. It was known that this loss of charge had to be due to the attraction by the leaves of charged particles (called ions) in the surrounding air. The leaves would attract those ions having a charge opposite to that of the leaves, due to the fact that opposite charges attract; eventually the accumulation of ions in this way would neutralize the charge that had been acquired by the leaves, and they would cease to repel each other. Scientists wanted to know where these ions came from. It was thought that they must be the result of radiation emanating from Earth's crust, since it was known that radiation could produce ions in the air. This led scientists to predict that there would be fewer ions present the further one traveled away from Earth's surface. Hess's experiments, in which he took electroscopes high above Earth's surface in a **balloon**, showed that this was not the case. At high altitudes, the electroscopes lost their charge even faster than they had on the ground, showing that there were more ions in the air and thus that the radiation responsible for the presence of the ions was stronger at higher altitudes. Hess concluded that there was a radiation coming into our atmosphere from outer space.

As physicists became interested in cosmic radiation, they developed new ways of studying it. The Geiger-Muller counter consists of a wire attached to an **electric circuit** and suspended in a gaseous chamber. The passage of a cosmic ray through the chamber produces ions in the gas, causing the counter to discharge an electric pulse. Another instrument, the cloud chamber, contains a gas which condenses into vapor droplets around ions when these are produced by the passage of a cosmic ray. In the decades following Hess's discovery, physicists used instruments such as these to learn more about the nature of cosmic radiation.

The nature and origin of cosmic rays

An atom of a particular element consists of a nucleus surrounded by a cloud of electrons, which are nega-

tively charged particles. The nucleus is made up of protons, which have a positive charge, and neutrons, which have no charge. These particles can be further broken down into smaller constituents; all of these particles are known as **subatomic particles**. Cosmic rays consist of nuclei and of various subatomic particles. Almost all of the primary cosmic rays are nuclei of various **atoms**. The great majority of these are single protons, which are nuclei of **hydrogen** atoms. The next most common primary cosmic ray is the nucleus of the helium atom, made up of a **proton** and a **neutron**. Hydrogen and helium nuclei make up about 99% of the primary cosmic radiation. The rest consists of nuclei of other elements and of electrons.

When primary cosmic rays enter Earth's atmosphere, they collide with molecules of gases present there. These collisions result in the production of more high-energy subatomic particles of different types; these are the secondary cosmic rays. These include photons, neutrinos, electrons, positrons, and other particles. These particles may in turn collide with other particles, producing still more secondary radiation. If the **energy** of the primary particle that initiates this process is very high, this cascade of collisions and particle production can become quite extensive. This is known as a shower, air shower, or cascade shower.

The energy of cosmic rays is measured in units called **electron** volts (abbreviated eV). Primary cosmic rays typically have energies on the order of billions of electron volts. Some are vastly more energetic than this; a few particles have been measured at energies in excess of 1019 eV. This is in the neighborhood of the amount of energy required to lift a weight of 2.2 lb (1 kg) to a height of 3.3 ft (1 m). Energy is lost in collisions with other particles, so secondary cosmic rays are typically less energetic than primary ones. The showers of particles described above diminish as the energies of the particles produced decrease. The energy of cosmic rays was first determined by measuring their ability to penetrate substances such as gold or **lead**.

Since cosmic rays are mostly charged particles (some secondary rays such as photons have no charge), they are affected by magnetic fields. The paths of incoming primary cosmic rays are deflected by the **earth's magnetic field**, somewhat in the way that **iron** filings will arrange themselves along the lines of **force** emitted by a magnet. More energetic particles are deflected less than those having less energy. In the 1930s it was discovered that more particles come to **Earth** from the West than from the East. Because of the nature of Earth's magnetic field, this led scientists to the conclusion that most of the incoming cosmic radiation consists of positively charged particles. This was an important step towards the discovery that the primary cosmic rays are

KEY TERMS

. . . .

Electron—A negatively charged particle, ordinarily occurring as part of an atom. The atom's electrons form a sort of cloud about the nucleus.

Electron volt (eV)—The unit used to measure the energy of cosmic rays.

Electroscope—A device for demonstrating the presence of an electric charge, which may be positive or negative.

Ion—An atom or molecule which has acquired electrical charge by either losing electrons (positively charged ion) or gaining electrons (negatively charged ion).

Neutron—Particle found in the nucleus of an atom, possessing no charge.

Nucleus—The central mass of an atom. The nucleus is composed of neutrons and protons.

Primary cosmic ray—Cosmic ray originating outside Earth's atmosphere.

Proton—Positively charged particle composing part of the nucleus of an atom. Primary cosmic rays are mostly made up of single protons.

Secondary cosmic ray—Cosmic ray originating within Earth's atmosphere as a result of a collision between another cosmic ray and some other particle or molecule.

Shower (also air shower or cascade shower)—A chain reaction of collisions between cosmic rays and other particles, producing more cosmic rays.

mostly bare atomic nuclei, since atomic nuclei carry a positive charge.

The ultimate origin of cosmic radiation is still not completely understood. Some of the radiation is believed to have been produced in the "big bang" at the origin of the universe. Other cosmic rays are produced by our **Sun**, particularly during solar disturbances such as solar flares. Exploding stars, called supernovas, are also a source of cosmic rays.

The fact that cosmic ray collisions produce smaller subatomic particles has provided a great deal of insight into the fundamental structure of matter. The construction of experimental equipment such as particle **accelerators** has been inspired by a desire to reproduce the conditions under which high-energy radiation is produced, in order to gain better experimental control of collisions and the production of particles.

See also Particle detectors.

Resources

Books

Friedlander, Michael. *Cosmic Rays.* Cambridge: Harvard University Press, 1989.

Longair, M. S. *High Energy Astrophysics.* Cambridge: Cambridge University Press, 1981.

Millikan, Robert Andrews. *Electrons (+ and -), Protons, Photons, Neutrons, and Cosmic Rays.* Chicago: University of Chicago Press, 1935.

Periodicals

"Cosmic Rays: Are Air Crews At Risk?" *Occupational and Environmental Medicine* 59, no. 7 (2002): 428-432.

"Radiation Risk During Long-Term Spaceflight." *Advaces in Space Research* 30, no. 4 (2002): 989-994.

John Bishop

Cosmology

Cosmology is the study of the origin, structure and evolution of the universe.

The origins of cosmology predate the human written record. The earliest civilizations constructed elaborate myths and folk tales to explain the wanderings of the **Sun, Moon,** and stars through the heavens. Ancient Egyptians tied their religious beliefs to celestial objects and Ancient Greek and Roman philosophers debated the composition and shape of **Earth** and the Cosmos. For more than 13 centuries, until the Scientific Revolution of the sixteenth and seventeenth centuries, the Greek astronomer Ptolemy's model of an Earth-centered Cosmos composed of concentric crystalline spheres dominated the Western intellectual tradition.

Polish astronomer Nicolaus Copernicus's (1473–1543) reassertion of the once discarded heliocentric (Sun-centered) theory sparked a revival of cosmological thought and work among the astronomers of the time. The advances in empiricism during the early part of the Scientific Revolution, embraced and embodied in the careful observations of Danish astronomer Tycho Brahe (1546–1601), found full expression in the mathematical genius of the German astronomer Johannes Kepler (1571–1630) whose laws of planetary **motion** swept away the need for the errant but practically useful Ptolemaic models. Finally, the patient observations of the Italian astronomer and physicist Galileo, in particular his observations of moons circling **Jupiter** and of the phases of **Venus**, empirically laid to rest cosmologies that placed Earth at the center of the Cosmos.

English physicist and mathematician Sir Isaac Newton's (1642–1727), important *Philosophiae Naturalis Principia Mathematica* (Mathematical principles of natural philosophy) quantified the **laws of motion** and gravity and thereby enabled cosmologists to envision a clockwork-like universe governed by knowable and testable natural laws. Within a century of Newton's *Principia,* the rise of concept of a mechanistic universe led to the quantification of celestial dynamics, that, in turn, led to a dramatic increase in the observation, cataloging and quantification of celestial phenomena. In accord with the development of natural theology, scientists and philosophers argued conflicting cosmologies that argued the existence and need for a supernatural God who acted as "prime mover" and guiding force behind a clockwork universe. In particular, French mathematician, Pierre Simon de Laplace (1749–1827) argued for a completely deterministic universe, without a need for the intervention of God. Most importantly to the development of modern cosmology, Laplace asserted explanations for celestial phenomena as the inevitable result of time and statistical probability.

By the dawn of the twentieth century, advances in **mathematics** allowed the development of increasingly sophisticated cosmological models. Many of advances in mathematics pointed toward a universe not necessarily limited to three dimensions and not necessarily absolute in time. These intriguing ideas found expression in the intricacies of relativity and theory that, for the first time, allowed cosmologists a theoretical framework upon which they could attempt to explain the innermost workings and structure of the universe both on the scale of the subatomic world and on the grandest of galactic scales.

As direct consequence of German-American physicist Albert Einstein's (1879–1955) relativity theory, cosmologists advanced the concept that space-time was a creation of the universe itself. This insight set the stage for the development of modern cosmological theory and provided insight into the evolutionary stages of stars (e.g., **neutron** stars, pulsars, black holes, etc.) that carried with it an understanding of nucleosythesis (the formation of elements) that forever linked the physical composition of **matter** on Earth to the lives of the stars.

Twentieth-century progress in cosmology has been marked by corresponding and mutually beneficial advances in technology and theory. American astronomer Edwin Hubble's (1889–1953) discovery that the universe was expanding, Arno A. Penzias and Robert W. Wilson's observation of **cosmic background radiation**, and the detection of the elementary particles that populated the very early universe all proved important confirmations of the **Big Bang theory**. The Big Bang theory asserts that all matter and **energy** in the Universe, and the four dimensions of time and **space** were created from the primordial explosion of a singularity of enormous **density, temperature**, and **pressure**.

During the 1940s Russian-born American cosmologist and nuclear physicist George Gamow (1904–1968) developed the modern version of the big bang model based upon earlier concepts advanced by Russian physicist Alexander (Aleksandr Aleksandrovich) Friedmann (also spelled as Fridman, 1888–1925) and Belgian astrophysicist and cosmologist Abbé Georges Lemaître (1894–1966). Big bang based models replaced static models of the universe that described a homogeneous universe that was the same in all directions (when averaged over a large span of space) and at all times. Big bang and static cosmological models competed with each other for scientific and philosophical favor. Although many astrophysicists rejected the steady state model because it would violate the law of mass-energy conservation, the model had many eloquent and capable defenders. Moreover, the steady model was interpreted by many to be more compatible with many philosophical, social and religious concepts centered on the concept of an unchanging universe. The discovery quasars and of a permeating cosmic background **radiation** eventually tilted the cosmological argument in favor of big bang-based models.

Technology continues to expand the frontiers of cosmology. The **Hubble Space Telescope** has revealed gas **clouds** in the cosmic voids and beautiful images of fledgling galaxies formed when the universe was less than a billion years old. Analysis of these pictures and advances in the understanding of the fundamental constituents of nature continue to keep cosmology a dynamic discipline of **physics** and the ultimate fusion of human scientific knowledge and philosophy.

Evolution of cosmological thought

Using such instruments as the Hubble Space Telescope, modern cosmology is an attempt to describe the large scale structure and order of the universe. To that end, one does not expect cosmology to deal with the detailed structure such as planets, stars, or even galaxies. Rather it attempts to describe the structure of the universe on the largest of scales and to determine its past and future.

One of the earliest constraints on any description of the universe is generally attributed to Heinreich Olbers in 1826. However, more careful study traces the idea back to Thomas Digges in 1576 and it was thoroughly discussed by Edmond Halley at the time of Isaac Newton. The notion, still called Olbers' paradox, is concerned with why the night sky is dark. At the time of Newton it was understood that if the universe was finite then Newton's Law of Gravity would require that all the matter in the universe should pull itself together to that **point** equidistant from the boundary of the universe. Thus, the prevailing wisdom was that the universe was infinite in extent and therefore had no center. However, if this were true, then the extension of any line of sight should sooner or later encounter the surface of a **star**. The night sky should then appear to have the brightness of the average star. The sun is an average star, thus one would expect the sky to be everywhere as bright as the sun. It is not, so there must be a problem with the initial assumptions.

An alternative explanation, pointed out in 1964 by Ed Harrison, is that the universe had a finite beginning and the **light** from distant stars had not yet had time to arrive and that is why the night sky is dark.

At the turn of the twentieth century cosmology placed the sun and its **solar system** of planets near the center of the **Milky Way Galaxy** which comprised the full extent of the known Universe. However, early in the century information began to be compiled that would change the popular view that the universe was static and made of primarily stars. On April 26, 1920, there was a historic debate between H. D. Curtis and Harlow Shapley concerning the nature of some fuzzy clouds of light which were called nebulae. Shapley thought they were objects within the galaxy while Curtis believed them to be "island universes" lying outside the Milky Way.

Although at the time most agreed that Shapley had won the debate, science eventually proved that Curtis was right. Within a few years Edwin Hubble detected a type of star, whose **distance** could be independently determined, residing within several of these fuzzy clouds. These stars clearly placed the "clouds" beyond the limits of the Milky Way. While Hubble continued to use the term "island universe," more and more extragalactic nebulae were discovered, and they are now simply known as galaxies. In the span of a quarter of a century the scale for the universe had been grown dramatically.

The expanding universe

During the time of Hubble's work, V. M. Slipher at the Lowell Observatory had been acquiring spectra of these fuzzy clouds. By breaking the light of astronomical objects into the various "colors" or wavelengths which make up that light, astronomers can determine much about the composition, temperature, and pressure of the material that emitted that light. The familiar **spectrum** of **hydrogen** so common to so many astronomical objects did not fall at the expected wavelengths, but appeared shifted to longer wavelengths. We now refer to such a change in wavelength as a **redshift**. Slipher noted that the fainter galaxies seemed to have larger redshifts and he sent his collection of galactic spectra off to Hubble.

Hubble interpreted the redshift as being caused by the **Doppler effect**, and thus representing motion away from us. The increasing faintness was related to distance so that in 1929 Hubble turned Slipher's redshift-brightness relation into a velocity-distance relation and the concept of the expanding universe was born. Hubble found that the **velocity** of distant galaxies increased in direct proportion to their distance. The constant of proportionality is denoted by the symbol H_0 and is known as Hubble's constant.

Although the historical units of Hubble's constant are (km/s/mpc), both kilometers and megaparsecs are lengths so that the actual units are inverse time (i.e. 1/sec). The **reciprocal** of the constant basically gives a value for the time it took distant galaxies to arrive at their present day positions. It is the same time for all galaxies. Thus, the inverse of Hubble's constant provides an estimate for the age of the expanding Universe called the Hubble age.

Due to its importance, the determination of the correct value of the Hubble constant has been a central preoccupation of many astronomers from the time of Hubble to the present. Since the gravitational pull of the matter in the universe on itself should tend to slow the expansion, values of Hubble's constant determined by hypothetical astronomers billions of years ago would have yielded a somewhat larger number and hence a somewhat younger age. Therefore the Hubble age is an upper limit to the true **age of the Universe** which depends on how much matter is in the Universe.

The notion of a Heavens (i.e. universe) that was dynamic and changing was revolutionary for the age. Einstein immediately modified the equations of the General Theory of Relativity which he applied to the universe as a whole to deal with a dynamic universe. Willem de Sitter and quite independently Alexandre Friedmann expanded on Einstein's application of **General Relativity** to a dynamically expanding universe.

The concept of a universe that is changing in time suggests the idea of predicting the state of the universe at earlier times by simply reversing the present dynamics. This is much like simply running a motion picture backwards to find out how the movie began. A Belgian priest by the name of Georges Lemaitre carried this to its logical conclusion by suggesting that at one time the universe must have been a very congested place with matter so squeezed together that it would have behaved as some sort of primeval atom.

The big bang

The analysis of Einstein, de Sitter, Friedmann, Lemaitre, and others showed that the dynamic future of the expanding universe depended on the local density. Simply put, if the density of the universe were sufficiently high, then the gravitational pull of the matter in any given **volume** on itself would be sufficient to eventually stop the expansion. Within the description given by the General Theory of Relativity, the matter would be said to warp space to such an extent that the space would be called closed. The structure of such a universe would allow the expansion to continue until it filled the interior of a **black hole** appropriate for the **mass** of the entire universe at which point it would begin to collapse. A universe with less density would exhibit less space warping and be said to be open and would be able to expand forever.

There exists a value for the density between these extremes where the matter of the universe can just stop the expansion after an infinite time. Such a universe is said to be flat. One of the central questions for observational cosmology continues to be which of these three cases applies to our universe.

George Gamow concerned himself with the early phases of an expanding universe and showed that Lemaitre's primeval atom would have been so hot that it would explode. After World War II, a competing cosmology developed by Hermann Bondi, Thomas Gold, and Fred Hoyle was put forth in order to avoid the ultimate problem with the expanding universe, namely, it must have had an origin. The Steady State Cosmology of Bondi, Gold, and Hoyle suggested that the universe has existed indefinitely and that matter is continuously created so as to replace that carried away by the observed expansion.

This rather sophisticated cosmology replaced the origin problem of the expanding universe by spreading the creation problem out over the entire history of the universe and making it a part of its continuing existence. It is somewhat ironic that the current name for the expanding universe cosmology as expressed by Gamow is derived from the somewhat disparaging name, big bang, given to it by Fred Hoyle during a BBC interview.

Gamow and colleagues noted that a very hot primeval atom should radiate like a blackbody (i.e., a perfect thermal radiator), but that radiation should be extremely red-shifted by the expansion of the universe so that it would appear today like a very cold blackbody. That prediction, made in 1948, would have to wait until 1965 for its confirmation. In that year Arno Penzias and Robert Wilson announced the discovery of microwave radiation which uniformly filled the sky and had a blackbody temperature of about 2.7K (-454.5°F [-270.3°C]).

While Gamow's original prediction had been forgotten, the idea had been re-discovered by Robert Dicke and his colleagues at Princeton University. Subsequent observation of this background radiation showed it to fit all the characteristics required by radiation from the early stages of the big bang. Its discovery spelled the end to the elegant Steady State Cosmology which could not easily accommodate the existence of such radiation.

Implications of the big bang

After the Second World War, the science of nuclear physics developed to a point at which it was clear that nuclear reactions would have taken place during the early phases of the big bang. Again scientists ran the motion picture backwards through an era of nuclear physics, attempting to predict what elements should have been produced during the early history of the universe. Their predictions were then compared to the elemental abundances of the oldest stars and the agreement was amazingly good.

The detailed calculations depended critically on whether the calculated model was for an open or closed universe. If the universe were open, then the era of nuclear reactions would not last long enough to produce elements heavier than hydrogen and helium. In such models some **deuterium** is formed, but the amount is extremely sensitive to the initial density of matter. Since deuterium tends to be destroyed in stars, the current measured value places a lower limit on the initial amount made in the big bang. The best present estimates of primordial deuterium suggest that there is not enough matter in the universe to stop its expansion at any time in the future.

In the event of an open universe, scientists are pretty clear what the future holds. In 1997, researchers from the University of Michigan released a detailed projection of the four phases of the universe, including the ultimate end in the Dark Era, some 10^{100} years hence. Currently, the universe is in the Stelliferous Era, dominated by high-energy stars and filled with galaxies. Some 1000 trillion years from now, the universe will enter the Degenerate Era. Stars will have burned down into degenerate husks that can no longer support hydrogen burning reactions, and will exist as white dwarfs, red dwarfs, brown dwarfs, or neutron stars; some massive stars will have collapsed into black holes, which will consume the other star relics.

Next, the universe will progress into the Black Hole Era, about 100 trillion trillion trillion years from the present. At that time, black holes will have swallowed up the remaining bodies in the universe and will gradually leak radiation themselves, essentially evaporating away over trillions of years. Finally, the universe will reach the Dark Era, in which no matter will exist, only a soup of elementary particles like electrons, positrons, neutrinos, and other exotic particles.

In 1998, astronomers studying a certain group of supernovas discovered that the older objects were receding at a speed about the same as the younger objects. According to the theory of a closed universe, the expansion of the universe should slow down as it ages, and older supernovas should be receding more rapidly than the younger supernovas. The fact that observations have shown the opposite has led scientists to believe that the universe is either open or flat.

Cosmologists still battle over the exact nature of the universe. Most scientists agree that the age of the universe ranges between 13 and 15 billion years. The exact age, for instance, is a matter of great controversy between rival research teams at Carnegie Observatories and the Space Telescope Science Institute. When researchers recently used an analysis of polarized light to show that the universe is not isotropic, i.e., not the same in all directions, their findings were disputed almost as soon as they were published.

Trouble in paradise

In the last quarter of the twentieth century some problems with the standard picture of the big bang emerged. The extreme uniformity of the cosmic background radiation, which seemed initially reasonable, leads to a subtle problem. Consider the age of elements of the cosmic background radiation originating from two widely separated places in the sky. The distance between them is so great that light could not travel between them in an amount of time less than their age. Thus the two regions could never have been in contact during their existence. Why then, should they show the same temperature? How was their current status coordinated? This is known as the horizon problem.

The second problem has to do with the remarkable balance between the energy of expansion of the universe and the energy associated with the gravitational forces of the matter opposing that expansion. By simply counting the amount of matter we see in the universe, we can account for about 1% of the matter required to stop the expansion and close the universe. Because the expansion causes both the expansion energy and the energy opposing the expansion to tend to **zero**, the **ratio** of their difference to either one tends to get larger with time. So one can ask how good the agreement between the two was, say, when the cosmic background radiation was formed.

The answer is that the agreement must have been good to about 1 part in a million. If one extends the logic back to the nuclear era where our physical understanding is still quite secure, then the agreement must be good to about thirty digits. The slight departure between these two fundamental properties of the universe necessary to produce what we currently observe is called the "Flatness Problem." There is a strong belief among many cosmologists that agreement to 30 digits suggests perfect agreement and there must be more matter in the Universe than we can see.

This matter is usually lumped under the name **dark matter** since it escapes direct visible detection. It has become increasingly clear that there is indeed more matter in the universe than is presently visible. Its gravitational effect on the **rotation** of galaxies and their motion within clusters of galaxies suggests that we see perhaps only a tenth of the matter that is really there. However, while this amount is still compatible with the abundance of deuterium, it is not enough to close the Universe and solve the flatness problem.

Any attempt to run the motion picture further backwards before the nuclear era requires physics which, while less secure, is plausible. This led to a modification of the big bang by Alan Guth called inflation. Inflation describes an era of very rapid expansion where the space containing the matter-energy that would eventually become galaxies spread apart faster than the speed of light for a short period of time.

Inflation solves the horizon problem in that it allowed all matter in the universe to be in contact with all other matter at the beginning of the inflation era. It also requires the exact balance between expansion energy and energy opposed to the expansion, thereby solving the flatness problem. This exact balance requires that there be an additional component to the dark matter that did not take part in the nuclear reactions that determined the initial composition of the Universe. The search for such matter is currently the source of considerable effort.

Finally, one wonders how far back one can reasonably expect to run the movie. In the earliest microseconds of the universe's existence the conditions would have been so extreme that the very forces of nature would have merged together. Physical theories that attempt to describe the merger of the strong nuclear force with the electro-weak force are called Grand Unified Theories, or GUTs for short. There is currently much effort being devoted to testing those theories. At sufficiently early times even the force of gravity should become tied to the other forces of nature. The conditions which lead to the merging of the forces of nature are far beyond anything achievable on Earth so that the physicist must rely on predictions from the early Universe to test these theories.

Ultimately **quantum mechanics** suggests that there comes a time in the early history of the universe where all theoretical descriptions of the universe must fail. Before a time known as the Planck Time, the very notions of time and space become poorly defined and one should

KEY TERMS

. . . .

Grand Unified Theory—Any theory which brings the description of the forces of electromagnetism, weak and strong nuclear interactions under a single representation.

Hubble constant—The constant of proportionality in Hubble's Law which relates the recessional velocity and distance of remote objects in the universe whose motion is determined by the general expansion of the universe.

Inflation cosmology—A modification to the early moments of the big bang Cosmology which solves both the flatness problem and the horizon problem.

Megaparsec—A unit of distance used in describing the distances to remote objects in the universe. One megaparsec (i.e., a million parsecs) is approximately equal to 3.26 million light years or approximately ten trillion trillion centimeters.

Olbers' paradox—A statement that the dark night sky suggests that the universe is finite in either space or time.

Planck time—An extremely short interval of time (i.e., 10^{43} sec) when the conventional laws of physics no longer apply.

Primeval atom—The description of the very early expanding universe devised by Abbe Lemaitre.

Spectra—The representation of the light emitted by an object broken into its constituent colors or wavelengths.

Steady state cosmology—A popular cosmology of the mid twentieth century which supposed that the universe was unchanging in space and time.

not press the movie further. Beyond this time science becomes ineffective in determining the structure of the universe and one must search elsewhere for its origin.

See also Relativity, general; Relativity, special; String theory; Symmetry.

Resources

Books

Harrison, E.R. *Cosmology: The Science of the Universe.* Cambridge, England: Cambridge University Press, 1981.

Hawking, Stephen. W. *The Illustrated A Brief History of Time.* 2nd ed. New York: Bantam Books, 2001.

Kirshner, Robert P. *The Extravagant Universe: Exploding Stars, Dark Energy, and the Accelerating Cosmos.* Princeton, NJ: Princeton University Press, 2002.

Weinberg, S. *The First Three Minutes.* New York: Basic Books, 1977.

Periodicals

Glanz, James. "Evidence Points to Black Hole At Center of the Milky Way." *New York Times.* October 17, 2002.

Guth, A.H., and P.J. Steinhardt. "The Inflationary Universe," *Scientific American* 250, no. 5 (1984): 116–28.

Other

Cambridge University. "Cambridge Cosmology." <http://www.damtp.cam.ac.uk/user/gr/public/cos_home.html> (cited February 14, 2003).

National Air and Space Administration. "Cosmology: The Study of the Universe." <http://map.gsfc.nasa.gov/m_uni.html> (cited February 5, 2003).

K. Lee Lerner
George W. Collins, II

Cotingas

Cotingas are a highly diverse group of **birds** that make up the family Cotingidae. **Species** of cotingas occur widely in tropical **forests** of South and Central America. Cotingas are fly-catching birds, and are similar in many respects to species of **tyrant flycatchers** (family Tyrannidae), although these families are not closely related.

Species of cotingas are extremely variable in size, shape, **color**, **behavior**, and natural history, and the family is therefore difficult to characterize. As a result, estimates of the number of species range from about 70 to 80, depending on the taxonomic treatment that is consulted. Many of these cotingas have a highly local (or **endemic**) distribution in tropical rain forests, and many species are endangered.

The largest cotinga is the crow-sized, umbrella-bird (*Cephalopterus ornatus*). This is a slate-gray, 16 in (40 cm) long bird with a large crest over the top of the head, and an inflatable orange throat-sac, which is used to give **resonance** to its low-pitched, bellowing calls. The smallest species is the kinglet calyptura (*Calyptura cristata*), only 3 in (7.5 cm) long. Some cotingas are rather drab in color, while others are extraordinarily beautiful, with hues of deep red, orange, purple, and yellow occurring in some species.

The feeding habits of cotingas are also highly varied. Some cotingas are exclusively fruit-eaters, while others are insectivorous, but most have a mixed diet of both of these types of foods. The insect-hunting species tend to glean their **prey** from the surfaces of foliage or branches. Alternatively, they may "fly-catch," that is sit motionless while scanning for large, flying **insects**, which, when seen, are captured in the beak during a brief aerial sally.

Perhaps the most famous species in the cotinga family are the cocks-of-the-rock (*Rupicola* spp.). For example, males of the Guianan cock-of-the-rock (*Rupicola rupicola*) are colored a beautiful golden orange, with an extraordinary semi-circular, flattened crest over the entire top of the head, long plumes over the wings, and delicate black-and-white markings. Male cocks-of-the-rock have a spectacular **courtship** display in which several cocks gather at a traditional strutting ground. Each bird clears a small area, known as a "court," in which to perform his display. When a female appears, the cocks fly down to their individual court, where they assume a still pose, designed to maximize the visual impact of their charismatic, orange crest on the female. Although all of the cocks seem spectacularly attractive to any human observer, the female is able to discern one that is even more-so, and she chooses him as her mate.

Other cotingas are noted for their extremely loud calls, which can resonate through even the densest tropical **rainforest**. Male bell-birds (*Procnias* spp.) advertise themselves to females with their bell-like calls, while male pihas (*Lipaugus* spp.) make extremely loud, piercing sounds to proclaim their virility.

The only cotinga to occur in the United States is the rose-throated becard (*Platypsaris aglaiae*), which is present in local populations close to the Mexican border in Arizona, New Mexico, and Texas.

Cotton

Cotton is a fiber obtained from various **species** of plants, genus *Gossypium,* family Malvaceae (Mallow), and is the most important and widely used natural fiber in the world. Cotton is primarily an agricultural crop, but it can also be found growing wild. Originally cotton species were perennial plants, but in some areas cotton has been selectively bred to develop as an annual **plant**. There are more than 30 species of *Gossypium,* but only four species are used to supply the world market for cotton. *Gossypium hirsutum,* also called New World or upland cotton, and *G. barbadense,* the source of Egyptian cotton and Sea Island Cotton, supply most of the world's cotton fiber. *G. barbadense* was brought from Egypt to the United States around 1900. A **hybrid** of these two cotton species known as Pima cotton, is also an important source of commercial cotton. These species have relatively longer fibers and greater resistance to the boll weevil, the most notable insect pest of cotton plants. Asian cotton plants, *G. ar-*

boreum and *G. herbaceum* grow as small shrubs and produce relatively short fibers. Today, the United States produces one-sixth of the world's cotton. Other leading cotton producing countries are China (the world's biggest producer), India, Pakistan, Brazil, and Turkey. The world production of cotton in the early 1990s was about 18.9 million metric tons per year. The world's largest consumers of cotton are the United States and **Europe**.

History

Cotton was one of the first cultivated plants. There is evidence that the cotton plant was cultivated in India as long as 5,000 years ago. Specimens of cotton cloth as old as 5,000 years have been found in Peru, and scientists have found ancient specimens of the cotton plant dating 7,000 years old in caves near Mexico City. Cotton was one of the resources sought by Columbus, and while he did not manage to find a shorter route to India, he did find species of cotton growing wild in the West Indies.

Cotton plant

The cotton plant grows to a height of 3–6 ft (0.9–1.8 m), depending on the species and the region where it is grown. The leaves are heart-shaped, lobed, and coarse veined, somewhat resembling a maple **leaf**. The plant has many branches with one main central stem. Overall, the plant is cone or **pyramid** shaped.

After a cotton seed has sprouted (about four to five weeks after planting), two "seed" leaves provide food for the plant until additional "true" leaves appear. **Flower** buds protected by a fringed, leafy covering develop a few weeks after the plant starts to grow, and then bloom a few weeks later. The flower usually blooms in the morning and then withers and turns **color** within two to three days. The bloom falls off the plant, leaving a ripening seed pod.

Pollination must occur before the flower falls off. Pollen from the stamens (male part) is transferred to the stigma (female part) by **insects** and **wind**, and travels down the stigma to the ovary. The ovary contains ovules, which become **seeds** if fertilized. The ovary swells around the seeds and develops into a boll. The cotton boll is classified as a fruit because it contains seeds. As the bolls develop, the leaves on the plant turn red.

About four months are needed for the boll to ripen and split open. A cotton boll contains 27 to 45 seeds and each seed grows between 10,000 and 20,000 hairs or fibers. Each fiber is a single **cell**, 3,000 times longer than wide. The fibers develop in two stages. First, the fibers grow to their full length (in about three weeks). For the following three to four weeks, layers of **cellulose** are de-

Cotton plants in cultivation in North Carolina. *JLM Visuals. Reproduced by permission.*

posited in a crisscross fashion, building up the wall of the fiber. After the boll matures and bursts open, the fibers dry out and become tiny hollow tubes that twist up, making the fiber very strong. The seed hairs or fibers grow in different lengths. The outer and longer fibers grow to 2.5 in (6.4 cm) and are primarily used for cloth. These fibers are very strong, durable, flexible, and retain dyes well. The biological function of the long seed hairs is to help scatter the seeds around in the wind. The inner, short fibers are called linter.

Growing, harvesting, processing

Cotton requires a long growing season (from 180 to 200 days), sunny and warm weather, plenty of **water** during the growth season, and dry weather for harvest. Cotton grows near the equator in tropical and semitropical climates. The Cotton Belt in the United States reaches from North Carolina down to northern Florida and west to California. A crop started in March or April will be ready to harvest in September. Usually, cotton seeds are planted in rows. When the plants emerge, they need to be thinned. **Herbicides**, rotary hoes, or flame cultivators are used to manage weeds. **Pesticides** are also used to control bacterial and fungal diseases, and insect **pests**.

Harvesting

For centuries, harvesting was done by hand. Cotton had to be picked several times in the season because bolls of cotton do not all ripen at the same time. Today, most cotton is mechanically harvested. Farmers wait until all the bolls are ripe and then defoliate the plants with chemicals, although sometimes defoliation occurs naturally from frost.

Processing

Harvested cotton needs to be cleaned before going to the gin. Often, the cotton is dried before it is put through the cleaning equipment which removes leaves, dirt, twigs, and other unwanted material. After cleaning, the long fibers are separated from the seeds with a cotton gin and then packed tightly into bales of 500 lb (227 kg). Cotton is classified according to its staple (length of fiber), grade (color), and character (smoothness). At a textile mill, cotton fibers are spun into yarn and then woven or knitted into cloth. The seeds, still covered with linter, are sent to be pressed in an oil mill.

Cotton by-products

Cotton seeds are valuable by-products. The seeds are delinted by a similar process to ginning. Some linter is used to make candle wicks, string, cotton balls, cotton batting, **paper**, and cellulose products such as rayon, **plastics**, photographic film, and cellophane. The delinted seeds are crushed and the kernel is separated from the hull and squeezed. The cottonseed oil obtained from the kernels is used for cooking oil, shortening, soaps, and cosmetics. A semi-solid residue from the refining process is called **soap** stock or foots, and provides **fatty acids** for various industrial uses such as insulation materials, soaps, linoleum, oilcloth, waterproofing materials, and as a paint base. The hulls are used for fertilizer, plastics, and paper. A liquid made from the hulls called furfural is used in the chemical industry. The remaining mash is used for **livestock** feed.

See also Natural fibers.

Resources

Books

Basra, Amarjit. *Cotton Fibers: Developmental Biology, Quality Improvement, & Textile Processing.* Food Products Press, 2001.

Jenkins, Johnie N., and Sukumar Saha, eds. *Genetic Improvement of Cotton: Emerging Technologies.* Science Publishers, Inc., 2001.

Lewington, Anna. *Plants for People.* New York: Oxford University Press, 1990.

Stewart, J. M. *Biotechnology in Cotton Research and Production.* CABI Publishing, 2003.

Christine Miner Minderovic

Coulomb

A coulomb (abbreviation: C) is the standard unit of charge in the **metric system**. It was named after the French physicist Charles A. de Coulomb (1736-1806) who formulated the law of electrical **force** that now carries his name.

History

By the early 1700s, Sir Isaac Newton's law of gravitational force had been widely accepted by the scientific community, which realized the vast array of problems to which it could be applied. During the period 1760-1780, scientists began to search for a comparable law that would describe the force between two electrically charged bodies. Many assumed that such a law would follow the general lines of the gravitational law, namely that the force would vary directly with the magnitude of the charges and inversely as the **distance** between them.

The first experiments in this field were conducted by the Swiss mathematician Daniel Bernoulli around 1760. Bernoulli's experiments were apparently among the earliest quantitative studies in the field of **electricity**, and they aroused little interest among other scientists. A decade later, however, two early English chemists, Joseph Priestley and Henry Cavendish, carried out experiments similar to those of Bernoulli and obtained qualitative support for a gravitation-like relationship for electrical charges.

Conclusive work on this subject was completed by Coulomb in 1785. The French physicist designed an ingenious apparatus for measuring the relatively modest force that exists between two charged bodies. The apparatus is known as a torsion balance. The torsion balance consists of a non-conducting horizontal bar suspended by a thin fiber of **metal** or silk. Two small spheres are attached to opposite ends of the bar and given an electrical charge. A third ball is then placed adjacent to the ball at one end of the horizontal rod and given a charge identical to those on the rod.

In this arrangement, a force of repulsion develops between the two adjacent balls. As they push away from each other, they cause the metal or silk fiber to twist. The amount of twist that develops in the fiber can be measured and can be used to calculate the force that produced the distortion.

Coulomb's law

From this experiment, Coulomb was able to write a mathematical expression for the electrostatic force between two charged bodies carrying charges of q_1 and q_2 placed at a distance of r from each other. That mathematical expression was, indeed, comparable to the gravitation law. That is, the force between the two bodies is proportional to the product of their charges ($q_1 \times q_2$) and inversely proportional to the square of the distance between them ($1/r^2$). Introducing a proportionality constant of k, Coulomb's law can be written as: $q_1 \times q_2 \ F = kr^2$. What this law says is that the force between two charged bodies drops off rapidly as they are separated from each other. When the distance between them is doubled, the force is reduced to one-fourth of its original value. When the distance is tripled, the force is reduced to one-ninth.

Coulomb's law applies whether the two bodies in question have similar or opposite charges. The only difference is one of sign. If a positive value of F is taken as a force of attraction, then a **negative** value of F must be a force of repulsion.

Given the close relationship between **magnetism** and electricity, it is hardly surprising that Coulomb discovered a similar law for magnetic force a few years later. The law of magnetic force says that it, too, is an inverse square law. In other words: $p_1 \times p_2 \ F = kr^2$ where p_1 and p_2 are the strengths of the magnetic poles, r is the distance between them, and k is a proportionality constant.

Applications

Coulomb's law is absolutely fundamental, of course, to any student of electrical phenomena in **physics**. However, it is just as important in understanding and interpreting many kinds of chemical phenomena. For example, an atom is, in one respect, nothing other than a collection of electrical charges, positively charged protons, and negatively charged electrons. Coulombic forces exist among these particles. For example, a fundamental problem involved in a study of the atomic nucleus is explaining how the enormous electrostatic force of repulsion among protons is overcome in such a way as to produce a stable body.

Coulombic forces must be invoked also in explaining molecular and crystalline architecture. The four bonds formed by a **carbon** atom, for example, have a particular geometric arrangement because of the mutual force of repulsion among the four **electron** pairs that make up those bonds. In crystalline structures, one arrangement of ions is preferred over another because of the forces of repulsion and attraction among like-charged and oppositely-charged particles respectively.

Electrolytic cells

The coulomb (as a unit) can be thought of in another way, as given by the following equation: 1 coulomb = 1

---

ampere $\times$ 1 second. The ampere (amp) is the metric unit used for the measurement of electrical current. Most people know that electrical appliances in their home operate on a certain number of "amps." The ampere is defined as the flow of electrical charge per second of **time**. Thus, if one multiplies the number of amps times the number of seconds, the total electrical charge (number of coulombs) can be calculated.

This information is of significance in the field of electrochemistry because of a discovery made by the British scientist Michael Faraday in about 1833. Faraday discovered that a given quantity of electrical charge passing through an electrolytic cell will cause a given amount of chemical change in that cell. For example, if one **mole** of electrons flows through a cell containing **copper** ions, one mole of copper will be deposited on the **cathode** of that cell. The Faraday relationship is fundamental to the practical operation of many kinds of electrolytic cells.

See also Electric charge.

Resources

Books

Brady, James E., and John R. Holum. *Fundamentals of Chemistry*. 2nd edition. New York: John Wiley and Sons, 1984.

Holton, Gerald, and Duane H. D. Roller. *Foundations of Modern Physical Science*. Reading, MA: Addison-Wesley Publishing Company, 1958.

Shamos, Morris H., ed. *Great Experiments in Physics*. New York: Holt, Rinehart and Winston, 1959.

Wilson, Jerry D. *Physics: Concepts and Applications.* 2nd edition. Lexington, MA: D. C. Heath and Company, 1981.

David E. Newton

Countable

Every set that can be counted is countable, but this is no surprise. The interesting case for countable sets comes when we abandon finite sets and consider infinite ones.

An infinite set of numbers, points, or other elements is said to be "countable" (also called *denumerable*) if its elements can be paired one-to-one with the **natural numbers**, 1, 2, 3, etc. The term countable is somewhat misleading because, of course, it is humanly impossible actually to count infinitely many things.

The set of even numbers is an example of a countable set, as the pairing in Table 1 shows.

Of course, it is not enough to show the way the first eight numbers are to be paired. One must show that no matter how far one goes along the list of even natural numbers there is a natural number paired with it. In this case this is an easy thing to do. One simply pairs any even number 2n with the natural number n.

What about the set of **integers**? One might guess that it is uncountable because the set of natural numbers is a proper subset of it. Consider the pairing in Table 2.

Remarkably it works.

The secret in finding this pairing was to avoid a trap. Had the pairing been that which appears in Table 3, one would never reach the **negative** integers.

In working with infinite sets, one considers a pairing complete if there is a scheme that enables one to reach *any* number or element in the set after a finite number of steps. (Not everyone agrees that that is the same thing as reaching them *all*.) The former pairing does this.

Are the rational numbers countable? If one plots the rational numbers on a number line, they seem to fill it up. Intuitively one would guess that they form an uncountable set. But consider the pairing in Table 4, in which the rational numbers are represented in their **ratio** form.

The listing scheme is a two-step procedure. First, in each ratio, the denominator and numerator are added. All those ratios with the same total are put in a **group**, and these groups are listed in the order of increasing totals. Within each group, the ratios are listed in order of increasing size. Thus the ratio 9/2 will show up in the group whose denominators and numerators total 11, and within that group it will fall between 8/3 and 10/1. The

TABLE 1							
Even Nos.	2	4	6	8	10	12	14	16 ...
Nat. Nos.	1	2	3	4	5	6	7	8 ...

TABLE 2									
Integers	0	1	-1	2	-2	3	-3	4	...n	-n ...
Nat. Nos.	1	2	3	4	5	6	7	8	...2n	2n + 1 ...

TABLE 3								
Integers	0	1	2	3	4	5	6	7	8 ...
Nat. Nos.	1	2	3	4	5	6	7	8	9 ...

TABLE 4													
Rat. Nos.	0/1	1/1	1/2	2/1	1/3	2/2	3/1	1/4	2/3	3/2	4/1	1/5	2/4 ...
Nat. Nos.	1	2	3	4	5	6	7	8	9	10	11	12	13 ...

TABLE 5	
1	.31754007...
2	.11887742...
3	.00037559...
4	.39999999...
5	.14141414...
6	.44116798...

list will eventually include any positive **rational number** one can name.

Unfortunately, there are two flaws. The list leaves out the negative numbers and it pairs the same rational number with more than one natural number. Because 1/1, 2/2, and 3/3 have different numerators and denominators, they show up at different places in the list and are paired with different natural numbers. They are the same rational number, however. The first flaw can be corrected by interleaving the negative numbers with the positive numbers, as was done with the integers. The second flaw can be corrected by throwing out any ratio which is not in lowest terms since it will already have been listed.

Correcting the flaws greatly complicates any formula one might devise for pairing a particular ratio a/b with a particular natural number, but the pairing is nevertheless one-to-one. Each ratio a/b will be assigned to a group a + b, and once in the group will be assigned a specific place. It will be paired with exactly one natural number. The set of rational numbers is, again remarkably, countable.

Another countable set is the set of algebraic numbers. Algebraic numbers are numbers which satisfy polynomial equations with **integral** coefficients. For instance, $\sqrt{2}$, I, and $(-1 + \sqrt{5})/2$ are algebraic, satisfying $x^2 - 2 = 0$, $x^2 + 1 = 0$, and $x^2 + x - 1 = 0$, respectively.

Are all infinite sets countable?

The answer to this question was given around 1870 by the German mathematician George Cantor. He showed that the set of numbers between 0 and 1 represented by infinite decimals was uncountable. (To include the finite decimals, he converted them to infinite decimals using the fact that a number such as 0.3 can be represented by the infinite decimal 0.29—where the 9s repeat forever.) He used a *reductio ad absurdum* **proof**, showing that the assumption that the set *is* countable leads to a contradiction. The assumption therefore has to be abandoned.

He began by assuming that the set *was* countable, meaning that there was a way of listing its elements so that, reading down the list, one could reach any number in it. This can be illustrated with the example in Table 5.

From this list he constructs a new decimal. For its first digit, he uses a 1, which is different from the first digit in the first number in the list. For its second digit, he uses a 3, which is different from the second digit in the second number. For its third digit, he uses a 3 again, since it is different from the third digit in the third number.

He continues in this fashion, making the n-th digit in his new number different from the n-th digit in the n-th number in the list. In this example, he uses 1s and 3s, but he could use any other digits as well, as long as they differ from the n-th digit in the n-th number in the list. (He avoids using 9's because the finite decimal 0.42 is the same number as the infinite decimal 0.4199999... with 9's repeating.) The number he has constructed in this way is 0.131131... Because it differs from each of the numbers in at least one decimal place, it differs every number in the assumed complete list. If one chooses a number and looks for it in a listing of a countable set of numbers, after a finite number of steps he will find it (assuming that list has been arranged to demonstrate the countability of the set). In this supposed listing he will not. If he checks four million numbers, his constructed number will differ from them all in at least one of the first four million decimal places.

Thus the assumption that the set of infinite decimals between 0 and 1 is countable is a false assumption. The set has to be uncountable. The infinitude of such a set is different from the infinitude of the natural numbers.

Resources

Books

Friedrichs, K.O. *From Pythagoras to Einstein.* Washington, DC: Mathematical Association of America, 1965.

Gullberg, Jan, and Peter Hilton. *Mathematics: From the Birth of Numbers.* W.W. Norton & Company, 1997.

Zippin, Leo. *Uses of Infinity.* Washington, DC: The Mathematical Association of America, 1962.

J. Paul Moulton

Counting numbers *see* **Natural numbers**

Coursers and pratincoles

Coursers and the closely related pratincoles are 17 **species** of **birds** that comprise the family Glareolidae, in the order Charadriiformes, which also contains the **plovers**, **sandpipers**, and other families of waders and shorebirds. The pratincoles occur in southern **Europe** and **Asia**, **Africa**, Southeast Asia, and Australasia, but coursers only occur in Africa, the Middle East, and India.

Coursers and pratincoles breed in sandy or stony deserts, in grassy plains, or in savannas, but always near **water**. Both types of birds fly gracefully, using their long, pointed wings.

Coursers have relatively long legs, three toes on their feet, a square tail, and a relatively long, thin, somewhat down-curved beak. Coursers are nomadic during their non-breeding season, undertaking wanderings in unpredictable directions, as is the case of many other bird species that breed in deserts.

Pratincoles have shorter legs, four toes, a deeply forked tail, and a short bill with a wide gape. Pratincoles undertake long-distance migrations during the non-breeding season, usually in flocks. The sexes are similar in both of these types of birds.

Pratincoles largely predate on flying **insects**, much in the manner of swallows (an alternate common name for these birds is swallow-plover). They also feed on the ground, running after their **prey** with rapid, short bursts of speed. Coursers are also insectivorous, but they feed exclusively on terrestrial insects, which are caught on the run. Coursers will also eat **seeds** when they are available.

Pratincoles occur in groups and nest in large, loosely structured colonies. Coursers are less social than this and do not nest in colonies. The nests of coursers and pratincoles are simple scrapes made in the open. In most species there are two eggs in a clutch, which are incubated by both the female and the male parents. These birds mostly nest in hot habitats, so the purpose of incubation is often to keep the eggs cool, rather than warm as in most birds. Some species moisten their eggs to keep them cooler.

Young coursers and pratincoles are precocious, hatching with their eyes open and are able to walk one day after **birth**. However, they are led to a sheltering bush or other hiding place as soon as they are mobile, and they shelter there while the parents bring them food. The young birds are well camouflaged to blend in with their surroundings.

If a **predator** is near the nest or babies, adult pratincoles will perform injury-feigning distraction displays meant to lure the **animal** away. Both coursers and pratincoles also have startle displays that they deploy under these conditions, in which the wings and tail are raised suddenly to reveal bold patterns of coloration in an attempt to unnerve the predator.

Species of coursers

The double-banded courser (*Rhinoptilus africanus*) of Africa lays only one egg, and the nest is commonly located near antelope dung as an aid to camouflage. The nest of the three-banded courser (*R. cinctus*) of Africa is a relatively deep scrape in which the clutch of two eggs is two-thirds buried in **sand** during incubation. The Egyptian plover (*Pluvialis aegyptius*) is a courser that breeds along large **rivers** in central and northeastern Africa. This species also buries its eggs and even its babies, and it also regurgitates water to cool these down on especially hot afternoons.

Species of pratincoles

The collared pratincole, red-winged pratincole, or swallow plover (*Glareola pratincola*) is highly unusual in having migratory populations in both the southern and northern hemispheres. Northern birds breed in open steppes, savannas, and dry mudflats in southern Europe and southeastern Asia, and winter in Africa. Birds that breed in southern Africa, migrate to northern Africa to spend their non-breeding season. The black-winged pratincole (*G. nordmanni*) is a widespread species that breeds from southeastern Europe through central Asia, and winters on tropical shores in south and southeast Asia. The oriental pratincole (*G. maldivarum*) breeds in central and southern Asia, and migrates as far south as **Australia**.

The Australian pratincole (*Stiltia isabella*) breeds widely across much of that **island continent**, including the semi-arid interior. This species migrates north to spend its non-breeding season in the coastal tropics, from northern Australia to Indonesia. This is the only species in its genus, and it is rather intermediate in form to the coursers and pratincoles. Like the coursers, the Australian pratincole has a relatively long beak and long legs, a short tail, and no hind toe. In addition, this species does not have comb-like structures called pectinations on the claw of the middle toe, a characteristic that all other pratincoles and the coursers exhibit.

Resources

Books

Bird Families of the World. Oxford: Oxford University Press, 1998.

Brooke, M., and T. Birkhead. *The Cambridge Encyclopedia of Ornithology*. Cambridge, U.K.: Cambridge University Press, 1991.

Hayman, P., J. Marchant, and T. Prater. *Shorebirds: An Identification Guide to the Waders of the World*. London: Croom Helm, 1986.

Bill Freedman

Courtship

Courtship is a complex set of behaviors in animals that leads to mating. Courtship **behavior** communicates to each of the potential mates that the other is not a threat. It also reveals information to each **animal** that the **species**, gender, and physical condition of the other are suitable for mating. Pre-mating activities are for the most part ritualistic. They consist of a series of fixed action patterns that are species-specific. Each fixed action triggers an appropriate fixed reaction by the partner, with one action stimulating the next. Courtship allows one or both sexes to select a mate from several candidates. Usually, the females do the choosing. In some species of **birds**,

males display in a lek, a small communal area, where females select a mate from the displaying males. Males, generally, compete with each other for mates, and females pick the best quality male available. The danger of courtship is that it can attract predators instead of mates.

Several basic factors influence a female's choice of mate. First, if a female provides parental care, she chooses as competent a male as possible. For example, in birds such as the common tern, the female selects a good **fish** catcher. As part of courtship, the male birds display fish to the female, and may even feed them to her. This demonstrates his ability to feed the young. In addition, females tend to select males with resources such as food or shelter which help a mating pair to produce more offspring that survive. In the long-jawed longhorned beetle that lives in the Arizona **desert**, males battle each other for saguaro **cactus** fruit. The females mate in exchange for access to the fruit. A male endowed with large mandibles can defeat other males, take over the fruit, and thus attract females. Genetic fitness is another important factor in mate **selection**. In species that lack parental care, offspring rely for survival on qualities that they inherit from their parents. During courtship, energetic displays and striking appearance indicate good health. Vigorous, attractive parents generally pass immunities to their offspring. Attractiveness may depend on the intensity of secondary sex characteristics, which in birds, for example, include colorful plumage and long tails. Another advantage is that inherited attractive features make offspring desirable to mates.

Courtship in insects

Insect courtship is ritualistic and has evolved over **time**. Male balloon flies of the family Empididae spin oval balloons of silk. Then they fly in a swarm, carrying their courtship objects aloft. Females approach the swarm and select their mates. As a female pairs off with a male, she accepts his balloon. In some species of balloon flies, the male brings the female a dead insect to eat during copulation. This may prevent her from eating him. In other species, the male carries a dead insect inside a silk balloon. Apparently, in the course of **evolution**, the suitor's gift-giving began with "candy," then a "box of candy," and finally just the empty "box." Other courtship strategies in **insects** include female **moths** that release a scent signal (or pheromone) that males of the same species recognize. When a male detects the signal, he flies upstream to the female. In queen **butterflies**, courtship is complex and requires steps that must occur in the proper order. First, the female flaps her wings and draws the male's attention and pursues her. As he hovers nearby his hairpencils (brushlike organs) release a pheromone. Then the receptive female lands

A male peacock displays his plumage as part of the courtship ritual. *Photograph by Norbert Wu. Stock Market. Reproduced by permission.*

on a nearby **plant**. Next, the male brushes his hairpencils on her antennae. She responds by closing her wings. This signals the male to land on her and begin mating. In other courtship behavior, male **crickets** rub their forewings together and produce a pulsed courtship song. In fireflies, the male's flashing **light** and the female's flashing answer is another type of courtship behavior. In fireflies, both sexes respond to a specific set of intervals between flashes.

Courtship in fish

In 1973, Niko Tinbergen won a Nobel Prize for his work on animal behavior. One of the topics he studied was courting in stickleback, small **freshwater** fish. At breeding time, the male stickleback changes **color** from dull brown, to black and blue above and red underneath. At this time, he builds a tunnel-shaped nest of **sand**. Females swollen with unfertilized eggs cruise in schools through the male territory. The male performs his zigzag courtship dance toward and away from the female fish. Attracted to the red color on the male's belly, a female ready to lay eggs displays her swollen abdomen. The male leads her to the nest. He pokes the base of her tail

with his snout, and the female lays her eggs and then swims away. The male enters the nest and fertilizes the eggs. In this manner, he may lead three or four females to his nest to lay eggs. Tinbergen showed in his studies that seeing the color red caused increased aggressiveness in males and attraction in females.

Courtship in birds

Adult birds generally return to their nesting grounds each mating season. A male claims a territory by singing a distinctive song. He then sings a song that attracts a female. Birds have different courtship rituals. Some use song, while others display colorful plumage. Woodcocks fly upward in a **spiral**, and **birds of paradise** do somersaults. Male frigatebirds—large birds with wings that spread wider than 6.6 ft (2 m)—breed in late winter on the coast of tropical islands in the western Atlantic and Pacific Oceans. The male perches in a low **tree** or brush and his red throat pouch inflates like a balloon. Its red color attracts females hovering overhead. Then the male spreads his wings, shakes them, and makes a whinnying sound. Finally, the pair come together, mate, and build a nest.

KEY TERMS

. . . .

Display—Showy exhibition by an animal that reveals information to others.

Lek—Communal area used by birds and insects for courtship and mate selection.

Pheromone—Chemical odorant that provides communication between animals.

Ritual—Species-specific behavior pattern or ceremony used for communication between animals.

Courtship in mammals

Mammals use various strategies in courtship. **Pheromones** act as sexual lures that bring members of the opposite sex together. These attractants are so powerful that a male dog can **smell** a female in estrus more than half a mile (1 km) away. The fact that at **puberty** humans begin to produce odorous sweat suggests the role of pheromones in primate courtship. Sex selection also exists in **primates**. Females usually choose their male partners, but sometimes the reverse occurs. Recent research reveals that male lion-tailed **macaques** remain aloof during the day. At night, however, they seek out sleeping estrous females for mating. Until this study, biologists thought that it was the females who initiated mating. In humans, various cultures determine the customs of courtship. For example, in some societies, marriages are arranged by relatives. In these cases, a woman is matched to a man with the appropriate resources. Just as other female animals select mates with resources, humans tend to select mates with wealth and status. Further, even if a woman has no say in the selection of her husband, she will help arrange the marriage of her offspring. This merely delays female mate choice by a generation.

See also Sexual reproduction.

Resources

Books

Batten, Mary. *Sexual Strategies*. New York: G.P. Putnam's Sons, 1992.

Chinery, Michael. *Partners and Parents*. Crabtree Publishing, 2000.

Otte, Jean Pierre. Marjolijn De Jager, trans. *The Courtship of Sea Creatures*. New York: George Braziller, 2001.

Periodicals

Davies, Nicholas B. "Backyard Battle of the Sexes." *Natural History* (April 1995).

Dennis, Jerry. "Mates for Life." *Wildlife Conservation* (May-June 1993).

Fernald, Russell D. "Cichlids in Love." *Sciences* (July-August 1993).

Hancock, Leah. "Whose Funny Valentine?" *Natural Wildlife* (February-March 1995).

Jackson, Robert. "Arachnomania." *Natural History* (March 1995).

Robert, Daniel, and Ronald R. Hoy. "Overhearing Cricket Love Songs." *Natural History* (June 1994).

Wilson, J. F., et al. "Genetic Evidence for Different Male and Female Roles During Cultural Transitions in the British Isles." *Proceedings of the National Academy of Sciences of the United States of America* 98 (2001): 5078-5083.

Other

Films for the Humanities and Sciences. *Mating Signals*. Princeton, 1994-5.

Films for the Humanities and Sciences. *The Rituals of Courtship*. Princeton, 1994-5.

Bernice Essenfeld

Covalent bond *see* **Chemical bond**

Coyote *see* **Canines**

Coypu

The coypu or nutria (*Myocastor coypu*) is a **species** of semi-aquatic, dog-sized rodent in the family Capromyidae. These animals are native to central and southern **South America**, but they have become widely established elsewhere, mostly as a result of animals that have escaped from fur farms or that have been deliberately released.

Coypus have a stout, 17-25 in (43-64 cm) long body, with a roughly triangular-shaped head, and a round, scaly, sparsely-haired tail, which is 10-17 in (25-43 cm) long. Adult animals weight 15-20 lb (7-9 kg), with males being somewhat larger than females. The eyes and ears of coypus are small, and the legs are short. The forelegs are much larger than the hind and have four webbed toes as well as a single free toe used for grooming the fur. The toes have prominent claws. The fur is soft, dense, and lustrous, consisting of long guard hairs over a velvety underfur. The **color** of the guard hairs ranges from yellow-brown to red-brown, while the underfur is blackish.

Coypus are semi-aquatic animals, and are excellent swimmers. They typically live in the vicinity of slow-moving **rivers** and streams, or near the edges of shallow lakes, marshes, and other **wetlands**. Coypus mostly live in **freshwater** habitats, but in some places they occur in **brackish** and **saltwater** wetlands as well. Coypus den in burrows dug into banks, or in mounds of reedy vegetation that are constructed when ground suitable for digging is not available. Coypus live in pairs or small family groups and sometimes in larger colonies.

Coypus typically forage during twilight hours over a **distance** of up to several hundred yards, travelling along well-worn pathways through the usually grassy **habitat**. Coypus are shy and wary when foraging, and flee quickly to their burrow when any hint of danger is perceived. Coypus are mainly vegetarian in their feeding, although they will also eat molluscs.

The fur of coypus is a valued product, and this species has been introduced as a fur-bearing species into various parts of the United States and **Europe**. Coypus are also cultivated on fur farms, where they will breed continuously, and can be quite productive. Cultivated coypus can have a white or yellowish fur, in addition to the darker colors of the wild animals.

Coypus are considered to be **pests** in many of the places where they have become naturalized because they damage **irrigation** ditches and earthen **dams** with their burrows, compete with native fur-bearers, and, when abundant, can significantly deplete the abundance of forage.

See also Beavers.

Crabs

Crabs are some of the best known arthropods—a terms that means jointed foot (Greek: *arthron*, joint; *pous*, foot). They are among the most successful of all living **species** (about 4,500 species have been described), with members adapted to living on land and in **water**; some species even succeed in living in both habitats. The majority, however, live in the marine environment. Unlike **lobsters** (to which they are closely related), which have a long and cylindrical body with an extended abdomen, crabs have a broad, flattened body and a short, symmetrical abdomen—adaptations that enable them to squeeze beneath **rocks** and into crevices for feeding purposes as well as concealment.

The bulk of the body is taken up by the abdomen. Attached to this is a small head which bears long **eye** stalks that fit into special sockets on the carapace. There are also several pairs of antennae of unequal length and feeding mouthparts known as maxillipeds. The first pair of walking legs are large in comparison with the remainder of the body and end in pinching claws. These are usually referred to as chelipeds. In most species, the tips of the remaining four pairs of legs terminate in pointed tips. When feeding, food is picked up by the chelipeds, torn apart, and passed to the maxillipeds in small portions,

from where it is pushed towards the pharynx. While some species are active predators of small **fish**, others are detritus feeders and scoop large volumes of mud towards the mouth region using the chelipeds as spades. These species then filter out any food particles and reject the remainder of the materials. Some species of burrowing crabs, which remain concealed in the soft sea bed, create a water current down into their burrows and filter out food particles in a similar manner. Their chelipeds are also fringed with tiny hair-like structures known as setae, which help extract the largest unwanted materials from the water current before other parts are ingested.

When moving on land or on the sea bed, crabs usually move in a sideways manner: the leading legs pull the body forward and those on the opposite side assist by pushing. Some species may use just two or three pairs of legs when moving quickly, stopping occasionally to turn around and reverse the order in which the legs move. Contrary to popular opinion, few crabs actually swim. One group of specialized swimming crabs (the family Portunidae) have an oval shaped body and the last pair of walking legs are flattened and act as paddles that propel the **animal**. Examples of these swimming crabs include the common blue crab (*Callinectes sapidus*), the green crab (*Carcinides maenas*) and the lady, or calico crab (*Ovalipes ocellatus*).

The remaining species, the "true crabs" vary considerably in size and **behavior**. Some of the largest of these are the spider crabs (family Maiidae). These are all marine species that live in the littoral zone, frequently skulking around on the sea bed in harbors and estuaries. This group contains the largest known arthropod, the giant Japanese crab (*Macrocheira kaempferi*) which can measure up to 13 ft (4 m) in diameter when fully extended. Most members of this family are scavenging animals. Many elect to carry a range of small **sponges** and other marine organisms on their outer carapace for concealment.

The aptly named fiddler crabs (family Ocyopidae), are easily recognized by the massively enlarged front claw of the male. The claw is usually carried horizontally in front of the body and has been likened to a fiddle; the smaller opposing claw is known as the bow. When males are trying to attract females, they wave these large claws two and fro; crabs with larger claws seem to attract more suitors than those with tiny claws. These crabs are usually a **light** brown **color** with mottled purple and darker brown patches on the carapace—a pattern that helps to conceal them on the dark sands and mud flats on which they live.

Unlike all other crabs, the tiny hermit crab has a soft body which is inserted in the shell of a marine snail for protection. Hermit crabs never kill the original occupant

A hermit crab. *Photograph by Thomas Dimock. Stock Market. Reproduced by permissions.*

of the shell and frequently change "homes" as they grow, slipping out of one shell and into another. House hunting is a demanding task, and hermit crabs spend considerable time inspecting new prospective pieces of real estate, checking for size and weight. The shell is held on through a combination of modified hind limbs, which grasp some of the internal rings of the shell, and the **pressure** of the body against the shell wall. When resting, the crab can withdraw entirely inside the shell, blocking the opening with its claws. Hermit crab shells are commonly adorned with **sea anemones** and hydroids, the reason seeming to be that these provide some protection against small predators due to the battery of specialized stinging cells that these organisms possess. In return for this service, the anemones and hydroids may benefit from the guarantee that they will always be in clean water and the possibility of obtaining food scraps from the crab when it is feeding. The importance of this relationship for the crab is seen when a hermit crab changes its shell, as they usually delicately remove the anemones and hydroids from their former home to their new abode.

Of the terrestrial species, one of the most distinctive groups are the robber, or coconut, crabs which live in deep burrows above the high water mark. These crabs rarely venture into the sea, apart from when they lay their eggs. They have overcome the problem of obtaining **oxygen** by converting their gill chambers to modified chambers lined with moisture, enabling them to breathe atmospheric oxygen. Closely related to the hermit crab, robber crabs have developed a toughened upper surface on their abdomen which means that have no need of a shell for protection. Coconut crabs—so called because of their habit of digging in the soft soils of coconut plantations—occasionally climb trees and sever the stems attaching young coconuts, on which they feed.

Crabs have a complicated **life history**. Mating is usually preceded by a short period of **courtship**. The eggs are laid shortly after copulation and are retained on the female's body until the larvae emerge. The tiny "zoea" larvae, as they are known, are free-living and grow through a series of body molts to reach a stage known as the "megalops" larvae, at which stage the first resemblance to the parent crabs is visible. Further development leads to the immature and mature adult form.

Cranberry *see* **Heath family (Ericaceae)**

Crane

The crane is an invention of ancient origin that is used to move heavy weights in both the vertical and horizontal directions, to load and unload heavy objects, and to construct tall buildings. Cranes can move objects weighing up to several hundred tons, depending on their design capacity, and they can be powered by human or animal power, water power, steam, internal combustion engines (gasoline or diesel), or electric power. One common forerunner of the crane was the *shaduf*, prevalent in Egypt and India around 1500 B.C. Employed by a single person for lifting water, the *shaduf* consisted of a vertical support, a long, pivoting beam, and a counterweight.

The first true cranes, founded on the principles of levers and counterweights, used a pulley system fixed to a single mast or boom. Lifting power was provided by humans or draft animals operating a treadmill or large wheel. Eventually, a second mast and guy wires were added to increase the strength and stability of this early form of crane.

One of the most significant developments in crane design, which probably occurred during medieval times with the advent of Gothic architecture, was the jib crane, which features a pivoting horizontal arm (a jib) that projects outward from the top of the boom. The addition of hinged movement to the outermost section of the jib allows for even further versatility and movement.

Jib cranes are also known as derrick cranes. Derrick is the term originally applied to gallows structures when Englishman Godfrey Derrick was a well-known hangman. Today, the derrick is a large hoisting machine similar in most respects to the crane, except for its typically stationary foundation. Oil derricks, for example, are specialized steel towers used for raising and lowering equipment for drilling oil wells. One of the most powerful cranes, a barge derrick, is a double-boomed structure capable of lifting and moving ships weighing up to 3,000 tons (2,700 metric tons).

Other cranes with specialized uses include the cantilever crane featuring a suspended horizontal boom and used in shipyards; the overhead traveling crane, also called a bridge crane, that is guided by rails and a trolley-suspended pulley system and used for indoor work; the gantry crane is a specialized bridge crane that is suspended between legs and moves laterally on ground rails; and the tractor-mounted crawler crane, which is a hydraulic-powered crane with a telescoping boom. A simple example of a small-scale crane is the fork-lift truck. Like its much larger relatives, the fork-lift is limited not so much by the size of its hoisting apparatus as by the force of its rear counterweight.

Cranes

Cranes are tall, wading birds known for their beauty, elaborate courtship dances, and voices that boom across their wetland habitat. Their family Gruidae, is among the oldest on Earth. Today 15 crane species are found throughout the world, except in South America and Antarctica. Two species, the whooping crane (*Grus americana*) and the sandhill crane (*G. canadensis*) are found in North America. Cranes belong to order Gruiformes, which also includes rails, coots, trumpeters, and the limpkin.

Cranes have long legs, a long neck, and a narrow, tapered bill. Most of them have a featherless spot on the top of the head that exposes colored skin. Some have wattles, or flaps of flesh, growing from the chin. The wattled crane (*Bugeranus carunculatus*) of eastern and southern Africa has very large wattles. This bird looks more like a chunky stork than a typical crane. Although cranes do look rather like storks and herons, those birds have a toe that points backward, enabling them to grasp a tree branch while perching. Cranes have a backward toe, but in all except the crowned crane (*Balearica*) it is raised up off the ground and of no use in grasping branches. Crowned cranes are able to perch in trees like storks.

Most cranes migrate fairly long distances to their nesting sites. Their large, strong wings allow them, once airborne, to glide on air currents, taking some of the strain out of the long trip. When flying, cranes stretch their neck and legs straight out, making a long straight body-line. They can cruise at speeds of about 45 mph (72 kph).

Cranes are highly vocal birds. They make many different sounds, from a low, almost purring sound, apparently of contentment, to a loud, high-pitched call that announces to other birds one is about to take flight. Mating pairs of cranes will often point their beaks to the sky and make long, dramatic calls that have been called unison calls or bonding calls. Cranes have a long windpipe that gives volume to their calls.

Cranes eat grains, especially liking waste corn and wheat found in harvested fields. They also eat invertebrates they catch in the water. Both in water and on land, cranes often stand on one foot, tucking the other under a wing.

Dancing and mating

Cranes are noted for their amazing dances that bond individual males and females together. These dances are elaborate and ballet-like, and are among the most intricate and beautiful in the animal kingdom. The cranes bow, leap high into the air, and twirl with their wings

Whooping cranes. © *U.S. Fish & Wildlife Service. Reproduced by permission*

held out like a skirt. However, this wonderful dance is not only a courtship dance because once a pair has bonded, they are mated for life. Cranes dance at other times, too, possibly as a way of relieving frustration, which might otherwise erupt into aggression. They also appear to dance for pleasure. Very young cranes start dancing with excitement.

A pair of cranes construct a raised nest of dried **grasses** by water. Either parent might start incubating as soon as the first egg is laid, although they usually lay two eggs. Crowned cranes often lay three eggs. The eggs hatch after about 28-31 days of incubation. Both parents feed the chicks and take care of them long past the time that they grow their adult feathers. The young are yellowish tan or grayish and very fuzzy. Once they reach adult size, their parents drive them away to establish lives of their own. The life spans of cranes vary considerably. Sandhill cranes rarely live more than 20 years. One Siberian crane (*Bugeranus leucogeranus*) was known to live 82 years, but 30-40 years is more usual.

While at their nesting site, most cranes go through a period of molting, or losing their feathers. Some of them have a period of up to a month during which so many feathers have been shed they are flightless.

Species of cranes

The largest crane, and the rarest Asian crane, is the red-crowned, or Japanese, crane (*Grus japonicus*). This bird can weigh up to 25 lb (11.4 kg). It has vivid red feathers on the top of its head, but its body is snowy white. It appears to have a black tail, but actually, these feathers are the tips of its wings. Although formerly widespread, the red-crowned crane is now reduced to very small populations in eastern **Asia** (this species breeds in Russia, and winters in China, Japan, North Korea, and South Korea). In 1952, this crane became Japan's national bird. There are fewer than about 1,200 of these birds left in the wild.

The smallest crane is the demoiselle crane (*Anthropoides virgo*) of **Europe** and North Africa. It has white **ear** tufts that stretch backward from its eyes and hang off the back of the head. Demoiselle cranes live on drier ground than other cranes. The blue crane (*A. paradisea*) of Africa has wingtip feathers that reach backward and to the ground, like a bustle. These two cranes do not nest by water, but in **grasslands** or even semiarid land. The blue crane is the national bird of South Africa. It has the surprising ability when excited, of puffing out its cheeks until its head looks frightening.

The tallest crane is the sarus crane (*G. antigone*) of India, Cambodia, Nepal, Vietnam, and northern **Australia**. Standing 6 ft (2 m) tall, it is a gray bird with a head and throat of vivid red. The red **color** ends abruptly in a straight line around the white neck. This species is among the least social of cranes, and it becomes aggressive when nesting.

A frequent resident of zoos is the crowned crane (*Balearica pavonina*) of Africa, which has a beautiful puff of golden feathers coming from the back of the head. It has a red wattle beneath its black and white head, a light gray neck, dark gray back and tail, and white, sometimes yellowish, wings. The West African subspecies has a black neck instead of gray and lacks the red wattle.

The rare black-necked, or Tibetan crane (*Grus nigricollis*) of the Himalayas breeds on the high plateau of Tibet. It migrates to the valleys of southwest China and Bhutan to spend the winter. The black-necked crane is a medium-sized crane with a stocky appearance; it has a larger body and shorter neck and legs than related species, perhaps as an **adaptation** to the cold climate of the Tibetan plateau. This crane has a black neck and head, plus a striking black trailing edge to its wings. A golden circle around the **eye** makes the eye look enormous against the black feathers. It is estimated that about 5,500 black-necked cranes survive in the wild.

The seriously endangered Siberian crane is as beautiful as it is rare, with long reddish pink legs, a red-orange face, and a snowy white body with black areas on its wings. This crane breeds only at two locations in Siberia and winters in China, India, and Iran. Only a few thousand of these birds are left.

Whooping crane

The whooping crane (*Grus americana*) is the rarest crane and the tallest American bird. This crane stands 5 ft (1.5 m) tall and has a wingspan of 7 ft (2.1 m). Adolescent birds have a golden-yellow neck, back, and beak, with golden edges to the black-tipped wings. By adulthood, only the wing tips are black; the rest of the bird is white except for the crown of the head and cheeks, and part of the long, pointed beak, all of which are red.

The population of the whooping crane was down to only 16 in 1941. The population decline of this species was mostly caused by hunting and habitat loss. In 1937, the wintering area of the whooping crane on the Gulf coast of Texas was protected as Aransas National Wildlife Refuge. In the meantime, some of the few remaining whooping cranes had been taken into captivity. Each egg that was laid was carefully incubated, but few of the young survived. Then, a nesting area was discovered inside Wood Buffalo National Park in northwestern Canada. Whoopers generally lay two olive-colored eggs, but only one of them hatches. Canadian **wildlife** biologists began removing one egg from wild nests. Most of the eggs were hatched in an incubator, while others were placed into the nests of wild sandhill cranes, which served as foster parents. Captive-reared birds have been returned to the wild. This, coupled with careful protection of the wild birds, has allowed the numbers of whooping cranes to gradually increase. By 1995, several hundred whooping cranes existed. However, the species remains perilously endangered.

Sandhill crane

Sandhill cranes are smaller than whooping cranes. They are generally light-gray in color, with a red crown, black legs, and white cheeks. There are large breeding populations in Siberia, Alaska, and northern Canada east of James Bay, as well as in the western and central United States. There are six subspecies of sandhill crane. Three of them—the greater, lesser, and Canadian sandhill cranes—are migratory birds. The other three—Flori-

da, Mississippi, and Cuban— do not migrate. The lesser sandhill, which is the smallest subspecies (less than 4 ft (1.2 m) tall and weighing no more than 8 lb (3.6 kg)) migrates the greatest **distance**. Many birds that winter in Texas and northern Mexico nest in Siberia.

The populations of sandhill cranes were greatly reduced by hunting and habitat loss in the 1930s and 1940s. However, wherever protected, they have been making a good comeback. One population in Indiana increased from 35 to 14,000 over a 40-year period. In one of the most amazing sights in nature, perhaps half a million sandhill cranes land on the sandbars of the Platte River in Nebraska while they are migrating.

Resources

Books

Forshaw, Joseph. *Encyclopedia of Birds.* New York: Academic Press, 1998.
Friedman, Judy. *Operation Siberian Crane: The Story Behind the International Effort to Save an Amazing Bird.* New York: Dillon Press, 1992.
Grooms, Steve. *The Cry of the Sandhill Crane.* Minocqua, WI: NorthWord Press, 1992.
Horn, Gabriel. *The Crane.* New York: Crestwood House, 1988.
Katz, Barbara. *So Cranes May Dance: A Rescue from the Brink of Extinction.* Chicago: Chicago Review, 1993.

Jean Blashfield

Crayfish

Crayfish are **freshwater** crustaceans of the order Decapoda, which includes **crabs**, shrimps, **lobsters**, and hermit crabs. Crayfish are nocturnally active, live in shallow freshwater habitats, and feed on aquatic **plant** and **animal** life, as well as dead organic **matter**. Their natural predators include **fish**, **otters**, **turtles**, and wading **birds**. Crayfish are particularly vulnerable to predation during their periodic molts, when their hard exoskeleton is shed to permit body growth. Aggressive **behavior** among crayfish often occurs over access to resources such as **habitat**, food, or mates. Crayfish are used by humans as live fish bait, and are also a popular culinary delicacy.

History and habitat

Crayfish evolved from marine ancestors dating back some 280 million years. There are more than 300 **species** of crayfish worldwide, which are classified into three families: the Astacidae, the Cambridae (found only in the Northern Hemisphere), and the Parastacidae (indigenous

A crayfish at the Fish Point State Wildlife Area, Michigan.
Photograph by Robert J. Huffman. Field Mark Publications. Reproduced by permission.

to the Southern Hemisphere). A few species have adapted to tropical habitats, but most live in temperate regions. None occur in **Africa** or the Indian subcontinent, although one species is found in Madagascar. Crayfish live in **water**, hiding beneath **rocks**, logs, **sand**, mud, and vegetation. Some species dig burrows, constructing little chimneys from moist **soil** excavated from their tunnel and carried to the surface. Some terrestrial species spend their whole life below ground in burrows, emerging only to find a mate. Other species live both in their tunnels as well as venturing into open water. Many species live mostly in open water, retreating to their burrows during pregnancy and for protection from predators and cold weather.

Appearance

Crayfish are usually colored in earth tones of muted greens and browns. The body has three primary sections: the cephalothorax (the fused head and thorax), which is entirely encased by a single shell; the carapace, a six-segmented abdomen; and a five-sectioned, fan-shaped tail, the telson. Five pairs of strong, jointed, armored legs (pereiopods) on the cephalothorax are used for walking and digging. The first pair of legs, known as the chelipeds, end in large pincers (chelae), which are used for defense and food gathering. Two pairs of small antennae (the antennae and antennules) are specialized chemical detectors used in foraging and finding a mate. The antennae project on either side of the tip of the rostrum, which is a beak-like projection at the front of the head. A third and longer pair of antennae are tactile, or **touch** receptors. Two compound eyes provide excellent **vision**, except in some cave-dwellers that live in perpetual dark and are virtually blind. Below the rostrum are two pairs of mandibles (the jaws) and three pairs of maxillipeds, which are small appendages that direct food to the mouth. The second pair of maxillipeds facilitate gill ventilation by swishing water through the banks of gills located at the base of each pereiopod on the sides of the carapace in the gill chambers. The strong, long, muscular abdomen has ten tiny appendages (the pleopods) which aid in swimming movements. When threatened, the crayfish propels itself backward quickly with strong flips of the telson, located at the tip of the abdomen.

Breeding habits

Crayfish usually mate in the fall. Females excrete **pheromones** which are detected by the antennules of males. The openings of the sex organs are located on the front end of the abdomen just below the thorax. In the male, the first two pairs of abdominal pleopods are used as organs of sperm transfer. Using his first set of pleopods, the male deposits sperm into a sac on the female's abdomen. The female stays well hidden as ovulation draws near, lavishly grooming her abdominal pleopods which will eventually secure her eggs and hatchlings to her abdomen. Strenuous abdominal contractions signal the onset of egg extrusion. Cupping her abdomen, the female crayfish collects her eggs as they are laid (they can number 400 or more), securing them with her pleopods, fastidiously cleaning them with thoracic appendages, and discarding any diseased eggs.

Young crayfish hatchlings emerge from the eggs in the spring, and closely resemble adult crayfish in form (although much smaller). The hatchlings cling tightly to their mother's pleopods, eventually taking short foraging forays and scurrying back for protection at the slightest disturbance. During this time, the mother remains relatively inactive and is extremely aggressive in protecting her young from other, non-maternal crayfish, which would cannibalize the young. The mother and young communicate chemically via pheromones; however, the

KEY TERMS

Antennule—Small antenna on the front section of the head.

Carapace—Shell covering the cephalothorax.

Cephalothorax—The head and thorax (upper part of the body) combined.

Chela—Pinchers on first pair of legs used for defense and food gathering.

Chelipeds—First pair of pereiopods ending with large pinchers.

Mandibles—Jaws.

Maxillipeds—Small leg-like appendages beneath the cephalothorax which aid in feeding.

Pereiopods—Ten jointed, armor-plated legs attached to the cephalothorax.

Pleopods—Small, specialized appendages below the abdomen which aid in swimming.

Rostrum—Beak-like projection at the front of the head.

Telson—Fan-shaped tail composed of five segments.

young cannot differentiate between their own mother and another maternal female.

See also Crustacea; Shrimp.

Resources

Books

Felgenhauer, Bruce E., Les Watling, and Anne B. Thistle. *Functional Morphology of Feeding and Grooming in Crustacea.* Rotterdam/Brookfield: A.A. Balkema, 1989.

Hobbs, Horton H., Jr. *The Crayfishes of Georgia.* Washington: Smithsonian Institution Press, 1981.

Marie L. Thompson

Cream of tartar *see* **Potassium hydrogen tartrate**

Crestfish

Crestfish, also called unicornfish, are a small family (Lophotidae) of deepwater, marine **bony fish** in the order Lampridiformes. These rare **fish** have unusual boxlike heads with protruding foreheads and ribbon-shaped silvery bodies with crimson fins. The prominent dorsal fin extends from the tip of the head to beyond the tail; the first rays of this fin form a crest above the head, giving these fish their common name. Crestfish also have a short anal fin, but the pelvic fin is absent. At least one **species** of crestfish (*Lophotus capellei*) has no scales, and another species (*L. lacepede*) has tiny, cycloid scales that are easily rubbed off.

Some crestfish can be quite large. Specimens as large as 6 ft (1.8 m) long have been found off the coast of southern California, although smaller fish (about 3.5 ft/1 m) are more typical. Crestfish are related to ribbon-fish (family Trachipteridae) and oarfish (family Regalecidae), but are distinguished from these related families by having high foreheads and anal fins.

See also Opah.

Cretaceous-Tertiary event *see* **K-T event (Cretaceous-Tertiary event)**

Creutzfeldt-Jakob disease

Creutzfeldt-Jakob **disease** is a rare encephalopathy, or **brain** disease, that causes a swift, progressive **dementia** and neuromuscular changes. It was first described by German psychiatrist Alfons Maria Jakob (1884–1931) in 1921. He gave credit to Hans Gerhard Creutzfeldt (1885–1964), also a German psychiatrist, for describing the **syndrome** first without realizing he had stumbled onto a repeating set of symptoms that constitute a syndrome. Although it is now known that what Creutzfeldt described was not the same syndrome that Jakob had discovered, the **disease** retains its compound name.

Creutzfeldt-Jakob disease is the more common of two known human spongiform encephalopathies, the other being **kuru**. Both encephalopathies are thought to be caused by **prions**, infectious agents made up of gene-lacking **proteins**.

The early stages of Creutzfeldt-Jakob disease include symptoms similar to those of **Alzheimer disease**. Disturbances in **vision** and other senses, confusion, and inappropriate **behavior** patterns are the usual early signs. During the following months the patient will invariably progress first to dementia and then into a **coma**. Jerking movements of the arms and legs are common, and convulsions are less common. Muscle spasms and rigidity occur in the late stages of the disease. Usually, the patient will deteriorate and die in less than a year, although some will live for as long as two years.

Creutzfeldt-Jakob disease occurs throughout the world, usually equally among men and women at about age 60. Rarely do young people get the disease. Three forms of the disease have been identified, classified by virtue of transmission. About 10% of Creutzfeldt-Jakob patients in the United States have a positive family history of Creutzfeldt-Jakob disease (hereditary form). About 85% have no identifiable risk factors for the disease, and yet have developed Creutzfeldt-Jakob disease (sporadic form). Only 1% are thought to have contracted the disease through exposure to infective materials (acquired form).

There are no laboratory tests to help diagnose Creutzfeldt-Jakob disease. Many tests which are performed when a patient is suspected of having Creutzfeldt-Jakob disease are done to make sure that the patient does not have some other, potentially treatable disease. A brain scan using **magnetic resonance imaging (MRI)** will usually show the degeneration of the cerebrum and enlargement of the brain ventricles (fluid-filled openings in the center of the brain) which characterize encephalopathy. A definitive **diagnosis** can be made only when a specimen (biopsy) of the brain **tissue** is stained and studied under a **microscope**. Biopsying the brain, however, is a highly invasive procedure and is rarely done to diagnose Creutzfeldt-Jakob disease, as information gained from the biopsy does not lead to any change in treatment. Other neurological measurements are less directly correlated to diagnosing the disease. The electroencephalogram (which monitors the pattern of electric brain waves) may show certain repeated signs, but not always. The cerebrospinal fluid (which fills spaces in the brain and surrounds the spinal cord) may have an elevated level of protein when tested; this measurement is currently being investigated to determine whether the presence or elevation of specific proteins is diagnostic of this disease. The changes seen in brain tissue are not found in any **organ** outside the central **nervous system**. The etiologic agent that causes the disease can be found in other organs, but evidently it has no effect on their structure or function.

The etiologic agent of Creutzfeldt-Jakob disease has unique characteristics. The agent can withstand **heat**, **ionizing radiation**, and ultraviolet light—immunities that documented viruses do not possess. Most scientists consider the agent to be a unique, nonviral pathogen that does not contain any DNA or RNA, the nucleic acids containing the reproductive code for any **organism**. Thus, a new theory emerged and gained widespread acceptance and the infectious agent is now though to be an unconventional proteinaceous particle called a "prion", ("proteinaceous infectious agent").

Prion proteins are believed to be proteins that are able to exist in two forms: a normal form, performing some unknown but presumably necessary physiological function in the host, and an abnormal, infectious form, which can cause the disease. When a substantial amount of the normal host protein is converted to the abnormal form (by processes which are not understood fully to date) prion disease can ensue.

The means of transmitting this disease is unknown. A case has been documented in which a patient who received a transplanted cornea from the **eye** of a person with Creutzfeldt-Jakob disease also developed the disease. Another person contracted the disease when electrodes previously used on an infected patient were used in an electroencephalogram. Even the spread of the disease within families is the result of a mysterious mechanism. It is not always seen and may be the result of more than one family member having a genetic predisposition to the disease.

There is no cure for Creutzfeldt-Jakob disease. Symptoms are treated to make the patients more comfortable, but nothing has been found that will interfere with the progress of the disease or kill the causal agent. In fact, very special precautions are required around these patients because the means of spreading the disease is not known. Only three means of sterilization of instruments or other objects that have touched the patient are known—lye, **bleach**, and intense heat. It is recommended that clothing and bed linens of the patient be burned, as washing will not kill the etiologic agent.

Researchers are very interested in more clearly identifying the infective agent, and defining its characteristics and the methods of transmission. Concern has been raised regarding the chance that the disease could potentially be passed through contaminated **blood** transfusions. No such case has been identified, but the issue has been raised. Contaminated beef has been held resonsible for a related disease called bovine spongiform encephalopathy, leading researchers to question whether Creutzfeldt-Jakob disease could also be spread through contaminated food products.

Recently in the United Kingdom, there have been several unusual cases of CJD in younger people. These cases surfaced in 1994 and 1995, with a disease that differed significantly from classical sporadic CJD and has been termed variant CJD (vCJD). The patients were all under the age of 42, with an average age of 28, as opposed to the typical age of 63 for classical CJD. The average course of the duration of vCJD was 13 months, in contrast to the average four to six month duration for classical CJD. The electroencephalographic (EEG) electrical activity in the brain for vCJD was different from

KEY TERMS

Cerebrum—The upper, main part of the human brain, it is made up of the two hemispheres, and is the most recent brain structure to have evolved. It is involved in higher cognitive functions such as reasoning, language, and planning.

Encephalopathy—Any abnormality in the structure or function of the brain.

Etiologic agent—The cause of a disease. An etiologic agent could be a bacterium, a virus, a parasite, or some other factor.

classical CJD. While the brain **pathology** of the vCJD cases was identifiable as CJD, the pattern differed because of the presence of large aggregates of prion protein plaques. Scientists ascertained a likely connection between this new form of CJD and bovine spongiform encephalopathy (BSE), a prion disease in cattle, which reached **epidemic** proportions in the 1980s in the United Kingdom. There have now been over 100 cases of vCJD, most of which have occurred in the United Kingdom, and the numbers are still rising. A few cases of vCJD have been found in France, Ireland, Italy, and Hong Kong. Because the incubation period between exposure to the agent and the onset of symptoms may be as long as 40 years, it is uncertain whether these vCJD cases may signal the beginning of an epidemic or whether the incidence of vCJD will remain low.

Creutzfeldt-Jakob disease remains a medical mystery, since scientists are not yet certain about its means of transmission, its cure, or its prevention. Much work is being done to collect brain, tissue, and body fluid samples from all sufferers, in order to advance research into the disease.

See also Neuroscience.

Resources

Books

Margulies, Phillip. *Creutzfeldt-Jacob Disease (Epidemics).* New York: Rosen Publishers, 2003.

Prusiner, Stanley B. *Prion Biology and Diseases.* Cold Spring Harbor, NY: Cold Spring Harbor Laboratory Press, 1999.

Ratzan, Scott C., ed. *Mad Cow Crisis: Health and the Public Good* New York: New York University Press, 1998.

Organizations

Creutzfeldt-Jakob Disease Foundation, Inc. P.O. Box 5312 Akron, Ohio 44334. <http://www.cjdfoundation.org>.

Larry Blaser

Crickets

Crickets (order Orthoptera, family Grillidae) are found throughout the world except for the polar regions. More than 900 **species** have been described. Often heard, but more seldom seen, at first glance crickets are quite similar to **grasshoppers** and bush crickets—also known as long-horned grasshoppers or katydids—but may be distinguished from these **insects** by their much longer, thread-like antennae. Crickets can also be easily identified by long hindlegs used for jumping. Most species are black or brown in appearance, which helps to conceal them and therefore reduce the risk of detection from predatory small **mammals** and **birds**.

A cricket's body is divided into three sections: the head, which bears the large eyes, antennae, and mouthparts; the thorax, which is separated into three segments (each of which bears one pair of segmented legs and efficient claws) and supports the wings; and the abdomen. There are typically two pairs of wings, of which the front pair are thicker. At rest, the wings are held flat over the insect's back with the edges bent down along the sides of the body. Some species, however, may just have one pair of wings, while others still have lost all power of flight. Among the latter are some of the smallest members of the Orthoptera, the ant-loving crickets that measure less than an inch (3–5mm) and live in underground ant nests. Despite the presence of wings on some species, none of the crickets are exceptionally good fliers; most rely on a combination of short flights and jumps to move to new feeding patches.

Crickets are active during the day and night, depending on the species; some prefer the warmth of full sunlight, others prefer the twilight hours of dusk, while quite a few species are only active at nighttime. Most species live in open grassland or woodlands, but others such as the mole crickets (family Gryllotalpidae) spend much time underground, only emerging occasionally to fly from one site to another. Mole crickets are easily distinguished from all other insects by their broad, spade-like front legs which have evolved as powerful digging tools. Armed with these and strong blade-like teeth, the mole cricket is well equipped to dig shallow tunnels in moist soils.

Crickets are versatile insects and are capable of feeding off a wide range of organisms. Largely vegetarian, they eat a great variety of young leaves, shoots, and stems but may also eat other insects (adults and larvae), as well as a range of dead or decaying **matter**, including household wastes. Like other orthopterans such as grasshoppers and locusts, their mouthparts are designed for biting and chewing.

Sound is important to all orthopterans, and crickets have specialized **hearing** organs (called tympanum) on their front legs, by which they are able to detect vibrations. Living in dense **forests** or rugged, tall **grasslands** can pose certain problems for small insects when it comes to finding a suitable mate. Crickets have solved this problem through an elaborate system of singing, which advertises their presence. Only male crickets are able to "sing" by a process known as stridulation. When he is ready to perform, the male chooses his auditorium carefully and, with both wings held above the body, begins to slowly rub one against the other. The right forewing bears a special **adaptation** which has been described as a toothed rib, or file; when this is rubbed against the hind margin of the left forewing, it produces a musical sound. In order to avoid confusion between different species, each cricket has its own distinct song, which varies not only in duration, but also in pitch. Among the songs commonly used are those used to attract females from a wide area around the male, while others serve as a **courtship** song once the amorous male has succeeded in gaining the interest of a passing female.

As the courtship ritual proceeds—either within a short burrow or amongst vegetation—the male deposits a number of small capsules (spermatophores) containing sperm cells. These are then collected by the female and used to fertilize her eggs. Female crickets lay their eggs singly in the ground and in **plant** tissues, sometimes using their long, needle-like ovipositors, specialized egg-laying organs, to make an incision in the plant stem. When the eggs hatch, a small nymph-like replica of the adult cricket emerges and immediately begins to feed. As the nymphs grow, they molt, casting off their outer skeleton, perhaps as many as 10 times before they finally reach adult size. Depending on when the eggs were laid, the young nymphs or eggs themselves may have to spend the winter in a dormant state underground, emerging the following spring and summer to develop and breed.

Among the exceptions to this general pattern of reproduction is the parental nature of the mole crickets. Females may lay up to 300 eggs in an underground nest where, unlike most other insects, she guards them from potential predators and remains with the young nymphs for several weeks after they hatch, sometimes until just before they begin to disperse. As with other crickets, these nymphs then pass through many stages of growth and do not reach full adult size until the year after hatching.

Although widely conceived as major crop **pests**, most crickets are thought to cause relatively little harm. Some species, such as the house cricket (*Acheta domestica*) which has unwittingly been transported all around the world by humans, are considered a pest and health hazard because of their liking for garbage heaps. In contrast, mole crickets are now being increasingly viewed as beneficial to many horticulturalists as they feed on a wide range of soil-dwelling larvae that cause considerable damage to **crops**. They, and other grassland species, have, however, suffered heavily through hanging agricultural practices, as well as the increased and often excessive use of **pesticides** and inorganic fertilizer which reduces the natural diversity of other insects (many of which are eaten by crickets) and plants in agricultural areas.

David Stone

Critical habitat

All **species** have particular requirements for their ecological **habitat**. These specific needs are known as critical habitat, and they must be satisfied if the species is to survive. Critical habitat can involve specific types of food, a habitat required for breeding (as is the case of species that nest in **tree** cavities), or some other crucial environmental requirement.

Some critical habitat features are obvious, and they affect many species. For example, although some species that live in **desert** regions are remarkably tolerant of **drought**, most have a need for regular access to **water**. As a result, the few moist habitats that occur in desert landscapes sustain a relatively great richness of species, all of which are dependent on this critical habitat feature. Such moist habitats in the desert are called oases, and in contrast to more typical, drier landscapes, they are renowned for the numbers of species that can be supported at relatively large population densities.

Salt licks are another critical habitat feature for many large species of mammalian herbivores. Herbivores consume large quantities of **plant** biomass, but this food source is generally lacking in **sodium**, resulting in a significant nutritional deficiency and intense craving for salt. Salt licks are mineral-rich places to which large herbivores may gravitate from a very wide area, sometimes undertaking long-distance movements of hundreds of miles to satisfy their need for sodium. Because of their importance in alleviating the scarcity of **minerals**, salt licks represent a critical habitat feature for these animals.

Many species of shorebirds, such as **sandpipers** and **plovers**, migrate over very long distances between their wintering and breeding habitats. Many species of the Americas, for example, breed in the arctic **tundra** of **North America** but winter in **South America**, some as far south as Patagonia. For some of these species, there are a few places along their lengthy **migration** route that provide critical staging habitats where the animals stop for a short time to feed voraciously. These critical habitats allow the animals to re-fuel after an arduous, energetically demanding part of their migration and to prepare for the next, similarly formidable stage.

For example, parts of Chesapeake Bay and the Bay of Fundy are famous critical habitats for migrating shorebirds on the east coast of North America. Chesapeake Bay is most important for **birds** migrating north to their breeding grounds, because at that time **horseshoe crabs** (*Limulus polyphemus*) spawn in the bay, and there is an enormous, predictable abundance of their nutritious eggs available for the shorebirds to eat. The Bay of Fundy is most important during the post-breeding, southern migration, because at that time its tidal mudflats support a great abundance of small crustaceans that can be eaten by these birds, which can occur there in flocks of hundreds of thousands of individuals. There are additional places on the east coast of North America that provide important staging habitat for these birds during their migrations, but none support such large populations as Chesapeake Bay and the Bay of Fundy. Therefore, these represent critical habitats that must be preserved if the shorebirds are to be sustained in large populations.

Another critical habitat feature for many species that live in **forests** is large dimension, dead **wood**, either lying on the ground as logs or as standing snags. Many species of birds, **mammals**, and plants rely on dead wood as a critical habitat feature that may provide cavities for nesting or resting, a feeding substrate, places from which to sing or survey the surroundings for food and enemies, or in the case of some plants, suitable places for seedlings to establish. **Woodpeckers**, for example, have an absolute need for snags or living trees with heart-rotted interiors in which they excavate cavities that they use for breeding or roosting. Many other animals are secondary users of woodpecker cavities.

Sometimes, different species develop highly specific relationships to the degree that they cannot survive without their obligate partners, which therefore represent critical, biological components of their habitat. For example, the dodo (*Raphus cucullatus*) was a flightless, turkey-sized bird that used to live on the **island** of Mauritius in the Indian Ocean. The dodo became extinct by overhunting and introduced predators soon after its island was discovered by Europeans. However, this bird lived in an intimate relationship with a species of tree, the tambalacoque (*Calvaria major*), that, like the dodo, only occurs on Mauritius. The large, tough **fruits** of this tree were apparently an important food source for dodos. The dodos found these fruits on the forest floor, ingested them whole, ground up the outer, fleshy coat in their muscular gizzard, and digested and absorbed the **nutri-**

ents in their alimentary tract. However, the hard, inner **seeds** were not digested and were defecated by the dodos, prepared for **germination** in a process that botanists call scarification. Dodos were the only animals on Mauritius that could perform this function. Since the dodo became extinct in the early 1600s, no seeds of tambalacoque were able to germinate since then, although some mature trees managed to persist. In the 1980s, the need of tambalacoque for this sort of scarification was discovered, and seeds can now be germinated after they have been eaten and scarified by passage through a domestic turkey. Seedlings of this species are now being planted in order to preserve this unique species of tree.

Because so many species and natural ecosystems are now endangered by human influences, it is very important that critical habitat needs be identified and understood. This knowledge will be essential to the successful preservation of those many **endangered species** that now must depend on the goodwill of humans for their survival. An important aspect of the strategy to preserve those species will be the active management and preservation of their critical habitat.

See also Symbiosis.

Resources

Books

Freedman, B. *Environmental Ecology*. 2nd ed. San Diego: Academic Press, 1995.

Bill Freedman

Critical mass *see* **Nuclear fission**

Crocodiles

The crocodile order (Crocodylia) consists of several families of large, unmistakable, amphibious **reptiles**: the crocodiles (Crocodylidae), gavials (Gavialidae), and the alligators and caimans (Alligatoridae). Although these animals look superficially like lizards, they are different in many important respects, and are believed by biologists to be the most highly evolved of the living reptiles.

Crocodilians are amphibious animals, spending most of their time in **water** but emerging onto land to bask in the **sun** and lay their eggs. Their usual **habitat** is in warm tropical or subtropical waters. Most **species** occur in **freshwater**, with only the **saltwater** crocodile being partial to marine habitats. **Fish** are the typical food of most adult crocodilians, but the biggest species will

An American crocodile (*Crocodylus acutus*). *Photograph by Tom & Pat Leeson. The National Audubon Society Collection/ Photo Researchers, Inc. Reproduced by permission.*

also eat large **mammals**, including humans. Younger crocodilians eat **invertebrates** and small fish.

Crocodilians are economically important for their thick, attractive hide, which can be used to make fine leather for expensive consumer goods, such as shoes, handbags, and other luxury items. Wild crocodilians are hunted for their hide wherever they occur, and in some areas they are also raised on ranches for this purpose. Crocodilian meat is also eaten, often as a gourmet food.

Most populations of wild crocodilians have been greatly reduced in size because of overhunting and habitat loss, and numerous species are endangered.

Biology of crocodilians

Among the more distinctive characteristics of the crocodilians are their almost completely four-chambered **heart**, teeth that are set into sockets in the jaw, a palate that separates the mouth from the nasal chambers, and spongy lungs. These animals also have a protective covering of partially calcified, horny plates on their back. The plates are not connected with each other, and are set into the thick, scaly skin, allowing a great freedom of movement. Crocodilians have a heavy body with squat legs and a large, strong, scale-ridged tail.

Sinusoidal motions of the powerful tail are used to propel the **animal** while swimming. The tail is also a formidable weapon, used to subdue **prey** and also for defense. Although they appear to be ungainly and often

spend most of their time lying about, crocodilians can actually move quite quickly. Some crocodilians can even lift their body fully above the ground, and run quickly using all four legs; a human cannot outrun a crocodile over a short **distance** on open land.

Crocodilians have numerous adaptations for living in water. They have webbed feet for swimming slowly and their nostrils, eyes, and ears are set high on the head so they can be exposed even while most of the body and head are below the surface. When a crocodilian is totally submerged, its **eye** is covered by a semi-transparent nictitating **membrane**, and flaps of skin seal its nostrils and ears against water inflow. Crocodilians often float motionless in the water, commonly with the body fully submerged and only the nostrils and eyes exposed. To accomplish this **behavior**, crocodilians regulate their body **density** by varying the amount of air held in the lungs. Also, the stomach of most adult crocodiles contain stones, to as much as 1% of the animal's total body weight. The stones are thought to be used as buoyancy-regulating ballast.

Crocodilians are poikilothermic, meaning they do not regulate their body **temperature** by producing and conserving metabolic **heat**. However, these animals are effective at warming themselves by basking in the sun, and they spend a great deal of time engaged in this activity. Crocodilians commonly bask through much of the day, often with their mouths held open to provide some cooling by the **evaporation** of water. Only when the day is hottest will these animals re-enter the water to cool down. Most species of crocodilians are nocturnal predators, although they will also hunt during the day if prey is available.

Stories exist of **birds** entering the open mouths of crocodiles to glean leeches and other **parasites**. This phenomenon has not been observed by scientists, although it is well known that crocodiles will tolerate certain species of birds picking external parasites from their skin, but not necessarily inside of their mouth.

Male crocodilians are territorial during the breeding season, and chase other males away from places that have good nesting, basking, and feeding habitat. The male animals proclaim their territory by roaring loudly, and sometimes by snapping their jaws together. Intruders are aggressively chased away, but evenly matched animals may engage in vicious fights. Territory-holding males do not actively assemble females into a harem. Rather, they focus on chasing other males away from a territory. Females will enter the defended territory if they consider it to be of high quality.

All crocodilians are predators, and they have large, strong jaws with numerous sharp teeth for gripping their prey. Crocodiles do not have cutting teeth. If they capture prey that is larger than they can eat in a single gulp, it is dismembered by gripping strongly with the teeth and rolling their body to tear the carcass. Some crocodiles will opportunistically cooperate to subdue a large mammal, and then to tear it into bits small enough to be swallowed. However, extremely large animals with tough skin, such as a dead hippopotamus, will be left to rot for some time until the carcass softens and can be torn apart by the crocodiles.

All crocodilians are **oviparous**, laying hard, white-shelled eggs. The nest may be a pit dug into a beach above high water, or it may be made of heaps of aquatic vegetation, which help to incubate the eggs through heat produced during **decomposition**. Often, a number of females will nest close to each other, but each builds a separate nest. The nesting grounds are used by the same females, year after year. In many species the female carefully guards and tends her nest. Fairly open, sandy beaches are generally preferred as sites upon which to build the nest or nest mound.

Typically, 20–40 eggs are laid at a time, but this varies with species and the size of the female. Incubation time varies with species and temperature, but ranges, in the case of the Nile crocodile, from 11 to 14 weeks. Predators as diverse as **monitor lizards**, **mongooses**, dogs, **raccoons**, and even **ants** seek out crocodile nests to eat the eggs and newly hatched young.

An infant crocodilian has a small, so-called "egg tooth" at the end of its snout, which helps it to break out of the shell when ready to hatch. All of the baby crocodilians hatch within a short time of each other, synchronized in part by the faint peeping noises they make during the later stages of incubation. In some crocodilians, the mother assists her babies in hatching, by gently taking eggs into her mouth and cracking them with her teeth. The mother also may guard her offspring for some time after hatching, often allowing them to climb onto her body and head. Female crocodiles are very aggressive against intruders while their eggs are hatching and newborn babies are nearby, and under these circumstances they will even emerge from the water to chase potential predators away. Nevertheless, young crocodilians are vulnerable to being eaten by many predators, and this is a high-risk stage of the life cycle.

Potentially, crocodilians are quite long-lived animals. Individuals in zoos have lived for more than 50 years, and the potential longevity of some species may be as great as a century.

Species of crocodilians

The gavial or gharial (family Gavialidae) is a single species, *Gavialus gangeticus*, which lives in a number of

The endangered false gavial (*Tomistoma schlegelii*). *Photograph by A. Cosmos Blank. The National Audubon Society Collection/Photo Researchers, Inc. Reproduced by permission.*

sluggish, tropical **rivers** in India, Bangladesh, and Indochina. Gavials have a long, slender snout, and are almost exclusively fish eaters, catching their prey with sideways sweeps of the open-mouthed head. Gavials can attain a length of about 20 ft (6 m). Gavials are considered holy in the Hindu religion, and this has afforded these animals a measure of protection in India. Unfortunately, this is not sufficiently the case anymore, and gavials have become severely endangered as a result of overhunting for their hide.

The true crocodiles (family Crocodylidae) include about 16 species that live in tropical waters. Crocodiles are large, stout animals, with a much heavier snout than that of the gavial. The main food of crocodiles is fish, but some species can catch and subdue large mammals that venture close to their aquatic habitat, or attempt to cross rivers in which the crocodiles are living. Perhaps the most famous species is the Nile crocodile (*Crocodylus niloticus*) of **Africa**, which can grow to a length of 23 ft (7 m). This crocodile can be a **predator** of unwary humans, although its reputation in this respect far exceeds the actual risks, except in certain places. This species used to be very abundant and widespread in Africa, but unregulated hunting and, to a lesser degree, habitat loss, have greatly reduced its population.

The most dangerous crocodilian to humans is the estuarine or salt-marsh crocodile (*Crocodylus porosus*), which lives in salt and **brackish** waters from northern **Australia** and New Guinea, through most of Southeast

Asia, to southern India. This species can achieve a length of more than 23 ft (7 m). Individuals of this crocodile species sometimes occur well out to sea.

Other species are the mugger crocodile (*Crocodylus palustris*) of India, Bangladesh, and Ceylon; the Australian crocodile (*C. johnsoni*) of northern Australia; and the New Guinea crocodile (*C. novaeguineae*) of New Guinea and parts of the Philippines.

The American crocodile (*Crocodylus acutus*) is a rare and **endangered species** of brackish estuaries in southern Florida, occurring more widely in central and northwestern **South America** and the Caribbean. This species can achieve a length of 20 ft (6 m). The Orinoco crocodile (*C. intermedius*) occurs in the Orinoco and Amazon Rivers of South America.

The false gavial (*Tomistoma schlegelii*) is a slender-snouted species of Southeast Asia.

The alligators and caimans (family Alligatoridae) are seven species that occur in fresh water, with a broader head and more rounded snout than crocodiles. The American alligator (*Alligator mississipiensis*) can achieve a length of 13 ft (4 m), and occurs in the southeastern United States as far north as South Carolina and Alabama. This species was endangered by unregulated hunting for its hide. However, strict **conservation** measures have allowed for a substantial recovery of the species, and it is now the subject of a regulated hunt.

The Chinese alligator (*Alligator sinensis*) occurs in the lower reaches of the Yangtze and Kiang Rivers in southern China, and it is the only member of its family to occur outside of the Americas. The Chinese alligator can grow as long as 6.5 ft (2 m).

Caimans are animals of freshwater habitat in South and Central America. The black caiman (*Melanosuchus niger*) can achieve a length of 16 ft (5 m). The spectacled caiman (*Caiman crocodilus*) and the broad-nosed caiman (*C. latisrostris*) of eastern Brazil can both grow somewhat longer than 6.5 ft (2 m). The dwarf caiman (*Paleosuchus palpebrosus*) and the smooth-fronted caiman (*P. trigonatus*) live in more swiftly flowing streams and rivers and are relatively small species that do not exceed about 5 ft (1.5 m) in length.

Crocodilians and people

The larger species of crocodilians are fierce predators. In particular, crocodiles have posed a long-standing risk to domestic **livestock** that try to drink from their aquatic habitat as well as to unwary humans. For this reason, crocodiles are commonly regarded as dangerous **pests**, and they are sometimes killed to reduce the risks associated with their presence.

Because some species of crocodilians are dangerous, they are greatly feared in many places. This fear is justified in some cases, at least in places where large human-eaters are abundant. In some cultures, the deep fear and revulsion that people have for dangerous crocodilians has transformed into an attitude of reverence. For example, a pool near Karachi, Pakistan, contains a number of large mugger crocodiles, which are venerated as priest-like entities and worshipped by pilgrims. In other places, human sacrifices have been made to crocodiles to pacify animist spirits.

The skins of crocodilians can be used to make a very tough and beautiful leather. This valuable product is widely sought for use in making expensive shoes, handbags, wallets, belts, suitcases, and other items. Crocodilians are readily hunted at night, when they can be found using searchlights that reflect brightly off their eyes. In almost all parts of the range of crocodilians, they have been hunted to endangerment or extirpation. Almost all species in the crocodile family are endangered to some degree.

In a few places, however, strict conservation regulations have allowed the populations of crocodilians to increase from historically depleted lows. This has been particularly true of the American alligator, which was considered to be an endangered species only a few decades ago, but has now recovered sufficiently to allow for a carefully regulated sport and market hunt.

Some species of crocodilians are also ranched, usually by capturing young animals in the wild and feeding them in confinement until they reach a large enough size to slaughter for their hide. The meat of crocodilians is also a saleable product, but it is secondary in importance to the hide.

Crocodilians are sometimes used to entertain people, and some species are kept as pets. The Romans, for example, sometimes displayed Nile crocodiles in their circuses. To amuse the masses of assembled people, the crocodiles would be killed by humans, or alternatively, humans would be killed by the crocodiles.

In more modern times, alligator wrestling has been popular in some places, for example, in parts of Florida. The key to successful alligator wrestling is to hold the jaws of the animal shut, which can be accomplished using only the hands (but watch out for the lashing tail!). Crocodilians have very powerful muscles for closing their jaws, but the muscles to open their mouth are quite weak.

Resources

Books

Alderton, D. *Crocodiles and Alligators of the World.* U.K.: Blandford Press, 1991.

Cogger, H.G., and R.G. Zweifel. *Encyclopedia of Reptiles and Amphibians.* San Diego: Academic Press, 1998.

KEY TERMS

Endangered—Refers to species or populations of organisms that are so small that there is a likelihood of imminent local extirpation, or even global extinction over its entire range.

Extirpated—The condition in which a species is eliminated from a specific geographic area of its habitat.

Nictitating membrane—An inner eyelid.

Overhunting—Hunting of an animal at a rate that exceeds its productivity, so that the population size decreases, often to the point of endangerment.

Oviparous—This refers to an animal that lays eggs, from which the young hatch after a period of incubation.

Poikilotherm—Refers to animals that have no physiological mechanism for the regulation of their internal body temperature. These animals are also known, less accurately, as "cold-blooded." In many cases, these animals bask in the sun or engage in other behaviors to regulate their body temperature.

Dudley, K. *Alligators and Crocodiles.* Raintree/Steck Vaughan, 1998.

Dudley. K. *Crocodiles and Alligators of the World.* Sterling Publications, 1998.

Grenard, S. *Handbook of Alligators and Crocodiles.* New York: Krieger, 1991.

Halliday, T. R., and K. Adler. *The Encyclopedia of Reptiles and Amphibians.* New York: Facts on File, 1986.

Messel, H., F. W. King, and J. P. Ross, eds. *Crocodiles: An Action Plan for Their Conservation.* Gland, Switzerland: International Union for the Conservation of Nature (IUCN), 1992.

Webb, G., S. Manolis, and P. Whitehead, eds. *Crocodiles and Alligators.* Australia: Surrey Beatty, 1988.

Zug, George R., Laurie J. Vitt, and Janalee P. Caldwell. *Herpetology: An Introductory Biology of Amphibians and Reptiles.* 2nd ed. New York: Academic Press, 2001.

Bill Freedman

Crocus *see* **Lily family (Liliaceae)**

Crop rotation

Crop rotation is a method of maintaining **soil** fertility and structure by planting a particular parcel of agricultural land with alternating **plant** species. Most crop rotation schedules require that a field contain a different crop each year, and some schemes incorporate times when the field remains uncultivated, or lies fallow. Farmers rotate **crops** to control **erosion**, promote soil fertility, contain **plant diseases**, prevent insect infestations, and discourage weeds. Crop rotation has been an important agricultural tool for thousands of years. Modern organic farmers depend on crop rotation to maintain soil fertility and to fight **pests** and weeds, functions that conventional farmers carry out with chemical **fertilizers** and **pesticides**. Crop rotation can also prevent or correct some of the problems associated with monocultures (single crop farms), including persistent weeds, insect infestations, and decreased resistance to plant diseases.

History

For 2,000 years, since the Romans spread their farming practices throughout the Roman Empire, European farmers followed a Roman cropping system called "food, feed, and fallow." Farmers divided their land into three sections, and each year planted a food grain such as **wheat** on one section, **barley** or oats as feed for **livestock** on another, and let the third plot lie fallow. On this schedule, each section lay fallow and recovered some of its **nutrients** and organic **matter** every third year before it was again sown with wheat. Farmers following the "food, feed, fallow" system typically only harvested six to ten times as much seed as they had sown, and saved a sixth to a tenth of their harvest to sow the following year. Low yields left little grain for storage; crops failed and people often starved during years of flood, **drought**, or pest infestation.

The size of agricultural allotments in **Europe** gradually increased beginning in the fifteenth century, allowing farmers more space to experiment with different crop rotation schedules. By 1800, many European farmers had adopted a four-year rotation cycle developed in Holland and introduced in Great Britain by Viscount Charles "Turnip" Townshend in the mid-1700s. The four-field system rotated wheat, barley, a root crop like turnips, and a nitrogen-fixing crop like clover. Livestock grazed directly on the clover, and consumed the root crop in the field. In the new system, fields were always planted with either food or feed, increasing both grain yields and livestock productivity. Furthermore, adding a nitrogen-fixing crop and allowing manure to accumulate directly on the fields improved soil fertility; eliminating a fallow period insured that the land was protected from soil erosion by stabilizing vegetation throughout the cycle.

Subsistence farmers in tropical **South America** and **Africa** followed a less orderly crop rotation system called "slash and burn" agriculture. Slash and burn rota-

Adjacent fields of rice and wheat, Sacramento Valley, California. *JLM Visuals. Reproduced by permission.*

tion involves cutting and burning nutrient-rich tropical vegetation in place to enhance a plot of nutrient-poor tropical soil, then planting crops on the plot for several years, and moving on to a new plot. Slash and burn agriculture is a successful strategy as long as the agricultural plots remain small in relation to the surrounding **rainforest**, and the plot has many years to recover before being cultivated again. Large-scale slash and burn agriculture results in permanent destruction of rainforest ecosystems, and in complete loss of agricultural productivity on the deforested land.

Crop rotation fell out of favor in developed nations in the 1950s, when farmers found they could maintain high-yield **monoculture** crops by applying newly developed chemical fertilizers, pesticides, and weed killers to their fields. Large-scale commercial agriculture that requires prescribed chemical treatments has become the norm in most developed nations, including the United States. However, substantial concerns about the effect of agricultural chemicals on human health, and damage to soil structure and fertility by monoculture crops, have led many farmers to return to more "natural" practices like crop rotation in recent decades. So-called conventional farmers use crop rotation in concert with chemical treat-

ments. Organic farmers, who cannot by definition use chemical treatments, rely entirely upon methods like crop rotation to maintain soil health and profitable crop yields.

Current crop rotation practices

Because climate, soil type, extent of erosion, and suitable cash crops vary around the globe, rotation schemes vary as well. The principles of crop rotation, however, are universal: to maintain soil health, combat pests and weeds, and slow erosion farmers should alternate crops with different characteristics—sod-base crops with row crops, weed-suppressing crops with those that do not suppress weeds, crops susceptible to specific **insects** with those that are not, and soil-enhancing crops with those that do not enhance soils.

Farmers use cover crops to stabilize soils during the off-season when a cash crop has been harvested. Cover crops are typically grown during dry or cold **seasons** when erosion and nutrient depletion threaten exposed soil. Slow-starting legume crops like sweet clover, red clover, crimson clover, and vetch can be planted during the cash crop's growing season. These nitrogen-fixing **legumes** also restore **nitrogen** to depleted soils during

KEY TERMS

. . . .

Fallow—Cultivated land that is allowed to lie idle during the growing season so that it can recover some of its nutrients and organic matter.

Nutrients—The portion of the soil necessary to plants for growth, including nitrogen, potassium, and other minerals.

Organic matter—The carbonaceous portion of the soil that derives from once living matter, including, for the most part, plants.

the off-season, which will benefit the next cash crop. Farmers typically plant fast-growing crops like rye, oats, ryegrass, and Austrian winter peas after harvesting the cash crop. Cover crops are plowed into the soil as "green manure" at the end of the season, a practice that increases soil organic content, improves structure, and increases permeability.

Increasing the number of years of grass, or forage, crops in a rotation schedule usually improves soil stability and permeability to air and **water**. Sloping land may experience excessive soil loss if row crops like corn, or small-grain crops like wheat, are grown on it for too many years in a row. Rotation with sod-based forage crops keeps soil loss within tolerable limits. Furthemore, forage crops can reverse the depletion of organic nutrients and soil compaction that occur under corn and wheat.

Crop rotation also works to control infestations of crop-damaging insects and weeds. Crop alternation interrupts the reproductive cycles of insects preying on a specific plant. For example, a farmer can help control cyst nematodes, **parasites** that damage soybeans, by planting soybeans every other year. Crop rotation discourages weeds by supporting healthier crop plants that out compete wild species for nutrients and water, and by disrupting the weed-friendly "ecosystems" that form in long-term monocultures. Rotation schedules that involve small fields, a large variety of rotated crops, and a long repeat interval contain insect infestations and weeds most successfully. Complex rotations keep weeds and insects guessing, and farmers can exert further control by planting certain crops next to each other. For example, a chinch bug infestation in a wheat field can be contained by planting soybeans in the next field instead of a chinch bug host like forage **sorghum**.

Farmers undertaking crop rotations must plan their planting more carefully than those who plant the same crop year after year. Using a simple principle, that there are the same number of fields or groups of fields as there

are years in the rotation, farmers can assure that they produce consistent amounts of each crop each year even though the crops shift to different fields.

Resources

Books

Bender, Jim. *Future Harvest: Pesticide Free Farming.* Lincoln, NB: University of Nebraska, 1994.
Pollan, Michael. *Botany of Desire: A Plant's Eye View of the World.* New York: Random House, 2001.
Troeh, Frederick R., and Louis M. Thompson. *Soils and Soil Fertility.* New York: Oxford University Press, 1993.

Organizations

Organic Farming Research Foundation, P.O. Box 440 Santa Cruz, CA, 95061. (831) 426-6606. <http://www.ofrf.org/index.html.>

Beth Hanson
Laurie Duncan

Crops

Crops are any organisms that humans utilize as a source of food, materials, or **energy**. Crops may be utilized for subsistence purposes, to barter for other goods, or to sell for a cash profit. They may be harvested from wild ecosystems, or they may be husbanded and managed, as occurs with domesticated **species** in agriculture.

In general, the purpose of management is to increase the amount of crop productivity that is available for use by humans. There is, however, a continuum in the intensity of crop-management systems. Species cultivated in agriculture are subjected to relatively intensive management systems. However, essentially wild, free-ranging species may also be managed to some degree. This occurs, for example, in **forestry** and in the management of **ocean** fisheries and certain species of hunted animals. Most crops are **plant** species, but animals and **microorganisms** (e.g., **yeast**) can also be crops.

In general, unmanaged, free-ranging crops are little modified genetically or morphologically (in form) from their non-crop ancestors. However, modern, domesticated crops that are intensively managed are remarkably different from their wild ancestors. In some cases a non-domesticated variety of the species no longer exists.

Hunting and gathering; crops obtained from unmanaged ecosystems

Human beings can only be sustained by utilizing other species as sources of food, material, and energy.

Humans have thus always had an absolute requirement for the goods and services provided by other species, and this will always be the case. Although direct genetic modification has in the last few years been added (controversially) to artificial **selection** as a technique for adjusting plant and **animal** species to our needs, there is no prospect of a time when technology will make it possible to meet our needs by applying energy directly to raw materials. Humans have always needed and always will need to harvest other living things.

Prior to the discovery of the first agricultural techniques about 9,000–11,000 years ago, almost all human societies were sustained by gathering edible or otherwise useful products of wild plants, and by hunting wild animals. The plant foods that were gathered as crops by these early peoples included starchy tubers, **fruits**, and **seeds**. Other plants were harvested as sources of fuel or to provide materials such as **wood** and **bark** for the construction of shelters, canoes, and other tools or weapons. Animals, meanwhile, were hunted for their meat, hide, and bones. Prior to the development of agriculture—the breeding and deliberate nurturing of useful plants and animals—these activities were undertaken in essentially natural ecosystems—ecosystems not intensively modified by the hunting and gathering activities of people.

Humans have been rather eclectic in their choice of crops, selecting a wide range of useful species of plants, animals, and microorganisms from among the vast diversity of species available in most places or regions. Humans are considered omnivorous, because they feed at all levels of ecological food webs: on plants, animals, and other types of organisms, both living and dead.

Human societies that subsist only by the hunting and gathering of wild crops are now virtually extinct. However, modern human societies continue to obtain important plant and animal crops from essentially unmanaged ecosystems.

Plants

Among plant crops that humans continue to obtain from natural ecosystems, trees are among the most notable. In the parlance of forestry, the terms "virgin" and "primary" are used to refer to older, natural **forests** from which wild, unmanaged trees have not yet been harvested by humans. "Secondary" forests have sustained at least one intensive harvest of their resource of trees in the past, and have since grown back. In general, the ecological characteristics of secondary forests are quite different from those of the more natural, primary forests that may have once occurred on the same site.

Trees have always been an important crop for humans, being useful as sources of fuel, food (edible fruits and nuts), and wood for tools, structures, furniture, **paper**, and vehicles (boats, wagons, etc.). Even today, trees harvested from natural forests are important crops in most countries. **Tree biomass** is an essential source of energy for cooking and space heating for more than one-half of the world's people, almost all of whom live in relatively poor, tropical countries. Trees are also an important crop for people living in richer countries, mostly as a source of lumber for the construction of buildings and furniture and as a source of pulpwood for paper manufacture.

In many places, the primary natural forest has been depleted by the cutting of trees, which has often been followed by conversion of the land to agriculture and urban land-uses. This pattern of forest loss has been common in **North America**, **Europe**, and elsewhere. To compensate to some degree for the loss of natural forests in those regions, efforts have been made to establish managed forests, so that tree crops will continue to be available.

Trees are not the only plant crops gathered from wild, unmanaged ecosystems. In particular, people who live a subsistence lifestyle in tropical forests continue to obtain much of their food, medicine, and materials from wild plants, usually in combination with hunting and subsistence agriculture. Even in North America, small crops of a few wild food plants continue to be gathered. Some examples include harvests of wild **rice** (*Zizania aquatica*), strawberry (*Fragaria virginiana*), low-bush blueberry (*Vaccinium angustifolium*), and fiddleheads (from the ostrich fern, *Matteucia struthiopteris*). Other minor wild crops include various species of edible **mushrooms** and marine **algae**.

Terrestrial animals

Wild animals have always been an important source of food and useful materials for humans. Most people who live in rural areas in poorer countries supplement their diet with meat obtained by hunting wild animals. Hunting is also popular as a sport among many rural people in wealthier countries. For example, each year in North America millions of **deer** are killed by hunters as food and a source of hide, especially white-tailed deer (*Odocoileus virginianus*), mule deer (*O. hemionus*), and elk (*Cervus canadensis*). There are also large hunts of upland game **birds**, such as ruffled **grouse** (*Bonasa umbellus*) and of wild **ducks** and **geese**, especially the mallard (*Anas platyrhynchos*), Canada goose or honker (*Branta canadensis*), and snow goose (*Chen hyperboreus*).

Aquatic animals

Aquatic animals have also provided important wild-meat crops for people living in places where there is access to the natural bounties of streams, **rivers**, lakes, and marine shores. Important food crops harvested from **freshwater** ecosystems of North America include species of **salmon**, trout, and other **fish**, as well as **crayfish**, freshwater mussels, and other **invertebrates**. Most of the modern marine fisheries also rely on harvesting the productivity of unmanaged populations of fish, invertebrates (e.g., shellfish), **seals**, and whales as wild crops.

Agriculture; crops from managed ecosystems

As considered here, agricultural crops are managed relatively intensively for the sustained productivity of food and materials useful to humans. In this sense, agricultural systems can involve the cultivation of plants and **livestock** on farms, as well as the cultivation of fish and invertebrates in aquaculture and the growing of trees in agroforestry plantations.

Agricultural systems can vary tremendously in the intensity of their management practices. For example, species of terrestrial crop plants may be grown in mixed populations, a system known as *polyculture*. These systems are often not weeded or fertilized very intensively. Mixed-cropping systems are common in nonindustrial agriculture, for example, in subsistence agriculture in many tropical countries.

In contrast, some monocultural systems in agriculture attempt to grow crops in single-species populations. Such intensively managed systems usually rely heavily on the use of **fertilizers**, **irrigation**, and **pesticides** such as **insecticides**, **herbicides**, and fungicides. Of course, heavy, sophisticated, energy-requiring machinery is also required in intensive agricultural systems to plow the land, apply **agrochemicals**, and harvest the crops. The intensive-management techniques are used in order to substantially increase the productivity of the species of crop plants. However, these gains are expensive in of both terms money and environmental damage. The U.S. Environmental Protection Agency estimates that **pollution** from agricultural runoff—sediment, animal wastes, pesticides, salts, and nutrients—account for 70% of the river miles in the United States that are impaired by pollution, 49% of the impaired fresh-water **lake** acreage, and 27% of the impaired square miles of marine estuary (inlets where rivers mingle with the ocean). According to the U.S. Department of Agriculture, an average of 11,600 lbs (5270 kg) of **soil** is being eroded from every acre of cultivated U.S. cropland; even this is a notable improvement over soil-loss rates of 20 years ago. Soil-loss rates are even higher in poorer parts of the world.

Agricultural plants

Hundreds of species of plants are cultivated by humans under managed agricultural conditions. However,

most of these species are tropical crops of relatively minor importance—minor in terms of their contribution to the global production of all agricultural plants. In fact, a small number of plant species contribute disproportionately to the global harvest of plant crops in agricultural systems. Ranked in order of their annual production (measured in millions of metric tons per year), the world's 15 most-important food crops are: 1) sugar cane (740 million tonnes per year); 2) **wheat** (390); 3) rice (370); 4) corn or maize (350); 5) white **potato** (300); 6) **sugar beet** (260); 7) **barley** (180); 8) **sweet potato** (150); 9) cassava (110); 10) **soybean** (80); 11) wine **grapes** (60); 12) tomato (45); 13) **banana** (40); 14) beans and peas (40); 15) orange (33).

Some care should be taken in interpreting these data in terms of the yield of actual foodstuffs. The production data for sugar cane, for example, reflect the entire harvested plant, and not just the refined sugar that is the major economic product of this crop. In contrast, the data for wheat and other grain crops reflect the actual harvest of seeds, which are much more useful nutritionally, pound for pound, than whole sugar cane (or even refined sugar).

Most agricultural crops are managed as annual plants, meaning that they are cultivated over a cycle of one year or less, with a single rotation involving sowing, growth, and harvesting. This is true of all the grains and **legumes** and most **vegetables**. Other agricultural species are managed as perennial crops, which are capable of yielding crops on a sustained basis once established. This is typically the manner in which tree-fruit crops such as oranges are managed and harvested, as are certain tropical species such as oil-palm (*Elaeis guineensis*) and para rubber (*Hevea brasiliensis*).

Some extremely valuable crops are not utilized for food, raw material, or energy. Instead, these crops are used for the production of important medicines, as is the case of the rosy periwinkle (*Catharanthus roseus*), which produces several chemicals that are extremely useful in treatment of certain types of cancers. Other crops are used to produce very profitable but illegal drugs. Examples of these sorts of crops include **marijuana** (*Cannabis sativa*), **cocaine** (from *Erythroxylon coca*), and the opium poppy (*Papaver somniferum*).

Agricultural animals

Enormous numbers of domesticated animals are cultivated by people as food crops. In many cases the animals are used to continuously produce some edible product that can be harvested without killing them. For example, milk can be collected daily from various species of **mammals**, including cows, **goats**, **sheep**, **horses**, and camels. Similarly, chickens can produce eggs regularly. All of the above animals, plus many other domesticated species, are also routinely slaughtered for their meat.

The populations of some of these domesticated animals are very large. In addition to approximately six billion people, the world today supports about 1.7 billion sheep and goats (*Ovis aries* and *Capra hircus*), 1.3 billion cows (*Bos taurus* and *B. indica*), 0.9 billion **pigs** (*Sus scrofa*), and 0.3 billion horses, camels, and **water** buffalo (*Equus caballus*, *Camelus dromedarius*, and *Bubalus bubalis*). In addition, there are about 10–11 billion domestic fowl, most of which are chickens (*Gallus gallus*).

These populations of domesticated animals are much larger than those maintained by any wild large animals. For example, no wild mammals of a comparable size to those listed above have populations greater than about 50 million, which is equivalent to less than 1% of a typical domestic-livestock population.

Aquaculture

Aquaculture is an aquatic analogue of terrestrial agriculture. In aquaculture, animals or seaweeds are cultivated under controlled, sometimes intensively managed conditions, to be eventually harvested as food for humans. Increasingly, aquaculture is being viewed as an alternative to the exploitation of wild stocks of aquatic animals and seaweeds.

The best opportunities to develop aquaculture occur in inland regions where there are many ponds and small lakes, and in protected coastal locations on the oceans. Fresh-water aquaculture is especially important in **Asia**, where various species of fish are cultivated in artificial ponds, especially **carp** (*Cyprinus carpio*) and tilapia (*Aureochromis niloticus*). In North America, various species of fish are grown in inland aquaculture in small ponds, most commonly rainbow trout (*Salmo gairdneri*) and **catfish** (*Ictalurus* spp.).

Aquaculture is also becoming increasingly important along sheltered marine coastlines in many parts of the world. In the tropics, extensive areas of mangrove forest are being converted into shallow ponds for the cultivation of prawns (*Penaeus monodon*) and giant prawns (*Macrobrachium rosenbergii*). Negative impacts of this practice include the destruction of the mangrove forests themselves (and of other coastal environments)—already endangered by other types of coastal development—and the nonselective mass harvesting of fish and other sea life supply feed for the **shrimp**. In North America and Western Europe, the cultivation of Atlantic salmon (*Salmo salar*) has become an important industry in recent decades, using pens floating in shallow, coastal embayments, and sometimes in the open ocean. Research is being undertaken into the potential domestication of other marine crops, including species of fish and seaweeds that are now harvested from unmanaged ecosystems.

Agroforestry

Agroforestry is a forest-related analogue of agriculture. In agroforestry, trees are usually cultivated under intensively managed conditions, to eventually be harvested as a source of lumber, pulpwood, or fuelwood. In many regions, this sort of intensive forestry is being developed as a high-yield alternative to the harvesting of natural forests.

The most important trees grown in plantations in the temperate zones as agroforestry crops are species of **pines** (*Pinus* spp.), spruces (*Picea* spp.), larch (*Larix* spp.), and poplar (*Populus* spp.). Depending on the species, site conditions, and economic product that is desired, intensively managed plantations of these trees can be harvested after a growth period of only 10 to 40–60 years, compared with 60 to more than 100 years for natural, unmanaged forest in the same regions. Increasingly, these temperate species of trees are being selectively bred and hybridized to develop high-yield varieties, in **parallel** with the ways in which food crops have been culturally selected from their wild progenitors during the process of domestication.

Unfortunately, there is a downside to such practices. Large monocultures of genetically uniform crops, whether of trees, corn, or any other plant, are by their very nature more vulnerable to climate variation, **pests**, and **disease** than are more diverse plant communities. The large-scale replacement of managed natural forests with uniform high-yield varieties has led to the appearance of pseudoforests in places such as Sweden and parts of the United States: large tracts of land covered by uniform specimens of one species of tree, supporting little **wildlife**. **Habitat** loss from monocultural tree cropping has placed over a thousand forest-dwelling species on the endangered list in Sweden alone.

Fast-growing, high-yield species of trees are also being grown under agroforestry systems in the tropics, for use locally as a source of fuelwood, and also for animal fodder, lumber, and pulpwood. Various tree species are being grown in this way, including species of pine, eucalyptus (*Eucalyptus* spp.), she-oak (*Casuarina* spp.), and tree-legumes (such as *Albizia procera* and *Leucaena leucocephala*). Plantations of slower-growing tropical hardwoods are also being established for the production of high-value lumber, for example, of **mahogany** (*Swietenia mahogani*) and teak (*Tectona grandis*).

See also Livestock.

Resources

Books

Conger, R. H. M., and G. D. Hill. *Agricultural Plants.* 2nd ed. Cambridge, U.K.: Cambridge University Press, 1991.

Freedman, B. *Environmental Ecology.* 2nd ed. San Diego Academic Press, 1994.

Klein, R. M. *The Green World. An Introduction to Plants and People.* New York: Harper & Row, 1987.

Other

U.S. Department of Agriculture. "1997 National Resources Inventory." Natural Resources Conservation Service. December 2000 [cited October 18, 2002]. <http://www.nrcs.usda.gov/technical/NRI/>.

U.S. Environmental Protection Agency. "National Management Measures to Control Nonpoint Source Pollution from Agriculture." 2000 [cited October 18, 2002]. <http://www.nrcs.usda.gov/technical/NRI/1997/summary_report/body.html#revised>.

Bill Freedman

Cross multiply

If two fractions are equal, say

$$\frac{a}{b} = \frac{c}{d}$$

then it is always true that the products of the numbers given by

$$\frac{a}{b} \bowtie \frac{c}{d}$$

are also equal or $ad = bc$. This is the most common form of cross **multiplication**. That

$$\frac{a}{b} = \frac{c}{d}$$

implies $ad = bc$ can be shown by multiplying both sides of

$$\frac{3}{x} = \frac{7}{21}$$

by the common denominator bd and canceling.

Cross multiplying is a common first step in solving proportions.

$$\frac{a}{b} \bowtie \frac{c}{d} \qquad \text{has the sum} \qquad \frac{ad + bc}{bd}$$

is equivalent to

$3 \times 21 = x \times 7$ or $7x = 63$.

Therefore, $x = 9$.

Adding fractions can also be done with cross multiplication.

$$\frac{a}{b} = \frac{c}{d}$$

Note that the cross product of two vectors does not involve cross multiplying.

Cross section

In solid **geometry**, the cross section of a three-dimensional object is a two-dimensional figure obtained by slicing the object **perpendicular** to its axis and viewing it end on. Thus, a sausage has a circular cross section, a 4×4 fence post has a **square** cross section, and a football has a circular cross section when sliced one way and an elliptical cross section when sliced another way. More formally, a cross section is the **locus** of points obtained when a **plane** intersects an object at right angles to one of its axes, which are taken to be the axes of the associated rectangular coordinate system. Since we are free to associate a coordinate system relative to an object in any way we please, and because every cross section is one dimension less than the object from which it is obtained, a careful choice of axes provides a cross section containing nearly as much information about the object from which it is obtained, a careful choice of axes provides a cross section containing nearly as much information about the object as a full-dimensional view.

Often choosing an axis of **symmetry** provides the most useful cross section. An axis of symmetry is a line segment about which the object is symmetric, defined as a line segment passing through the object in such a way that every line segment drawn perpendicular to the axis having endpoints on the surface of the object is bisected by the axis. Examples of three-dimensional solids with an axis of symmetry include: right parallelepipeds (most ordinary cardboard boxes), which have rectangular cross sections; spheres (basketballs, baseballs, etc.), which have circular cross sections; and pyramids with square bases (such as those found in Egypt), which have square cross sections.

Other times, the most useful cross section is obtained by choosing an axis **parallel** to the axis of symmetry. In this case, the plane that intersects the object will contain the axis of symmetry. This is useful for picturing such things as fancy parfait glasses in two dimensions.

Finally, there are innumerable objects of interest that have no axis of symmetry. In this case, care should be taken to choose the cross section that provides the most detail.

The great usefulness of a properly chosen cross section comes in the representation of three-dimensional objects using two-dimensional media, such as **paper** and pencil or flat computer screens. The same idea helps in the study of objects with four or more dimensions. A three-dimensional object represents the cross section of one or more four-dimensional objects. For instance, a cube is the cross section of a four-dimensional hypercube. In general, one way to define the cross section of any N-dimensional object as the locus of points obtained when any (N-1) dimensional" surface" intersects an N-dimensional "solid" perpendicular to one of the solid's axes. Again, the axes of an N-dimensional object are the N axes of the associated rectangular coordinate system. While this concept is impossible to represent geometrically, it is easily dealt with algebraically, using vectors and matrices.

Crows and jays

The members of the crow family (Corvidae) are among the world's most intelligent **birds**. The family has recently undergone taxonomic expansion, brought about by evidence gathered through **genetic testing**, and now includes such diverse **species** as birds-of-paradise, **orioles**, and **drongos**. Crows and jays belong to the subfamily Corvinae. The corvids comprise 113 species in 25 genera, which include ravens, crows, jays, magpies, rooks, nutcrackers, and jackdaws. Corvids are passerine or perching birds and count among their numbers the largest of the passerines. Corvids vary considerably in size, ranging from the tiny Hume's ground jay, which is 7.5 in (19 cm) long and weighs 1.5 oz (45 g), to the thick-billed raven, which is 25 in (64 cm) long and weighs 53 oz (1.5 kg).

Corvids originated in the northern and tropical areas of the Old World. From there they spread to their current geographic range; they are found around the world and are found everywhere except in the high Arctic, Antarctic, the southern part of **South America**, and New Zealand. Corvids vary widely in size and appearance, particularly those species found in the **forests** of Central and South America and Southeast Aisa. There are 27 oriental genera, which experts consider strong evidence that the corvids first evolved in **Asia** and then spread north and east, across the Bering land-bridge, into **North America** and, eventually, to Central and South America.

General characteristics

Crows are large to very large, robustly built birds, with tails that are short or medium length. The tail and primary feathers are stiff. The bill varies in shape from species to species, but is relatively long, although it can be stout or slender. The feet and legs are very strong, with scales on the front of the toes and smooth skin on the back. Among the crows the plumage is black, black and white, black and gray or sooty brown, while jays can also be green, blue, gray, and chestnut. The males and females appear similar; that is, there is not the sexual dimorphism found in other birds such as **pheasants** and **ducks**, where the male is brilliantly colored and the female has dull plumage. Some species of jays and magpies have crests. Common to all the corvids is a tuft of bristles at the base of the beak, just above the nostrils and there are more bristles fringing the mouth.

The personality of crows and jays can be described as aggressive, intelligent, quarrelsome, and sometimes playful. The voice of a corvid, once heard, is not easily forgotten. They produce an astounding range of harsh or more musical calls, which are part of languages, researchers also discovered. The repertoire of the blue jay (*Cyanocitta cristata*), for example, includes high-pitched shrieks (usually aimed at intruders such as **cats**, **owls**, or humans; a cry of *jeer-jeer*; a ringing, bell-like *tull-ull*; a call that sounds like the word "teacup'; a rapid clicking call; a soft, lisping song; a sound like a rusty gate; and imitations of the red-tailed hawk, the black-capped chickadee, the northern oriole, the grau catbird, the American goldfinch, and eastern wood pewee. Some species can even imitate human **speech**.

This range of imitative ability is common to several members of the crow family, and is both evidence of the family's intelligence and part of the reason these birds figure so prominently in human fiction. Experiments with captive common, or American, crows (*Corvus brachyrhynchos*) have proved the birds have excellent puzzle-solving abilities, can count up to three or four, have good memories, and can quickly learn to equate certain sounds or symbols with food. Caged jackdaws that were exposed to a 10-second burst of **light** accompanied by a four-minute recording of jackdaw distress calls, followed by two minutes of silence and darkness, soon learned to peck a key that shut off the light and the recording. A captive blue jay used a tool (a scrap of newspaper from the bottom of its cage) to reach a pile of food pellets that lay outside the cage, just out of reach of its beak. Most interesting, several other captive jays who watched this one method of problem solving soon used it too.

Although these captive experiments provide much information, observations of wild corvids provide even better evidence of the corvids' intelligence. In Norway and Sweden, springtime fishermen make holes in the **ice** and drop their fishing lines through them into the **water**. Hooded crows have been seen picking up the line and walking backward as far as they can, pulling the line out of the hole. The crow will do this as often as it needs to bring the end of the line to the surface, as well as the bait or the hooked fish—which the crow then devours.

The corvids are wary as well as smart. One bird researcher noted that blue jays' intelligence is the key to their not falling victim to the prowling cats that kill so many other species of bird. Crows also show signs of coming to one another's aid; the cawing of a common crow will bring crows from everywhere around, ready to mob the **predator**.

Crows live in varied habitats, including forests, **grasslands**, deserts, steppes, farms, and urban areas. They are mostly tree-dwelling, but the ground-jays have adapted to a life on the ground so much that they will run from a threat rather than fly.

They are highly gregarious birds. A flock may consist of as few as six to as many as a few hundred birds. Within the flock is a social hierarchy, particularly among the crows, the pinyon jays, scrub jays, and Mexican jays. However, mated pairs nest on their own. Corvids are generally aboreal nesters, building a nest of twigs lined with soft materials, although some species nest in holes or build domed nests. The female incubates the eggs (two to eight, depending on the species) alone for 16 to 22 days (again, the length of incubation depends on the species), and her mate feeds her while she does so and helps feed the young after they are born. Corvids do not carry food in their beaks, but rather in their throat or in a small pouch within the chin, under the tongue. Although the members of the crow family are known to be raucous, they become secretive near their nests, drawing as little attention as possible to themselves and their nestlings.

Young common crows fledge between 28 and 35 days old; among the family the nestling period ranges from 20 to 45 days. Although captive crows have been known to live 20 years or more, most wild corvids do not live that long.

A Steller's jay (*Cyanocitta stelleri*) in California. *Photograph by Robert J. Huffman. Field Mark Publications. Reproduced by permission.*

The diet of crows and jays is varied both among and within species. The American, or common, crow eats **insects**, spiders, crustaceans, **snails**, **salamanders**, earthworms, **snakes**, **frogs**, the eggs and chicks of other birds, and carrion. The crows will crack the shells of clams, mussels, and other bivalve **mollusks** by picking them up, flying with them to a height, and then dropping them to **rocks** below (herring **gulls** and crows were seen practicing this tactic at the same time, but the gulls dropped the mollusks onto the mud; the crows figured out much sooner that aiming for the rock was a better, more certain, method). The corvids are not solely carnivorous, however; the blue jay eats about three times as much vegetable matter—including acorns, corn, berries, currants, sorrel, and even cultivated cherries—as it does **animal** matter. Blue jays have been known to eat **mice**, small **fish**, and even **bats**. Another common North American crow, the fish crow of the eastern United States, also eats **shrimp**, fiddler **crabs**, **crayfish**, and turtle eggs.

Most wild corvids that have been studied have been seen hiding food for future use. Small **prey** items, such as insects and earthworms, are not usually hidden, but unexpected "bonuses" are hidden away in small holes or under fallen leaves, although hiding places in trees or buildings will also be used. The Canada jay would be unable to recover ground-buried food during the harsh northern winter, so instead hides food in pine and fir trees, sticking it to the branches with saliva (this species has developed accordingly large salivary **glands** for this task).

Ravens and crows have both been reported to hide some of a large amount of food before settling down to eat the remainder. The apparently excellent memories of crows serve them well in rediscovering these food caches, although success varies among species. Many of the acorns hidden by blue jays in the fall are never recovered. The nutcrackers, on the other hand, have been known to recover 70% of the **seeds** they store.

Their natural enemies include owls, **eagles**, and **buzzards**, and they have had a long-running battle with human beings. In the United States, common crows are fond of corn and other cultivated **crops**, and as a result have been shot at and poisoned. The house crow of India—a tremendously successful commensal species which has tied its life so tightly with that of man that its survival alone would be unlikely—has been destroyed in several places because its large flocks caused it to be considered a health threat.

Despite this animosity, most cultures have tales to tell about the corvids. Ravens figure prominently in Inuit legend. Two ravens, Huginn and Munnin, were the companions of the Norse god Odin. A legendary Celtic warrior god named Bran was also accompanied by a raven, and the bird is known by his name (Cigfran) in Celtic Welsh, Cornish, and Breton. In Cornwall, legend has it that the raven and another corvid, the red-billed chough, hold the spirit of King Arthur, and woe to he who harms either of these birds! From far back the raven has been associated with death, particularly with foretelling it—perhaps because of its close association with the Vikings, this raven-death association was particularly strong in western **Europe**.

And the legends are not all just in the past: even in the late twentieth century captive ravens are kept in the Tower of London. The belief is that when the last raven leaves it, the monarchy will fall.

Despite the success of the family as a whole, at least 22 species of corvids are endangered, including the Hawaiian crow (*Corvus tropicus*) and the Marianas crow (*C. kubaryi*).

F. C. Nicholson

Crustacea

The crustacea (subphylum Mandibulata, class Crustacea) are a diverse group of animals. This class includes some of the more familiar **arthropods**, including **barnacles**, **copepods**, **crabs**, prawns, **lobsters**, and wood **lice**. More than 30,000 **species** have been identified, the majority of which are marine-dwelling. Terrestrial species such as woodlice and pill bugs are believed to have evolved from marine species. Most crustaceans are free-living but some species are parasitic—some even on other crustaceans. Some species are free-swimming, while others are specialized at crawling or burrowing in soft sediments.

Despite such an extraordinary diversity of species, many crustaceans have a similar structure and way of life. The distinctive head usually bears five pairs of appendages: two pairs of antennae that play a sensory role in detecting food as well as changes in **humidity** and **temperature**; a pair of mandibles that are used for grasping and tearing food; and two pairs of maxillae that are used for feeding purposes. The main part of the body is taken up with the thorax and abdomen, both of which are often covered with a toughened outer skeleton, or exoskeleton. Attached to the trunk region are a number of other appendages which vary both in number and purpose in different species. In crabs, for example, one pair of appendages may be modified for swimming, another for feeding, another for brooding eggs and yet another for catching **prey**.

Crustacea exhibit a wide range of feeding techniques. The simplest of these are those species that practice filter feeding such as the copepods and tiny shrimps. Feeding largely on **plankton** and suspended materials, the **animal** creates a mini **water** current towards the mouth by the rhythmic beating of countless number of fine setae that cover the specialized feeding limbs of these species. Food particles are collected in special filters and then transferred to the mouth. Larger species such as crabs and lobsters are active hunters of small **fish** and other organisms, while some species adopt a scavenging role, feeding on dead animals or plants and other waste materials.

Apart from the smaller species, which rely on gas exchange through the entire body surface, most crustaceans have special gills that serve as a means of obtaining **oxygen**. Simple excretory organs ensure the removal of body wastes such as **ammonia** and **urea**. Most crustaceans have a series of well-developed sensory organs that include not only eyes, but also a range of chemical and tactile receptors. All crustaceans are probably capable of detecting a **light** source but in some of the more developed species, definite shapes and movements may also be detected.

Breeding strategies vary considerably amongst the crustacea. Most species are dioecious (being either male or female), but some, such as the barnacles, are **hermaphrodite**. **Fertilization** is usually internal through direct copulation. The fertilized eggs then mature either in a specialized brood chamber in some part of the female's body, or attached directly to some external appendage such as a claw. Most aquatic species hatch into a free-swimming larvae that progresses through a series of body molts until finally arriving at the adult size.

See also Zooplankton.

Cryobiology

Cryobiology is the scientific study of the effects of freezing and sub-freezing temperatures on biological fluids, cells, and tissues. It is an extension of **cryogenics**, which is the study of the properties of **matter** at very low temperatures. Cryobiological techniques have application in genetic research, **livestock** breeding, **infertility** treatment, and **organ** transplantation. A related field, cryogenics, is devoted to the study of low temperatures effects. Of real and important economic benefit to livestock breeding, there are those who ignore the devastating effects of freezing on subcellular structures to argue that cryonic preservation of humans after death might allow reanimation at a later date. The possibility of such reanimation via cryonics is presently rejected by mainstream scientists.

The terms cryobiology and cryogenics are derived from the Greek *kryos*, meaning icy cold. Temperatures used in cryogenics range from -148°F (-100°C) to near **absolute zero** -459.67°F (-273.15°C). Such ultra-low temperatures can be achieved by the use of super -cooled gases. The study of these gases dates back to 1877, when Swiss physicist Raoul Pictet (1846–1929) and French physicist/chemist Louis Cailletet (1832–1913) first learned how to liquefy **oxygen**. Although they worked independently and used different methods, both men discovered that oxygen could be liquefied at -297.67°F (-183.15°C). Soon after, other researchers liquefied **nitrogen** at -321.07°F (-196.15°C). Other breakthroughs in cryogenics included James Dewar's invention of the **vacuum** flask in 1898. Dewar's double- walled vacuum storage vessel allowed liquefied gases to be more readily studied. In the last 100 years, a variety of other methods for insulating super-cooled fluids have been developed.

In the twentieth century, scientists began applying cryogenic techniques to biological systems. They explored methods for treating **blood**, semen, **tissue**, and organs with ultra-low temperatures. In the last few decades, this research has resulted in advances in genetic research, livestock breeding, infertility treatment, and organ transplantation.

In the area of genetic research, cryobiology has provided an inexpensive way to freeze and store the embryos of different strains of research laboratory animals, such as **mice**. Maintaining a breeding colony of research animals can be expensive, and cryogenic storage of embryos can reduce cost by 75%. When the animals are needed, they can be thawed and implanted.

In agriculture, cryopreservation allows livestock breeders to mass produce the embryos of genetically desirable cattle. For example, hundreds of eggs can be harvested from a single prize dairy cow and frozen for later implantation in other mothers. Using similar techniques, **pigs** that a re too **fat** to reproduce on their own can be artificially implanted with embryos developed from frozen eggs and sperm. In addition, cryobiologists are examining the possibility of increasing buffalo herds by freezing **bison** embryos and later implanting them into cows to give **birth**.

Cryobiology has met with great success in the treatment of human infertility. The use of frozen sperm, eggs, and embryos increases the success **rate** of fertility treatments because it allows doctors to obtain a large number of samples that can be stored for future **fertilization**. Techniques for freezing sperm were relatively easy to develop, and in 1953, the first baby fertilized with previously frozen sperm was born. The process for freezing embryos is much more complicated, however. It involves removing **water** from the cells and replacing it with an organic antifreeze that prevents the formation of **ice** crystals that can cause cells to burst. Advances over the last few decades have made this technique highly successful, and in 1984, the first baby was born from a previously frozen embryo. Freezing eggs is an even more difficult challenge because the fragile **membrane** structure that surrounds the eggs make them difficult to freeze without causing severe damage. However, scientists working at Reproductive Biology Associates in Atlanta, Georgia, have successfully frozen eggs using a chemical solution similar to the ovaries' natural fluids. They have also learned to collect eggs at a certain point in the hormone cycle to increase the eggs' chances of surviving the freeze-thaw process. Although still experimental, their technique has been used to freeze eggs, which were later thawed and used to impregnate a woman. In 1997, the first birth resulting from frozen eggs was recorded.

Organ storage is another important area of cryobiological research. Using conventional methods, organs can only be stored for very short periods of time. For example, a kidney can be kept for only three days, a liver for no more than 36 hours, and hearts and lungs for no more than six hours. If these organs could be frozen without subcellular damage, storage times would be lengthened almost indefinitely. Although researchers have made great advances, they have not yet perfected the process of freezing and reviving organs. The problem they face is that the formation of ice crystals can damage fragile tissue. However, researchers at South Africa's H. F. Verwoerd Hospital have devised a way around this problem using a cryopreservant liquid that protects the organs during the freezing process. Boris Rubinsky, another important researcher in the field, has discovered an antifreeze protein that has been successfully used to freeze and revive rat livers. Rubinsky's **proteins** are derived from **fish** living in the Arctic that have evolved to survive in very cold water. These proteins alter the structure of ice **crystal** in order to be less damaging to cells. Another experimental new technique, called vitrification, is used to cool organs so quickly that their molecules do not have time to form damaging ice crystals. Continued success in these areas may one day lead to reliable methods of freezing and storing organs for future transplant.

Public awareness of cryonic preservation reached a high point in 2002 with the alleged cryonic freezing of American baseball player, and Hall of Fame member, Ted Williams. Although cryonic freezing has lost its novel popularity over the last decade—allegedly forcing some cryonic firms to disband, thaw, and bury previously stored remains—some long established firms continue to offer cryonic storage of bodies (in some cases only the detached heads of the deceased are frozen and kept in storage in hope that one day they might be reattached to a functional body).

See also Embryo and embryonic development; Embryo transfer; Embryology.

Cryogenics

Cryogenics is the science of producing and studying low-temperature environments. The word cryogenics comes from the Greek word "kryos," meaning cold; combined with a shortened form of the English verb "to generate," it has come to mean the generation of temperatures well below those of normal human experience.

More specifically, a low-temperature environment is termed a cryogenic environment when the **temperature** range is below the point at which permanent gases begin to liquefy. Among others, they include **oxygen**, **nitrogen**, **hydrogen**, and helium. The origin of cryogenics as a scientific discipline coincided with the discovery by nineteenth century scientists, that the permanent gases can be liquefied at exceedingly low temperatures. Consequently, the term cryogenic applies to temperatures from approximately -148°F (-100°C) down to **absolute zero**.

The temperature of a **sample**, whether it be a gas, liquid, or solid, is a measure of the **energy** it contains, energy that is present in the form of vibrating **atoms** and moving molecules. Absolute zero represents the lowest attainable temperature and is associated with the complete absence of atomic and molecular **motion**. The existence of absolute zero was first pointed out in 1848 by William Thompson (later to become Lord Kelvin), and is now known to be -459°F (-273°C). It is the basis of an

absolute temperature scale, called the Kelvin scale, whose unit, called a Kelvin rather than a degree, is the same size as the Celsius degree. Thus, -459°F corresponds to -273°C corresponds to 0K (note that by convention the degree symbol is omitted, so that 0K is read "zero Kelvin"). Cryogenics, then, deals with producing and maintaining environments at temperatures below about 173K (-148°F [-100°C]).

In addition to studying methods for producing and maintaining cold environments, the field of cryogenics has also come to include studying the properties of materials at cryogenic temperatures. The mechanical and electrical properties of many materials change very dramatically when cooled to 100K or lower. For example, rubber, most **plastics**, and some metals become exceedingly brittle, and nearly all materials contract. In addition, many metals and **ceramics** lose all resistance to the flow of **electricity**, a phenomenon called superconductivity. Very near absolute zero (2.2K) liquid helium undergoes a transition to a state of superfluidity, in which it can flow through exceedingly narrow passages with no **friction**.

History

The development of cryogenics as a low temperature science is a direct result of attempts by nineteenth century scientists to liquefy the permanent gases. One of these scientists, Michael Faraday, had succeeded, by 1845, in liquefying most of the gases then known to exist. His procedure consisted of cooling the gas by immersion in a bath of **ether** and dry **ice** and then pressurizing the gas until it liquefied. Six gases, however, resisted every attempt at liquefaction and were thus known at the time as permanent gases. They were oxygen, hydrogen, nitrogen, **carbon monoxide**, methane, and nitric oxide. The noble gases, helium, neon, argon, krypton, and xenon, not yet discovered.

Of the known permanent gases, oxygen and nitrogen, the primary constituents of air, received the most attention. For many years investigators labored to liquefy air. Finally, in 1877, Louis Cailletet in France and Raoul Pictet in Switzerland, succeeded in producing the first droplets of liquid air, and in 1883 the first measurable quantity of liquid oxygen was produced by S. F. von Wroblewski at the University of Cracow. Oxygen was found to liquefy at 90K (-297°F [-183°C]), and nitrogen at 77K (-320°F [-196°C]).

Following the liquefaction of air, a race to liquefy hydrogen ensued. James Dewar, a Scottish chemist, succeeded in 1898. He found the **boiling point** of hydrogen to be a frosty 20K (-423°F [-253°C]). In the same year, Dewar succeeded in freezing hydrogen, thus reaching the lowest temperature achieved to that time, 14K

TABLE 1			
Cryogen	**Boiling Point**		
	°F	°C	K
Oxygen	-297	-183	90
Nitrogen	-320	-196	77
Hydrogen	-423	-253	20
Helium	-452	-269	4.2
Neon	-411	-246	27
Argon	-302	-186	87
Krypton	-242	-153	120
Xenon	-161	-107	166

(-434°F [-259°C]). Along the way, argon was discovered (1894) as an impurity in liquid nitrogen, and krypton and xenon were discovered (1898) during the fractional **distillation** of liquid argon. Fractional distillation is accomplished by liquefying a mixture of gases each of which has a different boiling point. When the mixture is evaporated, the gas with the highest boiling point evaporates first, followed by the gas with the second highest boiling point, and so on. Each of the newly discovered gases condensed at temperatures higher than the boiling point of hydrogen, but lower than 173K (-148°F [-100°C]).

The last element to be liquefied was helium gas. First discovered in 1868 in the **spectrum** of **Sun**, and later on **Earth** (1885), helium has the lowest boiling point of any known substance. In 1908, the Dutch physicist Heike Kamerlingh Onnes finally succeeded in liquefying helium at a temperature of 4.2K (-452°F).

Methods of producing cryogenic temperatures

There are essentially only four physical processes that are used to produce cryogenic temperatures and cryogenic environments: **heat** conduction, evaporative cooling, cooling by rapid expansion (the Joule-Thompson effect), and adiabatic demagnetization. The first two are well known in terms of everyday experience. The third is less well known but is commonly used in ordinary refrigeration and air conditioning units, as well as cryogenic applications. The fourth process is used primarily in cryogenic applications and provides a means of approaching absolute zero.

Heat conduction is familiar to everyone. When two bodies are in contact, heat flows from the higher temperature body to a lower temperature body. Conduction can occur between any and all forms of **matter**, whether gas, liquid, or solid, and is essential in the production of cryogenic temperatures and environments. For example, samples may be cooled to cryogenic temperatures by immersing them directly in a cryogenic liquid or by placing them in an atmosphere cooled by cryogenic refrigeration. In either case, the sample cools by conduction of heat to its colder surroundings.

The second physical process with cryogenic applications is evaporative cooling, which occurs because atoms or molecules have less energy when they are in the liquid state than when they are in the vapor, or gaseous, state. When a liquid evaporates, atoms or molecules at the surface acquire enough energy from the surrounding liquid to enter the gaseous state. The remaining liquid has relatively less energy, so its temperature drops. Thus, the temperature of a liquid can be lowered by encouraging the process of **evaporation**. The process is used in cryogenics to reduce the temperature of liquids by continuously pumping away the atoms or molecules as they leave the liquid, allowing the evaporation process to cool the remaining liquid to the desired temperature. Once the desired temperature is reached, pumping continues at a reduced level in order to maintain the lower temperature. This method can be used to reduce the temperature of any liquid. For example, it can be used to reduce the temperature of liquid nitrogen to its freezing point, or to lower the temperature of liquid helium to approximately 1K (-458°F [-272°C]).

The third process makes use of the Joule-Thompson effect, and provides a method for cooling gases. The Joule-Thompson effect involves cooling a pressurized gas by rapidly expanding its **volume**, or, equivalently, creating a sudden drop in **pressure**. The effect was discovered in 1852 by James P. Joule and William Thompson, and was crucial to the successful liquefaction of hydrogen and helium.

A valve with a small orifice (called a Joule-Thompson valve) is often used to produce the effect. High pressure gas on one side of the valve drops very suddenly, to a much lower pressure and temperature, as it passes through the orifice. In practice, the Joule-Thompson effect is used in conjunction with the process of heat conduction. For example, when Kamerlingh Onnes first liquefied helium, he did so by cooling the gas through conduction to successively lower temperatures, bringing it into contact with three successively colder liquids: oxygen, nitrogen, and hydrogen. Finally, he used a Joule-Thompson valve to expand the cold gas, and produce a mixture of gas and liquid droplets.

Today, the two effects together comprise the common refrigeration process. First, a gas is pressurized and cooled to an intermediate temperature by contact with a colder gas or liquid. Then, the gas is expanded, and its temperature drops still further. Ordinary household refrigerators and air conditioners work on this principle, using freon, which has a relatively high boiling point. Cryogenic refrigerators work on the same principle but use cryogenic gases such as helium, and repeat the process in stages, each stage having a successively colder gas until the desired temperature is reached.

The fourth process, adiabatic demagnetization, involves the use of paramagnetic salts to absorb heat. This phenomenon has been used to reduce the temperature of liquid helium to less than a thousandth of a degree above absolute zero in the following way. A paramagnetic salt is much like an enormous collection of very tiny magnets called magnetic moments. Normally, these tiny magnets are randomly aligned so the collection as a whole is not magnetic. However, when the salt is placed in a magnetic field by turning on a nearby electromagnet, the north poles of each magnetic moment are repelled by the north pole of the applied magnetic field, so many of the moments align the same way, that is, opposite to the applied field. This process decreases the **entropy** of the system.

Entropy is a measure of randomness in a collection; high entropy is associated with randomness, zero entropy is associated with perfect alignment. In this case, randomness in the alignment of magnetic moments has been reduced, resulting in a decrease in entropy. In the branch of **physics** called **thermodynamics**, it is shown that every collection will naturally tend to increase in entropy if left alone. Thus, when the electromagnet is switched off, the magnetic moments of the salt will tend to return to more **random** orientations. This requires en-

ergy, though, which the salt absorbs from the surrounding liquid, leaving the liquid at a lower temperature. Scientists know that it is not possible to achieve a temperature of absolute zero, however, in their attempts to get ever closer, a similar process called nuclear demagnetization has been used to reach temperatures just one millionth of a degree above absolute zero.

Laser cooling and Bose-Einstein condensate

Scientists have demonstrated another method of cooling via the reduction of energy, using lasers instead of electromagnets. **Laser** cooling operates on the principle that temperature is really a measure of the energy of the atoms in a material. In laser cooling, the **force** applied by a laser beam is used to slow and nearly halt the motion of atoms. Slowing the atoms reduces their energy, which in turn reduces their temperature.

In a laser cooling setup, multiple lasers are aimed from all directions at the material to be cooled. Photons of **light**, which carry **momentum**, bombard the atoms from all directions, slowing the atoms a bit at a time. One scientist has likened the process to running through a hailstorm—the hail hits harder when you are running, no matter which direction you run, until finally you just give up and stop. Although laser-cooled atoms do not completely stop, they are slowed tremendously—normal atoms move at about 1,000 mi (1,600 km) per hour; laser-cooled atoms travel at about 3 ft (0.9 m) per hour.

Using laser cooling and magnetic cooling techniques, scientists have cooled rubidium atoms to 20 billionths of a degree above absolute zero, creating a new state of matter called Bose-Einstein condensate, in which the individual atoms condense into a superatom that acts as a single entity. Predicted many years before by Albert Einstein and Satyendra Nath Bose, Bose-Einstein condensate is has completely different properties from any other kind of matter, and does not naturally exist in the Universe.

The researchers first used laser cooling to bring the temperature of the rubidium atoms down to about 10 millionths of a degree above absolute zero, which was still too warm to produce Bose-Einstein condensate. The atoms held in the laser light trap were then subjected to a strong magnetic field that held them in place. Called a time-averaged orbiting potential trap, the magnetic trap had a special twist that allowed the scientists to remove the most energetic (hottest) atoms, leaving only the very cold atoms behind.

Since the initial demonstration in 1995, scientists have continued to work with Bose-Einstein condensate, using it to slow light to less than 40 MPH (64 km/h), and even to produce an atom laser which produces bursts of atoms with laser-like properties. Multiple studies are underway to help researchers understand the properties of this baffling material.

Applications

Following his successful liquefaction of helium in 1908, Kamerlingh Onnes turned his attention almost immediately to studying the properties of other materials at cryogenic temperatures. The first property he investigated was the **electrical resistance** of metals, which was known to decrease with decreasing temperature. It was presumed that the resistance would completely disappear at absolute zero. Onnes discovered, however, that for some metals the resistance dropped to zero very suddenly at temperatures above absolute zero. The effect is called superconductivity and has some very important applications in today's world. For example, superconductors are used to make magnets for particle **accelerators** and for **magnetic resonance imaging (MRI)** systems used in many hospitals.

The discovery of superconductivity led other scientists to study a variety of material properties at cryogenic temperatures. Today, physicists, chemists, material scientists, and biologists study the properties of metals, as well as the properties of insulators, semiconductors, plastics, composites, and living **tissue**. In order to chill their samples they must bring them into contact with something cold. This is done by placing the sample in an insulated container, called a dewar, and cooling the inner space, either by filling it with a cryogenic liquid, or by cooling it with a cryogenic refrigerator.

Over the years, this research has resulted in the identification of a number of useful properties. One such property common to most materials that are subjected to extremely low temperatures is brittleness. The **recycling** industry takes advantage of this by immersing recyclables in liquid nitrogen, after which they are easily pulverized and separated for reprocessing. Still another cryogenic material property that is sometimes useful is that of thermal contraction. Materials shrink when cooled. To a point (about the temperature of liquid nitrogen), the colder a material gets the more it shrinks. An example is the use of liquid nitrogen in the assembly of some **automobile** engines. In order to get extremely tight fits when installing valve seats, the seats are cooled to liquid nitrogen temperatures, whereupon they contract and are easily inserted in the engine head. When they warm up, a perfect fit results.

Cryogenic liquids are also used in the space program. For example, cryogens are used to propel rockets into space. A tank of liquid hydrogen provides the fuel to be burned and a second tank of liquid oxygen is provided

---

KEY TERMS

Absolute zero—Absolute zero is the lowest temperature possible. It is associated with the absence of molecular motion and is equal to 0K (-459°F [-273°C]).

Boiling point—The boiling point of a liquid is the temperature at which it boils, also the temperature at which its vapor condenses.

Bose-Einstein condensate—A material state in which a collection of supercooled atoms fall into the same quantum state, essentially acting like a single superatom.

Cryogen—A cryogen is a liquid that boils at temperatures below about 173K (-148°F [-100°C]).

Entropy—The measurement of a tendency towards increased randomness and disorder.

Kelvin temperature scale—The Kelvin temperature scale is an absolute temperature scale with the same size unit, called the Kelvin, as the Celsius scale, but shifted so that zero Kelvin (0K) corresponds to absolute zero.

Superconductivity—Superconductivity is the ability of a material to conduct electricity without loss, that is, without electrical resistivity, at a temperature above absolute zero. The phenomenon occurs in certain materials when their electrical resistance drops suddenly and completely to zero at a specific cryogenic temperature, called the critical temperature.

Thermodynamics—Thermodynamics is the study of energy in the form of heat and work, and the relationship between the two.

---

for **combustion**. A more exotic application is the use of liquid helium to cool orbiting infrared telescopes. Any object warmer than absolute zero radiates heat in the form of infrared light. The infrared sensors that make up a telescope's **lens** must be cooled to temperatures that are lower than the equivalent temperature of the light they are intended to sense, otherwise the **telescope** will be blinded by its own light. Since temperatures of interest are as low as 3K (-454°F [-270°C]), liquid helium at 1.8K (-456°F [-271°C]) is used to cool the sensors.

Cryogenic preservation has even extended to a hotly debated topic: in vitro **fertilization**. In vitro fertilization is a technique that improves the chances of a woman being pregnant by removing an egg and fertilizing it with a sperm, cultivating the zygote until it becomes an embryo and implanting it into the uterus of a female recipient. A recent report by clinical geneticists at the Georgia Reproductive Specialists LLC, in Atlanta published results from the first conception and delivery of non-identical twins after a successful transfer of human blastocyts (a stage during the early development of the embryo shortly after fertilization) that were cryogenically preserved at day six and seven after fertilization. Controversy in the medical community surrounds the value of cryogenically preserving later stage human blastocytes. Although using cryogenic preservation for maintaining embryo may lead to bioethical complications, the medical potential for preserving tissues or organs that thawed maitain viability has tremendous medical benefits.

Finally, the production of liquefied gases has itself become an important cryogenic application. Cryogenic liquids and gases such as oxygen, nitrogen, hydrogen, helium, and argon all have important applications. Shipping them as gases is highly inefficient because of their low densities. This is true even at extremely high pressures. Instead, liquefying cryogenic gases greatly increases the weight of cryogen that can be transported by a single tanker.

See also Cryobiology; Faraday effect; Gases, liquefaction of; Gases, properties of; Temperature regulation.

Resources

Books

Asimov, Isaac. *Asimov's Chronology of Science and Discovery.* New York: Harper and Row, 1989.

Baluchandran, B., D. Gubser, and K.T. Hartwig, eds. *Advances in Cryogenic Engineering: Proceedings of the International Cryogenic Materials Conference Icmc (AIP Conference Proceedings, 614)* American Institute of Physics, 2002.

Periodicals

Sills, E.S., Sweitzer CL, P.C., Morton, M. Perloe, CR Kaplan, M.J. Tucker. "Dizygotic Twin Delivery Following In Vitro Fertilization and Transfer of Thawed Blastocysts Cryopreserved at Day 6 and 7." *Fertility and Sterility* (2003): 79 (February): 424–7.

J. R. Maddocks

Cryptography, encryption, and number theory

Cryptography is a division of applied **mathematics** concerned with developing schemes and formula to enhance the privacy of communications through the use of codes. Cryptography allows its users, whether governments, military, businesses, or individuals, to maintain privacy and confidentiality in their communications. Encryption is any form of coding, ciphering, or secret writing. Encryption of data, therefore, includes any and all attempts to conceal, scramble, encode, or encipher any information. In the modern world, however, the term data usually implies digital data, that is, information in the form of binary digits ("bits," most often symbolized as 1s and 0s).

The goal of encryption is to be "crack proof" (i.e, only able to be decoded and understood by authorized recipients). Cryptography is also a means to ensure the integrity and preservation of data from tampering. Modern cryptographic systems rely on functions associated with advanced mathematics, including a specialized branch of mathematics termed **number theory** that explores the properties of numbers and the relationships between numbers.

Although cryptography has a long history of use and importance in military and diplomatic affairs, the importance of cryptography increased during the later half of the twentieth century. Increasing reliance on electronic communication and data storage increased demand for advancements in cryptologic science. The use of cryptography broadened from its core diplomatic and military users to become of routine use by companies and individuals seeking privacy in their communications.

In addition to improvements made to cryptologic systems based on information made public from classified government research programs, international scientific research organizations devoted exclusively to the advancement of cryptography (e.g., the International Association for Cryptologic Research (IACR)), began to apply applications of mathematical number theory to enhance privacy, confidentiality, and the security of data. Applications of number theory were used to develop increasingly involved algorithms (i.e., step-by-step procedures for solving a mathematical problems). In addition, as commercial and personal use of the Internet grew, it became increasingly important, not only to keep information secret, but also to be able to verify the identity of message sender. Cryptographic use of certain types of algorithms called "keys" allow information to be restricted to a specific and limited audiences whose identities can be authenticated.

In some cryptologic systems, encryption is accomplished, for example, by choosing certain **prime numbers** and then products of those prime numbers as basis for further mathematical operations. In addition to developing such mathematical keys, the data itself is divided into blocks of specific and limited length so that the information that can be obtained even from the form of the message is limited. Decryption is usually accomplished by following an elaborate reconstruction process that itself involves unique mathematical operations. In other cases, decryption is accomplished by performing the inverse mathematical operations performed during encryption.

Although it often debated as to whether what was to become known as the RSA **algorithm** was, at least it part, developed earlier by government intelligence agencies, in August 1977, Ronald Rivest, Adi Shamir, and Leonard Adleman published an algorithm destined to become a major advancement in cryptology. The RSA algorithm underlying the system derives its security from the difficulty in factoring very large composite numbers. As of 2003, the RSA algorithm became the most commonly used encryption and authentication algorithm in the world. The RSA algorithm was used in the development of Internet web browsers, spreadsheets, data analysis, email, and word processing programs.

Because digital data are numerical, their efficient encryption demands the use of ciphering rather than coding. A cipher is a system of rules for transforming any message text (the plaintext) into an apparently **random** text (the ciphertext) and back again. Digital computers are ideal for implementing ciphers; virtually all ciphering today is performed on digital data by digital computers.

See also Computer languages; Computer memory, physical and virtual memory; Computer software; Internet and the World Wide Web.

Resources

Other

National Institute of Standards and Technology. "Advanced Encryption Standard: Questions and Answers." Computer Resource Security Center. March 5, 2001 (cited March 26, 2003) <http://csrc. nist.gov/encryption/aes/round2/aesfact. html>.

Nechvatal, James, et al. "Report on the Development of the Advanced Encryption Standard." National Institute of Standards and Technology. October 2, 2000. (cited March 26, 2003) <http://csrc.nist.gov/encryption/aes/round2/r2report. pdf>.

K. Lee Lerner
Larry Gilman

Crystal

A crystal is a solid in which the particles that make up the solid take up a highly ordered, definite, geometric arrangement that is repeated in all directions within the crystal.

Crystals have always attracted the curiosity of humans. Archaeologists have unearthed shells, claws, teeth, and other crystalline solids dating to 25,000 B.C. that have holes, as though worn as necklaces, and that are engraved with symbols of magic. The treasures of the ancient Egyptian king, Tutankhamen, abound with crystals in the forms of gems and jewels. These were not only intended for personal adornment, but were designed in symbolic fashion and believed to possess mystical and religious powers. Healers used crystals in their magical rites and cures.

In ancient Greece, Archimedes made a study of regular solids, and Plato and Aristotle speculated on the relationship between regular solids and the elements. In the sixteenth century, the German naturalist, Giorgius Agricola, classified solids by their external forms, and Johannes Kepler observed that snowflakes were always six-sided (circa 1611), commenting on geometrical shapes and arrangements that might produce this effect. In the seventeenth century, noted philosophers and mathematicians, including René Descartes, Robert Hooke, and Christiaan Huygens followed and expanded Kepler's postulates.

In 1671, an English translation of a study by a Danish-born scientist, Nicolaus Steno, was published in London. It described his investigative work on crystals of quartz, which consists of silicon andoxygen. An Italian scientist, Domenico Guglielmini, developed a structural theory of crystals over the years 1688-1705. Later, measurements of crystals by the French scientist, Jean Baptiste Louis Romé Delisle, were published between 1772-1783. In 1809, the British scientist, William Hyde Wollaston, described an improved goniometer instrument for making accurate measurements on small crystals.

The study of crystals has led to major advances in our understanding of the **chemistry** of biological processes. In 1867, Louis Pasteur discovered two types of **tartaric acid** crystals which were related as the left hand is to the right; that is, one was the mirror image of the other. This led to the discovery that most biomolecules, molecules upon which living systems are based, exhibit this same type of "handedness." In fact, scientists have speculated on the possibility of life having evolved from crystals.

Detailed analyses of crystal structures are carried out by x-ray **diffraction**. In 1912, Max von Laue predicted that the spacing of crystal layers is small enough to cause diffraction (breaking of **light**, when it hits an opaque surface, into colored bands). William Henry Bragg and his son, William Lawrence Bragg, were awarded the Nobel Prize in chemistry (1915) for their development of crystal structure analysis using x-ray diffraction. In 1953, James Watson and Francis Crick deduced the **double helix** structure of DNA (deoxyribonucleic acid, one of the nucleic acids which controls heredity in living organisms) partly from the results of x-ray diffraction analysis of DNA. In recognition of this advancement in the study of the processes of life, they were awarded the Nobel Prize in 1962. Throughout the twentieth century the study of crystalline molecules has continued to expand our knowledge by providing detailed structures of vitamins, **proteins** (enzymes, myoglobin, bacterial membranes), **liquid crystals**, polymers, and organic and inorganic compounds.

Today, crystals are still worn for decorative purposes in the form of gems and jewels; there are still believers in the mystical powers of crystals, but there is no scientific basis for any of the many claims made for them by "New Age" promoters. Crystals are used in modern technological applications, such as lasers.

Common classes of crystalline solids

The standard classes of crystalline solids are the metals, ionic compounds, molecular compounds, and network solids.

The metals are those elements occurring on the left side of the **periodic table** (a classification of elements based on the number of protons in their nuclei), up to the diagonal that connects boron and astatine. The nuclei of **metal atoms** take up highly ordered, crystalline arrangements; the atomic electrons are relatively free to move throughout the metal, making metals good conductors of **electricity**.

When a metallic element combines with a **nonmetal** (an element which is on the right side of the boron-astatine diagonal) an ionic compound is obtained. Ionic compounds do not consist of molecules, but are made up of ordered arrays of ions. An ion is a charged atom or **molecule**; a positive ion (or **cation**) is produced when an atom gives up an **electron**, and a **negative** ion (or **anion**) is the result when an atom gains an electron. The attraction of opposite charges of cations and anions (electrostatic attraction) keeps them in close proximity to each other. In compounds, the ions assume the ordered arrangements characteristic of crystals. The strong electrostatic forces between oppositely charged ions make it very difficult to separate the ions and break down the crystal structure; thus, ionic compounds have very high melting points (generally higher than 1,742°F [950°C]). Because the electrons in ionic compounds are not free to move throughout the crystal, these compounds do not conduct electricity unless the ions themselves are released by heating to high temperatures or by dissolving the compound.

When nonmetallic elements combine in reactions, the resulting compound is a molecular compound. With-

in such compounds the atoms are linked by shared electrons, so that ions are not present. However, partial charges arise in individual molecules because of uneven distribution of electrons within each molecule. Partial positive charges in one molecule can attract partial negative charges in another, resulting in ordered crystalline arrangements of the molecules. The forces of attraction between molecules in crystals of covalent compounds are relatively weak, so these compounds require much less **energy** to separate the molecules and break down the crystals; thus, the melting points of covalent compounds are usually less than 572°F (300°C). Because charged particles are not present, covalent compounds do not conduct electricity, even when the crystals are broken down by melting or by dissolving.

Network solids are substances in which atoms are bonded covalently to each other to form large networks of molecules of nondefinite size. Examples of network solids include **diamond** and graphite, which are two crystalline forms of **carbon**, and silicates, such as **sand**, rock, and **minerals**, which are made up of silicon and **oxygen** atoms. Because the atoms occupy specific bonding sites relative to each other, the resulting arrangement is highly ordered and, therefore, crystalline. Network solids have very high melting points because all the atoms are linked to their neighbors by strong covalent bonds. Thus, the melting point of diamond is 6,332°F (3,500°C). Such solids are insoluble because the energy required to separate the atoms is so high.

Internal structures of metallic crystals

A complete description of the structure of a crystal involves several levels of detail. Metallic crystals are discussed first for simplicity, because the atoms are all of the same type, and can be regarded as spherical in shape. However, the basic concepts are the same for all solids.

If the spheres are represented by points, then the pattern of repeating points at constant intervals in each direction in a crystal is called the lattice. Fourteen different lattices can be obtained geometrically (the Bravais lattices). If lines are drawn through analogous points within a lattice, a three-dimensional arrangement of structural units is obtained. The smallest possible repeating structural unit within a crystal is called the unit cell, much like a **brick** is the smallest repeating unit (the unit cell) of a brick wall.

The 14 unit cell types are based on seven types of crystal systems. These are the cubic, triclinic, monoclinic, orthorhombic, trigonal, tetragonal, and hexagonal systems. The specific crystal system and type of unit cell observed for a given solid is dependent on several factors. If the particles that make up the solid are approximately spherical, then there is a tendency for them to pack together with maximum efficiency. Close-packed structures have the maximum packing efficiency, with 74% of the crystal **volume** being occupied by the particles. Close-packing occurs in two different ways: cubic close-packing (ccp), which gives rise to cubic unit cells (the face-centered cube), and hexagonal close-packing (hcp), which gives hexagonal unit cells.

The placement of atoms that produces each of these arrangements can be described in terms of their layering. Within each layer, the most efficient packing occurs when the particles are staggered with respect to one another, leaving small triangular spaces between the particles. The second layer is placed on top of the first, in the depressions between the particles of the first layer. Similarly, the third layer lies in the depressions of the second. Thus, if the particles of the third layer are also directly over depressions of the first layer, the layering pattern is ABCABC, in which the fourth layer is a repeat of the first. This is called cubic close-packing, and results in the face-centered cubic unit cell. Such close-packed structures are common in metals, including **calcium**, strontium, **aluminum**, rhodium, iridium, nickel, palladium, platinum, **copper**, silver, and gold. If the third layer particles are also directly over particles of the first, the repeating layer pattern is ABAB. This is called hexagonal close-packing, and produces the hexagonal unit cell. This packing arrangement also is observed for many metals, including beryllium, **magnesium**, scandium, **yttrium**, lanthanum, **titanium**, zirconium, hafnium, technetium, rhenium, rubidium, osmium, cobalt, zinc, and cadmium.

Other layering patterns in which the particles are not close-packed occur frequently. For example, particles within a layer might not be staggered with respect to one another. Instead, if they align themselves as in a **square** grid, the spaces between the particles also will be square. The second layer fits in the depressions of the first; the third layer lies in depressions of the second, and over particles of the first layer, giving the layering pattern (ABAB) with a space-filling efficiency of 68%. The resulting unit cell is a body-centered cube. Metals which have this arrangement of atoms include the **alkali metals**, **barium**, vanadium, niobium, tantalum, chromium, molybdenum, tungsten, manganese, and **iron**.

Common internal structures of crystals of ionic solids

Although ionic solids follow similar patterns as described above for metals, the detailed arrangements are more complicated, because the positioning of two different types of ions, cations and anions, must be considered. In general, it is the larger ion (usually, the anion) that determines the overall packing and layering, while the smaller ion fits in the holes (spaces) that occur throughout the layers.

Two types of holes occupied by cations exist (in close-packed ionic structures.). These are named tetrahedral and octahedral. An ion in a tetrahedral site would be in contact with four ions of opposite charge, which, if linked by imaginary lines, produces a **tetrahedron**. An ion in an octahedral site would be in contact with six ions of opposite charge, producing an octahedron. The number of oppositely charged ions in contact with a given ion is called its coordination number (CN). Therefore, an ion in a tetrahedral site has a coordination number of four; an ion in an octahedral site has a coordination number of six. The total number of octahedral holes is the same as the number of close-packed ions, whereas there are twice as many tetrahedral holes as close-packed atoms. Because tetrahedral holes are smaller, they are occupied only when the **ratio** of the smaller ion's radius to the radius of the larger ion is very small. As the radiusratio of the smaller ion to the larger ion becomes greater, the smaller ion no longer fits into tetrahedral holes, but will fit into octahedral holes.

These principles can be illustrated by several examples. The repeating structural unit of crystalline **sodium chloride** (table **salt**) is the face-centered cubic unit cell. The larger chloride ions are cubic close-packed (ABCABC layering pattern). The radius ratio of **sodium** ion to chloride ion is about 0.6, so the smaller sodium ions occupy all the octahedral sites. Chloride and sodium ions both have coordination numbers of six. This structure occurs frequently among ionic compounds and is called the sodium chloride or rock salt structure.

In the sphalerite (or zinc blende) crystalline form of zinc sulfide, the larger sulfide ions are cubic close-packed (ABCABC layering), giving a face-centered cubic unit cell. The small zinc ions occupy tetrahedral sites. However, the number of tetrahedral holes is twice the number of sulfide ions, whereas the number of zinc ions is equal to the number of sulfide ions. Therefore, zinc ions occupy only half of the tetrahedral holes. In the wurtzite structure, another crystalline form of zinc sulfide, the sulfide ions are hexagonally close-packed, (ABAB layering), giving a hexagonal unit cell. Again, the zinc ions occupy half the tetrahedral sites.

Another common structure, the fluorite structure, is often observed for ionic compounds which have twice as many anions as cations, and in which the cations are larger than the anions. The structure is named after the compound, calcium fluoride, in which the calcium ions are cubic close-packed, with fluoride in all the tetrahedral sites.

TABLE 1. COMMON CRYSTAL STRUCTURES OF IONIC COMPOUNDS

Compound	Structure name	Radius ratio and C.N. of cation and anion	Packing and layering
halides of lithium, sodium, potassium, rubidium; ammonium halides; silver halides; oxides and sulfides of magnesium, calcium, strontium, and barium	sodium chloride	0.41 to 0.75 6:6	chloride ccp, sodium in every octahedral hole
zinc sulfide, copper(I) chloride, cadmium (II) sulfide, mercury (II) sulfide	sphalerite	0.23 to 0.41 4:4	sulfide ccp, zinc in half the tetrahedral holes
zinc sulfide, zinc oxide, beryllium oxide, manganese (II) sulfide, silver iodide, silicon carbide, ammonium fluoride	wurtzite	0.23 to 0.41 4:4	sulfide hcp, zinc in half the tetrahedral holes
calcium fluoride, barium chloride, mercury (II) fluoride, lead (IV) oxide, barium fluoride, strontium fluoride	fluorite	0.72 and up 8:4	calcium ccp, fluoride in all tetrahedral holes
cesium chloride, calcium sulfide, cesium cyanide	cesium chloride	0.72 and up 8:8	chloride in primitive cubes, cesium at the centers

As discussed for metals, many compounds have structures that do not involve close-packing. For example, in the cesium chloride structure, the larger chloride ions are arranged in primitive cubes, with cesium ions occupying positions at the cube centers.

Many other structures are observed for ionic compounds. These involve similar packing arrangements as described above, but vary in number and types of occupied holes, and the distribution of ions in compounds having more than two types of cation and/or anion.

Crystal structures of molecular compounds and network solids

The molecules that make up molecular compounds may not be approximately spherical in shape. Therefore, it is difficult to make detailed generalizations for molecular compounds. They exhibit many crystal structures that are dependent on the best packing possible for a specific molecular shape.

The most common network solids are diamond, graphite, and silicates. Diamond and graphite are two crystalline forms of carbon. In diamond, each carbon atom is covalently bonded to all four of its nearest neighbors in all directions throughout the network. The resulting arrangement of atoms gives a face-centered cubic unit cell. In graphite, some of the covalent bonds are double bonds, forcing the carbon atoms into a planar arrangement of fused six-membered rings, like a chicken-wire fence. Sheets of these fused rings of carbon lie stacked upon one another.

Silicates, present in sand, clays, minerals, **rocks**, and gems, are the most common solid inorganic materials. In the arrays, four oxygen atoms bond to one silicon atom to give repeating tetrahedral units. Silicate units can share oxygen atoms with adjacent units, giving chain silicates, sheet silicates, and framework silicates.

Crystallinity in macromolecules

Macromolecules are giant **polymer** molecules made up of long chains of repeating molecular units and bonded

Name	Composition	Impurity	Common color	Crystal system
Diamond	carbon		colorless and other	cubic
Ruby	aluminum oxide	chromium	red	hexagonal
Sapphire	aluminum oxide	titanium, iron	blue and other	hexagonal
Emerald	beryllium-aluminum silicate	chromium	green	hexagonal
Jade	calcium-magnesium-iron silicate	iron	green and other	monoclinic
Opal	silicon oxide hydrates	(scattered light)	various	none
Topaz	aluminum fluoride-hydroxide-silicate	unknown	colorless and other	orthorhombic
Turquoise	copper-aluminum-hydroxide-phosphate	copper	blue and other	none
Zircon	zirconium silicate	iron	colorless and other	tetragonal

covalently to one another. Macromolecules occur widely in nature as carbohydrates, proteins, and nucleic acids. Polymers, **plastics**, and rubber also are macromolecules.

Macromolecules may be likened to a plate of spaghetti, in which the individual strands of the macromolecules are entangled. Notably, there is a lack of order in this system, and a lack of crystallinity. However, a marked degree of order does exist in certain regions of these entanglements where segments of neighboring chains may be aligned, or where chain folding may promote the alignment of a chain with itself. Regions of high order in macromolecules, called crystallites, are very important to the physical and chemical properties of macromolecules. The increased forces of attraction between chains in these regions give the polymer strength, impact resistance, and resistance to chemical attack. Polymers can be subjected to some form of **heat** treatment followed by controlled cooling and, sometimes, stretching, in order to promote greater alignment of chains, a higher degree of crystallinity, and a consequent improvement in properties.

Crystal defects and growth of crystals

The growth and size of a crystal depends on the conditions of its formation. **Temperature**, **pressure**, the presence of impurities, etc., will affect the size and perfection of a crystal. As a crystal grows, different imperfections may occur, which can be classified as either point defects, line defects (or dislocations), and **plane** defects.

Point defects occur: a) if a particle site is unoccupied (a Schottky defect); b) if a particle is not in its proper site (which is vacant) but is in a space or hole (a Frenkel defect); or c) if an extra particle exists in a space or hole, with no corresponding vacancy (an anti-Schottky defect). Line defects occur: a) if an incomplete layer of particles occurs between other, complete layers (an edge dislocation); or b) if a layer of particles is not planar, but is out of alignment with itself so that the crystal grows in a **spiral** manner (a screw dislocation). Plane defects occur; a) if two crystallites join to form a larger crystal in which the rows and planes of the two crystallites are mismatched (a grain boundary); or b) if a layer in an ABCABC pattern occurs out of sequence (a stacking fault).

Sometimes, imperfections are introduced to crystals intentionally. For example, the conductivity of silicon and germanium can be increased by the intentional addition of arsenic or antimony impurities. This procedure is called "doping," and is used in materials, called semiconductors, that do not conduct as well as metals under normal conditions. The additional electrons provided by arsenic or antimony impurities (they have one more electrons in their outermost shells than do silicon or germanium) are the source of increased conductivity.

Experiments in decreased gravity conditions aboard the space shuttles and in Spacelab I demonstrated that proteins formed crystals rapidly, and with fewer imperfections, than is possible under regular gravitational conditions. This is important because macromolecules are difficult to crystallize, and usually will form only crystallites whose structures are difficult to analyze. Protein analysis is important because many diseases (including Acquired Immunity Deficiency Syndrome, **AIDS**) involve enzymes, which are the highly specialized protein catalysts of **chemical reactions** in living organisms. The analysis of other biomolecules may also benefit from these experiments. It is interesting that similar advantages in crystal growth and degree of perfection have also been noted with crystals grown under high gravity conditions.

Gemstones

Although the apparent perfection of gems is a major source of their attraction, the rich colors of many gemstones are due to tiny impurities of colored metal ions within the crystal structure. Table 2 lists some common gemstones and their crystalline structures.

The value and desirable properties of crystals promote scientific attempts to synthesize them. Although methods of synthesizing larger diamonds are expensive, diamond films can be made cheaply by a method called chemical vapor deposition (CVD). The technique involves methane and **hydrogen** gases, a surface on which the film can deposit, and a microwave oven. Energy from microwaves breaks the bonds in the gases, and, after a series of reactions, carbon films in the form of diamond are produced. The method holds much promise for: a) the tool and cutting industry (because diamond is the hardest known substance); b) **electronics** applications (because diamond is a conductor of heat, but not electricity); and c) medical applications (because it is tissue-compatible and tough, making it suitable for joint replacements, **heart** valves, etc.).

See also Diffraction.

KEY TERMS

. . . .

Close-packing—The positioning of atoms, ions, or molecules in a crystal in such a way that the amount of vacant space is minimal.

Covalent bond—A chemical bond formed when two atoms share a pair of electrons with each other.

Diffraction—A wave-like property of light: when a ray of light passes through a tiny opening it spreads out in all directions, as though the opening is the light source.

Electrostatic attraction—The force of attraction between oppositely charged particles, as in ionic bonding.

Ionic compound—A compound consisting of positive ions (usually, metal ions) and negative ions (nonmetal ions) held together by electrostatic attraction.

Lattice—A pattern obtained by regular repetition of points in three dimensions.

Liquid crystal—A compound consisting of particles which are highly ordered in some directions, but not in others.

Macromolecule—A giant molecule consisting of repeating units of small molecules linked by covalent bonds.

Periodic Table—A classification of the known elements, based upon their atomic numbers (the numbers of protons in the nuclei).

Unit cell—The simplest three-dimensional repeating structure in a crystal lattice.

Resources

Books

Hall, Judy. *The Illustrated Guide To Crystals.* London: Sterling Publications, 2000.

Hankin, Rosie. *Rocks, Crystals & Minerals: Complete Identifier.* New York: Knickerbocker Press, 1999.

Knight, Sirona. *Pocket Guide to Crystals and Gemstones.* Berkeley, CA: Crossing Press, 1998.

Lima-de-Faria, J., ed. *Historical Atlas of Crystallography.* Published for The International Union of Crystallography by Dordrecht: Boston: Kluwer Academic Publishers, 1990.

Massimo D. Bezoari

Cube root *see* **Radical (math)**

Cubic equations

A cubic equation is one of the form $ax^3 + bx^2 + cx + d = 0$ where a,b,c and d are **real numbers**. For example, $x^3-2x^2-5x+6 = 0$ and $x^3 - 3x^2 + 4x - 2 = 0$ are cubic equations. The first one has the real solutions, or roots, -2, 1, and 3, and the second one has the real root 1 and the complex roots 1+i and 1-i.

Every cubic equation has either three real roots as in our first example or one real root and a pair of (conjugate) complex roots as in our second example.

There is a formula for finding the roots of a cubic equation that is similar to the one for the quadratic equation but much more complicated. It was first used by Geronimo Cardano in 1545, even though he had obtained the formula from Niccolo Tartaglia under the promise of secrecy.

Resources

Books

Bittinger, Marvin L., and Davic Ellenbogen. *Intermediate Algebra: Concepts and Applications.* 6th ed. Reading, MA: Addison-Wesley Publishing, 2001.

Roy Dubisch

Cuckoos

Cuckoos, coucals, anis, malkohas, and roadrunners are approximately 127 **species** of **birds** that make up the family Cuculidae. These birds are mostly tropical in distribution, but some species also breed in the temperate zones. Many species are parasitic breeders, laying their eggs in the nests of other species of birds. Species of the cuckoo family occupy a great diversity of habitats, ranging from **desert** to temperate and tropical **forests**.

The cuckoos vary greatly in size, with the range of body length being about 6–27.5 in (16–70 cm). These birds tend to have an elongated body, a rather long neck, a long tail, rounded wings, and a stout, down-curved beak. The basal coloration of the body is generally a brown, grey, or black hue, often with barring of the underparts or a white breast. Males and females are similarly colored, but juveniles are generally different.

A large number of species in the cuckoo family are nest-parasites. Instead of constructing their own nests, these parasitic birds seek out and discover nests of other species, and then lay an egg inside. If the host is of a similar size as the parasitizing cuckoo, then several eggs may be laid in the nest, but on separate days. Only one

A greater roadrunner. *Maslowski/Photo Researchers, Inc. Reproduced by permission.*

egg is laid if the cuckoo is substantially larger than the host, as is often the case. The female cuckoo may also remove any pre-existing eggs of the host species.

The host birds commonly do not recognize the foreign egg, and incubate it as if it was their own. The host then cares for the parasitic hatchling until it fledges, and often afterwards as well. In most cases, the host species is much smaller than the parasite, and it is quite a chore to feed the voracious young cuckoo. The young cuckoo commonly hatches quite quickly and ejects the unhatched eggs of the host from the nest, or it ejects or otherwise kills the babies of the host. Once their nest is discovered by a female cuckoo, the parasitized hosts are rarely successful in raising any of their own young under these sorts of circumstances.

Male cuckoos maintain a breeding territory, largely using a loud and distinctive, often bell-like call. Interestingly, females of the nest-parasitic species of cuckoos also maintain a territory, independent of that of males of their species. In this case, the defended area involves foraging **habitat** for the discovery of nests of other species, rather than for access to females, as in the case of the male cuckoos.

Many species of cuckoos that breed in the temperate zones undertake a long-distance **migration** between their breeding and non-breeding ranges. This is true of species breeding in the Northern Hemisphere, which winter to the south, and also of species breeding in the Southern Hemisphere, which winter to the north. For example, the shining cuckoo (*Chalcites lucidus*) of temperate New Zealand migrates across open waters of the Pacific Ocean, to winter in tropical habitats of the Bismarck Archipelago and Solomon Islands off New Guinea.

Most species in the cuckoo family feed mostly on **insects** and other **arthropods**. Some of the smaller species of cuckoos will eat the hairy caterpillars of certain types of **moths** and **butterflies**. Hairy caterpillars are often an abundant type of food, in part because they are rejected by most other types of birds, which find the hairs to be irritating and distasteful. Some of the larger species of cuckoos will also feed on lizards, **snakes**, small **mammals**, and other birds.

Species of cuckoos

The best-known species in the Cuculidae is the Eurasian cuckoo (*Cuculus canorus*), which breeds widely in forests and thickets of **Europe** and **Asia**. This species is the best-studied of the nest-parasites, laying single eggs in the nests of a wide range of smaller species. Although the egg of the Eurasian cuckoo is usually larger than those of the parasitized host, it is often colored in a closely similar way to the host species. **Individual** Eurasian cuckoos are known to have laid single eggs in as many as 20 nests of other species in one season. The call of the male Eurasian cuckoo is the famous, bi-syllabic: "cuck-coo," a sound that has been immortalized in literature and, of course, in cuckoo-clocks. Northern populations of this species migrate to **Africa** or southern Asia to spend their non-breeding season.

Two familiar cuckoos of **North America** are the yellow-billed cuckoo (*Coccyzus americanus*) and the black-billed cuckoo (*C. erythrophthalmus*). Both of these species breed in open woodlands and brushy habitats. The yellow-billed cuckoo ranges over almost all of the United States, southern Ontario, and northern Mexico, and winters in **South America**. The black-billed cuckoo ranges over southeastern North America, and winters in northwestern South America. This species is most abundant in places where there are local outbreaks of caterpillars. Both of these species build their own nests and raise their two to four babies. However, both species are occasional nest-parasites on other species, including each other.

A much larger American species is the greater roadrunner (*Geococcyx californianus*), a terrestrial bird of dry habitats in the southwestern United States and Central America. The greater roadrunner is the largest cuculid in North America. This species commonly feeds on lizards and snakes, including poisonous rattlesnakes. The greater roadrunner is not a nest-parasite. Roadrunners are fast runners, although not so fast and intelligent as the one that always gets the better of Wile E. Coyote in the famous Warner Bros. cartoons.

Two species of anis breed in North America, the smooth-billed ani (*Crotophaga ani*) of southern Florida,

and the groove-billed ani (*C. sulcirostris*) of southern Texas. These species also occur widely in Central and South America, and on many Caribbean islands. Anis build communal, globular, stick-nests in trees. Each of the several cooperating pairs of anis has its own nesting chamber, and incubate their own eggs. Both parents share in the brooding of the eggs and raising of the young, although there is some degree of cooperative feeding of young birds within the commune. Anis have home ranges, but because of their communal nesting, they do not appear to defend a territory.

The coucals are relatively large birds of Africa, South and Southeast Asia, and Australasia. Rather weak flyers, coucals are skulking birds that occur near the edges of scrubby and wooded habitats. The greater coucal or crow-pheasant (*Centropus sinensis*) of southern and southeastern Asia is a large (20 in [53 cm] body length), black, widespread species. Coucals build their own large globular nest of **grasses** and leaves near the ground in dense vegetation. The male and female share the incubation and rearing of the three to five babies.

Resources

Books

Bird Families of the World. Oxford: Oxford University Press, 1998.
Meinzer, W. *The Roadrunner.* Austin: Texas Tech University Press, 1993.

Bill Freedman

Cucumber *see* **Gourd family (Cucurbitaceae)**

Curare

Curare (pronounced cue-rah'-ree) is a general term for certain chemical substances found in different plants throughout the world's rainforests. These plants produce a harmless sap which for centuries the natives of the rainforests have refined into a deadly poison. The way of refining and delivering the poison from certain types of plants is similar for natives occupying equatorial regions from **South America**, **Africa**, and Southeast **Asia**. Animals are hunted with blowguns loaded with darts that have been prepared with lethal doses of the curare preparations.

The word curare is derived from *woorari*, a word of native American origin from the Amazon and Orinoco basins meaning poison. There are different plants used to produce the poisons for the tips of the darts used in hunting. The blowgun is particularly effective against arboreal animals, such as **monkeys** and **birds**. The hunters final curare preparation is composed of "curares" or poisons from various plants. Curares from these plants share the same chemical composition. They are all alkaloids. An **alkaloid** is an organic compound containing **nitrogen** and usually **oxygen**. They are found in seed plants and are usually colorless and bitter like **codeine** or **morphine**.

The **plant** *Strychnos toxifera* produces the strongest type of curare for the hunters of the rainforests. Other curare type plants, however, have been used in western medicine as anesthetics after it was discovered that curares can have non-lethal effects such as skeletal **muscle relaxants**. Tubocurarine, an anesthetic muscle relaxant introduced into medical practice in the early 1940s contains a curare alkaloid from the chondrodendron plant family.

History

Early eighteenth and nineteenth century researchers studied the effects of curare. In 1780 Abbe Felix Fontana found that its action was not on the nerves and **heart** but on the ability of the voluntary muscles to respond to stimuli. In British experiments, several English researchers showed that animals injected with curare would recover if their **respiration** was artificially continued. Laboratory experiments were continued throughout the nineteenth century using curare to find out more about the relationship between the nervous and skeletal muscle system. In 1850 Claude Bernard using curare identified the neuromuscular junction where the curare interferes with the acceptance of the neural impulse. Earlier in that century Squire Waterton had conjectured that curare could be used in the treatment of **tetanus**.

The first use of curare in **surgery** was in 1912. A German physician and physiologist, Arthur Lawen, wrote about his use of curare in surgery on a patient. He was able to relax the patients abdominal muscles with a small amount of regular **anesthesia** after administering curare. In order to control the curare he also learned how to intubate (insert a tube into the patient's trachea) and then ventilate the lungs, that is add air through the tube to control breathing. His reports, which were published only in German, were ignored largely because anesthesi-

ologists at that time had not learned the techniques of intubation and ventilation.

In 1938 Richard and Ruth Gill returned from a trip to South America to New York with a large stock of crude curare. They collected these plants from their Ecuadorian ranch for the Merck Company. At that time there was some interest in using curare for the treatment of a friend who had multiple sclerosis. Merck lost interest in the project, but some of the Gill's curare stock passed on to Squibb & Co. Early use of the drug for anesthetic purposes, however, were not successful, and interest was dropped at that time for further clinical experimentation.

Interest in curare resumed in 1939 when psychiatrists from the American midwest began to use it to treat certain categories of patients. Children with spastic disorders were injected with curare but when no long range improvement was observed, these psychiatric researchers passed it on to those who were using Metrazol, a drug that was a precursor to electroconvulsive therapy (ECT), formerly referred to as shock treatment. The curare appeared to absorb some of the intense muscle responses or seizures, thus helping to avoid seizure induced fractures to the bones. Other psychiatrists began to experiment with the drug after its successful application to ECT.

Shortly afterwards a Canadian physician, Harold Griffith, began to prepare to use curare for surgery after he saw the positive results of its use in psychiatric patients. He first utilized curare in an operation on January 23, 1942. Then he reported on the successful use of curare as a muscle relaxant for this operation, which was an appendectomy. He administered the curare after the patient's trachea was anesthetized and intubated early in the operation. The muscles of the abdominal wall became relaxed by the curare to help in the performance of the operation. Twenty-five other patients received similar treatment. After Griffith's report of his work, the use of curare and other synthetic type curare muscle relaxants became the standard practice for surgical procedures requiring muscle relaxation.

Tubocurarine

Since 1942 there have been about 50 different relaxants used in clinical anesthesia. Tubocurarine, whose chemical structure was determined 1935, is the prototype of a muscle relaxant that still contains the alkaloid constituent of curare and produces a similar physiological effect. Another semisynthetic derivative of tubocurarine is even more potent. It is given intravenously since it is not active when taken orally.

Anesthetic muscle relaxants block nerve impulses between the junctions of the nerve and muscle. It is be-

KEY TERMS

Electroconvulsive therapy (ECT)—Formerly known as "shock treatment." The administration of a low dose electric current to the head in conjunction with muscle relaxants to produce convulsions. A treatment method whose underlying action is still not fully understood, it has proven effective in relieving symptoms of some severe psychiatric disorders for which no other treatment has been effective, for example, severe depression.

Intubation—The insertion of a tube through the throat and trachea as part of the anesthetic procedure using a muscle relaxant.

Muscle relaxant—A drug used to relax the voluntary muscles during an operation.

Tubocarine—One of the many muscle relaxants used with curarelike effects.

Ventilation—Keeping air flowing through the lungs after the breathing muscles have been relaxed during an operation.

lieved they accomplish this task preventing the acceptance of **acetylcholine**, which is a chemical **neurotransmitter**, by the muscle fiber. In addition to the main clinical use of curare is as an accessory drug in surgical anesthesia to obtain relaxation of skeletal muscle, it is also used to facilitate diagnostic procedures, such as laryngoscopy and **endoscopy**. It is also used in cases of tetanus and myasthenia gravis, an autoimmune disorder.

Resources

Books

Barash, Paul G., Bruce F. Cullen, and Robert K. Stoelting. *Clinical Anesthesia.* Philadelphia, Lippincott, 1992.
Dripps, Robert D., James E. Eckenhoff, and Leroy D. Vandam. *Introduction to Anesthesia.* Philadelphia: Saunders, 1988.
Gold, Mark, and Michael Boyette. *Wonder Drugs: How They Work.* New York: Simon & Schuster, 1987.

Jordan P. Richman

Curium *see* **Element, transuranium**

Curlews

Curlews are large, brownish shorebirds (family Scolopacidae) with long legs and lengthy, downward curving bills, adapted for probing into sediment and **soil** for their food of **invertebrates**.

Although neither **species** of North American curlew is common, the most abundant ones are the long-billed curlew (*Numenius americanus*) and the whimbre, or Hudsonian curlew (*N. phaeopus*). The long-billed curlew breeds in wet meadows and grassy habitats in the western United States and southwestern Canada, and winters on mud flats and beaches in southern California and parts of the Gulf of Mexico. This species appears to be declining in abundance, likely as a result of the loss of most of its natural **habitat**, and possibly because of damage caused by **pesticides**.

The whimbrel breeds further to the north in two subarctic populations, one in coastal Alaska and northwestern Canada, and the other around the west coast of Hudson Bay. The whimbrel also breeds in northern Eurasia. The winter range of this species is very broad, ranging from the southern coastal United States, to the coasts of Central and **South America**, and some Pacific islands.

The bristle-thighed curlew (*N. tahitiensis*) is a rare species with a total population of fewer than 10,000 individuals. The bristle-thighed curlew breeds in montane habitat in western Alaska, and migrates directly south, to winter on widely scattered islands of the Pacific Ocean, including the Hawaiian Islands. The 5,000–5,600 mi (8,000–9,000 km) **migration** of this species is an extraordinary feat of non-stop flight while navigating over trackless **water**, in search of its scattered wintering islands.

The Eskimo curlew (*N. borealis*) is the smallest of the North American species, only 11 in (28 cm) in body length. This species was once abundant during its migrations. However, the Eskimo curlew was decimated by market hunting during the nineteenth century, and is now exceedingly rare, and on the verge of **extinction** (in fact, some biologists believe it is already extinct). The Eskimo curlew is one of many examples of once abundant species that have become extinct or endangered as a result of uncontrolled, unscrupulous exploitation by humans. Such tragedies represent lessons to be learned, so that similar calamities of **biodiversity** can be avoided in the future.

See also Sandpipers; Shore birds.

Currants *see* **Saxifrage family**
Current *see* **Electric current**

Currents

Currents are steady, smooth movements of **water** following a specific course; they proceed either in a cyclical pattern or as a continuous stream. In the Northern Hemisphere, currents generally move in a clockwise direction, while in the Southern Hemisphere they move counterclockwise. There are three basic types of **ocean** currents: surface currents; currents produced by long wave movements or **tides**; and deep water currents. Furthermore, turbidity currents play a role in shaping underwater topography. Measured in a variety of ways, currents are responsible for absorbing solar **heat** and redistributing it throughout the world.

Surface currents

Perhaps the most obvious type of current, surface currents are responsible for the major surface circulation patterns in the world's oceans. They are the result of the **friction** caused by the movements of the atmosphere over water; they owe their existence to the winds that form as a result of the warming of air masses at the sea surface near the equator and in temperate areas. When **wind** blows across the water's surface, it sets the water in **motion**. If the wind is constant and strong enough, the currents may persist and become permanent components of the ocean's circulation pattern; if not, they may be merely temporary. Surface currents can extend to depths of about 656 ft (200 m). They circle the ocean basins on both sides of the Equator in elliptical rotations.

There are several forces that affect and sustain surface currents, including the location of land masses, wind patterns, and the **Coriolis effect**. Located on either side of the major oceans (including the Atlantic, Indian, and Pacific), land masses affect currents because they act as barriers to their natural paths. Without land masses, there would be a uniform ocean movement from west to east at intermediate latitudes and from east to west near the equator and at the poles. The Antarctic Circumpolar Current can illustrate the west to east movement. Because no land barriers obstruct the prevailing current traveling between the southern tips of **South America** and **Africa** and the northern coast of **Antarctica**, the Antarctic Circumpolar Current consistently circles the globe in a west to east direction. Interestingly, this current is the world's greatest, flowing at one point at a **rate** of 9.5 billion cubic feet per second.

In addition to the presence of land barriers, two other factors work together to affect the surface currents—wind patterns and the Coriolis effect. The basic wind patterns that drive the currents in both hemispheres are the trade winds and the westerly winds. The Coriolis effect is a **force** that displaces particles, such as water, traveling on a rotating **sphere**, such as **Earth**. Thus, currents develop as water is deflected by the turning of Earth. At the equator, the effect is nonexistent, but at greater latitudes the

Coriolis effect has a stronger influence. As the trades and the westerlies combine with the Coriolis effect, elliptical circulating currents, called gyres, are formed. There are two large subtropical gyres dominating each side of the equator. In the Northern Hemisphere, the gyre rotates in a clockwise direction; in the Southern Hemisphere, it rotates counterclockwise. At the lower latitudes of each hemisphere, there are smaller, tropical gyres which move in the opposite direction of the subtropical gyres.

A good illustration of a surface current is the Gulf Stream, also called the Gulf Current. This current is moved by the trade winds in the Atlantic Ocean near the equator flowing in a northwesterly direction. Moving along the coasts of South and **North America**, the Gulf Stream circles the entire Atlantic Ocean north of the equator. Currents similar to this exist in the Pacific Ocean and in the Atlantic south of the equator.

One of the major consequences of surface currents is their ability to help moderate the earth's temperatures. As surface currents move, they absorb heat in the tropical regions and release it in colder environments. This process is referred to as a net poleward **energy transfer** because it moves the solar **radiation** from the equator to the Poles. As a result, places like Alaska and Great Britain are warmer than they otherwise would be.

Tidal currents

Tidal currents are horizontal water motions associated with the sea's changing tides. Thus, in the ocean, wave tides cause continuous currents that change direction 360 degrees every tidal cycle, which typically lasts six to 12 hours. These tides can be very strong—reaching speeds of 6 in (15 cm) per second and moving sediment long distances—or they can be weak and slow. Of interest to swimmers, rip currents are outward-flowing tidal currents, moving in narrow paths out to sea. The flow is swift in order to balance the consistent flow of water toward the beach brought by waves. In general, tidal currents are of minimal effect beyond the **continental shelf**.

Deep water (or density) currents

Deep water currents move very slowly, usually around 0.8-1.2 in (2-3 cm) per second. They dominate approximately 90% of the oceans' circulation. Water circulation of this type is called thermohaline circulation. Basically, these currents are caused by variations in water **density**, which is directly related to **temperature** and **salt** level, or salinity. Colder and saltier water is heavier than warmer, fresher water. Water gets denser in higher latitudes due to (1) the cooling of the atmosphere and (2) the increased salt levels, which result from the freezing of surface water. (Frozen water normally contains mostly **freshwater**, leaving higher concentrations of salt in the water that remains liquid.) Differences in water density generate slow moving currents, due to the sinking of the colder, saltier water into deeper parts of the oceans' basins and the displacement of lighter, fresher currents.

Turbidity currents

Turbidity currents are local, rapid-moving currents that travel along the ocean floor and are responsible for shaping its landscape. These currents result from water, heavy with suspended sediment, mixing with lighter, clearer water. Causes of turbidity currents are earthquakes or when too much sediment piles up on a steep underwater slope. They can move like avalanches. Turbidity currents often obscure the visibility of the ocean floor.

Measuring currents

Oceanographers measure currents in a variety of ways using a variety of equipment, yielding results that range from crude to sophisticated. Currents can be measured directly, by clocking the water movement itself, or indirectly, by looking at some characteristic closely related to water movement. Two common direct ways to measure currents are the lagrangian and the eulerian methods. Lagrangian measurements monitor the movement of water by watching objects that are released into the current. These objects are monitored and recollected at a later time. Eulerian measurements look at the movement of water past a designated fixed location and usually include an anchored current meter.

Ocean currents and climate

The oceans cover over 70% of Earth's surface. In the tropics, ocean water absorbs heat from the atmosphere. As the warmed water is carried north by surface currents, immense amounts of stored **energy**, in the form of heat, are transferred from one part of the world to another, contributing to **weather** and climate patterns. For example, the Gulf Stream carries warm water far up the eastern coast of North America, and then swings east, towards **Europe**. The warm water of the Gulf Stream heats the air above it, creating a warmer climate for Iceland and western Europe than would otherwise exist. Thermohaline currents also carry stored heat from the tropics to the mid-latitudes.

Oceanographers and climatologists are still exploring the important relationships between the oceans and their currents and ongoing **global climate** change due to greeenhouse gases. Much of the heat resulting from **global warming** is being stored in the oceans, according

KEY TERMS

Coriolis effect—Generically, this force affects particles traveling on a rotating sphere. As it pertains to currents, it is a deflection of water caused by the turning of the earth. At the equator, the effect is nonexistent but it gets stronger toward the poles. Water tends to swirl to the right in the Northern Hemisphere and to the left in the Southern Hemisphere.

Gyre—Typically elliptical in shape, a gyre is a surface ocean current that results from a combination of factors, including: the Coriolis effect, the earth's rotation, and surface winds.

Rip currents—Narrow areas in the ocean where water flows rapidly out to sea. The flow is swift in order to balance the consistent flow of water toward the beach brought by waves.

Thermohaline circulation—The flow of water caused by variations in water density rather than caused by the wind. In certain situations, colder water from the sea floor mixes upward with the warmer water. As it does this, it rotates faster, moving toward the two poles.

Turbidity currents—Local, rapid-moving currents that result from water heavy with suspended sediment mixing with lighter, clearer water. Causes of turbidity currents are earthquakes or when too much sediment piles up on a steep underwater slope. They can move like avalanches.

to scientists, thus delaying part of the surface warming global climate change theorists have expected to see as the result of human-induced increases in greenhouse gases in the atmosphere. The heat being stored by ocean waters will contribute to warming trends throughout the world as the water is circulated by oceanic currents.

Resources

Books

Davis, Richard A., Jr. *Oceanography, An Introduction to the Marine Environment.* Dubuque, IA: William C, Brown Publishers, 1991.

Goudie, Andrew, ed. *The Encyclopaedic Dictionary of Physical Geography.* New York: Blackwell Reference, 1985.

Groves, Donald G., and Lee M. Hunt. *Ocean World Encyclopedia.* New York: McGraw-Hill Book Company, 1980.

Hendrickson, Robert. *The Ocean Almanac.* Garden City, New York: Doubleday and Company, 1984.

Ocean Science. San Francisco: W. H. Freeman and Company, 1977.

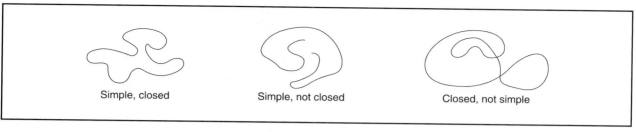

Simple, closed   Simple, not closed   Closed, not simple

Figure 1. *Illustration by Hans & Cassidy. Courtesy of Gale Group.*

Pinet, Paul. *Invitation to Oceanography.* 2nd ed. Boston: Jones & Bartlett Pub., 1999.
Thurman, Harold V., and Alan P. Trujillo. *Essentials of Oceanography.* 7th ed. Englewood Cliffs, NJ: Prentice Hall, 2001.

Curve

Informally, one can picture a curve as either a line, a line segment, or a figure obtained from a line or a line segment by having the line or line segment bent, stretched, or contracted in any way. A **plane** curve, such as a **circle**, is one that lies in a plane; a curve in three dimensional space, such as one on a **sphere** or cylinder, is called a skew curve.

Plane curves are frequently described by equations such as $y = f(x)$ or $F(x,y) = 0$. For example, $y = 3x + 2$ is the equation of a line through (1,5) and (2,8) and $x^2 + y^2 - 9 = 0$ is the equation of a circle with center at (0,0) and radius 3. Both of the curves described by the equations $y = 3x + 2$ and $x^2 + y^2 - 9 = 0$ are examples of algebraic curves. On the other hand, a curve described by an equation such as $y = \cos x$ is an example of a transcendental curve since it involves a transcendental **function**.

Another way of describing a curve is by means of parametric equations. For example, the **parabola** $y = x^2$ can be described by the parametric equations $x = t$, $y = t^2$ and the helix by $x = a \cos \theta$, $y = a \sin \theta$, $I = b\theta$.

A closed curve is a curve with no endpoints such as a circle or a figure eight. A curve that does not cross itself is a simple curve. So a circle is a simple closed curve whereas a figure eight is closed but not simple. Figure 1 shows another simple closed curve, a curve that is simple but not closed, and a curve that is closed but not simple.

Cushing syndrome

In the early twentieth century, Harvey Cushing, (1869–1939), an American neurosurgeon, described a set of symptoms that he identified as a **syndrome** later called Cushing syndrome or **disease**. The cause of the syndrome at the time was unknown, but since then, a better understanding of the underlying causes of Cushing syndrome have been elucidated.

Cushing syndrome is a disorder that is caused by excessive exposure of the body's tissues to a hormone called cortisol for long periods of time. It is a rare disease that effects adults ages 20–50, with roughly 10–15 million people affected each year. People that take steroidal **hormones** such as prednisone for **arthritis** or **asthma** may develop symptoms similar to Cushing syndrome. Overproduction of cortisol also causes the disease. Cortisol is produced in the cortex of the adrenal gland. The adrenal cortex is the outer layer of the gland. The adrenal **glands** rest like limp, triangular caps atop each kidney. The adrenal glands release cortisol in response to stimulation by another hormone, adrenocorticotropin (ACTH), that is produced by the pituitary gland. The pituitary gland is located at the base of the **brain** and is the control center for the other glands. ACTH is only produced when there are inadequate levels of cortisol in the bloodstream. Dysregulated cortisol production can be due to abnormalities in the adrenal glands, the pituitary gland, or abnormal regulation of ACTH production.

Cortisol has a variety of functions throughout the body. It is important in the regulation of **blood pressure** and serves as an anti-inflammatory mediator. It also regulates **insulin metabolism** as well as plays a role in protein, **carbohydrate**, and **lipid** metabolism. High levels of cortisol can cause **sodium** and **water** retention. Excessive cortisol production also can affect **heart** functions, muscle movements, blood **cell** development as well as other necessary bodily functions. Perhaps one of the most important functions of cortisol is to help the body respond to **stress**.

Certain tumors and other conditions can upset the balance between the pituitary gland and the adrenal gland, resulting in excessive cortisol production. There are several types of tumors related to overproduction of cortisol. Pituitary adenomas are non-cancerous tumors that produce increased amounts of ACTH and cause the greatest number of cases of Cushing syndrome with a five-fold higher risk for women. Cancerous tumors that develop outside the pituitary can produce ACTH, a condition also known as ectopic ACTH syndrome. Other cases of Cushing syndrome come from adrenal tumors. A non-cancerous **tumor** in the adrenal **tissue** can result in excessive cortisol release. Symptoms rapidly develop in cancerous tumors of the adrenal glands due to high levels of hormone production. Familial Cushing syndrome is condition that involves an inherited susceptibility to developing endocrine gland tumors but only accounts for a small population of patients.

Cushing syndrome patients can develop a rounded, moon-shaped face, a large **fat** pad below the neck and on the back between the shoulders (called a buffalo hump), and an accumulation of fat on the abdomen. The **volume** of abdominal fat develops dramatically, often hanging over the beltline. Sometimes vertical purplish stripes or striations will appear on the abdomen. Weakness and wasting away of the muscles can occur. The skin bruises easily and wounds heal slowly. Women develop brittle bones, or **osteoporosis**, rendering them vulnerable to fractures, especially in the pelvis and spinal areas of the body. The beginning stages of diabetes can develop and include glucose intolerance. Psychiatric symptoms, excess hair growth in women, and high blood pressure can all be part of the symptomatology. Excessive cortisol production in children can result in growth retardation.

The first step in treating Cushing syndrome patients is to identify the cause of the excessive ACTH production. It may be the result of abnormalities in the function of the pituitary gland or the adrenal cortex, exposure to cortisone for unrelated medical treatment purposes, or a tumor that stimulates ACTH or cortisol production. Cushing syndrome symptoms in patients that are receiving treatment of cortisol can reverse the symptoms by refraining from taking these medications. A tumor of the pituitary gland may require surgical removal if other treatments such as chemotherapy, **radiation**, or cortisol inhibitors prove unsuccessful. Removal of the adrenal gland is necessary if the tumor lies within it. Removal of one adrenal gland will not affect the endocrine balance since the other gland will naturally compensate. If both glands are removed, however, the patient must be given cortisol and other hormones to compensate for the lack of adrenal gland function.

A better understanding of the contribution of cortisol or ACTH and the cause of altered hormonal regulation has lead to better diagnostic tests as well as improved therapies. Current research is underway to identify the specific genetic defects associated with developing the disease.

See also Endocrine system; Genetic disorders; Genetics.

Resources

Books

Nussbaum, Robert L., Roderick R. McInnes, and Huntington F. Willard. *Genetics in Medicine*. Philadelphia: Saunders, 2001.

Rimoin, David L. *Emery and Rimoin's Principles and Practice of Medical Genetics*. New York: Churchill Livingstone, 2002.

Periodicals

McKusick, Victor A. "The Cardiovascular Aspects of Marfan's Syndrome." *Circulation* 11 (1955): 321–342.

Bryan Cobb

Cuttlefish

Cuttlefish are squid-like cephalopod **mollusks** of the family Sepiidae, in the order Sepioidea. Cephalopod literally means "head-footed animal" and is the name given to advanced mollusks (such as cuttlefish, **squid** and **octopus**) whose heads are circled with tentacles. Cuttlefish have a relatively well-developed **brain**, sensitive organs of **smell** and **hearing**, highly developed eyes, and a relatively advanced reproductive system.

There are more than 100 **species** of cuttlefish common in the warmer waters of the Mediterranean, the European Atlanticcoast, and abundant in the Indian Ocean, and western Pacific. Cuttlefish are found in shallow, sandy, coastal waters, where they feed upon their usual diet of

Sepia latimanus, **a cuttlefish, off Mota Island, Vanuatu.** *© Fred McConnaughey, National Audubon Society Collection/Photo Researchers, Inc. Reproduced with permission.*

shrimp. Cuttlefish are not found in the oceans around the United States. The smallest species of cuttlefish (*Hemisepies typicus*), grows to about 3 in (7.5 cm) long, while the largest species (*Sepia latimanus*) can reach up to 5.5 ft (1.6 m) in length. The best known species of cuttlefish *Sepia officinalis*, or common cuttlefish, grows up to about 3 ft (91 cm) long, including its tentacles.

Cuttlefish have ten tentacles (decapod), eight of which are short and have rows of suckers at their ends. The other two tentacles are longer and are retractable tentacles that can be used to catch **prey**. These tentacles have club-shaped ends with suckers, which can catch prey faster than the tongue of a lizard or frog, and can retract into sockets beside each **eye**. The cuttlefish mouth bears a strong beak-like structure that can bite and tear the prey, and cuttlefish salivary **glands** can secrete an immobilizing poison with the saliva.

The skin of a cuttlefish has pigment cells (chromatophores) that are under nervous and hormonal control, which enables the **animal** to become red, orange, yellow, brown, or black. Cuttlefish are often colored brownish green with white, irregular stripes that provide a perfect camouflage among seaweed. Cuttlefish also have a purple ribbon-like fin running the length of the body, and they are iridescent in sunlight. Cuttlefish can change their **color** and pattern at will, in a fraction of a second, a **behavior** which is thought to be a form of communication. They can become invisible by taking on the colors and designs of their surrounding environment, including other cuttlefish.

The body of a cuttlefish is a flattened oval and supported by a shield shaped, internal, calcareous (**calcium**) shell that contains a number of tiny, gas filled chambers. The cuttlebone has a hydrostatic function—it can change the proportion of gas and liquid it contains, thus controlling its specific gravity. Cuttlebones are used in bird cages as a good source of calcium, and to help keep the bird's beak trimmed. Cuttlefish bone is also pulverized and used in polish.

These mollusks swim by undulating their side fins and by using a funnel in the mantle cavity to maintain a stationary position in the **water** and to propel itself backward with a great deal of speed if necessary. The cuttlefish can control the direction of the funnel, and so control the **force** with which the water is expelled. Another defense capability of the funnel is a brownish black ink (called sepia) that is ejected when danger is sensed. The pigment in the ink is made of **copper** and **iron**, which are extracted from the cephalopod's **blood**. The sepia ink is the original India ink and is used by artists and industries as a pigment for paints, inks, and dyes.

Cybernetics

Cybernetics is a term that was originated by American mathematician Norbert Wiener (1894–1964) in the late 1940s. Based on common relationships between humans and machines, cybernetics is the study and analysis of control and communication systems. As Wiener explains in his 1948 book, *Cybernetics: or Control and Communication in the Animal and the Machine*, any machine that is "intelligent" must be able to modify its **behavior** in response to feedback from the environment.

This theory has particular relevance to the field of computer science. Within modern research, considerable attention is focused on creating computers that emulate the workings of the human mind, thus improving their performance. The goal of this research is the production of computers operating on a neural network. During the late 1990s work has progressed to the point that a neural network can be run, but, unfortunately, it is generally a **computer software** simulation that is run on a conventional computer. The eventual aim, and the continuing area of research in this field, is the production of a neural computer. With a neural computer, the architecture of the **brain** is reproduced. This system is brought about by transistors and resistors acting as neurons, axons, and dendrites. By 1998 a neural network had been produced on an **integrated circuit**, which contained 1024 artificial neurons. The advantage of these neural computers is that they are able to grow and adapt. They can learn from past experience and recognize patterns, allowing them to operate intuitively, at a faster **rate**, and in a predictive manner.

Another potential use of cybernetics is one much loved by science fiction authors, the replacement of ailing body parts with artificial structures and systems. If a structure, such as an **organ**, can take care of its own functioning, then it need not be plugged into the human **nervous system**, which is a very difficult operation. If the artificial organ can sense the environment around itself and act accordingly, it need only be attached to the appropriate part of the body for its correct functioning. An even more ambitious future for the cybernetics industry is the production of a fully autonomous life form, something akin to the robots often featured in popular science fiction offerings. Such an artificial life form with **learning** and deductive powers would be able to operate in areas that are inhospitable to human life. This could include long-term **space** travel or areas of high radioactivity.

Cycads

The cycads are a relatively small phylum (Cycadophyta) in the **plant** kingdom Plantae. The cycads are considered to be gymnosperms, because they bear their **seeds** naked on modified leaves called sporophylls. In contrast, the evolutionarily more recent angiosperms (flowering plants) bear their seeds inside of ovaries. Cycads grow in tropical and subtropical regions of the world. Cycads are sometimes referred to as "living fossils" because they are very similar to extinct **species** that were much more abundant several hundreds of million years ago. The foliage of many species of cycads resembles that of palm trees, and plants in the genus *Cycas* are commonly called "Sago palms." However, cycads are only distantly related to the **palms**, and their similarity is superficial.

General characteristics

Many cycad species are shrub-sized in stature, but some species are 20-60 ft (6-18 m) tall at maturity. The cycads typically have an unbranched central stem, which is thick and scaly. Most species grow relatively slowly and have a large, terminal rosette of leaves. The leaves of most species are compound, in that they are composed of numerous small leaflets. Cycad leaves remain green for 3-10 years, so they are considered evergreen. Many cycad species, though short in stature, have a thick tap root which can extend as much as 30-40 ft (9-12 m) beneath the **soil** surface. The function of the tap root is to take up **water** from deep beneath the surface. Cycads also produce coralloid (coral-like) roots, which grow near the surface and are associated with symbiotic cyanobacteria. In a process known as **nitrogen fixation**, the cyanobacteria take in atmospheric **nitrogen** gas (N_2) and transform it to **ammonia** (NH_3), a chemical form that can be used by the plant as a nutrient. In reciprocation, the plant provides **habitat** and carbohydrates to the cyanobacteria. The cycads are the only gymnosperms known to form symbiotic relationships with cyanobacteria.

There are about 200 species of cycads in the world. They are **endemic** to tropical and subtropical regions, and are found in Central America, **South America**, **Africa**, **Asia**, and **Australia**. The greatest richness of cycad species is in Mexico and Central America. *Zamia integrifolia* is the only species of cycad native to the United States and is found in Florida and Georgia. Several foreign cycad species are grown as ornamental plants in Florida and elsewhere in the southern United States.

The stems and seeds of most cycads are very rich in starch. In earlier times, the Seminole Indians of Florida used *Zamia* as an important food source. They dried and then ground up the starchy stem of *Zamia* to make a flour which they called "coontie." In India, the stem of another cycad, *Cycas circinalis*, is still used to make Sago flour. However, cycads are of little economic importance today, except as ornamental plants.

Life cycle

Cycads, like all seed-producing plants, have a dominant diploid sporophyte phase in their life cycle—this is the large, familiar, green plant seen in nature. Cycads and other gymnosperms do not have true flowers and their seeds are borne naked. In the more evolutionarily recent angiosperms (flowering plants), the seed is enveloped by a coat or fruit which originates from the ovary.

All species of cycads are dioecious, meaning the male and female reproductive structures are borne on separate plants. The male reproductive structure, known as an androstrobilus, superficially looks like a large pine cone, though it is much simpler in structure. It consists of many densely packed, modified leaves, known as microsporophylls. Each microsporophyll produces a large quantity of pollen grains on its dorsal surface. The pollen grain is the small, multicellular, male haploid gametophyte phase of the cycad life cycle. The pollen is dispersed by **wind** or by **insects** to the gynostrobilus, or the female reproductive structure.

The gynostrobilus of cycads also looks like a large pine cone, but it has a morphology different from the androstrobilus. When a pollen grain lands on the gynostrobilus, it germinates and grows a pollen tube, a long tubular **cell** that extends to deep within the multicellular, female haploid gametophyte. Then a sperm cell of the pollen grain swims through the pollen tube using its whip-like tail, or **flagella**, and fertilizes the egg to form a zygote. The zygote eventually develops into an embryo, and then a seed. Cycad seeds are rich in starch and have a pigmented, fleshy outer layer known as the sarcotesta. The seeds are often dispersed by **birds** or **mammals**, which eat them for the nutritious sarcotesta, and later defecate the still-viable seed.

It is significant that the cycads have flagellated sperm cells, which is considered a primitive (i.e., ancient) characteristic. Other evolutionarily ancient plants, such as mosses, liverworts, and **ferns**, also have flagellated sperm cells. More evolutionarily recent plants, such as the flowering plants, do not have flagellated sperm cells. In fact, other than the cycads, only one species of

A cycad in Hawaii. *JLM Visuals. Reproduced with permission.*

gymnosperm, the gingko, or maidenhair **tree** (*Ginkgo biloba*), has flagellated sperm cells. In other gymnosperms and angiosperms, the sperm is transported directly to the female ovule by a sperm tube.

Evolution

The earliest cycad fossils are from the Permian period (about 300 million years ago). Paleobotanists believe that cycads evolved from the **seed ferns**, a large group of primitive, seed-bearing plants with fern-like leaves. The seed ferns originated at least 350 million years ago, and became extinct more than 200 million years ago.

Although cycads are considered to be gymnosperms, because they bear naked seeds which are not enclosed by a fruit, fossil evidence suggests they are not closely related to other gymnosperms, such as the conifers. Therefore, many paleobotanists consider the gymnosperms to be an unnatural grouping of unrelated plants.

Cycads were particularly abundant and diverse during the Mesozoic era, so paleobotanists often refer to the

KEY TERMS

. . . .

Cyanobacteria (singular, cyanobacterium)—Photosynthetic bacteria, commonly known as blue-green algae.

Diploid—Nucleus or cell containing two copies of each chromosome, generated by fusion of two haploid nuclei.

Gametophyte—The haploid, gamete-producing generation in a plant's life cycle.

Haploid—Nucleus or cell containing one copy of each chromosome.

Rosette—A radial cluster of leaves, often on a short stem.

Sporophyll—An evolutionarily modified leaf which produces spores.

Sporophyte—The diploid, spore-producing generation in a plant's life cycle.

Mesozoic as "the age of cycads." This is also the era during which dinosaurs were the dominant animals, so zoologists refer to this as "the age of dinosaurs." Consequently, museum drawings and dioramas which depict re-creations of **dinosaur** life typically show cycads as the dominant plants.

The cycads are no longer a dominant group of plants, and there are only about 200 extant (surviving) species. The flowering plants essentially replaced the cycads as ecologically dominant species on land more than 100 million years ago.

See also Paleobotany.

Resources

Books

Jones, D. C. *Cycads of the World.* Washington, DC: Smithsonian Institute Press, 1993.

Margulis, L., and K.V. Schwartz. *Five Kingdoms.* New York: W. H. Freeman and Company, 1988.

Peter A. Ensminger

Cyclamate

Cyclamate (chemical formula $C_6H_{13}NO_3S$) is an artificial, noncaloric sweetener with approximately 30 times the sweetness of ordinary table sugar. It is currently sold in more than 50 countries. In the United States,

however, the Food and Drug Administration (FDA) has not allowed its sale since 1970.

University of Illinois graduate student Michael Sveda first synthesized cyclamate in 1937. Some say that he discovered its sweet **taste** by chance when he accidentally got some on the cigarette he was smoking. The university eventually transferred patent rights to Abbott Laboratories, which brought the sweetener to market in 1950.

Most cyclamate sales were as a 10-1 mixture with saccharin, marketed under the brand name Sucaryl®. (Since saccharin is about 10 times as sweet as cyclamate, each compound contributed roughly half the mixture's sweetening power.) The mixture was attractive because the two compounds together are sweeter and better-tasting than either alone. Cyclamate alone becomes relatively less sweet as its **concentration** increases—that is, raising the concentration ten-fold increases the total sweetness only six-fold. Thus, if cyclamate were used alone in very sweet products such as soft drinks, manufacturers would have to use large amounts. Besides cost, this risks development of the "off" flavors sometimes encountered at high cyclamate concentrations.

Another reason for combining saccharin with cyclamate is that the sweet taste of cyclamate develops slowly, although it lingers attractively on the tongue. On the other hand, saccharin has a bitter aftertaste that is much less noticeable in the mixture than when saccharin is used alone. Indeed, cyclamate is better than sugar at masking bitter flavors.

Unlike more recent low-calorie sweeteners, cyclamate is extremely stable. It can be used in cooking or baking and in foods of any level of acidic or basic character. Scientists have found no detectable change in Sucaryl tablets stored for seven years or more.

Regulatory controversy

Cyclamate's regulatory problems began in 1969, when a small number of **rats** fed very large amounts of Sucaryl for two years (virtually their entire lives) developed bladder **cancer**. This led the FDA to ban use of cyclamate—but not of saccharin, the mixture's other ingredient—the following year. The issue was far from settled, however. In 1973, Abbott Laboratories filed what the FDA calls a Food Additive Petition—that is, a request to allow use of cyclamate in foods. (A fine legal point is that this was the first such request for cyclamate. The law requiring FDA permission for use of food additives was not passed until 1958, so cyclamate and other additives used before that time were exempt.) This request was accompanied by a number of additional studies supporting the compound's safety. The FDA considered and debated this petition for seven years before finally rejecting it in 1980.

In 1982, Abbott Laboratories filed another Food Additive Petition, this time joined by an industry group called the Calorie Control Council. As of 1995, the FDA still has not acted. With the passage of so many years, however, the issue has become almost purely one of principle: Since the patent on cyclamate has expired, few believe Abbott Laboratories would manufacture and market the sweetener if allowed to do so. Possibly, though, another company might choose to offer it.

Does cyclamate cause cancer?

The idea that cyclamate may cause cancer rests on one study: When scientists fed 80 rats a cyclamate/saccharin mixture at a level equal to 5% of their diets, 12 of them developed bladder cancer within two years. Since then, there have been more than two dozen studies in which animals were fed similar levels of cyclamate for their entire lives; none has given any indication that the sweetener causes cancer.

As a result, the Cancer Assessment Committee of the FDA's Center for Food Safety and Applied Nutrition concluded in 1984 that, "the collective weight of the many experiments... indicates that cyclamate is not carcinogenic (not cancer causing)." The results of the 1969 study that led to banning of cyclamate, the committee says, "are... not repeatable and not explicable." The following year, the National Academy of Sciences added that, "the totality of the evidence from studies in animals does not indicate that cyclamate... is carcinogenic by itself." A joint committee of the World Health Organization (WHO) and the Food and Agriculture Organization (FAO) has similarly concluded that cyclamate is safe for human consumption. Unlike the two United States groups, the WHO/FAO panel addressed issues of genetic damage as well as cancer.

One of the most peculiar aspects of the entire regulatory situation is that, although the apparently incriminating study used a mixture of cyclamate and saccharin, only cyclamate was banned. We now know-although we did not in 1969—that saccharin itself produces occasional bladder cancers. So if the rats' diets did indeed cause their cancers (which some scientists doubt), most people today would assume that the saccharin was at fault.

Despite strong evidence for cyclamate's safety—the WHO/FAO committee commented, "one wonders how may common foodstuffs would be found on such testing to be as safe as that"—future United States use of the sweetener remains uncertain on both regulatory and economic grounds. Nevertheless, many people hope that the FDA will soon clear this 25-year-old case from its docket. Whether manufacture of cyclamate will then resume remains to be seen.

Resources

Books

Klaassen, Curtis D. *Casarett and Doull's Toxicology.* 6th ed. Columbus: McGraw-Hill, Inc., 2001.

Periodicals

Lecos, Chris W. "Sweetness Minus Calories = Controversy." *FDA Consumer* (February 1985): 18.23.

W. A. Thomasson

Cycloalkane *see* **Hydrocarbon**

Cyclone and anticyclone

The terms cyclone and anticyclone are used to describe areas of low and high **atmospheric pressure**, respectively. Air flowing around one or the other of these areas is said to be moving cyclonically in the first case and anticyclonically in the second. In the northern hemisphere, cyclonic winds travel in a counterclockwise direction and anticyclonic winds, in a clockwise direction. When a cyclone or anticyclone is associated with a wave front, it is called a wave, a frontal, or a mid-latitude cyclone or anticyclone.

Vertical air movements are associated with both cyclones and anticyclones. In the former case, air close to the ground is forced inward, toward the center of a cyclone, where **pressure** is lowest, and then begins to rise upward. At some height, the rising air begins to diverge outward away from the cyclone center.

In an anticyclone, the situation is reversed. Air at the center of an anticyclone is forced away from the high pressure that occurs there and is replaced by a downward draft of air from higher altitudes. That air is replaced, in turn, by a convergence of air from higher altitudes moving into the upper region of the anticyclone.

Distinctive **weather** patterns tend to be associated with both cyclones and anticyclones. Cyclones and low pressure systems are generally harbingers of rain, **clouds**, and other forms of bad weather, while anticyclones and high pressure systems are predictors of fair weather.

One factor in the formation of cyclones and anticyclones may be the development of irregularities in a **jet stream**. When streams of air in the upper atmosphere begin to meander back and forth along an east-west axis, they mayadd to cyclonic or anticyclonic systems that already exist in the lower troposphere. As a result, relatively stable cyclones (or anticyclones) or families of cyclones (or anticyclones) may develop and travel in an easterly or northeasterly direction across the **continent**.

On relatively rare occasions, such storms may pick up enough **energy** to be destructive of property and human life. Tornadoes and possibly hurricanes are examples of such extreme conditions.

See also Tornado; Wind.

Cyclosporine

Cyclosporines are drugs used in the field of immunosuppressant medicine to prevent the rejection of transplanted organs. They were discovered by Jean F. Borel in 1972. The cyclosporine used for transplant **surgery** is called cyclosporine A (CsA) and in 1984 it was added to the group of medicines used to prevent transplant rejection. Cyclosporine A is the most common form of the Norwegian fungus *Tolypocladium inflatum.*

The discovery of cyclosporine has led to a significant rise in the number of **organ** transplant operations performed as well as the **rate** of success. Cyclosporine has increased both the short- and long-term survival rates for transplant patients, especially in **heart** and liver operations. The rejection of grafted tissues occurs when white **blood** cells (lymphocytes) called T-helper cells stimulate the activity of cell-destroying (cytotoxic) T-killer cells. It is believed that cyclosporine interferes with the signals sent by the T-helper cells to the T-killer cells. These **T cells**, along with other white blood cells like monocytes and macrophages, cause the **tissue** rejection of the implanted organs.

Cyclosporine has proven to be the most effective medicine used to combat the body's own **immune system**, which is responsible for the rejection of transplanted organs. In addition to curtailing the activity of T-helper cells, cyclosporine is also able to fight the infectious illnesses that often occur after a transplant operation and can lead to death.

Cyclopsporine must be administered very carefully, since it can produce a number of toxic side effects, including kidney damage. Many clinical trials have been conducted using other drugs in combination with cyclosporine in an effort to reduce these side effects.

Immunosuppression

There are two types of immunosuppression: specific suppression and nonspecific suppression. In specific sup-

pression, the blocking agent restricts the immune system from attacking one or a specific number of antigens (foreign substances). In nonspecific immunosuppression, the blocking agent prevents the immune system from attacking any antigen. Nonspecific immunosuppression, therefore, breaks down the ability of the body to defend itself against infections.

In the case of organ transplants, the recipient's body responds to the donor's organ tissues as if they were infecting foreign tissues. A drug that is a specific suppressing agent could block the immune system's antigenic response to the newly implanted organ. While specific suppression has been accomplished in **animal** transplants, it has not succeeded in human trials. So far, all the drugs used to suppress the immune system after an organ transplant are nonspecific suppressants.

In administering nonspecific immunosuppressants, a balance has to be maintained between the need for protecting the new organ from the immune system's attack (rejection) and the immune system's normal functioning, which protects the **individual** from infectious diseases. As **time** passes after the transplant operation, the body slowly begins to accept the new organ and the amount of immunosuppressant drugs can be decreased. If, however, the immunosuppressant is suddenly decreased shortly after the operation, larger doses may have to be given several years later to avoid rejection of the transplanted organ.

All the immunosuppressant drugs have side effects and individuals react to them in different ways. One strategy often used is to create a mix of the different drugs—usually azathioprine, cyclosporine, and prednisone—for the transplant patient. For example, a patient whose blood **pressure** is elevated by cyclosporine could take less of that drug and be introduced instead to prednisone. If an adverse reaction takes place with azathioprine, cyclosporine could replace it.

Administration

Cyclosporine can be taken either orally or by intravenous injection. While the injection reduces the amount of a dose by two-thirds, it can also produce side effects like kidney damage and other neural disturbances. As an oral preparation, it can be taken mixed with other liquids or in capsules. It is most effective when taken with a meal, since the digestive process moves it toward the smaller intestine, where it is absorbed.

In order to prevent rejection, many doses of cyclosporine have to be taken, usually starting with high dosages and then reducing them over time. The size of the dose is determined by the weight of the individual. Dosages also vary from one individual to another de-

KEY TERMS

Antigen—A molecule, usually a protein, that the body identifies as foreign and toward which it directs an immune response.

Donor organ—An organ transplanted from one person (often a person who is recently deceased) into another.

Macrophage—A large cell in the immune system that engulfs foreign substances to dispose of them.

Organ recipient—A person into whom an organ is transplanted.

T-helper cells—Immune system cells that signal T killer cells and macrophages to attack a foreign substance.

pending on the patient's ability to withstand organ rejection. Frequent blood tests are done on the patient to monitor all the factors that go into successful drug therapy.

Another problem for transplant patients is the cost of cyclosporine; a year's supply can cost as much as $6,000. Although medical insurance and special government funds are available to pick up the cost of this drug, the expense of the medication still poses a problem for many transplant patients.

Side effects

Aside from potential damage to the kidneys, there are a number of other side effects of cyclosporine. They include elevated blood pressure, a raise in potassium levels in the blood, an increase in hair growth on certain parts of the body, a thickening of the gums, problems with liver functioning, tremors, seizures, and other neurological side effects. There is also a small risk of **cancer** with cyclosporine as well as with the other immunosuppressant drugs.

See also Antibody and antigen; Transplant, surgical.

Resources

Books

Auchinloss, Hugh, Jr., et al. *Organ Transplants: A Patient's Guide.* Cambridge: H. F. Pizer, 1991.
Barrett, James T. *Textbook of Immunology.* St. Louis: Mosby, 1988.
Joneja, Janice V., and Leonard Bielory. *Understanding Allergy, Sensitivity, and Immunity.* New Brunswick: Rutgers University Press, 1990.
Sell, Stewart. *Basic Immunology.* New York: Elsevier, 1987.

Weiner, Michael A. *Maximum Immunity.* Boston: Houghton Mifflin, 1986.

Periodicals

Kiefer, D. M. "Chemistry Chronicles: Miracle Medicines." *Today's Chemist* 10, no. 6 (June 2001): 59–60.

Jordan P. Richman

Cyclotron

A cyclotron is a type of particle accelerator designed to accelerate protons and ions to high velocities and then release them so as to strike a target. Observations of such collisions yield information about the nature of atomic particles. In contrast to the enormous particle **accelerators** used in particle **physics** today, the first cyclotron, built in 1930 by U.S. physicist E. O. Lawrence (1901–1958), measured just 4.5 in (12 cm) in diameter.

A charged particle moving at right angles to a magnetic field is subject to a **force** that is at right angles both to the field and to the charged particle's direction of **motion** at instant; this force follows the particle to follow a spiraling path. In a cyclotron, a pair of hollow, D-shaped pieces of **metal** are mounted above a powerful electromagnet, with their flat sides facing one another. One of the Ds is given a **negative** charge and the other is given a positive charge.

A charged particle, say a **proton**, is injected into this environment. With its positive charge, the proton is attracted by the negative D and repelled by the positive D; these forces start it into motion toward the negatively charged D. Once the particle is moving, the magnetic field deflects it into a curved path, back toward the positive D. Before the positive D can repel the proton, it is switched to a negative charge, thus attracting the proton rather than repelling it. Thus, the magnetic field keeps the particle on a circular path, while the alternating positive and negative charges on the D-shaped pieces of metal keep the proton chasing a negatively charged target indefinitely. As the proton circles inside the cyclotron it gains speed and thus **energy**; for a fixed magnetic-field strength, the size of the **circle** it travels increases correspondingly. Ultimately, before it can strike either of the metal Ds, it is propelled out of the cyclotron by a bending magnet and directed toward a target.

The cyclotron was a revolutionary device for its time, but has since been outmoded for particle-physics research purposes as cyclotrons are not capable of accelerating particles to the high speeds required for today's experiments in subatomic physics. High speeds are required

This domed building houses a cyclotron at the Lawrence Berkeley Laboratory at the University of California, Berkeley. *Photograph by John Spragens Jr. Photo Researchers, Inc. Reproduced by permission.*

for such research because, as Einstein proved, **mass** is proportional to energy. When an particle moves at high speed, say in a cyclotron, it has considerable energy of motion and its mass is therefore greatly increased. One way to boost the speed of the particle further is to switch the electrical polarities of the Ds at a gradually lower **frequency**. A more sophisticated version of the cyclotron, the synchrocyclotron, includes the complicated **electronics** necessary to do this. However, the most efficient method of compensating for the increased mass of high-energy particles is to increase the applied magnetic field as the particle speed increases. The class of device that does this is called a synchrotron, and includes the most powerful particle accelerators in existence today. These installations have rings more than 1.2 mi (2 km) in diameter, a far cry from Lawrence's first cyclotron.

Cyclotron-type warping of charged-particle paths occurs in nature as well as in cyclotrons: wherever charged particles move through a magnetic field (e.g., when charged particles from the **Sun** encounter the magnetic field of a **planet**), they are forced to follow spiraling paths. Since **acceleration** of a charged particle—any change in the particle's direction or velocity—causes it to emit electromagnetic **radiation**, charged particles encountering magnetic fields in **space** emit radiation. This radiation, termed cyclotron radiation, can reveal the interactions of particles and magnetic fields across the cosmos, and is of importance in **astronomy**.

Human-built cyclotrons of the fixed-field type are not used in physics research any more, but are increasingly important in medicine. Proton-beam therapy is a recent innovation in radiosurgery (**surgery** using radiation) in which protons accelerated by a cyclotron are beamed at a target in the human body, such as a **tumor** at the back of the **eye**. The energy of these protons can be carefully controlled, and their stopping **distance** inside living **tissue** (i.e., the depth at which they deposit their energy) precisely predicted. These features mean that tumors inside the body can be targeted while minimizing damage to healthy tissues.

Resources

Other

De Martinis, C., et al. "Beam Tests on a Proton Linac Booster for Hadronotherapy." Proceedings of European Particle Accelerator Conference, Paris, France. 2002 (cited Feb. 6, 2003) <http:accelconf.web.cern.ch/AccelConf/e02/PAPERS/MOPRI095. pdf>.

Cypress *see* **Swamp cypress family (Taxodiaceae)**

Cystic fibrosis

Cystic fibrosis (CF) is genetic **disease** characterized by defects in the transport of a **molecule** called chloride. Abnormalities in CF have been described in several organs and tissues, including the airways, pancreas, bile ducts, gastrointestinal tract, sweat **glands**, and male reproductive. Lung function is often normal at **birth**; however, airway obstruction and **inflammation** as well as bacterial colonization are characteristically seen in the CF airways. The pathophysiological consequences that follow are believed to stem from repetitive cycles of bacterial **infection**, which contributes to a progressive deterioration in lung function.

In the United States, the disease affects about one in every 3,900 babies born annually, and it is estimated that approximately 30,000 Americans are afflicted with this disease. The genetic defect that causes CF is most common in people of northern European descent. It is estimated that one in 25 of these individuals are carriers of a defective **gene** that causes CF. Currently, there is no cure for CF and the disease can be fatal. In the past, individuals with CF would die sometime during childhood. With pharmacological intervention due to drug discovery from many years of research, the age of survival has increased 31 years.

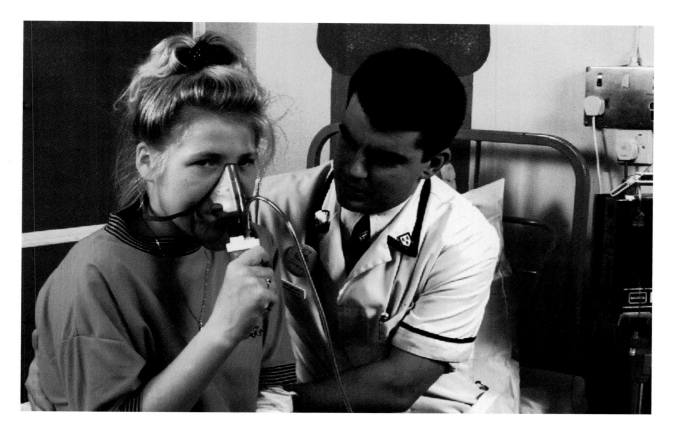

A cystic fibrosis patient using a nebulizer to loosen and subsequently expectorate the build up of thick mucus in the lungs.
© Simon Fraser, National Audubon Society Collection/Photo Researchers, Inc. Reproduced with permission.

The genetic basis of CF

CF is a homozygous recessive genetic disorder. In this type of disorder, two defective copies of the gene, one from each parent, must combine to produce the disease. If two people who each carry the defective copy of the gene have a child, chances are that one in four of their offspring will have CF.

In 1989, a team of researchers located the defective CF gene, which was found to be located on **chromosome 7**. Genes are segments of **deoxyribonucleic acid (DNA)** that code for certain **proteins**. If the sequence of DNA mutates in a gene, the protein for which it encodes also can change. In CF, a change or **mutation** in the DNA sequence of the gene can lead to the production of a defective version of an important protein. This protein is called the cystic fibrosis transmembrane conductance regulator, or CFTR. The protein works as an ion pump within the **cell membrane** and regulates the movement of **sodium** and chloride (electrolytes that makeup **salt**) into and out of cells. In people with CF, this pump does not work properly. As a result, **water** is retained within the cells. A dry, sticky mucus builds up in the tissues that are affected.

Clinical manifestations in CF

Most of the symptoms of CF are related to the sticky mucus that clogs the lungs and pancreas. People with CF have difficulty breathing and are highly susceptible to bacterial infections of the lungs. Normally, **bacteria** are expelled from the lungs by coughing and the movement of mucus up the airways to the throat where the bacteria can be expelled. In people with CF, the mucus is too thick to be removed and bacteria are able to inhabit the lungs and cause infection.

In addition to the airways, other tissues are affected in CF. The abnormalities found in these tissues are characterized by abnormally thick and dehydrated secretions, which appear to cause obstruction resulting in **organ** dysfunction. For example, the pancreatic ducts are obstructed resulting in **tissue** degeneration, fibrosis (scarring), and organ dysfunction. The pancreas secretes enzymes during digestion that break food into smaller pieces so that the body can absorb **nutrients**. Enzymes speed up **chemical reactions** and the enzymes in the pancreas are important for digestion of foods. Failure of the pancreas to function normally results in pancreatic

enzyme insufficiency, which is observed in approximately 85% of CF patients. Without treatment, enzyme deficiency results in protein and fats being poorly digested, which can lead to severe **malnutrition**.

In the gastrointestinal tract (the organ that digests and processes broken down food), accumulation of mucous secretions also occur. Dehydrated intestinal contents combined with abnormal mucous secretions are thought to predispose patients to bowel obstruction, which is a characteristic symptom in 10–20% of CF newborns. The bile ducts of CF patients can also be obstructed, producing gall bladder disease and elevations in liver function enzymes, occasionally leading to liver failure.

Greater than 95% of males with CF are infertile due to structural alterations in the reproductive tract that results in the sperm being incapable of **fertilization**, or azoospermia. These structures include the vas deferens and seminal vesicles, which are both an important part of the male reproductive tract and contribute to transportation of the sperm. If the vas deferens is absent at birth, it is a condition called **congenital** bilateral absence of the vas deferens (CBAVD). CBAVD is characteristic in male CF patients. Reduced fertility has also been noted in females with CF and may be related in part to abnormal mucous composition in their reproductive tract.

In the sweat gland, a characteristically detectable salty sweat represents the traditional gold standard test for diagnosing CF. Testing for CF involves analyzing sweat for elevated levels of salt. The ducts of the sweat glands normally function to reabsorb sodium and chloride across the water impermeable tissues. In CF, failure to reabsorb chloride ions in the ducts results in sodium and chloride, or salts, to be concentrated in sweat. Clinical manifestations include a predisposition to dehydration.

Despite the multi-organ involvement of the disease, respiratory failure is the primary cause of death in more than 90% of CF patients. Current hypotheses suggest that, in the CF airways, defective **electrolyte** transport results in alterations in the **volume** of liquid that covers the airways, the salt content, and/or mucus composition, which leads to thick mucus secretions that cannot be easily cleared out of the airways. The resulting airway microenvironment is conducive to chronic bacterial colonization and infection by specific bacterial **pathogens**, including *Streptococcus pneumoniae*, *Hemophilus pneumoniae*, and *Staphylococcus aeureus*. These bacteria typically infect the lungs of CF children. Adults are most susceptible to *Pseudomonas aeurginosa*. A rare type of bacteria called *Pseudomonas cepacia* currently infects people with CF at alarming rates. *Pseudomonas cepacia* causes a severe infection and hastens lung damage leading to an earlier death. These infections, coupled with an abnormal inflammatory response, leads to airway destruction and death in the vast majority of CF patients.

The CF gene was identified by researchers in 1989. There are many mutations (over 1,000) that cause CF. Some of these mutations cause a less severe disease; others cause a more severe disease. However, the same **gene mutation** in different people will not always have the same clinical manifestations.

Treating CF

Currently, no cure for CF exists. Treatment of the disease mainly involves alleviating symptoms caused by the build-up of mucus. To combat the lung infections, many patients are given large doses of **antibiotics** to prevent a severe, life-threatening infection. Some people undergo a course of antibiotics four times a year, on a predetermined schedule. Mucus in the lungs also can be broken down by drugs called mucolytic agents. These agents can be orally. Other drugs are inhaled as **aerosols**.

A drug called Pulmozyme is an enzyme which breaks down the excess DNA present in the mucus of CF patients that accumulates as a result of the inflammatory process. Pulmozyme helps to thin the mucus, allowing it to be more easily expelled. Clearing the thick mucus from the lungs can also be accomplished by physiotherapy. Physiotherapy includes breathing exercises and percussion, the administration of blows to the back and chest to loosen the mucus.

To control the malabsorption of nutrients, many people with CF take pancreatic enzymes in pill form with every meal. A diet high in **fat**, protein, and carbohydrates is also recommended to increase the nutrient intake. Multivitamins can also help prevent deficiencies of certain vitamins. When these methods do not result in adequate weight gain some people supplement their diets with feeding tubes, or a nutrient-rich **solution** infused through a tube placed in the stomach. Newer advances in the types of pancreatic enzymes and nutritional supplements offered to CF patients are helping such patients avoid malnutrition.

A number of other recent therapies are available for CF patients. These include an inhaled form of the antibiotic called tobramycin. Previously given intravenously to treat infections, inhaled tobramycin appears to improve lung function, while avoiding some of the detrimental side effects associated with IV tobraymycin. There are also other exciting development drugs that are aimed at improving the function of the protein defective in CF.

Gene therapy for CF

Researchers hoped that by discovering of the gene responsible for CF, a genetic approach to curing the disease will be developed. In **gene therapy**, a normal gene is inserted into cells to replace the defective gene. In most gene therapy experiments, cells from an affected organ are removed from the body and infected with a **virus** that has been modified to carry the normal gene. The newly infected cells are then put back into the body. In CF, this method has not yet been successful. The primary reason is that the lungs are equipped with a complex barrier, preventing successful penetration and delivery of the normal gene.

In 1994, researchers successfully transferred a virus containing the normal CFTR gene into four CF patients. The patients inhaled the virus into the nasal passages and lungs. An adenovirus, the virus used to carry the gene, is considered to be relatively safe but can cause several undesirable side effects. Nevertheless, one patient in this experiment developed side effects including headache, fatigue, and fever. It is unclear whether the experiment improved mucus clearance in the lungs or if the corrected gene produced adequate amounts of the corrected protein.

Before gene therapy can be considered, researchers must overcome several obstacles. The most important obstacle is the use of viruses as carriers for the normal genes. Some scientists feel that viruses are too dangerous, especially for patients who already have a chronic disease. Current studies are underway to investigate the use of liposomes, or small microscopic spheres consisting of a fatty-substance called a **lipid**, to transport the corrected gene.

A test for the CF gene

Recently researchers have located a number of defects on particular genes that appear to be responsible for the majority of CF cases. Knowing the location of these gene mutations makes it possible to test for carriers of the disease (individuals who have only one defective gene copy, and therefore have no symptoms themselves). Currently the test detects 85% of all CF gene mutations, which can be determined by extracting DNA from a persons **blood**, cheek cells, or saliva. Some researchers feel that this detection **rate** is still too low and that testing should be performed only on persons who have a familial history of CF or are Northern European by descent. Others argue that because the test is relatively inexpensive and easy to perform the test should be offered to everyone. At this time, testing for the gene responsible for CF remains controversial. In particular, testing parents prior to or during a pregnancy to determine their carrier status has resulted in controversy.

See also Genetic disorders; Respiratory diseases; Respiratory system.

KEY TERMS

. . . .

Allele—Any of two or more alternative forms of a gene that occupy the same location on a chromosome.

CF transmembrane conductance regulator—A protein that acts as a pump in the cell membranes of airway and pancreas cells. The pump regulates the transport of sodium and chloride into and out of cells.

Homozygous recessive disorder—Genetic disorder caused by the presence of two defective alleles.

Percussion—A technique in which blows are administered to the back and chest to loosen mucus in the respiratory tract.

Resources

Books

Kepron, Wayne. *Cystic Fibrosis: Everything You Need to Know (Your Own Personal Health)*. Toronto: Firefly, 2003.

Orenstein, David M. *Cystic Fibrosis: Medical Care*. Philadelphia: Lippincott Williams & Wilkins Publishers, 2000.

Shale, Dennis J. *Cystic Fibrosis*. BMJ Books, 2002.

Periodicals

"Gene therapy for CF reaches Human Lungs." *Science News* 146 (September, 3 1994): 149.

Johnson, Larry G. "Gene Therapy for CF." *Chest* 107 (February 1995): 775–815.

Other

Cystic Fibrosis Foundation. "About Cystic Fibrosis." (cited January 15, 2003) <http//www.cff.org/about_cf/what_is_cf.cfm>.

Bryan Cobb
Kathleen Scogna

Cytochrome

Cytochromes are electron-transporting protein pigments concerned with **cell respiration** that contain an iron-containing **molecule** called heme, allied to that of hemoglobin. When the **iron** of heme accepts an **electron**, it changes from the oxidized ferric (Fe III) state to the reduced ferrous (Fe II) state. The oxidation of cytochromes to molecular **oxygen** and their subsequent reduction by oxidizable substances in the cell is the main way in which atmospheric oxygen enters into the **metab-**

olism of the cell. About 90% of all oxygen consumed is mediated by the cytochromes.

Cytochromes make up two of the three large **enzyme** complexes that together comprise the electron transport or respiratory chain. This chain represents the end of oxidative phosphorylation, the process by which many organisms synthesize the energy-rich molecules of **adenosine triphosphate** (ATP) needed for life processes.

The source of the electrons to drive the respiratory chain is from the metabolic breakdown (**catabolism**) of food molecules. Two major pathways of metabolism-glycolysis and the Krebs cycle-break down glucose molecules and provide the electrons for the third pathway, the respiratory chain.

Glycolysis is the preliminary process during which the 6-carbon sugar molecule glucose is split into 3-carbon products, a process that renders only a few electrons for the respiratory chain. The more efficient **Krebs cycle**, which uses the 2-carbon products of glycolysis as raw materials for a cyclic series of enzymatic reactions, produces many more electrons.

The electron transport chain extends from the initial electron donor, nicotinamide adenine dinucleotide (NADH), to oxygen, the final electron acceptor.

The exchange of electrons begins at the NADH dehydrogenase **complex**, which passes electrons to ubiquinone (coenzyme Q). Ubiquinone, in turn, passes electrons to the cytochrome b-c$_1$ complex, which is composed of cytochromes and iron-sulfur **proteins**. The last cytochrome in this complex (cytochrome c) passes electrons to the cytochrome oxidase complex, composed of both cytochromes and **copper atoms**. Finally, the cytochrome oxidase complex passes electrons to oxygen.

The exchange of electrons along the respiratory chain generates a gradient of protons across the **membrane** in which the chain is located. When the protons flow back across the membrane, they activate the enzyme ATP synthetase, which produces ATP from **adenosine diphosphate** (ADP).

Cells that use the respiratory chain produce most of the supply of high-energy molecules of ATP needed for life. Many **bacteria** do not use oxygen (i.e., they are **anaerobic**), and consequently lack respiratory chain enzymes. These bacteria must rely on the less efficient glycolysis to produce ATP.

Cytochromes occur in organisms as varied as bacteria, **yeast**, humans, and **insects**. Indeed, beginning in 1925, researcher David Keilin made the first observations of cytochrome activity by studying the change in the wavelengths of **light** absorbed by cytochromes of flight muscles of living insects as the cytochromes underwent oxidation and reduction. He correctly postulated that these pigments underwent reduction and oxidation as they accepted electrons and then transferred them along the chain to the final electron acceptor, oxygen.

The heme group of cytochromes consists of a carbon-based ring called porphyrin, in which the iron atom is tightly bound by **nitrogen** atoms at each corner of a **square**. Related porphyrin molecules include hemoglobin, the oxygen-carrying molecule in **blood**, and **chlorophyll**, the green pigment of photosynthetic plants.

Cytology

Cytology is the branch of **biology** that studies cells, the building blocks of life. The name for this science is translated from kytos, the Greek term for "cavity." Cytology's roots travel back to 1665, when British botanist Robert Hooke, examining a cross-section of **cork**, gave the spaces the name "cells," meaning "little rooms" or "cavities."

Cytology's **beginnings** as a science occurred in 1839 with the first accurately conceived **cell** theory. This theory maintains that all organisms plants and animals alike are comprised of one or more like units called cells. Each of these units individually contain all the properties of life, and are the cornerstone of virtually all living organisms. Further, cell theory states that hereditary traits are passed on from generation to generation via **cell division**. Cell division generally has a regular, timed cyclical period during which the cell grows, divides, or dies. Virtually all cells perform biochemical functions, generating and transmitting **energy**, and storing genetic data carried down to further generations of cells. Cytology differs from its cousin, **pathology**, in that cytology concentrates on the structure, function and **biochemistry** of normal and abnormal living cells. Pathology pursues changes in cells caused by decay and death.

Cells can vary dramatically in size and shape from **organism** to organism. While **plant** and animals cell diameters generally average between 10–30 micrometers (0.00036–0.00108 inches), sizes can range from a few thousand atomic diameters for single-celled **microorganisms**, all the way up to 20–in (50–cm) diameters for the monocellular ostrich egg. Cell structures also differ between advanced single-celled and multicellular organisms (plants and animals) and more primitive prokaryotic cells (e.g., **bacteria**). Plant cells are the most representative of a prototypical cell, as they have a nucleus, cell **membrane** and cell wall. **Animal** cells, on the other hand, lack a formalized cell wall, although they contain the former two. **Prokaryote** cells (e.g., bacteria) are unique in that they lack a nucleus and possess no membrane-enclosed organelles. Exceptions to the cell theory include syncytial organisms (e.g., certain **slime molds** and microscopic **flatworms**) without cellular partitions; however, they are derived secondarily from organisms with cells via the breakdown of cellular membranes. Finally, the number of cells within an organism can range from one for organisms like an **amoeba**, to 100 trillion cells for a human being.

Cytology has greatly benefitted from the **electron microscope**, which reveals internal and external cell dynamics too small to be monitored by traditional optical microscopes. Also, **fluorescence** or contrast **microscopy** with more traditional visual observation equipment enables the cell substance to be revealed when a specific cell material is stained with a chemical compound to illuminate specific structures within the cells. For example, basic dyes (e.g., hematoxylin) illuminates the nucle-us, while acidic dyes (e.g., eosin) stain the cytoplasm (the cellular material within the membrane (excluding the nucleus). Finally, newer techniques including radioactive isotopes and high-speed centrifuges have helped advance cytology.

Cytological techniques are beneficial in identifying the characteristics of certain hereditary human diseases, as well as in plant and **animal breeding** to help determine the chromosonal structure to help design and evaluate breeding experiments. A far more controversial discussion deals with the role of cytology as it relates to cloning.

Over **time**, cytology's prominence as a separate science has diminished, integrating into other disciplines to create a more comprehensive biological-chemical approach. Associated disciplines include cytogenetics (study of **behavior** of chromosomes and genes relating to heredity) and cytochemistry (study of chemical contents of cells and tissues).

D

Dacron *see* **Artificial fibers**

Daffodil *see* **Amaryllis family (Amaryllidaceae)**

Daisy *see* **Composite family**

Dams

Dams are structures designed to restrict the flow of a stream or river, thus forming a pond, **lake**, or reservoir behind the wall. Dams are used for flood control, for production of hydroelectric power, to store and distribute **water** for agriculture and human populations, and as recreation sites.

Classification of dams

Dams may be classified according to the general purpose for which they are designed. These include storage, diversion, and detention.

Storage dams are built to provide a reliable source of water for short or long periods of **time**. Small dams, for example, are often built to capture spring runoff for use by **livestock** in the dry summer months. Storage dams can be further classified by the specific purpose for which the water is being stored, such as municipal water supply, recreation, hydroelectric power generation, or **irrigation**.

Diversion dams are typically designed to raise the elevation of a water body to allow that water to be conveyed to another location for use. The most common applications of diversion dams are supplying irrigation canals and transferring water to a storage reservoir for municipal or industrial use.

Detention dams are constructed to minimize the impact of **flooding** and to restrict the flow **rate** of a particular channel. In some cases, the water trapped by a detention dam is held in place to recharge the subsurface groundwater system. Other detention dams, called debris dams, are designed to trap sediment.

Large dams frequently serve more than one of these purposes and many combine aspects of each of the three main categories. Operation of such multipurpose dams is complicated by sometimes opposing needs. In order to be most effective, storage behind a flood control dam should be maintained at the lowest level possible. After a detention event occurs, water should be released as quickly as possible, within the capacity of the downstream channel. Conversely, the efficient and economic operation of storage and diversion dams requires that water levels be maintained at the highest possible levels. Releases from these reservoirs should be limited to the intended user only, such as the power generating turbines or the municipal water user. Operators of multipurpose dams must balance the conflicting needs of the various purposes to maintain the reliability, safety, and economic integrity of the dam. Operators must use a variety of information to predict the needs of the users, the expected supply, and the likelihood of any abnormal conditions that might impact the users or the dam itself. Failure to do so can threaten even the largest of dams.

During the El Niño of 1983, climate and hydrologic forecasts failed to predict abnormally heavy spring runoff in the Rocky Mountains. Dam operators along the Colorado River maintained high water storage levels, failing to prepare for the potential of the flooding. By the time operators began to react, water was bypassing the dams via their spillways and wreaking havoc throughout the system. Ultimately, the Glen Canyon dam in Arizona was heavily impacted with flood flows eroding large volumes of rock from within the canyon walls that support the dam. Fortunately, the flooding peaked and control was regained before the dam was breached.

Dam construction

There are four main types of dams: arch, buttress, gravity, and embankment dams. The type of construction

Hoover dam from downstream looking toward the east at the face of the dam. The Colorado River is in the foreground. The dam is 725 ft (221 m) high and impounds enough water in Lake Mead to cover the state of New York up to 1 ft (0.3 m) deep. *Photograph by Roy Morsch. Stock Market. Reproduced by permission.*

for each dam is determined by the proposed use of the structure, qualities of the intended location, quantity of water to be retained by the structure, materials available for construction, and funding limitations.

Arch dams use an upstream-facing arch to help resist the **force** of the water. They are typically built in narrow canyons and are usually made of **concrete**. Good contact between the concrete and the **bedrock** are required to prevent leakage and ensure stability. A dome dam is a special variant with curves on the vertical and horizontal planes, while the arch dam is only curved on the horizontal **plane**. In addition, dome dams are much thinner than arch dams.

A buttress dam is characterized by a set of angled supports on the downstream side that help to oppose the force of the water. This design can be employed in wide valleys where a solid bedrock foundation is not available. Because of the **steel** framework and associated labor needed for construction, these dams are no longer economically viable.

The gravity dam withstands the force of the water behind it with its weight. Made of cement or masonry, this type of dam normally utilizes a solid rock foundation but can be situated over unconsolidated material if provisions are made to prevent the flow of water beneath the structure. The solid, stable nature of this dam is favored by many and often incorporated into the spillway designs of embankment dams.

An embankment dam uses the locally available material (**rocks**, gravel, **sand**, clay, etc.) in construction. Just as with gravity dams, the weight of embankment dams is used to resist the force of the water. The permeability of the materials that make up these dams allows water to flow into and through the dam. An impervious **membrane** or clay core must be built into them to counteract the flow and protect the integrity of the structure. Because the materials are locally available and the construction of these dams is relatively simple, the cost of construction for this type of dam is much lower than the other types. Embankment dams are the most common.

Impact of dams

Dams have long been acknowledged for providing **electricity** without the **pollution** of other methods, for flood protection, and for making water available for agriculture and human needs. Within recent decades, however, the environmental impacts of dams have been debated. While dams do perform important functions, their effects can be damaging to the environment. People have begun to question whether the positive contributions of some dams are outweighed by those negative effects.

The damming of a river will have dramatic consequences on the nature of the environment both upstream and downstream of the dam. The magnitude of these effects are usually directly related to the size of the dam.

Prior to dam construction, most natural **rivers** have a flow rate that varies widely throughout the year in response to varying conditions. Of course once constructed, the flow rate of the river below a dam is restricted. The dam itself and the need to control water releases for the various purposes of the particular dam result in a flow rate that has a smaller range of values and peaks that occur at times related to need rather than the dictates of nature. In cases where the entire flow has been diverted for other uses, there may no longer be any flow in the original channel below the dam.

Because water is held behind the dam and often released from some depth, the **temperature** of the water below the dam is usually lower than it would be prior to dam emplacement. The temperature of the water flow is often constant, not reflecting the natural seasonal variations that would have been the case in the free-flowing river. Similarly, the **chemistry** of the water may be altered. Water exiting the lake may be higher in dissolved salts or have lower **oxygen** levels than would be the case for a free-flowing river.

Impoundments increase the potential for **evaporation** from the river. Because the surface area of a lake is so great when compared to the river that supplies it, the loss of water to evaporation must be considered. In some **desert** areas, potential annual evaporation can be greater than 7 ft (2.1 m), meaning that over the course of one year, if no water flowed into or out of the system, the reservoir would drop in elevation by 7 ft (2.1 m). At Lake Mead on the Colorado River in Arizona and Nevada, evaporation losses in one year can be as great as 350 billion gal (1.3 trillion l).

The impoundment of water behind a dam causes the **velocity** of the water to drop. Sediment carried by the river is dropped in the still water at the head of the lake. Below the dam, the river water flows from the clear water directly behind the dam. Because the river no longer carries any sediment, the erosive potential of the river is increased. **Erosion** of the channel and banks of the river below the dam will ensue. Even further downstream, sediment deprivation affects shoreline processes and biological productivity of coastal regions.

This problem has occurred within the Grand Canyon below Glen Canyon Dam. After the construction of the dam was completed in 1963, erosion of the sediment along the beaches began because of the lack of incoming sediment. By the early 1990's, many beaches were in danger of disappearing. In the spring of 1996, an experimental controlled flood of the river below Glen Canyon Dam was undertaken to attempt to redistribute existing sediments along the sides of the channel. While many of the beaches were temporarily rebuilt, this redistribution of sediments was short lived. Research on this issue is continuing, however, the fundamental problem of the lack of input sediment for the river downstream of the dam remains unresolved.

The environmental changes described above create a new environment in which native **species** may or may not be able to survive. New species frequently invade such localities, further disrupting the system. Early photographs of rivers in the southwest desert illustrate the dramatic modern invasion of non-native plants. Entire lengths of these rivers and streams have been trans-

formed from native desert plants to a dense riparian environment. Native species that formerly lived in this zone have been replaced as a result of the changes in river flow patterns. The most commonly cited species affected by the presence of dams is the **salmon**. Salmon have been isolated from their spawning streams by impassable dams. The situation has been addressed through the use of **fish** ladders and by the use of barges to transport the fish around the obstacles, but with only limited success.

David B. Goings

KEY TERMS

Arch dam—A thin concrete dam that is curved in the upstream direction.

Buttress dam—A dam constructed of concrete and steel that is characterized by angled supports on the downstream side.

El Niño—The phase of the Southern Oscillation characterized by increased sea water temperatures and rainfall in the eastern Pacific, with weakening trade winds and decreased rain along the western Pacific.

Embankment dam—A simple dam constructed of earth materials.

Gravity dam—A massive concrete or masonry dam that resists the force of the water by its own weight.

Impoundment—The body of water, a pond or lake, that forms behind a dam.

Permeability—The capacity of a geologic material to transmit a fluid, such as water.

Spillway—A passage for water to flow around a dam.

Resources

Books

Dunar, Andrew J., and Dennis McBride. *Building Hoover Dam: An Oral History of the Great Depression.* New York: Twayne Pub, 1993.

Fradkin, Philip L. *A River No More: The Colorado River and the West.* Berkeley: University of California Press, 1996.

High Country News. *Western Water Made Simple.* Washington, DC: Island Press, 1987.

Keller, Edward. *Environmental Geology.* Upper Saddle River: Prentice-Hall, Inc., 2000.

Pearce, Fred. *The Dammed: Rivers, Dams, and the Coming World Water Crisis.* London: Bodley Head, 1992.

Reisner, Marc. *Cadillac Desert: The American West and its Disappearing Water.* New York: Penguin Books Ltd., 1993.

Periodicals

"Water: The Power, Promise, and Turmoil of North America's Fresh Water." *National Geographic* Special Edition. November 1993.

Other

"Colorado River Watershed." Arizona Department of Water Resources. [cited October 16, 2002.] <http://www.adwr.state.az.us/AZWaterInfo/OutsideAMAs/UpperColoradoRiver/Watersheds/coloradoriver.html>.

"Lower Colorado River Operations." United States Bureau of Reclamation. [cited October 16, 2002]. <http://www.lc.usbr.gov/lcrivops.html>.

"Controlled Flooding of the Colorado River in Grand Canyon." United States Geological Survey. February 14, 1999 [cited October 16, 2002]. <http://az.water.usgs.gov/flood.html>.

Damselflies

Damselflies are the smaller and more delicate members of the insect order Odonata, which includes the **dragonflies**. The damselfly suborder Zygoptera is characterized by similar fore and hind wings, which are both narrow at the base. Most damselflies can be easily distinguished from their larger and heavier dragonfly relatives in the field by their fluttering flight, and when at rest by their holding their wings up vertically or in a V-position when at rest.

Damselflies are usually found sitting on overhanging branches or other objects near **water**. They feed on small flying **insects** such as **mosquitoes** and gnats, which they catch in flight.

Although most damselflies are small and very slender, many have brightly colored bodies. The males are usually more colorful than the females, and often have spots or markings of vivid blue, green, or yellow.

Damselflies have a worldwide distribution. One of the larger and more conspicuous **species** in **North America**, found on shaded bushes overhanging small streams, is the black-winged damselfly (*Calopterix maculata*). The male of this species has all-black wings and a metallic-green body, whereas the female has gray wings with a small white dot (stigma) near the tip.

Damselflies mate on the wing in the same unusual fashion as dragonflies, and lay their eggs in the water. The eggs hatch into wingless larvae, called naiads, that remain on the bottom of the pond or stream. The damselflies larvae feed on smaller insect larvae and other aquatic animals. Damselfly larvae resemble dragonfly larvae except for the three leaf-like gills at the end of the body.

These beautiful, delicate animals neither sting nor bite. Indeed, damselflies help to control the disease-carrying mosquitoes and biting midges.

Dandelion *see* **Composite family**

Dark matter

Dark matter is the term astronomers use to describe material in the Universe that is non-luminous—that is, material that does not emit or reflect **light** and that is therefore invisible. Everything seen when looking through a **telescope** is visible because it is either emitting or reflecting light; stars, nebulae, and galaxies are examples of luminous objects. However, luminous matter appears to make up only a small fraction of all the matter in the Universe, perhaps only a few **percent**. The rest of the matter is cold, dark, and hidden from direct view.

Because dark matter is invisible, it can only be detected through indirect means, primarily by analyzing its effect on visible material. Although dark matter does not shine, it still exerts a gravitational **force** on the matter around it. For example, it is possible to measure the velocities of many stars in our **galaxy** and in other galaxies. The measured velocities do not agree, in general, with those calculated on the assumption that the visible material of the galaxies (i.e., their stars and **clouds** of glowing gas) constitute all or even most of their **mass**. Additional, unseen mass, therefore, must exist in the vicinity of the galaxies, tugging on their stars. Such data seem to indicate the presence of massive "halos" of dark matter surrounding the galaxies that would account for most of their mass. Recent observations of the shapes formed by galaxies clumping together throughout the Universe have confirmed that dark matter does not pervade **space** uniformly or form structures independently of the galaxies, but is concentrated around the galaxies.

The identity of the Universe's dark matter remains a subject of research and dispute among physicists. A number of possibilities have been proposed: (1) Astronomers hold that supermassive black holes exist at the centers of most galaxies, contributing several hundred million or even on the order of a billion solar masses to each galaxy. (One solar mass is a quantity of matter equal to the mass of the Sun.) In 2001, observations of x-ray bursts from the center of our galaxy confirmed the presence of a large **black hole** there. Such black holes supply invisible mass and count as "dark matter." (2) Multitudes of non-luminous brown dwarfs or machos

A composite image of a cluster of galaxies NGC 2300 (seen at optical wavelengths) and the recently discovered gas cloud (seen in x-ray emission) in which they are embedded. The cloud is considered to be strong evidence for the existence of dark matter because the gravitational pull of the cluster is not strong enough to hold it together. Some astronomers have suggested that dark matter (so-called because it does not emit detectable radiation) is preventing the cloud from dispersing into space. *U.S. National Aeronautics and Space Administration (NASA).*

(massive compact halo objects)—dim blobs of gas not massive enough to initiate fusion reactions at their centers and thereby become stars—may **orbit** each galaxy. Such objects have been detected using gravitational lensing, but not in sufficient numbers to account for the amount of dark matter that is believed to exist. (3) The **subatomic particles** known as neutrinos, which pervade the universe in very great numbers, were shown in 1998 to have a small mass, ending a decades-long dispute among physicists about whether they are massless. It had been thought that neutrinos, if they have mass, might account for the Universe's dark matter; however, calculations now show that each neutrino's mass is so small that neutrinos can account for at most a fifth of the dark mat-

ter in the Universe. (4) Particles of some unknown kind, generically termed wimps (weakly-interacting massive particles), may permeate the space around the galaxies, held together in clouds by gravity.

Dark matter, which may turn out to be a combination of such factors, has long been thought to play a crucial role in determining the fate of the Universe. The most widely accepted theory regarding the origin and evolution of the universe is the **big bang theory**, which provides an elegant explanation for the well-documented expansion of the universe. One question is whether the universe will expand forever, propelled by the force of the big bang, or eventually stop expanding and begin to contract under its own gravity, much as a ball thrown into the air eventually turns around and descends. The deciding factor is the amount of mass in the universe: the more mass, the more overall gravity. There is a critical mass threshold above which the universe will eventually turn around and begin to contract (a "closed" universe). Below this threshold the expansion will continue forever (an "open" universe). It turns out that the luminous material currently observed throughout the Universe does not amount to nearly enough mass to halt the expansion. But what if there is a huge quantity of unseen mass out there, invisible but with a profound gravitational effect on the Universe? Dark matter, it was long thought, might supply the "missing gravity" necessary to halt the Universe's expansion.

Debate over this question persisted for decades, but has probably been resolved by observations made during the last five years that indicate that the expansion of the Universe, far from slowing down, is accelerating. If this result is confirmed, then the fate of the Universe is at last definitely known: it will expand forever, becoming darker, colder, and more diffuse.

To account for the observed **acceleration**, physicists have postulated a "dark energy," still mysterious in origin, that pervades the Universe and actually helps to push things apart rather than keep them together. Since **energy** (even "dark energy") and matter are interchangeable, some of the Universe's dark matter may thus turn out to be not matter at all, but energy.

Resources

Periodicals

Glanz, James. "Evidence Points to Black Hole At Center of the Milky Way." *New York Times,* September 6, 2001.

Glanz, James. "Photo Gives Weight to Einstein's Thesis of Negative Gravity." *New York Times,* April 3, 2001.

Overbye, Dennis. "Dark Matter, Still Elusive, Gains Visibility." *New York Times* January 8, 2002.

Wilford, John Noble. "Constructing a More Plausible Universe With 'Warm Dark Matter.'" *New York Times,* January 2, 2001.

Other

Elgaroy, Oystein, Jaqueline Mitton, and Peter Bond. "Neutrinos Only Account for One Fifth of Dark Matter." *Daily University Science News.* April 10, 2002 [cited January 9, 2003]. <http://unisci.com/stories/20022/0410026.htm>.

Dates *see* **Palms**

Dating techniques

Dating techniques are procedures used by scientists to determine the age of a specimen. Relative dating methods tell only if one **sample** is older or younger than another; absolute dating methods provide a date in years. The latter have generally been available only since 1947. Many absolute dating techniques take advantage of **radioactive decay**, whereby a radioactive form of an element is converted into a non-radioactive product at a regular **rate**. Others, such as **amino acid** racimization and cation-ratio dating, are based on chemical changes in the organic or inorganic composition of a sample. In recent years, a few of these methods have come under close scrutiny as scientists strive to develop the most accurate dating techniques possible.

Relative dating

Relative dating methods determine whether one sample is older or younger than another. They do not provide an age in years. Before the advent of absolute dating methods, nearly all dating was relative. The main relative dating method is **stratigraphy**.

Stratigraphy

Stratigraphy is the study of layers of **rocks** or the objects embedded within those layers. It is based on the assumption (which nearly always holds true) that deeper layers were deposited earlier, and thus are older, than more shallow layers. The sequential layers of rock represent sequential intervals of **time**. Although these units may be sequential, they are not necessarily continuous due to erosional removal of some intervening units. The smallest of these rock units that can be matched to a specific time interval is called a bed. Beds that are related are grouped together into members, and members are grouped into formations. Stratigraphy is the principle method of relative dating, and in the early years of dating studies was virtually the only method available to scientists.

Seriation

Seriation is the ordering of objects according to their age. It is a relative dating method. In a landmark study, archaeologist James Ford used seriation to determine the chronological order of American Indian pottery styles in the Mississippi Valley. Artifact styles such as pottery types are seriated by analyzing their abundances through time. This is done by counting the number of pieces of each style of the artifact in each stratigraphic layer and then graphing the data. A layer with many pieces of a particular style will be represented by a wide band on the graph, and a layer with only a few pieces will be represented by a narrow band. The bands are arranged into battleship-shaped curves, with each style getting its own **curve**. The curves are then compared with one another, and from this the relative ages of the styles are determined. A limitation to this method is that it assumes all differences in artifact styles are the result of different periods of time, and are not due to the immigration of new cultures into the area of study.

Faunal dating

The term faunal dating refers to the use of **animal** bones to determine the age of sedimentary layers or objects such as cultural artifacts embedded within those layers. Scientists can determine an approximate age for a layer by examining which **species** or genera of animals are buried in it. The technique works best if the animals belonged to species which evolved quickly, expanded rapidly over a large area, or suffered a **mass extinction**. In addition to providing rough absolute dates for specimens buried in the same stratigraphic unit as the bones, faunal analysis can also provide relative ages for objects buried above or below the fauna-encasing layers.

Pollen dating (palynology)

Each year seed-bearing plants release large numbers of pollen grains. This process results in a "rain" of pollen that falls over many types of environments. Pollen that ends up in **lake** beds or peat bogs is the most likely to be preserved, but pollen may also become fossilized in arid conditions if the **soil** is acidic or cool. Scientists can develop a pollen chronology, or calendar, by noting which species of pollen were deposited earlier in time, that is, residue in deeper sediment or rock layers, than others.

The unit of the calendar is the pollen zone. A pollen zone is a period of time in which a particular species is much more abundant than any other species of the time. In most cases, this tells us about the climate of the period, because most plants only thrive in specific climatic conditions. Changes in pollen zones can also indicate changes in human activities such as massive **deforestation** or new types of farming. Pastures for grazing **livestock** are distinguishable from fields of grain, so changes in the use of the land over time are recorded in the pollen history. The dates when areas of **North America** were first settled by immigrants can be determined to within a few years by looking for the introduction of ragweed pollen.

Pollen zones are translated into absolute dates by the use of radiocarbon dating. In addition, pollen dating provides relative dates beyond the limits of radiocarbon (40,000 years), and can be used in some places where radiocarbon dates are unobtainable.

Fluorine is found naturally in ground **water**. This water comes in contact with skeletal remains under ground. When this occurs, the fluorine in the water saturates the bone, changing the mineral composition. Over time, more and more fluorine incorporates itself into the bone. By comparing the relative amounts of fluorine composition of skeletal remains, one can determine whether the remains were buried at the same time. A bone with a higher fluorine composition has been buried for a longer period of time.

Absolute dating

Absolute dating is the term used to describe any dating technique that tells how old a specimen is in years. These are generally analytical methods, and are carried out in a laboratory. Absolute dates are also relative dates, in that they tell which specimens are older or younger than others. Absolute dates must agree with dates from other relative methods in order to be valid.

Amino acid racimization

This dating technique was first conducted by Hare and Mitterer in 1967, and was popular in the 1970s. It requires a much smaller sample than radiocarbon dating, and has a longer range, extending up to a few hundred thousand years. It has been used to date coprolites (fossilized feces) as well as fossil bones and shells. These types of specimens contain **proteins** embedded in a network of **minerals** such as **calcium**.

Amino acid racimization is based on the principle that amino acids (except glycine, which is a very simple amino acid) exist in two mirror image forms called stereoisomers. Living organisms (with the exception of some microbes) synthesize and incorporate only the L-form into proteins. This means that the **ratio** of the D-form to the L-form is **zero** (D/L=0). When these organisms die, the L-amino acids are slowly converted into D-amino acids in a process called racimization. This occurs because protons (H^+) are removed from the amino acids by acids or bases present in the burial environment. The protons are quickly replaced, but will return to either side of the amino acid, not necessarily to the side from which they came. This may form a D-amino acid instead

Dendrochronology is a dating technique that makes use of tree growth rings. *Photograph by Charles Krebs. The Stock Market. Reproduced by permission.*

of an L-amino acid. The reversible reaction eventually creates equal amounts of L- and D-forms (D/L=1.0).

The rate at which the reaction occurs is different for each amino acid; in addition, it depends upon the moisture, **temperature**, and **pH** of the postmortem conditions. The higher the temperature, the faster the reaction occurs, so the cooler the burial environment, the greater the dating range. The burial conditions are not always known, however, and can be difficult to estimate. For this reason, and because some of the amino acid racimization dates have disagreed with dates achieved by other methods, the technique is no longer widely used.

Cation-ratio dating

Cation-ratio dating is used to date rock surfaces such as stone artifacts and cliff and ground drawings. It can be used to obtain dates that would be unobtainable by more conventional methods such as radiocarbon dating. Scientists use cation-ratio dating to determine how long rock surfaces have been exposed. They do this by chemically analyzing the varnish that forms on these surfaces. The varnish contains cations, which are positively-charged **atoms** or molecules. Different cations move throughout the environment at different rates, so the ratio of different cations to each other changes over time. **Cation** ratio dating relies on the principle that the cation ratio $(K^+ + Ca^{2+})/Ti^{4+}$ decreases with increasing age of a sample. By calibrating these ratios with dates obtained from rocks from a similar microenvironment, a minimum age for the varnish can be determined. This technique can only be applied to rocks from **desert** areas, where the varnish is most stable.

Although cation-ratio dating has been widely used, recent studies suggest it has many problems. Many of the dates obtained with this method are inaccurate due to improper chemical analyses. In addition, the varnish may not actually be stable over long periods of time. Finally, some scientists have recently suggested that the cation ratios may not even be directly related to the age of the sample.

Thermoluminescence dating

Thermoluminescence dating is very useful for determining the age of pottery. Electrons from quartz and other minerals in the pottery clay are bumped out of their normal positions (ground state) when the clay is exposed to **radiation**. This radiation may come from radioactive substances such as **uranium**, present in the clay or burial medium, or from cosmic radiation. When the ceramic is heated to a very high temperature (over 932°F [500°C]), these electrons fall back to the ground state, emitting **light** in the process and resetting the "clock" to zero. The longer the exposure to the radiation, the more electrons that are bumped into an excited state, and the more light that is emitted upon heating. The process of displacing electrons begins again after the object cools. Scientists can determine how many years have passed since a ceramic was fired by heating it in the laboratory and measuring how much light is given off. Thermoluminescence dating has the advantage of covering the time interval between radiocarbon and potassium-argon dating, or 40,000–200,000 years. In addition, it can be used to date materials that cannot be dated with these other two methods.

Optically stimulated **luminescence** (OSL) has only been used since 1984. It is very similar to thermoluminescence dating, both of which are considered "clock setting" techniques. Minerals found in sediments are sensitive to light. Electrons found in the sediment grains leave the ground state when exposed to light, called recombination. To determine the age of a sediment, scientists expose grains to a known amount of light and compare these grains with the unknown sediment. This technique can be used to determine the age of unheated sediments less than 500,000 years old. A disadvantage to this technique is that in order to get accurate results, the sediment to be tested cannot be exposed to light (which would reset the "clock"), making sampling difficult.

Tree-ring dating

This absolute dating method is also known as dendrochronology. It is based on the fact that trees produce one growth ring each year. Narrow rings grow in cold and/or dry years, and wide rings grow in warm years with plenty of moisture. The rings form a distinctive pattern, which is the same for all members in a given species and geographical area. The patterns from trees of different ages (including ancient **wood**) are overlapped, forming a master pattern that can be used to date timbers thousands of years old with a resolution of one year. Timbers can be used to date buildings and archaeological sites. In addition, **tree** rings are used to date changes in the climate such as sudden cool or dry periods. Dendrochronology has a range of 1-10,000 years or more.

Radioactive decay dating

As previously mentioned, radioactive decay refers to the process in which a radioactive form of an element is converted into a nonradioactive product at a regular rate. Radioactive decay dating is not a single method of absolute dating but instead a group of related methods for absolute dating of samples.

Potassium-argon dating

When volcanic rocks are heated to extremely high temperatures, they release any argon gas trapped in them. As the rocks cool, argon-40 (^{40}Ar) begins to accumulate. Argon-40 is formed in the rocks by the radioactive decay of potassium-40 (^{40}K). The amount of ^{40}Ar formed is proportional to the decay rate (**half-life**) of ^{40}K, which is 1.3 billion years. In other words, it takes 1.3 billions years for half of the ^{40}K originally present to be converted into ^{40}Ar. This method is generally only applicable to rocks greater than three million years old, although with sensitive instruments, rocks several hundred thousand years old may be dated. The reason such old material is required is that it takes a very long time to accumulate enough ^{40}Ar to be measured accurately. Potassium-argon dating has been used to date volcanic layers above and below fossils and artifacts in east **Africa**.

Radiocarbon dating

Radiocarbon is used to date charcoal, wood, and other biological materials. The range of conventional radiocarbon dating is 30,000–40,000 years, but with sensitive instrumentation this range can be extended to 70,000 years. Radiocarbon (^{14}C) is a radioactive form of the element **carbon**. It decays spontaneously into nitrogen-14 (^{14}N). Plants get most of their carbon from the air in the form of **carbon dioxide**, and animals get most of their carbon from plants (or from animals that eat plants). Atoms of ^{14}C and of a non-radioactive form of carbon, ^{12}C, are equally likely to be incorporated into living organisms—there is no discrimination. While a **plant** or animal is alive, the ratio of ^{14}C/^{12}C in its body will be nearly the same as the ^{14}C/^{12}C ratio in the atmosphere. When the **organism** dies, however, its body stops incorporating new carbon. The ratio will then begin to change as the ^{14}C in the dead organism decays into ^{14}N. The rate at which this process occurs is called the half-life. This is the time required for half of the ^{14}C to decay into ^{14}N. The half-life of ^{14}C is 5,730 years. Scientists can tell how many years have elapsed since an organism died by comparing the ^{14}C/^{12}C ratio in the remains with the ratio in the atmosphere. This allows us to determine how much ^{14}C has formed since the death of the organism.

A problem with radiocarbon dating is that diagenic (after death) **contamination** of a specimen from soil, water, etc. can add carbon to the sample and affect the measured ratios. This can lead to inaccurate dates. Another problem lies with the assumptions associated with radiocarbon dating. One assumption is that the ^{14}C/^{12}C ratio in the atmosphere is constant though time. This is not completely true. Although ^{14}C levels can be measured in tree rings and used to correct for the ^{14}C/^{12}C ratio in the atmosphere at the time the organism died, and can even be used to calibrate some dates directly, radiocarbon remains a more useful relative dating technique than an absolute one.

Uranium series dating

Uranium series dating techniques rely on the fact that radioactive uranium and thorium isotopes decay into a series of unstable, radioactive "daughter" isotopes; this process continues until a stable (non-radioactive) lead **isotope** is formed. The daughters have relatively short half-lives ranging from a few hundred thousand years down to only a few years. The "parent" isotopes have half-lives of several thousand million years. This provides a dating range for the different uranium series of a few thousand years to 500,000 years. Uranium series have been used to date uranium-rich rocks, deep-sea sediments, shells, bones, and teeth, and to calculate the ages of ancient lake beds. The two types of uranium series dating techniques are daughter deficiency methods and daughter excess methods.

In daughter deficiency situations, the parent radioisotope is initially deposited by itself, without its daughter (the isotope into which it decays) present. Through time, the parent decays to the daughter until the two are in equilibrium (equal amounts of each). The age of the **deposit** may be determined by measuring how much of the daughter has formed, providing that neither isotope has entered or exited the deposit after its initial formation. Carbonates may be dated this way using, for example, the daughter/parent isotope pair protactinium-231/uranium-235 (^{231}Pa/^{235}U). Living **mollusks** and corals will only take up dissolved compounds such as isotopes of uranium, so they will contain no protactinium, which is insoluble. Protactinium-231 begins to accumulate via the decay of ^{235}U after the organism dies. Scientists can determine the age of the sample by measuring how much ^{231}Pa is present and calculating how long it would have taken that amount to form.

In the case of a daughter excess, a larger amount of the daughter is initially deposited than the parent. Non-uranium daughters such as protactinium and thorium are insoluble, and precipitate out on the bottoms of bodies of water, forming daughter excesses in these sediments. Over time, the excess daughter disappears as it is converted back into the parent, and by measuring the extent to which this has occurred, scientists can date the sample. If the radioactive daughter is an isotope of uranium, it will dissolve in water, but to a different extent than the parent; the two are said to have different solubilities. For example, ^{234}U dissolves more readily in water than its parent, ^{238}U, so lakes and oceans contain an excess of this daughter isotope. This excess is transferred to organisms such as mollusks or corals, and is the basis of ^{234}U/^{238}U dating.

Fission track dating

Some volcanic minerals and glasses, such as obsidian, contain uranium-238 (^{238}U). Over time, these substances become "scratched." The marks, called tracks, are the damage caused by the fission (splitting) of the uranium atoms. When an atom of ^{238}U splits, two "daughter" atoms rocket away from each other, leaving in their wake tracks in the material in which they are embedded. The rate at which this process occurs is proportional to the decay rate of ^{238}U. The decay rate is measured in terms of the half-life of the element, or the time it takes for half of the element to split into its daughter atoms. The half-life of ^{238}U is 4.47×10^9 years.

When the mineral or **glass** is heated, the tracks are erased in much the same way cut marks fade away from hard candy that is heated. This process sets the fission track clock to zero, and the number of tracks that then form are a measure of the amount of time that has passed since the heating event. Scientists are able to count the tracks in the sample with the aid of a powerful **microscope**. The sample must contain enough ^{238}U to create enough tracks to be counted, but not contain too much of the isotope, or there will be a jumble of tracks that cannot be distinguished for counting. One of the advantages of fission track dating is that it has an enormous dating range. Objects heated only a few decades ago may be dated if they contain relatively high levels of ^{238}U; conversely, some meteorites have been dated to over a billion years old with this method.

See also Pollen analysis; Strata.

Resources

Books

Geyh, Mebus A., and Helmut Schleicher. *Absolute Age Determination. Physical and Chemical Dating Methods and Their Application.* New York: Springer-Verlag, 1990.

Göksu, H.Y., M. Oberhofer, and D. Regulla, eds. *Scientific Dating Methods.* Boston: Kluwer Academic Publishers, 1991.

Lewis, C.L.E. *The Dating Game* Cambridge: Cambridge University Press, 2000.

Wagner, Günther, and Peter Van Den Haute. *Fission-Track Dating.* Boston: Kluwer Academic Publishers, 1992.

Periodicals

Hofreiter, M., et al. "Ancient DNA." *Nature Reviews Genetics* 2 (2001): 353-359.

Kathryn M. C. Evans

DDT (Dichlorodiphenyl-trichloroacetic acid)

Dichlorodiphenyl-trichloroacetic acid (or DDT) is a chlorinated **hydrocarbon** that has been widely used as an insecticide. DDT is virtually insoluble in **water**, but is freely soluble in oils and in the **fat** of organisms. DDT is also persistent in the environment. The combination of persistence and **lipid solubility** means that DDT biomagnifies, occurring in organisms in preference to the non-living environment, especially in predators at the top of ecological food webs. Environmental **contamination** by DDT and related chemicals is a widespread problem, including the occurrence of residues in **wildlife**, in drinking water, and in humans. Ecological damage has included the poisoning of wildlife, especially avian predators.

DDT and other chlorinated hydrocarbons

Chlorinated hydrocarbons are a diverse group of synthetic compounds of **carbon**, **hydrogen**, and **chlorine**, used as **pesticides** and for other purposes. DDT is a particular chlorinated hydrocarbon with the formula 2,2-bis-(*p*-chlorophenyl)-1,1,1-trichloroethane.

The insecticidal relatives of DDT include DDD, aldrin, dieldrin, heptachlor, and methoxychlor. DDE is a related non-insecticidal chemical, and an important, persistent, metabolic-breakdown product of DDT and DDD that accumulates in organisms. Residues of DDT and its relatives are persistent in the environment, for example, having a typical **half-life** of 5-10 years in **soil**.

A global contamination with DDT and related chlorinated hydrocarbons has resulted from the combination of their persistence and a tendency to become widely dispersed with wind-blown dusts. In addition, their selective partitioning into fats and lipids causes these chemicals to bioaccumulate. Persistence, coupled with **bioaccumulation**, results in the largest concentrations of these chemicals occurring in predators near or at the top of ecological food webs.

Uses of DDT

DDT was first synthesized in 1874. Its insecticidal qualities were discovered in 1939 by Paul Muller, a Swiss scientist who won a Nobel Prize in medicine in 1948 for his research on the uses of DDT. The first important use of DDT was for the control of insect vectors of human diseases during and following World War II. At about that time the use of DDT to control insect **pests** in agriculture and **forestry** also began.

The peak production of DDT was in 1970 when 386 million lb (175 million kg) were manufactured globally. The greatest use of DDT in the United States was 79 million lb (36 million kg) in 1959, but the maximum production was 199 million lb (90 million kg) in 1964, most of which was exported. Because of the discovery of a widespread environmental contamination with DDT and its breakdown products, and associated ecological damage, most industrialized countries banned its use in the early 1970s. The use of DDT continued elsewhere, however, mostly for control of insect vectors of human and **livestock** diseases in less-developed, tropical countries. Largely because of the **evolution** of resistance to DDT by many pest **insects**, its effectiveness for these purposes has decreased. Some previously well-controlled diseases such as **malaria** have even become more common recently in a number of countries (the reduced effectiveness of some of the prophylactic pharmaceuticals used to threat malaria is also importance in the resurgence of this **disease**). Eventually, the remaining uses of DDT will probably be curtailed and it will be replaced by other **insecticides**, largely because of its increasing ineffectiveness.

Until its use was widely discontinued because of its non-target, ecological damages, DDT was widely used to kill insect pests of **crops** in agriculture and forestry and to control some human diseases that have specific insect vectors. The use of DDT for most of these pest-control purposes was generally effective. To give an indication of the effectiveness of DDT in killing insect pests, it will be sufficient to briefly describe its use to reduce the incidence of some diseases of humans.

In various parts of the world, **species** of insects and ticks are crucial as vectors in the transmission of disease-causing **pathogens** of humans, livestock, and wild animals. Malaria, for example, is a debilitating disease caused by the protozoan *Plasmodium* and spread to people by **mosquitoes**, especially species of *Anopheles*. **Yellow fever** and related viral diseases such as **encephalitis** are spread by other species of mosquitoes. The incidence of these and some other important diseases can be greatly reduced by the use of insecticides to reduce the abundance of their arthropod vectors. In the case of mosquitoes, this can be accomplished by applying DDT or another suitable insecticide to the aquatic breeding **habitat**, or by applying a persistent insecticide to walls and ceilings which serve as resting places for these insects. In other cases, infestations of body **parasites** such as the human louse can be treated by dusting the skin with DDT.

The use of DDT has been especially important in reducing the incidence of malaria, which has always been an important disease in warmer areas of the world. Malaria is a remarkably widespread disease, affecting more than 5% of the world's population each year during the 1950s. For example, in the mid-1930s an **epidemic** in Sri Lanka affected one-half of the population, and 80,000 people died as a result. In **Africa**, an estimated 2-5 million children died of malaria each year during the early 1960s.

The use of DDT and some other insecticides resulted in large decreases in the incidence of malaria by greatly reducing the abundance of the mosquito vectors. India, for example, had about 100 million cases of malaria per year and 0.75 million deaths between 1933 and 1935. In 1966, however, this was reduced to only 0.15 million cases and 1,500 deaths, mostly through the use of DDT. Similarly, Sri Lanka had 2.9 million cases of malaria in 1934 and 2.8 million in 1946, but because of the effective use of DDT and other insecticides there were only 17 cases in 1963. During a vigorous campaign to control malaria in the tropics in 1962, about 130 million lb (59 million kg) of DDT was used, as were 7.9 million lb (3.6 million kg) of dieldrin and one million lb (0.45 million kg) of lindane. These insecticides were mostly sprayed inside of homes and on other resting habitat of mosquitoes, rather than in their aquatic breed-

DDT sperulites magnified 50 times. *Photograph by David Malin. Photo Researchers, Inc. Reproduced by permission.*

ing habitat. More recently, however, malaria has resurged in some tropical countries, largely because of the development of insecticide resistance by mosquitoes and a decreasing effectiveness of the pharmaceuticals used to prevent the actual disease.

Environmental effects of the use of DDT

As is the case with many actions of environmental management, there have been both benefits and costs associated with the use of DDT. Moreover, depending on socio-economic and ecological perspectives, there are large differences in the perceptions by people of these benefits and costs. The controversy over the use of DDT and other insecticides can be illustrated by quoting two famous persons. After the successful use of DDT to prevent a potentially deadly plague of **typhus** among Allied troops in Naples during World War II, Winston Churchill praised the chemical as "that miraculous DDT powder." In stark contrast, Rachael Carson referred to DDT as the "elixir of death" in her ground-breaking book *Silent Spring*, which was the first public chronicle of the ecological damage caused by the use of persistent insecticides, especially DDT.

DDT was the first insecticide to which large numbers of insect pests developed genetically based resistance. This happened through an evolutionary process involving **selection** for resistant individuals within large populations of pest organisms exposed to the toxic pesticide. Resistant individuals are rare in unsprayed populations, but after spraying they become dominant because the insecticide does not kill them and they survive to reproduce and pass along their genetically based tolerance. More than 450 insects and **mites** have populations that are resistant to at least one insecticide. Resistance is most common in the **flies** (Diptera), with more than 155 resistant species, including 51 resistant species of malaria-carrying mosquito, 34 of which are resistant to DDT.

As mentioned previously, the ecological effects of DDT are profoundly influenced by certain of its physical/chemical properties. First, DDT is persistent in the environment because it is not readily degraded to other chemicals by **microorganisms**, sunlight, or **heat**. Moreover, DDE is the primary breakdown product of DDT, being produced by enzymatic **metabolism** in organisms or by inorganic de-chlorination reactions in alkaline environments. The persistences of DDE and DDT are similar, and once released into the environment these chemicals are present for many years.

Another important characteristic of DDT is its insolubility in water, which means that it cannot be "diluted" into this ubiquitous solvent, so abundant in Earth's environments and in organisms. In contrast, DDT is highly soluble in fats (or lipids) and oils, a characteristic shared with other chlorinated hydrocarbons. In ecosystems, most lipids occur in the tissues of living organisms. Therefore, DDT has a strong affinity for organisms because of its high lipid solubility, and it tends to biomagnify tremendously. Furthermore, top predators have especially large concentrations of DDT in their fat, a phenomenon known as food-web accumulation. In ecosystems, DDT and related chlorinated hydrocarbons occur in extremely small concentrations in water and air. Concentrations in soil may be larger because of the presence of organic **matter** containing some lipids. Larger concentrations occur in organisms, but the residues in plants are smaller than in herbivores, and the highest concentrations occur in predators at the top of the food web, such as humans, predatory **birds**, and marine **mammals**. For example, DDT residues were studied in an estuary on Long Island where DDT had been sprayed onto **salt** marshes to kill mosquitoes. The largest concentrations of DDT occurred in fish-eating birds such as ring-billed gull (76 ppm), and double-crested cormorant, red-breasted merganser, and herring gull (range of 19-26 ppm).

Lake Kariba, Zimbabwe, is a tropical example of food-web bioconcentration of DDT. Although Zimbabwe banned DDT use in agriculture in 1982, it is still used to control mosquitoes and tsetse fly (a vector of diseases of cattle and other large mammals). The **concentration** of DDT in water of Lake Kariba was extremely small, less than 0.002 ppb, but larger in sediment of the lake (0.4 ppm). **Algae** contained 2.5 ppm, and a filter-feeding mussel contained 10 ppm in its lipids. Herbivorous **fish** contained 2 ppm, while a bottom-feeding species of fish contained 6 ppm. The tigerfish and cormorant (a bird) feed on small fish, and these contained 5 ppm and 10

ppm, respectively. The top **predator** in Lake Kariba is the Nile crocodile, and it contained 34 ppm. Lake Kariba exhibits a typical pattern for DDT and related chlorinated hydrocarbons; a large bio-concentration from water, and to a lesser degree from sediment, as well as a food-web magnification from herbivores to top predators.

Global contamination with DDT

Another environmental feature of DDT is its ubiquitous distribution in at least trace concentrations everywhere in the **biosphere**. This global contamination with DDT, and related chlorinated hydrocarbons such as PCBs, occurs because they enter into the atmospheric cycle and thereby become very widely distributed. This results from: (1) a slow **evaporation** of DDT from sprayed surfaces; (2) off-target drift of DDT when it is sprayed; and (3) entrainment by strong winds of DDT-contaminated dust into the atmosphere.

This ubiquitous contamination can be illustrated by the concentrations of DDT in animals in **Antarctica**, very far from places where it has been used. DDT concentrations of 5 ppm occur in fat of the southern polar skua, compared with less than 1 ppm in birds lower in the food web of the Southern Ocean such as the southern fulmar and species of penguin.

Much larger concentrations of DDT and other chlorinated hydrocarbons occur in predators living closer to places where the chemicals have been manufactured and used. The concentration of DDT in **seals** off the California coast was as much as 158 ppm in fat during the late 1960s. In the Baltic Sea of **Europe** residues in seals were up to 150 ppm, and off eastern Canada as much as 35 ppm occurred in seals and up to 520 ppm in porpoises.

Large residues of DDT also occur in predatory birds. Concentrations as high as 356 ppm (average of 12 ppm) occurred in bald **eagles** from the United States, up to 460 ppm in western **grebes**, and 131 ppm in herring **gulls**. White-tailed eagles in the Baltic Sea have had enormous residues—as much as 36,000 ppm of DDT and 17,000 ppm PCBs in fat, and eggs with up to 1,900 ppm DDT and 2,600 ppm PCBs.

Ecological damage

Some poisonings of wildlife were directly caused by exposure to sprays of DDT. There were numerous cases of dying or dead birds being found after the spraying of DDT, for example, after its use in residential areas to kill the beetle vectors of Dutch **elm** disease in **North America**. Spray rates for this purpose were large, about 1.5-3.0 lb (1.0-1.5 kg) of DDT per **tree**, and resulted in residues in earthworms of 33-164 ppm. Birds that fed on DDT-laced **invertebrates** had intense exposures to DDT, and many were killed.

Sometimes, detailed investigations were needed to link declines of bird populations to the use of organochlorines. One such example occurred at Clear Lake, California, an important waterbody for recreation. Because of complaints about the nuisance of a great abundance of non-biting aquatic insects called midges, Clear Lake was treated in 1949 with DDD at 1 kg/ha. Prior research had shown that this dose of DDD would achieve control of the midges but would have no immediate effect on fish. Unfortunately, the unexpected happened. After another application of DDD in 1954, 100 western grebes were found dead as were many intoxicated birds. Eventually, the breeding population of these birds on Clear Lake decreased from about 2,000 to none by 1960. The catastrophic decline of grebes was linked to DDD when an analysis of the fat of dead birds found residues as large as 1,600 ppm. Fish were also heavily contaminated. The deaths of birds on Clear Lake was one of the first well documented examples of a substantial mortality of wildlife caused by organochlorine insecticides.

Damage to birds also occurred in places remote from sprayed areas. This was especially true of raptorial (that is, predatory) birds, such as **falcons**, eagles, and **owls**. These are top predators, and they food-web accumulate chlorinated hydrocarbons to high concentrations. Declines of some species began in the early 1950s, and there were extirpations of some breeding populations. Prominent examples of predatory birds that suffered population declines from exposure to DDT and other organochlorines include the bald eagle, golden eagle, **peregrine falcon**, **prairie falcon**, osprey, brown pelican, double-crested cormorant, and European sparrowhawk.

Of course, birds and other wildlife were not only exposed to DDT. Depending on circumstances, there could also be significant exposures to other chlorinated hydrocarbons, including DDD, aldrin, dieldrin, heptachlor, and PCBs. Scientists have investigated the relative importance of these chemicals in causing the declines of predatory birds. In Britain, the declines of **raptors** did not occur until dieldrin came into common use, and this insecticide may have been the primary cause of the damage. However, in North America DDT use was more common, and it was probably the most important cause of the bird declines there.

The damage to birds was mainly caused by the effects of chlorinated hydrocarbons on reproduction, and not by direct toxicity to adults. Demonstrated effects of these chemicals on reproduction include: (1) a decrease in clutch size (i.e., the number of eggs laid); (2) the production of a thin eggshell which might break under the incubating parent; (3) deaths of embryos, unhatched chicks, and nestlings; and (4) pathological parental **behavior**. All of these effects could decrease the numbers of young successfully raised. The reproductive **pathology** of chlorinated hydrocarbons caused bird populations to decrease because of inadequate recruitment.

This **syndrome** can be illustrated by the circumstances of the peregrine falcon, a charismatic predator whose decline attracted much attention and concern. Decreased reproductive success and declining populations of peregrines were first noticed in the early 1950s. In 1970, a North American census reported almost no successful reproduction by the eastern population of peregrines, while the arctic population was declining in abundance. Only a local population in the Queen Charlotte Islands of western Canada had normal breeding success and a stable population. This latter population is non-migratory, inhabiting a region where pesticides are not used and feeding largely on non-migratory seabirds. In contrast, the eastern peregrines bred where chlorinated hydrocarbon pesticides were widely used, and its **prey** was generally contaminated. Although the arctic peregrines breed in a region where pesticides are not used, these birds winter in sprayed areas in Central and **South America** where their food is contaminated, and their prey of migratory **ducks** on the breeding grounds is also contaminated. Large residues of DDT and other organochlorines were common in peregrine falcons (except for the Queen Charlottes). Associated with those residues were eggshells thinner than the pre-DDT condition by 15-20% and a generally impaired reproductive **rate**.

In 1975, another North American survey found a virtual extirpation of the eastern peregrines, while the arctic population had declined further and was clearly in trouble. By 1985 there were only 450 pairs of arctic peregrines, compared with the former abundance of 5,000-8,000. However, as with other raptors that suffered from the effects of chlorinated hydrocarbons, a recovery of peregrine populations has begun since DDT use was banned in North America and most of Europe in the early 1970s. In 1985, arctic populations were stable or increasing compared with 1975, as were some southern populations, although they remained small. This recovery has been enhanced by a captive-breeding and release program over much of the former range of the eastern population of peregrine falcons.

It is still too soon to tell for certain, but there are encouraging signs that many of the severe effects of DDT and other chlorinated hydrocarbons on wildlife are becoming less severe. Hopefully, in the future these toxic damages will not be important.

See also Biomagnification.

Resources

Books

Freedman, B. *Environmental Ecology.* 2nd ed. San Diego: Academic Press, 1994.

Smith, R.P. *A Primer of Environmental Toxicology.* Philadelphia: Lea & Febiger, 1992.

Bill Freedman

de Broglie wavelength *see* **Quantum mechanics**

Deafness and inherited hearing loss

Deafness is the lack of functional sense of **hearing** in both ears. Loss of hearing can result from environmental or genetic causes and it can be temporary or permanent.

Environmental loss of hearing results from occupational noise, **noise pollution**, accidents, or intake of certain drugs. Inherited loss of hearing can be caused by mutations in any of over a hundred of genes known to affect hearing, and can affect various cells of the inner **ear**, the main hearing center.

The inner ear contains the **organ** of Corti, which has hair cells that are responsible for converting sounds to neuronal signals. These cells possess receptors responsive to mechanical movement, stereocilia, which are long, hair like-extensions. Stereocilia are surrounded by a liquid (endolymph) containing high **concentration** of potassium ions, while the main channels of the inner ear are filled with a liquid (perilymph) rich in **sodium** ions. Changes in **pressure** resulting from a sound **stimulus**, are passed through the perilymph and result in a mechanical distortion of the stereocilia. This induces transfer of potassium ions into the hair cells and stimulation of the neurons.

Hearing loss in the inherited cases of deafness can be syndromic or non-syndromic. Deafness or profound hearing loss is often associated with other genetic syndromes (for example Waardenburg or Pendred **syndrome**). These account for about 30% of all inherited hearing loss. In these patients the mutations can affect the inner ear (sensineuronal) or sound transmission through the outer and middle ear (conductive), or can affect both. The remainder of the inherited cases are non-syndromic (patients do not exhibit any other symptoms except for the loss of hearing).

Knowledge about the role of particular genes and general metabolic and signal transduction pathways in the cells of the inner ear is very limited. As a result, the genes causing the non-syndromic hearing loss are divided into artificial groups based on the mode of transmission: autosomal recessive (77%), autosomal dominant (22%), X **chromosome** linked (around 1%) and mitochondrial (less than 1%).

Hearing loss in case of recessive mutations is usually **congenital** as the patient receives two copies of the same **mutation** from the parents, resulting in complete absence of a functional protein. In contrast, the dominant mutations can be transmitted by only one parent, resulting in one good and one bad copy of the **gene** (two copies can be obtained as well if both parents are affected and have the same mutation in the same gene). Such patients usually exhibit progressive hearing loss, as initially there is some functional protein.

Defects that result in deafness or progressive hearing loss can affect different parts of the inner ear, hair cells, non-sensory cells, and the tectorial **membrane**.

A large number of mutations causing deafness affect the **proteins** essential for proper function of the organ of Corti. The changes found in the hair cells affect mainly the motor proteins, such as myosins (myosin 7A, 6 and 15), or interacting with actin filaments (espin), but also potassium (KCNQ4) and **calcium** ion transporters, cadherins (vezatin and harmonin), and a transcription factor (POU4F3). This results in the changed structure of the hair cells, loss of their mechanical ability to stretch and distort, and affects ion-dependent signalling. Another mechanical part of the inner ear is the tectorial membrane. This gelatinous structure composed of proteins of the extracellular matrix is bound to the stereocilia of some hair cells. Mutations that affect it are found mainly in the proteins of the extracellular matrix (**collagen** 11 or tectoin). These changes are thought to affect the mechanical and structural properties of the membrane.

Even the hearing of infants can be tested. *Photograph by James King-Holmes. Photo Researchers, Inc.*

The non-sensory cells are an important part of the inner ear, forming tight barriers preventing mixing of the potassium and sodium rich liquids, and supporting the hair cells. The mutations in these cells affect mainly the gap and tight junction proteins (connexin 26, 30, 31 and claudin 14) responsible for cell-to-cell contact, but also ion transporters (calcium and potassium), and a transcription factor (POU3F4).

Other genes that are mutated have not been yet localised to a particular **cell** type. Moreover, it has been suggested that there are certain genes (modifiers) increasing the susceptibility of age-related or noise-induced hearing loss. This could also explain why the mutations in the same gene result in very different severity of hearing loss.

See also Genetic disorders.

Resources

Periodicals

Avraham, Karen B. "Modifying with Mitochondria." *Nature Genetics* (February 2001): 136-137.

Petit, Christine, Jacqueline Levilliers, and Jean-Pierre Hardelin. "Molecular Genetics of Hearing Loss." *Annual Review of Economics and Human Genetics.* (2001): 589.646.

Steel, Karen P., and Corne J. Kros, "A Genetic Approach to Understanding Auditory Function." *Nature Genetics* (February 2001): 143-149.

Other

Atlantic Coast Ear Specialists, P.C. "Anatomical Tour Of The Ear." [cited January 22, 2003] <http://www.earaces.com/anatomy.htm>.

Van Camp, Guy, and Richard Smith. *The Hereditary Hearing Loss Homepage* [cited January 22, 2003]. <http://www.uia.ac.be/dnalab/hhh>.

Agnieszka Lichanska

Decimal fraction

A decimal fraction is a numeral that uses the numeration system, based on 10, to represent numbers that are not necessarily whole numbers. The numeral includes a dot, called the decimal point.

The digits to the right of the decimal point extend the place-values to tenths, hundredths, thousandths, and so forth. For example, the decimal fraction 5.403 means "5 ones, 4 tenths, 0 hundredths, and 3 thousandths." The same number can also be represented by a common fraction, such as 5403/1000, or as a mixed number, 5 403/1000.

See also Fraction, common.

Fungus creates circular patterns on decomposing maple leaves. *CORBIS/Gary Braasch. Reproduced by permission.*

Decomposition

Decomposition is the natural process by which large organic materials and molecules are broken down into simpler ones. The ultimate products of decomposition are simple molecules, such as **carbon dioxide** and **water**. Sometimes misunderstood as being undesirable, decomposition is actually an extremely vital ecological process. Living organisms are composed of cells and tissues, which are in turn made of complex organic molecules, including many that are large by molecular standards. Such large molecules are termed, macromolecules. Examples of macromolecules include **cellulose**, which comprises **plant cell** walls, triglyceride fats within **animal** cells, and **proteins** of plants, animals, **fungi**, protozoans, and **bacteria**. While alive, cells and whole organisms constantly maintain and add to the macromolecules necessary for life, in effect counteracting decomposition. Upon death, however, such maintenance and growth functions cease. Decomposition, then, begins the process of breaking cells down into their component molecules, and macromolecules into simpler organic and inorganic molecules. If decomposition did not occur, the world would be overcome with mountainous piles of dead **biomass**.

Decomposition is a process that recycles **nutrients** back to the **soil** from formerly living organisms. The process can involve soil organisms breaking-down large pieces of organic **matter** into smaller ones. Earthworms, **insects**, and **snails** are examples of animals involved in the initial stages of the decomposition process. Detritus is the term given to the disintegrated organic material produced by these animals. Earthworms are examples of detritivores, or organisms that consume detritus for **energy**. After larger particles are broken down, **microorganisms** further the decomposition process by secreting chemicals that digest organic material in detritus. The most prominent organisms that do this are bacteria and fungi. Bacteria and fungi that thrive in soil and feed upon dead organic matter are called saprophytes. Detritivores and saprophytes are essential in the **recycling** and disintegration processes of decomposition. The partially digested organic material left in soil, called **humus**, is then available for plants to use.

Humus is a soil component essential for plant growth. Found largely in topsoil, humus is created from dead living material in a process called humification. Humification of dead plant matter, for example, involves not only decomposition processes of detritivores, but also the physical action of **weathering** such as freezing, thawing, drying, and **erosion**. Humus is a major source of nutrients for plants. Essential **minerals** slowly leach from humus into the surrounding soil water, which are then absorbed by plant roots. Acting somewhat like a sponge, humus also helps retain soil moisture, while simultaneously keeping soil aerated by preventing compaction.

Humans can also make use of the natural process of decomposition. **Composting** is the gathering of waste organic material, most often plant material, into an aerated pile to facilitate partial decomposition into humus. The organic humus can then be used as a soil conditioner and fertilizer for gardens or on agricultural land. Normally, the process of decomposition can take a long time. However, in compost piles, the decomposition of organic matter is accelerated by turning to enhance **oxygen** availability, and by the build-up of **heat** and moisture. In compost piles, the action of saprophytes creates heat, which helps accelerate the entire process. The center of the compost pile is insulated from the exterior by the outer debris, and retains moisture well.

Terry Watkins

Deer

Deer are members of the order Artiodactyla, the even-toed **ungulates**. This order also includes the antelopes, bovines, and giraffes. Deer are generally slender and long-legged, and their most striking characteristic is the presence of antlers, which are often used to differentiate **species**.

The deer family, Cervidae, includes about 45 species, which are divided among 17 genera and five subfamilies: the Hydropotinae, the Chinese water deer; the Muntiacinae, the muntjacs of **Asia**; the Cervinae, the Eurasian deer; the Odocoleinae, the New World deer, **moose**, and **caribou**; and the Moschinae, the musk deer of China, Southeast Asia, and the Himalayas. Some taxonomists argue that the Moschinae should not be considered a subfamily of the Cervidae, but an entirely separate family (Moschidae), based on the relatively pronounced differences between *Moschus* and other deer. Unlike other deer, *Moschus* has a gall bladder, and where the females of other species have two pairs of teats, *Moschus* has only one pair.

Deer have short hair ranging in **color** from yellowish to dark brown. The underbelly and throat are lighter colored, and many species have a distinctive rump patch, an area of light hair fringed with darker hair. (A startled deer will lift its tail and flash the white of its rump patch as an alarm to other deer nearby.) The head of deer is angular, with the eyes set well on the side. The ears are oblong and the nose is usually covered with soft hair. The senses of **hearing** and **smell** are excellent. **Vision** is less so, as far as giving the **animal** an accurate picture of the world around it. Although a deer cannot accurately perceive form at distances greater than about 200 ft (60 m), it can detect slight movements up to 1,000 ft (300 m) away.

Besides the flash of the rump patch, deer communicate through sound and smell, and they produce a variety of vocalizations, from the roar of the red deer to the **bark** of the muntjac. Deer also have scent **glands** near their eyes, which they use to mark their territory on branches and twigs. Dung is also used as a territorial marker. Males will sniff a female's urine to learn if she is in estrus.

The legs of deer are long and slender, well-suited for fast running to escape their many predators. During evolutionary **time**, the leg bones of deer became longer and the weight of the animal became supported entirely on the third and fourth toes, eventually resulting in the **evolution** of cloven hooves. The second and fifth toes are short and positioned up, as dewclaws. The first digit has vanished and the bones of the palm (metacarpals and metatarsals) have been forged into a single bone, the cannon bone. Similar evolutionary changes have occurred in other herbivores that run to escape predators, such as **horses**.

Deer range in size from the *Pudu* (two species, standing 10-17 in [25-43 cm] at the shoulder and weighing 13-29 lb [6-13 kg]) to *Alces*, the moose, which stands 56-94 in (140-235 cm) at the shoulder and weighs 440-1,900 lb (200-850 kg). Most species of deer have antlers, which are usually found only on the males. However, in *Rangifer*, the caribou, both sexes have antlers. Other species, such as the Chinese water deer and the tufted deer, have tusks. Tusked deer are considered to be more evolutionarily ancient than those with antlers, because tusks are characteristic of the primitive chevrotains, or mouse deer. More recent species of deer are generally considered to have larger bodies, larger and more complex antlers, and a more gregarious social system.

Deer originated in Eurasia in the Oligocene, and were present in **North America** in the Miocene, and in **South America** in the Pleistocene. Perhaps the most well-known fossil species is the Irish elk (*Megaloceros gigantus*). Although not as large as the modern moose, *Megaloceros* carried a rack of antlers that had a spread of 6 ft (1.8 m) and weighed more than the rest of the animal's skeleton. Analysis of fossil specimens suggests that the Irish elk was well-suited for life in the open and able to run quickly for long distances. The common name is misleading, for *Megaloceros* was neither an elk nor exclusively Irish, although the first specimens were found in Irish peat bogs. (Long before *Megaloceros* came to the attention of science, the Irish were using its great antlers as gateposts and, in County Tyrone, even as a temporary bridge.)

Deer occur naturally throughout most of the world, with the exception of sub-Saharan **Africa**, **Australia**, and **Antarctica**. As an **introduced species**, deer have thrived in Australia, New Zealand, Cuba, New Guinea, and other places. For large herbivores, they are remarkably adaptable. Although most typically fond of wooded areas, some deer have adapted to semi-aquatic habitats (the Chinese water deer and moose), open **grasslands** (the Pampas deer of South America), and the arctic **tundra** (the caribou). Slowly, deer are returning to areas frequented by humans; in suburban America, white-tailed deer are becoming a common backyard sight, and throughout the **mountains** of New Hampshire and Maine roadsigns warn of moose crossing.

Deer are herbivores. Lacking upper incisors, they bite off forage by pressing their lower incisors against a callous pad on the upper gum. Their teeth have low crowns, well-suited to their diet that, depending on the species, includes twigs, leaves, aquatic plants, **fruits**, **lichens**, and grass. During hard winters, white-tailed deer may strip and eat the bark from trees. Furthermore, some temperate species, such as the white-tailed deer, actually alter their **metabolism** during the winter, lessening their need for food and therefore decreasing the likelihood of starvation. So strong is this natural **adaptation**, that even in captivity these species will eat less in the winter, even if the amount of food available remains constant.

Like other artiodactyls, deer are ruminants with a four-chambered stomach. When food is first eaten, it is stored in the first chamber, the rumen, where **bacteria** begin to break it down. It is later regurgitated into the mouth and, as a cud, is chewed again and mixed with saliva. When swallowed a second time, the food bypasses the rumen and goes into the second stomach chamber, the reticulum, and then passes into the third chamber, the omasum, and then into the fourth chamber, the abomasum. Food then moves into the small intestine, where **nutrients** are absorbed. Although this entire process takes about 80 hours, it converts about 60% of the **cellulose** in the food into usable sugars and thus is remarkably effective.

Deer vary their diet depending on the seasonal availability of forage and their nutritional needs. Fallow deer, for instance, eat a great deal of grass; it comprises about 60% of their diet during the summer. In the fall, there is less grass but more fruit is available, such as acorns. As the grass proportion of their diet declines, the deer turn to fruit, which at the height of fall makes up almost 40% of their food intake. The winter diet consists of browse, that is, woody stems of shrubs such as ivy and holly.

A three-year study of moose living on Isle Royale, Michigan, determined three major limiting factors on what moose could eat to satisfy their nutritional requirements. First was the capacity of the rumen, second was the time available for feeding, and third was the need for **sodium**, an important nutrient that is difficult to obtain on this glacier-scrubbed **island** in Lake Superior. Researchers calculated that to meet their needs, the moose would have to eat particular amounts of both terrestrial plants and higher-sodium aquatic plants each day. Remarkably, the observed diet of the moose in the study matched the scientists' predictions.

Like all herbivores, deer must spend a great deal of time eating in order to obtain sufficient **nutrition** from their food. Studies of wild red deer in Scotland found that females without calves spent 9.8 hours foraging each summer day, while the larger males spent 10.4 hours. Lactating females with calves spent 11.1 hours per day feeding. Spending large amounts of time feeding makes deer vulnerable to predators, but the tendency to herd, and the ability to eat fast and store food in the rumen, help make them less vulnerable.

In North America, predators of adult deer include the brown bear, bobcat, cougar, coyote, wolf, **wolverine**, and packs of roving domesticated dogs, while golden **eagles** sometimes take young deer. In South America, deer are taken by jaguars. Eurasian deer must deal with dholes (wild dogs), tigers, and wolves. One reptilian **predator**, the Komodo dragon of Indonesia, depends largely on the Timor hog deer (*Cervus rusak timoensis*). Deer have long been hunted by humans as well. Other causes of death include fighting between males, **automobile** and train accidents, falling through **ice** and drowning, becoming entangled in fences or stuck in the crotches of trees when reaching high for browse, being caught in forest fires, becoming stuck in swampy areas, and falling over snow-covered banks or cliffs. Many of the deer shot by hunters escape only to die of their wounds later. Particularly harsh winters also decimate deer populations.

Deer antlers are found primarily in the males and are a social and sexual symbol as well as a weapon. The huge antlers of the Irish elk were the long-term result of sexual **selection**, whereby females consistently bred with males that had the largest antlers. This can explain how the **gene** for larger and larger antlers was passed down through generations until the tremendous 6 ft (1.8 m) span was reached.

Antlers differ from horns. Horns are a permanent outgrowth of the skull and are covered by a layer of keratin.

A wapiti (*Cervus elaphus*), or red deer, in Yellowstone National Park, Wyoming. Wapiti is a Native American word for "white" and refers to the light colored rump of this species of deer. *Photograph by Robert J. Huffman. Field Mark Publications. Reproduced by permission.*

Antlers, on the other hand, are grown and shed annually. They consist of a bare bony core supported on bony disks, called pedicles, that are part of the skull. There is a tremendous investment of **energy** in the regrowth of antlers, which regrow to be more elaborate with each year as the deer ages, adding more prongs, or "points." Antlers are often damaged during mating-season fights, which seriously curtail a male's reproductive success. A study of red deer males with damaged antlers showed that they had less mating success than did males with intact racks. However, the experimental removal of the antlers of a dominant male showed it to be only a temporary setback; despite the loss, the male retained his status and his females.

The antlers of temperate-region species of deer begin to grow during early summer. The growing antlers are covered by a thin layer of skin covered by short, fine hairs. Aptly called velvet, this skin nourishes the antlers with a plentiful supply of **blood** until they reach full growth in late summer. The **blood supply** then ceases, and the velvet dries up. The deer rubs off the velvet to reveal fresh new and, for a short time after shedding the velvet, gory antlers.

Those species of deer considered to be more evolutionarily recent are generally more gregarious, but the diet of these species may also be related to social organization. Those deer that primarily browse, such as roe deer, live in small groups or alone, for their food is generally found only in small patches. On the other hand, caribou, which graze on lichens and **sedges** over extensive open areas, may occur in large herds of several thousand animals. Such grazers may find an extra benefit in their herding **behavior**, with extra eyes and ears alert for predators.

Mating strategies

During the mating season of temperate species, males use one of three strategies to obtain access to receptive females. They may defend a territory that overlaps the ranges of females, as does the muntjac. They may defend a single doe against all suitors, as does the white-tailed deer. Or they may attempt to assemble and hold a harem of females, as does the red deer (*Cervus elaphus*). The males and females of this gregarious species spend most of the year in single-sex herds, which generally have particular ranges. Come September, the females gather in rutting areas, and are soon joined by the males, which compete for the females through displays of roaring, spraying urine, and fighting.

Fighting begins when the challenger appears, and he and the holder of the harem roar at each other. After several minutes of vocalizing, they walk **parallel** to each other, tense and alert, until one of them turns toward the other and lowers his antlers. They lock antlers and begin shoving each other. When one succeeds in pushing the other backwards, the loser runs off.

The fights are dangerous. Almost a quarter of the males in a Scottish study were injured during the rut, 6% permanently. A male between ages of seven to 10 has the best chance of winning such an encounter, which a harem holder must face about five times during the mating season. There is another danger besides injury; young males often lurk at the fringes of a harem, waiting until the harem holder is distracted and then spiriting away a female or two. The apparent benefits of holding a harem are deceiving; although there may be as many as 20 females in the harem, the male will father only about four or five calves.

Females of tropical species of deer come into estrus several times a year. Gestation lasts from 176 days in the Chinese water deer to 294 days in the roe deer. The female deer delivers from one to six young (six in *Hydropotes*), but one or two is the norm. The young of most deer are born spotted.

The males of the Cervinae (such as the red deer) are called stags, the females, hinds, and the young, calves. Among the Odocoileinae, the male deer are called bucks, the females does, and the young fawns. Exceptions are *Alces* (moose) and *Rangifer* (caribou), in which the males are bulls, the females, cows, and the young, calves.

Besides the moose, other North American species include the white-tailed deer (*Odocoileus virginianus*) found from southern Canada, throughout most of the United States, Mexico, and down to Bolivia and northeastern Brazil. The white-tailed deer may be the most abundant species of wild large mammal, with a population of about 60 million individuals. The mule deer (*O. hemionus*), named for its large ears, ranges from the southern Yukon and Manitoba to northern Mexico. The tiny Key deer (*O. v. clavium*) is an endangered subspecies of the white-tailed deer; only about 250 remain in the western Florida Keys.

F. C. Nicholson

Deer mouse

The deer mouse (*Peromyscus maniculatus*) is a small, native rodent with an almost ubiquitous distribution in **North America**. The deer mouse ranges from the subarctic boreal forest, through wide areas of more southern **conifer** and mixed-wood **forests**, to drier habitats as far south as some regions of Mexico.

The deer mouse is highly variable in size and **color** over its range. Its body length ranges from 2.8 to 3.9 in (7 to 10 cm), the tail 2.0-5.1 in (5-13 cm), and the body weight 0.6-1.2 oz (18-35 g). Many geographic variants of the deer mouse have been described as subspecies. The color of the deer mouse ranges from greyish to reddish brown, with the body being dark above and white beneath. The bicolored coat of these **mice** gives rise to its common name, a reference to a superficial resemblance to the coloration of white-tailed and mule deer (*Odocoileus* spp.). The deer mouse can be difficult to distinguish from some closely related **species**, such as the white-footed mouse (*P. leucopus*), another widely distributed, but more eastern species.

The deer mouse occurs in a very wide range of **habitat** types. This species occurs in deserts, prairies, and forests, but not in **wetlands**. The deer mouse is quite tolerant of certain types of disturbance, and its populations are little affected by light wildfires or the harvesting of trees from its habitat.

The deer mouse nests in burrows dug in the ground, or in stumps or rotting logs. This species also sometimes nests in buildings. Deer mice can climb well, and they do so regularly in certain habitats. The deer mouse is a nocturnal feeder on a wide range of nuts and **seeds**, and when this sort of food is abundant it is stored for leaner times, because deer mice are active all winter. Deer mice also feed on **insects** when they are available.

The home range of deer mice can vary from about 0.5 to 3 acres (0.2 to 1.2 hectares), but this varies with habitat quality and also over **time**, because the abundance of these **rodents** can be somewhat irruptive if food supply is unusually large. Within any year, deer mice are generally most abundant in the late autumn, and least so in the springtime. Deer mice are quite tolerant of each other, and during winter they may **sleep** huddled in a cozy group to conserve **heat**. The typical longevity of a wild **animal** is two years, but deer mice can live for eight years in captivity.

Depending on latitude, the deer mouse breeds from February to November, raising as many as four litters per year of typically three to five young each. Young deer mice are capable of breeding once they are older than five or six weeks. Adult males often assist with rearing their progeny.

When they are abundant, deer mice are important **prey** for a wide range of small predators, such as foxes, **weasels**, **hawks**, **owls**, **snakes**, and other species. In this sense, deer mice and other small **mammals** are critical links in ecological food webs.

Deer mice are sometimes a problem in **forestry**, in situations where they eat large quantities of **tree** seeds and thereby inhibit the natural regeneration of harvested stands. However, deer mice also provide a service to forestry, by eating large numbers of potentially injurious insects, such as sawflies and budworms.

Deer mice may also be considered **pests** when they occur in homes, because they raid stored food and may shred fabrics and furnishings to get material with which to build their nests. In some regions, exposure to the feces of deer mice may result in people developing a potentially deadly **disease** caused by a microorganism known as hantavirus.

However, when closely viewed, deer mice prove to be inquisitive and interesting creatures. Deer mice are readily tamed, and they make lively pets.

Bill Freedman

Deforestation

Deforestation refers to a longer-term conversion of forest to some other kind of **ecosystem**, such as agricultural or urbanized land. Sometimes, however, the term is used in reference to any situation in which **forests** are disturbed, for example by clear-cut harvesting, even if another forest subsequently regenerates on the site. Various human activities result in net losses of forest area and therefore contribute to deforestation. The most important causes of deforestation are the creation of new agricultural land and unsustainable harvesting of trees. In recent decades, deforestation has been proceeding most rapidly in underdeveloped countries of the tropics and subtropics.

The most important ecological consequences of deforestation are: the depletion of the economically important forest resource; losses of **biodiversity** through the clearing of tropical forests; and emissions of **carbon dioxide** with potential effects on **global climate** through an enhancement of Earth's **greenhouse effect**. In some cases, indigenous cultures living in the original forest may be displaced by the destruction of their **habitat**.

Historical deforestation

Ever since the development of agriculture and settlements, humans have converted forest into agroecosystems of various sort, or into urban land. There are numerous references in historical, religious, and anthropo-

logical literature to forests that became degraded and were then lost through overharvesting and conversion. For example, extensive forests existed in regions of the Middle East that are now almost entirely deforested. This can be evidenced by reference in the Bible to such places as the Forest of Hamath, the Wood of Ziph, and the Forest of Bethel, the modern locations of which are now **desert**. The cedars of Lebanon were renowned for their abundance, size, and quality for the construction of buildings and ships, but today they only survive in a few endangered groves of small trees. Much of the deforestation of the Middle East occurred thousands of years ago. However, even during the Crusades of the eleventh century through the thirteenth century, extensive pine forests stretched between Jerusalem and Bethlehem, and some parts of Lebanon had cedar-dominated forests into the nineteenth century. These are all now gone.

Similar patterns of deforestation have occurred in many regions of the world, including most of the Mediterranean area, much of **Europe**, south **Asia**, much of temperate North and **South America**, and, increasingly, many parts of the sub-tropical and tropical world.

Deforestation today

From earliest times to the present, the global extent of deforestation has been about 12%. This loss included a 19% loss of closed forest in temperate and boreal latitudes, and a 5% loss of tropical and subtropical forests.

However, in recent decades the dynamics of deforestation have changed greatly. The forest cover in wealthier countries of higher latitudes has been relatively stable. In fact, regions of western Europe, the United States, and Canada have experienced an increase in their forest cover as large areas of poorer-quality agricultural land have been abandoned and then regenerated to forest. Although these temperate regions support large forest industries, post-harvest regeneration generally results in new forests, so that ecological conversions to agriculture and other non-forested ecosystems do not generally occur.

In contrast, the **rate** of deforestation in tropical regions of Latin America, **Africa**, and Asia have increased alarmingly in recent decades. This deforestation is driven by the rapid growth in size of the human population of these regions, with the attendant needs to create more agricultural land to provide additional food, and to harvest forest **biomass** as fuel. In addition, increasing globalization of the trading economy has caused large areas of tropical forest to be converted to agriculture to grow **crops** for an export market in wealthier countries, often to the detriment of local people.

In 1990, the global area of forest was 4.23 billion acres (1.71 billion ha), equivalent to 91% of the forest

Deforestation in the jungle in Brazil. *Photograph by Ulrike Welsch. National Audubon Society Collection/Photo Researchers, Inc. Reproduced by permission.*

area existing in 1980. This represents an annual rate of change of about -0.9% per year, which if projected into the future would result in the loss of another one-half of Earth's remaining forest in only 78 years. During this period of time deforestation (indicated as percent loss per year) has been most rapid in tropical regions, especially West Africa (2.1%), Central America and Mexico (1.8%), and Southeast Asia (1.6%). Among nations, the most rapid rates of deforestation are: Côte d'Ivoire (5.2%/year), Nepal (4.0%), Haiti (3.7%), Costa Rica (3.6%), Sri Lanka (3.5%), Malawi (3.5%), El Salvador (3.2%), Jamaica (3.0%), Nicaragua (2.7%), Nigeria (2.7%), and Ecuador (2.3%).

These are extremely rapid rates of national deforestation. A rate of forest loss of 2% per year translates into a loss of one-half of the woodland area in only 35 years, while at 3%/year the **half-life** is 23 years, and at 4%/year it is 18 years.

Past estimates of global deforestation have been criticized as unreliable. These surveys of forest changes, compiled by the United Nations Food and Agriculture Organization (FAO), have been considered to be problematic because of inconsistencies in the collection methodology. Problems included potentially biased information (frequently from agencies within the country itself), inconsistent definitions of **land use**, and data gathering techniques that changed from survey to survey. These issues are being addressed through the use of **remote sensing** techniques. **Satellite** imaging of the forests is now being used to produce consistent and verifiable information on a global scale. Scientists and policy-makers involved with the issue of deforestation rely on dependable and accurate data. This reliable information permits them to monitor changes and accurately determine the extent of the forest.

Loss of a renewable resource

Potentially, forests are a renewable natural resource that can be sustainably harvested to gain a number of economically important products, including lumber, pulp for the manufacture of **paper**, and fuelwood to produce **energy**. Forests also provide habitat for game **species** and also for the much greater diversity of animals that are not hunted for sport or food. In addition, forests sustain important ecological services related to clean air and **water** and the control of **erosion**.

Any loss of forest area detracts from these important benefits and represents the depletion of an important natural resource. Forest harvesting and management can be conducted in ways that encourage the regeneration of another forest after a period of recovery. However, this does not happen in the cases of agricultural conversion and some types of unsustainable forest harvesting. In such cases, the forest is "mined" rather than treated as a renewable natural resource, and its area is diminished.

Deforestation and biodiversity

At the present time, most of Earth's deforestation involves the loss of tropical forests, which are extremely rich in species. Many of the species known to occur in tropical forests have local (or **endemic**) distributions, so they are vulnerable to **extinction** if their habitat is lost. In addition, tropical forests are thought to contain millions of additional species of plants, animals, and **microorganisms** as yet undiscovered by scientists.

Tropical deforestation is mostly caused by various sorts of conversions, especially to subsistence agriculture, and to market agriculture for the production of export commodities. Tropical deforestation is also caused by unsustainable logging and fuelwood harvesting. Less important causes of tropical deforestation include hydroelectric developments that flood large reservoirs and the production of charcoal as an industrial fuel. Because these extensive conversions cause the extinction of innumerable species, tropical deforestation is the major cause of the global biodiversity crisis.

Deforestation and the greenhouse effect

Mature forests contain large quantities of organic **carbon**, present in the living and dead biomass of plants, and in organic **matter** of the forest floor and **soil**. The quantity of carbon in mature forests is much larger than in younger, successional forests, or in any other type of ecosystem, including human agroecosystems. Therefore, whenever a mature forest is disturbed or cleared for any purpose, it is replaced by an ecosystem containing a much smaller quantity of carbon. The difference in carbon content of the ecosystem is balanced by an **emission** of carbon dioxide (CO_2) to the atmosphere. This CO_2 emission always occurs, but its rate can vary. The CO_2 emission is relatively rapid, for example, if the biomass is burned, or much slower if resulting timber is used for many years and then disposed into an **anaerobic landfill**, where biological **decomposition** is very slow.

Prior to any substantial deforestation caused by human activities, Earth's vegetation stored an estimated 990 billion tons (900 billion metric tons) of carbon, of which 90% occurred in forests. Mostly because of deforestation, only about 616 billion tons (560 billion metric tons) of carbon are presently stored in Earth's vegetation, and that quantity is diminishing further with time. It has been estimated that between 1850 and 1980, CO_2 emissions associated with deforestation were approximately equal to emissions associated with the **combustion** of **fossil fuels**. Although CO_2 emissions from the use of fossil fuels has been predominant in recent decades, continuing deforestation is an important source of releases of CO_2 to the atmosphere.

The CO_2 **concentration** in Earth's atmosphere has increased from about 270 ppm prior to about 1850, to about 360 ppm in 1999, and it continues to increase. Many atmospheric scientists hypothesize that these larger concentrations of atmospheric CO_2 will cause an increasing intensity of an important process, known as the greenhouse effect, that interferes with the rate at which **Earth** cools itself of absorbed solar **radiation**. If this theory proves to be correct, then a climatic warming could result, which would have enormous implications for agriculture, natural ecosystems, and human civilization.

Causes of deforestation

Any solution to the problem of deforestation must first address the social and economic reasons for the activity. While population growth and social unrest have been cited as causes, the most important reasons for deforestation are economic. The average annual income of many people in those countries most heavily impacted by deforestation is at extremely low levels. These people are forced to survive by any means necessary including subsistence agriculture and utilization of wood for cooking and heating (about two thirds of tropical people use wood fuels as their major source of energy, particularly poorer people). They often follow new logging roads to the only available property, forest land. The search for valuable hardwoods, such as **mahogany** and teak, is the other major source of forest clearing. Cattle ranching, plantations, and **mining** also provide considerable economic incentive for the destruction of forests. The social and economic value of these activities is much greater for the people involved than any perceived environmental value of the forests.

The Earth Summits of 1992 and 2002 attempted to address the linkage of social issues, economics, and the environment, though little agreement among nations was achieved. Some **conservation** organizations have shown that the economic wealth of one country can be traded for the environmental riches of another. A plan known as a "debt-for-nature" swap is one possible method for addressing both economic and environmental issues associ-

KEY TERMS

Conversion—A longer-term change in character of the ecosystem at some place, as when a natural forest is harvested and the land developed into an agroecosystem.

ated with deforestation. Many countries in which deforestation is rampant are relatively poor and are often in debt to more developed nations. Under the plan, the debt is bought (at a significant discount) in exchange for a pledge by the country to protect some portion of its valuable biologic resources, or to fund the activities of local conservation organizations within its borders. Agreements of this type have been emplaced in Bolivia, Madagascar, Zambia and other countries.

See also Rainforest; Slash-and-burn agriculture.

Resources

Books

Wood, Charles H. and Roberto Porro, eds. *Deforestation and Land Use in the Amazon.* Gainesville: University of Florida Press, 2002.

Periodicals

Stokstad, Erik. "U.N. Report Suggests Slowed Forest Losses." *Science* (March 23, 2001): 2294.

Other

National Aeronautic and Space Administration. "Better Monitoring of National and Global Deforestation Possible with Satellites." May 30, 2001 [cited January 31, 2003]. <http://earthobservatory.nasa. gov/Newsroom/MediaAlerts/2001/200105304788.html>.

Bill Freedman

Degree

The word "degree" as used in **algebra** refers to a property of **polynomials**. The degree of a polynomial in one **variable** (a monomial), such as $5x^3$, is the **exponent**, 3, of the variable. The degree of a monomial involving more than one variable, such as $3x^2y$, is the sum of the exponents; in this case, $2 + 1 = 3$. The degree of a polynomial with more than one **term** is the highest degree among its monomial terms. Thus the degree of $5x^2y + 7x^3y^2z^2 + 8x^4y$ is $3 + 2 + 2 = 7$.

The degree of a polynomial equation is the highest degree among its terms. Thus the degree of the equation $5x^3 - 3x^2 = x + 1$ is 3.

Dehydroepiandrosterone (DHEA)

Dehydroepiandrosterone (DHEA) is one of the androgens secreted by the adrenal cortex. An androgen is a hormone that stimulates masculine characteristics and is present in both males and females. The adrenal **glands** are small structures located at the tops of the kidneys. The adrenal medulla is the central portion of the adrenal gland and the adrenal cortex is the outer portion. The adrenal glands produce **hormones** that are involved in **metabolism** and **stress** reactions. These hormones are all produced from **cholesterol**. Three types of hormones are synthesized by the adrenal glands, glucocorticoids, mineralocorticoids, and sex hormone precursors. DHEA is one of the sex hormone precursors, which means it is eventually converted into the male sex hormone testosterone.

DHEA, along with its derivative dehydroepiandrosterone sulfate (DHEAS) are the most abundant steroids produced by the adrenal glands. Despite the high concentrations of DHEA in both the **blood** and the **brain**, no receptors have been found for the hormone and scientists have not determined its function in the body. The only scientifically proven fate of DHEA in the body is that it is eventually converted into the sex hormones. Many scientists have reported numerous beneficial effects of DHEA on the body. This hormone is marketed as a "wonder drug" that can be used to boost immune function, metabolism, endocrine function, as well as neurological functions. These claims are the results of recent studies involving DHEA supplementation and are quite preliminary. It is unknown whether these effects are directly due to the DHEA, or if they are the result of one of the metabolites of the hormone. In other words, the action of DHEA in the body is unknown.

DHEA as a neurosteroid

DHEA is different from other sex hormone precursors in that it is also a neurosteroid. A neurosteroid is a steroid that accumulates in the **nervous system** independently of its production in the endocrine glands. This means that DHEA found in the nervous system was not produced by the adrenal glands. DHEA has been found in the brains of humans, **rodents**, rabbits, **monkeys**, and dogs in relatively high concentrations. Recent studies have suggested that the hormone acts directly on the brain. Although the hormone itself has been found in the adult brain, the **enzyme** needed for its production is only found in the brains of fetuses.

KEY TERMS

. . . .

Anabolic steroid—Any of a group of synthetic steroid hormones sometimes abused by athletes in training to temporarily increase the size of their muscles.

Antioxicant—Any substance that prevents oxidation from occurring.

Atrophy—Decreasing in size or wasting away of a body part or tissue.

Autoimmune—Misdirected immune response in which lymphocytes mount an attack against normal body cells.

Endocrine system—A system of glands and other structures that secrete hormones to regulate certain body functions such as growth and development of sex characteristics.

Enzyme—Biological molecule, usually a protein, which promotes a biochemical reaction but is not consumed by the reaction.

Exogenous—Produced by factors outside the organism or system.

Interleukin—One of a variety of communication signals that drive immune responses.

Lupus—An autoimmune disease characterized by skin lesions.

Metabolite—A product of the controlled, enzyme-mediated chemical reactions by which cells acquire and use energy.

Morphology—Dealing with the form and structure of organisms.

Oxidation—Loss of electrons by a compound during a certain type of chemical reaction called a redox reaction.

Retrovirus—A type of virus that inserts its genetic material into the chromosomes of the cells it infects.

Seminiferous tubules—Tubes lining the testes which produce sperm.

Steroids—A group of organic compounds that belong to the lipid family and that include many important biochemical compounds including the sex hormones, certain vitamins, and cholesterol.

Testes—Male gonads, primary reproductive organs in which male gametes and sex hormones are produced.

Because the enzyme needed for its production is found only in the brains of fetuses, it is thought that the hormone is somehow related to the organization and development of the brain. When DHEA is added to cultures of developing neurons from the brains of mouse embryos, it causes morphological changes such as increasing axon length. These studies suggest that certain developmental neurological disorders may actually be the result of lower than normal concentrations of DHEA in the fetal brain.

Actions of DHEA

The level of DHEA in the blood declines with age and also during times of stress or illness. Concentrations of this hormone in humans peak at about age 20, after which they steadily decline. By the time a person is 80 years old, their DHEA levels are only about 20% what they were at their peak. The level of DHEA also declines with age in the brain. It has been suggested that this decline may play a role in some age-related illnesses. DHEA has been shown to act as an antioxidant, to enhance **memory**, and also to serve as a neuroprotector. Certain age-related diseases of the central nervous system, such as **Alzheimer disease**, are thought to be a re-

sult of oxidative stress in the brain. Because DHEA has been shown to demonstrate antioxidant properties in the brain, it has been hypothesized that it can be used to treat these age-related disorders. Although its action in these cases is still unclear, it is thought that it acts by protecting regions of the hippocampus from oxidative stress. It may also work by affecting the production of interleukin-6 (IL-6), which is believed to play a role in the progression of these diseases.

DHEA and DHEAS have been found to be useful in the treatment of certain autoimmune diseases. In one study, when **mice** were treated with DHEA at a young age, the onset of autoimmune **disease** was delayed and the mice lived longer. Once the mice had already shown signs of the disease, however, the hormone had no effect on disease progression. Another study demonstrated that DHEA supplementation helped reduce some of the effects of a **retrovirus infection** that caused acquired immune-deficiency syndrome (**AIDS**) in mice, such as **vitamin** E loss and **lipid** oxidation. DHEA boosted immune function and increased vitamin E levels in healthy mice as well. DHEA and DHEAS were also found to boost immune function in humans afflicted with **Cushing syndrome** and to delay the onset of lupus in mice.

It has also been suggested that DHEA is involved in stimulating bone mineral content and **density**. In **rats**, DHEA supplementation increased both lumbar spine and total body bone mineral density. This implies DHEA could be used to treat bone loss in aging patients or those suffering from certain diseases. Patients suffering from anorexia nervosa demonstrate severe bone loss as a side effect of this disease. DHEA levels in these patients are much lower than normal. DHEA supplementation in anorexic patients not only increased their bone mineral density, but also resulted in the resumption of menstrual cycles in many cases. Systemic lupus erythematosus patients also demonstrate severe bone loss. Preliminary clinical trials of DHEA supplements in these patients have suggested that this hormone could be used to treat bone loss in lupus sufferers as well.

There have been countless other claims of the benefits of DHEA in the body. DHEA may lower serum cholesterol levels. It may also protect against bacterial infections. DHEA has been found to decrease allergic responses in mice. Research is being conducted regarding the role of DHEA as a possible treatment for **tuberculosis**. This hormone has also been found to decrease **anxiety** in rats. The list of theories and proposals for the actions of DHEA on the body and the brain goes on and on.

Marketing

DHEA is currently available as a nutritional supplement. The Food and Drug Administration (FDA) has not approved its use for specific disorders. Because it is classified as a supplement, it can be purchased without a prescription. Since DHEA is a precursor to the production of testosterone, it could be considered an exogenous androgenic or anabolic steroid. Use of these steroids to enhance performance is banned from sports organizations and the military. At low doses, DHEA has a minimal effect on urine testosterone levels (the test used to screen for use of these drugs); however, at high doses this hormone would result in a "positive" test.

Side effects

The majority of the available research investigates the benefits of DHEA supplementation, but few studies discuss the possible adverse side effects associated with this hormone. One study found that prolonged DHEA treatment in rats induced liver tumors, especially in females. In male rats, sustained delivery DHEA and DHEAS treatments caused atrophy of the seminiferous tubules and testes. The application of the studies on the benefits of DHEA is also limited. Much research has been conducted on rats and mice, but few clinical trials on humans have actually been performed. More research is needed on the toxicity and morphological effects of DHEA and DHEAS, as well as on its specific action on humans, before its widespread use.

Resources

Books

Solomon, Eldra Pearl. *Biology.* Orlando: Saunders College Publishing, 1999.

Starr, Cecie. *Biology—Concepts and Applications.* Belmont, CA: Wadsworth Publishing Company, 1997.

Periodicals

Gordon, C. "Changes in Bone Turnover Markers and Menstrual Function after Short-term DHEA in Young Women with Anorexia Nervosa." *Journal of Bone and Mineral Research* 14, no. 1 (1999).

Prasad, A. "Dehydroepiandrosterone DecreasesBehavioral Dispair in High- but not Low-anxiety Rats." *Physiology & Behavior* 62, no. 5 (1997).

Jennifer McGrath

Delta

A delta is a low-lying, almost flat **landform**, composed of sediments deposited where a river flows into a **lake** or an **ocean**. Deltas form when the **volume** of sediment deposited at a river mouth is greater than what waves, **currents**, and **tides** can erode. Deltas extend the coastline outward, forming new land along the shore. However, even as the delta is constructed, waves, currents, or tidal activity may redistribute sediment. Although they form in lakes, the largest deltas develop along seashores. Deltas are perhaps the most complex of all sedimentary environments. The term delta comes from the resemblance between the outline of some deltas and the fourth letter in the Greek alphabet—delta—which is shaped like a triangle.

Some areas of the delta are influenced more by river processes, while marine (or lake) activities control other parts. Deltas do not form if wave, current, or tide activity is too intense for sediment to accumulate. The degree of influence by river, wave, current and tide activity on delta form is often used to classify deltas. Among the many factors that determine the characteristics of a delta are the volume of river flow, sediment load and type, coastal topography and **subsidence rate**, amount and character of wave and current activity, tidal range, **storm** frequency and magnitude, **water** depth, **sea level** rise or fall, and climate.

Delta construction

Delta plain

As a river flows toward the sea or a lake, it occupies a single, large, relatively straight channel known as the main distributary channel. The main distributary may soon branch off, like the roots of a tree, into many separate smaller distributaries. The number of branches formed depends on many different factors such as river flow, sediment load, and shoreline slope. Large sand-filled distributary channels occupy the delta plain, the nearly level, landward part of the delta, which is partly subaerial (above lake or sea level).

The natural levees that flank large distributary channels are another element of the delta plain. Natural levees are mounds of **sand** and silt that form when flood waters flow over the banks of the distributary and **deposit** their sediment load immediately adjacent to the channel. Unlike natural levees on **rivers**, delta levees do not grow especially large. Therefore, they are easily broken through by flood waters—a process called avulsion. This forms small channels, or crevasses, that flow away from the distributaries, like the little rootlets from the larger roots of a tree. The fan-shaped deposits formed during breeching of the distributaries are called crevasse splays.

Between the distributary channels, a variety of shallow, quiet water environments form, including **freshwater** swamps and lakes, and **saltwater** marshes. It is in these wetland basins that large volumes of organic **matter** and fine-grained sediment accumulate.

Delta front

When sediment-laden river water flows into the standing water at the mouth of a distributary, the river water slows and deposits its load. This forms a sediment body called a distributary mouth bar, or bar finger sand—so named because the distributary channels look a bit like the fingers on your hand. Distributary mouth bars form on the delta front, the gently seaward-sloping, marine-dominated part of the delta that is all subaqueous, or below water level.

Subaqueous levees may extend out from the natural levees of the delta front onto the delta plain. These help confine the water flow seaward of the distributary mouth to a relatively narrow path, so that the delta continues growing outward at a rapid pace. The area of the delta front between the distributary mouth bars is called the interdistributary bay; the salinity here is **brackish** to marine.

Water **velocity** slows consistently outward from the distributary mouth; the river water consequently deposits finer and finer sediment as it flows seaward. Eventually, a point is reached where the average grain size decreases

to clay-sized sediment with only minor silt. This is the prodelta area, where the bottom generally has a very low slope. On occasion, large blocks of sediment break free from the delta plain, slide down the steeper delta front, and become buried in prodelta muds.

Delta morphology

Vertical character

Ideally, if a delta is growing seaward, or prograding, as deltas typically do, a thick deposit with three stacked sediment sequences develops. The lower sequence, or bottomset beds, contain flat-lying silt and clay layers of the prodelta. The middle sequence, or foreset beds, contain seaward-inclined layers of sand and silt produced by distributary mouth bar sedimentation. The upper sequence, or topset beds, consists of flat-lying sand deposits formed in distributary channels and natural levees, interlayered, or interbedded, with fine-grained interdistributary bay deposits. A variety of factors, such as marine influence, changes in sediment supply or sea level, tend to complicate this picture; the actual sediment distribution in a deltaic sequence is typically very complex.

Surface character

The landward to seaward transect—from distributary channel sands to prodelta muds—outlined above is typical of deltas, like the Mississippi River delta, which experience minimal marine influence. These lobate, or "bird's foot" deltas, migrate farther and farther out into the ocean, as a result of multiple distributary channels and bar sands, each building seaward-extending lobes. On coastlines where waves or currents erode and redistribute much of the delta's sand, this lobate form becomes highly modified.

In areas where wave power significantly modifies the delta, the sands of distributary mouth bars, and to a lesser degree, distributary channels and natural levees, are reworked into shore-parallel sand bodies known as **barrier islands**, or shore-attached beaches. These commonly form along coasts exposed to powerful waves and where water depth rapidly increases seaward. This allows waves to erode both the delta plain and delta front, and produces a delta with a smoother, more regular, convex shape. The Niger Delta located along the west coast of **Africa** is an example of a wave-dominated delta.

In locations where tidal range—the difference between high and low tide—is fairly high, strong tidal currents sweep across the delta front and up the channels of the delta plain. These reversing currents erode the delta and redistribute the deposits into large sand bodies oriented **perpendicular** to shore. This tends to make the

Aerial view of Mississippi River Delta (bird's foot delta). *National Oceanic and Atmospheric Administration.*

coastline concave and also gives the delta a smoother, more regular shape. The Ganges-Brahmaputra Delta at the border between India and Bangladesh is an example of a tide-dominated delta.

Delta abandonment

As the river that formed a delta inevitably goes through changes upstream, a particular lobe may be abandoned. This usually occurs because a crevasse forms upstream by avulsion, and provides a more direct route or a steeper slope by which water can reach the sea or lake. As a result, the crevasse evolves into a new distributary channel and builds up a new delta lobe, a process called delta switching. The old distributary channel downstream is filled in by fine-grained sediment and abandoned. Over the last 5,000 years, the Mississippi River has abandoned at least six major lobes by avulsion.

An even larger-scale redirection of flow threatens to trigger abandonment of the entire Mississippi River delta. The Atchafalaya River, which follows an old course of the Mississippi, has a steeper slope than the modern Mississippi. At a point where the Atchafalaya River flows directly adjacent to the Mississippi, it is possible for the Atchafalaya to capture, or pirate, the flow of the Mississippi. This could permanently redirect the Mississippi away from New Orleans, and into Atchafalya Bay to the west. Since the 1950s, the U.S. Army Corps of Engineers has controlled the flow of the Mississippi in this area. In the 1980s, when the Mississippi River had several **seasons** of unusually high water levels, additional efforts were necessary to avert this disaster, but it may threaten again in the future.

Delta destruction

Abandoned delta lobes experience rapid subsidence primarily due to sediment compaction. As water depth increases accordingly, enhanced wave and current attack contribute to rapid **erosion** of old delta deposits—a process called delta retreat—and the loss of vast tracts of ecologically important **wetlands** and barrier islands. Modern flood controls, such as channelization and levee construction, sharply reduce avulsion and delta switching. As a result, little or no new sediment is contributed to the delta plain outside of the large channels. Consequently, compaction, subsidence and erosion continue unabated. This enhanced effect results in the loss of up to 15,000 acres of wetlands per year in inactive areas of the Mississippi River delta plain. **Global warming** could accelerate this effect by triggering higher rates of global sea level rise and resulting in more rapid increases in water depth. Increased hurricane incidence in the 1990s and beyond also takes a toll.

Construction of **dams** upstream impacts deltas too. Dams not only trap water, they trap sediment as well. This sediment load would normally contribute to delta progradation, or at least help stabilize delta lobes. Construction of dams instead results in significant delta retreat. For example, construction of Egypt's Aswan High Dam on the Nile River in 1964 lead to rapid erosion of the delta plain with loss of both wetlands and agricultural lands.

Deltas and human activity

Deltas have been important centers of human activity throughout history, in part because of the fertility of the land and easy access to transportation. Many early human civilizations developed on deltas. For example, the Nile River delta has hosted Egyptian cultures for over seven thousand years.

Deltas contain large expanses of wetlands where organic matter rapidly accumulates. Consequently, delta muds are very rich organic in organic materials and make good **hydrocarbon** source **rocks** when buried to appropriate depths. Not surprisingly, deltaic deposits contain extensive supplies of **coal**, oil, and gas. Deltaic sand bodies are also excellent reservoir rocks for mobile hydrocarbons. This combination of factors makes deltas perhaps the most important hydrocarbon-bearing environment on **Earth**. Due to this economic bonanza, modern and ancient deltas have probably been more throughly studied than any other **sedimentary environment**.

Deltas are very low relief; most areas are rarely more than a few feet above sea level. Therefore, they contain freshwater, brackish, and saltwater basins with

KEY TERMS

. . . .

Delta front—The seaward, gently sloping part of a delta, which is below water level.

Delta plain—The landward, nearly level part of a delta, some of which is below sea or lake level and some above.

Delta retreat—Landward migration of a delta due to erosion of older delta deposits.

Distributary channel—A large channel within a delta, which delivers water and sediment into an ocean or a lake.

Grain size—The size of a sediment particle; for example, gravel (greater than 2mm), sand (2–1/16 mm), silt (1/16–1/256 mm) and clay (less than 1/256 mm).

Sediment load—The amount of sediment transported by wind, water, or ice.

Sedimentary environment—An area on the earth's surface, such as a lake or stream, where large volumes of sediment accumulate.

Tidal range—Vertical distance between high tide and low tide during a single tidal cycle.

correspondingly diverse, complex ecologies. Minor changes in the elevation of the delta surface can flood areas with water of much higher or lower salinity, so delta **ecology** is easily impacted by human activities. As indicated above, humans have significantly altered deltas and will continue to do so in hopes of curbing **flooding**. As a result, we will continue to see accelerated delta retreat, and wetlands destruction, unless humans develop new flood control technologies or new methods for wetlands protection.

Resources

Books

Leeder, Mike. *Sedimentology and Sedimentary Basins: From Turbulence to Tectonics.* London: Blackwell Science. 1999.

Selby, M.J. *Earth's Changing Surface.* London: Oxford University Press.1985.

Skinner, Brian J., and Stephen C. Porter. *The Dynamic Earth: An Introduction to Physical Geology.* 4th ed. John Wiley & Sons, 2000.

Thurman, Harold V., and Alan P. Trujillo. *Essentials of Oceanography.* 7th ed. Englewood Cliffs, NJ: Prentice Hall, 2001.

Clay Harris

Dementia

Dementia is a decline in a person's ability to think and learn. To distinguish true dementia from more limited difficulties due to localized **brain** damage, the strict medical definition requires that this decline affect at least two distinct spheres of mental activity; examples of such spheres include **memory**, verbal fluency, calculating ability, and understanding of **time** and location.

Some definitions of dementia also require that it interfere with a person's work and social life. However, this may be difficult to show when a person's work and social life is already limited, either by choice or by another mental or physical disorder. As a result, the most recent and most authoritative definition (that developed jointly by the National Institute for Neurological and Communicative Disorders and Stroke—part of the National Institutes of Health—and the Alzheimer's Disease and Related Disorders Association) does not include this criterion. The NINCDS-ADRDA definition focuses strictly on a decline from a previously higher level of mental function.

The term dementia goes back to antiquity, but was originally used in the general sense of being "out of one's mind." Identification specifically with difficulties in thinking and **learning** occurred in the late eighteenth and early nineteenth centuries. Even then, however, the term was used for almost any sort of thinking, learning, or memory problem, whether temporary or permanent and without regard to cause. The most typical picture was of a young adult suffering from insanity or a disease affecting the brain.

This picture changed later in the nineteenth century, as psychiatrists (then called alienists) sought to group disorders in ways that would help reveal their causes. Temporary stupor, dementia associated with insanity, and memory problems resulting from damage to a specific area of the brain were all reclassified. The central core of what was left was then senile dementia: the substantial, progressive loss of mental function sometimes seen in older people and now recognized as resulting from one or more specific, identifiable diseases. Current definitions still recognize the existence of dementia in younger people, however.

Diagnosis

The first step in diagnosing dementia is to show that the person's ability to think and learn has in fact declined from its earlier level. His or her current ability in different spheres of mental activity can be measured by any of a variety of mental status tests. The difficulty comes in comparing these current ability levels with those at earlier times. A patient's own reports cannot be relied upon, since memory loss is typically part of dementia. Frequently, however, family members' descriptions of what the person once could do will establish that a decline has occurred. In other cases, comparison with what a person has accomplished throughout his or her life is enough to show that a decline has occurred. If neither source of information provides a clear answer, it may be necessary to readminister the mental status test several months later and compare the two results.

Is any decline, no matter how small, sufficient to establish a **diagnosis** of dementia? The answer is not entirely clear. Research has shown that most older people suffer a small but measurable decrease in their mental abilities. For example, one recent study followed 5,000 people, some for as many as 35 years. This study found that scores on tests of mental abilities did not change between ages 25 and 60, but declined about 10% between ages 60 and 70. More significantly, people in their late eighties had scores more than 25% below those seen earlier.

Since none of the people tested were considered demented, one might assume that these declines are normal. It is still possible, however, that some tested individuals were in the early stages of dementia; these people's results may then have pulled down the average scores for the group as a whole and created a false impression of a sizable "normal" drop in IQ. This ambiguity is particularly unfortunate because it has significant implications at the **individual** level: No one knows whether, if an older person's mental sharpness starts to decline, this a normal part of aging or a possible signal of approaching dementia.

Once the existence of dementia has been established, the next question is: What is causing the condition? Alzheimer's disease is by far the most common cause of dementia, especially in older adults. One recent study found that it directly caused 54% of dementias in people over 65, and may have been partially responsible for up to 12% more.

Unfortunately, there is no direct way to diagnose Alzheimer's disease in a living person; only microscopic examination of the brain after death can conclusively establish that a person had this disorder. The same is true for the second most common cause, multi-infarct dementia. Both diagnoses are made by excluding other causes of dementia.

It is particularly crucial to exclude causes for which appropriate treatment might prove helpful. Among the most common and important of these are side effects of medications an individual may be taking—for example, sleeping pills, antidepressants, certain types of high **blood pressure** medications, or others to which a person may be particularly sensitive. Medications are particularly likely to be responsible when the affected person is not only confused and forgetful, but also is not alert to what is going on around him or her.

Older individuals—the group most likely to suffer dementia from other causes—are particularly likely to be taking multiple drugs for their various disorders. Sometimes these drugs interact, producing side effects such as dementia that would not occur with any single drug at the same dosage. Drug side effects, including dementia, may also be more common in older people because their body's ability to eliminate the drug often declines with age. Reduced speed of elimination calls for a corresponding reduction in dosage that does not always occur.

Another common, but treatable, cause of dementia, or of what looks like dementia, is **depression**. Some psychiatrists refer to the slowed thinking and confusion sometimes seen in people with depression as pseudodementia because of its psychological origin. Others believe the distinction does not reflect a meaningful difference. In any case, effective treatment of the depression will relieve the dementia it has produced.

Causes

Dementia can result from a wide variety of disorders and conditions. Some are quite rare, while others are moderately common. In some cases (measles, for example) dementia may be a rare complication of an otherwise common disease; in other cases, such as **infection** with Human Immunodeficiency Virus (HIV), an impact on mental function well known to medical specialists may not be widely recognized by the general public.

Non-Alzheimer degenerative dementias

In addition to **Alzheimer disease**, dementia may result from several other conditions characterized by progressive degeneration of the brain. The three most common of these are Pick's disease, **Parkinson disease**, and **Huntington disease** (Huntington's chorea).

Like Alzheimer disease, Pick's disease affects the brain's cortex—that is, the outer part where most of the higher mental functions take place. In other respects, however, the disorders are quite different. In Pick's disease, for example, microscopic examination of the brain reveals dense inclusions (Pick bodies) within the nerve cells, while the cells themselves are inflated like blown-up balloons. This does not at all resemble the neurofibrillary tangles and beta-amyloid plaques seen in Alzheimer disease. However, since microscopic examination of a living person's brain is rare, symptoms are used to distinguish the two diseases in practice.

Typically, Pick's disease affects different parts of the cortex than does Alzheimer disease. This influences the order in which symptoms appear. The earliest symptoms of Pick's disease include personality changes such as loss of tact and concern for others, impaired judgment, and loss of the ability to plan ahead. Loss of language skills occurs later, while memory and knowledge of such things as where one is and the time of day are preserved until near the end. In contrast, memory and time-space orientation are among the first things lost in Alzheimer disease, while personality changes and loss of language skills are late symptoms.

Both Parkinson disease and Huntington chorea initially affect deeper brain structures, those concerned with motor functions (that is, movement of the voluntary muscles). Indeed, most descriptions of Parkinson disease focus on the muscular rigidity that the disorder produces. In the later stages, however, nearly all patients with the disease will develop some degree of dementia as well.

Shortly after appearance of the choreiform movements that typify Huntington disease, most patients will begin to have trouble thinking clearly and remembering previous events. By the time they die, Huntington patients are intellectually devastated.

Vascular dementias

Although degenerative disorders account for the majority of dementia cases, a respectable minority result from interference with blood flow in or to the brain. Most such cases are due to a series of small strokes. Each **stroke** in the series may be unnoticeable, but the long-term result is a continuing and eventually severe decline in mental function.

(A stroke, known technically as an *infarct*, is a failure of blood flow beyond a certain point in an artery. Usually this is due to a blood clot at that point, but sometimes it results from a break in the artery allowing much or all of the blood to escape. Although the fundamental causes are almost diametrically opposite—a clot at the wrong place versus no clot where one is needed—the effects are virtually the same.) Unlike the degenerative dementias, which follow a relatively predictable course, vascular dementias can be quite variable. When and precisely where the next stroke occurs will determine both how quickly the dementia progresses and the extent to which different mental abilities are affected. Typically, however, vascular dementias are characterized by sudden onset, step-wise progression, and occurrence of motor symptoms early in the disorder. High blood pressure is usually present as a predisposing factor. Most, but not all, physicians believe that other **heart** attack risk factors, such as diabetes, cigarette smoking, and high **cholesterol**, also increase the risk of developing vascular dementia.

Traditionally, physicians have distinguished two major types of vascular dementia. In multiple-infarct dementia, examination of the brain after death shows a number of small but individually identifiable areas where strokes have destroyed the brain **tissue**. In Binswanger's disease, individual areas of destruction cannot be identified. Almost the entire "white matter" of the brain—the portion occupied primarily by axons rather than nerve **cell** bodies—is affected to some degree. There is no sharp line between the two disorders, however, just as there is none between multiple-infarct dementia and the effect of two or three large strokes.

Dementia may also result from a reduction in blood flow to the brain as a whole. The most common cause is a severe narrowing of the carotid **arteries** in the neck. This may be considered analogous to partial plugging of an automobile's fuel line, whereas the local damage resulting from a stroke is more like knocking out a piston. Most other dementias similarly represent damage to the engine itself. (Alzheimer disease might perhaps be likened to cylinder-wall deposits causing the pistons to stick, although we do not know enough about the origin of the disease to be sure this analogy is entirely accurate.)

Infectious dementias

Many infections either attack the brain as their primary target or can spread to it. If enough brain tissue is destroyed as a result, the outcome may be dementia. Brain infections can be due to viruses, **bacteria**, **fungi**, or **parasites**. For example, the measles **virus** will occasionally attack the brain, producing a condition known as subacute sclerosing panencephalitis that causes dementia and eventually death. The herpes (cold sore) virus can also cause dementia if it attacks the brain.

Infection by mosquito-borne **encephalitis** virus may leave survivors with significantly reduced mental function. The frequency with which this occurs depends, however, both on the particular virus involved and the age of the individual. Dementia is rare following infection with Western encephalitis virus, but is found in more than half those under five years of age who survive an Eastern encephalitis virus attack. Similarly, equine encephalitis virus produces severe illness, often leading to serious brain damage or death, in children under 15; in older people, however, the disease is typically quite mild and causes no lasting problems.

Nevertheless, serious viral infections of the brain are relatively uncommon. The one exception is infection with the human immunodeficiency virus (HIV)—the virus that causes **AIDS**. This is the most common infectious cause of dementia in the United States today, and the number of people affected continues to grow.

Although popular accounts of HIV infection focus on the damage it causes to the **immune system**, the virus also typically attacks the brain. Nearly all HIV-infected people will develop dementia at some time during their illness. However, how soon this dementia occurs and how severe it may become varies widely. About 20% of people with HIV infection develop dementia before they develop the opportunistic infections that define progression to full-blown AIDS.

Over the past half-century, **antibiotics** have greatly reduced the threat from bacterial infection of the brain or the meninges that surround it. In one respect, however, the situation may be said to have worsened. Formerly, 95% of those with acute bacterial **meningitis** died; today, most survive, but the disease often leaves them with reduced mental capacities or other central **nervous system** problems.

On the other hand, tuberculous meningitis, which once accounted for up to 5% of children's hospital admissions, has now been almost eliminated. There has also been a major reduction in syphilis of the central nervous system, a disease whose effects once resulted in 15–30% of mental hospital admissions. Unfortunately, both diseases are undergoing resurgences. The incidence of **tuberculosis** has increased 20% since 1985, while estimates suggest that 50,000 undetected and untreated new cases of syphilis occur each year. In the absence of treatment, 25–30% of syphilis cases will spread to the brain or meninges and result, over the course of years, in paralysis and dementia.

Fungal infections of the brain and meninges are generally rare except in people with weakened immune systems. Parasitic infections are also rare in this country. Elsewhere, however, the well-known African **sleeping sickness**, spread by a type of biting fly found only in equatorial **Africa**, is due to a parasite known as a trypanosome. **Malaria**, a mosquito-born parasitic disease, may also at times attack the brain and result in dementia.

Two infectious dementias that are quite rare but of tremendous scientific interest are **Creutzfeldt-Jakob disease** and **kuru**. The probable cause of these diseases are **prions**, infectious agents made up of gene-lacking **proteins**.

Miscellaneous causes

The dementia that can result from medication side effects or overdoses has been discussed in connection with diagnosis. Certain **vitamin** deficiencies may also cause dementia. The only one that is not extremely rare in developed countries, however, is **Korsakoff's syn-**

KEY TERMS

Pick's disease—A degenerative brain disorder causing progressive dementia.

Vascular dementia—Loss of mental function due to a number of small, individually unnoticeable, strokes or to some other problem with the blood vessels in or supplying blood to the brain.

drome. This results from thiamine deficiency produced by intense, prolonged **alcohol** abuse. Yet another potential cause of dementia is deficiency of thyroid hormone; unlike many other dementias, this is usually reversible once adequate amounts of the hormone are available.

In yet other cases, diseases of the kidney or liver may lead to build-up of toxic materials in the blood; dementia then becomes one symptom that these materials have reached poisonous levels. Chronic hypoglycemia (low bloodsugar), often due to disorders of the pancreas, may also impair mental function.

Although both head injuries and brain tumors usually affect only a single **sphere** of mental activity (and thus, by definition, do not produce dementia) this is not always the case. Prize fighters in particular are likely to have experienced multiple blows to the head, and as a result often suffer from a generalized dementia. Conditions such as near-drowning, in which the brain is starved of **oxygen** for several minutes, may also result in dementia.

Almost 3% of dementia cases are due to **hydrocephalus** (literally "water on the brain;" more precisely, an accumulation within the brain of abnormal amounts of cerebrospinal fluid). This usually results from an injury that makes it difficult for the fluid to reach the areas where it is supposed to be reabsorbed into the bloodstream. In the most common form, and the one most easily overlooked, pressure within the brain remains normal despite the fluid build-up. The extra fluid nevertheless distorts the shape of the brain and impairs its function. Installing shunts that allow the fluid to reach its proper place usually cures the dementia.

Resources

Books

Cummings, Jeffrey L., and Frank D. Benson, eds. *Dementia: A Clinical Approach*. 2nd ed. Boston: Butterworth-Heinemann, 1992.

Safford, Florence. *Caring for the Mentally Impaired Elderly: A Family Guide*. New York: Henry Holt, 1987.

Whitehouse, Peter J., ed. *Dementia*. Philadelphia: F.A. Davis Company, 1993.

Periodicals

Hyman, S.E. "The Genetics of Mental Illness: Implications for Practice." *Bulletin of the World Health Organization* 78 (April 2000): 455-463.

W. A. Thomasson

Dengue fever

Dengue fever is an illness caused by four closely related viruses (DEN-1, DEN-2, DEN-3, and DEN-4). Even though these viruses are closely related, they are recognized by the **immune system** as being different from each other. Thus, an **infection** with one **virus** does not provide immune protection against infections with the remaining three viral types. A person can have four bouts of dengue fever in his/her lifetime.

Dengue fever is a tropical **disease**. This is mainly because the virus is carried by a mosquito called *Aedes aegypti*. This mosquito is a normal resident of tropical climates. When the mosquito bites a human to obtain its **blood** meal, the virus can be transmitted to humans.

The reverse is true as well; the virus can be sucked up into a mosquito if the mosquito feeds on a dengue-infected human. The virus can subsequently be spread to another person. In this way, a large number of dengue fever cases can appear in a short time. Furthermore, if more than one of the dengue viruses is in circulation at the same time, more than one dengue fever **epidemic** can occur simultaneously.

Like **mosquitoes**, the dengue viruses have been around for centuries. The first written records of dengue fever epidemics date back to the late eighteenth century. The last major global epidemic began just after World War II. This epidemic is ongoing and has grown worse since 1990.

Dengue, also called breakbone or dandy fever (because of the severe joint and muscle **pain** that can result from the infection), is **endemic** to the tropics, meaning that the infection is always present. The dengue viruses belong to the arbovirus group. The term arbovirus is a derivative for arthropod borne, reflecting the fact that the viruses are transmitted by an insect.

The incubation period for dengue fever is usually five to eight days, but may be as few as three or as many as 15 days. Once the virus has had a sufficient incubation, the onset of the disease is sudden and dramatic. The first symptom is usually sudden chills. A headache follows and the patient feels pain with **eye** movement. Within hours the patient is debilitated by extreme pain in the legs and joints. Body **temperature** may rise to 104°F (40°C). A pale rash may appear, usually on the face, but it is transient and soon disappears.

These symptoms persist for up to 96 hours, followed by a rapid loss of fever and profuse sweating. The patient begins to feel better for about a day, and then a second bout of fever occurs. This temperature rise is rapid, but peaks at a lower level than the first episode. A rash appears on the extremities and spreads rapidly to the trunk and face. Palms of the hands and soles of the feet may turn bright red and swollen.

There is no cure for dengue. Treatment is palliative, that is, intended to ease the symptoms of the disease. Individuals usually recover completely from dengue after a convalescent period of several weeks with general weakness and lack of **energy**.

Recently, attenuated forms of all four dengue viruses have been developed in Thailand. These weakened viruses may be candidates for use in vaccines, even a **vaccine** that would simultaneously protect someone from all four versions of dengue fever. However, such a vaccine is still years from production.

No vaccine currently exists to prevent the disease. The only preventive measure that can be taken is to eradicate the aedes mosquito or to reduce exposure to them. Patients who have the disease should be kept under mosquito netting to prevent mosquito bites, which can then spread the virus.

See also Insecticides.

Resources

Books

Gubler, D. J., G. Kuno, and K. Gubler. *Dengue and Dengue Hemorrhagic Fever.* New York: CABI Publishing, 1997.

Richman, D. D., R. J. Whitley, and F. G. Hayden. *Clinical Virology.* 3rd ed. Washington, DC: American Society for Microbiology Press, 2002.

Periodicals

Kuhn, R. J., et al. "Structure of Dengue Virus: Implications for Flavivirus Organization, Maturation, and Fusion." *Cell* 108 (March 2002): 717–725.

Other

Centers for Disease Control and Prevention. Division of Vector-Borne Infectious Diseases. PO Box 2087, Fort Collins, CO 80522. <http://www.cdc.gov/ncidod/dvbid/dengue//>.

Brian Hoyle

Denitrification

Denitrification is a microbial process by which fixed **nitrogen** is lost from **soil** or aquatic systems to the atmosphere. This loss occurs when **bacteria** convert nitrogen-containing molecules, in particular, nitrate (NO_3^-) and nitrite (NO_2^-), to gaseous nitrous oxide (N_2O) and dinitrogen (N_2).

The biology of denitrification

Respiration is a chemical process in which **energy** is released when electrons are passed from a donor **molecule** to an acceptor molecule. In addition to energy being released, the respiratory process results in the donor molecule being converted to an oxidized molecule, meaning it has lost electrons, and the acceptor molecule being converted to a reduced molecule, meaning it has gained electrons. Typically, the acceptor molecule is **oxygen**, but in **anaerobic** environments, which lack oxygen, bacteria may reduce molecules other than oxygen that have high reduction potentials or ability to accept electrons in a process known as anaerobic respiration. Denitrification occurs when bacteria reduce nitrate or nitrite by this process. In a sequence of four reductions, nitrate is converted to dinitrogen gas, the molecular form in which nitrogen escapes from soils and aquatic systems. The four-step sequence is: 1.) Nitrate is reduced to nitrite. 2.) Nitrite is reduced to nitric oxide. 3.) Nitric oxide (NO) is reduced to nitrous oxide. 4.) Nitrous oxide is reduced to dinitrogen. Depending on its physiological capabilities, a single **organism** may carry out all of these reductions, or it may carry out only a few.

In addition to dinitrogen, small amounts of nitrous oxide leave aquatic and soil systems. This happens because not all bacteria that produce nitrous oxide can subsequently reduce it to dinitrogen. Therefore, some nitrous oxide can leak out of cells into the atmosphere, if it is not first reduced to dinitrogen by other organisms. Nitric oxide is also a gas, but organisms that have the ability to reduce nitrite to nitric oxide always have the ability to reduce nitric oxide to nitrousoxide. For this reason, nitric oxide is not an important product of denitrification.

In aquatic and soil systems fixed nitrogen primarily exists as a component of three inorganic molecules; nitrate, nitrite, and ammonium (NH_4^+), and in the **proteins** and other types of organic molecules that comprise living and dead organisms. Although only nitrogen from the molecules nitrate and nitrite is converted to a gaseous form and removed from these systems, nitrogen from proteins and ammonium can also be removed if it is first oxidized to nitrate or nitrite. This conversion begins in a process termed **ammonification**, when nitrogen is released from the **biomass** of dead organisms, which produces ammonium. Ammonium can then be converted to nitrate in an **aerobic** respiratory reaction called **nitrifica-**

tion, in which the ammonium serves as an **electron** donor, and oxygen an electron acceptor.

Importance

Along with dinitrogen fixation, ammonification, and nitrification, denitrification is a major component of the **nitrogen cycle**. Estimates of nitrogen fluxes from terrestrial and marine ecosystems to the atmosphere as a result of microbial denitrification range from 90×10^{12} to 243×10^{12} grams per year for terrestrial systems and 25×10^{12} to 179×10^{12} grams per year for marine systems. Scientist generally agree that less than 10% of these fluxes occur with nitrogen as a component of nitrous oxide. The range in these estimates reflects the difficulty researchers face in measuring denitrification and extrapolating the measurements to a global scale.

Humans are primarily interested in denitrification because this process is responsible for fixed nitrogen being removed from sewage and lost from cropland. Environmentally harmful nitrate concentrations in sewage discharge can be reduced by storing wastes under denitrifying conditions before releasing them into the environment. Although denitrification is a beneficial process in **sewage treatment**, it is considered a problem in agriculture. Farmers increase their crop yields by applying nitrogen containing **fertilizers** to their land. As a result of denitrification, crop yields may be reduced because much of the added nitrogen is lost to the atmosphere. This loss of fixed nitrogen may have global consequences. Increased denitrification from cropland is responsible for increased amounts of nitrous oxide in the atmosphere. Although nitrous oxide is not the major end product of denitrification, it is highly reactive and may contribute to the depletion of **ozone** in the stratosphere.

Resources

Books

Kupchella, Charles, and Margaret Hyland. *Environmental Science: Living Within the System of Nature.* 2nd ed. Boston: Allyn and Bacon, 1989.

Prescott, L., J. Harley, and D. Klein. *Microbiology.* 5th ed. New York: McGraw-Hill, 2002.

Steven MacKenzie

Density

The density of an object is defined simply as the **mass** of the object divided by the **volume** of the object. For a concrete example, imagine you have two identical boxes. You are told that one is filled with feathers and the other is filled with cement. You can tell when you pick up the boxes, without looking inside, which is the box filled with cement and which is the box filled with feathers. The box filled with cement will be heavier. It would take a very large box of feathers to equal the weight of a small box of cement because the box of cement will always have a higher density.

Density is a property of the material that does not depend on how much of the material there is. One pound of cement has the same density as one ton of cement. Both the mass and the volume are properties that depend on how much of the material an object has. Dividing the mass by the volume has the effect of canceling the amount of material. If you are buying a piece of gold jewelry, you can tell if the piece is solid gold or gold plated **steel** by measuring the mass and volume of the piece and computing its density. Does it have the density of gold? The mass is usually measured in kilograms or grams and the volume is usually measured in cubic meters or cubic centimeters, so the density is measured in either kilograms per cubic meter or in grams per cubic centimeter.

The density of a material is also often compared to the density of **water** to give the material's specific gravity. Typical **rocks** near the surface of **Earth** will have specific gravities of 2 to 3, meaning they have densities of two to three times the density of water. The entire Earth has a density of about five times the density of water. Therefore the center of Earth must be a high density material such as nickel or **iron**. The density provides an important clue to the interior composition of objects, such as Earth and planets, that we can't take apart or look inside.

Dentistry

Dentistry is the medical activity focused on treating the teeth, the gums and the oral cavity. This includes treating teeth damaged due to accidents or **disease**, filling teeth damaged due to tooth decay, and replacing damaged or injured teeth with replacement teeth. Major disciplines of dentistry include orthodontics, which focuses on the correction of tooth problems such as gaps between the teeth, crowded teeth and irregular bite; and periodontics, which addresses gum problems. Dentistry is considered an independent medical art, with its own licensing procedure. Medical doctors are not licensed to treat teeth; likewise dentists are not licensed to treat other parts of the body.

Skill and superstition

Ancient, medieval, and early Renaissance dental practice can be seen as a stew of the sensible and the outrageous. In each era, stories of practitioners with wisdom and skill coexist with outrageous tales of superstition and myth connected to teeth. In the ancient and Islamic worlds, doctors often performed dental work. The cleaning and extracting of teeth was often performed by individuals with little or no medical training.

Ancient men and women worked hard to alleviate dental **pain**. As early as 1550 B.C., the ancient Egyptians documented their interest in dentistry in the *Ebers Papyrus*, a document discovered in 1872. The *Papyrus* listed various remedies for toothache, including such familiar ingredients as dough, honey, onions, incense, and fennel **seeds**.

The Egyptians also turned to superstition for help preventing tooth pain. The mouse, which was considered to be protected by the **Sun** and capable of fending off death, was often used by individuals with a toothache. A common remedy involved applying half of the body of a dead mouse to the aching tooth while the body was still warm.

The Greeks offered a variety of conventional and unconventional dental therapy. One of the more illustrious dental pioneers was Hippocrates (460–375 B.C.), whose admonition to do no harm continues to be a central goal of medical practice. Hippocrates said that food lodged between teeth was responsible for tooth decay, and suggested pulling teeth that were loose and decayed.

Hippocrates also offered advice for bad breath. He suggested a mouth wash containing oil of anise seed and myrrh and white wine. Other ancient Greeks took a more superstitious approach, with some depending on the mythical power of the mouse to protect their teeth. A recipe for bad breath from the fifth century B.C. called for a range of ingredients including the bodies of three **mice**, including one whose intestines had been removed, and the head of a hare. The ingredients were burned and mixed with dust and **water** before consumption.

The Etruscans, who lived in Tuscany, Italy, between approximately 1000 and 400 B.C., also made great advances in dentistry. They are often cited for the sophistication of their gold crowns and bridges. One bridge which has been preserved included three artificial teeth attached to gold bands, which hooked around the natural teeth. The artificial teeth were actually real teeth taken from an immature calf, then divided in two.

The Romans built upon the Etruscan knowledge of dentistry and took seriously the challenge of keeping teeth healthy. Celsus, a Roman writer who lived about 100 B.C., wrote about toothache, dental abscesses and other dental ailments. For toothache, which he called "among the worst of tortures," he suggested the use of hot poultices, mouthwash, and steam. He also suggested using pain-killers such as opium. The Romans also made bridgework.

Clean teeth were valued by the Romans, and affluent families had slaves clean their mouths using small sticks of **wood** and tooth powder. Such powders could include burned eggshell, bay-leaves and myrrh. These powders could also include more unusual ingredients, such as burned heads of mice and lizard livers. **Earth** worms marinated in vinegar were used for a mouth wash, and urine was thought of as a gum strengthener.

The Romans, like individuals in many other cultures, believed that worms in the teeth caused pain. A vast well of superstition can also be found concerning the premature appearance of teeth. Babies born with one or more teeth were considered dangerous in **Africa**, Madagascar, and India, and were at one point killed. In contrast, the ancient Romans considered children born with teeth to be special, and children were often given a name, "Dentatus" in reference to their early dental development.

Non-western advances

Cultures outside Western civilization also focused on the teeth. The Chinese were the first to develop a silver amalgam filling, which was mentioned in medical texts as early as A.D. 659. The Chinese also developed full dentures by the twelfth century A.D. and invented the toothbrush model for our contemporary toothbrushes in the fifteenth century. Dental advances also flourished in the Islamic culture, which emerged around the spiritual and political power of Muhammad (570-632) and his followers. Innovators drew from the translated works of Aristotle, Plato, and Hippocrates, whose work was translated by Egyptians with links to Greece.

Mohammed's teaching called explicitly for the maintenance of clean teeth. Clean teeth were seen as a way of praising God, and he was reported to say "a prayer which is preceded by the use of the toothpick is worth 75 ordinary prayers." Dental powders, mouth wash, and polishing sticks were used to keep teeth clean.

Dental **surgery** advanced greatly with the teaching of Albucasis (936-1013), a surgeon whose extensive writing about surgery in the *Al-Tasrif* influenced Islamic and medieval European medical practitioners. He described surgery for dental irregularities, the use of gold wire to make teeth more stable, and the use of artificial teeth made of ox-bone. Albucasis also was one of the first to document the size and shape of dental tools, including drawings of dental saws, files, and extraction forceps in his book.

As the Islamic world moved ahead in dentistry, European dental practice was overwhelmed by the superstition, ignorance, and religious fervor of the Middle Ages. Scientific research was discouraged during the medieval era, which stretched from the fifth to the fifteenth century. Suffering and illness were widely considered to be punishment from God. Knowledge of dental **anatomy** and treatment did not advance during the Middle Ages, though the range of superstitious dental treatments flowered.

One fourteenth century therapy called for eating the brains of a hare to make lost teeth grow again. Charms made of stone, wood, or **paper** devoted to a religious figure were believed to ward off disease. Religious officials suggested prayer as the best protector.

The practice of dentistry during the Middle Ages was generally limited to the pulling of teeth that were decayed or destroyed. This task initially fell to barbers, who also performed minor surgery in England in the fifteenth century and were called barber-surgeons. Transient tooth-pullers, who traveled from place to place, also made money extracting teeth.

From counting teeth to replacing them

By the end of the fifteenth century, the emphasis on obedience to authority was changing, in part under the influence of advances such as the discovery of the **printing** press in 1436. Dentistry benefitted from the new spirit of inquiry. Contemporary thinkers, such as anatomist Andreas Vassalius (1514-1564) challenged classical ideas about dentistry. One indication of the stagnation of independent thinking was Vassalius's successful challenge of Aristotle's belief that men had more teeth than women.

Ambrose Pare (1510-1590), a Frenchman trained as a barber surgeon, gained fame as one of the great med-

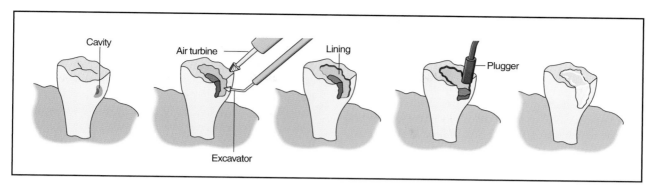

Teeth damaged by dental cavities can be excavated and filled with amalgam. *Illustration by Hans & Cassidy. Courtesy of Gale Group.*

ical and dental surgeons of the era. His work resembled the work of a contemporary oral surgeon, focusing on the removal of teeth, the setting of fractured jaws and the draining of dental abscesses. He published extensively, documenting methods for transplanting teeth and for creating devices that held artificial teeth made of bone in place using silver or gold wire.

The eighteenth century saw many significant advances in dentistry, many of them inspired by the work of Pierre Fauchard (1678-1761). By the year 1700, Parisian dentists such as Fauchard were considered members of a distinct profession, complete with an examining board for new dentists. Fauchard's work is best known through his writing about the profession in the 1728, two-volume, *Le Chirurgien Dentiste*, a 863-page tome. In the book, Fauchard explained how to fill teeth with **lead** or gold leaf tin foil, and various types of dentures. He also told how to make crowns from ivory or human teeth, how to straighten teeth, and how to protect teeth against periodontal damage.

Fauchard also took aim at some of the dental superstitions of the day, which included the erroneous belief that worms in the mouth played a role in tooth decay. His information was not all accurate, however, and Fauchard did suggest the use of urine as a mouth wash.

Another great eighteenth century finding was the development of porcelain, glazed white clay, as a substance for false teeth. Prior to this time, ivory was commonly used. Carving ivory was time consuming and difficult. The first porcelain teeth were developed by M. DeChateau, a French druggist, and M. Dubois De Chamant, a dentist.

DeChateau was frustrated that his teeth had discolored due to the chemicals he tasted while mixing substances for customers. After noticing that the chemicals never discolored his porcelain mortar and pestle, DeChateau decided that porcelain teeth would save him embarrassment and unhappiness. Gaining the help of

DeChamant, the two men discovered a way to effectively fit and create a pair of false teeth made of porcelain, gaining a patent on the teeth in 1788.

The nineteenth century saw the development of many dental tools and practices that would be the **bedrock** for twentieth century dentistry. Many of the great advances were made by Americans, who emerged as great dental innovators. The world's first dental school, the Baltimore College of Dentistry, opened in 1847, providing an organized curriculum to replace the apprenticeship system.

At the start of the century, false teeth were available to only the affluent. They were made of porcelain, which was not expensive. But they needed to be fastened to plates made of gold or silver, which were costly. The successful **vulcanization** of rubber in 1830 by American Charles Goodyear brought cheap false teeth to the masses. Now false teeth could be attached to vulcanized rubber, and dental laboratories emerged everywhere to keep up with the demand.

The development of **anesthesia** in the United States was a technological breakthrough which revolutionized surgical and dental practice. Many innovators experimented with the use of gases in the eighteenth and nineteenth centuries. Joseph Priestley, a British cleric, invented nitrous oxide, or laughing gas, in 1772. The substance caused euphoria, then sedation and unconsciousness.

Though researchers explored the application of nitrous oxide and **ether** in the early nineteenth century, the gases were not used for anesthetic purposes until the 1840s. Physician Crawford Williamson Long, a Georgia physician, first used ether to remove a **tumor** from a patient in 1842. Dentist Horace Wells used nitrousoxide on patients having their teeth pulled in 1844.

But dentist William Thomas Green Morton is widely credited with the first public display of anesthesia, in part because of the great success of his public demonstration and in part because of his canny alliance with influential physicians. Morton successfully extracted a tooth from a patient anesthetized with ether in 1846 in Boston.

Ether, nitrous oxide and **chloroform** were all used successfully during tooth extraction. But these gases were not effective for many other procedures, particularly those which took a long period of time to complete.

A breakthrough came in the form of the drug **cocaine**, an addictive drug derived from **coca** leaves which was highly valued in the nineteenth and early twentieth centuries for its pain-killing power. In 1899, cocaine was first used in New York as a local anesthetic to prevent pain in the lower jaw. Cocaine was effective but habit-forming and sometimes harmful to patients. The development of procaine, now known as novocaine, in 1905 provided dentists with a safer anesthetic than cocaine. Novocaine could be used for tooth grinding, tooth extraction and many other dental procedures.

Development of a drill powered by a footpedal in 1871 and the first electric drill in 1872 also changed the practice of dentistry.

Another major discovery of the era was the x ray by William Conrad Roentgen of Germany in 1895. The first x ray of the teeth was made in 1896. At the time, there was some skepticism about **x rays**. The *Pall Mall Gazette* of London railed in 1896 about the "indecency" of viewing another person's bones. William Herbert Rollins of New England reported as early as 1901 that x rays could be dangerous and should be housed properly to prevent excess exposure. Contemporary dentists continue to use x rays extensively to determine the condition of the teeth and the roots.

Modern dentistry

Cavities and fillings

The great nineteenth century advances in dentistry provided dentists with the tools to repair or remove damaged teeth with a minimum of pain. The hallmarks of dentistry in the twentieth century have been advances in the preservation of teeth.

The success of these efforts can be seen in the fact that more older Americans retain their teeth. For example, the number of Americans without teeth was 18 million in 1986, according to the Centers for Disease Control. By 1989, the number had dropped to 16.5 million. Children also have fewer dental caries, the technical name for cavities. While nearly three quarters of all 9-year-olds had cavities in the early 1970s, only one-third of 9-year-olds had cavities in the late 1980s, according to the Centers for Disease Control.

But many dental problems and challenges still exist. The two most common types of oral disease are dental caries and periodontal disease, Rowe reports. Dental caries stem from the destruction of the tooth by microbial activity on the surface. Dental caries occur when **bacteria** forms a dental plaque on the surface of the tooth. Plaque is a deposit of bacteria and their products which is sticky and colorless. After the plaque is formed, food and the bacteria combine to create acids that slowly dissolve the substance of the tooth. The result is a hole in the tooth which must be filled or greater damage may occur, including eventual loss of the tooth.

Many different strategies exist to prevent dental caries. These include the reduction of sugar consumption. While some foods, such as starches, do not digest completely in the mouth, other foods, such as sugars, break down quickly in the mouth and are particularly harmful. Tooth brushing also helps reduce plaque. Other preventive techniques, such as the use of fluoride and sealants, are also helpful.

Fluoride was recognized as early as 1874 as a protector against tooth decay. Great controversy surrounded the addition of fluoride to the public water supply in many communities in the 1950s and 60s, as concerns were raised about the long-term health affects of fluoride. While controversy on the issue remains in some areas, public health experts suggest that fluoride has greatly improved dental health in young and old people. The number of cavities are reduced 65% in areas in which water is fluoridated.

Another advance was the development of sealants for children in the late 1960s. These sealants, made of a clear plastic material, are typically added to an etched tooth surface to protect the tooth from decay. They can protect teeth from cavities for up to 15 years. They are generally used on the chewing surfaces of back teeth, which are most prone to tooth decay. Sealants are currently recommended for all children by the American Dental Association.

Regular dental check-ups are used to monitor teeth and prevent dental caries from growing large. Contemporary dentists typically examine teeth using dental equipment to poke and probe teeth and x rays to see potential dental caries before they can be seen easily without aid. To detect problems, x-ray beams are focused on special photographic film placed in the mouth. The x rays create a record of the tooth, with the film documenting dental cavities or other problems in the tooth.

The process of fixing dental caries can be a short procedure depending on the size of the cavity. Small cavities may require no anesthesia and minimal drilling, while extensive dental caries may require novocaine or nitrous oxide to dull the pain and extensive drilling. Typically the process of filling a cavity begins with the dentist using a drill or a hand tool to grind down the part of the tooth surrounding the dental carry. The dentist then shapes the cavity, removes debris from the cavity, and dries it off. At this point a cement lining is added as to insulate the inside of the tooth. The cavity is filled by inserting an amalgam or some other substance in small increments, compressing the material soundly.

Teeth are usually filled with an amalgam including silver, **copper**, tin, mercury, indium, and palladium. Other materials may be used for front teeth where metallic fillings would stand out. These include plastic composite material, which can be made to match tooth **color**.

Controversy about the possible safety hazards of mercury in amalgam fillings led some Americans to have their amalgam fillings removed in the early 1990s. While mercury is a proven toxic chemical, there is no proof that mercury in amalgam fillings causes disease, according to the American Dental Association. Still, some experts suggest that dentists seek alternatives to mercury to combat potential problems and fear linked to mercury exposure.

Tooth replacement

Teeth that have large cavities, are badly discolored, or badly broken often are capped with a crown, which covers all or part of the crown, or visible portion, of the tooth. This can be made of gold or dental porcelain. Dental cement is used to keep the crown in place.

Bridges are created when individuals need some tooth replacement but not enough to warrant dentures, which offer more extensive tooth replacement. These devices clasp new teeth in place, keep decayed teeth strong, and support the teeth in a proper configuration. Missing or damaged teeth may lead to difficulty speaking and eating. Like bridges for **rivers** or streams, dental bridges can be constructed many different ways, depending on the need and the area that needs bridging. There are **cantilever** dental bridges and many other types. Some are removable by the dentist, and may be attached to the mouth by screw or soft cement. Others, called fixed bridges, are intended to be permanent.

Dentures, a set of replacement teeth, are used when all or a large part of the teeth must be replaced. New teeth can be made of acrylic resin or porcelain. Creating a base to set the teeth in is an ambitious undertaking, requiring the development of an impression from the existing teeth and jaws and the construction of a base designed to fit the mouth exactly and not add errors. Contemporary dentists generally use acrylic **plastics** as the base for dentures. Acrylic plastic is mixed as a dough, heated, molded, and set in shape.

Gum disease and bad breath

Gum disease is an immense problem among adults. The more common gum diseases, gingivitis, can be found in about 44% of all employed Americans 18–64. Periodontitis can be found in at least 14% of this group, though it and gingivitis is far more common among older people. Gingivitis is the **inflammation** of gum **tissue**, and is marked by bleeding, swollen gums. Periodontitis involves damage to the periodontal ligament, which connects each tooth to the bone. It also involves damage to the alveolar bone to which teeth are attached.

Untreated periodontal disease results in exposure of tooth root surfaces and pockets between the teeth and supporting tissue. This leaves teeth and roots more susceptible to decay and tooth loss.

Periodontitis and gingivitis are caused primarily by bacterial dental plaque. This plaque includes bacteria which produce destructive enzymes in the mouth. These enzymes can damage cells and **connective tissue**. To prevent gum disease from forming, experts suggest regular brushing, flossing and removal of bacterial plaque using various dental tools. Regular mechanical removal of plaque by a dentist or hygienist is also essential.

Periodontal surgery is necessary when damage is too great. During this procedure, gums are moved away from bone and teeth temporarily to allow dentists to clean out and regenerate the damaged area.

Another less serious dental problem is halitosis, or bad breath. Bad breath can be due to normal body processes or to illness. Halitosis early in the morning is normal, due to the added amount of bacteria in the mouth during **sleep** and the reduced secretion of saliva, which cleanses the mouth. Another normal cause of bad breath is when one is hungry. This occurs because the pancreatic juice enters the intestinal tract when one has not eaten for some time, causing a bad **smell**. Certain foods also cause bad breath, such as garlic, **alcohol**, and fatty meat, which causes halitosis because the **fatty acids** are excreted through the lungs.

Halitosis can also be caused by a wealth of illnesses, ranging from diabetes to kidney failure and chronic lung disease. Dental problems such as plaque and dental caries can also contribute to bad breath. Treatment for the condition typically involves treating the illness, if that is causing the problem, and improving oral hygiene. This means brushing the tongue as well as the teeth.

Orthodontics: the art of moving teeth

The practice of orthodontics depends on the fact that the position of teeth in the mouth can be shaped and changed gradually using **pressure**. Orthodontia is used to correct problems ranging from a bite that is out of alignment, to a protruding jaw, to crowded teeth. Typically orthodontia begins when individuals are in their early teens, and takes about two years. However, with the development of clear plastic braces, adults are increasingly likely to turn to orthodontia to improve their appearance, and make eating and talking more comfortable.

The process may require some teeth to be pulled. The removal of teeth allows for the growth of other teeth to fill the newly-vacant area. Braces are made up of a network of wires and bands made of stainless **steel** or clear plastic. The tubes are often anchored on the molars and the wires are adjusted to provide steady pressure on the surface of the teeth. This pressure slowly moves the tooth to a more desirable location in the mouth and enables new bone to build up where it is needed. Orthodontia can also be used to help move the jaw by anchoring wires to the opposing jaw.

A look forward

Laser beams are already used in dentistry and in medical practice. But lasers are currently not used for everyday dentistry, such as the drilling of teeth. In the future, as laser technology becomes more refined, lasers may take the place of many conventional dental tools. Using lasers instead of dental tools would cut down on the opportunity to be exposed to blood-borne illness, and reduce damage to surrounding tissue.

Researchers also are exploring new ways to treat periodontal disease, such as more specific antibacterial therapy and stronger antibacterial agents. Many researchers also see a stronger role for fluoride in the future, in addition to its current presence in many public water supplies. Some dentists advocate the use of fluoride in sealants. A 1991 study reported that a sealant including fluoride reduced tooth irritation for some individuals with sensitive teeth.

While dentistry has made immense progress since days when a dead mouse was considered high dental technology, there is still progress to be made. Future challenges for the dental profession include continuing to reduce tooth loss and decay due to neglect and the aging process.

Resources

Books

Aschheim, Kenneth W., and Barry G. Dale. *Esthetic Dentistry: A Clinical Approach to Techniques and Materials.* 2nd ed. St. Louis: Mosby, Inc., 2001.

Hupp, James, and Larry J. Peterson. *Contemporary Oral and Maxillofacial Surgery.* 4th ed. St. Louis: Mosby, Inc., 2002.

KEY TERMS

Abscess—An enclosed collection of liquefied tissue, known as pus, somewhere in the body.

Bridge—Replacement for a missing tooth or teeth which is supported by roots or natural teeth.

Gingivitis—Gum inflammation

Vulcanization—A process in which sulfur and raw latex are combined at a high temperature to make rubber more durable.

Proffit, William, and Henry W. Fields. *Contemporary Orthodontics.* 3rd ed. Chicago: Year Book Medical Publishing, 2000.

Roderick, Cawson, and William Binnie, Anderw Barrett, and John Wright. *Oral Disease.* 3rd ed. St. Louis: Mosby, Inc., 2001.

Periodicals

Gift, Helen C., Stephen B. Corbin, and Ruth E. Nowjack-Raymer. "Public Knowledge of Prevention of Dental Disease." *Public Health Reports* 109, 397. (May-June 1994).

"What Will the Future Bring?" *Journal of the American Dental Association* 123. (April 1992): 40-46.

Patricia Braus

Denumberable *see* **Countable**

Deoxyribonucleic acid (DNA)

Deoxyribonucleic acid (DNA), "the master molecule," is a natural **polymer** which encodes the genetic information required for the growth, development, and reproduction of an **organism**. Found in all cells, it consists of chains of units called nucleotides. Each nucleotide unit contains three components: the sugar deoxyribose, a phosphate group, and a nitrogen-containing amine or base with a ring-type structure. The base component can be any of four types: adenine, cytosine, guanine or thymine.

DNA molecules are very long and threadlike. They consist of two polymeric strands twisted about each other into a **spiral** shape known as a **double helix**, which resembles two intertwined circular staircases. DNA is found within the **cell** nucleus in the chromosomes, which are extremely condensed structures in which DNA is associated with **proteins**. Each **species** contains a characteristic number of chromosomes in their cells. In hu-

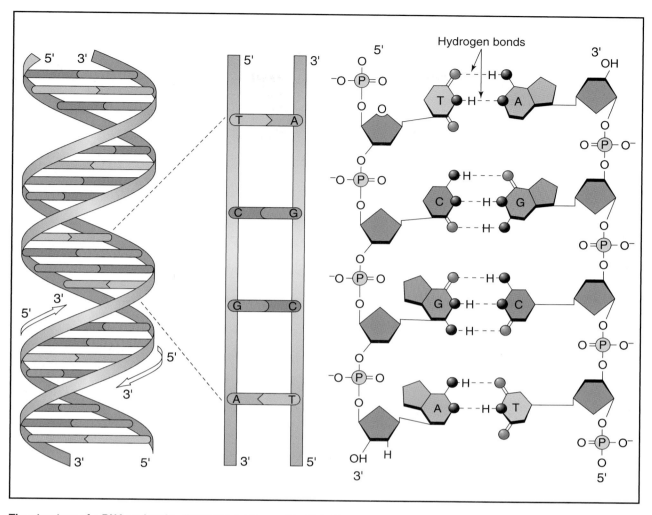

The structure of a DNA molecule. *Illustration by Argosy. The Gale Group.*

mans, every cell contains 46 chromosomes (except for egg and sperm cells which contain only 23). The total genetic information in a cell is called its **genome**.

The fundamental units of heredity are genes. A **gene** is a segment of a DNA **molecule** that encodes the information necessary to make a specific protein. Proteins are the "workhorses" of the cell. These large, versatile molecules serve as structural components: they transport molecules in and out of cells, catalyze cellular reactions, and recognize and eliminate invaders. Imagine a community in which the trash collectors, goods distributors, manufacturers, and police are all on strike, and you get an idea of the importance of proteins in the life of a cell.

DNA not only encodes the "blueprints" for cellular proteins but also the instructions for when and where they will be made. For example, the **oxygen** carrier hemoglobin is made in red **blood** cells but not in nerve cells, though both contain the same total genetic content.

Thus, DNA also contains the information necessary for regulating how its genetic messages are used.

Human cells are thought to contain between 50,000 and 100,000 genes. Except in the case of identical twins, a comparison of the genes from different individuals always reveals a number of differences. Therefore, each person is genetically unique. This is the basis of DNA "fingerprinting," a forensic procedure used to match DNA collected from a crime scene with that of a suspect.

Through the sum of their effects, genes direct the function of all organs and systems in the body. Defects in the DNA of just one gene can cause a genetic disorder which results in **disease** because the protein encoded by the defective gene is abnormal. The abnormal hemoglobin produced by people afflicted with **sickle cell anemia** is an example. Defects in certain genes called oncogenes, which regulate growth and development, give rise to **cancer**. Only about 100 genes are thought to be oncogenes. Therefore, defects in DNA can affect the two kinds of genetic information it carries, messages directing the manufacture of proteins and information regulating the expression, or carrying out, of these messages.

History

Prior to the discovery of the nucleic acids, the Austrian monk Gregor Mendel (1822-1884) worked out the laws of inheritance by the selective breeding of pea-plants. As early as 1865 he proposed that "factors" from each parent were responsible for the inheritance of certain characteristics in plants. The Swiss biochemist Friedrich Miescher (1844-1895) discovered the nucleic acids in 1868 in nuclei isolated from pus cells scraped from surgical bandages. However, research on the chemical structure of nucleic acids lagged until new analytical techniques became available in the mid twentieth century. With the advent of these new methods came evidence that the **nucleic acid** we now know as DNA. DNA was present in the nuclei of all cells and evidence about the chemical structure of its nucleotide components.

Despite knowledge of the chemical structure of nucleotides and how they were linked together to form DNA, the possibility that DNA was the genetic material was regarded as unlikely. As late as the mid twentieth century, proteins were thought to be the molecules of heredity because they appeared to be the only cellular components diverse enough to account for the large variety of genes. In 1944, Oswald Avery (1877-1955) and his colleagues showed that non-pathogenic strains of *pneumococcus*, the bacterium that causes pneumonia, could become pathogenic (disease-causing) if treated with a DNA-containing extract from heat-killed pathogenic strains. Based on this evidence, Avery concluded that DNA was the genetic material. However, widespread acceptance of DNA as the bearer of genetic information did not come until a report by other workers in 1952 that DNA, not protein, enters a bacterial cell infected by a **virus**. This showed that the genetic material of the virus was contained in its DNA, confirming Avery's hypothesis.

Shortly afterwards in 1953, James Watson (1928-) and Francis Crick (1916-) proposed their double helix model for the three-dimensional structure of DNA. They correctly deduced that the genetic information was encoded in the form of the sequence of nucleotides in the molecule. With their landmark discovery began an era of molecular **genetics** in **biology**. Eight years later investigators cracked the genetic code. They found that specific trinucleotide sequences—sequences of three nucleotides—are codes for each of 20 amino acids, the building blocks of proteins.

In 1970 scientists found that **bacteria** contained restriction enzymes molecular "scissors" that recognize a particular sequence of 4-8 nucleotides and will always cut DNA at or near that sequence to yield specific (rather than **random**), consistently reproducible DNA fragments. Two years later it was found that the bacterial **enzyme** DNA ligase could be used to rejoin these fragments. This permitted scientists to construct "recombinant" DNA molecules; that is, DNA molecules composed of segments from two different sources, even from different organisms. With the availability of these tools, **genetic engineering** became possible and **biotechnology** began.

By 1984 the development of **DNA fingerprinting** allowed forensic chemists to compare DNA samples from a crime scene with that of suspects. The first conviction using this technique came in 1987. Three years later doctors first attempted to treat a patient unable to produce a vital immune protein using **gene therapy**. This technique involves inserting a portion of DNA into a patient's cells to correct a deficiency in a particular function. The **Human Genome Project** also began in 1990. The aim of this project is to determine the nucleotide sequence in DNA of the entire human genome, which consists of about three billion nucleotide pairs. In 2001, researchers announced the completion of the sequencing of a human genome, promising refinement by 2003.

Structure

Deoxyribose, the sugar component in each nucleotide, is so called because it has one less oxygen atom than ribose, which is present in **ribonucleic acid (RNA)**. Deoxyribose contains five **carbon** atoms, four of which lie in a ring along with one oxygen atom. The fifth carbon atom is linked to a specific carbon atom in the ring. A phosphate group is always linked to deoxyribose via a **chemical bond** between an oxygen atom in the phosphate group and the carbon atom in deoxyribose by a chemical bond between a **nitrogen** atom in the base and a specific carbon atom in the deoxyribose ring.

The nucleotide components of DNA are connected to form a linear polymer in a very specific way. A phosphate group always connects the sugar component of a nucleotide with the sugar component of the next nucleotide in the chain. Consequently, the first nucleotide bears an unattached phosphate group, and the last nucleotide has a free hydroxyl group. Therefore, DNA is not the same at both ends. This directionality plays an important role in the replication of DNA.

DNA molecules contain two polymer chains or strands of nucleotides and so are said to be double-stranded. (In contrast, RNA is typically single-stranded.) Their shape resembles two intertwined spiral staircases in which the alternating sugar and phosphate groups of the nucleotides compose the sidepieces. The steps con-

Diagramatic representations of the chemical stuctures of the nitrogenous bases that comprise the rungs of the twisted DNA helical ladder are shown above. The dashed lines represent the potential hydrogen bonds that link Adenine with Thymine (A-T base pairing) or Cytosine with Guanine (C-G) base pairing. The specific base sequence becomes the fundamental element of the genetic code. *Illustration by Argosy. The Gale Group.*

sist of pairs of bases, each attached to the sugars on their respective strands. The bases are held together by weak attractive forces called **hydrogen** bonds. The two strands in DNA are antiparallel, which means that one strand goes in one direction (first to last nucleotide from top to bottom) and the other strand goes in the opposite direction (first to last nucleotide from bottom to top).

Because the sugar and phosphate components which make up the sidepieces are always attached in the same way, the same alternating phosphate-sugar sequence repeats over and over again. The bases attached to each sugar may be one of four possible types. Because of the **geometry** of the DNA molecule, the only possible base pairs that will fit are adenine (A) paired with thymine (T), and cytosine (C) paired with guanine (G).

The DNA in our cells is a masterpiece of packing. The double helix coils itself around protein cores to form nucleosomes. These DNA-protein structures resemble beads on a string. Flexible regains between nucleosomes allows these structures to be wound around themselves to produce an even more compact fiber. The fibers can then be coiled for even further compactness. Ultimately, DNA is paced into the highly condensed chromosomes. If the DNA in a human cell is stretched, it is approximately 6 ft (1.82 m) long. If all 46 chromosomes are laid end-to-end, their total length is still only about eight-

thousandths of an inch. This means that DNA in chromosomes is condensed about 10,000 times more than that in the double helix. Why all this packing? The likely answer is that the fragile DNA molecule would get broken in its extended form. Also, if not for this painstaking compression, the cell might be mired in its own DNA.

Function

DNA directs a cell's activities by specifying the structures of its proteins and by regulating which proteins and how much are produced, and where. In so doing, it never leaves the nucleus. Each human cell contains about 6 ft (2 m) of highly condensed DNA which encodes some 50,000–100,000 genes. If a particular protein is to be made, the DNA segment corresponding to the gene for that protein acts as a template, a pattern, for the synthesis of an RNA molecule in a process known as transcription. This messenger RNA molecule travels from the nucleus to the cytoplasm where it in turn acts as the template for the construction of the protein by the protein assembly apparatus of the cell. This latter process is known as translation and requires an adaptor molecule, transfer RNA, which translates the genetic code of DNA into the language of proteins.

Eventually, when a cell divides, its DNA must be copied so that each daughter cell will have a complete set of genetic instructions. The structure of DNA is perfectly suited to this process. The two intertwined strands unwind, exposing their bases, which then pair with bases on free nucleotides present in the cell. The bases pair only in a certain combination; adenine (A) always pairs with thymine (T) and cytosine (C) always pairs with guanine (G). The sequence of bases along one strand of DNA therefore determines the sequence of bases in the newly forming complementary strand. An enzyme then joins the free nucleotides to complete the new strand. Since the two new DNA strands that result are identical to the two originals, the cell can pass along an exact copy of its DNA to each daughter cell.

Sex cells, the eggs and sperm, contain half the number of chromosomes as other cells. When the egg and sperm fuse during **fertilization**, they form the first cell of a new **individual** with the complete complement of DNA—46 chromosomes. Each cell (except the sex cells) in the new person carries DNA identical to that in the fertilized egg cell. In this way the DNA of both parents is passed from one generation to the next. Thus, DNA plays a crucial role in the propagation of life.

Replication of DNA

DNA replication, the process by which the double-stranded DNA molecule reproduces itself, is a complicated process, even in the simplest organisms. DNA synthesis—making new DNA from old—is complex because it requires the interaction of a number of cellular components and is rigidly controlled to ensure the accuracy of the copy, upon which the very life of the organism depends. This adds several verification steps to the procedure. Though the details vary from organism to organism, DNA replication follows certain rules that are universal to all.

DNA replication (duplication, or copying) is always semi-conservative. During DNA replication the two strands of the parent molecule unwind and each becomes a template for the synthesis of the complementary strand of the daughter molecule. As a result both daughter molecules contain one new strand and one old strand (from the parent molecule), hence the term semi-conservative. The replication of DNA always requires a template, an intact strand from the parent molecule. This strand determines the sequence of nucleotides on the new strand. Wherever the nucleotide on the template strand contains the base A, then the nucleotide to be added to the daughter strand at that location must contain the base T. Conversely, every T must find an A to pair with. In the same way, Gs and Cs will pair with each other and with no other bases.

Replication begins at a specific site called the replication origin when the enzyme DNA helicase binds to a portion of the double stranded helix and "melts" the bonds between base pairs. This unwinds the helix to form a replication fork consisting of two separated strands, each serving as a template. Specific proteins then bind to these single strands to prevent them from re-pairing. Another enzyme, DNA polymerase, proceeds to assemble the daughter strands using a pool of free nucleotide units which are present in the cell in an "activated" form.

High fidelity in the copying of DNA is vital to the organism and, incredibly, only about one error per one trillion replications ever occurs. This high fidelity results largely because DNA polymerase is a "self-editing" enzyme. If a nucleotide added to the end of the chain mismatches the complementary nucleotide on the template, pairing does not occur. DNA polymerase then clips off the unpaired nucleotide and replaces it with the correct one.

Occasionally errors are made during DNA replication and passed along to daughter cells. Such errors are called mutations. They have serious consequences because they can cause the insertion of the wrong **amino acid** into a protein. For example, the substitution of a T for an A in the gene encoding hemoglobin causes an amino acid substitution which results in sickle cell **anemia**. To understand the significance of such mutations requires knowledge of the genetic code.

The genetic code

Genetic information is stored as nucleotide sequences in DNA (or RNA) molecules. This sequence specifies the identity and position of the amino acids in a particular protein. Amino acids are the building blocks of proteins in the same way that nucleotides are the building blocks of DNA. However, though there are only four possible bases in DNA (or RNA), there are 20 possible amino acids in proteins. The genetic code is a sort of "bilingual dictionary" which translates the language of DNA into the language of proteins. In the genetic code the letters are the four bases A, C, G, and T (or U instead of T in RNA). Obviously, the four bases of DNA are not enough to code for 20 amino acids. A sequence of two bases is also insufficient, because this permits coding for only 16 of the 20 amino acids in proteins. Therefore, a sequence of three bases is required to ensure enough combinations or "words" to code for all 20 amino acids. Since all words in this DNA language, called **codons**, consist of three letters, the genetic code is often referred to as the triplet code.

Each codon specifies a particular amino acid. Because there are 64 possible codons (for example $4^3 = 64$ different 3-letter "words" can be generated from a 4-letter "alphabet") and only 20 amino acids, several different codons specify the same amino acid, so the genetic code is said to be degenerate. However, the code is unambiguous because each codon specifies only one amino acid. The sequence of codons are not interrupted by "commas" and are always read in the same frame of reference, starting with the same base every time. So the "words" never overlap.

Since DNA never leaves the nucleus, the information it stores is not transferred to the cell directly. Instead, a DNA sequence must first be copied into a messenger RNA molecule, which carries the genetic information from the nucleus to protein assembly sites in the cytoplasm. There it serves as the template for protein construction. The sequences of nucleotide triplets in messenger RNA are also referred to as codons.

Four codons serve special functions. Three are stop codons that signal the end of protein synthesis. The fourth is a start codon which establishes the "reading frame" in which the message is to be read. For example, suppose the message is PAT SAW THE FAT RAT. If we overshoot the reading frame by one "nucleotide," we obtain ATS AWT HEF ATR AT, which is meaningless.

The genetic code is essentially universal. This means that a codon which specifies the amino acid tryptophan in bacteria also codes for it in man. The only exceptions occur in mitochondria and chloroplasts and in some **protozoa**. (Mitochondria and chloroplasts are subcellular compartments which are the sites of **respiration** in animals and plants, respectively, and contain some DNA.) The structure of the genetic code has evolved to minimize the effect of mutations. Changes in the third base of a codon do not necessarily result in a change in the specified amino acid during protein synthesis. Furthermore, changes in the first base in a codon generally result in the same or at least a similar amino acid. Studies of amino acid changes resulting from mutations have shown that they are consistent with the genetic code. That is, amino acid changes resulting from mutations are consistent with expected base changes in the corresponding codon. These studies have confirmed that the genetic code has been deduced correctly by demonstrating its relevance in actual living organisms.

Expression of genetic information

Genetic information flows from DNA to RNA to protein. Ultimately, the linear sequence of nucleotides in DNA directs the production of a protein molecule with a characteristic three dimensional structure essential to its proper function. Initially, information is transcribed from DNA to RNA. The information in the resulting messenger RNA is then translated from RNA into protein by small transfer RNA molecules.

In some exceptional cases the flow of genetic information from DNA to RNA is reversed. In retroviruses, such as the **AIDS** virus, RNA is the hereditary material. An enzyme known as reverse transcriptase makes a copy of DNA using the virus' RNA as a template. In still other viruses which use RNA as the hereditary material, DNA is not involved in the flow of information at all.

Most cells in the body contain the same DNA as that in the fertilized egg. (Some exceptions to this are the sex cells, which contain only half of the normal complement of DNA, as well as red blood cells which lose their nucleus when fully developed.) Some "housekeeping" genes are expressed in all cells because they are involved in the fundamental processes required for normal function. (A gene is said to be expressed when its product, the protein it codes for, is actively produced in a cell.) For example, since all cells require **ribosomes**, structures which function as protein assembly lines, the genes for ribosomal proteins and ribosomal RNA are expressed in all cells. Other genes are only expressed in certain cell types, such as genes for antibodies in certain cells of the **immune system**. Some are expressed only during certain times in development. How is it that some cells express certain genes while others do not, even though all contain the same DNA? A complete answer to this question is still in the works. However, the main way is by controlling the start of transcription. This is accomplished by the interaction of proteins called transcription factors with DNA sequences near the gene. By binding to these sequences transcription factors may turn a gene on or off.

Another way is to change the **rate** of messenger RNA synthesis. Sometimes the stability of the messenger RNA is altered. The protein product itself may be altered, as well as its transport or stability. Finally, gene expression can be altered by DNA rearrangements. Such programmed reshuffling of DNA is the means of generating the huge assortment of antibody proteins found in immune cells.

Genetic engineering and recombinant DNA

Restriction enzymes come from **microorganisms**. Recall that they recognize and cut DNA at specific base pair sequences. They cleave large DNA molecules into an assortment of smaller fragments ranging in size from a few to thousands of base pairs long, depending on how often and where the cleavage sequence appears in the original DNA molecule. The resulting fragments can be separated by their size using a technique known as **electrophoresis**. The fragments are placed at the top of a porous gel surrounded by a **solution** which conducts **electricity**. When a voltage is applied, the DNA frag-

KEY TERMS

Codon—The base sequence of three consecutive nucleotides on DNA (or RNA) that codes for a particular amino acid or signals the beginning or end of a messenger RNA molecule.

Cytoplasm—All the protoplasm in a living cell that is located outside of the nucleus, as distinguished from *nucleoplasm,* which is the protoplasm in the nucleus.

Gene—A discrete unit of inheritance, represented by a portion of DNA located on a chromosome. The gene is a code for the production of a specific kind of protein or RNA molecule, and therefore for a specific inherited characteristic.

Genetic code—The blueprint for all structures and functions in a cell as encoded in DNA.

Genetic engineering—The manipulation of the genetic content of an organism for the sake of genetic analysis or to produce or improve a product.

Genome—The complete set of genes an organism carries.

Nucleotide—The basic unit of DNA. It consists of deoxyribose, phosphate, and a ring-like, nitrogen-containing base.

Nucleus—A compartment in the cell which is enclosed by a membrane and which contains its genetic information.

Replication—The synthesis of a new DNA molecule from a pre-existing one.

Transcription—The process of synthesizing RNA from DNA.

Translation—The process of protein synthesis.

---

ments move towards the bottom of the gel due to the **negative** charge on their phosphate groups. Because it is more difficult for the large fragments to pass through the pores in the gel, they move more slowly than the smaller fragments.

DNA fragments isolated from a gel in this way can be joined with DNA from another source, either of the same or a different species, into a new, **recombinant DNA** molecule by enzymes. Usually, such DNA fragments are joined with DNA from subcellular organisms—"parasites" that live inside another organism but have their own DNA. Plasmids and viruses are two such examples. Viruses consist only of nucleic acids encapsulated in a protein coat. Though they can exist outside the cell, they are inactive. Inside the cell, they take over its metabolic machinery to manufacture more virus particles, eventually destroying their host. Plasmids are simpler than viruses in that they never exist outside the cell and have no protein coat. They consist only of circular double-stranded DNA. Plasmids replicate their DNA independently of their hosts. They are passed on to daughter cells in a controlled way as the host cell divides.

Cells that contain the same recombinant DNA fragment are clones. A clone harboring a recombinant DNA molecule that contains a specific gene can be isolated and identified by a number of techniques, depending upon the particular experiment. Thus, recombinant DNA molecules can be introduced into rapidly growing microorganisms, such as bacteria or **yeast**, to produce large quantities of medically or commercially important proteins normally present only in scant amounts in the cell. For example, human **insulin** and interferon have been produced in this manner.

In recent years a technique has been developed which permits analysis of very small samples of DNA without repeated cloning, which is laborious. Known as the polymerase chain reaction, this technique involves "amplifying" a particular fragment of DNA by repeated synthesis using the enzyme DNA polymerase. This method can increase the amount of the desired DNA fragment by a million-fold or more.

See also Chromosome; Enzyme; Genetics; Meiosis; Mitosis; Mutation; Nucleic acid.

Resources

Books

Berg, Paul, and Maxine Singer. *Dealing with Genes—The Language of Heredity.* Mill Valley, CA: University Science Press, 1992.

Blueprint for Life. Journey Through the Mind and Body series. Alexandria, VA: Time-Life Books, 1993.

Lee, Thomas F. *Gene Future.* New York: Plenum Publishing Corporation, 1993.

Rosenfeld, Israel, Edward Ziff, and Borin Van Loon. *DNA for Beginners.* New York: Writers and Readers Publishing Cooperative Limited, 1983.

Sofer, William H. *Introduction to Genetic Engineering.* Stoneham, MA: Butterwoth-Heineman, 1991.

Patricia V. Racenis

Deposit

A deposit is an accumulation of **Earth** materials, usually loose sediment or **minerals**, that is laid down by a natural agent. Deposits are all around you—the **sand** on the beach, the **soil** in your backyard, the **rocks** in a mountain stream. All of these consist of earth materials transported and laid down (that is, deposited) by a natural agent. These natural agents may include flowing **water**, **ice**, or gusts of **wind** (all operating under the influence of gravity), as well as gravity acting alone. For example, gravity alone can cause a rock fall along a highway, and the rock fall will form a deposit at the base of the slope. The agents of transport and deposition mentioned above are mechanical in nature and all operate in the same way. Initially, some **force** causes a particle to begin to move. When the force decreases, the **rate** of particle **motion** also decreases. Eventually particle motion ceases and mechanical deposition occurs.

Not all deposits form by mechanical deposition. Some deposits form instead by chemical deposition. As you may know, all naturally occurring water has some minerals dissolved in it. Deposition of these minerals may result from a variety of chemical processes; however, one of the most familiar is **evaporation**. When water evaporates, dissolved minerals remain behind as a solid residue. This residue is a chemical deposit of minerals.

Ocean water is very rich in dissolved minerals—that is why ocean water tastes salty. When ocean water evaporates, a deposit containing a variety of minerals accumulates. The mineral halite (that is, table **salt**) would make up the bulk of such a deposit. Large, chemically derived mineral deposits, which formed by the evaporation of ancient saline lakes, are currently being mined in several areas of the western United States. The Bonneville Salt Flats in Utah is a good example of an "evaporite" mineral deposit. Due to the arid climate, evaporite minerals are still being deposited today at Great Salt Lake in Utah.

The term "deposit" generally applies only to accumulations of earth materials that form at or near the earth's surface, that is, to particles, rocks, or minerals that are of sedimentary origin. However, **ore** deposits are an exception to this generality. The phrase "ore deposit" applies to any valuable accumulation of minerals, no matter how or where it accumulates. Some ore deposits do form by mechanical or chemical deposition (that is, they are of sedimentary origin).

For example, flowing streams deposit gold-bearing sand and gravel layers, known as placers. Placers, therefore, form by mechanical deposition. Some **iron** ores, on the other hand, form when subsurface waters chemically deposit iron in porous zones within sediments or rocks. However, many ore deposits do not form by either mechanical or chemical deposition, and so are not of sedimentary origin.

See also Sediment and sedimentation.

Depositional environment *see* **Sediment and sedimentation**

Depression

Depression is a psychoneurotic disorder characterized by lingering sadness, inactivity, and difficulty in thinking and concentration. A significant increase or decrease in appetite and time spent sleeping, feelings of dejection and hopelessness, and sometimes suicidal tendencies may also be present. It is one of the most common psychiatric conditions encountered, and affects up to 25% of women and 12% of men. Depression differs from grief, bereavement, or mourning, which are appropriate emotional responses to the loss of loved persons or objects.

Depression has many forms and is very responsive to treatment. Dysthymia, or minor depression, is the presence of a depressed mood for most of the day for two years with no more than two months' freedom from symptoms. Bipolar disorder (manic-depressive disorder) is characterized by recurrent episodes of **mania** and major depression. Manic symptoms consist of feelings of inflated self-esteem or grandiosity, a decreased need for **sleep**, unusual loquacity, an unconnected flow of ideas, distractibility, or excessive involvement in pleasurable activities that have a high potential for painful consequences, such as buying sprees or sexual indiscretions. Cyclothymia is a chronic mood disturbance and is a milder form of bipolar disorder.

Chemically speaking, depression is apparently caused by reduced quantities or reduced activity of the monoamine neurotransmitters serotonin and norepinephrine within the **brain**. Neurotransmitters are chemical agents released by neurons (nerve cells) to stimulate neighboring neurons, thus allowing electrical impulses to be passed from one **cell** to the next throughout the **nervous system**. They transmit signals between nerve cells in the brain.

Introduced in the late 1950s, **antidepressant drugs** have been used most widely in the management of major mental depression. All antidepressants accomplish their task by inhibiting the body's re-absorption of these neurotransmitters, thus allowing them to accumulate and remain in contact longer with their receptors in nerve cells. These changes are important in elevating mood and relieving depression.

Antidepressants are typically one of three chemical types: a tricyclic antidepressant (so called because its molecules are composed of three rings), a monoamine oxidase (MAO) inhibitor, or a serotonin reuptake inhibitor. The tricyclic antidepressants act by inhibiting the inactivation of norepinephrine and serotonin within the brain.

The MAOs apparently achieve their effect by interfering with the action of monoamine oxidase, an **enzyme** that causes the breakdown of norepinephrine, serotonin, and **dopamine** within nerve cells.

In the 1980s a new type of antidepressant called a serotonin reuptake inhibitor proved markedly successful. Its chemical name is fluoxetine, and it apparently achieves its therapeutic effect by interfering solely with the reabsorption of serotonin within the brain, thus allowing that **neurotransmitter** to accumulate there. Fluoxetine often relieves cases of depression that have failed to respond to tricyclics or MAOs, and it also produces fewer and less serious side effects than those drugs. It had thus become the single most widely used antidepressant by the end of the twentieth century. The most commonly used serotonin reuptake inhibitors are Prozac, Paxil, and Zoloft.

Medical experts agree that antidepressants are only a part of the therapeutic process when treating depression. Some form of psychotherapy is also needed in order to reduce the incidence for chronic recidivism of the illness.

Depth perception

Depth **perception** is the ability to see the environment in three dimensions and to estimate the spatial distances of objects from ourself and from each other. Depth perception is vital for our survival, being necessary to effectively navigate around and function in the world. Without it we would be unable to tell how far objects are from us, and thus how far we would need to move to reach or avoid them. Moreover, we would not be able to distinguish between, for instance, stepping down a stair from stepping off of a tall building.

Our ability to perceive depth encompasses space perception, or the ability to perceive the differential distances of objects in space. While researchers have discovered much about depth perception, numerous interesting questions remain. For instance, how are we able to perceive the world in three dimensions when the images projected onto the retina are basically two-dimensional and flat? And how much of a role does **learning** play in depth perception? While depth perception results primarily from our sense of **vision**, our sense of **hearing** also plays a role. Two broad classes of cues used to aid visual depth perception have been distinguished-the monocular (requiring only one **eye**), and the binocular (requiring both eyes working together.)

Monocular cues

The following cues require only one eye for their perception. They provide information that helps us estimate spatial distances and to perceive in three dimensions.

Interposition

Interposition refers to objects appearing to partially block or overlap one another. When an object appears partially blocked by another, the fully visible object is perceived as being nearer, and this generally corresponds to reality.

Shading and lighting

In general, the nearer an object is to a **light** source, the brighter its surface appears to be, so that with groups of objects, darker objects appear farther away than brighter objects. And in looking at single objects, the farther parts of an object's surface are from the source of light, the more shadowed and less bright they will appear. Varying shading and lighting then provide information about distances of objects from the source of light, and may serve as a cue to the **distance** of the object from the observer. In addition, some patterns of lighting and shading seem to provide cues about the shapes of objects.

Aerial perspective

Generally, objects having sharp and clear images appear nearer than objects with blurry or unclear images. This occurs because light is scattered or absorbed over long distances by particles in the atmosphere such as **water** vapor and dust which to a blurring of objects' lines. This is why on clear days, very large objects such as **mountains** or buildings appear closer than when viewed on hazy days.

Elevation

This cue, sometimes referred to as "height in the plane" or "relative height," describes how the horizon is seen as vertically higher than the foreground. Thus objects high in the visual field and closer to the horizon line are perceived as being farther away than objects lower in the visual field and farther away from the horizon line.

Above the horizon line this relationship is reversed, so that above the horizon, objects that are lower and nearer to the horizon line appear farther away than those up higher and at a greater distance from the horizon line.

Texture gradients

Textures that vary in complexity and **density** are a characteristic of most object surfaces and they reflect light differentially. Generally, as distance increases, the size of elements making up surface texture appear smaller and the distance between the elements also appears to decrease with distance. Thus if one is looking at a field of grass, the blades of grass will appear smaller and arranged more closely together as their distance increases. Texture gradients also serve as depth and distance cues in groupings of different objects with different textures in the visual field, as when looking at a view of a city. Finally, abrupt changes in texture usually indicate an alteration in the direction of an object's surface and its distance from the observer.

Linear perspective

Linear perspective is a depth cue based on the fact that as objects increase in distance from the observer their images on the retina are transformed so that their size and the space separating them decrease until the farthest objects meet at what is called the vanishing **point**. It is called the vanishing point because it is the point where objects get so small that they are no longer visible. In addition, physically **parallel** lines such as those seen in railroad tracks are perceived as coming closer together until they meet or converge at the vanishing point.

Motion parallax

Whenever our eyes move (due to eye movement alone, or head, or body movement) in relation to the spatial environment, objects at varying distances move at different rates relative to their position and distance from us. In other words, objects at different distances relative to the observer are perceived as moving at different speeds. **Motion parallax** refers to these relatively perceived object motions which we use as cues for the perception of distance and motion as we move through the environment.

As a rule, when the eyes move, objects close to the observer seem to move faster than objects farther away. In addition, more distant objects seem to move smaller distances than do nearer objects. Objects that are very far away, such as a bright **star** or the **moon**, seem to move at the exact same **rate** as the observer and in the same direction.

The amount and direction of movement are relative to the observer's fixation point or where they are fo-

A baby puts to use his/her depth perception by reaching out and touching a toy overhead. *Photograph by Gabe Palmer. Stock Market. Reproduced by permission.*

cussing. For instance, if you were travelling on a train and focussing on the middle of a large field you were passing, any objects closer to you than your fixation point would seem to be moving opposite to your direction of movement. In addition, those objects beyond your fixation point would appear to be moving in the same direction as you are moving. Motion parallax cues provide strong and precise distance and depth information to the observer.

Accommodation

Accommodation occurs when curvature of the eye **lens** changes differentially to form sharp retinal images of near and far objects. To focus on far objects the lens becomes relatively flat and to focus on nearer objects the lens becomes more curved. Changes in the lens shape are controlled by the ciliary muscles and it seems that feedback from alterations in ciliary muscle tension may furnish information about object distance.

Retinal size

As an object's distance from the viewer increases, the size of its image on the retina becomes smaller. And, generally, in the absence of additional visual cues, larger objects are perceived as being closer than are smaller objects.

Familiarity

While not exactly a visual cue for perceiving space or depth as are the previous ones discussed, our familiarity with spatial characteristics of an object such as its size or shape due to experience with the object may contribute to estimates of distance and thus spatial perception. For instance, we know that most cars are taller or higher than children below the age of five, and thus in the absence of other relevant visual cues, a young child seen in front of a car who is taller than the car would be perceived as being closer than the car.

Binocular cues

Monocular cues certainly provide a great deal of spatial information, but depth perception also requires binocular functioning of the eyes, that is, both eyes working together in a coordinated fashion. Convergence and retinal disparity are binocular cues to depth perception.

Convergence

Convergence refers to the eyes' disposition to rotate inward toward each other in a coordinated manner in order to focus effectively on nearby objects. With objects that are farther away, the eyes must move outward toward one's temples. For objects further than approximately 20 ft (6 m) away no more changes in convergence occur and the eyes are essentially parallel with each other. It seems that feedback from changes in muscular tension required to cause convergence eye movements may provide information about depth or distance.

Retinal disparity and stereopsis

Retinal disparity refers to the small difference between the images projected on the two retinas when looking at an object or scene. This slight difference or disparity in retinal images serves as a binocular cue for the perception of depth. Retinal disparity is produced in humans (and in most higher **vertebrates** with two frontally directed eyes) by the separation of the eyes which causes the eyes to have different angles of objects or scenes. It is the foundation of stereoscopic vision.

Stereoscopic vision refers to the unified three-dimensional view of objects produced when the two dif-

ferent images are fused into one (binocular fusion). We still do not fully understand the mechanisms behind stereopsis but there is evidence that certain cells in some areas of the **brain** responsible for vision are specifically responsive to the specific type of retinal disparity involving slight horizontal differences in the two retinal images. This indicates that there may be other functionally specific cells in the brain that aid depth perception. In sum, it seems that we use numerous visual depth cues, binocular vision, and functionally specific cells in the **nervous system** to make accurate depth judgements.

Auditory depth cues

Auditory depth cues are used by everyone but are especially important for the blind. These include the relative loudness of familiar sounds, the amount of reverberation of sounds as in echoes, and certain characteristics of sounds unique to their **frequency**. For instance, higher frequency sounds are more easily absorbed by the atmosphere.

Development of depth perception

A theme running throughout the study of perception in general since the time of the ancient Greeks has been whether perceptual processes are learned (based on past experience) or innate (existent or potential at **birth**). In terms of depth perception, research using the visual cliff with animals and human infants too young to have had experience with depth perception indicates that humans and various **species** of animals are born with some innate abilities to perceive depth.

The visual cliff is one the most commonly used methods of assessing depth perception. It is an apparatus made up of a large box with a clear or see-through panel on top. One side of the box has a patterned surface placed immediately under the clear surface, and the other side has the same patterned surface placed at some distance below the clear surface. This latter side gives the appearance of a sharp drop-off or cliff. The subject of the study will be placed on the **glass** and consistent movement toward the shallow side is seen as an indication of depth perception ability. Newborn infants who cannot crawl commonly show much distress when placed face down over the "cliff" side.

Research with animals raised without opportunities to see (for example if reared in the dark) sustain long-lasting deficits in their perceptual abilities. Indeed, such deprivation may even affect the weight and **biochemistry** of their brains. This research indicates that while humans and some **animal** species have innate mecha-

KEY TERMS

Accommodation—Changes in the curvature of the eye lens to form sharp retinal images of near and far objects.

Aerial-perspective—A monocular visual cue referring to how objects with sharp and clear images appear nearer than objects with blurry or unclear images.

Binocular cues—Visual cues that require the coordinated use of both eyes.

Convergence—The tendency of the eyes to rotate toward each other in a coordinated manner in order to focus effectively on nearby objects.

Elevation—A monocular visual cue referring to an object's placement in relation to the horizon.

Interposition—A monocular cue referring to how when objects appear to partially block or overlap with each other, the fully visible object is perceived as being nearer.

Linear perspective—A monocular depth cue involving the apparent convergence of parallel lines

in the distance, as well as the perceived decrease in the size of objects and the space between them with increasing distance from the observer.

Monocular cues—Visual cues that one eye alone can perceive.

Motion parallax—The perception of objects moving at different speeds relative to their distance from the observer.

Retina—An extremely light-sensitive layer of cells at the back part of the eyeball. Images formed by the lens on the retina are carried to the brain by the optic nerve.

Stereoscopic vision—The unified three-dimensional view of objects produced when the two slightly different images of objects on the two retinas are fused into one.

Texture gradient—A monocular visual cue referring to how changes in an object's perceived surface texture indicate distance from the observer and changes in direction of the object.

nisms for depth perception, these innate abilities require visual experience in order to develop and become fully functioning. This research also suggests that animals and humans may have developmentally sensitive periods in which visual experience is necessary or permanent perceptual deficits may occur.

Current research/future developments

In sum, while environmental cues, binocular vision, and physiological aspects of the nervous system can account for many aspects of depth perception, numerous questions remain. Advances in understanding the physiological basis of vision have been great since the 1950s and this has greatly influenced research and theorizing in perception in general, and depth perception in particular. Researchers are eagerly looking at the structure of the nervous system to see if it might explain further aspects of depth perception. In particular, researchers continue to explore the possibility that additional fine tuned detector cells may exist that respond to specific visual stimuli. Finally, some psychologists have begun using certain basic principles of associative learning theory to explain a number of well-known yet poorly understood elements of perceptual learning. Both of these approaches show great potential for furthering our understanding of many processes in perception.

Resources

Books

Coren, S., L.M. Ward, and J.T. Enns. *Sensation and Perception.* 4th Ed. Fort Worth, TX: Harcourt Brace Jovanovich, 1994.

Masin, S.C., ed. *Foundations of Perceptual Theory.* New York: Elvesier Science, Inc., 1993.

Ono, T., et al., eds. *Brain Mechanisms of Perception and Memory: From Neuron to Behavior.* New York: Oxford University Press, 1993.

Schiffman, H.R. *Sensation and Perception: An Integrated Approach.* 3rd Ed. New York: John Wiley & Sons, 1990.

Marie Doorey

Derivative

In **mathematics**, the derivative is the exact **rate** at which one quantity changes with respect to another. Geometrically, the derivative is the slope of a **curve** at a **point** on the curve, defined as the slope of the tangent to the curve at the same point. The process of finding the derivative is called differentiation. This process is central to the branch of mathematics called differential **calculus**.

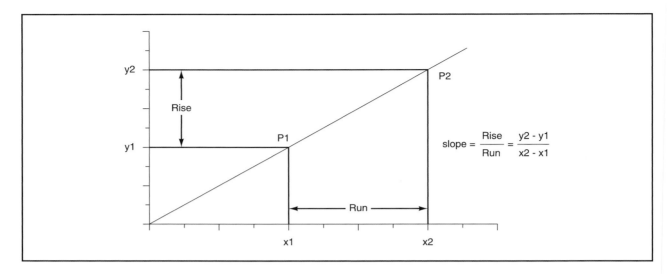

Figure 1. *Illustration by Hans & Cassidy. Courtesy of Gale Group.*

History and usefulness

Calculus was first invented by Sir Isaac Newton around 1665. Newton was a physicist as well as a mathematician. He found that the mathematics of his time was not sufficient to solve the problems he was interested in, so he invented new mathematics. About the same time another mathematician, Goltfried Leibnez, developed the same ideas as Newton. Newton was interested in calculating the **velocity** of an object at any instant. For example, if you sit under an apple **tree**, as legend has it Newton did, and an apple falls and hits you on the head, you might ask how fast the apple was traveling just before impact. More importantly, many of today's scientists are interested in calculating the rate at which a satellite's position changes with respect to time (its rate of speed). Most investors are interested in how a stock's value changes with time (its rate of growth). In fact, many of today's important problems in the fields of **physics**, **chemistry**, **engineering**, economics, and **biology** involve finding the rate at which one quantity changes with respect to another, that is, they involve finding the derivative.

The basic concept

The derivative is often called the "instantaneous" rate of change. A rate of change is simply a comparison of the change in one quantity to the simultaneous change in a second quantity. For instance, the amount of money your employer owes you compared to the length of time you worked for him determines your rate of pay. The comparison is made in the form of a **ratio**, dividing the change in the first quantity by the change in the second quantity. When both changes occur during an infinitely short period of time (in the same instant), the rate is said to be "instantaneous," and then the ratio is called the derivative.

To better understand what is meant by an instantaneous rate of change, consider the graph of a straight line (see Figure 1).

The line's slope is defined to be the ratio of the rise (vertical change between any two points) to the run (simultaneous horizontal change between the same two points). This means that the slope of a straight line is a rate, specifically, the line's rate of rise with respect to the horizontal axis. It is the simplest type of rate because it is constant, the same between any two points, even two points that are arbitrarily close together. Roughly speaking, arbitrarily close together means you can make them closer than any positive amount of separation. The derivative of a straight line, then, is the same for every point on the line and is equal to the slope of the line.

x_1	x_2	t_1	t_2	x_2-x_1	t_2-t_1	$(x_2-x_1)/(t_2-t_1)$
0	8	0	0.707106781	8	0.707106781	11.3137085
1	7	0.25	0.661437828	6	0.411437828	14.58300524
3	5	0.433012702	0.559016994	2	0.126004292	15.87247514

TABLE 1

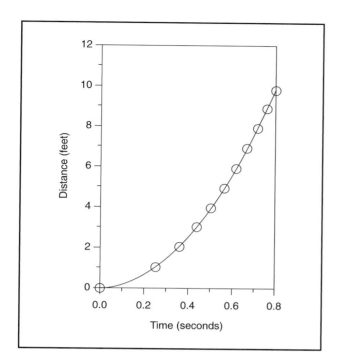

Figure 2. *Illustration by Hans & Cassidy. Courtesy of Gale Group.*

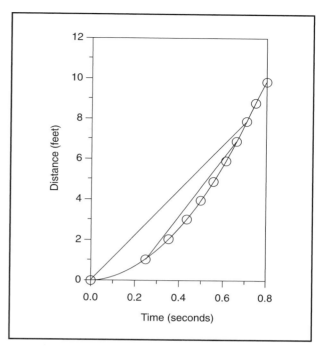

Figure 3. *Illustration by Hans & Cassidy. Courtesy of Gale Group.*

Determining the derivative of a curve is somewhat more difficult, because its instantaneous rate of rise changes from point to point (see Figure 2).

We can estimate a curve's rate of rise at any particular point, though, by noticing that any section of a curve can be approximated by replacing it with a straight line. Since we know how to determine the slope of a straight line, we can approximate a curve's rate of rise at any point, by determining the slope of an approximating line segment. The shorter the approximating line segment becomes, the more accurate the estimate becomes. As the length of the approximating line segment becomes arbitrarily short, so does its rise and its run. Just as in the case of the straight line, an arbitrarily short rise and run can be shorter than any given positive pair of distances. Thus, their ratio is the instantaneous rate of rise of the curve at the point or the derivative. In this case the derivative is different at every point, and equal to the slope of the tangent at each point. (A tangent is a straight line that intersects a curve at a single point.)

A concrete example

A fairly simple, and not altogether impractical example is that of the falling apple. Observation tells us that the apple's initial speed (the instant before letting go from the tree) is **zero**, and that it accelerates rapidly. Scientists have found, from repeated measurements with various falling objects (neglecting **wind** resistance), that the **distance** an object falls on the **earth** (call it S) in a specified time period (call it T) is given by the following equation (see Figure 2):

$$(1)\ S = 16\,T^2$$

Suppose you are interested in the apple's speed after it has dropped 4 ft (1.2 m). As a first **approximation**, connect the points where $Sl_1=0$ and $Sl_2=8$ (see Figure 3 and line 1 of Table 1).

Using equation (1), find the corresponding times, and calculate the slope of the approximating line segment (use the formula in Figure 1). Repeat this process numerous times, each time letting the two points get closer together. If a **calculator** or computer spreadsheet is available this is rather simple. Table 1 shows the result for several approximating line segments.

The line segments corresponding to the first two entries in the table are drawn in Figure 3. Looking at Figure 3, it is clear that as the approximating line gets shorter, its slope approximates the rate of rise of the curve more accurately.

Resources

Books

Allen, G.D., C. Chui, and B. Perry. *Elements of Calculus.* 2nd ed. Pacific Grove, CA: Brooks/Cole Publishing Co, 1989.

Boyer, Carl B. *A History of Mathematics*. 2nd ed. Revised by Uta C. Merzbach. New York: John Wiley and Sons, 1991.

Downing, Douglas. *Calculus the Easy Way*. 2nd ed. Hauppauge, NY: Barron's Educational Services, Inc., 1988.

Periodicals

McLaughlin, William I. "Resolving Zeno's Paradoxes." *Scientific American* 271 (1994): 84-89.

J. R. Maddocks

Desalination

Desalination, also called desalting, is the removal of **salt** from seawater. It provides essential **water** for drinking and industry in **desert** regions or wherever the local water supply is **brackish**. In 1991, about 3.5 billion gallons of desalinated water were produced in about 4,000 desalination plants worldwide. Most of this water was produced through **distillation**. However, other methods, including reverse **osmosis** and electrodialysis, are becoming increasingly important.

At its simplest, distillation consists of boiling the seawater to separate it from dissolved salt. The water vapor rises to a cooler region where it condenses as pure liquid water. **Heat** for distillation usually comes from burning **fossil fuels**. To reduce costs and **pollution**, desalination plants are designed to use as little fuel as possible. Many employ flash distillation, in which heated seawater is pumped into a low **pressure** chamber. The

low pressure causes the water to vaporize, or "flash," even though it is below its boiling **temperature**. Therefore, less heat is required. Multi-stage flashing passes the seawater through a series of chambers at successively lower pressures. For even greater efficiency, desalination plants can be linked with electrical power plants. Heat from the hot gasses that turn the generators is recycled to warm the incoming seawater. Distillation is widely used in the Middle East, where fossil fuel is plentiful but fresh water is scarce.

Reverse osmosis uses high pressure to force pure water out of **saltwater**. Normal osmosis occurs when pure water and saltwater are separated by a semi-permeable **membrane**, which permits only water to flow through. Under these conditions, the pure water will move into the saltwater side, but if the saltwater is squeezed under high enough pressure, **freshwater** moves out of it. Pressures on the order of 60 atmospheres (800-1,200 psi) are required to push pure water out of seawater. Reverse osmosis is widely used to desalinate brackish water, which is less salty than seawater and therefore requires pressures only about half as great.

Like reverse osmosis, electrodialysis is presently best suited for desalinating brackish water. Salts consist of ions, which are **atoms** that have acquired electrical charge by losing or gaining electrons. Because of their charge, ions are attracted to oppositely charged electrodes immersed in the saltwater. They move toward the electrodes, leaving a region of pure water behind. Special membranes prevent the ions from drifting back into the purified water as it is pumped out.

Ongoing research seeks to improve existing desalination methods and develop new ones. The costs of distillation could be greatly reduced if clean, renewable **energy** were used to heat the water. Solar, geothermal, and oceanic temperature differences are among the energy sources being studied. Reverse osmosis could be used on a larger scale, and with saltier water, through development of semi-permeable membranes able to withstand higher pressures for longer times. All desalination methods leave extremely salty residues. New methods for disposing of these must be developed as the world's use of desalination grows.

Desert

A desert is an arid land area where more **water** is lost through **evaporation** than is gained from **precipitation**. Deserts include the familiar hot, dry desert of rock and **sand** that is almost barren of plants, the semiarid deserts of scattered trees, scrub, and **grasses**, coastal deserts, and the deserts on the **polar ice caps** of the Antarctic and Greenland.

Most desert regions are the result of large-scale climatic patterns. As the **earth** turns on its axis, large air swirls are produced. Hot air rising over the equator flows northward and southward. The air currents cool in the upper regions and descend as high **pressure** areas in two subtropical zones. North and south of these zones are two more areas of ascending air and low pressures. Still farther north and south are the two polar regions of descending air. As air rises, it cools and loses its moisture. As it descends, it warms and picks up moisture, drying out the land. This downward movement of warm air masses over the earth have produced two belts of deserts. The belt in the northern hemisphere is along the Tropic of Cancer and includes the Gobi Desert in China, the Sahara Desert in North **Africa**, the deserts of southwestern **North America**, and the Arabian and Iranian deserts in the Middle East. The belt in the southern hemisphere is along the Tropic of Capricorn and includes the Patagonia Desert in Argentina, the Kalihari Desert of southern Africa, and the Great Victoria and Great Sandy Deserts of **Australia**.

Coastal deserts are formed when cold waters move from the Arctic and Antarctic regions toward the equator and come into contact with the edges of continents. The cold waters are augmented by upwellings of cold water from **ocean** depths. As the air currents cool as they move across cold water, they carry **fog** and mist, but little rain. These types of currents result in coastal deserts in southern California, Baja California, southwest Africa, and Chile.

Mountain ranges also influence the formation of deserts by creating rain shadows. As moisture-laden air currents flow upward over windward slopes, they cool and lose their moisture. Dry air descending over the leeward slopes evaporates moisture from the **soil**, resulting in deserts. The Great Basin Desert was formed from a rain shadow produced by the Sierra Nevada **mountains**. Desert areas also form in the interior of continents when prevailing winds are far from large bodies of water and have lost much of their moisture.

Desert plants have evolved methods to conserve and efficiently use available water. Some flowering desert plants are ephemeral and live for only a few days. Their **seeds** or bulbs can lie dormant in the soil for years, until a heavy rain enables them to germinate, grow, and bloom. Woody desert plants can either have long root systems to reach deep water sources or spreading shallow roots to take up moisture from dew or occasional rains. Most desert plants have small or rolled leaves to reduce the surface area from which **transpiration** of water can take place, while others drop their leaves during dry periods. Often leaves have a waxy coating that prevents water loss. Many desert plants are succulents, which store water in leaves, stems, and roots. Thorns and spines of the **cactus** are used to protect a plant's water supply from animals.

Desert animals have also developed protective mechanisms to allow them to survive in the desert environment. Most desert animals and **insects** are small, so they can remain in cool underground burrows or hide under vegetation during the day and feed at night when it is cooler. Desert **amphibians** are capable of dormancy during dry periods, but when it rains, they mature rapidly, mate, and lay eggs. Many **birds** and **rodents** reproduce only during or following periods of winter rain that stimulate vegetative growth. Some desert rodents (e.g., the North American kangaroo rat and the African gerbil) have large ears with little fur to allow them to sweat and cool down. They also require very little water. The desert camel can survive nine days on water stored in its stomach. Many larger desert animals have broad hooves or feet to allow them to move over soft sand. Desert **reptiles** such as the horned toad can control their metabolic **heat** production by varying their **rate** of heartbeat and the rate of body **metabolism**. Some **snakes** have developed a sideways shuffle that allows them to move across soft sand. Deserts are difficult places for humans to live, but people do live in some deserts, such as the Aborigines in Australia and the Tuaregs in the Sahara.

Desert soils are usually naturally fertile since little water is available to leach **nutrients**. **Crops** can be grown on desert lands with **irrigation**, but evaporation of the irrigation water can result in the accumulation of salts on the soil surface, making the soil unsuitable for further crop production. Burning, **deforestation**, and overgrazing of lands on the semiarid edges of deserts are enabling deserts to encroach on the nearby arable lands in a process called **desertification**. Desertification in combination with shifts in global **atmospheric circulation** has resulted in the southern boundary of the Sahara Desert advancing 600 mi (1,000 km) southward. A desertification study conducted for the United Nations in 1984 determined that 35% of the land surface of the earth was threatened by desertification processes.

Desertification

Desertification is the gradual degradation of productive arid or semi-arid land into biologically unproductive land. The French botanist, André Aubreville, coined the term in 1949 to describe to the transformation of productive agricultural land in northern **Africa** to desert-like, uncultivable fallowland. Loss of biological and ecological viability occurs when natural variations, like extended **drought** related to climate change, and unsustainable human activities such as over-cultivation and poor **irrigation** practices, strip drylands of their stabilizing vegetation, **soil nutrients**, and natural **water** distribution systems. The earth's arid and semi-arid regions are, by definition, areas with scarce **precipitation**; even very small changes can quickly destroy the fragile ecosystems and soil horizons that remain productive in areas with very little water. Desertification does not, per se, result in the development of a **desert**. Though desertified land and deserts are both dry, the barren, gullied wastelands left by desertification barely resemble the subtle biological productivity of healthy desert ecosystems. In some cases, careful land stewardship has successfully reversed desertification, and has restored degraded areas to a more productive condition. In the worst cases, however, semi-desert and desert lands have lost their sparse complement of plants and animals, as well as their ability to support agriculture.

Desertification is a particularly pressing social and environmental issue in regions where natural dryness and human poverty coincide. The earth's deserts and semi-arid **grasslands** occur in the subtropical bands between 15° and 30° north and south where extended periods of high **pressure** persist under the trade winds. Northern and southern Africa, the Arabian **peninsula**, southern **Asia**, the **continent** of **Australia**, southern **South America** and the North American Southwest lie in the subtropical zones. Desertification is usually discussed in the context of dry regions and ecosystems, but it can also affect prairies, savannas, **rainforests**, and mountainous habitats. **Global climate** change can alter the boundaries of these naturally dry regions, and change the precipitation patterns within them. Arid and semi-arid regions with large, impoverished populations of subsistence farmers, like northern Africa, or with large-scale commercial agriculture, like the American Southwest, are particularly susceptible to destructive desertification.

Sometimes desertification is the result of purely natural processes. Long-term changes in climatic conditions have led to decreased precipitation in a number of regions. The northern Sahara, for example, has experienced numerous fluctuations between arid and wet conditions over the past 10,000 years, as have the basins of the American West. **Radar** images collected aboard the **space shuttle** *Endeavor* show extensive river systems buried beneath more recent Saharan sands, and preserved fossil vegetation and **lake** shorelines suggest that forests surrounded filled lakes in northern Nevada and Utah.

Cyclical atmospheric and oceanographic variations, like the El Niño phenomenon in the southern Pacific, may also trigger extended regional drought. Environmental scientists warn that anthropogenic (human-induced) global climate change could also lead to bring desertification to previously unaffected regions. Until the twentieth century, humans were able to simply move their agricultural activity away from land rendered unusable by desertification. However, rapid twentieth century population growth, and a corresponding need for high agricultural productivity, has rendered that strategy untenable.

The Sahelian drought and United Nations convention to combat desertification

Desertification first captured major international attention in the 1970s when a decade of severe drought in the Sahel region of Africa brought starvation to the impoverished populations of countries along the southern border of the Sahara Desert. Sparse stabilizing vegetation died, and the Saharan sands encroached, covering depleted agricultural land. Reduced water **volume** in **rivers** compromised irrigation and hydroelectric generation. Dust storms brought health problems and ruined equipment and buildings. **Livestock** died of starvation. More than 100,000 sub-Saharan Africans died of thirst and starvation between 1972 and 1984, and more than 750,000 people in Mali, Mauritania, and Niger were completely dependent on international food aid during 1974.

The United Nations convened the Conference on Desertification (UNOCD) in 1977 in Nairobi, Kenya in response to the Sahelian Crisis. The conference brought together 500 delegates from 94 countries. The delegates approved 28 recommendations for slowing or reversing desertification, with the hope of stabilizing the area of land degraded by desertification, and preventing further crises. Scientists advising the 1977 convention estimated that desertification threatened 11.6 million mi^2 (30 million km^2), or 19% of the earth's land surface, in 150 nations worldwide. Despite the action undertaken in the years following the UNOCD, and numerous examples of "local successes," the United Nations Environment Programme (UNEP) concluded in 1991 that the threats and effects of desertification had worsened worldwide. The 1994 United Nations Convention to Combat Desertification (UNCCD) was proposed at the 1992 the United Nations Conference on Desertification (UNCOD)in Rio de Janeiro. By March 2002, 179 nations had agreed to the UNCCD treaty which advocates a program of locally-implemented efforts to reverse and/or prevent the desertification that directly affects 250 million people, and threatens at least one billion.

The burning of forests, like this one in the Amazon Basin of Brazil, contributes to the process of desertification. *National Center for Atmospheric Research. Reproduced by permission.*

Desertification in North America

Arid lands in parts of **North America** are among those severely affected by desertification; almost 90% of such habitats are considered to be moderately to severely desertified. The arid and semi-arid lands of the western and southwestern United States are highly vulnerable to this kind of damage. The perennial **grasses** and shrubs that dominate arid-land vegetation provide good forage for cattle, but overstocking leads to overgrazing and degradation of the natural vegetation cover, which, in turn, contribute to **erosion** and desertification. In addition, excessive withdrawals of **groundwater** to irrigate **crops** and supply cities is exceeding the ability of the aquifers to replenish, resulting in a rapid decline in height of the water table. Groundwater depletion by overpumping of the sandstone Ogalalla **aquifer** in the Southwestern United States has contributed to desertification in Nebraska, Kansas, Oklahoma and Texas. More-

over, the salts left behind on the soil surface after the irrigation water has evaporated results in land degradation through salinization, creating toxic conditions for crops and contaminating groundwater. Salination resulting from decades of heavy irrigation has compromised the soil quality in California's San Joaquin Valley, which produces much of the produce sold in the United States.

Studies of pre-industrial, aboriginal people in the western and southwestern United States suggest even small numbers of people could induce long-lasting ecological changes, including desertification. For example, native Americans reliant on mesquite beans for food planted mesquite throughout the Chihuahuan Desert of Arizona, New Mexico, Texas, and northern Mexico. Stands of mesquite developed around campsites and watering holes, and replaced the local grasses and other vegetation. The Pueblan culture, which flourished in the southwestern United States beginning around A.D. 800, used the meager stands of trees for housing material, resulting in local **deforestation**.

Processes of desertification

Desertification is a process of continuous, gradual **ecosystem** degradation, during which plants and animals, and geological resources such as water and soil, are stressed beyond their ability to adjust to changing conditions. Because desertification occurs gradually, and the processes responsible for it are understood, it can often be avoided by planning or reversed before irreparable damage occurs. The physical characteristics of land undergoing desertification include progressive loss of mature, stabilizing vegetation from the ecosystem, or loss of agricultural crop cover during periods of drought or economic infeasibility, and a resulting loss of unconsolidated topsoil. This process is called deflation. Erosion by **wind** and water then winnows the fine-grained silt and clay particles from the soil; dramatic dust storms like those observed during the 1930's Dust Bowl in the American mid-west, and in northern Africa, were essentially composed of blowing topsoil. Continued irrigation of desertified land increases soil salinity, and contaminates groundwater, but does little to reverse the loss of productivity. Finally, ongoing wind and water erosion leads to development of gullies and **sand** dunes across the deflated land surface.

The forces causing these physical changes to occur may be divided into natural, human or cultural, and administrative causes. Among the natural forces are wind and water erosion of soil, long-term changes in rainfall patterns, and other changes in climatic conditions. The role of drought is variable and related in part to its duration; a prolonged drought accompanied by poor land management may be devastating, while a shorter drought might not have lasting consequences. As such, drought thus stresses the ecosystem without necessarily degrading it permanently. Rainfall similarly plays a variable role that depends on its duration, the seasonal pattern of its occurrence, and its spatial distribution.

The list of human or cultural influences on desertification includes vegetation loss by overgrazing, depletion of groundwater, surface runoff of rainwater, frequent burning, deforestation, the influence of invasive non-native **species**, physical compaction of the soil by livestock and vehicles, and damage by strip-mining. Desertification caused by human influences has a long historical record; there is evidence of such damage caused around the Tigris and Euphrates rivers in ancient Mesopotamia. Administrative influences contributing to desertification include encouragement of the widespread cultivation of a single crop for export, particularly if irrigation is required, and the concentration of dense human populations in arid lands. Poor economic conditions, like the Great Depression in United States in the 1930s, also contribute to degradation of croplands. During that crisis, American farmers were simultaneously confronted with bankruptcy and a decade-long drought, and they left millions of acres of plowed, bare cropland unplanted. According to the 1934 *Yearbook of Agriculture,* "Approximately 35 million acres of formerly cultivated land have essentially been destroyed for crop production.... 100 million acres now in crops have lost all or most of the topsoil; 125 million acres of land now in crops are rapidly losing topsoil."

Considering these factors together, desertification can be viewed as a process of interwoven natural, human, and economic forces causing continuous degradation over **time**. Therefore, ecosystem and agricultural degradation caused by desertification must be confronted from scientific, social and economic angles. Fortunately, scientists believe that severe desertification, which renders the land irreclaimable, is rare. Most desertified areas can be ecologically reclaimed or restored to agronomic productivity, if socioeconomic and cultural factors permit restoration.

Land management

Land management measures that combat desertification focus on improving sustainability and long-term productivity. Though damaged areas cannot always be restored to their pre-desertified conditions, they can often be reclaimed by designing a new state that can better withstand cultural and climatic stresses. Specific measures include developing a resilient vegetation cover of mixed trees, shrubs, and grasses suitable to local conditions that protects the soil from wind and water erosion and compaction. Redistribution of water resources, and redesign of water delivery systems, can reduce the effects of salination, groundwater depletion, and wasteful water use. Finally, limiting the agricultural demands made on drought-prone arid and semi-arid lands can be accomplished by encouraging farmers to grow drought-tolerant plants, and to move water-hungry crops, like **cotton** and **rice**, to more suitable climates.

Land management methods that halt or reverse desertification have been known in the United States since the end of the nineteenth century, but they have only recently have they been put into widespread use. United States federal policies do support **soil conservation** in croplands and rangelands, but only about one third of such land is under federal protection. The United States government has strongly supported the autonomy of private landowners and corporations who, in turn, have often traded sustainable land-use practices for short-term profits. Even the ravages of the dust bowl did not result in widespread anti-desertification measures.

In the developing world, particularly in Africa, where poverty and political unrest are common, progress toward mitigation of desertification and its devastating social costs has been slow. Many of the world's poorest people, who live in countries with the weakest and most corrupt governments, rely on unsustainable agriculture and nomadic grazing to subsist. Many African countries, including Niger, Mali, and Senegal, have experienced positive results with implementation of a system of local self-regulation. This strategy, encouraged by the United Nations, involves a pastoral association or a community assuming responsibility for maintaining a water source and its surrounding **rangeland**, while receiving free veterinary and health services. By these and similar means, cultural habits, subsistence needs, economic concerns, and ecological **conservation** can be addressed in a single, integrated program. Furthermore, such programs help local communities reduce their dependence on ineffective or corrupt centralized governments, and on the international aid community. Such comprehensive anti-desertification programs have been very successful on a limited scale, but progress has been slow because of the extreme poverty and sociopolitical powerlessness of the communities involved. The key to the success of any anti-desertification program is the need to adapt to local conditions, including those associated with climate, **ecology**, culture, government, and historical land-use.

See also Ecological integrity; Ecological monitoring; Ecological productivity; Crop rotation; Deforestation; Forestry; Land use; Water conservation.

Resources

Books

Mainguet, M. *Aridity: Droughts and Human Development.* Springer Verlag, 1999.

Press, Frank, and Raymond Siever. *Understanding Earth. Chapter 14: Winds and Deserts.* New York: W.H. Feeman and Company, 2001.

Sheridan, David. *Desertification of the United States.* Washington, DC: Council on Environmental Quality, U.S. Government Printing Office, 1981.

Other

National Atmospheric and Space Administration. "From the Dust Bowl to the Sahel." Distributed Active Archive Centers Earth Observatory. May 18, 2001. [cited October 24, 2002] <http://earthobservatory.nasa.gov/Study/DustBowl/>.

Public Broadcasting System/WGBH. "Surviving the Dust Bowl." The American Experience. 1999 [cited October 24, 2002]. <www.pbs.org/ wgbh/amex/dustbowl/>.

United Nations Convention to Combat Desertification. "Text and Status of the Convention to Combat Desertification."

February 14, 2002 [cited October 24, 2002]. <http://www.unccd.int/convention/menu.php>.

Marjorie Pannell
Laurie Duncan

Determinants

A determinant, signified by two straight lines ‖, is a **square** array of numbers or symbols that has a specific value. For a square **matrix**, say, A, there exists an associated determinant, |A|, which has elements identical with the corresponding elements of the matrix. When matrices are not square, they do not possess corresponding determinants.

In general, determinants are expressed as shown in Figure 1, in which a_{ij}s are called elements of the determinant, and the horizontal and vertical lines of elements are called rows and columns, respectively. The sloping line

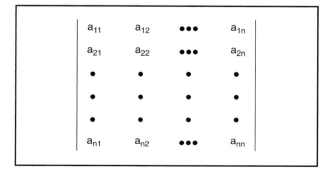

Figure 1. *Illustration by Hans & Cassidy. Courtesy of Gale Group.*

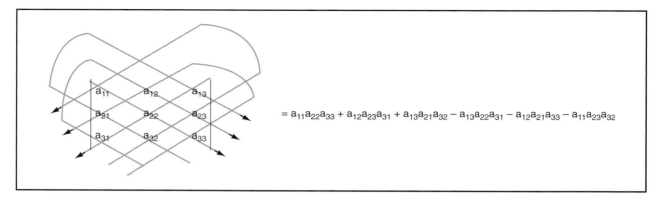

$$= a_{11}a_{22}a_{33} + a_{12}a_{23}a_{31} + a_{13}a_{21}a_{32} - a_{13}a_{22}a_{31} - a_{12}a_{21}a_{33} - a_{11}a_{23}a_{32}$$

Figure 2. *Illustration by Hans & Cassidy. Courtesy of Gale Group.*

(a)

$$a_{11} \begin{vmatrix} a_{22} & a_{23} \\ a_{32} & a_{33} \end{vmatrix} - a_{21} \begin{vmatrix} a_{12} & a_{13} \\ a_{32} & a_{33} \end{vmatrix} + a_{31} \begin{vmatrix} a_{12} & a_{13} \\ a_{22} & a_{23} \end{vmatrix}$$

or

(b)

$$- a_{12} \begin{vmatrix} a_{21} & a_{23} \\ a_{31} & a_{33} \end{vmatrix} + a_{22} \begin{vmatrix} a_{11} & a_{13} \\ a_{31} & a_{33} \end{vmatrix} - a_{32} \begin{vmatrix} a_{11} & a_{13} \\ a_{21} & a_{23} \end{vmatrix}$$

Figure 3. *Illustration by Hans & Cassidy. Courtesy of Gale Group.*

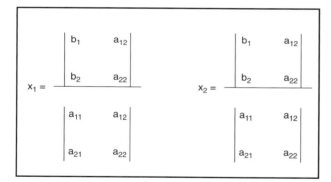

Figure 4. *Illustration by Hans & Cassidy. Courtesy of Gale Group.*

consisting of a_{ii} elements is called the principal diagonal of the determinant. Sometimes, determinants can be written in a short form, $|a_{ij}|$. The n value, which reflects how many n^2 quantities are enclosed in $||$, determines the order of a determinant.

For determinants of third order, that is, $n = 3$, or three rows of elements, we can evaluate them as illustrated in Figure 2.

By summing the products of terms as indicated by the arrows pointing towards the right-hand side and subtracting the products of terms as indicated by the arrows pointing towards the left-hand side, we can obtain the value of this determinant. The determinant can also be evaluated in terms of second-order determinants (two rows of elements), as in Figures 3(a) or 3(b).

Each of these second-order determinants, multiplied by an element a_{ij}, is obtained by deleting the ith row and the jth column of elements in the original third-order determinant, and it is called the "minor" of the element a_{ij}. The minor is further multiplied by $(-1)^{i+j}$, which is exactly the way we determine either the "+" or "-" sign for each determinant included in Figures 3 as shown, to become the "cofactor," C_{ij}, of the corresponding element.

Determinants have a variety of applications in **engineering mathematics**. Now, let's consider the system of two linear equations with two unknowns x_1 and x_2: $a_{11}x_1 + a_{12}x_2 = b_1$ and $a_{21}x_1 + a_{22}x_2 = b_2$.

We can multiply these two equations by a_{22} and $-a_{12}$, respectively, and add them together. This yields ($a_{11}a_{22}$ -

KEY TERMS

Cofactors—By deleting the *i*th row and the *j*th column from a determinant, say, A, of order n, an (n 1)th order determinant is obtained as the minor of the element a_{ij}. The minor multiplied by $(1)^{I+j}$ is called the cofactor of a_{ij} in A.

Cramer's rule—A method, proposed by Swiss mathematician G. Cramer (1704–1752), has been widely used to find solutions to simultaneous linear equations via the operation of determinants and by including the coefficients of variables in simultaneous equations as the elements of determinants.

$a_{12}a_{21})x_1 = b_1a_{22} - b_2a_{12}$, i.e., $x_1 = (b_1a_{22} - b_2a_{12})/(a_{11}a_{22} - a_{12}a_{21})$. Similarly, $x_2 = (b_1a_{21} - b_2a_{11})/(a_{12}a_{21} - a_{11}a_{21})$ can be obtained by adding together the first equation multiplied by a_{21} and second equation multiplied by $-a_{11}$. These results can be written in determinant form as in Figure 4.

This is generally called Cramer's rule. Notice that in Figure 4, elements of the determinant in the denominator are the same as the coefficients of x_1 and x_2 in the two equations. To solve for x_1 (or x_2), we then replace the elements that correspond to the coefficients of x_1 (or x_2) of the determinant in the numerator with two constant terms, b_1 and b_2. When b_1 and b_2 both are equal to **zero**, the system defined by the two equations is said to be homogeneous. In this case, it will have either only the trivial solution $x_1 = 0$ and $x_2 = 0$ or additional solutions if the determinant in the denominator in figure 5 is zero. When at least b_1 or b_2 is not zero (that is, a nonhomogeneous system) and the denominator has a value other than zero, the solution to the system is then obtained from figure 4. Cramer's rule is also applicable to systems of three linear equations. Therefore, determinants, along with matrices, have been used for solving simultaneous linear and differential equations involved in various systems, such as reactions in chemical reactors, stiffness of spring-connected masses, and currents in an electric network.

Pang-Jen Kung

Deuterium

Deuterium is an **isotope** of **hydrogen** with atomic **mass** of 2. It is represented by the symbols ^2H or D. Deuterium is also known as heavy hydrogen. The nucle-

us of the deuterium atom, consisting of a **proton** and a **neutron**, is known as a deuteron and is represented in nuclear equations by the symbol d.

Discovery

The possible existence of an isotope of hydrogen with atomic mass of two was suspected as early as the late 1910s after Frederick Soddy had developed the concept of isotopes. Such an isotope was of particular interest to chemists. Since the hydrogen atom is the simplest of all atoms—consisting of a single proton and a single electron—it is the model for most atomic theories. An atom just slightly more complex—one that contains a single neutron—also could potentially contribute valuable information to existing atomic theories.

Among those who sought for the heavy isotope of hydrogen was Harold Urey, at the **time** professor of **chemistry** at Columbia University. Urey began his work with the realization that any isotope of hydrogen other than hydrogen-1 (also known as protium) must exist in only minute quantities. The evidence for that fact is that the **atomic weight** of hydrogen is only slightly more than 1.000. The fraction of any isotopes with mass greater than that value must, therefore, be very small. Urey designed an experiment, therefore, that would allow him to detect the presence of heavy hydrogen in very small concentrations.

Urey's search for deuterium

Urey's approach was to collect a large **volume** of liquid hydrogen and then to allow that liquid to evaporate very slowly. His hypothesis was that the lighter and more abundant protium isotope would evaporate more quickly than the heavier hydrogen-2 isotope. The volume of liquid hydrogen remaining after **evaporation** was nearly complete, then, would be relatively rich in the heavier isotope.

In the actual experiment, Urey allowed 4.2 qt (4 l) of liquid hydrogen to evaporate until only 0.034 oz (1 ml) remained. He then submitted that **sample** to analysis by **spectroscopy**. In spectroscopic analysis, **energy** is added to a sample. **Atoms** in the sample are excited and their electrons are raised to higher energy levels. After a moment at these higher energy levels, the electrons return to their ground state, giving off their excess energy in the form of **light**. The bands of light emitted in this process are characteristics for each specific kind of atom.

By analyzing the spectral pattern obtained from his 0.034 oz (1 ml) sample of liquid hydrogen, Urey was able to identify a type of atom that had never before been detected, the heavy isotope of hydrogen. The new isotope was soon assigned the name deuterium. For his discovery of the isotope, Urey was awarded the 1934 Nobel Prize in chemistry.

Properties and preparation

Deuterium is a stable isotope of hydrogen with a relative atomic mass of 2.014102 compared to the atomic mass of protium, 1.007825. Deuterium occurs to the extent of about 0.0156% in a sample of naturally occurring hydrogen. Its melting point is -426°F (-254°C)—compared to -434°F (-259°C) for protium—and its **boiling point** is -417°F (-249°C)–compared to -423°F (-253°C) for protium. Its macroscopic properties of **color**, odor, **taste**, and the like are the same as those for protium.

Compounds containing deuterium have slightly different properties from those containing protium. For example, the melting and boiling points of heavy **water** are, respectively, 38.86°F (3.81°C) and 214.56°F (101.42°C). In addition, deuterium bonds tend to be somewhat stronger than protium bonds. Thus **chemical reactions** involving deuterium-containing compounds tend to go more slowly than do those with protium.

Deuterium is now prepared largely by the **electrolysis** of heavy water, that is, water made from deuterium and **oxygen** (D_2O). Once a great rarity, heavy water is now produced rather easily and inexpensively in very large volumes.

Uses

Deuterium has primarily two uses, as a tracer in research and in thermonuclear fusion reactions. A tracer is any atom or group of atoms whose participation in a physical, chemical, or biological reaction can be easily observed. Radioactive isotopes are perhaps the most familiar kind of tracer. They can be tracked in various types of changes because of the **radiation** they emit.

Deuterium is an effective tracer because of its mass. When it replaces protium in a compound, its presence can easily be detected because it weights twice as much as a protium atom. Also, as mentioned above, the bonds formed by deuterium with other atoms are slightly different from those formed by protium with other atoms. Thus, it is often possible to figure out what detailed changes take place at various stages of a chemical reaction using deuterium as a tracer.

Fusion reactions

Scientists now believe that energy produced in the **sun** and other stars is released as the result of a series of thermonuclear fusion reactions. The term fusion refers to

the fact that two small nuclei, such as two hydrogen nuclei, fuse—or join together—to form a larger nucleus. The term thermonuclear means that such reactions normally occur only at very high temperatures, typically a few millions of degrees Celsius. Interest in fusion reactions arises not only because of their role in the manufacture of stellar energy, but also because of their potential value as sources of energy here on **Earth**.

Deuterium plays a critical role in most thermonuclear fusion reactions. In the solar process, for example, the fusion sequence appears to begin when two protium nuclei fuse to form a single deuteron. The deuteron is used up in later stages of the cycle by which four protium nuclei are converted to a single helium nucleus.

In the late 1940s and early 1950s scientists found a way of duplicating the process by which the sun's energy is produced in the form of thermonuclear fusion weapons, the so-called hydrogen bomb. The detonating device in this type of weapon was **lithium** deuteride, a compound of lithium **metal** and deuterium. The detonator was placed on the casing of an ordinary fission ("atomic") bomb. When the fission bomb detonated, it set off further nuclear reactions in the lithium deuteride which, in turn, set of fusion reactions in the larger hydrogen bomb.

For more than four decades, scientists have been trying to develop a method for bringing under control the awesome fusion power of a hydrogen bomb for use in commercial power plants. One of the most promising approaches appears to be a process in which two deuterons are fused to make a proton and a triton (the nucleus of a hydrogen-3 isotope). The triton and another deuteron then fuse to produce a helium nucleus, with the release of very large amounts of energy. So far, the technical details for making this process a commercially viable source of energy have not been completely worked out.

See also Nuclear fusion; Radioactive tracers.

Resources

Books

Asimov, Isaac. *Asimov's Biographical Encyclopedia of Science and Technology.* 2nd revised edition. Garden City, NY: Doubleday & Company, Inc., 1982.

Greenwood, N.N., and A. Earnshaw. *Chemistry of the Elements.* 2nd ed. Oxford: Butterworth-Heinneman Press, 1997.

Joesten, Melvin D., David O. Johnston, John T. Netterville, and James L. Wood. *World of Chemistry.* Philadelphia: Saunders, 1991.

Thomson, John F. *Biological Effects of Deuterium.* New York: Macmillan, 1963.

David E. Newton

Developmental processes

Developmental processes are the series of biological changes associated with information transfer, growth, and differentiation during the life cycle of organisms. Information transfer is the transmission of DNA and other biological signals from parent cells to daughter cells. Growth is the increase in size due to **cell** expansion and **cell division**. Differentiation is the change of unspecialized cells in a simple body pattern to specialized cells in more complex body pattern. While nearly all organisms, even single-celled **bacteria**, undergo development of some sort; the developmental process of complex multicellular organisms is emphasized here. In these organisms, development begins with the manufacture of male and female sex cells. It proceeds through **fertilization** and formation of an embryo. Development continues following **birth**, hatching, or **germination** of the embryo and culminates in **aging and death**.

History

Until the mid-1800s, many naturalists supported a theory of development called epigenesis, which held that the eggs of organisms were undifferentiated, but had a developmental potential which could be directed by certain external forces. Other naturalists supported a theory of development called preformationism, which held that the entire complex morphology of mature **organism** is present in miniature form in the egg, a developmental form called the homunculus.

These early theories of development relied on little experimental evidence. Thus, biologists often criticized the original theory of epigenesis because it seemed to propose that mystical forces somehow directed development, a view clearly outside the realm of science. Biologists also rejected preformationism, since studies of **cytology** and **embryology** clearly showed that development is much more than the simple growth of a preformed organism.

The modern view is that developmental processes have certain general features of both preformationism and epigenesis. Thus, we know that the simple cells of an egg are preformed in the sense that they contain a preformed instruction set for development which is encoded in their genes. Similarly, we know that the egg is relatively formless, but has the potential to develop into a complex organism as it grows. Thus modern developmental **biology** views development as the expression of a preformed genetic program which controls the epigenetic development of an undifferentiated egg into a morphologically complex adult.

Evolutionary aspects

People have long been interested in the connection between the development of an organism, its ontogeny, and the evolutionary ancestry of the **species**, its **phylogeny**. Anaximander, a philosopher of ancient Greece, noted that human embryos develop inside fluid-filled wombs and proposed that human beings evolved from **fish** as creatures of the **water**.

This early idea was a progenitor to recapitulation theory, proposed in the 1800s by Ernst Haeckel, a German scientist. Recapitulation theory is summarized by the idea that the embryological development of an **individual** is a quick replay of its evolutionary history. As applied to humans, recapitulation theory was accepted by many evolutionary biologists in the 1800s. It also influenced the intellectual development of other disciplines outside of biology, including philosophy, politics, and **psychology**.

By the early 1900s, developmental biologists had disproven recapitulation theory and had shown that the relationship between ontogeny and phylogeny is more complex than proposed by Haeckel. However, like Haeckel, modern biologists hold that the similarities in the embryos of closely related species and the transient appearance of certain structures of mature organisms early in development of related organisms indicates a connection between ontogeny and phylogeny. One modern view is that new species may evolve when **evolution** alters the timing of development, so that certain features of ancestral species appear earlier or later in development.

Information transfer

Nearly every multicellular organism passes through a life cycle stage where it exists as a single undifferentiated cell or as a small number of undifferentiated cells. This developmental stage contains molecular information which specifies the entire course of development encoded in its many thousands of genes. At the molecular level, genes are used to make **proteins**, many of which act as enzymes, biological catalysts which drive the thousands of different biochemical reactions inside cells.

Adult multicellular organisms can consist of one quadrillion (a one followed by 15 zeros) or more cells, each of which has the same genetic information. (There are a few notable exceptions, such as the red **blood** cells of **mammals**, which do not have DNA, and certain cells in the unfertilized eggs of **amphibians**, which undergo **gene** amplification and have multiple copies of some genes.) F. C. Steward first demonstrated the constancy of DNA in all the cells of a multicellular organism in the 1950s. In a classical series of experiments, Steward separated a mature carrot **plant** into individual cells and showed that each cell, whether it came from the root, stem, or **leaf**, could be induced to develop into a mature carrot plant which was genetically identical to its parent. Although such experiments cannot typically be done with multicellular animals, animals also have the same genetic information in all their cells.

Many developmental biologists emphasize that there are additional aspects of information transfer during development which do not involve DNA directly. In addition to DNA, a fertilized egg cell contains many proteins and other cellular constituents which are typically derived from the female. These cellular constituents are often asymmetrically distributed during cell **division**, so that the two daughter cells derived from the fertilized egg have significant biochemical and cytological differences. In many species, these differences act as biological signals which affect the course of development. There are additional spatial and temporal interactions within and among the cells of a developing organism which act as biological signals and provide a form of information to the developing organism.

Growth

Organisms generally increase in size during development. Growth is usually allometric, in that it occurs simultaneously with cellular differentiation and changes in overall body pattern. Allometry is a discipline of biology which specifically studies the relationships between the size and morphology of an organism as it develops and the size and morphology of different species.

A developing organism generally increases in complexity as it increases in size. Moreover, in an evolutionary line, larger species are generally more complex that the smaller species. The reason for this correlation is that the **volume** (or weight) of an organism varies with the cube of its length, whereas gas exchange and food assimilation, which generally occur on surfaces, vary with the square of its length. Thus, an increase in size requires an increase in cellular specialization and morphological complexity so that the larger organism can breathe and eat.

Depending on the circumstances, natural **selection** may favor an increase in size, a decrease in size, or no change in size. Large size is often favored because it generally makes organisms faster, giving them better protection against predators, and making them better at dispersal and food gathering. In addition, larger organisms have a higher **ratio** of volume to surface area, so they are less affected by environmental variations, such as **temperature** variation. Large organisms tend to have a prolonged development, presumably so they have more **time** to develop the morphological complexities needed to support their large size. Thus, evolutionary selection for large size leads to a prolongation of development as well as morphological complexity.

Sometimes the coordination between growth and differentiation goes awry, resulting in a developmental abnormality. One such abnormality is an undifferentiated **mass** of cells called a **tumor**. A tumor may be benign, in which case it does not invade adjacent cells; alternatively, it may be malignant, or cancerous, in which case the proliferating cells invade their neighbors. Cancers often send colonies of tumor cells throughout the body of an individual, a process called metastasis.

Cancers can be caused by damaging the DNA, the molecular information carrier, of a single cell. This damage may be elicited by a variety of factors such as carcinogenic chemicals, viral **infection**, or ultraviolet **radiation**. In addition, some cancers may arise from unprovoked and spontaneous damage to DNA. Basic studies of the different developmental processes may lead to a better understanding of **cancer** and how it might be prevented or cured.

Differentiation

Differentiation is the change of unspecialized cells in a simple body pattern to specialized cells in a more complex body pattern. It is highly coordinated with growth and includes morphogenesis, the development of the complex overall body pattern.

Below, we emphasize molecular changes in organisms which lead to development. However, this does not imply that external factors have no role in development. In fact, external factors such as changes in **light**, temperature, or nutrient availability often elicit chemical changes in developing organisms which profoundly influence development.

The so-called "Central Dogma of Biology" says that spatial and temporal differences in gene expression cause cellular and morphological differentiation. Since DNA makes RNA, and RNA makes protein, there are basically three levels where a cell can modulate gene expression: 1) by altering the transcription of DNA into RNA; 2) by altering the translation of RNA into protein; and 3) by altering the activity of the protein, which is usually an **en-**

zyme. Since DNA and RNA are themselves synthesized by proteins, the gene expression patterns of all cells are regulated by highly complex biochemical networks.

A few simple calculations provide a better appreciation of the complexity of the regulatory networks of gene expression which control differentiation. Starting with the simplifying assumption that a given protein (gene product) can be either absent or present in a cell, there are at least ten centillion (a one followed by 6,000 zeros) different patterns of gene expression in a single typical cell at any time. Given that a multicellular organism contains one quadrillion or more cells, and that gene expression patterns change over time, the number of possible gene expression patterns is enormous.

Perhaps the central question of developmental biology is how an organism can select the proper gene expression pattern among all these possibilities. This question has not yet been satisfactorily answered. However, in a 1952 **paper**, Alan Turing showed that simple chemical systems, in which the component chemicals diffuse and react with one another over time, can create complex spatial patterns which change over time. Thus, it seems possible that organisms may regulate differentiation by using a Turing-like reaction-diffusion mechanism, in which proteins and other molecules diffuse and interact with one another to modulate gene expression. Turing's original model, while relatively simple, has been a major impetus for research about pattern development in biology.

Lastly, aging must also be considered a phase of development. Many evolutionary biologists believe that all organisms have genes which have multiple effects, called pleiotropic genes, that increase reproductive success when expressed early in development, but cause the onset of old age when expressed later in development. In this view, natural selection has favored genes which cause aging and death because the early effects of these genes outweigh the later effects.

Resources

Books

Beurton, Peter, Raphael Falk, and Hans-Jörg Rheinberger., eds. *The Concept of the Gene in Development and Evolution.* Cambridge, UK: Cambridge University Press, 2000.

Bonner, J. T. *First Signals: The Evolution of Multicellular Development.* Princeton, NJ: Princeton University Press, 2000.

Emde, Robert N., and John K. Hewitt, eds. *Infancy to Early Childhood: Genetic and Environmental Influences on Developmental Change.* New York: Oxford University Press, 2001.

Hall, B.K. *Evolutionary Developmental Biology.* Chapman and Hall, Inc., 1992.

Kugrens, P. *Developmental Biology.* Kendall-Hunt Publishing Co., 1993.

KEY TERMS

Differentiation—Developmental change of unspecialized cells in a simple body pattern to specialized cells in a more complex body pattern.

Gene expression—Molecular process in which a gene is transcribed into a specific RNA (ribonucleic acid), which is then translated into a specific protein.

Morphogenesis—Development of the complex overall body form of an organism.

Ontogeny—Entire developmental life history of an organism.

Phyologeny—Evolutionary history or lineage of an organism or group of related organisms.

Periodicals

Hayflick, L. "The Future of Aging." *Nature* no. 408 (2000): 103-4.

Peter Ensminger

Dew point

The dew point is that **temperature** below which the **water** vapor in a body of air cannot all remain vapor. When a body of air is cooled to its dew point or below, some fraction of its water vapor shifts from gaseous to liquid phase to form **fog** or cloud droplets. If a smooth surface is available, vapor condenses directly onto it as drops of water (dew). The dew point of a body of air depends on its water vapor content and **pressure**. Increasing the fraction of water vapor in air (i.e., its relative **humidity**) raises its dew point; the water molecules are more crowded in humid air and thus more likely to coalesce into a liquid even at a relatively warm temperature. Decreasing the pressure of air lowers its dew point; lowering pressure (at constant temperature) increases the average **distance** between molecules and makes water vapor less likely to coalesce.

If the dew point of a body of air is below 32°F (0°C), its water vapor will precipitate not as liquid water but as **ice**. In this case, the dew point is termed the frost point.

Air at ground level often deposits dew on objects at night as it cools. In this case, the dew point of the air remains approximately constant while its temperature drops. When the dew point is reached, dew forms. Ground mist and fog may also form under these conditions.

The dew point can be measured using a dew-point hygrometer. This instrument, invented in 1751, consists essentially of a **glass** with a **thermometer** inserted. The glass is filled with ice water and stirred. As the temperature of the glass drops, the air in contact with it is chilled; when it reaches its dew point, water condenses on the glass. The temperature at which condensation occurs is recorded as the dew point of the surrounding air.

See also Atmosphere, composition and structure; Atmospheric pressure; Atmospheric temperature; Clouds; Evaporation; Evapotranspiration; Precipitation; Weather forecasting.

Diabetes mellitus

Diabetes mellitus is a group of diseases characterized by high levels of glucose in the **blood** resulting from defects in **insulin** production (insulin deficiency), insulin action (insulin resistance), or both. Insulin is a hormone produced by the pancreas. When eaten, foods are converted to a type of sugar called glucose that enters the bloodstream. Insulin is needed to move glucose into the body cells where it is used for **energy**, and excesses are stored in the liver and **fat** cells. Insufficient amounts of working insulin cause blood sugar levels to rise and large amounts of glucose are excreted in the urine. Consistently high levels of glucose in the bloodstream damage the nerves and blood vessels, and can lead to **heart disease**, **stroke**, high blood **pressure**, blindness, kidney disease, amputations, and dental disease.

The exact cause of diabetes is unknown, although **genetics** and environmental factors such as **obesity** and lack of **exercise** appear to play roles. Diabetes can be associated with serious complications and death, but people with diabetes can take an active role in controlling the disease and lowering the risk of complications.

History of diabetes

The history of diabetes mellitus dates back to ancient Egypt, where its symptoms were described around 2000 B.C. The Greeks later gave the disease its name in the first century A.D. The word diabetes means siphon, which describes a major symptom of the condition, frequent urination. Mellitus means honey, and depicts one of the early signs of diabetes, sugar in the urine.

Incidence of diabetes

Over 17 million people in the United States, or 6.2% of the population, have diabetes. More than one third of diabetes victims are unaware that they have the disease. Higher rates of diabetes occur in certain populations: 13% of African Americans, 10.2% of Latino Americans, and 15.1% of Native Americans have diabetes. Prevalence of diabetes increases with age. Approximately 151,00 people less than 20 years of age have diabetes, but nearly 20.1% of the U.S. population age 65 and older has diabetes. In the United States, 8.9% of all women and 8.3% of all men have diabetes. Over one million people are newly diagnosed with diabetes each year. Greater than 450,000 deaths each year in the United States are attributed to diabetes.

Types of diabetes

There are three major types of diabetes: type 1, type 2, and gestational diabetes. Type 1 diabetes was previously called insulin-dependent diabetes or juvenile-onset diabetes. Type 1 diabetes develops when the body's **immune system** destroys pancreatic beta cells, the only cells that produce insulin. The body, in turn, produces little or no insulin, resulting in insulin deficiency. Without insulin, the body is unable to use glucose for energy and begins to break down fats for fuel. Ketones are formed when fat is burned for energy. Excess ketones build up in the blood and lower the blood **pH** value leading to ketoacidosis.

Symptoms of type 1 diabetes usually appear suddenly and include increased thirst, frequent urination, increased hunger, tiredness, and weight loss. Risk factors for type 1 diabetes include autoimmune, genetic, and environmental factors. Although it usually begins when people are under the age of 30, type 1 diabetes may occur at any age. Almost 10% of the United States diabetes population has type 1 diabetes.

Type 2 diabetes was previously called noninsulin-dependent or adult onset diabetes. It begins as insulin resistance, a disorder in which normal to excessive amounts of insulin is made by the body, but the cells cannot use insulin properly. The ability to make insulin gradually decreases with **time** due to the progressive nature of the disease. In its early stages, type 2 diabetes often has no symptoms. When they do occur, symptoms may initiate gradually and include fatigue, dry skin, numbness or tingling in hands or feet, frequent infections, slow healing of cuts and sores, problems with sexual function, and increased hunger and thirst. With type 2 diabetes, hyperosmolar **coma** can develop from blood glucose levels (often referred to as blood sugar) becoming dangerously high. If the elevated blood sugar is not adequately controlled, it can cause severe dehydration, a serious condition requiring immediate treatment. Type 2 diabetes is associated with obesity, family history of diabetes, prior history of gestational diabetes, impaired glucose tolerance, physical inactivity, and race/ethnicity. Type 2 diabetes is diagnosed in children and adolescents in increasing numbers. About 85% of the U.S. diabetes population has type 2 diabetes.

Gestational diabetes occurs during pregnancy and affects 4% of all pregnant women. During pregnancy, the placenta supplies the baby with glucose and **water** from the mother's blood. **Hormones** made by the placenta are needed for pregnancy, but can keep the mother's insulin from functioning effieciently. As the pregnancy continues, more of these hormones are manufactured. When the mother is not able to make enough insulin to compensate for the increased hormone levels and to maintain normal blood glucose, gestational diabetes develops. Treatment is required to normalize maternal blood glucose levels to avoid complications in the fetus. After pregnancy, up to 10% of women with gestational diabetes are found to have type 2 diabetes. Women who have had gestational diabetes have a 20%–50% chance of developing diabetes in the next 5–10 years.

Pre-diabetes

Before type 2 diabetes fulminates (fully develops), people with diabetes usually have a pre-diabetic condition in which blood glucose levels are higher than normal, but not yet high enough for a **diagnosis** of diabetes. At least 16 million people in the US ages 40–74 have pre-diabetes. Pre-diabetes is sometimes referred to as impaired glucose tolerance or impaired fasting glucose. With pre-diabetes, a person is likely to develop diabetes and may already be experiencing the adverse health effects Research has shown that long term damage to the heart and **circulatory system** may already be occurring during pre-diabetes. Diet, increased activity level, and medication may help to prevent or delay type 2 diabetes from developing. If untreated, most people with pre-diabetes develop type 2 diabetes within 3–10 years.

Tests for diabetes

There are three test methods used to diagnose diabetes and each must be confirmed, on a subsequent day, by any one of the three methods. The first method includes symptoms of diabetes (increased urination, increased thirst, unexplained weight loss) plus a casual **plasma** glucose **concentration** (blood test taken any time of day without regard to time since last meal) of equal to or greater than 200 mg. The second test method is a fasting plasma glucose (no caloric intake for at least eight hours) of equal to or greater than 126 mg. The third method is a two-hour after meal blood sugar of equal to or greater than 200 mg during an oral glucose tolerance test. Testing for diabetes should be considered in all individuals at age 45

years and above (particularly if overweight), and if normal, should be repeated every three years. Testing should be considered at a younger age or carried out more frequently in individuals who are overweight and who have additional risk factors among the following:

- first-degree relative with diabetes
- habitually physically inactive lifestyle
- member of high-risk ethnic population (African-American, Hispanic-American, Native American, Asian American, Pacific Islander)
- previous delivery of baby weighing greater than 9 lb (4.1 kg) or history of gestational diabetes
- high blood pressure
- HDL **cholesterol** less than 35 mg or a triglyceride level greater than 250 mg.
- PCOS (polycystic ovarian syndrome)
- impaired glucose tolerance or impaired fasting glucose
- history of vascular disease.

Other tests used in the management of diabetes include c-peptide levels and hemoglobin A1c levels. C-Peptide levels determine if the body is still producing insulin. C-Peptide is the connecting peptide portion of the insulin **molecule** that is produced in the pancreas. C-Peptide and insulin are secreted into the bloodstream in equal amounts by the pancreas. Measurement of C-Peptide is a reliable indicator of the amount of insulin produced by the person's pancreas. HbA1c (hemoglobin A1c) measures the average blood sugar control over a 2–3 month period. The A1c goal recommended by the American Diabetes Association is <7%, which correlates with average blood sugars of less than 150 mg.

Treatment for diabetes

Diabetes is treated with meal planning, exercise, medication, and blood glucose monitoring. Meal planning involves eating the right amount of food at the right time. Carbohydrates have the greatest impact on blood sugars. Keeping track of carbohydrates and spreading them throughout the day helps to control blood sugars. Exercise helps to reduce **stress**, control blood pressure and blood fats, and improves insulin resistance.

Diabetes medications include oral agents and insulin. There are several classes of oral medications. Sulfonylureas and meglitinides help the pancreas to produce more insulin. Alpha-glucosidase inhibitors slow down the digestion and absorption of starches and sugars. Biguanides stop the liver from releasing extra sugar when it is not needed. Thiozolidinediones treat insulin resistance.

Various types of insulin are available and have different action times designed to match to physiological needs

KEY TERMS

. . . .

Autoimmune response—Misdirected immune response in which the body's immune system accidentally recognizes the body's own cells as foreign and destroys them. Type 1 diabetes results from an autoimmune response in which the body destroys the beta cells in the pancreas.

Gestational diabetes—A type of diabetes that occurs in pregnancy.

Glucose—simple sugar made from other carbohydrates that is circulated in the blood at a narrow limit of concentration. Also known as blood sugar.

Hyperosmolar coma—A coma related to high levels of glucose in the blood and requiring emergency treatment. Ketones are not present in the urine; can occur in Type 2 diabetes that is out of control.

Impaired fasting glucose—a condition in which fasting glucose levels are >110 mg, but <126 mg. Now known as pre-diabetes.

Impaired glucose tolerance—a condition in which blood glucose levels rise after meals to levels that are higher than normal. Now called pre-diabetes.

Insulin deficiency—A condition in which little or no insulin is produced by the body.

Insulin resistance—Inability to use the insulin made by the body

Ketoacidosis—formation of ketones (acetones) in the blood from lipid (fat) metabolism and a high blood acid content. Occurs in uncontrolled Type 1 diabetes.

Ketones—acids indicating insufficient insulin that converts fat into glucose in the blood.

Type 1 diabetes—A condition in which the body makes little or no insulin (insulin deficiency). People with this type of diabetes must take injections of insulin.

Type 2 diabetes—A condition in which the body makes insulin but the cells cannot use it well (insulin resistance). It is treated with diet, exercise, and diabetes medication.

---

of the body for persons who no longer make enough insulin. The body requires a continuous, low level of insulin acting to meet baseline needs. Long-acting insulins provide the baseline or basal insulin needs. The body also requires insulin to cover carbohydrates eaten. Short-acting insulins provide coverage for meal boluses. With the wide variety of diabetes medications, the physician can determine a treatment plan that works best for the **individual**.

Blood glucose monitoring serves as the cornerstone tool for measuring the effects of food, exercise, and diabetes medications. Patients can check their blood sugars at various times of the day to keep track of how well the current treatment plan is keeping the sugars are under control. Results of tests are recorded and taken to MD office visits for the MD to evaluate trends and adjust the treatment plan.

Additional management of diabetes is geared toward prevention of complications. **Eye** problems may have no symptoms in their early, treatable stages; therefore annual dilated eye exams are needed. Urine should be checked annually for the protein microalbumin. Poor circulation, nerve damage, and difficulty fighting infections can make foot problems serious considerations for people with diabetes. Daily self-foot exams and foot exams at each physician visit can help identify problems early. Blood fat (lipids–cholesterol and triglyceride) levels should be checked annually.

See also Acids and bases; Metabolic disorders; Metabolism.

Resources

Periodicals

Davidson, Mayer B. MD. "American Diabetes Association: Clinical Practice Recommendations 2003." *Diabetes Care* (2003): 26 Supp1.

Organizations

American Diabetes Association. [cited March 15, 2003]. <http://www.diabetes.org/main/application/commercewf>.

Margaret Meyers
Phyllis Tate

Diagnosis

Diagnosis, from *gnosis*, the Greek word for knowledge, is the process of identifying a **disease** or disorder in a person by examining the person and studying the results of medical tests.

The diagnosis begins when the patient is presented to the doctor with a set of symptoms or perceived abnormalities such as **pain**, nausea, fever, or other untoward feeling. Often the diagnosis is relatively simple, and the physician can arrive at a clinical conclusion and prescribe the proper treatment. At other times, the symptoms may be subtle and seemingly unrelated, making the diagnosis difficult to finalize and requiring laboratory work.

The diagnosis is based on data the physician obtains from three sources, the first being the patient. This includes the patient's **perception** of his or her symptoms, medical history, family history, occupation, and other relevant facts. The physician then narrows the diagnosis with a second set of information obtained from the physical examination of the patient. The third source is the data obtained from medical tests, such as a **blood** test, x ray, or an electrocardiogram.

Patient information

The physician begins the examination by asking about the patient's symptoms. The patient may be asked to describe the symptoms and how long he or she has been experiencing them. If the patient is in pain, information is collected about the location, type, and duration of the pain. Other symptoms that may be present but may not have been noticed by the patient must be explored.

The patient's occupation may have a bearing on his or her illness. Perhaps he or she works around chemicals that may cause illness. A job of repetitive bending and lifting may result in muscle strain or back pain. A police officer or fire fighter may have periods of boredom interrupted by periods of **stress** or fear.

The physician must learn when the symptoms first appeared and whether they have worsened over **time** or remained the same in intensity. If the patient has more than one symptom, the physician must know which appeared first and in what order the others appeared. The doctor will also ask if the symptoms are similar to ones the patient has experienced in the past or if they are entirely new.

The medical history of the patient's family also may be helpful. Some diseases are hereditary and some, though not hereditary, are more likely to occur if the patient's parent or other close relative has had such a disease. For example, the person whose father has had a **heart** attack is more likely to have a heart attack than is a person whose family has been free of heart disease.

Personal habits, such as smoking or drinking large amounts of **alcohol**, also contribute to disease. Lack of **exercise**, lack of **sleep**, and an unhealthy diet are all involved in bringing about symptoms of disease.

The physical examination

In addition to exploring the patient's clinical history, the physician will carry out a physical examination to further narrow the list of possible conditions. The patient's **temperature**, blood **pressure**, and **rate** of **respiration** will be measured. He or she will be weighed and his or her height measured. The physician will use an otoscope to examine the eardrums and to look into the throat for signs of **inflammation**, **infection**, or other abnormal conditions.

The heart and lungs can be examined superficially using a stethoscope. Abnormalities in the heartbeat or in the functioning of the heart valves can be heard in this way, and the presence of **water** or other fluid in the lungs can be heard as noises called rales. The physician also can study the sounds made by the intestines by listening to them through the stethoscope.

Using his fingers, a technique called palpation, the physician probes the abdomen for signs of pain or an abnormal lump or growth. He also feels the neck, the axillary area (armpit) and other locales to locate any enlarged lymph nodes, a sign of an infection. Such probing also may bring to light the presence of a tender area previously unknown to the patient.

If the patient is complaining of an injury, the physician can carefully palpate around the injury to determine its size. He can bend an leg or arm to assess the integrity of the joint. Using other maneuvers, he can determine whether a ligament has been torn and if it may need surgical correction.

The laboratory examination

Having learned the patient's clinical history and made his physical examination, the physician may then decide to submit specimens from the patient to a laboratory for testing. Fluids such as blood, urine, stomach fluid, or spinal fluid can be collected.

Basic laboratory tests of blood include a count of the number of white and red blood cells. An elevated number of white blood cells indicates an infection is present, but does not pinpoint the location of the infection. Blood also carries **hormones** and other components that are directly affected by disease or inflammation.

Far from the laboratory of the 1960s, the modern clinical laboratory is one of **automation** and high technology. Whereas before the laboratory technician was required to mix together the chemicals for each test, newer technology requires only that a blood specimen be placed in one end of a machine. The blood is carried through the machine and minute amounts of the chemicals are added as needed and the results printed out. This technology also enables the measurement of blood or urine components in amounts much smaller than previous technology allowed—often at microgram levels. A microgram is one millionth of a gram. To measure such a minute amount, the **chemistry** involved is precise and the reading of the results is beyond the capability of the human **eye**.

Both blood and urine may contain evidence of alcohol, illicit drugs, or toxic substances that the patient has taken. Infectious organisms from the blood or urine can be grown in culture dishes and examined to determine what they are. **Bacteria** in blood or urine are often too sparsely distributed to be seen under the **microscope**, but bacteria in a blood specimen wiped across a plate of culture medium will grow when the plate is placed in an incubator at body temperature.

The physician also may want to obtain **x rays** of an injured area to rule out the possibility of a fractured bone. The presence of a heart condition can often be determined by taking an **electrocardiogram (ECG)**, which measures the electrical activity of the heart. Changes in the ECG can indicate the presence of heart disease or give evidence of a past heart attack. CAT (**computerized axial tomography**) scans use x rays to produce images of one layer of hard or soft **tissue**, a procedure useful in detecting small tumors. **Magnetic resonance imaging (MRI)** uses **radio waves** in a magnetic field to generate images of a layer of the **brain**, heart, or other **organ**. Ultrasound waves are also sometimes used to detect tumors.

Physicians can collect other kinds of information by injecting substances into the patient. Injection of radiopaque liquids, which block the passage of x rays, allow x-ray examination of soft tissues, such as the spinal cord, that are normally undetectable on x-ray photographs. **Metabolic disorders** can sometimes be pinpointed using a procedure called scintigraphy, in which a radioactive **isotope** is circulated through the body. A gamma camera is then used to record the **concentration** of the isotope in various tissues and organs.

Other laboratory specimens can be obtained by invasive techniques. If the physician finds a suspicious lump or swelling and needs to know its nature, he can remove part of the lump and send it to the laboratory to be examined. The surgical removal of tissue for testing is called a biopsy. In the laboratory, the specimen is sliced very thin, dyed to accentuate differences in tissues, and examined under the microscope. This enables the physician to determine whether the lump is malignant (cancerous) or benign (noncancerous). If it is **cancer**, further tests can determine if it is the primary **tumor** or if it has grown (metastasis) as a result of being spread from the primary tumor. Other tests can determine what kind of cancer it is.

The method of actually looking into the body cavity used to mean a major surgical procedure called a laparotomy. In that procedure, an incision was made in the abdomen so the physician could look at each organ and other internal structure in order and determine the presence of disease or parasite. Now the laparotomy is carried out using a flexible scope called a laparoscope, which is inserted into the body through a small incision. The scope is attached to a **television** monitor that gives the physician an enlarged view of the inside of the body. The flexibility of the scope allows it to be guided around the organs, and a light attached to the scope helps the physician see each organ. Also, the laparoscope is equipped with the means to collect biopsy specimens or suction blood out of the abdomen. Minor **surgery** can also be carried out to stop a bleeding blood vessel or remove a small growth from an organ.

Once the above steps the physician deems necessary have been carried out, he or she will then study the evidence collectively and arrive at a diagnosis. Once having determined the diagnosis, he or she can prescribe the proper treatment.

Larry Blaser

Dialysis

Dialysis is a process by which small molecules in a **solution** are separated from large molecules. The principle behind the process was discovered by the Scottish chemist Thomas Graham in about 1861. Graham found that the **rate** at which some substances, such as inorganic salts, pass through a semipermeable **membrane** is up to 50 times as great as the rate at which other substances, such as **proteins**, do so. We now know that such rate differences depend on the fact that the openings in semipermeable membranes are very nearly the size of **atoms**, ions, and small molecules. That makes possible the passage of such small particles while greatly restricting the passage of large particles.

In a typical dialysis experiment, a bag made of a semipermeable membrane is filled with a solution to be dialyzed. The bag is then suspended in a stream of running

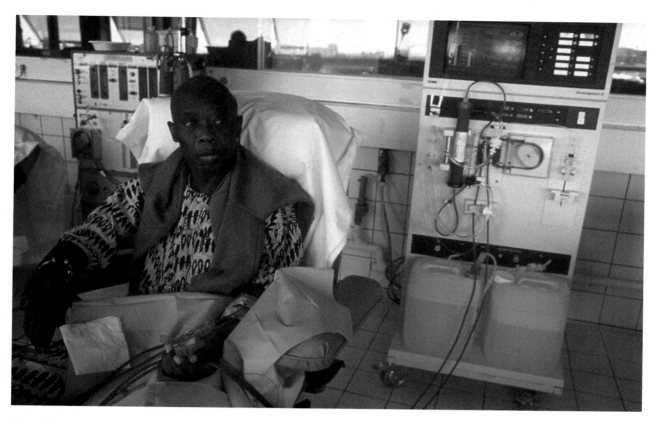

Man with damaged kidneys on a dialysis machine. The blood is removed from his arm, passed through the machine, and returned to his body. © *Giovaux Communication/Phototake NYC Reproduced with permission.*

water. Small particles in solution within the bag gradually diffuse across the semipermeable membrane and are carried away by the running water. Larger molecules are essentially retained within the bag. By this process, a highly efficient separation of substances can be achieved.

The kidney is a dialyzing **organ**. By the process described above, it filters waste products such as **urea** out of the **blood** and forces them into the urine, in which they are excreted from the body. Proteins and other important large molecules are retained in the blood.

A person whose kidneys have been damaged by **disease** or physical injury requires some artificial method for cleansing her or his blood. A device for carrying out this task–the artificial kidney machine–was developed in the early 1910s largely through the efforts of John J. Abel and his colleagues at the Johns Hopkins University. In the kidney machine, blood is removed from a person's arm, passed through a dialyzing system, and then returned to the patient. The machine functions much as a natural kidney does with one important exception. A natural kidney has a mechanism known as reverse dialysis for returning to the body certain small molecules (primarily glucose) that should not be excreted. The kidney machine is unable to do so, and glucose that it removes must be replaced by intravenous injection.

Electrodialysis is a form of dialysis in which the separation of ions from larger molecules is accelerated by the presence of an electrical field. In one arrangement, the solution to be dialyzed is placed between two other solutions, each containing an electrode. Cations within the middle solution are attracted to one electrode and anions to the other. Any large molecules in the middle solution remain where they are.

One possible application of electrodialysis is the **desalination** of water. In this procedure, **sodium** ions from seawater migrate to the **cathode** and chloride ions to the **anode** of an electrodialysis apparatus. Relatively pure water is left behind in the central compartment.

See also Osmosis.

Diamond

Diamond is a mineral with the same **carbon** composition as graphite, but with different structure.

Diamonds are formed in the compression of coal. *Photograph by Rick Gayle. Stock Market. Reproduced by permission.*

Diamonds are a globally traded commodity used for a variety of industrial and artistic purposes. In December 2000, the United Nations General Assembly unanimously adopted a resolution articulating the role of diamonds in fuelling international conflict and dedicated to breaking the link between the illicit transaction of rough diamonds and armed conflict. These "conflict diamonds" are facing increasing import-export trade restrictions.

The **atoms** making up a mineral may be arranged either randomly, or in an orderly pattern, if—as with diamonds—a mineral's atoms show long-range organization, the mineral is termed a crystalline mineral. The objects commonly called crystals are crystalline **minerals** of relatively large size that happen to have developed smooth faces. Diamonds are the hardest mineral (10 on the **Mohs' scale**), with the highest refractive index of 2.417 among all transparent minerals, and has a high dispersion of 0.044. Diamonds are brittle. Under UV **light**, the diamond frequently exhibits **luminescence** with different colors. It has a **density** of 3.52 g/cm³. The **mass** of diamonds is measured in carats; 1 carat = 0.2 grams. Diamonds rarely exceed 15 carats. Diamonds are insoluble in acids and alkalis, and may burn in **oxygen** at high temperatures.

Nitrogen is the main impurity found in diamonds, and influences its physical properties. Diamonds are divided into two types, with type I containing 0.001–0.23% nitrogen, and type II containing no nitrogen. If nitrogen exists as clusters in type I diamonds, it does not affect the **color** of the stone (type Ia), but if nitrogen substitutes carbon in the **crystal** lattice, it causes a yellow color (Ib). Stones of type II may not contain impurities (IIa), or may contain boron substituting carbon, producing a blue color and semiconductivity of the diamond.

Diamonds form only at extremely high **pressure** (over 45000 atmospheres) and temperatures over 2012°F (1100°C) from liquid ultrabasic magmas or peridotites. Diamonds, therefore, form at great depths in the Earth's crust. They are delivered to the surface by explosive volcanic phenomena with rapid cooling rates, which preserve the diamonds from transformation. This process happens in kimberlites (a peridotitic type of breccia), which constitutes the infill of diamond- bearing pipes. Also found with diamonds are olivine, serpentine, carbonates, pyroxenes, pyrope garnet, magnetite, hematite, graphite and ilmenite. Near the surface, kimberlite weathers, producing yellow loose mass called yellow ground, while deeper in **Earth**, it changes to more dense blue ground. Diamonds are extremely resistive to **corrosion**, so they can be fond in a variety of secondary deposits where they arrived after several cycles of **erosion** and sedimentation (alluvial diamond deposits, for example). Even in diamond-bearing rock, the diamond **concentration** is 1 g in 8–30 tons of rock.

Most diamonds are used for technical purposes due to their hardness. Gem quality diamonds are found in over 20 counties, mainly in **Africa**. The biggest diamond producer is South Africa, followed by Russia. Usually, diamonds appear as isolated octahedron crystals. Sometimes they may have rounded corners and slightly curved faces. Microcrystalline diamonds with irregular or globular appearance are called Bort (or boart), while carbonado are roughly octahedral, cubic or rhombic dodecahedral, blackish, irregular microcrystalline aggregates. Both are valued for industrial applications because they are not as brittle as diamond crystals. Frequently, diamonds have inclusions of olivine, sulfides, chrome-diopside, chrome-spinels, zircon, rutile, disthene, biotite, pyrope garnet and ilmenite. Transparent crystals are usually colorless, but sometimes may have various yellowish tints. Rarely, diamonds may be bright yellow, blue, pale green, pink, violet, and even reddish. Some diamonds are covered by translucent skin with a stronger color. Diamonds become green and radioactive after **neutron** irradiation, and yellow after further heating. They become blue after irradiation with fast electrons. Diamonds have different hardnesses along their different faces. Diamonds from different deposits also have different hardnesses. This quality allows for the polishing of faceted diamonds by diamond powder.

Most diamond gems are faceted into brilliant cuts. Due to the high reflective index, all light passing through the face of such facetted diamonds is reflected back from the back facets, so light is not passing through the stone. This can be used as a diagnostic property, because most simulants (except cubic zirconia) do not have this property. Diamonds do have many simulants, including zircon, corundum, phenakite, tourmaline, topaz, beryl, quartz, scheelite, sphalerite, and also synthetic gemstones such as cubic zirconia, Yttrium-aluminum garnet, strontium titanate, rutile, spinel, and litium niobate. Diamonds have high thermal conductivity, which allows it to be readily and positively distinguished from all simulated gemstones. The most expensive diamonds are those with perfect structure and absolutely colorless or slightly bluish-white color. Yellow tint reduces the price of the diamond significantly. Bright colored diamonds are extremely rare, and have exceptionally high prices.

In January 2003, a number of international concerns came to a preliminary consensus on the Kimberley Process Certification Scheme to curtail international trade in what are termed conflict diamonds.

As of early 2003, nearly 50 countries agreed to use and require standardized, tamper-proof packaging and official certificates attesting to the source of the enclosed diamonds when shipping rough uncut diamonds. Such controls are designed to stem illegal trade in diamonds and to reduce the ability of despotic regimes to exploit diamond trade to perpetuate their political and or military power (e.g., the protocols prohibit trade in contraband diamonds from rebel sources in Sierra Leone). Without proper certification many nations and industrial sources are agreed to import or purchase contraband diamonds.

See also Mineralogy.

Yavor Shopov

Resources

Books

Hart, Matthew. *Diamond: A Journey to the Heart of an Obsession.* New York: Walker & Co., 2001.
Klein, Cornelis. *Manual of Mineral Science.* 22nd. ed. New York: John Wiley & Sons, 2001.
Schumann, Walter. *Gemstones of the World.* London: Sterling Publications, 2000.

Diatoms

Algae are a very diverse group of simple, nucleated, plant-like aquatic organisms that are primary producers. Primary producers are able to utilize **photosynthesis** to create organic molecules from sunlight, **water**, and **carbon dioxide**. Ecologically vital, algae account for roughly half of photosynthetic production of organic material on **earth** in both **freshwater** and marine environments. Algae exist either as single cells or as multicellular organizations. Diatoms are microscopic, single-celled algae that have intricate glass-like outer **cell** walls partially composed of silicon. Different **species** of diatom can be identified based upon the structure of these walls. Many diatom species are planktonic, suspended in the water column moving at the mercy of water **currents**. Others remain attached to submerged surfaces. One bucketful of water may contain millions of diatoms. Their abundance makes them important food sources in aquatic ecosystems. When diatoms die, their cell walls are left behind and sink to the bottom of bodies of water. Massive accumulations of diatom-rich sediments compact and solidify over long periods of **time** to form rock rich in fossilized diatoms that is mined for use in **abrasives** and filters.

Diatoms belong to the taxonomic phylum *Bacillariophyta*. There are approximately 10,000 known diatom species. Of all algae phyla, diatom species are the most

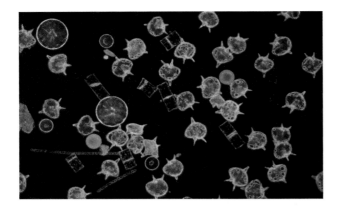

Diatom plankton. *Photograph by Dougals P. Wilson. COR-BIS/Dougals P. Wilson; Frank Lane Picture Angency. Reproduced by permission.*

numerous. The diatoms are single-celled, eukaryotic organisms, having genetic information sequestered into subcellular compartments called nuclei. This characteristic distinguishes the group from other single-celled photosynthetic aquatic organisms, like the blue-green algae that do not possess nuclei and are more closely related to **bacteria**. Diatoms also are distinct because they secrete complex outer cell walls, sometimes called skeletons. The skeleton of a diatom is properly referred to as a frustule.

Diatom frustules are composed of very pure hydrated silica within a layer of organic, **carbon** containing material. Frustules are really comprised of two parts: an upper and lower frustule. The larger upper portion of the frustule is called the epitheca. The smaller lower piece is the hypotheca. The epitheca fits over the hypotheca like the lid fits over a shoe box. The singular algal diatom cell lives protected inside the frustule halves like a pair of shoes snuggled within a shoe box.

Frustules are very ornate, having intricate designs delineated by patterns of holes or pores. The pores that perforate the frustules allow gases, **nutrients**, and metabolic waste products to be exchanged between the watery environment and the algal cell. The frustules themselves may exhibit bilateral **symmetry** or radial symmetry. Bilaterally symmetric diatoms are like human beings, having a single **plane** through which halves are mirror images of one another. Bilaterally symmetric diatoms are elongated. Radially symmetric diatom frustules have many mirror image planes. No matter which diameter is used to divide the cell into two halves, each half is a mirror image of the other. For example, apple pie is radially symmetric. No matter which diameter one chooses in slicing the pie into halves, each half is the mirror image of the other. Similarly, radially symmetric diatoms are round and flattened, like apple pies. The combination of symmetry and perforation patterns of di-

atom frustules make them very beautiful biological structures that also are useful in identifying different species. Because they are composed of silica, a very inert material, diatom frustules remain well preserved over vast periods of time within geologic sediments.

Diatom frustules found in **sedimentary rock** are microfossils. Because they are so easily preserved, diatoms have an extensive fossil record. Specimens of diatom algae extend back to the Cretaceous Period, over 135 million years ago. Some kinds of rock are formed nearly entirely of fossilized diatom frustules. Considering the fact that they are microscopic organisms, the sheer numbers of diatoms required to produce rock of any thickness is staggering. Rock that has rich concentrations of diatom fossils is known as diatomaceous earth, or diatomite. Diatomaceous earth, existing today as large deposits of chalky white material, is mined for commercial use in abrasives and in filters. The fine abrasive quality of diatomite is useful in cleansers, like bathtub scrubbing powder. Also, many toothpaste products contain fossil diatoms. The fine porosity of frustules also makes refined diatomaceous earth useful in fine water filters, acting like microscopic sieves that catch very tiny particles suspended in **solution**.

Fossilized diatom collections also tell scientists a lot about the environmental conditions of past eras. It is known that diatom deposits can occur in layers that correspond to environmental cycles. Certain conditions favor mass deaths of diatoms. Over many years, changes in diatom deposition rates in sediments, then, are preserved as diatomite, providing clues about prehistoric climates.

Diatom cells within frustules contain chloroplasts, the organelles in which photosynthesis occurs. Chloroplasts contain **chlorophyll**, the pigment **molecule** that allows plants and other photosynthetic organisms to capture solar **energy** and convert it into usable chemical energy in the form of simple sugars. Because of this, and because they are extremely abundant occupants of freshwater and **saltwater** habitats, diatoms are among the most important **microorganisms** on Earth. Some estimates calculate diatoms as contributing 20-25% of all carbon fixation on Earth. Carbon fixation is a term describing the photosynthetic process of removing atmospheric carbon in the form of carbon dioxide and converting it to organic carbon in the form of sugar. Due to this, diatoms are essential components of aquatic food chains. They are a major food source for many microorganisms, aquatic **animal** larvae, and grazing animals like **mollusks** (**snails**). Diatoms are even found living on land. Some species can be found in moist **soil** or on mosses. Contributing to the abundance of diatoms is their primary mode of reproduction, simple asexual **cell division**. Diatoms divide asexually by **mitosis**. During division, diatoms construct new frustule cell walls. After a cell divides, the epitheca and hypotheca separate, one remaining with each new daughter cell. The two cells then produce a new hypotheca. Diatoms do reproduce sexually, but not with the same frequency.

Resources

Books

Round, F. E., et al. *Diatoms: Biology and Morphology of the Genera.* Cambridge University Press, 1990.
Stoermer, Eugene F., and John P. Smol. *The Diatoms: Applications for the Environmental and Earth Sciences.* Cambridge University Press, 1999.

Terry Watkins

Dielectric materials

Dielectric materials are substances that have very low conductivity. That is, they are electrical insulators through which an electrical current flows only with the greatest of difficulty. Technically, a dielectric can be defined as a material with **electrical conductivity** of less than one millionth of a mho (a unit of electrical conductance) per centimeter.

In theory, dielectrics can include solids, liquids, and gases, although in practice only the first two of these three **states of matter** have any practical significance. Some of the most commonly used dielectrics are various kinds of rubber, **glass**, **wood**, and polymers among the solids; and **hydrocarbon** oils and silicone oils among the liquids.

The dielectric constant

A common measure of the dielectric properties of a material is the dielectric constant. The dielectric constant can be defined as the tendency of a material to resist the flow of an electrical current across the material. The lower the value of the dielectric constant, the greater its resistance to the flow of an electrical current.

The standard used in measuring the dielectric constant is a **vacuum**, which is assigned the value of one. The dielectric constants of some other common materials are as follows: dry air (at one atmosphere of **pressure**): 1.0006; **water**: 80; glass: 4-7; wax: 2.25; amber: 2.65; mica: 2.5-7; **benzene**: 2.28; **carbon tetrachloride**: 2.24; and methyl **alcohol**: 33.1. Synthetic polymers are now widely used as dielectrics. The dielectric constants for these materials range from a low of about 1.3 for polyethylene and 2.0 for polytetrafluoroethylene (Teflon) to a high of about 7.2-8.4 for a melamine-formaldehyde resin.

Uses

Almost any type of electrical equipment employs dielectric materials in some form or another. Wires and cables that carry electrical current, for example, are always coated or wrapped with some type of insulating (dielectric) material. Sophisticated electronic equipment such as rectifiers, semiconductors, transducers, and amplifiers contain or are fabricated from dielectric materials. The insulating material sandwiched between two conducting plates in a **capacitor** is also made of some dielectric substance.

Liquid dielectrics are also employed as electrical insulators. For example, **transformer** oil is a natural or synthetic substance (mineral oil, silicone oil, or organic esters, for example) that has the ability to insulate the coils of a transformer both electrically and thermally.

Synthetic dielectrics

A number of traditional dielectric materials are still widely used in industry. For example, **paper** impregnated with oil is often still the insulator of choice for coating wires that carry high-voltage current. But synthetic materials have now become widely popular for many applications once filled by natural substances, such as glass and rubber. The advantage of synthetic materials is that they can be designed so as to produce very specific properties for specialized uses. These properties include not only low dielectric constant, but also strength, hardness, resistance to chemical attack, and other desirable qualities.

Among the polymers now used as dielectrics are the polyethylenes, polypropylenes, polystyrenes, polyvinyl chlorides, polyamides (Nylon), polymethyl methacrylates, and polycarbonates.

Breakdown

When a dielectric material is exposed to a large electrical field, it may undergo a process known as breakdown. In that process, the material suddenly becomes conducting, and a large current begins to flow across the material. The appearance of a spark may also accompany breakdown. The point at which breakdown occurs with any given material depends on a number of factors, including **temperature**, the geometric shape of the material, and the type of material surrounding the dielectric. The ability of a dielectric material to resist breakdown is called its intrinsic electric strength.

Breakdown is often associated with the degradation of a dielectric material. The material may oxidize, physically break apart, or degrade in some other way that will make conductance more likely. When breakdown does occur, then, it is often accompanied by further degradation of the material.

See also Electronics; Oxidation state.

Resources

Books

Scaife, B.K. *Principles of Dielectrics.* New York: Oxford University Press, 1989.

Periodicals

Gridnev, S. A. "Electric Relaxation In Disordered Polar Dielectrics." *Ferroelectrics* 266, no. 1 (2002): 171-209.

David E. Newton

Diesel engine

Diesel engines are a class of **internal combustion engine** in which the fuel is burned internally and the **combustion** products are used as the working fluid. Unlike the spark-ignited (SI) engines found in the majority of today's automobiles in which the premixed fuel-air mixture is ignited by an electric spark, diesel engines are characterized by a spontaneously initiated combustion process where the ignition is brought about by very high **temperature** compressed air. A small amount of diesel fuel is injected at the end of the compression stroke into the cylinder where the fuel auto-ignites. Because of their higher actual operating efficiencies, as compared with SI engines that require pre-ignition, diesel engines are primarily used in heavy-duty vehicles such as trucks, ships, locomotives, etc.

Diesel engines were first developed by Rudolf Diesel (1858-1913) in the late nineteenth century. The original concept was to build a multifuel engine and to use **coal** as a primary fuel. However, for some reason, coal-fueled diesel engines so far have gained only occasional interest from the industry (e.g., when fuel-oil prices are high), most of the diesel engines currently being used rely on **petroleum** fuels. They are four-stroke cycle engines, and operate from several hundred up to around one thousand rpm. In addition to pistons, cylinders, crankshaft, and various valves, diesel engines are also equipped with controlled fuel injection systems, exhaust systems, cooling systems, and so on. Sufficient lubrication is required to prevent excessive wear of various parts in engines. Since pre-ignition is not required, the compression step can be continued to reach a higher **pressure** or a higher compression **ratio** than that in SI engines. This results in compressed air with a tempera-

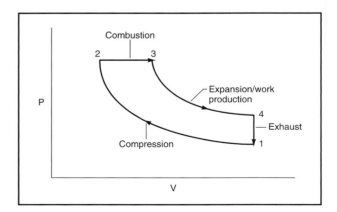

Figure 1. A typical pressure-volume (P-V) diagram for air-standard cycle of diesel engines. *Illustration by Hans & Cassidy. Courtesy of Gale Group.*

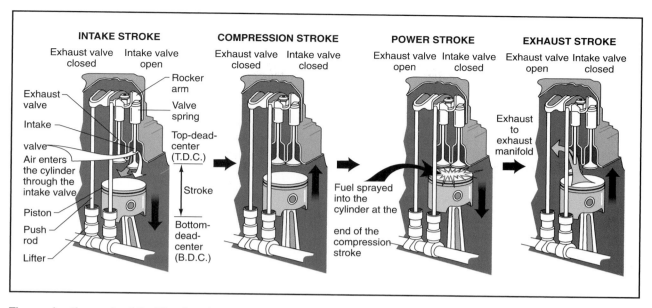

INTAKE STROKE

Exhaust valve closed   Intake valve open

Exhaust valve

Intake

valve

Air enters the cylinder through the intake valve

Piston

Push rod

Lifter

Rocker arm

Valve spring

Top-dead-center (T.D.C.)

Stroke

Bottom-dead-center (B.D.C.)

COMPRESSION STROKE

Exhaust valve closed   Intake valve closed

POWER STROKE

Exhaust valve open   Intake valve closed

Fuel sprayed into the cylinder at the

end of the compression stroke

EXHAUST STROKE

Exhaust valve open   Intake valve closed

Exhaust to exhaust manifold

The combustion cycle of the Diesel engine. *Illustration by Hans & Cassidy. Courtesy of Gale Group.*

ture exceeding the ignition point of the injected fuel for auto-ignition. To achieve high combustion efficiency, the fuel jets must draw in and mix well with air, ignite, and burn, all within less than one millisecond, when they impact on the cold combustion chamber walls. Engine performance is closely related to compression ratio, piston speed, supercharging, turbo-charging, etc., and engine size is normally in terms of power rating (i.e., horsepower; for instance, 20,000 hp applicable for ship propulsion). In principle, the same engine frame can be designed for different output by varying the number of cylinders (10, 12, 16 cylinders, and so on).

In reality, because the fuel-air mixture is burned and the products of combustion are emitted, the process for **work** production via combustion in diesel engines is complex and not cyclical. However, in order to analyze it, the actual operation is frequently represented approximately by a cyclical process, called "Diesel cycle." From the point of view of **thermodynamics**, the working fluid is assumed to be air, the compression and expansion stages are assumed to be adiabatic (without the loss or gain of **heat**) and reversible, and the combustion and exhaust strokes are replaced by constant-pressure heat-absorption and constant-volume heat-rejection stages. As shown in Figure 1, a typical pressure-volume (P-V) diagram for air-standard diesel engine operation, after the intake, air is compressed adiabatically along the path 1-2 and its temperature is increased substantially. At point 2 where the piston begins to reverse its **motion**, the fuel is injected and added slowly so that combustion is initiated and sustained at constant pressure following the path 2-3. After completion of the combustion, there is the work

stroke, i.e., along the path 3-4 where the high-temperature and high-pressure products of combustion are expanded to produce mechanical work. Then the exhaust valve is opened, the spent combustion products and waste heat are exhausted, and the pressure is rapidly reduced as the path 4-1. This, therefore, completes typical four strokes in each cycle of engine operation. The thermal efficiency of the cycle can be obtained from the net work produced divided by the heat absorbed during the entire cyclical process.

Overall, diesel engines can be viewed as a piston-and-cylinder assembly and the work-producing machine. Their operation cycle is similar to that in SI engines which are based on the Otto (after the German inventor for the first internal-combustion engine produced in the mid 1860s) cycle; however, the latter require an external combustion initiator and have combustion occurring under an almost constant-volume condition, which is different from the path 2-3 as shown in Figure 1. In these engines, the chemical (molecular) **energy** of the fuel (hydrocarbons) is released by a combustion process. Energy is evolved as heat and part of the heat is subsequently converted into useful work or mechanical energy. Because of the loss of heat during the process, research and development efforts have been made constantly in chamber design, new coatings for rings and liner, **emission** control, alternative fuels, and associated compressor and **turbine** technologies to improve the conversion efficiency. As can be expected, diesel engines will continue finding a variety of applications in the future, such as power generation as well as land, marine, and **aircraft** transport.

KEY TERMS

. . . .

Adiabatic—A process during which no heat is transferred between the system and surroundings is described as "adiabatic."

Heat engine—A device converts heat to mechanical work in a periodic process.

Reversible—A process occurs in such a way that both the system and its surroundings can be returned to their initial states.

Supercharging—Methods to increase the inlet manifold air pressure above ambient pressure so that power output in engines is increased.

Thermal efficiency—The ratio of net work to thermal energy input.

Turbocharging—An approach to utilizing high-temperature exhaust gas by expanding it through a turbine for driving the supercharging compressor.

Resources

Books

Haddad, S.D., ed. *Advanced Diesel Engineering and Application.* New York: John Wiley & Sons, 1988.

Periodicals

Rutland, Christopher. "Probability Density Function Combustion Modeling of Diesel Engine." *Combustion Science and Technology* 174, no. 10 (2002): 19-54.

Ryan III, T.W. " Coal-Fueled Diesel Development: A Technical Review." *Journal of Engineering for Gas Turbines and Power* (1994).

Other

"Combustion Modelling For Direct Injection Diesel Engines." *Proceedings Of The Institution Of Mechanical Engineers* 215, no. 5 (2001): 651-663.

Pang-Jen Kung

Diet *see* **Nutrition**

Diethylstilbestrol (DES)

The substance diethylstilbestrol (DES) is a synthetic, nonsteroidal estrogen, which was first made in 1938. Initially the substance was seen as a great scientific breakthrough, drawing on research that documented the importance of naturally occurring estrogen in women. Wide-scale use of DES by pregnant women to prevent miscarriage beginning in the 1940s ended in 1971 when researchers discovered that some daughters of women who took DES had developed a rare **cancer**, called clear-cell adenocarcinoma of the vagina. Researchers have since found that daughters of women who took DES face a higher risk of certain cancers and of structural abnormalities in the genital area. The example of DES, used by two to three million American women, has been used to dramatize the risk of improperly tested medicine during pregnancy.

Medical breakthrough

The development of DES by British scientist Edward Dodds (1899-1973) was one in a long line of twentieth century medical advances which reflected new understanding of the female **reproductive system**. While doctors had observed pregnancy, childbirth, miscarriages, and **infertility** for centuries, they did not gain an understanding of the hormonal functions behind these processes until the twentieth century. Through a series of discoveries, researchers learned that the process of pregnancy required a complicated series of hormonal triggers to occur successfully. They also learned that **hormones** were critical for the development of sexual characteristics in both men and women.

An early breakthrough was the successful isolation in 1923 of estrogen, the female sex hormone produced in men and women. Initially, the natural form of estrogen was extracted from **animal** ovaries, a process that was time-consuming and expensive. By 1936, the first synthetic estrogen was manufactured from **plant** steroids. In 1942, DES was approved by the U.S. Food and Drug Administration for use by menopausal women and for several other purposes. The drug was not approved for use by pregnant women until 1947, following reports that the drug could reduce the incidence of miscarriage.

At the time DES was first used for pregnant women, the substance was seen as a new weapon against infertility, stillbirths, and prematurity, according to a 1991 account by Edith L. Potter, a pathologist who specialized in obstetrics and **gynecology** during the time DES was prescribed. George and Olive Smith, the Harvard Medical School researchers who promoted the use of DES during pregnancy, believed that stillbirths and premature births were caused by a failure of the placenta to produce sufficient quantities of progesterone, Potter observes. They thought DES would relieve this condition, so that the pregnancies could be carried to term.

In the observation of practicing physicians, the substance appeared to cause no harm. Potter examined more than 10,000 infants who died of unrelated causes during this time, and she writes of seeing no apparent abnormalities due to DES. Problems did not become apparent until the DES infants were no longer infants.

Contemporary critics have faulted studies promoting DES for pregnant women because they did not include control groups—individuals given a **placebo** to help gauge the true effectiveness of DES. Without control groups, there was no scientific way to tell if the use of DES made women less likely to miscarry.

Signs of trouble

The largest number of DES prescriptions were ordered in 1953. By the middle 1950s, a series of studies suggested that DES did not actually help prevent miscarriages. Yet the drug continued to be given to pregnant women throughout the 1960s. Then, in the late 1960s, doctors noticed a series of cases of vaginal cancer in teenage girls and women in their twenties. This was troubling, because vaginal cancer had previously been seen primarily in much older women. By 1971, researchers had definitely linked DES to the vaginal cancer cases, and a report of the association appeared in the influential *New England Journal of Medicine.* That same year, the FDA prohibited the use of DES during pregnancy.

A total of about 600 cases of cancer of the cervix and vagina have been diagnosed in DES daughters. Daughters also are at higher risk of structural abnormalities of the reproductive tract and of poor pregnancy outcome. About one-half of all DES daughters experience an ectopic pregnancy, a premature **birth**, or a miscarriage. In addition, DES daughters are at a higher risk of infertility than women whose mothers did not take the drug.

The most common health problem reported by DES daughters is adenosis of the vagina. Adenosis is the abnormal development of glandular **tissue**. This occurs in 30% or more of DES daughters. About 25% of DES daughters have physical abnormalities of the cervix or vagina. Vaginal cancer occurs in less than 1 per 1,000 females whose mothers took DES.

Daughters of women who took DES are not the only ones at higher risk of health problems. Mothers who took DES face a slightly increased risk of breast cancer, and anecdotal evidence suggests that DES sons have a higher risk of testicular and semen abnormalities. Infertility and a higher risk of some types of cancer have also been reported among some DES sons.

Effects on the developing embryo

Various theories to account for the effects of DES have been presented. What is clear is that the diseases associated with DES derive from structural damage of the fetus caused by the drug. The drug is most damaging when taken early in pregnancy, when the reproductive organs are formed. (Researchers have found that daughters of mothers who took DES in their eighteenth week of pregnancy or later had fewer abnormalities.) One explanation suggests that DES exposure causes abnormal development of the Mullerian ducts, paired structures present in the early embryo. During a normal pregnancy, the Mullerian ducts form the female reproductive tract, including the uterus, the fallopian tubes, the vagina, and the cervix. DES causes the persistence of a type of glandular, or secreting, epithelial **cell** in the vagina. During normal development, this type of cell is transformed to a squamous, or flattened, cell. The persistence of this type of cell, researchers speculate, could make affected women more susceptible to a cancer-promoting factor. They have also suggested that vaginal cancer does not develop until after menstruation begins because this susceptible tissue reacts to estrogens released naturally in women who menstruate.

To explain the higher **rate** of premature deliveries and infertility in DES daughters, scientists point to the abnormal development of the cervix or endometrium in the embryo. Another possible explanation is that DES somehow causes defects in the **connective tissue** of the fetal cervix and uterus, so that these organs cannot develop normally.

There is also a possibility that the abnormal cells in the vagina and cervix of DES daughters will become malignant later in life. While this is unusual, physicians are advised to examine DES daughters regularly to monitor their condition. Arthur L. Herbst reports that only 16 of the hundreds of thousands of DES daughters who have been examined have had tumors develop from vaginal adenosis or a related condition of the cervix, cervical ectropion.

Predictions that the number of DES daughters with cancer would continue to grow dramatically have fortunately proven false. But researchers warn that DES daughters, sons, and mothers may face other health complications as they get older. The legacy of DES has been long-lasting and troubling, a reminder of the power of medicine to hurt as well as help.

Resources

Books

Herbst, Arthur L. "Problems of Prenatal DES Exposure," in *Comprehensive Gynecology.* Edited by Arthur L. Herbst, Daniel R. Mishell Jr., Morton A. Stenchever, and William Droegemueller. St. Louis: Mosby Year Book, 1992.

Periodicals

Colton, Theodore, et al. "Breast Cancer in Mothers Prescribed Diethylstilbestrol in Pregnancy: Further Follow-up." *Journal of the American Medical Association* 269 (April 28, 1993): 2096.

Henderson, Charles. "DES Registries in Need of Update." *Cancer Weekly* (June 22, 1992): 10.

Kushner, Susan. "In the Graveyard of Western Medicine." *East West Natural Health* (July-August 1992): 144.

Potter, Edith L. "A Historical View: Diethylstilbestrol Use During Pregnancy: A 30-Year Historical Perspective." *Pediatric Pathology* 11 (1991): 781-89.

Patricia Braus

Diffraction

Diffraction is the deviation from a straight path that occurs when a wave such as **light** or sound passes around an obstacle or through an opening. The importance of diffraction in any particular situation depends on the relative size of the obstacle or opening and the wavelength of the wave that strikes it. The **diffraction grating** is an important device that makes use of the diffraction of light to produce spectra. Diffraction is also fundamental in other applications such as x-ray diffraction studies of crystals and holography.

Fundamentals

All waves are subject to diffraction when they encounter an obstacle in their path. Consider the shadow of a flagpole cast by the **Sun** on the ground. From a **distance** the darkened zone of the shadow gives the impression that light traveling in a straight line from the Sun was blocked by the pole. But careful observation of the shadow's edge will reveal that the change from dark to light is not abrupt. Instead, there is a gray area along the edge that was created by light that was "bent" or diffracted at the side of the pole.

When a source of waves, such as a light bulb, sends a beam through an opening or aperture, a diffraction pattern will appear on a screen placed behind the aperture. The diffraction pattern will look something like the aperture (a slit, **circle**, **square**) but it will be surrounded by some diffracted waves that give it a "fuzzy" appearance.

If both the source and the screen are far from the aperture the amount of "fuzziness" is determined by the wavelength of the source and the size of the aperture. With a large aperture most of the beam will pass straight through, with only the edges of the aperture causing diffraction, and there will be less "fuzziness." But if the size of the aperture is comparable to the wavelength, the diffraction pattern will widen. For example, an open window can cause **sound waves** to be diffracted through large angles.

Fresnel diffraction refers to the case when either the source or the screen are close to the aperture. When both source and screen are far from the aperture, the term Fraunhofer diffraction is used. As an example of the latter, consider starlight entering a **telescope**. The diffraction pattern of the telescope's circular mirror or **lens** is known as Airy's disk, which is seen as a bright central disk in the middle of a number of fainter rings. This indicates that the image of a **star** will always be widened by diffraction. When optical instruments such as telescopes have no defects, the greatest detail they can observe is said to be diffraction limited.

Applications

Diffraction gratings

The diffraction of light has been cleverly taken advantage of to produce one of science's most important tools—the diffraction grating. Instead of just one aperture, a large number of thin slits or grooves—as many as 25,000 per inch—are etched into a material. In making these sensitive devices it is important that the grooves are **parallel**, equally spaced, and have equal widths.

The diffraction grating transforms an incident beam of light into a **spectrum**. This happens because each groove of the grating diffracts the beam, but because all the grooves are parallel, equally spaced and have the same width, the diffracted waves mix or interfere constructively so that the different components can be viewed separately. Spectra produced by diffraction gratings are extremely useful in applications from studying the structure of **atoms** and molecules to investigating the composition of stars.

KEY TERMS

Airy's disk—The diffraction pattern produced by a circular aperture such as a lens or a mirror.

Bragg's law—An equation that describes the diffraction of light from plane parallel surfaces.

Diffraction limited—The ultimate performance of an optical element such as a lens or mirror that depends only on the element's finite size.

Diffraction pattern—The wave pattern observed after a wave has passed through a diffracting aperture.

Diffractometer—A device used to produce diffraction patterns of materials.

Fresnel diffraction—Diffraction that occurs when the source and the observer are far from the diffraction aperture.

Interference pattern—Alternating bands of light and dark that result from the mixing of two waves.

Wavelength—The distance between two consecutive crests or troughs in a wave.

X-ray diffraction—A method using the scattering of x rays by matter to study the structure of crystals.

X-ray diffraction

X rays are light waves that have very short wavelengths. When they irradiate a solid, **crystal** material they are diffracted by the atoms in the crystal. But since it is a characteristic of crystals to be made up of equally spaced atoms, it is possible to use the diffraction patterns that are produced to determine the locations and distances between atoms. Simple crystals made up of equally spaced planes of atoms diffract x rays according to Bragg's Law. Current research using x-ray diffraction utilizes an instrument called a diffractometer to produce diffraction patterns that can be compared with those of known crystals to determine the structure of new materials.

Holography

When two **laser** beams mix at an **angle** on the surface of a photographic plate or other recording material, they produce an **interference** pattern of alternating dark and bright lines. Because the lines are perfectly parallel, equally spaced, and of equal width, this process is used to manufacture holographic diffraction gratings of high quality. In fact, any hologram (*holos*—whole: *gram*—message) can be thought of as a complicated diffraction grating. The recording of a hologram involves the mixing of a laser beam and the unfocused diffraction pattern of some object. In order to reconstruct an image of the object (holography is also known as wavefront reconstruction) an illuminating beam is diffracted by **plane** surfaces within the hologram, following Bragg's Law, such that an observer can view the image with all of its three-dimensional detail.

See also Hologram and holography; Wave motion.

John Appel

Diffraction grating

A **diffraction** grating is an optical device consisting of many closely spaced **parallel** lines or grooves. In a transmission type of grating, **light** passes through the narrow transparent slits that lie between the dark lines on a **glass** or plastic plate. In a reflecting grating, light is reflected by the many parallel, narrow, smooth surfaces and absorbed or scattered by the lines cut in the reflecting surface of the grating.

During the 1870s, Henry Rowland, a **physics** professor at Johns Hopkins University, developed a machine that used a fine **diamond** point to rule glass gratings with nearly 15,000 parallel lines per inch. Today, there are carefully ruled gratings that have as many as 100,000 lines per inch. On the other hand, you can obtain inexpensive replica gratings reproduced on film with 13,400 lines per inch. To diffract very short electromagnetic waves, such as **x rays**, the **distance** between the lines in the grating must be comparable to the distance between **atoms**. Gratings with these small separations are obtained by using the regularly arranged rows of closely spaced ions found in the lattice structure of **salt** crystals.

Like a **prism**, a diffraction grating separates the colors in white light to produce a **spectrum**. The spectrum, however, arises not from refraction but from the diffraction of the light transmitted or reflected by the narrow lines in the grating. When light passes through a narrow opening, it is diffracted (spread out) like **water** waves passing through a narrow barrier. With a transmission type diffraction grating, light waves are diffracted as they pass through a series of equally spaced narrow openings. (A similar effect takes place if light is reflected from a reflecting grating.) The beam formed by the combination of diffracted waves from a number of openings in a transmission grating forms a wave front that travels in the same direction as the original light beam. This beam is often referred to as the central maximum.

If the light is not monochromatic, the direction of the diffracted beams will depend on the wavelength. The first order beam for light of longer wavelength, such as red light, will travel at a greater **angle** to the central maximum than the first order beam for light of a shorter wavelength, such as blue light. As a result, white light diffracted by the grating will form a spectrum along each ordered beam. If light from a glowing gas, such as mercury vapor, passes through a diffraction grating, the separate **spectral lines** characteristic of mercury will appear.

Knowing the distance between the slits in the grating and the **geometry** of the **interference** pattern produced by the diffracted light, it is possible to measure the wavelength of the light in different parts of the spectrum. For this reason, diffraction gratings are often used in spectroscopes to examine the spectral lines emitted by substances undergoing chemical analysis.

An ordinary LP record or CD, when held at a sharp angle to a light source, will produce a spectrum characteristic of a reflection grating. The narrow, closely spaced grooves in the disc diffract the reflected light and produce the interference pattern that separates light into colors. A simple transmission grating can be made by looking at the light from a showcase filament with your eyes nearly closed. Light passing through the narrow openings between your eyelashes will be diffracted and give rise to an interference pattern with its characteristic bright and dark bands.

See also Wave motion.

Diffusion

Diffusion is the movement of molecules along a **concentration** gradient, from an area of high concentration to one of low concentration. Diffusion proceeds until the two concentrations are equal. Diffusion occurs in both gases and liquids.

Concentration gradients

Molecules always diffuse from areas of high concentration to areas of low concentration. The difference between the concentration of a substance in one area compared to another area is the concentration gradient. For example, placing ink on the surface of **water** establishes a concentration gradient in which the surface of the water has a high concentration of ink, and the rest of the water has a low concentration. As the ink diffuses, it moves from the area of high concentration to the area of low concentration, eventually resulting in a **solution** with equal concentrations of ink.

The importance of diffusion

Diffusion takes place not only in liquid solutions, but in gases. The odor of bread wafting through a house is an example of the diffusion of bread-smell chemicals from a high concentration to a lower concentration.

Diffusion in cells

Cells are bounded by a double **membrane** composed of lipids. This membrane is punctured intermit-

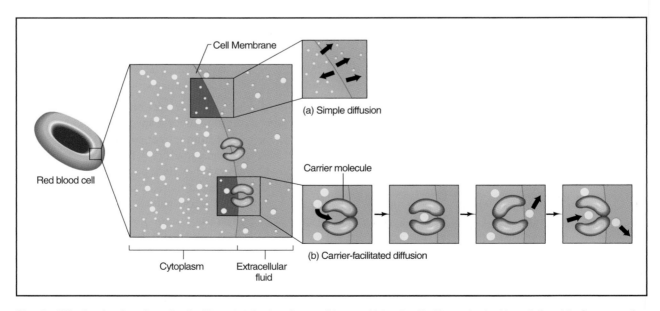

Simple diffusion (top) and carrier-facilitated diffusion (bottom) in a red blood cell. *Illustration by Hans & Cassidy. Courtesy of Gale Group.*

tently with tiny pores. The membrane of a **cell** is thus selectively permeable: it keeps out certain substances but lets others pass through. The substances that pass through move in either direction, either into or out of the cell, depending on the concentration gradient. For example, very small ions pass through the **lipid** membrane through tiny pores in the membrane. Ions move down the concentration gradient that exists between the cytoplasm of the cell and the environment outside the cell, called the extracellular fluid. The extracellular fluid usually contains less ions than the highly concentrated cytoplasm, so ions tend to move from the cytoplasm, down the concentration gradient, into the extracellular fluid. This process is called simple diffusion.

Substances such as glucose or **urea** cannot pass easily into the cell because their molecules are too large, or because they are electrically charged. In these cases, the substances need assistance in getting across the membrane. Special molecules called carrier molecules, situated within the cell membrane, bind to glucose and other substances and bring about their passage into the cell. Because these substances are moving down a concentration gradient, but are assisted by carrier molecules, this type of diffusion is called carrier-facilitated diffusion.

Water diffusion: osmosis

The special case of diffusion of water into and out of cells is called **osmosis**. Because osmosis is the diffusion of water, it is the movement of water from an area with a high concentration of water molecules to an area with a low concentration of water molecules; that is, water diffuses from an area in which water is abundant to an area in which water is scarce. Osmosis in cells is usually defined in different terms, however. It is the movement of water from a low concentration of salts to an area with a high concentration of salts, across a semi-permeable membrane.

Resources

Books

Byrne, John H. *An Introduction to Membrane Transport and Bioelectricity: Foundations of General Physiology and Electrochemical Signaling.* 2nd ed. New York: Raven Press, 1994.
Denny, Mark. *Air and Water: The Biology and Physics of Life's Media.* Princeton: Princeton University Press, 1993.
Yeagle, Philip. *The Membrane of Cells.* 2nd ed. San Diego: Academic Press, 1993.

Kathleen Scogna

Diffusion (of gases) *see* **Gases, properties of**

Digestive system

The digestive system is a group of organs responsible for the conversion of food into absorbable chemicals which are then used to provide **energy** for growth and repair. The digestive system is also known by a number of other names, including the gut, the digestive tube, the alimentary canal, the gastrointestinal (GI) tract, the intestinal tract, and the intestinal tube. The digestive system consists of the mouth, esophagus, stomach, and small and large intestines, along with several **glands**, such as the salivary glands, liver, gall bladder, and pancreas. These glands secrete digestive juices containing enzymes that break down the food chemically into smaller, more absorbable molecules. In addition to providing the body with the **nutrients** and energy it needs to function, the digestive system also separates and disposes of waste products ingested with the food.

Food is moved through the alimentary canal by a wavelike muscular **motion** known as peristalsis, which consists of the alternate contraction and relaxation of the smooth muscles lining the tract. In this way, food is passed through the gut in much the same manner as toothpaste is squeezed from a tube. *Churning* is another type of movement that takes place in the stomach and small intestine, which mixes the food so that the digestive enzymes can break down the food molecules.

Food in the human diet consists of carbohydrates, **proteins**, fats, vitamins, and **minerals**. The remainder of the food is fiber and **water**. The majority of minerals and vitamins pass through to the bloodstream without the need for further digestive changes, but other nutrient molecules must be broken down to simpler substances before they can be absorbed and used.

Ingestion

Food taken into the mouth is first prepared for digestion in a two step process known as mastication. In the first stage, the teeth tear and break down food into smaller pieces. In the second stage, the tongue rolls these pieces into balls (boluses). Sensory receptors on the tongue (**taste** buds) detect taste sensations of sweet, **salt**, bitter, and sour, or cause the rejection of bad-testing food. The olfactory nerves contribute to the sensation of taste by picking up the aroma of the food and passing the sensation of **smell** on to the **brain**.

The sight of the food also stimulates the salivary glands. Altogether, the sensations of sight, taste, and smell cause the salivary glands, located in the mouth, to produce saliva, which then pours into the mouth to soften the food. An **enzyme** in the saliva called amylase begins the break down of carbohydrates (starch) into simple sugars, such as maltose. Ptyalin is one of the main amylase enzymes found in the mouth; ptyalin is also secreted by the pancreas.

The bolus of food, which is now a battered, moistened, and partially digested ball of food, is swallowed, moving to the throat at the back of the mouth (pharynx). In the throat, rings of muscles force the food into the esophagus, the first part of the upper digestive tube. The esophagus extends from the bottom part of the throat to the upper part of the stomach.

The esophagus does not take part in digestion. Its job is to get the bolus into the stomach. There is a powerful muscle (the esophageal sphincter), at the junction of the esophagus and stomach, which acts as a valve to keep food, stomach acids, and bile from flowing back into the esophagus and mouth.

Digestion in the stomach

Chemical digestion begins in the stomach. The stomach, a large, hollow, pouched-shaped muscular **organ**, is shaped like a lima bean. When empty, the stomach becomes elongated; when filled, it balloons out.

Food in the stomach is broken down by the action of the gastric juice containing hydrochloric acid and a protein-digesting enzyme called *pepsin*. Gastric juice is secreted from the linings of the stomach walls, along with *mucus*, which helps to protect the stomach lining from the action of the acid. The three layers of powerful stomach muscles churn the food into a fine semiliquid paste called *chyme*. From time to time, the chyme is passed through an opening (the pyloric sphincter), which controls the passage of chyme between the stomach and the beginning of the small intestine.

Gastric juice

There are several mechanisms responsible for the secretion of gastric juice in the stomach. The stomach begins its production of gastric juice while the food is still in the mouth. Nerves from the cheeks and tongue are stimulated and send messages to the brain. The brain in turn sends messages to nerves in the stomach wall, stimulating the secretion of gastric juice before the arrival of the food. The second signal for gastric juice production occurs when the food arrives in the stomach and touches the lining. This mechanism provides for only a moderate addition to the amount of gastric juice that was secreted when the food was in the mouth.

Gastric juice is needed mainly for the digestion of protein by pepsin. If a hamburger and bun reach the stomach, there is no need for extra gastric juice for the bun (**carbohydrate**), but the hamburger (protein) will require a much greater supply of gastric juice. The gastric juice already present will begin the break down of the large protein molecules of the hamburger into smaller molecules: polypeptides and peptides. These smaller molecules in turn stimulate the cells of the stomach lining to release the hormone *gastrin* into the bloodstream.

Gastrin then circulates throughout the body, and eventually reaches the stomach, where it stimulates the cells of the stomach lining to produce more gastric juice. The more protein there is in the stomach, the more gastrin will be produced, and the greater the production of gastric juice. The secretion of more gastric juice by the increased amount of protein in the stomach represents the third mechanism of gastric juice secretion.

Alexis St. Martin's stomach

An understanding of the complex mechanisms of gastric juice secretion began with an American army doctor, William Beaumont (1785-1853). He was able to directly observe the process of digestion in the stomach from the wound of a soldier named Alexis St. Martin.

In 1822, Beaumont treated the soldier for an accidental gunshot wound. This wound left a large hole in the left side of St. Martin's body, tearing away parts of the ribs, muscles, and stomach wall. When the wound healed, the stomach wall had grown to the outer body wall, leaving a permanent hole from the outer body to the interior of the stomach. When St. Martin ate, bandages were needed to keep the food in place. For the first time in medical history, a physician was able to study the inner workings of the stomach. Beaumont's observations and experiments on St. Martin's stomach extended over 11 years.

In that time, he observed the secretion of gastric juice and placed the fluid from St. Martin's stomach on a piece of meat. There he could observe the digestion of protein. He was also able to observe the churning movements of the stomach when food entered it. Beaumont's investigation of St. Martin's stomach laid the groundwork for later investigations into the complexities of the digestive process.

Digestion and absorption in the small intestine

While digestion continues in the small intestine, it also becomes a major site for the process of absorption, that is, the passage of digested food into the bloodstream, and its transport to the rest of the body.

The small intestine is a long, narrow tube, about 20 ft (6 m) long, running from the stomach to the large in-

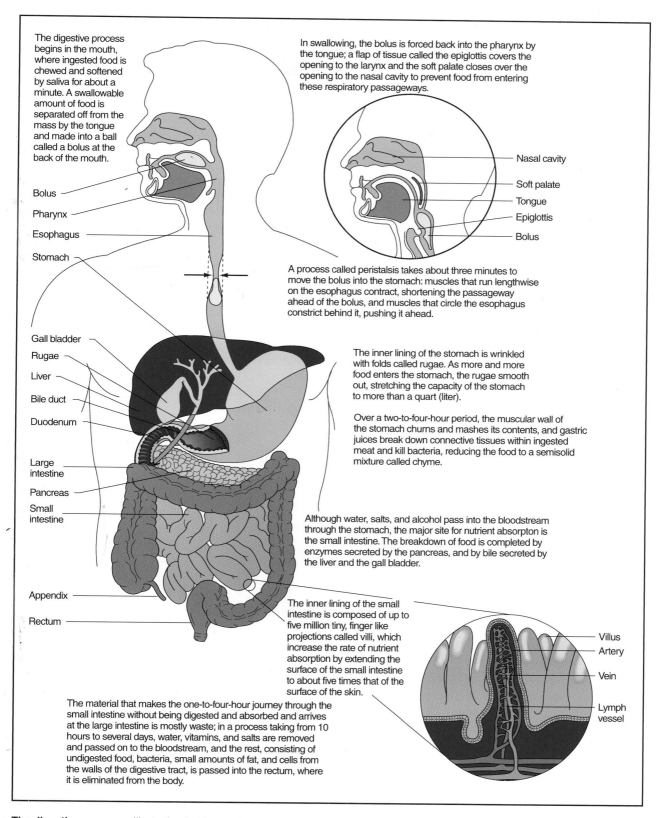

The digestive process begins in the mouth, where ingested food is chewed and softened by saliva for about a minute. A swallowable amount of food is separated off from the mass by the tongue and made into a ball called a bolus at the back of the mouth.

Bolus

Pharynx

Esophagus

Stomach

Gall bladder

Rugae

Liver

Bile duct

Duodenum

Large intestine

Pancreas

Small intestine

Appendix

Rectum

In swallowing, the bolus is forced back into the pharynx by the tongue; a flap of tissue called the epiglottis covers the opening to the larynx and the soft palate closes over the opening to the nasal cavity to prevent food from entering these respiratory passageways.

Nasal cavity

Soft palate

Tongue

Epiglottis

Bolus

A process called peristalsis takes about three minutes to move the bolus into the stomach: muscles that run lengthwise on the esophagus contract, shortening the passageway ahead of the bolus, and muscles that circle the esophagus constrict behind it, pushing it ahead.

The inner lining of the stomach is wrinkled with folds called rugae. As more and more food enters the stomach, the rugae smooth out, stretching the capacity of the stomach to more than a quart (liter).

Over a two-to-four-hour period, the muscular wall of the stomach churns and mashes its contents, and gastric juices break down connective tissues within ingested meat and kill bacteria, reducing the food to a semisolid mixture called chyme.

Although water, salts, and alcohol pass into the bloodstream through the stomach, the major site for nutrient absorption is the small intestine. The breakdown of food is completed by enzymes secreted by the pancreas, and by bile secreted by the liver and the gall bladder.

The inner lining of the small intestine is composed of up to five million tiny, finger like projections called villi, which increase the rate of nutrient absorption by extending the surface of the small intestine to about five times that of the surface of the skin.

Villus

Artery

Vein

Lymph vessel

The material that makes the one-to-four-hour journey through the small intestine without being digested and absorbed and arrives at the large intestine is mostly waste; in a process taking from 10 hours to several days, water, vitamins, and salts are removed and passed on to the bloodstream, and the rest, consisting of undigested food, bacteria, small amounts of fat, and cells from the walls of the digestive tract, is passed into the rectum, where it is eliminated from the body.

The digestive process. *Illustration by Hans & Cassidy. Courtesy of Gale Group.*

testine. The small intestine occupies the area of the abdomen between the diaphragm and hips, and is greatly coiled and twisted. The small intestine is lined with muscles that move the chyme toward the large intestine. The mucosa, which lines the entire small intestine, contains millions of glands that aid in the digestive and absorptive processes of the digestive system.

The small intestine, or small bowel, is sub-divided by anatomists into three sections, the duodenum, the jejunum, and the ileum. The duodenum is about 1 ft (0.3 m) long and connects with the lower portion of the stomach. When fluid food reaches the duodenum it undergoes further enzymatic digestion and is subjected to pancreatic juice, intestinal juice, and bile.

The pancreas is a large gland located below the stomach that secretes pancreatic juice into the duodenum via the pancreatic duct. There are three enzymes in pancreatic juice which digest carbohydrates, lipids, and proteins. Amylase, (the enzyme found in saliva) breaks down starch into simple sugars such as maltose. The enzyme maltase in intestinal juice completes the break down of maltose into glucose.

Lipases in pancreatic juice break down fats into **fatty acids** and **glycerol**, while proteinases continue the break down of proteins into amino acids. The gall bladder, located next to the liver, secretes bile into the duodenum. While bile does not contain enzymes; it contains bile salts and other substances that help to emulsify (dissolve) fats, which are otherwise insoluble in water. Breaking the **fat** down into small globules allows the lipase enzymes a greater surface area for their action.

Chyme passing from the duodenum next reaches the jejunum of the small intestine, which is about 3 ft (0.91 m) long. Here, in the jejunum, the digested breakdown products of carbohydrates, fats, proteins, and most of the vitamins, minerals, and **iron** are absorbed. The inner lining of the small intestine is composed of up to five million tiny, finger-like projections called villi. The villi increase the **rate** of absorption of the nutrients into the bloodstream by extending the surface of the small intestine to about five times that of the surface area of the skin.

There are two transport systems that pick up the nutrients from the small intestine. Simple sugars, amino acids, glycerol, and some vitamins and salts are conveyed to the liver in the bloodstream. Fatty acids and vitamins are absorbed and then transported through the **lymphatic system**, the network of vessels that carry lymph and white **blood** cells throughout the body. Lymph eventually drains back into the bloodstream and circulates throughout the body.

The last section of the small intestine is the ileum. It is smaller and thinner-walled than the jejunum, and it is the preferred site for **vitamin** B_{12} absorption and bile acids derived from the bile juice.

Absorption and elimination in the large intestine

The large intestine, or colon, is wider and heavier then the small intestine, but much shorter—only about 4 ft (1.2 m) long. It rises up on one side of the body (the ascending colon), crosses over to the other side (the transverse colon), descends (the descending colon), forms an s-shape (the sigmoid colon), reaches the rectum, and anus, from which the waste products of digestion (feces or stool), are passed out, along with gas. The muscular rectum, about 5 in (13 cm) long, expels the feces through the anus, which has a large muscular sphincter that controls the passage of waste matter.

The large intestine extracts water from the waste products of digestion and returns some of it to the bloodstream, along with some salts. Fecal matter contains undigested food, **bacteria**, and cells from the walls of the digestive tract. Certain types of bacteria of the large intestine help to synthesize the vitamins needed by the body. These vitamins find their way to the bloodstream along with the water absorbed from the colon, while excess fluids are passed out with the feces.

Liver

The liver is the largest organ in the body and plays a number of vital roles, including metabolizing the breakdown products of digestion, and detoxifying substances that are harmful to the body. The liver also provides a quick source of energy when the need arises and it produces new proteins. Along with the regulation of stored fats, the liver also stores vitamins, minerals, and sugars. The liver controls the excretion and production of **cholesterol** and metabolizes **alcohol** into a mild toxin. The liver also stores iron, maintains the hormone balance, produces immune factors to fight infections, regulates blood clotting, and produces bile.

The most common liver disorder in the United States and other developed countries is **cirrhosis** of the liver. The main cause for this **disease** is **alcoholism**. Cirrhosis is characterized by the replacement of healthy liver cells by fibrous **tissue**. The replacement process is gradual and extends over a period of 2-10 years to complete. There is no cure for the disease. Symptoms may not be noticed in its early development, but in its advanced stages there are a number of symptoms and the condition can lead to **coma**. Close medical attention is required to treat the disease.

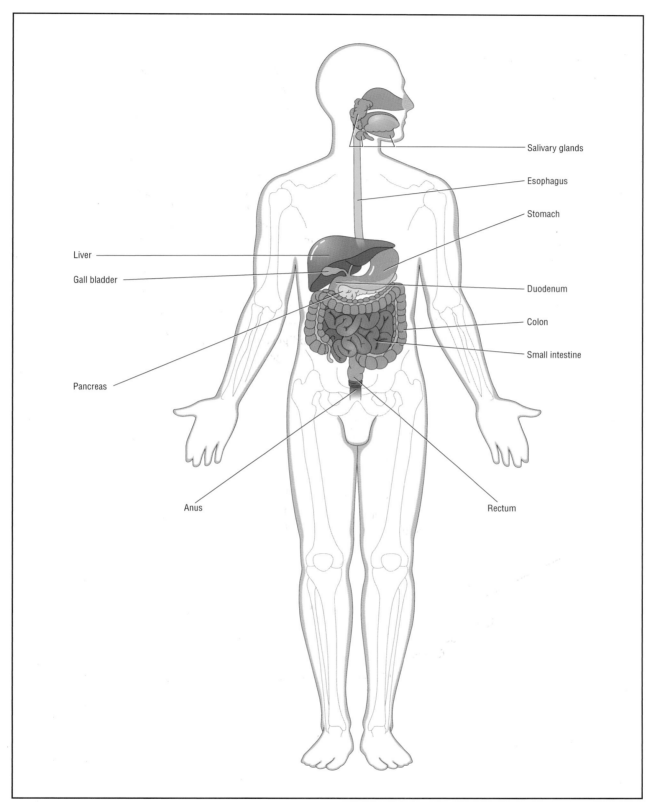

Salivary glands

Esophagus

Stomach

Liver

Gall bladder

Duodenum

Colon

Small intestine

Pancreas

Anus

Rectum

The human digestive system. *Illustration by Argosy. The Gale Group.*

Another common liver disorder is **hepatitis**. It is an **inflammation** of the liver caused by viruses. The most noticeable symptom of this disease is **jaundice**, that is, the skin, eyes, and urine turn yellow. The nine viruses known to cause hepatitis include Hepatitis A, B, C, D, and E; the recently discovered F and G viruses; and two herpes viruses (Epstein-Barr and cytomegalovirus).

Gallbladder

The gallbladder lies under the liver and is connected by various ducts to the liver and the duodenum. The gallbladder is a small hollow organ resembling a money pouch. Its main function is to store bile until it is concentrated enough to be used by the small intestine. The gall bladder can store about 2 oz (57 g) of bile. Bile consists of bile salts, bile acids, and bile pigments. In addition, bile contains cholesterol dissolved in the bile acids. If the amount of cholesterol in the bile acids increases or the amount of acid decreases, then some of the cholesterol will settle out of the acid to form gallstones that accumulate and block the ducts to the gallbladder.

Infection in the gallbladder can be another cause for gallstones. Gallstones may be in the gallbladder for years without giving any signs of the condition, but when they obstruct the bile duct they cause considerable **pain** and inflammation. Infection and blockage of the bile flow may follow. Surgical removal of the gallbladder may be necessary to treat this condition. Since the liver both produces and stores sufficient amounts of bile, the loss of the gallbladder does not interfere with the digestive process provided fat intake in the diet is regulated.

If the gallstones contain mainly cholesterol, drug treatment for gallstones may be possible. But if there is too much other material in the gallstones, **surgery** may still be necessary. Even after the condition has been treated successfully by drugs and diet, the condition can return. The drug treatment takes years to dissolve the gallstones.

Appendix

The appendix is a hollow finger-like projection that hangs from the occum at the junction between the small intestine and the large intestine. The appendix does not function in humans; however, in some animals, such as rabbits, the appendix is rather large and helps in the digestion of **cellulose** from **bark** and **wood**, which rabbits eat. The appendix in humans is therefore a vestigial organ, which may have had uses for earlier types of ancestral human digestive processes before the **evolution** of *Homo sapiens*.

If food gets trapped in the appendix, an irritation of its membranes may occur leading to swelling and inflammation, a condition known as appendicitis. If the condition becomes serious, removal of the appendix is necessary to avoid a life-threatening condition if it were to rupture.

Pancreas

When food reaches the small intestine, the pancreas secretes pancreatic juices. When there is no food in the small intestine, the pancreas does not secrete its juices. The economy of this process puzzled researchers who wondered what the mechanism for this control might be. In 1902, William Bayliss and Ernest Starling, two British physiologists, conducted experiments to find the answer. They reasoned that the same mechanism that initiated gastric juices when food first enters the mouth might be the same mechanism for releasing the flow of pancreatic juices.

These researchers made an extract from the lining of the small intestine and injected it into an experimental **animal**. The extract caused the animal to secrete large amounts of pancreatic juice. They concluded that the extract from the intestinal lining must have some substance responsible for the flow, which they named secretin. The experiment gave the first real **proof** for the existence of **hormones**, substances secreted by one group of cells that travel around the body which target other groups of cells.

Insulin is another important hormone secreted by a group of cells within the pancreas called the islets of Langerhans, which are part of the **endocrine system** rather than the digestive system. Insulin released into the bloodstream targets liver and muscle cells, and allows them to take excess sugar from the blood and store it in the form of glycogen. When the pancreas does not produce sufficient insulin to store dietary sugar, the blood and urine levels of sugar reach dangerous levels. **Diabetes mellitus** is the resultant disease. Mild cases can be controlled by a properly regulated diet, but severe cases require the regular injection of insulin.

Disorders of the digestive system

Several disorders of the esophagus are esophagitis, esophageal spasm, and esophageal **cancer**. Esophagitis (heartburn) is an inflammation of the esophagus usually caused by the reflux of gastric acids into the esophagus and is treated with (alkalis) antacid. Esophageal spasm is also caused by acid reflux and is sometimes treated with nitroglycerine placed under the tongue. Esophageal cancer can be caused by smoking and is generally fatal.

Disorders of the stomach include hiatal **hernia, ulcers**, and gastric cancer. A hiatal hernia occurs when a portion of the stomach extends upwards into the thorax

KEY TERMS

Amylase—A digestive enzyme found in saliva and the pancreas that breaks down carbohydrates to simple sugars.

Bile—A greenish yellow liquid secreted by the liver and stored in the gall bladder that aids in the digestion of fats and oils in the body.

Gastric juice—Digestive juice in produced by stomach wall that contains hydrochloric acid and the enzyme pepsin.

Gastrin—A hormone produced by the stomach lining in response to protein in the stomach that produces increased gastric juice.

Helicobacter pylori—Recently discovered bacteria that live in gastric acids and are believed to be a major cause of most stomach ulcers.

Lower esophageal sphincter—A strong muscle ring between the esophagus and the stomach that keeps gastric juice, and even duodenal bile from flowing upwards out of the stomach.

Lymphatic system—The transport system linked to the cardiovascular system that contains the immune system and also carries metabolized fat and fat soluble vitamins throughout the body.

Mucosa—The digestive lining of the intestines.

Nutrients—Vitamins, minerals, proteins, lipids, and carbohydrates needed by the body.

Peristalsis—The wavelike motion of the digestive system that moves food through the digestive system.

Villi—Finger-like projections found in the small intestine that add to the absorptive area for the passage of digested food to the bloodstream and lymphatic system.

through a large opening in the diaphragm. It is a condition that commonly occurs to people over the age of 50. Stomach ulcers are sores that form in the lining of the stomach. They may vary in size from a small sore to a deep cavity, surrounded by an inflamed area, sometimes called ulcer craters. Stomach ulcers and ulcers that form in the esophagus and in the lining of the duodenum are called peptic ulcers because they need stomach acid and the enzyme pepsin to form. Duodenal ulcers are the most common type. They tend to be smaller than stomach ulcers and heal more quickly. Ulcers that form in the stomach lining are called gastric ulcers. About four million people have ulcers and 20% of those have gastric ulcers. Those people who are at most risk for ulcers are those who smoke, middle-age and older men, chronic users of alcohol, and those who take anti-inflammatory drugs, such as aspirin and ibuprofen.

Until 1993, the general belief in the medical community concerning the cause of stomach ulcers was that there were multiple factors responsible for their development. By 1993 there was mounting evidence that an S-shaped bacterium, *Helicobacter pylori*, could be one of the factors causing ulcers. *Helicobacter pylori* live in the mucous lining of the stomach near the surface cells and may go undetected for years. Researchers argued that irritation to the stomach caused by the bacteria weakened the lining, making it more susceptible to damage by acid and resulting in the formation of ulcers.

Barry Marshall, an Australian gastroenterologist, was the chief proponent of the theory that stomach ulcers are caused by *H. pylori* infections, rather than a multiple factor explanation, such as **stress** or poor diet. Although Marshall was discouraged by his colleagues from pursuing this line of research, he demonstrated his hypothesis by swallowing a mixture containing *H. pylori*. Marshall soon developed gastritis, which is the precursor condition to ulcers.

The treatment of ulcers has undergone a radical change with Marshall's discovery that stomach ulcers are caused by *H. pylori* infections. Ulcer patients today are being treated with **antibiotics** and antacids rather than special diets or expensive medicines. It is believed that about 80% of stomach ulcers may be caused by the bacterial infection, while about 20% may be from other causes, such as the use of anti-inflammatory medicines.

Resources

Books

Maryon-Davis, Alan, and Steven Parker. *Food and digestion*. London; New York: F. Watts, 1990.

Peikin, Steven R. *Gastrointestinal Health*. New York: Harper-Collins, 1991.

Jordan P. Richman

Digital and analog *see* **Digital Recording**

Digital audio tape *see* **Magnetic recording/audiocassette**

Digital recording

Digital recording is a technique for preserving audio signals and video or visual images as a series of pulses that can be stored on magnetic tapes, optical discs (compact discs), or computer diskettes. These pulses are stored in the form of a series of binary digits (that is, zeros and ones). To make the recording, an analog-to-digital converter transforms the sound signal or visual image into digital information (a complex series of zeros and ones) that is recorded on high- speed magnetic tape or on disc or diskette. The system that plays back or reads out the sound or image translates the binary code back into analog (line like) signals using a digital-to-analog converter. Tape players, **compact disc** players, video disc players, and CD-ROM (Compact Disc-Read-Only Memory) players in home computers are examples of digital-to-analog converters used to play back audio and video codes in our homes.

Analog versus digital recording

Analog recordings were the only ones made until the digital revolution of the 1970s, and used a variety of methods that are now considered outdated like long-playing (LP) records, eight-track tapes (on either **metal** or magnetic tape), and home movies. In the history of sound and **video recording** from early in the twentieth century until the 1970s, the analog system seemed ideally suited to recording because sound and **light** both have linear properties; however, with the technological revolution that occurred late in the century, high speed and other characteristics of digital processing made digital recording possible, as did steadily falling costs due to **mass production** of computers, plastic coated discs, **laser** players, and other devices. The word "digital" means that numbers are used and refers to the encoding of signals as strings of zeros or ones. Digital recording has higher fidelity in sound and video because it provides a wide range of dynamics and low levels of distortion, which make analog recording less like the source sound or image.

Digital recording formats

Digital recording itself produces truer audio and video, and the systems developed for playing them also help eliminate **interference**. The most familiar audio system, the compact disc, uses a laser beam player to read the digital information coded on the disc. Digital audio tape (DAT) became available in the late 1980s. It uses magnetic tape and a specialized DAT recorder with a microprocessor to convert audio signals to digital data

during recording and to switch the data back to analog signals for playback. DAT systems are available to the average consumer but are used extensively by professionals. Digital compact cassette (DCC) recorders can play both DAT tapes and the analog tape cassettes that are more common.

Video systems parallel the audio methods. Compact discs for video recordings were initially considered impractical because of the complexities of carrying both images and sound, so larger diameter laser discs that also use a laser beam player to decode information on the discs were one system of playing back video. Laser discs (also called videodiscs) store audio information in digital form and video as analog data. Video tapes in Beta and VHS formats (both analog forms) were easier to mass produce at smaller cost. Another portion of the video problem was that video could be recorded on compact discs, but the level of fidelity of the disc was better than any **television** could reproduce. The video recording industry had to wait for televisions to catch up. In the late 1990s, high-density television (HDTV) became available, and the digital video disc (**DVD**) and DVD players rapidly became more popular in anticipation of better television technology. The DVD (also called the digital versatile disc) can accommodate all the sound and light needed for a *Star Wars* movie, for example, because it holds almost five billion bytes of data and may soon hold over eight billion data bytes; a typical CD-ROM for home computer use stores only 650 million bytes. Digital cameras were also introduced in 1997, the same year that DVD players were first widely sold. Improvement of HDTV was given a push by government; a phase-out of nondigital TV signals is to occur over 10 years (beginning in 1998) to be replaced by the digital images from satellites, digital network broadcasts, and DVD sources. Cable systems are also converting to digital signals.

Advances in the home computer industry are closely linked with audio and video digital recording systems. First, home computers have increasingly included audio and video playback systems. Second, the mergers of audio and video giants with Internet firms have shown that all these services may soon be provided directly to our homes through one cable, phone line, or other shared system. And third, the technology for putting more and better information on a compact disc has made the disc the leading medium for sound recordings (as the compact disc), video (in the form of DVDs), and information (CD-ROMs and recordable and erasable CDs for data, sound, and video). Erasable and recordable compact discs are called CD-Es and CD-Rs, respectively; following their introduction in the late 1990s, the equipment for using them (with home sound systems and computers) quickly became affordable. The DVD also has a close relative for computer data storage called the DVD disc drive that replaces the CD-ROM in some personal computers (PCs). Eventually, technology may produce a single type of disc that can be encoded and played back by computer, audio recorder/players, and video recorder/players (depending, of course, on the information on the disc).

Advantages and complexities of digital recording

Recording, particularly of music and video images, consumes massive amounts of digital memory. High-density discs are ideal for these types of recording. Direct digital recordings can play back recorded or modulated sound. Tape recording, though convenient and easy, could not store digital data until the development of the DAT tape. Carrier signals in digital recording are always pulse waves that alternate between voltages (analog signals). Consequently, the modulation method in most digital systems for music is pulse code modulation (PCM). On CDs and other disc formats read by laser, the physical structure uses islands or raised points and pits or low points as the zeros and ones; in pulse modulation, the high ends of the pulses are easily represented as the islands, and the low pulses are the pits. Pulse code modulation is actually an old development in the history of recordings. It was developed in 1939 by A. H. Reeves, but it took technology many years to find practical uses for Reeves' invention.

The conversion of sampling **frequency** from analog to digital is critical to sound recording. The frequency is measured or sampled many times per second and then averaged to produce the single piece of data for the digital input. Increasing the sampling frequency improves the sound quality but decreases the storage economy especially on tapes. Resolution is another important specification and describes the number of bits used to represent the amplitude of an instant on the recording. Each bit doubles the possibility for representing instantaneous amplitude levels. Typically, 14-bit resolution is used to give a range of 16,384 possibilities for representing instantaneous amplitude values. Quantization is the process that converts the collection of values into the amplitude the listener hears.

Recording media are all imperfect, thanks to specks of dust or other **contamination** that prevents equipment from imprinting the data on the medium. Data on CD and DAT tape are even more tightly compressed than those on analog tapes, so loss of data is effectively magnified. To fix this, special error correction codes are built into the data stream to weave the **sample** values throughout the data. Some of these error correction codes can be very complex, and, of course, they also consume valuable storage space on the CD or tape. The recording engineer must compromise the number of error codes to make enough storage for the sound data.

Analog systems also have the disadvantage that, when a recording is played back and rerecorded, distortion is increased by about 0.5 %. Each subsequent copy will be worse. Analog discs and tape are also nonlinear and do not record all sounds equally, leading to inaccurate reproduction. In a digital recording system, this distortion does not occur. The master recording may have minimal quantization errors, but these do not compound when copies are made. In this case, the absolute zero-or-one character of the digital world works to an advantage because the copy is equally absolute unless the digital recording is reconverted to an analog signal. Thousands of copies can be made from a digital master without distortion; similarly, digital media on CDs can be played back thousands of times without distortion.

The future of digital recording

Recordable and erasable CDs are giving the compact disc greater versatility. Compact disc recorders allow the user to record audio from various sources on CDs. The recorders require attentive use because the recording procedure depends on the type, quality, and input device of the source material. If the source is a CD that can be played on a machine with digital optical output, it can be connected directly to the CD recorder as input and be dubbed like an audio tape. The recorder evaluates the sonic range of the original and digitally synchronizes it; if tracks are recorded from several CDs, the recorder must resynchronize with each track.

Erasable CDs followed recordable CDs quickly. Erasable CDs or CD-Es can be overwritten when the data on them becomes obsolete. High-density CD-Rs and CD-Es are also being developed and are anticipated by the music industry because they can store music detail more completely. Enhanced audio CDs include music videos, lyrics, scores that the home musician can play, and interviews with the musicians. Enhanced audio CDs can be played on a CD-ROM drive and viewed on a monitor or connected television set. High Definition Compatible Digitals or HDCDs are also being marketed on a limited basis. They produce more realistic sound but require a CD player with a built-in decoder. Tapes in a high-density format are also on the horizon; analysts expect HDTV videocassette recorders to be less expensive than conventional VCRs eventually because the design and concept are simpler while producing higher quality video reproduction. Ultimately, the changes in computer architecture and the uniting of computers with communications systems will bring all

KEY TERMS

Bit and byte—A bit is the smallest element representing data in a computer's memory. A byte consists of eight bits that the computer processes as a unit.

High density—In recording systems, the ability to store large audio, video, or information files in a small space.

types of recordings of the best digital quality into the home from many media sources.

Resources

Books

Evans, Brian. *Understanding Digital TV: The Route to HDTV.* New York: IEEE Press, 1995.

Horn, Delton E. *DAT: The Complete Guide to Digital Audio Tape.* Blue Ridge Summit, PA: TAB Books, 1991.

Leathers, David. *Pro Tools Bible: The Complete Guide to Digital Recording.* McGraw-Hill, 2003.

Watkinson, John. *An Introduction to Digital Audio.* 2nd ed. Focal Press, 2002.

Periodicals

Alldrin, Loren. "Little Bits of Audio: All About Turning Your Audio into Little Ones and Zeros." *Videomaker* 12, no. 10 (April 1998): 24.

"A New Spin." *Time International* 150, no. 48. (July 27, 1998): 34.

Gallagher, Mitch. "Who's Afraid of Hard Disk?" *Guitar Player* 33, no. 16 (June 1999): 115.

Houkin, K. "Digital Recording in Microsurgery." *Journal of Neurosurgery* 92 no. 1 (2000): 176-180.

O'Malley, Chris. "A New Spin: Digital Recorders Finally Give Music Lovers a Way to Make Bit-perfect Copies of their Favorite CDs. A Guide to Three Leading Formats." *Time* 152, no. 8 (August 24, 1998): 64.

Other

"CD-E: The Future of Compact Disc Technology." *The CD Erasable Page.* <http://home.cdacrchive.com/info/cd_erasable.htm>.

Gillian S. Holmes

Digital signals *see* **Analog signals and digital signals**

Digitalis

Digitalis is a drug that has been used for centuries to treat **heart disease**. The active ingredient in the drug is glycoside, a chemical compound that contains a sugar **molecule** linked to another molecule. The glycoside compound can be broken down into a sugar and nonsugar compound. Though current digitalis drugs are synthetic, that is, man-made, early forms of the drug were derived from a **plant**.

Digitalis is a derivative of the plant *Digitalis purpurea*, or purple foxglove. The plant's name, Digitalis (from the Latin *digit*, finger) describes the finger-shaped purple flowers it bears. The effects of the plant extract on the heart were first observed in the late eighteenth century by William Withering, who experimented with the extract in fowls and humans. Withering reported his results in a treatise entitled, "The Foxglove and an Account of its Medical Properties, with Practical Remarks on Dropsy." His explanations of the effects of foxglove on the heart have not stood up to the test of time, but his prediction that it could be "converted to salutary ends" certainly has. Indeed, digitalis remains the oldest drug in use for the treatment of heart disease, as well as the most widespread, in use today.

The digitalis drugs come in many forms, differing in their chemical structure. As a group they are classified as cardiac inotropes. Cardiac, of course, refers to the heart. An inotrope is a substance that has a direct effect on muscle contraction. Positive inotropism is an increase in the speed and strength of muscle contraction, while **negative** inotropism is the opposite. Digitalis has a positive inotropic effect on the heart muscle.

How digitalis is used

Digitalis is used to bolster the ailing heart in congestive heart failure. In this condition, the heart muscle has stretched while straining to pump **blood** against a back **pressure**. The back pressure may be caused by high blood pressure, or it may be the result of a leak caused by a faulty aortic valve or a hole in the wall (septum) dividing the right and left halves of the heart. When these conditions occur, the heart muscle, or myocardium, must exert greater and greater pressure to **force** blood through the body against the resistant force. Over time the strain will stretch the heart muscle, and the size of the heart increases. As the heart muscle changes in these ways, its pumping action becomes less and less effective. Congestive heart failure occurs when the myocardium has been stretched too far. At this juncture the patient must have a heart transplant or he will die.

The administration of digitalis, however, can forestall the critical stage of the disease. Digitalis has a direct and immediate effect on the myocardium. By a mechanism not well understood, digitalis increases the levels of intracellular **calcium**, which plays an important role in the contraction of the muscles. Almost as soon as the drug has been administered, the heart muscle begins to contract faster and with greater force. As a result, its pumping efficiency increases and the supply of blood to the body is enhanced. Digitalis also tends to bring about a decrease in the size of the ventricles of the failing heart as well as a reduction in wall tension.

In addition to its immediate effect on the heart muscle, the drug affects the autonomic **nervous system**, slowing the electrical signal that drives the heartbeat. As heart contractions become more efficient, the heart **rate** slows. For this reason, the drug is said to have a negative chronotropic effect (the prefix *chrono-* refers to time).

As digitalis stabilizes the myocardium, appropriate steps can also be taken to correct the original cause of the disease, if possible. The patient's blood pressure can be lowered with medications, or heart **surgery** can be performed to replace a faulty valve or patch a hole in the septum. When it is not possible to improve cardiac function by other means, the patient can be maintained on digitalis for many years.

Risks and side effects

The effect of digitalis is dose related. The higher the dose, the more pronounced the cardiac reaction. It is this immediate and direct effect of the drug that dictates that the physician closely monitor his patient and adjust the digitalis dosage as needed to provide the corrective effect, while being careful not to institute a toxic reaction. Digitalis is a very potent and active drug and can quickly create an overdose situation if the patient is not closely watched. In the case of an overdose, the patient's heart will begin to beat out of rhythm (arrhythmia) and very rapidly (tachycardia). In addition, the drug may affect the nervous system and cause headaches, **vision** problems such as blurring and **light** sensitivity, and sometimes convulsions.

Withering already recognized the toxicity of digitalis and warned against the careless administration of the drug in too high a dose. Despite Withering's warnings, physicians in the early nineteenth century often overdosed their patients. As a consequence, the drug was considered too dangerous for the greater part of the nineteenth century and was used little. Later in the same century, however, the beneficial properties of digitalis were reassessed, and the drug became an essential element in the cardiologist's pharmacopeia.

Other drugs to treat diseases have been developed over time, of course, but none has replaced digitalis as the standard therapy for heart failure. A drug of ancient lineage, digitalis remains one of the most reliable and most used medicines.

KEY TERMS

. . . .

Aortic valve—The one-way valve that allows blood to pass from the heart's main pumping chamber, the left ventricle, into the body's main artery, the aorta.

Cardiologist—A physician who specializes in the diagnosis and treatment of heart disease.

Myocardium—The heart muscle.

Oxygenation—The process, taking place in the lungs, by which oxygen enters the blood to be transported to body tissues.

Septum—The wall that divides the right side of the heart (which contains "used" blood that has been returned from the body) from the left side of the heart (which contains newly oxygenated blood to be pumped to the body).

Resources

Books

The Complete Drug Reference: United States Pharmacopeia. Yonkers, NY: Consumer Reports Books, 1992.

Larson, David E., ed. *Mayo Clinic Family Health Book.* New York: William Morrow, 1996.

Larry Blaser

Dik-diks

Dik-diks (genus *Madoqua*) are small (dog-sized) African antelopes belonging to the family of Bovidae, which includes cattle, **sheep**, and **goats**, as well as antelopes, **gazelles**, and impalas. Like all bovids, dik-diks have even-toed hooves, horns, and a four-chambered stomach. There are five **species** of dik-dik—Kirk's (the largest), Günther's, Salt's, Red-bellied, and Swayne's (the smallest), as well as 21 subspecies.

Dik-diks belong to the tribe Neotragini, the **dwarf antelopes**. These small animals weigh only up to 12 lb (6 kg), stand a little over 1 ft (40 cm) in height at the shoulders, and are less than 2 ft (67 cm) in length. Dik-diks are found in the Horn of **Africa**, East Africa, and in some parts of southwest Africa. In spite of their small size, dik-diks are heavily hunted for their skin, which is used to make gloves. Dik-diks have big eyes, a pointed snout, and a crest of erect hair on their forehead. These antelopes can withstand prolonged high temperatures because of their ability to cool down by nasal panting.

A Kirk's dik-dik. *Photograph by Renee Lynn. The National Audubon Society Collection/Photo Researchers, Inc. Reproduced by permission.*

Habitat and diet

Dik-diks live in arid bush country and eat a diet of fallen leaves, green leaves, and fruit. This diet is digested with the aid of **microorganisms** in the dik-dik's four-chambered stomach and by the regurgitation and rechewing of food (chewing the cud). Because of their small size, the dik-dik's **rumination** process is much faster than in larger hoofed animals. With the reduction of forest **habitat** in Africa over the past 12 million years, it is believed that the small size of animals like the dik-dik has been favorable to their survival.

The dik-dik, like all cud-chewing animals, has a specialized jaw and tooth structure that is adaptable to its diet. The front part of the dik-dik jaw is large compared to the **brain** area of its skull. The jaws come together elongated, and there are no teeth at the end of the upper jaw. The overall structure functions like a shovel that can tear off great quantities of food at a fast pace and then chop it up for the rumination process.

Social organization

A male and a female dik-dik form a permanent pair bond and together they occupy a territory 12-75 acres (5-30 ha) in size. The female is slightly larger than the male, which reflects her greater role in caring for her offspring. Dik-diks give **birth** twice a year (coinciding with the rainy **seasons**) to one offspring at a time. For the first few weeks after its birth, the young dik-dik lies hidden in the bush. Its mother makes contact by bleating sounds which are answered by the offspring.

Like other dwarf antelopes, dik-diks have efficient scent **glands** that are used to mark their territory. These

glands are located in the front part of the eyes (suborbital glands) and on their hooves. Dik-diks are therefore able to mark both the ground and bushes of their territory with their scents.

Territorial behavior

Another distinctive aspect of dik-dik territorial **behavior** is a ritual that accompanies defecation and urination. The female urinates first, then defecates on a pile of dung that marks their territory. The male waits behind her while she squats during this activity. He then sniffs, scrapes, squats, and deposits his urine and feces over the female's. Some scent marking of neighboring plants is also part of this ritual. There can be between 6-13 such locations around a dik-dik pair's territory.

The male dik-dik defends the territory from both male and female intruders. Generally, conflicts over territory are infrequent. While rival males will engage in a rushing ritual, they rarely attack one another physically. The offspring of a dik-dik pair is allowed to remain in the territory until it reaches maturity, which is about six months for females and twelve months for male offspring. The male dik-dik usually intervenes when the mature male offspring tries to approach the mother. The adult male challenge leads to submissive behavior by the younger male. Eventually, the male or female offspring are driven from the territory but they quickly bond with another young dik-dik in an unclaimed territory.

Resources

Books

Estes, Richard D. *Behavior Guide to African Mammals.* Berkeley: University of California, 1991.

The Safari Companion. Post Mills, VT: Chelsea Green, 1993.

Vita Richman

Dingo *see* **Canines**

Dinosaur

Dinosaurs are a group of now-extinct, terrestrial **reptiles** in the order Dinosauria that lived from about 225 million years ago to 66 million years ago, during the Mesozoic era. **Species** of dinosaurs ranged from chicken-sized creatures such as the 2 lb (1 kg) **predator** *Compsognathus* to colossal, herbivorous animals known as "sauropods," which were larger than any terrestrial animals that lived before or since. Some dinosaurs were enormous, awesomely fierce predators, while others were mild-mannered herbivores, or **plant** eaters, that reached an immense size. The word "dinosaur" is derived from two Greek words, meaning "terrible lizard." The term refers to some of the huge and awesome predatory dinosaurs—the first of these extinct reptiles to be discovered that were initially thought to be lizard-like in appearance and **biology**. But Richard Owen (1804-1892), the British expert in **comparative anatomy**, also coined the word in awe of the complexity of this wide variety of creatures that lived so long ago and yet were so well-adapted to their world.

Dinosaurs were remarkable and impressive animals but are rather difficult to define as a zoological group. They were terrestrial animals that had upright legs, rather than legs that sprawled outward from the body. Their skulls had two temporal openings on each side (in addition to the opening for the eyes), as well as other common and distinctive features. The dinosaurs were distinguished from other animals, however, by distinctive aspects of their **behavior**, **physiology**, and ecological relationships. Unfortunately, relatively little is known about these traits because we can only learn about dinosaurs using their fossil traces, which are rare and incomplete. It is clear from the available evidence that some species of dinosaurs were large predators, others were immense herbivores, and still others were smaller predators, herbivores, or scavengers. Sufficient information is available to allow paleontologists to assign scientific names to many of these dinosaurs and to speculate about their evolutionary and ecological relationships.

Although they are now extinct, the dinosaurs were among the most successful large animals ever to live on **Earth**. The dinosaurs arose during the interval of **geologic time** known as the Mesozoic (middle life) era, often called the "golden age of reptiles" or "the age of dinosaurs." Radiometric dating of volcanic **rocks** associated with dinosaur fossils suggests they first evolved 225 million years ago, during the late Triassic Period and became extinct 66 million years ago, at the end of the Cretaceous period. Dinosaurs lived for about 160 million years and were the dominant terrestrial animals on Earth throughout the Jurassic and Cretaceous periods—a span of over 100 million years.

Interestingly, mammal-like animals co-existed almost continuously with the dinosaurs and obviously prospered after the last of the dinosaurs became extinct. Although they co-existed in time with dinosaurs, **mammals** were clearly subordinate to these reptiles. It was not until the disappearance of the last dinosaurs that an adaptive radiation of larger species of mammals occurred, and they then became the dominant large animals on Earth.

It is not known exactly what caused the last dinosaurs to become extinct. It must be stressed, however, that dinosaurs were remarkably successful animals. These creatures were dominant on Earth for an enormously longer length of time than the few tens of thousands of years that humans have been a commanding species.

Biology of dinosaurs

The distinguishing characteristics of the dinosaurs include the structure of their skull and other bones. Dinosaurs typically had 25 vertebrae, plus three vertebrae that were fused to form their pelvic bones. The dinosaurs displayed an enormous range of forms and functions, however, and they filled a wide array of ecological niches. Some of the dinosaurs were, in fact, quite bizarre in their shape and, undoubtedly, their behavior.

The smallest dinosaurs were chicken-like carnivores that were only about 1 ft (30 cm) long and weighed 5-6 lb (2-3 kg). The largest dinosaurs reached a length of over 100 ft (30 m) and weighed 80 tons (73 metric tons) or more—more than any other terrestrial **animal** has ever achieved. The largest blue whales can weigh more than this, about 110 tons (100 metric tons), representing the largest animals ever to occur on Earth. The weight of these aquatic animals is partially buoyed by the **water** that they live in; whales do not have to fully support their immense weight against the forces of gravity, as the dinosaurs did. When compared with the largest living land animal, the African **elephant**, which weighs as much as 7.5 tons (6.8 metric tons), the large species of dinosaurs were enormous creatures.

Most species of dinosaurs had long tails and long necks, but this was not the case for all species. Most of the dinosaurs walked on all four legs, although some species were bipedal, using only their rear legs for locomotion. Their forelegs were greatly reduced in size and probably were used only for grasping. The tetrapods that walked on four legs were all peaceful herbivores. In contrast, many of the bipedal dinosaurs were fast-running predators.

The teeth of dinosaur species were highly diverse. Many species were exclusively herbivorous, and their teeth were correspondingly adapted for cutting and grinding vegetation. Other dinosaurs were fierce predators, and their teeth were shaped like serrated knives, which seized and stabbed their **prey** and cut it into smaller pieces that could be swallowed whole.

Until recently, it was widely believed that dinosaurs were rather stupid, slow-moving, cold-blooded (or poikilothermic) creatures. Some scientists now believe, however, that dinosaurs were intelligent, social, quick-moving, and probably warm-blooded (or homoiothermic) animals. This is a controversial topic, and scientific consensus has not been reached on whether or not some of the dinosaurs were able to regulate their body **temperature** by producing **heat** through metabolic reactions. It is absolutely undeniable that dinosaurs were extremely capable animals. This should not be a surprise to us, considering the remarkable evolutionary successes that they attained.

Fossils and other evidence of the dinosaurs

Humans never co-existed with dinosaurs, yet a surprising amount is known about these remarkable reptiles. Evidence about the existence and nature of dinosaurs is entirely indirect; it has been gleaned from fossilized traces that these animals left in sediment deposits.

The first indications suggesting the existence of the huge, extinct creatures that we now know as dinosaurs were traces of their ancient footprints in sedimentary rocks. Dinosaurs left their footprints in soft mud as they moved along marine shores or riverbanks. That mud was subsequently covered over as a new layer of sediment accumulated, and it later solidified into rock. Under very rare circumstances, this process preserved traces of the footprints of dinosaurs. Interestingly, the footprints were initially attributed to giant **birds** because of their superficial resemblance to tracks made by the largest of the living birds, such as the ostrich and emu.

The first fossilized skeletal remains to be identified as those of giant, extinct reptiles were discovered by miners in western **Europe**. These first discoveries were initially presumed to be astonishingly gigantic, extinct lizards. Several naturalists recognized substantial anatomical differences between the fossil bones and those of living reptiles, however, and so the dinosaurs were "discovered." The first of these finds consisted of bones of a 35-50 ft (10-15 m) long **carnivore** named *Megalosaurus*; this was the first dinosaur to be named scientifically. A large **herbivore** named *Iguanodon* was found at about the same time in sedimentary rocks in mines in England, Belgium, and France.

Discoveries of fantastic, extinct mega-reptiles in Europe were soon followed by even more exciting finds of dinosaur fossils in **North America** and elsewhere. These events captured the fascination of both naturalists and the general public. Museums started to develop extraordinary displays of re-assembled dinosaur skeletons, and artists prepared equally extraordinary depictions of dinosaurs and their hypothesized appearances and habitats.

This initial hey-day of dinosaur fossil discoveries occurred in the late nineteenth and early twentieth centuries. During this period, many of the most important finds were made by North American paleontologists who discovered and began to mine rich deposits of fossils in the prairies. There was intense scientific interest in these American discoveries of fossilized bones of gargantuan, seemingly preposterous animals, such as the awesome predator *Tyrannosaurus* and the immense herbivore *Apatosaurus* (initially known as *Brontosaurus*). Unfortunately, the excitement and scientific frenzy led to competition among some of the paleontologists, who wanted to be known for discovering the biggest, fiercest, or weirdest dinosaurs. The most famous rivals were two American scientists, Othniel C. Marsh and Edwin Drinker Cope.

Other famous discoveries of fossilized dinosaur bones were made in the Gobi Desert of eastern **Asia**. Some of those finds include nests with eggs that contain fossilized embryos used to study dinosaur development. Some nests contain hatchlings, suggesting that dinosaur parents cared for their young. In addition, the clustering of the nests of some dinosaurs suggests social behavior including communal nesting, possibly for mutual protection against marauding predatory dinosaurs. In the valley of Ukhaa Tolgod, Mongolia, the skeleton of an adult oviraptor was found hunched over her nest of eggs, just like any incubating bird.

A find of dinosaur eggs in an Argentinian desert in 1998 is one of the largest collections ever discovered. It consists of hundreds of 6 in (15 cm) eggs of Titanosaurs, 45 ft (13.7 m) long relatives of the Brontosaurus. The eggs were laid 70-90 million years ago, and skeletons of about 36 15-in (38-cm) long babies were also found in the mudstone. The paleontologists named the site "Auca Maheuvo," after a local **volcano** and the Spanish words for "more eggs." They hope to assemble an "ontological series" of eggs and embryos from the fossils to show all the stages of baby dinosaur development. Other scientists have speculated that this type of dinosaur gave **birth** to live young, and the discovery of the egg bonanza resolves that question.

Fossilized dinosaur bones have been discovered on all continents. Discoveries of fossils in the Arctic and in **Antarctica** suggest that the climate was much warmer when dinosaurs roamed Earth. It is also likely that polar dinosaurs were migratory, probably traveling to high latitudes to feed and breed during the summer and return-

Dinosaur bones being excavated near Kauchanaburi, Thailand. *Photograph by Bill Wassman. Stock Market. Reproduced by permission.*

ing to lower latitudes during the winter. These migrations may have occurred mostly in response to the lack of sunlight during the long polar winters, rather than the cooler temperatures.

Although the most important fossil records of dinosaurs involve their bones, there is other evidence as well. In addition to footprints, eggs, and nests, there have also been finds of imprints of dinosaur skin, feces (known as coprolites), rounded gizzard stones (known as gastroliths), and even possible stomach contents. Fossilized imprints of feathers associated with dinosaurs called Sinosauropteryx and Protarchaeiopteryx found in the Liaoning Province of China show not only long flight and tail feathers but downy under feathers. In addition, fossilized plant remains are sometimes associated with deposits of dinosaur fossils, and these can be used to infer something about the habitats of these animals. Inferences can also be based on the geological context of the locations of fossils, such as their proximity to **ocean** shores or geographical position for polar dinosaurs. These types of information have been studied and used to infer the shape, physiology, behavior, and ecological relationships of extinct dinosaurs.

Major groups of dinosaurs

There is only incomplete knowledge of the evolutionary relationships of dinosaurs with each other and with other major groups of reptiles. This results from the fact that dinosaurs, like any other extinct **organism**, can only be studied through their fossilized remains, which are often rare and fragmentary. Nevertheless, some dinosaur species bear clear resemblances to each other but are also obviously distinct from certain other dinosaurs.

The dinosaurs evolved from a group of early reptiles known as *thecodonts*, which arose during the Permian period (290-250 million years ago) and were dominant throughout the Triassic (250-208 million years ago). It appears that two major groups of dinosaurs evolved from the thecodonts, the ornithischian (bird hips) dinosaurs and the saurischian (lizard hips) dinosaurs. These two groups are distinguished largely on the basis of the anatomical structure of their pelvic or hip bones.

Both of these dinosaur lineages originated at about the same time. Both evolved into many species that were ecologically important and persisted until about 66 million years ago. Both groups included quadrupeds that walked

A dinosaur foot print in Tuba City, Arizona. *JLM Visuals. Reproduced by permission.*

on all four legs, as well as bipeds that walked erect on their much-larger hind legs. All of the ornithischians had bird-like beaks on their lower jaws and all were herbivores. Most of the carnivorous, or predatory, dinosaurs were saurischians, as were some of the herbivorous species. Interestingly, despite some resemblance between ornithischian dinosaur and bird physiology, it appears that the first birds actually evolved from saurischians.

Carnivorous dinosaurs

The carnosaurs were a group of saurischian predators, or theropods, that grew large and had enormous hind limbs but tiny fore limbs. *Tyrannosaurus rex* was the largest carnivore that has ever stalked Earth; its scientific name is derived from Greek words for "tyrant reptile king." This fearsome, bipedal predator of the Late Cretaceous could grow to a length of 45 ft (14 m) and may have weighed as much as 9 tons (8.2 metric tons). *Tyrannosaurus rex* (*T. rex*) had a massive head and a mouth full of about 60 dagger-shaped, 6-in-long (15-cm-long), very sharp, serrated teeth, which were renewed throughout the life of the animal. This predator probably ran in a lumbering fashion using its powerful hind legs, which may also have been wielded as sharp-clawed, kicking weapons. It is thought that *T. rex* may have initially attacked its prey with powerful head-butts and then torn the animal apart with its enormous, 3 ft long (1 m long) jaws. Alternatively, *T. rex* may have been a **scavenger** of dead dinosaurs. The relatively tiny fore legs of *T. rex* probably only had little use. The long and heavy tail of *T. rex* was used as a counter-balance for the animal while it was running and as a stabilizing prop while it was standing.

Albertosaurus was also a large theropod of the Late Cretaceous. *Albertosaurus* was similar to *Tyrannosaurus*, but it was a less massively built animal at about 25 ft (8 m) long and 2 tons (1.8 metric tons) in weight. *Albertosaurus* probably moved considerably faster than *Tyrannosaurus*.

Allosaurus was a gigantic, bipedal predator of the Late Jurassic. *Allosaurus* could grow to a length of 36 ft (12 m) and a weight of 2 tons (1.8 metric tons). The jaws of *Allosaurus* were loosely hinged, and they could detach to swallow large chunks of prey.

Spinosaurus was a "fin-back" (or "sail-back") dinosaur of the Late Cretaceous period that was distantly related to *Allosaurus*. *Spinosaurus* had long, erect, skin-covered, bony projections from its vertebrae that may have been used to regulate body temperature or perhaps for behavioral displays to impress others or attract a mate. *Spinosaurus* could have achieved a length of 40 ft (13 m) and a weight of 7 tons (6.3 metric tons). These animals had small, sharp teeth and were probably carnivores. *Dimetrodon* and *Edaphosaurus*, early Permian pelycosaurs (mammal-like reptiles, not dinosaurs), are sometimes confused with *Spinosaurus* as they also had sail-like back spines.

Not all of the fearsome dinosaurian predators or theropods were enormous. *Deinonychus*, for example, was an Early Cretaceous dinosaur that grew to about 10 ft (3 m) and weighed around 220 lb (100 kg). *Deinonychus* was one of the so-called running lizards, which were fast, agile predators that likely hunted in packs. As a result, *Deinonychus* was probably a fearsome predator of animals much larger than itself. *Deinonychus* had one of its hind claws enlarged into a sharp, sickle-like, slashing weapon, which was wielded by jumping on its prey and then kicking, slashing, and disemboweling the victim. The scientific name of *Deinonychus* is derived from the Greek words for terrible claw.

The most infamous small theropod is *Velociraptor*, or "swift plunderer," a 6-ft-long (2-m-long) animal of the Late Cretaceous. Restorations of this fearsome, highly intelligent, pack-hunting, killing machine were used in the movie *Jurassic Park*.

Oviraptosaurs (egg-stealing reptiles) were relatively small, probably highly intelligent theropods that were fast-running hunters of small animals, and some are believed to have also been specialized predators of the nests of other dinosaurs. The best known of these animals is Late Cretaceous *Oviraptor*. *Ingenia*, a somewhat smaller oviraptorsaur, was about 6 ft (2 m) long, weighed about 55 lb (25 kg), and also lived during the Late Cretaceous. *Microvenator* of the early Cretaceous was less than 3 ft (1 m) long and weighed about 12 lb (6 kg).

Herbivorous dinosaurs

The sauropods were a group of large saurischian herbivores that included the world's largest-ever terrestrial animals. This group rumbled along on four, enormous, pillar-like, roughly equal-sized legs, with a long tail trailing behind. Sauropods also had very long necks, and their heads were relatively small, at least in comparison with the overall **mass** of these immense animals. The teeth were peg-like and were mostly used for grazing, rather than for chewing their diet of plant **matter**. Digestion was probably aided by large stones in an enormous gizzard, in much the same way that modern, seed-eating birds grind their food. The sauropods were most abundant during the Late Jurassic. They declined afterwards and were replaced as dominant herbivores by different types of dinosaurs, especially the hadrosaurs.

Apatosaurus (previously known as *Brontosaurus* or the ground-shaking "thunder lizard") was a large sauropod that lived during the Late Jurassic and reached a length of 65 ft (20 m) and a weight of 30 tons (27 metric tons). *Diplodocus* was a related animal of the Late Jurassic, but it was much longer in its overall body shape. A remarkably complete skeleton of *Diplodocus* was found that was 90 ft (27 m) long overall, with a 25 ft (8 m) neck and a 45 ft (14 m) tail, and an estimated body weight of 11 tons (10 metric tons). In comparison, the stouter-bodied *Apatosaurus* was slightly shorter but considerably heavier. *Brachiosaurus* also lived during the Late Jurassic and was an even bigger herbivore, with a length as great as 100 ft (30 m) and an astonishing weight that may have reached 80 tons (73 metric tons), although conservative estimates are closer to 55 tons (50 metric tons). *Supersaurus* and *Ultrasaurus* were similarly large. *Seismosaurus* may have been longer than 160 ft (50 m), and *Argentinosaurus* (recently discovered in Patagonia, **South America**) may set a new weight record of 100 tons (91 metric tons).

Stegosaurus was a 30 ft long (9 m long), Late Jurassic tetrapod with a distinctive row of triangular, erect, bony plates running along its back. These may have been used to regulate heat. *Stegosaurus* had sharp-spiked projections at the end of its tail, which were lashed at predators as a means of defense. *Dacentrurus* was a 13-ft-long (4-m-long), Jurassic-age animal related to *Stegosaurus*, but it had a double row of large spikes along the entire top of its body, from the end of the tail to the back of the head.

The ceratopsians were various types of "horned" dinosaurs. *Triceratops* was a three-horned dinosaur and was as long as 33 ft (10 m) and weighed 6 tons (5.4 metric tons). *Triceratops* lived in the late Cretaceous, and it had a large bony shield behind the head with three horns projecting from the forehead and face, which were used as defensive weapons. *Anchiceratops* was a 7-ton (6.3-metric-ton) animal that lived somewhat later. It was one of the last of the dinosaurs and became extinct 66 million years ago at the end of the Cretaceous period. There were also rhinoceros-like, single-horned dinosaurs, such as the 20-ft-long (6-m-long), 2-ton (1.8-metric-ton) *Centrosaurus* of the late Cretaceous. Fossilized skeletons of this animal

have been found in groups, suggesting that it was a herding dinosaur. The horned dinosaurs were herbivores, and they had parrot-like beaks useful for eating vegetation.

Ankylosaurus was a late Cretaceous animal that was as long as 36 ft (11 m) and weighed 5 tons (4.5 metric tons). *Ankylosaurus* was a stout, short-legged, lumbering herbivore. This animal had very heavy and spiky body armor and a large bony club at the end of its tail that was used to defend itself against predators.

The duck-billed dinosaurs or hadrosaurs included many herbivorous species of the Cretaceous period. Hadrosaurs are sometimes divided into groups based on aspects of their head structure; they could have a flattish head, a solid crest on the top of their head, or an unusual, hollow crest. Hadrosaurs were the most successful of the late Cretaceous dinosaurs in terms of their relative abundance and wide distribution.

Hadrosaurs apparently were social animals; they lived in herds for at least part of the year and migrated seasonally in some places. Hadrosaurs appear to have nested communally, incubated their eggs, and brooded their young. Hadrosaurs had large hind legs and could walk on all four legs or bipedally if more speed was required—these animals were probably very fast runners.

Hadrosaurus was a 5-ton (4.5-metric-ton), late Cretaceous animal and was the first dinosaur to be discovered and named in North America-in 1858 from fossils found in New Jersey. *Corythosaurus* was a 36 ft-long (11 m-long), 4 ton (3.6 metric ton), Late Cretaceous herbivore that had a large, hollow, helmet-like crest on the top of its head. *Parasaurolophus* of the late Cretaceous was similar in size, but it had a curved, hollow crest that swept back as far as 10 ft (3 m) from the back of the head. It has been suggested that this exaggerated helmet may have worked like a snorkel when this animal was feeding underwater on aquatic plants; however, more likely uses of the swept-back helmet were in species recognition and resonating the loud sounds made by these hadrosaurs. *Edmontosaurus* was a large, non-helmeted hadrosaur that lived in the Great Plains during the late Cretaceous and was as long as 40 ft (13 m) and weighed 3 tons (2.7 metric tons). *Anatosaurus* was a 3 ton (2.7 metric-ton) hadrosaur that lived as recently as 66 million years ago and was among the last of the dinosaurs to become extinct. The hadrosaurs probably were a favorite prey for some of the large theropods, such as *Tyrannosaurus rex*.

Other extinct orders of Mesozoic-age reptiles

Several other orders of large reptiles lived at the same time as the dinosaurs and are also now extinct.

The pterosaurs (order Pterosauria) were large, flying reptiles that lived from the late Triassic to the late Cretaceous. Some species of pterosaurs had wingspans as great as 40 ft (12 m), much wider than any other flying animal has ever managed to achieve. Functional biologists studying the superficially awkward designs of these animals have long wondered how they flew. Some species of pterosaurs are thought to have fed on **fish**, which were scooped up as the pterosaur glided just above the water surface.

The ichthyosaurs (Ichthyosauria), plesiosaurs (Plesiosauria), and mosasaurs (Mososauria) were orders of carnivorous marine reptiles that became extinct in the Late Cretaceous. The ichthyosaurs were shark-like in form, except that their vertebral column extended into the lower part of their caudal (or tail) fin, rather than into the upper part like the **sharks**. Of course, ichthyosaurs also had well-developed, bony skeletons, whereas sharks have a skeleton composed entirely of cartilage rather than bone. The plesiosaurs were large animals reaching a length as great as 45 ft (14 m). These marine reptiles had paddle-shaped limbs, and some species had very long necks. Mosasaurs were large lizards that had fin-shaped limbs and looked something like a cross between a crocodile and an eel; but they grew to lengths of more than 30 ft (9 m).

Theories about the extinction of dinosaurs

There are many theories about what caused the **extinction** of the last of the dinosaurs, which occurred at the end of the Cretaceous period, about 66 million years ago. Some of the more interesting ideas include: the intolerance of these animals to rapid climate change, the emergence of new species of dominant plants that contained toxic chemicals the herbivorous dinosaurs could not tolerate, an inability to compete successfully with the rapidly evolving mammals, insatiable destruction of dinosaur nests and eggs by mammalian predators, and widespread **disease** to which dinosaurs were not able to develop immunity. All of these hypotheses are interesting, but the supporting evidence for any one of them is not enough to convince most paleontologists.

Interestingly, at the time of the extinction of the last of the dinosaurs, there were also apparently mass extinctions of other groups of organisms. These included the reptilian order Pterosauria, along with many groups of plants and **invertebrates**. In total, perhaps three quarters of all species and one half of all genera may have become extinct at the end of the Cretaceous. A popular hypothesis for the cause of this catastrophic, biological event was the impact of a meteor hitting Earth. The im-

pact of an estimated 6 mi-wide (10 km-wide) meteorite could have spewed an enormous quantity of fine dust into the atmosphere, which could have caused climate changes that most large animals and other organisms could not tolerate. As with the other theories about the end of the dinosaurs, this one is controversial. Many scientists believe the extinctions of the last dinosaurs were more gradual and were not caused by the shorter-term effects of a rogue meteorite.

Another interesting concept concerns the fact that dinosaurs share many anatomical characteristics with Aves, the birds, a group that clearly evolved from a dinosaur ancestor. In fact, there are excellent fossil remains of an evolutionary link between birds and dinosaurs. The 3 ft-long (1 m-long), Late Jurassic fossil organism *Archaeopteryx* looked remarkably like *Compsognathus* but had a feathered body and could fly or glide. Moreover, some of the living, **flightless birds** such as emus and ostriches and recently extinct birds such as elephant birds and moas bear a remarkable resemblance to certain types of dinosaurs. Because of the apparent continuity of anatomical characteristics between dinosaurs and birds, some paleontologists believe that the dinosaurs did not actually become extinct. Instead, the dinosaur lineage survives today in a substantially modified form, as the group Aves, the birds.

See also Evolution; Fossil and fossilization; Paleontology.

Resources

Books

Carpenter, K., and P.J. Currie. *Dinosaur Systematics. Approaches and Perspectives.* Cambridge, UK: Cambridge University Press, 1990.

Cowen, R. *History of Life.* London: Blackwell Scientific Publishing, 1995.

Palmer, Douglas. *The Marshall Illustrated Encyclopedia of Dinosaurs & Prehistoric Animals: A Comprehensive Color Guide to over 500 Species.* New York: Todtri, 2002.

Prothero, Donald R. *Bringing Fossils To Life: An Introduction To Paleobiology.* Columbus: McGraw-Hill Science/Engineering/Math, 1997.

Weishampel, D.B., ed. *The Dinosauria.* Berkeley, CA: University of California Press, 1990.

Bill Freedman

Diode

A diode is an electronic device that has two electrodes arranged in such a way that electrons can flow in only one direction. Because of this ability to control the flow of electrodes, a diode is commonly used as a rectifier, a device that connects alternating current into direct current. In general, two types of diodes exist. Older diodes were **vacuum** tubes containing two **metal** components, while newer diodes are solid state devices consisting of one n-type and one p-type semiconductor.

The working element in a **vacuum tube** diode is a metal wire or cylinder known as the **cathode**. Surrounding the cathode or placed at some **distance** from it is a metal plate. The cathode and plate are sealed inside a **glass** tube from which all air is removed. The cathode is also attached to a heater, which when turned on, causes the cathode to glow. As the cathode glows, it emits electrons.

If the metal plate is maintained at a positive potential difference compared to the cathode, electrons will flow from the cathode to the plate. If the plate is **negative** compared to the cathode, however, electrons are repelled and there is no electrical current from cathode to plate. Thus, the diode acts as a rectifier, allowing the flow of electrons in only one direction, from cathode to plate.

One use of such a device is to transform alternating current to direct current. Alternating current is current that flows first in one direction and then the other. But alternating current fed into a diode can move in one direction only, thereby converting the current to a one-way or direct current.

Newer types of diodes are made from n-type semiconductors and p-type semiconductors. N-type semicon-ductors contain small impurities that provide an excess of electrons with the capability of moving through a system. P-type semiconductors contain small impurities that provide an excess of positively charged "holes" capable of moving through the system.

A semiconductor diode is made by joining an n-type semiconductor with a p-type semiconductor through an external circuit containing a source of electrical current. The current is able to flow from the n-semiconductor to the p-semiconductor, but not in the other direction. In this sense, the n-semiconductor corresponds to the cathode and the p-semiconductor to the plate in the vacuum tube diode. The semiconductor diode has most of the same functions as the older vacuum diode, but it operates much more efficiently and takes up much less space than does a vacuum diode.

See also Electrical conductivity; Electric current.

Dioxin

Chlorinated dioxins are a diverse group of organic chemicals. TCDD, or 2,3,7,8-tetrachlorodibenzo-*p*-dioxin, is a particular dioxin that is toxic to some **species** of animals in extremely small concentrations. As such, TCDD is the most environmentally controversial of the chlorinated dioxins, and the focus of this entry.

TCDD and other dioxins

Dioxins are a class of organic compounds, with a basic structure that includes two **oxygen atoms** joining a pair of **benzene** rings. Chlorinated dioxins have some amount of substitution with **chlorine** for **hydrogen** atoms in the benzene rings. A particular chemical, 2,3,7,8-tetrachlorodibenzo-*p*-dioxin (abbreviated as TCDD, or as 2,3,7,8-TCDD), is one of 75 chlorinated derivatives of dibenzo-*p*-dioxin. There is a very wide range of toxicity within the larger group of dioxins and chlorinated dioxins, but TCDD is acknowledged as being the most poisonous dioxin compound, at least to certain species of animals.

Dioxins have no particular uses. They are not manufactured intentionally, but are synthesized incidentally during some industrial processes. For example, under certain conditions relatively large concentrations of dioxins are inadvertently synthesized during industrial reactions involving 2,4,5-trichlorophenol. A well known case of this phenomenon is in the manufacturing of the phenoxy herbicide, 2,4,5-T (2,4,5-trichlorophenoxy **acetic acid**). This chemical (which is no longer used) was once manufactured in large amounts, and much of the material was badly contaminated by TCDD. Concentrations in the range of 10-50 parts per million (ppm, or mg per liter) occurred in 2,4,5-T manufactured for use during the Vietnam War. However, there was a much smaller **contamination** (less than 0.1 ppm) in 2,4,5-T manufactured after 1972, in accordance with regulations enacted by the U.S. Environmental Protection Agency.

TCDD is also a trace contaminant of other products manufactured from trichlorophenol, including hexachlorophene, once commonly used as a topical antibacterial treatment. TCDD is incidentally synthesized when **wood** pulp is bleached using chlorine-based oxidants. The low-temperature **combustion** of chlorine-containing organic materials (for example, in cigarettes, burning garbage dumps, and barbecues) also produce dioxins, including TCDD. The **incineration** of municipal waste synthesizes small quantities of TCDD, although the relatively high temperatures reduce the yield of dioxins compared with the smoldering kinds of combustion just mentioned. Dioxins are also synthesized naturally in trace quantities, mostly during forest fires.

TCDD is a persistent chemical in the environment, and because it is virtually insoluble in **water**, but highly soluble in fats and oils, it strongly biomagnifies and occurs in especially large concentrations in predators at the top of the ecological food web. Moreover, TCDD is globally distributed, meaning that any chemical analysis of a biological **tissue**, especially of the **fat** of an **animal**, will detect residues of this dioxin (assuming that the analytical **chemistry** is sensitive enough).

Toxicity

TCDD is the most toxic of the chlorinated dioxins, while octachlorodioxin may be the least so. However, there are large differences in the susceptibility of species to suffering toxicity from TCDD. The guinea pig, for example, is extremely sensitive to TCDD, thousands of times more so than the hamster.

Short-term or acute toxicity is often indicated by a laboratory assay known as LD_{50}, or the dose of a chemical required to kill one-half of a test population of organisms over a period of several days. Guinea pigs have a LD for TCDD in food of only 0.0006 mg/kg (that is, 0.0006 mg of TCDD per kg body weight). In comparison, **rats** have a LD_{50} for TCDD in food of 0.022-0.045 mg/kg, while **hamsters** have a LD_{50} of 1 mg/kg. Clearly, TCDD is a highly toxic chemical, although species vary greatly in sensitivity.

Depending on the dose and biological sensitivity, the symptoms of TCDD toxicity in **mammals** can include severe weight loss, liver damage, lesions in the vascular system, stomach **ulcers**, a persistent **acne** known as chloracne, **birth defects**, and ultimately, death.

Much of what is known about the toxicity of TCDD to humans has come from studies of: (1) industrial exposures of chemical workers, (2) people living near a toxic waste dump at Times Beach, Missouri, and (3) an accidental event at Seveso, Italy, in 1976. The latter case involved an explosion at a chlorophenol **plant** that released an estimated 2.2-11 lb (1-5 kg) of TCDD to the surroundings, and caused residues as large as 51 ppm to occur in environmental samples. This accident caused the deaths of some **livestock** within 2-3 days, but remarkably it was not until 2.5 weeks had passed that about 700 people were evacuated from the severely contaminated residential area near the factory. The exposure of humans to TCDD at Seveso caused 187 diagnosed cases of chloracne, but there were apparently no statistically detectable increases in the rates of other human diseases, or of deformities of children born to exposed women.

Overall, studies of humans suggest that they are among the least-sensitive mammals to suffering toxicity from TCDD. While chloracne is a common symptom of an acute human exposure to TCDD, the evidence showing increased rates of TCDD-related **disease**, mortality, **cancer**, or **birth** defects are equivocal, and controversial. Some scientists believe there is no evidence that a human has ever died from an acute exposure to TCDD. However, there is unresolved scientific controversy about the possible effects of longer-term, chronic exposures of humans to TCDD, which might result in increased rates of developmental abnormalities or cancers. Unless large, these effects would be difficult to detect, because of the great environmental and genetic variations that must be overcome in epidemiological studies of humans.

TCDD in Vietnam

To deprive their enemy of food and cover during the Vietnam War, the U.S. military sprayed large quantities of **herbicides**. More than 547,000 sq mi (1.4 million ha) of terrain were sprayed at least once. The most commonly used herbicide was a 50:50 mixture of 2,4,5-T and 2,4-D, known as **Agent Orange**. More than 46 million lb (21 million kg) of 2,4,5-T and 55 million lb (25 million kg) of 2,4-D were sprayed during this extensive military program.

An important aspect of the military use of herbicides in Vietnam was contamination of the 2,4,5-T by TCDD. A **concentration** as large as 45 ppm was measured in Agent Orange, but the average concentration was about 2 ppm. In total, 243-375 lb (110-170 kg) of TCDD was sprayed with herbicides onto Vietnam.

Because TCDD is known to be extremely toxic to some laboratory animals, there has been tremendous

KEY TERMS

. . . .

Acute toxicity—A poisonous effect produced by a single, short-term exposure to a toxic chemical, resulting in obvious tissue damage, and even death of the organism.

Chronic toxicity—This is a poisonous effect that is produced by a long period of exposure to a moderate, sub-acute dose of some toxic chemical. Chronic toxicity may result in anatomical damages or disease, but it is not generally the direct cause of death of the organism.

Epidemiology—The study of the incidence, control, and transmission of diseases in large populations.

controversy over the possible short- and long-term effects of exposure of soldiers and civilians to TCDD in Vietnam. Although claims have been made of effects in exposed populations, the studies have not been convincing to many scientists, and there is still controversy. The apparent, mainstream opinion from the most rigorous epidemiological studies suggests that large toxic effects have not occurred, which is encouraging. It is also likely that the specific effects of TCDD added little to the very substantial ecological effects caused by the use of military use of herbicides, and other weapons of **mass** destruction, during the Vietnam War.

See also Biomagnification.

Resources

Books

Freedman, B. *Environmental Ecology.* 2nd ed. San Diego: Academic Press, 1995.

Periodicals

Harris, W.E. "Dioxins—An Overview." *Tappi Journal* (April 1990): 267-69.

Bill Freedman

Diphtheria

Diphtheria is a serious **disease** caused by the bacterium *Corynebacterium diptheriae*. Usually, the **bacteria** initially infect the throat and pharynx. During the course of the **infection**, a membrane-like growth appearing on the throat can obstruct breathing. Some strains of this bacterium release a toxin, a substance that acts as a poison in the body. This toxin, when released into the

bloodstream, travels to other organs of the body and can cause severe damage.

Diphtheria was first formally described as a disease in 1826. In 1888, *Corynebacterium diptheriae* was identified as the cause of the disease. A few years later, researchers discovered the antitoxin, or antidote, to the diphtheria toxin. If the antitoxin is given to a person with diphtheria in the early stages of the infection, the antitoxin neutralizes the toxin. This treatment, along with an aggressive vaccination program, has virtually eliminated the disease in the United States. Other countries that do not have an aggressive vaccination program, have numerous cases of diphtheria, many of which end in death.

Incidence of diphtheria

Since most children in the United States are vaccinated against diphtheria, the domestic incidence of the disease is very low. When diphtheria does occur, it tends to strike adults, because fewer adults than children have been immunized against the disease. In developing countries, where less than 10% of the children are vaccinated against diphtheria, about one million deaths are caused each year by this disease. Diphtheria is highly contagious. The disease is prevalent in densely-populated areas, especially during the winter months when more people crowd together indoors. Transmission of the bacteria occurs when an infected person sneezes or coughs and a susceptible person breathes in the saliva or mucus droplets form the air.

Diphtheria toxin

Interestingly, diphtheria toxin is produced by strains of *Corynebacterium diptheriae* that have themselves been infected with a special type of **virus** called a **bacteriophage**. The particular bacteriophage that infects *C. diptheriae* carries with it the **gene** that produces the diphtheria toxin. Strains of *C. diptheriae* without the bacteriophage do not produce the toxin.

The diphtheria toxin consists of two subunits, A and B. The B subunit binds to the **plasma membrane** of a **cell**. Once it is bound to the membrane, it pulls the A subunit into the cell. The A subunit is the active segment of the toxin, producing most of the effects. Once inside the cell, the A subunit disrupts protein synthesis; once this mechanism is disrupted, the cell cannot survive for long. Diphtheria toxin thus kills cells. Cells in the throat and respiratory tract are killed first; if the toxin spreads in the bloodstream to other organs—such as the **heart**, kidney, and brain—severe and even fatal damage can result.

Symptoms

The incubation period—the **time** from exposure to the bacteria to the first symptoms—is one to seven days.

The first symptoms of diphtheria are fatigue, a low-grade fever, and a sore throat. As the disease progresses, the throat swells, sometimes so much that the patient has noticeable neck swelling. The bacteria infect the throat first before spreading to the larynx (voice box) and trachea (windpipe). At the site of infection, the throat is red and sore. In reaction to the infection, the throat tissues release a discharge containing fibrous material and immune cells. This discharge covers the throat tissues and appears as a grayish, membrane-like material. The throat and trachea continue to swell; if not relieved, the swelling may obstruct the airway, leading to death by suffocation.

Sometimes diphtheria bacteria infect the skin first. When this type of infection occurs, skin lesions appear. For reasons that are not clear, the diphtheria characterized by infection is more contagious than the disease characterized by respiratory infection. The skin-type of diphtheria is more common in tropical and sub-tropical countries.

Treatment

Diphtheria is treated with an antitoxin that can only neutralize the toxin that has not yet bound to a cell membranes; it cannot neutralize the toxin that has already bound to and penetrated a cell. For this reason, antitoxin must be administered early in infection. In fact, some experts recommend giving doses of antitoxin if diphtheria is even suspected, since the additional time spent waiting for confirming lab results allows for more of the toxin to spread and penetrate the cells.

Vaccine

The diphtheria **vaccine** consists of a small amount of the toxin that has been altered so as not to cause toxic effects. The vaccine works by prompting the body's **immune system** to make antitoxin against the altered vaccine toxin. The diphtheria toxin is combined with the **tetanus** toxin and the pertussis (**whooping cough**) toxin in one vaccine, abbreviated DPT. The DPT is given in four doses. In the United States, infants are given their first DPT dose at about six to eight weeks of age. If all four doses are administered before age four, the child should have a DPT "booster" before beginning kindergarten. This shot "boosts" the immunity to the disease.

A person can be tested for their immunity to diphtheria by the Schick test, which demonstrates the presence of antitoxin within the body. In this test, a small amount of diphtheria toxin is placed under the skin of the forearm. If the site develops a reaction—such as redness or swelling—the person has not developed the antitoxin from a previous infection or a vaccine, and is therefore susceptible to diphtheria. If no reaction is present,

KEY TERMS

. . . .

Antitoxin—A antidote to a toxin that neutralizes its poisonous effects.

Bacteriophage—A virus that infects bacteria. When a bacteriophage that carries the diphtheria toxin gene infects diphtheria bacteria, the bacteria produce diphtheria toxin.

Schick test—A test that checks for the presence of diphtheria antitoxin in the body.

Toxin—A poisonous substance.

the person had already developed the antitoxin. The Schick test is useful for adults who cannot find their immunization records or cannot remember if they had diphtheria in childhood.

See also Childhood diseases.

Resources

Periodicals

Kleinman, Lawrence C. "To End an Epidemic: Lessons From the History of Diphtheria." *New England Journal of Medicine* 326 (March 12, 1992): 773.

"Misfiring Magic Bullets (Report on Adverse Effect from Diphtheria-Pertussis-Tetanus and Rubella Vaccines)." *Science News* 140 (July 20, 1991): 45.

Peter, Georges. "Childhood Immunizations." *New England Journal of Medicine* 327 (December 7, 1992): 25.

Kathleen Scogna

Diplopia *see* **Vision disorders**

Dipole

Dipole, literally, means "two poles," two electrical charges, one **negative** and one positive. Dipoles are common in **atoms** whenever electrons (-) are unevenly distributed around nuclei (+), and in molecules whenever electrons are unevenly shared between two atoms in a covalent bond.

When a dipole is present, the atom or covalent bond is said to be polarized, or divided into negative and positive regions. This is indicated by the use of partial negative (δ-) and partial positive (δ+) signs. The magnitude and direction of the electrical charge separation is indicated by using an arrow, drawn from the positive pole in a **molecule** to the negative pole.

In covalent bonds, permanent dipoles are caused when two different atoms share their electrons unevenly. The atom that is more electronegative—the one that holds electrons more tightly—pulls the electrons closer to itself, creating a partial negative charge there. The less electronegative atom becomes partially positive as a result because it has lost partial possession of the electrons. The electric strength of a dipole generally increases as the electronegativity difference between the atoms in the bond increases. This strength, called a dipole moment, can be measured experimentally. The size of a dipole moment is expressed in Debye units in honor of the Dutch chemist, Peter Debye (1884-1966).

The dipole moments of a series of molecules are listed below:

Molecule	Dipole Moment (in Debye units, D)
HF	1.91 D
HCl	1.03 D
HBr	0.78 D
HI	0.38 D

The measurement of dipole moments can help determine the shape of a molecule. The net dipole moment of a **water** molecule (H_2O) represents the overall electrical charge distribution in that molecule. (See Figure 1.)

The H_2O molecule is bent. Its dipole vectors do not cancel. The water molecule has a net resultant dipole moment of 1.87 D. If the molecule were linear, the measured dipole moment would be **zero**. Its individual dipoles in the two oxygen-hydrogen covalent bonds would have cancelled each other out.

Individual atoms (and ions) will be naturally polarized if their electrons happen to move irregularly about their nuclei creating, at least temporarily, lopsided looking atoms with δ+ and δ- portions. Natural collisions occurring between atoms can induce this temporary deformity from an atom's normal spherical, symmetric shape. Larger atoms are considered to be "softer" than smaller, "harder" atoms. Larger atoms are then more likely to be polarized or to have stronger dipoles than smaller atoms.

The presence of dipoles helps to explain how atoms and molecules attract each other. Figure 2 shows how the electrically positive side of one xenon atom (Xe) lines up and pulls towards the negative side of another xenon atom. Likewise, the positive side of one H-Cl molecule is attracted to the negative side of another H-Cl molecule. When many atoms and molecules are present in **matter**, these effects continue on indefinitely from atom to atom and molecule to molecule.

Dipole forces tend to organize matter and pull it together. Atoms and molecules most strongly attracted to

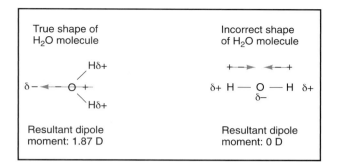

Figure 1. *Illustration by Hans & Cassidy. Courtesy of Gale Group.*

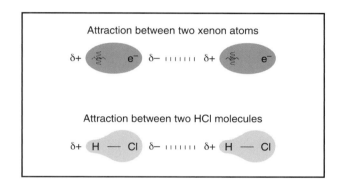

Figure 2. *Illustration by Hans & Cassidy. Courtesy of Gale Group.*

each other will tend to exist as solids. Weaker interactions tend to produce liquids. The gaseous state of matter will tend to exist when the atoms and molecules are nonpolar, or when virtually no dipoles are present.

Direct current *see* **Electric current**

Direct variation

If one quantity increases (or decreases) each **time** another quantity increases (or decreases), the two quantities are said to vary together. The most common form of this is direct variation in which the **ratio** of the two amounts is always the same. For example, speed and **distance** traveled vary directly for a given time. If you travel at 4 mph (6.5 kph) for three hours, you go 12 mi (19.5 km), but at 6 mph (9.5 kph) you go 18 mi (28.5 km) in three hours. The ratio of distance to speed is always 3 in this case.

The common ratio is often written as a constant in an equation. For example, if s is speed and d is distance, the **relation** between them is direct variation for d = ks, where k is the constant. In the example above, k = 3, so the equation becomes d = 3s. For a different time interval, a different k would be used.

Often, one quantity varies with respect to a power of the other. For example, of $y = kx^2$, then y varies directly with the square of x. More than two variables may be involved in a direct variation. Thus if z = kxy, we say that z is a joint (direct) variation of z with x and y. Similarly, if $z = kx^2/y$, we say that z varies directly with x^2 and inversely with y.

Directrix *see* **Conic sections**

Disaccharide *see* **Carbohydrate**

Disease

Disease can be defined as a change in the body processes that impairs its normal ability to function. Every day the **physiology** of the human body demands that oxygenation, acidity, salinity, and other functions be maintained within a very narrow spectrum. A deviation from the norm can be brought about by **organ** failure, toxins, heredity, **radiation**, or invading **bacteria** and viruses.

Normally the body has the ability to fight off or to neutralize many pathogenic organisms that may gain entrance through an opening in the skin or by other means. The **immune system** mobilizes quickly to rid the body of the offending alien and restore or preserve the necessary internal environment. Sometimes, however, the invasion is one that is beyond body's resistance, and the immune system is unable to overcome the invader. A disease may then develop. When the internal functions of the body are affected to the point that the **individual** can no longer maintain the required normal parameters, symptoms of disease will appear.

The **infection** brought about by a bacterium or **virus** usually generates specific symptoms, that is, a series of changes in the body that are characteristic of that invading **organism**. Such changes may include development of a fever (an internal body **temperature** higher than the norm), nausea, headache, copious sweating, and other readily discernable signs.

Much more important to the physician, though, are the internal, unseen changes that may be wrought by such an invasion. These abnormalities may appear only as changes from the norm in certain chemical elements of the **blood** or urine. That is the reason patients are asked to contribute specimens for analysis when they are ill, especially when their symptoms are not specific to a given disease. The function of organs such as the liver, kidneys, thyroid gland, pancreas, and others can be determined by the levels of various elements in the blood **chemistry**.

For a disease that is considered the result of a pathogenic invasion, the physician carries out a bacterial culture. Certain secretions such as saliva or mucus are collected and placed on a thin plate of culture material. The bacteria that grow there over the next day or so are then analyzed to determine which **species** are present and thus, which antibiotic would be most effective in eradicating them.

Viruses present special challenges, since they cannot be seen under a **microscope** and are difficult to grow in cultures. Also, viruses readily adapt to changes in their environment and become resistant to efforts to treat the disease they cause. Some viral diseases are caused by any number of forms of the same virus. The common cold, for example, can be caused by any one of some 200 viruses. For that reason it is not expected that any **vaccine** will be developed against the cold virus. A vaccine effective against one or two of the viruses will be completely useless against the other 198 or 199 forms.

The agents that cause a disease, the virus or bacterium, are called the etiologic agents of the disease. The etiologic agent for strep throat, for example, is a bacterium within the *Streptococcus* genus. Similarly, the tubercle bacillus is the etiologic agent of **tuberculosis**.

Modern medicine has the means to prevent many diseases that plagued civilization in the recent past. Polio, a crippling disease brought about by the **poliomyelitis** virus, was neither preventable nor curable until the middle 1950s. Early in that decade an outbreak of polio affected an abnormally large number of young people. Research into the cause and prevention of polio immediately gained high priority, and by the middle of the decade Dr. Jonas Salk had developed a vaccine to prevent polio. Currently all young children in developed countries can be vaccinated against the disease.

Similar vaccines have been developed over the years to combat other diseases that previously were lethal. **Whooping cough**, **tetanus**, **diphtheria**, and other diseases that at one time meant certain death to victims, can be prevented. The plague, once a dreaded killer of thousands, no longer exists among the human population. An effective vaccine has eradicated it as a dread disease.

The resistance to disease is called immunity. A few people are naturally immune to some diseases, but most have need of vaccines. This type of immunity, attained by means of a vaccine, is called artificial immunity. Vaccines are made from dead bacteria and are injected into the body. The vaccine causes the formation of antibodies, which alert the immune system in the event a live bacterium invades.

The body's immune system, responsible for guarding against invading **pathogens**, may itself be the cause of disease. Conditions such as rheumatoid **arthritis** and Lupus are considered to be the result of the immune system mistaking its own body for foreign **tissue** and organizing a reaction to it. This kind of disease is called an autoimmune disease—auto, meaning one's own, and immune referring to the immune system. Scientists have found that little can be done to combat this form of disease. The symptoms can be treated to ease the patient's discomfort or preserve his life, but the autoimmune reaction seldom can be shut down.

See also Epidemic; Epidemiology; Etiology; Syndrome.

Larry Blaser

Dissociation

Dissociation is the process by which a **molecule** separates into ions. It may also be called ionization, but because there are other ways to form ions, the term dissociation is preferred. Substances dissociate to different degrees, ranging from substances that dissociate very slightly, such as **water**, to those that dissociate almost completely, such as strong **acids and bases**. The extent to which a substance dissociates is directly related to its ability to conduct an **electric current**. A substance that dissociates only slightly (as in the case of a weak acid like vinegar) is a weak **electrolyte**, as it conducts **electricity** poorly. A substance that is almost completely dissociated (such as table **salt**, NaCl, or hydrochloric acid, HCl) conducts electricity very well. The ability to conduct electricity is based on the ionic makeup of a substance. The more ions a substance contains, the better it will conduct electricity.

Dissociation of water

Pure water dissociates only slightly. About one water molecule out of every 10 million is dissociated and the rest remain in non-dissociated (or molecular) form. This ionization of water (sometimes called self- or auto-ionization) can be summarized by the following formula. Pure water produces very few ions from its dissociation and so is a poor electrolyte, or conductor of electricity.

The following equation describes the process in which a water molecule ionizes (separates into ions) to form a **hydrogen** ion (**proton**) and a hydroxide ion.

$$H_2O \longleftrightarrow H^+ + OH^-$$

Another way to describe the dissociation of water is as follows:

$$H_2O + H_2O \longleftrightarrow H_3O^+ + OH^-$$

where two water molecules form a hydronium ion (essentially a water molecule with a proton attached) and a hydroxide ion.

Dissociation of acid and bases

Acids are molecules that can donate protons (hydrogen or H^+ ions) to other molecules. An alternate view is that an acid is a substance that will cause an increase in the **concentration** of hydrogen ions in a **solution**.

The dissociation of a strong acid (such as hydrochloric acid, HCl) is essentially 100%.

$$HCl \longrightarrow H^+ + Cl^-$$

In this case, nearly every HCl molecule is dissociated (separated into ions). When any substance dissociates, both positive and **negative** ions will be formed. In this case, the positive ion (**cation**) is a proton, and the negative ion (**anion**) is the chloride ion. A strong acid is a strong electrolyte and a good conductor of an electric current. In the case of a strong base, nearly 100% of the molecules are dissociated as well, and strong bases (such as **sodium hydroxide**, NaOH) are also strong electrolytes.

$$NaOH \longrightarrow Na^+ + OH^-$$

A weak acid, such as hydrofluoric acid is only slightly dissociated. Many more of the molecules exist in the molecular (undissociated or unionized) form than in the ionized form. Since it forms fewer ions, a weak acid will be a weak electrolyte.

$$HF \longleftrightarrow H^+ + F^-$$

In the case of a weak base, such as **aluminum hydroxide**, $Al(OH)_3$, only a small **percent** of molecules ionize, producing few ions, and making weak bases weak electrolytes as well.

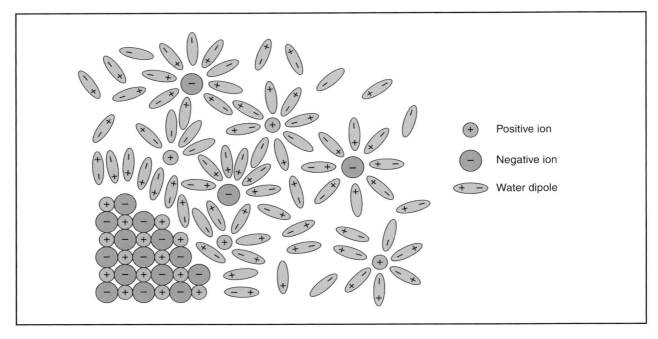

Figure 1. Illustration of the solvation process, in which the negative end of a water molecule faces the positive sodium ion and the positive end faces the negative ion. *Illustration by Hans & Cassidy. Courtesy of Gale Group.*

$$Al(OH)_3 \longleftrightarrow Al^{+3} + 3OH^-$$

In any dissociation reaction, the total charges will mathematically cancel each other out. The case above has a positive three charge on the **aluminum** ion and a negative one charge on each of the three hydroxide ions, for a total of **zero**.

Dissociation of salts

Salts are the product of the **neutralization** reaction between an acid and a base (the other product of this neutralization reaction being water). Salts that are soluble in water dissociate into their ions and are electrolytes. Salts that are insoluble or only slightly soluble in water form very few ions in solution and are non-electrolytes or weak electrolytes. **Sodium chloride**, NaCl, is a water-soluble salt that dissociates totally in water.

$$NaCl \rightarrow Na^+ + Cl^-$$

The process by which this takes place involves the surrounding of each positive **sodium** ion and each negative chloride ion by water molecules. Water molecules are polar and have two distinct ends, each with a partial positive or negative charge. Since opposite charges attract, the negative end of the water molecule will face the positive sodium ion and the positive end will face the negative ion. This process, illustrated in Figure 1, is known as solvation.

Resources

Periodicals

Carafoli, Ernest, and John Penniston. "The Calcium Signal." *Scientific American* 253 (November 1985).

Ezzell, Carol. "Salt's Technique for Tickling the Taste Buds." *Science News* 140 (November 2, 1991).

Louis Gotlib

Distance

Distance has two different meanings. It is a number used to characterize the shortest length between two geometric figures, and it is the total length of a path. In the first case, the distance between two points is the simplest instance.

The absolute distance between two points, sometimes called the displacement, can only be a **positive number**. It can never be a **negative** number, and can only be **zero** when the two points are identical. Only one straight line exists between any two points P_1 and P_2. The length of this line is the shortest distance between P_1 and P_2.

In the case of **parallel** lines, the distance between the two lines is the length of a **perpendicular** segment connecting them. If two figures such as line segments, triangles, circles, cubes, etc. do not intersect, then the necting straight line (P_1, P_2) which is the shortest distance between the two points, is given by the equation d = RADIC(x_1 MINUS x_2)^2 + (y_1 - y_2)^2 in many types of **physics** and **engineering** problems, for example, in tracking the trajectory of an atomic particle, or in determining the lateral **motion** of a suspension bridge in the presence of high winds.

Path length

The other meaning of distance is the length of a path. This is easily understood if the path consists entirely of line segments, such the perimeter of a pentagon. The distance is the sum of the lengths of the line segments that make up the perimeter. For curves that are not line segments, a continuous path can usually be approximated by a sequence of line segments. Using shorter line segments produces a better **approximation**. The limiting case, when the lengths of the line segments go to zero, is the distance. A common example would be the circumference of a **circle**, which is a distance.

Kristin Lewotsky

distance between them is the shortest distance between any pair of points, one of which lies on one figure, one of which lies on the other.

To determine the distance between two points, we must first consider a coordinate system. An *xy* coordinate system consists of a horizontal axis (*x*) and vertical axis (*y*). Both axes are infinite for positive and negative values. The crossing **point** of the lines is the origin (O), at which both *x* and *y* values are zero.

We define the coordinates of point P_1 as (x_1, y_1), and point P_2 by (x_2, y_2). The distance, the length of the con-

Distillation

Distillation is one of the most important processes for separating the components of a **solution**. The solution is heated to form a vapor of the more volatile com-

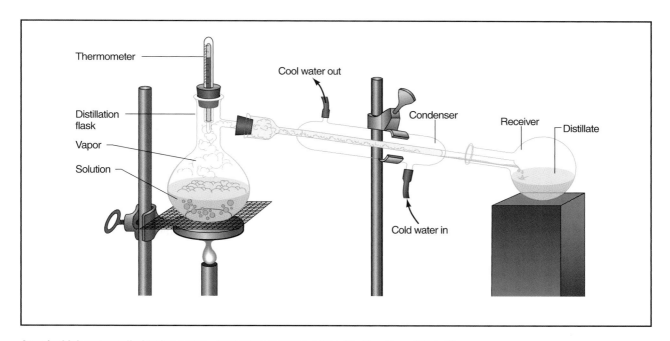

A typical laboratory distillation setup. *Illustration by Hans & Cassidy. Courtesy of Gale Group.*

ponents in the system, and the vapor is then cooled, condensed, and collected as drops of liquid. By repeating vaporization and condensation, individual components in the solution can be recovered in a pure state. Whiskey, essences, and many pure products from the oil refinery industry are processed via distillation.

General principles

Distillation has been used widely to separate volatile components from nonvolatile compounds. The underlying mechanism of distillation is the differences in **volatility** between individual components. With sufficient **heat** applied, a gas phase is formed from the liquid solution. The liquid product is subsequently condensed from the gas phase by removal of the heat. Therefore, heat is used as the separating agent during distillation. Feed material to the distillation apparatus can be liquid and/or vapor, and the final product may consist of liquid and vapor. A typical apparatus for simple distillation used in **chemistry** laboratories is one in which the still pot can be heated with a **water**, steam, or oil bath. When liquids tend to decompose or react with **oxygen** during the course of distillation, the working **pressure** can be reduced to lower the boiling points of the substances and hence the **temperature** of the distillation process.

In general, distillation can be carried out either with or without reflux involved. For the case of single-stage differential distillation, the liquid mixture is heated to form a vapor that is in equilibrium with the residual liquid. The vapor is then condensed and removed from the system without any liquid allowed to return to the still pot. This vapor is richer in the more volatile component than the liquid removed as the bottom product at the end of the process. However, when products of much higher purity are desired, part of the condensate has to be brought into contact with the vapor on its way to the condenser and recycled to the still pot. This procedure can be repeated for many times to increase the degree of separation in the original mixture. Such a process is normally called "rectification."

Applications

Distillation has long been used as the separation process in the chemical and **petroleum** industries because of its reliability, simplicity, and low-capital cost. It is employed to separate **benzene** from toluene, methanol or **ethanol** from water, **acetone** from **acetic acid**, and many multicomponent mixtures. Fractionation of crude oil and the production of **deuterium** also rely on distillation.

Today, with 40,000 distillation towers in operation, distillation makes about 95% of all current industrial

---

KEY TERMS

Activity coefficient—The ratio of the partial pressure of a component in the gas phase to the product of its mole fraction in the liquid phase and its vapor pressure as a pure liquid which is an important factor encountered in many vapor-liquid separation processes.

Bubble point—For a saturated liquid, because any rise in temperature will form bubbles of vapor, the liquid is said at its bubble point.

Dew point—The point at which air or a gas begins to condense to a liquid.

Differential distillation—During distillation, only a very small portion of the liquid is flashed each time and the vapor formed on boiling the liquid is removed at once from the system.

Distillate—The product withdrawn continuously at the top of the distillation column.

Reflux—Part of the condensate from the condenser is returned to the top tray of the distillation column as reflux to provide liquid flow above the feed point for increasing separation efficiency.

---

separation processes; however, distillation systems also have relatively high **energy** consumption. Significant effort, therefore, has been made to reduce the energy consumption and to improve the efficiency in distillation systems. This includes incorporating new analytical sensors and reliable hardware into the system to achieve advanced process control, using heat rejected from a condenser of one column to reboil other columns, and coupling other advanced process such as adsorption and crystallization with distillation to form energy-saving hybrid operation systems.

Pang-Jen Kung

Distributive property

The distributive property states that the **multiplication** "distributes" over **addition**. Thus a $\times$ (b + c) = a $\times$ b + a $\times$ c and (b + c) $\times$ a = b $\times$ a + c $\times$ a for all real or **complex numbers** a, b, and c.

The distributive property is behind the common multiplication **algorithm**. For example, 27 $\times$ 4 means 4 $\times$ (2 tens + 7 ones). To complete the multiplication, you

---

A forest in Homestead, Florida, that was destroyed by Hurricane Hugo. *JLM Visuals. Reproduced by permission.*

use the distributive property: $4 \times (20 + 7) = (4 \times 20) + (4 \times 7) = 80 + 28 = 108$.

We use the distributive property more than once in carrying out such computations as $(3x + 4)(x + 2)$. Thus $(3x + 4)(x + 2) = (3x + 4)x + (3x + 4)2$ where $3x + 4$ is "distributed over" $x + 2$ and then $(3x + 4)x + (3x + 4)2 = 3x^2 + 4x) + (6x + 8) = 3x^2 + 10x + 8$ where x and 2 are "distributed" over $3x + 4$.

Disturbance, ecological

In the ecological context, disturbance is regarded as an event of intense environmental stress occurring over a relatively short period of **time** and causing large changes in the affected **ecosystem**. Disturbance can result from natural causes or from the activities of humans.

Disturbance can be caused by physical stressors such as volcanic eruptions, hurricanes, tornadoes, earthquakes, and over geological time, glacial advance, and retreat. Humans can also cause physical disturbances, for example, through construction activities. **Wildfire** is a type of chemical disturbance caused by the rapid **combustion** of much of the **biomass** of an ecosystem and often causing mortality of the dominant **species** of the community such as trees in the case of a forest fire. Wildfires can ignite naturally, usually through a **lightning** strike, or humans can start the blaze. Sometimes fires are set deliberately as a management activity in **forestry** or agriculture. Events of unusually severe **pollution** by toxic chemicals, **nutrients**, or **heat** may also be regarded as a type of disturbance if they are severe enough to result in substantial ecological damages. Disturbance can also be biological, as when a severe infestation of defoliating **insects** causes substantial mortality of trees in a forest, or of **crops** in agriculture. The harvesting of **forests** and other ecosystems by humans is another type of biological disturbance.

Ecologic disturbance can occur at a variety of spatial scales. The most extensive disturbances involve landscape-scale events, such as glaciation, which can affect entire continents. Tornadoes, hurricanes and wildfires can also affect very large areas; sometimes wildfires extend over millions of acres.

Some disturbances, however, are much more local in their effects. For example, the primary disturbance regime in **old-growth forests** is associated with the death of **individual**, large trees caused by **disease**, insect attack, or a lightning strike. This sort of microdisturbance event results in a gap in the otherwise closed forest canopy, which lets direct **light** reach the forest floor. This encourages the development of a different set of **plant** and **animal** communities than those usually found on the dark, moist forest floor. Further ecological changes occur when the dead **tree** falls to the ground and slowly rots. Diverse processes of ecological recovery occur in response to the within-stand patch dynamics associated with the deaths of large trees in old-growth forests.

Whenever an ecosystem is affected by a substantial disturbance event, individuals and even entire species may be weakened or killed off. Other ecological damages can also occur, such as changes in hydrologic processes or **soil contamination**. However, once the actual disturbance event is finished, a process known as **succession** begins, which may eventually produce a similar ecosystem to the one that existed prior to the disturbance.

In a number of regions around the world, human activities are producing dramatic ecological disturbances. Clear-cutting of tropical rainforests, the damming or polluting of **rivers** and streams, the introduction of various chemicals and particulates into the atmosphere from industrial facilities are all human processes that have major effects on many ecosystems. In cases where the ecological disturbance is ongoing, succession is forestalled, and the damaged ecosystems may fail to recover their complex and sophisticated functions.

See also Stress, ecological.

Diurnal cycles

Diurnal cycles refer to patterns within about a 24-hour period that typically reoccur each day. Most daily cycles are caused by the **rotation** of **Earth**, which spins once around its axis about every 24 hours. The term diurnal comes from the Latin word *diurnus,* meaning daily. Diurnal cycles such as **temperature** diurnal cycles, diurnal **tides**, and solar diurnal cycles affect global processes.

A temperature diurnal cycle is composed of the daily rise and fall of temperatures. The daily rotation of Earth causes the progression of daytime and nighttime, and the amount of solar insolation (i.e., the amount of sunlight falling on a given area). Insolation fluctuations result in changes in both air and surface temperatures. Except in unusual terrain, the daily maximum temperature generally occurs between the hours of 2 P.M. and 5 P.M. and then continually decreases until sunrise the next

day. The **angle** of the **Sun** to the surface of Earth increases until around noon when the angle is the largest (i.e., the sunlight most direct). The intensity of the Sun increases with the Sun's angle, so that the Sun is most intense around noon. However, there is a **time** difference between the daily maximum temperature and the maximum intensity of the Sun, called the lag of the maximum. This discrepancy occurs because air is heated predominantly by reradiating **energy** from Earth's surface. Although the Sun's intensity decreases after 12 P.M., the energy trapped within Earth's surface continues to increase into the afternoon and supplies **heat** to Earth's atmosphere. The reradiating energy lost from Earth must surpass the incoming solar energy in order for the air temperature to cool.

Diurnal tides are the product of one low tide and one high tide occurring roughly within a 24-hour period.

Earth experiences varying hours of daylight due to the solar diurnal cycle. Solar diurnal cycles occur because Earth's axis is tilted 23.5 degrees and is always pointed towards the North **Star**, Polaris. The tilt of Earth in conjunction with the **Earth's rotation** around the Sun affects the amount of sunlight Earth receives at any location on Earth.

Diurnal cycles are of increasing interest to biologists and physicians. A number of physiological and behavioral functions are correlated to diurnal cycles. For example, the release of the cortical **hormones** is controlled by adrenocorticotropic (ACTH) from the anterior pituitary gland. The level of ACTH has a diurnal periodicity, that is, it undergoes a regular, periodic change during the 24-hour time period. ACTH **concentration** in the **blood** rises in the early morning, peaks just before awaking, and reaches its lowest level shortly before **sleep**.

Division

Division is the mathematical operation that is the inverse of **multiplication**. If one multiplies 47 by 92 then divides by 92, the result is the original 47. In general, (ab)/b = a. Likewise, if one divides first then multiplies, the two operations nullify each other: (a/b)b = a. This latter relationship can be taken as the definition of division: a/b is a number which, when multiplied by b, yields a.

In the real world using ordinary **arithmetic**, division is used in two basic ways. The first is to partition a quantity of something into parts of a known size, in which case the quotient represents the number of parts, for example, finding how many three-egg omelets can be made from a dozen eggs. The second is to share a quantity among a known number of shares, as in finding how many eggs will be available per omelet for each of five people. In the latter case the quotient represents the size of each share. For omelets, if made individually you could use two eggs each for the five people and still have two left over, or you could put all the eggs in one bowl, so each person would get 2⅔ eggs.

The three components of a division situation can represent three distinct categories of things. While it would not make sense to add dollars to earnings-per-share, one can divide dollars by earnings-per-share and have a meaningful result (in this case, shares). This is true, too, in the familiar rate-distance-time relationship R = D/T. Here the categories are even more distinct. **Distance** is measured with a tape; **time** by a clock; and **rate** is the result of these two quantities.

Another example would be in preparing a quarterly report for share holders; a company treasurer would divide the total earnings for the quarter by the number of shares in order to compute the earnings-per-share. On the other hand, if the company wanted to raise $6,000,000 in new capital by issuing new shares, and if shares were currently selling for $18⅛, the treasurer would use division to figure out how many new shares would be needed, i. e., about 330,000 shares.

Division is symbolized in two ways, with the symbol ÷ and with a bar, horizontal or slanted. In a/b or a ÷ b, a is called the dividend; b, the divisor; and the entire expression, the quotient.

Division is not commutative; 6/4 is not the same as 4/6. It is not associative; (8 ÷ 4) ÷ 2 is not the same as 8 ÷ (4 ÷ 2). For this reason care must be used when writing expressions involving division, or interpreting them. An expression such as

$$\frac{\frac{3}{4}}{7}$$

is meaningless. It can be given meaning by making one bar noticably longer than the other

$$\frac{\frac{3}{4}}{7}$$

to indicate that 3/4 is to be divided by 7. The horizontal bar also acts as a grouping symbol. In the expressions

$$\frac{14 - 7}{8 + 2} \qquad \frac{3/4}{7} \qquad \frac{x^2 - 1}{x + 1}$$

The division indicated by the horizontal bar is the last operation to be performed.

In computing a quotient one uses an **algorithm**, which finds an unknown multiplier digit by digit or term by term.

$$4\overline{\smash{\big)}\,3.00} \qquad x + 1\overline{\smash{\big)}\,x^2 \quad -1}$$

In the algorithm on the left, one starts with the digit 7 (actually 0.7) because it is the biggest digit one can use so that 4×7 is 30 or less. That is followed by 5 (actually 0.05) because it is the biggest digit whose product with the divisor equals what remains of the dividend, or less. Thus one has found (.7 + 0.05) which, multiplied by 4 equals 3. In the algorithm on the right, one does the same thing, but with **polynomials**. One finds the polynomial of biggest **degree** whose product with the divisor is equal to the dividend or less. In the case of polynomials, "less" is measured by the degree of the polynomial remainder rather than its numerical value. Had the dividend been $x^2 - 4$, the quotient would still have been x - 1, with a remainder of -3, because any other quotient would have left a remainder whose degree was greater than or equal to that of the divisor.

These last two examples point out another way in which division is a less versatile operation than multiplication. If one is working with **integers**, one can always multiply two of them an have an integer for a result. That is not so with division. Although 3 and 4 are integers, their quotient is not. Likewise, the product of two polynomials is always a polynomial, but the quotient is not. Occasionally it is, as in the example above, but had one tried to divide $x^2 - 4$ by x + 1, the best one could have done would have been to find a quotient and remainder, in this case a quotient of x - 1 and a remainder of -3. Many sets that are closed with respect to multiplication (i.e., multiplication can always be completed without going outside the set) are not closed with respect to division.

One number that can never, ever be used as a divisor is **zero**. The definition of division says that (a/b)b = a, but the multiplicative property of zero says that $(a/b) \times 0 = 0$. Thus, when one tries to divide a number such as 5 by zero, one is seeking a number whose product with 0 is 5. No such number exists. Even if the dividend were zero as well, division by zero would not work. In that case one would have (0/0)0 = 0, and 0/0 could be any number whatsoever.

Unfortunately division by zero is a trap one can fall into without realizing it. If one divides both sides of the equation $x^2 - 1 = 0$ by x - 1, the resulting equation, x + 1 = 0, has one root, namely -1. The original equation had two roots, however, -1 and 1. Dividing by x - 1 caused one of the roots to disappear, specifically the root that made x - 1 equal to zero.

The division algorithm shown above on the left converts the quotient of two numbers into a decimal, and if the division does not come out even, it does so only approximately. If one uses it to divide 2 by 3, for instance, the quotient is 0.33333... with the 3s repeating indefinitely. No matter where one stops, the quotient is a little too small. To arrive at an exact quotient, one must use fractions. Then the "answer," a/b, looks exactly like the "problem," a/b, but since we use the bar to represent both division and the separator in the **ratio** form of a **rational number**, that is the way it is.

The algorithm for dividing rational numbers and leaving the quotient in ratio form is actually much simpler. To divide a number by a number in ratio form, one simply multiplies by its **reciprocal**. That is, (a/b) ÷ (c/d) = (a/b)(d/c).

Resources

Books

Bittinger, Marvin L., and Davic Ellenbogen. *Intermediate Algebra: Concepts and Applications.* 6th ed. Reading, MA: Addison-Wesley Publishing, 2001.

Eves, Howard Whitley. *Foundations and Fundamental Concepts of Mathematics.* NewYork: Dover, 1997.

Grahm, Alan. *Teach Yourself Basic Mathematics.* Chicago: Mc-Graw-Hill Contemporary, 2001.

Weisstein, Eric W. *The CRC Concise Encyclopedia of Mathematics.* New York: CRC Press, 1998.

J. Paul Moulton

DNA *see* **Deoxyribonucleic acid (DNA)**

DNA fingerprinting

Genetic, genomic, or DNA fingerprinting is the term applied to a range of techniques that are used to show similarities and dissimilarities between the DNA present in different individuals.

Genetic fingerprinting is an important tool in the arsenal of forensic investigators. Genetic fingerprinting allows for positive identification, not only of body remains, but also of suspects in custody. Genetic fingerprinting can also link suspects to physical evidence.

Sir Alec Jeffreys at the University of Leicester developed DNA fingerprinting in the mid 1980s. The sequence of nucleotides in DNA is similar to a fingerprint, in that it is unique to each person. DNA fingerprinting is used for identifying people, studying populations, and forensic investigations.

The mechanics of genetic fingerprinting

The nucleus of every **cell** in the human body contains deoxyribonucleic acid or DNA, a biochemical **molecule** that is made up of nearly three-billion nucleotides. DNA consists of four different nucleotides, adenine (A), thymine (T), guanine (G), and cytosine (C), which are strung together in a sequence that is unique to every **individual**. The sequence of A, T, G, and C in human DNA can be found in more combinations or variations than there are humans. The technology of DNA fingerprinting is based on the assumption that no two people have the same DNA sequence.

The DNA from a small **sample** of human **tissue** can be extracted using biochemical techniques. Then the DNA can be digested using a series of enzymes known as restriction enzymes, or restriction endonucleases. These molecules can be thought of as chemical scissors, which cut the DNA into pieces. Different endonucleases cut DNA at different parts of the nucleotide sequence. For example, the endonuclease called SmaI cuts the sequence of nucleotides CCCGGG between the third cytosine (C) and the first guanine (G).

After being exposed to a group of different restriction enzymes, the digested DNA undergoes gel **electrophoresis**. In this biochemical analysis technique, test samples of digested DNA are placed in individual lanes on a sheet of an agarose gel that is made from seaweed. A separate lane contains control samples of DNA of known lengths. The loaded gel is then placed in a liquid bath and an **electric current** is passed through the system. The various fragments of DNA are of different sizes and different electrical charges. The pieces move according to their size and charge with the smaller and more polar ones traveling faster. As a result, the fragments migrate down the gel at different rates.

After a given amount of **time**, the electrical current in the gel electrophoresis instrumentation is shut off. The gel is removed from the bath and the DNA is blotted onto a piece of nitrocellulose **paper**. The DNA is then visualized by the application of radioactive probe that can be picked up on a piece of x-ray film. The result is a film that contains a series of lines showing where the fragments of DNA have migrated. Fragments of the same size in different lanes indicate the DNA has been broken into segments of the same size. This demonstrates a similarity between the sequences under test.

Different enzymes produce different banding patterns and normally several different endonucleases are used in conjunction to produce a high definition banding pattern on the gel. The greater the number of enzymes used in the digestion, the finer the resultant resolution.

In genetic or DNA fingerprinting, scientists focus on segments of DNA in which nucleotide sequences vary a great deal from one individual to another. For example, 5–10% of the DNA molecule contains regions that repeat the same nucleotide sequence many times, although the number of repeats varies from person to person. Jeffreys targeted these long repeats called variable number of tandem repeats (VNTRs) when he first developed DNA fingerprinting. The DNA of each person also has different restriction fragment sizes, called restriction fragment length polymorphisms (RFLPs), which can be used as markers of differences in DNA sequences between people. Today, technicians also use short tandem repeats (STRs) for DNA fingerprinting. STRs are analyzed using polymerase chain reaction or **PCR**, a technique for mass-producing sequences of DNA. PCR allows scientists to work with degraded DNA.

Genetic fingerprinting as a forensic tool

Genetic fingerprinting is now an important tool in the arsenal of forensic chemists. It is used in forensics to examine DNA samples taken from a crime scene and compare them to those of a suspect. Criminals almost always leave evidence of their identity that contains DNA at the crime scene—hair, **blood**, semen, or saliva. These materials can be carefully collected from the crime scene and fingerprinted

Although DNA fingerprinting is scientifically sound, the use of DNA fingerprinting in courtrooms remains controversial. There are several objections to its use. Lawyers who misrepresent the results of DNA fingerprints may confuse jurors. DNA fingerprinting relies on the probability that individuals will not produce the same banding pattern on a gel after their DNA has been fingerprinted. Establishing this probability relies on population **statistics**. Each digested fragment of DNA is given a probability value. The value is determined by a formula relating the combination of sequences occurring in the

population. There is concern that not enough is known about the distribution of banding patterns of DNA in the population to express this formula correctly. Concerns also exist regarding the data collection and laboratory procedure associated with DNA fingerprinting procedures. For example, it is possible that cells from a laboratory technician could be inadvertently amplified and run on the gel. However, because each person has a unique DNA sequence and this sequence cannot be altered by **surgery** or physical manipulation, DNA fingerprinting is an important tool for solving criminal cases.

Historical uses of genetic fingerprinting

Jeffreys was first given the opportunity to demonstrate the power of DNA fingerprinting in March of 1985 when he proved a boy was the son of a British citizen and should be allowed to enter the country. In 1986, DNA was first used in forensics. In a village near Jeffreys' home, a teenage girl was assaulted and strangled. No suspect was found, although body fluids were recovered at the crime scene. When another girl was strangled in the same way, a 19-year-old caterer confessed to one murder but not the other. DNA analysis showed that the same person committed both murders, and the caterer had falsely confessed. Blood samples of 4,582 village men were taken, and eventually the killer was revealed when he attempted to bribe someone to take the test for him.

The first case to be tried in the United States using DNA fingerprinting evidence was of African-American Tommie Lee Edwards. In November 1987, a judge did not permit population **genetics** statistics that compared Edwards to a representative population. The judge feared the jury would be overwhelmed by the technical information. The trial ended in a mistrial. Three months later, Andrews was on trial for the assault of another woman. This time the judge did permit the evidence of population genetics statistics. The prosecutor showed that the probability that the chance that Edwards' DNA would not match the crime evidence was one in 10 billion. Edwards was convicted.

DNA fingerprinting has been used repeatedly to identify human remains. In Cardiff, Wales, skeletal remains of a young woman were found, and a medical artist was able to make a model of the girl's face. She was recognized by a social worker as a local run-away. Comparing the DNA of the femur of the girl with samples from the presumptive parents, Jeffreys declared a match between the identified girl and her parents. In Brazil, Wolfgang Gerhard, who had drowned in a boating accident, was accused of being the notorious Nazi of Auschwitz, Josef Mengele. Disinterring the bones, Jeffreys and his team used DNA fingerprinting to conclude that the man actually was the missing Mengele.

In addition to forensics, Genetic fingerprinting has been used to unite families. In 1976, a military junta in a South American country killed over 9,000 people, and the orphaned children were given to military couples. After the regime was overthrown in 1983, Las Abuelas (The Grandmothers) determined to bring these children to their biological families. Using DNA fingerprinting, they found the families of over 200 children.

DNA has been used to solve several historical mysteries. On July 16, 1918, the czar of Russia and his family were shot, doused with **sulfuric acid**, and buried in a mass grave. In 1989, the site of burial was uncovered, and bone fragments of nine skeletons were assembled. Genetic fingerprinting experts from all over the world pieced together the puzzle that ended in a proper burial to the Romanov royal family in Saint Petersburg in 1998.

See also Amino acid; Gene; Genetic engineering; Forensic science.

Resources

Books

Griffiths, A., et al. *Introduction to Genetic Analysis.* 7th ed. New York: W.H. Freeman and Co., 2000.

Jorde, L.B., J. C. Carey, M. J. Bamshad, and R. L. White. *Medical Genetics.* 2nd ed. Mosby-Year Book, Inc., 2000.

Klug, W., and M. Cummings. *Concepts of Genetics.* 6th ed. Upper Saddle River: Prentice Hall, 2000.

Watson, J.D., et al. *Molecular Biology of the Gene.* 4th ed. Menlo Park, CA: The Benjamin/Cummings Publishing Company, Inc., 1987.

Other

The University of Washington. "Basics of DNA fingerprinting." [cited March 4, 2003] <http://www.biology.washington.edu/fingerprint/ dnaintro.html>.

DNA replication

DNA, short for deoxyribonucleic acid, is a double-stranded, helical **molecule** that forms the molecular basis for heredity. For DNA replication to occur, this molecule must first unwind, or "unzip," itself to allow the information-encoding bases to become accessible. The base pairing within DNA is of a complementary nature and, consequently, when the molecule unzips, due to the action of enzymes, two strands are temporarily produced, each of which acts as a template. A replication fork is first made—the DNA molecule separates at a small region and then the **enzyme** DNA polymerase adds complementary nucleotides to each side of the freshly separated strands. The DNA polymerase adds nucleotides only to one end of the DNA. As a result, one strand (the leading strand) is replicated continuously, while the other strand (the lagging strand) is replicated discontinuously, in short bursts. Each of these small sections is finally joined to its neighbor by the action of another enzyme, DNA ligase, to give a complete strand. This whole process gives rise to two completely new and identical daughter strands of DNA.

In the semi-conservative method, two strands of the parent molecule unwind and each becomes a template for the synthesis of the complementary strand of the daughter molecule. A competing hypothesis, which would eventually be disproved, was the conservative hypothesis that states no unzipping occurs and a new DNA molecule is formed alongside the original parent molecule. Consequently, of the two molecules of DNA produced after a round of replication, one of them is the intact parent molecule. By using radioactively labeled **nitrogen** to produce new DNA over several generations of **cell** replication by a bacillus **species**, all of the DNA in the daughter cells contained labeled nitrogen. The bacilli were then placed in media containing unlabeled nitrogen. After a further round of DNA replication the DNA was examined and it was found to contain equal amounts of labeled and unlabeled nitrogen. In the second generation two types of DNA were found—half was identical to the DNA from the first generation and the remaining half was found to consist of entirely unlabeled nitrogen. These results are consistent with the zip fastener model of the semi-conservative hypothesis, but not at all consistent with the conservative hypothesis. Thus, it was shown that DNA replication proceeds via the semi-conservative replication method.

This method of replication, known as the semi-conservative hypothesis, was proposed from the outset of the discovery, with the description of the structure of DNA by biochemists James D. Watson and Francis Harry Compton Crick in 1953. In 1957, biochemist Arthur Kornberg first produced new DNA from the constituent parts and a parent strand, forming synthetic but not biologically active molecules of DNA outside the cell. However, not until the work of Matthew Meselson and Franklin W. Stahl in the late 1950s was the semi-conservative hypothesis conclusively proven true.

DNA synthesis

Deoxyribonucleic acid (DNA) synthesis is a process by which strands of nucleic acids are created. In a **cell**, DNA synthesis takes place in a process known as replication. Using **genetic engineering** and **enzyme chemistry**, scientists have also developed man-made methods for synthesizing DNA.

The DNA **molecule** was discovered by Francis Crick, James Watson, and Maurice Wilkins. In 1953, Watson and Crick used **x-ray crystallography** data from Rosalind Franklin to show that the structure of DNA is a **double helix**. For this work, Watson, Crick, and Wilkins received the Nobel Prize for **physiology** or medicine in 1962.

To understand how DNA is synthesized, it is important to understand its structure. DNA is a long chain **polymer** made up of chemical units called nucleotides. It is the genetic material in most living organisms that carries information related to protein synthesis. Typically, DNA exists as two chains of nucleotides that are chemically linked following base pairing rules. Each nucleotide is made up of a deoxyribose sugar molecule, a phosphate **group**, and one of four **nitrogen** containing bases. The bases include the purines adenine (A) and guanine (G), and the pyrimidines thymine (T) and cytosine (C). In DNA, adenine generally links with thymine and guanine with cytosine. The chains are arranged in a double helical structure that is similar to a twisted ladder or **spiral** staircase. The sugar portion of the molecule makes up the sides of the ladder and the bases compose the rungs. The phosphate group holds the whole structure together by connecting the sugars. The order in which the nucleotides are linked is known as the sequence that is determined by DNA **sequencing**.

In a eukaryotic cell, DNA is synthesized prior to **cell division** by a process called replication. At the start of replication the two strands of DNA are separated by various enzymes. Each strand then serves as a template for producing a new strand. Replication is catalyzed by an enzyme known as DNA polymerase. This molecule brings complementary nucleotides to each of the DNA strands. The nucleotides connect to form new DNA strands, which are exact copies of the original strand known as daughter strands. Since each daughter strand contains half of the parent DNA molecule, this process is known as semi-conservative replication. The process of replication is important because it provides a method for cells to transfer an exact duplicate of their genetic material from one generation of cell to the next.

After the nature of DNA was determined, scientists began to examine the cellular genes. When a certain **gene** was isolated, it became desirable to synthesize copies of that molecule. One of the first ways in which a large amount of a specific DNA was synthesized was though genetic **engineering**. Genetic engineering begins by combining a gene of interest with a bacterial plasmid. A plasmid is a small stretch of DNA that is found in many **bacteria**. The resulting **hybrid** DNA is called **recombinant DNA**. This new recombinant DNA plasmid is then injected into a bacterial cell. The cell is then cloned by allowing it to grow and multiply in a culture. As the cells multiply so do the copies of the inserted gene. When the bacteria has multiplied enough, the multiple copies of the inserted gene can then be isolated. This method of DNA synthesis can produce billions of copies of a gene in a couple of weeks.

In 1985, researchers developed a new process for synthesizing DNA called polymerase chain reaction (**PCR**). This method is much faster than previous known methods producing billions of copies of a DNA strand in just a few hours. It begins by putting a small section of double stranded DNA in a **solution** containing DNA polymerase, nucleotides, and primers. The solution is heated to separate the DNA strands. When it is cooled, the polymerase creates a copy of each strand. The process is repeated every five minutes in an automated machine until the desired amount of DNA is produced.

DNA technology

DNA technology has revolutionized modern science. **Deoxyribonucleic acid (DNA)**, or an organism's genetic material—inherited from one generation to the next—holds many clues that have unlocked some of the mysteries behind human **behavior**, **disease**, **evolution**, and aging. As technological advances lead to a better understanding of DNA, new DNA-based technologies will emerge. Recent advances in DNA technology including cloning, **PCR, recombinant DNA** technology, **DNA fingerprinting**, **gene therapy**, DNA microarray technology, and DNA profiling have already begun to shape medicine, forensic sciences, environmental sciences, and national security.

In 1956, the structure and composition of DNA was elucidated and confirmed previous studies more than a decade earlier demonstrating DNA is the genetic material that is passed down from one generation to the next. A novel tool called PCR (polymerase chain reaction) was developed not long after DNA was descovered. PCR represents one of the most significant discoveries or inventions in DNA technology and it lead to a 1993 Nobel Prize award for American born Kary Mullis (1949–).

PCR is the amplification of a specific sequence of DNA so that it can be analyzed by scientists. Amplification is important, particularly when it is necessary to analyze a small sequence of DNA in quantities that are large enough to perform other molecular analyses such as DNA **sequencing**. Not long after PCR technology was developed, **genetic engineering** of DNA through recombinant DNA technology quickly became possible. Recombinant DNA is DNA that has been altered using bacterial derived enzymes called restriction endonucleases that act like scissors to cut DNA. The pattern that is cut can be matched to a pattern cut by the same enzymes from a different DNA sequence. The sticky ends that are created bind to each other and a DNA sequence can therefore be inserted into another DNA sequence.

Restriction endonucleases are also important in genetic fingerprinting. In this case, enzymes that recognize specific DNA sequences can produce fragments of DNA by cutting different parts of a long strand of DNA. If there are differences in the sequence due to inherited variation—meaning that there are extra DNA or specific sequences altered such that the restriction enzymes no longer recognize the site, variable patterns can be produced. If these patterns are used to compare two different people, they will have a different fragment pattern or fingerprint. Genetic fingerprinting can be used to test for paternity. In forensics, genetic fingerprinting can be used to identify a criminal based on whether their unique DNA sequence matches to DNA extracted from a crime scene. This technology can also allow researchers to produce genetic maps of chromosomes based on these restriction **enzyme** fingerprints. Becasue there are many different enzymes, many different fingerprints can be ascertained.

Recombinant DNA technology can also be applied to splicing genes into molecular devices that can transport these genes to various cellular destinations. This technique, also called **gene** therapy, has been used to deliver corrected genes into individuals that have defective genes that cause disease. **Gene splicing** has also been applied to the environment as well. Various **bacteria** have been genetically modified to produce **proteins** that break down harmful chemical contaminants such as DDT. Currently, scientists are investigating the application of this technology to produce genetically engineered plants and **crops** that can produce substances that kill **insects**. Similarly, **fruits** can be engineered to have genes that produce proteins that slow the ripening process in an effort to extend their shelf life.

DNA microarray technology, also known as the DNA chip, is the latest in **nanotechnology** that allows researchers the have ability to study the **genome** in a high throughput manner. It can be used for gene expression profiling which gives scientists insights into what genes are being up or down-regulated. Various genetic profiles can be determined in order to estimate **cancer** risk or to identify markers that may be associated with disease. It has the ability only to detect changes in gene expression that are large enough to be detected above a baseline

KEY TERMS

Cells—The smallest living units of the body which together form tissues.

Deoxyribonucleic acid (DNA)—The genetic material in a cell.

DNA—Deoxyribonucleic acid; the genetic material in a cell.

Enzyme—Biological molecule, usually a protein, which promotes a biochemical reaction but is not consumed by the reaction.

Gene—A discrete unit of inheritance, represented by a portion of DNA located on a chromosome. The gene is a code for the production of a specific kind of protein or RNA molecule, and therefore for a specific inherited characteristic.

Genome—The complete set of genes an organism carries.

Protein—Macromolecules made up of long sequences of amino acids.

Recombinant DNA—DNA that is cut using specific enzymes so that a gene or DNA sequence can be inserted.

level. Therefore, it does not detect subtle changes in gene expression that might cause disease or play a role in the development of disease. It can also be used for genotyping, although clinical diagnostic genotyping using microarray technology is still being investigated.

Genes from other **species** can also be used to add new traits to a particular **organism**. For example, bacteria, **mice**, and plants have all had luminescent (**light** glowing) genes from jelly fish added to their genomes. Another reason for adding genes to a foreign organism is to manufacture various nutritional or pharmaceutical products. Some cows have been modified so that they can produce human **insulin** or vitamins in their milk in bulk. **Pigs** have been modified to overcome a number of transplantation problems so that some limited transplantation of organs can be carried out from pigs to humans, also called xenotransplation.

DNA technology is a relatively new area of research with enormous controversy. It will likely continue to be a large part of public debate and have an impact on every aspect of medical diagnostics, therapeutics, forensics, and genetic profiling.

See also Chromosome; Genetics; DNA replication; DNA synthesis; DNA technology; Molecular biology.

Resources

Books

Nussbaum, Robert L., Roderick R. McInnes, and Huntington F. Willard. *Genetics in Medicine.* Philadelphia: Saunders, 2001.

Rimoin, David L., *Emery and Rimoin's Principles and Practice of Medical Genetics.* London; New York: Churchill Livingstone, 2002.

Bryan Cobb

DNA vaccine

The use of a **vaccine** constructed of a protein has traditionally been to induce the formation of an antibody to the particular protein. Antibodies are crucial to an organism's attempt to stop an **infection** caused by a microorganism.

In the early 1990s, scientists observed that plasmid DNA (DNA that is present in **bacteria** that is not part of the main body of DNA) could affect test animals. Work began on constructing vaccines that were made of DNA instead of protein.

Instead of injecting a protein into the body to induce antibody formation, the stretch of DNA that codes for the antigen is injected into the body. When the DNA is expressed in the body, the antigen is produced and antibody formation occurs.

Another approach that can be taken with DNA vaccines is the use DNA that codes for a vital component of the disease-causing microorganism. Antibodies that form to that protein will also attack the infecting **microorganisms**.

The DNA can be injected in a **salt** solution using a hypodermic syringe. Alternatively, the DNA can be coated onto gold beads, which are then propelled at high speed into the body using an apparatus dubbed the **gene** gun. The production of the actual protein that stimulates antibody formation inside the body eliminates the chances of infection, as can occur with some viral vaccines that use intact and living viruses.

DNA vaccines have several advantages over the traditional vaccine methods. The immune response produced is very long, as the inserted DNA continues to be expressed and code for the production of protein. So, the use of booster injections to maintain immunity is not required. Second, a single injection can contain multiple DNA sequences, providing multiple vaccines. This advantage is especially attractive for childhood vaccinations, which currently require up to 18 visits to the physician over a decade. DNA vaccines are stable, even after a long **time** at room **temperature**. Finally, as the

proteins that are crucial to diseases are deciphered, the genes for those proteins can be used in DNA vaccines.

The injection of DNA does not damage the host's genetic material. So far, the presence of the added DNA has not stimulated an immune response of a host against its own components. This reaction, termed an autoimmune response, could be possible if some host DNA coded for a protein that was very similar to the protein coded for by the added DNA.

As encouraging as these results are, the technology does have limitations. For example, so far immunity can develop only to protein. Many infections in the body occur within a covering of sugary material, which makes the underlying protein components of the infecting microbe almost invisible to the **immune system**.

Despite the limitations, DNA vaccines against diseases show promise. As of 2002, a vaccine against infectious hematopoietic necrosis **virus** in **salmon** and trout is being tested. The viral infection causes the deaths of large numbers of commercially raised salmon and trout, and is an economic problem for commercial fisheries. Humans have displayed immune responses to diarrhea-causing viruses, malarial **parasites** and **tuberculosis** using DNA vaccines. Indeed, it is hoped that a DNA based **malaria** vaccine will be available by 2005. Recently, DNA vaccines against measles and **rabies** have been shown to be protective against the two microbiological diseases. Protection of **monkeys** against HIV has been achieved, and human trials were scheduled to begin by 2003 in Nairobi on a DNA **AIDS** vaccine.

See also Nucleic acid.

Resources

Books

Paterson, Y. *Intracellular Bacterial Vaccine Vectors: Immunology, Cell Biology, and Genetics.* New York: Wiley-Liss, 1999.

Snyder, L., and W. Champness. *Molecular Genetics of Bacteria.* 2nd ed. Washington, DC: American Society for Microbiology Press, 2002.

Periodicals

Arvin, A.M., "Measles Vaccine—a Positive Step Toward Eradicating a Negative Strand." *Nature Medicine* 6 (July 2000): 744.

Epstein, S.L., T.M. Tumpey, J.A. Misplon, et al. "DNA Vaccine Expressing Conserved Influenzae Virus Proteins Protective Against H5N1 Challenge Infection in Mice." *Emerging Infectious Diseases* 8 (August 2002): 796–801.

Dobsonflies

Dobsonflies are **species** of medium- to large-sized **insects** in the order Neuroptera, family Corydalidae.

The life cycle of dobsonflies is characterized by a complete **metamorphosis**, with four developmental stages: egg, larva, pupa, and adult. Adult dobsonflies are usually found near **freshwater**, especially streams, either resting on vegetation or engaged in an awkward, fluttering flight. Sometimes adult dobsonflies can be abundant at night around lights, even far from **water**. The immature stages of dobsonflies are aquatic and are usually found beneath stones or other debris in swiftly flowing streams.

Dobsonflies have rather soft bodies. The adults of North American species generally have body lengths of 0.75-1.5 in (2-4 cm) and wing spans of 2 in (5 cm) or greater. These insects have four wings with distinctive, many-veined membranes. The wings are held tent-like over the back when the insect is at rest. Dobsonflies have piercing mouthparts. Male dobsonflies have large mandibles about three times longer than the head and projecting forward. Female dobsonflies have much smaller mandibles. Adult dobsonflies are active at night and are not believed to feed, so the function of the exaggerated mandibles of the male insects are unknown. Dobsonflies lay their eggs on vegetation near water, and the larvae enter the water soon after hatching.

Larval dobsonflies are sometimes known as hellgrammites and are predators of other aquatic **invertebrates**. Larval dobsonflies are quite large, often longer than 3 in (8 cm) or more, with distinctive, tracheal gills projecting from the segments of their large abdomen. Larval dobsonflies are sometimes used as bait for trout fishing.

Various species of dobsonflies occur in **North America**. The species *Corydalus cornutus* is common but not abundant in eastern parts of the **continent**, while the genus *Dysmicohermes* is widespread in western regions.

Dog *see* **Canines**

Dogwood tree

Dogwood refers to certain **species** of trees and shrubs in the dogwood family (Cornaceae). The dogwoods are in the genus *Cornus*, which mostly occur in temperate and boreal **forests** of the Northern Hemisphere.

Species in the dogwood family have seasonally deciduous foliage. The leaves are simple, usually untoothed, and generally have an opposite arrangement on the twig. The flowers of dogwoods develop in the early springtime, often before the leaves. The flowers are small and greenish, and are arranged in clusters at the terminus of twigs. The flowers are sometimes surrounded by whitish leaves that are modified as petal-like, showy bracts, giving the overall impression of a single, large **flower**. The fruit is a drupe, that is, a hard-seeded structure surrounded by an edible pulp.

Several North American species of dogwood achieve the size of small trees. The flowering dogwood (*Cornus florida*) of the eastern United States can grow as tall as 43 ft (13 m), and is a species of rich hardwood forests. The flowering dogwood is an attractive species often cultivated for its large and showy, white-bracted inflorescences, its clusters of scarlet **fruits**, and the purplish coloration of its autumn foliage. The Pacific dogwood (*C. nuttallii*) is a component of conifer-dominated rainforests of the Pacific coast, and is also a popular ornamental species, for reasons similar to the flowering dogwood. Other tree-sized dogwoods of the western United States include western dogwood (*C. occidentalis*) and black-fruited dogwood (*C. sessilis*), while stiff cornel dogwood (*C. stricta*) occurs in the east.

Many other species of dogwoods are shrub sized, including alternate-leaved dogwood (*C. alternifolia*) and roughleaf dogwood (*C. drummondii*) of eastern **North America**. The widespread red-osier dogwood (*C. stolonifera*) is sometimes cultivated for its attractive, red twigs, which contrast well with the snows of winter.

The bunchberry or dwarf cornel (*C. canadensis*) is a diminutive species of dogwood that grows in the ground vegetation of northern forests.

The **wood** of tree-sized dogwoods is very hard, and has a few specialized uses, for example, in the manufacturing of shuttles for fabric mills, golf-club heads, and other uses where a very durable material is required. However, the major economic benefit of dogwoods is through their attractive appearance, which is often exploited in **horticulture**.

Dogwood stems are an important food for wild animals such as rabbits, hares, and **deer** that browse on woody plants during the winter. In addition, many species of **birds** and **mammals** feed on the fruits of various species of dogwoods.

Anthracnose disease

Dogwood anthracnose is a **disease** that affects the flowering and Pacific dogwood (*Cornus florida* and *C. nuttallii*). Pacific dogwood infections have been reported in Washington, Oregon, Idaho, and British Columbia. In the eastern states, flowering dogwood infections have been reported in Massachusetts, Connecticut, New York, New Jersey, Pennsylvania, and Delaware. More recently, anthracnose has been detected in Maryland, Virginia, West Virginia, North and South Carolina, Tennessee, and Georgia.

The only way the disease can be effectively controlled is if the disease is detected before extensive dieback has occurred. The removal of diseased twigs and branches helps to reduce potential sources of infection. It may also help to remove any fallen leaves. Succulent growth, which is encouraged by high **nitrogen fertilization**, can lead to trunk canker formation. To encourage trees to grow, a balanced fertilizer may be applied in early spring.

Applications of the fungicides chlorothalonil and mancozeb in the spring protect against leaf infections. When conditions favor development of the disease later in the growing season, additional applications of fungicides may be beneficial.

Randall Frost

Dolphins *see* **Cetaceans**

Domain

The domain of a **relation** is the set that contains all the first elements, x, from the ordered pairs (x,y) that make up the relation. In **mathematics**, a relation is defined as a set of ordered pairs (x,y) for which each y depends on x in a predetermined way. If x represents an element from the set X, and y represents an element from the set Y, the Cartesian product of X and Y is the set of all possible ordered pairs (x,y) that can be formed with an element of X being first. A relation between the sets X and Y is a subset of their Cartesian product, so the domain of the relation is a subset of the set X. For example, suppose that X is the set of all men and Y is the set of all women. The Cartesian product of X and Y is the set of all ordered pairs having a man first and women second. One of the many possible relations between these two sets is the set of all ordered pairs (x,y) such that x and y are married. The set of all married men is the domain of this relation, and is a subset of X. The set of all second elements from the ordered pairs of a relation is called the range of the relation, so the set of all married women is the range of this relation, and is a subset of Y. The **variable** associated with the domain of the relation is called the independent variable. The variable associated with the range of a relation is called the dependent variable.

Many important relations in science, **engineering**, business and economics can be expressed as functions of **real numbers**. A **function** is a special type of relation in which none of the ordered pairs share the same first element. A real-valued function is a function between two

A dogwood in bloom, Georgia. *JLM Visuals. Reproduced with permission.*

The disease is caused by the anthracnose fungus, *Discula sp.* **Infection** most often occurs during cool, wet spring and fall weather, but may occur throughout the growing season. Trees weakened by **drought** or cold are most likely to suffer severely from the disease. When heavy infection occurs for several years, woodland and ornamental dogwoods frequently die.

The origin of anthracnose disease has not been determined. It may have been introduced, or it may have resulted from an altered host/parasite relationship that transformed an innocuous fungus into a significant pathogen.

Cultivated dogwoods that are well cared for are better able to withstand anthracnose during years of widespread infestation. To build up their resistance, dogwoods should be watered during periods of drought. Mulching may help conserve **water**, as well as protect trees from physical injury. Overhead watering may contribute to **leaf** infections.

sets X and Y, both of which correspond to the set of real numbers. The Cartesian product of these two sets is the familiar Cartesian coordinate system, with the set X associated with the x-axis and the set Y associated with the y-axis. The graph of a real-valued function consists of the set of points in the **plane** that are contained in the function, and thus represents a subset of the Cartesian plane. The x-axis, or some portion of it, corresponds to the domain of the function. Since, by definition, every set is a subset of itself, the domain of a function may correspond to the entire x-axis. In other cases the domain is limited to a portion of the x-axis, either explicitly or implicitly.

Example 1. Let X and Y equal the set of real numbers. Let the function, f, be defined by the equation $y = 3x^2 + 2$. Then the variable x may range over the entire set of real numbers. That is, the domain of f is given by the set $D = \{x| -\infty \leq x \geq \infty\}$, read "D equals the set of all x such that **negative infinity** is less than or equal to x and x is less than or equal to infinity."

Example 2. Let X and Y equal the set of real numbers. Let the function f represent the location of a falling body during the second 5 seconds of descent. Then, letting t represent **time**, the location of the body, at any time between 5 and 10 seconds after descent begins, is given by $f(t) = \frac{1}{2}gt^2$. In this example, the domain is explicitly limited to values of t between 5 and 10, that is, $D = \{t| 5 \leq t \geq 5\}$.

Example 3. Let X and Y equal the set of real numbers. Consider the function defined by $y = PIx^2$, where y is the area of a **circle** and x is its radius. Since the radius of a circle cannot be negative, the domain, D, of this function is the set of all real numbers greater than or equal to **zero**, $D = \{x| x \geq 0\}$. In this example, the domain is limited implicitly by the physical circumstances.

Example 4. Let X and Y equal the set of real numbers. Consider the function given by $y = 1/x$. The variable x can take on any real number value but zero, because **division** by zero is undefined. Hence the domain of this function is the set $D = \{x| x \text{ NSIME } 0\}$. Variations of this function exist, in which values of x other than zero make the denominator zero. The function defined by $y = \frac{1}{2}-x$ is an example; $x=2$ makes the denominator zero. In these examples the domain is again limited implicitly.

See also Cartesian coordinate plane.

Resources

Books

Allen, G.D., C. Chui, and B. Perry. *Elements of Calculus.* 2nd ed. Pacific Grove, CA.: Brooks/Cole Publishing Co., 1989.

Bittinger, Marvin L., and Davic Ellenbogen. *Intermediate Algebra: Concepts and Applications.* 6th ed. Reading, MA: Addison-Wesley Publishing, 2001.

Grahm, Alan. *Teach Yourself Basic Mathematics.* Chicago: McGraw-Hill Contemporary, 2001.

Swokowski, Earl W. *Pre Calculus, Functions, and Graphs, 6th ed.* Boston: PWS-KENT Publishing Co., 1990.

J. R. Maddocks

Donkeys

Domestic donkeys, members of the order Perissodactyla, are large single-hoofed horse-like **mammals** with elongated heads. Donkeys usually stand between 9.5 and 11 hands high measured at the withers, that is, 38-44 in (95-110 cm) tall. Because of the large amount of interbreeding among different donkey **species**, donkeys differ markedly in appearance. They can be brown, gray, black, roan (a mixture of white and usually brown hair), or broken colored (a combination of brown or black and white markings). Also known as **asses**, donkeys originated in **Africa** and are very well suited to hot dry climates, but are sensitive to the cold. Donkeys are intelligent, calm, and require little food in relation to the amount of work they are able to perform.

Members of the order Perissodactyla, the odd-toed **ungulates**, are medium-sized to very large animals. The third digit of their limbs is the longest, and all four of their limbs are hoofed. These fast-running herbivores have a life expectancy of around 40 years. Today, there are only three families of odd-toed ungulates, the **tapirs**, the **rhinoceros**, and the **horses**.

Donkeys belong the horse family Equidae, which has only one genus (*Equus*) and six species. The African wild ass (*Equus africanus*) is thought to be the ancestor of donkeys.

The domestic donkey can be traced back to three subspecies of African wild ass. The first subspecies, the Nubian wild ass, used to be found throughout Egypt and the Sudan. Today, these asses are very rare; in fact, only a few survive in zoos. The Nubian wild ass is a small yellow-gray **animal** with a dark strip across its shoulder and one long stripe down its back. The two stripes together are known as the cross, and can be found in many of its ancestors. The second subspecies, the North African wild ass, is now extinct. The third species, the Somali wild ass, is larger and taller than the Nubian wild ass. These asses are grayish with a pink hue and have stripes on their legs. Their manes are very dark and stand upright, and these asses have a nearly black tassel on their tails. Like their cousins, they are declining in numbers. Indeed, there are only a few hundred of these animals in Somalia and a few thousand in Ethiopia.

In around 4000 B.C., inhabitants of the Nile Valley of Egypt first domesticated descendants of the donkey, specifically, Nubian wild asses. Thus, donkeys were domesticated a long time before horses were. Later, Nubian wild asses were domesticated in Arabia and throughout Africa as well.

Eventually, the Somali wild asses were also domesticated, and the two subspecies of asses were mixed. Human use of donkeys as pack animals during wartime and for transporting tradable goods during times of peace accelerated the breeding of various subspecies of donkeys. Today's donkeys have characteristics of both Nubian and Somali wild asses.

It is thought that the Etruscans, traveling from Turkey to Italy, brought the first donkeys to **Europe** in around 2000 B.C., and that donkeys were brought to Greece by way of Turkey. In Greece, donkeys were commonly used for work in vineyards because of their sure-footedness. Soon, people throughout the Mediterranean used donkeys to help cultivate **grapes**. The Romans used donkeys throughout their empire, for pack animals and for grape cultivation, which they promoted as far north as France and Germany. The Romans also brought donkeys to Britain when they invaded.

Donkeys and horses existed in **North America** before the last **ice** age, over 10,000 years ago, but then became extinct. These did not reappear in North America until the Spanish brought them on their explorations in the 1600s. One hundred years later, the Spanish brought donkeys to **South America**.

Donkeys have a long work history. Aside from being used as pack animals and for grape cultivation, donkeys have been used to draw wagons, pull **water** from wells, and help grind grain. These animals were popular because of their efficiency and hardiness. In fact, few domestic animals require as little food as donkeys for the amount of work they can accomplish. Their diet is relatively simple; they survive very well on grass and hay. Furthermore, they can work into their old age, about 40 years. Contrary to popular myth, donkeys are cautious, brave, and very intelligent. Like their ancestors, donkeys can be aggressive if the need arises. When they are attacked, they form a circle and fend off predators by kicking and biting.

Interestingly, the name "donkey" is the word that most English speaking people have for asses. It is derived from the Old English word "dun," referring to the animals' gray-brown **color**, and the "ky," a suffix for small. Thus, the early English people used the word "dunky" to describe the pony-sized dun-colored animal.

Resources

Books

Svendsen, Elisabeth D., MBE. *The Professional Handbook of the Donkey.* Devon, England: The Sovereign Printing Group, 1989.

Kathryn Snavely

Dopamine

Dopamine is a **neurotransmitter** (a chemical used to send signals between nerve cells) in the same family as epinephrine (adrenaline). Dopamine is one of the primary neurotransmitters and it affects motor functions (movement), emotions, **learning**, and **behavior**. It was originally identified as the **brain** chemical associated with pleasure. A decrease in the amount of dopamine in specific sections of the brain has been implicated as a possible cause of Parkinson's **disease**, while an excess of dopamine in some regions of the brain has been suggested as a possible cause of **schizophrenia**. Dopamine is also thought to play a role in **depression**, attention deficit hyperactivity disorder, high **blood pressure**, and drug **addiction**. Recently, dopamine has been used as a treatment for victims of **heart** attacks.

Basic definitions and chemical information

Dopamine is one of a group of chemicals known as catecholamine neurotransmitters. Catecholamines are a group of chemicals that include epinephrine (adrenalin); **histamine**, which is responsible for many of the symptoms of allergies; and serotonin, a **molecule** that has been suggested as aiding in **sleep**. This group of compounds is sometimes collectively known as the biogenic amines. Neurotransmitters are chemicals used by the body to signal or send information between nerve cells or nerve and muscle cells. The chemical structure of dopamine is shown below. The NH_2 group on the molecule is the amine group in the term biogenic amines.

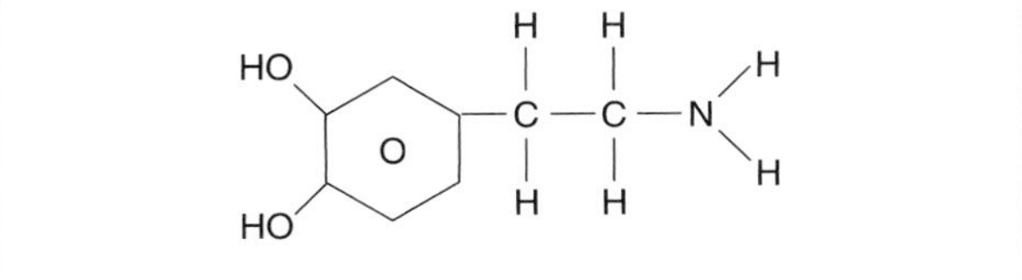

Figure 1. The chemical structure of dopamine. *Illustration by Hans & Cassidy. Courtesy of Gale Group.*

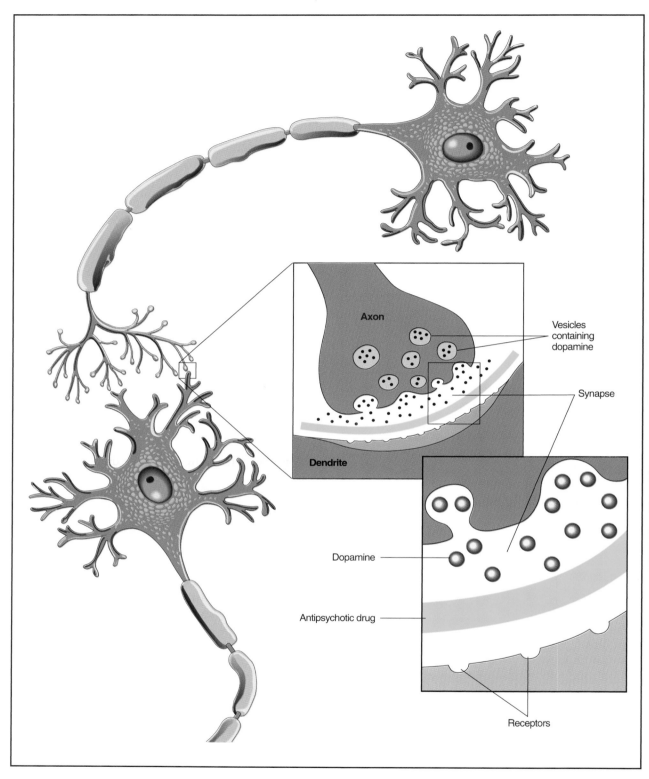

A representation of how an antipsychotic drug inhibits dopamine released by presynaptic neurons from reaching receptors on postsynaptic neurons. *Illustration by Hans & Cassidy. Courtesy of Gale Group.*

This entire group of chemicals has been implicated in depression and general moods.

Dopamine and Parkinsons disease

Parkinson disease is a disorder of the **nervous system** that is characterized by slow movements and difficulty initiating movements, a shuffle when walking, and increased muscular rigidity. It is estimated to affect as many as one million Americans and is far more prevalent in the elderly. The main cause of Parkinson disease is thought to be a lack of dopamine in a region of the brain known as the substantia nigra. Whether the cells in that area do not produce enough dopamine or whether there are too few of the dopamine-producing cells is a matter of debate and active research. A chemical known as Levodopa or L-dopa, which our bodies rapidly metabolize to dopamine, is the main treatment. Levodopa reduces the symptoms of the disease, but does not stop the progression of the disease. A lack of dopamine in some areas of the brain also has been implicated in depression.

Dopamine and schizophrenia

Schizophrenia is a form of **psychosis** or loss of contact with reality. It is estimated to affect about 1% of the population, or over 2.5 million Americans. A great deal of research is being done on the origins of schizophrenia. One widely accepted theory is that it is caused by an excess of dopamine or dopamine receptors. Receptors are **proteins** on the surfaces of cells that act as signal acceptors for the cells. They allow cells to send information, usually through neurotransmitter molecules. This hypersensitivity to dopamine (the prefix "hyper" means over or excessive) is treated by using chemicals that block (or inactivate) the receptors for the dopamine signals. However, there are a number of different types of dopamine receptors and there are many differences among individuals in the structures of these receptors. Drugs usually block all of the receptors, not just the ones related to schizophrenic symptoms, resulting in many side effects. Other approaches to the treatment of schizophrenia have focused on decreasing the amounts of dopamine in the brain. In doing so, however, symptoms of Parkinson disease often result, since less dopamine (or the ability to respond to dopamine) is present. The origins of schizophrenia are unclear; dopamine excess is probably not the sole cause of the disease as strong evidence for genetic and environmental factors exists as well. To date, treatments that focus on excess dopamine sensitivity have been the most successful.

Dopamine as heart medicine

Since dopamine can increase blood pressure, it is used as a treatment for shock (low blood pressure

throughout the body) which carries the risk of damage to major organs in patients who have suffered serious heart attacks. Dopamine raises the blood pressure and causes small blood vessels to constrict, thus raising the blood pressure throughout the body. Chemically related molecules such as adrenaline act similarly and both are often used to help patients.

Dopamine and attention deficit hyperactivity disorder

Attention deficit hyperactivity disorder (ADHD), a **syndrome** that affects as many as 3.5 million American children, and many adults as well, is characterized by an inability to pay attention, over-activity, and impulsive behaviors. ADHD has been associated with certain forms of the dopamine D4 receptor, and with **individual** differences in the **gene** that encodes the dopamine transporter, a molecule that binds and carries dopamine. ADHD often is treated with stimulatory drugs such as Ritalin, which increase the availability of dopamine in the brain.

Dopamine and drug addiction

Alcohol, **nicotine**, and a variety of other drugs including **marijuana**, **cocaine**, **amphetamines**, and heroin all appear to raise the level or the availability of dopamine in different parts of the brain. Pathways of nerve cells that produce dopamine and contain dopamine receptors are affected by all of these drugs. There is evidence that certain forms of the dopamine D4 receptor may predispose a person to drug addiction. Based on this information, researchers are attempting to develop drugs to treat addictions.

Dopamine and aging

Although individuals vary greatly in the amount of dopamine activity in their brains, in general dopamine appears to decline with age in those parts of the brain responsible for thinking. In particular, as people age, the number of dopamine D2 receptors decreases significantly. Thus, dopamine may be involved with the age-related loss of intellectual skills.

Resources

Books

Ackerman, S. *Discovering the Brain.* Washington, DC: National Academy Press, 1992.

Restak, Richard M. *Receptors.* New York: Bantam Press, 1994.

Periodicals

Bower, Bruce. "The Birth of Schizophrenia: A Debilitating Mental Illness May Take Root in the Fetal Brain." *Science News* (May 29, 1993): 346.

Concur, Bruce. "A Dangerous Pathway." *New Scientist* (July 5, 1997).

Miller, Susan. "Picking up Parkinson's Pieces." *Discover* (May 1991): 22.

Zamula, Evelyn. "Drugs Help People with Parkinson's Disease." *FDA Consumer* (January-February 1992): 28.

Louis J. Gotlib

Doppler effect

The Doppler effect was named after Johann Christian Doppler (1803-1853). This Austrian physicist observed and explained the changes in pitch and **frequency** of sound and **light** waves, as well as all other types of waves, caused by the **motion** of moving bodies. The general rule of the Doppler effect is that the wave frequencies of moving bodies rise as they travel toward an observer and fall as they recede from the point of observation.

While Doppler, in 1842, demonstrated the phenomenon named after him in the area of **sound waves**, in the same year he also predicted that light waves could be shown to exhibit the same response to the movement of bodies similar to those of sound waves.

Doppler effect in sound waves

The response of sound waves to moving bodies is illustrated in the example of the sounding of the locomotive whistle of a moving train. When the train blows its whistle while it is at rest in the station, stationary listeners who are either ahead of the engine or behind it will hear the same pitch made by the whistle, but as the train advances, those who are ahead will hear the sound of the whistle at a higher pitch. Listeners behind the train, as it pulls further away from them, hear the pitch of the whistle begin to fall.

The faster the train moves the greater will be the effect of the rising and falling of the pitch. Also, if the train remains at rest but the listeners either move toward the sounding train whistle or away from it, the effect will be the same. Those who move toward the train will hear a higher pitch, while those who travel away from the train will hear a lower pitch.

When the train is at rest it is the center of the sound waves it generates in circles around itself. As it moves forward, it ceases to be the center of the sound waves it produces. The sound waves move in the same direction of the train's motion. The train is chasing or crowding its waves up front, compressing them, so that the listener in front of the direction of its movement hears more waves per second, thus producing the effect of a higher frequency. The listener standing behind the train hears a lower pitch because the waves have spread out behind the forward motion of the train. Thus, there are fewer waves per second. The listener is now **hearing** a lower frequency than is actually being produced by the whistle.

In 1845, the Doppler effect received further confirmation in an elaborate experiment devised by a Dutch meteorologist, Christopher Heinrich Buys-Ballot. He placed a band of trumpet players on an open railroad flatcar and had it ride by listeners with perfect pitch who recorded their impressions of the notes produced by the whistle. Their written recordings of the pitches clearly demonstrated the Doppler wave effect.

Doppler effects in light waves

The Doppler effect in light waves can be observed by the spectral analysis of light emitted by luminous objects.

The light from a stationary distant object whose chemical composition is known is refracted at a specific band of light on a **spectroscope**. That band is known as its index of refraction. If the light, instead, appears at another frequency band in the spectroscope, it can be inferred from the Doppler effect that the body is in motion. When the light appears at a higher frequency band, then the body is no longer stationary but moving toward the observer. The Doppler effected light wave is displaced toward the higher frequency band, which is the blue end of the spectroscope. If the known body's light waves appear at a lower frequency band of the spectroscope, towards the red end, then the body is now in motion away from the observer.

With the use of the spectroscope, astronomers have been able to deduce the chemical composition of the stars. The Doppler effect enables them to determine their movements. In our own **galaxy**, all stars will be shifted either to the blue or red end because of a slight Doppler effect, indicating either a small movement toward or

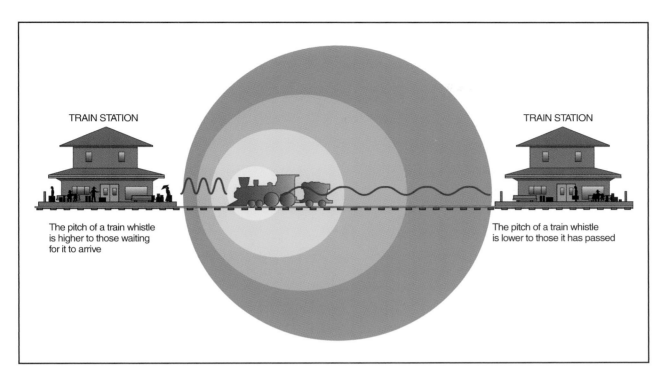

TRAIN STATION

The pitch of a train whistle is higher to those waiting for it to arrive

TRAIN STATION

The pitch of a train whistle is lower to those it has passed

The Doppler effect. *Illustration by Hans & Cassidy. Courtesy of Gale Group.*

away from **Earth**. In 1923, however, Edwin Hubble, an American astronomer, found that the light from all the galaxies outside our own were shifted so much toward the red as to suggest that they were all speeding away from our own at very great velocities. At the same time he saw that the recession of galaxies nearer to us was much less than those further away.

In 1929, Hubble and Milton Humason established a mathematical relationship that enables astronomers to determine the **distance** of galaxies by determining the amount of the galaxy's red shifts. This mathematical relationship is known as Hubble's law or Hubble's constant. Hubble's law shows that the greater the **velocity** of recession, the further away from the earth the galaxy is.

The concept of the expanding universe along with the corollary idea of the "big bang," that is, the instant creation of the universe from a compressed state of **matter**, owes much of its existence to Hubble's work, which in turn is an important development of the Doppler effect in light waves. While some recent research challenges the red shift phenomenon for galaxies, most astronomers continue to accept Hubble's findings.

Other uses of the Doppler effect

In addition to its uses in science, the Doppler effect has many practical applications. In maritime navigation, **radio waves** are bounced off orbiting satellites to mea-

KEY TERMS

Hubble's law—The law that states a galaxy's redshift is directly proportional to its distance from Earth.

Index of refraction—The ratio of the speed of light as it travels through any two media.

Redshift—The lengthening of the frequency of light waves as they travel away from an observer caused by the Doppler effect.

Spectroscope—An instrument for forming and examining light as it is refracted through an optical spectrum.

sure shifts which indicate changes in location. In highway traffic speeding detection, **radar** employs the Doppler effect to determine **automobile** speeds. There are also a number of medical applications of the Doppler effect found in ultrasonography, echocardiography, and **radiology**, all of which employ ultrasonic waves.

See also Wave motion.

Resources

Periodicals

Bruning, David. "Seeing a Star's Surface." *Astronomy* (October 1993): 34.

Powell, C. "The Redshift Blues." *Scientific American* (January 1990).

Stroh, Michael. "Gust Work: Meteorologists Decipher the Winds with Radar." *Science News* (July 11, 1992): 28-29.

Jordan P. Richman

Dories

Dories are **bony fish** in the family Zeidae. A dory has an oval body with a back that rises so that the main part of the body is higher than the head. The body itself is relatively thin and compressed and appears oval in side view.

Another distinguishing mark of the dories is a dark spot on each side of the body surrounded by a yellow ring. Dories typically are found in the middle depths of the seas where they live. Dories have extensible jaws, which can be extended outward as they capture their **prey**. This ability may explain why dories have relatively small teeth for a **carnivore**. Dories move slowly toward their prey, then display a burst of speed while extending their jaws to finish the kill.

The best known of the dories is *Zeus faber*, the john dory, which is found in the Atlantic from northern **Europe** to the tip of **Africa**, in the Mediterranean, and in the Pacific. The tasty john dory is a popular target for commercial fishermen. Specimens of *Zeus faber* can reach sizes of up to 3.3 ft (1 m). In **Australia**, the silver dory (*Cyttus australis*) is also a popular commercial **fish**. Despite the usual aversion people have to the "fishy" taste and **smell** of seafood that is not freshly caught, the john dory is said to taste better when aged several days after being caught.

The American john dory, *Zeus ocellata*, can be found all along the Atlantic coast of **North America** reaching 2.3 ft (0.7 m). Unlike its European counterpart, it is not a popular commercial **species**.

Fishermen often call the john dory "St. Peter's fish," a name that refers to the Apostle Peter who was at one time a fisherman. Two dark spots appear on the john dory as on other dories, one on each side of the body, which are said to be St. Peter's fingerprints.

Dormouse

Dormice are approximately ten **species** of **rodents** that make up the family Gliridae. Dormice typically live in trees, bushes, hedgerows, gardens, and rock piles.

Dormice have a superficial resemblance to **squirrels** (family Sciuridae), but they are smaller and differ in many other anatomical and behavioral characters.

Dormice have soft fur, and a long, bushy tail. Their forefeet have four digits, the hindfeet have five, and all have claws that aid in climbing. If a **predator** grabs a dormouse by the tail, that appendage is shed, giving the dormouse a chance to escape. Dormice are nocturnal animals, mostly foraging on **plant** materials, but also opportunistically taking **arthropods** and the contents of bird nests. In fact, in some places predation by dormice is believed to cause significant reductions in the populations of breeding **birds**.

Dormice become quite fat by the end of the autumn, approximately doubling their normal summer weight. Dormice spend most of the winter sleeping in their nest, except for relatively warm and sunny days, when they awake and eat food that they have stored.

The fat dormouse (*Glis glis*) is a rather arboreal species that occurs widely from Spain and France to western Russia, and has been introduced to southern England. The usual natural **habitat** is **angiosperm** and mixedwood **forests**. However, the fat dormouse also occurs in proximity to rural and suburban humans, and often nests in buildings. The fat dormouse is sometimes considered an important agricultural pest, especially in orchards where they may eat large quantities of valuable **crops** such as walnuts, or take small bites out of large numbers of softer **fruits**, making the produce unsalable. They are particularly regarded as a problem in Britain, where the populations of these animals are not well controlled by natural predators.

The fat dormouse is the largest of the dormice and it is sometimes eaten by people, some of whom consider the flesh of this **animal** to be a delicacy. In some respects, this epicurean taste for the fat dormouse is a leftover from the cuisine of the ancient Romans, who used to breed this dormouse in special pens for consumption when the animals were at their fattest.

The hazel mouse, or common dormouse (*Muscardinus avellanarius*) is the smallest species in this family, and occurs through much of **Europe**, **Asia** Minor, and western Russia. The usual habitat of these arboreal animals is forests and hedgerows, especially if there is a dense canopy of shrubs.

The garden, or orchard dormice (*Eliomys quercinus* and *E. melanurus*) occur in Europe, western Russia, Asia Minor, and northern **Africa**. These animals live in forests, swamps, and rocky habitats.

The tree, or forest dormouse (*Dryomys nitedula*) occurs in forests and shrubby habitats of much of Europe and

Asia Minor. The Japanese dormouse (*Glirulus japonicus*) only occurs in montane forests on the islands of Japan in eastern Asia. The mouse-like dormouse (*Myomimus personatus*) is a rare species that is only known from a few specimens collected in central Asia and Asia Minor. The African dormice (*Graphiurus* spp.) are three species that occur in a wide range of habitats in sub-Saharan Africa.

Double-blind study

New drugs undergo double-blind testing to determine whether they are effective. The test is called double-blind because neither the doctor who is administering the medication nor the patient who is taking it knows whether the patient is getting the experimental drug or a neutral substitute, called a **placebo**.

Getting a new drug approved is a long, complex process in order to ensure the drug is safe and effective and does what the manufacturer says it will do. The testing is done to satisfy the United States Food and Drug Administration (FDA), the government bureau that administers laws affecting the purity and safety of food and drugs.

A new medication first undergoes testing to insure that it is safe for humans to consume, that is, nontoxic. The drug then is tested to make certain it is effective against a specific **disease** in humans. Early testing must have shown it was effective against a bacterium or **virus** in the test tube (in vitro testing), but conditions are different in the human body. After passing toxicity and efficacy testing, the drug is placed in a double-blind study to compare it to other drugs used for similar purposes.

Thousands of patients in medical centers throughout the nation are assigned to the experimental group, which receives the new drug, or the control group, which receives the placebo or other older medication. Neither the patient nor his or her doctor know which group the patient is in.

The patient receives medication as stipulated by the doctor. The placebos are made to resemble the new drug's appearance and **taste** to make it difficult to tell the difference. These studies may take years to complete, so that a sufficient number of patients can be analyzed.

A safety committee oversees the study and determines which group each patient is in. If they notice that many patients become ill with the new drug, they can stop the test. If the new drug is proving exceptionally effective, they can stop the test to allow the drug to be given to all patients.

Double bond *see* **Chemical bond**

Double helix

The double helix refers to DNA's "spiral staircase" structure, consisting of two right-handed helical polynucleotide chains coiled around a central axis. Genes, which are specific regions of DNA, contain the instructions for synthesizing every protein. Because life cannot exist without **proteins**, the discovery of DNA's structure unveiled the secret of life: protein synthesis. In fact, the "central dogma" of **molecular biology** is that DNA is used to build **ribonucleic acid (RNA)**, which is used to build proteins, which in turn play a role in building DNA and RNA.

The discovery of the double-helix molecular structure of **deoxyribonucleic acid (DNA)** in 1953, one of the major scientific events of the twentieth century, and some would say in the history of **biology**, marked the culmination of an intense search involving many scientists. But ultimately, credit for the discovery and the 1962 Nobel Prize in Physiology or Medicine went to James Dewey Watson (who was an American postdoctoral student from Indiana University at the time) and Francis Harry Compton Crick, a researcher at the Cavendish Laboratory in Cambridge University, England. Their work, conducted at Cavendish Laboratory, significantly impacted the emerging field of molecular biology.

Prior to Watson and Crick's discovery, it had long been known that DNA contained four kinds of nucleotides, which are the building blocks of nucleic acids, such as DNA and RNA. A nucleotide contains a five-carbon sugar called deoxyribose, a phosphate group, and one of four nitrogen-containing bases: adenine (A), guanine (G), thymine (T), and cytosine (C). Thymine and cytosine are smaller, single-ringed structures called pyrimidines; adenine and guanine are larger, double-ringed structures called purines. Watson and Crick drew upon this and other scientific knowledge in concluding that DNA's structure possessed two nucleotide strands twisted into a double helix, with bases arranged in pairs such as A T, T A, G C, C G. Along the entire length of DNA, the double-ringed adenine and guanine nucleotide bases were probably paired with the single-ringed thymine and cytosine bases. Using **paper** cutouts of the nucleotides, Watson and Crick shuffled and reshuffled combinations. Later, they used wires and **metal** to create their model of the twisting nucleotide strands that form the double-helix structure. According to Watson and Crick's model, the diameter of the double helix measures 2.0 nanometers (nm). Each turn of the helix is 3.4 nm long, with 10 bases in each chain making up a turn.

Before Watson and Crick's discovery, no one knew how hereditary material was duplicated prior to **cell di-**

A model of the DNA double helix structure discovered by James Watson and Francis Crick. *Paul Seheult; Eye Ubiquitous/Corbis. Reproduced by permission.*

vision. Using their model, it is now understood that enzymes can cause a region of a DNA **molecule** to "unwind" one nucleotide strand from the other, exposing bases that are then available to become paired up with free nucleotides stockpiled in cells. A half-old, half-new DNA strand is created in a process that is called "semiconservative replication." When free nucleotides pair up with exposed bases, they follow a base-pairing rule which requires that A always pairs with T, and G always with C. This rule is constant in DNA for all living things, but the order in which one base follows another in a nucleotide strand differs from **species** to species. Thus, Watson and Crick's double-helix model accounts for both the sameness and the immense variety of life.

It is fair to say that Watson and Crick's discovery of the double helix would not have been possible without significant prior discoveries. In his 1968 book, *The Double Helix, A Personal Account of the Discovery of the Structure of DNA*, Watson wrote that the "race" to unveil the mystery of DNA was chiefly "a matter of five people:" Maurice Wilkins, Rosalind Franklin, Linus Pauling, Crick, and Watson. Wilkins, an Irish biophysicist who shared the 1962 Nobel Prize in Physiology or Medicine with Crick and Watson, extracted DNA gel fibers and analyzed them using x ray **diffraction**. The diffraction showed a helical molecular structure, and Crick and Watson used that information in constructing their double-helix model. Franklin, working in Wilkins' laboratory, between 1950 and 1953, produced improved x ray data using purified DNA samples, and through her work confirmed that each helix turn is 3.4 nm. Although her work suggested DNA might have a helix structure, she did not **postulate** a definite model. Pauling, an American chemist and twice Nobel laureate, in 1951 discovered the three-dimensional shape of the protein **collagen**. Pauling discovered that each collagen polypeptide or **amino acid** chain twists helically, and that the helical shape is held by **hydrogen** bonds. With Pauling's discovery, scientists worldwide began "racing" to discover the structure of other biological molecules, including the DNA molecule.

Down syndrome

Down **syndrome** is the most common cause of mental retardation. It can be caused by the presence of an extra **chromosome**. Chromosomes contain sequences of DNA called genes, which represent the genetic information that exists within almost every **cell** of the body. Twenty-three distinctive pairs, or 46 total chromosomes, are located within the nucleus (central DNA-containing structure) of each cell. When a sperm cell fertilizes an egg cell, the newly created zygote receives 23 chromosomes from each parent, for a total of 46 chromosomes. An aberration that occurs during division of the sperm or egg cell can cause this cell to have an extra chromosome for a total of 24 chromosomes instead of 23. This event occurs during **cell division** and is referred to as nondisjunction, or the failure of all chromosomes to separately properly resulting in retention of one of the chromosomes in one of the two new daughter cells and the loss of a chromosome in the other. Loss of a chromosome usually means that the embryo is nonviable. An embryo with an extra chromosome in all its cells (for a total of 47 chromosomes) is usually inconsistent with life. If the extra chromosome is number 21, and the fetus survives to term, the baby with have Down syndrome. This form of Down Syndrome is also called Trisomy 21 and

An older sibling plays with her younger sister who has Down syndrome. *Photograph by A. Sieveing. A. Sieveing/Petit Format/Photo Researchers, Inc. Reproduced by permission.*

accounts for approximately 95% of all Down syndrome patients.

In a very rare number of Down syndrome cases, the original egg and sperm cells begins with the correct number of chromosomes but shortly after **fertilization** during the phase where cells are dividing rapidly, a single cell can divide abnormally, creating a line of cells with an extra chromosome 21. This is called a cell line mosaicism. The **individual** with this type of Down syndrome has two types of cells: some with 46 chromosomes (the normal number), and some with 47 chromosomes (causing Down syndrome symptomatology). Individuals who are mosaic for Trisomy 21 typically have less severe signs and symptoms of the disorder.

Another relatively rare genetic abnormality that can cause Down syndrome is called a chromosome translocation. This is an event that unlike the numerical abnor-

mality causing Trisomy 21, there is a structural abnormality. Exchange of material from two different chromosomes during the production of sex cells can take place such that there is a whole chromosome 21 attached to another chromosome but the chromosome number is normal. These types of translocations, involving chromosome 21, occur in about 3–4% of cases of Down syndrome.

Down syndrome occurs in about one in every 800 liveborns. It affects an equal number of male and female babies. The majority of cases of Down syndrome occur due to a nondisjunction event that occurs in the maternal sex cells. As the maternal age increases, the risk of having a Down syndrome baby increases significantly. For example, at younger ages, the risk is about one in 1,250. By the time the woman is 35 years old, the risk increases to one in 385; by age 40 the risk increases to one in 100; and by age 45 the risk is one in 25. There is no known maternal age-related increased risk if down syndrome results from either a cell line mosaicism or translocation.

Causes and symptoms

While Down syndrome is a chromosomal disorder, a baby is usually identified at **birth** through observation of a set of common physical characteristics. Babies with Down syndrome tend to be overly quiet, less responsive, with weak, floppy muscles. Furthermore, a number of physical signs may be present. These include:

- flat appearing face
- small head
- flat bridge of the nose
- smaller than normal, low-set nose
- small mouth, with a protruding tongue
- upward slanting eyes
- extra folds of skin located at the inside corner of each **eye**, near the nose (called epicanthal folds)
- small, outwardly rotated ears
- small, wide hands
- an unusual, deep crease across the center of the palm (called a simian crease)
- a malformed fifth finger
- a wide space between the big and the second toes
- unusual creases on the soles of the feet
- shorter than normal height

Other types of defects often accompany Down syndrome. About one third of all children with Down syndrome have **heart** defects. These heart defects are characteristic of Down syndrome, including abnormal openings (or holes) in the walls which separate the heart's chambers (atrial septal defect, ventricular septal defect). These defects result in abnormal patterns of **blood** flow within the heart, resulting in inefficient **oxygen** delivery.

Malformations of the gastrointestinal tract are present in about 5–7% of children with Down syndrome. The most common malformation is a narrowed, obstructed duodenum (the part of the intestine into which the stomach empties). This disorder, called duodenal atresia, interferes with the baby's milk or formula leaving the stomach and entering the intestine for digestion. The baby often vomits forcibly after feeding, and cannot gain weight appropriately until the defect is surgically repaired.

Other medical conditions occurring in patients with Down syndrome include an increased chance of developing infections, especially **ear** infections and **pneumonia**; certain kidney disorders; thyroid **disease**; **hearing** loss; **vision** impairment requiring glasses (corrective lenses); and a 15-fold increased risk for developing **leukemia**.

Developmental milestones in a child with Down syndrome are delayed. Due to weak, floppy muscles (hypotonia), babies learn to sit up, crawl, and walk much later than their normal peers. Talking is also delayed. The extent of delayed **brain** development is considered to be mild-to-moderate. Most people with Down syndrome can learn to perform regular tasks and can have relatively easy, jobs (with supervision).

As people with Down syndrome age, they face an increased risk of developing **Alzheimer disease**, a degenerative disease that affects the brain. This occurs several decades earlier than the risk of developing Alzheimers disease in the general population. As people with Down syndrome age, they also have an increased chance of developing a number of other illnesses, including cataracts, thyroid problems, diabetes, leukemia, and seizure disorders.

Treatment

There is no cure for Down syndrome. However, some of the clinical manifestations can be treated. For example, heart defects and duodenal atresia can often be corrected with surgical repair. It was common only a few decades ago to enforce involuntary sterilization of individuals with Down syndrome. Additionally, it used to be common for these patients to be institutionalized. Today, involuntary sterilizations are illegal and most patients reside with their families. Many community support groups exist to help families deal with the emotional effects and to help plan for the affected individuals future. In general, Down syndrome people tend to be easy going and good natured.

Prognosis

The prognosis in Down syndrome is variable, depending on the types of complications (heart defects, susceptibility to infections, leukemia) of each affected individual. The severity of the developmental delay also varies. Without the presence of heart defects, about 90% of children with Down syndrome survive past their teenage years. In fact, people with Down syndrome can live until they are 50 years old.

The prognosis for a baby born with Down syndrome has improved compared to previous years. Modern medical treatments, including **antibiotics** to treat infections, and **surgery** to treat heart defects and duodenal atresia has greatly increased their life expectancy.

Diagnosis and prevention

Down syndrome can be diagnosed at birth, when the characteristic physical signs of Down syndrome are

KEY TERMS

Chromosomes—The structures that carry genetic information in the form of DNA. Chromosomes are located within every cell and are responsible for directing the development and functioning of all the cells in the body. The normal number of chromosomes in humans is 46 (23 pairs).

Developmental delay—A condition where an individual has a lower-than-normal IQ, and thus is developmentally delayed.

Egg cell—The female's reproductive sex cell.

Embryo—A stage in development after fertilization.

Karyotype—The specific chromosomal makeup of a particular cell.

Mosaic—A term referring to a genetic situation in which an different cells do not have the exact same composition of chromosomes. In Down syndrome, this may mean that some of the individual's cells have the normal 46 number of chromosomes, while other cells have an abnormal number, or 47 chromosomes.

Nondisjunction—A genetic term referring to an event that takes place during cell division in which one of the newly created cells has 24 chromosomes and the other cell has 22, rather than the normal 23.

Sperm—Substance secreted by the testes during sexual intercourse. Sperm includes spermatozoon, the mature male cell which is propelled by a tail and has the ability to fertilize the female egg.

Translocation—A genetic term referring to a situation during cell division in which a piece of one chromosome breaks off and sticks to another chromosome.

Trisomy—The condition of having three identical chromosomes, instead of the normal two.

Zygote—The cell resulting from the fusion of male sperm and the female egg. Normally the zygote has double the chromosome number of either gamete, and gives rise to a new embryo.

---

noted and chromosome analysis can also be performed to confirm the **diagnosis** and determine the recurrence risks.

At-risk pregnancies are referred for genetic counseling and prenatal diagnosis. Screening tests are available during a pregnancy to determine if the fetus has Down syndrome. During 14–17 weeks of pregnancy, a substance called AFP (alpha-fetoprotein) can be measured. AFP is normally found circulating in a pregnant woman's blood, but may be unusually high or low with certain disorders. Carrying a baby with Down syndrome often causes AFP to be lower than normal. This information alone, or along with measurements of two other **hormones**, is considered along with the mother's age to calculate the risk of the baby being born with Down syndrome.

A common method to directly determine whether the fetus has Down syndrome, is to test **tissue** from the fetus. This is usually done either by **amniocentesis**, or **chorionic villus sampling (CVS)**. In amniocentesis, a small amount of the fluid in which the baby is floating is withdrawn with a long, thin needle. In chorionic villus sampling, a tiny tube is inserted into the opening of the uterus to retrieve a small **sample** of the chorionic villus (tissue that surrounds the growing fetus). Chromosome analysis follow both amniocentesis and CVS to determine whether the fetus is affected.

Once a couple has had one baby with Down syndrome, they are often concerned about the likelihood of future offspring also being born with the disorder. In most cases, it is unlikely that the risk is greater than other woman at a similar age. However, when the baby with Down syndrome has the type that results from a translocation, it is possible that one of the two parents is a carrier of that defect. When one parent is a carrier of a particular type of translocation, the chance of future offspring having Down syndrome is increased. The specific risks can be estimated by a genetic counselor.

Resources

Books

Nussbaum, R.L., Roderick R. McInnes, and Huntington F. Willard. *Genetics in Medicine.* Philadelphia: Saunders, 2001.

Rimoin, David, L. *Emery and Rimoin's Principles and Practice of Medical Genetics.* London; New York: Churchill Livingstone, 2002.

Periodicals

"Medical and Surgical Care for Children with Down Syndrome." *The Exceptional Parent* 25 no. 11 (November 1995): 78+.

Rosalyn S. Carson-DeWitt
Bryan R. Cobb

Dragonflies

Dragonflies are large flying **insects** in the order Odonata. Dragonflies can be as large as 3 in (7.5 cm) in length, with a wing span of up to 8 in (20 cm). The fossilized remains of a huge dragonfly-like insect that had a wingspread of more than 2 ft (70 cm) is known from the Carboniferous period, some 300 million years ago.

Dragonflies are very distinctive insects, with large eyes that almost cover the entire head, a short thorax, a long slender abdomen, and glassy membranous wings. Dragonflies are classified in the suborder Anisoptera since their hindwings are larger than their forewings, and the wings are habitually held straight out when at rest. They feed on other insects, which they catch in flight.

Dragonflies are usually found around streams and ponds, where they feed, mate, and lay their eggs. The mating habits of dragonflies are conspicuous and unusual. The male generally sets up a territory over a part of a stream or pond which he patrols for most of the day. When a newly emerged female flies into the territory, the male flies above her and lands on her back, bends his abdomen far forward and deposits sperm on the underside of his second abdominal segment, which is the site of his penis. Then, grasping the female behind the head with a pair of forceps-like structures at the end of his abdomen, he flies off with her in tandem. When she is ready to mate, she curls her abdomen down and forward to place its end under the male's second abdominal segment, which has structures to hold it in place while the sperm are transferred to her reproductive tract. The pair may fly around in this unusual "wheel" configuration for several

A spotted skimmer dragonfly taking a drink. *Photograph by J.H. Robinson. The National Audubon Society Collection/Photo Researchers. Reproduced by permission.*

KEY TERMS

Globe-skimmer—One of the most widely distributed of all dragonflies.

Naiad—The aquatic larval stage of dragonflies.

Thorax—The body region of insects that supports the legs and wings.

minutes. Egg-laying begins within a short **time**, with the male either continuing to hold the female while she dips her abdomen into the **water** to lay the eggs, or waiting above her and then regrasping her after each egg-laying session. The eggs hatch into aquatic larval form (naiad) after a few days.

Like the adults, the wingless naiads feed on insects and other small aquatic animals. The lower lip (labium) of the larvae is retractable with jaws that can be thrust out in front of the head to catch and pull the **prey** back to the chewing mandibles. The naiads have gills in the last segments of the abdomen and ventilate the gills by pumping water in and out. The contraction of the pumping muscles also allows the larvae to "jet" forward rapidly out of harm's way. During the winter, the larvae live in the water, where they grow, shedding the external skeleton (molting) several times. In the spring, the larvae climb out of the water, molt again, and the newly-transformed adult dragonflies emerge and unfurl their wings.

Some 5,000 **species** of dragonflies are known, living in every **continent** except **Antarctica**, and on most islands as well. The principal families of dragonflies are the high-flying darners, the Aeshnidae, and the skimmers, the Libellulidae.

Resources

Books

Borror, D.J., and R.E. White. *A Field Guide to the Insects of America North of Mexico.* Boston: Houghton Mifflin, 1980.

Borror, D.J., D.M. Delong, and C.A. Triplehorn. *An Introduction to the Study of Insects.* 4th ed. New York: Holt, Reinhart & Winston, 1976.

Carde, Ring, and Vincent H. Resh, eds. *Encyclopedia of Insects.* San Diego: Academic Press, 2003.

d'Aguilar, J., J.L. Dommanget, and R. Prechac. *A Field Guide to the Dragonflies of Britain, Europe and North Africa.* London: Collins, 1986.

Herndon G. Dowling

Dream *see* **Sleep**

Drift net

Drift nets are lengthy, free-floating, 26-49 ft (8-15 m) deep nets, each as long as 55 mi (90 km). Drift nets are used to snare **fish** by their gills in pelagic, open-water situations. Because drift nets are not very selective of **species**, their use results in a large by-catch of non-target fish, **sharks**, **turtles**, seabirds, and marine **mammals**, which are usually jettisoned, dead, back to the **ocean**. Drift nets are an extremely destructive fishing technology.

Ecological damage caused by drift nets

Drift-net fisheries have been mounted in all of the world's major fishing regions, and unwanted by-catch is always a serious problem. This has proven true for pelagic fisheries for **swordfish**, **tuna**, **squid**, **salmon**, and other species. One example is the drift-net fishery for swordfish in the Mediterranean, 90% of which is associated with Italian fishers. This industry kills excessive numbers of striped dolphin and sperm whale, and smaller numbers of fin whale, Cuvier's beaked whale, long-finned pilot whale, and Risso's, bottlenose, and common dolphins, along with other non-target marine **wildlife**. As a result of concerns about the excessive by-catch in this swordfish fishery during the early 1990s, the European Union banned the use of drift nets longer than 1.5 mi (2.5 km) (prior to this action, the average set was 26 mi [12 km] in length). However, some fishing nations have objected to this regulation and do not enforce it. It remains to be seen whether this length restriction will prove to be useful in preventing the non-target, drift-net mortality in this fishery.

There are few monitoring data that actually demonstrate the non-target by-catch by drift nets. One measurement was made during a one-day monitoring of a typical drift-net set of 11 mile/day (19 km/day) in the Caroline Islands of the south Pacific. That single net, in one day, entangled 97 dolphins, 11 larger **cetaceans**, and 10 sea turtles. World-wide during the late 1980s, pelagic drift nets were estimated to have annually killed as many as one million dolphins, porpoises, and other cetaceans, along with millions of seabirds, tens of thousands of **seals**, thousands of sea turtles, and untold numbers of sharks and other large, non-target fish.

Although there are no hard data to verify the phenomenon, there are anecdotal reports of substantial reductions in the abundance of some of these groups of animals in regions that have experienced a great deal of drift netting. Consequently, the drift net by-catch is perceived (by proponents of this type of fishing technology) to be less of a problem than formerly, because the unintended by-catches are apparently smaller. However, this really reflects the likelihood that this rapacious fishing practice has created marine deserts, that only support sparse populations of large animals.

In addition, great lengths of drift nets and other fishing nets are lost at sea every year, especially during severe storms. Because the nets are manufactured of synthetic materials that are highly resistant to degradation, they continue to snare fish, sharks, mammals, **birds**, turtles, and other creatures for many years, as so-called ghost nets. Little is known about the magnitude of this problem, but it is undoubtedly an important cause of mortality of marine animals and other creatures.

In response to mounting concerns about unsustainable by-catches of non-target species of marine animals, which in some cases are causing population declines, the United Nations in 1993 banned the use of drift nets longer than 1.5 mi (2.5 km). Although this regulation would not eliminate the by-catches associated with drift netting, it would greatly reduce the amount of this unintended mortality, possibly by as much as two-thirds. Unfortunately, there has been a great deal of resistance from the fishing industry and certain fishing nations to the implementation of even this minimal regulation and have continued to allow the use of much longer nets. In addition, some illegal, or "pirate" fishers continue to use the extremely destructive, older-style drift nets.

Clearly, non-selective by-catches associated with drift nets cause an unacceptable mortality of non-target animals, some of which are threatened by this practice. A rapid improvement of this unsatisfactory state of environmental affairs could be achieved by using shorter drift nets, or by banning their use altogether. Unfortunately, because of economic self-interest of the world's nations and the fishing industry, this seemingly obvious and simple betterment has not yet proved possible.

Resources

Books

Berrill, M., and D. Suzuki. *The Plundered Seas: Can the World's Fish be Saved?* Sierra Club Books, 1997.

> **KEY TERMS**
>
> **By-catch**—A harvest of species of animals that are not the target of the fishery, caught during fishing directed towards some other, commercially desirable species.
>
> **Gill net**—A net that catches fish by snaring their gill covering.

Freedman, B. *Environmental Ecology.* 2nd ed. San Diego: Academic Press, 1995.

LaBudde, S. *Stripmining the Seas. A Global Perspective on Drift Net Fisheries.* Honolulu: Earthtrust, 1989.

Bill Freedman

Drongos

Drongos are 20 **species** of handsome **birds** that make up the family of perching birds known as Dicruridae. Drongos occur in **Africa**, southern and southeastern **Asia**, and Australasia. Their usual habitats are open **forests**, savannas, and some types of cultivated areas with trees.

Drongos are typically black colored with a beautiful, greenish or purplish iridescence. The wings of these elegant, jay-sized birds are relatively long and pointed, and the tail is deeply forked. The tail of some species is very long, with the outer feathers developing extremely long filaments with a "racket" at the end. The beak is stout and somewhat hooked, and is surrounded by short, stiff feathers known as rictal bristles, a common feature on many fly-catching birds other than drongos. The sexes are identical in **color** and size.

Drongos are excellent and maneuverable fliers, though not over long distances. They commonly feed by catching **insects** in flight, having discovered their **prey** from an exposed, aerial perch. Some species follow large **mammals** or **monkeys**, feeding on insects that are disturbed as these heavier animals move about.

Drongos sing melodiously to proclaim their territory, often imitating the songs of other species. They are aggressive in the defense of their territory against other drongos as well as potential predators. Some other small birds deliberately nest close to drongos because of the relative protection that is afforded against crows, **hawks**, and other predators.

Drongos lay three to four eggs in a cup-shaped nest located in the fork of a branch. The eggs are mostly incubated by the female, but both sexes share in the feeding and caring of the young.

The greater racket-tailed drongo (*Dicrurus paradiseus*) of India, Malaya, and Borneo has a very long tail, which is about twice the length of the body of the bird. More than one-half of the length of the tail is made up of the extended, wire-like shafts of the outer-two tail feathers, which end in an expanded, barbed surface—the racket. These seemingly ungainly tail-feathers flutter gracefully as these birds fly, but do not seem to unduly interfere with their maneuverability when hunting flying insects. The greater racket-tailed drongo is also famous for its superb **mimicry** of the songs of other species of birds.

Another well-known species is the king-crow (*Dicrurus macrocercus*) of India, so-named because of its aggressive dominance of any crows that venture too closely, and of other potential predators as well. Like other drongos, however, the king-crow is not a bully—it only chases away birds that are potentially dangerous.

Drosophila melanogaster

Throughout the last century, the fruit fly *Drosophila melanogaster,* has been the workhorse for genetic studies in eukaryotes. These studies provided the basis of much of scientists' understanding of fundamental aspects of eukaryotic **genetics**. Cloned fruit fly genes have led to the identification of mammalian cognates. Discoveries have shown that the conservation between the fruit fly and **mammals** is much greater than ever expected, from structure **proteins** to the higher order processes such as development, **behavior**, **sleep**, and other physiological responses.

Drosophila melanogaster, is a tiny fly, only 0.08–0.12 in (2–3 mm) in length and is often found around **grapes** and rotten bananas. They reproduce frequently, furnishing a new generation in less than two weeks; each generation includes hundreds of offspring. They are easy and inexpensive to maintain and easy to examine. Stable mutants will appear after a period of culture in the laboratory. All these characteristics make the fruit fly an ideal model for genetic studies.

In 1903, T. H. Morgan started his work on heredity and chromosomes using the fruit fly. In 1910, Morgan published his famous paper "Sex Limited Inheritance in *Drosophila*" in the journal *Science* that described a white-eyed male fruit fly mutant he observed and the crossing experiments he conducted in his laboratory. The research of Morgan and his associates demonstrated that genes for specific traits were located on separate chromosomes. Genes were arranged in a linear order and the relative **distance** of genes could be determined experimentally. These studies in the first third of the twentieth century established the **chromosome** theory. Morgan was awarded the Nobel Prize in 1933 for his discoveries on the research of the fruit fly.

The larval stage salivary gland chromosomes of the fruit fly are called polytene chromosomes. They are unique morphologically. The size and length of the chromosomes are greatly increased due to numerous rounds

Fruit fly (*Drosophila melanogaster*) resting on a piece of ripe fruit. *Photograph by Oliver Meckes. Photo Researchers, Inc. Reproduced by permission.*

of replication. This can be seen easily under the **microscope**. In 1934, T. S. Painter of the University of Texas published the first drawing of the fruit fly polytene chromosomes, which included the chromosomal localization of several genes. In 1935 and 1938, C. B. Bridges published the fruit fly polytene maps. His maps were so accurate that they are still used even today. These are the pioneer work on physical **gene** mapping.

The fruit fly **genome** sequence is the second and the largest **animal** genome sequenced. The fruit fly has four pairs of chromosomes. The whole genome is about 180 million base pairs. There are about 14,000 genes in the genome. Since the release of an initial draft sequence in 2000, scientists at the Berkeley Drosophila Genome Project (BDGP)and Celera (a private genomic company) continue to release improved versions of the Drosophila genome sequence. The latest version, (Release 3) was released in July 2002.

The conservation of biological processes from **flies** to mammals extends the influence of the fruit fly research to human health. When an uncharacterized fruit fly homologue of important human gene is isolated, the genetic techniques in the fruit fly system can be applied to its characterization. The identified fruit fly cognates of the human **disease** genes will also greatly expedite the progress of human disease research.

See also Clone and cloning; Genetic engineering; Genomics (comparative).

Resources

Periodicals

Adams, M.D., et al. "The Genome Sequence of *Drosophila melanogaster.*" *Science* 287 (2000): 2185–2195.

Celniker, S.E., et al. " Finishing a Whole Genome Shotgun: Release 3 of the *Drosophila melanogaster* Euchromatic Genome Sequence." *Genome Biology,* no. 3 (12.) (2002).

Hoskins, RA., et al. "Heterochromatic Sequences in a *Drosophila* Whole Genome Shotgun Assembly." *Genome Biology,* no. 3 (12.) (2002).

Rubin, G.M., and E.B. Lewis. "A Brief History of *Drosophila*'s Contributions to Genome Research." *Science* 287 (2000): 2216–2218.

Xiaomei Zhu

Drought

Drought is characterized by low **precipitation** compared to the normal amount for the region, low **humidity**, high temperatures, and/or high **wind** velocities. When these conditions occur over an extended period of **time**, drought causes low **water** supplies that are inadequate to support the demands of plants, animals, and people.

Drought is a temporary condition that occurs in moist climates. This is in contrast to the conditions of normally arid regions, such as deserts, that normally experience low average rainfall or available water. Under both drought and arid conditions, **individual** plants and animals may die, but the populations to which they belong survive. Both drought and aridity differ from desiccation, which is a prolonged period of intensifying drought in which entire populations become extinct. In **Africa** and **Australia**, periods of desiccation have lasted two to three decades. The loss of **crops** and cattle in these areas caused widespread suffering. Extensive desiccation may lead to **desertification** in which most **plant** and **animal** life in a region is lost permanently, and an arid **desert** is created.

Unlike a **storm** or a flood, there is no specific time or event that constitutes the beginning or end of a drought. Hydrologists evaluate the frequency and severity of droughts based on measurements of river basins and other water bodies. Climatologists and meteorologists follow the effects of **ocean** winds and volcanoes on **weather** patterns that cause droughts. Agriculturalists measure a drought's effects on plant growth. They may notice the onset of a drought long before hydrologists are able to record drops in underground water table levels. By observing weather cycles, meteorologists may be able to predict the occurrence of future droughts.

Severe drought in Botswana, Africa, was the result of an El Niño weather pattern that began in 1989. *Photograph by Tom Nebbia. Stock Market. Reproduced by permission.*

In addition to its duration, the intensity of a drought is measured largely by the ability of the living things in the affected vicinity to tolerate the dry conditions. Although a drought may end abruptly with the return of adequate rainfall, the effects of a drought on the landscape and its inhabitants may last for years.

Many factors affect the severity of a drought's impact on living organisms. Plants and animals are vulnerable to drought when stored water cannot replace the amount of moisture lost to **evaporation**. Plants and animals have several mechanisms that enable them to tolerate drought conditions. Many desert annuals escape drought altogether by having a short life span. Their life cycle lasts only a few weeks during the desert's brief, moist periods. The rest of the time they survive as dessication-resistant **seeds**. Some plants, such as cacti, evade drought by storing water in their tissues, while others, like mesquite trees, become dormant. Still others, such as the creosote bush, have evolved adaptations such as reduced **leaf** size and a waxy coating over the leaves that protect against water loss. Many animals that live in areas prone to drought like **snakes** and lizards forage and hunt at night, avoiding the desiccating effects of the sun's rays. Other animals have adaptations that allow them to survive without drinking, obtaining all of the water that they need from their food sources.

History

Studies of **tree** rings in the United States have identified droughts occurring as early as 1220. The thickness of annual growth rings of some tree **species**, such as red cedar and yellow pine, indicates the wetness of each season. The longest drought identified by this method began in 1276 and lasted 38 years. The tree ring method identified 21 droughts lasting five or more years during the period from 1210 to 1958. The earliest drought recorded and observed in the United States was in 1621. The most well-known American drought was the Dust Bowl on the Great Plains from 1931 to 1936. The years 1934 and 1936 were the two driest years in the recorded history of U.S. climate. The Dust Bowl encompassed an area approximately 399 mi (644 km) long and 298 mi (483 km wide) in parts of Colorado, New Mexico, Kansas, Texas, and Oklahoma. More recently, the United States experienced severe to extreme drought in over half of the country during 1987–89. This drought was the subject of national headlines when it resulted in the extensive fires in Yellowstone National Park in 1988.

Droughts have also had enormous impact in other regions of the world. A drought in northern China in 1876 dried up crops in an extensive region. Millions of people died from lack of food. Russia experienced severe droughts in 1890 and 1921. The 1921 drought in the Volga River **basin** caused the deaths of up to five million people—more than had died during World War I, which had just ended. India normally receives most of its rain during the **monsoon** season, which lasts from June to September. Winds blowing in from the Indian Ocean bring most of the country's rainfall during this season. The monsoon winds did not come during two droughts in 1769 and 1865. An estimated 10 million people died in each of those droughts, many from diseases like **smallpox**, which was extremely contagious and deadly because people were already weakened from lack of food. More recent severe droughts occurred in England (1921, 1933–34, and 1976), Central Australia (1945–72), and the Canadian prairies (1983–5).

Almost the entire **continent** of Africa suffered from droughts in the last quarter of the twentieth century. Ethiopia, usually considered the breadbasket of eastern Africa, was hit by a brutal drought in the early 1980's. A dry year in 1981 resulted in low crop yields. Three years later, another dry year led to the deaths of nearly a million people. Drought conditions again threatened eastern Africa in 2002. An estimated 15 million people in Ethiopia, three million in Kenya, 1.5 million in Eritrea and three million in Sudan could face starvation as a result of drought. According to the World Health Organization, drought is the cause of death for about half the people who are killed by natural disasters.

Between 1968 and 1973, Sahel (a region in east Burkina, a country in western Africa) suffered a great drought. An estimated 50,000–200,000 people died as a result. While the causes of these great African droughts are unknown, research is beginning to indicate that droughts could result from a combination of global and local climate patterns. **Satellite** imagery links **El Niño** and vertical ocean mixing patterns to dry weather in Sahel. In addition, desertification may be a positive feedback mechanism driving the climate towards drought conditions. In regions where there are few surface water reservoirs such as Sahel, the major source of precipitation is **transpiration** from plants. As plants become sparse in drought conditions, this source of water for precipitation is diminished. The diminished precipitation further decreases the growth of vegetation.

Since 1994, drought and famine in North Korea have been worsened by the political situation. The government of North Korea has allowed its people to starve rather than continue talks with South Korea and famine relief organizations. Over 50% of the children in North Korea are suffering from **malnutrition** and lack of water, and countless numbers have perished. The drought may accelerate political problems in the region if starving refugees from North Korea flee to China and South Korea.

Drought management

Because drought is a natural phenomenon, it cannot be eradicated. Consequently, methods to mitigate its devastating effects are crucial. Crop and **soil** management practices can increase the amount of water stored within a plant's root zone. For example, **contour plowing** and **terracing** decreases the steepness of a hillside slope and thus, reduces the amount and **velocity** of water runoff. Vegetation protects the soil from the impact of raindrops, which causes both **erosion** and soil crusting (hardening of the soil surface that prevents rain from percolating into the soil where it is stored). Both living plants and crop residues left by minimum tillage reduce soil crusting so the soil remains permeable and can absorb rainfall.

Other farming practices that lessen the impact of drought on crop production include strip cropping, windbreaks, and **irrigation**. Windbreaks or shelterbelts are

---

KEY TERMS
. . . .

Aquifer—A formation of soil or rock that holds water underground.

Arid climate—A climate that receives less than 10 in (25 cm) of annual precipitation and generally requires irrigation for agriculture.

Precipitation—Water particles that are condensed from the atmosphere and fall to the ground as rain, dew, hail or snow.

---

strips of land planted with shrubs and trees **perpendicular** to the prevailing winds. Windbreaks prevent soil, with its moisture-retaining properties, from being blown away by wind. Plants can also be specifically bred to adapt to the effects of weather extremes. For example, shorter plants encounter less wind and better withstand turbulent weather. Plants with crinkled leaves create small pockets of still air that slow evaporation.

From a social standpoint, drought severity is influenced by the vulnerability of an area or population to its effects. Vulnerability is a product of the demand for water, the age and health of the population affected by the drought, and the efficiency of water supply and **energy** supply systems. Drought's effects are more pronounced in areas that have lost **wetlands** that recharge aquifers, are dependent on agriculture, have low existing food stocks, or whose governments have not developed drought-response mechanisms.

See also Hydrologic cycle; Water conservation.

Resources

Books

Climate Change 2001: Impacts, Adaptation and Vulnerability. Intergovernmental Panel on Climate Change, 2001.

Defreitas, Stan. *The Water-Thrifty Garden.* Dallas: Taylor Publishing Company, 1993.

White, Donald A., ed. *Drought: A Global Assessment.* New York: Routledge Publishers, 2000.

Karen Marshall

Ducks

Ducks are waterfowl in the order Anseriformes, in the family Anatidae, which also includes **geese** and **swans**. Ducks occur on all continents except **Antarctica**, and are widespread in many types of aquatic habitats.

Almost all ducks breed in **freshwater** habitats, especially shallow lakes, marshes, and swamps. Most **species** of ducks also winter in these habitats, sometimes additionally using grain fields and other areas developed by humans. Some species of sea ducks breed on marine coasts, wintering in near-shore habitats. Most species of ducks undertake substantial migrations between their breeding and wintering grounds, in some cases flying thousands of miles, twice each year.

Ducks are well adapted to aquatic environments and are excellent swimmers with waterproof feathers, short legs, and webbed feet. The feathers are waterproofed by oil transferred from an oil gland at the base of the tail by the bill. Ducks eat a wide range of aquatic plants and animals, with the various species of ducks having long necks, wide bills and other attributes that are specialized for their particular diets. Most ducks obtain their food by either dabbling or diving. Dabbling ducks feed on the surface of the **water**, or they tip up to submerge their head and feed on reachable items in shallow water. Diving ducks swim underwater to reach deeper foods. Ducks have great economic importance as the targets of hunters and several species have been domesticated for agriculture. In general, duck populations have greatly declined world-wide, as a combined result of overhunting, **habitat** loss, and **pollution**.

Dabbling ducks

Dabbling ducks (subfamily Anatinae) are surface-feeding **birds** that eat vegetation and **invertebrates** found in shallow water they can reach without diving. Their **plant** foods include colonial **algae**, small vascular plants such as **duckweed** (e.g., *Lemna minor*), roots and tubers of aquatic plants, and the **seeds** of pondweed (*Potamogeton* spp.), smartweed (*Polygonum* spp.), wild **rice** (*Zizania aquatica*), **sedges** (*Carex* spp.), and bulrushes (*Scirpus* spp.). Dabbling ducks also eat aquatic invertebrates, and in fact these are the most important foods of rapidly growing ducklings.

Two widespread species of dabbling duck are mallards (*Anas platyrhynchos*) and pintails (*A. acuta*). These ducks range throughout the Northern Hemisphere, occurring in both **North America** and Eurasia. Other North American species include black ducks (*A. rubripes*), American widgeons (*Mareca americana*), shovelers (*Spatula clypeata*), blue-winged teals (*A. discors*), and wood ducks (*Aix sponsa*).

Bay and sea ducks

Bay and sea ducks (Aythyinae) are diving ducks that swim beneath the surface of the water in search of aquatic animals. Some species also eat plants, but this is generally less important than in the herbivorous dabbling ducks. Some bay and sea ducks, for example, common goldeneye (*Bucephala clangula*), ring-necked duck (*Aythya collaris*), and hooded merganser (*Lophodytes cucullatus*), eat mostly **arthropods** occurring in the water column. Other species including oldsquaws (*Clangula hyemalis*), lesser scaups (*Aythya affinis*), surf scoters (*Melanitta perspicillata*), and common eiders (*Someteria mollissima*) specialize on bottom living invertebrates. Some of these species are remarkable divers, descending as deep as 246 ft (75 m) in the case of oldsquaw ducks.

Tree or whistling ducks

Tree ducks (Dendrocygninae) are long-legged birds, and are much less common than most dabbling or diving ducks. Tree ducks tend to be surface feeders in aquatic habitats, but they also forage for nuts and seeds on land. Tree ducks have a generally southern distribution in North America. The most common North American species is the fulvous tree duck (*Dendrocygna bicolor*).

Stiff-tailed ducks

Stiff-tailed ducks (Oxyurinae) are small diving ducks with distinctive, stiffly-erect tails. This group is represented in North America by the ruddy duck (*Oxyura jamaicensis*).

Mergansers

Mergansers (Merginae) are sleek, diving ducks specialized for feeding on small **fish**, and have serrated bills, which apply a firm grip on their slippery **prey**. The most abundant species are the common merganser (*Mergus merganser*) and the red-breasted merganser (*M. serrator*).

Economic importance of ducks

Wild ducks have long been hunted for food, and more recently for sport. In recent decades, hunters kill about 10-20 million ducks each year in North America, shooting about 20% in Canada, and the rest in the United States. Duck hunting has a very large economic impact, because of the money hunters spend on travel, license fees, private hunting fees, and on firearms, ammunition, and other paraphernalia.

Prior to the regulation of the hunting of ducks and other game animals, especially before the 1920s, the killing of ducks was essentially uncontrolled. In areas where ducks were abundant, there were even commercial hunts to supply ducks to urban markets. The excessive

A lone mallard (*Anas platyrhynchos*) amid a group of black ducks in Castalia, Ohio. *Photograph by Robert J. Huffman. Field Mark Publications. Reproduced by permission.*

hunting during these times caused tremendous decreases in the populations of ducks and other waterfowl, as well as in other species of edible birds and **mammals**. Consequently, governments in the United States and Canada began to control excessive hunting, to protect breeding habitat, and to provide a network of habitat refuges to provide for the needs of waterfowl during **migration** and wintering. These actions have allowed subsequent increases in the populations of most species of waterfowl, although the numbers of some species still remain much smaller than they used to be.

A relatively minor but interesting use of ducks concerns the harvesting of the down of wild common eiders. The female of this species plucks down from her breast for use in lining the nest, and this highly insulating material has long been collected in northern countries, and used to produce eiderdown quilts and clothing.

Several species of ducks have been domesticated, and in some areas they are an important agricultural commodity. The common domestic duck is derived from the mallard, which was domesticated about 2,000 years ago in China. Farm mallards are usually white, and are sometimes called Peking duck. The common domestic

muscovy duck (*Cairina moschata*) was domesticated by aboriginal South Americans prior to the European colonization of the Americas.

Ducks are being increasingly used in a nonconsumptive fashion. For example, bird watchers often go to great efforts to see ducks and other birds, trying to view as many species as possible, especially in natural habitats. Like hunters, birders spend a great deal of money while engaging in their sport, to travel, to purchase binoculars and books, and to belong to birding, natural-history, and **conservation** organizations.

Factors affecting the abundance of ducks

The best aquatic habitat for ducks and other waterfowl are those with relatively shallow water, with very productive vegetation and large populations of invertebrates. Those habitats with a large **ratio** of shoreline to surface area, favors the availability of secluded nesting sites. These sorts of habitat occur to some degree in most regions, and are primarily associated with **wetlands**, especially marshes, swamps, and shallow, open water. In North America and elsewhere during the past century, extensive areas of these types of wetlands have been lost or degraded, mostly because they have been drained or filled in for agricultural, urban, or industrial use. Wetlands have also been degraded by **eutrophication** caused by excessive nutrient inputs, and by pollution by toxic chemicals and organic materials. These losses of habitat, in combination with overhunting, have caused large decreases in the populations of ducks throughout North America, and in most other places where these birds occur. Consequently, there are now substantial efforts to preserve or restore the wetlands required as habitat by ducks and other **wildlife**, and to regulate hunting of these animals.

The most important breeding habitats for ducks in North America occur in the fringing marshes and shallow open-water wetlands of small ponds in the prairies, known as "potholes." The marshy borders of potholes provide important breeding habitat for various species of dabbling ducks such as mallard, pintail, widgeon, and blue-winged teal, while deeper waters are important to lesser scaup, canvasbacks (*Aythya valisneria*), redheads (*Aythya americana*), and ruddy ducks. Unfortunately, most of the original **prairie** potholes have been filled in or drained to provide more land for agriculture. This extensive conversion of prairie wetlands has increased the importance of the remaining potholes as breeding habitat for North America's declining populations of ducks, and for other wildlife. As a result, further conversions of potholes are resisted by the conservation community, although agricultural interests still encourage the drainage of these important wetlands.

In years when the prairies are subject to severe **drought**, many of the smaller potholes are too dry to allow ducks to breed successfully, and ponds and wetlands farther to the north in Canada become relatively important for breeding ducks. Another important source of natural mortality of ducks and other waterfowl are infectious **disease**, such as avian **cholera**, which can sweep through dense staging or wintering populations, and kill tens of thousands of birds in a short period of **time**. When an **epidemic** of avian cholera occurs, wildlife managers attempt to manage the problem by collecting and burning or burying as many carcasses as possible, in order to decrease the exposure of living birds to the pathogen.

Lead shot is an important type of toxic pollution that kills large numbers of ducks and other birds each year. Lead shot from spent shotgun pellets on the surface mud and sediment of wetlands where ducks feed, may be ingested during feeding and retained in the duck's gizzard. There the shot is abraded, dissolved by acidic stomach fluids, absorbed into the **blood**, and then transported to sensitive organs, causing toxicity. An estimated 2-3% of the autumn and winter duck population of North America (some 2-3 million birds) dies each year from lead toxicity. As few as one or two pellets retained in the gizzard can be enough to kill a duck. Fortunately, **steel** shot is rapidly replacing lead shot, in order to reduce this unintended, toxic hazard to ducks and other wildlife.

Ducks and other aquatic birds may also be at some risk from acidification of surface waters as a result of **acid rain**. Although it is unlikely that acidification would have direct, toxic effects on aquatic birds, important changes could be caused to their habitat, which might indirectly affect the ducks. For example, fish are very sensitive to acidification, and losses of fish populations would be detrimental to fish-eating ducks such as mergansers. However, in the absence of the predation pressure exerted by fish in acidic lakes, aquatic invertebrates would become more abundant, possibly benefitting other species of ducks such as common goldeneye, ring-necked duck, and black duck. These scenarios are inevitably speculative, for not much is known about the effects of acid rain on ducks.

Ducks can also be affected by eutrophication in aquatic habitats, a condition characterized by large increases in productivity caused by large nutrient loads from sewage dumping or from the runoff of agricultural **fertilizers**. Moderate eutrophication often improves duck habitat by stimulating plant growth and their invertebrate grazers. However, intense eutrophication kills fish and severely degrades the quality of aquatic habitats for ducks and other wildlife.

Some species of ducks nest in cavities in trees, a **niche** that has become increasingly uncommon because of **forestry** and losses of woodlands to agriculture and urbanization. Together with overhunting, the loss of natural cavities was an important cause of the decline of the wood duck and hooded merganser (*Lophodytes cucullatus*) in North America. Fortunately, these species will nest in artificial cavities provided by humans, and these ducks have recovered somewhat, thanks in part to widespread programs of nest box erection in wetland habitats.

Agencies and actions

Because the populations of ducks and other waterfowl have been badly depleted by overhunting and habitat loss, conservation has become a high priority for governments and some private agencies. In North America, the U.S. Fish and Wildlife Service and the Canadian Wildlife Service have responsibilities for waterfowl at the federal level, as do states and provinces at the regional level. Ducks Unlimited is a non-governmental organization whose central concern is the conservation of duck populations. The Ducks Unlimited mandate is mostly pursued by raising and spending money to increase duck productivity through habitat management, with an aim of providing more birds for hunters. Other organizations have a non-consumptive mandate that is partly relevant to ducks, for example, the World Wildlife Fund, The Nature Conservancy, and the Nature Conservancy of Canada. On the international stage, the *Convention on Wetlands of International Importance, Especially as Waterfowl Habitat* is a treaty among national governments intended to facilitate worldwide cooperation in the conservation of wetlands, thereby benefiting ducks and other wildlife.

All of these agencies are undertaking important activities on behalf of ducks, other animals, and natural ecosystems. However, duck populations are still much smaller than they used to be, and some species are endangered. Much more must be done to provide the ducks of North America and the world with the protection and habitat that they require.

Status

- Mallard (*Anas platyrhynchos*). Often poisoned by lead bullets. Has been poisoned in the West by foraging in temporarily damp **lake** beds. One of the most abundant ducks in the world today. The population on the Great Plains seems to have been permanently diminished from historical levels. The status of wild mallards is unclear due to the large feral populations.

- Mottled duck (*Anas fulvigula*). Has suffered more from encroaching human habitation (draining and destruction of marshland) and agriculture than from hunting. Interbreeding with feral mallards threatens the genetic purity of the species.

- American black duck (*Anas rubripes*). The population has decreased in response to aerial spraying for **spruce** budworm, destruction of habitat, acid rain, overhunting, clearing of **forests**, and **competition** with the mallard (with which it hybridizes).

- Gadwall (*Anas strepera*). This bird's range has been eastward. Settlement of the northern Great Plains took a relatively large toll on this species. Current populations vary each year, but the population does not seem to be diminishing.

- Green-winged teal (*Anas crecca*). Audubon wrote in 1840 that hunters in the West shot six dozen of these birds upon their first migratory arrival. Today the population appears stable.

- American wigeon (*Anas americana*). Population apparently stable. Since the 1930s, the breeding range has expanded into eastern Canada and the northeastern United States.

- Northern pintail (*Anas acuta*). This bird's nests in fields are often plowed up. It has also suffered lead shot

poisoning. There is some indication of a decline in population since the 1960s, but the species is widespread and abundant today. Droughts on the northern plains may drastically reduce nesting success there.

- Northern shoveler (*Anas clypeata*). Population apparently stable.

- Blue-winged teal (*Anas discors*). Population apparently stable. Because this bird usually winters in Latin America, international cooperation is required to protect it.

- Cinnamon teal (*Anas cyanoptera*). Although the population has suffered by encroaching human habitation and agriculture (draining of wetlands and diverting water for **irrigation**), the current numbers appear stable.

- Ruddy duck (*Oxyura jamaicensis*). Current population is much less than historical levels, due mainly to shooting in the early 1900s and loss of breeding habitat.

- Masked duck (*Oxyura dominica*). Uncommon everywhere, but wide ranging in the tropics. Its secretive and nomadic **behavior** makes it hard to estimate this duck's population, or to protect it.

- Fulvous whistling duck (*Dendrocygna bicolor*). Population has declined in the Southwest in recent decades, but increased in the Southeast. There is some controversy over its effect on rice cultivation; some say it damages **crops**, others that it eats the weeds in the fields.

- Black-bellied whistling duck (*Dendrocygna autumnalis*). This bird is hunted mainly in Mexico. It will use nesting boxes. The population in the United States has increased greatly since the 1950s. Rare in Arizona before 1949, this bird in now a common nesting bird in that state.

- Wood duck (*Aix sponsa*). This duck has been hunted for its plumage, as a food source, and for its eggs. Development and forestry practices contributed to its decline. By the early 1900s, this bird was on the verge of **extinction**, but has since made a comeback. The wood duck readily uses nesting boxes.

- Canvasback (*Aythya valisineria*). The fact that breeding grounds have been reduced due to draining and cultivating of prairie potholes and freshwater marshes is probably responsible for the observed decline in numbers.

- Redhead (*Aythya americana*). Current population is well below historical levels, probably due to loss of nesting areas.

- Ring-necked duck (*Aythya collaris*). This duck's breeding range expanded eastward in the mid-1900s. The population suffered from lead shot poisoning. Since the 1930s, this bird has become a widespread breeder in eastern Canada and northern New England.

- Greater scaup (*Aythya marila*). Abundant. The fact that this bird congregates in large numbers in coastal bays in winter has caused concern that the species may be vulnerable to **oil spills** and other **water pollution**.

- Lesser scaup (*Aythya affinis*). Abundant, with relatively small fluctuations from year to year.

- Common eider (*Somateria mollissima*). Down from this duck, collected during incubation, is commercially valued. The taking of down, however, usually does not result in desertion of the nest. The populations have been increasing and stabilizing since 1930. Today this species is abundant, with a population estimated at several million. Local populations may be threatened by oil spills and other water pollution.

- King eider (*Somateria spectabilis*). Commercially valued as a source of down. Abundant in the Far North, with a population estimated to be several million.

- Spectacled eider (*Somateria fischeri*). Threatened or endangered. The population in the Yukon-Kuskokwim **delta** of western Alaska declined by 96% from 1970 to 1993. The status of the Siberian population is not well known.

- Steller's eider (*Polysticta stelleri*). Population in Alaska has declined significantly in recent decades.

- Labrador duck (*Camptorhynchus labradorius*). Extinct. The last known specimen was shot in 1875 on Long Island. Never abundant, this duck had a limited breeding range. Its extinction probably resulted from loss of habitat and hunting.

- Black scoter (*Melanitta nigra*). Population apparently stable. Birds at sea vulnerable to oil and other forms of pollution.

- White-winged scoter (*Melanitta fusca*). Population has declined in parklands and boreal forest of Canada, possibly due to advancing the hunting season to 2-3 weeks before some of the young can fly. Population today is apparently stable.

- Surf scoter (*Melanitta perspicillata*). Population apparently declined greatly in the early 1900s, but is now stable. Wintering populations are vulnerable to oil and other forms of pollution.

- Harlequin duck (*Histrionicus histrionicus*). The population appears stable in the Northwest. In the eastern part of North America, there has been a substantial decline over the past century.

- Oldsquaw (*Clangula hyemalis*). Abundant, with a population estimated to be in the millions. The tendency to congregate in large numbers makes it vulnerable to oil spills in northern seas. Large numbers of these birds are sometimes caught and killed in fishing nets.

- Barrow's goldeneye (*Bucephala islandica*). Population apparently stable. Will use nesting boxes.

- Common goldeneye (*Bucephala clangula*). Population apparently stable. Readily uses nesting boxes.

- Bufflehead (*Bucephala albeola*). Fairly common and widespread. Today less numerous than historically, due to unrestricted shooting in the early 1900s, and loss of habitat. Uses nesting boxes when cavities in trees are scarce.

- Common merganser (*Mergus merganser*). Population apparently stable in North America, possibly increasing in **Europe**.

- Red-breasted merganser (*Mergus serrator*). Population apparently stable.

- Hooded merganser (*Lophodytes cucullatus*). Population has declined due to loss of nesting habitat (large trees near water). Will use nesting boxes and cavities set up for wood ducks. Today the population appears to be increasing.

- Mandarin duck (*Aix galericulata*). Exotic. A native of **Asia**, this duck occasionally escapes, ending up in the wild.

- Spot-billed duck (*Anas poecilorhyncha*). An Alaskan stray. Native of Asia.

- Tufted duck (*Aythya fuligula*). A Western stray. The Eurasian counterpart of the North American Ring-necked duck. This bird occasionally reaches Alaska and the Pacific Coast from Asia, or the Northeast from Europe and Iceland.

See also Eutrophication.

Resources

Books

Bellrose, F.C. *Ducks, Geese, and Swans of North America.* Harrisburg, PA: Stackpole Books, 1976.

Forshaw, Joseph. *Encyclopedia of Birds.* New York: Academic Press, 1998.

Freedman, B. *Environmental Ecology.* 2nd ed. San Diego: Academic Press, 1995.

Godfrey, W.E. *The Birds of Canada.* Toronto: University of Toronto Press, 1986.

Johnsgard, P.A. *Ducks in the Wild. Conserving Waterfowl and Their Habitats.* Swan Hill Press, 1992.

Owen, M., and J.M. Black. *Waterfowl Ecology.* London: Blackie Pub., 1990.

Peterson, Roger Tory. *North American Birds.* Houghton Miflin Interactive (CD-ROM), Somerville, MA: Houghton Miflin, 1995.

Bill Freedman
Randall Frost

Duckweed

Duckweeds are small, floating to slightly submerged **species** of flowering plants in the genus *Lemna*. The simple body is leaf-like, generally flat on top and convex below, lacks stems or leaves, is oval to tear-dropped in shape, and has one unbranched root that lacks vascular (conducting) **tissue**. The upper surface of the **plant** is covered with waxy compounds so as to shed **water**.

Duckweeds are abundant throughout the world in **freshwater** ponds, lakes, and backwaters where the water is still, with the exception of the Arctic. Plants range in size from 0.05-0.8 in (1.5-20 mm) in length. One of the most widely distributed species, *Lemna minor,* typically grows to a length of 0.05-0.15 in (1.5-4 mm).

Reproduction in duckweeds is almost exclusively asexual, occurring as outgrowths from one end breaks off, often resulting in the development of a dense, green mat on the surface of the water. **Individual** bodies are generally short-lived, five to six weeks for *Lemna minor.* **Sexual reproduction** is rare in duckweeds, and appears to occur mostly in warmer regions. Flowers are unisexual and extremely simple, consisting of only one stamen in male and one pistil in female flowers. Each **flower** arises from a pouch in the body and is covered by a small, highly modified **leaf** called a spathe.

The watermeal (*Wolffia*) is a close relative of duckweed, and is the smallest flowering plant. Some species of watermeal consist of only a globular, rootless body, as small as 0.02 in (0.5 mm). Duckweeds and watermeals are an important food for waterfowl, which feed on these plants on the water surface.

Duikers

Duikers are small African antelopes in the large family of Bovidae. This family of hoofed animals includes antelope, **gazelles**, cattle, **sheep**, and **goats**. Like all bovids, duikers have even-toed hooves, horns, and a four-chambered stomach structure that allows them to digest a diet of plants. Duikers are found throughout sub-Saharan **Africa**. These small antelopes range in size from 22 in (55 cm) to as much as 57 in (1.45 m) in length, and weigh from as little as 3 lb (1.35 kg) to as much as 176 lb (80 kg).

There are 17 **species** of forest-dwelling duikers (*Cephalophus*). These are the blue, yellow-backed, bay, Maxwell's, Jentink's, black-fronted, red-flanked, Abbot's, banded or zebra, black, red, Ader's duiker, Peter's, Harvey's or Zanzibar duiker, bay or black-striped, Gabon or white-bellied, and Ogilby's duiker. There is only one species of **savanna** duiker, the grey (or Grimm's) duiker (*Sylvicapra grimmia*), which is found in thin forest and savanna woodlands. Duikers are heavily hunted for their meat, and many of the forest species are threatened or endangered.

Adaptation

Duikers have not been studied to any great extent in the wild because they live in dense **rainforest** habitats and are difficult to observe. They are, nonetheless, much sought after for their meat. The number of species of duikers increases with the size of the rainforests they inhabit.

The word duiker means divers or those that duck in Afrikaans. When duikers are alarmed they dive for cover into thickets. The front legs of duikers are shorter than the powerful hind legs. Duikers have a relatively big head, with a wide mouth, small ears, short backward-slanting horns, and a crest of erect hair on the forehead. Female duikers are on average a little larger than males and also possess horns.

Some species are active only during daylight hours (diurnal), some are active only at night (nocturnal), while others are active during both times of day. Duikers are browsers, that is, animals that eat the tender shoots, twigs, and leaves of bushy plants, rather than grazing on grass. Some species of duikers also eat buds, **seeds**, **bark**, and fruit, and even small **rodents** and **birds**. The moisture content of leaves is usually sufficient to satisfy their **water** needs during the rainy season, when duikers do not drink. Duikers are preyed upon by leopards, large predatory birds, and even **baboons**.

Social life

Duikers are not social animals and are usually seen alone or in pairs. Like other small browsing antelopes, duikers are territorial and monogamous (they mate for life). The size of the territory of a pair of duikers is between 5-10 acres (2-4 ha), and both sexes defend it from intrusion by other members of their species.

The care of the young is done mainly by the females. The first born offspring leaves its parents before a younger sibling is born. Within the territory, the male and female duikers rest and feed at different times, and often wander away from one another, which may be why they are seen alone so often.

The **courtship** ceremony of duikers includes close following of the female by the male, circling and turning by the female, hiding, moaning and snorting by both, mutual scenting, and then mating. Females give **birth** to

KEY TERMS

. . . .

Bovidae (Bovids)—A family of animals characterized by having even-toed hooves, horns, and a four-chambered stomach and by chewing its cud as part of its digestive process.

Browsers—Mammals that feed primarily on leaves of plants, as opposed to grazing on grass.

Diurnal—Refers to animals that are mainly active in the daylight hours.

Facultative monogamy—Among pairs of mated animals where the care of the young is left primarily to the female member and the offspring leaves the parents before the next sibling is born.

Nocturnal—Animals that are mainly active in the nighttime.

Savanna—A treeless plain of high grasses found in tropical climates.

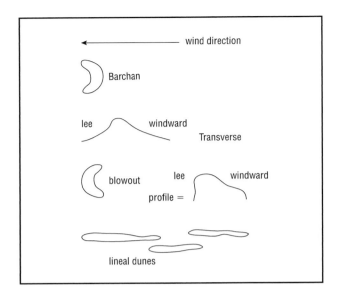

Dunes form different characteristic shapes depending on the amount of sand, amount of moisture, and the strength and prevailing direction of the wind (i.e., windward to leeward). *Illustration by Argosy. The Gale Group.*

one offspring at a time and may have from one to three young a year. Pregnancy usually lasts between four to five months or longer. The newborn duiker lies concealed for the first few weeks of its life when it is nursed by its mother.

Male duikers have scent **glands** underneath their eyes and on their hooves. The glands under their eyes extend downward and secrete through a series of pores rather than through one opening, as in other antelopes. Duikers in captivity have been seen to mark their territory as frequently as six times within ten minutes. Male duikers also use their scent glands to mark their opponents in battle, as well as engaging in mutual marking with their mates and their offspring.

Resources

Books

Estes, Richard D. *Behavior Guide to African Mammals.* Berkeley: University of California, 1991.
The Safari Companion. Post Mills, VT: Chelsea Green, 1993.
Spinage, C. A. *The Natural History of Antelopes.* New York: Facts on File, 1986.

Vita Richman

Dune

A dune is a wind-blown pile of **sand**. Over **time**, dunes become well-sorted deposits of materials by **wind**

or **water** that take on a characteristic shape and that retain that general shape as material is further transported by wind or water.

Desert dunes classifications are based upon shape include barchan dunes, relic dunes, transverse dunes, lineal dunes, and parabolic dunes. Dunes formed by wind are common in desert area and dunes formed by water are common in coastal areas. Dunes can also form on the bottom of flowing water (e.g., stream and river beds).

When water is the depositing and shaping agent, dunes are a bedform that are created by saltation and deposition of particles unable to be carried in suspension. Similar in shape to ripples—but much larger in size—dunes erode on the upstream side and extend via deposition the downstream or downslope side. Regardless of whether deposited by wind or water, dunes themselves move or migrate much more slowly than any individual deposition particle.

The sediments that accumulate on the windward slope are called topset deposits. When they reach the crest, they form an unstable and temporary surface called the brink. When enough sediments are captured on the brink they eventually tumble over the edge onto the slipface. This **motion** provides the advancement of the dune as it migrates in the direction of the wind. A temporary halt in dune movement can make a thin layer of sediments that become slightly bonded to one another. This layer becomes visible in side view and is even more recognizable in ancient deposits.

Distant view of red sand dunes in Namibia. *Photograph by Jan Halaska/Photo Researchers, Inc. Reproduced by permission.*

The sand forming dunes is usually composed of the mineral quartz eroded from **rocks**, deposited along streams or oceans or lakes, picked up by the wind, and redeposited as dunes. Sand collects and dunes begin to form in places where the wind speed drops suddenly, behind an obstacle such as a rock or bush, for example, and can no longer transport its load of sand.

Dunes move as wind bounces sand up the dune's gently-sloping windward side (facing the wind) to the peak of the slope where the wind's speed drops and sends sand cascading down the steeper lee side (downwind). As this process continues, the dune migrates in the direction the wind blows. The steeper lee side of the dune, called the slip face, maintains a 34° **angle** (called the angle of repose), much greater than the flatter (10°-12°) windward side. The sand may temporarily build up to an angle greater than 34°, but eventually it avalanches back to the angle of repose. Given enough sand and time, dunes override dunes to thicknesses of thousands of feet, as in the Sahara Desert, or

as in the fossilized dunes preserved in the sandstone of Zion and Arches National Parks in Utah. In the famous Navajo Sandstone in Zion National Park, crossbeds (sloping bedding planes in the rock) represent the preserved slip faces of 180 million year old former dunes.

Three basic dune shapes—crescent, linear, and star—range in size up to 330 ft (100 m) high and up to 1,000 ft (300 m) long and wide. Barchan and parabolic dunes are crescent-shaped like the letter C. Barchans form where the sand supply is minimal. The two ends of the barchan's crescent point downwind toward the direction the dune moves. In contrast, the pointed ends of a parabolic dune stab into the wind, a mirror image of a barchan. Bushes or some other obstruction anchor the tips of a parabolic dune.

Transverse and longitudinal dunes form as long, straight, or snake-like ridges. Transverse ridges run **perpendicular** to a constant wind direction, form with an abundance of sand available, and are asymmetric in **cross section** (the windward side gently-sloped, the slip face steep). The ridges of longitudinal dunes, however, run **parallel** to a slightly varying wind direction and are symmetrical in cross section—they have slip faces on either side of the ridge. Longitudinal dunes are also known as linear or seif (Arabic for sword) dunes.

Dune fields are large features of eolian or arid environments. They are associated with hot climate deserts such as the Sahara. Dune fields are not, however, exclusively restricted to these types of environments. Many dune fields are found in temperate climates where the processes of aridity in an arid climate combine to form dunes, but at a much slower **rate** than hot, arid climates.

Dune fields themselves are complex environments. Within the field, there are many microenvironments that lie between the dunes and at the bottom of dune valleys. Moisture may even accumulate and form small ponds. Scientists continue to study dunes and dune fields. They are one of the least understood structures in **geology** because of the difficulty in studying them. However, dune fields occur over about 30% of Earth's surface and certainly command more attention.

The formation and movements of dune fields are also of great interest to extraterrestrial or planetary geologists. Analysis of **satellite** images of **Mars**, for example, allow calculation of the strength and direction of the Martian winds and provide insight into Martian atmospheric dynamics. Dunes fields are a significant Martian **landform** and many have been observed to have high rates of **migration**.

See also Desertification; Erosion; Sediment and sedimentation; Sedimentary environment; Sedimentary rock.

Resources

Books

Goudie, Andrew S., et. al. *Aeolian Environments, Sediments, and Landforms.* Hoboken, NJ: John Wiley & Sons, 2000.

Tack, Francis, and Paul Robin. *Dunes.* Paris: Vilo International, 2002.

Other

United States Geological Survey. "Types of Dunes." (cited February 24, 2003) <http://pubs.usgs.gov/gip/deserts/dunes/>.

Brook Hall
K. Lee Lerner

Duplication of the cube

Along with squaring the **circle** and trisecting an **angle**, duplication of a cube is considered one of the three "unsolvable" problems of mathematical antiquity.

According to tradition, the problem of duplication of the cube arose when the Greeks of Athens sought the assistance of the oracle at Delos in order to gain relief from a devasting **epidemic**. The oracle told them that to do so they must double the size of the altar of Apollo which was in the shape of a cube.

Their first attempt at doing this was a misunderstanding of the problem: They doubled the length of the sides of the cube. This, however, gave them eight times the original **volume** since $(2x)^3 = 8\ x^3$.

In modern notation, in order to fulfill the instructions of the oracle, we must go from a cube of side x units to one of y units where $y^3 = 2x^3$, so that y = cube-root-2 x.

Thus, essentially, given a unit length, they needed to construct a line segment of length cuberoot-2 units. Now there are ways of doing this but not by using only a compass and an unmarked straight edge-which were the only tools allowed in classical Greek **geometry**.

Thus there is no solution to the Delian problem that the Greeks would accept and, presumably, the epidemic continued until it ran its accustomed course.

Resources

Books

Stillwell, John. *Mathematics: Its History.* Springer-Verlag, 1991.

Roy Dubisch

Dust devil

A dust devil is a relatively small, rapidly rotating **wind** that stirs up dust, **sand**, leaves, and other material as it moves across the ground. Dust devils are also known as whirlwinds or, especially in **Australia**, willy-willys. In most cases, dust devils are no more than 10 ft (3 m) in width and less than 300 ft (100 m) in height.

Dust devils form most commonly on hot dry days in arid regions such as a **desert**. They originate when a layer of air lying just above the ground is heated and begins to rise. Cooler air then rushes in to fill the **space** vacated by the rising column of warm air.

At some point, the rising column of air begins to spin. Unlike much larger storms such as hurricanes and tornadoes, dust devils may rotate either cyclonically or anticyclonically. Their size is such that the **earth's rotation** appears to have no effect on their direction of spin, and each direction occurs with approximately equal frequency. The determining factor as to the direction any one dust devil takes appears to be the local topography in which the

A dust devil in Kenya. *JLM Visuals. Reproduced by permission.*

storm is generated. The presence of a small hill, for example, might direct the storm in a cyclonically direction.

Some large and powerful dust devils have been known to cause property damage. In the vast majority of cases, however, such storms are too small to pose a threat to buildings or to human life.

DVD

In 1995, Philips and Sony introduced the digital video disc (DVD), which had the same dimensions as a standard compact disk (CD), but was able to store up to 4.7 gigabytes of data, such as high-definition digital video files. This is more than three times the capacity of a CD. DVD players use a higher-power **laser** than that used for CDs, which enables smaller pits (0.4 micrometre) and separation tracks (0.74 micrometre) to be used.

To record a DVD, semiconductor red lasers, with wavelengths of 630 nm, "burn" grooves into the medium. The peaks and valleys created are interpreted as binary numbers by the computer. To increase the amount of information stored, data is compressed using a form of lossy compression such as MPEG-2, an industry standard for audio and video sanctioned by a governing body called the Moving Pictures Experts Group. This technology strips unnecessary or redundant data from videos. The compression allows 135 minutes on a single side of an optical disc. Some DVD discs use two sides of the disc for longer movies, while others put a wide-screen version on one side of the disc and a standard 4:3 version on the other. But this was simply the first incarnation of DVD. Some discs now feature a dual-layer technology, meaning that a single side of the disc actually holds two separate MPEG video streams, like an upstairs/downstairs apartment. This allows a single side of the disc to hold 4-1/2 hours of video. Soon, there will be dual-sided/dual-layer discs, which will double the capacity to nine hours.

Most DVDs are write-once read-only disks. DVD-RAMs are now available for home use, allowing the user to record data on the DVD. These DVD-RAM drives can copy 315 megabytes (MB) of data in about 12 minutes; read and write DVD-RAM discs (as large as 5.2 gigabytes [GB]); and can read CD-ROM, CD-R, CD-RW, and DVD-ROM discs.

New technology pioneered by several major companies, such as Pioneer, introduced a rewriting method known as phase-change recording. To re-record, the disk is reheated with a laser at a different phase than the initial recording. The laser **light** hits the disk at a slightly different angle, changing the shape of the grooves. The data layer does not experience wear and tear because DVDs are read by laser light and never physically touched by mechanics. The data layer is coated with a protective plastic substrate.

The DVD defines the capability to display movies in three different ways. The wide-screen format provides a special anamorphic video signal that, when processed by a wide-screen **television** set, fills the entire screen and delivers optimum picture quality. Pan and Scan fills the screen of traditional 4:3 television sets with an entire picture, much like watching network movies. The Letterbox mode provides horizontal bands at the top and bottom to, in essence, create a wide-screen picture in a traditional television set.

DVD-Video supports multiple aspect ratios. Video stored on a DVD in 16:9 format is horizontally squeezed to a 4:3 (standard TV) aspect **ratio**. On wide-screen TVs, the squeezed image is enlarged by the TV to an aspect ratio of 16:9. DVD video players output wide-screen video in three different ways: letterbox (for 4:3 screens), pan and scan (for 4:3 screens), anamorphic or unchanged (for wide screens).

At the moment, DVD players resemble VCRs, however the race is on to make DVD players smaller and less expensive. Several companies developed a DVD player the size of a personal CD player with an integrated liquid **crystal** display viewing panel and speakers. A Chinese manufacturer incorporated the DVD player with the TV itself. Higher capacity and higher definition is not far off.

At the January 2000 Consumer Electronics Show, Pioneer showed off its high-definition DVD Recorder. Analysts say that high-definition DVD (HD-DVD) will become a reality when blue lasers, emitting wavelengths of 400-430 nm are perfected. The HD-DVD will require a storage capacity of 15 GB per side, more than three times that of current DVDs. A shorter wavelength laser increases the amount of information stored on an optical disk. Each time the wavelength is halved, the corresponding storage media can contain four times more data.

In order to promote the technology and establish a consensus on format standards, ten companies organized the DVD Consortium in 1995. These companies included: Hitachi, Ltd., Matsushita Electronic Industrial Co., Ltd., Mitsubishi Electric Corp., Philips Electronics N.V., Pioneer Electronics Corp., Sony Corp., THOMSON multimedia, Time Warner Inc., Toshiba Corp. and Victor Company of Japan Ltd. Today, the Forum boasts 122 member companies, including electronics manufacturers, software firms, and media companies worldwide.

Resources

Periodicals

Allingham, Philip. "DVDs Not Just Movies Anymore." *ZDNet AnchorDesk* (January 8, 2000).
Goldberg, Ron. "Digital Video Primer, One Pus Zero Equals a Whole Lot of Fun." *Home Theater* (September 1997).
Toupin, Laurie. "The Home of the Future." *Design News* (February 18, 1999).

Laurie Toupin

Dwarf antelopes

These small antelopes belong to the ruminant family Bovidae, and are grouped with the **gazelles** in the subfamily Antilopinae. The 13 **species** of dwarf antelopes are in the tribe Neotragini. Dwarf antelopes range from extremely small (3.3-4.4 lb or 1.5-2 kg) hare-sized royal antelopes and **dik-diks** to the medium-sized oribi and beira weighing from 30-50 lb (10-25 kg). Dwarf antelopes engage in territorial scent marking and possess highly developed scent **glands**. They are browsers, consuming a diet of young green leaves, fruit, and buds. Dwarf antelopes are also usually not dependent upon regular supplies of drinking **water** for their survival.

The food of the herbivorous dwarf antelope is digested by means of the four-chambered ruminant stomach. Dwarf antelopes browse or graze, consuming vegetation that is nutritionally rich. They lightly chew their food as they tear leaves from branches. After the food is swallowed, it enters the rumen of the stomach. Digestion is then aided by the process of **bacteria** breaking down **nutrients**. The food pulp is then regurgitated and chewed as cud to further break down the food before being swallowed and digested more completely.

Habitat

Dwarf antelopes are found in various terrains throughout the sub-Saharan regions of **Africa**. The klipspringer (*Oreotragus oreotragus*) is found in rocky areas in eastern and southern Africa. Four species of dik-diks are found in dry bush country in the horn of Afria, that is, Somalia and Ethiopia, as is the beira antelope (*Dorcatragus megalotis*). Oribis (*Ourebia ourebi*) are found in the **savanna** country, from West to East Africa and in parts of southern Africa.

Steenbok (*Raphicerus campestris*) inhabit bushy plains or lightly wooded areas in southern Africa, while the two species of grysbok (*R. melanotis* and *R. sharpei*) are found in stony, hilly areas and scrubby flat country in east-central Africa and the extreme south of the **continent**. The royal antelope (*Neotragus pygmaeus*) is found in dense **forests** in West Africa, and Bates' pygmy antelope (*N. batesi*) is found in the forests of the Zaire. Suni antelopes (*N. moschatus*) live in forests along the southeastern edge of Africa.

Characteristics

The horns of dwarf antelopes are short, straight spikes found only in the males, although klipspringer females sometimes have horns. Colorations are usually pale, varying from yellow to gray or brown with a white rump patch, while the steenbok is brick-colored. All dwarf antelopes have well-developed scent glands, particularly preorbital glands which can be easily seen on most species as dark slits beneath the eyes. Dwarf antelopes generally have narrow muzzles, prominent ears, and their nostrils are either hairy or bare.

Dwarf antelopes are territorial and many are in lifetime monogamous relationships. They tend to be solitary even though a mated pair shares the same territory. Terri-

KEY TERMS

Monogamy—Mating relationship where a male and female tend to become permanently paired.

Preorbital scent glands—Glands located below the eyes that are used to mark territory.

Scent-mark—To spread urine, feces, or special fluid from a body gland along a trail to let competitive animals know that the territory is taken.

tories can range in size from several hundred square feet to tens of acres depending upon the nature of the territory and the density of the group's population. Some monogamous pairs may have a second female, usually a female offspring that has not left the parental territory. Some dwarf antelope males may have two or more females within a small territory.

Scenting **behavior** among dwarf antelopes maintains the mating bond and protects the territory from intruders. Males mark their territory with the scent glands found under their eyes (preorbital glands), and on their hooves (pedal glands). They can mark both the ground of their territory, as well as branches and bushes. Additionally, males will scent their mates, which strengthens the ties between them. Ceremonial behavior in dunging is also seen. A pair will follow one another and deposit urine and feces on the same pile.

Males can be aggressive in defending their territories. They have been known to use their sharp horns to wound intruders. Usually, however, male rivals for females will more often only display aggressive behavior to one another before one retreats. The display of aggressive behavior can include pawing the ground, horning, alarm calls, chasing, and pretending to attack.

Parenting

Dwarf antelope females give **birth** to one offspring at a time, coinciding with seasonal rains. The gestation period is around six months, depending on the species. Infants hide in the grass for several weeks and the mother returns to feed them twice a day. As the fawn grows, it begins to follow the mother. Young females mature by the age of 6-10 months, while males reach maturity around 14 months. Somewhere between 9-15 months, young dwarf antelopes leave the territory to establish themselves on their own.

In klipspringer families, the pair are found close together, on the average 12-45 ft (4-15 m) apart, and the male assumes the role of lookout while the female cares

for the offspring. The male may even become involved with feeding the young klipspringer.

See also Antelopes and gazelles.

Resources

Books

Estes, Richard D. *Behavior Guide to African Mammals.* Berkeley: University of California, 1991.
Estes, Richard D. *The Safari Companion.* Post Mills, Vermont: Chelsea Green, 1993.
Haltenorth, T., and H. Diller. *A Field Guide to the Mammals of Africa.* London: Collins, 1992.

Vita Richman

Dwarf and mouse lemurs *see* **Lemurs**

Dyes and pigments

Color scientists use the term "colorant" for the entire **spectrum** of coloring materials, including dyes and pigments. While both dyes and pigments are sources of color, they are different from one another. Pigments are particles of color that are insoluble in **water**, oils, and **resins**. They need a binder or to be suspended in a dispersing agent to impart or spread their color. Dyes are usually water soluble and depend on physical and/or **chemical reactions** to impart their color. Generally, soluble colorants are used for coloring **textiles**, **paper**, and other substances while pigments are used for coloring paints, inks, cosmetics and **plastics**. Dyes are also called dyestuffs. The source of all colorants is either organic or inorganic.

Colorants are classified according to their chemical structure or composition (organic or inorganic), method of application, hue, origin (natural or synthetic), dyeing properties, utilization, and, sometimes, the name of the manufacturer and place of origin. The Society of Dyers and Colourists and the American Association of Textile Chemists and Colorists have devised a classification system, called the Color Index, that consists of the common name for the color, and a five-digit identification number.

Organic and inorganic colorants

Organic colorants are made of **carbon** atoms and carbon-based molecules. Most organic colors are soluble dyes. If an organic soluble dye is to be used as a pigment, it must be made into particle form. Some dyes are insoluble and must be chemically treated to become soluble.

Vegetable-based organic colorants are produced by obtaining certain extracts from the plants. An example of a dye that is not water soluble is indigo. Indigo is derived from plants of the genus *Indigofera*. By an oxidation process where the **plant** is soaked and allowed to ferment, a blue-colored, insoluble solid is obtained. To get the indigo dye into **solution**, a reducing agent (usually an alkaline substance such as caustic soda) is used. The blue dye, after reduction, turns a pale yellow. Objects dyed with indigo react with the air, oxidize, and turn blue. The imparted color is not always that of the dye itself. Animals are another, rather interesting, source of organic colorants. Royal purple, once worn only by royalty as the name suggests, is obtained from the Murex snail. Sepia is obtained from **cuttlefish**, and Indian yellow is obtained from the urine of cows that have been force-fed mango leaves.

Organic sources of color often have bright, vivid hues, but are not particularly stable or durable. Dyes that are not affected by **light** exposure and washing are called colorfast, while those that are easily faded are called fugitive. Most organic natural dyes need a fixing agent (mordant) to impart their color.

Inorganic colorants are insoluble, so by definition, they are pigments. This group of colorants is of mineral origin—elements, oxides, gemstones, salts, and complex salts. The **minerals** are pulverized and mixed with a dispersing or spreading agent. Sometimes heating the minerals produces different hues.

Synthetic colorants

Organic and inorganic colorants can be produced synthetically. Synthetic organic and inorganic colorants are copies of vegetable, **animal**, and mineral-based colorants, and are made in a laboratory. Until the nineteenth century, all colorants were of natural origin. The first synthetically made commercial colorant, mauve, was developed from aniline, a **coal** tar derivative, by William Henry Perkins in 1856. Today, chemists arrange and manipulate complex organic compounds to make dyes of all colors. Synthetic dyes, made in a controlled atmosphere, are without impurities and the colors are more consistent from batch to batch. Natural dyes still have some commercial value to craftspeople, but synthetic colorants dominate the manufacturing industry.

Pigments

The color of a pigment is deposited when the spreading agent dries or hardens. The physical property of a pigment does not change when it is mixed with the agent. Some organic dyes can be converted into pigments. For example, dyes that have **salt** groups in their chemical structure can be made into an insoluble salt by replacing the **sodium molecule** with a **calcium** molecule. Dyes that depend on chemical treatment to become soluble such as

indigo, can also be used as pigments. Pigments are also classified, in addition to classification mentioned above, by their color—white, transparent, or colored.

Pigments are also used for other purposes than just coloring a medium. Anticorrosive pigments, such oxides of **lead**, are added to paint to prevent the rusting of objects made of **iron**. Metallic pigments such as **aluminum**, bronze, and nickel are added to paints and plastics for decorative, glittery effects. Pulverized mica produces a sparkle effect and bismuth oxychloride gives a pearlescent appearance to paints and cosmetics.

Luminous pigments have the ability to radiate visible light when exposed to various **energy** sources. The luminous pigments that emit light after exposure to a light source and placed in the dark are called phosphorescent or commonly, glow-in-the-dark. Phosphorescent pigments are made from zinc or calcium sulfides and other mineral additives that produce the effect. Another good example of how dyes are made into pigments are some of the fluorescent pigments. Fluorescent pigments are those that are so intense that they have a glowing effect in daylight. These pigments are added to various resins, ground up, and used as a pigment. Some fluorescent pigments are illuminated by an ultraviolet light source (black light).

Dyes

Dyes are dissolved in a solution and impart their color by staining or being absorbed. What makes one organic source a dye and another not depends on a particular groups of **atoms** called chromophores. Chromophores include the azo group, thio group, nitroso group, **carbonyl group**, nitro group, and azoxy group. Other groups of atoms called auxochromes donate or accept electrons and attach to the dye molecule, enhance the color and increase **solubility**. Auxochrome groups include amino, hydroxyl, sulfonic, and substituted amino groups.

Other than chemical structure, dyes are classified by their dyeing properties. There are a great number of dyes and a greater number of fibers and materials that incorporate colorants in their manufacture. Certain dyes are used for specific materials depending on the chemical properties of the dye and the physical properties of the material to be dyed, or dyeing properties. Dyeing properties are categorized as basic or cationic, acid and premetalized, chrome and mordant, direct, **sulfur**, disperse, vat, azoic, and reactive dyes.

Utilization

Every manufactured object is colored by a dye or pigment. There are about 7,000 dyes and pigments, and new ones are patented every year. Dyes are used extensively in the textile industry and paper industry. Leather and **wood** are colored with dyes. Food is often colored with natural dyes or with a synthetic dye approved by a federal agency. Petroleum-based products such as waxes, lubricating oils, polishes, and gasoline are colored with dyes. Plastics, resins, and rubbers are usually colored by pigments. Dyes are used to stain biological samples, fur, and hair. Special dyes are added to photographic emulsions for color photographs.

Resources

Books

Gottsegen, Mark D. *The Painter's Handbook.* New York: Watson-Guptill Publications, 1992.

Gutcho, M. H., ed. *Inorganic Pigments: Manufacturing Processes.* 1980.

Lyttle, Richard B. *Paints, Inks, and Dyes.* New York: Holiday House, 1974.

Christine Miner Minderovic

Dynamics *see* **Newton's laws of motion**

Dysentery

Dysentery is an infectious **disease** that has ravaged armies and prisoner-of-war camps throughout history. The disease still is a major problem in tropical countries with primitive sanitary facilities. Refugee camps in **Africa** resulting from many civil wars are major sinks of infestation for dysentery.

Shigellosis

The acute form of dysentery, called *shigellosis* or bacillary dysentery, is caused by the bacillus (bacterium) of the genus *Shigella*, which is divided into four subgroups and distributed worldwide. Type A, *Shigella dysenteriae,* is a particularly virulent **species**. **Infection** begins from the solid waste from someone infected with the bacterium. **Contaminated soil** or **water** that gets on the hands of an **individual** often is conveyed to the mouth, where the person contracts the infection. **Flies** help to spread the bacillus.

Young children living in primitive conditions of overcrowded populations are especially vulnerable to the disease. Adults, though susceptible, usually will have less severe disease because they have gained a limited resistance. Immunity as such is not gained by infection, however, since an infected person can become reinfected by the same species of *Shigella*.

Once the bacterium has gained entrance through the mouth it travels to the lower intestine (colon) where it penetrates the mucosa (lining) of the intestine. In severe cases the entire colon may be involved, but usually only the lower half of the colon is involved. The incubation period is one to four days, that is the **time** from infection until symptoms appear.

Symptoms may be sudden and severe in children. They experience abdominal **pain** or distension, fever, loss of appetite, nausea, vomiting, and diarrhea. **Blood** and pus will appear in the stool, and the child may pass 20 or more bowel movements a day. Left untreated, he will become dehydrated from loss of water and will lose weight rapidly. Death can occur within 12 days of infection. If treated or if the infection is weathered, the symptoms will disappear within approximately two weeks.

Adults experience a less severe course of disease. They will initially feel a griping pain in the abdomen, develop diarrhea, though without any blood in the stool at first. Blood and pus will appear soon, however, as episodes of diarrhea recur with increasing frequency. Dysentery usually ends in the adult within four to eight days in mild cases and up to six weeks in severe infections.

Shigella dysenteriae brings about a particularly virulent infection that can be fatal within 12-24 hours. The patient has little or no diarrhea, but experiences delirium, convulsions, and lapses into a **coma**. Fortunately infection with this species is uncommon.

Treatment of the patient with dysentery usually is by fluid therapy to replace the liquid and electrolytes lost in sweating and diarrhea. **Antibiotics** may be used, but some *Shigella* species have developed resistance to them, so they may be relatively ineffective. Fluid therapy should be tendered with great care because patients often are very thirsty and will overindulge in fluids if given access to them. A hot water bottle may help to relieve abdominal cramps.

Some individuals can harbor the bacterium without having symptoms. Like those who are convalescent from the disease, the carriers without symptoms can spread the disease. This may occur by someone with improperly washed hands preparing food, which becomes infected with the **organism**.

Amebic dysentery

Another form of dysentery called amebic dysentery or intestinal amebiasis is spread by a protozoan, *Entamoeba histolytica*. The protozoan occurs in an active form, that which infects the bowel, and an encysted form, that which forms the source of infection. If the patient develops diarrhea the active form of **amoeba** will pass from the bowel and rapidly die. If no diarrhea is present the amoeba will form a hard cyst about itself and pass from the bowel to be picked up by another victim. Once ingested it will lose its shell and begin the infectious cycle. Amebic dysentery can be waterborne, so anyone drinking infested water that is not purified is susceptible to infection.

Amebic dysentery is common in the tropics and relatively rare in temperate climates. Infection may be so subtle as to be practically unnoticed. Intermittent bouts of diarrhea, abdominal pain, flatulence, and cramping mark the onset of infection. Spread of infection may occur with the organisms entering the liver, so abdominal tenderness may occur over the area of the liver. Because the amoeba invades the lining of the colon, some bleeding may occur, and in severe infections the patient may require blood transfusions to replace that which is lost.

Treatment again is aimed at replacement of lost fluids and the relief of symptoms. Microscopic examination of the stool will reveal the active protozoan or its cysts. Special medications aimed at eradicating the infectious organism may be needed.

An outbreak of amebic dysentery can occur seemingly mysteriously because the carrier of the amoeba may be without symptoms, especially in a temperate zone. This is the person with inadequate sanitation who can spread the disease through food that he has handled. Often the health officials can trace a disease outbreak back to a single kitchen and then test the cooks for evidence of amebic dysentery.

Before the idea of the spread of infectious agents was understood, dysentery often was responsible for more casualties among the ranks of armies than was actual combat. It also was a constant presence among prisoners who often died because little or no medical assistance was available to them. It is still a condition present throughout the world that requires vigilance. Prevention is the most effective means to maintain the health of populations living in close quarters. Hand washing, especially among food preparation personnel, and water purification are the most effective means of prevention. Adequate latrine facilities also help to contain any infectious human waste. A carefully administered packet of water and electrolytes to replace those lost can see a child through the infection.

See also Digestive system.

Larry Blaser

Dyslexia

Dyslexia is a disorder that falls under the broad category of **learning** disabilities. It is often described as a neurological **syndrome** in which otherwise normal people have difficulty reading and writing. Frequently, dyslexia is defined by what it is not—dyslexia is not mental retardation, a psychiatric or emotional disorder, or a **vision** problem. Dyslexia is not caused by poverty, psychological problems, lack of educational opportunities, or laziness; those who are identified as dyslectic have normal or above-normal intelligence, normal eyesight, and tend to come from average families.

There are dozens of symptoms associated with dyslexia. In reading and writing, those with dyslexia may skip words, reverse the order of letters in a word (for instance, writing or reading "was" for "saw"), or drop some letters from a word (for example, reading "run" instead of "running"). They may concoct strange spellings for common words, have difficulty remembering and following sequences (like reciting the alphabet in order), and have cramped, illegible handwriting. There is often a gap between what the person seems to be capable of doing and performance; it's not unusual for a student with dyslexia to earn straight As in science and fail English.

Reading and the brain

Sigmund Freud, the father of modern **psychiatry**, wrote in 1900 that painful childhood experiences or hatred of one or both parents caused dyslexia. He reasoned that children who could not openly rebel against a harsh mother or father defied their parents by refusing to learn to read. Freud recommended **psychoanalysis** to resolve such emotional problems.

Today, experts reject the psychoanalytic explanation of dyslexia. Sophisticated **brain** imaging technology known as functional magnetic **resonance** imaging (fMRI) clearly shows inactivity in a large area that links the angular gyrus—the visual cortex and visual association areas where print or writing is interpreted, to areas in the superior temporal gyrus (Wernicke's area) where language and phonetics are interpreted. Also, during phonologic reading tasks, the area associated with spoken language (Broca's area) showed activation in dyslexic readers where it did not in normal readers. Researchers believe this area may attempt to compensate for impairments in Wernicke's area.

Investigators have studied those with brain lesions (abnormal growths such as tumors) in Wernike's area. Although they had no reading difficulties before the lesion was large enough to detect, patients with brain lesions developed reading problems identical to those associated with dyslexia. Those with dyslexia also tend to have rapid, jerky, hard-to-control **eye** movements when they read—another indication of a misfire in the brain.

How we read

Reading is a complicated chain of events coordinated in the brain. Imagine a busy, computerized railroad yard: trains pull in on hundreds of tracks from all directions. The cargo of each train is documented in a central tower, then matched with a destination and assigned to one of dozens of tracks. Some trains may be sent to a holding area until their cargo is needed; others may be routed so they can make multiple stops. The computer system must analyze hundreds of pieces of information for each train pulling in, each train pulling out—so even a brief power failure can clog the railroad yard with thousands of trains, blocked from reaching their destination.

Scientists suspect a similar power failure in the brain is the cause of dyslexia. In normal reading, the eye sends pictures of abstract images (the printed word) to the brain. Each symbol is routed to various portions of the brain for processing or storage, symbols are interpreted and combined in combinations that make sense, then transferred to other portions of the brain that recognize the importance of the messages. Sometimes the messages are relayed to the lips, tongue, and jaw—reading aloud—or the fingers and hands—writing.

Investigators have identified three major "tracks" for sending written messages to the brain for interpretation. The phonic route recognizes individual letters and, over **time**, builds a list of groups of letters that generally appear together. The direct route is a "mental dictionary" of whole words recognized as a unit; the lexical route breaks strings of letters into a base word, prefixes, and suffixes. The lexical route might, for example, break the word "together" into "to-get-her." The areas of the brain responsible for channeling words along these different routes, processing them, then moving them along as a message that makes sense, coordinate thousands of pieces of information in normal reading. These bits of information are moved through the brain over neurons, the roadways of the **nervous system**, on neurotransmitters, naturally occurring chemicals that make it possible for messages to travel from one nerve **cell** to the next. In dyslexia, something jams the signals in the brain and interferes with the interpretation of the written word.

Causes of dyslexia

Researchers generally agree that **genetics** play a role in dyslexia. Studies of twins show that if one twin is dyslexic, the other is far more likely to have the disorder. Other studies show that dyslexia, which affects about 8% of the population, tends to run in families. It's common for a child with dyslexia to have a parent or other close relative with the disorder.

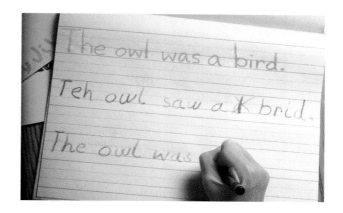

A student with dyslexia has difficulty copying words. © Will & Deni McIntyre/Science Source, National Audubon Society Collection/ Photo Researchers, Inc. Reproduced with permission.

Because dyslexia affects males far more often than females (the **rate** is about three to one), investigators are exploring the relationship of male **hormones** to dyslexia. Several studies indicate that an excess of the male hormone testosterone prior to **birth** may slow the development of the left side of the fetus's brain. Other researchers argue, however, that those with dyslexia rarely have problems with spoken language, which is also controlled by the left side of the brain and depends on some of the same areas that control reading and writing.

Treating dyslexia

Those with mild cases of dyslexia sometimes learn to compensate on their own, and many with dyslexia reach remarkable levels of achievement. Leonardo Da Vinci, the famous Renaissance inventor and artist who painted the Mona Lisa, is thought to have been dyslexic; so was Albert Einstein.

The severity of the disorder, early **diagnosis**, and prompt treatment seem to be the keys to overcoming the challenges of dyslexia. Linguistic and reading specialists can help those with dyslexia learn how to break reading and writing into specific tasks, how to better remember and apply reading skills, and how to independently develop reading and writing skills. Studies with community college students indicate that intensive sessions with a specialist significantly increase a student's reading and writing skills, and experts believe earlier intervention is even more effective.

Although dyslexia occurs independently, it can spark social, behavioral, and emotional problems. Children with dyslexia may be frustrated by their inability to understand and embarrassed by their "failure" in the classroom. They may perceive themselves as "stupid" and develop problems with self-esteem and motivation.

Future developments

Researchers are exploring the use of various drugs known to affect chemical activity in the brain. Although fMRI can not yet be used as a diagnostic tool, its use has proven the neurobiologic root to dyslexia and may help in devising methods of treatment.

Resources

Books

Selikowitz, Mark. *Dyslexia and Other Learning Difficulties: The Facts.* New York: Oxford University Press, 1993.

Ziminsky, Paul C. *In a Rising Wind: A Personal Journey Through Dyslexia.* Lanham: University Press of America, 1993.

Other

Facts About Dyslexia. National Institutes of Health, National Institutes of Child Health and Human Development. Washington, DC: U.S. Government Printing Office, 1993 (U.S. Government Printing Office NIH Publication 93-0384-P).

A. Mullig

Dysplasia

Dysplasia is a combination of two Greek words; *dys*, which means difficult or disordered; and *plassein*, to form. In other words, dysplasia is the abnormal or disordered formation of certain structures. In medicine, dysplasia refers to cells that have acquired an abnormality in their form, size, or orientation with respect to each other.

Dysplasia may occur as the result of any number of stimuli. Sunburned skin, for example, is dysplastic, but will correct itself as the sunburned skin heals itself. Any source of irritation causing **inflammation** of an area will result in temporary dysplasia. If the source of irritation is removed the dysplasia will rectify itself, and **cell** structure and organization will return to normal.

Unfortunately, dysplasia can become permanent. This can occur when a source of irritation to a given area cannot be identified and corrected, or for completely unknown reasons. The continually worsening area of dysplasia can develop into an area of malignancy (**cancer**). A tendency toward dysplasia can be genetic and/or can result from exposure to irritants or toxins, such as **cigarette smoke**, viruses, or chemicals.

The Pap smear, a medical procedure commonly performed on women, is a test for cervical dysplasia. The degree of dysplasia present in cervical cells can indicate progression to a cancerous condition.

Dysprosium *see* **Lanthanides**

Dystrophinopathies

Dystrophinopathies are progressive hereditary degenerative diseases (often called muscular dystrophies) of skeletal muscles due to an absence or deficiency of the protein dystrophin.

Dystrophin and the associated **proteins** form a complex system that connects the intracellular cytoskeleton to the extracellular matrix. The normal operation of this system is critical for maintaining the integrity of the delicate, elastic muscle **membrane** (sarcolemma) and the muscle fiber. The responsible **gene** is located on the short arm of the X **chromosome** at **locus** Xp21. It is an extremely large gene, comprising more than 2.5 million base pairs and 79 exons. The dystrophin gene produces several isoforms (alternative forms of a protein) of dystrophin. Seven distinct promoters have been identified, each driving a tissue-specific dystrophin.

The most common mechanisms of **mutation** are deletions and duplications largely clustered in the "hot spot," a DNA sequence associated with an abnormally high frequency of mutation or recombination, between exons 44 and 49. Whether the deletion is in the reading frame or out of frame determines whether dystrophin is absent from the muscle or present in a reduced, altered form. This has an important clinical significance because the former is usually associated with the severe Duchenne's variety of the **disease** (DMD), whereas the latter situation may cause the milder Becker's variant

(BMD). Thirty to 40% of DMD/BMD cases are associated with point mutations in the dystrophin gene.

In approximately two-thirds of cases, the dystrophin gene defect is transmitted to affected boys by carrier females following the pattern of Mendelian X-linked recessive inheritance. However, in one-third of cases, the defect arises as a result of a new mutation in the germ cells of parents or during very early embryogenesis.

DMD occurs at a frequency of one per 3,500 live births. The child who appears healthy at **birth** develops the initial symptoms around age five. Symptoms include clumsy gait, slow running, difficulty in getting up from the floor, difficulty in climbing stairs, and a waddling gait. Mental subnormality, if it occurs, is not progressive and is presumed to be due to the lack of **brain** dystrophin. The progression of muscle weakness and loss of function for the activities of daily living is relentless. In the typical situation, the child loses the ability to walk independently by age 9-12 years. Despite all therapeutic efforts, most patients die during the third decade. BMD is an allelic variant of DMD in which the mutation of the dystrophin gene produces a reduced amount of truncated dystrophin that is not capable of maintaining the integrity of the sarcolemma. However, the pace of muscle fiber loss is considerably slower than in DMD, which is reflected in a less severe clinical phenotype. The illness usually begins by the end of the first or at the beginning of the second decade. These boys, however, continue to walk independently past the age of 15 years and may not have to use a wheelchair until they are in their twenties or even later. In some people carrying a mutation in the dystrophin gene, no muscular symptoms are present at all

Confirmation of clinical **diagnosis** of DMD/BMD is largely based on deletion analysis of DNA. In the 30% of patients in whom a deletion is not found, a muscle biopsy is necessary to establish the absence of dystrophin by immunohistochemistry, or Western **blotting analysis**.

Treatment of dystrophinoptahies is palliative, aimed at managing the symptoms in an effort to optimize the quality of life. **Gene therapy** which is oriented towards replacement of the defective dystrophin gene with a wild-type one (or a functionally adequate one) or upregulation of the expression of the surrogate molecules such as utrophin is still in the research phase.

Prevention of dystrophinopathies is based on genetic counseling and molecular genetic diagnosis. Female carriers can namely opt for prenatal diagnosis (evaluation in the first trimester of pregnancy of whether the fetus has inherited the mutation) or even preimplantation genetic analysis of the embryo. In the last case, **fertilization** in vitro is followed by isolation and **genetic testing** of the single **cell** from the few-cell-stage embryo. The embryo is implanted into the uterus if no dystrophin gene defect is found.

See also Genetic disorders; Genetic engineering

Resources

Books

Barohn, R. J. "Muscular Dystrophies." In L. Goldman, and J.C. Bennett, eds. *Cecil Textbook of Medicine,* 21st ed. W.B. Saunders Co., 2002.

Brown, Susan C., ed. *Dystrophin.* Cambridge, UK: Cambridge University Press, 1997.

Periodicals

Muscular Dystrophy Association USA. "Progress in Utrophin Therapy for DMD." *Research Updates.* (February 7, no. 1 2000).

Organizations

Muscular Dystrophy Association. 3300 E. Sunrise Drive Tucson, AZ 85718. Tel. (800) 572-1717. <http://www.mdausa.org/>.

Other

National Institutes of Health. National Institute of Neurological Disorders and Stroke. "NINDS Muscular Dystrophy (MD) Information Page" May 29, 2001 [cited February 24, 2003]. <http://www.ninds.nih.gov/health_and_medical/disorders/md.htm>.

Borut Peterlin

E

e (number)

The number e, like the number **pi**, is a useful mathematical constant that is the basis of the system of natural **logarithms**. Its value correct to nine places is 2.718281828... The number e is used in complex equations to describe a process of growth or decay. It is therefore utilized in the **biology**, business, demographics, **physics**, and **engineering** fields.

The number e is widely used as the base in the exponential **function** $y = Ce^{kx}$. There are extensive tables for e^x, and scientific calculators usually include an e^x key. In **calculus**, one finds that the slope of the graph of e^x at any **point** is equal to e^x itself, and that the **integral** of e^x is also e^x plus a constant.

Exponential functions based on e are also closely related to sines, cosines, hyperbolic sines, and hyperbolic cosines: $e^{ix} = \cos x + i\sin x$; and $e^x = \cosh x + \sinh x$. Here i is the **imaginary number** RADIC-1. From the first of these relationships one can obtain the curious equation $e^{iPI} + 1 = 0$, which combines five of the most important constants in **mathematics**.

The constant e appears in many other formulae in **statistics**, science, and elsewhere. It is the base for natural (as opposed to common) logarithms. That is, if $e^x = y$, then $x = \ln y$. ($\ln x$ is the symbol for the natural logarithm of x.) $\ln x$ and e^x are therefore inverse functions.

The expression $(1 + 1/n)^n$ approaches the number e more and more closely as n is replaced with larger and larger values. For example, when n is replaced in turn with the values 1, 10, 100, and 1000, the expression takes on the values 2, 2.59..., 2.70..., and 2.717....

Calculating a decimal **approximation** for e by means of the this definition requires one to use very large values of n, and the equations can become quite complex. A much easier way is to use the Maclaurin series for e^x: $e^x = 1 + x/1! + x^2/2! + x^3/3! + x^4/4! + ...$ By letting x equal 1 in this series one gets $e = 1 + 1/1 + 1/2$ + 1/6 + 1/24 + 1/120 +.... The first seven terms will yield a three-place approximation; the first 12 will yield nine places.

Eagles

Eagles are large, diurnal **birds of prey** in the subfamily Buteonidae, which also includes **buzzards** and other broad-winged **hawks**. The buteonids are in the order Falconiformes, which also includes **falcons**, osprey, goshawks, and **vultures**.

Like all of these predatory **birds**, eagles have strong, raptorial (or grasping) talons, a large hooked beak, and extremely acute **vision**. Eagles are broadly distinguished by their great size, large broad wings, wide tail, and their soaring flight. Their feet are large and strong, armed with sharp claws, and are well-suited for grasping **prey**. Some **species** of eagles are uniformly dark-brown colored, while others have a bright, white tail or head. Male and female eagles are similarly colored, but juveniles are generally dark. Female eagles are somewhat larger than males.

Species of eagles occur on all of the continents, except for **Antarctica**. Some species primarily forage in terrestrial habitats, while others are fish-eating birds that occur around large lakes or oceanic shores. Eagles are fierce predators, but they also scavenge carrion when it is available.

North American eagles

The most familiar and widespread species of eagle in **North America** is the bald eagle (*Haliaeetus leucocephalus*). Mature bald eagles have a dark-brown body, and a white head and tail. Immature birds are browner and lack the bold white markings on the tail and head. They gradually develop the rich adult plumage, which is complete when the birds are sexually mature at four to five

The harpy eagle (*Harpia harpyja*) dwells in the forests of southeastern Mexico, Central America, and South America (as far south as Paraguay and northern Argentina), where it hunts monkeys, sloths, porcupines, reptiles, and large birds. Since it prefers virgin forest, its numbers have decreased wherever there is regular human access to forest habitat. *Photograph by Robert J. Huffman. Field Mark Publications. Reproduced by permission.*

years of age. The bald eagle mostly feeds on **fish** caught or scavenged in **rivers**, lakes, ponds, and coastal estuaries.

Bald eagles nest on huge platforms built of sticks, commonly located on a large **tree**. Because the nests are used from year to year, and new sticks are added each breeding season, they can eventually weigh several tons. Northern populations of bald eagles commonly migrate to the south to spend their non-breeding season. However, these birds are tolerant of the cold and will remain near their breeding sites as long as there is open **water** and a dependable source of fish to eat. Other birds winter well south of their breeding range.

The golden eagle (*Aquila chrysaetos*) is an uncommon species in North America, breeding in the northern

tundra, in mountainous regions, and in extensively forested areas. The golden eagle also breeds in northern **Europe** and **Asia**. This species has dark-brown plumage, and its wingspan is as great as 6.5 ft (2 m). These birds can predate on animals as large as young **sheep** and **goats**, but they more commonly take smaller **mammals** such as **marmots** and ground **squirrels**.

Golden eagles nest in a stick nest built on a large tree or a cliff. As with the bald eagle, the nest may be used for many years, and may eventually become a massive structure. Usually, two to three white-downed eaglets are hatched, but it is uncommon for more than one to survive and fledge. It takes four to five years for a golden eagle to become sexually mature.

Eagles elsewhere

The largest species of eagle is the harpy eagle (*Harpia harpyja*) of tropical **forests** of **South America**. This species mostly feeds on **monkeys** and large birds. The Philippine monkey-eating eagle (*Pithecophaga jefferyi*) and New Guinea harpy eagle (*Harpyopsis novaguineae*) are analogous species in Southeast Asia.

The sea eagle or white-tailed eagle (*Haliaeetus albicilla*) is a widespread species that breeds in coastal habitats from Greenland and Iceland, through Europe, to Asia. This species has a dark-brown body and white tail. Another fishing eagle (*H. vocifer*) breeds in the vicinity of lakes and large rivers in **Africa**.

The imperial eagle (*Aquila heliaca*) and spotted eagle (*A. clanga*) are somewhat smaller versions of the golden eagle, breeding in plains, steppes, and other open habitats from central Asia to Spain and northwestern Africa. These birds tend to eat smaller-sized mammals than the golden eagle.

The short-toed or snake eagle (*Circaetus gallicus*) breeds extensively in mountainous terrain in southern Europe and southwestern Asia. This species feeds on small mammals and **snakes**. Because it predates on large numbers of poisonous **vipers**, the short-toed eagle is highly regarded by many people living within its range.

The black eagle (*Ictinaetus malayensis*) is a species of tropical forest, ranging from India and southern China to the islands of Java and Sumatra in Indonesia.

Eagles and humans

Because of their fierce demeanor and large size, eagles have long been highly regarded as a symbol of power and grace by diverse societies around the world. Eagles have figured prominently in religion, mythology, art, literature, and other expressions of human culture.

KEY TERMS

. . . .

Diurnal—Refers to animals that are mainly active in the daylight hours.

Extirpated—The condition in which a species is eliminated from a specific geographic area of its habitat.

Raptor—A bird of prey. Raptors have feet adaptive for seizing, and a beak designed for tearing.

In North America, for example, the bald eagle is an important symbol in many Native American cultures. Many tribes believe that the feathers of this bird have powerful qualities, and they use these to ornament clothing and hats, or will hold a single feather in the hand as a cultural symbol and source of strength. Various tribes of the Pacific coast know the bald eagle as the "thunder bird," and they accord it a prominent place on totem poles.

Today, most North Americans regard the bald eagle as a valued species, and it is even a national symbol of the United States. However, some people consider eagles to be **pests**, believing the birds to be predators of domestic animals such as sheep, or of economically important fish. For these reasons, many eagles have been killed using guns, traps, and poison. Fortunately, these misguided attitudes about eagles are now in an extreme minority, and very few people still persecute these magnificent predators.

However, eagles and many other species of **raptors** are also damaged by other, less direct, human influences. These include the toxic effects of **insecticides** used in agriculture, some of which accumulate in wild animals and affect them or their reproduction. Eagles have also been poisoned by eating poisoned carcasses set out to kill other scavengers, such as coyotes or wolves. Eagles are also affected by ecological changes in their necessary breeding, migrating, and wintering habitats, especially damages caused by agriculture, urbanization, and **forestry**.

Because of these and other damaging effects of human activities, most of the world's species of eagles are much less abundant than they were a century or so ago. Many local populations of these magnificent birds have become endangered or have actually been extirpated. In more extreme cases of endangerment, some species are at risk of total biological **extinction**. The monkey-hunting harpy eagle, for example, is an extremely rare bird that requires extensive tracts of tropical **rainforest** in South America and is endangered because of its critically small and declining population, which has resulted mostly from **deforestation**.

Resources

Books

Forshaw, Joseph. *Encyclopedia of Birds.* New York: Academic Press, 1998.

Freedman, B. *Environmental Ecology.* 2nd ed. San Diego: Academic Press, 1995.

Gerrard, J., and G. Bortolotti. *The Bald Eagle.* Washington, DC: Smithsonian Press, 1988.

Johnsgard, P. A. *Hawks, Eagles, and Falcons of North America: Biology and Natural History.* Washington, DC: Smithsonian Press, 1990.

Savage, C. *Eagles of North America.* New York: Douglas & McIntyre, 1988.

Bill Freedman

Ear

The human ear is the anatomical structure responsible for **hearing** and balance. The ear consists of three parts: the outer, middle, and inner ears.

Outer ear

The outer ear collects sounds from the environment and funnels them through the auditory system. The outer ear is composed of three parts, the pinna (or auricle), the external auditory canal (or external auditory meatus), and the tympanic **membrane** (or eardrum).

Pinna

The two flap-like structures on either side of the head commonly called ears are actually the pinnas of the outer ear. Pinnas are skin-covered cartilage, not bone, and are therefore flexible. The lowest portion of the pinna is called the lobe or lobule and is the most likely site for earrings. The pinnas of most humans cannot move, but these structures are very mobile in other **mammals**, such as **cats** and dogs.

External auditory canal

The external auditory canal is a passageway in the temporal lobe of the skull that begins at the ear and extends inward and slightly upwards. In the adult human it is lined with skin and hairs and is approximately 1 in (2.5 cm) long.

The outer one-third portion of the canal is lined with a membrane containing ceruminous (ear wax producing) cells, and hair cells. The purpose of the cerumen and hairs is to protect the eardrum (which lies at the end of the canal) by trapping dirt and foreign bodies and keep-

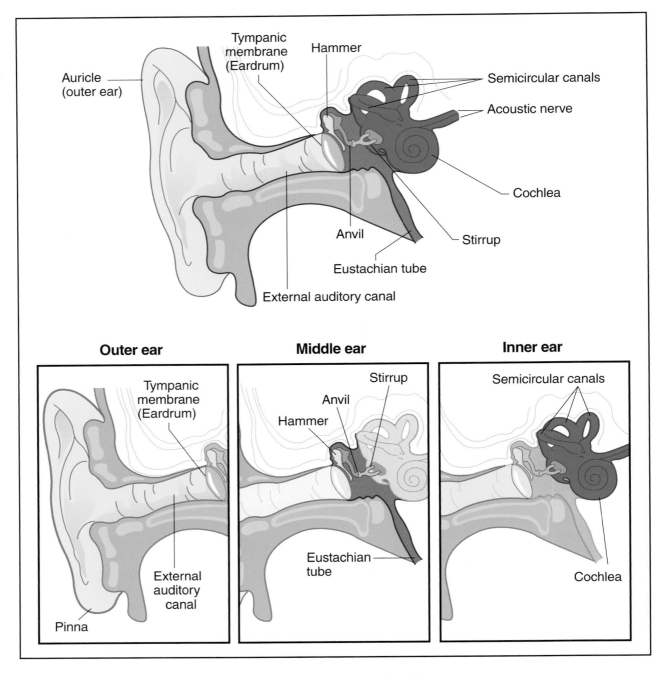

The anatomy of the human ear. *Illustration by Hans & Cassidy. Courtesy of Gale Group.*

ing the canal moist. In most individuals, cleaning of the external auditory canal (with Q-tips for example) is not needed. The inner two-thirds of the external auditory canal contains no **glands** or hair cells.

Tympanic membrane/eardrum

The human tympanic membrane or eardrum is a thin, concave membrane stretched across the inner end of the external auditory canal much like the skin covering the top of a drum. The eardrum marks the border between the outer ear and middle ear. The eardrum serves as a transmitter of sound by vibrating in response to sounds traveling down the external auditory canal, and beginning sound conduction in the middle ear.

In the adult human, the tympanic membrane has a total area of approximately 63 sq mm, and consists of three layers which contribute to the membrane's ability to vibrate while maintaining a protective thickness. The middle point of the tympanic membrane (the umbo) is attached to the stirrup, the first of three bones contained within the middle ear.

Middle ear

The middle ear transmits sound from the outer ear to the inner ear. The middle ear consists of an oval, air-filled space approximately 2 cubic cm in **volume**. The middle ear can be thought of as a room, the outer wall of which contains the tympanic membrane. The back wall, separating the middle ear from the inner ear, has two windows, the oval window and the round window. There is a long hallway leading away from the side wall of the room, known as the eustachian tube. The **brain** lies above the room and the jugular vein lies below. The middle ear is lined entirely with mucous membrane (similar to the nose) and is surrounded by the bones of the skull.

Eustachian tube

The eustachian tube connects the middle ear to the nasopharynx. This tube is normally closed, opening only as a result of muscle movement during yawning, sneezing, or swallowing. The eustachian tube allows for air **pressure** equalization, permitting the air pressure in the middle ear to match the air pressure in the outer ear. The most noticeable example of eustachian tube function occurs when there is a quick change in altitude, such as when a **plane** takes off. Prior to takeoff, the pressure in the outer ear is equal to the pressure in the middle ear. When the plane gains altitude, the air pressure in the outer ear decreases, while the pressure in the middle ear remains the same, causing the ear to feel "plugged." In response to this the ear may "pop." The popping sensation is actually the quick opening and closing of the eustachian tube, and the equalization of pressure between the outer and middle ear.

Bones/ossicles and muscles

Three tiny bones (the ossicles) in the middle ear form a chain which conducts **sound waves** from the tympanic membrane (outer ear) to the oval window (inner ear). The three bones are the hammer (malleus), the anvil (incus), and the stirrup (stapes). These bones are connected and move as a link chain might, causing pressure at the oval window and the transmission of **energy** from the middle ear to the inner ear. Sound waves cause the tympanic membrane to vibrate, which sets up vibrations in the ossicles, which amplify the sounds and transmits them to the inner ear via the oval windows.

In addition to bones, the middle ear houses the two muscles, the stapedius and the tensor tympani, which respond reflexively, that is, without conscious control.

Inner ear

The inner ear is responsible for interpreting and transmitting sound (auditory) sensations and balance (vestibular) sensations to the brain. The inner ear is small (about the size of a pea) and complex in shape, where its series of winding interconnected chambers, has been compared to (and called) a labyrinth. The main components of the inner ear are the vestibule, semicircular canals, and the cochlea.

Vestibule

The vestibule, a round open space which accesses various passageways, is the central structure within the inner ear. The outer wall of the vestibule contains the oval and round windows (which are the connection sites between the middle and inner ear). Internally, the vestibule contains two membranous sacs, the utricle and the saccule, which are lined with tiny hair cells and attached to nerve fibers, and serve as the vestibular (balance/equilibrium) sense organs.

Semicircular canals

Attached to the utricle within the vestibular portion of the inner ear are three loop-shaped, fluid filled tubes called the semicircular canals. The semicircular canals are named according to their location ("lateral," "superior," and "posterior") and are arranged **perpendicular** to each other, like the floor and two corner walls of a box. The semicircular canals are a key part of the vestibular system and allow for maintenance of balance when the head or body rotates.

Cochlea

The cochlea is the site of the sense organs for hearing. The cochlea consists of a bony, snail-like shell that contains three separate fluid-filled ducts or canals. The upper canal, the scala vestibuli, begins at the oval window, while the lower canal, the scala tympani, begins at the round window. Between the two canals lies the third canal, the scala media. The scala media is separated from the scala vestibuli by Reissner's membrane and from the scala tympani by the basilar membrane. The scala media contains the **organ** of Corti, (named after the nineteenth century anatomist who first described it). The organ of Corti lies along the entire length of the basilar membrane. The organ contains hair cells and is the site of the conversion of sound waves into nerve impulses, which are sent to the brain, for auditory interpretation along cranial nerve VIII, also known as the auditory nerve.

KEY TERMS

. . . .

Auricle—Also called pinna or external ear, is the flap-like organ on either side of the head.

Cerumen—Also known as ear wax, is an oily, fatty fluid secreted from glands within the external auditory canal.

Cochlea—A snail-shaped structure in the inner ear which contains the anatomical structures responsible for hearing.

Eustachian tube—A passageway leading from the middle ear to the nasopharynx or throat.

External auditory canal—Also called a meatus, is the tunnel or passageway which begins from the external ear and extends inward towards the eardrum.

Organ of Corti—A structure located in the scala media of the cochlea, contains hair cells responsible for hearing.

Ossicles—Three tiny, connected bones located in the middle ear.

Stapedius muscle—A muscle located in the middle ear which reflexively contracts in response to loud sounds.

Tympanic membrane—Also known as the eardrum, is a thin membrane located at the end of the external auditory canal, separates the outer ear from the middle ear.

Vestibular system—System within the body that is responsible for balance and equilibrium.

Resources

Books

Mango, Karin. *Hearing Loss.* New York: Franklin Watts, 1991.

Martin, Frederick. *Introduction to Audiology.* 4th ed. Upper Saddle River, NJ: Prentice Hall, 1991.

Moller, Aage R. *Sensory Systems: Anatomy and Physiology.* New York: Academic Press, 2002.

Periodicals

Mestel, Rosie. "Pinna to the Fore." *Discover* 14 (June 1993): 45–54.

Kate Glynn

Earth

Earth is our home **planet**. Its surface is mostly **water** (about 70%), and it has a moderately dense nitro-

gen-and-oxygen atmosphere that supports life—the only known life in the Universe. Rich in **iron** and nickel, Earth is a dense, rocky ball orbiting the **Sun** with its only natural **satellite**, the **Moon**. A complete revolution of the earth around the Sun takes about one year, while a **rotation** on its axis takes one day. The surface of Earth is constantly changing, as the continents slowly drift about on the turbulent foundation of partially molten rock beneath them. Collisions between landmasses build **mountains**; **erosion** wears them down. Slow changes in the climate cause equally slow changes in the vegetation and animals inhabiting a place.

Physical parameters of Earth

Earth is the third of nine planets in our **solar system**. It orbits the Sun at a **distance** of about 93,000,000 mi (150,000,000 km), taking 365.25 days to complete one revolution. Earth is small by planetary standards; with a diameter of 7,921 mi (12,756 km), it is only one-tenth the size of **Jupiter**. Its **mass** is 2.108×10^{26} oz (about six trillion kg), and it must speed this huge bulk along at nearly 19 mi (30 km) per second to remain in a stable **orbit**. The **mean** density of our planet is 5.5 grams per cubic centimeter. Unlike the outer planets, which are composed mainly of light gases, Earth is made of heavy elements such as iron and nickel, and is therefore much more dense. These characteristics—small and dense—are typical of the inner four planets, or terrestrial planets.

It was not until 1957, when the first man-made satellite was launched, that humans could see Earth as a beautiful whole. It seemingly floats in empty **space**, a world distinguished first by its vast oceans and only secondarily by its landmasses, everywhere draped in white swirls of **clouds**. This is a planet in the most fragile ecological balance, yet resilient to repeated catastrophe. It is our home, and it bears close examination.

The formation of Earth

About 4.5 billion years ago, our Sun was born from a contracting cloud of interstellar gas. The cloud heated as it shrank, until its central part blazed forth as the mature, stable **star** that exists today. As the Sun formed, the surrounding gas cloud flattened into a disk. In this disk the first solid particles formed and then grew as they accreted additional **matter** from the surrounding gas. Soon sub-planetary bodies, called planetesimals, built up, and then they collided and merged, forming the planets. The high temperatures in the inner solar system ensured that only the heavy elements, those that form rock and **metal**, could survive in solid form.

Thus were formed the small, dense terrestrial planets. Hot at first due to the collisions that formed it, Earth

The coastlines of Africa, Antarctica, and Arabia are visible in this photo of Earth taken in December, 1972, by Apollo 17. *U. S. National Aeronautics and Space Administration (NASA).*

began to cool. Its components began to differentiate, or separate themselves according to their density, much as the ingredients in a bottle of salad dressing will separate if allowed to sit undisturbed. To Earth's core went the heavy abundant elements, iron and nickel. Outside the core were numerous elements compressed into a dense but pliable substance called the mantle. Finally, a thin shell of cool, silicon-rich rock formed at Earth's surface: the crust, or **lithosphere**. Formation of the crust from the initial molten blob took half a billion years.

Earth's atmosphere formed as a result of outgassing of **carbon dioxide** from its interior, and accretion of gases from space, including elements brought to Earth by **comets**. The lightest elements, such as helium and most of the **hydrogen**, escaped to space, leaving behind an early atmosphere consisting of hydrogen compounds such as methane and **ammonia** as well as water vapor and nitrogen- and sulfur-bearing compounds released by

volcanoes. Carbon dioxide was also plentiful, but was soon dissolved in **ocean** waters and deposited in carbonate **rocks**. As the gases cooled, they condensed, and rains inundated the planet. The lithosphere was uneven, containing highlands made of buoyant rock such as granite, and basins of heavy, denser basalt. Into these giant basins the rains flowed, forming the oceans. Eventually life forms appeared, and over the course of a billion years, plants enriched the atmosphere with **oxygen**, finally producing the nitrogen-oxygen atmosphere we have today.

Earth's surface

Land

The lands of our planet are in a constant, though slow, state of change. Landmasses move, collide, and break apart according to a process called **plate tectonics**. The lithosphere is not one huge shell of rock; it is composed of several large pieces called plates. These pieces are constantly in **motion**, because **Earth's interior** is dynamic, with its core still molten and with large-scale convective currents in the upper mantle. The giant furnace beneath all of us moves our land no more than a few centimeters a year, but this is enough to have profound consequences.

Consider **North America**. The center of the **continent** is the magnificent expanse of the Great Plains and the Canadian Prairies. Flat and wide is the land around Winnipeg, Topeka, and Amarillo. On the eastern edge, the rolling folds of the Appalachian Mountains grace western North Carolina, Virginia, and Pennsylvania. In the west, the jagged, crumpled Rockies thrust skyward, tall, stark, and snow-capped.

These two great ranges represent one of the two basic land-altering processes: mountain building. Two hundred million years ago, North America was moving east, driven by the restless engine beneath it. In a shattering, slow-motion collision, it rammed into what is now **Europe** and North **Africa**. The land crumpled, and the ancient Appalachians rose. At that time, they were the mightiest mountains on Earth. A hundred million years later, North America was driven back west. Now the western edge of the continent rumbled along over the Pacific plate, and about 80 million years ago, a massive spate of mountain building formed the Rockies.

During the time since the Appalachians rose, the other land-altering process, erosion, has been hard at work on them. Battered by **wind** and water, their once sheer flanks have been worn into the low, rolling hills of today. Eventually they will be gone—and sometime long after that, so will the Rockies.

Mountain building can be seen today in the Himalayas, which are still rising as India moves northward into the underbelly of **Asia**, crumpling parts of Nepal and Tibet nearly into the stratosphere. Erosion rules in Arizona's Grand Canyon, which gradually is deepening and widening as the Colorado river slices now into ancient granite two billion years old. In time, the Canyon too will be gone.

This unending cycle of mountain building (caused by movement of the crustal plates) and erosion (by wind and water) has formed every part of Earth's surface today. Where there are mountains, as in the long ranks of the Andes or the Urals, there is subterranean conflict. Where a crustal plate rides over another one, burying and melting it in the hot regions below the lithosphere, volcanoes rise, dramatically illustrated by Mt. St. Helens in Washington and the other sleeping giants that loom near Seattle and Portland. Where lands lie wide and arid, they are sculpted into long, scalloped cliffs, as one sees in the deserts of New Mexico, Arizona, and Utah. Without ever being aware of it, we humans spend our lives on the ultimate roller coaster.

Water

Earth is mostly covered with water. The mighty Pacific Ocean covers nearly half the earth; from the proper vantage point in space one would see nothing but water, dotted here and there with tiny islands, with only **Australia** and the coasts of Asia and the Americas rimming the edge of the globe.

The existence of oceans implies that there are large areas of the lithosphere that are lower than others. This is because the entire lithosphere rides on a pliable layer of rock in the upper mantle called the **asthenosphere**. Parts of the lithosphere are made of relatively light rocks, while others are made of heavier, denser rocks. Just as corks float mostly above water while **ice** cubes float nearly submerged, the less dense parts of the lithosphere ride higher on the asthenosphere than the more dense ones. Earth therefore has huge basins, and early in the planet's history these basins filled with water condensing and raining out of the primordial atmosphere. Additional water was brought to Earth by the impacts of comets, whose nuclei are made of water ice.

The atmosphere has large circulation patterns, and so do the oceans. Massive streams of warm and cold water flow through them. One of the most familiar is the Gulf Stream, which brings warm water up the eastern coast of the United States.

Circulation patterns in the oceans and in the atmosphere are driven by **temperature** differences between adjacent areas and by the rotation of Earth, which helps create circular, or rotary, flows. Oceans play a critical role in the overall **energy** balance and **weather** patterns of our planet. Storms are ultimately generated by moisture in the atmosphere, and **evaporation** from the oceans is the prime source of such moisture. Oceans respond less dramatically to changes in energy input than land does, so the temperature over a given patch of ocean is far more stable than one on land.

Earth's atmosphere and weather

Structure of the atmosphere

Earth's atmosphere is the gaseous region above its lithosphere, composed of **nitrogen** (78% by number), oxygen (21%), and other gases (1%). It is only about 50 mi (80 km) from the ground to space: on a typical, 12-in (30 cm) globe the atmosphere would be less than 2 mm thick. The atmosphere has several layers. The most dense and significant of these is the troposphere; all weather occurs in this layer, and commercial jets cruise near its upper boundary, 6 mi (10 km) above Earth's surface. The stratosphere lies between 6 and 31 mi (10 and 50 km) above, and it is here that the **ozone** layer lies. In the mesosphere and the thermosphere one finds aurorae occurring after eruptions on the Sun; **radio** communications "bounce" off the ionosphere back to Earth, which is why you can sometimes pick up a Memphis AM radio station while you are driving through Montana.

The atmosphere is an insulator of almost miraculous stability. Only 50 mi (80 km) away is the cold of outer space, but the surface remains temperate. **Heat** is stored by the land and the atmosphere during the day, but the resulting heat **radiation** (infrared) from the surface is prevented from radiating away by gases in the atmosphere that trap infrared radiation. This is the well-known **greenhouse effect**, and it plays an important role in the atmospheric energy budget. It is well for us that Earth's climate is this stable. A global temperature decrease of two degrees could trigger the next advance of the current ice age, while an increase of three degrees could melt the **polar ice caps**, submerging every coastal city in the world.

Weather

Despite this overall stability, the troposphere is nevertheless a turbulent place. It is in a state of constant circulation, driven by **Earth's rotation** as well as the constant heating and cooling that occurs during each 24-hour period.

The largest circulation patterns in the troposphere are the Hadley cells. There are three of them in each hemisphere, with the middle, or Ferrel cell, lying over the latitudes spanned by the continental United States. Northward-flowing surface air in the Ferrel cell is deflected toward the east by the Coriolis **force**, with the result that winds—and weather systems—move from west to east in the middle latitudes of the northern hemisphere.

Near the top of the troposphere are the jet streams, fast-flowing currents of air that **circle** Earth in sinuous paths. If you have ever taken a commercial **plane** flight, you have experienced the **jet stream**: eastbound flights get where they are going much faster than westbound flights.

Circulation on a smaller scale appears in the cyclones and anticyclones, commonly called low and high **pressure** cells. Lows typically bring unsettled or stormy weather, while highs mean sunny skies. Weather in most areas follows a basic pattern of alternating pleasant weather and storms, as the endless progression of highs and lows, generated by Earth's rotation and temperature variation, passes by. This is a great simplification, however, and weather in any given place may be affected, or even dominated, by local features. The climate in Los Angeles is entirely different from that in Las Vegas, though the two cities are not very far apart. Here, local features—specifically, the mountains between them—are as important as the larger circulation patterns.

Beyond the atmosphere

Earth has a magnetic field that extends tens of thousands of kilometers into space and shields Earth from most of the **solar wind**, a stream of particles emitted by the Sun. Sudden enhancements in the solar wind, such as a surge of particles ejected by an eruption in the Sun's atmosphere, may disrupt the magnetic field, temporarily interrupting long-range radio communications and creating brilliant displays of aurorae near the poles, where the magnetic field lines bring the charged particles close to the earth's surface.

Farther out, at a mean distance of about 248,400 mi (400,000 km), is Earth's only natural satellite, the Moon. Some scientists feel that the earth and the Moon should properly be considered a "double planet," since the Moon is larger relative to our planet than the satellites of most other planets.

Life

The presence of life on Earth is, as far as we know, unique. Men have walked on the Moon, and it seems certain there is no life on our barren, airless satellite. Unmanned spacecraft have landed on **Venus** and **Mars** and have flown close to every other planet in the solar system except **Pluto**. The most promising possibility, Mars, yielded nothing to the automated experiments performed by the Viking spacecraft that touched down there.

The **origin of life** on Earth is not understood, but a promising experiment was performed in 1952 that may hold the secret. Stanley Miller and Harold Urey simulated conditions in Earth's early oceans, reproducing the surface and atmospheric conditions thought to have existed more than three billion years ago. A critical element of this experiment was simulated **lightning** in the form of an **electric arc**. Miller and Urey found that under these conditions, amino acids, the essential building blocks of life, had formed in their primitive "sludge." Certainly this was a long way from humans—or even an amoeba—but the experiment proved that the early Earth may have been a place where organic compounds, the compounds found in living creatures, could form.

Life has existed on dry land only for the most recent 10% of Earth's history, since about 400 million years ago. Once life got a foothold beyond the oceans, however, it spread rapidly. Within 200 million years **forests**

KEY TERMS

. . . .

Core—The innermost layer of Earth's interior. The core is composed of molten iron and nickel, and it is the source of Earth's magnetic field.

Erosion—One of the two main processes that alter Earth's surface. Erosion, caused by water and wind, tends to wear down surface features such as mountains.

Lithosphere—The outermost layer of Earth's interior, commonly called the crust. The lithosphere is broken into several large plates that move slowly about. Collisions between the plates produce mountain ranges and volcanism.

Mantle—The thick layer of Earth's interior between the core and the crust.

Mountain-building—One of the two main processes that alter Earth's surface. Mountain-building occurs where two crustal plates collide and crumple, resulting in land forms thrust high above the surrounding terrain.

Terrestrial planets—Planets with Earth-like characteristics relatively close to the Sun. The terrestrial planets are Mercury, Venus, Earth, and Mars.

Troposphere—The layer of air up to 15 mi (24 km) above the surface of the Earth, also known as the lower atmosphere.

spread across the continents and the first **amphibians** evolved into dinosaurs. **Mammals** became dominant after the demise of the dinosaurs 65 million years ago, and only in the last two million years have humans come onto the scene.

See also Antarctica; Cartography; Continental drift; Earth science; Earth's magnetic field; Earthquake; Geologic time; Geology; Hydrologic cycle; Paleontology; South America; Volcano.

Resources

Books

Beatty, J., and A. Chaikin. *The New Solar System.* Cambridge: Cambridge University Press, 1991.

Cater, John. *Key to the Future: The History of Earth Science.* New York: Routledge, 2002.

Hamblin, W. K., and E.H. Christiansen. *Earth's Dynamic Systems.* 9th ed. Upper Saddle River: Prentice Hall, 2001.

Hancock, P. L. and B. J. Skinner, eds. *The Oxford Companion to the Earth.* Oxford: Oxford University Press, 2000.

Press, F., and R. Siever. *Understanding Earth.* 3rd ed. New York: W. H Freeman and Company, 2001.

Periodicals

"The Dynamic Earth." *Scientific American* 249 (special issue, September 1983): 46–78+.

Haneberg, William C. "Determistic and Probabilistic Approaches to Geologic Hazard Assessment." Environmental & Engineering Geoscience 6, no. 3 (August 2000): 209–226.

Hoehler, T. M., B. M. Bebout, and D. J. Des Marais. "The Role of Microbial Mats in the Production of Reduced Gases on the Early Earth." *Nature* 412 (July 2001): 324–327.

Jeffrey C. Hall

Earth science

Befitting a dynamic **Earth**, the study of Earth science embraces a multitude of subdisciplines. At the **heart** of Earth science is the study of **geology**. Literally meaning "to study the Earth," traditional geological studies of **rocks**, **minerals**, and local formations have within the last century, especially in the light of the development of plate tectonic theory, broadened to include studies of **geophysics** and **geochemistry** that offer sweeping and powerful explanations of how continents move, to explanations of the geochemical mechanisms by which **magma** cools and hardens into a multitude of **igneous rocks**.

At the heart of Earth science is the study of geology. Literally meaning "to study the Earth," traditional geological studies of rocks, minerals, and local formations have within the last century, especially in the light of the development of plate tectonic theory, broadened to include studies of geophysics and geochemistry that offer sweeping and powerful explanations of how continents move, to explanations of the geochemical mechanisms by which magma cools and hardens into a multitude of igneous rocks.

Earth's formation and the **evolution** of life upon its fragile outer crust was dependent upon the conditions established during the formation of the **solar system**. The **Sun** provides the **energy** for life and drives the turbulent atmosphere. A study of Earth science must, therefore, not ignore a treatment of Earth as an astronomical body in **space**.

At the opposite extreme, deep within **Earth's interior**, **radioactive decay** adds to the **heat** left over from the condensation of Earth from cosmic dust. This heat drives the forces of **plate tectonics** and results in the tremendous variety of features that distinguish Earth. To understand Earth's interior structure and dynamics, seismologists probe the interior structure with seismic shock waves.

It does not require the spectacular hurricane, **tornado**, landslide, or volcanic eruption to prove that Earth's atmosphere and seas are dynamic entities. Forces that change and shape Earth appear on a daily basis in the form of **wind** and **tides**. What Earth scientists, including meteorologists and oceanographers seek to explain—and ultimately to quantify—are the physical mechanisms of change and the consequences of those changes. Only by understanding the mechanisms of change can predictions of **weather** or climatic change hope to achieve greater accuracy.

The fusion of disciplines under the umbrella of Earth science allows a multidisciplinary approach to solving complex problems or multi-faceted issues of resource management. In a addition to hydrogeologists and cartographers, a study of ground **water** resources could, for example, draw upon a wide diversity of Earth science specialists.

Although modern earth science is a vibrant field with research in a number of important and topical areas (e.g., identification of energy resources, waste disposal sites, etc), the span of geological process and the enormous expanse of **geologic time** make critical the study of ancient processes (e.g., paleogeological studies). Only by understanding how processes have shaped Earth in the past—and through a detailed examination of the geological record—can modern science construct meaningful predictions of the potential changes and challenges that lie ahead.

Earth, air, and water

Earth science is the study of the physical components of Earth—its water, land, and air—and the processes that influence them. Earth science can also be thought of as the study of the five physical spheres of Earth: atmosphere (gases), **lithosphere** (rock), pedosphere (**soil** and sediment), **hydrosphere** (liquid water), and cryosphere (**ice**). As a result, Earth scientists must consider interactions between all three states of matter—solid, liquid, and gas—when performing investigations. The subdisciplines of Earth science are many, and include the geosciences, **oceanography**, and the atmospheric sciences.

The geosciences involve studies of the solid part of Earth and include geology, geochemistry, and geophysics. Geology is the study of Earth materials and processes. Geochemistry examines the composition and interaction of Earth's chemical components. Geophysicists study the dynamics of Earth and the nature of interactions between its physical components.

Oceanography involves the study of all aspects of the oceans: **chemistry**, water movements, depth, topography, etc. Considerable overlap exists between oceanography and the geosciences. However, due to the special tools and techniques required for studying the oceans, oceanography and the geosciences continue to be thought of as separate disciplines.

The atmospheric sciences, **meteorology** and climatology, involve the study of the atmosphere. Meteorology is the study of the **physics** and chemistry of the atmosphere. One of the primary goals of meteorology is the analysis and prediction of short-term weather patterns. Climatology is the study of long-term weather patterns, including their causes, variation, and distribution.

Due to the interactions between the different spheres of Earth, scientists from these different subdisciplines often must work together. Together, Earth scientists can better understand the highly involved and interrelated systems of Earth and find better answers to the difficult questions posed by many natural phenomena. In addition, due to the interwoven nature of the biotic (living) and abiotic (nonliving) parts of Earth's environment, Earth scientists sometimes work with life scientists (i.e., biologists, ecologists, agronomists, etc.) who study Earth's **biosphere**.

Earth science research focuses on solving the many problems posed by increasing human populations, decreasing natural resources, and inevitable natural hazards. Computer and **satellite** technologies are increasingly utilized in the search for and development of Earth's resources for present and future use.

See also Astronomy; Atmosphere, composition and structure; Atmospheric circulation; Atmospheric optical phenomena; Atmospheric pressure; Atmospheric temperature; Biochemistry; Earth's magnetic field; Earth's rotation; Fossil and fossilization; Fossil fuels; Gravity and gravitation; Latitude and longitude; Mineralogy; Sediment and sedimentation.

Resources

Books

Press, Frank, and Raymond Siever. *Understanding Earth.* New York: W. H. Freeman and Company, 2001.

Tarbuck, Edward. D., Frederick K. Lutgens, and Tasa Dennis. *Earth: An Introduction to Physical Geology.* 7th ed. Upper Saddle River, NJ: Prentice Hall, 2002.

Periodicals

Hellfrich, George, and Wood, Bernard, "The Earth's Mantle." *Nature.* (August 2, 2001): 501–507.

Other

United States Geological Survey. "Science for a Changing World." [cited February 24, 2003]. <http://www.usgs.gov/>.

K. Lee Lerner
Jack Vine

Earth's core *see* **Earth's interior**

Earth's crust *see* **Earth's interior**

Earth's interior

It is approximately 3,950 mi (6,370 km) from Earth's surface to its center. Geologists understand the structure and composition of the surface by direct observation and by analysis of rock samples raised by drilling projects; however, the depth of drill holes and, therefore, the depth **limit** of scientists' ability to directly observe Earth's interior is severely limited. Even the deepest drill holes (7.5 mi [12 km]) penetrate less than 0.2% of the **distance** to Earth's center. Thus, we know more about the layers near Earth's surface than about the depths, and can only investigate conditions deeper in the interior through indirect means.

Geologists collect indirect information about the deep interior from several different sources. Some **rocks** found at the surface, such as kimberlite, originate deep in Earth's crust and in the mantle. These rocks provide geologists with samples of the composition of Earth's interior; however, their depth limit is still on the order of a few tens of miles. Another source of information, because of its ability to probe **Earth** to its very core, is more important: seismic waves. When an **earthquake** occurs anywhere on the **planet**, seismic waves—mechanical vibrations transmitted by the solid or liquid rock of Earth's interior—travel outward from the earthquake center. The speed, **motion**, and direction of seismic waves changes dramatically at depth different levels within Earth, and these are known as seismic transition zones. From such data, scientists have concluded that Earth is composed of three basic parts: the crust, the mantle, and the core.

The crust

The outermost layer of Earth is the crust, a thin shell of rock that covers the globe. There are two types of crust: (1) the continental crust, which consists mostly of light-colored rock of granitic composition and underlies the continents, and (2) the oceanic crust, which consists mostly of dark-colored rock of basaltic composition and underlies the oceans. The continents have an average elevation of about 2,000 ft (609 m) above **sea level**, while the average elevation (depth) of the **ocean** floor is 10,000 ft (3,048 m) below sea level. An important difference between continental and oceanic crust is their difference in **density**. Continental crust has a lower average density (2.6 g/cm^3) than does oceanic crust (3.0 g/cm^3). This density difference allows the continents to float perma-

nently on the upper mantle, persisting more or less intact for billions of years. Oceanic crust, in contrast, is barely able to float on the mantle (which has a density of about 3.3 g/cm^3). As oceanic crust ages, it accumulates a heavy underlayer of cooled mantle rock; the resulting two-layer structure eventually sinks of its own weight into the mantle, where it is melted down and recycled. Because of this recycling process, no oceanic crust older than about 200 million years exists on the surface of the earth. About 16% of the mantle consists of recycled oceanic crust; only about 0.3% consists of recycled continental crust.

Another difference between the oceanic crust and continental crust is their difference in thickness. The oceanic crust is 3–6 mi (5–10 km) thick, while the continental crust averages about 20 mi (35 km) in thickness and can reach 40 mi (70 km) in certain sections, particularly those found under recently elevated mountain ranges such as the Himalayas.

The bottom of the crust (both the oceanic and continental varieties) is determined by a distinct seismic transition zone termed the Mohorovičić discontinuity. The Mohorovičić discontinuity, commonly referred to as "the Moho" or the "M-discontinuity," is the transition or boundary between the bottom of the crust and the solid, uppermost layer of the mantle (the lithospheric mantle). As the thickness of the crust varies, the depth to the Moho varies, from 3–6 mi (5–10 km) under the oceans to 20–40 mi (35–70 km) under the continents.

The Moho was first discovered by the Croatian geophysicist Andrija Mohorovičić (1857–1936) in 1908. On October 8, 1908, Andrija Mohorovičić observed seismic waves from an earthquake in Croatia. He noticed that both the compressional (or primary [P]) waves and the shear (or secondary [S]) waves, at one **point** in their journey, picked up speed as they traveled farther from the earthquake. This suggested that the waves had been deflected. He noted that this increase in speed seemed to occur at a depth of about 30 mi (50 km). Since seismic waves travel faster through denser material, he reasoned that there must be an abrupt transition at that depth from the material of the crust to denser rocks below. This transition zone was later named for its discoverer. The Moho is a relatively narrow transition zone, estimated to be 0.1–1.9 mi (0.2–3 km) thick. It is defined by the level within the earth where P wave **velocity** increases abruptly from an average speed of 4.3 mi/sec (6.9 km/sec) to about 5.0 mi/sec (8.1 km/sec).

The mantle

Underlying the crust is the mantle, which comprises about 82% of Earth's **volume** and 65% of its **mass**. The uppermost section of the mantle, which is solid, is called the lithospheric mantle. This section extends from the

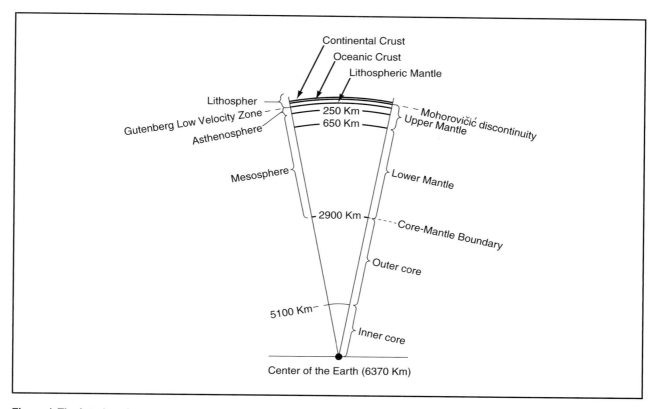

Figure 1. The interior of the earth. *Illustration by Hans & Cassidy. Courtesy of Gale Group.*

Moho down to an average depth of 40 mi (70 km), fluctuating between 30 and 60 mi (50–100 km). The density of this layer is greater than that of the crust, averaging 3.3 g/cm³. Like the crust, this section is solid, and is cool relative to the material below. The lithospheric mantle, combined with the overlying solid crust, is termed the **lithosphere**, a word derived from the Greek *lithos*, meaning rock. At the base of the lithosphere is another seismic transition, the Gutenberg low velocity zone. At this level, the velocity of S waves decreases dramatically, and seismic waves appear to be absorbed more strongly than elsewhere within the earth. Scientists interpret this to mean that the layer below the lithosphere is a "weak" or "soft" zone of partially melted material (1–10% molten material). This zone is termed the *asthenosphere*, from the Greek *asthenes*, meaning "weak." This transition between the lithosphere and the **asthenosphere** is named after German geologist Beno Gutenberg (1889–1960), who made several important contributions to our understanding of Earth's interior. It is at this level that some important Earth dynamics occur, affecting those of us here at the earth's surface. At the Gutenberg low velocity zone, the lithosphere is carried on top of the weaker, less-rigid asthenosphere, which seems to be in continual circulation. This circulatory motion creates stress in the rigid rock layers above it, and the slabs or

plates of the lithosphere are forced to jostle against each other like **ice** cubes floating in a bowl of swirling **water**. This motion of the lithospheric plates is known as *plate tectonics* (from the Greek *tektonikos*, meaning construction), and is responsible for many surface phenomena, including earthquakes, volcanism, mountain-building, and **continental drift**.

The asthenosphere extends to a depth of about 155 mi (250 km). Below that depth, seismic wave velocity increases, suggesting an underlying denser, solid phase.

The rest of the mantle, from the base of the asthenosphere at 155 mi (250 km) to the core at 1,800 mi (2,900 km), is called the mesosphere ("middle sphere"). Mineralogical and compositional changes are suggested by sharp velocity changes in the mesosphere. Notably, there is a seismic discontinuity at about 250 mi (410 km) of depth, attributed to a possible mineralogical change (presumably from an abundance of the mineral olivine to the mineral spinel), and another at about 400 mi (660 km), attributed to a possible increase in the **ratio** of **iron** to **magnesium** in mantle rocks. Except for these variations, down to 560 mi (900 km) the mesosphere seems to consist of predominantly solid material that displays a relatively consistent pattern of gradually increasing density and seismic wave velocity with increasing depth and

KEY TERMS

· ·

Continental crust—Layer of crust (about 35 km thick) that underlies the earth's continents; comprised of light-colored, relatively lightweight granitic rock.

Core—The part of Earth below 1,800 mi (2,900 km). Comprised of a liquid outer core and a solid inner core.

Gutenberg discontinuity—The seismic transition zone that occurs at 1,800 mi (2,900 km) and separates the lower mantle (solid) and the underlying outer core (liquid). Also known as the core-mantle boundary (CMB).

Gutenberg low velocity zone—The transition zone that occurs at 30–60 mi (50–100 km), between the rigid lithosphere and the underlying "soft" or partially melted asthenosphere.

Lithospheric mantle—The rigid uppermost section of the mantle, less than 60 mi (100 km) thick. This section, combined with the crust, constitutes the lithosphere, or the solid and rocky outer layer of Earth.

Mantle—The thick middle layer of the Earth that extends from the core to the crust, a thickness of almost 1,800 mi (2,900 km). The mantle is predominantly solid, although it includes the partially melted asthenosphere.

Mesosphere—The solid section of the mantle directly beneath the asthenosphere. Extends from 150 mi (250 km) down to 1,800 mi (2,900 km).

Mohorovičić discontinuity—The seismic transition zone indicated by an increase in primary seismic wave velocity that marks the transition from the crust to the uppermost section of the mantle.

Oceanic crust—Thin (3–6-mi [5–10-km] thick) crust that floors the ocean basins and is composed of basaltic rock: denser than continental crust.

P waves—Primary or compression waves that travel through Earth, generated by seismic activity such as earthquakes; can travel through solids or liquids.

S waves—Secondary or shear waves that travel through Earth, generated by seismic activity such as earthquakes; cannot travel through liquids (e.g., outer core).

Seismic transition zone—A layer in the Earth's interior where seismic waves undergo a change in speed and partial reflection; caused by change in composition, density, or both.

Seismic wave—A disturbance produced by compression or distortion on or within the earth, which propagates through Earth materials; a seismic wave may be produced by natural (e.g., earthquakes) or artificial (e.g., explosions) means.

---

pressure. Below the 560 mi (900 km) depth, the P and S wave velocities continue to increase, but the **rate** of increase declines with depth.

Although much of the mantle is "solid," the entire mantle actually convects or circulates like a pot of boiling water. Images produced by analysis of seismic waves show that dense slabs of oceanic crust plunge all the way through the mantle to the outer surface of the core, which indicates that the entire mantle is in motion, mixing thoroughly with itself over geological **time**.

The core

At a depth of 1,800 mi (2,900 km) there is another abrupt change in the seismic wave patterns, the **Gutenberg discontinuity** or core-mantle boundary (CMB). The density change at the CMB is greater than that at the interface of air and rock on the Earth's outer surface. At the CMB, P waves decrease while S waves disappear completely. Because S waves cannot be transmitted through liquids, it is thought that the CMB denotes a

phase change from the solid mantle above to a liquid outer core below. This phase change is believed to be accompanied by an abrupt **temperature** increase of 1,300°F (704°C). This hot, liquid outer core material is much denser than the cooler, solid mantle, probably due to a greater percentage of iron. It is believed that the outer core consists of a liquid of 80–92% iron, alloyed with lighter element. The composition of the remaining 8–20% is not well understood, but it must be a compressible element that can mix with liquid iron at these immense pressures. Various candidates proposed for this element include silicon, **sulfur**, or **oxygen**.

The actual boundary between the mantle and the outer core is a narrow, uneven zone that contains undulations on the order of 3–6 mi (5–8 km) high. These undulations are affected by heat-driven **convection** activity within the overlying mantle, which may be the driving force for **plate tectonics**. The interaction between the solid mantle and the liquid outer core is also important to Earth dynamics for another reason; eddies and currents in the iron-rich, fluid outer core are ultimately responsible for the **Earth's magnetic field**.

There is one final, deeper transition, evident from seismic wave data: Within Earth's core, at a depth of about 3,150 mi (5,100 km), P waves encounter yet another seismic transition zone. This indicates that the material in the inner core is solid. The immense pressures present at this depth probable cause a phase change, from liquid to solid. Density estimates are consist with the hypothesis that the solid, inner core is nearly pure iron.

The **heat** that keeps the whole interior of the Earth at high temperatures is derived from two sources: heat of formation and radioactive metals. As the Earth accreted from the original solar nebula, impacts of new material delivered sufficient **energy** to melt most or all of the forming planet's bulk. As most of the new Earth's iron sank its center through its bulkier, lighter elements (silicon, oxygen, etc.), further energy was released, sufficient to raise the temperature of the core by several thousand degrees Centigrade. Radioactive elements such as **uranium** and thorium, mostly resident in the mantle, have continued to supply the Earth's interior with heat in the billions of years since its formation; however, the Earth's interior continues to cool, steadily losing its primordial heat to **space** through the crust. As the core cools, its inner, solid portion grows at the expense of its outer, liquid portion. The current rate of thickening of the inner core is about 0.04 inch (1 mm) per year.

See also Magma.

Resources

Books

Magill, Frank N., ed. *Magill's Survey of Science: Earth Science*. Hackensack, NJ: Salem Press, Inc., 1990.

Tarbuck, Edward. D., Frederick K. Lutgens, and Tasa Dennis. *Earth: An Introduction to Physical Geology*. 7th ed. Upper Saddle River, NJ: Prentice Hall, 2002.

Winchester, Simon. *The Map That Changed the World: William Smith and the Birth of Modern Geology*. New York: Harper Collins, 2001.

Periodicals

Buffett, Bruce A. "Earth's Core and the Geodynamo." *Science*. (June 16, 2000): 2007–2012.

Hellfrich, George, and Bernard Wood "The Earth's Mantle." *Nature*. (August 2, 2001): 501–507.

Mary D. Albanese

Earth's magnetic field

Earth acts as a **dipole** magnet with the positive and **negative** magnetic poles near, but not aligned exactly with, the north and south geographic poles. Because of this difference, detailed maps commonly distinguish between true north and magnetic north. The difference, known as magnetic declination, must be taken into account when navigating with a magnetic compass. Its magnetic field also molds the configuration of Van Allen **radiation** belts, which are bands of high-energy charged particles around Earth, and helps to shield Earth from a stream of charged particles known as the **solar wind**.

The origin of Earth's magnetic field is not fully understood, but it is generally thought to be a result of electrical currents generated by movement of molten **iron** and nickel within Earth's outer core. This is known as the dynamo effect. The **Moon**, in contrast, has no magnetic field and is thought to have a small core that is only partially molten. A small portion of Earth's magnetic field is generated by movement of ions in the upper atmosphere.

Magnetic field lines produced by the dynamo effect radiate far into space. Unlike the field lines around a simple dipole magnetic, however, the field lines around Earth are asymmetric. Those nearest the **Sun** are compressed by the solar wind whereas those shielded from the Sun by Earth are highly elongated. The effect is similar to that of **water** passing around the bow of a ship. Charged particles become trapped along magnetic field lines, just as iron filings around a simple dipole magnet, and form Earth's **magnetosphere**.

Earth's magnetic field can change for short or long periods of **time**. The magnetic field can change quickly, within an hour, in magnetic storms. These short-term changes occur when the magnetic field is disturbed by **sunspots**, which send **clouds** of charged particles into Earth's atmosphere. The same particles excite **oxygen**, **nitrogen**, and **hydrogen atoms** in the upper atmosphere, causing the aurora borealis and aurora australis.

Long-term changes occur when Earth's magnetic field reverses itself, which occurs on average every 500,000 years. The last known reversal was about 700,000 years ago. Evidence of past reversals was discovered in the 1950s and 1960s, when magnetometers towed behind ships showed that the magnetic polarity of volcanic **rocks** comprising the oceanic crust alternated in strips **parallel** to mid-ocean ridges. Further research showed that the strips occur because oceanic plates grow outward from mid-ocean ridges. Grains of magnetite within volcanic rocks such as basalt, which constitutes most of the oceanic crust, align themselves with the prevailing magnetic field when the **temperature** of the rock falls below its Curie point (about 1,075°F [580°C] for basalt). Thus, the alternating strips represent changes in the polarity of Earth's magnetic field as oceans grow through time. The existence of magnetic strips was one of the key pieces of evidence that led to widespread acceptance of **plate tectonics** among scientists.

See also Earth's interior; Earth's rotation; Electromagnetic field; Magma; Magnetism; Volcano.

Earth's mantle *see* **Earth's interior**

Earth's rotation

All objects in the universe and our **solar system** move in **space**. The **earth** moves in two ways. It rotates like a top on its axis, an imaginary line through the north and south poles, and revolves in an **orbit** around the **Sun**. Centrifugal **force** results from the earth's **rotation**; without gravity, centrifugal force could cause objects to fly into space. Because the force of the earth's gravity is 289 times stronger than its centrifugal force, gravity prevents objects from leaving the surface. Centrifugal force causes the earth to bulge at the equator, making it slightly ovoid in shape.

The earth's counterclockwise rotation is in the opposite direction of the apparent movement of heavenly bodies. Thus, although the Sun and stars appear to move from east to west, the earth rotates from west to east.

The rotation of Earth can be proven in several ways. One is the Foucault experiment. This was first conducted in 1851, by the Frenchman Léon Foucault. He suspended a heavy **iron** ball from a 200 ft (61 m) wire, creating a pendulum, from the dome of the Pantheon in Paris. He put **sand** underneath the pendulum, and placed a pin on the bottom of the ball, so it would leave a mark on its swing from side to side. On each swing, over the course of 24 hours, the mark in the sand would move to the right. The direction in the path showed movement of the earth against the swing of the pendulum.

A more modern proof of the rotation is shown by the orbits of artificial satellites. A **satellite** is launched from the Kennedy Space Center at a 30 **degree** angle to an orbit 100 mi (161 km) above Earth. Its orbit stays at approximately the same **plane** in space. If Earth did not rotate, the satellite would pass over Cape Canaveral each time it completed an orbit, but it does not. As it completes the first orbit, it flies over Alabama and over Louisiana on the third. Each time the satellite passes over locations in the United States, it is 1,000 mi (1,609 km) farther to the west. Tracking stations have made this observation with hundreds of satellites.

Another way of proving rotation is through the prevailing winds. In the northern hemisphere, they move in a counter clockwise direction, while in the southern hemisphere, they blow clockwise.

The inner core of Earth rotates faster than the crust. Earth is made up of an inner core, an outer core, the mantle and the crust. The inner core, a solid **mass** of iron about 1,500 mi (2,400 km) in diameter, is suspended in the molten **metal** of the outer core, which is about 1,400 mi (2,240 km) thick. The mantle of the earth is a layer of mostly solid material about 1,700 mi (2,720 km) thick; the crust itself varies from 4 to 25 mi (6 to 40 km) thick. Although the inner core spins in the same direction as Earth's crust, the inner core rotates at a different speed. Scientists estimate that the inner core of Earth spins from one to three degrees a year faster than the crust of Earth. This means that while a point on the crust of Earth moves 360° in a year, a similar point on the inner core would move 361-363° in a year. Thus, over approximately 360 years, the inner core would make one complete rotation more than the crust of Earth.

Generally when we speak of Earth's rotation, we are talking about the rotation of Earth's surface. This rotation is used to define time. Since people began to measure time, it's been done by the movement of Sun and stars. One rotation of Earth makes up one 24 hour day. This is in contrast to the time of revolution around the Sun of 365 days, or one year. Because the earth's axis is not **perpendicular** to the equator, but leans at a 23.5° angle, the amount of daylight varies over the course of a year.

Over time the speed of Earth's rotation has slowed. Knowing how fast it spins, at any given time, is important to navigators and pilots in finding locations.

See also Gravity and gravitation.

Earthquake

An earthquake is the shaking or vibration of Earth's surface as the result of sudden movement along a **fault**, the movement of molten rock within the **Earth**, or human activities. The terms temblor and seism are often used as synonyms for earthquake. The location of an earthquake source within the Earth is known as its focus, and the point on the Earth's surface directly above the focus is known as the epicenter.

Earthquakes are common events. The United States Geological Survey estimates that more than three million earthquakes occur on Earth each year, which is equivalent to more than 9,000 earthquakes per day. Virtually all of these are too small to be noticed by humans and many occur in remote areas far from seismometers. Since

Aerial view of Interstate 5, a California highway damaged by the Northridge earthquake in January 1994. *Photograph by Robert A. Eplett. Governor's Office of Emergency Services.*

1900, there has been on average about 1 magnitude 8 earthquake, 18 magnitude 7.0 to 7.9 earthquakes, 120 magnitude 6.0 to 6.9 earthquakes, and 800 magnitude 5.0 to 5.9 earthquakes on Earth each year.

Earthquakes can range in severity from small events that are imperceptible to humans to devastating shocks that level cities and kill thousands. The world's most destructive earthquake, which occurred in China during the year 1556, killed 830,000 people. Twenty other earthquakes in **Europe**, **Asia**, and the Middle East are known to have resulted in more than 50,000 deaths each. The most devastating earthquake to strike the United States was the 1906 San Francisco earthquake, which killed about 3,000 people as a result of shaking and resulting fires. Modern **engineering** and construction methods have significantly reduced the danger posed by earthquakes in developed countries. In the United States, for example, only five earthquakes since 1950 have killed more than 60 people. The great Alaskan earthquake of 1964, the second largest earthquake ever recorded by seismologists, killed only 15 people. An additional 110 perished, however, in earthquake triggered tsunamis that struck coastal Alaska, Oregon, and California. Most of the fatal earthquakes occurring in the United States since 1950 have killed only one or two people, and the vast majority of earthquakes do not kill anyone.

The size of an earthquake is described by its magnitude, which reflects the amount of **energy** released by the temblor. There are many different ways of calculating earthquake magnitude, the most famous of which was proposed in the 1930s by the American seismologist Charles Richter (1900–1985). The Richter magnitude is the base 10 logarithm of the largest seismic wave amplitude recorded on a particular kind of **seismograph** located 62 mi (100 km) from the earthquake epicenter. Adjustments must be made if other kinds of seismographs are used or if they are located at a different **distance** from the epicenter. An earthquake of a given magnitude will produce waves 10 times as large as those from an earthquake of the next smaller magnitude. The energy released increases by a factor of about 30 from one magnitude to the next. The Richter scale is open-ended, meaning that it has no mathematical upper or lower limits. In reality, however, there are no faults on Earth large enough to produce a magnitude 10 earthquake. The two largest recorded earthquakes were the magnitude 9.5

Chilean earthquake of 1956 and the magnitude 9.2 Prince William Sound, Alaska, earthquake of 1964.

The effects of an earthquake are measured by its intensity. Unlike magnitude, earthquake intensity varies from place to place. The most common measure of intensity is the modified Mercalli scale, which ranges from an intensity of I (not felt except by a few people under especially favorable circumstances) to XII (total destruction, with objects thrown in the air and lines of sight distorted). Surveys and interviews after a large earthquake can be used to create an isoseismic **map**, which shows the distribution of reported earthquake intensities. Most isoseismic maps show a distorted bull's **eye** pattern of concentric rings of equal intensity area centered around the epicenter.

Causes of earthquakes

Tectonic plate movements

Some earthquakes occur in areas where the tectonic plates comprising Earth's **lithosphere** move horizontally past each other along large faults or zones of faults. Examples of this type include earthquakes along the San Andreas and Hayward faults in California. Earthquakes also occur in places where a continental plate subducts an oceanic plate, for example along the western coast of **South America**, the northwest coast of **North America** (including Alaska), and in Japan. If two continental plates collide but neither is subducted, as in Europe and Asia from Spain to Vietnam, earthquakes occur as the **rocks** are lifted to form mountain ranges.

In other parts of the world, for example the Basin and Range physiographic province of the western United States and the East African Rift, continental plates are being stretched apart by tectonic forces. The result is that some parts of the Earth's crust are lifted to form mountain ranges while neighboring blocks subside to form basins that collect sediment eroded from the **mountains**. Earthquakes can occur when movement occurs along faults developed as a result of the stretching.

Faults are planes of weakness, across which rock has moved in opposite directions, within the Earth's crust. They can range in size from continental scale features such as the San Andreas fault in California to small features across which only a few millimeters or centimeters of movement has occurred. Tectonic plate motions increase the level of stress within Earth's crust, which is accommodated as elastic strain energy, until the stress exceeds the strength of the fault. Then, the energy is suddenly released as the rocks on each side of the fault slip past each other to create an earthquake. This process is analogous to a rubber band snapping when it was been stretched to the breaking point. Because there is a fric-

tional resistance to movement along faults, rapid seismic slip can generate enough **heat** to melt the adjacent rocks and form a glassy rock known as pseudotachylyte. In other cases, the elastic strain energy is slowly and quietly dissipated through a process known as aseismic creep.

Magma movement

Rhythmic earthquakes known as harmonic tremors, which are caused by **magma** and volcanic gas moving through conduits in the Earth's crust just as air moves through a pipe **organ**, can foreshadow or accompany volcanic eruptions. Recent studies have also suggested that very large earthquakes, such as the magnitude 9.0 earthquake that affected the west coast of the United States in 1700, may trigger volcanic activity for several decades after their occurrence as the Earth's crust slowly adjusts to the initial movement. Seismologists can also use earthquake activity to infer the presence of magma that has not yet erupted and formed a **volcano**. Swarms of small earthquakes near Socorro, New Mexico, for example have helped scientists to locate a **mass** of molten rock about 12 mi (20 km) beneath the Earth's surface. Detailed measurements have shown that the surface is being lifted by about 2 mm per year in that area, but there are no obvious signs that a pool of molten rock lies beneath the surface.

Human activity

Explosions, especially from underground nuclear bomb testing, can produce small earthquakes. Earthquakes caused by explosions produce vibrations different than those caused by movement along faults, and seismic monitoring is an important part of nuclear test ban treaty verification. The implosive demolition of the Kingdome, a sports stadium in Seattle, in the year 2000 produced a magnitude 2.3 earthquake. Seismologists were able to deploy seismometers before the demolition and use the manmade earthquake to learn more about the **geology** of the area by studying how seismic waves were reflected and refracted beneath Earth's surface. Another well-known example of earthquakes due to human activity occurred at the Rocky Mountain Arsenal near Denver, Colorado, during the 1960s. The **pressure** of hazardous waste being injected deep into the Earth through disposal wells was large enough to trigger a series of earthquakes. A subsequent experiment in an oilfield near Rangely, Colorado, showed that earthquakes could be triggered at will be injecting **water** under pressure.

Seismic waves

Rapid slip along a fault generates waves in much the same way as does a pebble falling into a pool of water, and waves moving outward from an earthquake focus are reflected and refracted each time they encounter a different rock type. There are four different kinds of seismic waves, two of which are known as body waves and two of which are known as surface waves. Body waves travel deep through the Earth, whereas surface waves travel along the Earth's surface and generally cause the most damage.

The two types of body waves are P-waves and S-waves. P-waves, also known as primary waves, travel the fastest of the four types. They move by alternately compressing and stretching the rock through which they pass. P-wave **velocity** depends on the rock type and **density**, but it is generally about 6 km/s (4 mi/s). S-waves, also known as secondary waves, move by shearing or moving from side to side the rock through which they pass. S-waves move more slowly than P-waves and, depending on the type of rock, have a velocity of about 2 mi/s (3 km/s).

The two types of surface waves are known as Rayleigh and Love waves. They travel more slowly than either P- or S-waves, but often cause more damage than body waves because they travel along the Earth's surface and have a greater effect on buildings.

Seismologists can determine the epicenter of an earthquake by noting the times that seismic waves arrive at three or more different seismometers. **Multiplication** of the wave velocity by the travel time gives the distance to the epicenter, which is the radius of a **circle** with its center at the seismometer. The radii from at least three circles will intersect at a point that is the earthquake epicenter. In practice, seismologists first make a rough estimate of the epicenter and then refine their estimate as additional data become available, for example by using velocities corresponding to specific rock types rather than a general estimate.

Collapse of buildings

To construct a house or building under static conditions, the materials need only to be stacked up, attached to each other, and balanced. These kinds of buildings are not designed to accelerate rapidly and change directions like cars or airplanes. Buildings in seismically active areas, however, must be designed and built to withstand the dynamic **acceleration** that can occur during an earthquake. Large buildings and structures such as **bridges**, in particular, must be designed so that vibrations arising from earthquakes are damped and not amplified.

Because noticeable earthquakes are rare in most areas, people may not recognize that the objects and buildings around them represent potential hazards. It is not movement of the ground surface alone that kills peo-

ple. Instead, deaths from earthquakes result from the collapse of buildings and falling objects in them, fires, and tsunamis. The type of construction that causes the most fatal injuries in earthquakes is unreinforced **brick**, stone, or **concrete** buildings that tend not to be flexible and to collapse when shaken.

The most earthquake-resistant type of home is a low wooden structure that is anchored to its foundation and sheathed with thick plywood. Some of the traditional architecture of Japan approximates this shock-resistant design, including wooden buildings that are more than a thousand years old. Unfortunately, **wood** and **paper** houses can be easily ignited in the fires that are common after large earthquakes. Both unreinforced masonry and shock-resistant wood houses are used by different cultures in areas of high earthquake risk.

Active faults lie under many parts of the world that do not commonly experience earthquakes. The crust under such places as Italy, California, and Central America moves often enough that an earthquake there, although still unpredictable, is not entirely unexpected. But other populated areas, such as the East Coast and Mississippi Valley in the United States, periodically experience earthquakes just as big as those in any earthquake-prone part of the world, although far less frequently.

Earthquake-triggered landslide

Earthquakes can trigger landslides and rock falls many kilometers from their epicenters. Local governments can enact zoning regulations to prevent development in areas susceptible to landslides during earthquakes or heavy rainstorms. In other cases, potentially hazardous slopes can be excavated and regarded into a configuration that is able to resist the destabilizing effects of a large earthquake.

Seismically-triggered landslides can reshape the landscape. In 1959, an earthquake triggered a landslide that dammed the Madison River in Montana and created Hebgen Lake. To prevent this natural dam from washing out and causing catastrophic floods, the U.S. Army Corps of Engineers built an emergency spillway through the landslide material. This enabled them to control the release of the water from the new **lake**. Prehistoric landslides have dammed the Columbia River and could be the source of a legend of the Northwest Indians. In this legend, tribes walked across the Columbia River on a bridge of land to meet each other.

Liquefaction of soil

Seismic shaking can transform water-saturated **sand** into a liquid mass that will not support heavy loads such as buildings. This phenomenon, called liquefaction, causes much of the destruction associated with some earthquakes. Mexico City, for example, rests on the ancient lakebed of Lake Texcoco, which is a large basin filled with liquefiable sand and ground water. In the Mexico City earthquake of 1985, the wet sand beneath tall buildings liquefied and most of the 10,000 people who died were in buildings that collapsed as their foundations sank into liquefied sand.

Jets of sand sometimes erupt from the ground during an earthquake. These sand geysers or mud volcanoes occur when formations of soft, wet sand is liquefied and forcefully squeezed up through cracks in the ground. Despite these names, they have no relation to real geysers or volcanoes. Although they generally cause little damage, they are indications that more widespread liquefaction may have occurred or may be possible in the next earthquake.

Subsidence

Earthquakes can cause affected areas to increase or decrease in elevation by several feet, which can in turn lead to **flooding** in coastal areas. Port Royal, on the south shore of Jamaica, subsided several feet in an earthquake in 1692 and suddenly disappeared as the sea rushed into the new depression. Eyewitnesses recounted the seismic destruction of the infamous pirate anchorage as follows: "... in the space of three minutes, Port-Royall, the fairest town of all the English plantations, exceeding of its riches,... was shaken and shattered to pieces, sunk into and covered, for the greater part by the sea.... The earth heaved and swelled like the rolling billows, and in many places the earth crack'd, open'd and shut, with a **motion** quick and fast... in some of these people were swallowed up, in others they were caught by the middle, and pressed to death.... The whole was attended with... the noise of falling mountains at a distance, while the sky... was turned dull and reddish, like a glowing oven." Ships arriving later in the day found a small shattered remnant of the city that was still above the water. Charts of the Jamaican coast soon appeared printed with the words "Port Royall Sunk." During the New Madrid (Missouri) earthquake of 1811, a large area of land subsided around the bed of the Mississippi River in west Tennessee and Kentucky. The Mississippi was observed to flow backwards as it filled the new depression and created what is now known as Reelfoot Lake. The last great earthquake in the U.S. Pacific Northwest occurred two years before Port Royal sank in 1690. In the 300 years since then, no major earthquake has released the potential energy that has been building under the crust. Geologists have found buried **forests** and deposits indicating that coastal areas were periodically flooded, probably as the result of major earthquakes.

Tsunamis

An earthquake can create a large wave known as a **tsunami** (the Japanese term) or, seismic sea wave. A tsunami is barely detectable as it moves through deep water. Where the **ocean** becomes shallow near the shore, however, the fast-moving tsunami becomes a large wave that rises out of the sea and strikes the shore with unstoppable **force**. In a small, mountain-ringed bay, a tsunami can rush hundreds of meters up a sea-facing mountainside. A wall of water forms when a large tsunami enters a shallow bay or estuary, and it can move upriver for many miles. Sometimes tsunamis are mistakenly referred to as tidal waves, because they resemble a tide-related wave called a tidal bore.

The most destructive tsunamis in history have killed tens of thousands of people, many of them located great distances from the earthquake epicenter. The tsunami produced by a 1946 earthquake in the Aleutian Islands, Alaska, killed a total of 165 people. Of that number, 159 were in Hawaii, 5 were in Alaska, and 1 was in California. Coastal towns affected by tsunamis often have no topographic barriers between them and the sea and had no warning of the impending disaster. Building a breakwater to divert a tsunami and expend its energy is sometimes an option for otherwise unprotected coastal towns.

Secondary hazards: fire, disease, famine

Cities depend on networks of lifeline structures to distribute water, power, and food and to remove sewage and waste. These networks, whether power lines, water mains, or roads, are easily damaged by earthquakes. Elevated freeways collapse readily, as demonstrated by a section of the San Francisco Bay Bridge in 1989 and the National Highway Number 2 in Kobe, Japan, in 1995. The combination of several networks breaking down at once multiplies the hazard to lives and property. Live power lines fall into water from broken water mains, creating an electric shock hazard. Fires may start at ruptured gas mains or chemical storage tanks, but many areas may not be accessible to fire trucks and other emergency vehicles. Even if areas are accessible, there may not be water for fire-fighting. The great fire that swept San Francisco in 1906 could not be stopped by regular firefighting methods and entire blocks of buildings had to be demolished to halt the fire. Most of the 143,000 people killed in Tokyo and Yokohama because of the 1923 Kwanto perished in fires.

Famine and **epidemic disease** can quickly strike large displaced populations deprived of their usual food distribution system, sanitation services, and clean water.

KEY TERMS

Active fault—A fault where movement has been known to occur in recent geologic time.

Aftershock—A subsequent earthquake (usually smaller in magnitude) following a powerful earthquake that originates at or near the same place.

Epicenter—The location where the seismic waves of an earthquake first appear on the surface, usually almost directly above the *focus*.

Fault—A fracture in the earth's crust accompanied by a displacement of one side relative to the other.

Focus—The location of the seismic event deep within the earth's crust that causes an earthquake. Also called the earthquake's *hypocenter*.

Foreshock—A small earthquake or tremor that precedes a larger earthquake shock.

Modified Mercalli scale—A scale used to evaluate earthquake intensity based on effects felt and observed by people during the earthquake.

Richter scale—A scale used to compare earthquakes based on the energy released by the earthquake.

Seismic wave—A disturbance produced by compression or distortion on or within the earth, which propagates through Earth materials; a seismic wave may be produced by natural (e.g., earthquakes) or artificial (e.g., explosions) means. P waves, S waves, and surface waves are vibrations in rock and soil that transfer the force of the earthquake from the focus into the surrounding area.

Subsidence—A sinking or lowering of the earth's surface.

Furthermore, collapsed hospitals may be of no use to a stricken community that urgently needs medical services. After an earthquake, relief operations commonly offer inoculation against infectious diseases. In countries that do not have sufficient organization, trained personnel, or resources to handle an earthquake-generated refugee population, more people may die of secondary causes than the direct effects of seismic shaking. Even in the most prepared countries, the disruption of networks may prevent relief operations from working as planned. In the aftermath of the January, 1995, earthquake in Kobe, Japan, plans for emergency relief made before the disaster did not work as well as planned. Local residents, wary of the danger of aftershocks, had to live outdoors in winter without food, water, or power.

Historical incidence of earthquakes

Catastrophic earthquakes happened just as often in the past as they do today. Earthquakes shattered stone-walled cities in the ancient world, sometimes hastening the demise of civilizations. Knossos, Chattusas, and Mycenae, ancient capitals of countries located in tectonically active mountain ranges, fell to pieces and were eventually deserted. Scribes have documented earthquakes in the chronicles of ancient realms. An earthquake is recorded in the Book of Zachariah, and the Apostle Paul wrote that he got out of jail when the building fell apart around him in an earthquake. In the years before international news services, few people heard about distant earthquakes. Only a few handwritten accounts have survived, giving us limited knowledge of earthquakes in antiquity. Because of limited and lost data, earthquakes seem to have been less common in ancient times. In China, home of the first seismometer, the Imperial government has recorded earthquakes for over a thousand years. Their frequency has not changed through the ages.

See also Continental drift; Mass wasting; Plate tectonics.

Resources

Books

Reiter, L. *Earthquake Hazard Analysis.* New York: Columbia University Press, 1990.

Periodicals

Hill, D. P., F. Pollitz, and C. Hewhall. "Earthquake-Volcano Interactions." *Physics Today* 55, no. 11 (November 2002): 41–47.

Sykes, L. R. "Four Decades of Progress in Seismic Identification Help Verify the CTBT." *EOS, Transactions, American Geophysical Union* 83, no. 44 (October 29, 2002): 497–500.

Other

Spall, Henry. "NEIC: An Interview with Charles F. Richter." July 8, 2002 [cited November 8, 2002]. <http://neic.usgs.gov/neis/seismology/people/int_richter.html>.

U.S. Geological Survey. "Earthquake Image Glossary." July 29, 2002 [cited November 8, 2002]. <http://earthquake.usgs.gov/image_glossary/>.

U.S. Geological Survey. "EQ Facts and Lists." September 5, 2002 [cited November 8, 2002]. <http://earthquake.usgs.gov/bytopic/>.

U.S. Geological Survey. "USGS Earthquake Hazards Program." November 8, 2002 [cited November 8, 2002]. <http://earthquakes.usgs.gov/>.

Bill Hanneberg

Earthworms *see* **Segmented worms**

Earwigs

Earwigs are long-bodied **insects** with chewing mouthparts and many-jointed antennae in the order Dermaptera. Earwigs have small, vestigial forewings modified into a wing case, but their membranous hind-wings are large, folded, and functional, although they are not often used for flying. Earwigs hatch into nymphs which closely resemble the adults, only they are much smaller. **Metamorphosis** in earwigs is simple, with no radical changes in shape during development from the nymphal stages to the adult form.

The most readily distinguishing characteristic of earwigs is the pair of unjointed, forceps-like structures that terminate their abdomen. These unusual organs are modified from common insect structures known as cerci, and they differ between the sexes, those of females having less curvature. The pincers are brandished when earwigs are disturbed, and can give a significant pinch to the finger, so they are clearly useful in defense. The pincers may also have other uses, possibly in folding the rather complicated wings after a flight.

Earwigs are nocturnal animals, and they hide during the day in dark, damp places. Most **species** of the more than 1,200 species of earwigs are scavengers of a wide range of organic debris, including carrion. Some species are herbivorous, some are opportunistic predators of other insects, and a few, specialized species are **parasites** of **mammals**.

The most common native earwig in **Europe** is *Forficula auricularia*, a species that is now also widespread in **North America**, New Zealand, and elsewhere due to accidental introductions by humans. The European earwig is omnivorous, eating a wide range of dead organic **matter**, and also preying on other insects. The female of this species broods her eggs and young hatchlings. During summers when the European earwig is particularly abundant, it may be considered a pest because of its ubiquitous presence in **flower** gardens, under all manner of moist things, in basements and kitchens, and in laundry hanging on clotheslines. These earwigs may damage **vegetables** and flowers during their feeding, but they are not really an important pest. In fact, the European earwig may be beneficial in some respects, by cleaning up organic debris, and perhaps by preying on other, more important insect **pests**.

A total of 18 species of earwigs occur in North America. The seaside earwig (*Anisolabis maritima*) is a native species that occurs on both the Atlantic and Pacific coasts of North America. The red-legged earwig (*Euborellia annulipes*), striped earwig (*Labidura bidens*), and handsome earwig (*Prolabia pulchella*) occur in the southern United States. The toothed earwig (*Spongovostox apicedentatus*) occurs in dry habitats in the southwestern states. The little earwig (*Labia minor*) is another species that was introduced from Europe.

Some species of earwigs have relatively unusual, specialized lifestyles. *Arixenia* is a small earwig that is a viviparous breeder, giving **birth** to live young. This species is an ectoparasite of the Indian bat (*Cheiromeles torquatus*). *Hemimerus* is also a small, viviparous earwig, and a blind ectoparasite of the giant rat (*Cricetomys gambianus*) of west **Africa**.

Earwigs received their common name from the folk belief that these insects would sometimes crawl into the ears of people as they slept, seeking refuge in those dark, moist cavities. This may, indeed, sometimes occur, and it would certainly be disconcerting to have an earwig, or any other insect in one's **ear**. However, there is no evidence that earwigs in the ear are a common problem, except as very rare accidents.

Bill Freedman

Eating disorders

Eating disorders are psychological conditions that involve either overeating, voluntary starvation, or both. No one is sure what causes eating disorders, but researchers think that family dynamics, biochemical abnormalities, and society's preoccupation with thinness may all contribute. Eating disorders are virtually unknown in parts of the world where food is scarce and within less affluent socioeconomic groups in developed countries. Although these disorders have been known throughout history, they have gained attention in recent years, in part because some celebrities have died as a result of their eating disorders.

Young people are more likely than older people to develop an eating disorder—the condition usually begins before age 20. Although both men and women can develop the problem, it is more common in women. Only about 5% of people with eating disorders are male. In either males or females, eating disorders are considered serious and potentially deadly. Many large hospitals and psychiatric clinics have programs especially designed to treat these conditions.

Anorexia nervosa, anorexic bulimia, and **obesity** are the most well known types of eating disorders. The word anorexia comes from the Greek word meaning "lack of appetite." But the problem for people with anorexia is not that they are not hungry. They starve themselves out of fear of gaining weight, even when they are severely underweight. The related condition, anorexic bulimia, literally means being "hungry as an ox." People with this problem go on eating binges, often gorging on junk food. Then they force their bodies to get rid of the food, either by making themselves vomit or by taking large amounts of laxatives. A third type of eating disorder is obesity caused by uncontrollable overeating. Being slightly overweight is not a serious health risk. But being 25% or more over one's recommended body weight can lead to many health problems.

Anorexia

People with anorexia starve themselves until they look almost like skeletons. But their self-images are so distorted that they see themselves as **fat**, even when they are emaciated. Some refuse to eat at all; others nibble only small portions of fruit and **vegetables** or live on diet drinks. In addition to fasting, they may **exercise** strenuously to keep their weight abnormally low. No matter how much weight they lose, they always worry about getting fat.

This self-imposed starvation takes a heavy toll on the body. Skin becomes dry and flaky. Muscles begin to waste away. Bones stop growing and may become brittle. The **heart** weakens. With no body fat for insulation, it's hard to keep warm. Downy hair starts to grow on the face, back, and arms in response to lower body **temperature**. In women, menstruation stops and permanent **infertility** may result. Muscle cramps, dizziness, fatigue, even **brain** damage, kidney and heart failure are possible. An estimated 10–20% of people with anorexia die, either as a direct result of starvation or by suicide.

Researchers believe anorexia is caused by a combination of biological, psychological and social factors. They are still trying to pinpoint the biological factors, but they have zeroed in on some psychological and social triggers of the disorder. Many people with anorexia come from families in which parents are overprotective and have unrealistically high expectations of their children. The condition seems to run in families, which leads researchers to believe it may have a genetic basis. Anorexia often seems to develop after a young person goes through some stressful experience, such as moving to a new town, changing schools, or going through **puberty**. Low self-esteem, fear of losing control, and fear of growing up are common characteristics of anorectics (people with anorexia). The need for approval, combined with our culture's idealization of extreme thinness, also contributes.

The obvious cure for anorexia is eating, but that is the last thing a person with anorexia wants to do. It is unusual for the person himself or herself to seek treatment—usually a friend, family member, or teacher initiates the process. Hospitalization, combined with psy-

chotherapy and family counseling, is often needed to get the condition under control. Force feeding may be necessary if the person's life is in danger. Some 70% of anorexia patients who are treated for about six months return to normal body weight. About 15–20% can be expected to relapse, however.

Bulimia

Like anorexia, bulimia results in starvation. But there are behavioral, physical and psychological differences between the two conditions. Bulimia is much more difficult to detect because people who have it tend to be of normal weight or overweight, and they hide their habit of binge eating followed by purging, vomiting, or using laxatives. In fact, bulimia was not widely recognized, even among medical and mental health professionals, until the 1980s. Unlike anorectics, bulimics (people with bulimia) are aware that their eating patterns are abnormal, and they often feel remorse after a binge. For them, overeating offers an irresistible escape from **stress**. Many suffer from **depression**, repressed anger, **anxiety**, and low self esteem, combined with a tendency toward perfectionism. About 20% of bulimics also have problems with **alcohol** or drug **addiction**, and they are more likely than other people to commit suicide.

Many people occasionally overeat, but are not considered bulimic. According to the American Psychiatric Association's definition, a bulimic binges on enormous amounts of food at least twice a week for three months or more.

Bulimics plan their binges carefully, setting aside specific times and places to carry out their secret habit. They may go from restaurant to restaurant, to avoid being seen eating too much in any one place. Or they may pretend to be shopping for a large dinner party, when actually they intend to eat all the food themselves. Because of the expense of consuming so much food, some resort to shoplifting.

During a binge, bulimics favor high **carbohydrate** foods, such as doughnuts, candy, ice cream, soft drinks, cookies, cereal, cake, popcorn, and bread, consuming many times the amount of calories they normally would consume in one day. No matter what their normal eating habits, they tend to eat quickly and messily during a binge, stuffing the food in their mouths and gulping it down, sometimes without even tasting it. Some say they get a feeling of euphoria during binges, similar to the "runner's high" that some people get from exercise.

The self-induced vomiting that often follows eating binges can cause all sorts of physical problems, such as damage to the stomach and esophagus, chronic heartburn, burst **blood** vessels in the eyes, throat irritation, and erosion of tooth enamel from the acid in vomit. Excessive use of laxatives can be hazardous, too. Muscle cramps, stomach pains, digestive problems, dehydration, and even poisoning may result. Over **time**, bulimia causes **vitamin** deficiencies and imbalances of critical body fluids, which in turn can lead to seizures and kidney failure.

Some researchers believe that an imbalance in the brain chemical serotonin underlies bulimia, as well as other types of compulsive **behavior**. The production of serotonin, which influences mood, is affected by both **antidepressant drugs** and certain foods. But most research on bulimia focuses on its psychological roots.

Bulimia is not as likely as anorexia to reach life-threatening stages, so hospitalization is not usually necessary. Treatment generally involves psychotherapy and sometimes the use of antidepressant drugs. Unlike anorectics, bulimics usually admit they have a problem and want help overcoming it. Estimates of the rates of recovery from bulimia vary widely, with some studies showing low rates of improvement and others suggesting that treatment is effective. Even after apparently successful treatment, some bulimics relapse.

Obesity

Obesity is an excess of body fat. But the question of what constitutes an excess has no clear answer. Some doctors classify a person as obese whose weight is 20% or more over the recommended weight for his or her height. But other doctors say standard height and weight charts are misleading. They maintain that the proportion of fat to muscle, measured by the skinfold "pinch" test, is a better measure of obesity. A person who is overweight, they point out, is not necessarily obese. A very muscular athlete, for example, might have very little body fat, but still might weigh more than the recommended weight for his or her height.

The causes of obesity are complex and not fully understood. While compulsive overeating certainly can lead to obesity, it is not clear that all obesity results from overindulging. Recent research increasingly points to biological, as well as psychological and environmental factors that influence obesity.

In the United States, people with low incomes are more likely to be obese than are the wealthy. Women are almost twice as likely as men to have the problem, but both men and women tend to gain weight as they age.

In those people whose obesity stems from compulsive eating, psychological factors seem to play a large role. Some studies suggest that obese people are much more likely than others to eat in response to stress, loneliness, or depression. As they are growing up, some peo-

ple learn to associate food with love, acceptance, and a feeling of belonging. If they feel rejected and unhappy later in life, they may use food to comfort themselves.

Just as emotional **pain** can lead to obesity, obesity can lead to psychological scars. From childhood on, obese people are taunted and shunned. They may even face discrimination in school and on the job. The low self-esteem and sense of isolation that result may contribute to the person's eating disorder, setting up an endless cycle of overeating, gaining more weight, feeling even more worthless and isolated, then gorging again to console oneself.

People whose obesity endangers their health are said to be morbidly obese. Obesity is a risk factor in diabetes, high blood **pressure**, **arteriosclerosis**, angina pectoralis (chest pains due to inadequate blood flow to the heart), varicose **veins**, **cirrhosis** of the liver, and kidney **disease**. Obesity can cause complications during pregnancy and in surgical procedures. Obese people are about one and one half times more likely to have heart attacks than are other people. Overall, the death **rate** among people ages 20-64 is 50% higher for the obese than for people of normal weight.

Since compulsive eating patterns often have their beginnings in childhood, they are difficult to break. Some obese people get caught up in a cycle of binging and dieting—sometimes called yo-yo dieting—that never results in permanent weight loss. Research has shown that strict dieting itself may contribute to compulsive eating. Going without their favorite foods for long periods makes people feel deprived. They are more likely, then, to reward themselves by binging when they go off the diet. Other research shows that dieting slows the dieter's **metabolism**. When the person goes off the diet, he or she gains weight more easily.

The most successful programs for dealing with overeating teach people to eat more sensibly and to increase their physical activity to lose weight gradually without going on extreme diets. Support groups and therapy can help people deal with the psychological aspects of obesity.

Resources

Books

Epstein, Rachel. *Eating Habits and Disorders*. New York: Chelsea House Publishers, 1990.

Matthews, John R. *Eating Disorders*. New York: Facts On File, 1991.

Porterfield, Kay Marie. *Focus on Addictions*. Santa Barbara: ABC-CLIO, 1992.

Periodicals

Berry, Kevin. "Anorexia? That's a Girls' Disease." *Times Educational Supplement* (April 16, 1999): D8.

Dansky, Bonnie S., Timothy D. Brewerton, and Dean G. Kilpatrick. "Comorbidity of Bulimia Nervosa and Alcohol Use Disorders: Results from the National Women's Study." *The International Journal of Eating Disorders* 27, no. 2 (March 1, 2000): 180.

Nancy Ross-Flanigan

Ebola virus

The Ebola **virus** is one of a number of viruses that cause a devastating **disease** in humans and closely related **species** such as **monkeys**, **gorillas**, and **chimpanzees**. The disease is known as a hemorrhagic fever because of the massive internal bleeding caused by the viral **infection**. Most strains of Ebola hemorrhagic fever progresses quickly from the initial appearance of symptoms to resolution, which is often death.

The name of the Ebola virus comes from a river located in the Democratic Republic of the Congo, where the virus was discovered in 1976. At that time, the country was known as Zaire. The virus was designated Ebola Zaire. Over 300 cases were associated with this initial outbreak. That same year, another type of Ebola virus, named Ebola Sudan, caused an outbreak in another African country, Sudan. Outbreaks also occurred in 1979, 1995, and 1996. In the latest outbreak, which has been ongoing in Gabon since late 2001, 54 people have died as of February of 2002.

The Ebola virus is a species in one group in a collection of viruses classified as Filoviridae. They were originally classified as rhabdoviruses. However, **genome** sequencing revealed more of a similarity to paramyxoviruses. However, filoviruses are sufficiently distinct from the other nonsegmented negative-stranded RNA viruses to warrant taxonomic status as a separate virus family. This family of viruses causes a disease that is typified by copious internal bleeding and bleeding from various orifices of the body, including the eyes. The disease can be swiftly devastating and results in death in about 90% of cases.

As of 2003, four species of Ebola virus have been identified. The species differ slightly in their genetic sequences and in the immune reaction they elicit in those who become infected. A different immune reaction means that the protein components that cause the immune reaction (antigens) are different.

Three of the Ebola virus species cause disease in humans. These are Ebola-Zaire (isolated in 1976), Ebola-Sudan (also isolated in 1976), and Ebola-Ivory Coast (isolated in 1994). The fourth species, called Ebola-Reston, causes disease only in **primates**.

Ebola Reston is named for the United States military primate research facility where the virus was isolated, during a 1989 outbreak of the disease caused by infected monkeys that had been imported from the Philippines. Until the non-human involvement of the disease was proven, the outbreak was thought to be the first outside of **Africa**. This incident brought the virus and the disease it causes to prominent public attention, particularly in the United States.

The appearance of the Ebola virus only dates back to 1976. The explosive onset of the illness and the fact that the outbreaks tend to occur in underdeveloped and remote regions of Africa have complicated both the treatment of the disease and the tracking of its **habitat** and origin. Indeed, the source of the Ebola virus is still unknown. However, the related filovirus, which produces similar effects, establishes a latent infection (one that does not always produce symptoms, but which is infectious and capable of being spread) in African monkeys, **macaques**, and chimpanzees. A credible theory, therefore, is that the Ebola virus normally resides in an **animal** that lives in Africa. As of 2002, the Ebola virus has not been isolated from live primates in the wild. Chimpanzees found dead and infected with the Ebola virus, however, were responsible for the initiation of Ebola outbreaks in 1996 in two regions of Gabon when local residents ate infected meat from the dead chimpanzees.

Almost all confirmed cases of Ebola from 1976 to 2002 have been in Africa. In the past, one **individual** in Liberia presented immunological evidence of exposure to Ebola, but had no symptoms. As well, a laboratory worker in England developed Ebola fever as a result of a laboratory accident in which the worker was punctured by an Ebola-contaminated needle. No human case has been reported in **North America**. A suspected case in Hamilton, Ontario, Canada in early 2001 turned out not to be Ebola, although the exact cause of this illness remains unresolved.

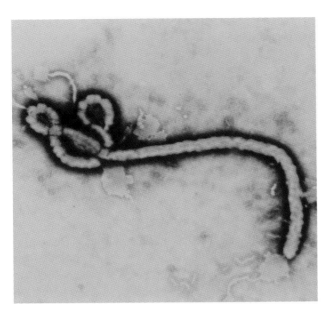

The Ebola virus. © *Corbis Sygma/Corbis. Reproduced by permission.*

Infection with the Ebola virus produces a high fever, headache, muscle aches, abdominal **pain**, tiredness and diarrhea within a few days. Some people will also display bloody diarrhea and vomit **blood**. At this stage of the disease some people do recover. But, for most of those who are infected, the disease progresses within days to produce copious internal bleeding, shock, and death.

The initial infection is presumably by contact between the person and the animal that harbors the **virus**. Subsequent person-to-person spread likely occurs by **contamination** with the infected blood or body tissues of an infected person in the home or hospital setting, or via contaminated needles. Following the initial human infection, spread occurs from direct contact with the blood and/or secretions of the infected person. Within a few days of infection, symptoms that typically develop include a fever, headache, muscle aches, fatigue, and diarrhea. Within one week of infection, hemorrhaging causes blood to flow from the eyes, nose, and ears. Death usually occurs within nine or 10 days of the onset of infection.

The severity and lethality of Ebola hemorrhagic fever effectively limit the outbreaks to a relatively few individuals. Outbreaks of infection with the Ebola virus appear sporadically, rapidly move through the population, and usually end just as suddenly. The high death rate associated with infection is likely the **limiting factor** in the infection's spread. Viruses require a host **cell** in which to reproduce. Killing the host eliminates the use of the host's cells as a reproductive factory for the virus. In other words, death is so rapid that the virus has a limited chance to spread to other victims.

That infected people tend to be in more under-developed regions, where even the health care facilities are not as likely to be equipped with isolation wards, furthers the risk of spread. The person-to-person passage is immediate. Unlike the animal host, people do not harbor the virus for lengthy periods of time.

Whether Ebola viruses can be transmitted through the air is not known. In the outbreak at the Reston facility, Ebola Reston may well have been transmitted from monkey to monkey via the air distribution system. Indeed, some of the monkeys that were infected were never in physical contact with the other infected monkeys. However, if the other species of the Ebola virus are capable of similar transmission has not yet been documented. Laboratory studies have shown that Ebola virus can remain infectious when aerosolized. But the current consensus is that airborne transmission, although possible, plays a minor role in the spread of the virus.

Similarly, the natural reservoir of the Ebola viruses is not known. Between the sporadic outbreaks, the Ebola virus probably is resident in this natural reservoir, whatever that reservoir may be.

Currently there is no cure for the infection caused by the Ebola virus. However, near the end of a 1995 outbreak in Kikwit, Africa, blood products from survivors of the infection were transfused into those actively experiencing the disease. Of those eight people who received the blood, only one person died. Whether or not the transfused blood conveyed a protective factor was not established, as resources were being devoted to bedside care rather than research. A detailed examination of this possibility awaits another outbreak. With the advent of **recombinant DNA** technology, the molecular structure of Ebola viruses, the details of their replication and interaction with the host are under intense study.

The molecular basis for the establishment of an infection by the Ebola virus is still also more theoretical than fact. One clue has been the finding of a glycoprotein in the circulating fluid of infected humans and monkeys that is a shortened version of a viral component. It may be that this protein acts as a decoy for the **immune system**, diverting the immune defenses from the actual site of viral infection. It has also been proposed that the virus suppresses the immune system, via the selective invasion and damage of the spleen and the lymph nodes (which are vital in the functioning of the human immune system).

The devastating infection caused by the Ebola virus is all the more remarkable given the very small size of the viral genome. Fewer than a dozen genes have been detected. How the virus establishes an infection and evades the host immune system with only the capacity to code for less than twelve **proteins** is unknown.

KEY TERMS

· ·

Epidemiologist—A physician or scientist who studies the distribution, sources, and control of diseases in populations.

See also Epidemiology.

Resources

Books

Peters, C. J., and M. Olshaker. *Virus Hunter: Thirty Years of Battling Hot Viruses Around the World.* New York: Doubleday, 1998.

Periodicals

Sanchez, A., M. P. Kiley, B. P. Holloway, et al. "Sequence Analysis of the Ebola Virus Genome: Organization, Genetic Elements, and Comparison with the Genome of Marburg Virus." *Virus Research* 29 (1993): 215–240.

Organizations

United States Centers for Disease Control and Prevention, Special Pathogens Branch. 1600 Clifton Road, Atlanta, GA 30333. (404) 639–3311. <http://www.cdc.gov/ncidod/dvrd/spb/mnpages/dispages/ebola.htm>.

Brian Hoyle

Ebony

Ebony (*Diospyros* spp., family Ebenaceae) are **species** of tropical hardwood trees favored for their hard and beautiful **wood**. Only the black or brown heartwood is used commercially. There are more than 300 species of ebony, ranging in size from shrubs to trees taller than 100 ft (30 m). The best commercial ebony comes from India, Madagascar, Nigeria, Zaire, and the Celebes Islands. Most species of ebony are found in the tropics, but some are found in warm temperate zones. The latter includes the American persimmon (*Diospyrus virginiana*), whose heartwood is not a full black and does not have the extreme density that is so desirable for carving and fine woodwork. Aggressive harvesting of ebony has rendered many species of ebony rare and endangered, and consequently, quite valuable.

Plants in the Ebenaceae family have simple, alternate, coriaceous (or leathery) leaves that are oblong or lanceolate, and vary in length according to species. The flowers are white or greenish-white, with at least four stamens. The globular **fruits** are sought by animals and humans alike because of their sweetness when ripe. Some indigenous tribes use the fruit to make beer. The leaves and other parts of the **tree** are used in traditional medicine to treat intestinal **parasites**, wounds, **dysentery**, and fever, but laboratory tests have not verified the efficacy of this medicinal usage.

The wood of the ebony is so dense, it rapidly dulls tools used for working, sawing, or turning it. Even **termites** will bypass a fallen ebony log. This density contributes to ebony's commercial appeal, as it results in a finish that will take a high polish, adding to its beauty. The properties, attributed to ebony through both fact and myth, have been recognized for many generations. It has long been a favorite material for carving in **Africa**. Some rulers in India had scepters made from it, and also used it for their drinking vessels as it was believed to neutralize poisons. Today, ebony is used for many purposes, including tool and knife handles, furniture, inlay work, wall paneling, golf club heads, and musical instruments. For many years ebony was used for the black keys on the piano, but increasing costs have necessitated the use of synthetic substitutes. Today, only the most expensive concert pianos are still made with ebony. Ebony is also used in stringed instruments for tension pegs and fingerboards.

Although there are many species of ebony, only a few provide commercial-grade wood, and the demand far exceeds the supply. Africa is the source of the most desirable, jet-black heartwood. It comes from the species *Diospyrus crassiflora*, commonly called African ebony. This ebony is prized for its intensely black core. With a wood-density of 64 lb/cu. ft (1,030 kg/cu. m), it has a specific gravity of 1.03 and will not float in **water**. It is found in Cameroon, Ghana, Nigeria, and Zaire.

Diospyrus macassar, commonly called Macassar ebony, is not as plentiful as the African species, but its greater density makes it even more useful in certain types of manufacturing. With a weight of 68 lb/cu ft (1,090 kg/cu m), it is even more dense than African ebony. It has a specific gravity of 1.09, and also does not float. Macassar ebony is found mostly in the Celebes Islands of Indonesia, with some minor growth in India. The heartwood is frequently streaked with lighter bands, and this type is favored by piano makers. Because they are so difficult to dry, the trees are usually girdled to kill them and then left standing for two years to dry out. After they are felled and cut into lumber, they must dry for another six months.

Diospyros mespiliformis, also known as the Jakkalsbessie (Jackal's berry), Transvaal ebony, or Rhodesian ebony, is a straight tree that grows 70 ft (21 m) tall with a trunk up to 4 ft (1.4 m) or more in diameter. It is more widespread and abundant than other ebonies, but the

KEY TERMS

Calyx—All the sepals of a flower, collectively.

Coriaceous—Leathery in texture, thicker than normal.

Dioecious—Plants in which male and female flowers occur on separate plants.

Lanceolate—Lance shaped.

Sepals—Usually outermost division of the calyx.

heartwood is more brown than black, limiting its appeal. Among many native cultures, it serves a medicinal purpose and concoctions derived from the leaves and **bark** are used to treat wounds, fevers, and intestinal parasites. **Color** aside, the density of Rhodesian ebony renders it desirable for furniture, knife handles, and flooring. The fruit is edible.

Diospyrus virginiana, the persimmon, or American ebony, is a native of the southeastern United States. It takes approximately 100 years to mature and grows to a height of 65 ft (20 m). Like the tropical ebonies, it has simple, alternate coriaceous leaves. The flowers are yellowish green and the fruit is yellow, globose, and somewhat larger than its tropical cousins (up to 2.5 in [6.4 cm] in diameter). The fruits are filled with many **seeds** and have a sweet, custard-like interior. Due to its hardness, the wood is used for handles, furniture, and golf club heads. Since there are no vast groves of persimmon, it is not of great economic importance. Persimmon weighs 53–55 lb/cu ft (826–904 kg/cu m).

The growing scarcity of all types of commercial ebony has steadily increased its value. All commercially valuable species are becoming rare, and some are endangered in their wild habitats. Many of the uses of ebony can be substituted by synthetic materials, such as hard **plastics**, although these do not have the aesthetic appeal of true ebony wood.

Resources

Books

Dale, Ivan R. *Kenya Trees and Shrubs*. London: Hatchards, 1991.

J. Gordon Miller

Echiuroid worms

Echiuroid worms, or echiurans, commonly called spoon worms, are soft-bodied, unsegmented, marine animals of worldwide distribution. The approximately 125 **species** in the phylum Echiura occur mostly in the shallow intertidal zone of oceans. Most burrow or form tubes in **sand** or mud. Some live in discarded shells of **sea urchins** and **sand dollars**. Others inhabit cracks and crevices in **rocks** or coral fragments. Body length varies from a fraction of an inch to 20 in (50 cm) or more. There are two body divisions: the body proper, or trunk; and a proboscis, which is highly mobile and extensible, but not retractable into the trunk. The trunk may be smooth, or it may have rows of small papillae or tubercles, giving it a superficially segmented appearance. At its anterior end, close to the base of the proboscis, it bears a pair of curved or hooked chitinous processes called setae. (Presence of setae is a major characteristic of the phylum Annelida, in which the setae are more numerous.) In some species there are additional setae near the posterior end of the trunk. The body wall is muscular, and a spacious, fluid-filled cavity separates it from the gut. This cavity does not extend into the proboscis. The gut is much longer than the trunk, and parts of it are coiled. It begins at the mouth at the base of the proboscis, and terminates at the anus at the opposite end.

The proboscis may be short, broad, and spoon-shaped, as in the common genus *Echiurus*, or it may be much longer than the trunk, narrow, and divided into two branches at the tip as in *Bonellia*. Its edges are rolled over ventrally to form a trough, which is lined by cilia and mucus-secreting cells. It is used in feeding to collect organic particles from the sandy or muddy substrate and to transport it to the mouth. Tube-dwelling echiurans, for example *Urechis*, collect their food by filter-feeding. This worm secretes a mucus funnel that strains out particles from **water** pumped into the tube. From time to time the mucus is transported to the mouth and ingested, and a new funnel is formed.

Respiratory gas exchange in echiuroid worms occurs between the body fluid and sea water, usually across the body wall. But at least in some species, water is pumped in and out of the lower part of the gut through the anus, and gas exchange takes place across the wall of the gut. Although a **circulatory system** of closed vessels is present, the **blood** is colorless and serves mainly to transport **nutrients**. Excretion is performed by tubular structures called nephridia whose number varies widely among the species. One end of each nephridium is funnel-shaped and ciliated, and opens into the body cavity. The other end is narrow and opens to the outside by means of a minute pore. The **nervous system** is simple, without a **brain** or specialized sense organs. Sexes are separate. Each **individual** has a single gonad, testis, or ovary, which develops from the lining of the body cavity. Immature gametes are shed into the body cavity. Upon maturation they are transported to the outside through the nephridial tubes. **Fertilization** is external, and a planktonic "trochophore" larva (found also in the **segmented worms**, phylum Annelida, and a few other groups) develops.

Although in most echiurans males and females of a particular species look alike, an interesting example of sexual dimorphism and sex differentiation is presented by the European form *Bonellia viridis*. In this species, the male, which is ciliated and lacks a proboscis, is about 0.4-0.8 in (1-2 mm) long. In contrast, the female's trunk alone is 2.3-3.1 in (6-8 cm), with an even longer proboscis. The male lives in the female's pharynx or occasionally in her body cavity. The trochophore larva in this species develops into a female if it settles (at **metamorphosis**) some **distance** away from an existing female. On the other hand, if it settles close to a female, it develops into a male. It has been suggested that females produce and release into the surrounding water a "hormone" which has a masculinizing effect on the developing trochophore.

Echo *see* **Acoustics**

Echolocation

In the **animal** kingdom, echolocation is an animal's determination of the position of an object by the interpretation of echoes of sounds produced by the animal. Echolocation is an elegant evolutionary **adaptation** to a low-light **niche**. The only animals that have come to exploit this unique sense ability are mammals—bats, dolphins, porpoises, and toothed whales. It is now believed that these animals use sound to "see" objects in equal or greater detail than humans can see with reflected **light**.

Echolocation is an adaptation to night life or to life in dark, cloudy waters. Long ago, **bats** that ate **insects** during the day might have been defeated in the struggle for survival by **birds**, which are agile and extremely sharp-sighted insectivores. Similarly, toothed whales, porpoises, and dolphins might have been quickly driven to **extinction** by **sharks**, which have a very keen sense of **smell**. These marine **mammals** not only compete with sharks for food sources, but have themselves been preyed upon by sharks. Echolocation helps them find food and escape from predators.

Bats

Echolocation in bats was first clearly described in 1945 in a seminal paper by Griffin and Galambos enti-

tled *Development of the Concept of Echolocation*. Bats that eat **frogs**, **fish**, and insects use echolocation to find their **prey** in total or near-total darkness. After emitting a sound, these bats can tell the **distance**, direction, size, surface texture, and material of an object from information in the returning echo. Although the sounds emitted by bats are at high frequencies, out of the range of human **hearing**, these sounds are very loud—as high as 100 decibels, which is as loud as a chainsaw or jackhammer. People may hear the calls as clicks or chirps. The fruit-eating and nectar-loving bats do not use echolocation. These daytime and dusk-active bats have strong eyes and noses for finding food.

Bats use echolocation to hunt for food and to avoid collisions. A group of insect-eating bats was trained to tell the difference between insect larvae with fuzzy bristles and larvae that had their bristles removed. Researchers injected the bristle-less larvae with a chemical that made them taste bitter, then offered both the normal and the nonfuzzy larvae to the bats; the bats could instantly distinguish between them. In another experiment, a group of bats was conditioned to detect very thin wires in total darkness.

One question that puzzled scientists is how a bat can hear the echo of one sound while it is emitting another sound; why is the bat not deafened or distracted by its own sounds? The answer is that the bat is deafened—but only for a moment. Every time a bat lets out a call, part of its middle **ear** moves, preventing sounds from being heard. Once the bat's call is made, this structure moves back, allowing the bat to hear the echo from the previous call.

One family of bats, the Vespertilionidae, emits ultrasonic sound pulses from their mouths in a narrow directed beam and uses their large ears to detect the returning echoes. Each sound pulse lasts 5-10 milliseconds and decreases in **frequency** from about 100,000 Hz at the beginning down to about 30,000 Hz. This change in frequency (or frequency modulation) is roughly equivalent to a human looking at an object under a range of different colors of light. When the bat is just "looking around" it puts out approximately 10 pulses per second. If it hears something interesting, it takes a closer look by increasing the number of pulses per second to approximately 200. This is roughly equivalent to a person shining a brighter light on an object under investigation.

Marine mammals

Echolocation may work better under **water** than it does on land because water is a more effective and efficient conveyer of **sound waves**. Echolocation may be more effective for detecting objects underwater than light-based **vision** is on land. Sound with a broad frequency range has a more complex interaction with the

KEY TERMS

. . . .

Frequency—For a sound wave, the number of waves that pass a given point per unit of time.

Hertz—A unit of measurement for frequency, abbreviated Hz. One hertz is one cycle per second.

Insectivore—Any animal that feeds on insects.

Ultrasonic vibrations—Acoustic vibrations with frequencies higher than the human threshold of hearing.

objects that reflect it than does light. For this reason, sound can convey more information than light.

Like bats, marine mammals such as whales, porpoises, and dolphins emit pulses of sounds and listen for the echo. Also like bats, these sea mammals use sounds of many frequencies and a highly direction-sensitive sense of hearing to navigate and feed. Echolocation provides all of these mammals with a highly detailed, three-dimensional image of their environment.

Whales, dolphins, and porpoises all have a weak sense of vision and of smell, and all use echolocation in a similar way. They first emit a frequency-modulated sound pulse. A large fatty deposit, sometimes called a melon, found in its head helps the mammal to focus the sound. The echoes are received at a part of the lower jaw sometimes called the acoustic window. The echo's vibration is then transmitted through a fatty **organ** in the middle ear where it is converted to neural impulses and delivered to the **brain**. The brains of these sea mammals are at least as large relative to their body size as is a human brain relative to the size of the human body.

Captive porpoises have shown that they can locate tiny objects and thin wires and distinguish between objects made of different metals and of different sizes. This is because an object's material, structure, and texture all affect the nature of the echo returning to the porpoise.

Like bats, the toothed whales have specially adapted structures in the head for using echolocation. Some **species** of toothed whales have a bony structure in the head that insulates the back of the skull where sounds are received from the front of the skull where sounds are produced. The middle ear cavity is divided into a complex sinus that may help to acoustically separate the right and the left ears. This would enable the whale to more easily glean information from the echoes it receives. Other structures help to reduce the confusion of transmitted and received sound throughout the skull.

See also Acoustics; Cetaceans; Radar.

Resources

Books

Campbell, N., J. Reece, and L. Mitchell. *Biology*. 5th ed. Menlo Park: Benjamin Cummings, Inc. 2000.

Harrison, Richard, and M. M. Bryden, eds. *Whales, Dolphins, and Porpoises*. New York: Facts on File, 1988.

John Henry Dreyfuss

Eclipses

It is a coincidence of nature that the apparent size of the **Sun** and the **Moon** in the sky are about the same. Thus on those rare occasions when the orbital **motion** of **Earth** and Moon cause them to align with the sun, as seen from points on Earth, the Moon will just cover the surface of the Sun and day will suddenly become night. Those who are located in the converging lunar shadow that just reaches Earth will see a *total eclipse* of the sun. The converging shadow cone, within which the sun is completely hidden by the moon, the umbral shadow of the moon.

One can imagine a diverging cone with the Moon at its apex in which only part of the sun is covered by the moon. This shadow is called the penumbra, or partially dark shadow. Folks on the earth located in this shadow will see the sun partially obscured or covered by the Moon.

Such an eclipse is called a *partial solar eclipse*. Because the base of this shadow cone is far larger than the umbral shadow, far more people see partial solar eclipses than see total solar eclipses. However, the impact on the observer of a total solar eclipse is far greater. Even a nearly total solar eclipse permits a small fraction of the solar surface to be visible, but covering the bright photosphere completely drops the light-level to a millionth of its normal value. During totality one can safely look directly at the sun and its corona, but this should not be done outside of totality during any partial or annular phases. The photospheric surface of the Sun is so bright, its focused image on the retina of the **eye** can do permanent damage to an individual's **vision**, including total blindness. Even viewing the sun through colored or smoked **glass** should be avoided, for the filter may pass infrared or ultraviolet **light** not obvious to the observer, but which can still do extensive damage. While specially designed "sun filters" may provide viewing safety, the safest approach to looking at the sun is projecting its image from a small **telescope** or monocular onto a screen. Direct viewing and photographs of the projected image can then be made in relative safety.

To the observer of a total solar eclipse many strange phenomena are apparent at the same time. The progressive coverage of the solar photosphere by the Moon reduces the solar heating of Earth, causing the local **temperature** to fall. The drop in temperature is accompanied by a rise in **humidity** and often a **wind** change. The covering of the central part of the sun's disk also brings about a subtle **color** shift toward the yellow. In the final seconds before totality the last bright regions of the sun's disk shine through the valleys at the limb of the moon, causing bright spots called "Baily's Beads." As the last of these disappear, the blood-red upper atmosphere of the sun called the *chromosphere* will briefly appear before it too is covered, revealing the winding, sheet-white corona that constitutes the outer atmosphere of the sun and is less bright than the full moon.

Birds fly to roost and animals behave as if night had truly arrived. All the senses are assaulted at once both by the changes in the local environment and the changes to the sun. A solar eclipse makes such an impression on people that it is said St. Patrick used one to convert the Celtic Irish to Christianity in the fifth century. The ancient historian Herodotus reported that a total solar eclipse that occurred during a battle between the Lydians and the Medes in 585 B.C. caused the soldiers to throw down their weapons and leave the field. Otherwise professional astronomers have been known to stand and stare at the phenomenon, forgetting to gather the data they have practiced for months and traveled thousands of miles to obtain.

Outside the narrow band traced across the earth by the tip of the moon's umbral shadow, part of the sun will be covered from those located in the expanding cone of the lunar penumbral shadow. An eclipse seen from such locations is said to be a partial solar eclipse. If the moon is near its farthest point from Earth, its dark umbral shadow does not quite reach Earth. Should this occur when the alignment for a solar eclipse is correct, the bright disk of the Sun will only be partially covered. At the middle of the eclipse a bright annulus of the solar photosphere will completely surround the dark disk of the moon. Such eclipses are called *annular eclipses* and may be considered a special case of a partial solar eclipse. Since part of the photosphere is always visible, one never sees the chromosphere or corona and the sky never gets as dark as during a total solar eclipse. However, there is a definite change in the color of the sunlight. Since the visible photosphere at the limb of the annularly eclipsed sun is cooler and more yellow than the center of the solar disk, the effect is for the daylight color to be shifted to the yellow. The effect is quite pronounced for eclipses occurring around local noon.

Because the area on the earth covered by the moon's umbra during a total eclipse is so small, it is quite rare for an **individual** to see one even though their **frequency**

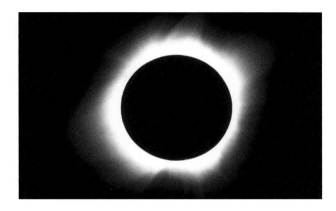

A total solar eclipse in La Paz, Baja California, Mexico, on July 11, 1991. *Photograph by Francois Gohier. Photo Researchers, Inc. Reproduced by permission.*

of occurrence is somewhat greater than lunar eclipses. Lunar eclipses occur when the moon passes into the shadow cast by the earth. During a lunar eclipse, the bright disk of the full moon will be progressively covered by the dark disk of the umbral shadow of the earth. If the eclipse is total, the moon will be completely covered by that shadow. Should the alignment between the sun, earth, and moon be such that the moon simply grazes the earth's umbra, the eclipse is called a partial. Lunar orbital paths that pass only through the penumbral shadow of the earth are called *penumbral lunar eclipses*. The dimming of the moon's light in these eclipses is so slight that it is rarely detected by the human eye so little notice of these eclipses is made.

Since the moon is covered by the shadow of the earth, any point on the earth from which the moon can be seen will be treated to a lunar eclipse. Thus they are far more widely observed than are total eclipses of the sun. However, because the sun is so much brighter than the full moon, the impact of a total lunar eclipse is far less than for a total solar eclipse. Unlike a solar eclipse where the shadow cast by the moon is totally dark, some light may be refracted by the earth's atmosphere into the earth's umbra so that the disk of the moon does not totally disappear during a total lunar eclipse. Since most of the blue light from the sun is scattered in the atmosphere making the sky blue, only the red light makes it into the earth's umbral shadow. Therefore, the totally eclipsed moon will appear various shades of red depending on the cloud cover in the atmosphere of the earth.

Lunar eclipses do not occur every time the moon is full, nor do solar eclipses happen each time the moon is new. Although the line-up between the sun, earth, and moon is close at these lunar phases, it is not perfect. The orbital **plane** of the moon is tipped about five degrees to

TABLE OF ECLIPSES 1995-2010

Date	Type of Eclipse	Time of Mid-Eclipse-EST*	Duration of Eclipse**	Total Length of Eclipse	Region of Visibility†
Apr. 15, 95	Lunar-Partial	7:19 AM	—	1h 12min	E. Hemisph.
Apr. 29, 95	Solar-Annular	1 AM	6min 38sec	—	Pacific S. America
Oct. 24, 95	Solar-Total	Midnight	2min 10sec	—	Asia, Borneo, Pacific Ocean
Apr. 3, 96	Lunar-Total	7:11 PM	86min	3h 36min	W. Hemisph.
Sep. 26, 96	Lunar-Total	9:55 PM	70min	3h 22min	W. Hemisph.
Mar. 8, 97	Solar-Total	8 PM	2min 50sec	—	Siberia
Mar. 23, 97	Lunar-Partial	11:41 PM	—	3h 22min	W. Hemisph.
Sep. 16, 97	Lunar-Total	1:47 PM	62min	3h 16min	E. Hemisph.
Feb. 26, 98	Solar-Total	Noon	4min 8sec	—	W. Pacific, S. Atlantic
Aug. 21, 98	Solar-Annular	9 PM	3min 14sec	—	Sumatra, Pacific Ocn.
Feb. 16, 99	Solar-Annular	2 AM	1min 19sec	—	Indian Ocn., Australia
Jul. 28, 99	Lunar-Partial	6:34 AM	—	2h 22min	Europe-Asia
Aug. 11, 99	Solar-Total	6 AM	2min 23sec	—	Atlantic Ocn., Europe-Asia
Jan. 20, 00	Lunar-Total	11:45 PM	76min	3h 22min	W. Hemisph.
Jul. 16, 00	Lunar-Total	8:57 AM	106min	3h 56min	E. Hemisph.
Jan. 9, 01	Lunar-Total	3:22 PM	60min	3h 16min	E. Hemisph.
Jun. 21, 01	Solar-Total	7 AM	4m 56sec	—	S. Atlantic, S. Africa
Jul. 5, 01	Lunar-Partial	9:57 AM		2h 38min	E. Hemisph.
Dec. 14, 01	Solar-Annular	4 PM	3min 54sec	—	Pacific Ocn., Cent. Amer.
Jun. 10, 02	Solar-Annular	7 PM	1min 13sec	—	Pacific Ocn.
Dec. 4, 02	Solar-Total	3 AM	2min 4sec	—	Indian Ocn., Australia
May 15, 03	Lunar-Total	10:41 PM	52min	3h 14min	W. Hemisph.
May 30, 03	Solar-Annular	11 PM	3min 37sec	—	Iceland &E. Arctic
Nov. 8, 03	Lunar-Total	8:20 PM	22min	3h 30min	W. Hemisph.
Nov. 23, 03	Solar-Total	6 PM	1min 57sec	—	Antarctica
May 4, 04	Lunar-Total	3:32 PM	76min	3h 22min	E. Hemisph.

TABLE OF ECLIPSES 1995-2010 (cont'd)

Date	Type of Eclipse	Time of Mid-Eclipse-EST*	Duration of Eclipse**	Total Length of Eclipse	Region of Visibility†
Oct. 27, 04	Lunar-Total	10:05 PM	80min	3h 38min	W. Hemisph.
Apr. 8, 05	Solar-Annular-Total	4 PM	42sec	—	N. Central, Pacific Ocn.
Oct. 3, 05	Solar-Annular	6 AM	4m 32sec	—	Atlantic Ocn., Spain, Africa
Oct. 17, 05	Lunar-Partial	7:04 AM	—	56min	E. Hemisph.
Mar. 29, 06	Solar-Total	5AM	4min 7sec	—	Atlantic Ocn., Africa, Turk.
Sep. 7, 06	Lunar-Partial	1:52 PM	—	1h 30min	E. Hemisph.
Sep. 22, 06	Solar-Annular	7 AM	7min 9sec	—	N.E. of S.Amer. Atlan.
Mar. 3, 07	Lunar-Total	6:22 PM	74min	3h 40min	W. Hemisph.
Aug. 28, 07	Lunar-Total	5:38 AM	90min	3h 32min	W. Hemisph.
Feb. 6, 08	Solar-Annular	11 PM	2min 14sec	—	S. Pacific, Antarctic
Feb. 20, 08	Lunar-Total	10:57 PM	50min	3h 24min	W. Hemisph.
Aug. 1, 08	Solar-Total	5 AM	2min 28sec	—	Arctic-Cand., Siberia
Aug. 16, 08	Lunar-Partial	4:11 PM	—	3h 8min	E. Hemisph.
Jan. 26, 09	Solar-Annular	3 AM	7min 56sec	—	S. Atlantic, Indian Ocn.
Jul. 21, 09	Solar-Total	10 PM	6min 40sec	—	East Asia, Pacific Ocn.
Dec. 31, 09	Lunar-Partial	2:24 PM	—	1h 00min	E. Hemisph.
Jan. 15, 10	Solar-Annular	2 AM	11min 10sec	—	Africa, Indian Ocn.
Jun. 26, 10	Lunar-Partial	6:40 AM	—	2h 42min	E. Hemisph.
Jul. 11, 10	Solar-Total	3 PM	5min 20sec	—	Pacific Ocn., S. America
Dec. 21, 10	Lunar-Total	3:18 AM	72min	3h 28min	W. Hemisph.

* Eastern Standard time is used for convience. Since the path of a solar eclipse spans a good part of the earth, only an approximate time to the nearest hour is given the mid-point of that path.

** The time of the eclipse duration is for maximum extend of totality, except for annular eclipses where it marks the maximum duration of the annular phase.

† The visible location of lunar eclipses is approximately half the globe where the Moon is visible. For convience, the globe has been split into eastern and western hemispheres. Depending on the time of mid-eclipse, more or less of the entire eclipse may be visible from the specified hemisphere.

the orbital plane of the earth. These two planes intersect in a line called the line of nodes. That line must be pointed at the sun in order for an eclipse to occur. Should the moon pass by the node between the earth and the sun while the line of nodes is aimed at the sun, the alignment between the sun, moon, and earth will be perfect and a solar eclipse will occur. If the moon passes through the node lying beyond Earth when the line of nodes is properly oriented, we see a lunar eclipse.

Except for slow changes to the Moon's **orbit**, the line of nodes maintains an approximately fixed orientation in **space** as it is carried about the sun by Earth's motion. Therefore, about twice a year the line of nodes is pointing straight at the Sun and eclipses can occur. If the alignment is closely maintained during the two weeks between new moon and full moon, a solar eclipse will be followed by a lunar eclipse. A quick inspection of the table of pending eclipses shows that 20 of the 47 listed eclipses occur within two weeks of one another, indicating that these are times of close alignment of the line of nodes with the sun. A further inspection shows that these pairs occur about 20 days earlier each year indicating that the line of nodes is slowly moving westward across the sky opposite to the annual motion of the Sun. At this **rate** it takes about 18.6 years for the nodes to complete a full circuit of the sky. Thus every 18-19 years eclipses will occur at about the same season of the year. After three of these seasonal cycles, or 56 years, the eclipses will occur on, or about, the same day. It is this long seasonal cycle that Gerald Hawkins associated with the 56 "Aubry Holes" at Stonehenge. He used this agreement to support his case that Stonehenge was used to predict eclipses and the Aubry Holes were used to keep track of the yearly passage of time between seasonal eclipses.

There are other cycles of eclipses that have been known since antiquity. It is a reasonable question to ask how long it will be before an eclipse will re-occur at the same place on the earth. The requirements for this to happen are relatively easy to establish. First, the moon must be at the same phase (i.e., either new or full depending on whether the eclipse in question is a solar or lunar eclipse). Secondly, the moon must be at the same place in its orbit with respect to the orbital node. Thirdly, the sun and moon must have the same **distance** from the earth for both eclipses. Finally, if the solar eclipses are to have similar paths across the earth, they must happen at the same time of the year. The first two conditions are required for an eclipse to happen at all. Meeting the third condition assures that the umbral shadow of the moon will reach the earth to the same extent for both eclipses. This means that the two eclipses will be of the same type (i.e., total or annular in the case of the sun). The last con-

dition will be required for solar eclipses to be visible from the same location on the earth.

The interval between successive phases of the moon is called the *synodic month* and is 29.5306 days long. Due to the slow motion of the line of nodes across the sky, successive passages of a given node, called the *nodal month*, occur every 27.2122 days. Finally, successive intervals of closest approach to Earth (i.e. perigee passage) are known as the *anomalistic* month, which is 27.55455 days long. For the first three conditions to be met, the moon must have traversed an **integral** number of synodic, nodical, and anomalistic months in a nearly integral number of days. One can write these constraints as equations whose solutions are **integers**. However, such equations, called Diophantine equations, are notably difficult to solve in general. The ancient Babylonians found that 223 synodic months, 242 nodical months, and 239 anomalistic months all contained about 6,585 1/3 days, which turns out to be just 11 days in excess of 18 years. They referred to the cycle as the Saros cycle, for it accurately predicted repeats of lunar eclipses of the same type and duration. However, the cycle missed being an integral number of days by about eight hours. Thus, solar eclipses would occur eight hours later after each Saros, which would be more than enough to move the path of totality away from any given site. After three such cycles, sometimes referred to as the Triple Saros, lasting 54 years and a month, even the same solar eclipses would repeat with fairly close paths of totality. Since the multiples of the various months do not exactly result in an integral number of days, the repetitions of the eclipses are not exactly the same, but they are close enough to verify the predictability and establish the cycles. The Babylonians were able to establish the Saros with some certainty. Their ability to do so supports Hawkins' notion that the people who built Stonehenge were also capable of establishing the seasonal eclipse cycle.

It is tempting to look for cycles of even longer duration in search of a set of synodic, nodical, and anomalistic months that would yield a more close number of days, but such a search would be fruitless. There are other subtle forces perturbing the orbit of the moon so that longer series of eclipses fail to repeat. Indeed, any series of lunar eclipses fails to repeat after about 50 Saros or about 870 years. Similar problems exist for solar eclipses.

See also Calendars.

Resources

Books

Arny, T. T. *Explorations: An Introduction to Astronomy.* St. Louis: Mosby, 1994.

KEY TERMS

....

Anomalistic month—The length of time required for the moon to travel around its orbit from its point of closest approach to the earth and back again.

Chromosphere—The bright red "color sphere" seen surrounding the sun as a narrow band when the photosphere is obscured.

Corona—A pearly white irregular shaped region surrounding the sun. It is visible only when the photosphere and chromosphere are obscured.

Node—The intersection of the lunar orbit with the plane of the earth's orbit about the sun.

Nodical month—The length of time required for the moon to travel around its orbit from a particular node and back again.

Penumbra—From the Greek term meaning "partially dark." Within the penumbral shadow part of the light source contributing to the eclipse will still be visible.

Photosphere—From the Greek term meaning "light-sphere." This is the bright surface we associate with sunlight.

Saros—A cycle of eclipses spanning 18 years and 11 days first recorded by the Babylonians.

Synodic month—The time interval in which the phases of the Moon repeat (from one Full Moon to the next), and averages 29.53 days.

Umbra—From the Greek meaning dark. Within the umbral shadow no light will be visible except in the case of the earth's umbral shadow where some red sunlight may be refracted by the atmosphere of the Earth.

Hawkins, G. S. *Stonehenge Decoded.* New York: Dell Publishing Co. Inc., 1965.

Periodicals

Schaefer, B. E. "Solar Eclipses that Changed the World." *Sky & Telescope* 87 (1984): 36-39.

George W. Collins II

Ecological economics

Conventional and ecological economics

Economics is conventionally considered to be a social science that examines the allocation of scarce resources among various potential uses that are in **competition** with each other. As such, economics attempts to predict and understand the patterns of consumption of goods and services by individuals and society. A core assumption of conventional economics is that individuals and corporations seek to maximize their profit within the marketplace.

In conventional economics, the worth of goods or services are judged on the basis of their direct or indirect utility to humans. In almost all cases, the goods and services are assigned value (that is, they are valuated) in units of tradable currency, such as dollars. This is true of: (1) manufactured goods such as televisions, automobiles, and buildings, (2) the services provided by people like farmers, doctors, teachers, and baseball players, and (3) all natural resources that are harvested and processed for use by humans, including nonrenewable resources such as metals and **fossil fuels**, and renewable resources such as agricultural products, **fish**, and **wood**.

Ecological economics differs from conventional economics in attempting to value goods and services in ways that are not only based on their usefulness to humans, that is, in a non-anthropocentric fashion. This means that ecological economics attempts to take into account the many environmental and social costs associated with the depletion of natural resources, as well as the degradation of ecological systems through **pollution**, **extinction**, and other environmental damages. Many of these important problems are associated with the diverse economic activities of humans, but the degradation is often not accounted for by conventional economics. From the environmental perspective, the most important problem with conventional economics has been that the marketplace has not recognized the value of important ecological goods and services. Therefore, their degradation has not been considered a cost of doing business. Ecological economics attempts to find ways to consider and account for the real costs of environmental damage.

Ecological goods and services

Humans have an absolute dependence on a continuous flow of natural resources to sustain their economic systems. There are two basic types of natural resources: nonrenewable and renewable. By definition, sustainable economic systems and sustainable human societies cannot be based on the use of nonrenewable resources, because these are always depleted by usage, a process referred to as "mining." Ultimately, sustainable systems can only be supported by the use of renewable resources, which if harvested and managed wisely, can be available forever. Because most renewable resources are the goods and services of ecosystems, economic and ecological systems are highly interdependent.

Potentially, renewable natural resources can sustain harvesting indefinitely. However, to achieve a condition of sustainable usage, the **rate** of harvesting must be smaller than the rate of renewal of the resource. For example, flowing **water** can be sustainably used to produce hydroelectricity or for **irrigation**, as long as the usage does not exceed the capacity of the landscape to yield water. Similarly, biological natural resources such as trees and hunted fish, waterfowl, and **deer** can be sustainably harvested to yield valuable products, as long as the rate of cropping does not exceed the renewal of the resource. These are familiar examples of renewable resources, partly because they all represent ecological goods and services that are directly important to human welfare, and can be easily valuated in terms of dollars.

Unlike conventional economics, ecological economics also considers other types of ecological resources to be important, even though they may not have direct usefulness to humans, and they are not valuated in dollars. Because the marketplace does not assign value to these resources, they can be degraded without conventional economic cost even though this results in ecological damage and ultimately harms society. Some examples of ecological resources that markets consider to be "free" goods and services include:

(1) non-exploited **species** of plants and animals that are not utilized as an economic resource, but are nevertheless important because they may have undiscovered uses to humans (perhaps as new medicines or foods), or are part of the aesthetic environment, or they have intrinsic value which exists even if they are not useful to humans;

(2) ecological services such as control over **erosion**, provision of water and nutrient cycling, and cleansing of pollutants emitted into the environment by humans, as occurs when growing vegetation removes **carbon dioxide** from the atmosphere and when **microorganisms** detoxify chemicals such as **pesticides**.

Use of renewable resources by humans

As noted above, sustainable economic systems can only be based on the wise use of renewable resources. However, the most common way in which humans have used potentially renewable resources is by "overharvesting," that is, exploitation that exceeds the capacity for renewal so that the stock is degraded and sometimes made extinct. In other words, most use of potentially renewable resources has been by **mining**, or use as if it were a nonrenewable resource.

There are many cases of the mining and degradation of potentially renewable resources, from all parts of the world and from all human cultures. In a broad sense, this **syndrome** is represented by extensive **de-forestation**, collapses of wild fisheries, declines of agricultural **soil** capability, and other resource degradations. The extinctions of the dodo, great auk, Steller's sea cow, and passenger pigeon all represent overhunting so extreme that it took potentially renewable resources beyond the brink of biological extinction. The overhunting of the American **bison** and various species of **seals** and whales all represent biological mining that took potentially renewable resources beyond the brink of economic extinction, so that it was no longer profitable to exploit the resource.

These and many other cases of the degradation of renewable resources occurred because conventional economics did not value resource degradation properly. Consequently, profit was only determined on the basis of costs directly associated with catching and processing the resource, and not on the costs of renewal and depletion. Similarly, conventional economics considers non-valuated goods and services such as **biodiversity**, **soil conservation**, erosion control, water and nutrient cycling, and cleansing air and water of pollutants to be free resources so that no costs are associated with their degradation.

Ecologically sustainable systems

The challenge of ecological economics is to design systems of resource harvesting and management that are sustainable, so that human society can be supported forever without degrading the essential, ecological base of support.

Ecologically sustainable systems must sustain two clusters of values: (1) the health of economically valuated, renewable resources, such as trees, fish, and agricultural soil capability, and (2) acceptable levels of ecological goods and services that are not conventionally valuated. Therefore, a truly sustainable system must be able to yield natural resources that humans need, and to provide that sustenance forever. However, the system must also provide services related to clean air and water and nutrient cycling, while also sustaining **habitat** for native species and their natural ecological communities.

To achieve this goal, ecologically sustainable systems will have to be based on two ways of managing ecosystems: (1) as working ecosystems, and (2) as ecological reserves (or protected areas). The "working ecosystems" will be harvested and managed to yield sustainable flows of valuated resources, such as forest products, hunted animals, fish, and agricultural commodities. However, some environmental costs will be associated with these uses of ecosystems. For example, although many species will find habitats available on working lands to be acceptable to their purposes, other native species and most natural communities will be at

KEY TERMS

. . . .

Anthropocentric—Considering the implications of everything from the perspective of utility to humans, and to human welfare.

risk on working landscapes. To sustain the ecological values that cannot be accommodated by working ecosystems, a system of ecological reserves will have to be developed. These reserves must be designed to ensure that all native species are sustained at viable population levels, that there are viable areas of natural ecosystems, and that ecosystems will be able to supply acceptable levels of important services, such as control of erosion, nutrient cycling, and cleansing the environment of pollution.

So far, ecologically sustainable systems of the sort described above are no more than a concept. None exist today. In fact, humans mostly exploit the potentially renewable goods and services of ecosystems in an non-sustainable fashion. Clearly this is a problem, because humans rely on these resources to sustain their economy. Ecological economics provides a framework for the design of better, ecologically sustainable systems of resource use. However, it remains to be seen whether human society will be wise enough to adopt these sustainable methods of organizing their economy and their interactions with ecosystems.

See also Alternative energy sources; Ecosystem; Sustainable development.

Resources

Books

Costanza, R. *Ecological Economics: The Science and Management of Sustainability.* New York: Columbia University Press, 1991.

Freedman, B. *Environmental Ecology.* 2nd ed. San Diego: Academic Press, 1995.

Jansson, A.M., M. Hammer, C. Folke, and R. Costanza, eds. *Investing in Natural Capital: The Ecological Economics Approach to Sustainability.* Washington, DC: Island Press, 1994.

Shortle, J. S., and Ronald C. Griffin, eds. *Irrigated Agriculture and the Environment.* Northampton, MA: Edward Elgar, 2001.

Periodicals

Hooke, Roger L. "On the History of Humans as Geomorphic Agents." *Geology,* vol. 28, no. 9 (September 2000): 843-846.

Bill Freedman

Ecological integrity

Ecological integrity is a relatively new concept that is being actively discussed by ecologists. However, a consensus has not yet emerged as to the definition of ecological integrity. Clearly, human activities result in many environmental changes that enhance some **species**, ecosystems, and ecological processes, while at the same time causing important damage to others. The challenge for the concept of ecological integrity is to provide a means of distinguishing between responses that represent improvements in the quality of ecosystems, and those that are degradations.

The notion of ecological integrity is analogous to that of health. A healthy **individual** is relatively vigorous in his or her physical and mental capacities, and is uninfluenced by **disease**. Health is indicated by diagnostic symptoms that are bounded by ranges considered to be normal, and by attributes that are regarded as desirable. Unhealthy conditions are indicated by the opposite, and may require treatment to prevent further deterioration. However, the metaphor of human and **ecosystem** health is imperfect in some important respects, and has been criticized by ecologists. This is mostly because health refers to individual organisms, while ecological contexts are much more complex, involving many individuals of numerous species, and both living and nonliving attributes of ecosystems.

Environmental stress is a challenge to ecological integrity

Environmental stress refers to physical, chemical, and biological constraints on the productivity of species and the development of ecosystems. When they increase or decrease in intensity, stressors elicit ecological responses. Stressors can be natural environmental factors, or they can be associated with the activities of humans. Some environmental stressors are relatively local in their influence, while others are regional or global in scope. Stressors are challenges to ecological integrity.

Species and ecosystems have some capacity to tolerate changes in the intensity of environmental stressors, an attribute known as resistance. However, there are limits to resistance, which represent thresholds of tolerance. When these thresholds are exceeded, substantial ecological changes occur in response to further increases in the intensity of environmental stress.

Environmental stressors can be categorized as follows:

Physical stress

Physical stress refers to brief but intense events of kinetic **energy**. Because of its acute, episodic nature, this is a type of disturbance. Examples include volcanic eruptions, windstorms, and explosions.

Wildfire

Wildfire is another disturbance, during which much of the **biomass** of an ecosystem combusts, and the dominant species may be killed.

Pollution

Pollution occurs when chemicals occur in concentrations large enough to affect organisms, and thereby cause ecological change to occur. Toxic pollution can be caused by gases such as **sulfur dioxide** and **ozone**, elements such as mercury and arsenic, and **pesticides**. **Nutrients** such as phosphate and nitrate can distort ecological processes such as productivity, causing a type of pollution known as **eutrophication**.

Thermal stress

Thermal stress occurs when releases of **heat** cause ecological responses, as occurs near natural, hot **water** vents in the **ocean**, or with industrial discharges of heated water.

Radiation stress

Radiation stress is associated with excessive loads of ionizing energy. This can be important on mountaintops, where there are intense exposures to ultraviolet radiation, and in places where there are uncontrolled exposures to **radioactive waste**.

Climatic stress

Climatic stress is caused by excessive or insufficient regimes of **temperature**, moisture, solar radiation, or combinations of these. **Tundra** and deserts are climatically stressed ecosystems, while tropical **rainforest** occurs in places where climate is relatively benign.

Biological stress

Biological stress is associated with the complex interactions that occur among organisms of the same or different species. Biological stress can result from **competition**, herbivory, predation, parasitism, and disease. The harvesting and management of species and ecosystems by humans is a type of biological stress.

Large changes in the intensity of environmental stress result in various types of ecological responses. For example, when an ecosystem is disrupted by an intense disturbance, there may be substantial mortality of its species and other damage, followed by recovery through

succession. In contrast, a longer-term intensification of environmental stress, possibly associated with chronic pollution or climate change, causes more permanent ecological adjustments to occur. Relatively vulnerable species are reduced in abundance or eliminated from sites that are stressed over the longer term, and their modified niches are assumed by organisms that are more tolerant. Other common responses include a simplification of species richness, and decreased rates of productivity, **decomposition**, and nutrient cycling. These changes represent an ecological conversion, or a longer-term change in the character of the ecosystem.

Components of ecological integrity

Many studies have been made of the ecological responses to disturbance and to longer-term changes in the intensity of environmental stress. These studies have examined stressors associated with, for example, pollution, the harvesting of species from ecosystems, and the conversion of natural ecosystems into managed agroecosystems. The commonly observed patterns of change in these sorts of stressed ecosystems are considered to represent some of the key elements of ecological integrity. Such observations can be used to develop indicators of ecological integrity, which are useful in determining whether this condition is improving or being degraded over time. It has been suggested that greater ecological integrity is displayed by systems with the following characteristics:

Resiliency and resistance

Ecosystems with greater ecological integrity are, in a relative sense, more resilient and resistant to changes in the intensity of environmental stress. In the ecological context, resistance refers to the capacity of organisms, populations, and communities to tolerate increases in stress without exhibiting significant responses. Resistance is manifest in thresholds of tolerance. Resilience refers to the ability to recover from disturbance.

Biodiversity

In its simplest interpretation, **biodiversity** refers to the number of species occurring in some ecological community or in a designated area, such as a park or a country. However, biodiversity is better defined as the total richness of biological variation, including genetic variation within populations and species, the numbers of species in communities, and the patterns and dynamics of these over large areas.

Complexity of structure and function

The structural and functional complexity of ecosystems is limited by natural environmental stresses associated with climate, **soil**, **chemistry**, and other factors, and by stressors associated with human activities. As the overall intensity of stress increases or decreases, structural and functional complexity responds accordingly. Under any particular environmental regime, older ecosystems will generally be more complex than younger ecosystems.

Presence of large species

The largest, naturally occurring species in any ecosystem generally appropriate relatively large amounts of resources, occupy a great deal of **space**, and require large areas to sustain their populations. In addition, large species are usually long-lived, and therefore integrate the effects of stressors over an extended time. Consequently, ecosystems that are subject to an intense regime of environmental stress cannot support relatively large species. In contrast, mature ecosystems of relatively benign environments are dominated by large, long-lived species.

Presence of higher-order predators

Because top predators are dependent on a broad base of **ecological productivity**, they can only be sustained by relatively extensive and/or productive ecosystems.

Controlled nutrient cycling

Recently disturbed ecosystems temporarily lose some of their capability to exert biological control over nutrient cycling, and they often export large quantities of nutrients dissolved or suspended in streamwater. Systems that are not "leaky" of their nutrient capital in this way are considered to have greater ecological integrity.

Efficient energy use and transfer

Large increases in environmental stress commonly result in community **respiration** exceeding productivity, so that the standing crop of biomass decreases. Ecosystems that are not degrading in their capital of biomass are considered to have greater integrity than those in which biomass is decreasing over time.

Ability to maintain natural ecological values

Ecosystems that can naturally maintain their species, communities, and other important characteristics, without interventions by humans through management, have greater ecological integrity. For example, if a rare species of **animal** can only be sustained through intensive management of its **habitat** by humans, or by management of its demographics, possibly by a captive-breeding and release program, then its populations and ecosystem are lacking in ecological integrity.

Components of a "natural" community

Ecosystems that are dominated by non-native, **introduced species** are considered to have less ecological integrity than those composed of native species.

The last two indicators involve judgements about "naturalness" and the role of humans in ecosystems, which are philosophically controversial topics. However, most ecologists would consider that self-organizing, unmanaged ecosystems have greater ecological integrity than those that are strongly influenced by human activities. Examples of the latter include agroecosystems, **forestry** plantations, and urban and suburban ecosystems. None of these systems can maintain themselves in the absence of large inputs of energy, nutrients, and physical management by humans.

Indicators of ecological integrity

Indicators of ecological integrity vary widely in their scale, complexity, and intent. For example, certain metabolic indicators can suggest the responses by individual organisms and populations to toxic stress, as is the case of assays of detoxifying **enzyme** systems that respond to exposure to persistent **chlorinated hydrocarbons**, such as DDT and PCBs. Indicators related to populations of **endangered species** are relevant to the viability of those species, as well as the integrity of their natural communities. There are also indicators relevant to processes occurring at the level of landscape. There are even global indicators, for example, relevant to climate change, depletion of stratospheric ozone, and **deforestation**.

Sometimes, relatively simple indicators can be used to integrate the ecological integrity of a large and complex ecosystem. In the western United States, for instance, the viability of populations of spotted **owls** (*Strix occidentalis*) is considered to be an indicator of the integrity of the types of old-growth forest in which this endangered bird breeds. If plans to harvest and manage those **forests** are judged to pose a threat to the viability of a population of spotted owls or the species, this would indicate a significant challenge to the integrity of the entire old-growth forest ecosystem.

Ecologists are also beginning to develop holistic indicators of ecological integrity. These are designed as composites of various indicators, analogous to certain economic indices such as the Dow-Jones Index of the stock market, the Consumer Price Index, and gross domestic product indices of economies. Composite economic indicators like these are relatively simple to design because all of the input data are measured in a common way, for example, in dollars. However, in **ecology** there is no common currency among the various indicators of ecological integrity, and it is therefore difficult to develop composite indicators that people will agree upon.

In spite of the difficulties, ecologists are making progress in their development of indicators of ecological integrity. This is an important activity for ecologists, because people and their larger society need objective information about changes in the integrity of species and ecosystems so that actions can be taken to prevent unacceptable degradations. It is being increasingly recognized that human economies can only be sustained over the longer term by ecosystems with integrity. These must be capable of supplying continuous flows of renewable resources, such as trees, **fish**, agricultural products, and clean air and water. There are also important concerns about the intrinsic value of native species and their natural ecosystems, all of which must be sustained along with humans. A truly sustainable economy can only be based on ecosystems with integrity.

See also Indicator species; Stress, ecological.

Resources

Books

Babaev, Agadzhan, and Agajan G. Babaev, eds. *Desert Problems and Desertification in Central Asia: The Researches of the Desert Institute.* Berlin: Springer Verlag, 1999.

Freedman, B. *Environmental Ecology.* 2nd ed. San Diego: Academic Press, 1995.

Hamblin, W. K., and E.H. Christiansen. *Earth's Dynamic Systems.* 9th ed. Upper Saddle River: Prentice Hall, 2001.

Woodley, S., J. Kay, and G. Francis, eds. *Ecological Integrity and the Management of Ecosystems.* Boca Raton, FL: St. Lucie Press, 1993.

Periodicals

Caballero, A., and M. A. Toro. "Interrelations Between Effective Population Size and Other Pedigree Tools for the Management of Conserved Populations." *Genetical Research* 75, no. 3 (June 2000): 331-43.

Karr, J. "Defining and Assessing Ecological Integrity: Beyond Water Quality." *Environmental Toxicology and Chemistry* 12 (1993): 1521-1531.

Bill Freedman

Ecological monitoring

Governments everywhere are increasingly recognizing the fact that human activities are causing serious environmental and ecological damage. To effectively deal with this environmental crisis, it is important to understand its dimensions and dynamics. What, specifically, are the damages, how are they changing over **time**, and what are the best means of prevention or mitigation? To develop answers to these important questions, longer-term programs of monitoring and research must be designed and implemented. These programs must be capable of detecting environmental and ecological change over large areas, and of developing an understanding of the causes and consequences of those changes.

Humans and their societies have always been sustained by environmental resources. For almost all of human history the most important resources have been potentially renewable, ecological resources. Especially important have been **fish** and terrestrial animals that could be hunted, edible plants that could be gathered, and the productivity of managed, agricultural ecosystems. More recently, humans have increasingly relied on the use of nonrenewable mineral resources that are mined from the environment, especially **fossil fuels** and metals.

However, the ability of ecosystems to sustain humans is becoming increasingly degraded. This is largely because of the negative consequences of two, interacting factors: (1) the extraordinary increase in size of the human population, which numbered about 6.0 billion in 1999, and (2) the equally incredible increase in the quantities of resources used by **individual** humans, especially people living in developed countries with an advanced economy, such as those of **North America** and Western **Europe**.

Environmental and ecological degradations are important for two reasons: (1) they represent decreases in the ability of Earth's ecosystems to sustain humans and their activities, and (2) they represent catastrophic damage to other **species** and to natural ecosystems, which have their own intrinsic value, regardless of their importance to humans. The role of programs of environmental and ecological monitoring is to detect those degradations, to understand their causes and consequences, and to find ways to effectively deal with the problems.

Monitoring, research and indicators

In the sense used here, environmental monitoring is an activity that involves repeated measurements of inorganic, ecological, social, and/or economic variables. This is done with a view to detecting important changes over time, and to predicting future change. Within this larger context, ecological monitoring deals with changes in the structure and functioning of ecosystems.

Monitoring investigates scientific questions that are rather uncomplicated, involving simple changes over time. However, the success of monitoring depends on: (1) the astute choice of a few, appropriate indicators to measure over time, from a diverse array of potential indicators, and (2) successful data collection, which can be expensive and difficult, and requires longer-term commitments because important changes may not detectable by short-term studies.

It is important to understand that monitoring programs must be integrated with research, which examines relatively complex questions about the causes and consequences of important environmental and ecological changes that may be detected during monitoring. The ultimate goals of an integrated program of ecological monitoring and research are to: (1) detect or forecast changes, and (2) determine the causes and implications of those changes.

Monitoring involves the repeated measurement of indicators, which are relatively simple measurements related to more complex aspects of environmental quality. Changes in indicators are determined through comparison with their historical values, or with a reference or control situation. Often, monitoring may detect changes in indicators, but the causes of those changes may not be understood because the base of environmental and ecological knowledge is incomplete. To discover the causes of those changes, research has to be undertaken.

For example, monitoring of **forests** might detect a widespread decline of some species of **tree**, or of an entire forest community. In many cases the causes of obvious forest declines are not known, but they are suspected to be somehow related to environmental stressors, such as **air pollution**, insect damage, climate change, or **forestry**. These possibilities must be investigated by carefully designed research programs. The ecological damages associated with forest declines are very complex, and are related, for example, to changes in: productivity, amounts of living and dead **biomass**, age-class structure of trees and other species, nutrient cycling, **soil erosion**, and **biodiversity** values. However, in an ecological monitoring program designed to study forest health, only a few well-chosen indicators would be measured. A sensible indicator of changes in the forest as an economic resource might be the productivity of trees, while a species of mammal or bird with specific **habitat** needs could be used as an indica-

tor of the **ecological integrity** of mature or older-growth forests.

Indicators can be classified according to a simple model of stressor-exposure-response:

(1) Stressors are the causes of environmental and ecological changes, and are associated with physical, chemical, and biological threats to environmental quality. Stressors and their indicators are often related to human activities, for example, emissions of **sulfur dioxide** and other air pollutants, concentrations of **secondary pollutants** such as **ozone**, the use of **pesticides** and other toxic substances, or occurrences of disturbances associated with construction, forestry, or agriculture. Natural stressors include wildfires, hurricanes, volcanic eruptions, and climate change.

(2) Exposure indicators are relevant to changes in the intensity of stressors, or to doses accumulated over time. Exposure indicators might only measure the presence of a stressor, or they might be quantitative and reflect the actual intensity or extent of stressors. For example, appropriate exposure indicators of ozone in air might be the **concentration** of that toxic gas, while disturbance could be indicated by the annual extent of habitat change caused by forest fires, agriculture, clear-cutting, or urbanization.

(3) Response indicators reflect ecological changes that are caused by exposure to stressors. Response indicators can include changes in the health of organisms, populations, communities, or ecoscapes (landscapes and seascapes).

Indicators can also take the form of composite indices, which integrate complex information. Such indices are often used in finance and economics, for example, stock-market indices such as the Dow-Jones, and consumer price indices. For reporting to the public, it is desirable to have composite indices of environmental quality, because complex changes would be presented in a simple manner. However, the design of composite indices of environmental quality or ecological integrity are controversial, because of difficulties in selecting component variables and weighing their relative importance. This is different from composite economic indicators, in which all variables are measured in a common currency, such as dollars.

Monitoring addresses important issues

Environmental monitoring programs commonly address issues related to changes in: (1) environmental stressors, for example, the chemical quality of **water**, air, and soil, and activities related to agriculture, forestry, and construction; (2) the abundance and productivity of economically important, ecological resources such as agricultural products, forests, and hunted fish, **mammals**, and **birds**; and (3) ecological values that are not economic resources but are nevertheless important, such as rare and **endangered species** and natural communities.

Monitoring programs must be capable of detecting changes in all of the above values, and of predicting future change. In North America, this function is carried out fairly well for categories (1) and (2), because these deal with economically important activities or resources. However, there are some important deficiencies in the monitoring of noneconomic ecological values. As a result, significant environmental issues involving ecological change cannot be effectively addressed by society, because there is insufficient monitoring, research, and understanding. A few examples are:

(1) Is a widespread decline of populations of migratory songbirds occurring in North America? If so, is this damage being caused by stressors occurring in their wintering habitat in Central and **South America**? Or are changes in the breeding habitat in North America important? Or both? What are the causes of these changes, and how can society manage the stressors that are responsible?

(2) What is the scope of the global biodiversity crisis that is now occurring? Which species are affected, where, and why? How are these species important to the integrity of the **biosphere**, and to the welfare of humans? Most of the extinctions are occurring because of losses of tropical forest, but how are people of richer countries connected to the biodiversity-depleting stressors in poorer countries?

(3) What are the biological and ecological risks of increased exposures to ultraviolet **radiation**, possibly caused by the depletion of stratospheric ozone resulting from emissions of chlorofluorocarbons by humans?

(4) What constitutes an acceptable exposure to potentially toxic chemicals? Some toxins, such as metals, occur naturally in the environment. Are there thresholds of exposure beyond which human emissions should not increase the concentrations of these chemicals? Is any increase acceptable for non-natural toxins, such as synthetic pesticides, TCDD, PCBs, and radionuclides?

These are just a small **sample** of the important ecological problems that have to be addressed by ecological monitoring, research, and understanding. To provide the information and knowledge needed to deal with environmental problems, many countries are now designing programs for longer-term monitoring and research in **ecology** and environmental science.

In the United States, for example, the Environmental Monitoring and Assessment Program (EMAP) of the En-

vironmental Protection Agency is intended to provide information on ecological changes across large areas, by monitoring indicators at a large number of sites spread across the entire country. Another program has been established by the National Science Foundation and involves a network of Long-Term Ecological Research (LTER) sites (21 in 1999), although these are mostly for fundamental ecological research and are not necessarily relevant to environmental problems. These are important programs in ecological monitoring and research, but they are still in their infancy and it is too soon to determine how well they will contribute to resolution of the environmental crisis.

State-of-the-environment reporting and social action

The information from programs of environmental monitoring and research must be reported to government administrators, politicians, corporations, and individuals. This information can influence the attitudes of these groups, and thereby affect environmental quality. Decision makers in government and industry need to understand the causes and consequences of environmental damage, and the costs and benefits of alternative ways of dealing with those changes. Their decisions are based on the balance of the perceived costs associated with the environmental damage, and the shorter-term, usually economic benefits of the activity that is causing the degradation.

Information from environmental monitoring and research is interpreted and reported to the public by the media, educational institutions, state-of-the-environment reporting by governments, and by non-governmental organizations. All of these sources of information help to achieve environmental literacy, which eventually influences public attitudes. Informed opinions about the environment will then influence individual choices of lifestyle, which has important, mostly indirect effects on environmental quality. Public opinion also influences politicians and government administrators to more effectively manage and protect the environment and ecosystems.

See also Ecosystem; Indicator species; Population, human; Stress, ecological.

Resources

Books

Freedman, B. *Environmental Ecology.* 2nd ed. San Diego: Academic Press, 1995.
Goldsmith, F.B., ed. *Monitoring for Conservation and Ecology.* London: Chapman and Hall, 1991.
Spellerberg, I.F. *Monitoring Ecological Change.* Cambridge: Cambridge University Press, 1991.

Periodicals

Hooke, Roger L. "On the History of Humans as Geomorphic Agents." *Geology*, vol. 28, no. 9 (September 2000): 843-846.

Bill Freedman

Ecological productivity

Ecological productivity refers to the primary fixation of solar **energy** by plants and the subsequent use of that fixed energy by plant-eating herbivores, animal-eating carnivores, and the detritivores that feed upon dead **biomass**. This complex of energy fixation and utilization is called a food web.

Ecologists refer to the productivity of green plants as primary productivity. Gross primary productivity is the total amount of energy that is fixed by plants, while net primary productivity is smaller because it is adjusted for energy losses required to support **plant respiration**. If the net primary productivity of green plants in an **ecosystem** is positive, then the biomass of vegetation is increasing over **time**.

Gross and net secondary productivities refer to herbivorous animals, while tertiary productivities refer to carnivores. Within food webs, a pyramid-shaped structure characterizes ecological productivity. Plants typically account for more than 90% of the total productivity of the food web, herbivores most of the rest, and carnivores less than 1%. Any dead plant or **animal** biomass is eventually consumed by decomposer organisms, unless ecological conditions do not allow this process to occur efficiently, in which case dead biomass will accumulate as peat or other types of non-living organic **matter**.

Because of differences in the availabilities of solar **radiation**, **water**, and **nutrients**, the world's ecosystems differ greatly in the amount of productivity that they sustain. Deserts, **tundra**, and the deep **ocean** are the least productive ecosystems, typically having an energy fixation of less than 0.5×10^3 kilocalories per square meter per year (thousands of kcal/m^2/yr; it takes one **calorie** to raise the **temperature** of one gram of water by 34°F [1°C] under standard conditions, and there are 1,000 calories in a kcal). **Grasslands**, montane and boreal **forests**, waters of the **continental shelf**, and rough agriculture typically have productivities of 0.5-3.0×10^3 kcal/m^2/yr. Moist forests, moist prairies, shallow lakes, and typical agricultural systems have productivities of 3-10×10^3 kcal/m^2/yr. The most productive ecosystems are fertile estuaries and marshes, coral reefs, terrestrial

vegetation on moist alluvial deposits, and intensive agriculture, which can have productivities of 10-25×10^3 kcal/m^2/yr.

See also Carnivore; Ecological pyramids; Food chain/web; Herbivore.

Ecological pyramids

Ecological pyramids are graphical representations of the trophic structure of ecosystems. Ecological pyramids are organized with the productivity of plants on the bottom, that of herbivores above the plants, and carnivores above the herbivores. If the **ecosystem** sustains top carnivores, they are represented at the apex of the ecological **pyramid** of productivity.

A fact of ecological energetics is that whenever the fixed **energy** of **biomass** is passed along a food chain, substantial energy losses occur during each transfer. These energy losses are a necessary consequence of the so-called second law of **thermodynamics**. This universal principle states that whenever energy is transformed from one state to another, the **entropy** of the universe must increase (entropy refers to the randomness of distributions of **matter** and energy). In the context of transfers of fixed biological energy along the trophic chains of ecosystems, increases in entropy are represented by losses of energy as heat (because energy is converted from a highly ordered state in biomass, to a much less-ordered condition as heat). The end result is that transfers of energy between organisms along food chains are inefficient, and this causes the structure of productivity in ecological food webs to always be pyramid shaped.

Ecological food webs

Ecological food webs are based on the productivity of green plants (or photoautotrophs), which are the only organisms capable of utilizing diffuse solar **radiation** to synthesize simple organic compounds from **carbon dioxide** and **water**. The fixed energy of the simple organic compounds, plus inorganic **nutrients**, are then used by plants in more complex metabolic reactions to synthesize a vast diversity of biochemicals. Plants utilize the fixed energy of their biochemicals to achieve growth and reproduction. On average, **plant photosynthesis** utilizes less than 1% of the solar radiation that is received at the surface of the **earth**. Higher efficiencies are impossible for a number of reasons, including the second law of thermodynamics, but also other constraining factors such as the availability of nutrients and moisture, appropriate temperatures for growth, and other environmental limita-

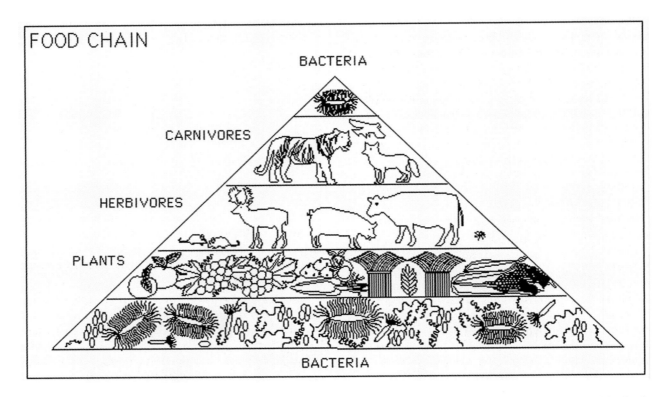

FOOD CHAIN

BACTERIA

CARNIVORES

HERBIVORES

PLANTS

BACTERIA

Ecological pyramids are based on the productivity of organisms. Plants account for 90% of the total productivity of the food web, and herbivores account for most of the rest. Carnivores are responsible for less than 1% of ecological productivity. *Courtesy of Gale Research.*

tions. However, even relatively fertile plant communities can only achieve conversion efficiencies of 10% or so, and only for relatively short periods of **time**.

The solar energy fixed by green plants in photosynthesis is, of course, the energetic basis of the productivity of all heterotrophic organisms that can only feed upon living or dead biomass, such as animals and **microorganisms**. Some of the biomass of plants is consumed as food by animals in the next trophic level, that of herbivores. However, herbivores cannot convert all of the energy of the vegetation that they eat into their own biomass. Depending on the digestibility of the food being consumed, the efficiency of this process is about 1-20%. The rest of the fixed energy of the plant foods is not assimilated by herbivores, or is converted into heat. Similarly, when carnivores eat other animals, only some of the fixed energy of the **prey** is converted into biomass of the **predator**. The rest is ultimately excreted, or is converted into heat, in accordance with the requirement for entropy to increase during any energy transformation.

Ecological pyramids

It is important to recognize that the second law of thermodynamics only applies to **ecological productivity**

(and to the closely related variable of energy flow). Consequently, only the trophic structure of productivity is always pyramid shaped. In some ecosystems other variables may also have a trophic structure that is pyramid shaped, for example, the quantities of biomass (also known as standing crop) present at a particular time, or the sizes or densities of populations. However, these latter variables are not pyramid shaped for all ecosystems.

One example of plants having a similar, or even smaller total biomass as the herbivores that feed upon them occurs in the open **ocean**. In that planktonic ecosystem the **phytoplankton** (or single-celled **algae**) typically maintain a similar biomass as the small animals (called **zooplankton**) that feed upon these microscopic plants. However, the phytoplankton cells are relatively short-lived, and their biomass is regenerated quickly because of the high productivity of these microorganisms. In contrast, the herbivorous zooplankton are longer lived, and they are much less productive than the phytoplankton. Consequently, the productivity of the phytoplankton is much larger than that of the zooplankton, even though at any particular time their biomasses may be similar.

In some ecosystems, the pyramid of biomass may be inverted, that is, characterized by a larger biomass of herbivores than of plants. This can sometimes occur in **grasslands**, where the dominant plants are relatively small, herbaceous **species** that may be quite productive, but do not maintain much biomass at any time. In contrast, the herbivores that feed on the plants may be relatively large, long-lived animals, and they may maintain a larger total biomass than the vegetation. Inverted biomass pyramids of this sort occur in some temperate and tropical grasslands, especially during the dry **seasons** when there can be large populations, and biomasses, of long-lived herbivores such as **deer**, **bison**, antelopes, **gazelles**, **hippopotamuses**, rhinos, elephants, and other big animals. Still, the annual productivity of the plants in grasslands is much larger than that of the herbivores.

Similarly, the densities of animals are not necessarily less than those of the plants that they eat. For example, **insects** are the most important herbivores in most **forests**, where they can maintain very large population densities. In contrast, the densities of **tree** populations are much smaller, because each **individual organism** is large and occupies a great deal of space. In such a forest, there are many more small insects than large trees or other plants, so the pyramid of numbers is inverted in shape. However, the pyramid of productivity in the forest is still governed by the second law of thermodynamics, and it is much wider at the bottom than at the top.

Sustaining top carnivores

Because of the serial inefficiencies of **energy transfer** along food chains, there are intrinsic, energetic limits to the numbers of top carnivores that ecosystems can sustain. If top predators such as lions or killer whales are to be sustained in some minimal viable productivity and population size, there must be a suitably large productivity of **animal** prey that these animals can exploit. Their prey must in turn be sustained by a suitably large productivity of appropriate plant foods. Because of these ecological constraints, only very productive or extensive ecosystems can sustain top predators.

African savannas and grasslands sustain more species of higher-order carnivores than any other existing terrestrial ecosystems. The most prominent of these top predators are lion, leopard, cheetah, **hyena**, and wild dog. Although these various species may kill each other during some aggressive interactions (lions and hyenas are well known for their mutual enmity), they do not eat each other, and each can therefore be considered a top predator. In this unusual case, a large number of top predators can be sustained because the ecosystem is very extensive, and also rather productive of vegetation in most years. Other, very extensive but unproductive ecosystems may only support a single species of top predator, as is the case of the wolf in the arctic **tundra**.

See also Autotroph; Carnivore; Food chain/web; Herbivore; Heterotroph; Trophic levels.

Resources

Books

Odum, E.P. *Ecology and Our Endangered Life Support Systems*. New York: Sinauer, 1993.

Ricklefs, R.E. *Ecology*. New York: W. H. Freeman, 1990.

Bill Freedman

Ecology

Ecology can be defined as the study of the relationships of organisms with their living and nonliving environment. Most ecologists are interested in questions involving the natural environment. Increasingly, however, ecologists are concerned about degradation associated with the ecological effects of humans and their activities. Ultimately, ecological knowledge will prove to be fundamental to the design of systems of resource use and management that will be capable of sustaining humans over the longer term, while also sustaining other **species** and natural ecosystems.

The subject matter of ecology

The subject matter of ecology is the relationships of organisms with their biological and nonliving environment. These are complex, **reciprocal** interactions; organisms are influenced by their environment, but they also cause environmental change, and are components of the environment of other organisms.

Ecology can also be considered to be the study of the factors that influence the distribution and abundance of organisms. Ecology originally developed from natural history, which deals with the richness and environmental relationships of life, but in a non-quantitative manner.

Although mostly a biological subject, ecology also draws upon other sciences, including **chemistry**, **physics**, **geology**, **mathematics**, computer science, and others. Often, ecologists must also deal with socioeconomic issues, because of the rapidly increasing importance of human impacts on the environment. Because it draws upon knowledge and information from so many disciplines, ecology is a highly interdisciplinary field.

The biological focus of ecology is apparent from the fact that most ecologists spend much of their time engaged in studies of organisms. Examples of common themes of ecological research include: (1) the physical and physiological adaptations of organisms to their environment, (2)

New land development, Germantown, Pennsylvania. *JLM Visuals. Reproduced by permission.*

patterns of the distribution of organisms in **space**, and how these are influenced by environmental factors, and (3) changes in the abundance of organisms over time, and the environmental influences on these dynamics.

Levels of integration within ecology

The universe can be organized along a **spectrum** of levels according to spatial scale. Ordered from the extremely small to the extremely large, these levels of integration are: **subatomic particles**, **atoms**, molecules, molecular mixtures, tissues, organs, **individual** organisms, populations of individuals, communities of populations, ecological landscapes, the **biosphere**, the **solar system**, the **galaxy**, and the universe. Within this larger scheme, the usual realm of ecology involves the levels ranging from (and including) individual organisms through to the biosphere. These elements of the ecological hierarchy are described below in more detail.

The individual

In the ecological and evolutionary contexts, an individual is a particular, distinct **organism**, with a unique complement of genetic information encoded in DNA. (Note that although some species reproduce by nonsexual means, they are not exceptions to the genetic uniqueness of evolutionary individuals.) The physical and physiological attributes of individuals are a function of (1) their genetically defined capabilities, known as the genotype, and (2) environmental influences, which affect the actual expression of the genetic capabilities, known as the phenotype. Individuals are the units that are "selected" for (or against) during **evolution**.

The population

A population is an aggregation of individuals of the same species that are actively interbreeding, or exchang-

ing genetic information. Evolution refers to changes over time in the aggregate genetic information of a population. Evolution can occur as a result of **random** "drift," as directional **selection** in favor of advantageous phenotypes, or as selection against less well-adapted genotypes.

The community

An ecological community is an aggregation of populations thar are interacting physically, chemically, and behaviorally in the same place. Strictly speaking, a community consists of all **plant**, **animal**, and microbial populations occurring together on a site. Often, however, ecologists study functional "communities" of similar organisms, for example, bird or plant communities.

The ecological landscape

This level of ecological organization refers to an aggregation of communities on a larger area of terrain. Sometimes, ecological units are classified on the basis of their structural similarity, even though their actual species may differ among widely displaced locations. A **biome** is such a unit, examples of which include alpine and arctic **tundra**, boreal forest, deciduous forest, **prairie**, **desert**, and tropical **rainforest**.

The biosphere

The biosphere is the integration of all life on **Earth**, and is spatially defined by the occurrence of living organisms. The biosphere is the only place in the universe known to naturally support life.

Energy and productivity

Less than 1% of the solar **energy** reaching Earth's surface is absorbed by green plants or **algae** and used in **photosynthesis**. However, this fixed solar energy is the energetic basis of the structure and function of ecosystems. The total fixation of energy by plants is known as gross primary production (GPP). Some of that fixed energy is used by plants to support their own metabolic demands, or **respiration** (R). The quantity of energy that is left over (that is, GPP MINUS R) is known as net primary production (NPP). If NPP has a positive value, then plant **biomass** accumulates over time, and is available to support the energy requirements of herbivorous animals, which are themselves available as food to support to carnivores. Any plant or animal biomass that is not directly consumed eventually dies, and is consumed by decomposers (or detritivores), the most important of which are **microorganisms** such as **bacteria** and **fungi**. The complex of ecological relationships among all of the plants, animals, and decomposers is known as a food web.

Environmental influences and biological interactions

Compared with the potential biological "demand," the environment has a limited ability to "supply" the requirements of life. As a result, the rates of critical ecological processes, such as productivity, are constrained by so-called limiting factors, which are present in the least supply relative to the biological demand. A limiting environmental factor can be physical or chemical in nature, and the factors act singly, but sequentially. For example, if a typical unproductive **lake** is fertilized with nitrate, there would be no ecological response. However, if that same lake was fertilized with phosphate, there would be a great increase in the productivity of single-celled algae. If the lake was then fertilized with nitrate, there would be a further increase of productivity, because the ecological requirement for phosphate, the primary **limiting factor**, had previously been satiated.

This example illustrates the strong influence that the environment has on rates of processes such as productivity, and on overall ecological development. The most complex, productive, and highly developed ecosystems occur in relatively benign environments, where climate and the supplies of **nutrients** and **water** are least limiting to organisms and their processes. Tropical **forests** and coral reefs are the best examples of well-developed, natural ecosystems of this sort. In contrast, environmentally stressed ecosystems are severely constrained by one or more of these factors. For example, deserts are limited by the availability of water, and tundra by a cold climate.

In a theoretically benign environment, with an unlimited availability of the requirements of life, organisms can maximize the growth of their individual biomass and of their populations. Conditions of unlimited resources might occur (at least temporarily), perhaps, in situations that are sunny and well supplied with water and nutrients. Population growth in an unlimited environment is exponential, meaning that the number of individuals doubles during a fixed time **interval**. For example, if a species was biologically capable of doubling the size of its population in one week under unlimited environmental conditions, then after one week of growth an initial population of N individuals would grow to $2N$, after two weeks $4N$, after three weeks $8N$, after four weeks $16N$, and after eight weeks it would be $256N$. A financial analogy will help to put this tremendous **rate** of population increase into perspective: an initial investment of $100 growing at that rate would be worth $25,600 after only 8 weeks.

Clearly, this is an enormous rate of growth, and it would never be sustainable under real-world ecological

(or economic) conditions. Before long, environmental conditions would become limiting, and organisms would begin to interfere with each other through an ecological process known as **competition**. In general, the more similar the ecological requirements of individuals or species, the more intense the competition they experience. Therefore, competition among similar-sized individuals of the same species can be very intense, while individuals of different sized species (such as trees and **moss**) will compete hardly at all.

Competition is an important ecological process, because it limits the growth rates of individuals and populations, and influences the kinds of species that can occur together in ecological communities. These ecological traits are also profoundly influenced by other interactions among organisms, such as herbivory, predation, and **disease**.

The goal of ecology

The larger objective of ecology is to understand the nature of environmental influences on individual organisms, their populations, and communities, on ecoscapes and ultimately at the level of the biosphere. If ecologists can achieve an understanding of these relationships, they will be well placed to contribute to the development of systems by which humans could sustainably use ecological resources, such as forests, agricultural **soil**, and hunted animals such as **deer** and **fish**. This is an extremely important goal because humans are, after all, completely reliant on ecologically goods and services as their only source of sustenance.

See also Biological community; Ecological productivity; Ecosystem; Stress, ecological.

KEY TERMS

· ·

Genotype—The genetic information encoded within the DNA of an organism (or a population, or a species). Not all of the information of the genotype is expressed in the phenotype.

Interdisciplinary—A field of investigation that draws upon knowledge from many disciplines.

Phenotype—The actual morphological, physiological, and behavioral expression of the genetic information of an individual, as influenced by its environment.

Population biology—Study of changes in the abundance of organisms over time, and of the causes of those changes.

Resources

Books

Begon, M., J.L. Harper, and C.R. Townsend. *Ecology: Individuals, Populations and Communities.* 3rd ed. London: Blackwell Sci. Pub., 1996.

Dodson, S.I., T.F.H. Allen, S.R. Carpenter, A.R. Ives, R.L. Jeanne, J.F. Kitchell, N.E. Langston, and M.G. Turner. *Ecology.* New York: Oxford University Press, 1998.

Freedman, B. *Environmental Ecology.* 2nd ed. San Diego: Academic Press, 1995.

Keller, E.A. *Introduction to Environmental Geology.* 2nd ed. Upper Saddle River: Prentice Hall, 2002.

Ricklefs, R.E. *Ecology.* New York: W.H. Freeman, 1990.

Bill Freedman

Ecosystem

The notion of ecosystem (or ecological system) refers to indeterminate ecological assemblages, consisting of communities of organisms and their environment. Ecosystems can vary greatly in size. Small ecosystems can be considered to occur in tidal pools, in a back yard, or in the rumen of an **individual** cow. Larger ecosystems might encompass lakes or stands of forest. Landscape-scale ecosystems comprise larger regions, and may include diverse terrestrial and aquatic communities. Ultimately, all of Earth's life and its physical environment could be considered to represent an entire ecosystem, known as the **biosphere**.

Often, ecologists develop functional boundaries for ecosystems, depending on the particular needs of their work. Depending on the specific interests of an ecologist, an ecosystem might be delineated as the shoreline vegetation around a **lake**, or perhaps the entire waterbody, or maybe the lake plus its terrestrial **watershed**. Because all of these units consist of organisms and their environment, they can be considered ecosystems.

Through biological productivity and related processes, ecosystems take sources of diffuse **energy** and simple inorganic materials, and create relatively focused combinations of these, occurring as the **biomass** of plants, animals, and **microorganisms**. Solar electromagnetic energy, captured by the **chlorophyll** of green plants, is the source of diffuse energy most commonly fixed in ecosystems. The most important of the simple inorganic materials are **carbon dioxide**, **water**, and ions or small molecules containing **nitrogen**, **phosphorus**, potassium, **calcium**, **magnesium**, **sulfur**, and some other **nutrients**.

Because diffuse energy and simple materials are being ordered into much more highly structured forms such as biochemicals and biomass, ecosystems (and life more generally) represent rare islands in which negative **entropy** is accumulating within the universe. One of the fundamental characteristics of ecosystems is that they must have access to an external source of energy to drive the biological and ecological processes that produce these localized accumulations of negative entropy. This is in accordance with the second law of **thermodynamics**, which states that spontaneous transformations of energy can only occur if there is an increase in entropy of the universe; consequently, energy must be put into a system to create negative entropy. Virtually all ecosystems (and life itself) rely on inputs of solar energy to drive the physiological processes by which biomass is synthesized from simple molecules.

To carry out their various functions, ecosystems also need access to materials—the nutrients referred to above. Unlike energy, which can only flow through an ecosystem, nutrients can be utilized repeatedly. Through biogeochemical cycles, nutrients are recycled from dead biomass, through inorganic forms, back into living organisms, and so on.

One of the greatest challenges facing humans and their civilization is understanding the fundamentals of ecosystem organization—how they function and how they are structured. This knowledge is absolutely necessary if humans are to design systems that allow a sustainable utilization of the products and services of ecosystems. Humans are sustained by ecosystems, and there is no tangible alternative to this relationship.

See also Biological community; Ecological productivity.

Bill Freedman

Ecotone

An ecotone is a zone of transition between distinct ecological communities or habitats. Usually, the word is used to refer to relatively sharp, local transitions, also known as edges.

Because many physical and chemical changes in the environment tend to be continuous, ecological transitions are often similarly gradual. For example, climate and **precipitation** change steadily across continents and up the slopes of **mountains**. Because these environmental changes are gradual, communities of plants and animals often intergrade through wide, continuous transitions.

Frequently, however, there are relatively sharp environmental interfaces associated with rapid changes occur-

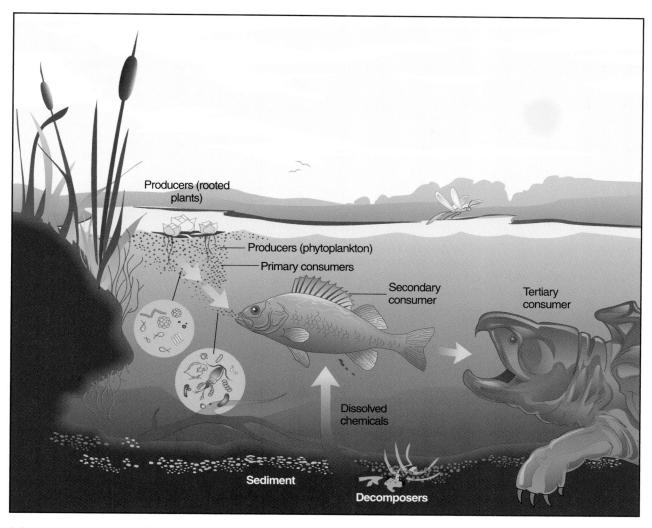

Producers (rooted plants)

Producers (phytoplankton)

Primary consumers

Secondary consumer

Tertiary consumer

Dissolved chemicals

Sediment

Decomposers

A freshwater ecosystem. *Illustration by Hans & Cassidy. Courtesy of Gale Group.*

ring naturally at the edges of major geological or **soil** discontinuities along the interface of aquatic and terrestrial habitats or associated with the boundaries of disturbances such as landslides and wildfires. These are the sorts of environmental contexts in which ecotones occur naturally. Human activities also favor the occurrence of many ecotones, for example, along the edges of clearcuts, agricultural fields, highways, and residential areas.

Disturbance-related ecotones exist in **space**, but often they eventually become indistinct as **time** passes because of the ecological process known as **succession**. For example, in the absence of an intervening disturbance, an ecotone between a forest and a field will eventually disappear if the field is abandoned and succession allows a mature forest to develop over the entire area.

The sharp ecological discontinuities at ecotones provide **habitat** for so-called "edge" **species** of plants and an-

imals. These have a relatively broad ecological tolerance and within limits can utilize habitat on both sides of the ecotone. Examples of edge plants include many shrubs and vines that are abundant along the boundaries of **forests** in many parts of the world. Some animals are also relatively abundant in edges and in habitat mosaics with a large **ratio** of edge to area. Some North American examples of edge animals include white-tailed and mule **deer**, snowshoe hare, cottontail rabbit, blue jay, and robin.

Because human activities have created an unnatural proliferation of ecotonal habitats in many regions, many edge animals are much more abundant than they used to be. In some cases this has resulted in important ecological problems. For example, the extensive range expansion of the brown-headed cowbird, a prairie-forest edge species, has caused large reductions in the breeding success of many small species of native **birds**, contributing to large declines in some of their populations. This has happened because the cowbird is a very effective social parasite which lays its eggs in the nests of other species who then rear the cowbird chick to the severe detriment of their own young.

See also Biological community.

Ecotourism

Ecotourism refers to outdoor recreation, sightseeing, and guided natural history studies in remote or fragile natural areas, or archeological and cultural sites. Ecotourism usually involves travel to engage in activities such as trekking and hiking, diving, mountaineering, biking, and paddling, while exploring a region's natural highlights, observing native animals, and **learning** about the area's natural history. Ecotourists may also visit local cultural and historical sites, and even participate in cultural activities. Many ecotours employ native guides and interpreters who can help visitors fully appreciate the natural and cultural significance of their experience.

Ecotourism and sustainable development

Ecotourism is touted as a successful tool for promoting sustainable economic practices in developing nations, and for encouraging environmental **conservation** worldwide. The guiding principle of **sustainable development** is to meet the needs and aspirations of a region's present generation of people without compromising those of future generations. Sustainable development policies also seek to develop economic systems that run with little or no net consumption of natural resources, and that avoid ecological damage. Ecotourism, like other successful sustainable development strategies, provides a strong economic incentive to protect natural resources. Economies that depend on ecotourism dollars have an obvious interest in preserving the natural and culture features that these amateur naturalists and explorers pay to see. Furthermore, the environmental impacts and resource needs of ecotourism, which include development of trail systems and access roads, use of fuel and vehicles for transportation to and from the wilderness, and establishment of campsites, are minimal, especially when compared to the **land use** practices that commercial nature travel often replaces. Finally, the firsthand experience of traveling in the wilderness, of observing natural complexity, and of reflecting on the fragility of ecosystems stressed by human uses often gives ecotourists and their local guides a new perspective on the value of environmental preservation and resource conservation.

A number of international organizations, including the United Nations Environment Programme (UNEP), and Conservation International, support ecotourism as a component of their sustainable development and environmental conservation strategies. In fact, 2002 was designated as the International Year of Ecotourism. While many governments and international non-governmental organizations (NGOs) promote ecotourism, they also caution that ecotourism must be practiced correctly in order to provide positive results for the region involved, and for the tour's participants. Many of the activities offered by ecotourism companies, including high-altitude mountaineering, whitewater paddling, diving, and travel in the remote wilderness, are inherently dangerous, and require highly skilled guides. Furthermore, some of the Earth's most remarkable natural features exist in politically unstable nations, where international visitors may be unwelcome, or even unsafe. Ecotourism, practiced incorrectly, can also cause significant environmental damage. A safari hunt for an endangered **animal** in a country that has lax **conservation laws**, for example, is not a sustainable ecotour. Finally, ecotourism enterprises that exploit another region's natural and cultural resources without contributing to the local economy do not meet the criteria for sustainable development. If none of the tourists' money goes to the local businesses or conservation agencies, then often-poorer countries bear the financial responsibility of providing protected natural and cultural sites for wealthy foreigners to visit, but receive none of the financial reward. Organizations like the World Tourism Organization (WTO) and the International Ecotourism Society (IES) investigate various ecotourism enterprises, and can provide potential ecotourists with valuable guidance in choosing a company to guide them on a safe, sustainable adventure.

Ecotourism enterprises

Many private companies offer a wide variety of ecotours, as do a number of development and conservation-related NGOs. These businesses often enlist the logistical and marketing assistance of government agencies in the countries where their tours take place. Ecotourism companies typically supply a number of services to their clients: transportation to and from remote venues, food and cooking, lodging, local guiding, outdoor skills training, and expert interpretation of natural and cultural features. These services promote in-depth exploration of the natural and cultural sites on the itinerary, minimize environmental impact, and allow clients to travel safely and comfortably in remote or environmentally fragile areas.

Ecotours are available to all types of potential adventures with all kinds of interests. Ecotourists can visit and explore all seven continents, and all four oceans. The National Geographic Society, for example, lists some of its top destinations for October, 2002: hiking Machu Picchu and Peru, cruising the Galapagos Islands, exploring the Alaskan Frontier, visiting the pyramids of Egypt, diving in the Caribbean, and photographing South African **wildlife**. Meanwhile, the Smithsonian Institution offers study trips to hundreds of locations including Patagonia, **Antarctica** and Falklands, the **rivers** of West **Africa**, Tahiti and Polynesia, Yellowstone, Baja California, **Australia**, and the Southern Amazon. Some ecotours are athletically strenuous, some are luxurious, and some are scientific. There are outfits that offer adventures for travelers on all types of budgets. There is also a wide range of ecotourism and outdoor education activities available to high school and college students. Some of these programs, including the National Outdoor Leadership School (NOLS), and Semester at Sea, offer high school and college credit for their courses. Other programs allow students to participate in international conservation efforts and natural science expeditions. Many schools and universities even offer their own off-campus programs to augment natural, environmental, and social science curricula.

See also Ecological economics; Ecological integrity; Ecological monitoring; Ecological productivity.

Resources

Books

Elander, M., and S. Widstrand. *Eco-Touring: The Ultimate Guide.* Buffalo, NY: Firefly Books, 1997.

Fennell, D.A. *Ecotourism: An Introduction.* New York: Routledge, 1999.

Harris, Rob, Ernie Heath, Lorin Toepper, and Peter Williams. *Sustainable Tourism. A Marketing Perspective.* Woburn, MA: Butterworth-Heinemann, 1998.

Honey, M. *Ecotourism and Sustainable Development: Who Owns Paradise?* Washington, DC: Island Press, 1998.

Organizations

Conservation International. 1919 M Street, NW, Suite 600, Washington, DC 20036. (800) 406–2306. <http://www.conservation.org/xp/CIWEB/home>

The International Ecotourism Society. P.O. Box 668, Burlington, VT 05402. (802) 651–9818. <http://www.ecotourism.org>

The National Outdoor Leadership School. 284 Lincoln St., Lander, WY 82520-2848. (307) 332–5300. <http://www.nols.org>

Semester at Sea—Institute for Shipboard Education. 811 William Pitt Union, Pittsburg, PA 15260. (412) 648–7490. <http://www.semesteratsea.com>

Other

National Geographic Institution. "National Geographic Expeditions" [cited October 28, 2002]. <http://www.nationalgeographic.com/ngexpeditions/>.

Smithsonian Institution. "Smithsonian Study Tours" [cited October 28, 2002]. <http://smithsonianstudytours.org/sst/start.htm.>.

United Nations Environmental Programme Production and Consumptin Unit, Tourism. "International Year of Ecotourism 2002." June 25, 2002 [cited October 28, 2002]. <http://www.uneptie.org/pc/tourism/ecotourism/iye.htm.>.

World Tourism Organisation. "The Leading Organization in the World of Tourism and Travel" [cited October 29, 2002]. <http:// www.world-tourism.org/.>.

Bill Freedman
Laurie Duncan

Ectoprocts *see* **Moss animals**

Edema

Edema is the accumulation of fluid in any given location in the body. Edema can result from trauma, as in a sprained ankle, or from a chronic condition such as **heart** or kidney failure. The word edema is from the Greek and means "swelling." The presence of edema can be an important diagnostic tool for the physician. A patient who is developing congestive heart failure often will develop edema in the ankles. Congestive heart failure means that the heart is laboring against very high **blood pressure** and the heart itself has enlarged to the point that it is not effectively circulating the blood. Excess fluid will leave the **circulatory system** and accumulate between the cells in the body. Gravity will pull the fluid to the area of the ankles and feet, which will swell. The physician can press on the swollen area and the depression left by his finger will remain after he lifts the pressure. The patient with congestive heart failure will develop edema in the lungs as well, and thus has a chronic cough.

Individuals who have liver failure, often because of excessive **alcohol** consumption over a period of years, will develop huge edematous abdomens. The collection of fluid in the abdomen is called ascites (ah-SITE-eez, from the Greek word for bag).

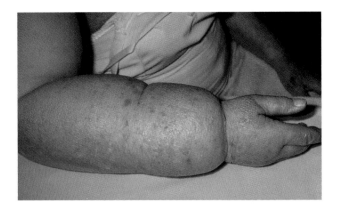

Gross lymphoedema in the arm of an elderly woman following radiotherapy treatment for breast cancer. *Photograph by Dr. P. Marazzi. National Audubon Society Collection/Photo Researchers, Inc. Reproduced by permission.*

The presence of edema is not a **diagnosis** in itself. It signifies a more serious clinical condition that requires immediate attention. The failing heart reaches a point when it can no longer cope with the huge load of fluid and will become an ineffective pump. At that point the only cure for the patient is to undergo a heart transplant. If the underlying problem is kidney failure, the patient can be placed on a **dialysis** machine several times a week to filter the excess **water** from the system along with any accumulated toxins.

Medications are available to help rid the body of excess fluid. These drugs are called diuretics and stimulate the kidneys to filter greater volumes of fluid which is eliminated as urine. These are potent medications, however, that require close monitoring by the physician.

EDTA *see* **Ethylenediaminetetra-acetic acid**

Eel grass

Eel **grasses** (or eelgrasses) are 18 **species** of herbaceous aquatic plants in the family Zosteraceae, 12 species of which are in the genus *Zostera*. However, some **plant** systematists have treated the eel grasses as a component of a much larger family, the pondweeds or Potamogetonaceae.

Eel grasses have long, strap-like leaves that emerge from a thin **rhizome** growing in the surface sediment of the shallow-water, estuarine or marine **wetlands** of the temperate zone where these plants grow. At the end of the growing season, the dead leaves and stems of eel grass break away from the perennating (living over from season to season) rhizomes of the plant and wash up on shores in large quantities.

The flowers of eel grasses are small, either male or female, and are aggregated into an inflorescence that may be unisexual or may contain flowers of both sexes. The fruit is a small seed.

Zostera marina is a common species of eel grass in **North America**. This species is widespread in estuaries and shallow, marine bays. It is eaten by many marine **invertebrates**, and by **swans**, **geese**, and **ducks** in estuaries and other marine wetlands. During the 1930s and 1940s, a mysterious **disease** affected eel grass beds over much of their range. The lack of forage was hard on some species of dependent **wildlife**, such as the Brant goose, which declined greatly in abundance. In fact, an invertebrate known as the eel grass limpet (*Lottia alvens*) became extinct at about this time. Fortunately, the eel grass beds have since recovered, and again provide **critical habitat** for many species of wildlife.

In the past, the large quantities of eel grass debris that often accumulate along shores in the autumn were collected and used for packing delicate objects and instruments for shipping, and for packing into the walls of houses as insulation. Today, the major economic importance of eel grasses is through the **habitat** and food they provide for aquatic wildlife.

Eels *see* **True eels**

Eggplant *see* **Nightshade**

Einsteinium *see* **Element, transuranium**

El Niño and La Niña

El Niño and La Niña are disruptions of the oceanic and atmospheric systems of the equatorial Pacific Ocean that have far reaching effects on Earth's **weather** patterns. El Niño and La Niña do not change with the regularity of the **seasons**; instead, they repeat about every two to seven years. They are the extremes in an aperiodic, or irregular, cycle called the El Niño Southern Oscillation (ENSO), during which warm surface waters from the western Pacific Ocean spread toward the South American coastline.

During normal, non-ENSO periods, southeasterly (east to west) Trade Winds push equatorial surface waters into western half of the Pacific, driving the northwest-flowing Southern Equatorial Current, and creating a mound of warm **water** around Indonesia. Sea surface temperatures near Indonesia are typically about 46°F

(8°C) higher than those near Ecuador, and the sea surface is about 1.6 ft (0.5 m) higher. The pool of warm water in the western Pacific warms the air above it, creating an upward current of moist air that rises to form rain **clouds**. The coastal areas and islands of the western Pacific typically enjoy abundant rainfall and support lush, biologically diverse rainforests, including those of Borneo and New Guinea. Meanwhile, along the coast of **South America**, cold, nutrient-rich waters from the deep **ocean** rise to the sea surface, since the warmer surface waters have been blown westward. The result is called an **upwelling**, which nourishes abundant **phytoplankton** and **zooplankton**, the tiny sea plants and animals that provide food for many other types of sea life. The South American upwelling is a very productive region for **fish** and the animals, including humans, who depend on fish for food. The cold water of an upwelling cools the air above it, creating high-pressure zones of sinking dry air. Regions near upwellings, like coastal Peru and Ecuador, tend to be arid (desert-like).

An ENSO event begins with a lessening of the trade winds in the equatorial Pacific, and a corresponding collapse of the sea surface slope between Indonesia and South America. The pile of warm water in the western Pacific sloshes toward the coast of South America, and shuts down the South American upwelling. A dramatic warming of the waters off of South America and a corresponding decline of marine productivity indicates the El Niño phase of the southern oscillation. La Niña, the opposite phase of an ENSO cycle, occurs when the southeast trade winds are particularly strong. La Niña events are accompanied by colder-than-normal temperatures off South America, and an intensification of the South American upwelling. La Niña events often, but not always, follow El Niño events.

Although El Niño and La Niña take place in a small portion of the southern tropical Pacific, the changes caused by ENSO events affect the weather in large parts of **Asia**, **Africa**, Polynesia, and North and South America. El Niño events occurred during 1982–1983, 1986–1987, 1991–1992, 1993, 1994, 1997–1998, and 2002–2003. (It is unusual to have two El Niños in row, as happed in 1993 and 1994.) Disruption of the vertical **atmospheric circulation** over the southern equatorial Pacific accompanies slackening of the Southern Equatorial Current during El Niño years. As a result, the intensity of the low-pressure system over the southwest Pacific lessens, as does the high-pressure system over the Andes, bringing abnormally dry conditions to Indonesia, and unusually wet weather to the west coast of South America. The more distant climatic effects of El Niños include dryer-than-normal conditions in eastern Africa and western South America; wetter-than-normal weather in equatorial Africa, southern South America, and the southern United States; and abnormally warm winters in Japan and northern **North America**. In the United States, the 1982–83 El Niño, was associated with record snowfall in parts of the Rocky Mountains, **flooding** in the southern United States, and heavy rain storms in southern California, which brought about floods and mud slides. The opposite effects, **drought** in the southern United States, and unusually cold winters in Japan, for example, often accompany La Niña episodes, which occurred recently in 1995–1996 and 1998–1999.

Discovery and study of the El Niño Southern Oscillation

The name El Niño comes from nineteenth century Peruvian and Ecuadorian fishermen. They noticed that each year, within a few months of the Christmas holiday, the seawater off the South American coast became warmer, the nearshore ocean currents assumed new patterns, and the fishing became poorer. Every few years, the changes were strong enough to wipe out a fishing season, and to bring significant, long-lasting changes in the weather. For example, normally dry areas on shore could receive abundant rain, turning deserts into lush **grasslands** for as long as these strong El Niños lasted. Because the phenomenon happened close to Christmas each year, the fishermen dubbed it El Niño, Spanish for "the boy child," after the Christ child. Only in the 1960s did scientists begin to realize that the strong El Niño events were more than a local South American phenomenon, and were rather one half of a multi-year atmospheric-oceanic cycle that affects the entire tropical Pacific Ocean. The other half of the ENSO cycle has been named La Niña, the girl child, or, less commonly, El Viejo, the old man.

The Southern Oscillation was detected, and named, in 1923 by Sir Gilbert Walker. Walker was the director of observatories in India, and was trying to understand the variations in the summer monsoons (rainy seasons) of India by studying the way **atmospheric pressure** changed over the Pacific Ocean. Based on meteorologists' previous **pressure** observations from many stations in the southern Pacific and Indian oceans, Walker established that, over the years, atmospheric pressure seesawed back and forth across the ocean. In some years, pressure was highest in the Indian Ocean near northern **Australia**, and lowest over the southeastern Pacific, near the **island** of Tahiti. In other years, the pattern was reversed. He also recognized that each pressure pattern had a specific related weather pattern, and the change from one phase to the other could mean the shift from rainfall to drought, or from good harvests to famine. In the late 1960s, Jacob Bjerknes, a professor at the University of California, first proposed that the Southern Oscillation and the strong El Niño sea warming were two aspects of the same atmosphere-ocean cycle, and explained the ENSO phenomenon in terms of physical mechanisms.

Regional and global effects of El Niño and La Niña

The atmosphere and the ocean form a coupled system, that is, they respond to each other. Changes in the ocean cause a response in the winds above it, and vice versa. For reasons not yet fully understood, an ENSO event begins with a change in the atmosphere-ocean system of the southern equatorial Pacific. The Southeasterly Trade Winds weaken, and they push less warm water to the western edge of the Pacific, causing far-reaching changes. Fewer rain clouds form over Indonesia, the Pacific Islands, Australia, and Southeast Asia. Lush rain **forests** dry out and become fuel for forest fires. The area of heavy rain shifts to the mid-southern Pacific, where heavy rains inundate usually arid islands. In the eastern Pacific, the surface water becomes warmer. Ocean upwelling is weakened, and the surface water runs low on the **nutrients** that support the ocean food chain. Many **species** of fish are driven elsewhere to find food; in severe El Niño years fish populations may be almost completely wiped out. Bird species that depend on fish must look elsewhere, and the human fishing population faces economic hardship. At the same time, the warmer waters offshore encourage the development of clouds and thunderstorms. Normally dry areas in western South America, such as Peru and Ecuador, may experience torrential rains, flooding, and mud slides during the El Niño phase.

The climatic effects of El Niño have long been noted in the tropical Pacific, and are now being studied around the world. The altered pattern of winds, ocean temperatures, and currents during an El Niño is believed to change the high-level winds, called the jet streams, that steer storms over North and South America. El Niños have been linked with milder winters in western Canada and the northern United States, as most severe storms are steered northward toward Alaska. As Californians saw in 1982–1983, and 1998–1999, an El Niño can cause extremely wet winters along the west coast, and bring torrential rains to the lowlands and heavy snows to the mountains. Alteration of the jet streams by El Niño can also contribute to **storm** development over the Gulf of Mexico, and to heavy rainfall in the southeastern United States. Similar changes

occur in countries of South America, such as Chile and Argentina, while droughts may affect Bolivia and parts of Central America.

El Niño also appears to affect monsoons, which are annual shifts in the prevailing winds that bring rainy seasons to India, southeast Asia, and portions of Africa. The rains of the **monsoon** are critical for agriculture; when the monsoon fails, millions of people risk starvation. It appears that while El Niños do not always determine monsoons, they contribute to weakened monsoons in India and southeastern Africa, and tend to strengthen those in eastern Africa.

In general, the effects of El Niño are reversed during the La Niña extreme of the ENSO cycle. During the 1998–1999 La Niña episode, for example, the central and northeastern United States experienced record snowfall and sub-zero temperatures, rainfall increased in the Pacific Northwest, and a record number of tornadoes plagued the southern states. Not all El Niños and La Niñas have equally strong effects on the **global climate** because every El Niño and La Niña event is of a different magnitude and duration.

Predicting El Niño and La Niña

The widespread weather impacts of the two extreme phases of the El Niño Southern Oscillation cycle make understanding and predicting ENSO events a high priority for atmospheric scientists and oceanographers. Researchers have developed computer models of the Southern Oscillation that mimic the behavior of the real atmosphere-ocean system, and predict future events. These computer simulations require the input of very large amounts of data about sea and **wind** conditions in the equatorial Pacific. A large and growing network of instruments, many of them owned and maintained by the National Atmospheric and Oceanographic Administration (NOAA, provides these data. Ocean buoys, permanently moored in a transect across the equatorial Pacific, constantly relay information on water **temperature**, wind, and air pressure to weather prediction stations around the world. The buoys are augmented by surface ships, island weather stations, and **Earth** observing satellites. Even with mounting data and improving computer models, El Niño, La Niña and the Southern Oscillation remain difficult to predict. However, the ENSO models, and analyses of past ENSO cycles, are now being used in several countries to help prepare for the next El Niño. Countries most affected by the variations in El Niño, such as Peru, Australia and India, presently use El Niño prediction to improve agricultural planning.

See also Air masses and fronts; Oceanography.

KEY TERMS

. . . .

Coupled system—A system with parts that are linked in such a way that they respond to changes in each other. The atmosphere and the ocean form a coupled system, so that changes in one will cause a response in the other, which will in turn cause another change in the first, etc.

El Niño—The phase of the Southern Oscillation characterized by increased sea water temperatures and rainfall in the eastern Pacific, with weakening trade winds and decreased rain along the western Pacific.

ENSO—Abbreviation for El Niño/Southern Oscillation.

Jet streams—High velocity winds that blow at upper levels in the atmosphere and help to steer major storm systems.

La Niña—The phase of the Southern Oscillation characterized by strong trade winds, colder sea water temperatures and dry weather in the eastern Pacific, with increased rainfall along the western Pacific.

Monsoon—An annual shift in the direction of the prevailing wind that brings on a rainy season and affects large parts of Asia and Africa.

Southern oscillation—A large scale variation in the winds, ocean temperatures and atmospheric pressure of the tropical Pacific Ocean which repeats about every three to four years.

Resources

Books

Open University Course Team. *Ocean Circulation.* Oxford: Pergamon Press, 1993.

Williams, Jack. *The Weather Book.* New York: Vintage Books, 1992.

Periodicals

Kerr, R.A. "A Successful Forecast of an El Niño Winter." *Science* (January 24, 1992): 402.

McPhaden, M.J. "TOGA-TAO and the 1991–93 El Niño-Southern Oscillation Event." *Oceanography* 6 (1993): 36–44.

Other

National Oceanographic and Atmospheric Administration. "El Niño theme page" [cited November 6, 2002]. <http://www.pmel.noaa.gov/tao/elnino/nino-home.html>.

James Marti
Laurie Duncan

Eland

Eland (*Taurotragus oryx*) are the largest African antelopes, weighing up to a 2,205 lb (1,000 kg) and standing 6.6 ft (2 m) at the shoulder. They belong to the family Bovidae in the order Artiodactyla, the even-toed hoofed **mammals**. Eland belong to the tribe Tragelaphini, a closely-related group of spiral-horned antelopes, whose members are not territorial. Both sexes posses long horns, and females are slightly smaller than males.

Characteristics

The horns of eland are about 2 ft (0.6 m) long, with one or two tight spirals. Eland have five or six white stripes on their bodies and white markings on their legs as well. The young are reddish brown, while older males are a bluish gray. Other distinctive markings include a crest running along their spines, a tuft of hair on the tail (like a cow tail), and a large loose flap of skin below the neck (the dewlap). This adds to the eland's bulky appearance.

Eland are not fast runners, but they can trot at a speed of 13 mph (21 kph) for long periods and can easily jump over a 6 ft (2 m) fence. They are gregarious, living in loosely structured herds where bonding is only evident between mothers and their calves. The size of herds can be as large as 500 with subgroups made up of eland of the same gender and age. Their home range areas can encompass more than 150 sq mi (389 sq km) and they travel over greater distances throughout the year.

Female eland reach maturity at three years, males at four or five years. Males continue to grow even after maturity. Eland mate every other year. The gestation period lasts about nine months, resulting in a single calf. The newborn calf lies concealed in the grass or undergrowth for about a month and is visited for nursing by its mother twice a day. After this, the calf joins other young calves, forming a nursery group, watched over by female eland who protect the young from predators.

Adaptation

Eland can adapt to a wide range of conditions. They can be found in arid regions, savannas, woodland and grassland areas, and in mountain ranges as high as 15,000 ft (4,570 m). Eland, like all bovids, are ruminants (cud-chewing animals) living on a diet of leaves, **fruits**, seed pods, flowers, tubers, and **bark**. They sometimes break down higher branches with their horns to feed on leaves of trees. Eland are adept at picking out high quality food from among poorer vegetation, a habit known as foliage gleaning. During rainy **seasons** eland graze on green grass.

A bull eland in eastern Africa. *Photograph by Christina Loke. Photo Researchers, Inc. Reproduced by permission.*

Eland are found in East **Africa** (Kenya, Malawi, and Mozambique) and in southern Africa (from Zimbabwe to South Africa). In West Africa (from Senegal to Sudan) a second **species**, the giant eland (*T. derbianus*) is found from Senegal in West Africa to southern Sudan and northern Uganda. Like other antelopes, eland are somewhat independent of drinking, since they are able to meet most of their needs from the **water** contained in plants they eat. Some of the strategies eland use in **water conservation** are common to all antelopes. Seeking shade during the hottest part of the day and feeding during the coolest part is one strategy. Other water-conservation strategies include the ability to concentrate urine, **heat** storage, the ability to allow body **temperature** to rise, and exhaling dry air by recovering water that would otherwise be lost.

Domestication and conservation

Rock paintings indicate a domestic relationship between eland and bushmen. In Natal, South Africa, eland have been domesticated for use as both dairy and draft animals, and for their tough hides. On their own and in low-density areas, they are endangered by agricultural development, which diminishes their range, and by hunting. Their meat is considered delicious and is prized as a source of protein. Ranched eland are susceptible to ticks. These antelope also died in large numbers during the rinderpest **epidemic** of 1896. Conservationists support planned domestication since it preserves species otherwise threatened by the encroaching **land use** of humans. The populations of eland today are much reduced. These formerly abundant antelope are now found mainly in reserves in South Africa and Botswana.

See also Antelopes and gazelles.

Resources

Books

Estes, Richard D. *Behavior Guide to African Mammals.* Berkeley: University of California, 1991.

KEY TERMS

Dewlap—A loose fold of skin that hangs from the neck.

Foliage gleaner—An animal that selects the most nutritious leaves for its diet.

Rinderpest—A contagious, often fatal, viral disease of cattle, sheep and goats, characterized by fever and the appearance of ulcers on the mucous membranes of the intestines.

Ruminant—A cud-chewing animal with a four-chambered stomach and even-toed hooves.

Tribe—A classification of animals that groups similar species exhibiting common features.

Estes, Richard D. *The Safari Companion.* Post Mills, Vermont: Chelsea Green, 1993.

Grzimek, Bernhard. *Encyclopedia of Mammals.* New York: McGraw-Hill, 1990.

Haltenorth, T., and H. Diller. *A Field Guide to the Mammals of Africa.* London: Collins, 1992.

MacDonald, David, and Sasha Norris, eds. *Encyclopedia of Mammals.* New York: Facts on File, 2001.

Nowak, Ronald M. *Walker's Encyclopedia of Mammals.* 5th ed. Baltimore: Johns Hopkins University Press, 1991.

Vita Richman

Elapid snakes

Elapid snakes are extremely venomous **snakes** such as cobras, mambas, kraits, tiger snakes, and coral snakes in the family Elapidae. The elapids are about 120 **species** in the subfamily Elapinae. The sea snakes (subfamily Hydrophiinae) and subfamily Laticaudinae make up the other two subfamilies in the Elapidae. Elapid snakes have a wide distribution from warm temperate climates to tropical climates, and are found on all continents except **Antarctica**.

Biology of elapid snakes

Elapid snakes have teeth on the front part of the upper jaw that are modified as paired fangs to inject venom into their victims. The fangs deliver the venom in much the same way that a hypodermic syringe delivers a drug, i.e., as a subcutaneous injection under **pressure** through narrow tubes. The fangs of elapid snakes are permanently erect, and when the mouth is closed they are enclosed within a pocket in the outer lip, outside of the lower mandible. At any one time, only two fanged teeth are functionally capable of delivering venom. However, there are a series of smaller, developing fangs available as replacements, should the primary ones be damaged, lost during use, or shed. Elapid snakes bite to subdue their **prey**, and when attempting to protect themselves from their own predators.

Three species of elapid snakes have the ability to deliver their venom through the air, by "spitting" rather accurately towards the eyes of a **predator**, in some species to a **distance** of up to 9.8 ft (3 m). This is primarily a defensive **behavior**, rather than one used for hunting. The spitting cobra (*Hemachatus hemachatus*) of South **Africa** is especially accurate, and can propel its venom as far as 6.5 ft (about 2 m). Other spitting cobras are the African black-necked cobra (*Naja nigricollis*) and a subspecies of the Asian cobra (*Naja naja sputatrix*). If the venom of a spitting cobra is not quickly washed from the eyes, blindness could occur.

When cobras feel threatened, they will raise the front of their body above the ground, and face the danger. At the same time, cobras use extensible neck ribs to spread their so-called "hood," as a further warning to the potential predator. The erect stance and spread hood of cobras is a warning display, used to caution predators about meddling with a dangerous snake.

Most elapid snakes are **oviparous**, meaning they lay eggs, that after a period of incubation hatch into young that are small replicas of the adult animals. Some species of elapid snakes, most commonly cobras, guard their eggs until they hatch. Some species, including the spitting cobra, are **ovoviviparous**, meaning the eggs are retained within the body of the female until they hatch, so that live snakes are born. Australian snakes in the genus *Denisonia* are viviparous, meaning true eggs are never formed by the female, and live young are born.

The greatest recorded longevity of an elapid snake was for the forest cobra (*Naja melanoleuca*), which lived 29 years in captivity.

Fish-eating sea snakes can reach a body length of 9.2 ft (2.8 m) and occur in tropical waters in eastern Africa and the Red Sea, **Asia**, **Australia**, and many Pacific islands. Sea snakes have very toxic venom, but most species are not aggressive, and they rarely bite humans. Sea snakes have a laterally compressed, paddle-shaped tail, well adaptive to swimming, and most species are ovoviviparous. Some species of sea snakes occasionally form mass aggregations, probably for breeding, and such gatherings have been estimated to contain several million individuals.

One especially seafaring species, the pelagic sea snake (*Pelamis platurus*), ranges from the east coast of

A Siamese cobra. *Photograph by Tom McHug. The National Audubon Society Collection/Photo Researchers, Inc. Reproduced by permission.*

Africa, through the Indo-Pacific region, and has even crossed the Pacific Ocean, occuring in tropical waters of western **South America**. Sea snakes are probably the basis of folk legends about sea serpents, although the living sea snakes do not closely resemble the fantastically large and aggressive serpents of folk lore.

Species of elapid snakes

Perhaps the world's most famous species of elapid snake is a subspecies of the Asian cobra (*Naja naja*) known as the Indian cobra (*N. n. naja*), which is the serpent that is most often used by snake charmers. Often, the cobra emerges from the urn or sack in which it is kept, and then assumes its warning stance of an erect fore-body and spread hood. In addition, the serpent "dances" sinuously in response to the movements of the flute, as it is waved about in front of the cobra. Actually,

the cobra is deaf to most of the music played by the charmer's flute—it is only responding to the movement of the instrument.

The world's longest venomous snake is the king cobra (*Ophiophagus hannah*), which can attain a length of 18 ft (5.5 m). This impressive but uncommon snake occurs in India and southeast Asia, and it feeds primarily on other species of snakes.

The mambas are four species of African elapids, of which the black mamba (*Dendroapsis polylepis*) is most feared, because it is relatively common and many people are bitten each year. This snake can grow to a length of 13 ft (4 m), and is probably the most swiftly moving of all snakes.

Elapid snakes are relatively diverse and abundant in Australia, where species of venomous snakes actually outnumber nonvenomous snakes by four to one. The largest, most dangerous species is the taipan (*Oxyuranus scutellatus*), an uncommon, aggressive, tropical species that can reach a length of 11.5 ft (3.5 m). However, several species of tiger snakes (*Notechis scutatus* and *N. ater*) are more common and widespread, and have particularly deadly venom. The death adders (*Acanthophis antarcticus* and *A. pyrrhus*) are viper-like elapids that are relatively common and widespread.

American elapids are represented by about 40 species of coral snakes, in the genera *Micrurus* and *Micruroides*. These snakes have extremely potent venom. However, coral snakes are not very aggressive, possessing relatively short fangs and a small mouth, so they cannot easily bite most parts of the human body, with fingers and toes being notable exceptions. Coral snakes are brightly colored with rings of black, red, and yellow.

The most widespread species in **North America** is the eastern coral snake (*Micrurus fulvius fulvius*), occurring widely in the southeastern United States from southern North Carolina to eastern Louisiana. The eastern coral snake likes to burrow, and is not often seen unless it is specifically looked for. This snake feeds almost entirely on **reptiles**, with **frogs** and small **mammals** also occasional prey. The eastern coral snake has brightly colored rings of red, yellow, and black on its body. These are a warning or aposematic coloration, intended to alert predators to the dangers of messing with this potentially dangerous, venomous snake.

However, in the coral snake the red and yellow rings occur adjacent to each other, unlike similarly colored but nonpoisonous species such as the scarlet kingsnake (*Lampropeltis triangulum*) and the scarlet snake (*Cemophora coccinea*). These latter snakes are mimics of the coral snake, which share aspects of its coloration to gain some measure of protection from predators. A folk saying was developed to help people remember the important differences in coloration between the coral snake and its harmless mimics: "Red touch yellow—dangerous fellow. Red touch black—venom lack." The Texas coral snake (*Micrurus fulvius tenere*) occurs in parts of the central and southwestern United States and Mexico.

Elapid snakes and humans

Species of elapid snakes are among the most feared of the serpents, and each year many people die after being bitten by these animals. This is especially true of certain tropical countries, particularly in India and tropical Asia, and in parts of Africa. For example, thousands of fatal snake bites occur each year in India alone. Wherever elapids and other poisonous snakes occur, there is a tangible risk of snake bite.

KEY TERMS

. . . .

Antivenin—An antitoxin that counteracts a specific venom, or a group of similar venoms. Antivenins are available for most types of snake venoms.

Aposematic—Refers to a bright coloration of an animal, intended to draw the notice of a potential predator, and to warn of the dangers of toxicity or foul taste.

However, in many places the magnitude of the risks of a snake bite are grossly overestimated by people. Except in the case of unusually aggressive species of snakes, it is extremely unlikely that a careful person will be bitten by a venomous snake, even where these animals are abundant. In the greater scheme of things, snake bites may be deadly, but in terms of actual risk, snakes are not usually very dangerous. This is especially true in North America, but somewhat less so in some tropical countries.

However, any bite by a poisonous snake should be treated as a medical emergency. First-aid procedures in the field can involve the use of a constriction band to slow the absorption of the venom into the general circulation, and perhaps the use of incision and suction to remove some of the poison. Antivenins are also available for the venoms of many species of poisonous snakes. Antivenins are commercially prepared serums that serve as antidotes to snake venoms if they are administered in time.

It is regrettable that so many poisonous snakes—and harmless snakes—are killed each year by people with fears that are essentially misguided and overblown. Snakes are a valuable component of natural ecosystems. Moreover, many species of snakes provide humans with useful services, for example, by preying on **rodents** that can potentially cause great damage in agriculture or serve as the vectors of human diseases.

Resources

Books

Cogger, Harold G., David Kirshner, and Richard Zweifel. *Encyclopedia of Reptiles and Amphibians.* 2nd ed. San Diego, CA: Academic Press, 1998.

Mattison, C. *Snakes of the World.* Poole, UK: Blandford Press, 1986.

Zug, George R., Laurie J. Vitt, and Janalee P. Caldwell. *Herpetology: An Introductory Biology of Amphibians and Reptiles.* 2nd ed. New York: Academic Press, 2001.

Bill Freedman

Elasticity

Elasticity is the ability of a material to return to its original shape and size after being stretched, compressed, twisted or bent. Elastic deformation (change of shape or size) lasts only as long as a deforming **force** is applied to the object, and disappears once the force is removed. Greater forces may cause permanent changes of shape or size, called plastic deformation.

In ordinary language, a substance is said to be "elastic" if it stretches easily. Therefore, rubber is considered a very elastic substance, and rubber bands are even called "elastics" by some people. Actually, however, most substances are somewhat elastic, including **steel**, **glass**, and other familiar materials.

Stress, strain, and elastic modulus

The simplest description of elasticity is Hooke's law, which states, "The stress is proportional to the strain." This **relation** was first expressed by the British scientist, Robert Hooke (1635-1702). He arrived at it through studies in which he placed weights on **metal** springs and measured how far the springs stretched in response. Hooke noted that the added length was always proportional to the weight; that is, doubling the weight doubled the added length.

In the modern statement of Hooke's law, the terms "stress" and "strain" have precise mathematical definitions. Stress is the applied force divided by the area the force acts on. Strain is the added length divided by the original length.

To understand why these special definitions are needed, first consider two bars of the same length, made of the same material. One bar is twice as thick as the other. Experiments have shown that both bars can be stretched to the same additional length only if twice as much weight is placed on the bar that is twice as thick. Thus, they both carry the same stress, as defined above.

The special definition of strain is required because, when an object is stretched, the stretch occurs along its entire length, not just at the end to which the weight is applied. The same stress applied to a long rod and a short rod will cause a greater extension of the long rod. The strain, however, will be the same on both rods.

The amount of stress required to produce a given amount of strain also depends on the material being stretched. Therefore, the **ratio** of stress to strain is a unique property of materials, different for each substance. It is called the elastic modulus (plural: moduli). It is also known as Young's modulus, after Thomas Young (1773-1829) who first described it. It has been measured

for thousands of materials. The greater the elastic modulus, the stiffer the material is. For example, the elastic modulus of rubber is about six hundred psi (pounds per square inch). That of steel is about 30 million psi.

Other elastic deformations

All deformations, no matter how complicated, can be described as the result of combinations of three basic types of stress. One is tension, which stretches an object along one direction only. Thus far, our discussion of elasticity has been entirely in terms of tension. Compression is the same type of stress, but acting in the opposite direction.

The second basic type of stress is shear stress. This results when two forces push on opposite ends of an object in opposite directions. Shear stress changes the object's shape. The shear modulus is the amount of shear stress divided by the **angle** through which the shape is strained.

Hydrostatic stress, the third basic stress, squeezes an object with equal force from all directions. A familiar example is the **pressure** on objects under **water** due to the weight of the water above them. Pure hydrostatic stress changes the **volume** only, not the shape of the object. Its modulus is called the bulk modulus.

Elastic limit

The greatest stress a material can undergo and still return to its original dimensions is called the elastic **limit**. When stressed beyond the elastic limit, some materials fracture, or break. Others undergo plastic deformation, taking on a new permanent shape. An example is a nail bent by excessive shear stress of a hammer blow.

Elasticity on the atomic scale

The elastic modulus and elastic limit reveal much about the strength of the bonds between the smallest particles of a substance, the **atoms** or molecules it is composed of. However, to understand elastic behavior on the level of atoms requires first distinguishing between materials that are crystalline and those that are not.

Crystalline materials

Metals are examples of crystalline materials. Solid pieces of metal contain millions of microscopically small crystals stuck together, often in **random** orientations. Within a single **crystal**, atoms are arranged in orderly rows. They are held by attractive forces on all sides. Scientists model the attractive force as a sort of a spring. When a spring is stretched, a restoring force tries to return it to its original length. When a metal rod is stretched in tension, its atoms are pulled apart slightly. The attractive force between the atoms tries to restore the original **distance**. The stronger the attraction, the more force must be applied to pull the atoms apart. Thus, stronger atomic forces result in larger elastic modulus.

Stresses greater than the elastic limit overcome the forces holding atoms in place. The atoms move to new positions. If they can form new bonds there, the material deforms plastically; that is, it remains in one piece but assumes a new shape. If new bonds cannot form, the material fractures.

The ball and spring model also explains why metals and other crystalline materials soften at higher temperatures. **Heat energy** causes atoms to vibrate. Their vibrations move them back and forth, stretching and compressing the spring. The higher the **temperature**, the larger the vibrations, and the greater the average distance between atoms. Less applied force is needed to separate the atoms because some of the stretching energy has been provided by the heat. The result is that the elastic modulus of metals decreases as temperature increases.

Elastomers

To explain the elastic behavior of materials like rubber requires a different model. Rubber consists of molecules, which are clusters of atoms joined by chemical bonds. Rubber molecules are very long and thin. They are polymers, long chain-like molecules built up by repeating small units. Rubber polymers consist of hundreds or thousands of atoms joined in a line. Many of the bonds are flexible, and can rotate. The result is a fine structure of kinks along the length of the **molecule**. The molecule itself is so long that it tends to bend and coil randomly, like a rope dropped on the ground. A piece of rubber, such as a rubber band, is made of vast numbers of such kinked, twisting, rope-like molecules.

When rubber is pulled, the first thing that happens is that the loops and coils of the "ropes" straighten out. The rubber extends as its molecules are pulled out to their full length. Still more stress causes the kinks to straighten out. Releasing the stress allows the kinks, coils and loops to form again, and the rubber returns to its original dimensions. Materials made of long, tangled molecules stretch very easily. Their elastic modulus is very small. They are called elastomers because they are very "elastic" polymers.

The "kink" model explains a very unusual property of rubber. A stretched rubber band, when heated, will suddenly contract. It is thought that the added heat provides enough energy for the bonds to start rotating again. The kinks that had been stretched out of the material return to it, causing the length to contract.

> ## KEY TERMS
>
> **Elastic deformation**—A temporary change of shape or size due to applied force, which disappears when the force is removed.
>
> **Elastic modulus**—The ratio of stress to strain (stress divided by strain), a measure of the stiffness of a material.
>
> **Plastic deformation**—A permanent change of shape or size due to applied force.
>
> **Strain**—The change in dimensions of an object, due to applied force, divided by the original dimensions.
>
> **Stress**—The magnitude of an applied force divided by the area it acts upon.

Sound waves

Elasticity is involved whenever atoms vibrate. An example is the movement of **sound waves**. A sound wave consists of energy that pushes atoms closer together momentarily. The energy moves through the atoms, causing the region of compression to move forward. Behind it, the atoms spring further apart, as a result of the restoring force.

The speed with which sound travels through a substance depends in part on the strength of the forces between atoms of the substance. Strongly bound atoms readily affect one another, transferring the "push" due to the sound wave from each atom to its neighbor. Therefore, the stronger the bonding force, the faster sound travels through an object. This explains why it is possible to hear an approaching railroad train by putting one's **ear** to the track, long before it can be heard through the air. The sound wave travels more rapidly through the steel of the track than through the air, because the elastic modulus of steel is a million times greater than the bulk modulus of air.

Measuring the elastic modulus

The most direct way to determine the elastic modulus of a material is by placing a **sample** under increasing stresses, and measuring the resulting strains. The results are plotted as a graph, with strain along the horizontal axis and stress along the vertical axis. As long as the strain is small, the data form a straight line for most materials. This straight line is the "elastic region." The slope of the straight line equals the elastic modulus of the material. Alternatively, the elastic modulus can be calculated from measurements of the speed of sound through a sample of the material.

Resources

Books

Goodwin, Peter H. *Engineering Projects for Young Scientists.* New York: Franklin Watts, 1987.

Periodicals

"A Figure Less Than Greek" *Discover* 13 (June 1992): 14.
Williams, Gurney, III. "Smart Materials." *Omni* (April 15, 1993): 42–44+.

Sara G. B. Fishman

Electric arc

An electric **arc** is a high-current, low-voltage electrical discharge between electrodes in the presence of gases. In an electric arc, electrons are emitted from a heated **cathode**. Arcs can be formed in high, atmospheric, or low pressures, and in various gases. They have wide uses as highly luminous lamps, as furnaces for heating, cutting and **welding**, and as tools for spectrochemical analysis.

Electrical conduction in gases

Gases consist of neutral molecules, and are, therefore, good insulators. Yet under certain conditions, a breakdown of the insulating property occurs, and current can pass through the gas. Several phenomena are associated with the electric discharge in gases; among them are spark, dark (Townsend) discharge, glow, corona, and arc.

In order to conduct **electricity**, two conditions are required. First, the normally neutral gas must create charges or accept them from external sources, or both. Second, an electric field should exist to produce the directional **motion** of the charges. A charged atom or **molecule**, or ion, can be positive or **negative**; electrons are negative charges. In electrical devices, an electric field is produced between two electrodes, called **anode** and cathode, made of conducting materials. The process of changing a neutral atom or molecule into an ion is called ionization. Ionized gas is called **plasma**. Conduction in gases is distinguished from conduction in solids and liquids in that the gases play an active role in the process. The gas not only permits free charges to pass though, but itself may produce charges. Cumulative ionization occurs when the original **electron** and its offspring gain enough **energy**, so each can produce another electron. When the process is repeated over and over, the resulting process is called an avalanche.

For any gas at a given **pressure** and **temperature** there is a certain voltage value, called *breakdown poten-*

tial, that will produce ionization. Application of a voltage above the critical value would initially cause the current to increase due to cumulative ionization, and the voltage is then decreased. If the pressure is not too low, conduction is concentrated into a narrow, illuminated, "spark" channel. By receiving energy from the current, the channel becomes hot and may produce shock-waves. Natural phenomena are the **lightning** and the associated thunder, that consist of high voltages and currents that cannot be artificially achieved.

An arc can be produced in high pressure following a spark. This occurs when steady conditions are achieved, and the voltage is low but sufficient to maintain the required current. In low pressures, the transient stage of the spark leads to the glow discharge, and an arc can later be formed when the current is further increased. In arcs, the *thermionic effect* is responsible for the production of free electrons that are emitted from the hot cathode. A strong electric field at the metallic surface lowers the barrier for electron **emission**, and provides a *field emission*. Because of the high temperature and the high current involved, however, some of the mechanisms of arcs cannot be easily studied.

Properties of the arc

The electric arc was first detected in 1808 by British chemist Humphry Davy. He saw a brilliant luminous flame when two **carbon** rods conducting a current were separated, and the **convection** current of hot gas deflected it in the shape of an arc. Typical characteristics of an arc include a relatively low potential gradient between the electrodes (less than a few tens of Volts), and a high current **density** (from 0.1 amperes to thousands amperes or higher). High gas temperatures (several thousands or tens of thousands degrees Kelvin) exist in the conducting channel, especially in high gas pressures. Vaporization of the electrodes is also common, and the gas contains molecules of the electrodes material. In some cases, a hissing sound may be heard, making the arc "sing." The potential gradient between the electrodes is not uniform. In most cases, one can distinguish between three different regions: the area close to the positive electrode, termed *cathode fall*; the area close to the negative electrode, or *anode rise*; and the main arc body. Within the arc body there is a uniform voltage gradient. This region is electrically neutral, where the cumulative ionization results in the number of positive ions equals the number of electrons or negative ions. The ionization occurs mainly due to excitation of the molecules and the gain of high temperature.

The cathode fall region is about 0.01 mm with a potential difference of less than about 10 Volts. Often thermionic emission would be achieved at the cathode.

The electrodes in this case are made of refractive materials like tungsten and carbon, and the region contains an excess of positive ions and a large **electric current**. At the cathode, transition is made from a metallic conductor in which current is carried by electrons, to a gas in which conduction is done by both electrons or negative ions and positive ions. The gaseous positive ions may reach the cathode freely and form a potential barrier. Electrons emitted from the cathode must overcome this barrier in order to enter the gas.

At the anode, transition is made from a gas, in which both electrons and positive ions conduct current, to the metallic conductor, in which current is carried only by electrons. With a few exceptions, positive ions do not enter the gas from the **metal**. Electrons are accelerated towards the anode and provide, through ionization, a supply of ions for the column. The electron current may raise the anode to a high temperature, making it a thermionic emitter, but the emitted electrons are returned to the anode, contributing to the large negative space charge around it. The melting of the electrodes and the introduction of their vapor to the gas adds to the pressure in their vicinities.

Uses of electric arcs

There are many types of arc devices. Some operate at **atmospheric pressure** and may be open, and others operate at low pressure and are therefore closed in a container, like **glass**. The property of high current in the arc is used in the mercury arc rectifiers, like the thyratron. An alternate potential difference is applied, and the arc transfers the current in one direction only. The cathode is heated by a filament.

The high temperature created by an electric arc in the gas is used in furnaces. *Arc welders* are used for welding, where a metal is fused and added in a joint. The arc can supply the **heat** only, or one of its electrodes can serve as the consumable parent metal. *Plasma torches* are used for cutting, spraying, and gas heating. Cutting may be done by means of an arc formed between the metal and the electrode.

Arc lamps provide high luminous efficiency and great brightness. The **light** comes from the highly incandescence (about 7,000°F [3,871°C]) electrodes, as in *carbon arcs*, or from the heated, ionized gases surrounded the arc, as in *flame arcs*. The carbon arc, where two carbon rods serve as electrodes, was the first practical commercial electric lighting device, and it is still one of the brightest sources of light. It is used in theater motion-picture projectors, large searchlights, and lighthouses. Flame arcs are used in **color photography** and in photochemical processes because they closely approximate

KEY TERMS

Artificial (hot) arc—An electric arc whose cathode is heated by an external source to provide thermionic emission, and not by the discharge itself.

Cold cathode arc—An electric arc that operates on low boiling-point materials.

Thermionic arc—An electric arc in which the electron current from the cathode is provided predominantly by thermionic emission.

natural sunshine. The carbon is impregnated with volatile chemicals, which become luminous when evaporated and driven into the arc. The color of the arc depends on the material; the material could be **calcium**, **barium**, **titanium**, or strontium. In some, the wavelength of the **radiation** is out of the visible **spectrum**. Mercury arcs produce ultraviolet radiation at high pressure. They can also produce visible light in a low pressure tube, if the internal walls are coated with **fluorescence** material such as phosphor; the phosphor emits light when illuminated by the ultraviolet radiation from the mercury.

Other uses of arcs include valves (used in the early days of the **radio**), and as a source of ions in nuclear **accelerators** and thermonuclear devices. The excitation of electrons in the arc, in particular the direct electron bombardment, leads to narrow **spectral lines**. The arc, therefore, can provide information on the composition of the electrodes. The spectra of metal alloys are widely studied using arcs; the metals are incorporated with the electrodes material, and when vaporized, they produce distinct spectra.

See also Electronics.

Ilana Steinhorn

Electric charge

Rub a **balloon** or styrofoam drinking cup against a wool sweater. It will then stick to a wall (at least on a dry day) or pick up small bits of **paper**. Why? The answer leads to the concept of electric charge.

Electromagnetic forces are one of the four fundamental forces in nature. The other three are gravitational, strong nuclear, and weak nuclear forces. The electromagnetic **force** unifies both electrical and magnetic forces. The magnetic forces occur whether the charges are moving or at rest. Electric charge is our way of measuring how much electric force an object can exert or feel.

Electric charge plays the same role in electric forces as **mass** plays in gravitational forces. The force between two electric charges is proportional to the product of the two charges divided by the **distance** between them squared, just as the force between two masses is proportional to the product of the two masses divided by the distance between them squared. These two force laws for electric and gravitational forces have exactly the same mathematical form.

There are however differences between the electric and gravitational forces. One difference is the electrical force is much stronger than the gravitational force. That is why the styrofoam cup mentioned above can stick to a wall. The electrical force pulling it to the wall is stronger than the gravitational force pulling it down.

The second major difference is that the gravitational force is always attractive. The electrical force can be either attractive or repulsive. There is only one type of mass, but there are two types of electric charge. Like charges will repel each other and unlike charges will attract. Most **matter** is made up of equal amounts of both types of charges, so electrical forces **cancel** out over long distances. The two types of charge are called positive and **negative**, the names given by Benjamin Franklin, the first American physicist. Contrary to what many people think, the terms positive and negative don't really describe properties of the charges. The names are completely arbitrary.

An important property of electric charges that was discovered by Benjamin Franklin is that charge is conserved. The total amount of both positive and negative charges must remain the same. Charge conservation is part of the reason the balloon and cup mentioned above stick to the wall. Rubbing causes electrons to be transferred from one object to another, so one has a positive charge and the other has exactly the same negative charge. No charges are created or destroyed; they are just transferred. The objects then have a net charge and electrical forces come into play.

Electric charge forms the basis of the electrical and magnetic forces that are so important in our modern electrical and electronic luxuries.

Electric circuit

An electric circuit is a system of conducting elements designed to control the path of **electric current** for a particular purpose. Circuits consist of sources of electric **energy**, like generators and batteries; elements that transform, dissipate, or store this energy, such as resistors, capacitors, and inductors; and connecting wires. Circuits often include a fuse or circuit breaker to prevent a power overload.

Devices that are connected to a circuit are connected to it in one of two ways: in series or in **parallel**. A series circuit forms a single pathway for the flow of current, while a parallel circuit forms separate paths or branches for the flow of current. Parallel circuits have an important advantage over series circuits. If a device connected to a series circuit malfunctions or is switched off, the circuit is broken, and other devices on the circuit cannot draw power. The separate pathways of a parallel circuit allows devices to operate independently of each other, maintaining the circuit even if one or more devices are switched off.

The first electric circuit was invented by Alessandro Volta in 1800. He discovered he could produce a steady flow of **electricity** using bowls of **salt solution** connected by **metal** strips. Later, he used alternating discs of **copper**, zinc, and cardboard that had been soaked in a salt solution to create his voltaic pile (an early **battery**). By attaching a wire running from the top to the bottom, he caused an electric current to flow through his circuit. The first practical use of the circuit was in **electrolysis**, which led to the discovery of several new chemical elements. Georg Ohm (1787-1854) discovered some conductors had more resistance than others, which affects their efficiency in a circuit. His famous law states that the voltage across a conductor divided by the current equals the resistance, measured in *ohms*. Resistance causes **heat** in an electrical circuit, which is often not wanted.

See also Electrical conductivity; Electrical power supply; Electrical resistance; Electronics; Integrated circuit.

Electric conductor

An electric conductor is any material that can efficiently conduct **electricity**, such as a **metal**, ionic **solution**, or ionized gas. Usually, this term refers to the current-carrier component of an **electric circuit** system.

Conduction of electricity

Conduction, the passage of charges in an electrical field, is done by the movement of charged particles in the conducting medium. Good *conductors* are materials that have available **negative** or positive charges, like electrons or ions. *Semiconductors* are less effective in conducting electricity, while most other materials are *insulators*.

In metals, the atomic nuclei form crystalline structures, where electrons from outer orbits are mobile, or "free." The *current* (the net transfer of **electric charge** per unit **time**) is carried by the free electrons. Yet the transfer of **energy** is done much faster than the actual movements of the electrons. Among metals at room **temperature**, silver is the best conductor, followed by **copper**. **Iron** is a relatively poor conductor.

In electrolytic solutions, the positive and negative ions of the dissolved salts can carry current. Pure **water** is a good insulator, and various salts are fair conductors; together, as sea water, they make a good conductor.

Gases are usually good insulators. Yet when they become ionized under the influence of strong electrical fields, they may conduct electricity. Some of the energy is emitted as **light** photons, with most spectacular effects are seen in **lightning**.

In semiconductors like germanium and silicon, a limited number of free electrons or holes (positive charges) are available to carry current. Unlike metals, the conductivity of semiconductors increases with temperature, as more electrons are becoming free.

Types of conductors

Electrical energy is transmitted by metal conductors. Wires are usually soft and flexible. They may be bare, or coated with flexible insulating material. In most cases, they have a circular cross-section. Cables have larger cross-sections than wires, and they are usually stranded, built up as an assembly of smaller solid conductors. Cords are small-diameter flexible cables that are usually insulated. Multi-conductor cable is an assembly of several insulated wires in a common jacket. Bus-bars are rigid and solid, made in shapes like rectangular, rods or tubes, and are used in switchboards.

Most conductors are made from copper or **aluminum**, which are both flexible materials. While copper is a better conductor, aluminum is cheaper and lighter. For overhead lines the conductors are made with a **steel** or aluminum-alloy core, surrounded by aluminum. The conductors are supported on insulators, which are usually ceramic or porcelain. They may be coated with rubber, polyethylene, **asbestos**, thermoplastic, and varnished cambric. The specific type of the insulating material depends on the voltage of the circuit, the temperature, and whether the circuit is exposed to water or chemicals.

Resistance to electrical energy

A perfect conductor is a material through which charges can move with no resistance, while in a perfect insulator it is impossible for charges to move at all. How-ever, all conducting materials have some resistance to the electrical energy, with several major effects. One is the loss of electrical energy that converts to **heat**; the other is that the heating of the conductors causes them to age. In addition, the energy loss within the conductors causes a reduction in the voltage at the load. The voltage drop needs to be taken into consideration in the design and operation of the circuit, since most utility devices are operating within a narrow range of voltage, and lower than desired voltage may not be sufficient for their operation.

Superconductors

Superconductors carry **electric current** without any resistance, therefore without energy loss. In addition, under the extremely high currents they are able to carry, superconductors exhibit several characteristics that are unknown in common conductors. For instance, they may repel external magnetic fields; magnets placed over superconducting materials will remain suspended in the air. While there is a great potential in using superconductors as carriers of electrical energy, and for frictionless means of transportation, currently their use is limited. One of the reasons is their relatively low operating temperature; mostly close to the **absolute zero**, some higher, up to 130K (-225°F [-143°C]).

See also Electronics.

Electric current

Electric current is the result of the relative **motion** of net **electric charge**. In metals, the charges in motion are electrons. The magnitude of an electric current depends upon the quantity of charge that passes a chosen reference point during a specified **time** interval. Electric current is measured in amperes, with one ampere equal to a charge-flow of one **coulomb** per second.

A current as small as a picoampere (one-trillionth of an ampere) can be significant. Likewise, artificial currents in the millions of amperes can be created for special purposes. Currents between a few milliamperes to a few amperes are common in **radio** and **television** circuits. An **automobile** starter motor may require several hundred amperes.

Current and the transfer of electric charge

The total charge transferred by an unvarying electrical current equals the product of current in amperes and the time in seconds that the current flows. If one ampere flows for one second, one coulomb will have moved in the conductor. If a changing current is graphed against time, the area between the graph's **curve** and the time axis will be proportional to the total charge transferred.

The speed of an electric current

Electrical currents move through wires at a speed only slightly less than the speed of **light**. The electrons, however, move from atom to atom more slowly. Their motion is more aptly described as a drift. Extra electrons added at one end of a wire will cause extra electrons to appear at the other end of the wire almost instantly. Individual electrons will not have moved along the length of the wire but the electric field that pushes the charge against charge along the conductor will be felt at the distant end almost immediately. To visualize this, imagine a cardboard mailing tube filled with ping-pong balls. When you insert an extra ball in one end of the tube, an identical ball will emerge from the distant end almost immediately. The original ball will not have traveled the length of the tube, but since all the balls are identical it will seem as if this has happened. This mechanical analogy suggests the way that charge seems to travel through a wire very quickly.

Electric current and energy

Heat results when current flows through an ordinary electrical conductor. Common materials exhibit an electrical property called resistance. **Electrical resistance** is analogous to **friction** in a mechanical system. Resistance results from imperfections in the conductor. When the moving electrons collide with these imperfections, they transfer kinetic **energy**, resulting in heat. The quantity of heat energy produced increases as the square of the current passing through the conductor.

Electric current and magnetism

A magnetic field is created in **space** whenever a current flows through a conductor. This magnetic field will exert a **force** on the magnetic field of other nearby current-carrying conductors. This is the principle behind the design of an **electric motor**.

An electrical **generator** operates on a principle similar to an electric motor. In a generator, mechanical energy forces a conductor to move through a magnetic field. The magnetic field forces the electrons in the conductor to move, which causes an electric current.

Direct current

A current in one direction only is called a direct current, or DC. A steady current is called pure DC. If DC varies with time it is called pulsating DC.

Alternating current

If a current changes direction repeatedly it is called an alternating current, or AC. Commercial electrical power is transported using alternating current because AC makes it possible to change the **ratio** of voltage to current with transformers. Using a higher voltage to transport electrical power across country means that the same power can be transferred using less current. For example, if transformers step up the voltage by a factor of 100, the current will be lower by a factor of 1/100. The higher voltage in this example would reduce the energy loss caused by the resistance of the wires to 0.01% of what it would be without the use of AC and transformers.

When alternating current flows in a circuit the charge drifts back and forth repeatedly. There is a transfer of energy with each current pulse. Simple electric motors deliver their mechanical energy in pulses related to the power line **frequency**.

Power lines in **North America** are based on AC having a frequency of 60 Hertz (Hz). In much of the rest of the world the power line frequency is 50 Hz. Alternating current generated aboard **aircraft** often has a frequency of 400 Hz because motors and generators can work efficiently with less **iron**, and therefore less weight, when this frequency is used.

Alternating current may also be the result of a combination of signals with many frequencies. The AC powering a loudspeaker playing music consists of a combination of many superimposed alternating currents with different frequencies and amplitudes.

Current flow vs. electron flow

We cannot directly observe the electrically-charged particles that produce current. It is usually not important to know whether the current results from the motion of positive or **negative** charges. Early scientists made an unfortunate choice when they assigned a positive polarity to the charge that moves through ordinary wires. It seemed logical that current was the result of positive charge in motion. Later it was confirmed that it is the negatively-charged **electron** that moves within wires.

The action of some devices can be explained more easily when the motion of electrons is assumed. When it is simpler to describe an action in terms of the motion of electrons, the charge motion is called electron flow. Current flow, conventional current, or Franklin convention current are terms used when the moving charge is assumed to be positive.

Conventional current flow is used in science almost exclusively. In **electronics**, either conventional current or electron flow is used, depending on which flow is most

KEY TERMS

Conventional current—Current assuming positive charge in motion.

Coulomb—The standard unit of electric charge, defined as the amount of charge flowing past a point in a wire in one second, when the current in the wire is one ampere.

Frequency—Number of times per unit of time an event repeats.

Hertz—A unit of measurement for frequency, abbreviated Hz. One hertz is one cycle per second.

Picoampere—One trillionth of an ampere or 10^{-12} amperes.

Speed of light—Speed of electromagnetic radiation, usually specified in a vacuum. Approximately 6.7×10^8 miles per hour (3×10^8 meters per second).

convenient to explain the operation of a particular electronic component. The need for competing conduction models could have been avoided had the original charge-polarity assignment been reversed.

See also Electronics.

Resources

Books

Hewitt, Paul. *Conceptual Physics.* Englewood Cliffs, NJ: Prentice Hall, 2001.

Hobson, Art. *Physics: Concepts and Connections.* Upper Saddle River, NJ: Prentice Hall, 1994.

Ostdiek, Vern J., and Donald J. Bord. *Inquiry Into Physics.* 3rd ed. St. Paul, MN: West Publishing Co., College & Schl. Div., 1995.

Donald Beaty

Electric motor

An electric motor is a machine used to convert electrical **energy** to mechanical energy. Electric motors are extremely important to modern-day life, being used in many different places, e.g., vacuum cleaners, dishwashers, computer printers, fax machines, video cassette recorders, **machine tools**, **printing** presses, automobiles, subway systems, **sewage treatment** plants and **water** pumping stations.

The major physical principles behind the operation of an electric motor are known as Ampère's law and

Faraday's law. The first states that an electrical conductor sitting in a magnetic field will experience a **force** if any current flowing through the conductor has a component at right angles to that field. Reversal of either the current or the magnetic field will produce a force acting in the opposite direction. The second principle states that if a conductor is moved through a magnetic field, then any component of **motion perpendicular** to that field will generate a potential difference between the ends of the conductor.

An electric motor consists of two essential elements. The first, a static component which consists of magnetic materials and electrical conductors to generate magnetic fields of a desired shape, is known as the *stator*. The second, which also is made from magnetic and electrical conductors to generate shaped magnetic fields which interact with the fields generated by the stator, is known as the *rotor*. The rotor comprises the moving component of the motor, having a rotating shaft to connect to the machine being driven and some means of maintaining an electrical contact between the rotor and the motor housing (typically, **carbon** brushes pushed against slip rings). In operation, the electrical current supplied to the motor is used to generate magnetic fields in both the rotor and the stator. These fields push against each other with the result that the rotor experiences a **torque** and consequently rotates.

Electrical motors fall into two broad categories, depending on the type of electrical power applied-direct current (DC) and alternating current (AC) motors.

The first DC electrical motor was demonstrated by Michael Faraday in England in 1821. Since the only available electrical sources were DC, the first commercially available motors were of the DC type, becoming popular in the 1880s. These motors were used for both low power and high power applications, such as electric street railways. It was not until the 1890s, with the availability of AC electrical power that the AC motor was developed, primarily by the Westinghouse and General Electric corporations. Throughout this decade, most of the problems concerned with single and multi-phase AC motors were solved. Consequently, the principal features of electric motors were all developed by 1900.

DC motor

The operation of a DC motor is dependent on the workings of the poles of the stator with a part of the rotor, or armature. The stator contains an even number of poles of alternating magnetic polarity, each pole consisting of an electromagnet formed from a pole winding wrapped around a pole core. When a DC current flows through the winding, a magnetic field is formed. The ar-

A cross section of a simple direct-current electric motor. At its center is the rotor, a coil wound around an iron armature, which spins within the poles of the magnet that can be seen on the inside of the casing. *Photograph by Bruce Iverson. Science Photo Library, National Audubon Society Collection/Photo Researchers, Inc. Reproduced by permission.*

mature also contains a winding, in which the current flows in the direction illustrated. This armature current interacts with the magnetic field in accordance with Ampère's law, producing a torque which turns the armature.

If the armature windings were to rotate round to the next pole piece of opposite polarity, the torque would operate in the opposite direction, thus stopping the armature. In order to prevent this, the rotor contains a commutator which changes the direction of the armature current for each pole piece that the armature rotates past, thus ensuring that the windings passing, for example, a pole of north polarity will all have current flowing in the same direction, while the windings passing south poles will have oppositely flowing current to produce a torque in the same direction as that produced by the north poles. The commutator generally consists of a split contact ring against which the brushes applying the DC current ride.

The **rotation** of the armature windings through the stator field generates a voltage across the armature which is known as the counter EMF (**electromotive force**) since it opposes the applied voltage: this is the consequence of Faraday's law. The magnitude of the counter EMF is dependent on the magnetic field strength and the speed of the rotation of the armature. When the DC motor is initially turned on, there is no counter EMF and the armature starts to rotate. The counter EMF increases with the rotation. The effective voltage across the armature windings is the applied voltage minus the counter EMF.

Types of DC motor

DC motors are more common than we may think. A car may have as many as 20 DC motors to drive fans,

seats, and windows. They come in three different types, classified according to the electrical circuit used. In the shunt motor, the armature and field windings are connected in **parallel**, and so the currents through each are relatively independent. The current through the field winding can be controlled with a field rheostat (variable resistor), thus allowing a wide variation in the motor speed over a large range of load conditions. This type of motor is used for driving machine tools or fans, which require a wide range of speeds.

In the series motor, the field winding is connected in series with the armature winding, resulting in a very high starting torque since both the armature current and field strength run at their maximum. However, once the armature starts to rotate, the counter EMF reduces the current in the circuit, thus reducing the field strength. The series motor is used where a large starting torque is required, such as in **automobile** starter motors, **cranes**, and hoists.

The compound motor is a combination of the series and shunt motors, having parallel and series field windings. This type of motor has a high starting torque and the ability to vary the speed and is used in situations requiring both these properties such as punch presses, conveyors and elevators.

AC motors

AC motors are much more common than the DC variety because almost all electrical supply systems run alternating current. There are three main different types of motor, namely polyphase induction, polyphase synchronous, and single phase motors. Since three phase supplies are the most common polyphase sources, most polyphase motors run on three phase. Three phase supplies are widely used in commercial and industrial settings, whereas single phase supplies are almost always the type found in the home.

Principles of three phase motor operation

The main difference between AC and DC motors is that the magnetic field generated by the stator rotates in the ac case. Three electrical phases are introduced through terminals, each phase energizing an individual field pole. When each phase reaches its maximum current, the magnetic field at that pole reaches a maximum value. As the current decreases, so does the magnetic field. Since each phase reaches its maximum at a different time within a cycle of the current, that field pole whose magnetic field is largest is constantly changing between the three poles, with the effect that the magnetic field seen by the rotor is rotating. The speed of rotation of the magnetic field, known as the synchronous speed, depends on the **frequency** of the power supply and the number of poles produced by the stator winding. For a standard 60 Hz supply, as used in the United States, the maximum synchronous speed is 3,600 rpm.

In the three phase induction motor, the windings on the rotor are not connected to a power supply, but are essentially short circuits. The most common type of rotor winding, the squirrel cage winding, bears a strong resemblance to the running wheel used in cages for pet **gerbils**. When the motor is initially switched on and the rotor is stationary, the rotor conductors experience a changing magnetic field sweeping by at the synchronous speed. From Faraday's law, this situation results in the induction of currents round the rotor windings; the magnitude of this current depends on the impedance of the rotor windings. Since the conditions for motor action are now fulfilled, that is, current carrying conductors are found in a magnetic field, the rotor experiences a torque and starts to turn. The rotor can never rotate at the synchronous speed because there would be no relative motion between the magnetic field and the rotor windings and no current could be induced. The induction motor has a high starting torque.

In squirrel cage motors, the motor speed is determined by the load it drives and by the number of poles generating a magnetic field in the stator. If some poles are switched in or out, the motor speed can be controlled by incremental amounts. In wound-rotor motors, the impedance of the rotor windings can be altered externally, which changes the current in the windings and thus affords continuous speed control.

Three-phase synchronous motors are quite different from induction motors. In the synchronous motor, the rotor uses a DC energized coil to generate a constant magnetic field. After the rotor is brought close to the synchronous speed of the motor, the north (south) pole of the rotor magnet locks to the south (north) pole of the rotating stator field and the rotor rotates at the synchronous speed. The rotor of a synchronous motor will usually include a squirrel cage winding which is used to start the motor rotation before the DC coil is energized. The squirrel cage has no effect at synchronous speeds for the reason explained above.

Single phase induction and synchronous motors, used in most domestic situations, operate on principles similar to those explained for three phase motors. However, various modifications have to be made in order to generate starting torques, since the single phase will not generate a rotating magnetic field alone. Consequently, split phase, **capacitor** start, or shaded pole designs are used in induction motors. Synchronous single phase motors, used for timers, clocks, tape recorders etc., rely on the reluctance or hysteresis designs.

KEY TERMS

....

AC—Alternating current, where the current round a circuit reverses direction of flow at regular intervals.

DC—Direct current, where the current round a circuit is approximately constant with time.

Rotor—That portion of an electric motor which is free to rotate, including the shaft, armature and linkage to a machine.

Stator—That portion of an electric motor which is not free to rotate, including the field coils.

Torque—The ability or force needed to turn or twist a shaft or other object.

Resources

Books

Anderson, Edwin P., and Rex Miller. *Electric Motors*. New York: Macmillan, 1991.

Periodicals

Gridnev, S. A. "Electric Relaxation In Disordered Polar Dielectrics." *Ferroelectrics* 266, no. 1 (2002): 171-209.

Iain A. McIntyre

Electric vehicles

Electric vehicles (EV), vehicles whose wheels are turned by electric motors rather than by a mechanical gasoline-powered drivetrain, have been long touted as saviors of the environment due to their low **pollution** and high fuel efficiency. However, they have yet to take over the highways and byways.

Thomas Davenport is credited with building the first practical EV in 1834, which was quickly followed by a two-passenger electric car in 1847, then an electric car in 1851 that could go 20 mph (32 km/h). The Edison Cell, a nickel-iron **battery**, was developed in 1900 and was a key factor in the development of early twentieth century electric vehicles. By 1900, electric vehicles had a healthy share of the pleasure car market. Of the 4,200 automobiles sold in the United States at the turn of the century, 38% were powered by **electricity**, 22% by gasoline, and 40% by steam. But by the 1920s, both electricity and steam had lost out to gasoline.

Automakers began working on electric vehicle batteries again during the 1960s as an offshoot of the U.S. **space** programs. Research continued during the oil embargo of the 1970s, through the 1980s and the 1990s. The results include a handful of commercially available electric automobiles capable of driving between 70-100 mi (113-160 km/h). As of the year 2000, a **sample** of the EVs on the market include:

- DaimlerChrysler EPIC Electric Minivan—Range: 80–90 mi (130–145 km/h); **Acceleration**: 0–60 mph (0–97 km/h) in 17 seconds; Maximum speed: 80 mph (130 km/h); Recharging time: four to five hours (220v)/ 30 minutes (440v); Battery: nickel-metal hydride.

- General Motors EV1 uses a lead-acid battery, changing to nickel-metal hydride battery—Range: with a **lead** acid battery 55–95 mi (90–153 km) and with a nickel-metal hydride battery 75–130 mi (120–210 km); Acceleration: 0–60 mph (0–97 km/h) in less than nine seconds; Maximum speed: 80 mph (130km/h); Recharging time: lead-acid battery 5.5–6 hours and nickel-metal hydride six to eight hours.

- Nissan Altra EV—Range: 120 mi (193 km); Acceleration: 0–50 mph (0–80 km/h) in 12 seconds; Maximum speed: 75 mph (120 km/h); Recharging time: five hours; Battery: lithium-ion battery.

- Toyota RAV4-EV-Range: 126 mi (203 km); Acceleration: 0–50 mph (0–80 km/h) in 12.8 seconds (or 0-60 mph [0–97 km/h] in about 18 seconds); Maximum speed: 79 mph (128 km/h); Recharging time: six to eight hours; Battery: Nickel-metal hydride.

Other auto manufacturers such as Ford, Honda, Mitsubishi, Daihatsu, BMW, Audi, Fiat, and Peugeot, are competing to produce a commercially viable vehicle for the masses.

The key components of an electric vehicle include **energy** storage cells, a power controller, and motors. Transmission of energy in electrical form eliminates the need for a mechanical drivetrain. A special braking design, called regenerative braking, uses the motor as a **generator**. This system feeds energy back to the storage system each time the brakes are used.

Batteries

As can be seen in the above list, the three main batteries employed today are lead-acid; nickel-metal hydride; and lithium-based batteries. Of these, experts predict nickel-metal hydride and lithium-based batteries have the greatest potential.

The lead acid battery uses lead oxide and spongy lead electrodes with **sulfuric acid** as an **electrolyte**. Generally, they consist of several cells put in series to form a battery, such as an **automobile** battery. The group of cells are generally in a polypropylene container. The advantages of the lead-acid battery are commercial avail-

ability, recyclability and low cost. The disadvantages are that they are heavy and the amount of energy stored per kilogram is less than other types of batteries.

Nickel-metal hydride operates by moving **hydrogen** ions between a nickel-metal hydride **cathode** and a nickel hydroxide **anode**. During discharge, hydrogen moves from cathode to anode. During charging, ions move in the opposite direction.

There are two types of **lithium** batteries, the lithium ion and the lithium **polymer**. A lithium ion type works by dissolving lithium ions, and transporting them between the anode and cathode. The battery has an anode made of lithium cobalt dioxide and a cathode from a non-graphitizing **carbon**. During operation, lithium ions move through a liquid electrolyte that contains a thin, microporous **membrane**. The lithium polymer uses lithium as an electrochemically active material and the electrolyte is a polymer or polymer-like material that conducts lithium ions.

Advantages

Electric vehicles are more efficient than internal **combustion** engines for several reasons. First, because the **electric motor** is directly connected to the wheels, it consumes no energy while the car is at rest or coasting. Secondly, the regenerative braking system can return as much as half an electric vehicle's kinetic energy to the storage cells. And thirdly, the motor converts more than 90% of the energy in its storage cells to motive **force**, whereas internal-combustion drives use less than 25% of the energy in a gallon (3.75 L) of gasoline.

One can recharge an EV for approximately one-third the cost to refuel a gasoline-powered car. The average monthly fuel cost for a typical EV driver is less than $15, while gasoline runs around an estimate $50. Although time of day and utility rates may minimally affect monthly total, the cost of recharging an electric car overnight costs less than a large cup of coffee.

The world land speed record for an electric car is just under 200 mph (320 km/h) as of 1996. Many operators of heavy vehicles, such as subway trains, locomotives and **mining** equipment, prefer electric motors because of the amount of instantaneous **torque** they offer; gasoline engines have to build power before they reach the peak rpm range that allows them to shift **gears**. Additionally, the average daily use of private vehicles in major U.S. cities is 40 mi (64 km); today's EVs can handle these trips with ease. An EV averages 40-100 mi (34-160 km) per charge.

Recognizing the need for alternative fuel vehicles (AFVs), U. S. President Bill Clinton issued Executive Order 13148 entitled "Greening the Government Through Federal Fleet and Transportation Efficiency" on the twen-

KEY TERMS

Battery—A battery is a container, or group of containers, holding electrodes and an electrolyte for producing electric current by chemical reaction and storing energy. The individual containers are called "cells." Batteries produce direct current (DC).

Cell—Basic unit used to store energy in a battery. A cell consists of an anode, cathode and the electrolyte.

Controller—Device managing electricity flow from batteries to motor(s), from "on-off" function to vehicle throttle control.

Direct current (DC)—Electrical current that always flows in the same direction.

Electrolyte—The medium of ion transfer between anode and cathode within the cell. Usually liquid or paste that is either acidic or basic.

Flywheels—Rapidly spinning wheel-like rotors or disks that store kinetic energy.

Hybrid vehicle—Vehicles having two or more sources of energy. There are two types of hybrid electric vehicles (HEVs), series and parallel. In a series hybrid, all of the vehicle power is provided from one source. For example, with an IC/electric series hybrid, the electric motor drives the vehicle from the battery pack and the internal combustion

engine powers a generator that charges the battery. In a parallel hybrid, power is delivered through both paths. In an IC/electric parallel hybrid, both the electric motor and the internal combustion engine power the vehicle.

Motor—Electromechanical device that provides power (expressed as horsepower and torque) to driveline and wheels of vehicle.

Regenerative braking—A means of recharging batteries using energy created by braking the vehicle. With normal friction brakes, a certain amount of energy is lost in the form of heat created by friction from braking. With regenerative braking, the motors act as generators. They reduce the energy lost by feeding it back into the batteries resulting in improved range.

Ultracapacitors—These are higher specific energy and power versions of electrolytic capacitors—devices that store energy as an electrostatic charge. They are electrochemical systems that store energy in a polarized liquid layer at the interface between an ionically conducting electrolyte and a conducting electrode.

Watt—The basic unit of electrical power equal to 1 joule per second.

ty fifth anniversary of **Earth** Day, April 21, 2000. The executive order sought to ensure that the federal government exercises leadership in the reduction of **petroleum** consumption through improvements in fleet fuel efficiency and the use of AFVs and alternative fuels. This includes procurement of innovative vehicles capable of large improvements in fuel economy such as **hybrid** electric vehicles. In 2003, President George W. Bush called for a federal institute to advance fuel cell development and use.

Hybrids

While pure electric vehicles are some time in the future, the world may be ready for a hybrid electric vehicle (HEV)—a vehicle that combines small internal-combustion engines with electric motors and electricity storage devices. These have the potential to reduce emissions almost as much as battery-powered electric vehicles as well as offering the extended range and rapid refueling that consumers expect from a conventional vehicle. Hybrid power systems were conceived as a way to compensate for the shortfall in battery technology. Because bat-

teries could supply only enough energy for short trips, an onboard generator, powered by an **internal combustion engine**, could be installed and used for longer trips.

The HEV is able to operate approximately twice as efficiently as conventional vehicles. Honda's Insight, the first hybrid car to be sold in the United States, is expected to go 700 mi (1,127 km) on a single tank of gas. The Toyota Prius is expected to go about 450 mi (724 km). For the driver, hybrids offer similar or better performance than conventional vehicles. More important, because such performance is available now, hybrids are a practical, technically achievable, alternative approach.

Essentially, a hybrid combines an energy storage system, a power unit, and a vehicle propulsion system. The primary options for energy storage include batteries, ultracapacitors, and flywheels.

Challenges still exist

Engineers still struggle with energy **density**. Average energy density in today's EV batteries is about 70W-hr/kg (one W-hr/kg is roughly one mile of range in a four-passenger sedan). In order to increase driving ranges, battery makers must find new alloys for cathodes and anodes. Merely placing more batteries per vehicle is not sufficient.

Cost remains a challenge, particularly battery costs. Automakers say they need to offer their customers $100-per-kilowatt-hour batteries. Today, the best long-term EV batteries cost $10,000–20,000 per kilowatt-hour.

Resources

Books

Question 349. *Science and Technology Desk Reference.* 2nd ed. Carnegie Library of Pittsburgh Science and Technology Department, 1996.

Periodicals

"The Case for Electric Vehicles." *Scientific American* (November 1996).
"Out of Juice." *Design News* (October 5, 1998).
"Waiting for the Supercar." *Scientific American* (April 1999).

Organizations

Electric Vehicle Association of the Americas. <http://www.evaa.org>.

Other

Office of Transportation Technologies. *Hybrid Electric Vehicle Program.* (2003). <http://www.newscientist.com/ns/970705/ndope.html>.

Laurie Toupin

Electrical conductivity

Conductivity is the term used to describe the ability of a material medium to permit the passage of particles or **energy**. Electrical conductivity refers to the movement of charged particles through **matter**. Thermal conductivity refers to the transmission of **heat** energy through matter. Together, these are the most significant examples of a broader classification of phenomena known as transport processes. In metals, electrical conductivity and thermal conductivity are related since both involve aspects of **electron motion**.

History

The early studies of electrical conduction in metals were done in the eighteenth and early nineteenth centuries. Benjamin Franklin (1706-1790) in his experiments with **lightning** (leading to his invention of the lightning rod), reasoned that the charge would travel along the metallic rod. Alessandro Volta (1745-1827) de-

rived the concept of electrical potential from his studies of static **electricity**, and then discovered the principle of the **battery** in his experiments with dissimilar metals in common contact with moisture. Once batteries were available for contact with metals, electric currents were produced and studied. Georg Simon Ohm (1787-1854) found the direct proportion relating current and potential difference, which became a measure of the ability of various metals to conduct electricity. Extensive theoretical studies of currents were carried out by André Marie Ampère (1775-1836).

To honor these scientists, the système internationale (SI) units use their names. The unit of potential difference is the volt, and potential difference is more commonly called voltage. The unit of **electrical resistance** is the ohm, and the unit of current is the ampere. The **relation** among these functions is known as **Ohm's law**.

Franklin is remembered for an unlucky mistake. He postulated that there was only one type of electricity, not two as others thought, in the phenomena known in his day. He arbitrarily called one form of static **electric charge** positive and attributed the opposite charge to the absence of the positive. All subsequent studies continued the convention he established. Late in the nineteenth century, when advancements in both electrical and **vacuum** technology led to the discovery of **cathode** rays, streams of particles issuing from a **negative** electrode in an evacuated tube, Sir Joseph John Thomson (1856-1940) identified these particles as common to all metals used as cathodes and negatively charged. The historical concept of a positive current issuing from an **anode** is mathematically self-consistent and leads to no analytical errors, so the convention is maintained but understood to be a convenience.

Materials

Electrical conduction can take place in a variety of substances. The most familiar conducting substances are metals, in which the outermost electrons of the **atoms** can move easily in the interatomic spaces. Other conducting materials include semiconductors, electrolytes, and ionized gases, which are discussed later in this article.

Metals

Metals are now known to be primarily elements characterized by atoms in which the outermost orbital shell has very few electrons with corresponding values of energy. The highest conductivity occurs in metals with only one electron occupying a state in that shell. Silver, **copper**, and gold are examples of high-conductivity metals. Metals are found mainly toward the left side of the **periodic table** of the elements, and in the transition columns. The electrons contributing to their conductivity are also the electrons that determine their chemical **valence** in forming compounds. Some metallic conductors are alloys of two or more **metal** elements, such as **steel**, brass, bronze, and pewter.

A piece of metal is a block of metallic atoms. In individual atoms the valence electrons are loosely bound to their nuclei. In the block, at room **temperature**, these electrons have enough kinetic energy to enable them to wander away from their original locations. However, that energy is not sufficient to remove them from the block entirely because of the potential energy of the surface, the outermost layer of atoms. Thus, at their sites, the atoms are ionized—that is, left with a net positive charge—and are referred to as ion cores. Overall, the metal is electrically neutral, since the electrons' and ion cores' charges are equal and opposite. The conduction electrons are bound to the block as a whole rather than to the nuclei.

These electrons move about as a cloud through the spaces separating the ion cores. Their motion is **random**, bearing some similarities to gas molecules, especially scattering, but the nature of the scattering is different. Electrons do not obey classical gas laws; their motion in detail must be analyzed quantum-mechanically. However, much information about conductivity can be understood classically.

A particular specimen of a metal may have a convenient regular shape such as a cylinder (wire) or a **prism** (bar). When a battery is connected across the ends of a wire, the electrochemical energy of the battery imparts a potential difference, or voltage between the ends. This electrical potential difference is analogous to a hill in a gravitational system. Charged particles will then move in a direction analogous to downhill. In the metal, the available electrons will move toward the positive terminal, or anode, of the battery. As they reach the anode, the battery injects electrons into the wire in equal numbers, thereby keeping the wire electrically neutral. This circulation of charged particles is termed a current, and the closed path is termed a circuit. The battery acts as the electrical analog of a pump. Departing from the gravitational analogy, in which objects may fall and land, the transport of charged particles requires a closed circuit.

Current is defined in terms of charge transport:

$$I = q/t$$

where I is current, q is charge, and t is **time**. Thus q/t is the **rate** of charge transport through the wire. In a metal, as long as its temperature remains constant, the current is directly proportional to the voltage. This direct proportion in mathematical terms is referred to as linear, because it can be described in a simple linear algebraic equation:

$$I = GV$$

In this equation, V is voltage and G is a constant of proportionality known as conductance, which is independent of V and remains constant at constant temperature. This equation is one form of Ohm's law, a principle applicable only to materials in which electrical conduction is linear. In turn, such materials are referred to as ohmics.

The more familiar form of Ohm's law is:

$$I = V/R$$

where R is 1/G and is termed resistance.

Conceptually, the idea of resistance to the passage of current preceded the idea of charge transport in historical development.

The comparison of electrical potential difference to a hill in gravitational systems leads to the idea of a gradient, or slope. The rate at which the voltage varies along the length of the wire, measured relative to either end, is called the electric field:

$$E = -(V/L)$$

The field E is directly proportional to V and inversely proportional to L in a linear or ohmic conductor. This field is the same as the electrostatic field defined in the article on electrostatics. The minus sign is associated with the need for a negative gradient to represent "downhill." The electric field in this description is conceptually analogous to the gravitational field near the earth's surface.

Experimental measurements of current and voltage in metallic wires of different dimensions, with temperature constant, show that resistance increases in direct proportion to length and inverse proportion to cross-sectional area. These variations allow the metal itself to be considered apart from specimen dimensions. Using a proportionality constant for the material property yields the relation:

$$R = \rho\,(L/A)$$

where ρ is called the resistivity of the metal. Inverting this equation places conduction rather than resistance uppermost:

$$G = \sigma\,(A/L)$$

where σ is the conductivity, the **reciprocal** ($1/\rho$) of the resistivity.

This analysis may be extended by substitution of equivalent expressions:

$$G = I/V$$
$$\sigma(A/L) = I/EL$$
$$\sigma = I/AE$$

Introducing the concept of current **density**, or current flowing per unit cross-sectional area:

$$J = I/A$$

yields an expression free of all the external measurements required for its actual calculation:

$$\sigma = J/E$$

This equation is called the field form of Ohm's law, and is the first of two physical definitions of conductivity, rather than mathematical.

The nature of conductivity in metals may be studied in greater depth by considering the electrons within the bulk metal. This approach is termed microscopic, in contrast to the macroscopic properties of a metal specimen. Under the influence of an internal electric field in the material, the **electron cloud** will undergo a net drift toward the battery anode. This drift is very slow in comparison with the random thermal motions of the individual electrons. The cloud may be characterized by the **concentration** of electrons, defined as total number per unit **volume**:

$$n = N/U$$

where n is the concentration, N the total number, and U the volume of metal (U is used here for volume instead of V, which as an algebraic symbol is reserved for voltage). The total drifting charge is then:

$$q = Ne = nUe$$

where e is the charge of each electron.

N is too large to enumerate; however, if as a first **approximation** each atom is regarded as contributing one valence electron to the cloud, the number of atoms can be estimated from the volume of a specimen, the density of the metal, and the atomic **mass**. The value of n calculated this way is not quite accurate even for a univalent metal, but agrees in order of magnitude. (The corrections are quantum-mechanical in nature; metals of higher valence and alloys require more complicated quantum-based corrections.) The average drift **velocity** of the cloud is the **ratio** of wire length to the average time required for an electron to traverse that length. Algebraic substitutions similar to those previously shown will show that the current density is proportional to the drift velocity:

$$J = nev_d$$

The drift velocity is superimposed on the thermal motion of the electrons. That combination of motions, in which the electrons bounce their way through the metal, leads to the microscopic description of electrical resistance, which incorporates the idea of a limit to forward motion. The limit is expressed in the term mobility:

$$\mu = v_d /E$$

so that mobility, the ratio of drift velocity to electric field, is finite and characteristic of the particular metal.

Combining these last two equations produces the second physical definition of conductivity:

$$\sigma = J/E = nev_d /E = ne\mu$$

The motion of electrons among vibrating ion cores may be analyzed by means of Newton's second law, which states that a net **force** exerted on a mass produces an **acceleration**:

$$F = ma$$

Acceleration in turn produces an increasing velocity. If there were no opposition to the motion of an electron in the **space** between the ion cores, the connection of a battery across the ends of a wire would produce a current increasing with time, in proportion to such an increasing velocity. Experiment shows that the current is steady, so that there is no net acceleration.

Yet the battery produces an electric field in the wire, which in turn produces an electric force on each electron:

$$F = eE$$

Thus, there must be an equal and opposite force associated with the behavior of the ion cores. The analogy here is the action of air molecules against an object falling in the atmosphere, such as a raindrop. This fluid **friction** generates a force proportional to the velocity, which reaches a terminal value when the frictional force becomes equal to the weight. This steady state, for which the net force is **zero**, corresponds to the drift velocity of electrons in a conductor. Just as the raindrop quickly reaches a steady speed of fall, electrons in a metal far more quickly reach a steady drift velocity manifested in a constant current.

Thus far, this discussion has required that temperature be held constant. For metals, experimental measurements show that conductivity decreases as temperature increases. Examination suggests that, for a metal with n and e fixed, it is a decrease in mobility that accounts for that decrease in conductivity. For moderate increases in temperature, the experimental variation is found to fit a linear relation:

$$\rho = \rho_0[1 + \alpha(T-T_0)]$$

Here the subscript "0" refers to initial values and a is called the temperature **coefficient** of resistivity. This coefficient is found to vary over large temperature changes.

To study the relationship between temperature and electron mobility in a metal, the behavior of the ion cores must be considered. The ion cores are arranged in a three-dimensional **crystal** lattice. In most common metals the structure is cubic, and the transport functions are not strongly dependent on direction. The metal may then be treated as isotropic, that is, independent of direction, and all the foregoing equations apply as written. For anisotropic materials, the orientational dependence of transport in the crystals leads to families of equations with sets of directional coefficients replacing the simple constants used here.

Temperature is associated with the vibrational kinetic energy of the ion cores in motion about their equilibrium positions. They may be likened to masses interconnected by springs in three dimensions, with their bonds acting as the springs. Electrons attempting to move among them will be randomly deflected, or scattered, by these lattice vibrations, which are quantized. The vibrational quanta are termed phonons, in an analogy to photons. Advanced conductivity theory is based on analyses of the scattering of electrons by phonons.

With the increase in vibrational energy as temperature is increased, the scattering is increased so that the drift motion is subjected to more disruption. Maintenance of a given current would thus require a higher field at a higher temperature.

If the ion cores of a specific metal were identical and stationary in their exact equilibrium lattice sites, the electron cloud could drift among them without opposition, that is, without resistance. Thus, three factors in resistance can be identified: (a) lattice vibrations, (b) ion core displacement from lattice sites, and (c) chemical impurities, which are wrong ion cores. The factors (a) and (b) are temperature-dependent, and foreign atoms contribute their thermal motions as well as their wrongness. Additionally, sites where ions are missing, or vacancies, also are wrong and contribute to scattering. Displacements, vacancies, and impurities are classed as lattice defects.

A direct extension of thermal behavior downward toward the **absolute zero** of temperature suggests that resistance should fall to zero monotonically. This does not occur because lattice defects remain wrong and vibrational energy does not drop to zero-quantum mechanics accounts for the residual zero-point energy. However, in many metals and many other substances at temperatures approaching zero, a wholly new phenomenon is observed, the sudden drop of resistivity to zero. This is termed superconductivity.

Semiconductors

Semiconductors are materials in which the conductivity is much lower than for metals, and widely variable through control of their composition. These substances are now known to be poor insulators rather than poor conductors, in terms of their atomic structure. Though some semiconducting substances had been identified and studied by the latter half of the nineteenth century, their properties could not be explained on the basis of classical **physics**. It was not until the mid-twentieth century, when modern quantum-mechanical principles were applied to the analysis of both metals and semiconductors,

that theoretical calculations of conductivity values agreed with the results of experimental measurements.

In a good insulator, electrons cannot move because nearly all allowed orbital states are occupied. Energy must then be supplied to remove an electron from an outermost bound position to a higher allowed state. This leaves a vacancy into which another bound electron can hop under the influence of an electric field. Thus, both the energized electron and its vacancy become mobile. The vacancy acts like a positive charge, called a hole, and drifts in the direction opposite to electrons. Electrons and holes are more generally termed charge carriers.

In good insulators the activation energy of charge carriers is high, and their availability requires a correspondingly high temperature. In poor insulators, that is, semiconductors, activation occurs at temperatures moderately above 80.6°F (27°C). Each substance has a characteristic value.

There are many more compounds than elements that can be classed as semiconductors. The elements are a few of those in column IV of the periodic table, which have covalent bonds: **carbon** (C), germanium (Ge), and silicon (Si). For carbon, only the graphite form is semiconducting; **diamond** is an excellent insulator. The next element down in this column, tin (Sn), undergoes a transition from semiconductor to metal at 59°F (15°C), below room temperature, indicative of an unusefully low activation energy. Other elements that exhibit semiconductor behavior are found in the lower portion of column VI, specifically selenium (Se) and tellurium (Te).

There are two principal groups of compounds with semiconducting properties, named for the periodic table columns of their constituents: III-V, including gallium arsenide (GaAs) and indium antimonide (InSb), among others; and II-VI, including zinc sulfide (ZnS), selenides, tellurides, and some oxides. In many respects these compounds mimic the behavior of column IV elements. Their chemical bonds are mixed covalent and ionic. There are also some organic semiconducting compounds, but their analysis is beyond the scope of this article.

A semiconductor is called intrinsic if its conductivity is the result of equal contributions from its own electrons and holes. The equation must then be expanded:

$$\sigma = n_e e \, \mu_e + n_h e \, \mu_h$$

In an intrinsic semiconductor, $n_e = n_h$, and e has the same numerical value for an electron (-) and the hole left behind (+). The mobilities are usually different. These terms add because the opposite charges move in opposite directions, resulting in a pair of like signs in each product.

For application in devices, semiconductors are rarely used in their pure or intrinsic composition. Under carefully controlled conditions, impurities are introduced which contribute either an excess or a deficit of electrons. Excess electrons neutralize holes so that only electrons are available for conduction. The resulting material is called n-type, n for negative carrier. An example of n-type material is Si with Sb, a column IV element with a column V impurity known as a donor. In n-type material, donor atoms remain fixed and positively ionized. When a column III impurity is infused into a column IV element, electrons are bound and holes made available. That material is called p-type, p for positive carrier. Column III impurities are known as acceptors; in the material acceptor atoms remain fixed and negatively ionized. An example of p-type material is Si with Ga. Both n-type and p-type semiconductors are referred to as extrinsic.

Thermal kinetic energy is not the only mechanism for the release of charge carriers in semiconductors. Photons with energy equal to the activation energy can be absorbed by a bound electron which, in an intrinsic semiconductor, adds both itself and a hole as mobile carriers. These photons may be in the visible range or in the near infrared, depending on E_G. In extrinsic semiconductors, photons of much lower energies can contribute to the pool of the prevailing carrier type, provided the material is cooled to cryogenic temperatures in order to reduce the population of thermally activated carriers. This behavior is known as photoconductivity.

Each separate variety of semiconductor is ohmic, with the conductivity constant at constant temperature. However, as the temperature is increased, the conductivity increases very rapidly. The concentration of available carriers varies in accordance with an exponential function:

$$n \propto \exp[-(E_G/kT)]$$

where E_G is the gap or activation energy, k is Boltzmann's constant (1.38×10^{23} joules/kelvin), T is absolute (kelvin) temperature, and the product kT is the thermal energy corresponding to temperature T. The increase in available charge carriers overrides any decrease in mobility, and this leads to a negative value for a. Indeed, a decrease in resistance with increasing temperature is a reliable indication that a substance is a semiconductor, not a metal. Graphite is an example of a conductor that appears metallic in many ways except for a negative ALPHA. The converse, a positive ALPHA, is not as distinct a test for metallic conductivity.

The Fermi level, E_F, can be shown differently for intrinsic, n-type, and p-type semiconductors. However, for materials physically connected, E_F must be the same for thermal equilibrium. This is a consequence of the laws of **thermodynamics** and energy conservation. Thus, the behavior of various junctions, in which the interior energy levels shift to accommodate the alignment of the Fermi level, is extremely important for the semiconductor devices.

Non-ohmic conductors

Non-ohmic conduction is marked by nonlinear graphs of current vs. voltage. It occurs in semiconductor junctions, electrolytic solutions, some ionic solids not in **solution**, ionized gases, and vacuum tubes. Respective examples include semiconductor p-n diodes, battery acid or alkaline solutions, alkali halide crystals, the ionized mercury vapor in a fluorescent lamp, and cathode ray tubes.

Ionic conductivities are much lower than electronic, because the masses and diameters of ions make them much less mobile. While ions can drift slowly in a gas or liquid, their motions through the interstices of a solid lattice are much more restricted. Yet, with their thermal kinetic energy, ions will diffuse through a lattice, and in the presence of an electric field, will wander toward the appropriate electrode. In most instances, both ionic and electronic conduction will occur, depending on impurities. Thus, for studies of ionic conductivity, the material must be a very pure solid.

In gases, the gas atoms must be ionized by an electric field sufficient to supply the ionization energy of the gas in the tube. For stable currents, the ratio of field to gas **pressure**, E/P, is a major parameter. Electrons falling back into bound states produce the characteristic **spectrum** of the gas, qualitatively associated with **color**, e.g., red for neon, yellow-orange for **sodium** vapor, or blue-white for mercury vapor.

The basic definition of a **plasma** in physics includes all material conductors, ohmic and non-ohmic. A plasma is a medium in which approximately equal numbers of opposite charges are present, so that the medium is neutral or nearly so. In a metal the negative electrons are separated from an equal number of positive ion cores. In a semiconductor there may be holes and electrons (intrinsic), holes and ionized acceptors (p-type), or electrons and ionized donors (n-type). In an electrolytic solution and in an ionic solid there are positive and negative ions. An ionized gas contains electrons and positive ions. A small distinction among these may be made as to whether the medium has one or two mobile carriers.

In contemporary usage, the term plasma usually refers to extremely hot gases such as those used in the Tokamak for **nuclear fusion** experiments. High-energy plasmas are discussed in the article on fusion as a means of generating electric power.

The remaining non-ohmic conduction category is the **vacuum tube**, in which a beam of electrons is emitted from either a heated cathode (thermionic) or a suitably illuminated cathode (photoelectric), and moves through evacuated space to an anode. The beam in its passage is subjected to electrostatic or magnetic fields for control. The evacuated space cannot be classed either as a material with a definable conductivity or as a plasma, since only electrons are present. However, there are relations of current and voltage to be analyzed. These graphs are generally nonlinear or linear over a limited range. But vacuum tubes are not called ohmic even in their linear ranges because there is no material undergoing the lattice behavior previously described as the basis for ohmic resistance.

Electrical conduction in the human body and other **animal** organisms is primarily ionic, since body fluids contain vital electrolytes subject to electrochemical action in organs. Further information is available in other articles, particularly those on the **heart**, the **brain**, and neurons.

See also Chemical bond; Electrolyte; Nonmetal.

Resources

Books

Halliday, David, Robert Resnick, and Kenneth Krane. *Physics.* 4th ed. New York: John Wiley and Sons, 1992.
Serway, Raymond A. *Physics for Scientists and Engineers.* 3rd ed. Philadelphia: W. B. Saunders Co.

Frieda A. Stahl

Electrical power supply

An electrical power supply is a device that provides the **energy** needed by electrical or electronic equipment to perform their functions. Often, that energy originates from a source with inappropriate electrical characteristics, and a power supply is needed to change the power to meet the equipment's requirements. Power supplies usually change alternating current into direct current, raise or lower the voltage as required, and deliver the electrical energy with a more constant voltage than the original source provides. Power supplies often provide protection against power source failures that might damage the equipment. They may also provide isolation from the electrical noise that is usually found on commercial power lines.

An electrical power supply can be a simple **battery** or may be more sophisticated than the equipment it supports. An appropriate power supply is an essential part of every working collection of electrical or electronic circuits.

The requirement for power supplies

Batteries could be used to supply the power for almost all electronic equipment if it were not for the high cost of the energy they provide compared to commercial power lines. Power supplies were once called battery eliminators, an apt name because they made it possible to use less expensive energy from a commercial power line where it is available. Batteries are still an appropriate and economical choice for portable equipment having modest energy requirements.

Batteries as power supplies

Two basic types of chemical cells are used in batteries that supply power to electronic equipment. Primary cells are normally not rechargeable. They are intended to be discarded after their energy reserve is depleted. Secondary cells, on the other hand, are rechargeable. The lead-acid secondary cell used in an automobile's battery can be recharged many times before it fails. Nickel-cadmium batteries are based on secondary cells.

Plug-in power supplies

The electrical energy supply for homes and businesses provided through the commercial power lines is delivered by an alternating current (AC). Electronic equipment, however, almost always requires direct-current power (DC). Power supplies usually change AC to DC by a process called rectification. Semiconductor diodes that pass current in only one direction are used to block the power line's current when its polarity reverses. Capacitors store energy for use when the diodes are not conducting, providing relatively constant voltage direct current as needed.

Power supply voltage regulation

Poor power line voltage regulation causes lights in a home to dim each time the refrigerator starts. Similarly, if a change in the current from a power supply causes the voltage to vary, the power supply has poor voltage regulation. Most electronic equipment will perform best when it is supplied from a nearly constant voltage source. An uncertain supply voltage can result in poor circuit performance.

Analysis of a typical power supply's performance is simplified by modeling it as a constant-voltage source in series with an internal resistance. The internal resistance is used to explain changes in the terminal voltage when the current in a circuit varies. The lower the internal resistance of a given power supply, the more current it can supply while maintaining a nearly-constant terminal voltage. An ideal supply for circuits requiring an unvarying voltage with changing load current, would have an internal resistance near **zero**. A power supply with a very-low internal resistance is sometimes called a "stiff" power supply.

An inadequate power source almost always compromises the performance of electronic equipment. Audio amplifiers, for example, may produce distorted sound if the supply voltage drops with each loud pulse of sound. There was a time when the pictures on **television** sets would shrink if the AC-line voltage fell below a minimum value. These problems are less significant now that voltage regulation has been included in most power supplies.

There are two approaches that may be used to improve the voltage regulation of a power supply. A simple power supply that is much larger than required by the average equipment demand will help. A larger power supply should have a lower effective internal resistance, although this is not an absolute rule. With a lower internal resistance, changes in the current supplied are less significant and the voltage regulation is improved compared to a power supply operated near its maximum capacity.

Some power supply applications require a higher internal resistance. High-power **radar** transmitters require a power source with a high internal resistance so that the output can be shorted each time the radar transmits a signal pulse without damaging the circuitry. Television receivers artificially increase the resistance of the very high voltage power supply for the picture tube by adding resistance deliberately. This limits the current that will be delivered should a technician inadvertently contact the high voltage which might otherwise deliver a fatal electrical shock.

Voltage-regulation circuits

Voltage-regulated power supplies feature circuitry that monitors their output voltage. If this voltage changes because of external current changes or because of shifts in the power line voltage, the regulator circuitry makes an almost instantaneous compensating adjustment.

Two common approaches are used in the design of voltage-regulated power supplies. In the less-common scheme, a shunt regulator connects in **parallel** with the power supply's output terminals and maintains a constant voltage by wasting current the external circuit, called the load does not require. The current delivered by the unregulated part of the power supply is always constant. The shunt regulator diverts almost no current when the external load demands a heavy current. If the external load is reduced, the shunt regulator current increases. The disadvantage of shunt regulation is that it dissipates the full power the supply is designed to deliver, whether or not the external circuit requires energy.

The more-common series voltage regulator design depends upon the variable resistance created by a **transistor** in series with the external circuit current. The transistor's voltage drop adjusts automatically to maintain a constant output voltage. The power supply's output voltage is sampled continuously, compared with an accurate reference, and the transistor's characteristics are adjusted automatically to maintain a constant output.

A power supply with adequate voltage regulation will often improve the performance of the electronic device it powers, so much so that voltage regulation is a very common feature of all but the simplest designs. Packaged integrated circuits are commonly used, simple three-terminal devices that contain the series transistor and most of the regulator's supporting circuitry. These "off the shelf" chips have made it very easy to include voltage regulation capability in a power supply.

Power supplies and load interaction

When a single power supply serves several independent external circuits, changes in current demand imposed by one circuit may cause voltage changes that affect the operation of the other circuits. These interactions constitute unwanted signal coupling through the common power source, producing instability. Voltage-regulators can prevent this problem by reducing the internal resistance of the common power source.

Ripple reduction

When an alternating current is converted to direct current, small voltage variations at the supply **frequency** are difficult to smooth out, or filter, completely. In the case of power supplies operated from the 60-Hz power line, the result is a low-frequency variation in the power supply's output called ripple voltage. Ripple voltage on the power supply output will add with the signals processed by electronic circuitry, particularly in circuits where the signal voltage is low. Ripple can be minimized by using more elaborate filter circuitry but it can be reduced more effectively with active voltage regulation. A voltage regulator can respond fast enough to **cancel** unwanted changes in the voltage.

Minimizing the effects of line-voltage changes

Power-line voltages normally fluctuate randomly for a variety of reasons. A special voltage-regulating **transformer** can improve the voltage stability of the primary power. This transformer's action is based on a coil winding that includes a **capacitor** which tunes the transformer's inductance into **resonance** at the power line frequency. When the line voltage is too high, the circulating current in the transformer's resonant winding tends to saturate the magnetic core of the transformer, reducing its efficiency and causing the voltage to fall. When the line voltage is too low, as on a hot summer day when air conditioners are taxing the capabilities of the generators and power lines, the circulating current is reduced, raising the efficiency of the transformer. The voltage regulation achieved by these transformers can be helpful even though it is not perfect. An early TV brand included resonant transformers to prevent picture-size variations that accompanied normal line-voltage shifts.

Resonant power transformers waste energy, a serious drawback, and they do not work well unless heavily loaded. A regulating transformer will dissipate nearly its full rated power even without a load. They also tend to distort the alternating-current waveform, adding **harmonics** to their output, which may present a problem when powering sensitive equipment.

Laboratory power supplies

Voltage-regulated power supplies are necessary equipment in scientific and technical laboratories. They provide an adjustable, regulated source of electrical power to test circuits under development.

Laboratory power supplies usually feature two programmable modes, a constant-voltage output over a selected range of load current and a constant-current output over a wide range of voltage. The crossover point where the action switches from constant voltage to constant current action is selected by the user. As an example, it may be desirable to limit the current to a test circuit to avoid damage if a hidden circuit **fault** occurs. If the circuit demands less than a selected value of current, the regulating circuitry will hold the output voltage at the selected value. If, however, the circuit demands more than the selected maximum current, the regulator circuit will decrease the terminal voltage to whatever value will maintain the selected maximum current through the load.

The powered circuit will never be allowed to carry more than the selected constant-current limit.

Simple transformer power supplies

Alternating current is required for most power lines because AC makes it possible to change the voltage to current **ratio** with transformers. Transformers are used in power supplies when it is necessary to increase or decrease voltage. The AC output of these transformers usually must be rectified into direct current. The resulting pulsating direct current is filtered to create nearly-pure direct current.

Switching power supplies

A relatively new development in power-supply technology, the switching power supply, is becoming popular. Switching power supplies are lightweight and very efficient. Almost all personal computers are powered by switching power supplies.

The switching power supply gets its name from the use of transistor switches, which rapidly toggle in and out of conduction. Current travels first in one direction then in the other as it passes through the transformer. Pulsations from the rectified switching signal are much higher frequencies than the power line frequency, therefore the ripple content can be minimized easily with small filter capacitors. Voltage regulation can be accomplished by varying the switching frequency. Changes in the switching frequency alter the efficiency of the power supply transformer enough to stabilize the output voltage.

Switching power supplies are usually not damaged by sudden short circuits. The switching action stops almost immediately, protecting the supply and the circuit load. A switching power supply is said to have stalled when excessive current interrupts its action.

Switching power supplies are light in weight because the components are more efficient at higher frequencies. Transformers need much less **iron** in their cores at higher frequencies.

Switching power supplies have negligible ripple content at audible frequencies. Variations in the output of the switching power supply are inaudible compared to the hum that is common with power supplies that operate at the 60-Hz AC-power line frequency.

The importance of power supplies

Electrical power supplies are not the most glamorous part of contemporary technology, but without them many electronic products that we take for granted would not be possible.

See also Electricity; Electronics.

Resources

Books

Cannon, Don L. *Understanding Solid-State Electronics.* 5th ed. Upper Saddle River, NJ: SAMS division of Prentice Hall Publishing Company, 1991.

Giancoli, Douglas C. *Physics: Principles With Applications.* 3rd ed. Upper Saddle River, NJ: Prentice Hall, 1991.

Donald Beaty

Electrical resistance

The electrical resistance of a wire or circuit is a way of measuring the resistance to the flow of an electrical current. A good electrical conductor, such as a **copper** wire, will have a very low resistance. Good insulators, such as rubber or **glass** insulators, have a very high resistance. The resistance is measured in ohms, and is related to the current in the circuit and voltage across the circuit by **Ohm's law**. For a given voltage, a wire with a lower resistance will have a higher current.

The resistance of a given piece of wire depends of three factors: the length of the wire, the cross-sectional area of the wire, and the resistivity of the material composing the wire. To understand how this works, think of **water** flowing through a hose. The amount of water flowing through the hose is analogous to the current in the wire. Just as more water can pass through a **fat** fire hose than a skinny garden hose, a fat wire can carry more current than a skinny wire. For a wire, the larger the cross-sectional area, the lower the resistance; the smaller the cross-sectional area, the higher the resistance. Now consider the length. It is harder for water to flow through a very long hose simply because it has to travel farther. Analogously, it is harder for current to travel through a longer wire. A longer wire will have a greater resistance. The resistivity is a property of the material in the wire that depends on the chemical composition of the material but not on the amount of material or the shape (length, cross-sectional area) of the material. Copper has a low resistivity, but the resistance of a given copper wire depends on the length and area of that wire. Replacing a copper wire with a wire of the same length and area but a higher resistivity will produce a higher resistance. In the hose analogy, it is like filling the hose with **sand**. Less water will flow through the hose filled with sand than through an identical unobstructed hose. The sand in effect has a higher resistivity to water flow. The total resistance of a wire is then the resistivity of the material composing the wire times the length of the wire, divided by the cross-sectional area of the wire.

Electrical resistivity survey methods *see* **Subsurface detection**

Electricity

Electricity is a natural phenomenon resulting from one of the most basic properties of **matter**, electrical charge. Our understanding of electrical principles has developed from a long history of experimentation. Electrical technology, essential to modern society for **energy** transmission and information processing, is the result of our knowledge about electrical charge at rest and electrical charge in **motion**.

Electrical charge

Electrical charge is a fundamental property possessed by a few types of particles that make up **atoms**. Electrical charge is found with either positive or **negative** polarity. Positive charge exactly neutralizes an equal quantity of negative charge. Charges with the same sign repel while unlike charges attract. The unit of electrical charge is the **coulomb**, named for Charles Coulomb, an early authority on electrical theory.

The most obvious sources of **electric charge** are the negatively-charged electrons from the outer parts of atoms and the positively-charged protons found in atomic nuclei. Electrical neutrality is the most probable condition of matter because most objects contain nearly equal numbers of electrons and protons. Physical activities that upset this balance will leave an object with a net electrical charge, often with important consequences.

Excess static electric charges can accumulate as a result of mechanical **friction**, as when someone walks across a carpet. Friction transfers charge between shoe soles and carpet, resulting in the familiar electrical shock when the excess charge sparks to a nearby person.

Many semiconductor devices used in **electronics** are so sensitive to static electricity that they can be destroyed if touched by a technician carrying a small excess of electric charge. Computer technicians often wear a grounded wrist strap to drain away an electrical charge that might otherwise destroy sensitive circuits they **touch**.

Electric fields

Charged particles alter their surrounding space to produce an effect called an electric field. An electric field is the concept we use to describe how one electric charge exerts **force** on another distant electric charge. Whether electric charges are at rest or moving, they are acted upon by a force whenever they are within an electric field. The **ratio** of this force to the amount of charge is the measure of the field's strength.

An electric field has vector properties in that it has both a unique magnitude and direction at every point in space. An electric field is the collection of all these values. When neighboring electrical charges push or pull each other, each interacts with the electric field produced by the other charge.

Electric fields are imagined as lines of force that begin on positive charges and end on negative charges. Unlike magnetic field lines, which form continuous loops, electric field lines have a beginning and an ending. This makes it possible to block the effects of an electric field. An electrically conducting surface surrounding a **volume** will stop an external electric field. Passengers in an **automobile** may be protected from **lightning** strikes because of this shielding effect. An electric shield enclosure is called a Faraday cage, named for Michael Faraday.

Electricity arcing over the surface of ceramic insulators.
Photograph by Robert Essel/Bikderberg. Stock Market. Reproduced with permissions.

Coulomb's law and the forces between electrical charges

Force, quantity of charge, and **distance** of separation are related by a rule called Coulomb's law. This law states that the force between electrical charges is proportional to the product of their charges and inversely proportional to the square of their separation.

The coulomb force binds atoms together to form chemical compounds. It is this same force that accelerates electrons in a TV picture tube, giving energy to the beam of electrons that creates the **television** picture. It is the electric force that causes charge to flow through wires.

The electric force binds electrons to the nuclei of atoms. In some kinds of materials, electrons stick tightly to their respective atoms. These materials are electrical insulators that cannot carry a significant current unless acted upon by an extremely strong electrical field. Insulators are almost always nonmetals. Metals are relatively good conductors of electricity because their outermost electrons are easily removed by an electric field. Some metals are better conductors than others, silver being the best.

Current

The basic unit of **electric current** is the ampere, named for the French physicist Andre Marie Ampere. One ampere equals 1 coulomb of charge drifting past a reference point each second.

Voltage

Voltage is the ratio of energy stored by a given a quantity of charge. Work must be performed to crowd same-polarity electric charges against their mutual repulsion. This work is stored as electrical potential energy, proportional to voltage. Voltage may also be thought of as electrical **pressure**.

The unit of voltage is the volt, named for Alessandro Volta. One volt equals one joule for every coulomb of electrical charge accumulated.

Resistance

Ordinary conductors oppose the flow of charge with an effect that resembles friction. This dissipative action is called resistance. Just as mechanical friction wastes energy as **heat**, current through resistance dissipates energy as heat. The unit of resistance is the ohm, named for Georg Simon Ohm. If 1 volt causes a current of 1 ampere, the circuit has 1 unit of resistance. It is useful to know that resistance is the ratio of voltage to current.

Mechanical friction can be desirable, as in automobile brakes, or undesirable, when friction creates unwanted energy loss. Resistance is always a factor in current electricity unless the circuit action involves an extraordinary low-temperature phenomenon called superconductivity. While superconducting materials exhibit absolutely no resistance, these effects are confined to temperatures so cold that it is not yet practical to use superconductivity in other than exotic applications.

Ohm's law

Ohm's law defines the relationship between the three variables affecting simple circuit action. According to Ohm's law, current is directly proportional to the net voltage in a circuit and current is inversely proportional to resistance.

Electrical power

The product of voltage and current equals electrical power. The unit of electrical power is the watt, named for James Watt. One watt of electrical power equals 1 joule per second. If 1 volt forces a 1-ampere current through a 1-ohm resistance, 1 joule per second will be wasted as heat. That is, 1 watt of power will be dissipated. A 100-watt incandescent lamp requires 100 joules for each second it operates.

Electricity provides a convenient way to connect cities with distant electrical generating stations. Electricity is not the primary source of energy, rather it serves as the means to transport energy from the source to a load. Electrical energy usually begins as mechanical energy before its conversion to electrical energy. At the load end of the distribution system the electrical energy is changed to another form of energy, as needed.

Commercial electrical power is transported great distances through wires, which always have significant resistance. Some of the transported energy is unavoidably wasted as heat. These losses are minimized by using very high voltage at a lower current, with the product of voltage and current still equal to the power required. Since the energy loss increases as the square of the current, a reduction of current by a factor of 1/100 reduces the power loss by a factor of 1/10,000. Voltage as high as 1,000,000 volts is used to reduce losses. Higher voltage demands bigger insulators and taller transmission towers, but the added expense pays off in greatly-reduced energy loss.

Alternating current and direct current

Direct current, or DC, results from an electric charge that moves in only one direction. A car's **battery**, for example, provides a direct current when it forces electrical charge through the starter motor or through the car's headlights. The direction of this current does not change.

Current that changes direction periodically is called alternating current, or AC. Our homes are supplied with alternating current rather than direct current because the use of AC makes it possible to step voltage up or down, using an electromagnetic device called a **transformer**. Without transformers to change voltage as needed, it would be necessary to distribute electrical power at a safer low voltage but at a much higher current. The higher current would increase the transmission loss in the powerlines. Without the ability to use high voltages, it would be necessary to locate generators near locations where electric power is needed.

Southern California receives much of its electrical power from hydroelectric generators in the state of Washington by a connection through an unusually long DC transmission line that operates at approximately one million volts. Electrical power is first generated as alternating current, transformed to a high voltage, then converted to direct current for the long journey south. The

KEY TERMS

. . . .

Ampere—A standard unit for measuring electric current.

Conductors—Materials that permit electrons to move freely.

Coulomb force—Another name for the electric force.

Electric field—The concept used to describe how one electric charge exerts force on another, distant electric charge.

Generator—A device for converting kinetic energy (the energy of movement) into electrical energy.

Insulator—An object or material that does not conduct heat or electricity well.

Joule—The unit of energy in the mks system of measurements.

Ohm—The unit of electrical resistance.

Semiconductor devices—Electronic devices made from a material that is neither a good conductor or a good insulator.

Volt—A standard unit of electric potential and electromotive force

Watt—The basic unit of electrical power equal to 1 joule per second.

direct-current power is changed back into AC for final distribution at a lower voltage. The use of direct current more than compensates for the added complexity of the AC to DC and DC to AC conversions.

Resources

Books

Asimov, Isaac. *Understanding Physics: Light, Magnetism, and Electricity.* Vol. 2. Signet Science Series. New York: NAL, 1969.

Giancoli, Douglas C. *Physics: Principles With Applications.* 3rd ed. Upper Saddle River, NJ: Prentice Hall, 1991.

Hewitt, Paul. *Conceptual Physics.* Englewood Cliffs, NJ: Prentice Hall, 2001.

Donald Beaty

Electrocardiogram (ECG)

The electrocardiogram, ECG or EKG, directly measures microvoltages in the **heart** muscle (myocardium) oc-

curring over specific periods of **time** in a cardiac, i.e., a heartbeat, otherwise known as a cardiac impulse. With each heartbeat, electrical currents called action potentials, measured in millivolts (mV), travel at predictable velocities through a conducting system in the heart. The potentials originate in a sinoatrial (SA) node which lies in the entrance chamber of the heart, called the right atrium. These currents also diffuse through tissues surrounding the heart whereby they reach the skin. There they are picked up by external electrodes which are placed at specific positions on the skin. They are in turn sent through leads to an electrocardiograph. A pen records the transduced electrical events onto special **paper**. The paper is ruled into mV against time and it provides the reader with a so-called rhythm strip. This is a non-invasive method to used evaluate the electrical counterparts of the myocardial activity in any series of heart beats. Careful observation of the records for any deviations in the expected times, shapes, and voltages of the impulses in the cycles gives the observer information that is of significant diagnostic value, especially for human medicine. The normal rhythm is called a sinus rhythm if the potentials begin in the sinoatrial (SA) node.

A **cardiac cycle** has a phase of activity called systole followed by a resting phase called diastole. In systole, the muscle **cell** membranes, each called a sarcolemma, allow charged **sodium** particles to enter the cells while charged potassium particles exit. These processes of **membrane** transfer in systole are defined as polarization. Electrical signals are generated and this is the phase of excitability. The currents travel immediately to all cardiac cells through the mediation of end-to-end high-conduction connectors termed intercalated disks. The potentials last for 200 to 300 milliseconds. In the subsequent diastolic phase, repolarization occurs. This is a period of oxidative restoration of **energy** sources needed to drive the processes. Sodium is actively pumped out of the fiber while potassium diffuses in. **Calcium**, which is needed to energize the **force** of the heart, is transported back to canals called endoplasmic reticula in the cell cytoplasm.

The action potentials travel from the superior part of the heart called the base to the inferior part called the apex. In the human four-chambered heart, a **pacemaker**, the SA node, is the first cardiac area to be excited because sodium and potassium interchange and energize both right and left atria. The impulses then pass downward to an atrioventricular (AV) node in the lower right atrium where their **velocity** is slowed, whereupon they are transmitted to a conducting system called the bundle of His. The bundle contains Purkinje fibers that transmit the impulses to the outer aspects of the right and left ventricular myocardium. In turn, they travel into the entire ventricular muscles by a slow process of **diffusion**. Repolarization of the myocardial cells takes place in a

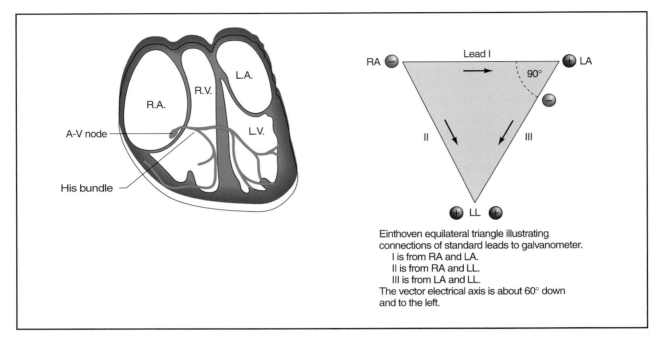

Figure 1. *Illustration by Hans & Cassidy. Courtesy of Gale Group.*

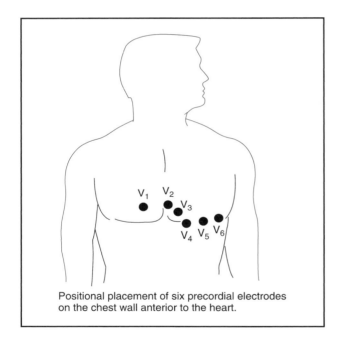

Figure 2. *Illustration by Hans & Cassidy. Courtesy of Gale Group.*

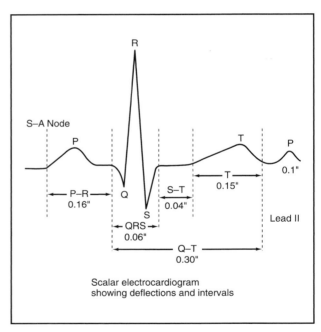

Figure 3. *Illustration by Hans & Cassidy. Courtesy of Gale Group.*

reverse direction to that of depolarization, but does not utilize the bundle of His.

The place where electrodes are positioned on the skin is important. In what are called standard leads, one electrode is fastened to the right arm, a second on the left arm, and a third on the left leg. They are labeled Lead I

(left arm to right arm), II (right arm to left leg), and III (left arm to left leg). These three leads form angles of an equilateral triangle called the Einthoven triangle (Figure 1). In a sense, the galvanometer is looking at the leads from three different points of view. The standard leads are in pairs called bipolar and the galvanometer mea-

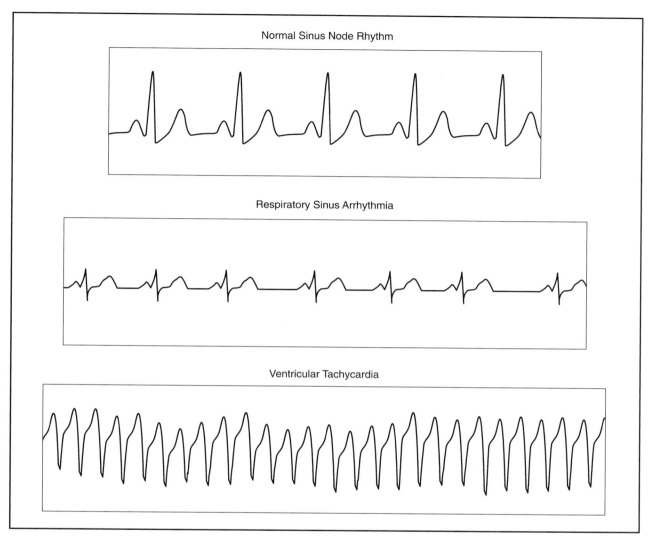

Figure 4. ECG trace recordings. *Illustration by Hans & Cassidy. Courtesy of Gale Group.*

sures them algebraically, not from **zero** to a finite value. The ECG record is called frontal, which is a record of events downward from base to apex.

The ECG displays a second set of leads called precordial. This means that they are positioned anterior to the heart at specific places on the skin of the chest. They measure electrical events, not in a frontal **plane** like the standard leads do, but tangentially, from anterior (ventral) to posterior (dorsal) or vice versa across the chest wall. They are numbered from their right to left positions as V1 through V6 (Figure 2). This allows them to sense impulses directly beneath the particular electrode put into the circuit. Events in these horizontal planes add significantly to **diagnosis**.

The ECG also shows a third set of leads which are three in number. These are called vectorial and are essen-

tial in obtaining vectorcardiograms because the transmission of action potentials in the heart is a directional or vector process. The direction of travel of the action potentials is found by vectorial analysis as it is in **physics**. The direction of travel of the action potentials is found by vectorial analysis as it is in physics. It takes two measurements of a completed record that are at right angles to one another to determine the resultant direction of all the potentials occurring at a given time. The resultant, computed as an arrow with a given length and direction, is considered to be the electrical axis of the heart. In the normal young adult it is predictably about minus 60 degrees below the horizontal isoelectric base line. The three vectorial leads are each 30 degrees away from the standard leads, appearing like spokes on a wheel. They explain why twelve leads appear in an ECG strip. In the recordings they are designated a VR, a VL and a VF. The lower case "a" means augmented, V is voltage and R, L and F are for the right arm, left arm and left foot.

The normal sinus ECG

A few selected examples of ECGs are displayed herein. In the normal ECG, as taken from standard lead II, there are three upward or positive deflections, P, R, and T and two downward **negative** deflections, Q and S. The P wave indicates atrial depolarization. The QRS complex shows ventricular activity. The S-T segment as well as the T wave indicate ventricular repolarization. There are atrial repolarization waves but they are too low in voltage to be visible (Figure 3).

The time line on the X axis is real time. The recording paper is read on this line as 0.04 seconds for each small vertical subdivision if the paper is running at 0.98 in (25 mm) per second. At the end of each group of five of these, which corresponds to 0.2 seconds, the vertical line is darker on the ruled paper. If the pulse **rate** is found to be 75 per minute, the duration of a cardiac cycle is 60/75 or 0.8 seconds. Variations in expected normal times for any part of a cycle indicate specific cardiac abnormalities. This is used to diagnose arrhythmias which have a basis in time deviation.

On the Y axis, every 0.4 in (10 mm) corresponds to 1 mV of activity in the heart. Although time on the X axis is real, the mV on the Y axis cannot always be taken literally. Voltages may partly lose significance in that a fatty person can to some extent insulate cardiac currents from reaching the skin.

Respiratory sinus arrhythmia

The young adult male, while resting, breathes about 12 times per minute. Each cycle takes five seconds, two for inspiration, and three for expiration. The ECG shows these differences graphically in every respiratory cycle and they are easily measurable between successive P waves. This is the only arrhythmia that is considered to be normal.

Ventricular tachycardia

The effect of the form of the wave on the ECG, as distinguished from the effect of the direction and force is illustrated in this disorder. Prominent signs include an extraordinary height of the waves and also the rapidity of the heart beat. Both X and Y axes must be examined.

Resources

Books

Eckert, Roger, David Randall, and George Augustine. *Animal Physiology.* 3rd ed. New York: W. H. Freeman and Company, 1988.

KEY TERMS

. . . .

Action potential—A transient change in the electrical potential across a membrane which results in the generation of a nerve impulse.

Depolarization—A tendency of a cell membrane when stimulated to allow charged (ionic) chemical particles to enter or leave the cell. This favors the neutralization of excess positive or negative particles within the cell.

Diastole—The phase of "rest" in a cardiac cycle, allowing reconstitution of energy needed for the phase of systole which follows.

Electrocardiogram (ECG)—A moving pen or oscilloscope record of the heart beats. Essentially, the measurement of microvolt changes by a galvanometer over time.

Precordial leads—These are six leads, labeled V1 through V6, which pass from electrodes on the ventral chest wall to the electrocardiograph. They are called unipolar in that the active electrode is placed on the chest while the second electrode is placed on an extremity, but only one transmits the action potentials originating in the heart.

Repolorization—The reverse passage of ions across the cell membranes following their depolarization. This reestablishes the resting differences, i.e., polarity, on either side of a cell membrane.

Standards (limb) leads—These are conductors connecting the electrocardiograph with the right arm, left arm, and left leg. They form a hypothetical Einthoven triangle. The leads are called bipolar in that each lead involves two electrodes placed RA and LA, RA and LL, LA and LL.

Vector—A quantity or term that can be expressed in terms of both magnitude (a number) and a direction.

Ganong, William F. *Review of Medical Physiology.* 16th ed. East Norwalk, CT: Appleton & Lange, 1993.

Guyton & Hall. *Textbook of Medical Physiology* 10th ed. New York: W. B. Saunders Company, 2000.

Guyton, Arthur C. *Human Physiology and Mechanisms of Disease.* 4th ed. Philadelphia: W.B. Saunders Co., 1987.

Harold M. Kaplan
Kathleen A. Jones

Electrochemical cell *see* **Cell, electrochemical**

Electrode *see* **Battery**

Electroencephalogram (EEG)

An electroencephalogram, usually abbreviated EEG, is a medical test that records electrical activity in the **brain**. During the test, the brain's spontaneous electrical signals are traced onto **paper**. The *electroencephalograph* is the machine that amplifies and records the electrical signals from the brain. The *electroencephalogram* is the paper strip the machine produces. The EEG changes with **disease** or brain disorder, such as **epilepsy**, so it can be a useful diagnostic tool, but usually must be accompanied by other diagnostic tests to be definitive.

To perform an EEG, electrodes, which are wires designed to detect electrical signals, are placed on the cranium either by inserting a needle into the scalp or by attaching the wire with a special **adhesive**. The electrodes are placed in pairs so that the difference in electric potential between them can be measured. The wires are connected to the electroencephalograph, where the signal is amplified and directed into pens that record the waves on a moving paper chart. The tracing appears as a series of peaks and troughs drawn as lines by the recording pens.

Basic alpha waves, which originate in the cortex, can be recorded if the subject closes his eyes and puts his brain "at rest" as much as possible. Of course, the brain is never still, so some brain activity is going on and is recorded in waves of about six to 12 per second, with an average of about 10 per second. The voltage of these waves is from five to 100 microvolts. A microvolt is one-one millionth of a volt. Thus, a considerable amount of amplification is required to raise the voltage to a discernable level.

The **rate** of the waves, that is, the number that occur per second, appears to be a better diagnostic indicator than does the amplitude, or strength. Changes in the rate indicating a slowing or speeding up are significant, and unconsciousness occurs at either extreme. **Sleep**, stupor, and deep **anesthesia** are associated with slow waves and grand mal seizures cause an elevated rate of brain waves. The only time the EEG line is straight and without any wave indication is at death. A person who is brain dead has a straight, flat EEG line.

The rates of alpha waves are intermediate compared with other waves recorded on the EEG. Faster waves, 14–50 waves per second, that are lower in voltage than alpha waves are called beta waves. Very slow waves, averaging 0.5-5 per second, are **delta** waves. The slowest brain waves are associated with an area of localized brain damage such as may occur from a **stroke** or blow on the head.

The **individual** at rest and generating a fairly steady pattern of alpha waves can be distracted by a sound or **touch**. The alpha waves then flatten somewhat, that is their voltage is less and their pattern becomes more irregular when the individual's attention is focused. Any difficult mental effort such as multiplying two four-digit numbers will decrease the amplitude of the waves, and any pronounced emotional excitement will flatten the pattern. The brain wave pattern will change to one of very slow waves, about three per second, in deep sleep.

Though the basic EEG pattern remains a standard one from person to person, each individual has his own unique EEG pattern. The same individual given two separate EEG tests weeks or months apart will generate the same alpha wave pattern, assuming the conditions of the tests are the same. Identical twins will both have the same pattern. One twin will virtually match the second twin to the extent that the two tracings appear to be from the same individual on two separate occasions.

Though the EEG is a useful diagnostic tool, its use in brain research is limited. The electrodes detect the activity of only a few neurons in the cortex out of the billions that are present. Electrode placement is standardized so the EEG can be interpreted by any trained neurologist. Also, the electrical activity being measured is from the surface of the cortex and not from the deeper areas of the brain.

The brain

The brain is the center of all human thought, feeling, emotion, movement, and touch, among other facilities. It consists of the prominent cerebrum, the cerebellum, and the medulla oblongata. The cerebral cortex, or outer layer, has specialized areas for sight, **hearing**, touch, **smell**, **taste**, and so on.

The basic **cell** of the brain is the **neuron**, which monitors information coming in to it and directs an appropriate response to a muscle or to another neuron. Each neuron is connected to other neurons through axons, which carry information away from a neuron, and dendrites, which carry information to the neuron. Thus, an axon from one neuron will end at a dendrite of another. The very tiny space between the two nerve endings is a **synapse**. The message is passed across the synapse by the release of certain chemical "messengers," from the axon which cross the space and occupy receptor areas in the dendrite. These chemicals are called neurotransmitters. Thus neurons are in constant electrical contact with other neurons, receiving and passing on information at the rate of billions of reactions a second.

Neuronal connections are established early in life and remain intact throughout one's lifetime. An interruption of those connections because of a stroke or accident results in their permanent loss. Sometimes, with great effort, alternative pathways or connections can be established to restore function to that area, but the original connection will remain lost.

Uses of the EEG

The electroencephalogram is a means to assess the **degree** of damage to the brain in cases of trauma, or to measure the potential for seizure activity. It is used also in sleep studies to determine whether an individual has a sleep disorder and to study brain wave patterns during dreaming or upon sudden awakening.

The EEG is also a useful second-level diagnostic tool to follow-up a computerized tomogram (CT) scan to assist in finding the exact location of a damaged area in the brain. The EEG is one of a **battery** of brain tests available and is seldom used alone to make a **diagnosis**. The EEG tracing can detect an abnormality but cannot distinguish between, for example, a **tumor** and a **thrombosis** (site of deposit of a **blood** clot in an artery).

Although persons with frequent seizures are more likely to have an abnormal EEG than are those who have infrequent seizures, EEGs cannot be solely used to diagnose epilepsy. Approximately 10% of epilepsy patients will have a normal EEG. A normal EEG, therefore, does not eliminate brain damage or seizure potential, nor does an abnormal tracing indicate that a person has epilepsy. Something as simple as visual stimulation or rapid breathing (hyperventilation) may initiate abnormal electrical patterns in some patients.

If the EEG is taken at the time the patient has a seizure, the pattern will change. A grand mal seizure will result in sharp spikes of higher voltage and greater **frequency** (25–30 per second). A petit mal seizure also is accompanied by sharp spikes, but at a rate of only three waves per second.

Also, the EEG is not diagnostic of mental illness. The individual who is diagnosed with **schizophrenia** or paranoia may have an EEG tracing interpreted as normal. Most mental illness is considered to be a chemical imbalance of some sort, which does not create abnormal electrical activity. However, an EEG may be taken of an individual who exhibits bizarre, abnormal behavior to rule out an organic source such as thrombosis as the cause.

KEY TERMS

. . . .

Electrode—A wire with a special terminal on it to attach to a part of the body to measure certain signals or transmit stimuli.

Patients being diagnosed for a brain disorder can be monitored on a 24-hour basis by a portable EEG unit. A special cap with electrodes is fitted onto the head where it will remain during the time the test is being run. The electroencephalograph is worn on the belt. A special attachment on the machine enables the patient to **telephone** the physician and transmit the data the machine has accumulated.

Resources

Books

Lerner, Brenda Wilmoth. "The Development of High-Tech Medical Diagnostic Tools." *Science and Its Times*. Vol. 7 Detroit: Gale Group, 2000.

Rosman, Isadore, ed. *Basic Health Care and Emergency Aid*. New York: Thomas Nelson, Inc., 1990.

Larry Blaser

Electrolysis

Electrolysis is the process of causing a chemical reaction to occur by passing an **electric current** through a substance or mixture of substances, most often in liquid form. Electrolysis frequently results in the **decomposition** of a compound into its elements. To carry out an electrolysis, two electrodes, a positive electrode (**anode**) and a **negative** electrode (**cathode**), are immersed into the material to be electrolyzed and connected to a source of direct (DC) electric current.

The apparatus in which electrolysis is carried out is called an *electrolytic cell*. The roots *-lys* and *-lyt* come from the Greek *lysis* and *lytos*, meaning to cut or decompose; electrolysis in an electrolytic cell is a process that can decompose a substance.

The substance being electrolyzed must be an **electrolyte**, a liquid that contains positive and negative ions and therefore is able to conduct **electricity**. There are two kinds of electrolytes. One kind is a ion compound **solution** of any compound that produces ions when it dissolves in **water**, such as an inorganic acid, base, or **salt**. The other kind is a liquefied ionic compound such as a molten salt.

In either kind of electrolyte, the liquid conducts electricity because its positive and negative ions are free to move toward the electrodes of opposite charge—the positive ions toward the cathode and the negative ions toward the anode. This transfer of positive charge in one direction and negative charge in the opposite direction constitutes an electric current, because an electric current is, after all, only a flow of charge, and it does not matter whether the carriers of the charge are ions or electrons. In an ionic solid such as **sodium chloride**, for example, the normally fixed-in-place ions become free to move as soon as the solid is dissolved in water or as soon as it is melted.

During electrolysis, the ions move toward the electrodes of opposite charge. When they reach their respective electrodes, they undergo chemical oxidation-reduction reactions. At the cathode, which is pumping electrons into the electrolyte, chemical reduction takes place-a taking-on of electrons by the positive ions. At the anode, which is sucking electrons out of the electrolyte, chemical oxidation takes place-a loss of electrons by the negative ions.

In electrolysis, there is a direct relationship between the amount of electricity that flows through the cell and the amount of chemical reaction that takes place. The more electrons are pumped through the electrolyte by the **battery**, the more ions will be forced to give up or take on electrons, thereby being oxidized or reduced. To produce one mole's worth of chemical reaction, one **mole** of electrons must pass through the cell. A mole of electrons, that is, 6.02×10^{23} of electrons, is called a *faraday*. The unit is named after Michael Faraday (1791-1867), the English chemist and physicist who discovered this relationship between electricity and chemical change. He is also credited with inventing the words *anode, cathode, electrode, electrolyte,* and *electrolysis.*

Various kinds of electrolytic cells can be devised to accomplish specific chemical objectives.

Electrolysis of water

Perhaps the best known example of electrolysis is the electrolytic decomposition of water to produce **hydrogen** and **oxygen**:

$$2H_2O \quad + \quad energy \quad \rightarrow \quad 2H_2 \quad + \quad O_2$$
water ⠀⠀⠀⠀⠀⠀⠀⠀⠀⠀⠀⠀⠀⠀⠀ hydrogen ⠀ oxygen
⠀⠀⠀⠀⠀⠀⠀⠀⠀⠀⠀⠀⠀⠀⠀⠀⠀⠀⠀ gas ⠀⠀⠀⠀ gas

Because water is such a stable compound, we can only make this reaction go by pumping **energy** into it—in this case, in the form of an electric current. Pure water, which does not conduct electricity very well, must first be made into an electrolyte by dissolving an acid, base, or salt in it. Then an anode and a cathode, usually

made of graphite or some non-reacting **metal** such as platinum, can be inserted and connected to a battery or other source of direct current.

At the cathode, where electrons are being pumped into the water by the battery, they are taken up by water molecules to form hydrogen gas:

$$4H_2O \quad + \quad 4e^- \quad \rightarrow \quad 2H_2 \quad + \quad 4OH^-$$
water ⠀⠀⠀⠀ electrons ⠀⠀⠀ hydrogen ⠀ hydroxide
⠀⠀⠀⠀⠀⠀⠀⠀⠀⠀⠀⠀⠀⠀⠀⠀⠀⠀ gas ⠀⠀⠀⠀ ions

At the anode, electrons are being removed from water molecules:

$$2H_2O \quad - \quad 4e^- \quad \rightarrow \quad O_2 \quad + \quad 4H^+$$
water ⠀⠀⠀⠀ electrons ⠀⠀⠀ oxygen ⠀⠀ hydrogen
⠀⠀⠀⠀⠀⠀⠀⠀⠀⠀⠀⠀⠀⠀⠀⠀⠀ gas ⠀⠀⠀⠀ ions

The net result of these two electrode reactions added together is

$$2H_2O \rightarrow 2H_2 + O_2.$$

(Note that when these two equations are added together, the four H^+ ions and four OH^- ions on the right-hand side are combined to form four H_2O molecules, which then **cancel** four of the H_2O molecules on the left-hand side.) Thus, every two molecules of water have been decomposed into two molecules of hydrogen and one **molecule** of oxygen.

The acid, base, or salt that made the water into an electrolyte was chosen so that its particular ions cannot be oxidized or reduced (at least at the voltage of the battery), so they do not react chemically and serve only to conduct the current through the water. **Sulfuric acid**, H_2SO_4, is commonly used.

Production of sodium and chlorine

By electrolysis, common salt, sodium chloride, NaCl, can be broken down into its elements, **sodium** and **chlorine**. This is an important method for the production of sodium; it is used also for producing other **alkali metals** and **alkaline earth metals** from their salts.

To obtain sodium by electrolysis, we will first melt some sodium chloride by heating it above its melting point of 1,474°F (801°C). Then we will insert two inert (non-reacting) electrodes into the melted salt. The sodium chloride must be molten in order to permit the Na^+ and Cl^- ions to move freely between the electrodes; in solid sodium chloride, the ions are frozen in place. Finally, we will pass a direct electric current (DC) through the molten salt.

The negative electrode (the cathode) will attract Na^+ ions and the positive electrode (the anode) will attract Cl^- ions, whereupon the following **chemical reactions** take place.

At the cathode, where electrons are being pumped in, they are being grabbed by the positive sodium ions:

$$Na^+ + e^- \rightarrow Na$$

sodium ion   electron   sodium atom

At the anode, where electrons are being pumped out, they are being ripped off the chloride ions:

$$Cl^- - e^- \rightarrow Cl$$

chloride ion   electron   chlorine atom

(The chlorine **atoms** immediately combine into diatomic molecules, Cl_2.) The result is that common salt has been broken down into its elements by electricity.

Production of magnesium

Another important use of electrolysis is in the production of **magnesium** from sea water. Sea water is a major source of that metal, since it contains more ions of magnesium than of any other metal except sodium. First, magnesium chloride, $MgCl_2$, is obtained by precipitating magnesium hydroxide from seawater and dissolving it in hydrochloric acid. The magnesium chloride is then melted and electrolyzed. Similar to the production of sodium from molten sodium chloride, above, the molten magnesium is deposited at the cathode, while the chlorine gas is released at the anode. The overall reaction is $MgCl_2 \rightarrow Mg + Cl_2$.

Production of sodium hydroxide, chlorine and hydrogen

Sodium hydroxide, NaOH, also known as lye and caustic soda, is one of the most important of all industrial chemicals. It is produced at the **rate** of 25 billion pounds a year in the United States alone. The major method for producing it is the electrolysis of brine or "salt water," a solution of common salt, sodium chloride in water. Chlorine and hydrogen gases are produced as valuable byproducts.

When an electric current is passed through salt water, the negative chloride ions, Cl^-, migrate to the positive anode and lose their electrons to become chlorine gas.

$$Cl^- + e^- \rightarrow Cl$$

chloride ion   electron   chlorine atom

(The chlorine atoms then pair up to form Cl_2 molecules.) Meanwhile, sodium ions, Na^+, are drawn to the negative cathode. But they do not pick up electrons to become sodium metal atoms as they do in molten salt, because in a water solution the water molecules themselves pick up electrons more easily than sodium ions do. What happens at the cathode, then, is

$$2H_2O + 2e^- \rightarrow H_2 + OH^-$$

water   electrons   hydrogen gas   hydroxide ions

The hydroxide ions, together with the sodium ions that are already in the solution, constitute sodium hydroxide, which can be recovered by **evaporation**.

This so-called *chloralkali* process is the basis of an industry that has existed for well over a hundred years. By electricity, it converts cheap salt into valuable chlorine, hydrogen and sodium hydroxide. Among other uses, the chlorine is used in the purification of water, the hydrogen is used in the **hydrogenation** of oils, and the lye is used in making **soap** and **paper**.

Production of aluminum

The production of **aluminum** by the Hall process was one of the earliest applications of electrolysis on a large scale, and is still the major method for obtaining that very useful metal. The process was discovered in 1886 by Charles M. Hall, a 21-year-old student at Oberlin College in Ohio, who had been searching for a way to reduce aluminum oxide to the metal. Aluminum was a rare and expensive luxury at that time, because the metal is very reactive and therefore difficult to reduce from its compounds by chemical means. On the other hand, electrolysis of a molten aluminum salt or oxide is difficult because the salts are hard to obtain in anhydrous (dry) form and the oxide, Al_2O_3, does not melt until 3,762°F (2,072°C).

Hall discovered that Al_2O_3, in the form of the mineral bauxite, dissolves in another aluminum mineral called cryolite, Na_3AlF_6, and that the resulting mixture could be melted fairly easily. When an electric current is passed through this molten mixture, the aluminum ions migrate to the cathode, where they are reduced to metal:

$$Al^{3+} + 3e^- \rightarrow Al$$

aluminum ion   electrons   molten aluminum metal

At the anode, oxide ions are oxidized to oxygen gas:

$$2O^{2-} - 2e^- \rightarrow O_2$$

oxide ion   electrons   oxygen gas

The molten aluminum metal sinks to the bottom of the cell and can be drawn off.

Notice that three moles of electrons (three faradays of electricity) are needed to produce each mole of aluminum, because there are three positive charges on each aluminum ion that must be neutralized by electrons. The production of aluminum by the Hall process therefore consumes huge amounts of electrical energy. The

recycling of beverage cans and other aluminum objects has become an important energy **conservation** measure.

Refining of copper

Unlike aluminum, **copper** metal is fairly easy to obtain chemically from its ores. But by electrolysis, it can be refined and made very pure—up to 99.999%. Pure copper is important in making electrical wire, because copper's **electrical conductivity** is reduced by impurities. These impurities include such valuable metals as silver, gold and platinum; when they are removed by electrolysis and recovered, they go a long way toward paying the electricity bill.

In the electrolytic refining of copper, the impure copper is made from the anode in an electrolyte bath of copper sulfate, $CuSO_4$, and sulfuric acid H_2SO_4. The cathode is a sheet of very pure copper. As current is passed through the solution, positive copper ions, Cu^{2+}, in the solution are attracted to the negative cathode, where they take on electrons and deposit themselves as neutral copper atoms, thereby building up more and more pure copper on the cathode. Meanwhile, copper atoms in the positive anode give up electrons and dissolve into the electrolyte solution as copper ions. But the impurities in the anode do not go into solution because silver, gold and platinum atoms are not as easily oxidized (converted into positive ions) as copper is. So the silver, gold and platinum simply fall from the anode to the bottom of the tank, where they can be scraped up.

Electroplating

Another important use of electrolytic cells is in the electroplating of silver, gold, chromium and nickel. Electroplating produces a very thin coating of these expensive metals on the surfaces of cheaper metals, to give them the appearance and the chemical resistance of the expensive ones.

In silver plating, the object to be plated (e.g., a spoon) is made from the cathode of an electrolytic cell. The anode is a bar of silver metal, and the electrolyte (the liquid in between the electrodes) is a solution of silver cyanide, AgCN, in water. When a direct current is passed through the cell, positive silver ions (Ag^+) from the silver cyanide migrate to the negative anode (the spoon), where they are neutralized by electrons and stick to the spoon as silver metal:

$$2H_2O \;+\; \text{energy} \;\rightarrow\; 2H_2 \;+\; O_2$$
water   hydrogen   oxygen
gas   gas

Meanwhile, the silver anode bar gives up electrons to become silver ions:

$$Ag \quad - \quad e^- \quad \rightarrow \quad Ag^\cdot$$
silver   electron   silver
atom   ion

Thus, the anode bar gradually dissolves to replenish the silver ions in the solution. The net result is that silver metal has been transferred from the anode to the cathode, in this case the spoon. This process continues until the desired coating thickness is built up on the spoon-usually only a few thousandths of an inch-or until the silver bar has completely dissolved.

In electroplating with silver, silver cyanide is used in the electrolyte rather than other compounds of silver such as silver nitrate, $AgNO_3$, because the cyanide ion, CN^-, reacts with silver ion, Ag^+, to form the complex ion $Ag(CN)_2^-$. This limits the supply of free Ag^+ ions in the solution, so they can deposit themselves only very gradually onto the cathode. This produces a shinier and more adherent silver plating. Gold plating is done in much the same way, using a gold anode and an electrolyte containing gold cyanide, AuCN.

Resources

Books

Chang, Raymond. *Chemistry*. New York: McGraw-Hill, 1991.
Sherwood, Martin, and Christine Sutton, eds. *The Physical World*. New York: Oxford, 1991.

Robert L. Wolke

Electrolyte

An electrolyte is a substance that will allow current to flow through the **solution** when dissolved in **water**.

Electrolytes promote this current flow because they produce positive and **negative** ions when dissolved. The current flows through the solution in the form of positive ions (cations) moving toward the negative electrode and negative ion (anions) moving the positive electrode.

Electrolytes can be classified as strong electrolytes and weak electrolytes. Strong electrolytes are substances that completely break apart into ions when dissolved. The most familiar example of a strong electrolyte is table **salt, sodium chloride**. Most salts are strong electrolytes, as are strong acids such as hydrochloric acid, **nitric acid**, perchloric acid, and **sulfuric acid**. Strong bases such as **sodium hydroxide** and **calcium** hydroxide are also strong electrolytes. Although calcium hydroxide is only slightly soluble, all of the compound which dissolves in completely ionized.

Weak electrolytes are substances which only partially dissociate into ions when dissolved in water. Weak acids such as **acetic acid**, found in vinegar, and weak bases such as **ammonia**, found in cleaning products, are examples of weak electrolytes. Very slightly soluble salts such as mercury chloride are also sometimes classified as weak electrolytes. Ligands and their associated **metal** ions can be weak electrolytes.

Not all substances that dissolve in water are electrolytes. Sugar, for example, dissolves readily in water, but remains in the water as molecules, not as ions. Sugar is classified as a non-electrolyte. Water itself ionizes slightly and is a very, very weak electrolyte.

Electromagnetic field

An electromagnetic field is an area in which electric and magnetic forces are interacting. It arises from electric charges in **motion**. Electromagnetic fields are directly related to the strength and direction of the **force** that a charged particle, called the "test" charge, would be subject to under the electromagnetic force caused by another charged particle or group of particles, called the source.

An electromagnetic field is best understood as a mathematical function or property of spacetime, but may be represented as a group of vectors, arrows with specific length and direction. For a static electric field, meaning there is no motion of source charges, the force $F\rightarrow$ on a test charge is $F\rightarrow = qE\rightarrow$, where q is the value of the test charge and $E\rightarrow$ is the vector electric field. For a static magnetic field (caused by moving charge inside an overall neutral group of charges, or a bar magnet, for example) the force is given by $F\rightarrow = qv\rightarrow \times B\rightarrow$, where $v\rightarrow$ is the charge **velocity**, $B\rightarrow$ is the

vector magnetic field, and the $\times$ indicates a cross-product of vectors.

A stationary charge produces an electric field, while a moving charge additionally produces a magnetic field. Since velocity is a relative concept dependent on one's choice of reference frame, **magnetism** and **electricity** are not independent, but linked together, hence the term **electromagnetism**.

Superposition of fields

Since all charges, moving or not, have fields associated with them, we must have a way to describe the total field due to all randomly distributed charges that would be felt by a positive charge, such as **proton** or positron, at any position and time. The total field is the sum of the fields produced by the individual particles. This idea is called the principle of superposition.

Let us look at arrow or vector representations of the fields from some particular charge distributions. There is an infinite number of possibilities, but we will consider only a few simple cases.

Electric fields

The field of a static point charge

According to Coulomb's law, the strength of the electric field from a nonmoving point charge depends directly on the charge value q and is inversely proportional to the **distance** from the charge. That is, farther from the source charge will be subject to the same strength of force regardless of whether it is above, below, or to the side of the source, as long as the distance is the same. A surface of the same radius all around the source will have the same field strength. This is called a surface of equipotential. For a **point source** charge, the surface of equipotential is a **sphere**, and the force F will push a positive charge radially outward. A test charge of **mass** m and positive charge q will feel a push away from the positive source charge with an **acceleration** (a) directly proportional to the field strength and inversely proportional to the mass of the test charge.

If a charge does not move because it is acted upon by the electromagnetic force equally from all directions, it is in a position of stable equilibrium.

The dipole field

Now let us consider the field from two charges, one positive and one **negative**, a distance d apart. We call this combination of charges a **dipole**. Remember, opposite charges attract, so this is not an unusual situation. A **hydrogen** atom, for example, consisting of an **electron** (negative charge) and a proton (positive charge) is a very small dipole, as these particles do not sit right on top of each other. According to the superposition principle mentioned above, we can just add the fields from each individual charge and get a rather complicated field. If we only consider the field at a position very far from the dipole, we can simplify the field equation so that the field is proportional to the product of the charge value and the separation of the two charges. There is also dependence on the distance along the dipole axis as well as radial distance from the axis.

The field of a line of charge

Next, we consider the field due to a group of positive charges evenly distributed along an infinite straight line, defined to be infinite because we want to neglect the effect of the endpoints as an unnecessary complication here. Just as the field of a point charge is directed radially outward in a sphere, the field of a line of charge is directed radially outward, but at any specific radius the surface of equipotential will be a cylinder.

Magnetic fields

Recall that the force of a stationary charge is $F\rightarrow = qE\rightarrow$, but if the charge is moving the force is $F\rightarrow = qE\rightarrow + qv\rightarrow \times B\rightarrow$. A steady (unchanging in time) current in a wire, generates a magnetic field. **Electric current** is essentially charges in motion. In an electrical conductor like **copper** wire, electrons move, while positive charges remain steady. The positive charge cancels the **electric charge** so the overall charge looks like **zero** when viewed from outside the wire, so no electric field will exist outside the wire, but the moving charges create a magnetic field from $F\rightarrow = qv\rightarrow \times B\rightarrow$ where $B\rightarrow$ is the magnetic field vector. The cross product results in magnetic field lines circling the wire. Because of this effect, solenoids (a current-carrying coil of wire that acts as a magnet) can be made by wrapping wire in a tight **spiral** around a metallic tube, so that the magnetic field inside the tube is linear in direction.

Relating this idea to Newton's first law of motion, which states that for every action there is an equal and opposite reaction, we see that an external magnetic field (from a bar magnet, for example) can exert a force on a current-carrying wire, which will be the sum of the forces on all the individual moving charges in the wire.

Electromagnetic fields

A simple example of a combination of electric and magnetic fields is the field from a single point charge,

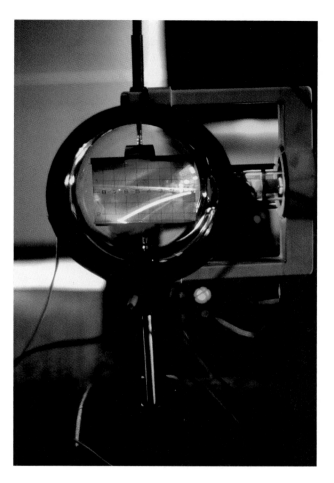

A demonstration of the effect of an electromagnetic field on an electron beam. The dark circles are the coils of wire of an electromagnet. Current flowing in the wire produces a magnetic field along the axis of the coils which deflects the electron beam in a direction perpendicular it. *Photograph by John Howard. National Audubon Society Collection/Photo Researchers, Inc. Reproduced by permission.*

say a proton, traveling through **space** at a constant speed in a straight line. In this case, the field vectors pointing radially outward would have to be added to the spiral magnetic field lines (circles extend into spirals because an individual charge is moving) to get the total field caused by the charge.

Maxwell's equations

A description of the field from a current which changes in time is much more complicated, but is calculable owing to James Clerk Maxwell (1831–1879). His equations, which have unified the laws of electricity and magnetism, are called Maxwell's equations. They are differential equations which completely describe the combined effects of electricity and magnetism, and are considered to be one of the crowning achievements of the nineteenth century. Maxwell's formulation of the theory of electromagnetic **radiation** allows us to understand the entire **electromagnetic spectrum**, from **radio waves** through visible **light** to gamma rays.

Electromagnetic induction

Electromagnetic induction is the generation of an **electromotive force** in a closed electrical circuit. It results from a changing magnetic field as it passes through the circuit. Some of the most basic components of electrical power systems—such as generators and transformers—make use of electromagnetic induction.

Fundamentals

The phenomenon of electromagnetic induction was discovered by the British physicist Michael Faraday in 1831 and independently observed soon thereafter by the American physicist Joseph Henry. Prior to that time, it was known that the presence of an **electric charge** would cause other charges on nearby conductors to redistribute themselves. Furthermore, in 1820 the Danish physicist Hans Christian Oersted demonstrated that an **electric current** produces a magnetic field. It seemed reasonable, then, to ask whether or not a magnetic field might cause some kind of electrical effect, such as a current.

An electric charge that is stationary in a magnetic field will not interact with the field in any way. Nor will a moving charge interact with the field if it travels **parallel** to the field's direction. However, a moving charge that crosses the field will experience a **force** that is **perpendicular** both to the field and to the direction of **motion** of the charge (Figure 1). Now, instead of a single charge, consider a rectangular loop of wire moving

through the field. Two sides of the loop will be subjected to forces that are perpendicular to the wire itself so that no charges will be moved. Along the other two sides charge will flow, but because the forces are equal the charges will simply bunch up on the same side, building up an internal electric field to counteract the imposed force, and there will be no net current (Figure 2).

How can a magnetic field cause current to flow through the loop? Faraday discovered that it was not simply the presence of a magnetic field that was required. In order to generate current, the magnetic flux through the loop must change with time. The term flux refers to the flow of the magnetic field lines through the area enclosed by the loop. The flux of the magnetic field lines is like the flow of **water** through a pipe and may increase or decrease with time.

To understand how the change in flux generates a current, consider a circuit made of many rectangular loops connected to a **light** bulb. Under what conditions will current flow and the light bulb shine? If the circuit is pulled through a uniform magnetic field there will be no current because the flux will be constant. But, if the field is non-uniform, the charges on one side of the loop will continually experience a force greater than that on the other side. This difference in forces will cause the charges to circulate around the loop in a current that

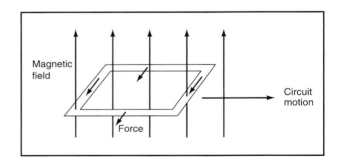

Figure 2. *Illustration by Hans & Cassidy. Courtesy of Gale Group.*

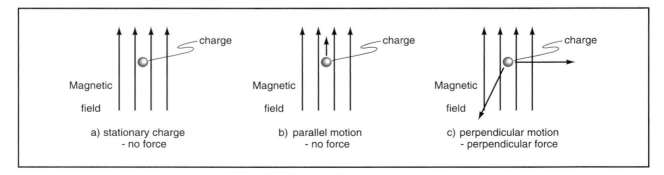

Figure 1. *Illustration by Hans & Cassidy. Courtesy of Gale Group.*

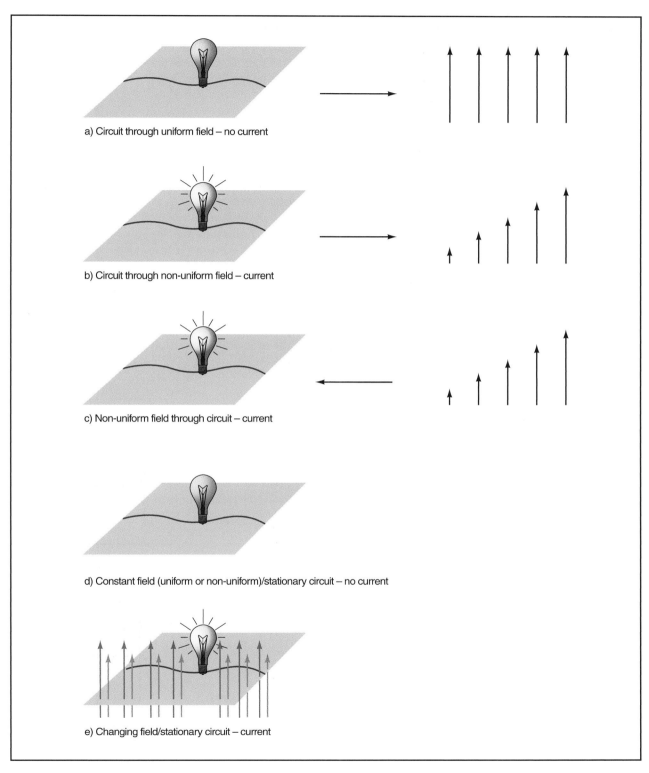

a) Circuit through uniform field – no current

b) Circuit through non-uniform field – current

c) Non-uniform field through circuit – current

d) Constant field (uniform or non-uniform)/stationary circuit – no current

e) Changing field/stationary circuit – current

Figure 3. *Illustration by Hans & Cassidy. Courtesy of Gale Group.*

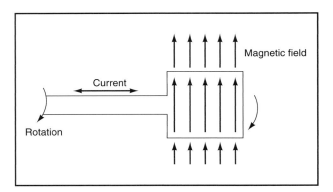

Figure 4. *Illustration by Hans & Cassidy. Courtesy of Gale Group.*

lights the bulb. The **work** done in moving each charge through the circuit is called the electromotive force or EMF. The units of electromotive force are volts just like the voltage of a **battery** that also causes current to flow through a circuit. It makes no difference to the circuit whether the changing flux is caused by the loop's own motion or that of the magnetic field, so the case of a stationary circuit and a moving non-uniform field is equivalent to the previous situation and again the bulb will light (Figure 3).

Yet a current can be induced in the circuit without moving either the loop or the field. While a stationary loop in a constant magnetic field will not cause the bulb to light, that same stationary loop in a field that is changing in time (such as when the field is being turned on or off) will experience an electromotive force. This comes about because a changing magnetic field generates an electric field whose direction is given by the right-hand rule—with the thumb of your right hand pointing in the direction of the change of the magnetic flux, your fingers can be wrapped around in the direction of the induced electric field. With an EMF directed around the circuit, current will flow and the bulb will light (Figure 3).

The different conditions by which a magnetic field can cause current to flow through a circuit are summarized by Faraday's law of induction. The variation in time of the flux of a magnetic field through a surface bounded by an electrical circuit generates an electromotive force in that circuit.

What is the direction of the induced current? A magnetic field will be generated by the induced current. If the flux of that field were to add to the initial magnetic flux through the circuit, then there would be more current, which would create more flux, which would create more current, and so on without limit. Such a situation would violate the conservation of **energy** and the tendency of physical systems to resist change. So the induced current will be generated in the direction that will create

KEY TERMS

Ampere—A standard unit for measuring electric current.

Faraday's law of induction—The variation in time of the flux of a magnetic field through a surface bounded by an electrical circuit generates an electromotive force in that circuit.

Flux—The flow of a quantity through a given area.

Generator—A device for converting kinetic energy (the energy of movement) into electrical energy.

Henry—A standard unit for measuring inductance.

Lenz's law—The direction of a current induced in a circuit will be such as to create a magnetic field which opposes the inducing flux change.

Mutual inductance—The ratio of the induced electromotive force in one circuit to the rate of change of current in the inducing circuit.

Right-hand rule (for electric fields generated by changing magnetic fields)—With the thumb of the right hand along the direction of change of magnetic flux, the fingers curl to indicate the direction of the induced electric field.

Self-inductance—The electromotive force induced in a circuit that results from the variation with time in the current of that same circuit.

Volt—A standard unit of electric potential and electromotive force.

magnetic flux which opposes the variation of the inducing flux. This fact is known as Lenz's law.

The **relation** between the change in the current through a circuit and the electromotive force it induces in itself is called the self-inductance of the circuit. If the current is given in amperes and the EMF is given in volts, the unit of self-inductance is the henry. A changing current in one circuit can also induce an electromotive force in a nearby circuit. The **ratio** of the induced electromotive force to the **rate** of change of current in the inducing circuit is called the mutual inductance and is also measured in henrys.

Applications

An electrical **generator** is an apparatus that converts mechanical energy into electrical energy. In this case the magnetic field is stationary and does not vary with time. It is the circuit that is made to rotate through the magnetic field. Since the area that admits the passage of magnetic field lines changes while the circuit rotates, the flux through the circuit will change, thus inducing a current (Figure 4). Generally, a **turbine** is used to provide the circuit's **rotation**. The energy required to move the turbine may come from steam generated by nuclear or **fossil fuels**, or from the flow of water through a dam. As a result, the mechanical energy of rotation is changed into electric current.

Transformers are devices used to transfer electric energy between circuits. They are used in power lines to convert high voltage **electricity** into household current. Common consumer **electronics** such as radios and televisions also use transformers. By making use of mutual inductance, the transformer's primary circuit induces current in its secondary circuit. By varying the physical characteristics of each circuit, the output of the **transformer** can be designed to meet specific needs.

John Appel

Electromagnetic radiation *see*
Electromagnetic spectrum

Electromagnetic spectrum

The electromagnetic **spectrum** encompasses a continuous range of frequencies or wavelengths of electromagnetic **radiation**, ranging from long wavelength, low **energy radio waves** to short wavelength, high **frequency**, high-energy gamma rays. The electromagnetic spectrum is traditionally divided into regions of **radio** waves, microwaves, infrared radiation, visible **light**, ultraviolet rays, **x rays**, and gamma rays.

Scottish physicist James Clerk Maxwell's (1831–1879) development of a set of equations that accurately described electromagnetic phenomena allowed the mathematical and theoretical unification of electrical and magnetic phenomena. When Maxwell's calculated speed of light fit well with experimental determinations of the speed of light, Maxwell and other physicists realized that visible light should be a part of a broader electromagnetic spectrum containing forms of electromagnetic radiation that varied from visible light only in terms of wavelength and wave frequency. Frequency is defined as the number of wave cycles that pass a particular point per unit **time**, and is commonly measured in Hertz (cycles per second). Wavelength defines the **distance** between adjacent points of the electromagnetic wave that are in equal phase (e.g., wavecrests).

Exploration of the electromagnetic spectrum quickly resulted practical advances. German physicist Henrich Rudolph Hertz regarded Maxwell's equations as a path to a "kingdom" or "great domain" of electromagnetic waves. Based on this insight, in 1888, Hertz demonstrated the existence of radio waves. A decade later, Wilhelm Röentgen's discovery of high-energy electromagnetic radiation in the form of x rays quickly found practical medical use.

At the beginning of the twentieth century, German physicist, Maxwell Planck, proposed that **atoms** absorb or emit electromagnetic radiation only in certain bundles termed quanta. In his work on the **photoelectric effect**, German-born American physicist Albert Einstein used the term **photon** to describe these electromagnetic quanta. Planck determined that energy of light was proportional to its frequency (i.e., as the frequency of light increases, so does the energy of the light). **Planck's constant**, $h = 6.626 \times 10^{-34}$ joule-second in the meter-kilogram-second system (4.136×10^{-15} eV-sec), relates the energy of a photon to the frequency of the electromagnetic wave and allows a precise calculation of the energy of electromagnetic radiation in all portions of the electromagnetic spectrum.

Although electromagnetic radiation is now understood as having both photon (particle) and wave-like properties, descriptions of the electromagnetic spectrum generally utilize traditional wave-related terminology (i.e., frequency and wavelength).

Electromagnetic fields and photons exert forces that can excite electrons. As electrons transition between allowed orbitals, energy must be conserved. This conservation is achieved by the **emission** of photons when an **electron** moves from a higher potential orbital energy to a lower potential orbital energy. Accordingly, light is emitted only at certain frequencies characteristic of every atom and **molecule**. Correspondingly, atoms and molecules absorb only a limited range of frequencies and wavelengths of the electromagnetic spectrum, and reflect all the other frequencies and wavelengths of light. These reflected frequencies and wavelengths are often the actual observed light or colors associated with an object.

The region of the electromagnetic spectrum that contains light at frequencies and wavelengths that stimulate the rod and cones in the human **eye** is termed the visible region of the electromagnetic spectrum. **Color** is the association the eye makes with selected portions of that visible region (i.e., particular colors are associated with specific wavelengths of visible light). A nanometer (10^{-9} m) is the most common unit used for characterizing the wavelength of visible light. Using this unit, the visible portion of the electromagnetic spectrum is located between 380 nm–750 nm and the component color regions of the visible spectrum are Red (670–770 nm), Orange (592–620 nm), Yellow (578–592 nm), Green (500–578 nm), Blue (464–500 nm), Indigo (444–464 nm), and Vio-

TABLE 1

Region	Frequency (Hz)	Wavelength (m)	Energy (eV)	Size Scale
Radio waves	$< 10^9$	> 0.3	$< 7 \times 10^{-7}$	Mountains, building
Microwaves	$10^9 - 3 \times 10^{11}$	$0.001 - 0.3$	$7 \times 10^{-7} - 2 \times 10^{-4}$	
Infrared	$3 \times 10^{11} - 3.9 \times 10^{14}$	$7.6 \times 10^{-7} - 0.001$	$2 \times 10^{-4} - 0.3$	
Visible	$3.9 \times 10^{14} - 7.9 \times 10^{14}$	$3.8 \times 10^{-7} - 7.6 \times 10^{-7}$	$0.3 - 0.5$	Bacteria
Ultraviolet	$7.9 \times 10^{14} - 3.4 \times 10^{16}$	$8 \times 10^{-9} - 3.8 \times 10^{-7}$	$0.5 - 20$	Viruses
X-rays	$3.4 \times 10^{16} - 5 \times 10^{19}$	$6 \times 10^{-12} - 8 \times 10^{-9}$	$20 - 3 \times 10^4$	Atoms
Gamma Rays	$> 5 \times 10^{19}$	$< 6 \times 10^{-12}$	$> 3 \times 10^4$	Nuclei

TABLE 2

Red	6300 - 7600 Å
Orange	5900 - 6300 Å
Yellow	5600 - 5900 Å
Green	4900 - 5600 Å
Blue	4500 - 4900 Å

let (400–446 nm). Because the energy of electromagnetic radiation (i.e., the photon) is inversely proportional to the wavelength, red light (longest in wavelength) is the lowest in energy. As wavelengths contract toward the blue end of the visible region of the electromagnetic spectrum, the frequencies and energies of colors steadily increase.

Like colors in the visible spectrum, other regions in the electromagnetic spectrum have distinct and important components. Radio waves, with wavelengths that range from hundreds of meters to less than a centimeter, transmit radio and **television** signals. Within the radio band, FM radio waves have a shorter wavelength and higher frequency than AM radio waves. Still higher frequency radio waves with wavelengths of a few centimeters can be utilized for **RADAR** imaging.

Microwaves range from approximately 1 ft (30 cm) in length to the thickness of a piece of **paper**. The atoms in food placed in a microwave oven become agitated (heated) by exposure to microwave radiation. Infrared radiation comprises the region of the electromagnetic spectrum where the wavelength of light is measured region from one millimeter (in wavelength) down to 400 nm. Infrared waves are discernible to humans as thermal radiation (**heat**). Just above the visible spectrum in terms of higher energy, higher frequency and shorter wavelengths is the ultraviolet region of the spectrum with light ranging in wavelength from 400 to 10 billionths of a meter. Ultraviolet radiation is a common cause of sunburn even when visible light is obscured or blocked by **clouds**. X rays are a highly energetic region of electromagnetic radiation with wavelengths ranging from about ten billionths of a meter to 10 trillionths of a meter. The ability of x rays to penetrate skin and other substances renders them useful in both medical and industrial radiography. Gamma rays, the most energetic form of electromagnetic radiation, are comprised of light with wavelengths of less than about ten trillionths of a meter and include waves with wavelengths smaller than the radius of an atomic nucleus (10^{15} m). Gamma rays are generated by nuclear reactions (e.g., **radioactive decay**, nuclear explosions, etc.).

Cosmic rays are not a part of the electromagnetic spectrum. Cosmic rays are not a form of electromagnetic radiation, but are actually high-energy charged particles with energies similar to, or higher than, observed gamma electromagnetic radiation energies.

Wavelength, frequency, and energy

The wavelength of radiation is sometimes given in units with which we are familiar, such as inches or centimeters, but for very small wavelengths, they are often given in angstroms (abbreviated Å). There are 10,000,000,000 angstroms in 3.3 ft (1 m).

An alternative way of describing a wave is by its frequency, or the number of peaks which pass a particular point in one second. Frequencies are normally given in cycles per second, or hertz (abbreviation Hz), after Hertz. Other common units are kilohertz (kHz, or thousands of cycles per second), megahertz (MHz, millions of cycles per second), and gigahertz (GHz, billions of cycles per second). The frequency and wavelength, when multiplied together, give the speed of the wave. For electromagnetic waves in empty **space**, that speed is the speed of light, which is approximately 186,000 miles per second (300,000 km per sec).

In addition to the wave-like properties of electromagnetic radiation, it also can behave as a particle. The energy of a particle of light, or photon, can be calculated from its frequency by multiplying by Planck's constant. Thus, higher frequencies (and lower wavelengths) have higher energy. A common unit used to describe the energy of a photon is the electron volt (eV). Multiples of this unit, such as keV (1000 electron volts) and MeV (1,000,000 eV), are also used.

Properties of waves in different regions of the spectrum are commonly described by different notation. Visible radiation is usually described by its wavelength, for example, while x rays are described by their energy. All of these schemes are equivalent, however; they are just different ways of describing the same properties.

Wavelength regions

The electromagnetic spectrum is typically divided into wavelength or energy regions, based on the characteristics of the waves in each region. Because the properties vary on a continuum, the boundaries are not sharp, but rather loosely defined.

Radio waves are familiar to us due to their use in communications. The standard AM radio band is at 540–1650 kHz, and the FM band is 88-108 MHz. This region also includes shortwave radio transmissions and television broadcasts.

We are most familiar with microwaves because of microwave ovens, which heat food by causing **water** molecules to rotate at a frequency of 2.45 GHz. In **astronomy**, emission of radiation at a wavelength of 8.2 in (21 cm) has been used to map neutral **hydrogen** throughout the **galaxy**. Radar is also included in this region.

The infrared region of the spectrum lies just beyond the visible wavelengths. It was discovered by William Herschel in 1800 by measuring the dispersing sunlight with a **prism**, and measuring the **temperature** increase just beyond the red end of the spectrum.

The visible wavelength range is the range of frequencies with which we are most familiar. These are the wavelengths to which the human eye is sensitive, and which most easily pass through Earth's atmosphere. This region is further broken down into the familiar colors of the rainbow, which fall into the wavelength intervals listed here.

A common way to remember the order of colors is through the name of the fictitious person ROY G. BIV (the I stands for indigo).

The ultraviolet range lies at wavelengths just short of the visible. Although humans do not use UV to see, it has many other important effects on **Earth**. The **ozone** layer high in Earth's atmosphere absorbs much of the UV radiation from the **sun**, but that which reaches the surface can cause suntans and sunburns.

We are most familiar with x rays due to their uses in medicine. X radiation can pass through the body, allowing doctors to examine bones and teeth. Surprisingly, x rays do not penetrate Earth's atmosphere, so astronomers must place x-ray telescopes in space.

Gamma rays are the most energetic of all electromagnetic radiation, and we have little experience with them in everyday life. They are produced by nuclear processes, for example, during radioactive decay or in nuclear reactions in stars or in space.

See also Electromagnetic field; Electromagnetic induction; Electromagnetism.

Resources

Books

Gribbin, John. *Q is for Quantum: An Encyclopedia of Particle Physics*. New York: The Free Press, 1998.

Griffiths, D.J. *Introduction to Quantum Mechanics.* Upper Saddle River, NJ: Prentice-Hall, Inc. 1995.

Jackson, J.D. *Classical Electrodynamics* New York: John Wiley and Sons, 1998.

Phillips, A.C. *Introduction to Quantum Mechanics.* New York: John Wiley & Sons, 2003.

Other

High Energy Astrophysics Science Archive Research Center, NASA. "Imagine the Universe. The Electromagnetic Spectrum" [cited February 24, 2003]. <http://imagine.gsfc.nasa.gov/docs/science/know_l1/emspectrum.html>.

K. Lee Lerner
David Sahnow

Electromagnetic waves *see* **Electromagnetic spectrum**

Electromagnetism

Electromagnetism is a branch of physical science that involves all the phenomena in which **electricity** and **magnetism** interact. This field is especially important to **electronics** because a magnetic field is created by an **electric current**. The rules of electromagnetism are responsible for the way charged particles of **atoms** interact.

Some of the rules of *electrostatics*, the study of electric charges at rest, were first noted by the ancient Romans, who observed the way a brushed comb would attract particles. It is now known that electric charges occur in two different types, called positive and **negative**. Like types repel each other, and differing types attract.

The **force** that attracts positive charges to negative charges weakens with **distance**, but is intrinsically very strong. The fact that unlike types attract means that most

of this force is normally neutralized and not seen in full strength. The negative charge is generally carried by the atom's electrons, while the positive resides with the protons inside the atomic nucleus. There are other less well known particles that can also carry charge. When the electrons of a material are not tightly bound to the atom's nucleus, they can move from atom to atom and the substance, called a conductor, can conduct electricity. On the contrary, when the **electron** binding is strong, the material is called an insulator.

When electrons are weakly bound to the atomic nucleus, the result is a semiconductor, often used in the electronics industry. It was not initially known if the electric current carriers were positive or negative, and this initial ignorance gave rise to the convention that current flows from the positive terminal to the negative. In reality we now know that the electrons actually run from the negative to the positive.

Electromagnetism is the theory of a unified expression of an underlying force, the so-called electromagnetic force. This is seen in the movement of **electric charge**, which gives rise to magnetism (the electric current in a wire being found to deflect a compass needle), and it was a Scotsman, James Clerk Maxwell, who published the theory unifying electricity and magnetism in 1865. The theory arose from former specialized work by Gauss, **Coulomb**, Ampère, Faraday, Franklin, and Ohm. However, one factor that did not contradict the experiments was added to the equations by Maxwell so as to ensure the conservation of charge. This was done on the theoretical grounds that charge should be a conserved quantity, and this addition led to the prediction of a wave phenomena with a certain anticipated **velocity**. **Light**, which has the expected velocity, was found to be an example of this electromagnetic **radiation**.

Light had formerly been thought of as consisting of particles (photons) by Newton, but the theory of light as particles was unable to explain the wave nature of light (**diffraction** and the like). In reality, light displays both wave *and* particle properties. The resolution to this duality lies in quantum theory, where light is neither particles or wave, but both. It propagates as a wave without the need of a medium and interacts in the manner of a particle. This is the basic nature of quantum theory.

Classical electromagnetism, useful as it is, contains contradictions (acausality) that make it incomplete and drive one to consider its extension to the area of quantum **physics**, where electromagnetism, of all the fundamental forces of nature, it is perhaps the best understood.

There is much **symmetry** between electricity and magnetism. It is possible for electricity to give rise to magnetism, and symmetrically for magnetism to give rise to electricity (as in the exchanges within an electric **transformer**). It is an exchange of just this kind that constitutes electromagnetic waves. These waves, although they do not need a medium of propagation, are slowed when traveling through a transparent substance.

Electromagnetic waves differ from each other only in amplitude, **frequency** and orientation (polarization). **Laser** beams are particular in being very coherent, that is, the radiation is of one frequency, and the waves coordinated in **motion** and direction. This permits a highly concentrated beam that is used not only for its cutting abilities, but also in electronic data storage, such as in CD-ROMs.

The differing frequency forms are given a variety of names, from **radio waves** at very low frequencies through light itself, to the high frequency x and gamma rays.

Many miracles depend upon the broad span of the **electromagnetic spectrum**. The ability to communicate across long distances despite intervening obstacles, such as the walls of buildings, is possible using the **radio** and **television** frequencies. **X rays** can see into the human body without opening it. These things, which would once have been labeled magic, are now ordinary ways we use the electromagnetic **spectrum**.

The unification of electricity and magnetism has led to a deeper understanding of physical science, and much effort has been put into further unifying the four forces of nature. The remaining known forces are the so called weak, strong, and gravitational forces. The weak force has now been unified with electromagnetism, called the electroweak force. There are proposals to include the strong force in a **grand unified theory**, but the inclusion of gravity remains an open problem.

The fundamental role of special relativity in electromagnetism

Maxwell's theory is in fact in contradiction with Newtonian mechanics, and in trying to find the resolution to this conflict, Einstein was lead to his theory of special relativity. Maxwell's equations withstood the conflict, but it was Newtonian mechanics that were corrected by relativistic mechanics. These corrections are most necessary at velocities, close to the speed of light. The many strange predictions about **space** and **time** that follow from special relativity are found to be a part of the real world.

Paradoxically, magnetism is a counter example to the frequent claims that relativistic effects are not noticeable for low velocities. The moving charges that compose an electric current in a wire might typically only be traveling at several feet per second (walking speed), and

the resulting Lorentz contraction of special relativity is indeed minute. However, the electrostatic forces at balance in the wire are of such great magnitude, that this small contraction of the moving (negative) charges exposes a residue force of real world magnitude, namely the magnetic force. It is in exactly this way that the magnetic force derives from the electric. Special relativity is indeed hidden in Maxwell's equations, which were known before special relativity was understood or separately formulated by Einstein.

Technological uses of electromagnetism

Before the advent of technology, electromagnetism was perhaps most strongly experienced in the form of **lightning**, and electromagnetic radiation in the form of light. Ancient man kindled fires which he thought were kept alive in trees struck by lightning.

Much of the magic of nature has been put to work by man, but not always for his betterment or that of his surroundings. Electricity at high voltages can carry **energy** across extended distances with little loss. Magnetism derived from that electricity can then power vast motors. But electromagnetism can also be employed in a more delicate fashion as a means of communication, either with wires (as in the **telephone**), or without them (as in radio communication). It also drives our electronics devices (as in computers).

Magnetism has long been employed for navigation in the compass. This works because **Earth** is itself a huge magnet, thought to have arisen from the great heat driven **convection** currents of molten **iron** in its center. In fact, it is known that Earth's magnetic poles have exchanged positions in the past.

Electromotive force

In an **electric circuit**, electromotive **force** is the **work** done by a source on an electrical charge. Because it is not really a force, the term is actually a misnomer; it is more commonly referred to by the initials EMF. EMF is another term for electrical potential, or the difference in charge across a **battery** or voltage source. For a circuit with no current flowing, the potential difference is called EMF.

Electrical sources that convert **energy** from another form are called seats of EMF. In the case of a complete circuit, such a source performs work on electrical charges, pushing them around the circuit. At the seat of EMF, charges are moved from low electrical potential to higher electrical potential.

Water flowing downhill in a flume is a good analogy for charges in an electric circuit. The water starts at the top of the hill with a certain amount of potential energy, just as charges in a circuit start with high electrical potential at the battery. As the water begins to flow downhill, its potential energy drops, just as the electrical potential of charges drops as they travels through the circuit. At the bottom of the hill, the potential energy is minimum, and work must be performed to pump it to the top of the hill to travel through the flume again. Similarly, in an electrical circuit, the seat of EMF performs work on the charges to bring them to a higher potential after their trip through the circuit.

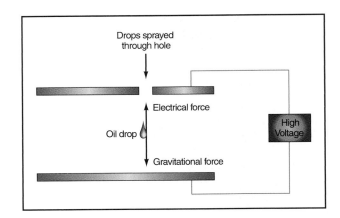

Figure 1. *Illustration by Hans & Cassidy. Courtesy of Gale Group.*

Electron

The electron is a negatively charged subatomic particle which is an important component of the **atoms** which make up ordinary **matter**. The electron is fundamental, in that it is not believed to be made up of smaller constituents. The size of the charge on the electron has for many years been considered the fundamental unit of charge found in nature. All electrical charges were believed to be **integral** multiples of this charge. Recently, however, considerable evidence has been found to indicate that particles classified as mesons and baryons are made up of objects called **quarks**, which have charges of either 2/3 or 1/3 the charge on the electron. For example, the neutrons and protons, which make up the nuclei of atoms, are baryons. However, scientists have never been able to observe an isolated quark, so for all practical purposes the charge on the electron can still be considered the fundamental unit of charge found in nature. The magnitude of this charge, usually designated by e, has been measured very precisely and is 1.602177×10^{-19} coulombs. The **mass** of the electron is small even by atomic standards and has the value 9.109389×10^{-31} kg ($0.5110 \ M_eV/c^2$, being only about 1/1836 the mass of the **proton**.

All atoms found in nature have a positively charged nucleus about which the negatively charged electrons move. The atom is electrically neutral and thus the positive electrical charge on the nucleus has the same magnitude as the **negative** charge due to all the electrons. The electrons are held in the atom by the attractive **force** exerted on them by the positively charged nucleus. They move very rapidly about the nucleus in orbits which have very definite energies, forming a sort of **electron cloud** around it. Some of the electrons in a typical atom can be quite close to the nucleus, while others can be at distances which are many thousands of times larger than the diameter of the nucleus. Thus, the electron cloud deter-

mines the size of the atom. It is the outermost electrons that determine the chemical behavior of the various elements. The size and shape of the electron clouds around atoms can only be explained utilizing a field of **physics** called **quantum mechanics**.

In metals, some of the electrons are not tightly bound to atoms and are free to move through the **metal** under the influence of an electric field. It is this situation that accounts for the fact that most metals are good conductors of **electricity** and **heat**.

Quantum theory also explains several other rather strange properties of electrons. Electrons behave as if they were spinning, and the value of the angular **momentum** associated with this spin is fixed; thus it is not surprising that electrons also behave like little magnets. The way electrons are arranged in some materials, such as **iron**, causes these materials to be magnetic. The existence of the positron, the **antiparticle** of the electron, was predicted by French physicist Paul Dirac in 1930. To predict this antiparticle, he used a version of quantum mechanics which included the effects of the theory of relativity. The positron's charge has the same magnitude as the electron's charge but is positive. Dirac's prediction was verified two years later when the positron was observed experimentally by Carl Anderson in a cloud chamber used for research on cosmic rays. The positron does not exist for very long in the presence of ordinary matter because it soon comes in contact with an ordinary electron and the two particles annihilate, producing a gamma ray with an **energy** equal to the energy equivalent of the two electron masses, according to Einstein's famous equation $E = mc^2$.

History

As has been the case with many developments in science, the discovery of the electron and the recognition

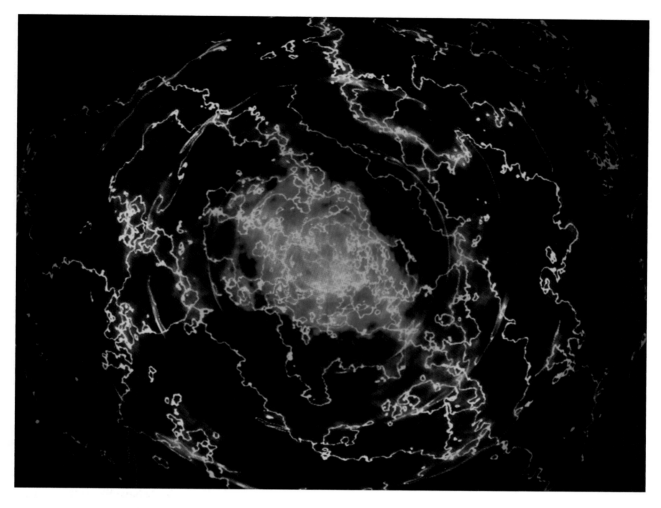

Electron cloud. *ArSciMed/Science Photo Library/Photo Researchers, Inc. Reproduced by permission.*

of its important role in the structure of matter evolved over a period of almost 100 years. As early as 1838, English physicist Michael Faraday found that when a charge of several thousand volts was applied between metal electrodes in an evacuated **glass** tube, an **electric current** flowed between the electrodes. It was found that this current was made up of negatively charged particles by observing their deflection in an electric field. Credit for the discovery of the electron is usually given to the English physicist J. J. Thomson. He was able to make quantitative measurements of the deflection of these particles in electric and magnetic fields and measure e/m, the **ratio** of their charge to mass.

Later, similar measurements were made on the negatively charged particles emitted by different **cathode** materials and the same value of e/m was obtained. When the same value of e/m was also obtained for "electrons" emitted by hot filaments (called thermionic

emission) and for photoelectrons emitted when **light** hits certain surfaces, it became clear that these were all the same type of particle, and the fundamental nature of the electron began to emerge. From these and other measurements it soon became known that the charge on the electron was roughly 1.6×10^{-19} coulombs. But the definitive experiment, which indicated that the charge on the electron was the fundamental unit of charge in nature, was carried out by Robert A. Millikan at the University of Chicago between 1907 and 1913. A schematic diagram of this famous "oil drop" experiment is shown in Figure 1. Charged oil drops, produced by an atomizer, were sprayed into the electric field maintained between two **parallel** metal plates. By measuring the terminal **velocity** of individual drops as they fell under gravity and again as they rose under an applied electric field, Millikan was able to measure the charge on the drops. He measured the charge on thousands of drops and was able to follow some drops for long periods of

Electron cloud

KEY TERMS

. . . .

Positron—The antiparticle of the electron. It has the same mass and spin as the electron but its charge, though equal in magnitude, is opposite in sign to that of the electron.

Quarks—Believed to be the most fundamental units of protons and neutrons.

Terminal velocity—Since the air resistance force increases with velocity, all objects falling in air reach a fixed or terminal velocity when this force directed upward equals the force of gravity directed down.

time and to observe changes in the charge on these drops produced by ionizing **x rays**. He observed many drops with only a single electronic charge and never observed a charge that was not an integral multiple of this fundamental unit. Millikan's original measurements gave a value of 1.591×10^{-19} coulombs. These results do not prove that nonintegral charges do not exist, but because many other different experiments later confirmed Millikan's result, he is generally credited with discovering the fundamental nature of the charge on the electron, a discovery for which he received the Nobel Prize in physics in 1923.

See also Neutron; Subatomic particles.

Robert L. Stearns

Electron cloud

The term **electron** cloud is used to describe the area around an atomic nucleus where an electron will probably be. It is also described as the "fuzzy" **orbit** of an atomic electron.

An electron bound to the nucleus of an atom is often thought of as orbiting the nucleus in much the same manner that a **planet** orbits a **sun**, but this is not a valid visualization. An electron is not bound by gravity, but by the **Coulomb force**, whose direction depends on the sign of the particles' charge. (Remember, opposites attract, so the **negative** electron is attracted to the positive **proton** in the nucleus.) Although both the Coulomb force and the gravitational force depend inversely on the square of the **distance** between the objects of interest, and both are central forces, there are important differences. In the classical picture, an accelerating charged particle, like

the electron (a circling body changes direction, so it is always accelerating) should radiate and lose **energy**, and therefore **spiral** in towards the nucleus of an atom... but it does not.

Since we are discussing a very small (microscopic) system, an electron must be described using quantum mechanical rules rather than the classical rules which govern planetary **motion**. According to **quantum mechanics**, an electron can be a wave or a particle, depending on what kind of measurement one makes. Because of its wave nature, one can never predict where in its orbit around the nucleus an electron will be found. One can only calculate whether there is a high probability that it will be located at certain points when a measurement is made.

The electron is therefore described in terms of its probability distribution or probability **density**. This probability distribution does not have definite cutoff points; its edges are somewhat fuzzy. Hence the term "electron cloud." This cloudy probability distribution takes on different shapes, depending on the state of the atom. At room **temperature**, most **atoms** exist in their lowest energy state or "ground" state. If energy is added-by shooting a **laser** at it, for example-the outer electrons can "jump" to a higher state (think larger orbit, if it helps). According to quantum mechanical rules, there are only certain specific states to which an electron can jump. These discrete states are labeled by *quantum numbers*. The letters designating the basic quantum numbers are n, l, and m, where n is the principal or energy **quantum number**, l relates to the orbital angular **momentum** of the electron, and m is a magnetic quantum number. The principal quantum number n can take integer values from 1 to **infinity**. For the same electron, l can be any integer from 0 to $(n - 1)$, and m can have any integer value from $-l$ to l. For example, if $n = 3$, we can have states with $l = 2$, 1, or 0. For the state with $n = 3$ and $l = 2$, we could have $m = -2, -1, 0, 1,$ or 2.

Each set of n, l, m quantum numbers describes a different probability distribution for the electron. A larger n means the electron is most likely to be found farther from the nucleus. For $n = 1$, l and m must be 0, and the electron cloud is spherical about the nucleus. For $n = 2$, $l = 0$, there are two concentric spherical shells of probability about the nucleus. For $n = 2$, $l = 1$, the cloud is more barbell-shaped. We can even have a daisy shape when $l = 3$. The distributions can become quite complicated.

Experiment has verified these distributions for one-electron atoms, but the wave function computations can be very difficult for atoms with more than one electron in their outer shell. In fact, when the motion of more than one electron is taken into account, it can take days for the largest computer to output probability distributions for even a low-lying state, and simplifying approximations must often be made.

Overall, however, the quantum mechanical wave equation, as developed by Schrödinger in 1926, gives an excellent description of how the microscopic world is observed to behave, and we must admit that while quantum mechanics may not be precise, it is accurate.

Electron microscope *see* **Microscopy**
Electronegativity *see* **Chemical bond**

Electronics

Electronics is a field of **engineering** and applied **physics** that grew out of the study and application of **electricity**. Electricity concerns the generation and transmission of power and uses **metal** conductors. Electronics manipulates the flow of electrons in a variety of ways and accomplishes this by using gases, materials like silicon and germanium that are semiconductors, and other devices like solar cells, light-emitting diodes (LEDs), masers, lasers, and microwave tubes. Electronics applications include **radio**, **radar**, **television**, communications systems and satellites, navigation aids and systems, control systems, **space** exploration vehicles, microdevices like watches, many appliances, and computers.

History

The history of electronics is a story of the twentieth century and three key components—the **vacuum tube**, the **transistor**, and the **integrated circuit**. In 1883, Thomas Alva Edison discovered that electrons will flow from one metal conductor to another through a **vacuum**. This discovery of conduction became known as the Edison effect. In 1904, John Fleming applied the Edison effect in inventing a two-element **electron** tube called a **diode**, and Lee De Forest followed in 1906 with the three-element tube, the triode. These vacuum tubes were the devices that made manipulation of electrical **energy** possible so it could be amplified and transmitted.

The first applications of electron tubes were in radio communications. Guglielmo Marconi pioneered the development of the wireless **telegraph** in 1896 and long-distance radio communication in 1901. Early radio consisted of either radio telegraphy (the transmission of Morse code signals) or radio telephony (voice messages). Both relied on the triode and made rapid advances thanks to armed forces communications during World War I. Early radio transmitters, telephones, and

telegraph used high-voltage sparks to make waves and sound. Vacuum tubes strengthened weak audio signals and allowed these signals to be superimposed on **radio waves**. In 1918, Edwin Armstrong invented the "super-heterodyne receiver" that could select among radio signals or stations and could receive distant signals. Radio broadcasting grew astronomically in the 1920s as a direct result. Armstrong also invented wide-band **frequency** modulation (FM) in 1935; only AM or amplitude modulation had been used from 1920 to 1935.

Communications technology was able to make huge advances before World War II as more specialized tubes were made for many applications. Radio as the primary form of education and entertainment was soon challenged by television, which was invented in the 1920s but didn't become widely available until 1947. Bell Laboratories publicly unveiled the television in 1927, and its first forms were electromechanical. When an electronic system was proved superior, Bell Labs engineers introduced the **cathode** ray picture tube and **color** television. But Vladimir Zworykin, an engineer with the Radio Corporation of America (RCA), is considered the "father of the television" because of his inventions, the picture tube and the iconoscope camera tube.

Development of the television as an electronic device benefitted from many improvements made to radar during World War II. Radar was the product of studies by a number of scientists in Britain of the reflection of radio waves. An acronym for *RAdio Detection And Ranging*, radar measures the **distance** and direction to an object using echoes of radio microwaves. It is used for **aircraft** and ship detection, control of weapons firing, navigation, and other forms of surveillance. Circuitry, video, pulse technology, and microwave transmission improved in the wartime effort and were adopted immediately by the television industry. By the mid-1950s, television had surpassed radio for home use and entertainment.

After the war, electron tubes were used to develop the first computers, but they were impractical because of the sizes of the electronic components. In 1947, the transistor was invented by a team of engineers from Bell Laboratories. John Bardeen, Walter Brattain, and William Shockley received a Nobel prize for their creation, but few could envision how quickly and dramatically the transistor would change the world. The transistor functions like the vacuum tube, but it is tiny by comparison, weighs less, consumes less power, is much more reliable, and is cheaper to manufacture with its combination of metal contacts and semiconductor materials.

The concept of the integrated circuit was proposed in 1952 by Geoffrey W. A. Dummer, a British electronics expert with the Royal Radar Establishment. Throughout the

1950s, transistors were mass produced on single wafers and cut apart. The total semiconductor circuit was a simple step away from this; it combined transistors and diodes (active devices) and capacitors and resistors (passive devices) on a planar unit or chip. The semiconductor industry and the silicon integrated circuit (SIC) evolved simultaneously at Texas Instruments and Fairchild Semiconductor Company. By 1961, integrated circuits were in full production at a number of firms, and designs of equipment changed rapidly and in several directions to adapt to the technology. Bipolar transistors and digital integrated circuits were made first, but analog ICs, large-scale integration (LSI), and very-large-scale integration (VLSI) followed by the mid-1970s. VLSI consists of thousands of circuits with on-and-off switches or gates between them on a single chip. Microcomputers, medical equipment, video cameras, and communication satellites are only examples of devices made possible by integrated circuits.

Electronic components

Integrated circuits are sets of electronic components that are interconnected. Active components supply energy and include vacuum tubes and transistors. Passive components absorb energy and include resistors, capacitors, and inductors.

Vacuum tubes or electron tubes are **glass** or ceramic enclosures that contain metal electrodes for producing, controlling, or collecting beams of electrons. A diode has two elements, a cathode and an **anode**. The application of energy to the cathode frees electrons which migrate to the anode. Electrons only flow during one half-cycle of an alternating (AC) current. A grid inserted between the cathode and anode can be used to control the flow and amplify it. A small voltage can cause large flows of electrons that can be passed through circuitry at the anode end.

Special purpose tubes use photoelectric **emission** and secondary emission, as in the television camera tube that emits and then collects and amplifies return beams to provide its output signal. Small amounts of argon, **hydrogen**, mercury, or neon vapors in the tubes change its current capacity, regulate voltage, or control large currents. The finely focused beam from a cathode-ray tube illuminates the coating on the inside of the television picture tube to reproduce images.

Transistors are made of silicon or germanium containing foreign elements that produce many electrons or few. N-type semiconductors produce a lot of electrons, and p-type semiconductors do not. Combining the materials creates a diode, and when energy is applied, the flow can be directed or stopped depending on direction. A triple layer with either n-p-n or p-n-p creates a triode, which, again can be used to amplify signals. The field-effect transistor or FET superimposes an electric field and uses that field to attract or repel charges. The field can amplify the current much like the grid does in the vacuum tube. FETs are very efficient because only a small field controls a large signal. A controlling terminal or gate is called a JFET or junction FET. Addition of metal semiconductors, metal oxides, or insulated gates produce other varieties of transistor that enhance different signal-transmitting aspects.

Integrated circuits

An integrated circuit consists of tens of thousands of transistors and other circuit elements that are fabricated in a substrate of inert material. That material can be ceramic or glass for a film-integrated circuit or silicon or gallium-arsenide for a semiconductor integrated circuit (SIC). These circuits are small pieces or chips that may be 0.08–0.15 sq in long (2–4 sq mm long). Designers are able to place these thousands of components on a chip by using photolithography to place the components and minute conducting paths in the proper patterns for the purpose of each type of circuit. Many chips are made simultaneously on a 4-sq-in (10-sq-cm) wafer.

Several methods are used to introduce impurities into the silicon in the planar process. A mask with some regions isolated is placed over the surface or **plane** of the wafer, and the surface of the silicon is altered or treated to modify its electrical character. Crystals of silicon are grown on the substrate in a process called epitaxy; another method, thermal oxidation, grows a film of silicon dioxide on the surface that acts as a gate insulator. During solid-state **diffusion**, impurities diffused as a gas or spread in a beam of ions are distributed or redistributed in regions of the semiconductor. The number of impurities diffused into the **crystal** can be carefully controlled so the movement of electrons through the chip will also be specific. Coatings can also be added by chemical vapor deposition, **evaporation**, and a method called sputtering used to deposit tungsten on the substrate; the results of all these methods are coatings on the substrate or disturbed surfaces of the substrate that are only **atoms** thick. Etching and other forms of **lithography** (using electron beams or **x rays**) are also used to pattern the wafer surface for the interconnection of the surface elements.

Resistors, capacitors, and inductors

Resistance to the flow of current can be controlled by the conductivity of the material, dimensions over which current flows, and the applied voltage. In electronic circuits, metal films, mixtures containing **carbon**, and resistance wire are used to make resistors. Capacitors have the ability to retain charge and voltage and to act as conductors, especially when currents change in flow. Inductors regulate rapid changes in signals and current intensity.

Sensors

Sensors are specialized electronic devices that detect changes in quantities such as **temperature**, electrical power levels, chemical concentrations, physical position, fluid flow, or mechanical properties like **velocity** or **acceleration**. When a sensor responds to change, it usually requires a **transducer** to convert the quantity the sensor has measured into electrical signals that are translated into printouts, electronic readouts, recordings, or information that is returned to the device to control the change measured. Specialized resistors and capacitors are sometimes used as combined sensors and transducers. Variable resistors respond to mechanical motions by changing them to electrical signals. The thermistor varies its resistance with temperature; a **thermocouple** also measures temperature changes in the form of small voltages as temperatures are measured at two different junctions on the thermocouple. Usually, sensors produce weak electronic signals, and added circuits amplify these. But sensors can be operated from a distance and in conditions such as extreme **heat** or cold or contaminated environments where working conditions are unpleasant or hazardous to humans.

Amplifiers

Amplifiers are electronic devices that boost current, voltage, or power. Audio amplifiers are used in radios, televisions, cassette recorders, sound systems, and citizens band radios. They receive sound as electrical signals, amplify these, and convert them to sound in speakers. Video amplifiers increase the strength of the visual information seen on the television screen by regulating the brightness of the image-forming **light**. Radio frequency amplifiers are used to amplify the signals of communication systems for radio or television broadcasting and operate in the frequency range from 100 kHz to 1 GHz and sometimes into the microwave-frequency range. Video amplifiers increase all frequencies equally up to 6 MHz; audio amplifiers, in contrast, usually operate below 20 kHz. But both audio and video amplifiers are linear amplifiers that proportion the output signal to the input received; that is, they do not distort signals. Other forms of amplifiers are nonlinear and do distort signals usually to some cutoff level. Nonlinear amplifiers boost electronic signals for modulators, mixers, oscillators, and other electronic instruments.

Oscillators

Oscillators are amplifiers that receive an incoming signal and their own output as feedback (that is, also as input). They produce radio and audio signals for precision signaling, such as warning systems, **telephone** electronics between individual telephones and central telephone stations, computers, alarm clocks, high-frequency communications equipment, and the high-frequency transmissions of broadcasting stations.

Power-supply circuits

Electronic equipment usually operates on direct current (DC) power supplies because these are more easily regulated. Power supplies in electrical outlets, however, are alternating currents (AC), so electronic equipment must be able to convert AC to DC. A team of devices is used for this conversion. The piece of equipment has an internal **transformer** that adjusts the voltage it receives from the outlet up or down to suit operation of the equipment. The transformer is also a ground, a type of insulation that reduces the possibility of electrical shock. A rectifier converts AC to DC, and a **capacitor** filters the converted voltage to level out any fluctuations. A voltage regulator may take the place of the capacitor, especially in more sophisticated equipment; modern voltage regulators are manufactured as integrated circuits.

Microwave electronics

Microwaves are the frequencies of choice for many forms of communications especially telephone and television signals that are transmitted long distances through overland methods, broadcast stations, and satellites. Microwave electronics are also used for radar.

Microwaves are within the frequency of 3 GHz to about 300 GHz; because of their high frequency **spectrum**, microwaves can carry large numbers of channels. They also have short wavelengths from 10 cm to 0.1 cm; wavelength dictates the size of **antenna** that can be used to transmit that particular wavelength, so the small antennae for microwave communications are very practical. They do require repeater stations to make long-distance links.

Electronic devices like capacitors, inductors, oscillators, and amplifiers were not usable with microwaves because their high frequency and the speeds of electrons are not compatible. This complication of component size was studied in detail in the 1930s. Finally, it was found that the velocity of the electrons could be modulated to the advantage of microwave applications. The modulating device, the klystron, was a tube that amplified the microwave signal in a resonating cavity. The klystron could amplify only a narrow range of microwave frequencies, but the traveling-wire tube (invented in 1934)—a similar velocity modulator—could amplify a wider frequency band using a wire helix instead of a resonating cavity.

High-powered and high-pulsed microwave use especially for radar required another device, the magnetron. The magnetron was perfected in 1939 and was a tube with multiple resonating cavities. While these devices were successful for their specialized uses, they were expensive and bulky (like other vacuum tubes); they have been replaced completely by semiconductors and integrated circuits with equally sophisticated and specialized solutions for handling the high frequencies of microwaves that fit much smaller spaces and can be mass produced economically.

Microwave electronics have also required adaptations of other parts of transmission systems. Conventional wires can not carry microwaves because of the energy they give off; instead, coaxial cables can carry microwaves up to 5 GHz in frequency because their self-shielding conductors prevent radiating energy. Waveguides are used for higher-frequency microwave transmission; waveguides are hollow metal tubes with a refractive interface that reflects energy back. Microstrips are an alternative to waveguides that connect microwave components and work by separating two conductors with dielectric material. Microstrips (also called striplines) can be manufactured using integrated circuit (IC) technology and are compatible with the small size of ICs.

Masers were first developed in 1954. The **Maser** (*M*icrowave *A*mplification by *S*timulated *E*mission of *R*adiation) can be used for amplifying and oscillating microwaves in signals from satellites, atomic clocks, spacecraft, and radio. Masers focus molecules in an excited energy state into a resonant microwave cavity which then emits them as stimulated emission of **radiation** through the microwave output.

Optical electronics

Optical electronics involve combined applications of optical (light) signals and electronic signals. Optoelectronics have a number of uses, but the three general classifications of these uses are to detect light, to convert solar energy to electric energy, and to convert electric energy to light. Like radio waves and microwaves, light is also a form of electromagnetic radiation except that its wavelength is very short. Photodetectors allow light to irradiate a semiconductor that absorbs the light as photons and converts these to electric signals. Light meters, burglar alarms, and many industrial uses feature photodetectors.

Solar cells convert light from the **sun** to electric energy. They use single-crystal doped silicon to reduce internal resistance and metal contacts to convert over 14% of the solar energy that strikes their surfaces to electrical output voltage. Cheaper, polycrystalline silicon sheets

and other lenses are being developed to reduce the cost and improve the effectiveness of solar cells.

Light-emitting diodes (LEDs) direct incoming voltage to gallium-arsenide semiconductors that, when agitated, emit photons of light. The wavelength of the emitted light depends on the material used to construct the semiconductor. The **LED** is used in many applications where illuminated displays are needed on instruments and household appliances. Liquid crystal displays (LCDs) use very low power levels to produce reflected or scattered light; they cannot be seen in the dark like LED displays because they do not produce light. Conductive patterns of electrodes overlie parallel-plate capacitors that hold large molecules of the liquid crystal material that works as the dielectric.

Like microwaves, optical electronics use waveguides to reflect, confine, and direct light. The most familiar form of optical waveguide is the optic fiber. These fine, highly specialized glass fibers are made of silica that has been doped with germanium dioxide.

Digital electronics

Digital electronics are the electronics that transformed our lives beginning in the 1970s. The personal computer is one of the best examples of this transformation because it has simplified tasks that were difficult or impossible for individuals to accomplish. Digital devices use simple "true-false" or "on-off" statements to represent information and to make decisions. In contrast, analog devices use a continuous system of values. Because digital devices only recognize one of two permissible signals, they are more tolerant to noise (unwanted electronic signals) and a range of components than analog devices. Digital systems are built of a collection of components that process, store, and transmit or communicate information. The basis of these components is the logic circuit that makes the true-false decision from what may be many true-false signals. The logic circuit is an integrated circuit from any one of a number of families of digital logic devices that use switches, transducers, and timing circuits to function. Digital logic gates are the most elementary inputs and outputs in a logic device. A logic gate is based on a simple operation in **Boolean algebra** (a form of **mathematics** that uses logic variables to express thought processes). For example, a logic gate may perform an "or," "and," or "not" function; to make it capable of a "nor" function, an "or" gate is followed by an inverter. By linking combinations of these gates, any decision is possible.

The most popular form of logic circuit is probably the transistor-transistor logic (TTL) circuit. High-speed systems use emitter coupled logic (ELC), and

KEY TERMS

· ·

Epitaxy—The growth of a crystalline substance on a substrate such that the crystals imitate the orientation of those in the substrate.

the complementary metal oxide semiconductor (CMOS) logic uses lower speeds to also lower power levels. Logic gates are also combined to make static-memory cells. These are combined in a rectangular array to form the random-access memory (RAM) familiar to home computer users. The binary digits that make up this memory are called "bits," and typical large-scale integrated (LSI) circuit memory chips have over 16,000 bits of static memory. Dynamic memory cells use capacitors to send memory to a selected cell or to "write" to that cell. Very-large-scale chips with 256,000 bits per chip were made beginning in the 1980s, and dynamic memory made these possible because of its high **density**.

Microprocessors have replaced combinations of switching and timing circuits. They are programmed to perform sets of tasks and a wider variety of logic functions. Electronic games and digital watches are examples of microprocessor systems. Digital methods have revolutionized music, library storage, medical electronics, and high definition television, among thousands of other tools that influence our lives daily. Future changes to so-called "computer architecture" are directed at greater speed; ultra-high-speed computers may operate by using superconducting circuits that operate at extremely cold temperatures, and integrated circuits that house hundreds of thousands of electronic components on one chip may be commonplace on our desktops.

Resources

Books

Boylstad, Robert, and Louis Nashalsky. *Electronics: A Survey.* Englewood Cliffs, NJ: Prentice Hall, 1985.
Houglum, Roger J. *Electronics: Concepts, Applications and History.* 2nd ed. Albany, NY: Delmar Publishers, 1985.
Patrick, Dale R., and Stephen W. Fardo. *Understanding Electricity and Electronics.* Upper Saddle River, NJ: Prentice Hall, 1989.
Riordan, Michael, and Lillian Hoddeson. *Crystal Fire: The Birth of the Information Age.* New York: W. W. Norton & Company, 1997.
Vergarra, William C. *Electronics in Everyday Life.* New York: Dover Publications, Inc., 1984.

Periodicals

Adler, Jerry. "Three Magic Wands." *Newsweek* (Winter 1997): 6+.

Bains, Sunny. "Double Helix Doubles as Engineer." *Science* (March 27, 1998): 2043+.
"Elephant Chips." *Discover* (July 1998): 62.

Gillian S. Holmes

Electrophoresis

Electrophoresis is a technique used for the separation of biological molecules based on their movement due to the influence of a direct **electric current**. The technique was pioneered in 1937 by the Swedish chemist Arne Tiselius for the separation of **proteins**. It has now been extended to the separation of many other different classes of biomolecules including nucleic acids, carbohydrates and amino acids.

Electrophoresis has become increasingly important in the laboratory for basic research, biomedical research and in clinical settings for the **diagnosis** of **disease**. Electrophoresis is not commonly used to purify proteins in large quantities because other methods exist which are simpler, faster, and more efficient. However, it is valuable as an analytical technique for detecting and quantifying minute traces of many biomolecules in a mixture. It is also useful for determining certain physical properties such as **molecular weight**, isoelectric point, and biological activity.

Electrophoretic theory

Electrophoretic separations are based upon the fact that the electrical **force** (F) on a charged particle (ion) in an electrical field (E) is proportional to the charge of the particle (q), or F = qE (Eq 1).

The **migration** of the charged particle in the electric field, called the electrophoretic mobility (μ), is defined as $\mu = v/E = q/f$ (Eq 2), where v is the **velocity** of the charged particle and f is a complex term called the frictional **coefficient**. The frictional coefficient relates to the size and the shape of the particle. From equation (2) it can be seen that electrophoretic mobility decreases for larger particles and increases with higher charge.

Methodology and applications

The electrophoresis equipment can have several designs. The simplest approach is the *moving boundary technique* (Figure 3). The charged molecules (i.e., proteins) to be separated are electrophoresed upward through a **buffer solution** toward electrodes immersed on either side of a U-shaped tube. The individual proteins are resolved because they have different mobilities as described above. This technique separates the biomolecules on the

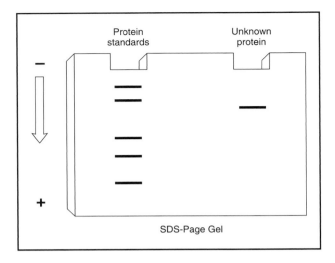

Figure 1. *Illustration by Hans & Cassidy. Courtesy of Gale Group.*

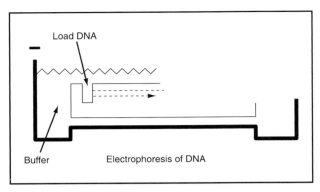

Figure 2. *Illustration by Hans & Cassidy. Courtesy of Gale Group.*

basis of their charges. Positively charged molecules migrate toward the **negative** electrode (**cathode**) and negatively charged particles move to the positive electrode (**anode**). The migration of the particles to the electrodes can be followed by instruments which measure the refractive index or absorption of **light** by the solution.

The most important class of electrophoresis is *zone electrophoresis*. In zone electrophoresis, the **sample** to be separated is applied to a solid **matrix** through which it migrates under the force of an applied electric potential. Two major classes of zone electrophoresis will be discussed below.

Gel electrophoresis

The sample is loaded into a gel matrix support. This has many important advantages:

1. The **density** and porosity of gels can be easily controlled and adjusted for different biomolecules or different experimental conditions. Since the gel pore size is on the order of the dimensions of the macromolecules, the separations are based on *molecular sieving* as well as electrophoretic mobility of the molecules. Large molecules are retarded relative to smaller molecules as they cannot pass as easily through the gel during electrophoresis.

2. Gels are easy to chemically modify, for related techniques such as affinity electrophoresis which separates biomolecules on the basis of biomolecular affinity or recognition.

3. Excellent separation power.

Gel electrophoresis of DNA

Highly purified agarose, a major component of seaweed, is used as the solid gel matrix into which the DNA

samples are loaded for electrophoresis. By varying the agarose **concentration** in the gel, DNA fragments in different size ranges can be separated.

The agarose is dissolved in **water**, heated and cast as a gel slab approximately 0.2 in (0.5 cm) in thickness. Wells are formed at one end of the gel for the loading of the DNA sample. The slab is then placed horizontally into the electrophoresis buffer chamber (Figure 2).

The DNA migrates in bands toward the positive electrode. The smaller molecules pass through the matrix more rapidly than the larger ones, which are restricted. The DNA bands are then stained using a fluorescent dye such as ethidium bromide. The stained gel is then viewed directly under ultraviolet light and photographed.

Gel electrophoresis of proteins

Because proteins are typically much smaller than DNA, they are run in gels containing polymers of much smaller pore size. The most common technique for protein separation is known as SDS-Polyacrylamide Gel Electrophoresis (SDS-PAGE) (Figure 1). In this approach the proteins are treated with a detergent (**sodium** dodecyl sulfate) which unfolds them and gives them similar shape and **ratio** of charge to **mass**. Thus, proteins, treated with SDS separate on the basis of mass, with the smaller proteins migrating more rapidly through the gel. The gel matrix is polyacrylamide, a synthetic copolymer which has excellent molecular sieving properties. Polyacrylamide gels are cast between **glass** plates to form what is called a "sandwich." Protein gels are generally run in a vertical fashion. The electrophoresis apparatus has an upper and lower buffer tank. The top and bottom of the sandwich is in contact with either buffer. Protein is loaded into the wells in the upper buffer tank and current is applied. The proteins thus migrate down through the gel in bands, according to their sizes. After electrophoresis, the polyacrylamide gel is removed from between the

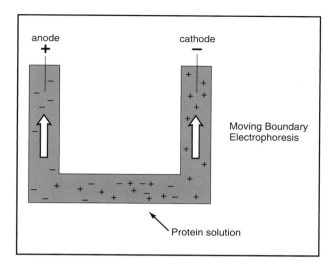

anode +

cathode −

Moving Boundary
Electrophoresis

Protein solution

Figure 3. *Illustration by Hans & Cassidy. Courtesy of Gale Group.*

glass plates and chemically stained to show protein bands which can then be studied. The SDS-PAGE technique allows researchers to study parts of proteins and protein-protein interactions. If a protein has different subunits they will be separated by SDS treatment and will form separate bands.

Paper electrophoresis

This technique is useful for the separation of small charged molecules such as amino acids and small proteins. A strip of filter **paper** is moistened with buffer and the ends of the strip are immersed into buffer reservoirs containing the electrodes. The samples are spotted in the center of the paper, high voltage is applied, and the spots migrate according to their charges. After electrophoresis, the separated components can be detected by a variety of staining techniques, depending upon their chemical identity.

Electrophoretic techniques have also been adapted to other applications such as the determination of protein isoelectric points. Affinity gels with biospecific properties are used to study binding sites and surface features of proteins. Continuous flow electrophoresis is applied to separations in free solution and has found very useful application in **blood** cell separation. Recently, High Performance Capillary Electrophoresis (HPCE) has been developed for the separation of many classes of biological molecules.

See also Bioassay; Biochemistry; Biotechnology; Proteomics.

Resources

Books

Lodish, H., et al. *Molecular Cell Biology.* 4th ed. New York: W. H. Freeman & Co., 2000.

Nelson, David L., and Michael M. Cox. *Lehninger Principles of Biochemistry.* 3rd ed. Worth Publishing, 2000.

Westheimer, Reiner. *Electrophoresis in Practice.* 3rd ed. New York: Springer Verlag, 2001.

Organizations

Human Proteome Organization. <http://www.hupo.org/.>

Leonard D. Holmes

Electrostatic devices

Electrostatics is the study of the behavior of electric charges that are at rest. The phenomenon of static **electricity** has been known for well over 2,000 years, and a variety of electrostatic devices have been created over the centuries.

The ancient Greek philosopher Thales (624-546 B.C.) discovered that when a piece of amber was rubbed, it could pick up light objects, a process known as triboelectrification. The Greek name for amber, *elektron*, gave rise to many of the words we use in connection with electricity. It was also noted that lodestone had the natural ability to pick up **iron** objects, although the early Greeks did not know that electricity and **magnetism** were linked.

In the late sixteenth century, William Gilbert (1544-1603) began experimenting with static electricity, pointing out the difference between static electric attraction

The Vivitron electrostatic particle accelerator under construction at the Centre des Recherches Nucleaires, Strasbourg, France. Vivitron, the largest Van de Graaff generator in the world, can generate a potential of up to 35 million volts. The accelerator will be used to fire ions of elements such as carbon at other nuclei. Under the right conditions this creates superdeformation, a relatively stable state in which the rotating nuclei have an elliptical form. Gamma rays given off by these nuclei reveal much about the internal structure of the nucleus. *Photograph by Philippe Plailly/Eurelios. National Audubon Society Collection/Photo Researchers, Inc. Reproduced by permission.*

and magnetic attraction. Later, in the mid-1600s, Otto von Guericke built the first electrostatic machine. His device consisted of a **sulfur** globe that was rotated by a crank and stroked by hand. It released a considerable static **electric charge** with a large spark.

A similar device was invented by Francis Hawkesbee in 1706. In his design, an iron chain contacted a spinning globe and conducted the electric charge to a suspended gun barrel; at the other end of the barrel another chain conducted the charge.

In 1745, the first electrostatic storage device was invented nearly simultaneously by two scientists working independently. Peter von Muschenbrock, a professor at the University of Leyden, and Ewald von Kleist of the Cathedral of Camin, Germany, devised a water-filled **glass** jar with two electrodes. A Leyden student who had been using a Hawkesbee machine to electrify the **water** touched the chain to remove it and nearly died from the electric shock. This device, known as the Leyden jar, could accumulate a considerable electric charge, and au-

diences willingly received electric shocks in public displays. One of these displays aroused the curiosity of Benjamin Franklin, who obtained a Leyden jar for study. He determined that it was not the water that held the electric charge but the glass insulator. This is the principle behind the electrical condenser (**capacitor**), one of the most important electrical components in use today.

Charles F. DuFay (1698-1739) discovered that suspended bits of **cork**, electrified with a statically charged glass rod, repelled each other. DuFay concluded that any two objects which had the same charge repelled each other, while unlike charges attracted. The science of electrostatics, so named by André Ampère (1775-1836), is based on this fact.

French physicist Charles **Coulomb** (1736-1806) became interested in the work of Joseph Priestley (1733-1804), who had built an electrostatic **generator** in 1769, and studied electrical repulsion. Coulomb used his torsion balance to make precise measurements of the **force** of attraction between two electrically charged spheres

A Van de Graaff generator is a device that is capable of building up a very high electrostatic potential. In this photo, the charge that has accumulated in the dome is leaking into the hair of a wig that has been placed on top of the generator. Because the charge is of one polarity, the hairs repel each other. © Adam Hart-Davis/Science Photo Library, National Audubon Society Collection/Photo Researchers, Inc. Reproduced with permission.

and found they obeyed an inverse square law. The mathematical relationship between the forces is known as Coulomb's law, and the unit of electric charge is named the coulomb in his honor.

Alessandro Volta invented a device in 1775 that could create and store an electrostatic charge. Called an electrophorus, it used two plates to accumulate a strong positive charge. The device replaced the Leyden jar, and the two-plate principle is behind the electrical condensers in use today.

Several other electrostatic machines have been devised. In 1765 John Reid, an instrument maker in London, built a portable static electric generating machine to treat medical problems. In 1783, John Cuthbertson built a huge device that could produce electrical discharges 2 ft (61 cm) in length. The gold leaf electroscope, invented in 1787, consists of two leaves which repel each other when they receive an electric charge. In 1881, British engineer James Wimshurst invented his Wimshurst machine, two glass discs with **metal** segments spinning opposite each other. Brushes touching the metal segments removed the charge created and conducted it to a pair of Leyden jars where it was stored for later use.

The most famous of all the electrostatic devices is the Van de Graaff generator. Invented in 1929 by Robert J. Van de Graaff, it uses a conveyor belt to carry an electric charge from a high-voltage supply to a hollow ball. It had various applications. For his experiments on properties of **atoms**, Van de Graaff needed to accelerate **subatomic particles** to very high **velocity**, and he knew that storing an electrostatic charge could result in a high po-

tential. Another generator was modified to produce **x rays** for use in the treatment of internal tumors. It was installed in a hospital in Boston in 1937. Van de Graaff's first generator operated at 80,000 volts, but was eventually improved to five million volts. It remains one of the most widely used experimental exhibits in schools and museums today.

Electroweak interactions *see* **Subatomic particles**

Element, chemical

A chemical element is a substance made up of only one kind of atom (**atoms** having the same **atomic number**). A compound, on the other hand, is made up of two or more kinds of atom combined together in specific proportions.

The atomic number of an element is the number of protons found in the nucleus of each atom of that element; the number of protons in the nucleus equals the number of electrons that can bind to the atom. (Since electrons and protons have equal but opposite electrical charges, atoms can bind as many electrons to themselves as they have protons in their nuclei.) Because the chemical properties of an atom—the ways in which it binds to other atoms—are determined by the number of electrons that can bind to its nucleus, every element has a unique set of chemical properties.

Some elements, such as the **rare gases**, exist as collections of single atoms; such a substance is monatomic. Others may exist as molecules that consist of two or more atoms of that element bonded together. For example, **oxygen** (O) can remain stable as either a diatomic (two-atom) **molecule** (O_2) or a triatomic (three-atom) molecule (O_3). (O_2 is the form of oxygen that we breathe; O_3 [ozone] is toxic to animals and plants, yet **ozone** in the upper atmosphere screens **Earth** from harmful solar radiation.) **Phosphorus** (P) is stable as a four-atom molecule (P_4), while **sulfur** (S) is stable as an eight-atom molecule (S_8).

Even though all atoms of a given element have the same number of protons in their nuclei, they may not have the same number of neutrons in their nuclei. Atoms of the same element having different numbers of neutrons in their nuclei are termed isotopes of that element. An **isotope** is named according to the sum of the number of protons and the number of neutrons in its nucleus. For example, 99% of all **carbon** (C), atomic number 6, has 6 neutrons in the nucleus of each atom; this isotope

TABLE 1. TWO DOZEN OF THE MOST COMMON AND/OR IMPORTANT CHEMICAL ELEMENTS.

Element	Symbol	Percent of all atoms[a]				Characteristics under ordinary room conditions
		In the universe	In the Earth's crust	In sea water	In the human body	
Aluminum	Al	—	6.3	—	—	A lightweight, silvery metal
Calcium	Ca	—	2.1	—	0.2	Common in minerals, seashells, and bones
Carbon	C	—	—	—	10.7	Basic in all living things
Chlorine	Cl	—	—	0.3	—	A toxic gas
Copper	Cu	—	—	—	—	The only red metal
Gold	Au	—	—	—	—	The only yellow metal
Helium	He	7.1	—	—	—	A very light gas
Hydrogen	H	92.8	2.9	66.2	60.6	The lightest of all elements; a gas
Iodine	I	—	—	—	—	A nonmetal; used as antiseptic
Iron	Fe	—	2.1	—	—	A magnetic metal; used in steel
Lead	Pb	—	—	—	—	A soft, heavy metal
Magnesium	Mg	—	2.0	—	—	A very light metal
Mercury	Hg	—	—	—	—	A liquid metal; one of the two liquid elements
Nickel	Ni	—	—	—	—	A noncorroding metal; used in coins
Nitrogen	N	—	—	—	2.4	A gas; the major component of air
Oxygen	O	—	60.1	33.1	25.7	A gas; the second major component of air
Phosphorus	P	—	—	—	0.1	A nonmetal; essential to plants
Potassium	K	—	1.1	—	—	A metal; essential to plants; commonly called "potash"
Silicon	Si	—	20.8	—	—	A semiconductor; used in electronics
Silver	Ag	—	—	—	—	A very shiny, valuable metal
Sodium	Na	—	2.2	0.3	—	A soft metal; reacts readily with water, air
Sulfur	S	—	—	—	0.1	A yellow nonmetal; flammable
Titanium	Ti	—	0.3	—	—	A light, strong, noncorroding metal used in space vehicles
Uranium	U	—	—	—	—	A very heavy metal; fuel for nuclear power

[a] If no number is entered, the element constitutes less than 0.1 percent.

TABLE 2. A WHO'S WHO OF THE ELEMENTS.		
Element	*Distinction*	*Comment*
Astatine (At)	The *rarest*	Rarest of the naturally occurring elements
Boron (B)	The *strongest*	Highest stretch resistance
Californium (Cf)	The *most expensive*	Sold at one time for about $1 billion a gram
Carbon (C)	The *hardest*	As diamond, one of its three solid forms
Germanium (Ge)	The *purest*	Has been purified to 99.99999999 percent purity
Helium (He)	The *lowest melting point*	-457.09°F (-271.72°C) at a pressure of 26 times atmospheric pressure
Hydrogen (H)	The *lowest density*	Density 0.0000899 g/cc at atmospheric pressure and 32°F (0°C)
Lithium (Li)	The *lowest-density metal*	Density 0.534g/cc
Osmium (Os)	The *highest density*	Density 22.57 g/cc
Radon (Rn)	The *highest-density gas*	Density 0.00973 g/cc at atmospheric pressure and 32°F (0°C)
Tungsten (W)	The *highest melting point*	6,188°F (3,420°C)

of carbon is called carbon 12 (^{12}C). An isotope is termed stable if its nuclei are permanent, and unstable (or radioactive) if its nuclei occasionally explode. Some elements have only one stable (nonradioactive) isotope, while others have two or more. Two stable isotopes of carbon are ^{12}C (6 protons, 6 neutrons) and ^{13}C (6 protons, 7 neutrons); a radioactive isotope of carbon is ^{14}C (6 protons, 8 neutrons). Tin (Sn) has ten stable isotopes. Some elements have no stable isotopes; all their isotopes are radioactive. All isotopes of a given element have the same outer **electron** structure and therefore the same chemical properties.

Ninety-two different chemical elements occur naturally on Earth; 81 of these have at least one stable isotope. Other elements have been made synthetically (artificially), usually by causing the nuclei of two atoms to collide and merge. Since 1937, when technetium (Tc, atomic number 43), the first synthetic element, was

made, the number of known elements has grown as nuclear chemists made new elements. Most of these synthetic elements have atomic numbers higher than 92 (i.e., more than 92 protons in their nuclei); since 92 is the atomic number of **uranium** (U), these artificial heavy elements are called "transuranium" (past-uranium) elements. The heaviest element so far is Element 114, whose synthesis was announced in January 1999. In June, 1999, scientists at Lawrence Berkeley National Laboratory in California announced the synthesis of elements 116 and 118; however, it was later revealed that these announcements had been based on fabricated (made-up) data, and the claim to have synthesized these elements was publicly retracted. The same researcher who falsified the data had participated in the work leading up to the announcements of elements 110 and 112 in 1994 and 1996, but later analysis confirmed that enough authentic evidence existed to support the announcement that 110 and 112 had been synthesized.

A survey of the elements

Of the 114 currently known elements, 11 are gases, two are liquids, and 101 are solids. (The transuranium elements are presumed to be solids, but since only a few atoms at a time can be synthesized it is impossible to be sure.) Many elements, such as **iron** (Fe), **copper** (Cu), and **aluminum** (Al), are familiar everyday substances, but many are unfamiliar, either because they are not abundant on Earth or because they are not used much by human beings. Less-common naturally occurring elements include dysprosium (Dy), thulium (Tm) and protactinium (Pa).

Every element (except a few of the transuranium elements) has been assigned a name and a one- or two-letter symbol for convenience in writing formulas and chemical equations; these symbols are shown above in parentheses. For example, to distinguish the four elements that begin with the letter c, **calcium** is symbolized as Ca, cadmium as Cd, californium as Cf, and carbon as C.

Many of the symbols for chemical elements do not seem to make sense in terms of their English names—Fe for iron, for example. Those are mostly elements that have been known for thousands of years and that already had Latin names before chemists began handing out the symbols. Iron is Fe for its Latin name, *ferrum*. Gold is Au for *aurum*, **sodium** is Na for *natrium*, copper is Cu for *cuprum*, and mercury is Hg for *hydrargyrum*, meaning liquid silver, which is exactly what it looks like, but is not.

Table 1 lists some of the most common and important chemical elements. Note that many of these are referred to in the last column as metals. In fact, 93 out of the 114 elements are metals; the others are nonmetals.

Notice that only two elements taken together—hydrogen and helium—make up 99.9% of the atoms in the entire universe. That is because virtually all the **mass** in the universe is in the form of stars, and stars are made mostly of H and He. Only H and He were produced in the big bang that began the universe; all other elements have been built up by nuclear reactions since that time, either naturally (in the cores of stars) or artificially (in laboratories). On Earth, only three elements—oxygen, silicon and aluminum—make up more than 87% of the earths's crust (the rigid, rocky outer layer of the **planet**, about 10.5 mi [17 km] under most dry land [less under the oceans]). Only six more elements—hydrogen, sodium, calcium, iron, **magnesium**, and potassium—account for more than 99% of Earth's crust.

The abundance of an element can be quite different from its importance to humans. Nutritionists believe that some 24 elements are essential to life, even though many are fairly rare and are needed in only tiny amounts.

History of the elements

Many substances now known as elements have been known since ancient times. Gold (Au) was found and made into ornaments during the late stone age, some 10,000 years ago. More than 5,000 years ago, in Egypt, the metals iron (Fe), copper (Cu), silver (Ag), tin (Sn), and **lead** (Pb) were also used for various purposes. Arsenic (As) was discovered around A.D. 1250, and phosphorus (P) was discovered around 1674. By 1700, about 12 elements were known, but they were not yet recognized as they are today.

The concept of elements—i.e., the theory that there are a limited number of fundamental pure substances out of which all other substances are made—goes back to the ancient Greeks. Empedocles (c. 495–435 B.C.) proposed that there are four basic "roots" of all materials: earth, air, fire, and **water**. Plato (c. 427–347 B.C.) referred to these four "roots" as *stoicheia* elements. Aristotle (384–322 B.C.), a student of Plato's, proposed that an element is "one of those simple bodies into which other bodies can be decomposed and which itself is not capable of being divided into others." Except for **nuclear fission** and other nuclear reactions discovered more than 2,000 years later, by which the atoms of an element can be decomposed into smaller parts, this definition remains accurate.

Several other theories were generated throughout the years, most of which have been dispelled. For example, the Swiss physician and alchemist Theophrastus Bombastus von Hohenheim (c. 1493–1541), also known as Paracelsus, proposed that everything was made of three "principles:" **salt**, mercury, and sulfur. An alchemist named van Helmont (c. 1577–c.1644) tried to explain everything in terms of just two elements: air and water.

Eventually, English chemist Robert Boyle (1627–1691) revived Aristotle's definition and refined it. In 1789, French chemist Antoine Lavoisier (1743–94) was able to publish a list of chemical elements that met Boyle's definition. Even though some of Lavoisier's "elements" later turned out to be compounds (combinations of actual elements), his list set the stage for the adoption of standard names and symbols for the various elements.

The Swedish chemist J. J. Berzelius (1779–1848) was the first person to employ the modern method of classification: a one- or two-letter symbol for each element. These symbols could be put easily together to show how the elements combine into compounds. For example, writing two Hs and one O together as H_2O would mean that the particles (molecules) of water consist of two **hydrogen** atoms and one oxygen atom, bonded together. Berzelius published a table of 24 elements, including their atomic weights, most of which are close to the values used today.

By the year 1800 only about 25 true elements were known, but progress was relatively rapid throughout the nineteenth century. By the time Russian scientist Dmitri Ivanovich Mendeleev (1834–1907) organized his **periodic table** in 1869, he had about 60 elements to reckon with. By 1900 there were more than 80. The list quickly expanded to 92, ending at uranium (atomic number 92). There it stayed until 1940, when synthesis of the transuranium elements began.

Organization of the elements

The task of organizing more than a hundred very different elements into some simple, sensible arrangement would seem difficult. Mendeleev's periodic table, however, is the answer. It even accommodates the synthetic transuranium elements without strain. In this encyclopedia, each individual chemical element is discussed under at least one of the following types of entry: (1) Fourteen particularly important elements are discussed in their own entries. They are aluminum, calcium, carbon, **chlorine**, copper, hydrogen, iron, lead, **nitrogen**, oxygen, silicon, sodium, sulfur, and uranium. (2) Elements that belong to any of seven families of elements—groups of elements that have similar chemical properties—are discussed under their family-name headings. These seven families are the **actinides**, **alkali metals**, **alkaline earth metals**, **halogens**, **lanthanides**, rare gases, and transuranium elements. (3) Elements that are not discussed either under their own name or as part of a family ("orphan elements") are discussed briefly below. Any element that is not discussed below can be found in the headings described above.

"Orphan" elements

Actinium. The metallic chemical element of atomic number 89. Symbol Ac, specific gravity 10.07, melting point 1,924°F (1,051°C), **boiling point** 5,788°F (3,198°C). All isotopes of this element are radioactive; the **half-life** of its most stable isotope, actinium-227, is 21.8 years. Its name is from the Greek *aktinos*, meaning ray.

Antimony. The metallic chemical element of atomic number 51. Symbol Sb, **atomic weight** 121.8, specific gravity 6.69, melting point 1,167°F (630.63°C), boiling point 2,889°F (1,587°C). One of its main uses is to **alloy** with lead in **automobile** batteries; actinium makes the lead harder.

Arsenic. The metallic chemical element of atomic number 33. Symbol As, atomic weight 74.92, specific gravity 5.73 in gray metallic form, melting point 1,503°F (817°C), sublimes (solid turns to gas) at 1,137°F (614°C). Arsenic compounds are poisonous.

Bismuth. The metallic chemical element of atomic number 83. Symbol Bi, atomic weight 208.98, specific gravity 9.75, melting point 520.5°F (271.4°C), boiling point 2,847.2°F (1,564°C). Bismuth oxychloride is used in "pearlized" cosmetics. Bismuth subsalicylate, an insoluble compound, is the major ingredient in Pepto-Bismol. The soluble compounds of bismuth, however, are poisonous.

Boron. The non-metallic chemical element of atomic number 5. Symbol B, atomic weight 10.81, specific gravity (amorphous form) 2.37, melting point 3,767°F (2,075°C), boiling point 7,232°F (4,000°C). Common compounds are borax, $Na_2B_4O_7 \cdot 10H_2O$, used as a cleansing agent and water softener, and **boric acid**, H_3BO_3, a mild antiseptic and an effective cockroach poison.

Cadmium. The metallic chemical element of atomic number 48. Symbol Cd, atomic weight 112.4, specific gravity 8.65, melting point 609.92°F (321.07°C), boiling point 1,413°F (767°C). A soft, highly toxic **metal** used in silver solder, in many other alloys, and in nickel-cadmium rechargeable batteries. Because it is an effect absorber of moving neutrons, it is used in control rods for nuclear reactors to slow the chain reaction.

Chromium. The metallic chemical element of atomic number 24. Symbol Cr, atomic weight 51.99, specific gravity 7.19, melting point 3,465°F (1,907°C), boiling point 4,840°F (2,671°C). A hard, shiny metal that takes a high polish. Used to electroplate **steel** for protection against **corrosion** and as the major ingredient (next to iron) in stainless steel. Alloyed with nickel, it makes Nichrome, a high-electrical-resistance metal that gets red hot when **electric current** passes through it; toaster and heater coils are made of Nichrome wire. Chromium is named from the Greek *chroma*, meaning **color**, because most of its compounds are highly colored. Chromium is responsible for the green color of emeralds.

Cobalt. The metallic chemical element of atomic number 27. Symbol Co, atomic weight 58.93. Cobalt is a grayish, hard, brittle metal closely resembling iron and nickel. These three metals are the only naturally occurring magnetic elements on Earth.

Gallium. The metallic chemical element of atomic number 31. Symbol Ga, atomic weight 69.72, melting point 85.6°F (29.78°C), boiling point 3,999°F (2,204°C). Gallium is frequently used in the **electronics** industry and in thermometers that measure a wide range of temperatures.

Germanium. The metallic chemical element of atomic number 32. Symbol Ge, atomic weight 72.59. In pure form, germanium is a brittle **crystal**. It was used to make the world's first **transistor** and is still used as a semiconductor in electronics devices.

Gold. The metallic chemical element of atomic number 79. Symbol Au, atomic weight 196.966. This most malleable of metals was probably one of the first elements known to humans. It is usually alloyed with harder metals for use in jewelry, coins, or decorative pieces.

Hafnium. The metallic chemical element of atomic number 72. Symbol Hf, atomic weight 178.49, melting point 4,040.6 ±S68°F (2,227 ±20°C), boiling point 8,315.6°F (4,602°C). Hafnium is strong and resistant to corrosion. It also absorbs neutrons well, making it useful in control rods of nuclear reactors.

Indium. The metallic chemical element of atomic number 49. Symbol In, atomic weight 114.82, melting point 313.89°F (156.61°C), boiling point 3,776°F (2,080°C). Indium is a lustrous, silvery metal that bends easily. It is often alloyed with other metals in solid-state-electronics devices.

Iridium. The metallic chemical element of atomic number 77. Symbol Ir, atomic weight 192.22. Iridium is an extremely dense metal that resists corrosion better than most others. In its pure state, it is often used in **aircraft** spark plugs.

Manganese. The metallic chemical element of atomic number 25. Symbol Mn, atomic weight 54.93. The biggest use of manganese is in steelmaking, where it is alloyed with iron. This element is required by all plants and animals, so it is sometimes added as manganese oxide to **animal** feed.

Mercury. The metallic chemical element of atomic number 80. Symbol Hg, atomic weight 200.59, melting point -37.96°F (-38.87°C), boiling point 673.84°F (356.58°C). Mercury is highly poisonous and causes irreversible damage to the nervous and excretory systems. This element was long used in thermometers because it expands and contracts at a nearly constant **rate**; however, mercury thermometers are being phased out in favor of alcohol-based and electronic thermometers because of mercury's high toxicity.

Molybdenum. The metallic chemical element of atomic number 42. Symbol Mo, atomic weight 95.94, melting point 4,753°F (2,623°C), boiling point 8,382°F (4,639°C). Molybdenum is used to make superalloyed metals designed for high-temperature processes. It is also found as a trace element in **plant** and animal tissues.

Nickel. The metallic chemical element of atomic number 28. Symbol Ni, atomic weight 58.71. Nickel is often mixed with other metals, such as copper and iron, to increase the alloy's resistance to **heat** and moisture.

Niobium. The metallic chemical element of atomic number 41. Symbol Nb, atomic weight 92.90, melting point 4,474.4 ±50°F (2,468 ±10°C), boiling point 8,571.2°F (4,744°C). Niobium is used to strengthen alloys used to make lightweight aircraft frames.

Osmium. The metallic chemical element of atomic number 76. Symbol Os, atomic weight 190.2. Osmium is hard and dense, weighing twice as much as lead. The metal is used to make fountain pen tips and electrical devices.

Palladium. The metallic chemical element of atomic number 46. Symbol Pd, atomic weight 106.42. Palladium is soft. It also readily absorbs hydrogen, and is therefore used to purify hydrogen gas.

Phosphorus. The nonmetallic chemical element of atomic number 15. Symbol P, atomic weight 30.97. Phosphorus is required by all plant and animal cells. Most of the phosphorus in human beings is in the bones and teeth. Phosphorus is heavily used in agricultural **fertilizers**.

Platinum. The metallic chemical element of atomic number 78. Symbol Pt, atomic weight 195.08, melting point 3,215.1°F (1,768.4°C), boiling point 6,920.6 ±212°F (3,827 ±100°C). Platinum withstands high temperatures well and is used in rocket and jet-engine parts. It is also used as a catalyst in **chemical reactions**.

Polonium. The metallic chemical element of atomic number 84. Symbol Po, atomic weight 209. Polonium is a product of uranium decay and is 100 times as radioactive as uranium.

Rhenium. The metallic chemical element of atomic number 75. Symbol Re, atomic weight 186.207, specific gravity 21.0, melting point 5,766.8°F (3,186°C), boiling point 10,104.8°F (5,596°C). Rhenium is used in chemical and medical instruments, as a catalyst for the chemical and **petroleum** industries, and in photoflash lamps.

Rhodium. The metallic chemical element of atomic number 45. Symbol Rh, atomic weight 102.91. This element is similar to palladium. Electroplated rhodium, which is hard and highly reflective, is used as a reflective material for optical instruments.

Ruthenium. The metallic chemical element of atomic number 44. Symbol Ru, atomic weight 101.07, specific gravity 12.5, melting point 4,233.2°F (2,334°C), boiling point 7,502°F (4,150°C). This element is alloyed with platinum and palladium to form hard, resistant contacts for electrical equipment that must withstand a great deal of wear.

Scandium. The metallic chemical element of atomic number 21. Symbol Sc, atomic weight 44.96, melting point 2,805.8°F (1,541°C), boiling point 5,127.8°F (2,831°C). Scandium is a silvery-white metal that develops a yellowish or pinkish cast when exposed to air. It has relatively few commercial applications.

Selenium. The nonmetallic chemical element of atomic number 34. Symbol Se, atomic weight 78.96. Selenium is able to convert **light** directly into **electricity**, and its resistance to electrical current decreases when it is exposed to light. Both properties make this element useful in photocells, exposure meters, and solar cells.

Silver. The metallic chemical element of atomic number 47. Symbol Ag, atomic weight 107.87. Silver has long been used in the manufacture of coins. It is also an excellent conductor of heat and electricity. Some compounds of silver are light-sensitive, making silver important in the manufacture of photographic films and papers.

Tantalum. The metallic chemical element of atomic number 73. Symbol Ta, atomic weight 180.95, melting point 5,462.6°F (3,017°C), boiling point of 9,797 ±212°F (5,425 ±100°C). Tantalum is a heavy, gray, hard metal that is used in alloys to pen points and analytical weights.

Technetium. The metallic chemical element of atomic number 43. Symbol Tc, atomic weight 98. Technetium was the first element to be produced synthetically; scientists have never detected the natural presence of this element on Earth.

Tellurium. The nonmetallic chemical element of atomic number 52. Symbol Te, atomic weight 127.60, melting point 841.1 ±32.54°F (449.5 ±0.3°C), boiling point 1,813.64 ±38.84°F (989.8 ±3.8°C). Tellurium is a grayish-white, lustrous, brittle metal. It is a semiconductor and is used in the electronics industry.

Thallium. The metallic chemical element of atomic number 81. Symbol Tl, atomic weight 204.38. Thallium is a bluish-gray metal that is soft enough to be cut with a knife. Thallium sulfate is used as a rodenticide and ant poison.

Tin. The metallic chemical element of atomic number 50. Symbol Sn, atomic weight 118.69. Tin is alloyed with copper and antimony to make pewter. It is also used as a soft solder and as coating to prevent other metals from corrosion.

Titanium. The metallic chemical element of atomic number 22. Symbol Ti, atomic weight 47.90, melting point 3,020 ±50°F (1,660 ±10°C), boiling point 5,948.6°F (3,287°C). This element occurs as a bright, lustrous brittle metal or dark gray powder. **Titanium** alloys are strong for their weight and can withstand large changes in **temperature**.

Tungsten. The metallic chemical element of atomic number 74. Symbol W, atomic weight 183.85, melting point 6,170 ±68°F (3,410 ±20°C). The melting point of tungsten is higher than that of any other metal. Its chief use is as a filament in electric light bulbs.

Vanadium. The metallic chemical element of atomic number 23. Symbol V, atomic weight 50.94. Pure vanadium is bright white. This metal finds its biggest use in strengthening steel.

Yttrium. The metallic chemical element of atomic number 39. Symbol Y, atomic weight 88.91, melting point 2,771.6 ±46.4°F (1,522 ±8°C), boiling point 6,040.4°F (3,338°C). **Yttrium** is a relatively active metal that decomposes in cold water slowly and in boiling water rapidly. Certain compounds containing yttrium have been shown to become superconducting at relatively high temperatures.

Zinc. The metallic chemical element of atomic number 30. Symbol Zn, atomic weight 65.39. Zinc, a brittle metal at room temperature, forms highly versatile alloys in industry. One zinc alloy is nearly as strong as steel, but has the malleability of plastic.

Zirconium. The metallic chemical element of atomic number 40. Symbol Zr, atomic weight 91.22, melting point 3,365.6 ±35.6°F (1,852 ±2°C), boiling point 7,910.6°F (4,377°C). Neutrons can pass through this metal without being absorbed; this makes it highly desirable as a construction material for the metal rods containing the fuel pellets in **nuclear power** plants.

See also Ammonia; Compound, chemical; Deuterium; Element, transuranium; Tritium; Valence.

Resources

Books

Lide, David R. *CRC Handbook of Chemistry and Physics.* 7th ed. Boca Raton, FL: CRC Press LLC, 1997.

Emsley, J. *The Elements.* 3rd ed. New York: Oxford Univ. Press, Inc., 1998.

Greenwood, N. N., and A. Earnshaw. *Chemistry of the Elements.* 2nd ed. Woburn, MA: Butterworth-Heinemann, 1997.

Periodicals

Seife, Charles. "Heavy-Element Fizzle Laid to Falsified Data." *Science* (July 19, 2002): 313–315.

Robert L. Wolke

Element, families of

A family of chemical elements usually consists of elements that are in the same group (the same column) on the **periodic table**. The term is also applied to certain closely related elements within the same period (row). Just as the **individual** members in a human family are all different but have common characteristics, like hair **color**, so to do the elements in a chemical family have certain properties in common, and others that make them unique.

The search for patterns among the elements

Johann Döbereiner (1780–1849) made one of the earliest attempts to organize the elements into families in 1829, when he observed that for certain groups of three elements, called triads, the properties of one element were approximately mid-way between those of the other two. However, because the number of elements known to Döbereiner was far less than it is today, the number of triads that he was able to find was very limited.

In 1864, John Newlands (1837–1898) noticed that when the known elements were arranged in order of increasing **atomic weight**, every eighth element showed similar properties. This observation, which was at first dismissed by the chemical community as being purely coincidental, is readily explicable using the modern periodic table and the concept of families of elements.

After organizing the elements known in 1869 so that those with similar properties were grouped together, Dmitri Mendeléev (1834–1907) predicted the existence and properties of several new elements. The subsequent discovery of these elements, and the **accuracy** of many of Mendeléev's predictions, fully justified the notion that the elements could be organized into families. Today, we recognize that the basis for this classification is the similarity in the electronic configurations of the **atoms** concerned.

The main-group families

For those families of elements found among the main-group elements, that is, elements in groups 1, 2, and 13 through 18 of the periodic table, each member of a given family has the same number of **valence** electrons. A detailed examination of the **electron** configurations of the elements in these families reveals that each family has its own characteristic arrangement of electrons. For example, each element in group 1, the **alkali metals**, has its valence electron in an s sublevel. As a result, all the elements in this family have an electron configuration which, when written in linear form, terminates with ns^1, where n is an integer representing the principal **quantum number** of the valence shell. Thus, the electron configuration of **lithium** is $1s^2 2s^1$;, that of **sodium** is $1s^2 2s^2 2p^6 3s^1$, potassium is $1s^2 2s^2 2p^6 3s^2 3p^6 4s^1$, and so on. In a similar way, the elements in group 2, the **alkaline earth metals**, each have two valence electrons and electron configurations that terminate in ns^2. For example, beryllium is $1s^2 2s^2$, **magnesium** is $1s^2 2s^2 2p^6 3s^2$, **calcium** is $1s^2 2s^2 2p^6 3s^2 3p^6 4s^2$. Because the s sub-level can only accommodate a maximum of 2 electrons, the members of group 13, which have

3 valence electrons, all have electron configurations terminating in $ns^2 np^1$; for example, **aluminum** is $1s^2 2s^2 2p^6 3s^2 3p^1$. The remaining main-group families, group 14 (the **carbon** family), group 15 (the pnicogens), group 16 (the chalcogens), group 17 (the **halogens**), and group 18 (the **rare gases**) have 4, 5, 6, 7, and 8 valence electrons, respectively. Of these valence electrons, two occupy an s sublevel and the remainder occupy the p sub-level having the same principal quantum number.

The similarity in electron configurations within a given main-group family results in the members of the family having similar properties. For example, the alkali metals are all soft, highly reactive elements with a silvery appearance. None of these elements is found uncombined in nature, and they are all willing to give up their single valence electron in order to form an ion with a charge of +1. Each alkali **metal** will react with **water** to give **hydrogen** gas and a **solution** of the metal hydroxide.

Characteristic patterns of **behavior** can also be identified for other main-group families; for example, the members of the carbon family all form chlorides of the type ECl_4 and hydrides of the type EH_4, and have a tendency towards *catenation*, that is, for identical atoms to join together to form long chains or rings. Similarly, although little is known about the heaviest, radioactive halogen, astatine, its congeners all normally exist as diatomic molecules, X_2, and show a remarkable similarity and predictability in their properties. All the members of this family are quite reactive-fluorine, the most reactive, combines directly with all the known elements except helium, neon and argon-and they all readily form ions having a charge of -1.

The family of elements at the far right of the periodic table, the rare gases, consists of a group of colorless, odorless gases that are noted for their lack of reactivity. Also known as noble gasses, the first compounds of these elements were not prepared until 1962. Even today there are only a limited number of krypton compounds known and still no known compounds of helium, neon, or argon.

Hydrogen: The elemental orphan

When the elements are organized into families, hydrogen presents a problem. In some of its properties, hydrogen resembles the alkali metals, but it also shows some similarities to the halogens. Many periodic tables include hydrogen in group 1; others show it in groups 1 and 17. An alternative approach is to recognize hydrogen as being unique and not to assign it to a family.

Other families of elements

In addition to the main-group families, other families of elements can be identified among the remaining elements of the periodic table.

The transition metals

The elements in groups 3 through 12, the transition metals or *d*-block elements, could be considered as one large family. Their characteristic feature, with some exceptions, is the presence of an incomplete *d* sublevel in their electron configurations. As with any large family, transition metals show considerable diversity in their behavior, although there are some unifying features, such as their ability to form ions with a charge of +2. Another similarity between these elements is that most of their compounds are colored.

The coinage metals and the platinum metals

At least two small family units can be identified within the larger transition-metal family. One of these small families, the coinage metals, consists of **copper**, silver and gold, the three elements in group 11. The other family, the platinum metals, includes elements from three groups: ruthenium and osmium from group 8; rhodium and iridium from group 9; and palladium and platinum from group 10.

The coinage metals are resistant to oxidation, hence their traditional use in making coins. Unlike the majority of the transition metals, the coinage metals each have a full *d* sublevel and one electron in an *s* sublevel, that is, an electron configuration that terminates in (n-1) d^{10} n s^2. One result of this electron configuration is that each of these metals will form an ion of the type M^+, although it is only for silver that this ion is relatively stable.

The platinum metals occur together in the same ores, are difficult to separate from one another, and are relatively unreactive.

The lanthanides and actinides

The **lanthanides** (or rare-earth elements) and **actinides** are two families that are related because they both result from electrons being added into an *f* sublevel. Both families have 14 members, the lanthanides consisting of the elements with atomic numbers 58 through 71, and the actinides including the elements with atomic numbers 90 through 103. However, it is sometimes convenient to consider lanthanum (**atomic number** 57) as an honorary member of the lanthanide family and to treat actinium (atomic number 89) in a similar manner with respect to the actinides.

The lanthanides are usually found together in the same ores and despite their alternative name of the rare-earth elements, they are not particularly rare. In contrast, only two of the actinides, thorium and **uranium**, occur in nature, the remainder having been synthesized by nu-

clear scientists. Members of both families form ions with a charge of +3, although other ions are also formed, particularly by the actinides.

See also Element, chemical; Element, transuranium.

Resources

Books

Emsley, John. *Nature's Building Blocks: An A-Z Guide to the Elements.* Oxford: Oxford University Press, 2002.
Norman, Nicholas C. *Periodicity and the s- and p-Block Elements.* Oxford Chemistry Primers, no. 51. New York: Oxford Univ. Press, 1997.
Silberberg, Martin. *Chemistry: The Molecular Nature of Matter and Change.* St. Louis: Mosby, 1996.

Arthur M. Last

Element, transuranium

A transuranium (beyond **uranium**) element is any of the chemical elements with atomic numbers higher than 92, which is the **atomic number** of uranium.

Ever since the eighteenth century when chemists began to recognize certain substances as chemical elements, uranium had been the element with the highest **atomic weight**; it had the heaviest **atoms** of all the elements that could be found on **Earth**. The general assumption was that no heavier elements could exist on this **planet**. The reasoning went like this: Heavy atoms are heavy because of their heavy nuclei, and heavy nuclei are unstable, or radioactive; they spontaneously transform themselves into other elements. Uranium and several even lighter elements—all those with atomic numbers higher than 83 (bismuth)—were already radioactive. Therefore, still heavier ones would probably be so unstable that they could not have lasted for the billions of years that Earth has existed, even if they were present when Earth was formed. In fact, uranium itself has a **half-life** that is just about equal to the age of Earth (4.5 billion years), so only one-half of all the uranium that was present when Earth was formed is still here.

If we could create atoms of elements beyond uranium, however, perhaps they would be stable enough to hang around long enough for us to study them. A few years, or even hours, would do. But in order to make an atom of an element with an atomic number higher than uranium which has 92 protons in its nucleus, we would have to add protons to its nucleus; one added **proton** would make an atom of element number 93, two added protons would make element 94, and so on. There was no way to add protons to nuclei, though, until the invention of the **cyclotron** in the early 1930s by Ernest Lawrence at the University of California at Berkeley. The cyclotron could speed up protons or ions (charged atoms) of other elements to high energies and fire them at atoms of uranium (or any other element) like machine-gun bullets at a target. In the resulting nuclear smashup, maybe some protons from the bullet nuclei would stick in some of the "hit" target nuclei, thereby transforming them into nuclei of higher atomic numbers. And that is exactly what happened. Shooting **light** atoms at heavy atoms has turned out to be the main method for producing even heavier atoms far beyond uranium.

Such processes are called *nuclear reactions*. Using nuclear reactions in cyclotrons and other "atom smashing machines," nuclear chemists and physicists over the years have learned a great deal about the atomic nucleus and the fundamental particles that make up the universe. Making new transuranium elements has been only a small part of it.

The road beyond uranium

Like any series of elements, the transuranium elements have similarities and differences in their chemical properties. Also like any other series of elements, they must fit into the **periodic table** in positions that match their atomic numbers and electronic structures. The

transuranium elements are often treated as a "family," not because their properties are closely related (although some of them are), but only because they represent the latest, post-1940 extension of the periodic table. Uniting them is their history of discovery and their radioactivity, more than their chemical properties.

Transuranium elements and the periodic table

We can think of the atomic numbers of the transuranium elements as mileposts along a Transuranium Highway that begins at uranium (milepost 92) and runs onward into transuranium country as far as milepost 110. As we begin our trip at 92, however, we realize that we are already three mileposts into another series of elements that began back at milepost 89: the *actinides*. Actinide Road runs from milepost 89 to 103, so it overlaps the middle of our 92-110 transuranium trip. (The road signs between 92 and 103 read both "Actinide Road" and "Transuranium Highway.")

The **actinides** and all of the transuranium elements fit in periods 6 and 7 on the Periodic Table. The names that go along with the symbols of the elements from 93 to 109 are: Np = neptunium, Pu = plutonium, Am = americium, Cm = curium, Bk = berkelium, Cf = californium, Es = einsteinium, Fm = fermium, Md = mendelevium, No = nobelium, and Lr = lawrencium. Proposed names for elements 104–109 are: Ru = rutherfordium, Db = dubnium, Sg = seaborgium, Bh = bohrium, Hs = hassium, and Mt = meitnerium. Element 110 has not yet been named.

The names of some of the transuranium elements and who discovered them have been the subjects of a raging battle among the world's chemists. In one corner of the name-game ring is the International Union of Pure and Applied **Chemistry** (IUPAC), a more-or-less official organization that among other things "makes the rules" about how new chemicals should be named. In another corner is the American Chemical Society (ACS) and most of the American and German scientists who discovered transuranium elements. The names listed above are the ACS recommendations.

History of the transuranium elements

In 1940, the first element with an atomic number higher than 92 was found, element number 93, now known as neptunium. This set off a search for even heavier elements. In the 15 years between 1940 and 1955 eight more were found, going up to atomic number 101 (mendelevium). Most of this work was done at the University of California laboratories in Berkeley, led by nuclear chemists Albert Ghiorso and Glenn Seaborg. Since 1955, the effort to find new transuranium elements has continued, although with rapidly diminishing returns. As of 1995 a total of 18 transuranium elements had been made, ranging up to atomic number 110. While the first nine transuranium elements were discovered within a 15 year period, discovering the last nine took almost 40 years. It was not that the experiments took that long; they had to await the development of more powerful cyclotrons and other ion-accelerating machines. This is because these transuranium elements do not exist on Earth; they have to be made artificially in the laboratory.

The transuranium story began when **nuclear fission** was discovered by Otto Hahn and Fritz Strassman in Germany in 1938. Chemists were soon investigating the hundreds of new radioactive isotopes that were formed in fission, which spews its nuclear products over half the periodic table. In 1940, E. M. McMillan and P. Abelson at the University of California in Berkeley found that one of those isotopes could not be explained as a product of nuclear fission. Instead, it appeared to have been formed by the radioactive transformation—rather than the fission—of uranium atoms, and that it had the atomic number 93.

When uranium was bombarded with neutrons, some uranium nuclei apparently had become radioactive and had increased their atomic number from 92 to 93 by emitting a (**negative**) beta particle. (It was already known that radioactive beta decay could increase the atomic number of an atom.) Because uranium had been named after **Uranus**, the seventh planet from the **sun** in our **solar system**, the discoverers named their "next" element *neptunium*, after the next (eighth) planet. When McMillan and other chemists at Berkeley, including G. Seaborg, E. Segrè, A. Wahl, and J. W. Kennedy, found that neptunium further decayed into the next higher element with atomic number 94, they named it plutonium, after the next (ninth) planet, **Pluto**. From there on, new transuranium elements were synthesized by using nuclear reactions in cyclotrons and other **accelerators**.

These experiments become more and more difficult as atomic numbers increase. For one thing, if you want to make the next higher transuranium element, you have to have some of the preceding one to use as a target, and the world's supply of that one may be only a few micrograms—a very tiny target indeed. It's worse than trying to hit a mosquito at 50 yards with a BB gun. While the probability of hitting one of these target nuclei with a "bullet" atom is incredibly small, the probability is even smaller that you will transform some of the nuclei you *do* hit into a particular higher atomic number nucleus, because once a "bullet" atom crashes into a target atom many different nuclear reactions can happen. To make matters even worse, the target element is likely to be

very unstable, with a half-life of only a few minutes. So it's not only an incredibly tiny target, it's a rapidly disappearing one. The **mosquitoes** are vanishing before your eyes while you're trying to shoot them.

The heaviest transuranium elements have therefore been made literally one atom at a time. Claims of discovery of new transuranium elements have often been based on the production of only half a dozen atoms. It is no wonder that the three major groups of discoverers, Americans, Russians, and Germans, have had "professional disagreements" about who discovered which element first. When organizations such as IUPAC and the ACS get into the act, trying to choose a fair name that honors the true discoverers of each element, the disagreements can get rather heated.

Cruising the transuranium highway

Following is a brief sketch of each of the transuranium elements. The chemical properties of these elements have in most cases been determined by nuclear chemists using incredibly ingenious experiments, often working with one atom at a time, and with radioisotopes that last only a few minutes. We will omit the chemical properties of these elements, however, because they are not available in sufficient quantities to be used in any practical chemical way; only their *nuclear* properties are important.

Neptunium (93)—Named after the planet **Neptune**, the next planet "in line" after Uranus, for which uranium (92) was named. Discovered in 1940 by McMillan and Abelson at the **Radiation** Laboratory of University of California, Berkeley (now called the Lawrence Radiation Laboratory), as a product of the **radioactive decay** of uranium after it was bombarded with neutrons. The neutrons produced uranium-239 from the "ordinary" uranium-238. The resulting uranium-239 has a half-life of 23.5 minutes, changing itself into to neptunium-239, which has a half-life of 2.35 days. Trace amounts of neptunium actually occur on Earth, because it is continually being formed in uranium ores by the small numbers of ever-present neutrons.

Plutonium (94)—First found in 1940 by Seaborg, McMillan, Kennedy, and Wahl at Berkeley as a secondary product of the radioactive decay of neutron-bombarded uranium. The most important **isotope** of plutonium is plutonium-239, which has a half-life of 24,390 years. It is produced in large quantities from the **neutron** bombardment of uranium-238 while ordinary **nuclear power** reactors are operating. When the reactor fuel is reprocessed, the plutonium can be recovered. This fact is of critical strategic importance because plutonium-239 is the major ingredient in **nuclear weapons**.

Americium (95)—Named after the Americas because europium, its just-above neighbor in the periodic table, had been named after **Europe**. Found by Seaborg, James, Morgan, and Ghiorso in 1944 in neutron-irradiated plutonium during the Manhattan Project (the atomic bomb project) in Chicago in 1944.

Curium (96)—Named after Marie Curie, the discoverer of the elements radium and polonium and the world's first nuclear chemist, and her husband, physicist Pierre Curie. First identified by Seaborg, James, and Ghiorso in 1944 after bombarding plutonium-239 with helium nuclei in a cyclotron.

Berkelium (97)—Named after Berkeley, California. Discovered in 1949 by Thompson, Ghiorso, and Seaborg by bombarding a few milligrams of americium-241 with helium ions. By 1962 the first visible quantity of berkelium had been produced. It weighed three billionths of a gram.

Einsteinium (99)—Named after Albert Einstein. Discovered by Ghiorso and his coworkers at Berkeley in the debris from the world's first large thermonuclear (**hydrogen** bomb) explosion, in the Pacific Ocean in 1952. About a hundredth of a microgram of einsteinium was separated out of the bomb products.

Fermium (100)—Named after physicist Enrico Fermi. Isolated in 1952 from the debris of a thermonuclear explosion in the Pacific by Ghiorso, working with scientists from Berkeley, the Argonne National Laboratory, and the Oak Ridge National Laboratory. Also produced by a group at the Nobel Institute in Stockholm by bombarding uranium with **oxygen** ions in a *heavy ion accelerator*, a kind of cyclotron.

Mendelevium (101)—Named after Dmitri Mendeleev, originator of the periodic table. Made by Ghiorso, Harvey, Choppin, Thompson, and Seaborg at Berkeley in 1955 by bombarding einsteinium-253 with helium ions. The discovery was based on the detection of only 17 atoms.

Nobelium (102)—Named after Alfred Nobel, Swedish discoverer of dynamite and founder of the Nobel prizes. Produced and positively identified in 1958 by Ghiorso, Sikkeland, Walton, and Seaborg at Berkeley, by bombarding curium with **carbon** ions. It was also produced, but not clearly identified as element 102, by a group of American, British, and Swedish scientists in 1957 at the Nobel Institute of **Physics** in Stockholm. IUPAC hastily named the element for the Swedish workers. The Berkeley chemists eventually agreed to the Swedish name, but not to the Swedes' credit for discovery. Ironically, in 1992 the International Unions of Pure and Applied Chemistry and of Pure and Applied Physics (IUPAC and IUPAP) credited the discovery of nobelium to a group of Russian scientists at the Joint Institute for Nuclear Research at Dubna, near Moscow.

Lawrencium (103)—Named for Ernest O. Lawrence, inventor of the cyclotron. Produced in 1961 by Ghiorso, Sikkeland, Larsh, and Latimer at Berkeley by bombarding californium with boron ions.

Elements 104 to 110—The identities of the true discoverers of these elements are tangled in an assortment of very difficult experiments performed by different groups of scientists at the American Lawrence Radiation Laboratory in Berkeley, the German Gesellschaft für Schwerionenforschung (Institute for Heavy-Ion Research) in Darmstadt, the Russian Joint Institute for Nuclear Research in Dubna, and the Swedish Nobel Institute of Physics in Stockholm.

The end of the road?

The Transuranium Highway would appear to be coming to a dead end for two reasons. Chemists do not have large enough samples of the heaviest transuranium elements to use as targets in their cyclotrons, and the materials are so radioactive anyway that they only last for seconds or at most a few minutes.

Element 110 has been made by a slightly different trick—shooting medium-weight atoms at each other. The nuclei of these atoms can fuse together and hopefully stick, to make a nucleus of a transuranium element. In November 1994, a group of nuclear chemists at the Heavy Ion Research Center at Darmstadt, Germany reported that by shooting nickel atoms (atomic number 28) at **lead** atoms (atomic number 82), they had made three atoms of element 110 (=28+82), which lasted for about a ten-thousandth of a second.

In spite of this gloomy picture, nuclear chemists are trying very hard to make *much* heavier, "superheavy" elements. There are theoretical reasons for believing that they would be more stable and would stick around much longer.

The Transuranium Highway may be still under construction.

Robert L. Wolke

Resources

Periodicals

Harvey, Bernard G. "Criteria for the Discovery of Chemical Elements." *Science* 193 (1976): 1271-2.

Hoffman, Darleane C. "The Heaviest Elements." *Chemical & Engineering News* (May 2, 1994): 14-34.

Seaborg, Glenn T., and Walter D. Loveland. *The Elements Beyond Uranium.* New York: John Wiley & Sons, Inc., 1990.

Elements, formation of

Elements are identified by the nuclei of the **atoms** of which they are made. For example, an atom having six protons in its nucleus is **carbon**, and one having 26 protons is **iron**. There are over 80 naturally occurring elements, with **uranium** (92 protons) being the heaviest (heavier nuclei have been produced in reactors on **Earth**). Nuclei also contain certain neutrons, usually in numbers greater than the number of protons.

Heavy elements can be formed from light ones by **nuclear fusion** reactions; these are nuclear reactions in which atomic nuclei merge together. The simplest reactions involve **hydrogen**, whose nucleus consists only of a single **proton**, but other fusion reactions, involving mergers of heavier nuclei, are also possible. When the universe formed in an initial state of very high **temperature** and **density** called the big bang, the first elements to exist were the simplest ones: hydrogen, helium (two protons), and little else. But we and the earth are made of much heavier elements, so a major question for scientists is how these heavier elements were created.

During the formation of the universe in the so-called big bang, only the lightest elements were formed: hydrogen, helium, **lithium**, and beryllium. Hydrogen and helium dominated; the lithium and beryllium were only made in trace quantities. The other 88 elements found in nature were created in nuclear reactions in the stars and in huge stellar explosions known as supernovas. Stars like the **Sun** and planets like Earth containing elements other than hydrogen and helium could only form after the first generation of massive stars exploded as supernovas, and scattered the atoms of heavy elements throughout the **galaxy** to be recycled.

History

The first indications that stars manufacture elements by nuclear reactions came in the late 1930s when Hans Bethe and C. F. von Weizsäcker independently deduced that the **energy** source for the Sun and stars was nuclear fusion of hydrogen in a process that formed helium. They received the Nobel prize in **physics** for this work.

George Gamow championed the **big bang theory** in the 1940s. Working with Ralph Alpher, he developed the theory that the elements formed during the big bang. With Robert Herman in the early 1950s, the pair used early computers to try to work out in detail how all the elements could have been formed during this cataclysmic period. The attempt was unsuccessful, but was one of the first large scientific problems to be tackled by computer.

Astronomers now realize that heavier elements could not have been formed in the big bang. The problem was that the universe cooled too rapidly as it expanded, and the extremely high temperatures required for nuclear reactions to occur did not last long enough for the creation of elements heavier than lithium or beryllium. By the time the universe had the raw materials to form the heavier elements, it was too cool.

In 1957, Margaret Burbidge, Geoffery Burbidge, William Fowler, and Fred Hoyle (referred to as B^{2}FH) published a monumental paper in which they outlined the specific nuclear reactions that occur in stars and supernovas to form the heavy elements. Fowler received the 1983 Nobel Prize in physics for his role in understanding nuclear processes in stars.

Formation of elements

During most of their lives, stars fuse hydrogen into helium in their cores, but the fusion process rarely stops at this point; most of the helium in the universe was made during the initial big bang. When the star's core runs out of hydrogen, the **star** begins to die out. The processes that occur during this period form the heavier elements.

The dying star expands into a **red giant star**. A typical red giant at the Sun's location would extend to roughly the earth's **orbit**. The star now begins to manufacture carbon atoms by fusing three helium atoms. Occasionally a fourth helium atom combines to produce **oxygen**. Stars of about the Sun's **mass** stop with this helium burning stage and collapse into white dwarfs about the size of the earth, expelling their outer layers in the process. Only the more massive stars play a significant role in manufacturing heavy elements.

Massive stars become much hotter internally than stars like the Sun, and additional reactions occur after all the hydrogen in the core has been converted to helium. At this point, massive stars begin a series of nuclear burning, or reaction, stages: carbon burning, neon burning, oxygen burning, and silicon burning. In the carbon burning stage, carbon undergoes fusion reactions to produce oxygen, neon, **sodium**, and **magnesium**. During the neon burning stage, neon fuses into oxygen and magnesium. During the oxygen burning stage, oxygen forms silicon and other elements that lie between magnesium and **sulfur** in the **periodic table**. These elements, during the silicon burning stage, then produce elements near iron on the periodic table.

Massive stars produce iron and the lighter elements by the fusion reactions described above, as well as by the subsequent **radioactive decay** of unstable isotopes. Elements heavier than iron are more difficult to make, however. Unlike nuclear fusion of elements lighter than iron, in which energy is released, nuclear fusion of elements heavier than iron requires energy. Thus, the reactions in a star's core stop once the process reaches the formation of iron.

Manufacturing heavy elements

How then are elements heavier than iron made? There are two processes, both triggered by the **addition** of neutrons to atomic nuclei: the s (slow) process and the r (rapid) process. In both processes, a nucleus captures a **neutron**, which emits an **electron** and decays into a proton, a reaction called a beta decay. One proton at a time, these processes build up elements heavier than iron. Some elements can be made by either process, but the s process can only make elements up to bismuth (83 protons) on the periodic table. Elements heavier than bismuth require the r process.

The s process occurs while the star is still in the red giant stage. This is possible because the reactions create excess energy, which keeps the star stable. Once iron has formed in the star's core, however, further reactions suck **heat** energy from the core, leading to catastrophic collapse, followed by rebound and explosion. The r process occurs rapidly when the star explodes.

During a **supernova**, the star releases as much energy as the Sun does in 10 billion years and also releases the large number of neutrons needed for the r process, creating new elements during the outburst. The elements that were made during the red giant stage, and those that are made during the supernova explosion, are spewed out into **space**. The atoms are then available as raw materials for the next generation of stars, which can contain elements that were not made during the big bang. These elements are the basic materials for life as we know it. During their death throes, massive stars sow the **seeds** for life in the universe.

See also Cosmology; Nuclear fission; Stellar evolution.

Resources

Books

Bacon, Dennis Henry, and Percy Seymour. *A Mechanical History of the Universe.* London: Philip Wilson Publishing, Ltd., 2003.

Emsley, John. *Nature's Building Blocks: An A-Z Guide to the Elements.* Oxford: Oxford University Press, 2002.

Riordan, Michael, and David N. Schramm. *The Shadows of Creation.* New York: Freeman, 1991.

KEY TERMS

. . . .

Beta decay—The splitting of a neutron into a proton and an electron.

Fusion—The conversion of nuclei of two or more lighter elements into one nucleus of a heavier element.

r process—Rapid process, the process by which some elements heavier than iron are made in a supernova.

Red giant—An extremely large star that is red because of its relatively cool surface.

s process—Slow process, the process by which some elements heavier than iron are made in a red giant.

Periodicals

Kirshner, Robert. "The Earth's Elements." *Scientific American* (October 1994): 59.

Paul A. Heckert

Elephant

Elephants are large, four-legged, herbivorous **mammals**. They have a tough, almost hairless hide, a long flexible trunk, and two ivory tusks growing from their upper jaw. Only two **species** of elephant exist today, the African elephant (*Loxodonta africana*) and the Asian or Indian elephant (*Elephas maximus*), both of which are threatened or endangered.

African elephants are the largest of all land animals, weighing up to 5 tons. There are two subspecies, the African bush elephant (*Loxodonta africana africana*) and the African forest elephant (*Loxodonta africana cyclotis*). Bush elephants inhabit grassland and **savanna**, while forest elephants live in tropical **rainforest**. Asian elephants are widely domesticated, with the few surviving wild elephants living mainly in forest and woodland. Field workers have differing opinions of the life span of elephants, some estimating between 60 and 80 years while others suggesting more than 100 years.

Evolution

Elephants are placed within the suborder Elephantoidea, in the order Proboscidea. The first identifiable ancestors of today's elephants were small beasts that lived

An African elephant (*Loxodonta africana*) in Amboseli National Park, Kenya. *Photograph by Malcolm Boulton. The National Audubon Society Collection/Photo Researchers, Inc. Reproduced by permission.*

50–70 million years ago and stood about 2 ft (0.75 m) tall. The suborder Elephantoidea originated in North **Africa** long before that region became extensively desertified, and from there elephants spread to every **continent** except **Australia** and **Antarctica**. The group once included three families, several genera, and hundreds of species. Today, however, the family Elephantidae includes only two living species: the Asian and the African elephant. Mammoths and mastodons also belonged to the suborder Elephantoidea, but these species become extinct about 10,000 years ago.

About 400,000 years ago Asian elephants inhabited a much wider range than they do today, including Africa. This species now survives only in southern **Asia**, from India to Sumatra and Borneo. The single species of Asian elephant has three subspecies: *Elephas maximus maximus* of Sri Lanka, *E. m. indicus* of India, Indochina, and Borneo, and *E. m. sumatranus* of Sumatra. African elephants

only ever existed in Africa, appearing in the fossil record about four million years ago. As recently as only a hundred years ago, some 10 million African elephants inhabited that continent. By 1999, however, their numbers were reduced by overhunting to only about 300,000.

Body

Asian and African elephants can be distinguished by the shape of their backs, the Asian having a convex, gently sloping back and the African a concave or saddle-shaped one. Male elephants (or bulls) are much larger than females (cows), being 20–40% taller and up to 70% heavier. The average African adult bull weighs about 5 tons and measures about 8 ft (2.4 m) to the shoulder. The largest elephant on record was a magnificent bull, now mounted as a specimen in the Smithsonian Museum in Washington, D.C., standing a massive 13 ft 2 in (4 m) at the shoulder.

Skin texture varies from the tough, thick, wrinkled, folds on the back and forehead, to the soft, thinner, pliable skin of the breast, ears, belly, and underside of the trunk. The tough skin bears a few, scattered, bristly hairs, while the thinner skin on the trunk, chin, **ear** rims, eyelids, knees, wrists, and tip of the tail has somewhat thicker hair. Daily skin care includes showers, dusting with **sand**, and full-bodied mud-packs which are later rubbed off against a **tree** or boulder, removing dead skin as well. These activities help to keep the skin moist, supple, protected from the **sun** and **insects**, and also aid in keeping the **animal** cool.

Limbs

Supporting the elephant's massive body are four sturdy, pillar-like legs. Although the back legs are slightly longer than the front legs, the high shoulder makes the forelimbs look longer. The back legs have knees with knee-caps, while the front leg joints are more like wrists. Elephants kneel on their "wrists," stand upright on their back legs, sit on their haunches, and can be trained to balance on their front feet. The feet have thick, sponge-like pads with ridged soles which act as shock-absorbers and climbing boots, helping these sure-footed animals to ascend embankments and negotiate narrow pathways with amazing dexterity. The African species has four toenails on its round front feet, and three on its oval-shaped back feet; the Asian species has five toes on the front feet and four on the back feet. In spite of their size, elephants can move quickly, but cannot make sustained runs, as all four feet are never off the ground at one time. Elephants often doze on their feet, but **sleep** lying down for about one to three hours at night.

Head

Elephants have a large skull which supports the massive weight of their tusks. The size and shape of the skull helps distinguish between African and Asian elephants and between females and males. Asian elephants have a high, dome-shaped forehead while African elephants display a lower, more gently angled forehead. Heads of the males are larger in both species. Also in both species, the neck is short, making the head relatively immobile. To see behind, elephants must move their entire body; they display excited, restless **behavior** and turn quickly when detecting unfamiliar sounds or smells from the rear.

Mouth and trunk

Elephants have a small mouth and a large, mobile tongue which cannot extend past the short lower lip.

Contributing to the elephant's unique appearance is its long, strong, flexible trunk, which is a fusion and elongation of the nose and upper lip. The trunk, with no bones and more than 100,000 muscles, is so strong and flexible it can coil like a snake around a tree and uproot it. At the end of this mighty "limb," which trails on the ground unless curled up at the end, are two nostrils and flexible finger-like projections. The tip is so sensitive and dexterous it can wipe a grain of sand from the elephant's **eye** and detect delicate scents blowing in the breeze. Using this remarkable appendage, an elephant can feed by plucking grass from the ground, or foliage from a tree, placing it in its mouth. **Water** drawn up the trunk may be squirted into the mouth for drinking, or sprayed over the body for bathing and cooling. Loud trumpeting sounds and soft, affectionate murmurs can echo through the trunk. The trunk is also used to tenderly discipline, caress, and guide young offspring, to stroke the mate, to fight off predators, and to push over trees during feeding. The trunk is clearly an essential **organ**. It is also sometimes the object of attack by an enemy, and damage to it causes extreme **pain** and can lead to death.

Teeth

The tusks of elephants begin as two front teeth which drop out after about a year. In their place grow ivory tusks which eventually protrude from beneath the upper lip. The tusks of female Asian elephants, however, remain short and are barely visible. Male African elephants grow the largest tusks, the longest recorded measuring approximately 137 in (348 cm) and weighing over 220 lb (100 kg) each. Today, however, tusks are much smaller in wild elephants because most of the older animals have been slaughtered for their ivory. Although there are variations, the long, cylindrical tusks grow in a gradual upward **curve**, somewhat resembling the sliver of a new **moon**. Elephants use their tusks as weapons in combat, and to dig up roots, strip **bark** off trees, lift objects, and (for females) to establish feeding dominance. Tusks continue to grow throughout an animal's life at an average of about 5 in (12.7 cm) a year; however, their length is not an accurate measure of the animals age, as the tips wear and break with daily use and during combat.

Elephants have large, grinding, molar teeth which masticate (chew and grind) their **plant** diet with a backward-forward jaw action. These teeth fall out when worn down, and are replaced by new, larger teeth. During its lifetime, an elephant may grow 24 of these large molar teeth, each weighing up to 9 lb (4 kg) in older animals. Only four teeth, two on each side of the jaw, are in use at any one time. As the teeth wear down, they move forward; the new teeth grow from behind and the worn teeth drop out. This pattern repeats up to six times over the

Wild asiatic elephants (*Elephas maximus*) in Yala National Park, Tissamaharama, Sri Lanka. *Photograph by Ben Simon. Stock Market. Reproduced with permissions.*

elephant's lifetime, and the most common method of determining an elephant's age is by tooth and jaw examination. Once all of its teeth have fallen out, an elephant can no longer chew its food, and will soon die.

Ears

One astute elephant observer noted that "the ears of Asian elephants are shaped like India, and African elephants like Africa!" The ears of African elephants are much larger than those of Asian elephants, and the ears of the African bush elephants are larger than those of the forest elephants. African elephants cool themselves off by fanning with their ears and, conversely, in extreme cold elephants must increase their activity level to produce enough body heat to prevent their ears being frostbitten. Elephants have a keen sense of **hearing**, and spread their ears wide to pick up distant sounds; the spread-out ears also intimidate enemies by making the elephants appear larger.

Eyes

The eyes of elephants are about the same size as a human's. The eyes are usually dark brown, with upper and lower lids, and long eyelashes on the upper lid. With one eye on either side of their head elephants have a wide visual field, although their eyesight is relatively poor, particularly in bright sunlight.

Social behavior

Few animals other than humans have a more complex social network than elephants, which field biologists are just beginning to decipher. These outgoing, emotionally demonstrative animals rarely fight among themselves and peacefully coexist with most other animals. Elephants give and receive love, care intensely for their young, grieve deeply for their dead, get angry, show fear, and are thought to be more intelligent than any other animals except the higher **primates**.

Group structure

Each elephant troop has its own home range, but territorial fights are rare even though ranges often overlap. While several hundred elephants may roam a similar range, small "kin groups" form between female relatives. The leader of each group is a respected old female with years of accumulated knowledge. This matriarch is the mother and grandmother of other members but sometimes allows her sisters and their offspring to join the group. Once a male reaches maturity, he is forced to leave. The entire group looks to the matriarch for guidance, particularly in the face of danger. Her actions, based on her superior knowledge, will determine whether the group flees or stands its ground. Young members learn from their elders how to find water and food during **drought**, when to begin travel and where to go, and many other survival skills. This knowledge is passed on from generation to generation.

Once a male elephant reaches sexual maturity at 12 years or older, the matriarch no longer tolerates him in the group. He will then live mostly alone or perhaps join a small, loosely-knit group of other males. Bull elephants seldom form long-term relationships with other males, but often one or two young males accompany an old bull, perhaps to learn from him. Bulls often spar with each other to establish a dominance hierarchy. Elephants have an excellent **memory**; once a social hierarchy is established, the same two elephants not only recognize each other, even after many years, but know which one is dominant. This way, they avoid fighting again to reestablish dominance. After about 25 years of age, male elephants experience annual periods of heightened sexuality called "musth," which lasts about a week in younger animals and perhaps three or four months as they near their 50s. During this time they aggressively search out females and challenge other bulls, sometimes even causing more dominant males to back down. Different bulls come into musth at different times of the year; however, two well-matched bulls in musth may fight to the death.

Mating

Female elephants come into estrus (heat), marking ovulation and the ability to get pregnant, for only a few days each year. Because the mating season is short, mature female elephants are never far from adult males. The scent of a female elephant in estrus attracts male bulls. A receptive female will hold her head high, producing a low, rumbling invitation as she leaves her group and runs quickly across the plains chased by the bulls. It appears she actually chooses her mate, for she seldom stops for a young bull but slows down for a larger, dominant male who, once she allows him to catch her, gently rests his trunk across her back in a caress. They may mate several times, and he may stay with her until the end of her estrus, warding off other bulls and fighting if necessary. She may, however, mate with others. Because males play no part in raising the young and are not needed to protect the mother or baby, their role appears to be purely reproductive.

At the end of estrus, the cow returns to her group and the male goes off in search of another mate. The gestation period of female elephants lasts for 22 months, longer than any other animal; pregnancies are spaced from three to nine years apart. There is usually only one offspring, but twin births do occur and both calves may survive under favorable conditions. There is much excitement in the group during a **birth**, and another female almost always tends to the birthing mother. An adult female and her sexually immature offspring are a "family unit" within the group. However, females assist each other in raising the young, with one mother even sometimes nursing the calf of another. In general, females reach sexual maturity between the age of 12 to 15 years and, over the course of 60 years, will bear from five to 15 offspring.

Communication

Elephants teach and learn by behavioral examples and "talk" with vocalized sounds that can be described as screams, trumpets, growls, and rumbles. Originating from the throat or head, these calls can signal danger and express anger, joy, sadness and sexual invitation. An animal separated from its family will make "contact rumbles," which are low, vibrating sounds that can be heard at great distances. Once reunited, the family engages in a "greeting ceremony," reserved strictly for close relatives, in which excited rumbling, trumpeting, touching of trunks, urinating, and defecating occurs. Vocal sounds range from high-pitched squeaks to extremely powerful infrasonic sounds of a **frequency** much lower than can be heard by the human ear.

Death

Elephants mourn deeply for their dead and often cover them with leaves, dirt, and grass. An animal will stand over the body of a dead loved one, gently rocking back and forth as other animals caress the mourner with their trunks. One field-biologist watching such a display wrote: "This isn't just a dead elephant; it is a living elephant's dead relative or friend."

Habitat and food

Because of their high intelligence level, elephants can adapt to and modify **habitat**, while their wide range of food choices permits habitation of a diverse range of ecosystems, including forest, woodland, savanna, grassy plains, swampy areas, and sparsely vegetated **desert**. Unfortunately, because of massive poaching for ivory and the destruction of much of the elephant's natural habitat, most African elephants are now restricted to the protection of national parks. Not so long ago, however, they freely followed age-old seasonal **migration** routes from one habitat to another.

Elephants need massive quantities of food, perhaps 300–350 lb (136–159 kg) a day, although proportional to their body-weight elephants eat less than **mice**. The diet of elephants includes roots, bark, grass, leaves, berries, seedpods, and other **fruits**. Elephants will uproot trees to obtain tasty treats from the top, or delicately pluck a single berry from a branch. Elephants never roam far from water, and will travel great distances in search of it. They may drink up to 50 gal (189 l) of water a day, and after drinking their fill, will splash themselves with water and mud, wash their young, and sometimes just frolic, tossing and squirting water about while their young splash, play, and roll in the mud. Surprisingly, populations of these water-loving creatures may inhabit desert areas, using their tusks and trunk to dig for water under dry river beds. The knowledge of where to find water is handed down from one generation to another.

The future

Only a few surviving elephant herds remain in the wild. In Asia, elephants are venerated. However, they are also highly valued as domestic animals for work and transport and most tamed animals must be captured from the wild (although there has been recent progress in captive breeding). One-third of the surviving 35,000 Asian elephants are now in captivity, and the survival of all wild herds is threatened.

The combination of habitat loss and ivory poaching have made the African elephants endangered. Ivory has been traded for thousands of years, but this commercial activity escalated dramatically after the middle of the twentieth century. During the 1980s, about 100,000 elephants were being slaughtered each year, their tusks ending up as billiard balls, piano keys, jewelry, and sculptures. The oldest males, bearing the biggest tusks, were killed first, but as their population diminished, younger males and females were also slaughtered, leaving young calves to grieve and usually die of starvation.

The unsustainable "elephant holocaust" was brought to public attention in the 1970s and bitter battles have ensued between government authorities, ivory traders, and conservationists. Not until 1989 was a ban imposed on the international trading in elephant ivory. This was enacted by the Convention on the International Trade in Endangered Species of Wild Flora and Fauna (or CITES). In the early 1990s, elephant kills in Kenya and other African countries dropped to almost **zero**, but by then the total surviving population of elephants had been reduced to an extremely low level. Today, elephants are worth more alive than dead in some regions, where local ivory prices have crashed from $30 a kilogram to $3, while tourists coming to see elephants and other **wildlife** bring hard currency to African governments, totaling more than $200 million a year.

However, the international ban on trading elephant ivory is extremely controversial, and there are strong calls to partially lift it. This is mostly coming from South Africa, Zimbabwe, and other countries of southern Africa. Effective **conservation** efforts in that region have resulted in the build-up of relatively large elephant populations in comparison with the natural areas that are available to support the herds, so that habitat damage is being caused. In fact, some of these countries engage in legal culls of some of their elephants, to ensure that the population does not exceed the **carrying capacity** of the available habitat. These countries also believe that they should be able to harvest their elephants at a sustainable **rate**, and to sell the resulting ivory in Asian markets, where the price for legal ivory is extremely high. This seems to be a sympathetic goal. In fact, CITES has decided to make limited trade exceptions; in April 1999, Zimbabwe held an auction for 20 tons of legal ivory, all of which was purchased by Japanese dealers. These relatively small, closely monitored sales of legal ivory from southern Africa will now probably occur periodically.

Unfortunately, it is extremely difficult to separate "legal" and "illegal" ivory in the international marketplace, and some people feel that this partial lifting of the ban could result in greater poaching activity in countries where elephant populations remain perilously small. The real problem, of course, is that the growing human population and its many agricultural and industrial activities are not leaving enough habitat for elephants and other wild creatures, resulting in the endangerment of many species.

Resources

Books

Douglas-Hamilton, Iain, and Oria Douglas-Hamilton. *Battle For the Elephant.* New York: Viking Penguin, 1992.

Hayes, Gary. *Mammoths, Mastodons, and Elephants—Biology, Behavior and the Fossil Record.* Cambridge: Cambridge University Press, 1991.

Moss, Cynthia. *Elephant Memories: Thirteen Years of Life in an Elephant Family.* New York: William Morrow and Company, Inc., 1988.
Redmond, Ian. *The Elephant Book.* Woodstock: The Overlook Press, 1991.
Scullard, H. H. *The Elephant in the Greek and Roman World.* Ithaca: Cornell University Press, 1974.

Marie L. Thompson

Elephant shrews

Elephant shrews are relatively small **mammals** in the family Macroscelididae, order Macroscelidea. Elephant shrews have a characteristic long, narrow snout that is broad at the base, and very sensitive and flexible but not retractile. This snout is movable in circular manner at the base, and has nostrils at the end. There are five genera with some 185 **species** of elephant shrews, living in continental **Africa** and on the **island** of Zanzibar. Elephant shrews live in thorn bush country, grassy plains, thickets, the undergrowth of **forests**, and on rocky outcrops.

The head and body length is 3.7–12.3 in (9.5–31.5 cm) and the tail 3.1–10.3 in (8–26.5 cm). The tail is usually slender and covered with bristles, which may be

An elephant shrew (*Elephantulus rozeti*). © *Tom McHugh, National Audubon Society Collection/Photo Researchers, Inc. Reproduced with permission.*

rough at the underside and terminating in knobs. There is a naked black musk gland under the tail which secretes a highly scented substance, especially in females. The body is covered with soft fur which is lacking on the rump. The two bones of the hind legs are joined. The feet have four or five toes, and the hands have five fingers. The females have two or three pairs of mammae.

The elephant shrews are active mainly in daytime, but in hot weather they may be nocturnal. These animals may hide during the day when harassed by diurnal predators, but are often seen **sun** bathing. They reside singly or in pairs in burrows, ground depressions, rock crevices, and in the crevices of termite mounds. Burrows of **rodents** are occasionally used. When running, elephant shrews leave runways with a broken appearance because of their jumping locomotion which is like a bouncing ball, running on their hind legs, with the tail extended upward.

The smaller species of elephant shrews feed on **ants**, **termites**, and slender shoots, roots, and berries, while some larger species prefer **beetles**. The members of the genus *Elephantulus* and *Rhynchocyon* produce squeaks, while *Petrodromus* make cricket-like calls. *Elephantulus* and *Petrodromus* rap their hind feet when nervous or to give an alarm. Species of *Petrodromus* and *Rhynchocyon* are known to beat their tails on the ground.

Gestation takes about two months and one or two young are born relatively large and well developed, fully covered with fur and with eyes open at **birth** or soon after. There is a short nursing period and elephant shrews become sexually mature at five to six weeks of age. The life span in the wild is probably 18 months or less, but captive North African elephant shrews lived 40 months. Larger elephant shrews of the genus *Petrodromus* of East Africa are snared and eaten by natives.

See also Shrews.

Elephant snout fish

Elephant snout fish belong to a diverse group of fishes that comprise the family Mormyridae. All are **freshwater species** that are confined to tropical parts of the African **continent**. Some 150 species have been described so far. The group takes its common or English name from the animals' extended snout. This **adaptation** is taken to the extreme in the genus *Gnathonemus* which has a pendulous, trumpet-shaped snout. This species, like many other in the genus, feeds almost exclusively on small crustaceans. In some species, the snout is so modified that it possesses only a tiny mouth equipped with just a few, but relatively large, teeth. In some species the "trunk" is pendulous, while in others it may be held straight out from the head. Although this adaptation may, at first sight appear at odds to a predatory **fish**, because these animals often frequent muddy waters, this slender, highly tactile snout is ideally suited for detecting and grasping small **prey** that hide in vegetation or amongst rubble or mud on the base of streams and lakes.

All mormyroid fish possess specialized electric organs, a feature that is not unusual in species living in either deep or gloomy waters. In the elephant snout fish, this feature is probably related to helping the fish to move around and avoid obstacles, as well as assisting with the location of prey. By emitting a series of short, pulsed electrical signals, the fish is able to detect and avoid obstacles. In the same way, it can detect and pinpoint living animals that also give off a small electric field. In this way they are able to identify potential food items and avoid conflict with other electric-producing snout fish. Members of the family Mormyridae probably constitute the most diverse group of electric fishes; most swim by synchronous movements of the opposite dorsal and anal fins, thereby keeping the electric organs arranged along the sides of the body in perfect alignment with the body.

Elephantiasis

Elephantiasis is an extreme symptom of human **infection** by a type of roundworm or nematode. It involves massive swelling of a limb or of the scrotum. The leg of an **individual** suffering from elephantiasis can become enlarged to two or three times normal diameter.

The actual name of the **disease** or infection which causes elephantiasis is lymphatic filariasis. Lymphatic filariasis is an important parasitic infection (a parasite is any **organism** which survives by living within another organism) in **Africa**, Latin America, the Pacific Islands, and **Asia**, and causes infection in about 250 million individuals (more than the number suffering from **malaria**). At one time, there was a small focus of infection that occurred in South Carolina, but this ended in the 1920s.

How lymphatic filariasis is spread

Lymphatic filariasis is caused by infestation by one of three nematodes (*Wucheria bancrofti*, *Brugia malayi*, or *Brugia timori*). These nematodes are spread to humans through the bite of **mosquitoes**. The mosquitoes are considered vectors, meaning that they spread the nematode, and therefore the disease. Humans are considered hosts, meaning that actual reproduction of the nematode occurs within the human body.

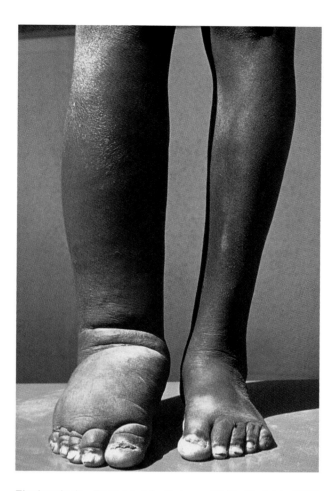

Elephantiasis. *Photograph by C. James Webb. Phototake NYC. Reproduced by permission.*

The nematode has a rather complicated lifecycle. The larval form lives within the mosquito, and it is this form which is transmitted to humans through the bite of an infected mosquito. The larvae pass into the human **lymphatic system**, where they mature into the adult worm. Adult worms living within the human body produce live offspring, known as microfilariae, which find their way into the bloodstream.

Microfilariae have interesting properties which cause them to be released into the bloodstream primarily during the night; this property is called nocturnal periodicity. Therefore, the vectors (carriers) of filariasis, which deliver the infective worm larvae to the human host, tend to be the more nocturnal (active at night) **species** of mosquito.

Symptoms and progression of filarial disease

The majority of the suffering caused by filarial nematodes occurs because of blockage of the lymphatic system. The lymphatic system is made up of a network of vessels which serve to drain **tissue** fluid from all the major organs of the body, including the skin, and from all four limbs. These vessels pass through lymph nodes on their way to empty into major **veins** at the base of the neck and within the abdomen. While it was originally thought that blockage of lymph flow occurred due to the live adult worms coiling within the lymphatic vessels, it is now thought that the worst obstructions occur after the adult worms die.

While the worm is alive, the human **immune system** attempts to rid itself of the foreign invader by sending a variety of cells to the area, causing the symptoms of **inflammation** (redness, heat, swelling) in the infected node and lymph channels. The skin over these areas may become thickened and rough. In some cases, the host will experience systemic symptoms as well, including fever, headache, and fatigue. This complex of symptoms lasts seven to ten days, and may reappear as often as ten times in a year.

After the worm's death, the inflammatory process is accelerated, and includes the formation of tough, fibrous tissue which ultimately blocks the lymphatic vessel. Lymph fluid cannot pass through the blocked vessel, and the back up of fluid results in swelling (also called **edema**) below the area of blockage. Common areas to experience edema are arms, legs, and the genital area (especially the scrotum). These areas of edema are also prone to infection with bacterial agents. When extreme, lymphatic obstruction to vessels within the abdomen and chest can lead to rupture of those vessels, with spillage of lymph fluid into the abdominal and chest cavities.

An individual who picks up filarial disease while traveling does not tend to experience the more extreme symptoms of elephantiasis that occur in people who live for longer periods of time in areas where the disease is common. It is thought that people who live in these areas receive multiple bites by infected mosquitoes over a longer period of time, and are therefore host to many, many more nematodes than a traveler who is just passing through. The larger worm load, as the quantity of nematodes present within a single individual is called, contributes to the severity of the disease symptoms suffered by that individual.

Diagnosis

An absolutely sure **diagnosis** (called a definitive diagnosis) of filarial disease requires that the actual nematode be identified within body tissue or fluid from an individual experiencing symptoms of infection. This is not actually easy to accomplish, as the lymph nodes and vessels in which the nematodes dwell are not easy to access. Sometimes, **blood** samples can be examined to reveal the presence of the microfilariae. Interestingly enough, because of the nocturnal periodicity of these microfilariae, the patient's blood must be drawn at night to increase the likelihood of the **sample** actually containing the parasite.

Many times, however, diagnosis is less sure, and relies on the patient's history of having been in an area where exposure to the nematode could have occurred, along with the appropriate symptoms, and the presence in the patient's blood of certain immune cells which could support the diagnosis of filarial disease.

Treatment

A drug called diethylcarbamazine (DEC) is quite effective at killing the microfilariae as they circulate in the blood, and injuring or killing some of the adult worms within the lymphatic vessels. An individual may require several treatment with DEC, as any adult worms surviving the original DEC treatment will go on to produce more microfilariae offspring.

As the nematodes die, they release certain chemicals which can cause an allergic-type reaction in the host, so many individuals treated with DEC will also need treatment with potent anti-allergy medications such as steroids and **antihistamines**. The tissue damage caused by elephantiasis is permanent, but the extreme swelling can be somewhat reduced by **surgery** or the application of elastic bandages or stockings.

Prevention

Prevention of lymphatic filariasis is very difficult, if not impossible, for people living in the areas where the causative nematodes are commonly found. Travelers to such areas can minimize exposure to the mosquito vectors through use of insect repellant and mosquito netting. Work is being done to determine whether DEC has any use as a preventive measure against the establishment of lymphatic filariasis.

See also Roundworms.

Resources

Books

Andreoli, Thomas E., et al. *Cecil Essentials of Medicine.* Philadelphia: W. B. Saunders Company, 1993.

Berkow, Robert, and Andrew J. Fletcher. *The Merck Manual of Diagnosis and Therapy.* Rahway, NJ: Merck Research Laboratories, 1992.

Cormican, M. G. and M. A. Pfaller. "Molecular Pathology of Infectious Diseases." In *Clinical Diagnosis and Management by Laboratory Methods.* 20th ed. Philadelphia: W. B. Saunders, 2001.

Isselbacher, Kurt J., et al. *Harrison's Principles of Internal Medicine.* New York: McGraw Hill, 1994.

KEY TERMS

Host—An animal or plant within which a parasite lives.

Microfilariae—Live offspring produced by adult nematodes within the host's body.

Nocturnal—Occurring at night.

Periodicity—The regularity with which an event occurs.

Vector—Any agent, living or otherwise, that carries and transmits parasites and diseases.

Mandell, Douglas, et al. *Principles and Practice of Infectious Diseases.* New York: Churchill Livingstone, 1995.
Prescott, L., J. Harley, and D. Klein. *Microbiology.* 5th ed. New York: McGraw-Hill, 2002.

Rosalyn Carson-DeWitt

Elevator

An elevator is an enclosed car that moves in a vertical shaft between the multi-story floors of a building carrying passengers or freight. All elevators are based on the principle of the counterweight, and modern elevators also use geared, electric motors and a system of cables and pulleys to propel them. The world's most often used means of mechanical transportation, it is also the safest. The elevator has played a crucial role in the development of the high-rise or skyscraper and is largely responsible for how our cities look today. It has become an indispensable factor of modern urban life.

History

Lifting loads by mechanical means goes back at least to the Romans who used primitive hoists operated by human, **animal**, or **water** power during their ambitious building projects. An elevator employing a counterweight is said to have been built in the seventeenth century by a Frenchman named Velayer, and it was also in that country that a passenger elevator was built in 1743 at the Versailles Palace for King Louis XV. By 1800, steam power was used to power such lift devices, and in 1830, several European factories were operating with hydraulic elevators that were pushed up and down by a plunger that worked in and out of a cylinder.

All of these lifting systems were based on the principle of the counterweight, by which the weight of one object is used to balance the weight of another object. For example, while it may be very difficult to pull up a heavy object using only a rope tied to it, this job can be made very easy if a weight is attached to the other end of the rope and hung over a pulley. This other weight, or counterweight, balances the first and makes it easy to pull up. Thus an elevator, which uses the counterweight system, never has to pull up the total weight of its load, but only the difference between the load-weight and that of the counterweight. Counterweights are also found inside the sash of old-style windows, in grandfather clocks, and in dumbwaiters.

Until the mid-nineteenth century, the prevailing elevator systems had two problems. The plunger system was very safe but also extremely slow, and it had obvious height limitations. If the plunger system was scrapped and the elevator car was hung from a rope to achieve higher speeds, the risk of the rope or cable breaking was an ever-present and very real danger. Safety was the main technical problem that the American inventor, Elisha Graves Otis (1811–1861) solved when he invented the first modern, fail-safe passenger elevator in 1853. In that year, Otis demonstrated his fail-safe mechanism at the Crystal Palace Exposition in London. In front of an astonished audience, he rode his invention high above the crowd and ordered that the cable holding the car be severed. When it was, instead of crashing to the ground, his fail-safe mechanism worked automatically and stopped the car dead.

The secret of Otis's success was a bow-shaped wagon spring device that would flex and jam its ends into the guide rails if tension on the rope or cable was released. What he had invented was a type of speed governor that translated an elevator's downward **motion** into a sideways, braking action. On March 23, 1857, Otis installed the first commercial passenger elevator in the Haughwout Department Store in New York, and the age of the skyscraper was begun. Until then, large city buildings were limited to five or six stories which was the maximum number of stairs people were willing to climb. When the iron-frame building was developed by architects in the 1880s, the elevator was ready to service them. By then, electric power had replaced the old steam-driven elevator, and the first commercial passenger elevator to be powered by **electricity** was installed in 1889 in the Desmarest Building in New York. In 1904, a "gearless" feature was added to the **electric motor**, making elevator speed virtually limitless. By 1915, automatic leveling had been introduced and cars would now stop precisely where they should.

Modern elevators

Today's passenger elevators are not fundamentally different from the Otis original. Practically all are elec-

KEY TERMS

Centrifugal force—The inertial reaction which causes a body to move away from a center about which it revolves.

Counterweight—The principle in which the weight of one object is used to balance the weight of another object; for an elevator, it is a weight which counterbalances the weight of the elevator car plus approximately 40% of its capacity load.

Hydraulic elevator—A power elevator where the energy is applied, by means of a liquid under pressure, in a cylinder equipped with a plunger or a piston; a direct-plunge elevator had the elevator car attached directly to the plunger or piston which went up and down a sunken shaft.

Microprocessor—The central processing unit of a microcomputer that contains the silicon chip which decodes instructions and controls operations.

Sheave—A wheel mounted in bearings and having one or more grooves over which a rope or ropes may pass.

Speed governor—A device that mechanically regulates the speed of a machine, preventing it from going any faster than a preset velocity.

trically propelled and are lifted between two guide rails by **steel** cables that loop over a pulley device called a *sheave* at the top of the elevator shaft. They still employ the counterweight principle. The safety mechanism, called the overspeed governor, is an improved version of the Otis original. It uses centrifugal **force** that causes a system of weights to swing outward toward the rails should the car's speed exceed a certain limit. Although the travel system has changed little, its control system has been revolutionized. Speed and **automation** now characterize elevators, with microprocessors gradually replacing older electromechanical control systems. Speeds ranging up to 1,800 ft (550 m) per minute can be attained. Separate outer and inner doors are another essential safety feature, and most now have electrical sensors that pull the doors open if they sense something between them. Most also have telephones, alarm buttons, and emergency lighting. Escape hatches in their roofs serve both for maintenance and for emergency use.

Modern elevators can also be programmed to provide the fastest possible service with a minimum number of cars. They can further be set to sense the weight of a car and to bypass all landing calls when fully loaded. In addition to regular passenger or freight elevators, today's specialized lifts are used in ships, **dams**, and even on rocket launch pads. Today's elevators are safe, efficient, and an essential part of our daily lives.

See also Building design/architecture.

Resources

Books

The First One Hundred Years. New York: The Otis Elevator Company, 1953.
Strakosch, George R. *Vertical Transportation: Elevators and Escalators.* New York: John Wiley & Sons, 1983.

Periodicals

Jackson, Donald Dale. "Elevating Thoughts from Elisha Otis and Fellow Uplifters." *Smithsonian* (November 1989): 211+.

Leonard C. Bruno

Ellipse

An ellipse is a kind of oval. It is the oval formed by the intersection of a **plane** and a right circular cone-one of the four types of **conic sections**. The other three are the **circle**, the **hyperbola**, and the **parabola**. The ellipse is symmetrical along two lines, called *axes*. The *major axis* runs through the longest part of the ellipse and its center, and the *minor axis* is **perpendicular** to the major axis through the ellipse's center.

Other definitions of an ellipse

Ellipses are described in several ways, each way having its own advantages and limitations:

1. The set of points, the sum of whose distances from two fixed points (the foci, which lie on the major axis) is constant. That is, P: $PF_1 + PF_2$ = constant.

2. The set of points whose distances from a fixed **point** (the focus) and fixed line (the directrix) are in a constant **ratio** less than 1. That is, P: PF/PD = e, where 0 < e < 1. The constant, e, is the eccentricity of the ellipse.

3. The set of points (x,y) in a Cartesian plane satisfying an equation of the form $x^2/25 + y^2/16 = 1$. The equation of an ellipse can have other forms, but this one, with the center at the origin and the major axis coinciding with one of the coordinate axes, is the simplest.

4. The set of points (x,y) in a Cartesian plane satisfying the parametric equations x = a cos t and y = b sin t, where a and b are constants and t is a **variable**. Other

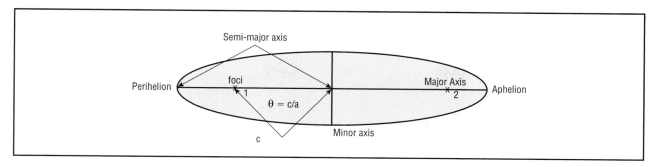

Earth travels around the Sun in a slighlty elliptical orbit. *Illustration by Argosy. The Gale Group.*

parametric equations are possible, but these are the simplest.

Features

In working with ellipses it is useful to identify several special points, chords, measurements, and properties:

The major axis: The longest chord in an ellipse that passes through the foci. It is equal in length to the constant sum in Definition 1 above. In Definitions 3 and 4 the larger of the constants a or b is equal to the semimajor axis.

The center: The midpoint, C, of the major axis.

The vertices: The end points of the major axis.

The minor axis: The chord which is perpendicular to the major axis at the center. It is the shortest chord which passes through the center. In Definitions 3 and 4 the smaller of a or b is the semiminor axis.

The foci: The fixes points in Definitions 1 and 2. In any ellipse, these points lie on the major axis and are at a **distance** c on either side of the center. If a and b are the semimajor and semiminor axes respectively, then $a^2 = b^2 + c^2$. In the examples in Definitions 3 and 4, the foci are 3 units from the center.

The eccentricity: A measure of the relative elongation of an ellipse. It is the ratio e in Definition 2, or the ratio FC/VC (center-to-focus divided by center-to-vertex). These two definitions are mathematically equivalent. When the eccentricity is close to **zero**, the ellipse is almost circular; when it is close to 1, the ellipse is almost a parabola. All ellipses having the same eccentricity are geometrically similar figures.

The **angle** measure of eccentricity: Another measure of eccentricity. It is the acute angle formed by the major axis and a line passing through one focus and an end point of the minor axis. This angle is the **arc** cosine of the eccentricity.

The area: The area of an ellipse is given by the simple formula PIab, where a and b are the semimajor and semiminor axes.

The perimeter: There is no simple formula for the perimeter of an ellipse. The formula is an elliptic **integral** which can be evaluated only by **approximation**.

The reflective property of an ellipse: If an ellipse is thought of as a mirror, any ray which passes through one focus and strikes the ellipse will be reflected through the other focus. This is the principle behind rooms designed so that a small sound made at one location can be easily heard at another, but not elsewhere in the room. The two locations are the foci of an ellipse.

Drawing ellipses

There are mechanical devices, called ellipsographs, based on Definition 4 for drawing ellipses precisely, but lacking such a device, one can use simple equipment and the definitions above to draw ellipses which are accurate enough for most practical purposes.

To draw large ellipses one can use the pin-and-string method based on Definition 1: Stick pins into the drawing board at the two foci and at one end of the minor axis. Tie a loop of string snugly around the three pins. Replace the pin at the end of the minor axis with a pencil and, keeping the loop taut, draw the ellipse. If string is used that does not stretch, the resulting ellipse will be quite accurate.

To draw small and medium sized ellipses a technique based on Definition 4 can be used: Draw two concentric circles whose radii are equal to the semimajor axis and the semiminor axis respectively. Draw a ray from the center, intersecting the inner circle at y and outer circle at x. From y draw a short horizontal line and from x a short vertical line. Where these lines intersect is a point on the ellipse. Continue this procedure with many different rays until points all around the ellipse have been located. Connect these points with a smooth curve. If this is done carefully, using ordinary drafting equipment, the resulting ellipse will be quite accurate.

Uses

Ellipses are found in both natural and artificial objects. The paths of the planets and some **comets** around the **Sun** are approximately elliptical, with the sun at one of the foci. The seam where two cylindrical pipes are joined is an ellipse. Artists drawing circular objects such as the tops of vases use ellipses to render them in proper perspective. In Salt Lake City the roof of the Mormon Tabernacle has the shape of an ellipse rotated around its major axis, and its reflective properties give the auditorium its unusual acoustical properties. (A pin dropped at one focus can be heard clearly by a person standing at the other focus.) An ellipsoidal reflector in a lamp such as those dentists use will, if the **light** source is placed at its focus, concentrate the light at the other focus.

Because the ellipse is a particularly graceful sort of oval, it is widely used for esthetic purposes, in the design of formal gardens, in table tops, in **mirrors**, in picture frames, and in other decorative uses.

Resources

Books

Finney, Thomas, Demana, and Waits. *Calculus: Graphical, Numerical, Algebraic.* Reading, MA: Addison Wesley Publishing Co., 1994.

J. Paul Moulton

Elm

Elms are trees (occasionally shrubs) of flowering plants in the genus *Ulmus*. Elm leaves possess stipules, and often have a nonsymmetrical **leaf**, that is, one half is larger than the other so that the bottom ends do not meet where they are attached to the mid-rib. Elms flower in the spring. Their flowers lack petals, form reddish brown clusters in the tops of the trees, appear before the leaves have fully expanded, and are pollinated by the **wind**. The **fruits** are samaras that are technically equivalent to the keys (fruits) of maple, although in elms the seed is surrounded entirely by a greenish, papery wing so that the oval or circular fruit (seed plus wing) resembles a fried egg.

There are about 30 **species** of elms in the world. Most occur in north temperate regions of **North America**, **Europe**, and **Asia**, and on **mountains** of tropical Asia. Elms rarely occur in large tracts in **forests**. Instead they are usually found interspersed among other deciduous trees.

Elm has, until recently, been an important ornamental and shade **tree**. In eastern North America there is hardly a single town without an Elm Street, named after the stately white or American elm (*Ulmus americana*), which is a tall tree reaching heights of 130 ft (40 m). American elm has elegant upswept limbs that arch down at the ends. Unfortunately, these beautiful trees are becoming rare because of a devastating **disease** called Dutch elm disease that was accidentally imported from Europe, probably on infected timber.

Dutch elm disease derives its name from the fact that it was first discovered on elm trees in Holland in 1921. There is no tree known as Dutch elm, and the disease originated in Asia. The disease appeared in the United States in 1930 and spread rapidly throughout the range of elm in eastern North America.

Dutch elm disease is caused by an ascomycete fungus (*Ophiostoma ulmi*) in partnership with an insect. Although the fungus causes the disease, an insect is generally necessary as an agent in spreading fungal spores from infected trees to healthy trees. The **bark beetles** *Scolytus multistriatus*, *S. scolytus*, and *Hylurgopinus rufipes* are the common agents of spread, although **birds** and, to a limited extent, **water** and wind can also spread the disease. The fungus can also spread from diseased trees to healthy trees by natural root grafts. Once a tree has been infected, the fungus initially grows within the water-conducting cells called vessels, where spores (different from those produced on coremia) are produced and released. Carried in the water of the vessels, these spores spread the **infection** to other parts of the tree. The symptoms shown by infected trees include: yellowing of the leaves, followed by wilting and their premature fall, death of branches causing the crown to appear sparse, and browning of the sapwood. If the infection spreads throughout the vascular system, death of the tree can occur within weeks of the infection, although many trees survive for several years before succumbing to the disease. In the later stages of the disease, the fungus moves out of the water-conducting cells into other tissues. There is no satisfactory method for managing Dutch elm disease. Not all species of elm have been affected by Dutch elm disease. Asiatic species, such as Siberian and Chinese elms, are generally resistant.

Elm **wood** is economically valuable. Most species produce fine timber with a distinctive pattern. The timber resists decay when waterlogged, thus making it quite useful in certain specialized uses, such as serving as underwater pilings. Before metalworking, elm was used in Europe in water pipes and water pumps; 200-year-old pipes are often dug up in London, England. The grain of elm wood is strongly interlocked so that it is very difficult to split. For this reason, elm wood is often used for certain kinds of furniture, such as the seats of chairs, since the driving in of legs and backs tends to split most other woods, and also for wheel-hubs and mallet heads. Elm wood has also been extensively used for coffin boards and shipping cases for heavy machinery.

The inner bark of elm, especially of roots, is fibrous and can be made into rope, and string for fishing-line, nets, or snares. The fruits are eaten by many birds and **squirrels**. Twigs and leaves are eaten by **deer** and rabbits. The leaves are quite nutritious and in ancient times in Europe, branches were cut off so that **livestock** could feed on the foliage. Elms have little value as food for people, although in times of famine, the ground bark, leaves, and fruits have been eaten by the Chinese, and bark ground into a meal and mixed with flour for bread has similarly been used in times of scarcity in Norway. The inner bark of slippery elm reputedly has some medicinal properties.

Les C. Cwynar

Embiids

Embiids are small, cylindrical, soft-bodied **insects** in the order Embioptera that spin tubular galleries of silk, an ability that gives them the common name web-spinners. They have chewing mouthparts, and undergo paurometabolism, or gradual **metamorphosis**, exhibiting a definite egg, nymph, and adult stage. In the **phylogeny**, or evolutionary history of the class Insecta, embiids are thought to be most closely related to the orders Dermaptera (the **earwigs**) and Plecoptera (the **stoneflies**).

A distinguishing morphological feature of the web-spinners is that the silk **glands** are in the tarsal segments of the front legs. Another unique characteristic of these insects is that the wings of the males are flexible when the insect is not in flight. Their wing **veins** are hollow, and fill with **blood** in order to stiffen the wing for flight.

Embiids are gregarious, or group-living, insects in which the wingless females live in silken galleries where

they care for their young. The males of this order are often, but not always winged, and do not feed as adults. Rather they die soon after finding and mating with a female. Individuals in all developmental stages have the ability to spin silk. They spin galleries in the **soil**, under **bark**, in dead **plant matter**, in rock crevices, and other such inconspicuous substrates. The females and nymphs living in the galleries are flightless but they are adapted to run backwards through the tunnel to flee potential predators that may discover the opening of the chamber.

These secretive insects are rare, with only 200 **species** known to exist, most of which are tropical. In the United States, there are ten species, all of which have a southern distribution. The main food source of Embioptera is dead plant matter, and this fact, as well as their relative rarity, make embiids of little known economic significance to humans.

Embolism

An embolism is the sudden blockage of a **blood** vessel by a blood clot that has been brought to that location by the bloodstream. The clot, called an embolus, from the Greek word meaning plug, is a blood clot that has formed inside the **circulatory** system and is floating in the bloodstream. It will remain on the move until it encounters a blood vessel too small for it to fit through, where it will plug the vessel and prevent any further circulation of blood through it.

A blood clot that forms in a given location and remains there is called a thrombus, from the Greek word for clot.

An embolism is named by the location in which the clot lodges. A pulmonary embolism is an embolus that has plugged a blood vessel, usually an artery, in one of the lungs. A coronary embolism is obscuring the channel in one of the coronary **arteries**, which feed the **heart** muscle. A cerebral embolism lodges in a blood vessel in the **brain** and perhaps precipitates a **stroke**.

When an embolus plugs a blood vessel, the tissues that are bathed by the blood in the vessel will die when the **blood supply** is cut off. Death of **tissue** resulting from the lack of blood is called an infarct. If the embolism is in a coronary artery and the infarct is in the heart muscle it is a heart attack. The seriousness of the attack is determined by which vessel the clot blocks and how much of the heart muscle is infarcted.

The same situation applies to other organs. A cerebral embolism can cause brain damage and bring about a stroke. A pulmonary embolism causes damage to the lung tissue that can be serious. Any of these embolisms can be fatal and must be treated quickly.

Drugs can be given to dissolve the clot and other drugs can be taken to prevent the formation of any more clots.

What causes most emboli to form is not known. Some may form after **surgery** if air gets into the bloodstream.

Embryo and embryonic development

An embryo is a stage directly after **fertilization** that signifies the early stages of growth and development of an **organism**. In humans, this stage ends during the third month of pregnancy, and is then called a fetus. Plants and invertebrate as well as vertebrate animals have an embryonic stage of development. For example, the embryo of the common North American leopard frog, *Rana pipiens* is from a **species** that has been studied extensively because it is common, relatively easy to induce ovulation in females, and developmental progress can be monitored in a **glass** dish due tot the fact that the embryo is neither within a shell (as with **reptiles** and **birds**) nor within the body of the mother (as with **mammals**). Much of what is known today about the stages of embryonic development and characterization of the embryo has been accomplished through research contributions related to the development of the frog. In fact, the technique of removing male and female sex cells to be used for fertilization and then replace the fertilized egg into the appropriate place in the female, also known as *in vitro* fertilization was first demonstrated in this species. This technique has revolutionized reproductive technologies in humans and has allowed certain reproductively challenged couples the option of having biologically-related children. Although there are many differences, development of humans is much like the frog in terms of the basic embryological terminology as well as the distinct stages of embryonic development. In the *Rana pipiens*, development is divided into an embryonic period which occurs prior to hatching from the jelly membranes that enclose the embryo, and larval development which is the period of the free-living feeding tadpole.

Embryonic development in the *Rana pipiens*

Embryonic development ordinarily is considered to begin with the formation of the fertilized egg called the zygote. In the case of North American leopard **frogs**, zygote formation occurs in breeding ponds or **wetlands** in the early spring. During winter, the frogs reside in the cold **water** of northern lakes. When the **ice** melts and the days grow longer, the frogs leave the cold lakes and seek shallow ephemeral bodies of water. Water has a high specific **heat**, which means that lakes resist a change in **temperature**. The shallow water of breeding ponds warms readily, which is essential for both ovulation (the release of eggs) and embryo development. Male frogs clasp mature female frogs and this may encourage egg release. The clasping male releases sperm as the eggs are extruded. A female may release as many as 3,000 eggs which, when fertilized, results in a potential of 3,000 embryos.

In vitro fertilization and stages of frog development

The processes leading to fertilization can be accomplished in the laboratory. Laboratory frogs are ordinarily kept cold until embryo formation is required. Female frogs brought to laboratory temperature (18°C or about 65° F) are injected with a pituitary extract combined with progesterone causing ovulation within 48 hours. The release of sperm is induced with an injection of **human chorionic gonadotropin** hormone. A zygote is formed when freshly ovulated ova are combined with freshly released sperm. The zygote cleaves (divides) into two cells within 3.5 hours called blastomeres. It divides again into four cells, then eight, and continues until a ball of cells forms known as a morula. With continued division, the morula gives rise to the blastula. The frog blastula is also a ball of cells but it is hollow. The mature blastula has about 3,000 cells, which can form within the first 24 hours in the laboratory.

Blastula cells vary in size depending on the amount of yolky granules they contain. Very yolky cells are large; cells with minimal yolk are tiny but clearly visible in the **microscope**. Regardless of size, the cells are considered to be developmentally equivalent. That means that they have not begun the process of specialization known as differentiation. Classic grafting experiments in the early 1900s demonstrated the lack of specialization by grafting small populations of blastula cells to a new site within an embryo with no effect on development. While 3,000 cells are present at the definitive blastula stage, there has been no net growth and, therefore, no increase in **mass**. A frog blastula has the same diameter as a zygote. Cleavage gives rise to an increasing number of cells which partitions the former zygote into ever smaller compartments (cells).

Gastrulation, which precedes the blastula stage, is a time of developmental change. Many living cells can migrate within a developing organism and the first migrations are observed during gastrulation. Some cells on the

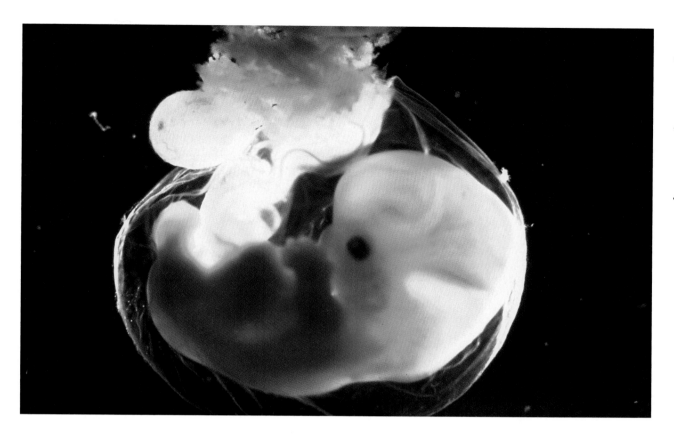

A human embryo at five to six weeks of development. *Photograph by Petit Fromat/Nestle. National Audubon Society Collection/Photo Researchers, Inc. Reproduced by permission.*

surface migrate to the interior. In the process of **migration**, the former population of equivalent and unspecialized cells becomes a structured **sphere** with a complicated interior with the potential to differentiate. The three primary germ layers may be detected at this time; the external ectoderm, the internal endoderm, and the intermediate mesoderm. The rearrangement of cells that result from migration forms an area that invaginates and cells move toward the interior. The site of movement inwards is known as the blastopore. The blastopore will eventually become the posterior opening of the **digestive system** and the positioning of the blastopore within the gastrula permits identification of an anterior and posterior axis as well as left and right sides.

Differentiation occurs during gastrulation. Cells begin their specialization, limiting their competence to differentiate into all of the different possible **cell** types. Thus, when a **graft** is made, which exchanges populations of cells from one area to another, the grafts develop in their new locations as if they had not been moved. For example, when cells destined to form nerve structures are grafted to a skin-forming area, they do not form skin but continue on their pathway to differentiate neural cells.

Tissue specific differentiation during embryogenesis

Specific **tissue** types form during embryo development. A portion of the ectoderm on the dorsal side of the embryo rolls up into a tube, which forms into the central **nervous system**. The anterior portion of the tube becomes the **brain** and the posterior portion becomes the spinal cord. Some mesoderm cells become specialized to form muscle and the muscle functions well before feeding occurs. This can be observed by muscular activity (bending and twisting of the embryo body) in the as yet unhatched embryo in its jelly membranes. Careful examination of the embryo at this time reveals a pulsation. The pulsation is the beating of the **heart** muscle, which begins to circulate embryonic **blood** cells. Embryonic gills are exposed on either side of the head. The structure of the gills is so delicate that blood cells, with a dissecting microscope, can be seen surging in synchrony with the beating heart. During embryonic development, the **excretory system** and the digestive system begin their formation. Within six days in the laboratory, all embryonic systems have begun developing. Hatching occurs at about this time, which marks the end of the embryonic period and the beginning of larval development. Larvae feed in about a week after fertilization. Embryonic development has been characterized by differentiation of **organ** systems but no increase in mass. Feeding begins and the tadpole grows enormously in size compared with its origin. Feeding and growing will continue until the larval tadpole begins its **metamorphosis** into a juvenile frog.

See also Embryo transfer; Embryology; Sexual reproduction.

Resources

Books

Larsen, William J. *Human Embryology.* 3rd. ed. Philadelphia: Elsevier Science, 2001.

Other

FreeEssays "Study on *Rana pipiens*" Research center database. September 2, 2002 [cited January 12, 2003]. <http://www.freeessays.cc/db/3/alw59.shtml>.

Herpnet "Northern Leopard Frog" Minnesota Herpetology. [cited January 12, 2003]. <http://www.herpnet.net/Minnesota-Herpetology/frogs_toads/NorthernLeopard_frog.html>.

Kimball, Dr. John W. "Frog Embryology" Online Biology Resource. September 2, 2002 [cited January 12, 2003]. <http://users.rcn.com/jkimball.ma.ultranet/BiologyPages/F/FrogEmbryology.html>.

UNSW "Embryology, Frog Development" UNSW Medical School. June 5, 2001 [cited January 12, 2003]. <http://anatomy.med.unsw.edu.au/cbl/embryo/OtherEmb/Frog.htm>.

Western Ecological Research Center. "Scientific Name: *Rana pipiens* Complex." USGS. April 12, 2002 [cited January 12, 2003]. <http://www.werc.usgs.gov/fieldguide/rapi.htm>.

Bryan H. Cobb

Embryo transfer

Developments in reproductive technology are occurring at a rapid **rate** in **animal** science as well as in human **biology**. *In vitro* **fertilization**, embryo culture, preservation of embryos by freezing (cryopreservation), and cloning technology yield embryos that are produced outside of the female **reproductive system**. Embryo transfer permits continued survival of embryos by insertion into the female reproductive system.

Traditionally, relatively little scientific attention has been focused on the technique of human embryo transfer, as it is considered an unimportant variable in the success of an *in vitro* fertilization cycle. Essentially, the characteristics of embryo transfer in the human have changed little during the twenty years since the British physician R. G. Edwards described the technique in the 1980s.

While **cell** culture *in vitro* has made remarkable strides, embryos can be sustained in culture for only a few days. Thus, their survival is dependent upon transfer to the hospitable and nurturing environment of the uterus of a foster mother. While embryo transfer may seem to be high technology, it actually got its start well over a century ago and over the past decade, **evolution** in culture conditions (as well as stage-specific or "sequential" complex media) have significantly increasing the viability of *in vitro* fertilized embryos.

Embryos may be transferred to the patient on day 1 (zygote stage) of development, on day two (two-cells to four-cells stage), on day three (six-cells to eight-cells stage), on day four (morula), or on days five to seven (different blastocyst stages). The process of embryo transfer includes several steps and the overall pregnancy rate is affected by several factors. One of the most difficult considerations of embryo transfer is the determination of which embryos are most suitable for transfer into the uterus. Recent advanced technologies such as micromanipulation yield increased successes in embryo implantation.

Selection of embryos is an important issue, because the higher the number of embryos transferred, the higher the pregnancy rate. The multiple pregnancy rate also increases, a condition that often is not desirable for the survival of all fetuses. To avoid this, it is important that embryos be selected prior to implantation. The embryo is selected for quality of viability and other important characteristics. Embryo scoring techniques have been developed to evaluate the potential for embryo implantation including cell number, fragmentation characteristics, cytoplasmic pitting, blastomere regularity, presence of vacuoles, and blastomere expansion. Cell stages at the time of transfer correlate to embryo survival because its **genome** activation occurs between the four-cell and eight-cell stages of pre-implantation development. Events used to score the embryos include embryo **metabolism** assessment and pre-implantation genetic **diagnosis**. Genetic abnormalities of the embryo as well as defects in uterine receptivity are also factors in the success of embryo transfer. Additional aspects of successful implantation include bed rest following embryo transfer, the absence of hydrosalpinx (blocked, fluid-filled fallopian tubes) that could lead to washing out of the implanted embryos by intermittent leakage of the hydrosalpingeal fluid. Should uterine contractions occur after transfer, the embryos could be expelled down through the cervix, or up into the fallopian tubes, instead of implanting in the uterus. Additionally, the presence of **blood** on the outside of the transfer catheter tip correlates with lower rates of successful implantation.

The majority of embryo transfers are currently carried out by cannulating (inserting a hollow tube into) the uterine cavity via the cervix (transcervical transfers). A catheter is inserted inside the uterus and the embryos are deposited at least one to two millimetres from the fundus. Disadvantages include possible cervical stenosis, the risk of **infection** from the introduction of **microorganisms**, and release of prostaglandins that may cause contractions. Interaction between the embryo and the endometrium is another important topic in determining embryo transfer success. Paracrine modulators are necessary for embryo-uterine interactions and they are more favorable during the so called " implantation window" that generally occurs between days 18 and 24 of the normal **menstrual cycle**. This time interval can be assessed either by endometrial biopsies or imaging techniques (ultrasound and magnetic **resonance** imaging).

See also Embryo and embryonic development; Embryology.

Resources

Periodicals

Ishihara, Baba K., et al. "Where Does the Embryo Implant After Embryo Transfer in Humans?" *Fertility and Sterility* (2000): 73, 123–5.

Lindheim, S.R., M.A. Cohen, and M.V. Sauer. "Ultrasound-guided Embryo Transfer Significantly Improves Pregnancy Rate in Women Undergoing Oocyte Donation." *International Journal of Gynecology and Obstetrics* (1999): 66, 281–4.

Tapanainen, Thomas C. and H. Martikainen. "The Difficulty of Embryo Transfer is an Independent Variable for Predicting Pregnancy in In-vitro Fertilization Treatments." *Fertility and Sterility* (1998): 70, Suppl 1, S433.

Antonio Farina

Embryology

Embryology is the study of the development of organisms. This is as true of plants as it is of animals.

Seed formation proceeds following **fertilization** in higher plants. The seed consists of the embryo, the seed coat, and another part sometimes called the endosperm. While plants are extraordinarily important for survival of **animal** life, animal embryology is described here.

The dictionary definition limits the meaning of the term "embryo" to developing animals that are unhatched or not yet born. Human embryos are defined as developing humans during the first eight weeks after conception. The reason that many embryologists have difficulty with this terminology is that it is purely arbitrary. It would be difficult indeed, if not impossible, to discriminate a human embryo nearing the end of the eighth week from

a developing human during the ninth week after conception. Correspondingly, there are no morphological events that distinguish a pre-hatching frog tadpole from a post-hatching tadpole (hatching never occurs synchronously in an egg mass—there are always those that hatch early and those larvae which are dilatory).

Embryologists consider development from a zygote to a multicellular **organism**. In the particular case of humans, development does not even stop at **birth**. Note that teeth continue to develop and sex **glands** with sexual differentiation mature long after birth. For a number of years, many embryologists have referred to their discipline as developmental **biology** to escape from the need to confine their studies to earlier stages. Embryology in the modern sense is the study of the **life history** of an animal and human embryology considers developmental aspects of life as a whole and not just the first eight weeks.

History of embryology as a science

The study of embryology, the science that deals with the formation and development of the embryo and fetus, can be traced back to the ancient Greek philosophers. Originally, embryology was part of the field known as "generation," a term that also encompassed studies of reproduction, development and differentiation, regeneration of parts, and **genetics**. Generation described the means by which new animals or plants came into existence. The ancients believed that new organisms could arise through **sexual reproduction, asexual reproduction,** or **spontaneous generation**. As early as the sixth century B.C., Greek physicians and philosophers suggested using the developing chick egg as a way of investigating embryology.

Aristotle (384–322 B.C.) described the two historically important models of development known as preformation and epigenesis. According to preformationist theories, an embryo or miniature **individual** preexists in either the mother's egg or the father's semen and begins to grow when properly stimulated. Some preformationists believed that all the embryos that would ever develop had been formed by God at the Creation. Aristotle actually favored the theory of epigenesis, which assumes that the embryo begins as an undifferentiated **mass** and that new parts are added during development. Aristotle thought that the female parent contributed only unorganized **matter** to the embryo. He argued that semen from the male parent provided the "form," or soul, that guided development and that the first part of the new organism to be formed was the **heart**.

Aristotle's theory of epigenetic development dominated the science of embryology until the work of physiologist William Harvey (1578–1657) raised doubts about

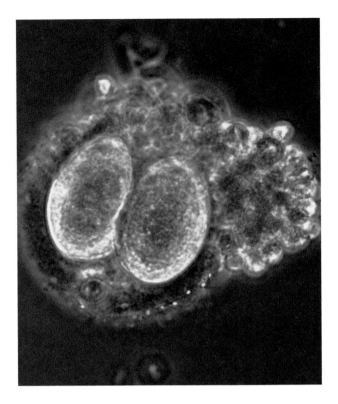

A human two-cell embryo 24 hours after fertilization. *Photograph by Richard G. Rawlins. Custom Medical Stock Photo. Reproduced by permission.*

many aspects of classical theories. In his studies of embryology, as in his research on the circulation of the **blood**, Harvey was inspired by the work of his teacher, Girolamo Fabrici (ca.1533–1619). Some historians think that Fabrici should be considered the founder of modern embryology because of the importance of his embryological texts: *On the Formed Fetus* and *On the Development of the Egg and the Chick*. Harvey's *On the Generation of Animals* was not published until 1651, but it was the result of many years of research. Although Harvey began these investigations in order to provide experimental proof for Aristotle's theory of epigenesis, his observations proved that many aspects of Aristotle's theory of generation were wrong.

Aristotle believed that the embryo essentially formed by coagulation in the uterus immediately after mating when the form-building principle of the male acted on the material substance provided by the female. Using **deer** that had mated, Harvey dissected the uterus and searched for the embryo. He was unable to find any signs of a developing embryo in the uterus until about six or seven weeks after mating had taken place. In addition to his experiments on deer, Harvey carried out systematic studies of the developing chick egg. His observa-

tions convinced him that generation proceeded by epigenesis, that is, the gradual addition of parts. Nevertheless, many of Harvey's followers rejected epigenesis and turned to theories of preformation.

Naturalists who favored preformationist theories of generation were inspired by the new mechanical philosophy and by the **microscope**, a device that allowed them to see the embryo at earlier stages of development. Some naturalists produced very unreliable observations of early embryos, but Marcello Malpighi (1628–1694) and Jan Swammerdam (1637–1680), two pioneers of **microscopy**, provided observations that seemed to support preformation. Based on Swammerdam's studies of **insects** and **amphibians**, naturalists suggested that embryos preexisted within each other like a nest of boxes. However, given such a theory, only one parent can serve as the source of the sequence of preformed individuals. At the time, the egg of many **species** was well known, but when the microscope revealed the existence of "little animals" in male semen, some naturalists argued that the preformed individuals must be present in the sperm.

Respected scientists of the time, including Albrecht von Haller (1708–1777), Charles Bonnet (1720–1793), Lazzaro Spallanzani (1729–1799), and René Antoine Ferchault de Reaumur (1683–1757), supported preformation. Bonnet's studies of **parthenogenesis** in **aphids** were regarded as strong support of ovist preformationism. Thus, some naturalists argued that the whole human race had preexisted in the ovaries of Eve, while others reported seeing homunculi (tiny people) inside spermatozoa. Other eighteenth century naturalists rejected both ovist and spermist preformationist views. One of the most influential was Casper Friedrich Wolff (1733–1794), who published a landmark article in the history of embryology, "Theory of Generation," in 1759. Wolff argued that the organs of the body did not exist at the beginning of gestation, but formed from some originally undifferentiated material through a series of steps. Naturalists who became involved in the movement known as nature philosophy found Wolff's ideas very attractive. During the nineteenth century, **cell** theory, the discovery of the mammalian ovum by Karl Ernst von Baer (1792–1876), and the establishment of experimental embryology by Wilhelm Roux (1850–1924) and Hans Driesch (1867–1941) transformed philosophical arguments about the nature of embryological development.

About a century ago, careful observations were made of a number of developing organisms. By this time, there was a cell theory and good microscopes were available. Next came a causal analysis. For instance, it was known that the dorsal ectoderm of all vertebrate embryos rolls up into a tube to form the central **nervous system**. What factors control the very regular appearance of the nervous system and subsequent differentiation into the various parts of the **brain** and the spinal cord? It was hypothesized that the underlying chordamesoderm cells of the gastrula signaled the ectoderm to become neural. The signal was referred to as induction. Other embryonic organs also seemed to appear as a result of induction. Chemical embryology sought to characterize the nature of inducing signals. Now, modern molecular embryology seeks to examine on the level of the **gene** what controls differentiation of specific **tissue** and cell typed of a developing organism.

There are practical considerations that drive some embryologists. The causes of developmental abnormalities (**congenital** malformations) in humans becomes more understandable with a consideration of embryology. The human embryo is extraordinarily vulnerable to drugs, viruses, and **radiation** during the first several months of development when many critical **organ** systems are developing.

See also Embryo and embryonic development; Embryo transfer; Clone and cloning.

Resources

Books

Gilbert, Scott F. *Developmental Biology*. 6th ed. Sunderland, MA: Sinauer Associates, Inc., 2000.
Larsen, William J. *Human Embryology*. 3rd. ed. Philadelphia: Elsevier Science, 2001.
Sadler, T.W., and Jan Langman. *Langman's Medical Embryology*, 8th ed. New York: Lippincott Williams & Wilkins Publishers, 2000.

Other

Intellimed, Inc. "Human Anatomy Online-Innerbody.com" [cited February 5, 2003]. <http://www.innerbody.com/htm/body.html>.

Lois Magner
K. Lee Lerner

Emission

In the context of **ecology** and environmental science, the word emission generally refers to a release of a

A nickel smelting facility emitting pollutants that carry for hundreds of miles in Russian Siberia. *Photograph by Josef Polleross. Stock Market. Reproduced by permission.*

substance or **energy** to the environment. Often, emissions refer to substances or energy that are ecological stressors, and can potentially cause deleterious changes.

Emission sources are of two broad types, point and diffuse. Point sources of emission are discrete and spatially localized. A **volcano** is a natural example of a point sources of emission of gases and particulates to the atmosphere. Some point sources associated with human activities are automobiles, chimneys of homes, smokestacks of power plants and smelters, and aquatic discharge pipes for chemical effluent and waste **heat** of factories. In contrast, diffuse emissions occur over large areas from many, often indistinct sources. For example, although each is a small **point source**, the large numbers of houses and vehicles in cities collectively represent a large, diffuse source of emissions of pollutants into the atmosphere.

Emitted energy can be of two types, heat or electromagnetic. Heat is a type of kinetic energy, involving the vibration of **atoms** or molecules. The more vigorously these are vibrating, the greater the heat content of a substance. Discharges of warm **water** from power plants and factories are examples of the emission of heat.

Electromagnetic energy is the energy of photons, which are entities that have properties of both particles and waves. Electromagnetic energy is divided into spectral components on the basis of wavelength, and includes **radio waves**, infrared, visible (so-called because it is perceptible by the human **eye**), ultraviolet, **x rays**, and gamma **radiation**. Electromagnetic energy is generally emitted by a point source, whose emission **spectrum** can involve one or more of the categories just noted.

Emissions of materials can be from diffuse or point sources. Human activities result in emissions of many chemicals that are important pollutants that cause environmental damages. Some of the most important of these emitted materials are gases such as **sulfur dioxide**, oxides of **nitrogen** (a mixture of nitric oxide and nitrogen dioxide), and **carbon dioxide**, particulates containing metals and organic compounds, and vapors of hydrocarbons, mercury, and other chemicals.

See also Non-point source; Stress, ecological.

Emphysema

Emphysema is an incurable lung **disease** that results in the destruction of air sacs in the lungs. It is brought about almost exclusively by smoking. In the past, the majority of its victims were male, but the disease has become more common in women as more women smoke.

Emphysema is also called chronic obstructive pulmonary (or lung) disease (COPD or COLD). Chronic **bronchitis**, that is, **inflammation** of the air tubes leading into the lungs, is closely associated with the development of emphysema. Some medical authorities consider emphysema and bronchitis to be the same disease.

Although emphysema is closely linked to smoking, a few patients—less than 1% of all cases—have a genetic lack of an **enzyme**, called alpha-1-antitrypsin, that leads to the development of the disease. Alpha-1-antitrypsin normally protects the elastic fibers in the walls of the small air sacs in the lungs. If the enzyme is not present, the air sacs are vulnerable to damage by **cigarette smoke**, chemical fumes, dust, and other substances that are inhaled.

The lungs

The lungs are two large, spongy sacs that lie on each side of the chest (thorax), separated by the **heart**. The right lung is divided into three lobes and the left into two lobes. Each lobe is further divided into two to five segments, which are divided by a **membrane**. Each segment is supplied with incoming air by a branching tube called a bronchiole. The bronchioles are connected to larger and larger bronchioles which in turn connect to large tubes called bronchi (singular: bronchus). The bronchus from each lung merges into a single tube called the trachea, which connects with the upper part of the **respiratory system** leading to the nose.

The lungs are made up of approximately 300 million tiny air sacs called alveoli (singular: alveolus). Each sac is surrounded by tiny **blood** vessels, and it is here that the **carbon dioxide** from the body is exchanged for fresh **oxygen** that has been inhaled into the lungs.

During **respiration**, the diaphragm, which forms the floor of the thorax, moves downward toward the abdomen and draws fresh air, inflating the lungs. When the diaphragm relaxes, it resumes its resting domed shape, which forces the air, now heavy with **carbon** dioxide, out of the lungs. The cycle then is repeated for the next breath. In the case of lung disease such as emphysema the declining number of alveoli means that less oxygen is absorbed into the blood with each breath. At first this may not be a discomfort, especially when the **individual** is at rest, but with **exercise** or as more of the alveoli are destroyed the patient will notice an increasing difficulty in getting his breath. He will tire easily and will need to sit down and gasp for air.

Emphysema

Not all smokers develop emphysema, but those who develop the condition become progressively worse over a period of years. Some scientists believe that cigarette smoke neutralizes the protective effects of alpha-1-antitrypsin so that the harmful elements in the smoke can damage the alveolar walls. This has yet to be proved, however.

Once destroyed, the alveoli cannot be repaired or replaced. Continued destruction of these air sacs leads to open, nonfunctional areas in the lungs, reducing the area through which oxygen and carbon dioxide can be exchanged.

In emphysema, the bronchi and bronchioles also become constricted, interfering with the free flow of air in and out of the lungs. The patient finds it harder and harder to breathe and will find himself straining to force air in and out of the lungs. The initial symptoms of emphysema are shortness of breath and a mild cough, both of which become more severe as the disease progresses.

In advanced stages of the disease, the patient will develop a barrel chest from the strain of breathing and probably will need to breathe pure oxygen with the aid of a machine. He or she will no longer have the lung power to blow out a candle held only a few inches from the mouth. The emphysema patient also is at increased risk of worsening the disease by catching **pneumonia** or even a common cold.

Treatment

There is no treatment that will reverse emphysema. The alveoli cannot be healed to restore normal respiration. Some patients may need to take medications to keep the bronchi as open as possible. Also, many emphysema patients require oxygen infusion to provide sufficient oxygen for the body functions. Emphysema patients are advised to avoid people who have colds or pneumonia or other contagious diseases. Also, the emphysema patient should be careful to wrap his face with a scarf when he goes out into cold air. The cold air will constrict the bronchi even more and increase the effort required to breathe. The patient should also avoid dusty areas, paint fumes, **automobile** exhaust, and other lung irritants. Above all, when the early symptoms of emphysema appear, the individual should stop smoking. This will prevent further lung damage and ease the burden of respiration. Continued cigarette smoking will worsen the condition and lead to an earlier death.

Special breathing techniques and respiratory exercises can strengthen the diaphragm, abdominal muscles, and chest muscles to make breathing easier. Oxygen tanks can be installed in the home and small, portable oxygen tanks can be obtained for travel. These may be so small that they can hang on the belt, or if larger supplies are needed, small packs containing tanks can be worn on the back or pushed in a cart.

KEY TERMS

Chronic—A disease or condition that devlops slowly and exists over a long period of time.

Diaphragm—The sheet-like muscle that separates the contents of the abdomen from the contents of the chest cavity. The diaphragm is a major muscle involved in breathing.

Pulmonary—Having to do with the lungs or respiratory system.

Individuals who have severe emphysema may be helped by a lung transplant. Until recently efforts to transplant only the lungs were met with little success. Now, however, improvement in medications and technology allow successful lung transplants.

Individuals who have a genetic lack of alpha-1-antitrypsin may benefit from having the enzyme infused into the lungs. This is an experimental procedure now being done that is showing promising results. It is necessary that the infusion be started in the early stages of the disease. Repeated infusions will be needed during the patient's lifetime.

See also Respiratory diseases.

Resources

Books

Griffith, H. Winter. *Complete Guide to Symptoms, Illness & Surgery for People over 50.* New York: The Body Press/Perigee Books, 1992.

Larson, David E., ed. *Mayo Clinic Family Health Book.* New York: William Morrow, 1996.

Larry Blaser

Emu *see* **Flightless birds**

Emulsion

An emulsion is a two phase system of immiscible liquids in which one liquid is dispersed in the other in the form of microscopic droplets. This dispersion is achieved through the use of emulsifying agents, known as surfactants, which act as chemical **bridges** between the two liquids. Emulsions provide a variety of benefits such as the ability to deliver **water** insoluble active materials from water based products, to improve control over a product's physical properties, and effectively dilute expensive or hazardous functional materials. These properties make emulsions useful in a variety of products including pharmaceuticals, cosmetics, paints and inks, **pesticides**, and foods.

Emulsions throughout history

Emulsions in one form or another, have been used for centuries. Early societies learned to take advantage of this natural phenomenon; for example, the ancient Egyptians used eggs, to emulsify berry extracts, with oils to form crude emulsified paints. In more recent times, emulsions have been developed to deliver medicine in the form of ointments. Today, many products including drugs, paints and inks, cosmetics, and even certain foods, are made possible through the use of emulsions.

Emulsions are created by surfactants

An emulsion can be described as a collection of tiny droplets of one liquid (e.g., an oil) dispersed in another liquid (e.g., water) in which it, the first liquid, is insoluble. Emulsions are formed by mixing these two liquids with a third substance, known as an emulsifier, which creates a uniform, stable dispersion of the first liquid in the second.

Emulsifiers belong to the class of chemicals known as surfactants, or surface active agents, which are able to reduce the **surface tension** of liquids. This ability is important because surface tension, (one of the physical properties which determines how liquids behave) must be overcome for the two liquids to effectively intermingle. Surfactants are able to create this effect by virtue of the dual nature of their molecular structure. One part of the **molecule** is soluble in water, the other in oil; therefore, when an emulsifying surfactant is added to a mixture of oil and water it acts as a "bridge" between the two immiscible materials. This bridging effect reduces the forces between the liquid molecules and allows them to be broken into, and maintained as, separate microscopic droplets.

In an emulsion millions of these tiny surfactant bridges surround the dispersed droplets, shielding them from the other liquid in which they are mixed. The dispersed drops are called the internal phase, the liquid that surrounds the drops is known as the external phase. Depending on the desired characteristics of the finished emulsion, either water or oil may comprise the external phase. The emulsion is further characterized by the type of charge carried by its emulsifiers: it can be anionic (containing a **negative** charge), cationic (containing a positive charge), or nonionic (containing no charge).

Characteristics of emulsions

The resulting emulsion has physical properties different from either of its two components. For example,

while water and oils are transparent, emulsions are usually opaque and may be designed to have a lustrous, pearlized appearance. While water and oil are thin free flowing liquids, emulsions can be made as thick creams which do not flow. Furthermore, the tactile and spreading properties of emulsions are different than the materials of which they are composed.

One of the most important characteristics of emulsions is their inherent instability. Even though the dispersed drops are small, gravity exerts a measurable **force** on them and over **time** they coalesce to form larger drops which tend to either settle to the bottom or rise to the top of the mixture. This process ultimately causes the internal and external phases to separate into the two original components. Depending on how the emulsion is formulated and the physical environment to which it is exposed, this separation may take minutes, months, or millennia.

Uses of emulsions

Many functional chemical ingredients (such as drugs) are not water soluble and require **alcohol** or other organic solvents to form solutions. These solvents may be costly, hazardous to handle, or toxic. Emulsions are useful because they allow ways to deliver active materials in water which is inexpensive and innocuous. A related advantage of emulsions is they allow dilution of these active ingredients to an optimal **concentration**. For example, in hair **conditioning** products the oils and other conditioning agents employed would leave hair limp and sticky if they were directly applied. Through emulsification, these materials can be diluted to an appropriate level which deposits on hair without negative side effects.

Emulsions are commonly used in many major chemical industries. In the pharmaceutical industry, they are used to make medicines more palatable, to improve effectiveness by controlling dosage of active ingredients, and to provide improved aesthetics for topical drugs such as ointments. Nonionic emulsions are most popular due to their low toxicity, ability to be injected directly into the body, and compatibility with many drug ingredients. Cationic emulsions are also used in certain products due to their antimicrobial properties.

In the agricultural industry, emulsions are used as delivery vehicles for **insecticides**, fungicides and pesticides. These water insoluble biocides must be applied to **crops** at very low levels, usually by spraying through mechanical equipment. Emulsion technology allows these chemicals to be effectively diluted and provides improved sprayability. Nonionic emulsions are often used in this regard due to their low foaming properties and lack of interaction with biocidal agents they are carrying.

In cosmetics, emulsions are the delivery vehicle for many hair and skin conditioning agents. Anionic and nonionic emulsions are used to deliver various oils and waxes which provide moisturization, smoothness and softness to hair and skin. Emulsions formed with cationic emulsifiers are themselves effective conditioning agents since their positive charge is attracted to the negative sites on the hair, thus allowing them to resist rinse off.

Many paints and inks are based on emulsions. Such products may be true liquid-in-liquid emulsions or they may be dispersions. Dispersions are similar to emulsions except that the dispersed phase is usually finely divided solid particles. The same surfactant technology used to formulate emulsions is used to create dispersions of pigments that are used in paints and inks. These dispersions are designed to dry quickly and form waterproof films, while not affecting the **color**. In this regard emulsions provide benefits over solvent containing systems because of reduced odor and flammability.

Many food products are in the form of emulsions. An example of a naturally occurring food emulsion is milk which contains globules of milk **fat** (cream) dispersed in water. The whiteness of milk is due to **light** scattering as it strikes the microscopic fat particles. Salad dressings, gravies and other sauces, whipped dessert toppings, peanut butter, and ice cream are also examples of emulsions of various edible fats and oils. In addition to affecting the physical form of food products, emulsions impact taste because emulsified oils coat the tongue, imparting "mouthfeel." Emulsions are useful tools in industries which directly impact many aspects of society. Although emulsions have been used for years, science still has much to learn. In part, this is due to the infinite number of combinations of emulsion systems and the task of fully characterizing their structure. New emulsion types are constantly being developed as new needs arise; for example a relatively recent advance in emulsion technology is the microemulsion, a special type of emulsion characterized by an extremely small particle size. Microemulsions are completely transparent and have enhanced stability as compared to conventional systems. As science continues to respond to the needs of industry, more unusual emulsion combinations will be developed resulting in improved medicines, cosmetics, pesticides, and dessert toppings.

Resources

Books

Garrett, H.E. *Surface Active Chemicals.* New York: Pergamon Press, 1972.

Hibbott, H.W., ed. *Handbook of Cosmetic Science.* New York: Macmillan, 1963.

Lissant, Kenneth J., ed. *Emulsions and Emulsion Technology.* New York: Marcel Dekker, 1974.

KEY TERMS

. . . .

Anionic—A type of surfactant characterized by a net negative charge on its water soluble portion.

Cationic—A type of surfactant characterized by a net positive charge on its water soluble portion.

Dispersion—A mixture composed of tiny particles suspended in a medium such as water.

Immiscible—Compounds which will not dissolve or form solutions with one another.

Microemulsions—A dispersion mixture which consists of particles so small that it is transparent.

Nonionic—A type of surfactant which has no net charge on its water soluble portion.

Surface tension—A force which causes a liquid to resist an increase in its surface area.

Surfactant—Chemical which has both water soluble and oil soluble portions and is capable of forming nearly homogenous mixtures of typically incompatible materials. Also known as an emulsifier.

Rosen, Milton J. *Surfactants and Interfacial Phenomena.* New York: Wiley, 1978.

Randy Schueller
Perry Romanowski

Encephalitis

Encephalitis is an inflammatory **disease** of the **brain**. It is caused by a **virus** that either has invaded the brain, or a virus encountered elsewhere in the body that has caused a sensitivity reaction in the brain. Most cases of encephalitis are considered secondary to a virus **infection** that stimulates an immune reaction.

An infection that involves the membranes associated with the spinal cord is called **meningitis**. This is a less-serious condition than encephalitis in that it usually has few long-term effects. Encephalitis is an infection of the brain **tissue** itself, not the surrounding protective covering (meninges), so the consequences are more serious.

Among the many forms of encephalitis are those that occur seasonally (Russian spring-summer encephalitis, for example), those that affect animals (bovine, fox, and equine), and a form that is carried by a mosquito. Viruses that have been directly implicated in causing encephalitis include the arbovirus, echovirus, poliovirus, and the herpes virus. Encephalitis occurs as a complication of, for example, **chickenpox**, polio, and vaccinia, which is a cowpox virus used in **smallpox** vaccinations, as well as the common flu virus. The herpes simplex virus, responsible for the common cold sore, eczema, and genital herpes; the measles (rubeola) virus; some of the 31 types of echoviruses that also cause a paralytic disease or an infection of the **heart** muscle; the coxsackievirus responsible for infections of the heart and its covering and paralysis; the mumps virus; the arboviruses that normally infect animals and can be spread by mosquito to humans—all have been implicated as causal agents in human encephalitis.

The virus responsible for the infection can invade the cranium and infect the brain via the **circulatory system**. The blood-brain barrier, a system that serves to protect the brain from certain drugs and other toxins, is ineffective against viruses. Once it has gained entrance into the brain the virus infects the brain tissue. The immediate reaction is an **inflammation** causing the brain to swell and activating the **immune system**. The tightly closed vault of the cranium leaves little room for the brain to enlarge, so when it does expand it is squeezed against the bony skull. This, with the active immune system can result in loss of brain cells (neurons), which can result in permanent postinfection damage, depending upon the location of the damage.

The **individual** who is developing encephalitis will have a fever, headache, and other symptoms that depend upon the affected area of the brain. He may fade in and out of consciousness and have seizures resembling epileptic seizures. He may also have rigidity in the back of the neck. Nausea, vomiting, weakness, and sore throat are common. Certain viruses may cause symptoms out of the **nervous system** as well. The mumps virus will cause inflammation of the parotid gland (parotitis), the spleen, and the pancreas as well as of the brain, for example. An infection by the herpes virus can cause hallucinations and bizarre **behavior**.

Treatment of encephalitis is difficult. It is important that the type of virus causing the infection be identified. Drugs are available to treat a herpes virus infection, but not others. Mortality (the death **rate**) can be as high as 50% among patients whose encephalitis is caused by the herpes virus. Infection by other viruses, such as the arbovirus, may have a mortality rate as low as 1%. Treatment is supportive of the patient. Reduction of fever, as well as treatment for nausea and headache are needed. Unfortunately, even those who survive viral encephalitis may have remaining neurologic defects and seizures.

Reye's syndrome

Reye's syndrome is a special form of encephalitis coupled with liver dysfunction seen in young children and those in their early teens.

KEY TERMS

Blood-brain barrier—A blockade of cells separating the circulating blood from elements of the central nervous system (CNS); it acts as a filter, preventing many substances from entering the central nervous system.

Meninges—The tough, fibrous covering of the brain and spinal cord.

Spinal cord—The long cord of nervous tissue that leads from the back of the brain through the spinal column, from which nerves branch to various areas of the body. Severing the cord causes paralysis in areas of the body below the cut.

Invariably, the individual who develops Reye's syndrome has had an earlier viral infection from which they seemingly have recovered. Hours or days later they will begin to develop symptoms such as vomiting, convulsions, delirium, and **coma**. A virus such as the **influenza** virus, varicella (measles), and coxsackie virus are responsible. For reasons unknown, giving a child aspirin tablets to reduce fever accompanying a cold or flu can trigger Reye's syndrome.

At the time the nervous system begins to show signs of infection, the liver is also being affected. Fatty deposits begin to replace functional liver tissue. Similar fatty tissue can be found in the heart muscle and the kidneys. The relationship between the viral effects on the brain and the **parallel** liver damage is not known.

Treatment is not specific to the virus, but is directed at relieving **pressure** on the brain and reducing symptoms. The head of the bed can be elevated and the room left very cool. Care is taken to maintain **blood** sugar level at normal and not let it drop. Other blood factors such as **sodium** and potassium also fall quickly and must be corrected.

The mortality rate for Reye's syndrome can be as high as 25-50%. Early **diagnosis** and initiation of treatment play an important part in keeping the mortality low. Other factors, including age and severity of symptoms, affect the outcome. Some children who survive Reye's syndrome will show signs of brain damage such as impaired mental capacity or seizures.

Thus, it is important that children who contract one of the common **childhood diseases** of viral origin, such as mumps, measles, or chickenpox be watched closely to insure they do not develop symptoms of a brain infection from the same virus.

Resources

Periodicals

Adams, R.M. "Meningitis and Encephalitis: Diseases that Attack the Brain." *Current Health* 21 (October, 1994): 27-29.

Larry Blaser

Endangered species

An endangered species of **plant**, **animal**, or microorganism is at risk of imminent **extinction** or extirpation in all or most of its range. Extinct **species** no longer occur anywhere on **Earth**, and once gone they are gone forever. Extirpated species have disappeared locally or regionally, but still survive in other regions or in captivity. Threatened species are at risk of becoming endangered in the foreseeable future. In the United States, the Endangered Species Act (ESA) of 1973 protects threatened and endangered species that meet specified criteria. Many nations have their own version of the ESA and, like the United States, are members of the World Conservation Union (IUCN), and signatories of the Convention on International Trade In Endangered Species of Wild Fauna and Flora (CITES).

Species have become extinct throughout geological history. Biological **evolution**, driven by natural climate change, catastrophic geologic events, and **competition** from better-adapted species, has involved extinction of billions of species since the advent of life on Earth about three billion years ago. In fact, according to the fossil record, only 2–4% of the species that have existed on Earth exist today. In modern times, however, species threatened by human activities are becoming extinct at a **rate** that far exceeds the pace of extinction throughout most of geologic history. (The mass species extinctions that occurred at the end of the Paleozoic and Mesozoic eras are noteworthy exceptions to this generalization. Geologic data show evidence that a cluster of very large meteorite impacts killed about 85% of the Earth's species, including the dinosaurs, about 65 million years ago at the end of the Mesozoic Cretaceous era.) Meteorite impacts notwithstanding, scientists approximate that present extinction rates are 1,000 to 10,000 times higher than the average natural extinction rate.

Human causes of extinction and endangerment

Human activities that influence the extinction and endangerment of wild species fall into a number of cate-

gories: (1) unsustainable hunting and harvesting that cause mortality at rates that exceed recruitment of new individuals, (2) **land use** practices like **deforestation**, urban and suburban development, agricultural cultivation, and **water** management projects that encroach upon and/or destroy natural habitat, (3) intentional or unintentional introduction of destructive diseases, **parasites**, and predators, (4) ecological damage caused by water, air, and **soil pollution**, and (5) anthropogenic (human-caused) **global climate** change. Alone or in combination, these stressors result in small, fragmented populations of wild flora and fauna that become increasingly susceptible to inbreeding, and to the inherent risks of small abundance, also called demographic instability. Without intervention, stressed populations often decline further, and become endangered.

Why are endangered species important?

Sociopolitical actions undertaken to preserve endangered species and their natural habitats often conflict with human economic interests. In fact, efforts to protect an endangered species usually require an economic sacrifice from the very business or government that threatened the plant or animal in the first place. It is necessary, therefore, to define endangered species in terms of their aesthetic, practical, and economic value for humans. Preservation of endangered species is important and practical for a number of reasons: (1) organisms other than humans have intrinsic moral and ethical value, and a natural right to exist, (2) many plants and animals have an established economic value, as is the case of domesticated species, and exploited **wildlife** like **deer**, **salmon**, and trees, (3) other species, including undiscovered medicinal plants and potential agricultural species, have as-yet unknown economic value, and (4) most species play a critical role in maintaining the health and integrity of their **ecosystem**, and are therefore indirectly important to human welfare. Such ecological roles include nutrient cycling, pest and weed control, species population regulation, cleansing chemical and organic pollution from water and air, **erosion** control, production of atmospheric **oxygen**, and removal of atmospheric **carbon dioxide**.

Rates of endangerment and extinction have increased rapidly in concert with human population growth. Though accelerating habit loss and extinction rates are hallmarks of the modern **biodiversity** crisis, the link between human enterprise and species extinction has existed for almost 100,000 years, during which time **Australia**, the Americas, and the world's islands lost 74–86% of their animals larger than 97 lb (44 kg). In North and **South America**, the disappearance of numerous large animals, including extraordinary species like mammoths, sabre-toothed **cats**, giant ground **sloths**, and armoured glyptodonts coincided with the arrival of significant population of humans between 11,000 and 13,000 years ago. More than 700 vertebrate animals, including about 160 species of **birds** and 100 **mammals**, have become extinct since A.D. 1600.

There is no accurate estimate of the number of endangered species. A thorough census of the earth's smallest and most numerous inhabitants—insects, marine **microorganisms**, and plants—has yet to be conducted. Furthermore, ecologists believe that a large percentage of the earth's as-yet uncataloged biodiversity resides in equatorial rainforests. Because human development is rapidly converting their tropical forest **habitat** into agricultural land and settlements, these multitudinous, unnamed species of small **invertebrates** and tropical plants are categorically endangered. There are perhaps several million endangered species, most of which are invertebrates living in tropical **forests**.

The large number of recorded threatened and endangered species is particularly disturbing given the small percentage of organisms evaluated during compilation of lists. The 2000 IUCN Red List of Threatened Species lists 11,046 species of plants and animals facing imminent extinction around the world. The Red List includes approximately one in four mammal species, and one in eight bird species. Only 4% of the world's named plant species were evaluated for Red List. In the United States, where species must meet a stringent set of criteria to be listed as endangered or threatened under the ESA, the 2002 list of threatened or endangered species included 517 animal species and 745 plant species. Another 35 species had been proposed for listing, and 257 species had been suggested as candidate species.

The United States Fish and Wildlife service has been thorough in its assessment of the nation's ecological resources, but even the ESA list is incomplete. Endangered species listing favors larger, more charismatic plants and animals, especially vertebrate animals and vascular plants; endangered species of **arthropods**, mosses, **lichens**, and other less-well known groups remain undercounted. The humid Southeast and the arid Southwest have the largest numbers of endangered species in the United States. These regions tend to have unique ecological communities with many narrowly-distributed **endemic** species, as well as extensive human urbanization and resource development that threaten them.

There are numerous examples of endangered species. In this section, a few cases are chosen that illustrate the major causes of extinction, the socioeconomic conflicts related to protection of endangered species, and some possible successful strategies for wildlife protection and conflict resolution.

Species endangered by unsustainable hunting

Overhunting and overfishing have threatened animal species since aboriginal Europeans, Australians, and Americans developed effective hunting technology thousands of years ago. The dodo, passenger pigeon, great auk, and Steller's sea cow were hunted to extinction. Unstainable hunting and fishing continue to endanger numerous animals worldwide. In the United States, many of the animals considered national symbols—bald eagle, grizzly bear, timber wolf, American **Bison**, bighorn **sheep**, Gulf of Mexico sea **turtles**—have been threatened by overhunting. (American bison, incidentally, are no longer considered threatened, but they exist mainly in managed herds, and have never repopulated their wide range across the American and Canadian west.)

The eskimo curlew is a large sandpiper that was abundant in **North America** in the nineteenth century. The birds were relentlessly hunted by market gunners during their **migration** from the prairies and coasts of Canada and the United States to their wintering grounds on the pampas and coasts of South America. The eskimo curlew became very rare by the end of the nineteenth century. The last observation of a curlew nest was in 1866, and the last "collection" of birds was in 1922. There have been a few reliable sightings of individuals in the Canadian Artic and small migrating flocks in Texas since then, but sightings are so rare that the species' classification changes to extinct between each one.

The Guadalupe fur seal was abundant along the coast of western Mexico in the nineteenth century, numbering as many as 200,000 individuals. This marine mammal was hunted for its valuable fur and almost became extinct in the 1920s. Fortunately, a colony of 14 **seals**, including pups, was discovered off Baja California on Guadalupe Island in 1950. Guadalupe Island was declared a pinnaped sanctuary in 1975; the species now numbers more than 1,000 animals, and has begun to spread throughout its former range. The Juan Fernandez fur seal of Chile had a similar history. More than three million individuals were killed for their pelts between 1797 and 1804, when the species was declared extinct. The Juan Fernandez seal was rediscovered in 1965; and its population presently numbers several thousand individuals.

Commercial whaling for meat and oil since the eighteenth century has threatened most of the world's baleen whale species, and several toothed whales, with extinction. (Baleen whales feed by straining microorganisms from seawater.) Faced with severe depletion of whale stock, 14 whaling nations formed the International Whaling Commission (IWC) in 1946. While the IWC was somewhat successful in restoring whale populations, it lacks authority to enforce hunting bans, and non-member

The endangered golden frog, Panama. *JLM Visuals. Reproduced by permission.*

nations often threaten to disregard IWC directives. The Marine Mammal Protection Act of 1972 banned all whaling in United States waters, the CITES treaty protects all whale species, and many whales have been protected by the ESA. In spite of these measures, only a few whale species have recovered to their pre-whaling populations, and a number of species remain on the brink of extinction. Seven baleen whales, and four toothed whales, remain on the ESA list and the IUCN Red List today: northern and southern right whales, bowhead whale, blue whale, fin whale, sei whale, humpback whale, sperm whale, vaquita, baiji, and Indus susu. The California gray whale is a rare success story. This species was twice hunted near extinction, but it has recovered its pre-whaling population of about 21,000 individuals. The gray whale was removed from the endangered species list in 1993.

Large predators and trophies

Many large predators are killed because they compete with human hunters for wild game like deer and elk, because they prey on domestic animals like sheep, or sometimes because they threaten humans. Consequently, almost all large predators whose former range has been developed by humans have become extirpated or endan-

gered. The list of endangered large predators in the United States includes most of the species that formerly occupied the top of the food chain, and that regulated populations of smaller animals and fishes: grizzly bear, black bear, gray wolf, red wolf, San Joaquin kit fox, jaguar, lynx, cougar, mountain lion, Florida panther, bald eagle, northern falcon, American alligator, and American crocodile.

A number of generally harmless species are, sadly, endangered because of their threatening appearance or reputation, including several types of **bats**, **condors**, non-poisonous **snakes**, **amphibians**, and lizards. Internationally, many endangered species face extinction because of their very scarcity. Though CITES agreements attempt to halt trade of rare animals and animal products, trophy hunters, collectors of rare pets, and traders of luxury animal products continue to threaten numerous species. International demand for products like **elephant** tusk ivory, rhino horn, aquarium fish, bear and cat skins, pet tropical birds, reptile leather, and tortoise shells have taken a toll on many of the earth's most extraordinary animals.

Endangerment caused by introduced species

In many places, vulnerable native species have been decimated by non-native species imported by humans.

Predators like domestic cats and dogs, herbivores like cattle and sheep, diseases, and broadly-feeding omnivores like **pigs** have killed, starved, and generally outcompeted native species after introduction. Some destructive species introductions, like the importation of **mongooses** to the Pacific islands to control snakes, are intentional, but most of the damage caused by exotic species and diseases is unintended.

For example, the native birds of the Hawaiian archipelago are dominated by a family of about 25 species known as **honeycreepers**. Thirteen species of honeycreepers have been rendered extinct by introduced predators and habitat loss since Polynesians discovered the islands, and especially since European colonization. The surviving 12 species of honeycreepers are all endangered; they continue to face serious threats from introduced diseases, like avian **malaria**, to which they have no immunity.

Deliberate introduction of the Nile **perch** caused terrible damage to the native fish population of Lake Victoria in eastern **Africa**. Fisheries managers stocked Lake Victoria, the world's second-largest **lake**, with Nile Perch in 1954. In the 1980s the perch became a major fishery resource and experienced a spectacular population increase that was fueled by predation on the lake's extremely diverse community of cichlid fishes. The collapse of the native fish community of Lake Victoria, which originally included more than 400 species, 90% of which only occurred in Lake Victoria, resulted in the extinction of about one-half of the earth's cichlid species. Today, most of the remaining cichlids are endangered, and many of those species exist only in captivity.

Species living on islands are especially vulnerable to introduced predators. In one case, the accidental introduction of the predatory brown **tree** snake to the Pacific **island** of Guam in the late 1940s caused a severe decline of native birds. Prior to the introduction of the snake there were 11 native species of birds on Guam, most of which were abundant. By the mid-1980s seven of the native species were extinct or extirpated on Guam, and four more were critically endangered. The Guam rail, a flightless bird, is now extinct in the wild, although it survives in captivity and will hopefully be captive-bred and released to a nearby, snake-free island.

Endangerment caused by habitat destruction

Many species have become extinct or endangered as their natural habitat has been converted for human land-use purposes. The American ivory-billed woodpecker, for example, once lived in mature, bottomland hardwood forests and cypress swamps throughout the southeastern United States. These habitats were heavily logged and/or converted to agricultural land by the early 1900s. There have been no reliable sightings of the American ivory-billed woodpecker since the early 1960s, and it is probably extinct in North America. A related subspecies, the Cuban ivory-billed woodpecker, is also critically endangered because of habitat loss, as is the closely related imperial woodpecker of Mexico.

The black-footed ferret was first discovered in the North American **prairie** in 1851. This small **predator** became endangered when the majority of its grassland habitat was converted to agricultural use. Farming in the American and Canadian plains also dramatically reduced the population of prairie dogs, the black-footed ferret's preferred food.

Furbish's lousewort is an example of a botanical species endangered by habitat destruction. This herbaceous plant only occurs along a 143-mi (230-km) reach of the St. John River in Maine and New Brunswick. It was considered extinct until a botanist "re-discovered" it in Maine in 1976. At that time, a proposed hydroelectric reservoir threatened the entire habitat of Furbish's lousewort. In the end, the controversial dam was not built, but the lousewort remains threatened by any loss of its habitat.

The northern spotted owl lives in the old-growth **conifer** forests of North America's Pacific Northwest. These small **owls** require large areas of uncut forest to breed, and became endangered when their habitat was greatly reduced and fragmented by heavy logging. The Environmental Species Act prescribes, and legally requires, preservation of large areas of extremely valuable timber land to protect the northern spotted owl. Upon receiving its status as an endangered species, the otherwise unremarkable owl became a symbol of the conflict between environmental preservation and commercial enterprise. For environmentalists, endangered classification of northern spotted owl brought the possibility of protecting the forests from all exploitation; for timber industry workers, the decision represented the government's choice to preserve a small bird instead of their livelihood. Small stores on the back roads of the Pacific Northwest expressed their resentment for the ESA by advertising such specialties as "spotted owl barbeque" and activities as "spotted owl hunts."

Like the northern spotted owl, the endangered red-cockaded woodpecker of the southeastern United States requires old-growth pine forest to survive. The woodpecker excavates nest cavities in heart-rotted trees, and younger plantation trees do not meet its needs. Suitable forests have been greatly diminished by conversion to agriculture, logging, and residential development. Natural disturbance like hurricanes and wildfires threaten the remaining diminished and fragmented populations of red-cockaded **woodpeckers**. The ESA has attempted to protect the red-cockaded woodpecker by establishing ecological reserves and non-harvested buffers around known nesting colonies outside the reserves. Also like the spotted owl, the red-cockaded woodpecker is maligned by farmers, loggers, and developers for its role in their economic restriction.

Tropical deforestation presents represents the single greatest threat to endangered species today, though destruction of coastal and shallow marine habitats associated with anthropogenic **global warming** may present an even larger challenges in the future. While there was little net change (-2%) in the total forest cover of North America between the 1960s and the 1980s, the global area of forested land decreased by 17% during that period. Conversion of species-rich tropical forests in Central America, South America, Africa, and the Pacific islands to unforested agricultural land accounts for most of the decline. (Ironically, tropical soils have such poor structure and nutrient content that they generally cannot support profitable agriculture once the forest **biomass** has been removed.)

In the mid-1980s, tropical rainforests were being cleared at a rate of 15–20 million acres (6–8 million hectares) per year, or about 6–8% of the total equatorial forest area. The causes of tropical deforestation include conversion to subsistence and market agriculture, logging, and harvesting of fuelwood. All of these activities represent enormous threats to the multitude of endangered species native to tropical countries. Recent efforts to slow the rate of deforestation have included international financial and scientific aid to help poorer tropical nations protect important ecosystems, and to adopt new, more sustainable, methods of profitable resource use.

Actions to protect endangered species

Numerous international agreements deal with issues related to the conservation and protection of endangered species. The scientific effort to more accurately catalog species and better define the scope of biodiversity has dramatically raised the number of recorded threatened and endangered species in recent years. In spite of these shocking **statistics** of endangerment, there is a good deal of evidence that national and international efforts to preserve endangered species have been very successful. Some of the most important international conventions are ratified by most of the world's nations, and have had significant power to enforce agreements in the decades since their introduction: (1) the 1971 Convention on Wetlands of International Importance that promotes wise use of **wetlands** and encourages designation of important wetlands as ecological reserves; (2) the 1972 Convention Concerning the Protection of the

World Cultural and Natural Heritage that designates of high-profile World Heritage Sites for protection of their natural and cultural values; (3) the 1973 Convention on International Trade in Endangered Species of Wild Fauna and Flora (CITES); (4) the 1979 Convention on the Conservation of Migratory Species of Wild Animals of 1979 that deals with species that regularly cross national boundaries or that occur in international waters; and (5) the 1992 Convention on Biological Diversity (CBD). The CBD was presented by the United Nations Environment Program (UNEP) at the United Nations Conference on Environment and Development at Rio de Janeiro, Brazil in 1992, and has been regularly updated since then; the most recent amendments to the CBD occurred at the 2002 United Nations Earth Summit in Johannesburg, South Africa. The CBD is a central element of another international program called the Global Biodiversity Strategy, a joint effort by the IUCN, UNEP, and the World Resources Institute to study and conserve biodiversity.

Many countries, like the United States, have also undertaken their own actions to catalog and protect endangered species and other elements of biodiversity. Many of these national conservation efforts, like the ESA, have and international component that deals with species migration and trade across borders, and that mesh with the international conventions. Another important aspect of endangered species protection is collaboration with non-governmental organizations like the World Wildlife Fund, the Nature Conservancy and the Ocean Conservancy. The United States, for example, has a network of **conservation** data centers (CDCs) that gather and compile information on the occurrence and abundance of biological species and ecosystems that was designed and established by The Nature Conservancy. The Nature Conservancy has also facilitated development of CDCs in Canada and in Central and South America.

International, national and non-governmental agencies attempting to conserve biodiversity and protect endangered species choose whether to pursue single-species approaches that focus on particular species, or to develop more comprehensive strategies that focus on larger ecosystems. Because there are so many endangered species, many of which have not even been discovered, the single-species approach has obvious limitations. While the method works well for charismatic, large animals like giant **pandas**, grizzly **bears**, whales, and whooping **cranes**, this approach fails to protect most endangered species. More effective strategies focus on entire natural ecosystems that include numerous, hidden elements of threatened biodiversity. Furthermore, more conservation policies are attempting to consider the social, political, and economic ramifications of a species or

KEY TERMS

. . . .

Endangerment—Refers to a situation in which a species is vulnerable to extinction or extirpation.

Endemic—Refers to species with a relatively local distribution, sometimes occurring as small populations confined to a single place, such as a particular oceanic island. Endemic species are more vulnerable to extinction than are more widespread species.

Extinction—The condition in which all members of a group of organisms have ceased to exist.

Extirpation—The condition in which a species is eliminated from a specific geographic area of its habitat.

environmental protection plan. As in the case of the northern spotted owl, policies that require large economic sacrifices and offer no immediate local benefits often alienate the very humans that could best help to preserve an endangered species or ecosystem. Modern environmental protection strategies attempt to present alternatives that permit sustainable human productivity.

See also Stress, ecological.

Resources

Books

Beacham, W. and K.H. Beetz. *Beacham's Guide to International Endangered Species.* Osprey, FL: Beacham Publications, 1998.

Burton, J.A., ed. *The Atlas of Endangered Species.* New York: Macmillan Library Reference, 1998.

Wilson, E.O. *The Diversity of Life.* Cambridge: Harvard University Press, 1992.

Other

International Union for Conservation of Nature and Natural Resources. "IUCN Home Page." October 17, 2002 [cited November 21, 2002]. <http://www.iucn.org/>.

United Nations Environment Programme. "Convention on International Trade in Endangered Species of Wild Fauna and Flora (CITES)." November 15, 2002 [cited November 19, 2002]. <http://www.cites.org/index.html>.

United States Fish and Wildlife Service. "The United States Endangered Species Program." [cited November 19, 2002]. <http://endangered.fws.gov/>.

World Wildlife Fund. "Endangered Species." October 17, 2002 [cited November 19, 2002]. <http://www.wwfus.org/species/species.cfm>.

Bill Freedman
Laurie Duncan

Endemic

Endemic is a biogeographic term referring to a distinct race or **species** that originated in a local place or region, and that has a geographically restricted distribution. Endemic species tend to occur in certain ecological contexts, being especially frequent in places that are ecologically isolated, and that have not been affected by a regional-scale, catastrophic disturbance for a very long time.

For example, islands situated in remote regions of the oceans are physically isolated from other landmasses. In any cases where such islands have continuously supported ecosystems for a long period of time (at least tens of thousands of years), their biota will mostly be composed of endemic species of plants and animals that are not found elsewhere. This attribute of island **biodiversity** occurs because, under highly isolated conditions, **evolution** over long periods of time proceeds towards the development of new species from older, founder species. In itself, this process leads to the evolution of distinctly different, endemic races and species on remote islands.

Moreover, the biotas of remote islands tend to be depauperate in numbers of species, compared with any continental area of similar size and **habitat** type. As a result, **competition** among similar species tends to not be very intense in some types of island habitats, and there may be relatively broad **niche** opportunities available to be exploited by plants and animals. Consequently, on the rare occasions when remote islands are successfully colonized by new species, these founders may be able to evolutionarily radiate, and develop a number of new species that are found nowhere else. Some of the more remarkable examples of evolutionary **radiation** of endemic species on oceanic islands include: (1) the 13 species of Darwin's **finches** (subfamily Geospizinae of the finch family, Fringillidae) on the Galápagos Islands of the Pacific Ocean; (2) the 25 species of **honeycreepers** (Drepanididae) of the Hawaiian Islands; and (3) the approximately 1,250 species of fruit flies (genus *Drosophila*) on the Hawaiian Islands. All of these diverse groups likely evolved by adaptive radiations into unfilled habitat opportunities of single, founder species.

Therefore, because of the evolutionary influences of isolation and adaptive radiation on islands, these places tend to have many endemic species. For example, almost 900 species of **birds**, about 90% of all bird species, are endemic to oceanic islands. The native **flora** of the Hawaiian archipelago is estimated to have originally contained 2,000 species of **angiosperm** plants, of which 94-98% were endemic. Similarly, 76% of the plants of the Pacific **island** of New Caledonia are endemic, as are 50% of those of Cuba, and 36% of those of Hispaniola.

Habitat "islands" can also occur on land, in the form of isolated communities located within a dominant **matrix** of another type of **ecosystem**. Large numbers of endemic species may evolve over long periods of time in terrestrial habitat islands. This could be part of the basis of the evolution of the extraordinary species richness of tropical rainforests, which may have been contracted into isolated fragments during past episodes of dry climate. It is likely that more than 90% of Earth's species live in tropical **forests**, and most of these are endemics.

Endemic species are rare in places that have "recently" been subject to some sort of catastrophic, regional-scale disturbance. (Note that in the sense used here, "recent" on the evolutionary time scale means within the most recent, several tens of thousands of years.) For example, much of **North America** was covered by glacial **ice** up until about 10,000-12,000 years ago. Although most of that region has since supported vegetation for about ten thousand years, this has not been enough time to allow many endemic species to develop. Consequently, countries like Canada, which were entirely glaciated, have virtually no endemic species.

Because endemic species have geographically restricted distributions, and they have often evolved in isolation from many types of diseases, predators, and competitors, they tend to be highly vulnerable to **extinction** as a result of the activities of humans. About three-quarters of the species of plants and animals that are known to have become extinct during the past few centuries were endemics that lived on islands. For example, of the 108 species of birds that have become extinct during that period, 90% lived on islands. About 97% of the endemic plants of Lord Howe Island are extinct or endangered, as are 96% of those of Rodrigues and Norfolk Islands, 91% of those of Ascension Island, and 81% of those of Juan Fernandez and the Seychelles Islands. Tremendous numbers of endemic species also occur in the tropical rainforests of continental regions. Most of these species are now endangered by the conversion of their habitat to agriculture, or by disturbances associated with **forestry** and other human activities.

See also Biological community.

Endive *see* **Composite family**

Bill Freedman

Endocrine system

The endocrine system is the body's network of nine **glands** and over 100 **hormones** which maintain and reg-

ulate numerous events throughout the body. The glands of the endocrine system include the pituitary, thyroid, parathyroids, thymus, pancreas, pineal, **adrenals**, and ovaries or testes: in addition, the hypothalamus, in the **brain**, regulates the release of pituitary hormones. Each of these glands secrete hormones (chemical messengers) into the **blood** stream. Once hormones enter the blood, they travel throughout the body and are detected by receptors that recognize specific hormones. These receptors exist on target cells and organs. Once a target site is bound by a particular hormone, a cascade of cellular events follows that culminates in the physiological response to a particular hormone.

The endocrine system differs from the exocrine system in that **exocrine glands** contain ducts which direct their hormones to specific sites; whereas endocrine hormones travel through blood until they reach their destination. The endocrine is also similar to the **nervous system**, because both systems regulate body events and communicate through chemical messengers with target cells. However, the nervous system transmits neurotransmitters (also chemical messengers) between neighboring neurons via nerve extension, and neurotransmitters do not generally enter the circulation. Yet, some overlap between hormones and neurotransmitters exists which gives rise to chemical signals called neurohormones which function as part of the neuroendocrine system. The endocrine system oversees many critical life processes involving **metabolism**, growth, reproduction, immunity, and **homeostasis**. The branch of medicine that studies endocrine glands and the hormones which they secrete is called endocrinology.

History of endocrinology

Although some ancient cultures noted biological observations grounded in endocrine function, modern understanding of endocrine glands and how they secrete hormones has evolved only in the last 300 years. Ancient Egyptian and Chinese civilizations castrated (removed the testicles of) a servile class of men called eunuchs. It was noted that eunuchs were less aggressive than other men, but the link of this **behavior** to testosterone was not made until recently.

Light was shed on endocrine function during the seventeenth and eighteenth centuries by a few significant advances. A seventeenth century English scientist, Thomas Wharton (1614-1673), noted the distinction between ductile and ductless glands. In the 1690s, a Dutch scientist named Fredrik Ruysch (1638-1731) first stated that the thyroid secreted important substances into the blood stream. A few decades later, Theophile Bordeu (1722-1776) claimed that "emanations" were given off by some body parts that influenced functions of other body parts.

One of the greatest early experiments performed in endocrinology was published by A. A. Berthold (1803-1861) in 1849. Berthold took six young male chickens, and castrated four of them. The other two were left to develop normally and used comparatively as control samples. Two of the castrated chickens were left to become chicken eunuchs. But what Berthold did with the other two castrated chickens is what really changed endocrinology. He transplanted the testes back into these two chickens at a distant site from where they were originally. The two castrated chickens never matured into roosters with adult combs or feathers. But the chickens who received transplanted testes did mature into normal adult roosters. This experiment revealed that hormones that could access the blood stream from any site would function correctly in the body and that hormones did, in fact, travel freely in the circulation.

The same year Berthold published his findings, Thomas Addison (1793-1860), a British scientist reported one of the first well documented endocrine diseases which was later named **Addison's disease** (AD). AD patients all had a gray complexion with sickly skin; they also had weak hearts and insufficient blood levels of hemoglobin necessary for **oxygen** transport throughout the body. On autopsy, each of the patients Addison studied were found to have diseased adrenal glands. This **disease** can be controlled today if it is detected early. President John F. Kennedy suffered from AD.

Basic endocrine principles

Most endocrine hormones are maintained at specific concentrations in the **plasma**, the non-cellular, liquid portion of the blood. Receptors at set locations monitor plasma hormonal levels and inform the gland responsible for producing that hormone if levels are too high or too low for a particular time of day, month, or other life period. When excess hormone is present, a **negative** feedback loop is initiated such that further hormone production is inhibited. Most hormones have this type of regulatory control. However, a few hormones operate on a positive feedback cycle such that high levels of the particular hormone will activate release of another hormone. With this type of feedback loop, the end result is usually that the second hormone released will eventually decrease the initial hormone's secretion. An example of positive feedback regulation occurs in the female **menstrual cycle**, where high levels of estrogen stimulate release of the pituitary hormone, luteinizing hormone (LH).

All hormones are influenced by numerous factors. The hypothalamus can release inhibitory or stimulatory hormones the determine pituitary function. And every physiological component that enters the circulation can effect some endocrine function. Overall, this system uses multiple bits of chemical information to hormonally maintain a biochemically balanced **organism**.

Endocrine hormones do not fall into any one chemical class, but most are either a protein (polypeptides, peptides, and glycoproteins are also included in this category) or steroids. Protein hormones bind cell-surface receptors and activate intracellular events that carry out the hormone's response. Steroid hormones, on the other hand, usually travel directly into the **cell** and bind a receptor in the cell's cytoplasm or nucleus. From there, steroid hormones (bound to their receptors) interact directly with genes in the DNA to elicit a hormonal response.

The pituitary

The pituitary gland has long been called "the master gland," because it secretes multiple hormones which, in turn, trigger the release of other hormones from other endocrine sites. The pituitary is roughly situated behind the nose and is anatomically separated into two distinct lobes, the anterior pituitary (AP) and the posterior pituitary (PP). The entire pituitary hangs by a thin piece of **tissue**, called the pituitary stalk, beneath the hypothalamus in the brain. The AP and PP are sometimes called the adenohypophysis and neurohypophysis, respectively.

The PP secretes two hormones, oxytocin and antidiuretic hormone (ADH), under direction from the hypothalamus. Direct innervation of the PP occurs from regions of the hypothalamus called the supraoptic and paraventricular nuclei. Although the PP secretes its hormones into the bloodstream through blood vessels that supply it, it is regulated in a neuroendocrine fashion. The AP, on the other hand, receives hormonal signals from the **blood supply** within the pituitary stalk.

AP cells are categorized according to the hormones that they secrete. The hormone-producing cells of the AP include: somatotrophs, corticotrophs, thyrotrophs, lactotrophs, and gonadotrophs. Somatotrophs secrete growth hormone; corticotrophs secrete adrenocorticotropic hormone (ACTH); thyrotrophs secrete thyroid stimulating hormone (TSH); lactotrophs secrete prolactin; and gonadotrophs secrete LH and follicle stimulatory hormone (FSH). Each of these hormones sequentially signals a response at a target site. While ACTH, TSH, LH, and FSH primarily stimulate other major endocrine glands, growth hormone and prolactin primarily coordinate an endocrine response directly on bones and mammary tissue, respectively.

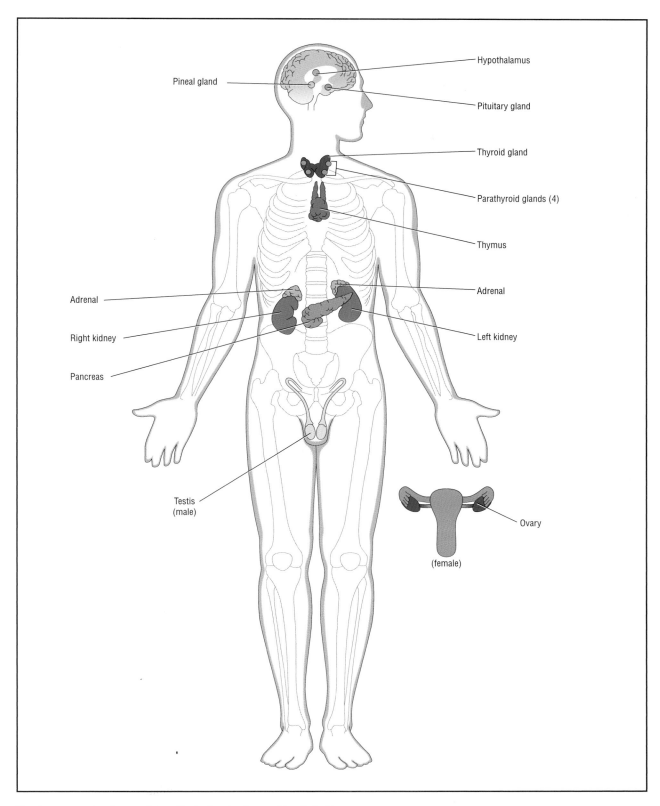

The human endocrine system. *Illustration by Argosy. The Gale Group.*

The pineal

The pineal gland is a small cone-shaped gland believed to function as a body clock. The pineal is located deep in the brain just below the rear-most portion of the corpus callosum (a thick stretch of nerves that connects the two sides of the brain). The pineal gland, also called the pineal body, has mystified scientists for centuries. The seventeenth century philosopher, Rene Descartes, speculated that the pineal was the seat of the soul. However, its real function is somewhat less grandiose than that.

The pineal secretes the hormone melatonin, which fluctuates on a daily basis with levels highest at night. Although its role is not well understood, some scientists believe that melatonin helps to regulate other diurnal events, because melatonin fluctuates in a 24-hour period. Exactly what controls melatonin levels is not well understood either; however, visual registration of light may regulate the cycle.

The thyroid

The thyroid is a butterfly-shaped gland that wraps around the back of the esophagus. The two lobes of the thyroid are connected by a band of tissue called the **isthmus**. An external covering of **connective tissue** separates each lobe into another 20-40 follicles. Between the follicles are numerous blood and lymph vessels in another connective tissue called stroma. The epithelial cells around the edge of the follicles produce the major thyroid hormones.

The major hormones produced by the thyroid are triiodothyronine (T3), thyroxine (T4), and calcitonin. T3 and T4 are iodine-rich molecules that fuel metabolism. The thyroid hormones play several important roles in growth, metabolism, and development. The thyroid of pregnant women often become enlarged in late pregnancy to accommodate metabolic requirements of both the woman and the fetus.

Thyroid hormones accelerate metabolism in several ways. They promote normal growth of bones and increase growth hormone output. They increase the **rate** of **lipid** synthesis and mobilization. They increase cardiac output by increasing rate and strength of **heart** contractions. They can increase **respiration**, the number of red blood cells in the circulation, and the amount of oxygen carried in the blood. In addition, they promote normal nervous system development including nerve branching.

The parathyroids

While most people have four small parathyroid glands poised around the thyroid gland, about 14% of the population have one or two additional parathyroid glands. Because these oval glands are so small, the extra space occupied by extra glands does not seem to be a problem. The sole function of these glands is to regulate **calcium** levels in the body. Although this may seem like a simple task, the maintenance of specific calcium levels is critical. Calcium has numerous important bodily functions. Calcium makes up 2-3% of adult weight with roughly 99% of the calcium in bones. Calcium also plays a pivotal role in muscle contraction and **neurotransmitter** secretion.

The thymus

In young children, the thymus extends into the neck and the chest, but after **puberty**, it begins to shrink. The size of the thymus in most adults is very small. Like some other endocrine glands, the thymus has two lobes connected by a stalk. The thymus secretes several hormones that promote the maturation of different cells of the **immune system** in young children. In addition, the thymus oversees the development and "education" of a particular type of immune system cell called a T lymphocyte, or T cell.

Although many details of thymal hormonal activity are not clear, at least four thymal products have been identified which control T cell and B cell (antibody-producing immune cells) maturation. The four products are: thymosin, thymic humoral factor (THF), thymic factor (TF), and thymopoietin. Because the viral disease, **AIDS**, is characterized by T cell depletion, some AIDS treatment approaches have tried administering tymosin to boost T cell production.

The pancreas

The pancreas is a large endocrine and exocrine gland situated below and behind the stomach in the lower abdomen. The pancreas is horizontally placed such that its larger end falls to the right and its narrower end to the left. Clusters of exocrine pancreatic cells called acini secrete digestive enzymes into the stomach; while endocrine cells secrete hormones responsible for maintaining blood glucose levels.

The endocrine cells of the pancreas are contained in the islets of Langerhans which are themselves embedded in a rich network of blood and lymph vessels. About one million islet cells make up about 2% of the total pancreas: the islet cells are quite small. The three major types of endocrine cells within the islets are alpha cells, beta cells, and **delta** cells. Beta cells make up about 70% of islet cells and secrete **insulin**. Alpha cells which secrete glucagon comprise roughly 20% of islet cells.

Delta cells, which comprise up to 8% of islet cells, secrete somatostatin. Another pancreatic hormone, called pancreatic polypeptide, is secreted by F cells in the islets and has just recently been characterized.

Insulin is secreted in response to high plasma glucose levels. Insulin facilitates glucose uptake into blood cells thus reducing plasma glucose levels. Glucagon has the opposite effect; low plasma glucose triggers the breakdown of stored glucogen in the liver and glucose release into the blood. By balancing these two hormones, the islets continually regulate circulating glucose levels. Both hormones are contained within secretory vesicles in the cells which release them. And monitoring of blood glucose concentrations are evaluated directly at the pancreas, as opposed to being mediated by another gland such as the pituitary. In addition, nerve endings at the islets contribute in regulating insulin and glucagon secretion. The hormone somatostatin is also released under conditions of increased blood glucose, glucagon, and amino acids. Somatostatin inhibits additional insulin release.

The overall effect of insulin is to store fuel in muscle and adipose tissue. It also promotes protein synthesis. Hence, insulin is said to be anabolic which means that it works to build up instead of break down energy-storing molecules. Glucagon, however, is metabolic, since it promotes glycogen breakdown. However, each of the pancreatic hormones is also considered to be paracrine, since they also modulate other pancreatic cells.

Diabetes mellitus (DM) is a very serious endocrine disorder caused by insulin dysfunctions. In type I DM, the beta cells are defective and can not produce enough insulin. Type II DM is caused by a lack of target cell-receptor responsiveness to available insulin. While type I requires regular insulin injections, type II can be controlled with diet.

The adrenals

One of the two adrenals sit atop each kidney and are divided into two distinct regions, the cortex and the medulla. The outer area makes up about 80% of each adrenal and is called the cortex. And the inner portion is called the medulla. The adrenals provide the body with important buffers against stress while helping it adapt to stressful situations.

The cells of the cortex form three distinct layers, the zona glomerulosa (ZG), the zona fasciculata (ZF), and the zona reticularis (ZR). All three layers secrete steroid hormones. The ZG secretes mineralocorticoids; the ZF secretes glucocorticoids; and the ZR secretes primary androgens. **Cholesterol** droplets are interspersed throughout the cortex for synthesis of cortical steroids. ACTH released from the anterior pituitary triggers glu-

cocorticoids release from the ZF. The major glucocorticoid in humans is cortisol, which fluctuates in **concentration** throughout the day: the highest levels exist in the morning around 8 a.m. with the lowest levels around midnight. Both physical and emotional stress can affect cortisol secretion. The major human mileralocorticoid is aldosterone. Aldosterone acts on the kidneys and is important in regulating fluid balance. It triggers increased extracellular fluid that leads to increased blood **pressure**.

Cells of the adrenal medulla, called chromaffin cells, secrete the hormones epinephrine (adrenaline) and non-epinephrine (nor-adrenaline). Chromaffin cells are neuroendocrine cells which function like some nerve fibers of the sympathetic nervous system. However, these cells are endocrine, because the neurohormones that they release target distant organs. Although the effects of these two medullary hormones are the same whether they originate in the endocrine or the nervous system, endocrine hormonal effects are prolonged, because they are removed more slowly from blood than from a nerve terminal. Both cortical and medullary hormones work together in emergencies, or stressful situations, to meet the physical demands of the moment.

The ovaries

The ovaries are located at the end of each fallopian tube in the female reproductive tract, and they produce the female reproductive hormones: estrogen, progesterone, and relaxin. Although the fluctuation of these hormones is critical to the female menstrual cycle, they are initially triggered by a hormone from the hypothalamus, called a releasing factor, that enables gonadotrophs in the pituitary to release LH and FSH that, in turn, regulate part of the menstrual cycle. All of these hormones work together as part of the endocrine system to ensure fertility. They are also important for the development of sexual characteristics during puberty.

Each month after puberty, females release a single egg (ovulation) when the pituitary releases LH. Prior to ovulation, the maturing egg releases increasing levels of estrogen that inform the pituitary to secrete LH. While an egg travels down the fallopian tube, progesterone is released which prevents another egg from beginning to mature. Once an egg is shed in the uterine lining (in menstruation), the cycle can begin again. During pregnancy, high levels of circulating estrogen and progesterone prevent another egg from maturing. Estrogen levels fall dramatically at **menopause**, signifying the end of menstrual cycling and fertility. Menopause usually occurs between the ages of 40 and 50.

Endocrine regulation of the female reproductive tract does not stop during pregnancy. In fact, more sex hormones are released during pregnancy than during any other phase of female life. At this time, a new set of conditions support and protect the developing baby. Even cells of the developing embryo begin to release some hormones that keep the uterine lining intact for the first couple of months of pregnancy. High progesterone levels prevent the uterus from contracting so that the embryo is not disturbed. Progesterone also helps to prepare breasts for lactation. Towards the end of pregnancy, high estrogen levels stimulate the pituitary to release the hormone, oxytocin, which triggers uterine contractions. Prior to delivery, the ovaries release the hormone, relaxin, a protein which causes the pelvic ligaments to become loose for labor.

The testes

The two testes are located in the scrotum, which hangs between the legs behind the penis. Most of the testes is devoted to sperm production, but the remaining cells, called Leydig cells, produce testosterone. Testosterone caries out two very important endocrine tasks in males: it facilitates sexual maturation, and it enables sperm to mature to a reproductively-competent form. Healthy men remain capable of fertilizing an egg throughout their post-pubertal life. However, testosterone levels do show a gradual decline after about the age of 40 with a total drop of around 20% by age 80.

Testosterone also has the important endocrine function of providing sexual desire in both men and women. Although reproduction can occur without maximal desire, this added incentive increases the likelihood of reproduction. Human sexual behavior is also influenced by several other factors including thoughts and beliefs.

Endocrine disorders

As much as 10% of the population will experience some endocrine disorder in their lifetime. Most endocrine disorders are caused by a heightened or diminished level of particular hormones. For example, excess growth hormone can cause giantism (unusually large stature). Tumors in endocrine glands are one of the major causes of hormone overproduction.

Hormone underproduction is often due to a **mutation** in the **gene** that codes for a particular hormone or the hormone's receptor; even if the hormone is normal, a defective receptor can render the hormone ineffective in eliciting a response. The distinction between the pancreatic endocrine disorders, diabetes mellitus types I and II, make this point very clearly. In addition, underproduction of **growth hormones** can lead to dwarfism. Insufficient calcitonin from the thyroid can also lead to cretinism which is also characterized by diminished stature due to low calcium availability for bone growth.

The importance of diet can not be overlooked in some endocrine disorders. For example, insufficient dietary iodine (required for T3 and T4 synthesis) can cause goiters. Goiters are enlarged thyroid glands caused by the thyroid's attempt to compensate for low iodine with increased size. Certain endocrine imbalances can also cause mental problems ranging from poor ability to concentrate to mental retardation (as in cretinism).

See also Chemoreception.

Resources

Books

Little, M. *The Endocrine System.* New York: Chelsea House Publishers, 1990.
Marieb, Elaine Nicpon. *Human Anatomy & Physiology.* 5th Edition. San Francisco: Benjamin/Cummings, 2000.

Louise Dickerson

Endoprocta

The phylum Endoprocta is a group of about 60 **species** that closely resemble **moss animals** or members of the phylum Bryozoa. With the exception of the genus *Urnatella* all endoprocts are marine-dwelling species. Like moss animals they are sessile, being attached to a wide range of submerged objects such as **rocks**, shells, **sponges**, corals, and other objects. These tiny animals—the largest measures just 0.2 in (5 mm) in length—may either live a solitary or colonial existence. All members of the phylum are filter feeders that extract **plankton** and other food particles from the **water** column.

Body form varies considerably in this phylum, with most species being an oval shape. Solitary species are usually positioned on a short stalk while colonial species may have a large number of **individual** animals all arising from a single, spreading stalk. The top of the **animal** is dominated by a ring of short tentacles which are produced directly from the body wall and which may be withdrawn when the animal is not feeding. The mouth is at the center of this ring, the anus at the opposite pole. As with other small, filter-feeding organisms, a mass of tiny cilia line the inner side of the tentacles. When they beat, they create a downward current drawing water and food particles towards the animal, in between the tentacles and towards the mouth. From here, additional cilia continue the downward movement of food items towards the stomach.

Asexual reproduction is common in both solitary and colonial species. Many endoprocts also produce male and female gametes. **Fertilization** is thought to be internal although the exact process underlying this is not known. Fertilized eggs develop into free-living and free-swimming larvae known as a trocophore, which, after a short period at sea, settles and attaches itself firmly to some substrate.

Endoscopy

Endoscopy is the use of a thin, lengthy, flexible scope that can be inserted into the body for the **diagnosis** and treatment of various conditions. Until the last third of the twentieth century, one of the limiting factors in the treatment of internal injuries or diseases was the need to perform open **surgery** on the patient. That meant putting him under **anesthesia**, carrying out the operation, sewing up the incision, and allowing the patient to recuperate from the procedure for several days in the hospital. In some instances, such as trauma, the need for open surgery only added to the time involved for the patient to be treated.

Surgeons for many years had attempted various means to penetrate the interior of the body without the need for a major incision.

The use of **x rays** allowed observation of bones and, with some use of enhancement, certain organs and the **blood** vessels. Although this procedure gave ample information about bone fractures and at times even bone **cancer**, x rays gave evidence of a **tumor** or other **disease** process without telling the physician what actually was the cause of the condition. Knowing the tumor or disease was present did not remove the necessity of surgical diagnosis. Clearly, a method was needed to look into the body to observe a pathologic condition as it existed rather than as a shadow on an x-ray plate.

As early as 1918, a physician named Takagi was attempting to use the technology of the day to examine the interior of joints. He used a cystoscope, an instrument used to examine the interior of the urinary bladder, but immediately came upon major problems. The cystoscope was a rigid tube with a **light** on it that was inserted in the urethra to examine the urinary bladder. Because it was rigid the instrument was not maneuverable or flexible enough to be guided around various anatomic structures. The light on the cystoscope was at the far end and could

be broken off easily inside the patient. Also the **heat** of the lamp, small as it was, soon heated the joint space to an unacceptable level, far too quickly for the doctor to carry out a thorough examination.

Technology available by the late 1970s, however, solved these problems and allowed a specialty to begin that today is widespread and beneficial. The space age brought on the science of **fiber optics**, long strands of **glass** that could carry light and **electricity** over long distances, around corners, in a small bundle compared to **copper** wires. Using fiber optics, the light source for the endoscope could be housed in the handle end of the scope so that the light itself never entered the body. Fiber optics also allowed the instrument to be flexible so that the doctor could steer the end into whatever area he wanted. The efficiency of fiber optics in carrying light and images meant that the diameter of the endoscope could be reduced considerably compared with the few scopes then available.

At first, the endoscope was called the arthroscope and used only to visualize the internal **anatomy** of joints such as the knee. Soon the scope was fitted with instruments such as scalpels and scissors to carry out surgery, a **vacuum** line to suck out any floating material that might interfere with the function of the joint, and a **television** camera so the physician, instead of peering through a small opening in the scope, could watch his progress on a larger television screen.

With these and other refinements the endoscope now can be used to penetrate nearly any area of the body, provide the physician with information on the condition of the area being examined, and provide the means for the physician to carry out surgical procedures through a tiny incision. The patient usually is in and out of the treatment facility the same day and the recovery from such minor surgery is rapid.

Like any other surgery, endoscopy is carried out in a sterile environment. The patient is positioned and appropriate anesthetic is administered. Often the area to be examined is filled with saline to expand the interior space and lift the overlying tissues from the area being examined. Saline is a mild **salt** solution. Through a small incision the tip of the endoscope is inserted into the joint space. The end of the endoscope being held by the surgeon has a "joy stick," a lever that protrudes and that can be used to guide the tip of the endoscope from one area into another. A second endoscope may be needed to assist with surgery, provide more light, maintain the saline environment, or for any of a number of other reasons. Using the lever the physician moves the tip of the endoscope with the TV camera from one area to another, examining the structures within the joint as he goes.

The use of the endoscope to penetrate joints is called arthroscopy. It is an especially useful procedure in sports medicine where athletes often suffer knee injuries as the result of running, jumping, or being tackled. The ligaments that hold the lower leg bone, the tibia, to the upper leg bone, the femur, can be ruptured if the knee is twisted or hit from the side. Arthroscopy allows the surgeon to examine the details of the injury, determine whether a more radical procedure is needed, and if not, to repair the torn ligaments and remove any loose material from the joint using the arthroscope. The athlete will require a few weeks of **physical therapy** to regain full strength, but the actual surgery can be completed in only a short time and he will be out of the hospital the same day he enters.

The laparoscope or peritoneoscope (so named because it penetrates the peritoneum, the lining of the abdominal cavity) is used to examine the interior of the abdomen. The tip of the scope with the TV camera attached can be guided around, above, and underneath the organs of the abdomen to determine the source of bleeding, the site of a tumor, or the probable cause of an illness. In the case of gallstones, which form in the gallbladder near the liver, the gallbladder can be removed using the surgical attachments. Suturing the stump of the gallbladder can also be accomplished using the attachments.

To examine the inner surface of the lower digestive tract the scope used is the sigmoidoscope. It can be passed into the colon to examine the walls of the intestine for possible cancer or other abnormal structures.

Obviously, the physician using the endoscope must be highly knowledgeable about anatomy. Human anatomy as it appears through the **lens** of the scope is considerably different from its appearance on the page of a book or during open surgery. The physician must be able to recognize structures, which may appear distorted through the lens of the endoscope, to determine the location of the tip of the endoscope and to know where to maneuver the scope next. Training is carried out under the guidance of a physician who has extensive experience with the endoscope.

Resources

Periodicals

Bechtel, S. "Operating Through a Keyhole." *Prevention* 45 (July 1993): 72+.

Frandzel, S. "The Incredible Shrinking Surgery." *American Health* 13 (April 1994): 80-84.

Larry Blaser

Endothermic

The term endothermic has two distinct meanings, depending on the context in which it is used. In **chemistry**, endothermic means that a chemical reaction or phase transition absorbs **heat**. (A phase transition is the transformation of **matter** from a gas, liquid, or solid into a different one of these states.) In **physiology**, the term endothermic refers to organisms which metabolically generate heat to maintain their body **temperature**.

Typically, the term endothermic describes a chemical reaction or phase transition that absorbs heat from the environment, in order to bring the reaction back to the temperature at which the reaction began. Heat is a form of **energy**, like kinetic energy and **radiation**. The units in which heat is expressed are known as joules or calories. The heat consumed or released in a chemical reaction is known as the enthalpy change, if measured at a constant **pressure**, or the energy change, if measured at a constant **volume**. The heat consumed may not be the same under these different conditions, because energy may be transferred to or from the reaction in the form of **work**, if it occurs at a constant pressure. A positive enthalpy or energy change indicates that a reaction is endothermic. The **evaporation** of **water** is a familiar, endothermic phase transition. When water moves from a liquid phase to a gaseous phase, it absorbs heat; that is why drying objects usually feel cool when touched.

Physiologists use the term endothermic when referring to organisms classified as endotherms, such as **mammals** or **birds**. Endothermic organisms maintain their body temperature above that of their surroundings by maintaining a high metabolic **rate**. (**Metabolism** refers to the material and energy transformations within an organism.) These transformations are highly exothermic of heat releasing, causing the endotherm's body temperature to remain high. The fact that endothermic organisms derive heat from exothermic **chemical reactions** can make this term confusing when used in physiology.

See also Temperature regulation.

Endothermic reaction *see* **Chemical reactions**

Energy

Energy is a state function commonly defined as the capacity to do **work**. Since work is defined as the movement of an object through a **distance**, energy can also be described as the ability to move an object through a distance. As an example, imagine that a bar magnet is placed next to a pile of **iron** filings (thin slivers of iron **metal**). The iron filings begin to move toward the iron bar because magnetic energy pulls on the iron filings and causes them to move.

Energy can be a difficult concept to understand. Unlike **matter**, energy can not be taken hold of or placed on a laboratory bench for study. We know the nature and characteristics of energy best because of the effect it has on objects around it, as in the case of the bar magnet and iron filings mentioned above.

Energy is described in many forms, including mechanical, **heat**, electrical, magnetic, sound, chemical, and nuclear. Although these forms appear to be very different from each other, they often have much in common and can generally be transformed into one another.

Over **time**, a number of different units have been used to measure energy. In the British system, for example, the fundamental unit of energy is the foot-pound. One foot-pound is the amount of energy that can move a weight of one pound a distance of one foot. In the **metric system**, the fundamental unit of energy is the joule (abbreviation: J), named after the English scientist James Prescott Joule (1818-1889). A joule is the amount of energy that can move a weight of one newton a distance of one meter.

Potential and kinetic energy

Every object has energy as a consequence of its position in **space** and/or its **motion**. For example, a baseball poised on a railing at the top of the observation deck on the Empire State Building has potential energy because of its ability to fall off the railing and come crashing down onto the street. The potential energy of the baseball—as well as that of any other object—is dependent on two factors, its **mass** and its height above the ground. The formula for potential energy is p.e. = m × g × h, where m stands for mass, h for height above the ground, and g for the gravitational constant (9.8 m per second per second).

Potential energy is actually a manifestation of the gravitational attraction of two bodies for each other. The baseball on top of the Empire State Building has potential energy because of the gravitational **force** that tends to bring the ball and **Earth** together. When the ball falls, both Earth and ball are actually moving toward each other. Since Earth is so many times more massive than the ball, however, we do not see its very minute motion.

When an object falls, at least part of its potential energy is converted to kinetic energy, the energy due to an object's motion. The amount of kinetic energy possessed by an object is a function of two variables, its mass and its **velocity**. The formula for kinetic energy is k.e. = $1/2m \times v^2$, where m is the mass of the object and v is its velocity. This formula shows that an object can have a lot of kinetic energy for two reasons. It can either be very heavy or it can be moving very fast. For that reason, a fairly light baseball falling over a very great distance and traveling at a very great speed can do as much damage as a much more massive object falling at a slower speed.

Conservation of energy

The sum total of an object's potential and kinetic energy is known as its mechanical energy. The total amount of mechanical energy possessed by a body is a constant. The baseball described above has a maximum potential energy and minimum kinetic energy (actually a **zero** kinetic energy) while at rest. In the fraction of a second before the ball has struck the ground, its kinetic energy has become a maximum and its potential energy has reached almost zero.

The case of the falling baseball described above is a special interest of a more general rule known as the law of **conservation** of energy. According to this law, energy can never be created or destroyed. In other words, the total amount of energy available in the universe remains constant and can never increase or decrease.

Although energy can never be created or destroyed, it can be transformed into new forms. In an electric iron, for example, an electrical current flows through metallic coils within the iron. As it does so, the current experiences resistance from the metallic coils and is converted into a different form, heat. A **television** set is another device that operates by the transformation of energy. An electrical beam from the back of the television tube strikes a thin layer of chemicals on the television screen, causing them to glow. In this case, electrical energy is converted into light. Many of the modern appliances that we use in our homes, such as the electric iron and the television set, make use of the transformation of energy from one form to another.

In the early 1900s, Albert Einstein announced perhaps the most surprising energy transformation of all. Einstein showed by mathematical reasoning that energy can be converted into matter and, vice versa, matter can be transformed into energy. He expressed the equivalence of matter and energy in a now famous equation, $E = m \times c^2$, where c is a constant, the speed of light.

Forms of energy

The operation of a **steam engine** is an example of heat being used as a source of energy. Hot steam is pumped into a cylinder, forcing a piston to move within the cylinder. When the steam cools off and changes back to **water**, the piston returns to its original position. The cycle is then repeated. The up-and-down motion of the piston is used to turn a wheel or do some other kind of work. In this example, the heat of the hot steam is used to do work on the wheel or some other object.

The source of heat energy is the motion of molecules within a substance. In the example above, steam is said to be "hot" because the particles of which it is made are moving very rapidly. When those particles slow down, the steam has less energy. The total amount of energy contained within any body as a consequence of particle motion is called the body's thermal energy.

One measure of the amount of particle motion within a body is **temperature**. Temperature is a measure of the average kinetic energy of the particles within the body. An object in which particles are moving very rapidly on average has a high temperature. One in which particles are moving slowly on average has a low temperature.

Temperature and thermal energy are different concepts, however, because temperature measures only the average kinetic energy of particles, while thermal energy measures the total amount of energy in an object. A thimbleful of water and a swimming pool of water might both have the same temperature, that is, the average kinetic energy of water molecules in both might be the same. But there is a great deal more water in the swimming pool, so the total thermal energy in it is much greater than the thermal energy of water in the thimble.

Electrical energy

Suppose that two ping pong balls, each carrying an electrical charge, are placed near to each other. If free to move, the two balls have a tendency either to roll toward each other or away from each other, depending on the charges. If the charges they carry are the same (both positive or both **negative**), the two balls will repel each other and roll away from each other. If they charges are opposite, the balls will attract each other and roll toward each other. The force of attraction or repulsion of the two balls is a manifestation of the electrical potential energy existing between the two balls.

Electrical potential energy is analogous to gravitational energy. In the case of the latter, any two bodies in the universe exert a force of attraction on each other that depends on the masses of the two bodies and the distance between them. Any two charged bodies in the universe, on the other hand, experience a force of attraction or repulsion (depending on their signs) that depends on the magnitude of their charges and the distance separating them. A **lightning** bolt traveling from the ground to a cloud is an example of electrical potential energy that has suddenly been converted to it "kinetic" form, an electrical current.

An electrical current is analogous to kinetic energy, that is, it is the result of moving electrical charges. An electrical current flows any time two conditions are met. First, there must be a source of electrical charges. A **battery** is a familiar source of electrical charges. Second, there must be a pathway through which the electric charges can flow. The pathway is known as a circuit.

An **electric current** is useful, however, only if a third condition is met—the presence of some kind of device that can be operated by electrical energy. For example, one might insert a **radio** into the circuit through which electrical charges are flowing. When that happens, the electrical charges flow through the radio and make it produce sounds. That is, electrical energy is transformed into sound energy within the radio.

Magnetic energy

A magnetic is a piece of metal that has the ability to attract iron, nickel, cobalt, or certain specific other kinds of metal. Every magnet contains two distinct regions, one known as the north pole and one, the south pole. As with electrical charges, unlike poles attract each other and like poles repel each other.

A study of magnets allows the introduction of a new concept in energy, the concept of a field. An energy field is a region in space in which a magnetic, electrical, or some other kind of force can be experienced. For example, imagine that a piece of iron is placed at a distance of 2 in (5 cm) from a bar magnet. If the magnet is strong enough, it may pull on the iron strongly enough to cause it to move. The piece of iron is said to be within the magnetic field of the bar magnet.

The concept of an energy field was, at one time, a very difficult one for scientists to understand and accept. How could one object exert a force on another object if the two were not in contact with each other? Eventually, it became clear that forces can operate at a distance from each other. Electrical charges and magnetic poles seem to exert their forces throughout a field along pathways known as lines of force.

One of the great discoveries in the history of **physics** was made by the English physicist James Clerk Maxwell (1831-1879) in the late nineteenth century. Maxwell found that the two major forms of energy known as **electricity** and **magnetism** are not really different from each other, but are instead closely associated with each other. That is, every electrical current has associated with it a magnetic field and every changing magnetic field creates its own electrical current.

As a result of Maxwell's work, it is often more correct to speak of electromagnetic energy, a form of energy that has both electrical and magnetic components. Scientists now know that a number of seemingly different types of energy are all actually forms of electromagnetic energy. These include **x rays**, gamma rays, ultraviolet light, visible light, infrared **radiation**, **radio waves**, and microwaves. These forms of electromagnetic energy differ from each other in terms of the wavelength and **frequency** of the energy wave on which they travel. The waves associated with x rays, for example, have very short wavelengths and very high frequencies, while the waves associated with microwaves have much longer wavelengths and much lower frequencies.

Sound, chemical, and nuclear energy

The fact that people can hear is a simple demonstration of the fact that sound is a form of energy. Sound is actually nothing other than the movement of air. When sound is created, **sound waves** travel through space, creating compressions in some regions and rarefactions in other regions. When these sound waves strike the human eardrum, they cause the drum to vibrate, creating the sensation of sound in the **brain**. Similar kinds of sound waves are responsible for the destruction caused by explosions. The sound waves collide with building, trees, people, and other objects, causing damage to them.

Chemical energy is a form of energy that results from the forces of attraction that hold **atoms** and other particles together in molecules. In water, for example, **hydrogen** atoms are joined to **oxygen** atoms by means of strong forces known as chemical bonds. If those are broken, the forces are released in the form of chemical energy. When a substance is burned, chemical energy is released. Burning (**combustion** or oxidation) is the process by which chemical bonds in a fuel and in oxygen molecules are broken and new chemical bonds are formed. The total energy in the new chemical bonds is less than it was in the original chemical bonds, and the difference is released in the form of chemical energy.

Nuclear energy is similar to chemical energy except that the bonds involved are those that hold together the particles of a nucleus, protons and neutrons. The fact that

KEY TERMS

Conservation of energy—A law of physics that says that energy can be transformed from one form to another, but can be neither created nor destroyed.

Field—A region in space in which a magnetic, electrical, or some other kind of force can be experienced.

Joule—The unit of measurement for energy in the metric system.

Kinetic energy—The energy possessed by a body as a result of its motion.

Magnetic pole—The region of a magnetic in which magnetic force appears to be concentrated.

Potential energy—The energy possessed by a body as a result of its position.

Temperature—A measure of the average kinetic energy of all the elementary particles in a sample of matter.

Thermal energy—The total amount of energy contained within any body as a consequence of the motion of its particles.

most atomic nuclei are stable is proof that some very strong nuclear forces exist. Protons are positively charged and one would expect that they would repel each other, blowing apart a nucleus. Since that does not happen, some kinds of force must exist to hold the nucleus together.

One such force is known as the strong force. If something happens to cause a nucleus to break apart, the strong force holding two protons together is released in the form of nuclear energy. That is what happens in an atomic (fission) bomb. A **uranium** nucleus breaks apart into two roughly equal pieces, and some of the strong force holding protons together is released as nuclear energy.

David E. Newton

Energy budgets

An **energy** budget describes the ways in which energy is transformed from one state to another within some defined system, including an analysis of inputs, outputs, and changes in the quantities stored. Ecological energy budgets focus on the use and transformations of energy in the **biosphere** or its components.

Solar electromagnetic **radiation** is the major input of energy to **Earth**. This external source of energy helps to **heat** the **planet**, evaporate **water**, circulate the atmosphere and oceans, and sustain ecological processes. Ultimately, all of the solar energy absorbed by Earth is re-radiated back to **space**, as electromagnetic radiation of a longer wavelength than what was originally absorbed. Earth maintains a virtually perfect energetic balance between inputs and outputs of electromagnetic energy.

Earth's ecosystems depend on solar radiation as an external source of diffuse energy that can be utilized by photosynthetic autotrophs, such as green plants, to synthesize simple organic molecules such as sugars from inorganic molecules such as **carbon dioxide** and water. Plants use the fixed energy of these simple organic compounds, plus inorganic **nutrients**, to synthesize an enormous diversity of biochemicals through various metabolic reactions. Plants utilize these biochemicals and the energy they contain to accomplish their growth and reproduction. Moreover, **plant biomass** is directly or indirectly utilized as food by the enormous numbers of heterotrophic organisms that are incapable of fixing their own energy. These organisms include herbivores that eat plants, carnivores that eat animals, and detritivores that feed on dead biomass.

Worldwide, the use of solar energy for this ecological purpose is relatively small, accounting for much less than 1% of the amount received at Earth's surface. Although this is a quantitatively trivial part of Earth's energy budget, it is clearly very important qualitatively, because this is the absorbed and biologically fixed energy that subsidizes all ecological processes.

Forms of energy

Energy is defined as the ability, or potential ability, of a body or system to do **work**. Energy can be measured in various units, such as the **calorie**, defined as the amount of energy required to raise the **temperature** of one gram of pure water from 59–61°F (15–16°C). (Note that the dietician's calorie is equivalent to one thousand of these calories, or one kilocalorie.) The Joule is another unit of energy, defined as the amount of work required to lift a weight of 1 kg by 10 cm, and equivalent to 0.24 calories.

Energy can exist in various states, all of which are interchangeable through various sorts of physical/ chemical transformations. The basic categories of energy are: electromagnetic, kinetic, and potential, but each of these can also exist in various states, as is described below:

(1) Electromagnetic energy is the energy of photons, or quanta of energy that have properties of both particles and waves, and that travel through space at a constant speed of 3×10^8 meters per second (that is, at the speed of **light**). The components of electromagnetic energy are characterized on the basis of wavelength ranges, which ordered from the shortest to longest wavelengths are known as: gamma, x ray, ultraviolet, light or visible, infrared, and **radio**. All bodies with a temperature greater than **absolute zero** (that is, -459°F [-273°C], or **zero** degrees on the kelvin scale) emit electromagnetic energy at a **rate** and spectral quality that is strictly determined by their surface temperature. Relatively hot bodies have much larger **emission** rates and their radiation is dominated by shorter wavelengths, compared with cooler bodies. The **Sun** has a surface temperature of about 11,000°F (6,093°C) and most of its radiation is in the wavelength range of visible light (0.4-0.7 æm or micrometers) and shorter-wave infrared (0.7-2 æm), while Earth has a surface temperature of about 77°F (25°C) and its radiation peaks in the longer-wave infrared range at about 10 æm.

(2) Kinetic energy is the energy of dynamic **motion**, of which there are two basic types, the energy of moving bodies, and that of vibrating **atoms** or molecules. The later is also known as thermal energy, and the more vigorous the vibration, the greater the heat content.

(3) Potential energy has the capacity to do work, but it must be mobilized to do so. Potential energy occurs in various forms, including the following: (a) Chemical potential energy is stored in the inter-atomic bonds of molecules. This energy can be liberated by so-called exothermic reactions, which have a net release of energy. For example, heat is released when the chemically reduced **sulfur** of sulfide **minerals** is oxidized to sulfate, and when crystalline **sodium chloride** is dissolved into water. All biochemicals also store potential energy, equivalent to 4.6 kilocalories per gram of **carbohydrate**, 4.8 Kcal/g of protein, and 6.0-9.0 Kcal/g of **fat**. (b) Gravitational potential energy is stored in **mass** that is elevated above some gravitationally attractive surface, as when water occurs above the surface of the oceans, or any object occurs above the ground surface. Unless obstructed, water spontaneously flows downhill, and objects fall downwards in response to gradients of gravitational potential energy. (c) Other types of potential energy are somewhat less important in terms of ecological energy budgets, but they include potential energies of compressed gases, electrical potential gradients associated with voltage differentials, and the potential energy of **matter**, which can be released by nuclear reactions.

Energy transformations and the laws of thermodynamics

As noted previously, energy can be transformed among its various states. For example: (a) electromagnetic energy can be absorbed by a dark object and converted to thermal kinetic energy, resulting in an increased temperature of the absorbing body; (b) gravitational potential energy of water high on a plateau is transformed into the kinetic energy of moving water and heat at a waterfall, or it can be mobilized by humans to drive a **turbine** and generate electrical energy; and (c) solar electromagnetic radiation can be absorbed by the **chlorophyll** of green plants, and some of the absorbed energy can be converted into the chemical potential energy of sugars, and the rest converted into heat.

All transformations of energy must occur according to certain physical principles, known as the laws of **thermodynamics**. These are universal laws, which means that they are always true, regardless of circumstances. The first law states that energy can undergo transformations among its various states, but it is never created or destroyed, so the energy content of the universe remains constant. A consequence of this law for energy budgets is that there must always be a zero balance between the energy inputs to a system, the energy outputs, and any net storage within the system.

The second law of thermodynamics states that transformations of energy can only occur spontaneously under conditions in which there is an increase in the **entropy** of the universe. (Entropy is related to randomness of the distributions of matter and energy). For example, Earth is continuously irradiated by solar radiation, mostly of visible and near-infrared wavelengths. Some of this energy is absorbed, which heats the surface of Earth. The planet cools itself in various ways, but ultimately this is done by radiating its own electromagnetic radiation back to space, as longer-wave infrared radiation. The transformation of relatively short-wave solar radiation into the longer-wave radiation emitted by Earth represents a degradation of the quality of the energy, and an increase in the entropy of the universe.

A corollary, or secondary proposition of the second law of thermodynamics is that energy transformations can never be completely efficient, because some of the initial content of energy must be converted to heat so that entropy can be increased. Ultimately, this is the reason why no more than about 30% of the energy content of gasoline can be converted into the kinetic energy of a moving **automobile**, and why no more than about 40% of the energy of **coal** can be transformed into **electricity** in a modern generating station. Similarly, there are upper limits to the efficiency by which green plants can photosynthetically convert visible radiation into biochemicals, even in ecosystems in which ecological constraints related to nutrients, water, and space are optimized.

Interestingly, plants absorb visible radiation emitted by the Sun, and use this relatively dispersed energy to fix simple inorganic molecules such as carbon dioxide,

water, and other nutrients into very complex and energy-dense biochemicals. The biochemicals of plant biomass are then used by heterotrophic organisms to synthesize their own complex biochemicals. Locally, these various biological syntheses represent energy transformations that substantially decrease entropy, rather than increase it. This is because relatively dispersed solar energy and simple compounds are focused into the complex biochemicals of living organisms. Are biological transformations not obeying the second law of thermodynamics?

This seeming physical paradox of life can be successfully rationalized, using the following logic: The localized bio-concentrating of **negative** entropy can occur because there is a constant input of energy into the system, in the form of solar radiation. If this external source of energy was terminated, then all of the negative entropy of organisms and organic matter would rather quickly be spontaneously degraded, producing heat and simple inorganic molecules, and thereby increasing the entropy of the universe. This is why life and ecosystems cannot survive without continual inputs of solar energy. Therefore, the biosphere can be considered to represent a localized **island**, in space and **time**, of negative entropy, fueled by an external (solar) source of energy. There are physical analogues to these ecological circumstances—if external energy is put into the system, relatively dispersed molecules of gases can be concentrated into a container, as occurs when a person blows energetically to fill a **balloon** with air. Eventually, however, the balloon pops, the gases re-disperse, the original energy input is converted into heat, and the entropy of the universe is increased.

Physical energy budgets

Physical energy budgets consider a particular, defined system, and then analyze the inputs of energy, its various transformations and storages, and the eventual outputs. This concept can be illustrated by reference to the energy budget of Earth.

The major input of energy to Earth occurs as solar electromagnetic energy. At the outer limits of Earth's atmosphere, the average rate of input of solar radiation is 2.00 calories per cm^2 per minute (this flux is known as the solar constant). About half of this energy input occurs as visible radiation, and half as near-infrared. As noted previously, Earth also emits its own electromagnetic radiation, again at a rate of 2.00 cal/cm^2/min, but with a spectrum that peaks in the longer-wave infrared, at about 10 æm. Because the rate of energy input equals the rate of output, there is no net storage of energy, and no substantial, longer-term change in Earth's surface temperature. Therefore, Earth represents a zero-sum, energy flow-through system. (Actually, over geological time there has been a small storage of energy, occurring as an accumulation of undecomposed biomass that eventually transforms geologically into **fossil fuels**. There are also minor, longer-term variations of Earth's temperature surface that represent climate change. However, these represent quantitatively trivial exceptions to the preceding statement about Earth as a zero-sum, flow-through system for energy.) Although the amount of energy emitted by Earth eventually equals the amount of solar radiation that is absorbed, there are some ecologically important transformations that occur between these two events. The most important ways by which Earth deals with its incident solar radiations are: (1) An average of about 30% of the incident solar energy is reflected back to outer space by Earth's atmosphere or its surface. This process is related to Earth's **albedo**, which is strongly influenced by the solar **angle**, the amounts of cloud cover and atmospheric particulates, and to a lesser **degree** by the character of Earth's surface, especially the types and amount of water (including **ice**) and vegetation cover. (2) About 25% of the incident energy is absorbed by atmospheric gases, vapors, and particulates, converted to heat or thermal kinetic energy, and then re-radiated as longer-wavelength infrared radiation. (3) About 45% of the incident radiation is absorbed at Earth's surface by living and non-living materials, and is converted to thermal energy, increasing the temperature of the absorbing surfaces. Over the longer term (that is, years) and even the medium term (that is, days) there is little or no net storage of heat. Virtually all of the absorbed energy is re-radiated by the surface as long-wave infrared energy, with a wavelength peak of about 10 æm. (4) Some of the thermal energy of surfaces causes water to evaporate from plant and non-living surfaces (see entry on **evapotranspiration**), or it causes ice or snow to melt. (5) Because of the uneven distribution of thermal energy on Earth's surface, some of the absorbed radiation drives mass-transport, distributional processes, such as winds, water currents, and waves on the surface of waterbodies. (6) A very small (averaging less than 0.1%) but ecologically critical portion of the incoming solar energy is absorbed by the chlorophyll of plants, and is used to drive **photosynthesis**. This photoautotrophic fixation allows some of the solar energy to be "temporarily" stored in the potential energy of biochemicals, and to serve as the energetic basis of life on Earth.

Certain gases in Earth's atmosphere absorb long-wave infrared energy of the type that is radiated by heated matter in dissipation mechanisms 2 and 3 (above). This absorption heats the gases, which then undergo another re-radiation, emitting even longer-wavelength infrared energy in all directions, including back to Earth's surface. The most important of the so-called radiatively active gases in the atmosphere are water and carbon dioxide, but the trace gases methane, nitrous oxide, **ozone**, and chlorofluorocarbons are also significant. This phenomenon, known as the **greenhouse effect**, significantly interferes with the rate of radiative cooling of Earth's surface.

If there were no greenhouse effect, and Earth's atmosphere was fully transparent to long-wave infrared radiation, surface temperatures would average about 17.6°F (-8°C), much too cold for biological processes to occur. Because the naturally occurring greenhouse effect maintains Earth's average surface temperature about 60 degrees warmer than this, at about 77°F (25°C), it is an obviously important factor in the habitability of our planet. However, human activities have resulted in increasing atmospheric concentrations of some of the radiatively active gases, and there are concerns that this could cause an intensification of Earth's greenhouse effect. This could lead to **global warming**, changes in the distributions of rainfall and other climatic effects, and severe ecological and socioeconomic damages.

Budgets of fixed energy

Ecological energetics examines the transformations of fixed, biological energy within communities and ecosystems, in particular, the manner in which biologically fixed energy is passed through the food web.

For example, studies of a natural oak-pine forest in New York found that the vegetation fixed solar energy equivalent to 11,500 kilocalories per hectare per year (10^3 Kcal/ha/yr). However, plant **respiration** utilized 6.5×10^3 Kcal/ha/yr, so that the actual net accumulation of energy in the **ecosystem** was 5.0×10^3 Kcal/ha/yr. The various types of heterotrophic organisms in the forest utilized another 3.0×10^3 Kcal/ha/yr to support their respiration, so the net accumulation of biomass by all of the organisms of the ecosystem was equivalent to 2.0×10^3 Kcal/ha/yr.

The preceding is an example of a fixed-energy budget at the ecosystem level. Sometimes, ecologists develop budgets of energy at the levels of population, and even for individuals. For example, depending on environmental circumstances and opportunities, **individual** plants or animals can optimize their fitness by allocating their energy resources into various activities, most simply, into growth of the individual or into reproduction.

However, biological energy budgets are typically much more complicated than this. For example, a plant can variously allocate its energy into the production of longer stems and more leaves to improve its access to sunlight, or it could grow longer and more roots to increase its access to **soil** nutrients, or more flowers and **seeds** to increase the probability of successful reproduction. There are other possible allocation strategies, including some combination of the preceding.

Similarly, a bear must makes decisions about the allocation of its time and energy into activities associated with resting, either during the day or longer-term **hibernation**, hunting for plant or **animal** foods, seeking a mate, taking care of the cubs.

See also Energy transfer; Food chain/web.

Resources

Books

Odum, E.P. *Ecology and Our Endangered Life Support Systems.* New York: Sinauer, 1993.
Ricklefs, R.E. *Ecology.* New York: W.H. Freeman, 1990.

Bill Freedman

Energy efficiency

Energy efficiency refers to any process by which the amount of useful **energy** obtained from some process is increased compared to the amount of energy put into that process. As a simple example, some automobiles can travel 40 mi (17 km) by burning a single gallon (liter) of gasoline, while others can travel only 20 mpg (8.5 km/l). The energy efficiency achieved by the first car is twice that achieved by the second car. In general, energy efficiency is measured in units such as mpg, lumens per watt, or some similar "output per input" unit.

History of energy concerns

Interest in energy efficiency is relatively new in the history of modern societies, although England's eighteenth century search for **coal** was prompted by the decline of the country's forest resources. For most of the

past century, however, energy resources seemed to be infinite, for all practical purposes. Little concern was expressed about the danger of exhausting the world's supplies of coal, oil, and **natural gas**, its major energy resources.

The turning point in that attitude came in the 1970s when the major oil-producing nations of the world suddenly placed severe limits on the amounts of **petroleum** that they shipped to the rest of the world. This oil embargo forced major oil users such as the United States, Japan, and the nations of Western **Europe** to face for the first time the danger of having insufficient petroleum products to meet their basic energy needs. Use of energy resources suddenly became a matter of national and international discussion.

Energy efficiency can be accomplished in a number of different ways. One of the most obvious is *conservation*; that is, simply using energy resources more carefully. For example, people might be encouraged to turn out lights in their home, to set their thermostats at lower temperatures, and to use bicycles rather than automobiles for transportation. Energy efficiency in today's world also means more complex and sophisticated approaches to the way in which energy is used in industrial, commercial, and residential settings.

Energy efficiency in buildings

Approximately one-third of all the energy used in the United States goes to **heat**, cool, and **light** buildings. A number of technologies have been developed that improve the efficiency with which energy is used in buildings. Some of these changes are simple; higher grades of insulation are used in construction, and air leaks are plugged. Both of these changes reduce the amount of heated or air-conditioned air (depending on the season) lost from the building to the outside environment.

Other improvements involve the development of more efficient appliances and construction products. For example, the typical gas furnace in use in residential and commercial buildings in the 1970s was about 63% efficient. Today, gas furnaces with efficiencies of 97% are readily available and affordable. Double-glazed windows with improved insulating properties have also been developed. Such windows can save more than 10% of the energy lost by a building in a year.

Buildings can also be designed to save energy. For example, they can be oriented on a lot to take advantage of solar heating or cooling. Many commercial structures also have computerized systems that automatically adjust heating and cooling schedules to provide a comfortable environment for occupants only when and in portions of the building that are occupied.

Entirely new technologies can be used also. For example, many buildings now depend exclusively on more efficient fluorescent lighting systems than on less efficient incandescent lights. In some situations, this single change can produce a greater savings in energy use than any other modification. The increasing use of solar cells is another example of a new kind of technology that has the potential for making room and **water** heating much more efficient.

Transportation

About one-third more of the energy used in the United States goes to moving people and goods from place to place. For more than two decades, governments have made serious efforts to convince people that they should use more energy-efficient means of transportation, such as bicycles or some form of **mass** transit (buses, trolleys, subways, light-rail systems, etc.). These efforts have had only limited success.

Another approach that has been more successful has been to encourage car manufacturers to increase the efficiency of **automobile** engines. In the 1970s, the average fuel efficiency of cars in the United States was 13 mpg (5.5 km/l). Over the next decade, efficiency improved nearly twice over to 25 mpg (10.6 km/l). In other nations, similar improvements were made. Cars in Japan, for example, increased from an average efficiency of 23 mpg (9.8 km/l) in 1973 to 30 mpg (12.8 km/l) in 1985.

Yet, even more efficient automotive engines appear to be possible. Many authorities believe that efficiencies approaching 50 mpg (21 km/l) should be possible by the year 2001. As of 1987, at least three cars with fuel efficiencies of more than 50 mpg (21 km/l) were already in production (the Ford Escort, Honda City, and Suzuki Sprint). Experimental cars with efficiencies close to 100 mpg (42.5 km/l) were also being tested; the Toyota AXV has achieved 98 mpg (41.7 km/l) on test tracks, and the Renault VESTA has logged 124 mpg (52.7 km/l).

To a large extent, automobile manufacturers have been slow to produce cars that have the maximum possible efficiencies because they question whether consumers will pay higher purchase prices for these cars. Improvements continue to be made, however, at least partly because of the legislative pressure for progress in this direction.

Energy efficiency in industry

The final third of energy use in the United States occurs in a large variety of industrial operations such as producing steam, heating plants, and generating **electricity** for various operations. Improvements in the efficiency with which energy is used in industry also depends on two principal approaches: the development of more effi-

cient devices and the invention of new kinds of technologies. More efficient motors are now available so that the same amount of **work** can be accomplished with a smaller amount of energy input. And, as an example of the use of new technologies, **laser** beam systems that can both heat and cut more efficiently than traditional tools are being given new applications. Industries are also finding ways to use computer systems to design and carry out functions within a plant more efficiently than traditional resource management methods.

One of the most successful approaches to improving energy efficiency in industry has been the development of **cogeneration** systems. Cogeneration refers to the process in which heat produced in an industrial operation (formerly regarded as "waste heat") is used to generate electricity. The plant saves money through cogeneration because it does not have to buy electrical power from local utilities.

Other techniques for increasing energy efficiency

Many other approaches are available for increasing the efficiency with which energy is used in a society. **Recycling** has become much more popular in the United States over the past few decades at least partly because it provides a way of salvaging valuable resources such as **glass**, **aluminum**, and **paper**. Recycling is also an energy efficient practice because it reduces the cost of producing new products from raw materials. Another approach to energy efficiency is to make use of packaging materials that are produced with less energy. The debate still continues over whether paper or plastic bags are more energy efficient, but at least the debate indicates that people are increasingly aware of the choices that can be made about packaging materials.

Government policies and regulations

Most governmental bodies were relatively unconcerned about energy efficiency issues until the OPEC (Organization of Oil Exporting Countries) oil embargo of 1973-74. Following that event, however, they began to search for ways of encouraging corporations and private consumers to use energy more efficiently. One of the first of many laws that appeared over the next decade was the Energy Policy and Conservation Act of 1975. Among the provisions of that act were: a requirement that new appliances carry labels indicating the amount of energy they use, the creation of a federal technical and financial assistance program for energy **conservation** plans, and the establishment of the State Energy Conservation Program. A year later, the Energy Conservation and Production Act of 1976 provided for the development of national mandatory Building Energy Performance Standards and the creation of the Weatherization Assistance Program to fund energy-saving retrofits for low-income households. Both of these laws were later amended and updated.

In 1991, the U.S. Environmental Protection Agency (EPA) established two voluntary programs to prevent **pollution** and reduce energy costs. The Green Lights Partnership provided assistance in installing energy-efficient lighting, and the Energy Star Buildings Partnership used Green Lights as its first of five stages to improve all aspects of building efficiency. The World Trade Center and the Empire State Building in New York City and the Sears Tower in Chicago (four of the world's tallest structures) joined the Energy Star Buildings Partnership as charter members and have reduced their energy costs by millions of dollars. The EPA also developed software with energy management aids for building operators who enlist in the partnership. By 1998, participating businesses had reduced their lighting costs by 40%, and whole-building upgrades had been completed in over 2.8 billion ft^2 (0.3 billion m^2) of building space. The EPA's environmental interest in the success of these programs comes not only from conserving resources but from limiting **carbon dioxide** emissions that result from energizing industrial plants and commercial buildings and that cause changes in the world's climate.

Green Market electric utilities

In 1998, restructuring of the electric power utilities opened the market place to "Green Power Markets" that offer environmental features along with power service. Green power sources will provide clean energy and improved efficiency based on technologies that rely on renewable energy sources. Educating the consumer and providing green power at competitive costs are seen as two of the biggest challenges to this new market. The Center for Resource Solutions in California pioneered the Green-e Renewable Electricity Branding Program that is a companion to green power and certifies electricity products that are environmentally preferred.

Results and the future

Efforts to increase public consciousness about energy efficiency issues have had some remarkable successes in the past two decades. Despite the increasing complexity of most developed societies and increased population growth in many nations, energy is being used more efficiently in almost every part of the world. Increased efficiency of energy use increased between 1973 and 1985 by as much as 31% in Japan, 23% in the United States, 20% in the United Kingdom, and 19% in Italy. At the beginning of this period, most experts had predicted that

---

KEY TERMS

Cogeneration—A process by which heat produced as a result of industrial processes is used to generate electrical power.

Mass transit—Any form of transportation in which significantly large numbers of riders are moved within the same vehicle at the same time.

Solar cell—A device by which sunlight is converted into electricity.

---

changes of this magnitude could be accomplished only as a result of the massive reorganization of social institutions; this has not been the case. Processes and inventions that continue to increase energy efficiency can be incorporated into daily life with minimal disruptions to personal lives and industrial operations.

Energy efficiency has a long way to go, however. In December 1997 in Kyoto, Japan, a **global warming** agreement was proposed to the nations of the world to cut **carbon** emissions, reduce levels of so-called "greenhouse gases" (methane, carbon dioxide, and nitrous oxide), and use existing technologies to improve energy efficiency. These technologies apply to all levels of society from governments and industries to the individual household. But experts acknowledge that the public must recognize the global warming problem as real and serious before existing technologies and a host of potential new products will be supported.

See also Alternative energy sources; Fluorescent light; Hydrocarbon.

Resources

Books

Flavin, Christopher, and Alan B. Durning. *Building on Success: The Age of Energy Efficiency.* Worldwatch Paper 82. Washington, DC: Worldwatch Institute, March 1988.

Hirst, Eric, et al. *Energy Efficiency in Buildings: Progress and Promise.* Washington, DC: American Council for an Energy-Efficient Economy, 1986.

Hoffmann, Peter, and Tom Harkin. *Tomorrow's Energy: Hydrogen, Fuel Cells, and Prospects for a Cleaner Planet.* Boston: MIT Press, 2001.

Meier, Alan K., et al. *Saving Energy through Greater Efficiency.* Berkeley: University of California Press, 1981.

Other

U.S. Congress, Office of Technology Assessment. *Building Energy Efficiency.* OTA-E-518. Washington, DC: U.S. Government Printing Office, May 1992.

David E. Newton

Energy transfer

Energy transfer describes the changes in energy (a state function) that occur between organisms within an **ecosystem**. Living organisms are constantly changing as they grow, move, reproduce, and repair tissues. These changes are fueled by energy. Plants, through **photosynthesis**, capture some of the Sun's radiant energy and transform it into chemical energy, which is stored as **plant biomass**. This biomass is then consumed by other organisms within the ecological **food chain/web**. A food chain is a sequence of organisms that are connected by their feeding and productivity relationships; a food web is the interconnected set of many food chains.

Energy transfer is a one-way process. Once potential energy has been released in some form from its storage in biomass, it cannot all be reused, recycled, or converted to waste **heat**. This means that if the **Sun**, the ultimate energy source of ecosystems, were to stop shining, life as we know it would soon end. Every day, the Sun provides new energy in the form of photons to sustain the food webs of **Earth**.

History of energy transfer research

In 1927, the British ecologist Charles Elton wrote that most food webs have a similar pyramidal shape. At the bottom, there are many photosynthetic organisms which collectively have a large biomass and productivity. On each of the following **trophic levels**, or feeding levels, there are successively fewer heterotrophic organisms, with a smaller productivity. The pyramid of biomass and productivity is now known as the Eltonian pyramid.

In 1942, Raymond L. Lindeman published a **paper** that examined food webs in terms of energy flow. Lindeman proposed that, by using energy as the currency of ecosystem processes, food webs could be quantified. This allowed him to explain that the Eltonian pyramid was a result of successive energy losses associated with the thermodynamic inefficiencies of energy transfer among trophic levels.

Current research in ecological energy transfer focuses on increasing our understanding of the paths of energy and **matter** within grazing and microbial food webs. Rather little is understood about such pathways because of the huge numbers of **species** and their complex interactions. This understanding is essential for proper management of ecosystems. The fate and effects of toxic chemicals within food webs must be understood if impacts on vulnerable species and ecosystems are to avoided or minimized.

The laws of thermodynamics and energy transfer in food webs

Energy transfers within food webs are governed by the first and second laws of **thermodynamics**. The first law relates to quantities of energy. It states that energy can be transformed from one form to another, but it cannot be created or destroyed. This law suggests that all energy transfers, gains, and losses within a food web can be accounted for in an energy budget.

The second law relates to the quality of energy. This law states that whenever energy is transformed, some of must be degraded into a less useful form. In ecosystems, the biggest losses occur as **respiration**. The second law explains why energy transfers are never 100% efficient. In fact, ecological efficiency, which is the amount of energy transferred from one trophic level to the next, ranges from 5-30%. On average, ecological efficiency is only about 10%.

Because ecological efficiency is so low, each trophic level has a successively smaller energy pool from which it can withdraw energy. This is why food webs have no more than four to five trophic levels. Beyond that, there is not enough energy to sustain higher-order predators.

Components of the food web

A food web consists of several components; primary producers, primary consumers, secondary consumers, tertiary consumers, and so on. Primary producers include green plants and are the foundation of the food web. Through photosynthesis, primary producers capture some of the Sun's energy. The net **rate** of photosynthesis, or net primary productivity (NPP), is equal to the rate of photosynthesis minus the rate of respiration of plants. In essence, NPP is the profit for the primary producer, after their energy costs associated with respiration are accounted for. NPP determines plant growth and how much energy is subsequently available to higher trophic levels.

Primary consumers are organisms that feed directly on primary producers, and these comprise the second trophic level of the food web. Primary consumers are also called herbivores, or plant-eaters. Secondary consumers are organisms that eat primary consumers, and are the third trophic level. Secondary consumers are carnivores, or meat-eaters. Successive trophic levels include the tertiary consumers, quaternary consumers, and so on. These can be either carnivores or omnivores, which are both plant- and animal-eaters, such as humans.

The role of the microbial food web

Much of the food web's energy is transferred to the often overlooked microbial, or decomposer, trophic level. Decomposers use excreted wastes and other dead biomass as a food source. Unlike the main, grazing food web, organisms of the microbial trophic level are extremely efficient feeders. Various species can rework the

KEY TERMS

. . . .

Biomass—Total weight, volume, or energy equivalent of all living organisms within a given area.

Ecological efficiency—Energy changes from one trophic level to the next.

First law of thermodynamics—Energy can be transformed but it cannot be created nor can it be destroyed.

Primary consumer—An organism that eats primary producers.

Primary producer—An organism that photosynthesizes.

Second law of thermodynamics—When energy is transformed, some of the original energy is degraded into less useful forms of energy.

same food particle, extracting more of the stored energy each time. Some waste products of the microbial trophic level re-enter the grazing part of the food web and are used as growth materials for primary producers. This occurs, for example, when earthworms are eaten by **birds**.

See also Ecological pyramids; Energy budgets.

Resources

Books

Bradbury, I. *The Biosphere*. New York: Belhaven Press, Pinter Publishers, 1991.

Incropera, Frank P., and David P. DeWitt. *Fundamentals of Heat and Mass Transfer*. 5th ed. New York: John Wiley & Sons, 2001.

Miller, G. T., Jr. *Environmental Science: Sustaining the Earth*. 3rd ed. Belmont, CA: Wadsworth Publishing Company, 1991.

Stiling, P. D. "Energy Flow in Ecosystems." In *Introductory Ecology*. Englewood Cliffs, NJ: Prentice-Hall, 1992.

Periodicals

Begon, M., J. L. Harper, and C. R. Townsend. "The Flux of Energy Through Communities." In *Ecology: Individuals, Populations and Communities*. 2nd ed. Boston: Blackwell Scientific Publications, 1990.

Jennifer LeBlanc

Engineering

Engineering is the art of applying science, **mathematics**, and creativity to solve technological problems.

The accomplishments of engineering can be seen in nearly every aspect of our daily lives, from transportation to communications, and entertainment to health care. And, although each of these applications is unique, the *process* of engineering is largely independent. This process begins by carefully analyzing a problem, intelligently designing a solution for that problem, and efficiently transforming that design solution into physical reality.

Analyzing the problem

Defining the problem is the first and most critical step of the problem analysis. To best approach a solution, the problem must be well-understood and the guidelines or design considerations for the project must be clear. For example, in the creation of a new **automobile**, the engineers must know if they should design for fuel economy or for brute power. Many questions like this arise in every engineering project, and they must all be answered at the very beginning if the engineers are to work efficiently toward a solution.

When these issues are resolved, the problem must be thoroughly researched. This involves searching technical journals and closely examining solutions of similar engineering problems. The purpose of this step is two-fold. First, it allows the engineer to make use of a tremendous body of work done by other engineers. And second, it ensures the engineer that the problem has not already been solved. Either way, the review allows him or her to intelligently approach the problem, and perhaps avoid a substantial waste of time or legal conflicts in the future.

Designing a solution

Once the problem is well-understood, the process of designing a solution begins. This process typically starts with brainstorming, a technique by which members of the engineering team suggest a number of possible general approaches for the problem. In the case of an automobile, perhaps conventional gas, solar, and electric power would be suggested to propel the vehicle. Generally, one of these is then selected as the primary candidate for further development. Occasionally, however, if time permits and several ideas stand out, the team may elect to pursue multiple solutions to the problem. More refined designs of these solutions/systems then "compete," and the best of those is chosen.

Once a general design or technology is selected, the work is sub-divided and various team members assume specific responsibilities. In the automobile, for example, the mechanical engineers in the group would tackle such problems as the design of the transmission and suspension systems. They may also handle air flow and climate-control concerns to ensure that the vehicle is both

Job Opportunities in Engineering and Technology '95. Princeton, NJ: Peterson's, 1994.

Newhouse, Elizabeth L., ed. *The Builders: Marvels of Engineering.* Washington, DC: National Geographic Society, 1992.

Scott Christian Cahall

KEY TERMS

Brainstorming—A process by which engineers discuss and suggest multiple possible approaches for solving a problem.

Design considerations—A set of guidelines or key issues that engineers must follow when they design a particular project.

Prototype—A preliminary working model of an engineering design, primarily used for testing and not for "show."

aerodynamic and comfortable to ride in. Electrical engineers, on the other hand, would concern themselves with the ignition system and the various displays and electronic gauges. They would also be responsible for the design of the communication system which links all of the car's sub-systems together. In any case, each of these engineers must design one aspect which operates in harmony with every other aspect of the system.

Bringing it to life

Once the design is complete, a prototype or preliminary working model is generally built. The primary function of the prototype is to demonstrate and test the operation of the device. For this reason, its cosmetics are typically of little concern, as they will likely change by the time the device reaches the market.

The prototype stage is where the device undergoes extensive testing to reveal any bugs or problems with the design. Especially with complex systems, it is often difficult to predict (on **paper**) where problems with the design may occur. If one aspect of the system happens to fail too quickly or does not function at all, it is closely analyzed and that sub-system is redesigned and retested (both on its own and within the complete system). This process is repeated until the entire system satisfies the design requirements.

Once the prototype is in complete working order and the engineers are satisfied with its operation, the device goes into the production stage. Here, details such as appearance, ease of use, availability of materials, and safety are given attention and generally result in additional final design changes.

Resources

Books

Adams, James L. *Flying Buttresses, Entropy, and O-Rings, The World of an Engineer.* Cambridge: Harvard University Press, 1991.

Engraving and etching

Engraving and etching are processes used to make intaglio prints. An intaglio print is made from a plate, usually a **metal** one, which has been had lines drawn into its surface. These lines trap ink when the ink is rolled across the surface of the plate. When the surface of the plate is wiped with a cloth, the lines retain their ink. A piece of damp **paper** is placed on the plate, and the two are run through a press, which forces them together. This process transfers the ink from the plate to the paper. In an etching, acids are used to draw into the plate. In an engraving, sharp tools are used to draw directly into the metal.

Engraving and etching have been used in **printing** for hundreds of years. Before the invention of modern, photographic-based techniques, they were the most commonly used method for reproducing images. Newspapers and printed advertisements formerly used engravings. Stamps and paper money are still printed using the engraving process because of its ability to reproduce fine lines and sharp details.

Origins and history of intaglio printing

Engraving first became popular in **Europe** during the fifteenth century, when paper became available far more widely than it had been previously. From the beginning, intaglio printing was used for both the sacred and the profane. Artists made engravings of religious scenes, while craftsmen used the new technique to make copies of famous paintings or decks of playing cards. In an age when the printing press and movable type were first being invented, this was the equivalent of today's mass-produced posters.

Propelled by the genius of artists like Albrecht Dürer (1471-1528), intaglio techniques quickly grew. Artists learned to create various kinds of shading through the use of dots, called stippling, and groups of **parallel** lines at various angles to each other, called cross-hatching. Engraving requires drawing a line while pressing into a plate the correct **distance** to create the desired shade and line width. Some artists rejected the tool used for this, called a burin, which has a square or diamond-shaped tip and shaft, and creates very clean

A laser etching machine used in computer chip production. *Photograph by Richard Nowitz. Phototake NYC. Reproduced by permission.*

lines. They preferred the softer look caused by cutting the plate with a needle, a technique called drypoint. A needle causes ridges of metal to rise next to the groove it cut. These ridges catch large amounts of printing ink, which produce rich, soft blacks when transferred to paper. Unfortunately, the ridges wear down rapidly, making drypoint unsuitable to large press runs. In engraving, any ridges caused by cutting the plate are removed with a sharp instrument called a scraper.

Etching became popular during the sixteenth century. It evolved from fifteenth century techniques for putting patterns on swords. Armorers would cover new swords with wax, then scratch through the wax, and put the sword in a weak acid until the acid bit a line in the metal. To make an etching, a metal plate is first covered with a ground, a thin layer of acid-resistant material. The artist then scratches gently into the ground with an etching needle, exposing parts of the plate. This process is much more like drawing with a pencil than engraving is, and many artists preferred it. After the ground is drawn on with the etching needle, the plate is put into acid, which eats at the plate where the needle has exposed it. The length of exposure to the acid determines the depth of the line.

Evolution of etching techniques

The same principals are employed today, though techniques have been refined and expanded. Etching grounds have evolved from an unpredictable waxy substance, to harder varnishes used by instrument makers, to **petroleum** derivatives like asphaltum. Early acids used on **copper** were made from a mixture of **sulfuric acid**, sodium nitrate, and alum, but they produced virulent fumes when used. Etchers switched to a weaker but safe mixture made from vinegar, **salt**, ammonium chloride,

and copper sulfate. This solution produced rich detail, and was used by Rembrandt. **Nitric acid** was often used beginning in the late eighteenth century, but because of its strength, it sometimes destroyed fine details. Around 1850 a popular mixture was discovered consisting of hydrochloric acid, potassium chloride, and **water**. Sometimes called Dutch Mordant, it is still used today, and is the most widely used copper-etching agent, along with ferric chloride.

Plate usage has also evolved. The first etchings were done on **iron** plates, which corroded quickly. Around 1520, the Dutch artist Lucas van Leyden (1494-1533) began using copper plates, and other printmakers swiftly followed. However, copper plates can hold a lot of detail and are not expensive, but copper is too soft for large editions. In van Leyden's day, people used to pound it with hammers, creating a **random** alignment of the copper molecules, and thus a stronger plate. Today, copper can be bought in an already-hardened form. Better still, after etching or engraving, copper plates can be strengthened by putting a microscopically thin layer of iron on them through electrodeposition. This process is called **steel** facing.

In electrodeposition, copper and iron plates are placed in an electricity-carrying solution. Direct-current **electricity** is run through the plates, giving the copper a **negative** charge, and the iron a positive charge. The current removes iron ions, carries them through the solution, and deposits them on the copper plate. This technique, invented in France in the mid-nineteenth century, quickly spread throughout the world.

Zinc is often used as an etching plate because it is less expensive than copper. It is softer than copper, however, and requires a two-stage process for steel facing that results in some loss of detail. Zinc's softness can be an advantage in etching, where the plate is sometimes changed by scraping and polishing. It is too soft for fine-line engraving, however.

Brass, an **alloy** of copper and zinc that is harder than either of them, is sometimes used for plates, and costs about as much as copper. Steel is the material of choice for engraving plates. Because it is so hard, very large editions can be printed. Steel also yields extremely fine detail. One disadvantage of steel is that it rusts easily. It must be stored carefully and protected from moisture with a coating of etching ground or oil.

Resources

Books

Ayres, Julia. *Printmaking Techniques.* New York: Watson-Guptill, 1993.

Scott M. Lewis

Enterobacteria

Enterobacteria are **bacteria** from the family Enterobacteriaceae, which are primarily known for their ability to cause intestinal upset. Enterobacteria are responsible for a variety of human illnesses, including urinary tract infections, wound infections, gastroenteritis, **meningitis**, septicemia, and **pneumonia**. Some are true intestinal **pathogens**; whereas others are merely opportunistic **pests** which attack weakened victims.

Most enterobacteria reside normally in the large intestine, but others are introduced in contaminated or improperly prepared foods or beverages. Several enterobacterial diseases are spread by fecal-oral transmission and are associated with poor hygienic conditions. Countries with poor **water** decontamination have more illness and death from enterobacterial **infection**. Harmless bacteria, though, can cause diarrhea in tourists who are not used to a geographically specific bacterial strain. Enterobacterial gastroenteritis can cause extensive fluid loss through vomiting and diarrhea, leading to dehydration.

Classification

Enterobacteria are a family of rod-shaped, **aerobic**, facultatively **anaerobic** bacteria. This means that while these bacteria can survive in the presence of **oxygen**, they prefer to live in an anaerobic (oxygen-free) environment. The Enterobacteriaceae family is subdivided into eight tribes including: Escherichieae, Edwardsielleae, Salmonelleae, Citrobactereae, Klebsielleae, Proteeae, Yersineae, and Erwineae. These tribes are further divided into genera, each with a number of **species**.

A transmission electron micrograph of *Escherichia coli.* *Photograph by Kari Lounatmaa. Photo Researchers. Reproduced by permission.*

Enterobacteria can cause **disease** by attacking their host in a number of ways. The most important factors are motility, colonization factors, endotoxin, and enterotoxin. Those enterobacteria that are motile have several **flagella** all around their perimeter (peritrichous). This allows them to move swiftly through their host fluid. Enterobacterial colonization factors are filamentous appendages, called fimbriae, which are shorter than flagella and bind tightly to the **tissue** under attack, thus keeping hold of its host. Endotoxins are the **cell** wall components which trigger high fevers in infected individuals. And enterotoxins are bacterial toxins which act in the small intestines and lead to extreme water loss in vomiting and diarrhea.

Pathology

A number of tests exist for rapid identification of enterobacteria. Most will ferment glucose to acid, reduce nitrate to nitrite, and test **negative** for **cytochrome** oxidase. These biochemical tests are used to pin-point specific intestinal pathogens. *Escherichia coli* (*E. coli*), *Shigella* species, *Salmonella*, and several *Yersinia* strains are some of these intestinal pathogens.

E. coli is indigenous to the GI tract and generally benign. However, it is associated with most hospital-acquired infections as well as nursery and travelers diarrhea. *E. coli* pathogenicity is closely related to the presence or absence of fimbriae on **individual** strains. Although most *E. coli* infections are not treated with **antibiotics**, severe urinary tract infections are.

The *Shigella* genus of the Escherichieae tribe can produce serious disease when its toxins act in the small intestine. *Shigella* infections can be entirely asymptomatic, or lead to severe **dysentery**. *Shigella* bacteria cause about 15% of pediatric diarrheal cases in the United States. However, they are a leading cause of infant mortality in developing countries. Only a few organisms are need to cause this fecal-orally transmitted infection. Prevention of the disease is achieved by proper sewage disposal and water **chlorination**, as well as personal hygiene such as hand-washing. Antibiotics are only used in more severe cases.

Salmonella infections are classified as nontyphoidal or typhoidal. Nontyphoidal infections can cause gastroenteritis and are usually due to contaminated food or water and can be transmitted by animals or humans. These infections cause one of the largest communicable bacterial diseases in the United States. They are found in contaminated **animal** products such as beef, pork, poultry, and raw chicken eggs. As a result, any food product which uses raw eggs, such as mayonnaise, homemade ice-cream, or Caesar salad dressing, could carry these bacteria. The best prevention when serving these dishes is to adhere strictly to refrigeration guidelines.

Typhoid *Salmonella* infections are also found in contaminated food and water, but humans are their only host. Typhoid Mary was a cook in New York from 1868 to 1914. She was typhoid carrier who contaminated much of the food she handled and was responsible for hundreds of typhoid cases. **Typhoid fever** is characterized by septicemia (**blood** poisoning), accompanied by a very high fever and intestinal lesions. Typhoid fever is treated with the drugs Ampicillin and Chloramphenicol.

Certain *Yersinia* bacteria cause the most notorious and fatal infections known to man. *Yersinia pestis* is the agent of **bubonic plague** and is highly fatal. The bubonic plague is carried by a rat flea and is thought to have killed at least 100 million people in the sixth century as well as 25% of the fourteenth century European population. This plague was also known as the Black Death, because it caused darkened hemorrhagic skin patches. The last widespread **epidemic** of *Y. pestis* began in Hong Kong in 1892 and spread to India and eventually San Francisco in 1900. The bacteria can reside in **squirrels**, prairie dogs, **mice**, and other **rodents** and are mainly found (in the United States.) in the Southwest. Since 1960, fewer than 400 cases have resulted in only a few deaths, due to rapid antibiotic treatment.

Two less severe *Yersinia* strains are *Y. pseudotuberculosis* and *Y. enterocolotica*. *Y. pseudotuberculosis* is

transmitted to humans by wild or domestic animals and causes a non-fatal disease which resembles appendicitis. *Y. enterocolotica* can be transmitted from animals or humans via a fecal-oral route and causes severe diarrhea with bloody stools.

Resources

Books

Koch, A.L. *Bacterial Growth and Form.* Dordrecht: Kluwer Academic Publishers, 2001.

Koneman, E., et al., eds. *Color Atlas and Textbook of Diagnostic Microbiology.* 4th ed. Philadelphia: J. B. Lippincott, 1992.

Prescott, L., J. Harley, and D. Klein. *Microbiology.* 5th ed. New York: McGraw-Hill, 2002.

Louise Dickerson

Entropy

Entropy is a physical quantity that is primarily a measure of the thermodynamic disorder of a physical system. Entropy has the unique property in that its global value must always increase or stay the same; this property is reflected in the second law of **thermodynamics**. The fact that entropy must always increase in natural processes introduces the concept of irreversibility, and defines a unique direction for the flow of **time**.

Entropy is a property of all physical systems, the **behavior** of which is described by the second law of thermodynamics (the study of **heat**). The first law of thermodynamics states that the total **energy** of an isolated system is constant; the second law states that the entropy of an isolated system must stay the same or increase. Note that entropy, unlike energy, is not conserved, but can increase. A

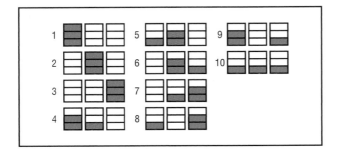

The 10 possible microstates for a system of three atoms sharing three units of energy. *Illustration by Argosy. The Gale Group.*

system's entropy can also decrease, but only if it is part of a larger system whose overall entropy does increase.

Entropy, first articulated in 1850 by the German physicist Rudolf Clausius (1822–1888), does not correspond to any property of **matter** that we can sense, such as **temperature**, and so it is not easy to conceptualize. It can be roughly equated with the amount of energy in a system that is not available for **work** or, alternatively, with the orderliness of a system, but is not precisely given by either of these concepts. A basic intuitive grasp of what entropy means can be given by **statistical mechanics**, as described below.

On a fundamental level, entropy is related to the number of possible physical states of a system, S = k log (Gamma), where S represents the entropy, k is Boltzmann's constant, and (Gamma) is the number of states of the system.

Consider a system of three independent **atoms** that are capable of storing energy in quanta or units of fixed size e$\dot{\epsilon}$. If there happens to be only three units of energy in this system, how many possible microstates—that is, distinct ways of distributing the energy among the atoms—are there? This question is most easily answered, for this example, by listing all the possibilities. There are 10 possible configurations.

If n_0 stands for the number of atoms in the system with 0ϵ energy, n_1 for the number with ϵ, n_2 for the number with 2ϵ, and n_3 for the number with 3ϵ. For example, in the microstates labeled 1, 2, and 3 in the figure that accompanies this article, $(n_0, n_1, n_2, n_3) = (2, 0, 0, 3)$; that is, two atoms have 0ϵ energy, no atoms have 1ϵ or 2ϵ, and one atom has 3ϵ.

Each class or group of microstates that corresponds to a distinct (n_0, n_1, n_2, n_3) distribution. There are three possible distributions, and where P stands for the number of microstates corresponding to each distribution, P can equal 3, 6 or 1. The three values of P can be verified by counting the microstates that themselves reflect the

energy distributions for a system of three atoms sharing three units of energy. Again, the number of possible microstates P corresponding to each distribution.

The distribution $P2$—representing the distribution $(n_0, n_1, n_2, n_3) = (1, 1, 1, 0)$—has the most possible microstates (six). If we assume that this system is constantly, randomly shifting from one microstate to another, that any microstate is equally likely to follow any other, and that we inspect the system at some randomly-chosen instant, we are, therefore, most likely to observe one of the microstates corresponding to distribution $P2$. Specifically, the probability of observing a microstate corresponding to distribution $P2$ is 0.6 (6 chances out of 10). The probability of observing distribution $P1$ is 0.3 (3 chances out of 10) and the probability of observing distribution $P3$ is 0.1 (1 chance out of 10).

The entropy of this or any other system, S, is defined as $S = k\ln(P_{max})$, where P_{max} is the number of microstates corresponding to the most probable distribution of the system ($P_{max} = 6$ in this example), k is the Boltzmann constant (1.3803×10^{-16} ergs per **degree** C), and $\ln(\dot{s})$ is the natural logarithm operation. Further inspection of this equation and the three-atom example given above will clarify some of the basic properties of entropy.

Entropy measures disorder

(1) Microstates 1, 2, and 3 of the three-atom system described above—those distributions in which the energy available to the system is segregated entirely to one atom—are in some sense clearly the most orderly or organized. Yet these three microstates (distribution 1) are also unlikely; their total probability of occurrence at any moment is only half that of distribution 2. Order is less likely than disorder.

Entropy is a probabilistic property

(2) Any system might transition, at any time, to one of its less probable states, because energy can, in a sense, go anywhere it wants to; it is simply much less likely to go some places than others. In systems containing trillions of atoms or molecules, such as a roomful of air, the probability that the system will transition to a highly organized state analogous to microstate 1 in the Figure—say, that all the energy in the room will concentrate itself in one **molecule** while all the rest cool to absolute zero—is extremely small; one would have to wait many trillions of years for such an even to happen.

Entropy is additive

If two systems with entropy S_1 and S_2, respectively, are brought into contact with each other, their combined entropy equals $S_1 + S_2$. This is assured by the logarithmic relationship between S and P_{max}, as follows: If the number of microstates corresponding to the most likely distribution of the first system is P_{max1} and that of the second system is P_{max2}, then the number of possible microstates of the most likely distribution of the *combined* system is simply $P_{max1}P_{max2}$. It is a fundamental properties of **logarithms** that $\ln(ab) = \ln(a) + \ln(b)$, so if S_{1+2} is the entropy of the combined system, then

$$S_{1PLUS2} = k\ln(P_{max1}P_{max2})$$
$$= k[\ln(P_{max1}) + \ln(P_{max2})]$$
$$= k\ln(P_{max1}) + k\ln(P_{max2})$$
$$= S_1 + S_2$$

Entropy is not conserved

All that is needed to increase the entropy of an isolated system is to increase the number of microstates its particles can occupy; say, by allowing the system to occupy more space. It is beyond the scope of this discussion to prove that the entropy of a closed system cannot ever decrease, but this can be made plausible by considering the first law of thermodynamics, which forbids energy to be created or destroyed. As long as a system has the same number of atoms and the same number of quanta of energy to share between them, it is plausible that the system possesses a minimum number of possible microstates—and a minimum entropy.

It is sometimes claimed that "entropy always increases," and that the second law requires that disorder must always increase when nature is left to its own devices. This is incorrect. Note that in the above example, a system of three *independent* atoms is stipulated; yet atoms rarely behave independently when in proximity to each other at low temperatures. They tend to form bonds, spontaneously organizing themselves into orderly structures (molecules and crystals). Order from disorder is, therefore, just as natural a process as disorder from order. At low temperatures, self-ordering predominates; at high temperatures, entropy effects dominate (i.e., order is broken down). Furthermore, any system that is not isolated can experience decreased entropy (increasing order) at the expense of increasing entropy elsewhere. **Earth**, which shares the **solar system** with the **Sun**, whose entropy is increasing rapidly, is one such non-isolated system. It is therefore an error to claim, as some writers do, that biological evolution—which involves spontaneously increasing order in natural molecular systems—contradicts thermodynamics. Entropy does not forbid molecular self-organization because entropy is only one property of matter; entropy does discourage self-organization, but other properties of matter encourage it, and in some circumstances (especially at relatively low temperatures, as on the surface of the earth) will prevail.

An alternative derivation of the entropy concept, based on the properties of heat engines (devices that turn heat flows partially into mechanical work) is often presented in textbooks. This method produces a definition of entropy that seems to differ from $S = k\ln(P_{max})$, namely, $dS = dQ_{rev}/T$, where dS is an infinitesimal (very small) change in a system's entropy at the fixed temperature T when an infinitesimal quantity of heat, Q_{rev}, flows reversibly into or out of the system. However, it can be shown that these definitions are exactly equivalent; indeed, the entropy concept was originally developed from the analysis of heat engines, and the statistical interpretation given above was not invented until later.

The entropy concept is fundamental to all science that deals with heat, efficiency, the energy of systems, **chemical reactions**, very low temperatures, and related topics. Its physical meaning is, in essence, that the amount of work the universe can perform is always declining as its "orderliness" declines, and must eventually approach **zero**. In other words, things are running down, and there is no way to stop them.

Resources

Books

Dugdale, J.S. *Entropy and Its Physical Meaning.* Cornwall, UK: Taylor & Francis, 1996.

Goldstein, M., and I. Goldstein. *The Refrigerator and the Universe: Understanding the Laws of Energy.* Cambridge, MA: Harvard University Press, 1993.

Lee, Joon Chang. *Thermal Physics: Entropy and Free Energies.* World Scientific Publishing Co., Inc., 2002.

K. Lee Lerner
Larry Gilman

Environmental ethics

Ethics is a branch of philosophy that primarily discusses issues dealing with human **behavior** and character. Ethics attempts to establish a basis for judging right from wrong and good from bad. Environmental ethics employs concepts from the entire field of philosophy, especially aesthetics, metaphysics, epistemology, philosophy of science, and social and political philosophy in an effort to relate moral values to human interactions with the natural world.

Aesthetics deals with perceptions of physical properties such as **color**, sound, **smell**, texture, and **taste**. Since environmental ethics is often involved with issues dealing with the protection of plants and animals, its appeal is often to aesthetic experiences of nature. Environmental ethics is also interconnected with political and social structures concerning the use of natural resources, so the field also touches the areas of social and political philosophy. In the struggle to conserve the environment, environmental ethicists also use the knowledge and theories of science, for example, in issues such as those dealing with **global warming** and **air pollution**.

Key issues

Just as philosophers try to answer questions about reality, environmental ethicists attempt to answer the questions of how human beings should relate to their environment, how to use Earth's resources, and how to treat other **species**, both **plant** and **animal**. Some of the conflicts that arise from environmental policies deal with the rights of individuals versus those of the state, and the rights of private property owners versus those of a community.

Methods of dealing with environmental issues vary among the organizations that are devoted to protecting the environment. An important milestone toward a national environmental movement was an event that first took place on many college campuses across the United States on April 22, 1970. This was called **Earth** Day, and it used social protest and demonstration as a way of raising awareness about environmental issues. Earth Day has since become an annual event. In the United States, such groups as the Audubon Society, the Sierra Club, the Wilderness Society, Greenpeace, the Environmental Defense Fund, and the National Wildlife Federation use education and the political arena to lobby Congress for laws to protect the environment. These groups sometimes also use the legal system as a method to change environmental actions and attitudes.

The call to conserve and protect the environment has resulted in the passage of many laws. Among them are the National Environmental Policy Act of 1969, the Clean Water Act of 1972, the Endangered Species Act of 1973, the National Forest Management Act of 1976, and the National Acid Precipitation Act of 1980. In 1970, the Environmental Protection Agency (EPA) was created to oversee federal environmental policies and laws. This increased governmental activity supports the belief of many environmental activists that more is accomplished by working through the political and social arenas than in theorizing about ethics. However, others maintain that the exploration of ideas in environmental ethics is an important springboard for social and political action.

Environmental issues are not universally supported. The conflicts between those who want to protect the natural environment and its species, and those for which this is a lesser concern, often center around economic issues. For example, environmentalists in the Pacific Northwest want to protect the **habitat** of the rare spotted owl, which inhabits **old-growth forests** on which the timber industry and many people depend for their livelihood. There is much controversy over who had the most "right" to use this forest. The **perception** of those who are economically affected by protection of the old-growth forest is that spotted **owls** have become more "important" than the needs of people. Environmentalists, on the other hand, believe that both are important and have legitimate needs.

Environmental attitudes

The cultural and economic background of a person is likely to influence his views with regard to the environment. While these views can vary significantly, they can generally be categorized into one of three positions: the development ethic, the preservation ethic, or the **conservation** ethic. Each of these attitudes represents a generalized moral code for interaction with the environment.

The development ethic considers the human race to be the master of nature and that the resources of Earth exist solely for human benefit. The positive nature of growth and development is an important theme within this philosophy. This view suggests that hard work and technological improvements can overcome any limitations of natural resources.

The preservation ethic suggests that nature itself has intrinsic value and deserves protection. Some preservationists assert that all life forms have rights equal to those of humans. Others seek to preserve nature for aesthetic or recreational reasons. Still other preservationists value the diversity represented by the natural environment and suggest that humans are unaware of the potential value of many species and their natural ecosystems.

The conservation ethic recognizes the limitations of natural resources on Earth and states that unlimited economic or population growth is not feasible. This philosophy seeks to find a balance between the availability and utilization of natural resources.

The fundamental differences between each of these attitudes are the basis for debate and conflict in environmental policies today. These views dictate the behavior of corporations, governments, and even individuals and the solution to any environmental issues will first require an acknowledgement and some consensus of attitudes.

Environmental ethics and the law

In 1970, the Environmental Protection Agency was established to oversee the various federal environmental laws that had been enacted. One of its major functions is to review the environmental impacts of highway projects, large-scale commercial and residential construction, power plants, and other large undertakings involving the federal government. A major tool of the EPA is its power to issue an "environmental impact statement" that evaluates a proposed project before it is undertaken. The advocates of this planning tool believe that is of great value in protecting the environment, particularly when a project is a potential threat to human health or the natural environment. However, others maintain that the agency and its work frustrate worthwhile projects and economic growth.

Lawyers who deal with environmental issues are key players in the issues raised by environmental ethics. They may bring court action against companies that, for example, leak toxic substances into the **groundwater** or emit harmful smoke from factories. Disasters like the 1989 *Exxon Valdez* oil spill in Alaska help fuel demands for better environmental controls, since cases like this clearly show the damage that can be caused to **fish**, **birds**, and the natural environment. The *Exxon Valdez* oil spill was also an economic loss to Alaskan fishermen, who blame the disaster for degrading their ability to fish and make a living. What is always being weighed legally and ethically is how much environmental damage to the environment and its inhabitants can be judged as reasonable, and how much is not.

Major contributors

Environmental ethicists trace the roots of modern American environmental attitudes to the idea of private ownership. During the European Middle Ages, a strong ideal of individual land ownership emerged, in contrast to control by a ruler or a governmental body. This became the basis of property rights in the American Colonies of Britain, as advocated by Thomas Jefferson. The strongly held belief of the right to hold private property remains a cornerstone of American thinking, and is often at odds with issues of environmental ethics.

Farming the land was not the only activity in the early history of the development of the North American **continent** by European settlers. Before there was significant development of farmland in the interior of the continent, explorers, trappers, and naturalists were looking over the landscape for other reasons. During the eighteenth and nineteenth centuries, the attitudes of naturalists and hunters about killing animals were similar. For the naturalist, it was the only way to examine new species up close. For the hunter, killing animals was a way of making a living, from sale of the meat, fur, or some other product, such as ivory.

It was not until Dr. Edwin James' expedition into the Rocky Mountains in 1819-1820 that it was suggested that some parts of the continent should be conserved for **wildlife**. One of the first to calculate the destruction of wildlife in **North America** was the artist George Catlin (1796-1872), who studied the Upper Missouri Indians. He was the first American to advocate a national park for people and animals alike. By 1872, it became clear that the plains buffalo had been massacred to the point of near **extinction**, and Congress established Yellowstone National Park as the first national park in the country.

Thomas Malthus

Thomas Malthus (1766-1834) had an enormous influence on the development of environmental ethics through his theory about population growth, which raised the primary question of how many human beings could be sustained by the ecosystems of Earth. Malthus was an English economist who published *An Essay on the Principle of Population as It Affects the Future Improvement of Society* in 1798, just as the **Industrial Revolution** was beginning in **Europe**. Malthus believed that if the natural forces of war, famine, and **disease** did not reduce the **rate** of growth of the human population, it would increase to the point where it could not be sustained by the natural resources that are available. The problem, as Malthus saw it, was that population increased geometrically, while resources could only grow arithmetically. In a later essay in 1803 he proposed late marriage and sexual abstinence as ways of restraining population growth. Malthus' ideas influenced other activists of the nineteenth century, including Robert Owen (1791-1858), who advocated **birth** control for the poor. However, there was a great deal of opposition to the ideas of Malthus from such social reformers as William Godwin, who took a more optimistic view of the benefits gained through "progress," and its possibilities for improving the lives of people.

While predicting a population explosion, Malthus did not foresee the technological changes that have increased the capacity of modern societies to sustain increasingly larger populations of people (at least for the time being). However, modern ecologists, social scientists, environmental ethicists, and politicians still must deal with the question of how large a population this **planet** can sustain without destroying its ecosystems, and subsequently much of the human population as well.

Theodore Roosevelt

President Theodore Roosevelt (1858-1919) was at the forefront of the conservation movement that developed in America from 1890 to 1920. By the end of the nineteenth century, the United States was an industrialized society. Many of the country's natural resources were already threatened, as its wildlife had been earlier by commercial hunting. Diminishing stocks of forest, **rangeland**, **water**, and mineral resources were all of great concern to Roosevelt during his presidency. In his government, Gifford Pinchot (1865-1946), the head of the Forest Service, and James R. Garfield, Secretary of the Interior, argued for a comprehensive policy to plan the use and development of the natural resources of the United States.

While there was political opposition in Congress and among industrial developers to such ideas, there was support among some segments of the public. John Muir (1838-1914) founded the Sierra Club in this atmosphere of support for the conservation of natural resources. Some American businesses that depended on renewable resources were also supportive. The goal of this emerging "conservation movement" was to sustain natural resources without causing undue economic hardship.

New federal agencies were formed, including the National Park Service and the Bureau of Reclamation. State **conservation laws** were also passed in support of the conservation movement. The historical periods of both World Wars and the economic depression of the 1930s slowed the conservation movement, but did not destroy its foundation.

Aldo Leopold

American conservationist Aldo Leopold (1887-1948) is credited as the founding father of wildlife **ecology**. In his tenure with the U.S. Forest Service, Leopold pioneered the concept of game management. He recognized the linkage between the available space and food supply and the number of animals that may be supported within an area. Leopold observed in his influential book *A Sand County Almanac* that the greatest impediment to achieving what he called a "land ethic" was the predominant view of the land as a commodity, managed solely in terms of its economic value. He argued that instead, humans need to view themselves as part of a community which also includes the land. In his effort to guide the change in philosophy, Leopold wrote, "A thing is right when it tends to preserve the integrity, stability, and beauty of the biotic community. It is wrong when it tends otherwise."

Rachel Carson

A reawakening of environmental issues took place when Rachel Carson (1907-1964) published her book *Silent Spring* in 1962. Carson was a biologist with the Fish and Wildlife Service, and had already published *The Sea Around Us* in 1951. In *Silent Spring*, she alerted the world to the dangers of harmful **pesticides** that were being used in agriculture, particularly DDT. Later American writers who carried the "environmental message" into the public arena include Paul Ehrlich and Barry Commoner.

In the decades following the publication of *Silent Spring*, the earlier conservation movement became transformed into a worldwide environmental movement. Evidence of this transformation includes the growth of organizations such as Greenpeace and the World Wildlife Fund, and an expansion of legislation designed to protect the environment, preserve species, and ensure the health of humans. Carson's writings made the world community aware of the interrelationships of humans and ecosystems. The ideas that **pollution** in one area can affect the environment in another, and that humans cannot live without the goods and services provided by ecosystems, are now commonly understood facts.

Concerns about **acid rain**, **deforestation**, global warming, and nuclear catastrophes like Chernobyl (1986) have helped the cause of those who argue for improved policies and laws to protect the whole environment and all of its inhabitants. In addition, activism has resulted in many non-governmental environmental organizations forming to tackle specific problems, some of a protective nature, others focused on conservation, and others designed to reclaim damaged areas.

High-profile Earth Summits sponsored by the United Nations met in Rio de Janeiro in 1992 and again in Johannesburg in 2002. The goal of these meetings was to develop plans for implementing social, economic, and environmental change for future development. The Kyoto Protocol of 1997 was an attempt to require the industrialized nations to reduce **emission** of "greenhouse gases." However, some environmentalists asssert that little of substance was accomplished in terms of getting specific commitments from governments around the world to undertake serious action. As in the past, economic pressures came into stark conflict with the philosophical premises of environmental ethics. The same questions of how much protection our environment and its resources need, and how much must this would interfere with economic progress, are still relevant today.

See also Ecology; Endangered species; Greenhouse effect

Resources

Books

Botzler, R.G., and S.J. Armstrong. *Environmental Ethics: Divergence and Convergence.* 2nd ed. New York: McGraw Hill Companies, 1997.

Miller, Peter, and Laura Westra. *Just Ecological Integrity: The Ethics of Maintaining Planetary Life.* Lanham, MD: Rowman & Littlefield, 2002.

KEY TERMS

Aesthetics—The branch of philosophy that deals with the nature of beauty.

Epistemology—The study of the nature of knowledge, its limits and validity.

Metaphysics—The study in philosophy of the nature of reality and being.

Philosophy of science—The study of the fundamental laws of nature as they are revealed through scientific investigation.

Political philosophy—The beliefs underlying a political structure within a society and how it should be governed.

Social philosophy—The beliefs underlying the social structure within a society and how people should behave toward one another.

VanDeVeer, Donald, and Christine Pierce. *The Environmental Ethics and Policy Book: Philosophy, Ecology, Economics.* Belmont, CA: Thomson/Wadsworth, 2003.

Other

University of Cambridge. "Environmental Ethics Resources on WWW" [cited January 28, 2003]. <http://www.ethics.ubc.ca/resources/environmental/>.

Vita Richman

Environmental impact statement

In the United States, the National Environmental Policy Act (NEPA) requires federal agencies to file an Environmental Impact Statement (EIS) for any major project or legislative proposal that may have significant environmental effects. According to the U.S. Environmental Protection Agency (EPA), "NEPA requires federal agencies to integrate environmental values into their decision making processes by considering the environmental impacts of their proposed actions and reasonable alternatives to those actions. To meet this requirement, federal agencies prepare a detailed statement known as an Environmental Impact Statement (EIS)." An EIS requires a federal agency to consider the environmental consequences of a proposed project, including air, **water**, and **soil pollution**; **ecosystem** disruption; overuse of shared water, fishery, agricultural, mineral, and forest resources; disruption of endangered species'

habitat; and threats to human health. While an EIS prompts an organization to disclose the environmental effects of their project before beginning construction or development, and to suggest ways to avoid adverse impacts, the document does not, in itself, prevent federal agencies from polluting, or otherwise endangering ecological and human health. An EIS is intended to help federal regulators and environmental compliance agencies determine whether the proposed development will violate environmental laws. United States environmental policy does not require corporations or individuals to file EIS documents, but it does expect private sector developments to comply with federal, state and local environmental and public health regulations.

Environmental impact assessment

In order to prepare an EIS, federal agencies must fully assess the environmental setting and potential effects of a project. Corporations and individuals must likewise determine whether their businesses and residences meet the environmental standards required by their national, state and local governments. Environmental impact assessment is a process that can be used to identify and estimate the potential environmental consequences of proposed developments and policies. Environmental impact assessment is a highly interdisciplinary process, involving inputs from many fields in the sciences and social sciences. Environmental impact assessments commonly examine ecological, physical/chemical, sociological, economic, and other environmental effects. Many nations require the equivalent of an EIS for projects proposed by their federal agencies, and almost all countries have environmental laws. Environmental assessment is a tool used by developers and civil planners throughout the international community. Governmental agencies, like the EPA in the United States, typically provide guidelines for risk assessment procedures and regulations, but the financial and legal responsibility to conduct a thorough environmental assessment rests with the agency, corporation, or **individual** proposing the project.

Environmental assessments may be conducted to review the potential effects of: (1) individual projects, such as the construction of a particular power plant, incinerator, airport, or housing development; (2) integrated development schemes, or proposals to develop numerous projects in some area. Examples include an industrial park, or an integrated venture to harvest, manage, and process a natural resource, such as a pulp mill with its associated wood-supply and forest-management plans; or (3) government policies that carry a risk of having substantial environmental effects. Examples include decisions to give national priority to the generation of **electricity** using nuclear reactors, or to clear large areas of natural forest to develop new lands for agricultural use. Government agencies and businesses proposing large projects usually hire an independent environmental consulting firm to conduct environmental risk assessments, and to preview the possible environmental, legal, health, safety, and civil **engineering** ramifications of their development plan. Environmental consultants also perform scientific environmental monitoring at existing sites, and, if necessary, recommend methods of cleaning up environmental **contamination** and damage caused by non-compliant projects.

Any project, scheme, or policy can potentially cause an extraordinary variety of environmental and ecological changes. Consequently, it is rarely practical to consider all a proposal's potential effects in an environmental impact assessment. Usually certain indicators, called "valued ecosystem components" (VECs), are selected for study on the basis of their importance to society. VECs are often identified through consultations with government regulators, scientists, non-governmental organizations, and the public. Commonly examined VECs include: (1) resources that are economically important, such as agricultural or forest productivity, and populations of hunted **fish** or game; (2) rare or **endangered species** and natural ecosystems; (3) particular **species**, communities, or landscapes that are of cultural or aesthetic importance; (4) resources whose quality directly affects human health, including drinking water, urban air, and agricultural soil; and (5) simple indicators of a complex of ecological values. The spotted owl (*strix occidentalis*), for example, is an indicator of the integrity of certain types of old-growth **conifer forests** in western **North America**. Proposed activities, like commercial **forestry**, that threaten a population of these **birds** also imperil the larger, old-growth forest ecosystem. Determination of VECs usually requires a site-specific scientific characterization and survey of the proposed development.

Conducting an environmental impact assessment

Risk assessment is the first phase of an environmental impact assessment. Environmental risk characterization is accomplished by predicting a proposed development's potential stressors at the site over **time**, and comparing these effects with the known boundaries of VECs. This is a preliminary reconnaissance **exercise** that may require environmental scientists to judge the severity and importance of potential interactions between stressors and VECs. It is highly desirable to undertake field, laboratory, or simulation research to determine the risks of interactions identified during preliminary screening. However, even the most thorough environmental assess-

ments usually present an incomplete characterization of a site's VECs and potential impacts. Inadequate time and funds often further constrain impact assessments.

Once potentially important risks to VECs are identified and studied, it is possible to consider various planning options. EIS documentation requires both environmental assessment and an explanation of planned responses to environmental impacts. During the planning stage of the impact assessment, environmental specialists provide objective information and professional opinions to project managers. There are three broad types of planning responses to environmental impacts:

(1) The predicted damages can be avoided by halting the development, or by modifying its structure. Avoidance is often disfavored by proponents of a development because irreconcilable conflicts with environmental quality can result in substantial costs, including canceled projects. Regulators and politicians also tend to have an aversion to this option, as lost socioeconomic opportunities and unfinished projects are generally unpopular.

(2) Mitigations can be designed and implemented to prevent or significantly reduce damages to VECs. For examples, acidification of a **lake** by industrial dumping could be mitigated by adding carbonate to the lake, a development that threatens the habitat of an endangered species could suggest moving the population to another site, or a proposed coal-fired generating station could offset its **carbon dioxide** emissions by planting trees. Mitigations are popular ways of resolving conflicts between project-related stressors and VECs. However, mitigations are risky because ecological and environmental knowledge is inherently incomplete, and many mitigation schemes fail to properly protect the VEC. Moving a population of animals from its home habit, for instance, usually results in mortality of the animals.

(3) A third planning response is to allow the projected environmental damages to occur, and to accept the degradation as an unfortunate cost of achieving the project's perceived socioeconomic benefits. This choice is common, because not all environmental damages can be avoided or mitigated, and many damaging activities can yield large, short-term profits.

It is nearly impossible to carry out large industrial or economic developments without causing some environmental damage. However, a properly conducted impact assessment can help decision makers understand the dimensions and importance of that damage, and to decide whether they are acceptable.

Environmental effects monitoring

Once an environmental assessment has been completed, and federal projects have filed an approved EIS,

KEY TERMS

. . . .

Environmental impact assessment—A process by which the potential environmental consequences of proposed activities or policies are presented and considered.

Mitigation—This is a strategy intended to reduce the intensity of environmental damages associated with a stressor. Mitigations are the most common mechanisms by which conflicts between stressors and valued ecosystem components are resolved during environmental impact assessments.

the project or policy may proceed. When the development is completed, the associated environmental stresses begin to affect species, ecosystems, and socioeconomic systems. The actual effects of a project must then be monitored to ensure compliance with environmental regulations, to observe and mitigate any unforeseen consequences, and to compare real damages to predicted ones. Monitoring might include measuring levels of chemical emissions into the air, water, or soil, observing the response of plants and animals at the site, or surveying the health of employees and nearby residents of the site. Socioeconomic impacts, like increased traffic congestion, or water availability for surrounding agricultural and municipal interests, are also including in appropriate monitoring schemes. Predictions set out in the pre-development environmental impact assessment usually determine the structure of monitoring programs. Monitoring programs must be flexible and adaptive, because "surprises," or unpredicted environmental changes, often occur. In this sense, environmental impact assessment is an ongoing process that extends over the lifetime of a development.

See also Air pollution; Ecological economics; Ecological integrity; Ecological monitoring; Ecological productivity; Environmental ethics.

Resources

Books

Canter, L.W. *Environmental Impact Assessment.* 2nd ed. Amsterdam: Kluwer, 1993.

Freedman, B. *Environmental Ecology.* 2nd ed. San Diego: Academic Press, 1994.

Wathern, P., ed. *Environmental Impact Assessment: Theory and Practice.* London: Unwin Hyman, 1998.

Other

United States Environmental Protection Agency. "Environmental Management." November 7, 2002 [cited November 8, 2002]. <http://www.epa.gov/ebtpages/environmentalman agement.html>.

United States Environmental Protection Agency. "National Center for Environmental Assessment." September 19, 2001 [cited November 8, 2002]. <http://cfpub1.epa.gov/ncea/cfm/nceahome.cfm>.

Laurie Duncan

Enzymatic engineering

The introduction of the techniques of modern **molecular biology**, beginning in the 1970s, have made possible the **genetic engineering** of **proteins**. New proteins can be created and existing proteins altered. Enzymes are proteins that function in **chemical reactions** by making the reaction occur more easily than if the **enzyme** was absent. Enzymatic **engineering** includes a wide variety of techniques related to the manipulation of molecules to create new enzymes, or proteins, that have useful characteristics. These techniques range from the modification of existing enzymes to the creation of totally new enzymes. Typically, this involves changing the protein on a genetic level by mutating a **gene**. Other ways that modifications to enzymes are made are by chemical reactions and synthesis.

Genetic engineering can improve the speed at which the enzyme-catalyzed chemical reaction proceeds, and can improve the amount of product that is generated by the chemical reaction. Furthermore, an enzyme that has a desirable activity, but which operates very slowly in its parent **organism**, can be taken out of that organism and installed in another organism where it performs faster. An example of this is the speeding up of the enzyme penicillin-G-amidase by slicing the enzyme into the genetic material of *Escherichia coli*.

Enzymes are relatively large, complex molecules that catalyze biochemical reactions. These include such diverse reactions as the digestion of sugars, replication of **deoxyribonucleic acid (DNA)**, or combating diseases. For this reason, the creation of improved enzymes has been an important field of study for the last 50 years. A variety of steps are involved in the creation of modified enzymes. These entail a determination of the molecular structure, function, and modifications to improve the effectiveness.

The key to creating new and improved enzymes and proteins is determining the relationship between the structure and function of the **molecule**. The structure of an enzyme can be described on four levels. The primary structure is related to the sequence of amino acids that are bonded together to form the molecule. The amino acids that make up the enzyme are determined by the

DNA sequence of the gene that codes for the enzyme. When enough amino acids are present, the molecule folds into secondary structures known as structural motifs. These structural motifs are further organized into a three dimensional, or tertiary structure. In many enzymes, multiple tertiary structures are combined to give the overall quaternary structure of the molecule. Scientists have found that the specific structural organization of amino acids in an enzyme are directly related to the function of that enzyme.

Structural information about enzymes is obtained by a variety of methods. The **amino acid** sequence is typically found using DNA sequencing. In this method, the pattern of nucleotides is determined for the gene that encodes the enzyme. This information is then compared to the genetic code to obtain the amino acid sequence. Other structural information can also be found by using **spectroscopy**, **chromatography**, magnetic **resonance**, and **x-ray crystallography**.

To investigate the function of an enzyme, it is helpful to be able to create large quantities of an enzyme. This is done by cloning a desired gene. Polymerase chain reaction (**PCR**), which is a method of making a large number of copies of a gene, is particularly helpful at this point. The cloned genes are then incorporated into a biological vector such as a bacterial plasmid or phage. Colonies of **bacteria** that have the specific gene are grown. Typically, these bacteria will express the foreign gene and produce the desired enzyme. This enzyme is then isolated for further study.

To create new enzymes, the DNA sequence of the gene can be modified. Modifications include such things as deleting, inserting, or substituting different nucleotides. The resulting gene will code for a slightly modified enzyme, which can then be studied for improved stability and functionality.

One of the major goals in enzymatic engineering is to be able to determine enzyme structure and functionality based on the amino acid sequence. With improvements in computer technology and our understanding of basic molecular interactions, this may someday be a reality.

Another promising area of enzymatic engineering is the engineering of proteins. By designing or modifying enzymes, the proteins that the enzymes help create can also be changed, or the amount of the protein that is made can be increased. Also, since enzymes are themselves proteins, the molecular tinkering with protein structure and genetic sequence can directly change the structure and function of enzymes.

See also Bioremediation; Biotechnology.

KEY TERMS

. . . .

Restriction enzymes—Enzymes that recognize certain sequences of DNA and cleave the DNA at those sites. The enzymes are used to generate fragments of DNA that can be subsequently joined together to create new stretches of DNA.

Site directed mutagenesis—The alteration of a specific site on a DNA molecule. This can be done using specialized viruses.

Resources

Periodicals

Laskin, A.I., G. Li, and Y.T. Yu. "Enzyme Engineering." *Annals of the New York Academy of Science* 864 (December 1998): 1–665.

Brian Hoyle

Enzyme

Enzymes are biological catalysts, agents which increase the **rate** of **chemical reactions** without being used up in the reaction. They are **proteins** which possess special binding sites for a certain biochemicals. Weak binding interactions with the biochemical allow enzymes to accelerate specific reaction rates millions of times. Enzyme kinetics is the study of enzyme reactions and mechanisms. Enzyme inhibitor studies have allowed researchers to develop therapies for treatment of **disease**.

Historical background of enzyme research

Louis Pasteur was among the first to study enzyme action. He incorrectly hypothesized that the conversion of sugar into **alcohol** by **yeast** was catalyzed by "ferments" that could not be separated from living cells. In 1897 the German biochemist Eduard Buchner (1860-1917) isolated the enzymes which catalyze alcoholic **fermentation** from living yeast cells, represented in the equation:

$$glucose \rightarrow ethanol + carbon\ dioxide$$

The early twentieth century saw dramatic advancement in enzyme studies. Emil Fischer (1852-1919) recognized the importance of substrate shape for binding by enzymes. Leonor Michaelis (1875-1949) and Maud Menten introduced a mathematical approach for quantifying enzyme-catalyzed reactions. James Sumner (1887-1955) and John Northrop (1891-1987) were among the first to produce highly ordered enzyme crystals and firmly establish

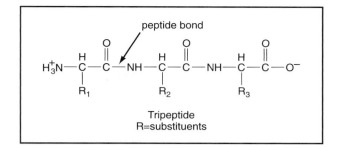

Figure 1. *Illustration by Hans & Cassidy. Courtesy of Gale Group.*

Figure 2. *Illustration by Hans & Cassidy. Courtesy of Gale Group.*

the protein nature of these biological catalysts. In 1937 Hans Krebs (1900-1981) postulated how a series of enzymatic reactions were coordinated in the **citric acid** cycle for the production of ATP from glucose metabolites. Today, enzymology is a central part of biochemical study.

Enzyme structure

Protein structural studies are a very active area of biochemical research, and it is known that the biological function (activity) of an enzyme is related to its structure. There are 20 common amino acids which make up the building blocks of all known enzymes. They have similar structures, differing mainly in their substituents. The organic substituent of an **amino acid** is called the R group. The structure of the amino acid alanine is depicted in Figure 1.

The biologically common amino acids are designated as L-amino acids because the amino group is on the left of the a alpha-carbon when the **molecule** is drawn as indicated. This is called Fischer projection. The amino acids are covalently joined via peptide bonds. A series of three amino acids is shown in Figure 2, illustrating the structure of a tripeptide. Enzymes have many peptide linkages and range in molecular **mass** from 12,000 to greater than one million.

Enzymes may also consist of more than a single polypeptide chain. Each polypeptide chain is called a subunit, and may have a separate catalytic function. Some enzymes have non-protein groups which are necessary for enzymatic activity. **Metal** ions and organic molecules called coenzymes are components of many enzymes. Coenzymes which are tightly or covalently attached to enzymes are termed prosthetic groups. Prosthetic groups contain critical chemical groups which allow the overall catalytic event to occur.

Enzymes bind their reactants (substrates) at special folds and clefts in their structures called "active sites." Because active sites have chemical groups precisely located and orientated for binding the substrate, they generally display a high **degree** of substrate specificity. The active site of an enzyme consists of two key regions. The

catalytic site, which interacts with the substrate during the reaction, and the binding site, the chemical groups of the enzyme which bind the substrate to allow chemical interaction at the catalytic site.

Although the details of enzyme active sites differ between different enzymes, there are common motifs. The active site represents only a small fraction of the total protein **volume**. The reason enzymes are so large is that many interactions are necessary for reaction. The crevice of the active site creates a microenvironment which is critical for catalysis. Environmental factors include polarity, hydrophobicity, and precise arrangement of **atoms** and chemical groups. In 1890 Emil Fischer compared the enzyme-substrate relationship to a "lock-and-key." Fischer postulated that the active site and the enzyme have complimentary three dimensional shapes. This model was extended by Daniel Koshland Jr. in 1958 by his "induced fit" model, which reasoned that actives sites are complimentary to substrate shape only after the substrate is bound (Fig. 3).

Enzyme function

Consider the simple, uncatalyzed chemical reaction reactant A → product B.

As the **concentration** of reactant A is increased, the rate of product B formation increases. Rate of reaction is defined as the number of molecules of B formed per unit **time**. In the presence of a catalyst, the reaction rate is accelerated. For reactant to be converted to product, a thermodynamic **energy** barrier must be overcome. This energy barrier is known as the activation energy (E_a). A catalyst speeds up a chemical process by lowering the activation energy which the reactant must reach before being converted to product. It does this by allowing a different chemical mechanism or pathway which has a lower activation energy (Fig. 4).

Enzymes have high catalytic power, high substrate specificity, and are generally most active in aqueous solvents at mild **temperature** and physiological **pH**. There

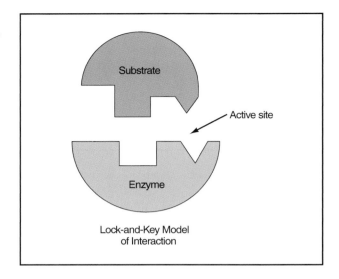

Figure 3. *Illustration by Hans & Cassidy. Courtesy of Gale Group.*

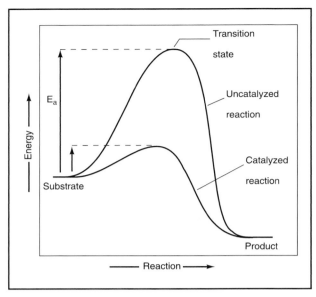

Figure 4. *Illustration by Hans & Cassidy. Courtesy of Gale Group.*

are thousands of known enzymes, but nearly all can be categorized according to their biological activities into six major classes: oxidoreductases, transferases, hydrolases, lyases, isomerases, and ligases. Most enzymes catalyze the transfer of electrons, atoms, or groups of atoms, and are assigned names according to the type of reaction.

Thousands of enzymes have been discovered and purified to date. The structure, chemical function, and mechanism of hundreds of enzymes has given biochemists a solid understanding of how they work. In the 1930s, J. B. S. Haldane (1892-1964) described the principle that interactions between the enzyme and its substrate can be used to distort the shape of the substrate and induce a chemical reaction. The energy released and used to lower activation energies from the enzyme-substrate interaction is called the binding energy. The chemical form of the substrate at the top of the energy barrier is called a transition state. The maximum number of weak enzyme-substrate interactions is attained when the substrate reaches its transition state. Enzymes stabilize the transition state, allowing the reaction to proceed in the forward direction. The rate-determining process of a simple enzyme catalyzed reaction is the breakdown of the **activated complex** between the transition state and the enzyme. In 1889 the Swedish chemist Svante Arrhenius (1859-1927) showed the dependence of the rate of reaction on the magnitude of the energy barrier (activation energy). Reactions which consist of several chemical steps will have multiple activated complexes. The over-all rate of the reaction will be determined by the slowest activated complex **decomposition**. This is called the rate-limiting step.

In addition to enhancing the reaction rate, enzymes have the ability to discriminate among competing sub-

strates. Enzyme specificity arises mainly from three sources: (1) Exclusion of other molecules from the active site because of the presence of incorrect functional groups, or the absence of necessary binding groups. (2) Many weak interactions between the substrate and the enzyme. It is known that the requirement for many interactions is one reason enzymes are very large compared to the substrate. The noncovalent interactions which bind molecules to enzymes are similar to the intramolecular forces which control enzyme conformation. Induced conformational changes shift important chemical groups on the enzyme into close proximity for catalysis. (3) Stereospecificity arises from the fact that enzymes are built from only L-amino acids. They have active sites which are asymmetric and will bind only certain substrates.

Regulation of enzyme activity

Enzyme activity is controlled by many factors, including environment, enzyme inhibitors, and regulatory binding sites on the enzyme.

Environment

Enzymes generally have an optimum pH range in which they are most active. The pH of the environment will effect the ionization state of catalytic groups at the active site and the ionization of the substrate. Electrostatic interactions are therefore controlled by pH. pH may also control the conformation of the enzyme through ionizable amino acids which are located distant from the active site, but are nevertheless critical for the three-dimensional shape of the macromolecule.

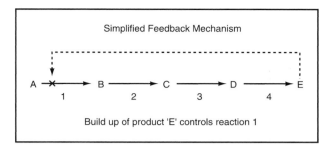

Simplified Feedback Mechanism

Build up of product 'E' controls reaction 1

Figure 5. *Illustration by Hans & Cassidy. Courtesy of Gale Group.*

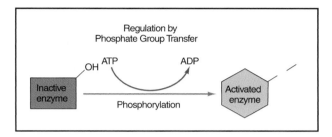

Regulation by
Phosphate Group Transfer

Figure 6. *Illustration by Hans & Cassidy. Courtesy of Gale Group.*

Enzyme inhibitors

Inhibitors diminish the activity of an enzyme by altering the way substrates bind. Inhibitor structure may be similar to the substrate, but they react very slowly or not at all. Chemically, inhibitors may be large organic molecules, small molecules or ions. They are important because they can be useful for chemotherapeutic treatment of disease, and for providing experimental insights into the mechanism of enzyme action. Inhibitors work in either a reversible or an irreversible manner. Irreversible inhibition is characterized by extremely tight or covalent interaction with the enzyme, resulting in very slow **dissociation** and inactivation. Reversible inhibition is displayed by rapid dissociation of the inhibitor from the enzyme. There are three main classes of reversible inhibition, competitive inhibition, noncompetitive inhibition, and mixed inhibition. They are distinguishable by how they bind to enzymes and response to increasing substrate concentration. Enzyme inhibition is studied by examining reaction rates and how rates change in response to changes in experimental parameters.

Regulatory binding sites

Regulatory enzymes are characterized by increased or decreased activity in response to chemical signals. Metabolic pathways are regulated by controlling the activity of one or more enzymatic steps along that path. Regulatory control allows cells to meet changing de-

KEY TERMS

Activation energy—The amount of energy required to convert the substrate to the transition state.

Active site—The fold or cleft of an enzyme which binds substrate and catalytically transforms it to product.

Hydrophobic—Nonpolar, insoluble in water.

Induced fit—Enzyme conformational change caused by substrate binding and resulting in catalytic activity.

Lock-and-key—Geometrically complementary shapes of the enzyme (lock) and the substrate (key) resulting in specificity.

Peptide bond—Chemical covalent bond formed between two amino acids in a polypeptide.

Polypeptide—A long chain of amino acids linked by peptide bonds.

Rate of reaction—The quantity of substrate molecules converted to product per unit time.

Transition state—The activated form of the reactant occupying the highest point on the reaction coordinate.

Weak interactions—Noncovalent interactions between the substrate and the enzyme, allowing enzyme conformational change and catalysis.

mands for energy and biomolecules. The scheme shown in Figure 5 illustrates the concept of a regulatory feedback mechanism.

Enzymatic activity is regulated in four general ways:

1. Allosteric control. Allosteric enzymes have distinct binding sites for effector molecules which control their rates of reaction.

2. Control proteins participate in cellular regulatory processes. For example, calmodulin is a Ca^{2+} binding protein which binds with high affinity and modulates the activity of many Ca^{2+}-regulated enzymes. In this way, Ca^{2+} acts as "second messenger" to allow crosstalk and feedback control in metabolic pathways.

3. Enzyme activity controlled by reversible covalent modification is carried out with a variety of chemical groups. A common example is the transfer of phosphate groups, catalyzed by enzymes known as kinases (Figure 6). Glucose **metabolism** is modulated by phosphate transfer.

4. Proteolytic activation. The above example is reversible in nature. Irreversible hydrolytic cleavage of peptide fragments from inactive forms of enzymes known as

zymogens converts them into fully active forms. Proteolytic enzymes are controlled by this mechanism.

See also Catalyst and catalysis; Krebs cycle.

Resources

Books

Branden, C., and J. Tooze. *Introduction to Protein Structure.* New York: Garland, 1991.

Lehninger, A.L., D.L. Nelson, and M.M. Cox. *Principles of Biochemistry.* 2nd ed. New York: Worth, 1993.

Periodicals

Kraut, J. "How Do Enzymes Work?" *Science* 242 (1988): 533-540.

Leonard D. Holmes

Epidemic

An epidemic is an outbreak of a **disease** among members of a specific population that exceeds the extent of occurrence of the disease normally found in that population. Epidemics affect those members of the population who do not have an acquired or inherent immunity to the disease. Although most epidemics are caused by infectious organisms, the term can be applied to an outbreak of any chronic disease, such as lung **cancer** or **heart** disease.

During an epidemic, organisms can spread in several ways. But in each case, there must be a continual source of disease organisms, that is, a reservoir of **infection**. The reservoir may be human (an infected food server), **animal** (bacteria-carrying rat), or an inanimate object (contaminated **water**).

For human diseases, the human body is the principal living reservoir of infectious organisms. Among the diseases spread by human carriers are **AIDS**, **typhoid fever**, **hepatitis**, gonorrhea, and streptococcal infections. While people who have signs and symptoms of a disease are obvious reservoirs of infections, some people may carry and spread the disease without showing any signs or symptoms. Still others may harbor the disease during the symptomless stage called the *incubation* period, before symptoms appear, or during the *convalescent period*, during which time they are recovering. This fuels the epidemic, since there is no apparent reason for people to take precautions to prevent transmission to others.

Animal reservoirs can also spread diseases. Those diseases that are spread from wild and domestic animals to humans are called *zoonoses*. **Yellow fever**, spread by the *Aedes* mosquito; **Lyme disease**, spread by ticks; **rabies**, spread by **bats**, **skunks**, foxes, **cats**, and dogs; and

bubonic plague, spread by **rats**, are examples of **zoonoses**.

Inanimate reservoirs, such as drinking water contaminated by the feces of humans and other animals, are a common environment for organisms causing gastrointestinal diseases.

Once infected, a person becomes a reservoir and can spread the disease in a variety of ways, called *contact transmission*. Direct contact transmission occurs when an infected host has close physical contact with a susceptible person. This person-to-person transmission can occur during touching, kissing, and sexual intercourse. Viral **respiratory diseases** such as the common cold and **influenza** are transmitted in this way, as are **smallpox**, hepatitis A, **sexually transmitted diseases** (genital herpes, gonorrhea, syphilis) and, in some cases, AIDS.

Indirect contact transmission occurs when the disease-causing **organism** is transmitted from the reservoir to a susceptible host by means of a inanimate carrier called a *fomite*, which can be a towel, drinking cup, or eating utensils. Hepatitis and AIDS epidemics can occur when contaminated syringes serve as fomites among intravenous drug users.

Droplet transmission of an epidemic occurs when microbes are spread in tiny bits of mucous called droplet nuclei that travel less than three feet from the mouth and nose during coughing, sneezing, laughing, or talking. Influenza, **whooping cough**, and **pneumonia** are spread this way.

Transmission of epidemics can occur through food, water, air, and **blood**, among other objects. Waterborne transmission occurs through contaminated water, a common means by which epidemics of **cholera**, waterborne shigellosis, and leptospirosis occurs. Foodborne poisoning in the form of staphylococcal **contamination** may occur when food is improperly cooked, left unrefrigerated, or prepared by an infected food handler.

Airborne transmission of viral diseases such as measles and **tuberculosis** occurs when the infectious organisms travel more than 3 ft (0.9 m) from the human reservoir to a susceptible host in a fine spray from the mouth and nose. Fungal infections such as histoplasmosis, coccidioidomycosis, and blastomycosis can be spread by airborne transmission as their spores are transported on dust particles.

Vectors, usually **insects**, are animals that carry **pathogens** from one host to another. These vectors may spread an epidemic by mechanical or biological transmission. When flies transfer organisms causing typhoid fever from the feces of infected people to food, the disease is spread through mechanical transmission. Biological transmission occurs when an arthropod bites a host and ingests infected blood. Once inside the arthropod vector, the disease-causing organisms may reproduce in the gut, increasing the number of **parasites** that can be transmitted to the next host. In some cases, when the host is bitten, the parasites are passed out of the vector and into a wound when the vector passes feces or vomits. The protozoan disease **malaria** is spread by the *Anopheles* mosquito vector.

After an epidemic is introduced into a population by one or more persons, the infection spreads rapidly if there are enough susceptible hosts. If so, the incidence of the disease increases over time, until it reaches a maximum, at which time it begins to subside. This subsidence of the epidemic is due mostly to the lack of susceptible individuals; most individuals already have the disease or have had the disease and gained an immunity to it.

After an epidemic subsides, there are too few susceptible individuals to support a new epidemic if the infection is reintroduced. This overall immunity of a host population to a potential epidemic disease is called *herd immunity*.

Herd immunity tends to disappear over time due to three factors: (1) deterioration of **individual** immunity; (2) death of immune individuals; (3) influx of susceptible individuals by **birth** or emigration into the area of the epidemic.

The rise in the number of infected individuals over time, followed by a fall to low levels, can be graphically depicted as an "epidemic curve," which usually represents a time period of days, weeks, or months. Some may even last years. For example, waterborne gastrointestinal infections may peak during the later summer months, reflecting the role of recreational swimming in areas where parasitic organisms exist.

Marc Kusinitz

Epidemiology

Epidemiology is the study of the occurrence, **frequency**, and distribution of diseases in a given population. As part of this study, epidemiologists—scientists who investigate epidemics (widespread occurrence of a **disease** that occurs during a certain time)—attempt to determine how the disease is transmitted, and what are the host(s) and environmental factor(s) that start, maintain, and/or spread the **epidemic**.

The primary focus of epidemiology are groups of persons, rather than individuals. The primary effort of epidemiologists is in determining the **etiology** (cause) of the disease and identifying measures to stop or slow its spread. This information, in turn, can be used to create strategies by which the efforts of health care workers and facilities in communities can be most efficiently allocated for this purpose.

In tracking a disease outbreak, epidemiologists may use any or all of three types of investigation: descriptive epidemiology, analytical epidemiology, and experimental epidemiology.

Descriptive epidemiology is the collection of all data describing the occurrence of the disease, and usually includes information about individuals infected, and the place and period during which it occurred. Such a study is usually retrospective, i.e., it is a study of an outbreak after it has occurred.

Analytical epidemiology attempts to determine the cause of an outbreak. Using the case control method, the epidemiologist can look for factors that might have preceded the disease. Often, this entails comparing a group of people who have the disease with a group that is similar in age, sex, socioeconomic status, and other variables, but does not have the disease. In this way, other possible factors, e.g., genetic or environmental, might be identified as factors related to the outbreak.

Using the cohort method of analytical epidemiology, the investigator studies two populations, one who has had contact with the disease-causing agent and another that has not. For example, the comparison of a group that received **blood** transfusions with a group that has not might disclose an association between blood transfusions and the incidence of a bloodborne disease, such as **hepatitis** B.

Experimental epidemiology tests a hypothesis about a disease or disease treatment in a group of people. This strategy might be used to test whether or not a particular antibiotic is effective against a particular disease-causing **organism**. One group of infected individuals is divided randomly so that some receive the antibiotic and others receive a placebo—a "false" drug that is not known to have any medical effect. In this case, the antibiotic is the variable, i.e., the experimental factor being tested to see if it makes a difference between the two otherwise similar groups. If people in the group receiving the antibiotic recover more rapidly than those in the other group, it may logically be concluded that the variable—antibiotic treatment—made the difference. Thus, the antibiotic is effective.

Although the sudden appearance of dreaded diseases has plagued humanity for millennia, it was not until the nineteenth century that the field of epidemiology can be said to have been born. In 1854, the British physician

John Snow (1813-1858) demonstrated the power of epidemiologic principles during an outbreak of **cholera** in London. Snow discovered that most of the victims of cholera he questioned obtained their drinking **water** from a well on London's Broad Street. Moreover, most of the people afflicted with the disease drank from the polluted section of the Thames River, which ran through London. Snow arranged to have the Broad Street pump closed, preventing people from drinking water from that well. Subsequently, the cholera epidemic subsided.

Since the days of Snow, epidemiology has grown into a very sophisticated science, which relies on **statistics** as well as interviews with disease victims. Today, epidemiologists study not only infectious diseases, such as cholera and **malaria**, but also noninfectious diseases, such as lung **cancer** and certain **heart** disorders.

In the process of studying the cause of an infectious disease, epidemiologists often view it in terms of the agent of **infection** (e.g., particular bacterium or **virus**), the environment in which the disease occurs (e.g., crowded slums), and the host (e.g., hospital patient). Thus, beta-hemolytic streptococci **bacteria** are the agent for acute **rheumatic fever**; but because not all persons infected with the organism develop the disease, the health of the host helps to determine how serious the disease will be for a particular person, or even, whether it will occur.

Among the important environmental factors that affect an epidemic of infectious diseases are poverty, overcrowding, lack of sanitation, and such uncontrollable factors as the season and climate.

Another way epidemiologists may view etiology of disease is as a "web of causation." This web represents all known predisposing factors and their relations with each other and with the disease. For example, a web of causation for myocardial infarction (heart attack) can include diet, hereditary factors, cigarette smoking, lack of **exercise**, susceptibility to myocardial infarction, and **hypertension**. Each factor influences and is influenced by a variety of other factors.

By identifying specific factors and how they are ultimately related to the disease, it is sometimes possible to determine which preventive actions can be taken to reduce the occurrence of the disease. In the case of myocardial infarction, for example, these preventive actions might include a change in diet, treatment for hypertension, eliminating smoking, and beginning a regular schedule of exercise.

Epidemiologic investigations are largely mathematical descriptions of persons in groups, rather than individuals. The basic quantitative measurement in epidemiology is a count of the number of persons in the group being studied who have a particular disease; for example, epidemiologists may find 10 members of a village in the African village of Zaire suffer from infection with **Ebola virus** infection; or that 80 unrelated people living in an inner city area have **tuberculosis**.

Any description of a group suffering from a particular disease must be put into the context of the larger population. This shows what proportion of the population has the disease. The significance of 10 people out of a population of 1,000 suffering tuberculosis is vastly different, for example, than if those 10 people were part of a population of one million.

Thus one of the most important tasks of the epidemiologist is to determine the prevalence rate—the number of persons out of a particular population who have the disease:

Prevalence **rate** = number of persons with a disease / total number in group. Prevalence rate is like a snapshot of a population at a certain point in time, showing how many people in that population suffer from a particular disease. For example, the number of people on March 15 suffering infection from the parasite cryptosporidium in a town with a polluted water supply might be 37 out of a population of 80,000. Therefore, the prevalence rate on March 15 is 37/80,000.

A prevalence rate can represent any time period, e.g., day or hour; and it can refer to an event that happens to different persons at different times, such as complications that occur after drug treatment (on day five for some people or on day two for others).

The incidence rate is the rate at which a disease develops in a group over a period of time. Rather than being a snapshot, the incidence rate describes a continuing process that occurs over a particular period of time.

Incidence rate = total number *per unit* developing a disease *over time* / total number of persons. For example, the incidence rate of prostate cancer among men in a particular country might be 2% per year; or the number of children getting measles in a town might be 3% per day. Once a person has developed a lifelong disease, such as **AIDS**, he or she cannot be counted in the denominator of the incidence rate, since these people cannot get the disease again. The denominator refers only to those in the population who have not yet developed the disease.

Period prevalence measures the extent to which one or all diseases affects a group during the course of time, such as a year.

Period prevalence = number of persons with a disease during a period of time / total number in group. In the case of a year, such as 1995, the period prevalence equals the prevalence at the beginning of 1995 plus the annual incidence during 1995.

Epidemiologists also measure attributable risk, which is the difference between two incidence rates of groups being compared, when those groups differ in some attribute that appears to cause that difference. For example, the lung cancer mortality rate among a particular population of non-smoking women 50 to 70 years old might be 20/100,000, while the mortality rate among woman in that age range who smoke might be 150/100,000. The difference between the two rates (150-20 = 130) is the risk that is attributable to smoking, if smoking is the only important difference between the groups regarding the development of lung cancer.

Epidemiologists arrange their data in various ways, depending on what aspect of the information they want to emphasize. For example, a simple graph of the annual occurrence of viral **meningitis** might show by the "hills" and "valleys" of the line in which years the number of cases increased or decreased. This might provide evidence of the cause and offer ways to predict when the incidence might rise again.

Bar graphs showing differences in rates among months of the year for viral meningitis might pinpoint a specific time of the year when the rate goes up, for example, in summertime. That, in turn, might suggest that specific summertime activities, such as swimming, might be involved in the spread of the disease.

One of the most powerful tools an epidemiologist can use is case reporting: reporting specific diseases to local, state and national health authorities, who accumulate the data. Such information can provide valuable leads as to where, when, and how a disease outbreak is spread, and help health authorities to determine how to halt the progression of an epidemic-one of the most important goals of epidemiology.

Resources

Books

Cohn, Victor. *News and Numbers. A Guide to Reporting Statistical Claims and Controversies in Health and Related Fields.* Ames: Iowa State University Press, 1989.

Nelson, K.E., C.M. Williams, and N.M.H. Graham. *Infectious Disease Epidemiology: Theory and Practice* Gaithersburg: Aspen Publishers, 2001.

Periodicals

Perera, F.P., and I.B. Weinstein. "Molecular Epidemiology: Recent Advances and Future Directions." *Carcinogenesis* 21 (2000): 517-524.

"Pidemics Of Meningococcal Disease. African Meningitis Belt, 2001." *Weekly Epidemiological Record / World Health Organization* 76, no. 37 (2001): 282-288.

Marc Kusinitz

Epilepsy

Epilepsy, from the Greek word for seizure, is a recurrent demonstration of a **brain** malfunction. The outward signs of epilepsy may range from only a slight smacking of the lips or staring into space to a generalized convulsion. It is a condition that can affect anyone, from the very young to adult ages, of both sexes and any race. Epilepsy was described by Hippocrates (c.460-c.377 B.C.), known as the "father of medicine." The incidence of epilepsy, that is, how many people have it, is not known. Some authorities say that up to one-half of 1% of the population are epileptic, but others believe this estimate to be too low. Many cases of epilepsy, those with very subtle symptoms, are not reported. The most serious form of epilepsy is not considered an inherited **disease**, though parents with epilepsy are more prone to have children with the disease. On the other hand, an epileptic child may have parents that show no sign of the condition, though they will have some abnormal brain waves.

Though the cause of epilepsy remains unknown, the manner in which the condition is demonstrated indicates the area of the brain that is affected. Jacksonian seizures, for example, which are localized twitching of muscles, originate in the frontal lobe of the brain in the motor cortex. A localized numbness or tingling indicates an origin in the parietal lobe on the side of the brain in the sensory cortex.

The recurrent symptoms, then, are the result of localized, excessive discharges of brain cells or neurons. These can be seen on the standard brain test called the **electroencephalogram (EEG)**. For this test electrodes are applied to specific areas of the head to pick up the electrical waves generated by the brain. If the patient experiences an epileptic episode while he is wired to the EEG, the abnormal brain waves can easily be seen and the determination made as to their origin in the brain. Usually, however, if the patient is not experiencing a seizure no abnormalities will be found in the EEG.

Grand mal seizures

Grand mal seizures are those that are characteristic of epilepsy in everyone's mind. Immediately prior to the seizure, the patient may have some indication that a seizure is immanent. This feeling is called an aura. Very soon after he has the aura the patient will lapse into unconsciousness and experience generalized muscle contractions that may distort the body position; these are clonic seizures. The thrashing movements of the limbs that ensue in a short time are caused by opposing sets of muscles alternating in contractions (hence, the other name for grand mal seizures: tonic-clonic seizures). The patient may also lose control of the bladder. When the seizures cease, usually after three to five minutes, the patient may remain unconscious for up to half an hour. Upon awakening, the patient may not remember having had a seizure and may be confused for a time.

Petit mal seizures

In contrast to the drama of the grand mal seizure, the petit mal may seem inconsequential. The patient interrupts whatever he is doing and for up to about 30 seconds may show subtle outward signs such as blinking his eyes, staring into space, or pausing in conversation. After the seizure he resumes his previous activity. Petit mal seizures are associated with heredity and they never occur in people over the age of 20 years. Oddly, though the seizures may occur several times a day, they do so usually when the patient is quiet and not during periods of activity. After **puberty** these seizures may disappear or they may be replaced by the grand mal type of seizure.

Status epilepticus

A serious form of seizure, status epilepticus indicates a state in which grand mal seizures occur in rapid succession with no period of recovery between them. This can be a life-threatening event because the patient has difficulty breathing and may experience a dangerous rise in **blood** pressure. This form of seizure is very rare, but can brought on if someone abruptly stops taking the medication prescribed for his epilepsy. It may also occur in **alcohol** withdrawal.

Treatment

A number of drugs are available for the treatment of epilepsy. The oldest is phenobarbital, which has the unfortunate side effect of being addictive. Other drugs currently on the market are less addictive, but all have the possibility of causing untoward side effects such as drowsiness or nausea or dizziness.

The epileptic patient needs to be protected from injuring himself during an attack. Usually for the patient having a petit mal seizure, little needs to be done. Occasionally these individuals may lose their balance and need to be helped to the ground to avoid hitting their head, but otherwise need little attention. The **individual** in a grand mal seizure should not be restrained, but may need to have some help to avoid his striking his limbs or head on the floor or nearby obstruction. If possible, roll the patient onto his side. This will maintain an open airway for him to breathe by allowing his tongue to fall to one side.

Epilepsy is a recurrent, lifelong condition that must be reckoned with. Medication can control seizures in a

substantial percentage of patients, perhaps up to 85% of those with grand mal manifestations. Some patients will experience seizures even with maximum dosages of medication. These individuals need to wear an identification bracelet to let others know of their condition. Epilepsy is not a reflection of insanity or mental retardation in any way. In fact, many who experience petit mal seizures are of above-average intelligence.

See also Anticonvulsants.

Resources

Books

Ziegleman, David. *The Pocket Pediatrician.* New York: Doubleday and Company, 1995.

Periodicals

Glanz, J. "Do Chaos-Control Techniques Offer Hope for Epilepsy?" *Science* 265 (August 26, 1994): 1174.

Larry Blaser

Episomes

An episome is a portion of genetic material that can exist independent of the main body of genetic material (called the **chromosome**) at some times, while at other times is able to integrate into the chromosome.

Examples of episomes include insertion sequences and transposons. Viruses are another example of an episome. Viruses that integrate their genetic material into the host chromosome enable the viral **nucleic acid** to be produced along with the host genetic material in a nondestructive manner. As an autonomous unit (i.e., existing outside of the chromosome) however, the viral episome destroys the host **cell** as it commandeers the host's replication apparatuses to make new copies of itself.

Another example of an episome is called the F factor. The F factor determines whether genetic material in the chromosome of one **organism** is transferred into another organism. The F factor can exist in three states that are designated as FPLUS, Hfr, and F prime.

FPLUS refers to the F factor that exists independently of the chromosome. Hfr stands for high frequency of recombination, and refers to a factor that has integrated into the host chromosome. The F prime factor exists outside the chromosome, but has a portion of chromosomal DNA attached to it.

An episome is distinguished from other pieces of DNA that are independent of the chromosome (i.e.,plasmids) by their large size.

Plasmids are different from episomes, as plasmid DNA cannot link up with chromosomal DNA. The plasmid carries all the information necessary for its independent replication. While not necessary for bacterial survival, plasmids can be advantageous to a bacterium. For example, plasmids can carry genes that confer resistance to **antibiotics** or toxic metals, genes that allow the bacterium to degrade compounds that it otherwise could not use as food, and even genes that allow the bacterium to infect an **animal** or **plant** cell. Such traits can be passed on to another bacterium.

Transposons and insertion sequences are episomes. These are also known as mobile genetic elements. They are capable of existing outside of the chromosome. They are also designed to integrate into the chromosome following their movement from one cell to another. Like plasmids, transposons can carry other genetic material with them, and so pass on resistance to the cells they enter. Class 1 transposons, for example, contain drug resistance genes. Insertion sequences do not carry extra genetic material. They code for only the functions involved in their insertion into chromosomal DNA.

Transposons and insertion sequences are useful tools to generate changes in the DNA sequence of host cells. These genetic changes that result from the integration and the exit of the mobile elements from DNA, are generically referred to as mutations. Analysis of the mobile element can determine what host DNA is present, and the analysis of the mutated host cell can determine whether the extra or missing DNA is important for the functioning of the cell.

See also Bacteria; Electrophoresis; Gene; Gene splicing; Recombinant DNA.

Resources

Books

Craig, N.L., R. Craigie, M. Gilbert, and A.M. Lambowitz. *Mobile DNA II*. Washington, DC: American Society for Microbiology Press, 2002.

Snyder, L., and W. Champness. *Molecular Genetics of Bacteria,* 2nd ed. Washington, DC: American Society for Microbiology Press, 2002.

Epsom salts *see* **Magnesium sulfate**

Epstein-Barr virus

Epstein-Barr **virus** (EBV) is part of the family of human herpes viruses. Infectious mononucleosis (IM) is the most common **disease** manifestation of this virus which, once established in the host, can never be completely eradicated. Very little can be done to treat EBV; most methods can only alleviate resultant symptoms. **Sleep** and rest—complete bedrest in severe cases—is still the best medicine for sufferers of this virus.

In addition to infectious mononucleosis, EBV has also been identified in association with—although not necessarily believed to cause—as many as 50 different illnesses and diseases, including chronic fatigue **syndrome**, rheumatoid **arthritis**, arthralgia (joint **pain** without **inflammation**), and myalgia (muscle pain). While studying aplastic **anemia** (failure of bone marrow to produce sufficient red **blood** cells), researchers identified EBV in bone marrow cells of some patients, suggesting the virus may be one causative agent in the disease. Also, several types of **cancer** can be linked to presence of EBV, particularly in those with suppressed immune systems, for example, suffering from **AIDS** or having recently undergone kidney or liver transplantation. The diseases include hairy **cell leukemia**, Hodgkin's and non-Hodgkin lymphoma, Burkitt's lymphoma (cancer of the **lymphatic system endemic** to populations in **Africa**), and nasopharyngeal carcinoma (cancers of the nose, throat, and thymus gland, particularly prevalent in East **Asia**). Very recently, EBV has been associated with malignant smooth-muscle **tissue** tumors in immunocompromised children. Such tumors were found in several children with AIDS and some who had received liver transplants. Conversely, it appears that immunosuppressed adults show no elevated rates of these tumors.

Discovery, disease, and research

EBV was first discovered in 1964 by three researchers—Epstein, Achong, and Barr—while studying a form of cancer prevalent in Africa called *Burkitt's lymphoma*. Later, its role in IM was identified. A surge of interest in the virus has now determined that up to 95% of all adults have been infected with EBV at some stage of their lives. In seriously immunocompromised individuals and those with inherited **immune system** deficiencies, the virus can become chronic, resulting in "chronic Epstein-Barr virus" which is extremely serious and can be fatal.

Age and the health of the immune system play important roles in EBV-related illnesses. Infectious mononucleosis is the most common illness resulting from primary EBV **infection**. In young children, it often takes on mild flu-like symptoms that improve with time, or the symptoms are so mild they go unnoticed. However, ill or older people, IM can become a debilitating infection, and complications can affect almost every **organ** of the body, including possible rupture of the spleen. In fewer than 1% of IM cases, neurologic complications develop, and the patient can develop **encephalitis**, **meningitis**, Guillian-Barre syndrome, or other serious conditions. Very rarely, IM causes complications like aplastic anemia, thrombocytopenia (reduced numbers of platelets, the clotting factor in blood, resulting in bleeding), and granulocytopenia (severe reduction in white cells in the blood allowing the potential for "superinfection"). More commonly, however, IM causes persistent fevers; swollen and tender lymph **glands** in the neck, groin, and armpits; sore throat and even severe **tonsillitis**; and sometimes a rash. Because the sore throat is so prominent, and because it may in fact become secondarily infected with strep, some patients are treated with ampicillin. Almost all of these individuals will break out in a rash which is often mistaken for evidence of a penicillin **allergy**. IM is almost always accompanied by fatigue and general malaise which can be quite severe initially, but bouts of which become less severe and prolonged over several weeks or months. Inflammation of the spleen and liver are also common in IM. The spleen in particular may grow very large, and may rupture spontaneously in about 0.5% of patients.

EBV is also implicated in another immune system-related illness called chronic fatigue syndrome (CFS). In several communities throughout the world, several epidemic-type outbreaks of what became known as CFS, including one in Lake Tahoe in 1985, baffled the medical community. There have been many conflicting theories in the ongoing search for the role of EBV in CFS. Initially, EBV was thought to be the causative factor; however, following intense studies performed during an outbreak of CFS in New York in 1985, it was determined that not everyone suffering from CFS developed antibodies to EBV. In cases where several children in one family all showed the same symptoms, only thee out of four showed infection by EBV. Further studies revealed that about one-third of the individuals affected by CFS had experienced primary infection with EBV years earlier and were once again fighting the virus. This syndrome became known as "active chronic EBV infection." In the other two-thirds of CFS sufferers, EBV levels were no different to those shown by the general population. It therefore seemed inaccurate to conclude EBV was the causative agent in CFS. Because some CFS patients experience reactivation of latent EBV, while in others EBV becomes active for the first time, it is possible that the effect of CFS on the immune system allows activation and reactivation of EBV. One theory about the interrelatedness of EBV and CFS is that, rather than causing CFS, EBV may, instead, trigger it.

Origin and development

EBV is restricted to a very few cells in the host. Initially, the infection begins with its occupation and replication in the thin layer of tissue lining the mouth, throat, and cervix, which allow viral replication. The virus then invades the B cells, which do not facilitate the virus's replication but do permit its occupation. Infected B cells may lie dormant for long periods or start rapidly producing new cells. Once activated in this way, the B cells often produce antibodies against the virus residing in them. EBV is controlled and contained by killer cells and suppressor cells known as CD4+ T lymphocytes in the immune system. Later, certain cytotoxic (destructive) CD8+ T lymphocytes with specific action against EBV also come into play. These cells normally defend the host against the spread of EBV for the life of the host.

A healthy body usually provides effective immunity to EBV in the form of several different antibodies, but when this natural defense mechanism is weakened by factors that suppress its normal functioning—factors such as AIDS, organ transplantation, bone marrow failure, chemotherapy and other drugs used to treat malignancies, or even extended periods of lack of sleep and overexertion—EBV escape from their homes in the B cells, disseminate to other bodily tissue, and manifest in disease.

Diagnostic blood tests cannot detect the virus itself. Infection is determined by testing for the antibodies pro-

KEY TERMS

Antibody—A molecule created by the immune system in response to the presence of an antigen (a foreign substance or particle). It marks foreign microorganisms in the body for destruction by other immune cells.

Arthralgia—Joint pain without inflammation.

B cell—Immune system white blood cell that produces antibodies.

Immunosuppressed/immunocompromised—Reduced ability of the immune system to fight disease and infection.

Myalgia—Muscular aches and pain.

Nasopharyngeal—Of the nose and throat.

T cells—Immune-system white blood cells that enable antibody production, suppress antibody production, or kill other cells.

duced by the immune system to fight the virus. The level of a particular antibody—the *heterophile* antibody—in the blood stream is a good indicator of the intensity and stage of EBV infection. Even though EBV proliferates in the mouth and throat, cultures taken from that area to determine infection are time-consuming, cumbersome, and usually not accurate.

Treatment is primarily supportive; encouraging rest and recuperation. When spleen enlargement is present, activities may need to be restricted to avoid the complication of splenic rupture. **Antibiotics** should be used only to treat documented bacterial infections which complicate the course of IM. Anti-viral medications do not seem to be helpful. Steroids should only be given in the rare instance of such complications as hemolytic anemia, severely swollen tonsils which are causing airway obstruction, or platelet destruction.

Disease transmission and prevention

Spread of the virus from one person to another requires close contact. Because of viral proliferation and replication in the lining of the mouth, infectious mononucleosis is often dubbed "the kissing disease." Also, because it inhabits cervical cells, researchers now suspect EBV may be sexually transmitted. Rarely is EBV transmitted via blood transfusion.

EBV is one of the latent viruses, which means it may be present in the body, lying dormant often for many years and manifesting no symptoms of disease.

The percentage of *shedding* (transmission) of the virus from the mouth is highest in people with active IM or who have become immunocompromised for other reasons. A person with active IM can prevent transmission of the disease by avoiding direct contact—such as kissing—with uninfected people. However, shedding has been found to occur in 15% of adults who test positive for antibodies but who show no other signs of infection, thus allowing the virus to be transmitted. Research efforts are directed at finding a suitable **vaccine**.

The prevalence of antibodies against EBV in the general population is high in developing countries and lower socioeconomic groups where individuals become exposed to the virus at a very young age. In developed countries, such as the United States, only 50% of the population shows traces of antibody by the age of five years, with an additional 12% in college-aged adolescents, half of whom will actually develop IM. This situation indicates that children and adolescents between the age of 10 to 20 years are highly susceptible to IM in developed countries, making it a significant health problem among young students and those in the military.

Resources

Books

Bell, David S. *The Doctor's Guide to Chronic Fatigue Syndrome: Understanding, Treating, and Living with CFIDS.* Reading, MA: Addison-Wesley, 1993.

Flint, S.J., et al. *Principles of Virology: Molecular Biology, Pathogenesis, and Control* Washington: American Society for Microbiology, 1999.

Richman, D.D., and R.J. Whitley. *Clinical Virology.* 2nd ed. Washington: American Society for Microbiology, 2002.

Periodicals

Liebowitz, David. "Epstein-Barr Virus—An Old Dog With New Tricks." *The New England Journal of Medicine* 332, no. 1 (5 January 1995): 55-57.

Marie L. Thompson

Equation, chemical

Chemical equations reveal the chemical species involved in a particular reaction, the charges and weight relationships among them, and how much **heat** of reaction results. Equations tell us the beginning compounds, called reactants, and the ending compounds, called products, and which direction the reaction is going. Equations are widely used in chemical **engineering**, they serve as the basis for chemical synthesis, reactor design, process control, and cost estimate. This allows chemical process engineers to prepare ahead of time for on-line production.

It is fairly difficult to take a few chemical compounds and derive chemical equations from them, because many variables need to be determined before the correct equations can be specified. However, to look at a chemical equation and know what it really means is not as difficult. To achieve this, there are certain conventions and symbols which we always have to keep in mind. Now let's start with a general chemical equation, $aA + bB \xrightarrow{\Delta} cC + dD\uparrow$, to explain those conventions and symbols, and few examples will then be given and discussed.

Conventions and symbols

In general, the reactants (A and B) are always placed on the left-hand side of the equation, and the products (C and D) are shown on the right. The symbol "$\rightarrow$" indicates the direction in which the reaction proceeds. If the reaction is reversible, the symbol "$\rightleftarrows$" should be used to show that the reaction can proceed in both the forward and reverse directions. Δ means that heat is added during the reaction, and not equal implies that D escapes while produced. Sometimes, Δ is replaced by "light" (to initiate reactions) or "flame" (for **combustion** reactions.) Instead of showing the symbol Δ, at the same place we may just indicate the operating **temperature** or what enzymes and catalysts are need to speed the reaction.

Each chemical species involved in an equation is represented by chemical formula associated with stoichiometric coefficients. For instance, a, b, c, and d are the stoichiometric coefficients for A, B, C, and D, respectively. Stoichiometric coefficients can be **integers**, vulgar fractions, (e.g. 3/4) or decimal fractions (e.g. 0.5). They define the **mole ratio** (not **mass** ratio) that permits us to calculate the moles of one substance as related to the moles of another substance in the chemical equation. In the present case, we know that a moles of A react with b moles of B to form c moles of C and d moles of D.

The chemical equation needs to be balanced, that is, the same number of **atoms** of each "element" (not compounds) must be shown on the right-hand side as on the left-hand side. If the equation is based on an **oxidation-reduction reaction** which involves **electron** transfer, the charges should also be balanced. In other words, the oxidizing agent gains the same number of electrons as are lost by the reducing agent. For this reason, we must know the oxidation numbers for elements and ions in chemical compounds. An element can also have more than one oxidation number, for instance, Fe^{2+} and Fe^{3+} for **iron**.

Under certain conditions, the information on phase, temperature, and **pressure** should be included in the equation. For instance, H_2O can exist as solid, liquid, and vapor (gas) that can be represented by $H_2O(s)$,

$H_2O(l)$, and $H_2O(g)$, respectively. If we have an infinitely dilute **solution**, say HCl, it can be denoted as HCl(aq). For **solubility** problems, A underlined ($\underline{A}$) means that A is a solid or precipitated phase. In many cases, the heat of reaction, ΔH, is also given; a **positive number** implies an **endothermic** reaction (where heat is absorbed), and a **negative** number implies an exothermic reaction (where heat is given off). Unless otherwise specified, the heat of reaction is normally obtained for all the chemical species involved in the reaction at the standard state of 77°F (25°C) and 1 atmosphere total pressure, the so-called "standard heat of reaction" and denoted by $\Delta H°$.

A few examples

NaOH + HCl $\rightarrow$ NaCl + H_2O means that (1) 1 mole of NaOH reacts with 1 mole of HCl to form 1 mole of NaCl and 1 mole of H_2O, (2) 40 g (that is, **molecular weight**) of NaOH react with 36.5 g of HCl to form 58.5 g of NaCl and 18 g of H_2O, or (3) 6.02×10^{23} molecules (1 mole) of NaOH react with 6.02×10^{23} molecules of HCl to form 6.02×10^{23} of NaCl and 6.02×10^{23} of H_2O. Notice that on both sides of the equation, we have one **chlorine** atom, two **hydrogen** atoms, one **oxygen** atom, and one **sodium** atom. This equation, then, is properly balanced.

For the reaction between permanganate (MnO_4) ion and ferrous (Fe^{2+}) ion in an acid solution, an expression is given like this, $KMnO_4 + FeSO_4 + H_2SO_4 \rightarrow Fe_2(SO_4)_3 + K_2SO_4 + MnSO_4 + H_2O$. Obviously this equation is not balanced. To remedy this, first, the equation can be rewritten as $MnO_4 + Fe^{2+} + H^+ \rightarrow Fe^{3+} + Mn^{2+} + H_2O$ if one recognizes that potassium (K^+) and sulfate (SO_4) ions do not enter into the reaction. Secondly, the oxidation number of manganese (Mn) is changed from +7 in MnO_4^- to +2 in Mn^{2+}, that is, Mn gains 5 electrons during the reaction. Similarly, one electron is lost from Fe^{2+} to Fe^{3+}. To make the number of electrons lost from one substance equal to the number of electrons gained by another in oxidation-reduction reactions, we need to use the least common multiple of 1 and 5, which is 5. So we have $MnO_4 + 5Fe^{2+} + H^+ \rightarrow 5Fe^{3+} + Mn^{2+} + H_2O$. Thirdly, the equation has to be balanced for the number of atoms of individual elements, too. Thus a final expression is obtained as $MnO_4^- + 5Fe^{2+} + 8H^+ \rightarrow 5Fe^{3+} + Mn^{2+} + H_2O$. Lastly we can add the potassium and sulfate back into the complete equation, $2KMnO_4 + 10FeSO_4 + 8H_2SO_4 \rightarrow 5Fe_2(SO_4)_3 + K_2SO_4 + 2MnSO_4 + 8H_2O$. At this stage, we do not have to worry about charge balances, but atom conservation needs to be checked again and corrected.

Derivation of equations for oxidation-reduction reactions sometimes can be simplified by using a series of

KEY TERMS

Compound—A pure substance that consists of two or more elements, in specific proportions, joined by chemical bonds. The properties of the compound may differ greatly from those of the elements it is made from.

Heat of formation, $\Delta H_f°$—The heat involved for the formation of 1 mole of a compound that is formed from the elements which make up the compound.

Oxidation-reduction reaction—A chemical reaction in which one or more atoms are oxidized, while one or more other atoms are reduced.

Standard potential—The electrochemical potential (volts) with respect to the standard state which is a measure for the driving force of a reaction.

Stoichiometry—Deals with combining weights of elements and compounds.

half reactions, whose expressions can be found in special tables of many textbooks. For example, with half-reactions of $Zn \rightarrow Zn^{2+} + 2e$ and $Fe^{2+} + 2e \rightarrow Fe$, by summing them up we can obtain the equation, $Zn + Fe^{2+} \rightarrow Zn^{2+} + Fe$. Since 2e is found both on the right and left sides of the equations and does not react with anything else, it can be dropped from the combined equation.

For those reactions in which we are interested for their heats of reaction, knowing how to derive the final equations from relevant formation reactions is very useful. For example, when the formation reactions at temperature of 77°F (25°C; 298K) are given as (1) $C(s) + O_2(g) \rightarrow CO_2(g)$, $\Delta H_f° = -94,501$ cal, (2) $C(s) + 0.5 O_2(g) \rightarrow CO(g)$, $\Delta H_f° = -26,416$ cal, and (3) $H_2(g) + 0.5 O_2(g) \rightarrow H_2O$, $\Delta H_f° = -57,798$ cal, we can obtain the equation, $CO_2(g) + H_2(g) \rightarrow CO(g) + H_2O(g)$, $\Delta H_f° = 9,837$ cal, by reversing (1), that is, $CO_2(g) \rightarrow C(s) + O_2(g)$, $\Delta H_f° = 94,501$ cal, and adding it to (2) and (3). Therefore, the result shows an endothermic reaction with the heat of reaction of 9,837 cal at 77°F (25°C; 298K).

Applications

Because the stoichiometric coefficients are unique for a given reaction, chemical equations can provide us with more information than we might expect. They tell us whether or not the conversion of specific products from given reactants is feasible. They also tell us that explosive or inflammable products could be formed if the reaction was performed under certain conditions.

Pang-Jen Kung

Equilibrium, chemical

Chemical equilibrium is the final condition of a chemical reaction after the reacting substances have completey reacted. Depending on the reaction, they may reach this condition quickly or slowly, but eventually they will come to a condition in which there are definite, unchanging amounts of all the relevant substances.

Chemical equilibrium is one of the most important features of chemical processes, involving everything from the dissolving of sugar in a cup of coffee to the vital reactions that carry essential **oxygen** to cells and poisonous **carbon dioxide** away from them.

How chemical equilibrium works

A chemical reaction between substance A and substance B. The reaction can be written as:

$$A + B \rightarrow C + D$$

If the reaction is not one-way, as indicated by the arrow, and if C and D can also react with one another to form A and B—the reverse of the process above, this process is indicated with the arrow going the other way:

$$A + B \leftarrow C + D$$

This is far from an unusual situation. In fact, almost all of the **chemical reactions** are reversible in they proceed in either direction, as long as A, B, C and D molecules are available. After all, a chemical reaction is a process in which **atoms** or electrons are rearranged, reshuffled, or transferred from one **molecule** to another. Chemists call a two-way reaction a reversible reaction, and they write the equation with a two-way arrow:

$$A + B \longleftrightarrow C + D$$

In adverse reaction A and B molecules collide to produce C and D molecules, just as we wrote in the first equation above. As they react, the number of remaining As and Bs will be going down. But as more and more Cs and Ds are produced, they will begin to collide and reproduce some of the As and Bs. The reaction is now going in both directions-to the right and to the left.

As more and more Cs and Ds build up, they will be bumping into each other more and more often, thereby reproducing As and Bs at a faster and faster clip. Pretty soon, the C+D (leftward) reaction will be going as fast as the A+B (rightward) reaction; As and Bs will be reappearing just as fast as they are disappearing. The result is that there will be no further change in the numbers of As and Bs. The system is at *equilibrium*.

A and B molecules continue to collide and produce C and D molecules. Likewise, we see that Cs and Ds are still madly bumping into each other to produce As and Bs. In other words, both reactions are still vigorous. But they're going on at exactly the same **rate**. If somebody is giving you money and taking it away at the same rate, the amount you have isn't changing. The total amounts of A, B, C and D aren't changing at all. It looks as if the reaction is over, and for all our practical purposes it is.

When the rate of leftward reaction is the same as the rate of rightward reaction, the system stays at equilibrium—no net movement in either direction.

Upsetting our equilibrium

"Can we influence an equilibrium reaction in any way? Are we stuck with whatever mixture is left after a reaction has come to its final equilibrium condition? That is, are we stuck with so much of each substance, A, B, C and D? Or can we tamper with those "final" equilibrium amounts? Can we force the reactions to give us more of what we want (for example, C and D) and less of what we don't want (for example, A and B)?"

Suppose we have a reaction that has already come to equilibrium:

$$A + B \longleftrightarrow C + D$$

There are certain equilibrium amounts of all four chemicals. But if the number of B molecules is increased, the As and Bs will continue to react because there are now more Bs for the As to collide into extra Cs and Ds are produced, and fewer As. The same thing would happen if we added some extra As instead of Bs: we'd still get an extra yield of Cs and Ds.

Why would we want to do this? Well, maybe we're a manufacturer of C or D, and the further we can push the reaction to produce more of them, the more we can sell. We have succeeded in pushing the equilibrium point in the direction we want—toward more C and D—by adding Bs alone; no extra As were needed. If we'd wanted to, we could have gotten the same result by adding some extra As instead of Bs, whichever one is cheaper.

We could also have accomplished the desired result—more C and D—not by speeding up the collision rate of As and Bs, but by slowing down the collision rate of Cs and Ds. That is, instead of adding extra As or Bs, we could keep removing some of the Cs or D's from the pot as they are formed. Then the A and B will keep colliding and producing more, unhindered by the building up of the backward process. We'd get a 100% reaction between A and B.

In both cases, it appears that when we changed the amount of any one of the chemicals (by adding or removing some) we imposed an unnatural stress upon the equilibrium system. The system then tried to relieve that stress by using up some of the added substance or by producing more of the removed substance. In 1884, French chemist Henry Le Châtelier (1850-1936) realized that these shifts in the equilibrium amounts of chemicals were part of a bigger principle, which now goes by his name. In general terms, Le Châtelier's principle can be stated this way: When any kind of stress is imposed upon a chemical system that has already reached its equilibrium condition, the system will automatically adjust the amounts of the various substances in such a way as to relieve that stress as much as possible.

Le Châtelier's principle is being applied every day in manufacturing industries to maximize efficiency of chemical reactions.

Resources

Books

Oxtoby, David W., et al. *The Principles of Modern Chemistry.* 5th ed. Pacific Grove, CA: Brooks/Cole, 2002.
Parker, Sybil, ed. *McGraw-Hill Encyclopedia of Chemistry.* 2nd ed. New York: McGraw Hill, 1999.
Umland, Jean B. *General Chemistry.* St. Paul: West, 1993.

Robert L. Wolke

Equinox

The Latin meaning of equinox is "equal night," the times of the year when day and night are equal in length. In **astronomy**, the equinox is the point at which the **Sun** appears to cross the equator as a result of **Earth's rotation** around the Sun. The vernal equinox, which occurs as the Sun moves from south to north across the equator, takes place around March 21 and marks the beginning of spring. On about September 23, the Sun moves from north to south across the equator, marking the autumnal equinox and beginning of autumn. It is important to realize that the Sun does not actually move; its apparent path is a reflection of Earth's orbital **rotation** about the sun, and the tilt of Earth's axis.

When you stand on **Earth** and gaze upward on a clear night, you see the sky as part of a giant **sphere** that surrounds Earth. Although we know it is Earth that rotates, it appears as though this star-bearing dome turns about us. Early astronomers thought the stars were attached to this giant sphere. Today, astronomers still find it useful to imagine a celestial sphere that surrounds Earth. The extension of Earth's north and south poles extend to the north and south celestial poles, and Earth's equator can be projected outward to the celestial equator. **Time** and horizontal angles are measured eastward from the vernal equinox—the point where the Sun crosses the celestial equator in March—and vertical angles are measured north or south of the celestial equator.

Earth's axis of rotation is tilted 23.5° to the **plane** of its **orbit**. This tilt causes the **seasons**, and (from our frame of reference) it makes the Sun and the planets, which have orbital planes **parallel** to Earth's appear to move north and south during the course of the year along a path called the ecliptic. Because the ecliptic is tipped relative to Earth's equator, it is also tipped relative to the celestial equator. The two points where the ecliptic intercepts the celestial sphere are the equinoxes. When the Sun reaches either equinox, it rises in a direction that is due east everywhere on Earth. After the vernal equinox, the sun continues to move northward along the ecliptic and rise a little farther north of east each day until it reaches the summer solstice—a point 23.5° above the equator—around June 22. The summer **solstice** marks the beginning of summer, after which, the Sun begins to move southward. It crosses the celestial equator (the autumnal equinox) and continues to move southward and rise a little farther south of east each day until it is 23.5° south of the celestial equator at the winter solstice around December 22. It then begins its northward movement back to the vernal equinox.

Erbium *see* **Lanthanides**

Erosion

Erosion is a group of processes that, acting together, slowly decompose, disintegrate, remove, and transport materials on the surface of **Earth**. Among geologists, there is no general agreement on what processes to include as a part of erosion. Some limit usage to only those processes that remove and transport materials. Other geologists also include **weathering** (**decomposition** and disintegration). This broad definition is used here.

Erosion is a sedimentary process. That is, it operates at the surface of the earth to produce, among other things, surficial materials. The material produced by erosion is called sediment (sedimentary particles, or grains). A thin layer of sediment, known as *regolith*, covers most of the earth's surface. Erosion of the underlying solid rock surface, known as **bedrock**, produces this layer of regolith. Erosion constantly wears down the earth's surface, exposing the **rocks** below.

Sources of erosional energy

The **energy** for erosion comes from five sources: gravity, the **Sun**, **Earth's rotation**, **chemical reactions**, and organic activity. These forces work together to break down and carry away the surficial materials of Earth.

Gravity exerts a **force** on all **matter**, Earth materials included. Gravity, acting alone, moves sediment down slopes. Gravity also causes **water** and **ice** to flow down slopes, transporting earth materials with them. Gravity and solar energy work together to create waves and some types of **ocean** currents. Earth's **rotation**, together with gravity, also creates tidal currents. All types of water movement in the ocean (waves and currents) erode and transport sediment.

Solar energy, along with gravity, produces **weather** in the form of rain, snow, **wind**, **temperature** changes, etc. These weather elements act on surface materials, working to decompose and disintegrate them. In addition, chemical reactions act to decompose earth materials. They break down and dissolve any compounds that are not stable at surface temperature and **pressure**. Organic activity, by both plants and animals, can also displace or disintegrate sediment. An example of this is the growth of a **tree** root moving or fracturing a rock.

Erosional settings

Erosion can occur almost anywhere on land or in the ocean. However, erosion does occur more rapidly in certain settings. Erosion happens faster in areas with steep slopes, such as **mountains**, and especially in areas where steep slopes combine with flowing water (mountain streams) or flowing ice (alpine **glaciers**). Erosion is also rapid where there is an absence of vegetation, which would stabilize the surficial materials. Some settings where the absence of vegetation helps accelerate erosion are deserts, mountain tops, or agricultural lands. Whatever the setting, water is a more effective agent of erosion than wind or ice—even in the **desert**.

Weathering

The first step in erosion is weathering. Weathering "attacks" solid rock, produces loose sediment, and makes the sediment available for transport. Weathering consists of a number of related processes that are of two basic types: mechanical or chemical.

Mechanical weathering

Mechanical weathering processes serve to physically break large rocks or sedimentary particles into smaller ones. That is, mechanical weathering disintegrates earth materials. An example of mechanical weathering is when water, which has seeped down into cracks in a rock, freezes. The pressure created by freezing and expanding of the water breaks the rock apart. By breaking up rock and producing sediment, mechanical weathering increases the surface area of the rock and so speeds up its **rate** of chemical weathering.

Severe soil erosion caused by overgrazing and clearance of vegetation in Eastern Province, Kenya, Africa. *Mark Boulton (1990). Photo Researchers, Inc. Reproduced by permission.*

Chemical weathering

Chemical weathering processes attack the **minerals** in rocks. Chemical weathering either decomposes minerals to produce other, more stable compounds or simply dissolves them away. Chemical weathering usually requires the presence of water. You may have noticed during a visit to a cemetery that the inscription on old marble headstones is rather blurred. This is because rainwater, which is a weak acid, is slowly dissolving away the marble. This dissolution of rock by rainwater is an example of chemical weathering.

Chemical weathering results in the formation of dilute chemical solutions (minerals dissolved in water) as well as weathered rock fragments. Chemical weathering, along with biological activity, contributes to the formation of soils. Besides surface area, both the temperature and the amount of moisture present in an environment control

the rate of chemical weathering. Chemical weathering usually happens fastest in warm, moist places like a tropical jungle, and slowest in dry, cold places like the Arctic.

Agents and mechanisms of transport

Transport of sediment occurs by one or more of four agents: gravity, wind, flowing water, or flowing ice. A simple principle controls transport; movement of sediment occurs only as long as the force exerted on the sediment grain by the agent exceeds the force that holds the grain in place (**friction** due to gravity). For example, the wind can only move a grain of **sand** if the force generated by the wind exceeds the frictional force on the bottom of the grain. If the wind's force is only slightly greater than the frictional force, the grain will scoot along on the ground. If the wind's force is much greater than the frictional force, the grain will roll or perhaps bounce along on the ground. The force produced by flowing air (wind), water, or ice is a product of its **velocity**.

When gravity alone moves rocks or sediment, this is a special type of transport known as **mass wasting** (or **mass** movement). This name refers to the fact that most mass wasting involves a large amount of sediment moving all at once rather than as **individual** grains. Along a highway, if a large mass of **soil** and rock from a hillside suddenly gives way and rapidly moves downhill, this would be a type of mass wasting known as a landslide. Mudflows and rockfalls are two other common types of mass wasting.

Products and impacts of erosion

As already mentioned, sediment grains and chemical solutions are common products of erosion. Wind, water, ice, or gravity transport these products from their site of origin and lay them down elsewhere in a process known as *deposition*. Deposition, which occurs in large depressions known as basins, is considered to be a separate process from erosion.

Soil is also an erosional product. Soil is formed primarily by chemical weathering of loose sediments or bedrock along with varying degrees of biological activity, and the addition of biological material. Some soil materials may also have undergone a certain amount of transport before they were incorporated into the soil.

Another important product of erosion is the landscape that is left behind. Erosional landscapes are present throughout the world and provide some of our most majestic scenery. Some examples are mountain ranges such as the Rocky Mountains, river valleys like the Grand Canyon, and the rocky sea cliffs of Northern California. Anywhere that you can see exposed bedrock, or where there is only a thin layer of regolith or soil covering bedrock, erosion has been at work creating a landscape. In some places one erosional agent may be responsible for most of the work; in other locations a combination of agents may have produced the landscape. The Grand Canyon is a good example of what the combination of flowing river water and mass wasting can do.

In addition to producing sediment, chemical solutions, soil, and landscapes, erosion also has some rather negative impacts. Two of the most important of these concern the effect of erosion on soil productivity and slope stability.

Soils are vital to both plants and animals. Without soils plants cannot grow. Without plants, animals cannot survive. Unfortunately, erosion can have a very negative impact on soil productivity because it decreases soil fertility. Just as erosion can lead to the deposition of thick lay-

Dust storms, like this one, occur when soils not properly anchored by vegetation are subjected to dry conditions and winds. *FMA Production. Reproduced by permission.*

ers of nutrient rich material, thereby increasing soil fertility, erosion can also remove existing soil layers. Soil forms very slowly—a 1-in (2.5-cm) thick soil layer typically takes 50-500 years to form. Yet an inch of soil or more can be eroded by a single rainstorm or windstorm, if conditions are right. Farming and grazing, which expose the soil to increased rates of erosion, have a significant impact on soil fertility worldwide, especially in areas where **soil conservation** measures are not applied. High rates of soil erosion can lead to crop loss or failure and in some areas of the world, mass starvation. On United States farmland, even with widespread use of soil **conservation** measures, the average rate of soil erosion is three to five times the rate of soil formation. Over **time**, such rates cut crop yields and can result in unproductive farmland.

Erosion is also a very important control on slope stability. Slopes, whether they are small hillsides or large mountain slopes, tend to fail (mass waste) due to a combination of factors. However, erosion is a significant contributor to nearly all slope failures. For example, in California after a **drought** has killed much of the vegetation on a hillside, the rains that signal the end of the drought lead to increased erosion. Eventually, due to the increased erosion, the slope may fail by some type of mass wasting, such as a mudflow or landslide.

Controls on erosion

The average rate of erosion at the surface of Earth is about 1 in (2.5 cm) per thousand years. However, the rate of erosion varies tremendously from place to place. Soil erosion in some areas exceeds one inch per year—one hundred times its rate of formation. This range in rates is dependent on several different controlling factors. These factors include the type and amount of **plant** cover and **animal** activity, the climate, the nature of surface materials, the slope **angle**, and human **land use**. However, many of these factors routinely help increase erosion in some ways, while decreasing it in others. In addition, a complex interplay between the different factors may exist. For example, a particular combination of surficial materials and plant cover may accelerate erosion in one climate, while decreasing it in another. The individual controls can be difficult to recognize and their effects difficult to discern as well.

Vegetation

Generally, plants tend to secure and stabilize sediment, but they may also be instrumental in helping to weather bedrock (for example, by prying open cracks during root growth). Animals may increase erosion by loosening soil, but they can also help stabilize it. An earthworm's sticky slime, for example, increases soil particle cohesion and helps the particles resist erosion.

Climate

As was mentioned above, warm, moist climates increase the rate of weathering and so speed up erosion as well. However, the plant cover in this setting usually helps decrease soil loss. Deserts tend to be very susceptible to erosion due to the limited amounts of vegetation. Fortunately, the low rainfall characteristic of deserts helps to limit erosional effects.

Surface material

Bedrock is more resistant to erosion than are sediments and soil. However, bedrock does display a range of susceptibility to erosion due to the different types of rock that may be present. Here again, the type of climate can have a major impact on erosion rates. In the desert, nearly all types of bedrock are very resistant to erosion, whereas in the tropics, nearly all types of rock weather rapidly.

Slope angle

The angle of a slope is one of the few consistent controls on erosion. The steeper the slope, when all other factors being equal, the more susceptible the slope will be to erosion.

Land use

Agriculture increases the likelihood of erosion by exposing soil to wind and rainfall. However, agriculture is not the only human land use that increases the likelihood of erosion. Logging, construction, landscaping, as well as many other activities make land more susceptible to erosion. Generally, any land use or activity that disturbs the natural vegetation or involves a change in slope, surface materials, etc., will increase the likelihood of erosion. There are some obvious exceptions, though. For example, pavement can temporarily halt erosion in almost all cases. However, nothing resists the erosive power of nature forever—all man-made structures will eventually weather and then fail.

Erosion and rejuvenation

Studies of erosion and the landscapes it leaves behind have been going on for over a century. This area of geologic inquiry, known as *geomorphology*, has long recognized that a balance exists between the erosion of land and its rejuvenation. If this were not the case, after a few tens to hundreds of millions of years, the earth's mountains would wear down to flat, relatively featureless plains and the basins would fill up with the sediment shed by the mountains. Instead, after billions of years of erosion, we still have mountains such as Mt. Everest in the Himalayas, which stands over 5.5 mi (8.8 km) above **sea level**, and ocean trenches such as the Marianas Trench, which reaches depths of more than 6.5 mi (10.4 km) below sea level.

The continued existence of rugged landscapes on the face of the earth is a result of a process of rejuvenation known as **plate tectonics**. Forces within the interior of the earth periodically re-elevate, or **uplift**, the earth's surface in various regions, while causing the lowering, or **subsidence**, of other regions. Plate **tectonics** therefore serves to maintain existing landscapes or build new ones. Currently, the Himalayas are an area of active uplift, but someday uplift will cease and erosion will slowly, but completely, wear them down. Someday the Marianas Trench may be filled in with sediment deposits. At the same time, new dramatic landscapes will be forming elsewhere on the Earth.

KEY TERMS

. . . .

Bedrock—The unweathered or partially weathered solid rock layer, which is exposed at the Earth's surface or covered by a thin mantle of soil or sediment.

Chemical weathering—The decomposition and decay of Earth materials caused by chemical attack.

Deposition—The accumulation of sediments after transport by wind, water, ice, or gravity.

Geomorphology—The study of Earth's land forms and the processes that produce them.

Mechanical weathering—The break up or disintegration of earth materials caused by the creation and widening of fractures. Also known as physical weathering.

Regolith—A thin layer of sediment that covers most of the earth's surface. Erosion of the underlying solid rock surface produces this layer.

Slope stability—The ability of the materials on a slope to resist mass wasting.

Soil productivity—The ability of a soil to promote plant growth.

Surficial material—Any type of loose Earth material, for example sediment or soil, found at the surface of Earth.

Erosion research

Research continues to focus on the factors that control erosion rates and ways to lessen the impact of land use on soil productivity. New methods of soil conservation are continually being developed and tested to decrease the impact of soil erosion on crop production.

Conventional tillage techniques leave fields bare and exposed to the weather for extended periods of time, which leaves them vulnerable to erosion. Conservation tillage techniques are planting systems that employ reduced or minimum tillage and leave 30% or more of the field surface protected by crop residue after planting is done. Leaving crop residue protects the soil from the erosive effects of wind and rain. Direct drilling leaves the entire field undisturbed. Specialized machines poke holes through the crop residue, and **seeds** or plant starts are dropped directly into the holes. No-till planting causes more disturbance to the crop residue on the field. Using this technique, the farmer prepares a seedbed 2 in (5 cm) wide or less, leaving most of the surface of the field undisturbed and still covered with crop residues. Strip rotary tillage creates a wider seed bed, 4-8 in (10-20 cm) wide, but still leaves crop residue between the seed beds. Conservation tillage techniques are particularly effective at reducing erosion from farm lands; in some cases reducing erosion by as much as 90%.

Other erosion research is focused on the factors that control mass wasting, especially where it is hazardous to humans. Stabilization of slopes in high risk areas is an increasingly important topic of study, as more people populate these areas every day.

See also Sediment and sedimentation.

Resources

Books

Dixon, D. *The Practical Geologist.* New York: Simon & Schuster, 1992.

Hamblin, W.K., and E.H. Christiansen. *Earth's Dynamic Systems.* 9th ed. Upper Saddle River: Prentice Hall, 2001.

Leopold, L.B. *A View of the River.* Cambridge: Harvard University Press, 1994.

Woodhead, James A. *Geology.* Boston: Salem Press, 1999.

Periodicals

Reganold, J. P., R. K. Papendick, and J.F. Parr. "Sustainable Agriculture." *Scientific American* (June 1990): 112-120.

Clay Harris

Error

Error is the amount of deviation in a physical quantity that arises as a result of the process of measurement or **approximation**. Another term for error is uncertainty.

Physical quantities such as weight, **volume, temperature**, speed, or **time** must all be measured by an instrument of one sort or another. No matter how accurate the measuring tool—be it an **atomic clock** that determines time based on atomic oscillation or a **laser** interferometer that measures **distance** to a fraction of a wavelength of **light** some finite amount of uncertainty is involved in the measurement. Thus, a measured quantity is only as accurate as the error involved in the measuring process. In other words, the error, or uncertainty, of a measurement is as important as the measurement itself.

As an example, imagine trying to measure the volume of **water** in a bathtub. Using a gallon bucket as a measuring tool, it would only be possible to measure the volume accurately to the nearest full bucket, or gallon. Any fractional gallon of water remaining would be added as an estimated volume. Thus, the value given for the volume would have a potential error or uncertainty of something less than a bucket.

Now suppose the bucket were scribed with lines dividing it into quarters. Given the resolving power of the human **eye**, it is possible to make a good guess of the measurement to the nearest quarter gallon, but the guess could be affected by factors such as viewing **angle**, **accuracy** of the scribing, tilts in the surface holding the bucket, etc. Thus, a measurement that appeared to be 6.5 gal (24.6 l) could be in error by as much as one quarter of a gallon, and might actually be closer to 6.25 gal (23.6 l) or 6.75 gal (25.5 l). To express this uncertainty in the measurement process, one would write the volume as 6.5 gallons +/-0.25 gallons.

As the resolution of the measurement increases, the accuracy increases and the error decreases. For example, if the measurement were performed again using a cup as the unit of measure, the resultant volume would be more accurate because the fractional unit of water remaining—less than a cup—would be a smaller volume than the fractional gallon. If a teaspoon were used as a measuring unit, the volume measurement would be even more accurate, and so on.

As the example above shows, error is expressed in terms of the difference between the true value of a quantity and its approximation. A positive error is one in which the observed value is larger than the true value; in a **negative** error, the observed value is smaller. Error is most often given in terms of positive and negative error. For example, the volume of water in the bathtub could be given as 6 gallons +/-0.5 gallon, or 96 cups +/-0.5 cup, or 1056 teaspoons +/-0.5 teaspoons. Again, as the uncertainty of the measurement decreases, the value becomes more accurate.

An error can also be expressed as a **ratio** of the error of the measurement and the true value of the measurement. If the approximation were 25 and the true value were 20, the relative error would be 5/20. The relative error can be also be expressed as a **percent**. In this case, the percent error is 25%.

Sources of error

Measurement error can be generated by many sources. In the bathtub example, error could be introduced by poor procedure such as not completely filling the bucket or measuring it on a tilted surface. Error could also be introduced by environmental factors such as **evaporation** of the water during the measurement process. The most common and most critical source of error lies within the measurement tool itself, however. Errors would be introduced if the bucket were not manufactured to hold a full gallon, if the lines indicating quarter gallons were incorrectly scribed, or if the bucket incurred a dent that decreased the amount of water it could hold to less than a gallon.

In electronic measurement equipment, various electromagnetic interactions can create electronic **interference**, or noise. Any measurement with a value below that of the electronic noise is invalid, because it is not possible to determine how much of the measured quantity is real, and how much is generated by instrument noise. The noise level determines the uncertainty of the measurement. Engineers will thus speak of the noise floor of an instrument, and will talk about measurements as being below the noise floor, or "in the noise."

Measurement and measurement error are so important that considerable effort is devoted to ensure the accuracy of instruments by a process known as **calibration**. Instruments are checked against a known, precision standard, and adjusted to be as accurate as possible. Even gas pumps and supermarket scales are checked periodically to ensure that they measure to within a predetermined error.

Nearly every country has established a government agency responsible for maintaining accurate measurement standards. In the United States, that agency is known as the National Institute of Standards and Technology (NIST). NIST provides measurement standards, calibration standards, and calibration services for a wide array of needs such as time, distance, volume, temperature, luminance, speed, etc. Instruments are referred to as "NIST traceable" if their accuracy, and measurement error, can be confirmed by one of the precision instruments at NIST.

Kristin Lewotsky

KEY TERMS

Calibration—A procedure for adjusting the performance of measuring equipment to ensure a precise, predetermined level of accuracy.

Electronic noise—Spurious signals generated in electrical measurement equipment that interfere with readings.

Uncertainty—The degree to which a measurement is unknown.

Escherichia coli

Escherichia coli is one of the most well-known and intensively studied **bacteria**. Often shortened to *E. coli*, the bacterium was discovered in 1885 by the German bacteriologist Dr. Theodor Escherich. Initially, the bacterium was termed *Bacterium coli,* but later the name was changed to honor Dr. Escherich.

Escherichia coli inhabits the intestinal tract of humans and other warm-blooded **mammals**. It constitutes approximately 0.1% of the total bacteria in the adult intestinal tract.

Escherich was also the first person to demonstrate that the bacterium could cause diarrhea and gastroenteritis (an **inflammation** of the intestinal tract) in infants. That the bacterium could cause intestinal maladies was at first confusing, given that it exists naturally in the intestinal tract. However, it is now known that certain types of *E. coli* exist that are more capable of causing **disease** than other types. If these types are present in **water** or food that is ingested, then an **infection** can result. The vast majority of the many types of *E. coli* are harmless to humans.

E. coli has been a popular bacterium for laboratory research almost since its discovery. This is because the **organism** can be grown quickly on simple and inexpensive lab growth media. Also, the organism can be used to study bacterial growth in the presence of **oxygen** (**aerobic** growth) and in the absence of oxygen (**anaerobic** growth). The ability of *E. coli* to grow aerobically and anaerobically classifies the bacterium as a facultative anaerobe.

A type (or strain) of *E. coli* that has been the workhouse of research is designated K12. K12's biochemical behavior and structure are well known. The huge amount of structural, biochemical, genetic, and behavioral information has made *E. coli* indispensable as a bacterial model system for numerous studies. Hospital laboratory scientists are also concerned with *E. coli,* as the bacterium is the primary cause of human urinary tract infections, as well as **pneumonia**, and traveler's diarrhea.

In its normal **habitat** of the intestinal tract, *E. coli* is beneficial. The bacteria participate in the digestion of food, and produce **vitamin** K and B—complex vitamins.

When *E. coli* is excreted from the intestinal tract, the bacteria are able to survive only a few hours. This characteristic of rapid death was recognized at the beginning of the twentieth century, when the bacterium began to be used as an indicator of fecal **pollution** of water. The presence of large numbers of *E. coli* in water is a strong indicator of recent fecal pollution, and so the possible presence of other intestinal bacteria that cause serious disease (i.e., Vibrio, **Salmonella**, Shigella). Even today, *E. coli* remains one of the important tests of water quality

In 1975, the United States Centers for Disease Control and prevention identified a new strain of *E. coli* that

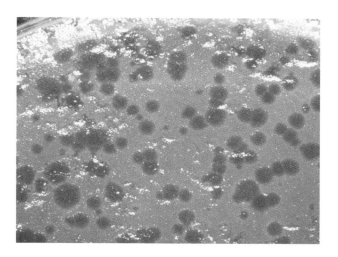

A culture of *E. coli*. © Lester V. Bergman/Corbis. Reproduced by permission.

was designated O157:H7. Strain O157:H7 was first linked to human disease in 1983, when it was shown to have caused two outbreaks of a severe gastrointestinal illness in the Unites States. This strain is capable of causing severe, even lethal infection. Those who recover sometimes have permanent kidney damage.

The origin of O157:H7 is not known for certain. The consensus among researchers, however, is that O157:H7 arose when a strain of *E. coli* living in the intestine and which was not disease causing became infected by a **virus**. The virus carried the genes coding for a powerful toxin called Shiga-like toxin. Thus, the *E. coli* acquired the ability to produce the toxin.

The toxin can destroy the cells that line the intestinal tract and can enter the bloodstream and migrate to the kidneys and liver. Severe damage to these organs can occur. The intestinal damage causes severe bleeding, which can be lethal in children and elderly people. During the summer of 2000, *E. coli* O157:H7 contaminated the drinking water of the town of Walkerton, Ontario, Canada. Over 2,000 people became ill and seven people died. The source of the strain was the intestinal tract of cattle, a known natural habitat of O157:H7.

E. coli can be spread to food by handling of the food with unwashed hands, particularly after using the bathroom. The solution to this spread of the bacterium is proper hand washing with **soap**. Other preventative measures include avoiding unpasteurized milk or apple cider, washing raw foods before consumption, and thorough cooking of ground meat (cattle carcasses can become contaminated with feces during slaughter and the bacterium can be passed on to the ground beef).

The **genome** sequences of several strains of *E. coli* have been obtained. The sequence of strain K12 has approximately 4300 protein coding regions making up about 88% of the bacterial **chromosome**. Most of these **proteins** function in getting **nutrients** into the **cell**. Much of the remainder of the genome is devoted to coding for proteins involved in processing of the nutrients to produce the **energy** needed for cell survival, growth, and division. The genome sequence of O157:H7 is very different from that of K12. Much of the genome of O157:H7 codes for unique proteins, over 1,300, some of which are necessary for infection. Many of these genes are thought to have been acquired from other **microorganisms**. Strain O157:H7 is designed to acquire genes and change quickly.

See also Water microbiology; Water treatment.

Resources

Books

Donnenberg, M. *Escherichia coli*: Virulence Mechanisms of a Versatile Pathogen. San Diego: Academic Press, 2002.

Kaper, J.B., and A.D. O'Brien. *Escherichia coli O157:H7 and Other Shiga Toxin-producing E.coli Strains*. Washington, DC: American Society for Microbiology Press, 1998.

Organizations

Centers for Disease Control and Prevention, 1600 Clifton Road, Atlanta, GA 30333. (404)639–3311. June 20, 2001 [cited November 17, 2002]. <http://www.cdc.gov/ncidod/dbmd/diseaseinfo.escherichiacoli_q.htm.>.

Brian Hoyle

Ester

Ester is an organic functional group that forms many sweet-smelling compounds. The chemical structure of an ester is represented by the general formula, R-CO-OR', where a central **carbon** atom that has two bonds to an **oxygen** atom (the **carbonyl group**), C=O, a single bond to another carbon atom represented by R, and a single bond to an oxygen atom connected to a carbon atom represented by R'. The R and R' groups can be the same or different. If the R and R' groups are bonded to each other, they form a ring and constitute a cyclic ester or lactone.

If an ester group is treated with **water (hydrolysis)**, it yields an organic acid and an **alcohol**. Esters are classified according to which acid and alcohol are produced by hydrolysis. For example, methyl acetate, CH_3-CO-OCH_3, forms methanol, CH_3OH, and **acetic acid** or vinegar, $HOCOCH_3$, when treated with water. It consists of a central carbon atom that has two bonds to an oxygen atom (the carbonyl group), C=O, a bond to a carbon atom connected to three **hydrogen atoms (methyl**

group), -CH_3, and a bond to an oxygen atom connected to a methyl group, -OCH_3. Low **molecular weight** esters that produce vinegar or acetic acid when hydrolyzed are used as solvents in coatings or paints on automobiles and inks. Larger esters can be used as a *plasticizer*, that is, to soften materials that are normally hard or brittle. Esters are important chemical compounds for various pharmaceutical and agricultural applications.

Esters having an acetic acid or vinegar base are called acetates. They are used extensively as solvents, due to their ability to dissolve various greases. Methyl acetate is a solvent for many oils and **resins** and is important in the manufacture of artificial leather. In the pharmaceutical and food processing industries, ethyl acetate is used as an extraction solvent. Butyl acetate does not discolor readily and dries at a reasonable **rate**; therefore it is an ideal solvent for inks. It is also used as a cleaning liquid for silicon wafers. Butyl acetate is a good solvent for the manufacture of lacquers, because it is a good medium for less compatible ingredients. Amyl acetates are utilized as extracting liquids in the isolation of penicillin and in the manufacture of polishes for leather goods.

Of all the chemical compounds approved by the Food and Drug Administration (FDA) for use as flavorings in food, esters make up the largest single portion of this group. The flavoring esters are a major constituent of essential oils which are used in foods as a natural flavoring. (An essential oil is a **solution** obtained from liquefying the vapors formed from heating the organic **matter** of a flavorful plant.) The American Indians used the dried leaves of a fern that commonly grows in the eastern United States, especially in North Carolina and Pennsylvania, to make **wintergreen** tea. When the essential oil of this **plant** was analyzed by scientists, it was made up almost entirely of the ester, methyl salicylate. This ester contains a **benzene** ring or **phenyl group** in its molecular structure and will produce methanol when heated with water. Methyl salicylate, or "sarsaparilla" flavor is also the major ingredient in "birch beer" or root beer. It also gives that wintergreen **taste** to toothpaste, chewing gum, and candy. Methyl salicylate can also be used to relieve muscular aches and pains and rheumatic conditions. Other aromatic esters are used as flavors or perfumes. Chocolate flavoring uses isoamyl phenylacetate, and strawberry is a mix of ethylmethylphenylglycinate and methyl cinnamate. The aromatic ester, benzyl acetate, is used largely in jasmine and gardenia fragrances, and benzyl benzoate, an ester with practically no odor, is used as a base in many perfumes. Some aromatic esters, such as Bitrex, are very bitter tasting, but are added to household products such as cleaning fluids, nail polish removers, and detergents to keep children from wanting to drink these poisonous substances.

Salicylic acid acetate, or aspirin, is only one of many esters used as medicines. Phenyl salicylate, a similar aromatic ester, is used in the treatment of rheumatic **arthritis**. Methyl phenidate, an ester which produces methanol when it is heated with water, is used to stimulate the central **nervous system**. The pharmaceutical industry has discovered that certain undesirable properties of drugs, such as bad taste or swelling of the skin at the spot of an injection, can be avoided by converting the original drug into an ester. The antibiotic dalactine (clindamycin), a bitter tasting drug, was converted to its palmitate ester in order to make its flavor less harsh.

The macrolides are large ring lactones or cyclic esters. Most of these unusual esters are isolated from **microorganisms** that grow in the **soil** and are being used as **antibiotics** in human and veterinary medicine. The well known antibiotic erythromycin is an example of a macrocylic lactone consisting of 12 carbon atoms and one oxygen atom bonded to a carbonyl group and the more potent roxithromycin contains 14 carbon atoms, an oxygen atom and a carbonyl group bonded as a cyclic ester.

See also Esterification.

Resources

Books

Kirk-Othmer Encyclopedia of Chemical Technology, Pigments to Powders, Handling. New York: Wiley, 1999.

Loudon, G. Mark. *Organic Chemistry.* Oxford: Oxford University Press, 2002.

Patai, S., ed. *Synthesis of Carboxylic Acids, Esters and Their Derivatives.* New York: Wiley, 1991.

Andrew J. Poss

Esterification

Esterification is the chemical process for making esters, which are compounds of the chemical structure R-CO-OR', where R and R' are either alkyl or aryl groups. The most common method for preparing esters is to **heat** a carboxylic acid, R-CO-OH, with an **alcohol**, R'-OH, while removing the **water** that is formed. A mineral acid catalyst is usually needed to make the reaction occur at a useful **rate**.

Esters can also be formed by various other reactions. These include the reaction of an alcohol with an acid chloride (R-CO-Cl) or an anhydride (R-CO-O-CO-R'). Early studies into the chemical mechanism of esterification, concluded that the **ester** product (R-CO-OR') is the union of the acyl group (R-C=O-) from the acid, R-CO-OH, with the alkoxide group (R'O-) from the alcohol, R'-OH rather than other possible combinations.

$$'R—CH \;+\; R—\overset{\displaystyle O}{\overset{\displaystyle \|}{C}}—CH \quad \xrightarrow[\text{catalyst}]{\text{mineral acid}} \quad R—\overset{\displaystyle O}{\overset{\displaystyle \|}{C}}—OR' \;+\; H_2O$$

alcohol   acid   ester   water

The chemical structure of the alcohol, the acid, and the acid catalyst used in the esterification reaction all effect its rate. Simple alcohols such as methanol (CH_3OH) and **ethanol** (CH_3CH_2OH) react very fast because they are relatively small and contain no **carbon** atom sidechains that would hinder their reaction. These differing rates of reaction were first reported by Nikolay Menschutkin (1842-1907) in 1879-83. He also noted that simple acids such as **acetic acid** or vinegar (CH_3CO_2H) form esters very easily. The most common acid catalysts are hydrochloric acid, HCl, and **sulfuric acid**, H_2SO_4, because they are very strong acids. At the end of the esterification reaction, the acid catalyst has to be neutralized in order to isolate the product. German chemists, during World War II, developed solid acid catalysts or **ion exchange resins** for use in the manufacture of esters. These solid catalysts work well with acid sensitive esters because they can be separated from the product by **filtration** and therefore, the catalyst does not spend very much time in contact with the acid unstable product.

The esterification process has a broad spectrum of uses from the preparation of highly specialized esters in the chemical laboratory to the production of millions of tons of commercial ester products. These commercial compounds are manufactured by either a batch or a continuous synthetic process. The batch procedure involves a single pot reactor that is filled with the acid and alcohol reactants. The acid catalyst is added and the water removed as the reaction proceeds. This method is most often used by chemists in the laboratory, but in a few cases, it is used by industry to make large quantities of esters. This batch process usually requires reactors that hold extremely large volumes of reactants. Butyl acetate is commonly prepared from butanol and acetic acid by this method. The continuous process for making esters was first patented in 1921 and has been used extensively in the manufacture of large quantities of esters. This procedure involves the mixing of streams of the reactants into a reaction chamber while the product is removed at the same time. Continuous esterification has the advan-

Patai, S., ed. *Synthesis of Carboxylic Acids, Esters and Their Derivatives*. New York: Wiley, 1991.

Andrew J. Poss

KEY TERMS

· ·

Acetates—Esters that are based on acetic acid or vinegar as their acid component.

---

tage that larger quantities of products can be prepared in shorter periods of time. This procedure can be run for days or weeks without interruption, but requires special equipment and special chemical **engineering** considerations. The continuous esterification process is used industrially to make methyl acetate from acetic acid and methanol and ethyl acetate from acetic acid and ethanol.

The alternative process of making esters from the reaction of an alcohol with an anhydride is important in the manufacture of drugs. This reaction gives an acid as a by-product.

$$
\begin{array}{cccc}
& \text{O} \quad \text{O} & \text{O} & \text{O} \\
& \| \quad \| & \| & \| \\
\text{'R--CH} & + \text{ R--C--O--C--R} & \text{R--C--OR'} & + \text{ R--C--CH} \\
\text{alcohol} & \text{anhydride} & \text{ester} & \text{acid}
\end{array}
$$

Acetic anhydride, $CH_3\text{-CO-O-CO-CH}_3$, a **derivative** of acetic acid, the acid in vinegar, is the most commonly used anhydride reactant. Phenyl acetate, one ester prepared industrially by this method, is an important intermediate in the synthesis of acetaminophen. Aspirin, or **acetylsalicylic acid**, is also prepared in large scale by the esterification reaction of an alcohol with acetic anhydride. This anhydride is important in the production of **cellulose** acetate by the esterification of cellulose or **cotton**. Cellulose acetate was first prepared in 1865 and was used extensively during World War I to coat airplane wings. Today, cellulose acetate finds its largest application as the fibrous material used in cigarette filters. It is also used in various yarns and **textiles** and to make the tips of felt-tip pens. Phthalic anhydride, an anhydride derivative of a **benzene** or phenyl ring, yields dimethyl phthalate when reacted with methanol in a reaction to that described for acetic anhydride. Dimethyl phthalate is used as a mosquito repellent and in the manufacture of certain polyesters. It is also a component in hair sprays and is added to **plastics** to soften them.

See also Carboxylic acids.

Resources

Books

Kirk-Othmer Encyclopedia of Chemical Technology. New York: Wiley, 1991.

Loudon, G. Mark. *Organic Chemistry*. Oxford: Oxford University Press, 2002.

Ethanol

Ethanol is an **alcohol** fuel that is manufactured by fermenting and distilling **crops** with a high starch or sugar content, such as grains, **sugarcane**, or corn. In the **energy** sector, ethanol can be used for **space** and **water** heating, to generate **electricity**, and as an alternative vehicle fuel, which has been its major use to date. Worldwide, ethanol is the mostly widely used alternative liquid fuel. Ethanol is also known as ethyl alcohol, drinking alcohol, and grain alcohol.

The United States produced 780 million gallons (3 billion l) of ethanol in 1986 and plans to increase this to 1.8 billion gal (7 billion l). This ethanol is mostly blended with conventional gasoline to make gasohol (90% gasoline and 10% ethanol). Gasohol accounts for 8% of national gasoline sales and 25-35% of sales in the farming states of Illinois, Iowa, Kentucky, and Nebraska, where much ethanol is manufactured from maize. Brazil, the world's largest producer and consumer of ethanol, uses a gasohol blend of 85-95% gasoline and 15-5% ethanol. Brazil produces 3 billion gal (12 billion l) of ethanol yearly, almost all of which is manufactured from sugar cane.

History

Interest in alternative fuels began with the realization that the supply of non-renewable fossil fuel is not infinite, a fact which has important economic and environmental consequences. For example, national dependence on foreign **petroleum** reserves creates economic vulnerabilities. In the United States, approximately 40% of the national trade deficit is a result of petroleum imports.

Environmentally, fossil fuel burning has negative consequences for local and global air quality. Locally, it causes high concentrations of ground-level **ozone**, **sulfur dioxide**, **carbon monoxide**, and particulates. Globally, fossil-fuel use increases concentrations of **carbon dioxide**, an important greenhouse gas.

Advantages of ethanol as an alternative fuel

Ethanol has many positive features as an alternative liquid fuel. First, ethanol is a renewable, relatively safe fuel that can be used with few engine modifications. Second, its energy **density** is higher than some other alterna-

---

tive fuels, such as methanol, which means less **volume** is required to go the same **distance**. The third benefit of ethanol is that it can improve agricultural economies by providing farmers with a stable market for certain crops, such as maize and sugar beets. Fourth, using ethanol increases national energy security because some use of foreign petroleum is averted.

Another benefit, though controversial, is that using ethanol might decrease emissions of certain emissions. Toxic, ozone-forming compounds are emitted during the **combustion** of gasoline, such as aromatics, olefins, and hydrocarbons, would be eliminated with the use of ethanol. The **concentration** of particulates, produced in especially large amounts by diesel engines, would also decrease. However, emissions of **carbon** monoxide and **nitrogen** oxides are expected to be similar to those associated with newer, reformulated gasolines. Carbon dioxide emissions might be improved (-100%) or worsened (+100%), depending the choice of material for the ethanol production and the energy source used in its production.

Disadvantages of ethanol as an alternative fuel

However, there are several problems with the use of ethanol as an alternative fuel. First, it is costly to produce and use. At 1987 prices, it cost 2.5-3.75 times as much as gasoline. The United States Department of the Environment (DOE) is funding a research program aimed at decreasing the cost to $0.60/gallon by the year 2000; in the last decade or so, the cost has dropped from $3.60/gallon to $1.27/gallon. There are also costs associated with modifying vehicles to use methanol or gasohol, but these costs vary, depending on the number of vehicles produced.

Another problem is that ethanol has a smaller energy density than gasoline. It takes about 1.5 times more ethanol than gasoline to travel the same distance. However, with new technologies and dedicated ethanol-engines, this is expected to drop to 1.25 times.

An important consideration with ethanol is that it requires vast amounts of land to grow the crops needed to generate fuel. The process for conversion of crops to ethanol is relatively inefficient because of the large water content of the **plant** material. There is legitimate concern, especially in developing countries, that using land for ethanol production will compete directly with food production.

Another problem is that ethanol burning may increase **emission** of certain types of pollutants. Like any combustion process, some of the ethanol fuel would come out the tailpipe unburned. This is not a major problem since ethanol emissions are relatively non-toxic. However, some

---

> **KEY TERMS**
> ....
> **Energy density**—The relative energy in a specific amount of fuel.
>
> **Enzymatic hydrolysis**—A low-cost, high-yield process of converting plant material into ethanol fuel using yeast.
>
> **Ethyl alcohol/drinking alcohol/grain alcohol**—Alternative terms for ethanol.
>
> **Gasohol**—A fuel mixture of gasoline and ethanol.
>
> **Swill**—Waste generated when ethanol fuel is produced.

of the ethanol will be only partially oxidized and emitted as acetylaldehyde, which reacts in air to eventually contribute to the formation of ozone. Current research is investigating means to reduce acetylaldehyde emissions by decreasing the engine warm-up period.

Finally, ethanol production, like all processes, generates waste products that must be disposed. The waste product from ethanol production, called swill, can be used as a **soil** conditioner on land, but is extremely toxic to aquatic life.

Current research and outlook

Current research is investigating ways to reduce the cost of ethanol production. DOE's research partners are developing a process, called enzymatic **hydrolysis**, that uses special strains of **yeast** to manufacture ethanol in an inexpensive, high-yield procedure. This project is also developing the use of waste products as a fuel for ethanol production.

Other researchers are investigating the problems of ethanol-fueled vehicles in prototype and demonstration projects. These projects are helping engineers to increase fuel efficiency and starting reliability. However, until the economic and technical problems are solved, ethanol will mostly be used in countries or areas that have limited access to oil reserves or have an excess of **biomass** for use in the ethanol production process.

See also Alternative energy sources.

Resources

Books

Miller, G.T., Jr. *Environmental Science: Sustaining the Earth.* 3rd ed. Belmont, CA: Wadsworth, 1991.
Poulton, M.L. *Alternative Fuels for Road Vehicles.* Boston: Computational Mechanics, 1994.

Jennifer LeBlanc

Ether

Ether is the common name of the organic compound whose chemical formula is $CH_3CH_2OCH_2CH_3$. Diethyl ether, as it is also known, is a very flammable liquid with a sweet smell. Ether has been prepared by reacting ethyl **alcohol** with strong acid, since the thirteenth century. Its properties were well known but its chemical structure was not determined until 1851 by Alexander William Williamson (1824-1904).

Ether was first used as an anesthetic to kill **pain** by W. T. G. Morton (1819-1868), a Boston dentist. Morton had learned about ether from a chemist named Charles T. Jackson (1805-1880). Eventually Morton convinced Dr. J. C. Warren (1778-1856) to let him use ether as an anesthetic on one of his patients. In Massachusetts General Hospital on October 16, 1846, Morton put a Mr. Abbott to **sleep** with ether, then called out, "Dr. Warren, your patient is now ready." The patient was asleep and relaxed. From this date on, ether became the most widely used anesthetic. It was eventually replaced, in about 1956, with newer anesthetics that are not flammable and have some more beneficial properties.

Ethyl ether has many uses in the chemical industry. It is an excellent solvent and is used for dissolving various waxes, fats, oils, and gums. Ether is an inert compound that is often used as the reaction medium for **chemical reactions** between very reactive **species**. When ether is mixed with **ethanol** the resulting **solution** has special solvent properties that are used to dissolve nitrocellulose the principal component in the manufacture of the explosive, guncotton. This solvent mixture is also used in the preparation of **plastics** and membranes. Ether can be purchased in aerosol cans to be used as a starting fluid for **automobile** engines.

See also Anesthesia.

Ethnoarchaeology

Ethnoarchaeology, a subfield of **archaeology**, is the study of contemporary cultures in order to interpret social organization within an archeological site. Traditionally, archaeology has been concerned with the identification, classification, and chronological ordering of remains. Archaeologists were able to describe a civilization according to its artifacts, but not to fully understand its culture. Archaeologists viewed artifacts as a means of adapting to an environment. Ethnoarchaeologists view artifacts as a possible means of communication or expression, and factor **random** human **behavior** into their models. In the last few decades, archaeologists have widened their perspective and adopted the methods, and the insights, of a variety of related scientific disciplines. Combined with traditional research methods, along with theoretical models and research methods borrowed from and shared with anthropologists and ethnologists, archaeologists have reconstructed various cultural activities from the past.

In his excavations of certain European sites of Paleolithic hunters-gatherers, the archaeologist Lewis Binford was puzzled by the fact that the remaining **animal** bones found on several sites showed significant variations in their assemblages. Were there cultural, or even particular, reasons for these variations? Binford therefore decided to turn to ethnology and study how a contemporary, hunter-gatherer culture kills animals and disposes of the remains. As reported in his 1978 study *Nunamiut Ethnoarchaeology*, Binford observed that a modern community of Nunamiut Eskimos left bone assemblages similar to those found on Paleolithic sites. This study might not provide an exact parallel for the prehistoric hunter-gatherers, but Binford could conclude that there were certain functions or actions that may be shared by all hunters and gatherers. Binford could also conclude that in both cultures, the variations in behavior reflected by the different manners of disposing animal remains, most likely pointed to the intrinsic variability of human behavior, and were not the result of cultural changes.

Ethnoarchaeologists explore how conscious and subconscious human behavior imposes itself on a culture's external, material world. The Garbage Project of Tucson, Arizona, is a good example. Organized by William L. Rathje, the Garbage Project involved collecting and sorting trash from a certain section of the city and a survey of its inhabitants was conducted. The inhabitants were questioned about, for example, what they buy, how much of a certain product they consume, and what they throw away. The garbage was carefully sorted in a laboratory, classified, identified, and compared with the survey. Using traditional archaeological methods and ethnoarchaeological models, project workers learned that what people say they do does not correspond to what they actually do. Of course, when studying an ancient site, archaeologists can only interpret cultures by what actually remains at a site—they cannot question the inhabitants about how artifacts were used. By studying the pattern of consumption in a modern urban environment, ethnoarchaeologists can deduce that prehistoric urban human behaviors were similar.

Ethnobotany

Ethnobotany is the study of the relationships between plants and people. Most often, however, the term is used in **relation** to the study of the use of plants by aboriginal people living relatively simple, pre industrial lifestyles.

Plants have always played a central role within indigenous cultures. **Plant** products are used as food, as sources of medicine, and as raw materials for the weaving of fabrics. In addition, **wood** is commonly used as a fuel for cooking and to keep warm, and as a material for the construction of homes and for manufacturing tools. The indigenous knowledge of particular **species** of plants useful for these purposes is a key cultural **adaptation** to having a successful life in local ecosystems.

Plants are also crucial to humans living in an advanced economy, such as that typical of **North America** and **Europe**. Almost all of the food consumed by these people is directly or indirectly (that is, through **livestock** such as cows and chickens) derived from plants. Similarly, most medicines are manufactured from plants, and wood is an important material in construction and manufacturing. Therefore, an intrinsic reliance on plant products is just as important for people living in modern cities, as for those living in tropical jungle.

It is true that substitutes are now available in advanced economies for many of the previous uses of plant products. For example, **plastics** and metals can be used instead of wood for many purposes, and synthetic fabrics such as nylon instead of plant-derived **textiles** such as **cotton**. Similarly, some medicines are synthesized by biochemists, rather than being extracted from plants (or from other organisms, such as certain microbes or animals). However, alternatives have not been discovered for many crucial uses of plants, so we continue to rely on domesticated and wild species as resources necessary for our survival.

Moreover, there is an immense diversity of potential uses of wild plants that scientists have not yet discovered. Continued bio-prospecting in tropical rainforests and other natural ecosystems will discover new, previously unknown uses of plants, as foods, medicines, and materials. A key element of this prospecting is the examination of already existing knowledge of local people about the utility of wild plants. This is the essence of the practice of ethnobotany.

The diversity of plants

All species are unique in their form and **biochemistry**, and all are potentially useful to people as sources of food, medicine, or materials. As we will see below, however, only a relatively few species are being widely used in these ways. This suggests that there are enormous opportunities of "undiscovered" uses of plants and other organisms.

There are about 255,000 species of plants, including 16,600 species of bryophytes (liverworts and mosses; phylum Bryophyta), 1,000 of **club mosses**, quillworts, and **horsetails** (Lycophyta and Sphenophyta), 12,000 of **ferns** (Pterophyta), 550 of conifers (Coniferophyta), and 235,000 of flowering plants (Anthophyta). Although **fungi** and **algae** are not plants, ethnobotanists also study their use. There are about 37,000 species of fungi, and 1,500 species of larger algae (such as seaweeds). These numbers refer only to those species that have been officially named by biologists there is also a huge number of species that have not yet been "discovered" by these scientists. Most of these undescribed species inhabit poorly known ecosystems, such as tropical rainforests. In many cases, however, they are well known to indigenous people, and are sometimes used by them as valuable resources. A major goal of ethnobotany is to examine the utilization of local species of plants by indigenous people, to see whether these uses might be exploited more generally in the larger economy.

Plants as food

Plant products are the most important sources of food for people. We eat plants directly, in the form of raw or cooked **vegetables** and **fruits**, and as processed foods such as bread. We also eat plants indirectly, as when we consume the meat of **pigs** or eggs of chickens that have themselves fed on plant foods.

Relatively few species of plants comprise most of the food that people eat. The world's most important **crops** are cereals, such as **barley**, maize, **rice**, **sorghum**, and **wheat**. **Tuber** crops are also important, such as cassava, **potato**, **sweet potato**, and turnip. However, remarkably few plants are commonly cultivated, amounting to only a few hundred species. Moreover, only 20 species account for 90% of global food production, and just three (wheat, maize, and rice) for more than half.

There is an enormously larger number of plants that are potentially edible (about 30,000 species), including about 7,000 species that are being utilized locally by indigenous peoples as nutritious sources of food. Potentially, many of these minor crops could be cultivated more widely, and thereby provide a great benefit to a much greater number of people.

A few examples of highly nutritious foods used by local cultures, which could potentially be cultivated to feed much larger numbers of people, include arrachacha (*Arracia xanthorhiza*), an Andean tuber; amaranths (three *Amaranthus* spp.), tropical American and Andean grains; Buffalo gourd (*Curcurbita foetidissima*), a Mexican tuber; maca (*Lepidium meyenii*), an Andean root vegetable; spirulina (*Spirulina platensis*), an African blue-green alga; wax gourd (*Benincasa hispida*), an Asian melon. These and many other local, traditional foods have been "discovered" by ethnobotanists, and could well prove to be important foods for many people in the future.

Plants as medicines and drugs

Similarly, a large fraction of the drugs used today in modern medicine are derived from plants. In North America, for example, about 25% of prescription drugs are derived from plants, 13% from **microorganisms**, and 3% from animals. In addition, most recreational drugs are products of plants or fungi. Several familiar examples are: **acetylsalicylic acid** (or ASA), a pain-killer originally derived from the plant meadowsweet (*Filipendula ulmaria*); **cocaine**, a local anaesthetic and recreational drug derived from the **coca** plant (*Erythroxylon coca*); **morphine**, a pain-killer derived from the opium poppy (*Papaver somniferum*); vinblastine and vincristine, two anti-cancer drugs extracted from the rosy periwinkle (*Catharantus roseus*); and taxol, an anti-cancer drug derived from the Pacific **yew** (*Taxus brevifolia*).

In almost all cases, the usefulness of medicinal plants was well known to local, indigenous people, who had long exploited the species in traditional, herbal medicinal practices. Once these uses became discovered by ethnobotanists, a much greater number of people were able to experience their medicinal benefits.

Conservation of ethnobotanical resources

The conservation of plant biodiversity

Most species of wild plants can only be conserved in their natural ecosystems. Therefore the enormous storehouse of potentially useful foods, medicines, and materials available from plant **biodiversity** can only be preserved if large areas of tropical **forests** and other natural ecosystems are conserved. However, **deforestation**, **pollution**, and other changes caused by humans are rapidly destroying natural ecosystems, causing their potentially invaluable biodiversity to be irretrievably lost through **extinction**. The **conservation** of natural ecosystems is of great importance for many reasons, including the fact that it conserves ethnobotanical resources.

The conservation of indigenous knowledge

Ethnobotanical studies conducted in many parts of the world have found that local cultures are fully aware of the many useful species occurring in their **ecosystem**. This is particularly true of people living a subsistence lifestyle, that is, they gather, hunt, or grow all of the food, medicine, materials, and other necessities of their lives. This local knowledge has been gained by trial and **error** over long periods of **time**, and in most cases has been passed across generations through oral transmission. Indigenous knowledge is an extremely valuable cultural resource that ethnobotanists seek to understand, because so many useful plants and other organisms are known to local people. Unfortunately, this local, traditional knowledge is often rapidly lost once indigenous people become integrated into modern, materialistic society. It is important that local indigenous peoples be given the opportunity to conserve their own culture.

Ethics in ethnobotanical research

Ethnobotanists working in remote regions have a number of special ethical responsibilities that they must fulfill. One of these is to ensure that their studies of indigenous cultures are not overly intrusive, and do not result in fundamental changes in local values and lifestyles. People living in remote areas should not be exposed to the full possibilities of modern lifestyles too quickly, or their local, often sustainable lifestyles could quickly disintegrate. This change is sometimes referred to as a kind of cultural genocide. Ethnobotanists have a responsibility to ensure that their studies do cause this to happen.

In addition, indigenous knowledge of useful plants (and other species) is an extremely valuable cultural resource of aboriginal people living in many parts of the world. The ethics of ethnobotanists must include actions to ensure that as this local knowledge is tapped to provide benefits for humans living throughout the world, the local people also benefit significantly from their generous sharing of valuable information about the uses of biodiversity. The benefits could take many forms, ranging from royalties on the sales of products, to the provision of hospitals and schools, and the legal designation of their land rights to places where they and their ancestors have lived for long periods of time.

Resources

Books

Balick, M.J., and P.A. Cox. *Plants, People, and Culture: The Science of Ethnobotany.* New York: W.H. Freeman and Co., 1997.

Balick, M.J., E. Elisabetcky, and S.A. Laird, eds. *Medicinal Resources of the Tropical Forest: Biodiversity and Its Importance to Human Health.* New York: Columbia University Press, 1995.

Chadwick, D. *Ethnobotany and the Search for New Drugs.* New York: John Wiley and Sons, 1994.

Cotton, C.M. *Ethnobotany: Principles and Applications.* New York: John Wiley and Sons, 1996.

Johnson, T. *CRC Ethnobotany Desk Reference.* Boca Raton, FL: Lewis Publishing Co., 1998.

KEY TERMS

. . . .

Biodiversity—The richness of biological variation, including that at the levels of genetics, species, and ecological communities.

Ethnobotany—The study of the relationships between plants and people, particularly in relation to preindustrial aboriginal cultures.

Indigenous knowledge—The understanding of local people of the usefulness of plants and other organisms occurring within their ecosystem.

Moerman, D.E. *Native American Ethnobotany.* Portland, OR: Timber Press, 1998.

Schultes, R.E., and S. Van Reis eds. *Ethnobotany: Evolution of a Discipline.* Portand, OR: Timber Press, 1995.

Other

Society for Economic Botany (2003). The New York Botanical Garden, Bronx, NY 10458-5126. (718) 817-8632. <http://www.econbot.org>.

Bill Freedman

Ethyl group

Ethyl group is the name given to the portion of an organic **molecule** that is derived from ethane by removal of a **hydrogen** atom (-CH$_2$CH$_3$). An ethyl group can be abbreviated in chemical structures as -Et. The ethyl group is one of the alkyl groups defined by dropping the -*ane* ending from the parent compound and replacing it with -*yl*. By this terminology, the ethyl group is derived from the parent alkane, ethane. Ethane has the **molecular formula** of CH$_3$CH$_3$. It is composed of two **carbon atoms** connected by a single bond, (C-C) and in turn, each carbon atom has three single bonds to hydrogen atoms (C-H). The ethyl group is the two carbon atom unit that is connected to a longer chain of carbon atoms or possibly a **benzene** ring.

Ethane is a gas at room **temperature** and burns very easily. The word ethane is derived from *aithein*, the Greek word for to blaze or to kindle. Ethane makes up about 15% of **natural gas** and can also be isolated from crude oil. It is used by industries for the production of ethylene and ethyl chloride.

Ethylene has the chemical formula of H$_2$C=CH$_2$ and is composed of two carbon atoms connected by a double bond (C=C), each carbon atom is also bonded to two hy-drogen atoms (C-H). Ethylene is made by heating a mixture of steam and ethane to very high temperature. Ethylene is used in the manufacture of the **polymer**, polyethylene, which as the name implies, consists of many two carbon atom units linked into a long chain. Polyethylene is the clear wrap that is used in grocery stores and homes for wrapping and preserving food. Polyethylene can also be colored and coated onto wires for insulation. Ethylene is converted into ethylene oxide by the addition of an **oxygen** atom with oxygen gas. Ethylene oxide is mixed with gaseous chlorofluorocarbons to make a gas blend used to sterilize medical equipment and materials that are sent into outer space.

When ethane and **chlorine** are heated to high temperatures, they undergo a chemical reaction that results in the product, ethyl chloride (CH$_3$CH$_2$Cl). Ethyl chloride is an alkyl halide that has of one of the hydrogen atoms of ethane replaced by a chlorine atom. Ethyl chloride is commonly used as a substrate for the addition of an ethyl group on to a benzene ring or another organic chemical. For example, two ethyl chloride molecules are used to make an important ingredient in sleeping pills called barbital or 5,5-diethylbarbituric acid. When ethyl chloride is rubbed onto the skin, it causes the skin to become numb or to loose all feeling. This local anesthetic property of ethyl chloride is used by members of the medical community to stop **pain** when lancing of boils and giving injections.

The incorporation of an ethyl group or chain of two carbon atoms into a molecule's structure can change the properties of the compound drastically. Butanoic acid is an acid composed of four carbon atoms. It is easily identified by its odor of rancid butter. When butanoic acid is converted into its ethyl **ester**, ethyl butyrate, it loses its disgusting smell. Ethyl butyrate has a broad **spectrum** of commercial applications from the perfume industry where it is routinely used for its rose fragrance to the food industry where its addition results in a natural pineapple flavor. **Benzoic acid**, another organic acid, is used as a food preservative because it does not interfere with the other flavors present. The corresponding ethyl ester of benzoic acid is ethyl benzoate. It is used as one of the primary ingredients of many kinds of fruit and berry flavored chewing gum. In the pharmaceutical industry, the addition of an ethyl group to a drug can have a profound effect on its properties. Barbital, or 5,5-diethylbarbituric acid, is used as a sedative and is a common component of many sleeping pills, whereas barbituric acid, the compound without the ethyl groups, has no calming or **sleep** inducing properties at all.

See also Anesthesia; Barbiturates.

Resources

Books

Arctander, S. *Perfume and Flavor Materials of Natural Origin.* Elizabeth, NJ: S. Arctander, 1960.

Carey, Francis A. *Organic Chemistry.* New York: McGraw-Hill, 2002.

Kirk-Othmer. *Encyclopedia of Chemical Technology.* New York: Wiley, 1991.

Ethylene *see* **Hydrocarbon**

Ethylene glycol

Ethylene **glycol** is an organic (**carbon** based) **molecule** most widely used as antifreeze in **automobile** engines and as an industrial solvent, a chemical in which other substances are dissolved. The addition of ethylene glycol to **water** raises the **boiling point** of the engine coolant and reduces the chances of a car's radiator "boiling over." The name ethylene glycol communicates much information about the chemical's structure. The "ethylene" portion of the name indicates that the molecules of ethylene glycol have two carbon **atoms** in them, and the "glycol" part of the name indicates that there are two hydroxy groups (OH units) attached on carbon atoms. Ethylene glycol has a freezing point of 8.6°F (-13°C) and a boiling point of 388°F (198°C), and is completely miscible with water.

Ethylene glycol is sweet tasting but highly toxic. It must therefore be kept away from children and pets. As little as 2 oz (56.7 g) can cause death in an adult, and much smaller doses can kill a child or a small **animal**. Ethylene glycol itself is metabolized by enzymes in the liver and becomes highly poisonous **oxalic acid**. Enzymes are protein catalysts, molecules that speed up the rates of **chemical reactions**. Pure ethylene glycol is colorless, syrupy liquid. The **color** of commercial antifreeze (usually green) is due to a dye that is added to help identify the source of a leak from a car.

Pure water has a boiling point of 212°F (100°C) at **sea level**. A special **pressure** cap is used on car radiators to make the boiling point higher, but even so, most cars would "boil over" were it not for the addition of other chemicals that raise the boiling point. Many chemicals can be used to raise the boiling point of water (including **salt** and sugar), but ethylene glycol is used because it does not damage parts of the car and is inexpensive and long-lasting. It also lowers the freezing point, thus requiring a lower **temperature** for the water to freeze. Ethylene glycol could just as well be called "anti-boil." In industries, ethylene glycol is used as a solvent (a substance that dissolves other chemicals) and as starting material for the production of Dacron and some types of polyurethane foam.

Ethylenediaminetetra-acetic acid

Ethylenediaminetetra-acetic acid, typically shortened to EDTA, is a chemical compound with the ability to form multiple bonds with **metal** ions, making it an important chemical to analytical scientists and industry alike.

The compounds used to create EDTA include ethylenediamine, formaldehyde, and **sodium** cyanide. When these compounds are mixed in an appropriate fashion, a series of **chemical reactions** take place. Formaldehyde reacts with the sodium cyanide to form formaldehyde cyanohydrin. In the presence of **sulfuric acid**, this compound then reacts with ethylenediamine forming an intermediate compound that eventually reacts with **water** to form EDTA.

Solid EDTA is readily dissolved in water where it can form multiple chemical bonds with many metal ions in a **solution**, in effect tying up the metal ions. A **molecule** such as EDTA which has at least one free pair of unbonded electrons and therefore can form chemical bonds with metal ions. These compounds are known as ligands. Ligands are typically classified by the number of available free **electron** pairs that they have for creating bonds. **Ammonia** (NH_3), for example, which has one pair of unbonded electrons, is known as a monodentate **ligand** (from Latin root words meaning one tooth). EDTA has six pairs of unbonded electrons and is called a hexadentate ligand. Ligands such as EDTA, which can form multiple bonds with a single metal ion, in effect surrounding the ion and caging it in, are known as chelating agents (from the Greek *chele,* meaning crab's claw). The EDTA molecule seizes the metal ion as if with a claw, and keeps it from reacting normally with other substances.

Chelating agents play an important roll in many products, such as food, soda, shampoo, and cleaners. All of these products contain unwanted metal ions which affect **color**, odor, and appearance. In addition to affecting the physical characteristics of these products, some metal ions also promote the growth of **bacteria** and other **microorganisms**. EDTA ties up metal ions in these products and prevents them from doing their damage.

EDTA is also used extensively by analytical chemists for titrations. A titration is one method for finding the amounts of certain metals in various samples based on a known chemical reaction. Since EDTA bonds with metal ions, the amount of metal in a **sample** can be calculated based on the amount of EDTA needed to react with it. In this way, chemists can determine the amount of such things as **lead** in drinking water or **iron** in **soil**.

Etiology

Etiology is study of the cause of **disease**. More specifically, etiology is the sum of knowledge regarding a disease, or knowledge of all pathological processes leading to a disease. There may be just one causative agent, but no one cause of disease. For example, as stated in Boyd's *Introduction To The Study Of Disease*, we know **tuberculosis** is caused by the tubercle bacillus. Many people are exposed to the tubercle bacillus yet only one may develop the disease. The bacilli can also lay dormant in the body for many years and became active as the result of an intercurrent **infection** (occurring during the course of an existing disease), prolonged **stress**, or starvation. While some diseases have no known cause, there are several factors that may contribute to, or predispose someone to a disease. Many variables must be considered when looking at the cause of a disease such as, gender, age, environment, lifestyle, heredity, **allergy**, immunity, and exposure.

Eubacteria

The eubacteria are the largest and most diverse taxonomic group of **bacteria**. Some regard this as an artifi-

cial assemblage, merely a group of convenience rather than a natural grouping. The eubacteria are all easily stained, rod-shaped or spherical bacteria. They are generally unicellular, but a small number of multicellular forms do occur. They can be motile or non-motile and the motile forms are frequently characterized by the presence of numerous flagellae. Many of the ecologically important bacteria responsible for the fixation of **nitrogen**, such as *Azotobacter* and *Rhizobium*, are found in this group.

The **cell** walls of all of these **species** are relatively thick and unchanging, thus shape is generally constant within groups found in the eubacteria. Thick cell walls are an evolutionary **adaptation** that allows survival in extreme situations where thinner walled bacteria would dry out. Some of the bacteria are gram positive whilst others are gram negative. One commonality that can be found within the group is that they all reproduce by transverse binary fission, although not all bacteria that reproduce in this manner are members of this group.

Eucalyptus tree *see* **Myrtle family (Myrtaceae)**

Eugenics

Eugenics is the study of improving the human race by selective breeding. Its rationale is to remove bad or deleterious genes from the population, increasing the overall fitness of humanity as a result. Campaigns to stop the criminal, the poor, the handicapped, and the mentally ill from passing on their genes were supported in the past by such people as British feminist Marie Stopes and Irish playwright George Bernard Shaw. In the United States in the early 1900s, enforced sterilization was carried out on those deemed unfit to reproduce.

One problem with practical eugenics is who decides what is a desirable characteristic and what is not. Both before and during the World War II, there was a eugenics program in place in Nazi Germany. Initially this was carried out on inmates of mental institutions, but very quickly the reasons for sterilization became more and more arbitrary. After the World War II, there was a backlash against eugenics and anthropologists such as Franz Boas and Ruth Benedict championed the opposite view, *tabula rasa* (blank slate). This view favored the theory that humans are born with an empty mind, which is filled by experience. American anthropologist Margaret Mead carried out work in Samoa that confirmed these ideas, demonstrating that people are programmed by their environment, not by their genes, and that providing a decent environment produces people who behave decently towards each other. Modern work suggests that it is a subtle interaction of both the genetic make-up and the environmental conditions that shape an **individual**.

Eugenics happens to a minor degree in modern society, most notably when a couple with family histories of **genetic disorders** decide not to have children or to terminate a pregnancy, based on genetic screening. In 1994 China passed restrictions on marriages which involved individuals with certain disabilities and diseases.

There is evidence that the practice of eugenics could never be truly effective. When one calculates the **frequency** of deleterious **alleles** in the population, it is found humans all carry at least 1% of alleles which if present in homozygous form would prove fatal. When scientists predict the effects that might be achieved by preventing all individuals possessing a given allele from breeding, it is found that the effect would be minimal. One problem associated with this prediction is the difficulty in detecting certain alleles when they are present in the heterozygous form.

Eukaryotae

Eukaryotae, or eukaryotic cells, are large and complex cells bounded by an outer **plasma membrane**. They contain many organelles within their cytoplasm and a nucleus separated from the cytoplasm by the nuclear membrane. Fossils of eukaryotic cells are present in **rocks** dated as 1.5 billion years old. All living things on **Earth**, except **bacteria** and blue-green **algae** (cyanobacteria), which are Prokaryotae, are composed of eukaryotic cells.

The nucleus of a eukaryotic **cell** is a membrane-bound compartment containing genetic information in the form of DNA organized into chromosomes. The nuclei of eukaryotic cells divide by **mitosis**, a process which results in two daughter nuclei that are identical to the parent cell. The cell's nucleus directs its overall functioning, while the membrane-bound organelles in the cytoplasm carry out a variety of specialized jobs in a coordinated fashion.

In plants, organelles called chloroplasts trap the **energy** from sunlight in a process called **photosynthesis**. Plants then use that energy to drive metabolic pathways. In both **animal** and **plant** eukaryotic cells, the cellular energy is generated by organelles called mitochondria. Other organelles, the lysosomes, are membrane-bound packages of digestive enzymes. These digestive en-

zymes, and other **proteins**, are manufactured in the **ribosomes** located on the rough endoplasmic reticulum, a kind of cellular highway. An organelle called the Golgi complex then moves the enzymes—and other proteins— into the membranes and distributes them.

Some eukaryotic cells have a flagellum, whip-like projection from the cell membrane that aids in the cell's locomotion. Others may have cilia, shorter, hair-like strands arranged around the perimeter of the cell in a characteristic way. The cilia of prokaryotic cells are less complex than those of eukaryotic cells.

The types and arrangement of a cell's organelles enable eukaryotic cells of multicellular organisms to perform specialized functions. In humans, the eukaryotic cells of a number of organs are highly specialized, but nevertheless maintain most of the defining features of the eukaryotic cell. For example, the cells of the **brain**, liver, bone, muscle of a growing baby divide by mitosis under the control of the DNA in the nucleus, with the liver cells producing more liver cells, and bone cells producing other bone cells.

See also Nucleus, cellular; Prokaryote.

Europe

The **continent** of Europe is a landmass bounded on the east by the Ural mountains, on the south by the Mediterranean Sea, and on the north and west by the Arctic and Atlantic Oceans. Numerous islands around this landmass are considered a part of Europe. Europe is also the westernmost part of the Eurasian supercontinent (several continental masses joined together).

Europe holds a unique place among the continents; much of it is "new," in geologic terms. Unlike other continents whose structures seem simple in comparison, Europe is a collection of all different kinds of geologic regions located side by side, many of which have little or nothing to do with each other. This should be kept in mind when reading about the regions in this article, which are not areas of similar origins, but rather areas of dissimilar origin that happen to be part of the same continent.

Forces that made Europe

Plate tectonics is the main **force** of nature responsible for the geologic history of Europe. This continent-building process may be simply explained:

The **earth** is covered by a thin, brittle layer called the lithosphere. This term is used here interchangeably with "the Earth's crust." Below the **lithosphere** is the asthenosphere, where solid rock stretches and flows. The lithosphere is composed of sections, called plates, and floats on top of the **asthenosphere**, because it is less dense than the asthenosphere. The **motion** of a tireless **heat** engine swirls and stirs the asthenosphere, moving the plates.

Over hundreds of millions of years, the plates have come together to form continents, including Europe. Other processes, such as sedimentation and **erosion**, modify the shape of the land that has been forged by plate **tectonics**.

Large-scale geologic elements of Europe

European geologic history, like that of all the continents, involves the formation of the following features as a result of plate tectonics:

Island arcs: When the edge of a plate of Earth's lithosphere runs over another plate, forcing the lower plate deep into the elastic interior, a long, curved chain of volcanic mountains usually erupts on the forward-moving edge of the upper plate. When this border between two plates forms in an **ocean**, the volcanic mountains constitute a string of islands (or archipelago). This is called an island arc. Italy's Appenine mountains originally formed as an island arc, then became connected into a single **peninsula** later.

Continental arcs: A continental arc is exactly like an island arc except that the volcanos erupt on a continent, instead of in the middle of an ocean. The chemical composition of the erupted rock is changed, because old continental **rocks** at the bottom of the lithosphere have melted and mixed with the **magma**. A clear-cut example of this kind of mountain chain no longer exists in Europe, but ancient continental arcs once played an important part in Europe's geologic past. Sicily's Mt. Aetna and Mt. Vesuvius on the Bay of Naples are good examples of the type of **volcano** that commonly make up a continental arc.

Sutures: A suture describes the place where two parts of a **surgery** patient's **tissue** are sewed or rejoined; it also describes the belts of mountains that form when two continents are shoved into each other, over tens of millions of years, to become one. The Alps and other ranges in southern Europe stand tall because of a continental collision between Europe and **Africa**. The Alps, and other European ranges, are the forerunners of what may be a fully developed suture uniting Europe and Africa. This suture would be a tall mountain range that stretches continuously from Iberia to easternmost Europe.

The collision of Africa with Europe is a continuous process that stretches tens of millions of years into the past and into the future. All the generations of humanity together have seen only a tiny increment of the continental movement in this collision. However, throughout history people have felt the collision profoundly, in earthquakes and volcanos, with all the calamities that attend them.

Rifts: Sometimes a continent is torn in pieces by forces moving in opposite directions beneath it. On the surface, this tearing at first makes a deep valley, which experiences both volcanic and **earthquake** activity. Eventually the valley becomes wide and deep enough that its floor drops below **sea level**, and ocean **water** moves in. This process, called rifting, is the way ocean basins are born on Earth; the valley that makes a place for the ocean is called a rift valley. Pieces of lithosphere, rifted away from Africa and elsewhere, have journeyed across the Earth's surface and joined with the edge of Europe. These pieces of lithosphere lie under southern England, Germany, France, and Greece, among other places.

Fault-block mountains: When a continent-sized "layer cake" of rock is pushed, the upper layers move more readily than the lower layers. The upper layers of rock are heavy-but easier to move than those beneath it (like a full filing cabinet is heavy-but when it's pushed, it moves more easily than the floor beneath it). Between near surface rocks and the deeper, more ancient crustal rocks, a flat-lying **fault** forms, also called a "detachment" fault (decollement in French). This horizontal crack, called a thrust fault, contains fluid (water, mostly). The same hydraulic force that makes hydraulic machines lift huge weights functions in this crack as well. The fluid is so nearly incompressible that a sizeable piece of a continent can slide on it when pushed. The fault block floats on fluid pressure between the upper and lower sections of the lithosphere like a fully loaded tractor trailer gliding effortlessly along a rain-slicked road. The mountains that are heaved up where the thrust fault reaches the surface are one kind of fault block mountains. Both the Jura mountains and the Carpathians are fault block mountains.

Another kind of fault block mountain comes from stretching of the earth's lithosphere. The lithosphere, like any other brittle material, develops cracks **perpendicular** to the direction of movement of the forces that are pulling it apart. In this case the force is lateral, so steep, nearly vertical faults form.

However they form, physical and chemical forces start wearing down mountain ranges while they are still rising. Europe has been criss-crossed by one immense range of mountains after another throughout its almost four-billion year history. Where did these mountains go?

If mountains are not continuously uplifted, they are worn down by erosion in a few million years. In Europe's geologic past, eroded particles from its mountains were carried by streams and dumped into the surrounding ocean or the continent's inland seas. Those particular rivers and seas are now gone from Earth, but the sediments that filled them remain, like dirt in a bathtub when the water is drained. The roots of all the mountain ranges that ever stood in Europe still exist, and much of the **sand** and clay into which the mountains were transformed still exists also, as rock or **soil** formations.

European geologic history

Unfortunately, the farther back in a continent's history we attempt to explore, the less we can know about it. Only a very incomplete record of ancient geologic events is preserved for study because most ancient rocks are subsequently destroyed. For the most distant **geologic time interval** in Europe's history, from 4.6 billion years ago (b.y.a.) to 3.8 b.y.a., the Hadean eon, almost nothing is known and so it is not discussed here. Enough is known however about the Archean eon (3.8-2.5 b.y.a.), the Proterozoic eon (2.5 b.y.a.-570 million y.a.), and the Phanerozoic eon (570 m.y.a.-present) to provide a fairly detailed picture. In fact, the events of the Phanerozoic eon are known well enough to discuss them with even smaller geologic **time** intervals. These from largest to smallest are era, period, and epoch. The older the events, the less detail is known, so only very recent geologic epochs are discussed.

Archean rocks in Europe

Europe was not formed in one piece, or at one time. Various parts of it were formed all over the ancient world, over a period of four billion years, and were slowly brought together and assembled into one continent by the processes of plate tectonics. What is now called Europe began to form more than 3 billion years ago, during the Archean eon.

Most geologists feel that prior to and during the creation of the oldest parts of Europe, Earth only superficially resembled the **planet** we live on today. Active volcanos and rifts abounded. The planet had cooled enough to have a solid crust and oceans of liquid water. The crust may have included hundreds of small tectonic plates, moving perhaps ten times faster than plates move today. These small plates, carrying what are now the earth's most ancient crustal rocks, moved across the surface of a frantic crazy-quilt planet. Whatever it truly was like, the oldest regions in Europe were formed in this remote world. These regions are in Finland, Norway (Lofoten Islands), Scotland, Russia, and Bulgaria.

Proterozoic Europe

The piece of Europe that has been in its present form for the longest time is the lithospheric crust underneath Scandinavia, the Baltic states, and parts of Russia,

Belarus, and Ukraine. This region moved around on its own for a long time, and is referred to as Baltica. It is the continental core, or craton, to which other parts were attached to form Europe.

In the two billion years of the Proterozoic eon (2.5 b.y.a to 570 m.y.a.), the geologic setting became more like the world as we know it. The cores of the modern continents were assembled, and the first collections of continents, or supercontinents, appeared. Life, however, was limited to **bacteria** and **algae**, and unbreatheable gases filled the atmosphere. Erosion clogged the rivers with mud and sand, because no land plants protected the earth's barren surface from the action of rain, **wind**, heat, and cold.

During the late Proterozoic an existing supercontinent began to break up. One of the better known episodes in this supercontinent's dismemberment was the Pan-African event. This was a period when new oceans opened up and mountains rose all around Africa. Over a period of a quarter-billion years, relatively small pieces of Africa tore away and were rafted along on the asthenosphere, colliding with Baltica. These blocks include: the London Platform (southern Britain); a strip of Ireland from between counties Wicklow and Kerry; the Iberian Peninsula (Spain and Portugal); Aquitaine, Armorica, and the Massif Central (France); the Inter-alpine region (Switzerland and Austria); the Austrian Alps (Carnic Alps); Bohemia; Moesia (Bulgaria and Romania); and the Rhodope region (Northern Greece, Bulgaria, and European Turkey, including Istanbul).

The result is that much of northern and central Europe, on a continental scale of reference, is made of blocks of Africa stuck together in a "paste" of **sedimentary rock**.

Most of the mountain belts created in the Proterozoic eon have long since worn away. The roots of these mountain belts are found in northern Europe and include:

The Karelian mountain belt, forming the middle third of Norway, Sweden, and Finland, including some of Karelia, is underlain by the worn-flat roots of the Karelian mountain range. This orogenic event (from *oros*, Greek for mountain, and *genesis*, Greek for origin) happened between 2 billion and 1.7 billion years ago.

The Svecofennian mountain belt, forming most of the southern third of Norway, Sweden, and Finland, are the roots of a Proterozoic mountain belt. They lie hidden beneath recent glacier-deposited sediments, and beneath dense coniferous **forests**. Because these factors make examination difficult, the Svecofennian mountain-building event is not yet well understood. It is presumed that the Svecofennian mountain belt marks where Baltica's edge was between 1.9 and 1.4 billion years ago. This moun-

Europe. *Illustration by Hans & Cassidy. Courtesy of Gale Group.*

tain belt rose in a time of worldwide mountain-building, and it may have been related to the assembly of North America's Canadian Shield.

The southernmost tips of Norway and Sweden were made in the SvecoNorwegian/Grenville orogenic event, between 1.2 and 0.9 billion years ago. This is the mountain chain that united Baltica with **North America**, before most of the western and southern parts of the European landmass existed.

The Ukrainian shield: At present, geologists do not agree on when or how the Ukrainian shield was put to-

gether. Parts of it were definitely in existence before the Proterozoic eon. During the Proterozoic it was assembled, and joined Baltica.

Paleozoic Europe

The sea washed over the oldest parts of Earth's surface many times during the three billion-plus years of Archean and Proterozoic history. Life flourished in the shallow tidewater for these 3,000 million years. Algae, one of the most ancient organisms, were joined later by worms and other soft-bodied animals. Little is known of early soft-bodied organisms, because they left no skeletons to become fossils. Only a handful of good fossils-preserved under very rare conditions-remain from the entire world's immense Precambrian, or combined Hadean, Archean and Proterozoic, rock records.

Then, about 570 million years ago, several unrelated lineages of shell-bearing sea animals appeared. This was the beginning of the Phanerozoic eon of earth history (from the Greek *phaneros*, meaning visible, and *zoe*, meaning life), which includes the time interval from 570 million years ago to the present day. The seas teemed with creatures whose bones and shells we have come to know in the fossil record. These include trilobites, **brachiopods**, the first fishes, and vast coral reefs. The first half of the Phanerozoic eon is called the Paleozoic era (from the Greek palaios, meaning ancient). For Europe, the Paleozoic era was largely a continuation of events that started in the late Proterozoic eon.

Remember, the **map** of the ancient Earth didn't look anything like a modern world map. As this new day of Earth's history began, Baltica sat near the south pole. Three oceans surrounded it, and each of these oceans shrank and closed during the Paleozoic. As plate tectonics shifted island arcs and already-formed blocks of crust around on Earth's surface, new rock came to be a part of the forming continent of Europe; underground masses of igneous rock accumulated near active margins of Baltica; and volcanos frequently erupted even as they do today in the Mediterranean region. Concurrent with this volcanic mountain-building, was the wearing-down of the volcanos, and the wearing-down of mountain cores, exposed by erosion. The sediment from the erosion of these **igneous rocks** accumulated at the borders of the continental blocks, and often became compressed into tightly folded rock masses only a few million years after they accumulated.

Baltica faced Greenland across the Iapetus Ocean. Early in the Paleozoic Era, Europe combined again with North America to form a giant continent called Laurussia (from *Laurentia*, meaning the core of North America, and Russia), or The North Continent. The line along which the continents were welded together runs along the length of Norway, through central Scotland, south-central Ireland, and southern Wales. In North America this line runs down the eastern coast of Greenland. As always happens when continents collide, there an associated mountain-building event. The resulting mountains are called the Caledonides, and they stood athwart the seam that welded Baltica and Laurentia together at that time. A continental arc developed, running from Denmark southeast down the ancient coast of Baltica, which is also considered to be a part of the Caledonian orogeny. Mountains in Norway, Scotland, Wales, and Ireland bear the shapes of folds made in the Caledonian orogeny. They shed sediment as they eroded, and a great **volume** of sand came to rest in the British Isles (then scattered to areas throughout the region). These sandstone masses attracted the attention of early British geologists, who named them the Old Red Sandstone. They proposed that there had been a continent to the west that had entirely worn away, which they called the Old Red Sandstone continent. This continent is now known to have been the Greenland coast of Laurentia.

The Uralian Ocean lapped at Baltica's northeastern coastline. Just before the end of the Paleozoic era, the Ural ocean that had washed the eastern shore of Baltica narrowed and closed forever. Two landmasses—Siberia and Kazakhstan—compressed and lifted the sediments of the Ural **ocean basin** into the Ural mountains. The Urals still stand today, and mark the eastern boundary of Europe within the Eurasian supercontinent.

Africa faced Baltica across the Tornquist Sea. (Two oceans that later lay between Africa and Europe, at different times, are called Tethys. Earlier versions of Tethys are called Paleo-Tethys, and later versions of it are called Neo-Tethys. No Tethys ocean exists today.) Near the close of the Paleozoic Era (245 m.y.a.), Laurussia united with the single southern supercontinent made up of much of the rest of the world's continents, named Gondwana. This created the Pangea (from the Greek *pan*, meaning all, and *Ge*, meaning Earth). The Central Pangean mountains arose from this mighty collision, and deformed the rocks laid down during the Paleozoic in central Europe. These mountains joined to the west with the Mauretanides of West Africa, and the Appalachians and Ouachitas of the eastern and southern United States to form a tremendous mountain range stretching along the equator for more than 4,200 mi (7000 km). The whole mountain range is called the Hercynian Mega-suture.

Piece by piece, continental fragments tore away from Africa's northern and western shores and were crushed into the jagged southern edge of Baltica. The same processes that transported these blocks of Europe also deformed them, often changing the crystalline fabric of the rock so much that the original igneous or sedimentary rock type can only be guessed. Between Africa and Europe, tectonic forces seem to have pushed miniature continental blocks in several directions at once-the reconstructions of this time are still unclear. The violently twisted line followed by the Variscan mountain belt anticipated the Alps and other ranges of modern Europe.

Many important rock bodies, called massifs, formed as parts of the Variscan mountains. Giant, bulbous masses of granite and other igneous rocks solidified to form England's Cornwall; Spain and Portugal's Cantabrian

The Matterhorn, 14,690 ft (4,480 m), in Switzerland was carved by alpine glaciers. *JLM Visuals. Reproduced by permission.*

Massif; France's Armorican Massif, Massif Central, and the Vosges; Germany's Black Forest; and the Erzgebirge Mountains on the German-Czech border. The Balkans also contain Variscan-age massifs.

Mesozoic and Cenozoic Europe

At the opening of the Mesozoic era, 245 million years ago, a sizeable part of western and southern Europe was squeezed up into the Central Pangean mountain system that sutured Laurussia and Gondwana together. Europe was almost completely landlocked, its southern regions part of a mountain chain that stretched from Kazakhstan to the west coast of North America.

The birth of a new ocean **basin**, the Atlantic, signaled the end of Pangaea. The Central Pangean Mountains, after tens of millions of years, had worn down to sea level and below. A new ocean basin, not the Mediterranean, but rather the Ligurean Ocean, began to open up between Africa and Europe. This formed a seaway between the modern North Atlantic and the Neo-Tethys Ocean (which no longer exists). Sea water began to leave deposits where high mountains had stood, and a layer cake of sediment-laid down on the shallow-sea bottom-began to accumulate throughout Europe.

Beginning at the close of the Mesozoic era (66 m.y.a.), and continuing through the Cenozoic era to the present day, a complex orogeny has taken place in Europe. The ocean basin of Tethys was entirely destroyed, or if remnants still exist, they are indistinguishable from the ocean crust of the Mediterranean Sea. Africa has shifted from west of Europe (and up against the United States' East Coast) to directly south of Europe, and their respective tectonic plates are now colliding.

As in the collision that made the Variscan mountain belt, a couple of dozen little blocks are being pushed sideways into southern Europe. The tectonic arrangement can be compared with a traffic jam in Rome or Paris, where numerous moving objects attempt to wedge into a **space** in which they can't all fit.

The Mediterranean desert

When sea level fell below the level of the Straits of Gibraltar around six million years ago, the western seawater passage from the Atlantic Ocean to the Mediterranean Sea closed, and water ceased to flow through this passage. At about the same time, northward-moving Arabia closed the eastern ocean passage out of the Mediterranean Sea and the completely landlocked ocean basin began to dry up. Not once, but perhaps as many as 30 times, all the water in the ancestral Mediterranean, Black, and Caspian seas completely evaporated, leaving a thick crust of crystallized sea **minerals** such as gypsum, sylvite, and halite (**salt**). It must have been a lifeless place, filled with dense, hot air like modern below-sea-level deserts such as Death Valley and the coast of the Dead Sea. The rivers of Europe, **Asia**, and Africa carved deep valleys in their respective continental slopes as they dropped down to disappear into the burning salt wasteland.

Many times, too, the entire basin flooded with water. A rise in global sea level would lift water from the Atlantic Ocean over the barrier mountains at Gibraltar. Then the waters of the Atlantic Ocean would cascade 2.4 mi (4 km) down the mountainside into the western Mediterranean basin. From Gibraltar to central Asia, the bone-dry basin filled catastrophically in a geological instant—a few hundred years. This "instant ocean" laid deep-sea sediment directly on top of the layers of salt. The widespread extent and repetition of this series of salt and deep-sea sediment layers is the basis for the theory of numerous catastrophic floods in the Mediterranean basin.

Pleistocene Europe

For reasons not yet fully understood, the Earth periodically experiences episodes of planet-wide, climatic cooling, the most recent of which is known as the Pleistocene Epoch. Large areas of the land and seas become covered with **ice** sheets thousands of feet thick that remain unmelted for thousands or hundreds of thousands of years. Since the end of the last ice age about eight to twelve thousand years ago, only Greenland and **Antarctica** remain covered with continent-sized **glaciers**. But during the last two million or so years, Europe's northern regions and its mountain ranges were ground and polished by masses of water frozen into miles-thick continental glaciers.

This ice age began in Europe when heavy snowfalls accumulated in Scandinavia and northern Russia. As the planet's climate cooled, summer's snowmelt did not remove all of winter's snowfall, and an increasingly thick layer of ice accumulated. Ice built up higher and higher in some areas, and eventually began flowing out from these ice centers. As the ice sheet spread over more of the continent, its brilliant surface reflected the sun's heat back out into space—cooling the climate even more. During several intervals, each lasting hundreds of thousands of years, the European ice sheet covered Scandinavia, northern Russia, all the lands around the Baltic Sea, and the British Isles. Ice caps covered the mountain ranges of Europe. Between these planetary deep-freezes were warm, or interglacial, intervals, some of them hundreds of thousands of years long.

Ice in glaciers is not frozen in the sense of being motionless. It is in constant motion, imperceptibly slow-but irresistible. Glaciers subject the earth materials beneath them to the most intense kind of scraping and scouring. An alpine glacier has the power to tear **bedrock** apart and move the shattered pieces miles away. These are the forces that shaped the sharp mountain peaks and u-shaped mountainvalleys of modern Europe. Many European mountain ranges bear obvious scars from alpine glaciation, and the flat areas of the continent show the features of a formerly glaciated plain.

Each time the cooling climate froze the equivalent of an ocean of water into glaciers, the global sea level fell drastically (100 ft [30 m] or more). Southern England became a western promontory of the main European landmass, and the North Sea's seabed lay exposed to the sky. The Adriatic Sea almost disappeared, and its seabed became an extension of the Po River Valley. Sardinia and Corsica were at that time one island, and Sicily became a peninsula of mainland Europe.

Unusual geographic conditions also followed the retreat of the ice sheets. Bare rock was exposed in many areas. Elsewhere unvegetated sediment, deposited by the melting glaciers, lay exposed to the elements. Windstorms removed tremendous amounts of the smaller sized grains in this sediment, and carried it far from where the glacier left it. The wind-blown sediment eventually settled out of the sky and formed layers of silt, called loess ("lurse"). Today, loess deposits form a broad belt in various regions from northern France to Russia, and beyond. These deposits originally must have covered almost all of central and eastern Europe.

Continental glaciation occurred several times during the last 2.2 million years. Geologists do not agree whether the ice will return again or not. However, even if the present climate is merely a warm period between glaciations, tens or hundreds of thousands of years may elapse before the next advance of the ice sheets.

Holocene Europe

Humans have lived in Europe for much of the Pleistocene epoch and the entire Holocene epoch (beginning at the end of the last ice age, about 10,000 years ago). During the past few thousand years, humans have been significantly altering the European landscape. **Wetlands** across Europe have been drained for agricultural use from the Bronze Age onward. The Netherlands is famous for its polders, below-sea-level lands made by holding back the sea with dikes. Entire volcanos (cinder cones) have been excavated to produce frost-resistant road fill.

Europe's heavily industrialized regions developed in their present locations as a result of geologic factors. The industrial districts are centered around places where transportation routes carved by rivers occurred in conjunction with **ore** deposits and **fossil fuels**.

Europe continues to change today. From the Atlantic coast of Iberia to the Caucasus, Europe's southern border is geologically active, and will remain so effectively forever, from a human frame of reference. Africa, Arabia, and the Iranian Plateau all continue to move northward, which will insure continued mountain-building in southern Europe.

Geologists are concerned about volcanic hazards, particularly under the Bay of Naples and in the Caucasus. Smaller earthquakes, floods, and other natural disasters happen every year or so. In historic times, in the Aegean Sea and at Pompeii, Herculaneum, and Lisbon, entire cities have been devastated or destroyed by volcanos, earthquakes, and seismic sea waves. These larger-scale natural disasters can and will continue to happen in Europe on an unpredictable schedule with predictable results.

Miscellaneous regions and events

Britain and Ireland

The northwest fringe of Europe is made up of the two very old islands, Great Britain and Ireland, and numerous smaller islands associated with them. Geologically, these islands are a part of the European continent, although culturally separate from it. Unlike many islands of comparable size, the British Isles do not result from a single group of related tectonic events. They are as complex as continents themselves, which in the last two centuries has provided plenty of subject matter for the new science of **geology**.

Scotland and Ireland are each made of three or four slices of continental crust. These slices came together around 400 million years ago like a deck of cards being put back together after shuffling.

Iberia

The Iberian Peninsula, occupied today by Portugal and Spain, is one of the pieces of lithosphere that was welded to Europe during the Variscan mountain-building event. Like Britain, it is an unusual "micro-continent" with a complex geologic history.

Alpine and related orogenies

Since the Paleozoic era, southern Europe has continued to acquire a jumbled mass of continental fragments from Africa. Even today, the rocks of Europe from the Carpathian mountains southwestward to the Adriatic and Italy are made up of "tectonic driftwood," and are not

resting on the type of solid, crystalline basement that underlies Scandinavia and Ukraine.

Since the late Mesozoic era, the widening Atlantic Ocean has been pushing Africa counterclockwise. All the blocks of lithosphere between Africa and Europe, including parts of the Mediterranean seafloor, will in all likelihood eventually become a part of Europe.

The Alps resulted from Europe's southern border being pushed by the northern edge of Africa. In Central Europe, freshly-made sedimentary rocks of early Mesozoic age, along with the older, metamorphosed, Variscan rocks below, were pushed into the continent until they had no other way to go but up. Following the path of least resistance, these rocks were shaped by powerful forces into complex folds called *nappes*, which means tablecloth in French. The highly deformed rocks in these mountains were later carved into jagged peaks by glaciers during the Pleistocene epoch.

The Jura Mountains, the Carpathians, and the Transylvanian Alps are made of stacks of flat-lying sedimentary rock layers. These mountain ranges were thrust forward in giant sheets out in front of the rising Alps.

A complex story of tectonic movement is recorded in the sea-floor rocks of the western Mediterranean. Corsica, Sardinia, Iberia, and two pieces of Africa called the "Kabylies"—formerly parts of Europe—moved in various directions at various speeds throughout the Cenozoic era.

On the western Mediterranean floor, new oceanic lithosphere was created. A subduction zone formed as an oceanic plate to the east sank below the western Mediterranean floor. The magma generated by this event gave rise to the Appenine mountains, which formed as an island arc on the eastern edge of the western Mediterranean oceanic plate. The Appenines began to rotate counterclockwise into their present position. The Tyrrhenian Sea formed as the crust stretched behind this forward-moving island arc. In the Balkans, blocks of lithosphere have piled into each other over tens of millions of years.

The Dinarides and Hellenides, mountains that run down the east coast of the Adriatic Sea, form the scar left after an old ocean basin closed. The compressed and deformed rocks in these mountain ranges contain pieces of ocean floor. Just east of these seacoast mountains is a clearly-recognized plate boundary, where the European and African plates meet. The boundary runs from the Pannonian Basin (in Hungary, Romania, and Yugoslavia), cuts the territory of the former Yugoslavia in half, and winds up in Greece's Attica, near Athens.

Further inland, the Pannonian Basin results from the lithosphere being stretched as the Carpathian mountains move eastward and northward.

KEY TERMS

Archean eon—The interval of geologic time from 3.8 billion years ago to 2.5 billion years ago.

Baltica—The oldest part of Europe, made of rocks formed between 3.8 and 2.5 billion years ago, that underlies Scandinavia, the Baltic states, Poland, Belarus, and much of western Russia and Ukraine.

Collisional mountain belt—A mountain range, like the Alps, caused by one continent running into another continent.

Continental arc—A volcanic mountain range, such as existed in Europe's geologic past, that forms on the tectonically active edge of a continent, over a subduction zone.

Craton—The part of a continent that has remained intact since Earth's earliest history, and which functions as a foundation, or basement, for more recent pieces of a continent. Baltica is considered to be a craton to which the rest of Europe is attached.

Fault block mountains—A mountain range formed by horizontal forces that squeeze a continent, fracturing its crust and pushing some crustal blocks

upward to form mountains while others drop down to form valleys.

Hadean eon—The interval of geologic time from the beginning of Earth (4.65 billion years ago) to 3.8 billion years ago.

Island arc—An curved row of islands of volcanic origin that develops where two lithospheric plates converge, usually near the edge of a continent, and associated with the formation of a deep trench parallel to the arc as oceanic crust is subducted.

Phanerozoic eon—The interval of geologic time beginning 570 million years ago, the rocks of which contain an abundant record of fossilized life.

Precambrian—The combined Hadean, Archean and Proterozoic eons, the first four billion years of Earth's history, and for which there is only a poor fossil record.

Proterozoic eon—The interval of geologic time beginning 2.5 billion years ago and ending 570 million years ago.

The Aegean Sea seems to have formed as continental crust has been stretched in an east-west direction. It is a submerged basin-and-range province, such as in the western United States. The Pelagonian Massif, a body of igneous and **metamorphic rock** that lies under Attica, Euboea, and Mount Olympus, forms part of the Aegean sea floor. The Rhodopian Massif, in northern Greece, Bulgaria, and Macedonia, also extends beneath the Aegean Sea. Faults divide the ridges from the troughs that lie between them. The faults indicate that the troughs have dropped into the crust between the high ridges.

The Balkan range in Bulgaria is thought to mark the crumpled edge of the European craton—the Proterozoic-age rocks extending north into Russia.

Europe is also host to isolated volcanos related to structural troughs within the continent. The Rhine river flows in a trough known as the Rhine Graben. Geologists believe the Rhine once flowed southward to join the Rhone river in France, but was diverted by upwarping of the crust around the Vogelsberg volcano. The Rhine then changed its course, flowing out to meet England's Thames river in the low-sea-level ice age.

Resources

Books

Hancock P. L., and B. J. Skinner, eds. *The Oxford Companion to the Earth.* Oxford: Oxford University Press, 2000.

Winchester, Simon, and Soun Vannithone. *The Map That Changed the World: William Smith and the Birth of Modern Geology.* New York: Perennial, 2002.

Periodicals

Hoffman, P., ed. "Stonehenge Blues." *Discover* 11 (October 1990): 10.

Other

Carius, Alexander, and Kurt M. Lietzmann, eds. *Environmental Change and Security: A European Perspective (Contributions to the International and European Environmental Policy).* Berlin: Springer Verlag, 2002.

Europium *see* **Lanthanides**

Eutrophication

The process of heightened biological productivity in a body of **water** is call eutrophication. The major factors controlling eutrophication in a body of water, whether large, small, warm, cold, fast-moving, or quiescent, are nutrient input and rates of primary production. Not all lakes experience eutrophication. Warmth and **light** increase eutrophication, (which in Greek means "well nourished") if nutrient input is high enough. Cold dark

The structure of a eutrophic lake. *Illustration by Hans & Cassidy. Courtesy of Gale Group.*

lakes may be high in **nutrients**, but if rates of primary production are low, eutrophication does not occur. Lakes with factors that limit **plant** growth are called oligotrophic. Lakes with intermediate levels of biological productivity are called mesotrophic.

Many lakes around developed areas experience cultural eutrophication, or an accelerated **rate** of plant growth, because additional nitrates and phosphates (which encourage plant growth) flow into the lakes from human activities. **Fertilizers**, **soil erosion** and **animal** wastes may run off from agricultural lands, while detergents, sewage wastes, fertilizers, and construction wastes are contributed from urban areas. These nutrients stimulate the excessive growth of green plants, including **algae**. Eventually these plants die and fall to the bottom of the **lake**, where decomposer organisms use the available **oxygen** to consume the decaying plants. With accelerated plant growth and subsequent death, these decomposers consume greater amounts of available oxygen in the water; other **species** such as **fish** and **mollusks** thus are affected. The water also becomes less clear as heightened levels of **chlorophyll** are released from the decaying plants. Native species may eventually be replaced by

those tolerant of **pollution** and lower oxygen levels, such as worms and **carp**.

While at least one-third of the mid-sized or larger lakes in the United States have suffered from cultural eutrophication at one time or another over the past 40 years, Lake Erie is the most publicized example of excessive eutrophication. Called a "dead" lake in the 1960s, the smallest and shallowest of the five Great Lakes was inundated with nutrients from heavily developed agricultural and urban lands surrounding it for most of the twentieth century. As a result, plant and algae growth choked out most other species living in the lake, and left the beaches unusable due to the smell of decaying algae that washed up on the shores. New pollution controls for **sewage treatment** plants and agricultural methods by Canada and the United States led to drastic reductions in the amount of nutrients entering the lake. Almost forty years later, while still not totally free of pollutants and nutrients, Lake Erie is a biologically thriving lake, and recreational swimming, fishing, and boating are strong components of the region's economy and aesthetic benefits.

Evaporation

Evaporation is a process that is commonly used to concentrate an aqueous **solution** of nonvolatile solutes and a volatile solvent. In evaporation, a portion of the solvent is vaporized or boiled away, leaving a thick liquid or solid precipitate as the final product. The vapor is condensed to recover the solvent or it can simply be discarded. A typical example is the evaporation of brine to produce **salt**.

Evaporation may also be used as a method to produce a liquid or gaseous product obtained from the condensed vapor. For instance, in desalinization processes, sea **water** is vaporized and condensed in a water-cooled **heat** exchanger and forms the fresh water product.

In general, evaporation processes can be expressed as:

The separating agent is heat, which is usually supplied by a low-pressure steam to provide the latent heat of vaporization. When the liquid, say, a water solution, is heated, boiling occurs at the heated surface and the liquid circulates. Because the dissolved solids (solutes) are less volatile than water (solvent), water will first escape gradually from the solution. As sufficient water is boiled off, the resulting liquid becomes saturated, and then the dissolved solids crystallize. To reduce **energy** consumption, via utilizing the latent heat of the generated vapor over and over again, heat introduced into one evaporator can be used in other evaporators involved in a multi-stage, or formally called multi-effect process.

A variety of approaches are employed to vaporize liquids or solutions. Liquids can flow as a thin-film layer on the walls of a heated tube, can be diffused on the heated surface, or can be spread in fine droplets into a hot, dry gas. Wet cloth or **paper** generally can be dried by evaporation of the moisture into a gas stream. For some heat-sensitive liquids, such as pharmaceutical products and foods, evaporation must be carried out under reduced **pressure**, in which the **boiling point** occurs at lower **temperature**. Alternatives to this approach are to increase heat-transfer area or to inject steam directly into the solution to heat it rapidly. Very often, fouling layers can build up next to heat-transfer surfaces and reduce the heat-transfer **coefficient** across the evaporator. In certain situations, material tends to foam during vaporization. Liquid can boil over into the vapor, resulting in the failure to separate components or concentrating solutions. Therefore, good evaporator designs and the understanding of liquid characteristics are very crucial in evaporation efficiency.

Evaporation is often employed to produce a concentrated liquid for industrial purposes. A slurry of crystals in a saturated liquid can be obtained by means of evapo-

ration. In the sugar industry, water is boiled off to produce sucrose prior to crystallization. Solutions such as caustic soda and organic colloids all can be concentrated by evaporation. Because of the their substantial thermal sensitivity, **vacuum** evaporation is normally applied for **concentration** of fruit juices. Concentrated juices are easily transported and during storage, are more resistant to environmental degradation than fresh juices.

See also Heat transfer; States of matter.

Pang-Jen Kung

Evapotranspiration

Evapotranspiration refers to the vaporization of **water** from both non-living and living surfaces on a landscape. Evapotranspiration is a composite of two words: **evaporation** and **transpiration**. Evaporation refers to the vaporization of water from surface waters such as lakes, **rivers**, and streams, from moist **soil** and **rocks**, and any other substrates that are non-living. Transpiration refers to the vaporization of water from any moist living surface, such as **plant** foliage, and the body or lung surfaces of animals.

Evapotranspiration is an important component of Earth's **energy** budget, accounting for a substantial part of the physical dissipation of absorbed solar **radiation**. In any moist **ecosystem**, air and surface temperatures would be much warmer than they actually are, if evapo-

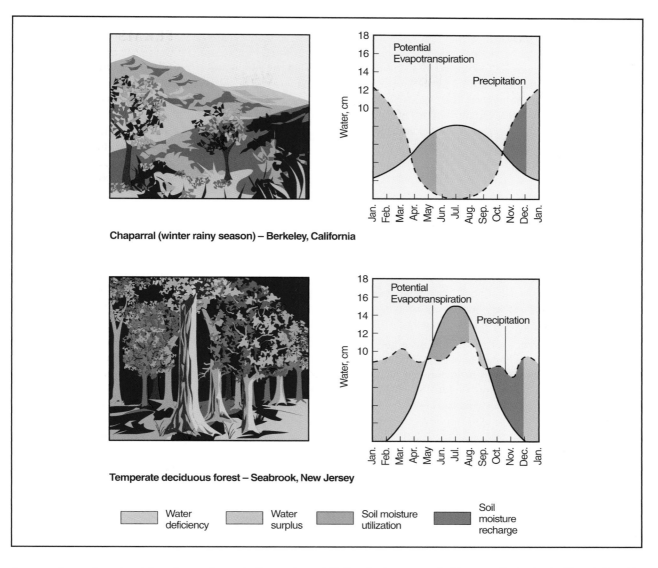

Chaparral (winter rainy season) – Berkeley, California

Temperate deciduous forest – Seabrook, New Jersey

Water deficiency   Water surplus   Soil moisture utilization   Soil moisture recharge

A comparison of yearly cycles of evapotranspiration and precipitation for two separate biomes. *Illustration by Hans & Cassidy. Courtesy of Gale Group.*

transpiration was not operating to consume thermal energy during the vaporization of water. Because of vigorous transpiration of water in moist **forests** during the day, air in and beneath the canopy is considerably cooler than it would be if this process did not occur.

Evapotranspiration from forests has a large influence on the water budget, both globally and locally. In the absence of forest transpiration, an equivalent quantity of water would have to leave the **watershed** as seepage to **groundwater** and streamflow. In temperate forests, evapotranspiration typically accounts for 10-40% of the total input of water by **precipitation**. However, there is a distinct seasonality in the patterns of evapotranspiration from forests. In a temperate region, evapotranspiration is largest during the growing season, when air temperatures are highest and the amount of plant foliage is at its annual maximum. During this season, evapotranspiration rates are often larger than the inputs of water by rainfall, and as a result groundwater is depleted, and sometimes the surface flow of forest streams disappears. Evapotranspiration is very small during the winter because of low temperatures, although some direct vaporization of water from snow or **ice** to vapor does occur, a process known as sublimation. Evapotranspiration is also relatively small during the springtime, because deciduous trees do not yet have foliage, and air temperatures are relatively cool.

The disturbance of forests disrupts their capacity to sustain evapotranspiration, by reducing the amount of foliage in the canopy. Consequently, clearcutting and **wildfire** often lead to an increase in streamflow, and sometimes to an increase in height of the water table. In general, the increase in streamflow is roughly proportional to the fraction of total foliage that is removed. In the first year after clear-cutting an entire watershed, the increase in streamflow can be as large as 40%, but most of this effect rapidly diminishes as vegetation re-grows after the disturbance.

See also Energy budgets.

Even and odd

An even number is an integer that is divisible by 2 and an odd number is an integer that is not divisible by 2. Thus 2, 4, 6, 8... are even numbers and 1, 3, 5, 7... are odd numbers. Any even number can be written as 2n where n is an integer and any odd number can be written as 2n + 1 or as 2n - 1, where n is an integer.

The sum, difference, or product of even numbers is an even number. On the other hand, the product of odd numbers is odd, but the sum or difference of two odd numbers is even.

Event horizon

In early 2001, data gathered by NASA's **Hubble Space Telescope** and the Chandra X-ray Observatory independently provided strong evidence of an event **horizon**, the observable boundary region surrounding an unobservable **black hole**.

The size of the event horizon surrounding a black hole is termed the Schwarzschild radius, named after the German astronomer Karl Schwarzschild (1873–1916), who studied the properties of geometric **space** around a singularity when warped according to general relativity theory. Gravitational fields bend **light**, and within a black hole the gravity is so strong that light is actually bent back upon itself. Another explanation in accord with the wave-particle duality of light is that inside the event horizon the gravitational attraction of the singularity is so strong that the required escape **velocity** for light is greater than the speed of light. As a consequence, because no object can exceed the speed of light, not even light itself can escape from the region of space within the event horizon.

More evidence gathered in 2002 convinced many scientists that in addition to **matter** spinning violently into the vortex of a black hole, the black hole itself rotates just like a **star**.

The event horizon is an observational boundary on the cosmic scale because no information generated within the black hole can escape. Processes occurring at or near the event horizon, however, are observable and offer important insights into the **physics** associated with black holes.

An **accretion disk** surrounding a black hole forms as matter accelerates toward the event horizon. The accelerating matter heats and emits strong highly energetic electromagnetic **radiation**, including **x rays** and gamma rays, that may be associated with some of the properties exhibited by quasars.

Although light and matter can not escape a black hole by crossing the event horizon, English physicist Stephen Hawking (1942–) formed an important concept later named Hawking radiation that offers a possible explanation to possible matter and **energy** leakage from black holes. Relying on quantum theories regarding **virtual particles** (particle-antiparticle pairs that exist so briefly only their effects (not their masses) can be measured), Hawking radiation occurs when a virtual particle crosses the event horizon and its partner particle cannot be annihilated and thus becomes a real particle with **mass** and energy. With Hawking radiation, mass can thus leave the black hole in the form of new particles created just outside the event horizon.

In general usage, an event horizon is a limiting boundary beyond which nothing can be observed or measured with regard to a particular phenomena (e.g., the cosmic event horizon beyond which nothing can be observed from **Earth**).

See also Astronomy; Astrophysics; Atomic models; Black hole; Blackbody radiation; Electromagnetic spectrum; Quantum mechanics; Quasar; Radio astronomy; Radio waves; Radio; Relativity, general; Relativity, special; Stellar evolution.

Evolution

Evolution refers to biological change. Biological evolution involves change in the genetic constitution of populations over **time** such that complexity is achieved due to the formation of new genes or gene-encoded functions rather than harmful mutations. These changes are passed on from parents to their offspring, but biological evolution does not involve **individual** organisms. Individuals develop, but they do not evolve. The study of evolution encompasses two major facets: deducing the history of descent among living things (phylogenetic systematics), and investigating **evolutionary mechanisms** as revealed by laboratory and field studies of living organisms and their fossilized remains. Evolutionary change occurs as a result of beneficial **mutation**, **migration**, genetic drift and natural **selection**, and it is ultimately a passive process, devoid of any purpose or goal. As a scientific theory, it is an interconnected series of hypotheses, corroborated by a large body of evidence. Thus, biologists accept the historical reality of evolution, even though many details that are unclear continue to be investigated.

Historical background

The birth of modern evolutionary theory can be traced to the mid-nineteenth century, and the publication of Charles Darwin's book, *The Origin of Species by Means of Natural Selection* in 1859. The book is considered by some to be the most influential in all of **biology** and by others an unsubstantiated theory. It was not, however, a new concept. Even in the late eighteenth century, the French scientist Maupertuis, the philosopher Diderot, and others entertained the notion that new life forms might arise by natural means, including **spontaneous generation**. Their definition of evolution more closely resembles our modern concept of development, since the course of evolution is unpredictable. Interestingly, Darwin did not use the word "evolution" in *The Origin of Species*, except in the form "evolved" which is the last word of the book.

French naturalist Jean-Baptiste Lamarck (1744–1829) was the first to clearly articulate the theory that species could change over time into new species, a process later called speciation. In his 1809 book *Philosophie zoologique*, he argued that living things progress inevitably toward greater perfection and complexity, through a process guided by the natural environment. A changing environment would alter the needs of living creatures, which would respond by using certain organs or body parts more or less. The use or disuse of these parts would consequently alter their size or shape, a change that would be inherited by the creatures offspring. Lamarck, although not the first, discussed this notion of "acquired characters." His contemporaries accepted. It is therefore unfortunate that Lamarck is famous today primarily because he was wrong in this belief, which Darwin upheld, in that it implies that bodily changes acquired during an individuals lifetime can affect the inherited material (now known as DNA) and be inherited. This hypothesis is not true as depicted in the example that a blacksmith's large muscles (an acquired character) are not passed on to the blacksmith's children.

Charles Robert Darwin (1809–1882) and his contemporary Alfred Russell Wallace (1823–1913) are cred-

ited with independently providing the first plausible theory for a mechanism to explain evolutionary change, which they called natural selection. However, Wallace did not present so great a synthesis, or consider all the ramifications of evolution, as Darwin did in *The Origin of Species* and his later works. One major difference between the two men was that Wallace did not believe that natural selection could have produced the human **brain**; Darwin rejected his colleague's contention that human intellect could only have been created by a higher power.

In *The Origin of Species*, Darwin concluded that some individuals in a species are better equipped to find food, survive **disease**, and escape predators than others. He reasoned that if the differences between the survivors and the doomed are passed from parents to their offspring, these differences will be inherited and gradually will characterize the whole population. Darwin knew that the greatest weakness in his theory was his lack of knowledge about the mechanics of inheritance; his attempt at a theory of inheritance, known as pangenesis, was later refuted. However, he is credited for his remarkable insight into the hidden variation in the form of traits that are genetically recessive and only manifest themselves when they occur as a genetic "double dose"— along. He also helped clarify the disctinction between **genotype and phenotype**. The genotype is a set of genetic instructions for creating an organism; the phenotype is the observable traits manifested by those instructions. The issue that explains the mechanism of inheritance, however, went unsolved for several decades later.

The modern synthesis

By the 1920s, Mendel's theory of heredity had been rediscovered, but the early Mendelians, such as William Bateson and Hugo de Vries, opposed Darwin's theory of gradual evolutionary change. These geneticists favored the position that evolution proceeds in large jumps, or macro-mutations, rather than the gradual, incremental changes proposed by Darwin. The dispute between the Mendelians and the Darwinians was reconciled by important theoretical work published in the early 1930s by independent population geneticists Ronald A. Fisher (1890–1962), J. B. S. Haldane (1892–1964), and Sewall Wright (1889–1988). Their classic work showed that natural selection was compatible with the laws of Mendelian inheritance.

This reconciliation gave rise to a period of prolific research in **genetics** as the basis of evolution. Theodosius Dobzhansky (1900–1975), a Russian emigrant to the United States, began his classic investigations of evolution in fruit fly populations, and in 1937 published a book entitled *Genetics and the Origin of Species*, which has been among the most influential books in modern genetics. Others, like E. B. Ford (1901–1988) and H. B. D. Kettlewell (1901–1979), helped pioneer the subject Ford called "ecological genetics." This was a fundamental concept because it brought together two formerly separate areas of investigation: ecological variation (the influence of different environmental conditions) and genetic variation (differences in genetic make-up) in natural populations.

Evidence of evolution

Darwin recognized that some of the best evidence of evolution remains hidden within the bodies of living creatures. He reasoned that, if organisms have a history, then their history can be deciphered from their remnants. In fact, virtually all living creatures possess vestigial or rudimentary features that were once functional traits of their ancestors. Fetal whales, still in their mothers' womb, produce teeth like all **vertebrates**, only to reabsorb them in preparation for a life filtering **plankton** from their **ocean habitat**. **Snakes**, whose vertebrate ancestors ceased walking on four legs millions of years ago, still possess vestigial hind limbs, with reduced hip and thigh bones.

In some cases the same structures may be adapted for new uses. The cat's paw, the dolphin's flipper, the bat's wing, and a human hand all contain counterparts of the same five bones forming the digits (fingers in humans). There is no known environmental or functional reason why there should be five digits. In theory, there could just as easily be four, or seven. The point is that the ancestor to all tetrapods (vertebrates with four legs) had five digits, and thus all living tetrapods have that number, although in a modified form. Such traits, when they reflect shared ancestry, are known as homologous structures or characters. Traits that have functional but not truly ancestral similarity, such as the wings of **insects** and the wings of **birds**, are known as analogous characteristics.

Further evidence of evolution involves the most basic homology of all, the genetic code, or the molecular "language" that is shared by all living things. Although the genes of each kind of organism may be represented by different "words," coding for each species unique structures, the molecular alphabet forming these words is the same for all living things on **Earth**. This is interpreted as evidence that the genetic code arose in a way that a single organism was the ancestor to all organisms that exist today.

Finally, there is the evidence from the fossil records that the remains of **invertebrates**, plants, and animals appear in the rocky layers of the earth's crust in the same order that their anatomical complexity suggests they should, with the more primitive organisms in the older layers and the more complex organisms in the more recent deposits. No one has ever found a flowering **plant** or a mammal in deposits from the Devonian age, for instance, because those organisms did not appear on Earth until much later.

Evolutionary mechanisms

To an evolutionary biologist, evolution is a change in the proportion of genes present in an existing population. Each individual in a breeding population possesses a unique genotype, the set of paired genetic alternatives that determines its physical attributes, or phenotype. According to a principle known as the Hardy-Weinberg equilibrium, if certain conditions prevail, the **frequency** of each genotype in the population will be constant over generations, unless evolution has occurred to cause a shift in the **gene** frequencies. What factors can cause the proportion of genotypes in a population to change?

Natural selection is one example. An organisms environment—including its habitat, diseases, predators, and others of its own kind—present numerous challenges to its survival and to its ability to reproduce. Individual organisms of the same species have slightly different phenotypes (observable attributes), and some of these individuals may be slightly better equipped to survive the hazards of existence than others. The survivors produce proportionally more offspring than do the others of their species, and so their traits will be better represented in subsequent generations. These traits lead to an outnumbered or skewed representation of the offspring compared to less successful individuals (in terms of reproductive capacities) with alternative genetic complements. The genotype that is better adapted to the prevailing conditions will spread throughout the population—at least until conditions change and new genotypes are favored.

Another means by which gene frequencies may change is genetic mutation. When organisms reproduce, a copy of the DNA from each parent is transmitted to the offspring. Normally, the parental DNA is copied exactly; but occasionally, errors occur during replication in their sex cells, the egg and the sperm. Many of these errors occur in non-coding regions of the DNA, and so may have no known effect on the offspring's phenotype; others may be lethal. In some cases, however, an offspring with a slightly modified genetic makeup survives. Mutation is thought to be an important source of new variation, particularly in rare cases where the mutation confers a selective survival advantage or a gain of function. After all, not all mutations are disadvantageous. In fact, natures method for ensuring the survival of a species is to rely on gene mutations that might enhance the function of the protein it encodes in such a way that there is a survival advantage. The determination of this advantage does not only rely on the type of mutation but the environment circumstances that prevail.

Evolution may also occur by genetic drift, a **random** process in which gene frequencies fluctuate due to chance alone. For example, if a certain genetic background is over-represented in one generation, perhaps because the region happened to be colonized by a small, homogeneous population, these genetic complements are more likely to remain high in the future generations. Genetic drift has more potent effects in small, isolated populations where some characteristic are not necessarily the result of natural selection and therefore do not represent a selective survival advantage.

Finally, the proportions of genotypes in a population may change as a result of migration, resulting in gene flow, or the movement of individuals (and their sex cells) into and out of the population.

Species diversity and speciation

Biologists have estimated that there are as many as 50 million species living on Earth today. Remarkably, it is believed that this figure is only 1% of the species that have ever lived. The great diversity of organisms, living and extinct, is the result of speciation, the splitting and divergence of biological lineages to create new species.

The term "species" is derived from the Latin word for kind, and species are considered to be the fundamental natural units of life which can and will evolve. Species have been defined in various ways over the years, but the most widely used definition at present is the biological species concept. Most often associated with the name of Ernst Mayr (1904-), it defines a species as a group of interbreeding populations that is reproductively separate from other such groups.

The crucial event for the origin of new species is reproductive separation. How do barriers to reproduction emerge? Perhaps a homogeneous population becomes separated into two distinct populations due to environmental factors. For example, when a river divides the species into two geographically distinct populations. Once they are separated, one or more evolutionary mechanisms may act on each in different ways over time, with the result that the two will diverge so they can no longer interbreed. As another example, perhaps a genetic variant arises in a population of interbreeding individuals by mutation or immigration; if the new variant spreads, bearers of the new variant may begin to breed preferentially with other bearers, preferring the new type. Along the way, further ecological, behavioral, and phenotypic changes may occur. Over many years, the organisms can diverge into two distinct species based on a

The long neck of the giraffe is an adaptation that allows it to survive competition for food resources. *Photograph by Stan Osolinski. Stock Market. Reproduced by permission.*

dually but mutually exclusive preferential mating selection process.

Phylogenetic systematics: Reconstructing evolutionary history

The process of classifying and reconstructing the evolutionary history, or **phylogeny**, of organisms is known as phylogenetic systematics. Its goal is to group species in ways that reflect a common ancestry. The members of each group, or taxon, share uniquely derived characteristics that have arisen only once. For instance, the taxon Amniota includes **amphibians**, **reptiles**, birds and **mammals**, all of which arose from a single common ancestor that possessed an amnion in the egg stage. Classifying species involves only one apect of phylogenetic systematics. Understanding the evolutionary interrelationships of organisms by investigating the mechanisms leading to diversification of life and the changes that take

KEY TERMS

. . . .

Genetic drift—Random change in gene frequencies in a population.

Genotype—The full set of paired genetic elements carried by each individual, representing the its genetic blueprint.

Hardy-Weinberg equilibrium—The principle that, if certain conditions prevail, the relative proportions of each genotype in a breeding population will be constant across generations.

Macromutation—Mutation having a pronounced phenotypic effect, one that produces an individual that is well outside the norm for the species as it existed previously.

Mutation—Alteration in the physical structure of the DNA, resulting in a genetic change that can be inherited.

Natural selection—The process of differential reproduction, in which some phenotypes are better suited to life in the current environment.

Nucleotide—Molecular unit that is the building block of DNA.

Phenotype—The outward manifestation of the genotype, including an organism's morphological, physiological, and many behavioral attributes.

Phylogeny—Branching tree diagram representing the evolutionary history of an organism, and its relationship to others via common ancestry.

Speciation—The divergence of evolutionary lineages, and creation of new species.

Systematics—The classification of organisms according to their evolutionary relationships, and shared genetic and phenotypic characteristics.

Taxon (taxa)—Group of related organisms at one of several levels, such as the family Hominidae, the genus Homo, or the species *Homo sapiens*.

Tetrapod—The group of vertebrates having four legs, including amphibians, reptiles, mammals and birds.

place over time encompasses phylogenetic systematics. Systematics exceeds **taxonomy** or naming groups within species by attempting to develop new theories to describe the possible mechanisms of evolution. Historical remnants leave residual clues that allow phylogeneticists to piece together using hypotheses and models to describe history and how organisms evolve.

Systematists gather as much evidence as they can concerning the physical, developmental, ecological, and behavioral traits of the species they wish to group, and the results of their analysis are one (or more) branching "tree" diagrams, representing the hypothetical relationships of these taxa. Analysis of the genetic material itself has become an increasingly valuable tool in deducing phylogenetic relationships. The approach is to determine the nucleotide sequence of one or more genes in the species and use this to contribute to determining the appropriate phylogeny. Comparison of the differences in these DNA sequences provides an estimate of the time elapsed since the divergence of the two lineages. With the ability to sequence an entire organisms **genome**, the remnants of various species DNA, extinct (depending on the quality of the DNA) and living, can be compared and contrasted. This has exciting implications in the analysis of evolutionary modeling and applications to phylogenetic systematics.

See also Adaptation; Competition; Extinction; Opportunistic species.

Resources

Books

Futuyma, Douglas J. *Science on Trial: The Case for Evolution.* New York: Random House, 1983.

Gould, Stephen J. *Ever Since Darwin.* New York: W. W. Norton, 1977.

Gould, Stephen J. *The Structure of Evolutionary Theory.* Cambridge, MA: Harvard University Press, 2002.

Ridley, Mark. *Evolution.* Cambridge, MA: Blackwell Scientific Publications, 1993.

Other

PBS. "Evolutionary Thought." 2001 [cited January 13, 2003]. <http://www.pbs.org/ wgbh/evolution/>.

University of California at Berkeley. "Welcome to the Evolution Wing." UCMP exhibit halls. November 15, 2002 [cited January 13, 2003]. <http://www.ucmp.berkeley.edu/ history/evolution.html>.

Susan Andrew

Evolution, convergent

Convergent **evolution** represents a phenomenon when two distinct **species** with differing ancestries evolve to display similar physical features. Environmental circumstances that require similar developmental or structural alterations for the purposes of **adaptation** can lead to convergent evolution even though the species differ in descent. These adaptation similarities that arise as a result of the same selective pressures can be misleading to scientists studying the natural evolution of a species. For example, the wings of all flying animals are very similar because the same laws of **aerodynamics** apply. These laws determine the specific criteria that govern the shape for a wing, the size of the wing, or the movements required for flight. All these characteristics are irrespective of the **animal** involved or the physical location. In various species of plants, which share the same pollinators, many structures and methods of attracting the pollinating species to the **plant** are similar. These particular characteristics enabled the reproductive success of both species due to the environmental aspects governing **pollination**, rather than similarities derived by being genetically related by descent.

One of the best examples of convergent evolution involves how **birds**, **bats**, and pterosaurs (all different species that evolved along distinct lineages at different times) learned to fly. Importantly, each species developed wings independently. These species did not evolve in order to prepare for future circumstances, but rather the development of flight was induced by selective pressure imposed by similar environmental conditions, even though they were at different points in **time**. The development potential of any species is not limitless, primarily due to inherent constraints in genetic capabilities. Only changes that are useful in terms of adaptation are preserved. Yet, changes in environmental conditions can lead toward less useful functional structures, such as the appendages that might have existed before wings. Another change in environmental conditions might result in alterations of the appendage to make it more useful, given the new conditions.

Understanding the reason why each different species developed the ability to fly relies on an understanding of the possible functional adaptations, based on the **behavior** and environmental conditions to which the species was exposed. Although only theories can be made about extinct species and flight since these behaviors can be predicted using by fossil records, these theories can often be tested using information gathered from their remains. Perhaps the wings of bird or bats were once appendages used for other purposes, such as gliding, sexual display, leaping, protection, or arms to capture **prey**. Convergent evolution is supported by the fact that these species come from different ancestors, which has been proven by DNA analysis. However, understanding the mechanisms that brings about these similarities in characteristics of a species, despite the differences in **genetics**, is more difficult.

Convergent evolution creates problems for paleontologists using evolutionary patterns in **taxonomy**, or the categorization and classification of various organisms based on relatedness. It often leads to incorrect relationships and false evolutionary predictions.

See also Competition; Extinction; Evolution, parallel; Evolution, evidence of.

Resources

Books

Merrell, David J. *The Adaptive Seascape: The Mechanism of Evolution.* Minneapolis: University of Minnesota Press, 1994.
Gould, Stephen Jay. *The Structure of Evolutionary Theory.* Cambridge, MA: Harvard University Press, 2002.

Periodicals

Berger, Joel, and Kaster, "Convergent Evolution." *Evolution* (1979): 33:511.

Other

PBS. "Convergent evolution." Evolution Library. 2001 [cited January 14, 2003]. <http://www.pbs.org/wgbh/evolution/library/01/4/l_014_01.html.>

Bryan Cobb

Evolution, divergent

Divergent **evolution** occurs when a group from a specific population develops into a new **species**. In order to adapt to various environmental conditions, the two groups develop into distinct species due to differences in the demands driven by the environmental circumstances. A good example of how divergent evolution occurs is in comparing how a human foot evolved to be very different from a monkey's foot, despite their common primate ancestry. It is speculated that a new species (humans) developed because there was no longer was a need for swinging from trees. Upright walking on the ground required alterations in the foot for better speed and balance. These differing traits soon became characteristics that evolved to permit movement on the ground. Although humans and **monkeys** are genetically similar, their natural **habitat** required different physical traits to evolve for survival.

If different selective pressures are placed on a particular **organism**, a wide variety of adaptive traits may result. If only one structure on the organism is considered, these changes can either add to the original function of the structure, or they can change it completely. Divergent evolution leads to speciation, or the development of a new species. Divergence can occur when looking at any group of related organisms. The differences are produced from the different selective pressures. Any genus of plants or animals can show divergent evolution. An example can involve the diversity of floral types in the orchids. The greater the number of differences present, the greater the divergence. Scientists speculate the greater that two similar species diverge indicates a longer length of **time** that the divergence originally took place.

There are many examples of divergent evolution in nature. If a freely-interbreeding population on an **island** is separated by a barrier, such as the presence of a new river, then over time, the organisms may start to diverge. If the opposite ends of the island have different pressures acting upon it, this may result in divergent evolution. Or, if a certain group of **birds** in a population of other bird of the same species varies from their migratory track due to abnormal **wind** fluctuations, they may end up in new environment. If the food source is such that only birds of the population with a variant beak are able to feed, then this trait will evolve by virtue of its selective survival advantage. The same species in the original geographical location and having the original food source do not require this beak trait and will, therefore, evolve differently.

Divergent evolution has also occurred in the red fox and the kit fox. While the kit fox lives in the **desert** where its coat helps disguise it from its predators, the red fox lives in **forests**, where the red coat blends into its surroundings. In the desert, the **heat** makes it difficult for animals to eliminate body heat. The ears of the kit fox have evolved to have greater surface area so that it can more efficiently remove excess body heat. Their different evolutionary fates are determined primarily on the different environmental conditions and **adaptation** requirements, not on genetic differences. If they were in the same environment, it is likely that they would evolve similarly. Divergent evolution is confirmed by DNA analysis where the species that diverged can be shown to be genetically similar.

See also Evolution, convergent; Extinction; Genetics; Opportunistic species.

Resources

Books

Merrell, David J. *The Adaptive Seascape: The Mechanism of Evolution.* Minneapolis: University of Minnesota Press, 1994.

Gould, Stephen Jay. *The Structure of Evolutionary Theory.* Cambridge, MA: Harvard University Press, 2002.

Ridley, Mark. *Evolution.* Cambridge, MA: Blackwell Scientific Publications, 1993.

Other

University of California at Berkeley. "Welcome to the Evolution Wing." UCMP exhibit halls. November 15, 2002 [cited January 13, 2003]. <http://www.ucmp.berkeley.edu/history/evolution.html>.

BioWeb. "Divergent and Convergent Evolution." Earlham. [cited January 14, 2003]. <http://bioweb.cs.earlham.edu/9-12/evolution/HTML/converge.html>.

Bryan Cobb

Evolution, evidence of

Evidence of **evolution** can be observed in numerous ways, including distribution of **species** (both geographically and through **time**), comparative anatomy, **taxonomy**, **embryology**, **cell biology**, **molecular biology**, and **paleontology**.

The English naturalist Charles Darwin (1809–1882) formulated the theory of evolution through natural **selection** in his ground breaking publication *The Origin of Species by Means of Natural Selection,* published in 1859. One of the first observations that Darwin made, prompting him to become a pioneer of evolutionary thinking, was on his journey aboard the HMS *Beagle* as a naturalist. Prior to the work of Darwin, most people accepted the biblical account of creation where all animals and plants were brought into the world, also called creationism. Darwin made extensive collections of the plants and animals that he came across wherever the ship stopped, and very soon he started to notice patterns within the organisms he studied.

Similarities emerged between organisms collected from widely differing areas. As well as the similarities, there were also striking differences. For example, **mammals** are present on all of the major landmasses; however, these mammals are different, even in similar habitats. One explanation of this is that in the past when the landmasses were joined, mammals spread over all of the available land. Subsequently, this land moved apart, and the animals became isolated. As time passed, **random** variation due to natural selection within the populations occurred. This process is known as adaptive **radiation**. From the same basic origin, many different forms have evolved. Each environment is slightly different, and slightly different forms are better suited to survive.

Evolution has historically been a subject of considerable debate mainly due to the difficulties in testing scientific hypotheses that are inherently associated with the topic. It is difficult to imagine how the magnanimous interspecies as well as intraspecies diversification could exist without evolutionary pathways. There are many critics that have alternative theories to describe this diversification, however, many lines of evidence exist that lend credence to the theory of evolution. This evidence helps to understand why, for example, islands in the Galapagos are inhabited be various organisms that are similar to those in the mainland, but belong to a different species, or why thousands of snail species and other **mollusks** are found only in Hawaii. A possible biological explanation is that the diversity and number of related species in these isolated habitats is a consequence of the **evolutionary mechanisms** that arose from a small number of common ancestors that were the original inhabitants. With the greatness in ecological diversity in different parts of the world and the absence of species competing for resources necessary for survival, diversification from evolutionary change was made possible.

If it is true that widely separated groups of organisms have ancestors in common, then intuitively it suggests they would have certain basic structures in common as well. The more structures they have in common, the more closely related they must be. The study of evolutionary relationships based on similarities and differences in the structural makeup of certain species is called comparative **anatomy**. What scientists look for are structures that may serve entirely different functions, but are basically similar. Such homologous (meaning the same) structures suggest a common ancestor. A classic example of this is the pentadactyl (five digits, as in the hand of humans) limb, which in suitably modified forms can be seen in all mammals. A even greater modified version of this can also be seen amongst **birds**.

Evolutionary relationships are reflected in taxonomy. Taxonomy is an artificial, hierarchical system showing categorical relationships between species based on specific defining characteristics. Each level within the taxonomic system denotes a greater degree of relatedness to a particular the **organism** if it is closer in the hierarchical scheme.

In embryology, the developing fetus is studied, and similarities with other organisms are observed. For example, annelids and mollusks are very dissimilar as adults. If, however, the embryo of a ragworm and a whelk are studied, one sees that for much of their development they are remarkably similar. Even the larvae of these two species are very much alike. This suggests that they both belong to a common ancestor. It is not, however, true that a developing organism replays its evolutionary stages as an embryo. There are some similarities with the more conserved regions, but embryonic development is subjected to evolutionary pressures as much as other areas of the life cycle.

The analysis of developmental stages of various related species, particularly involving reproduction suggests common descent and supports evolution. **Sexual reproduction** in both **apes** and humans, for example, is very similar. Molecular biology also produces evidences supporting evolution. Organisms such as fruit **flies** have similar **gene** sequences that are active in specific times during development and these sequences are very similar to sequences in **mice** and even humans that are activated in similar ways during development.

Even in cell biology, at the level of the individuals cell, there is evidence of evolution in that there are many similarities that can be observed when comparing various cells from different organisms. Many structures and pathways within the cell are important for life. The more important and basic to the functioning of the tissues in which cells contribute, the more likely it will be conserved. For example, the DNA code (the genetic material in the cell) is the very similar in comparing DNA from different organisms.

In molecular biology, the concept of a molecular clock has been suggested. The molecular clock related to the average **rate** in which a gene (or a specific sequence of DNA that encodes a protein) or protein evolves. Genes evolve at different rates the **proteins** that they encode. This is because gene mutations often do not change the protein. But due to mutations, the genetic sequence of a species changes over time. The more closely related the two species, the more likely they will have similar sequences of their genetic material, or DNA sequence. The molecular clock provides relationships between organisms and helps identify the point of divergence between the two species. Pseudogenes are genes that are part of an organisms DNA but that have evolved to no longer have important functions. Pseudogenes, therefore, represent another line of evidence supporting evolution, which is based on concepts derived from molecular **genetics**.

Perhaps the most persuasive argument that favors evolution is the fossil record. Paleontology (the study of fossils) provides a record that many species that are extinct. By techniques such as **carbon** dating and studying the placement of fossils within the ground, an age can be assigned to the fossil. By placing fossils together based on their ages, a gradual change in form can be identified, which can be carefully compared to species that currently exist. Although fossil records are incomplete, with many intermediate species missing, careful analysis of **habitat**, environmental factors at various timepoints, characteristics of extinct species, and characteristics of species that currently exist supports theories of evolution and natural selection.

See also Evolution, convergent; Evolution, divergent; Evolution, parallel; Evolutionary mechanisms; Extinction; Opportunistic species.

Resources

Books

Gould, Stephen Jay. *The Structure of Evolutionary Theory.* Cambridge, MA: Harvard University Press, 2002.

Merrell, David J. *The Adaptive Seascape: The Mechanism of Evolution.* Minneapolis: University of Minnesota Press, 1994.

Ridley, Mark. *Evolution.* Cambridge, MA: Blackwell Scientific Publications, 1993.

Other

Knowledge Matters, Ltd. "The Voyage of the Beagle" Charles Darwin. June 29, 1999 [cited January 17, 2003]. <http://www.literature.org/authors/darwin-charles/the-voyage-of-the-beagle/chapter-18.html>.

Bryan Cobb

Evolution, parallel

Parallel **evolution** occurs when unrelated organisms develop the same characteristics or adaptive mechanisms due to the nature of their environmental conditions. Or stated differently, parallel evolution occurs when similar environments produce similar adaptations. The morphologies (or structural form) of two or more lineages evolve together in a similar manner in parallel evolution, rather than diverging or converging at a particular point in **time**.

Parallel evolution is exemplified in the case of the tympanal and atympanal mouthears in hawkmoths, or Sphingidae **species**. These **insects** have developed a tympanum, or eardrum, similar to humans as a means to communicate through sound. Sounds induce vibrations of a **membrane** that covers the tympanum, known as the tympanic membrane. These vibrations are detected by small **proteins** at the surface of the tympanic membrane called auditory receptors. Within the Sphingidae species, two differing subgroups acquired **hearing** capability by developing alterations in their mouthparts by a distinctly independent evolutionary pathway.

Investigating the biomechanics of the auditory system reveals that only one of these subgroups has a tympanum. The other subgroup has developed a different mouthear structure that does not have a typanum, but has a mouthear with functional characteristics essentially the same as the subgroup with the tympanum. The evolutionary significance of how hearing capabilities developed in parallel in two different subgroups of a species reveals that distinct mechanisms can exist leading to similar functional capabilities with differing means for acquiring the same functional attribute. For both subgroups, hearing

---

KEY TERMS

Morphological—relates to the study of differences in shape or structure

Natural selection—The process of differential reproduction, in which some phenotypes are better suited to life in the current environment.

Speciation—The divergence of evolutionary lineages, and creation of new species.

Tympanum—The eardrum.

---

must have been an important characteristic for the species to survive given the environmental conditions.

Parallel speciation is a type of parallel evolution in which reproductive incompatibility in closely related populations is determined by traits that independently evolve due to **adaptation** to differing environments. These distinct populations are reproductively incompatible and only populations that live in similar environmental conditions are less likely to become reproductively isolated. In this way, parallel speciation suggests that there is good evidence for natural selective pressures leading to speciation, especially since reproductive incompatibility between to related populations is correlated with differing environmental conditions rather than geographical or genetic distances.

See also Evolution, convergent; Evolution, divergent; Evolutionary mechanisms; Survival of the fittest.

Resources

Books

Gould, Stephen Jay. *The Structure of Evolutionary Theory.* Cambridge, MA: Harvard University Press, 2002.

Other

Encyclopedia Britannica. "Parallel Evolution" Encyclopaedia Britannica Premium Service. [cited February 23, 2003]. <http://www.britannica.com/search?miid=1215119&query=parallel+evolution>.

Evolutionary change, rate of

Rates of **evolution** change vary widely over **time**, among characteristics, and among **species**. Evolutionary change can be estimated by examining fossils and species that are related to each other. The **rate** of change is governed by the life span of the species under examination, short-lived species are capable of changing more quickly than those that have a longer life span and reproduce less often. Yet, even short-lived species such as **bacteria**, which have generation times measured in minutes, do not manifest noticeable evolutionary changes in a humans lifetime.

One technique that has been used to examine the rate of evolutionary change is DNA analysis. This technology involves identifying the percentage of similarity between samples of DNA from two related organisms under study. The greater the similarity, the more recently the organisms are considered to have diverged from a common ancestor. The information that is obtained in this manner is compared to information obtained from other sources such as the fossil records and studies in comparative anatomy.

There are two competing hypotheses designed to explain the rate of evolutionary change. One is called the **punctuated equilibrium** hypotheses. This hypothesis states that there are periods of time in which the rate of evolutionary change is slow. These periods are interspersed with periods of rapid change. Rapid change usually prevails with the early establishment of a species, during which time the **organism** has not yet development environmental **adaptation**. For example, if a food source eventually becomes unavailable to a particular species, the rate of evolutionary change will increase rapidly. In this way, new characteristics will evolve so that the species can utilize a different food source or the species will become extinct. Once this trait is common within the species, the rate of evolutionary change will slow down. An example of punctuate equilibrium occurred millions of years ago when aquatics species inhabited the emerging land spaces. These species rapidly developed characteristics permitting life as land dwellers. Once these characteristics were developed and common among these species, the rate of evolutionary change slowed.

Another hypothesis is the gradual change hypothesis. This explanation of the evolutionary rate of change states that species evolve slowly over time. In this hypothesis, the rate of change is slow and species that do not change quickly enough to develop traits enabling them to survive will die. Although some species such as the **sequoia** (redwoods) or **crocodiles** have maintained distinct and similar characteristics over millions of years, some species such as the cichlids in the African rift lakes have rapidly change in appearance over thousands of years. These examples of slow changing species involve organisms that have an arrested rate of evolutionary change, but at one time, most likely endured a period of rapid evolutionary change resulting in tolerance to environmental changes.

There are several limiting factors that might control the evolutionary rate of change. Physical or biological changes influencing the environment can play a major role in the rate of evolutionary change. Another example is the **mutation** rate. The mutation rate in various species at different times can also limit the rate of evolutionary change such that a higher mutation rate correlates to a higher rate of change, assuming the environmental changes remain unchanged. Mutations do not seem to have major effects on limiting evolution because diversity in morphological evolution (evolution of physical characteristics) does not correlate well with DNA mutation rates. However, in some cases, evolution rates can depend on mutation rates. A good example is antibiotic resistance. Bacterial mutation rates can induce changes in the ability to become resistance to **antibiotics**.

If certain characteristic are more efficiently selected against in one species compared with a different species, then the rates of evolutionary change will vary between them. Selective pressures are greater in larger populations. Therefore, small populations might not be able to evolve rapidly enough in a rapidly changing environment. In this scenario, inefficient natural **selection** will be a **limiting factor** in the rate of evolution. Finally, constraints that occur when a mutation in a **gene** produces a beneficial characteristic for the species but impedes the function of other gene products can be a limiting factor on the rate of evolution. These architectural constraints can be bypassed if gene or whole genomes are duplicated. When this occurs, the extra genes can compensate for the negative effects on gene function imposed by the beneficial new function from the gene that is mutated. In a sense, these extra genes will speed up the rate of evolutionary change.

See also Evolution, convergent; Evolution, divergent; Evolution, parallel; Evolutionary mechanisms; Extinction; Genetics; Opportunistic species; Survival of the fittest.

Resources

Books

Gould, Stephen Jay. *The Structure of Evolutionary Theory.* Cambridge, MA: Harvard University Press, 2002.

Milligan, B.G. *Estimating Evolutionary Rates for Discrete Characters.* Clarendon Press; Oxford, England, 1994.

Ridley, Mark. *Evolution.* Cambridge, MA: Blackwell Scientific Publications, 1993.

Bryan Cobb

Evolutionary mechanisms

Evolution is the process of biological change over **time**. Such changes, especially at the genetic level are accomplished by a complex set of evolutionary mechanisms that act to increase or decrease genetic variation.

Evolutionary theory is the cornerstone of modern **biology**, and unites all the fields of biology under one theoretical umbrella to explain the changes in any given **gene** pool of a population over time. Evolutionary theory is theory in the scientific usage of the word. It is more than a hypothesis; there is an abundance of observational and experimental data to support the theory and its subtle variations. These variations in the interpretation of the role of various evolutionary mechanisms are due to the fact that all theories, no matter how highly useful or cherished, are subject to being discarded or modified when verifiable data demand such revision. Biological evolutionary theory is compatible with nucelosynthesis (the evolution of the elements) and current cosmological theories in **physics** regarding the origin and evolution of the universe. There is no currently accepted scientific data that is incompatible with the general postulates of evolutionary theory, and the mechanisms of evolution.

Fundamental to the concept of evolutionary mechanism is the concept of the syngameon, the set of all genes. By definition, a gene is a hereditary unit in the syngameon that carries information that can be used to construct **proteins** via the processes of transcription and translation. A gene pool is the set of all genes in a **species** or population.

Another essential concept, important to understanding evolutionary mechanisms, is an understanding that there are no existing (extant) primitive organisms that can be used to study evolutionary mechanism. For example, all eukaryotes derived from a primitive, common prokaryotic ancestral bacterium. Accordingly, all living eukaryotes have evolved as eukaryotes for the same amount of time. Additionally, no eukaryote **plant** or **animal cell** is more primitive with regard to the amount of time they have been subjected to evolutionary mechanisms. Seemingly primitive characteristics are simply highly efficient and conserved characteristics that have changed little over time.

Evolution requires genetic variation, and these variations or changes (mutations) can be beneficial, neutral or deleterious. In general, there are two major types of evolutionary mechanisms, those that act to increase genetic variation, and mechanisms that operate to decrease genetic mechanisms.

Mechanisms that increase genetic variation include **mutation**, recombination and gene flow.

Mutations generally occur via chromosomal mutations, point mutations, frame shifts, and breakdowns in DNA repair mechanisms. Chromosomal mutations include translocations, inversions, deletions, and **chromosome** non-disjunction. Point mutations may be nonsense mutations leading to the early termination of protein synthesis, missense mutations (a that results an a substitution of one **amino acid** for another in a protein), or silent mutations that cause no detectable change.

Recombination involves the re-assortment of genes through new chromosome combinations. Recombination occurs via an exchange of DNA between homologous chromosomes (crossing over) during **meiosis**. Recombination also includes linkage disequilibrium. With linkage disequilibrium, variations of the same gene (**alleles**) occur in combinations in the gametes (sexual reproductive cells) than should occur according to the rules of probability.

Gene flow occurs when individuals change their local genetic group by moving from one place to another. These migrations allow the introduction of new variations of the same gene (alleles) when they mate and produce offspring with members of their new group. In effect, gene flow acts to increase the gene pool in the new group. Because genes are usually carried by many members of a large population that has undergone **random** mating for several generations, random migrations of individuals away from the population or group usually do not significantly decrease the gene pool of the group left behind.

In contrast to mechanisms that operate to increase genetic variation, there are fewer mechanisms that operate to decrease genetic variation. Mechanisms that decrease genetic variation include genetic drift and natural **selection**.

Genetic drift results form the changes in the numbers of different forms of a gene (allelic **frequency**) that result from **sexual reproduction**. Genetic drift can occur as a result of random mating (random genetic drift) or be profoundly affected by geographical barriers, catastrophic events (e.g., natural disasters or wars that significantly affect the reproductive availability of selected members of a population), and other political-social factors.

Natural selection is based upon the differences in the viability and reproductive success of different genotypes with a population (differential reproductive success). Natural selection can only act on those differences in genotype that appear as phenotypic differences that affect the ability to attract a mate and produce viable offspring that are, in turn, able to live, mate and continue the species. Evolutionary fitness is the success of an entity in reproducing (i.e., contributing alleles to the next generation).

There are three basic types of natural selection. With directional selection an extreme phenotype is favored (e.g., for height or length of neck in giraffe). Stabilizing selection occurs when intermediate phenotype is fittest (e.g., neither too high or low a body weight) and for this reason it is often referred to a normalizing selection. Disruptive selection occurs when two extreme phenotypes are fitter that an intermediate phenotype.

Natural selection does not act with foresight. Rapidly changing environmental conditions can, and often do, impose new challenges for a species that result in **extinction**. In addition, evolutionary mechanisms, including natural selection, do not always act to favor the fittest in any population, but instead may act to favor the more numerous but tolerably fit. Thus, the modern understanding of evolutionary mechanisms does not support the concepts of social Darwinism.

The operation of natural evolutionary mechanisms is complicated by geographic, ethnic, religious, and social groups and customs. Accordingly, the effects of various evolution mechanisms on human populations are not as easy to predict. Increasingly sophisticated statistical studies are carried out by population geneticists to characterize changes in the human **genome**.

See also Chromosome mapping; Evolution, convergent; Evolution, divergent; Evolution, evidence of; Evolution, parallel; Evolutionary change, rate of; Genetic engineering; Genetic testing; Genotype and phenotype; Molecular biology.

Resources

Books

Beurton, Peter, Raphael Falk, Hans-Jörg Rheinberger., eds. *The Concept of the Gene in Development and Evolution.* Cambridge, UK: Cambridge University Press, 2000.

Bonner, J.T. *First Signals: The Evolution of Multicellular Development.* Princeton, NJ: Princeton University Press, 2000.

Lewin, B. *Genes.* 7th ed. New York, Oxford University Press Inc., 2000

Lodish, H., et. al. *Molecular Cell Biology.* 4th ed. New York: W. H. Freeman & Co., 2000.

Periodicals

Fraser, C.M., J. Eisen, R.D. Fleischmann, K.A. Ketchum, and S. Peterson. "Comparative Genomics and Understanding of Microbial Biology." *Emerging Infectious Diseases* 6, no. 5 (September-October 2000).

Veuille, E. "Genetics and the Evolutionary Process." *C. R. Acad. Sci III* 323, no.12 (December 2000):1155–65.

K. Lee Lerner

Excavation methods

Archeological excavation involves the removal of **soil**, sediment, or rock that covers artifacts or other evidence of human activity. Early excavation techniques involved destructive **random** digging and removal of objects with little or no location data recorded. Modern excavations often involve slow, careful extraction of sediments in very thin layers, detailed sifting of sediment samples, and exacting measurement and recording of artifact location.

About the time of the American Revolution, the then-future U.S. president Thomas Jefferson began excavating Indian burial mounds that had been constructed on his property in Virginia. His technique, which was to dig trenches and observe the successive **strata**, or layers of soil, anticipated the techniques of modern **archaeology**.

Between 1880 and 1890, General Pitt-Rivers initiated the practice of total site excavation, with emphasis on **stratigraphy** and the recording of the position of each object found. In 1904, William Mathew Flinders Petrie established principles of surveying and excavation that emphasized the necessity of not damaging the monuments being excavated, of exercising meticulous care when excavating and collecting artifacts, of conducting detailed and accurate surveys, and of publishing the findings as soon as possible following the excavation. In the same year, the archeologists R. Pumpelly and Hubert Schmidt, working in Turkestan, had already begun using sifting techniques to save small objects, and were recording the vertical and horizontal locations of even the smallest objects in each cultural layer.

Today, archeology is still equated with excavation in the popular mind. Most sites are no longer fully excavated unless in danger of destruction from building or **erosion**. Archaeologists leave a portion of some sites unexcavated to preserve artifacts and context for future research. Furthermore, there are now many alternatives to excavation for studying an archeological site, including surface archeology in which surface-exposed artifacts are detected and recorded; **remote sensing**; and the examination of soil and **plant** distributions. These techniques are nondestructive, and permit the archeologist to easily examine large areas. But even today, in almost all archeological projects, there comes a time when it is necessary to probe beneath the surface to obtain additional information about a site.

Before any excavation is begun, the site must be located. Techniques used to find a site may include remote sensing (for example, by aerial **photography**), soil surveys, and walk-through or surface surveys. The digging of shovel tests, augured core samples and, less commonly, trenches may also be used to locate archaeological sites. Soil samples may be collected from various sites and depths to determine whether any buried features are present.

When planning an archeological excavation, archeologists often use nondestructive techniques such as electrical resistivity meters and magnetometers to locate structures and artifacts on the site without digging. Soil testing may shed light on settlement patterns connected with the site. Aerial photography can also provide useful informa-

Dinosaur dig site, where it is thought that approximately 10,000 animals were buried by cretaceous volcanic ash flows. *Photograph by James L. Amos. Corbis. Reproduced by permission.*

tion for planning the excavation. In unobstructed fields, past human occupation of an area is evident through visible soil stains left by plowing, digging, and construction.

Before beginning the actual excavation, an archeologist prepares a topographical **map** of the site that includes such details as roads, buildings, bodies of **water**, and various survey points. This allows researchers to compare site location with natural landforms or regional terrain to establish settlement patterns, a theory about where people used to live and why they chose to live there.

Prior to excavating the site, a series of physical gridlines are placed over it to serve as points of reference. In staking out the grid, the archeologist essentially turns the site into a large piece of graph **paper** that can be used to chart any finds. The site grid is then mapped onto a sheet of paper. As objects are discovered in the course of the excavation, their locations are recorded on the site map, photographed in place, and catalogued.

Excavation strategies

Archeology has undergone radical changes since the time when an excavation was simply a **mining** of arti-

facts. Today, the removal of artifacts requires that the spatial relationships and context in which they are found be fully documented.

When an archeologist documents a find, he considers both vertical and horizontal relationships. Vertical relationships may yield information about the cultural history of the site, and horizontal relationships, about the way the site was used. In vertical excavation, the archeologist may use test units to identify and/or remove strata. Many archaeologists excavate sites in arbitrary levels, small increments of excavation, paying close attention to any changes in soil **color** or texture to identify various strata. In horizontal excavation, the archeologist may plow strips along the surface of the site to expose any objects lying near the surface. The excavation of a site proceeds by these methods until, layer by layer, the foundations of the site are uncovered. Often, excavation ends when sterile levels, strata without artifacts, are repeatedly uncovered.

Conventional excavation tools include, roughly in order of decreasing sensitivity, a magnifying **glass**, tape measure, pruning shears, bamboo pick, whiskbroom and dustpan, grapefruit knife, trowel, army shovel, hand pick, standard pick, shovel, and perhaps in some cases, even a bulldozer. Most of the excavation work is done with a shovel, but whenever fragile artifacts are encountered, the hand trowel becomes the tool of choice.

Mapping and recording

Archeologists record spatial information about a site with the aid of maps. Measuring tools range from simple tapes and plumb bobs to **laser** theodolites. The **accuracy** of a map is the degree to which a recorded measurement reflects the true value; the precision of the map reflects the consistency with which a measurement can be repeated.

In the course of an excavation, the archeologist carefully evaluates the sequential order that processes such as the collapse of buildings or the digging of pits contribute to the formation of a site. In addition, the archeologist typically notes such details as soil color and texture, and the presence and size of any stones.

The way the research proceeds at the site will depend on the goal of the excavation. If the purpose of the excavation is to document the placement of all retrieved artifacts and fragments for the purpose of piecing broken objects back together, the level of recording field data will be much finer than if the goal is simply to retrieve large objects. In cases where the goal of the site research is, for example, to recover flakes and chips of worked stone, digging at the site typically involves a trowel and whisk broom, and almost always, screening or sifting techniques. One-quarter-inch (6 mm) screens are usually fine enough for the recovery of most bones and artifacts, but finer meshes may be used in the recovery of **seeds**, small bones, and chipping debris. When a screen is used, shovels full of soil are thrown against or on the screen so that the dirt sifts through it, leaving any artifacts behind.

Another technique frequently utilized in the recovery of artifacts is water-screening. By using a water pump to hose down the material to be screened, the process of recovery is sped up and the loss of objects that might be missed or damaged in the course of being dry-screened can be avoided. A drawback in using this recovery technique is that it generates large quantities of mud that may cause environmental damage if dumped into a stream.

Elaborate flotation techniques may be used to recover artifacts, seeds, small bones, and the remains of charred plant material. In these techniques, debris from the site is placed in a container of pure or chemically treated water. The container is then shaken, causing small objects to float to the surface where they can be recovered. Archeologists have developed elaborate modifications of this technique, some involving multiple trays for the sorting of objects, to facilitate the recovery of artifacts. Many of these artifacts are analyzed under a **microscope** to look for clues about manufacture

KEY TERMS

. . . .

Electrical resistivity—A remote sensing technique that determines the character of subsurface sediments or the presence of objects based on variations in the resistance to an electrical current passing through the subsurface.

Magnetometer—An instrument designed to measure the strength of a magnetic field; magnetometers detect the presence of metallic objects that distort Earth's magnetic field.

Stratigraphy—The study of layers of rock or soil, based on the assumption that the oldest material will usually be found at the bottom of a sequence.

Theodolite—An optical instrument consisting of a small telescope used to measure angles in surveying, meteorology, and navigation.

Topographic map—A map illustrating the elevation or depth of the land surface using lines of equal elevation; also known as a contour map.

and use. The smallest artifacts, microartifacts such as pollen and seeds, require the use of a microscope for simple identification.

Prior to being sent to the laboratory for processing, artifacts are placed in a bag that is labeled with a code indicating where and in which stratigraphic layer the artifacts were found. All relevant information about an artifact is recorded in the field notes for the site.

Publication of findings

Because excavation permanently destroys at least a portion of a site as a source of archeological data for future generations, it is essential that the results of an excavation be promptly published in a form that is readily accessible. Current practice is to publish only portions of the complete field report, which is based on analyses of physical, biological, stratigraphic, and chronological data. But many archeologists are of the opinion that the public, which is widely viewed as having collective ownership of all matters relating to the past, has a right to view even unpublished field records and reports about a site.

See also Archeological mapping; Artifacts and artifact classification.

Further Reading

Books

Fagan, Brian M., ed. *The Oxford Companion to Archeology.* New York: Oxford University Press, 1996.

Maloney, Norah. *The Young Oxford Book of Archeology.* New York: Oxford University Press, 1997.

Nash, Stephen Edward, ed. *It's about Time: A History of Archaeological Dating in North America.* Salt Lake City, UT: University of Utah Press, 2000.

Sullivan, George. *Discover Archeology: An Introduction to the Tools and Techniques of Archeological Fieldwork.* Garden City, NY: Doubleday & Co., 1980.

Randall Frost

Exclusion principle, Pauli

The Pauli exclusion principle states that no two electrons in the same atom can have the same set of quantum numbers. The principle was first stated by the great Austrian-Swiss physicist Wolfgang Pauli (1900-1958) in 1925.

Historical background

The 1920s were a decade of enormous upheaval in atomic **physics**. Niels Bohr's model of the atom, proposed in 1913, had been a historic breakthrough in scientists' attempts to understand the nature of **matter**. But even from the outset, it was obvious that the **Bohr model** was inadequate to explain the fine points of atomic structure.

In some ways, the most important contribution made by Bohr was his suggestion that electrons can appear in only specific locations outside the atomic nucleus. These locations were designated as shells and were assigned **integral** "quantum" numbers beginning with 1. Electrons in the first shell were assigned the **quantum number** 1, those in the second **orbit**, the quantum number 2, those in the third shell, quantum number 3, and so on.

Eventually it became evident to scientists that additional quantum numbers would be needed to fully describe an **electron** in its orbit around the atomic nucleus. For example, the German physicist Arnold Sommerfeld (1868-1951) announced in 1915 that electrons traveled not in circles, but in ellipses around an atomic nucleus. The eccentricity of the elliptical orbit could be expressed, Sommerfeld said, by a second quantum number. By the mid-1920s, two additional quantum numbers, one defining the magnetic characteristics of an electron and one defining its spin, had been adopted.

The exclusion principle

In the early 1920s, Pauli reached an important insight about the nature of an electron's quantum numbers. Suppose, Pauli said, that an atom contains eight electrons. Then it should be impossible, he predicted, for any two of those electrons to have exactly the same set of quantum numbers.

As an example, consider an electron in the first orbit. All first-orbit electrons have a primary quantum number of 1. Then, mathematical rules determine the quantum numbers that are possible for any given primary quantum number. For example, Sommerfeld's secondary quantum number can be any integer (whole number) from 0 to one less than the primary quantum number, or, $l = 0 \rightarrow n - 1$. For an electron in the first shell (n = 1), l can only be 0. The third quantum number can have values that range from +l to -l. In this example, the third quantum number must also be 0. Finally, the fourth quantum number represents the spin of the electron on its own axis and can have values only of +1/2 or -1/2.

What Pauli's exclusion principle says about this situation is that there can be no more than two electrons in the first shell. One has quantum numbers of 1, 0, 0, +1/2 and the other, of 1, 0, 0, -1/2.

More variety is available for electrons in the second shell. Electrons in this shell have quantum number 2 (for second shell). Mathematically, then, the secondary quantum number can be either 1 or 0, providing more options for the value of the magnetic quantum number (+1, 0, or -1). If one writes out all possible sets of quantum numbers of the second shell, eight combinations are obtained. They are as follows:

2, 1, +1, +1/2

2, 1, 0, +1/2

2, 1, -1, +1/2

2, 1, +1, -1/2

2, 1, 0, - 1/2

2, 1, -1, -1/2

2, 0, 0, +1/2

2, 0, 0, -1/2

Electronic configurations

The Pauli exclusion principle is more than an intellectual game by which quantum number sets can be worked out for individual electrons. Beyond that, the principle allows one to determine the electronic configuration—the way the electrons are arranged—within any given atom. It shows that for an electron with 15 electrons, for example, two and only two can occupy the first shell, eight more (and only eight more) can occupy the second, leaving five electrons in a third shell.

The exclusion principle also demonstrates that, as with an onion, there are layers within layers. In the third shell of an atom, for example, there are three sub-divisions, known as *orbitals*. One orbital has the secondary quantum number of 0, one has the secondary quantum number of 1, and one, the secondary quantum number 2.

Rationalizing the periodic law

For more than half a century, chemists had known that the chemical elements display a regular pattern of properties, a discovery originally announced as the periodic law by the Russian chemist Dmitri Mendeleev (1834-1907) in about 1869. The Pauli exclusion principle provided important theoretical support for the periodic law. When a chart is made showing the electronic configuration of all the elements, an interesting pattern results. The elements have one, two, three, four (and so on) electrons in their outermost orbital in a regular and repeating pattern. All elements with one electron in their outermost orbital, for example, occur in column one of Mendeleev's **periodic table**. They have similar chemical and physical properties, it turns out, because they have similar electronic configurations.

Resources

Books

Miller, Franklin, Jr. *College Physics*. 5th ed. New York: Harcourt Brace Jovanovich, 1982.

David E. Newton

Excretory system

The excretory system removes cellular wastes and helps maintain the salt-water balance in an **organism**. In providing these functions, excretion contributes to the body's **homeostasis**, the maintenance of constancy of the internal environment. When cells break down **proteins**, they produce nitrogenous wastes, such as **urea**. The excretory system serves to remove these nitrogenous waste products, as well as excess salts and **water**, from the body. When cells break down carbohydrates during cellular respiration, they produce water and **carbon dioxide** as a waste product. The **respiratory system** gets rid of **carbon** dioxide every time we exhale. The **digestive system** removes feces, the solid undigested wastes of digestion, by a process called elimination or defecation. Organisms living in fresh and **salt** water have to adjust to the salt **concentration** of their aqueous environment, and terrestrial animals face the danger of drying out. It is the job of the excretory system to balance salt and water levels in addition to removing wastes. Different organisms have evolved a number of organs for waste removal. Protozoans, such as *Paramecium*, have specialized excretory structures; **flatworms**, which lack a **circulatory system**, have a simple excretory **organ**; earthworms, **grasshoppers**, and humans, have evolved an excretory system that works with the circulatory system.

Nitrogenous wastes

Nitrogenous waste products have their origin in the breakdown of proteins by cells. Cells catabolize amino acids to obtain **energy**. The first step of this process is deamination. During deamination, enzymes remove the amino group as **ammonia** (NH_3). Ammonia is toxic, even at low concentrations, and requires large amounts of water to flush it out of the body. Many animals, including humans, create a less poisonous substance, urea, by combining ammonia with carbon dioxide. An **animal** can retain urea for some time before excreting it, but it requires water to remove it from the body as urine. **Birds**, **insects**, land **snails**, and most **reptiles** convert ammonia into an insoluble substance, uric acid. This way, water is not required water to remove urea from the body. This method of ammonia excretion is particularly advantageous for animals, since they all lay eggs. If the embryo excreted ammonia inside the egg, it would rapidly poison its environment. Even urea would be dangerous. However, uric acid in solid form is a safe way to store nitrogenous wastes in the egg.

Excretion by organisms living in water

Some one-celled and simple multicellular aquatic organisms have no excretory organs; nitrogenous wastes simply diffuse across the **cell membrane** into the aqueous environment. In others, *Paramecium* for example, a specialized organelle, the contractile vacuole, aids in excretion by expelling excess water and some nitrogenous waste. In fresh water, the inside of cells has a higher concentration of salt than the surrounding water, and water constantly enters the cell by **osmosis**. Radiating canals in *Paramecium* collect excess water from the cell and deposit it in the contractile vacuole, which squeezes it out through a pore on the surface. This process requires energy supplied by ATP produced in the cell's mitochondria.

Saltwater-dwelling animals must survive in water that is has a higher salt concentration than in cells and body fluids. These animals run the risk of losing too much body water by osmosis or taking in too much salt by **diffusion**. Several adaptations protect them. The skin and scales of marine **fish** are relatively impermeable to salt water. In addition, the salts in the water they continually drink are excreted by special cells in their gills. In fact, marine fish excrete most of their nitrogenous waste as ammonia through the gills and only a little as urea, which conserves water. **Sharks** and other **cartilaginous fish**, on the other hand, store large amounts of urea in their **blood**. As a result, the concentration of their blood is slightly greater than the surrounding sea water, and water does not enter by osmosis. Special cells in the rectal gland of these fish excrete whatever excess salt does enter the system.

Excretion by land animals

As in aquatic animals, the excretory system in land animals removes nitrogenous waste and helps establish a balance between salt and water in the body. Terrestrial animals, however, also run the risk of drying out by **evaporation** from the body surface and the lungs. The elimination of feces and the excretion of urine also bring about water loss. Drinking, foods containing large amounts of water, and producing water during cellular respiration help overcome the loss. Animals that produce uric acid need less water than those excreting urine. Flame cells in flatworms, the nephridia in **segmented worms**, Malpighian tubules in insects, and kidneys in **vertebrates** are all examples of excretory systems.

Planarians and other flatworms have an excretory system that consists of two or more longitudinal branching tubules that run the length of the body. The tubules open to the outside of the animal through holes or pores in the surface. The tubules end in flame cells, bulb-shaped cells that contain cilia. The cilia create currents that move water and wastes through the canals and out the pores. Flatworms lack a circulatory system, so their flame cells excretory system picks up wastes directly from the body tissues.

The cells of segmented worms, such as earthworms, produce urea that is excreted through long tubules called nephridia, that work in conjunction with the earthworm's circulatory. Almost every segment of the earthworm's body contains a pair of nephridia. A nephridium consists of a ciliated funnel, a coiled tubule, an enlarged bladder, and a pore. The ciliated funnel collects wastes from the **tissue** fluid. The wastes travel from the funnel through the coiled tubule. Additional wastes from blood in the earthworm's **capillaries** enter the coiled tubule through its walls. Some of the water in the tubule is reabsorbed into the earthworm's blood. A bladder at the end of the nephridium stores the collected wastes. Finally the bladder expels the nitrogenous wastes through the pore to the outside.

Malpighian tubules are excretory organs that operate in association with the open circulatory system of grasshoppers and other insects. They consist of outpocketings of the digestive system where the midgut attaches to the hindgut. Blood in the open sinuses of the grasshoppers' body surrounds the Malpighian tubules. The ends of the tubules absorb fluid from the blood. As the fluid moves through the tubules, uric acid is precipitated. A lot of the water and other salts are reabsorbed into the grasshopper's blood. The remaining fluid plus uric acid passes out of the Malpighian tubule and enters the gut. Water is reabsorbed from the digestive tract. Finally, the uric acid is eliminated from the rectum as a dry mass.

The vertebrate excretory system works with circulatory system to remove wastes and water from blood, and convert them to urine. The urine is stored in a urinary bladder before it is expelled from the body. Kidneys are the main organs of excretion in vertebrates. Within the kidneys, working units called nephrons take in the liquid portion of the blood, filter out impurities, and return necessary substances to the blood stream. The remaining waste-containing portion is converted to urine and expelled from the body.

Excretion in humans

The main excretory system in humans is the urinary system. The skin also acts as an organ of excretion by removing water and small amounts of urea and salts (as sweat). The urinary system includes a pair of bean-shaped kidneys located in the back of the abdominal cavity. Each day, the kidneys filter about 162 qt (180 L) of blood, enough to fill a bathtub. They remove urea, toxins, medications, and excess ions and form urine. The kidneys also balance water and salts as well as **acids and bases**. At the same time, they return needed substances to the blood. Of the total liquid processed, about 1.3 qt (1.5 L) leaves the body as urine.

The size of an adult kidney is approximately 4 in (10 cm) long and 2 in (5 cm) wide. Urine leaves the kidneys in tubes at the hilus, a notch that occurs at the center of the concave edge. Blood vessels, lymph vessels, and nerves enter and leave the kidneys at the hilus. If we cut into a kidney, we see that the hilus leads into a **space** known as the renal sinus. We also observe two distinct kidney layers. There is the renal cortex, an outer reddish layer, and the renal medulla, a reddish brown layer. Within the kidneys, nephrons clear the blood of wastes, create urine, and deliver urine to a tube called a ureter, which carries the urine to the bladder. The urinary bladder is a hollow muscular structure that is collapsed when empty and pear-shaped and distended when full. The urinary bladder then empties urine into the urethra, a duct leading to outside the body. A sphincter muscle controls the flow of urine between the urinary bladder and the urethra.

Each kidney contains over one million nephrons, each of which consists of a tuft of capillaries surrounded by a capsule on top of a curving tube. The tuft of capillaries is called a glomerulus. Its capsule is cup-shaped and is known as Bowman's capsule. The glomerulus and Bowman's capsule form the top of a tube, the renal tubule. Blood vessels surround the renal tubule, and urine forms in it. The renal tubules of many nephrons join in collecting tubules, which in turn merge into larger tubes and empty their urine into the ureters in the renal sinus. The ureters exit the kidney at the hilus.

The job of clearing the blood of wastes in the nephrons occurs in three stages. They are **filtration**, reabsorption, and tubular secretion:

1. The first stage in clearing the blood is filtration, the passage of a liquid through a filter to remove impurities. Filtration occurs in the glomeruli. Blood **pressure** forces **plasma**, the liquid portion of the blood, through the capillary walls in the glomerulus. The plasma contains water, glucose, amino acids, and urea. Blood cells and proteins are too large to pass through the wall, so they stay in the blood. The fluid, now called filtrate, collects in the capsule and enters the renal tubule.

2. During reabsorption, needed substances in the filtrate travel back into the bloodstream. Reabsorption occurs in the renal tubules. There, glucose and other **nutrients**, water, and essential ions materials pass out of the renal tubules and enter the surrounding capillaries. Normally 100% of glucose is reabsorbed. (Glucose detected in the urine is a sign of **diabetes mellitus**, which is characterized by too much sugar in the blood due to a lack of insulin.) Reabsorption involves both diffusion and active transport, which uses energy in the form of ATP. The waste-containing fluid that remains after reabsorption is urine.

3. Tubular secretion is the passage of certain substances out of the capillaries directly into the renal tubules. Tubular secretion is another way of getting waste materials into the urine. For example, drugs such as penicillin and phenobarbital are secreted into the renal tubules from the capillaries. Urea and uric acid that may have been reabsorbed are secreted. Excess potassium ions are also secreted into the urine. Tubular secretions also maintain the **pH** of the blood.

The **volume** of the urine varies according to need. Antidiuretic hormone (ADH), released by the posterior pituitary gland, controls the volume of urine. The amount of ADH in the bloodstream varies inversely with the volume of urine produced. If we perspire a lot or fail to drink enough water, special nerve cells in the hypothalamus, called osmoreceptors, detect the low water concentration in the blood. They then signal neurosecretory cells in the hypothalamus to produce ADH, which is transmitted to the posterior pituitary gland and released into the blood, where it travels to the renal tubules. With ADH present, the kidney tubules reabsorb more water from the urine and return it to the blood, and the volume of urine is reduced. If we take in too much water, on the other hand, the osmoreceptors detect the overhydration and inhibit the production of ADH. Reabsorption of water is reduced, and the volume of urine is increased. **Alcohol** inhibits ADH production and therefore increases the output of urine.

The liver also plays an important role in excretion. This organ removes the ammonia and converts it into the less toxic urea. The liver also chemically changes and filters out certain drugs such as penicillin and erythromycin. These substances are then picked up by the blood and transported to the kidneys, where they are put into the execretory system.

The urinary system must function properly to ensure good health. During a physical examination, the physician frequently performs a urinalysis. Urine testing can reveal diseases such as diabetes mellitus, urinary tract infections, kidney stones, and renal **disease**. Urography, taking **x rays** of the urinary system, also helps diagnose urinary problems. In this procedure, an opaque dye is introduced into the urinary structures so that they show up in the x rays. Ultrasound scanning is another diagnostic tool. It uses high **frequency sound waves** to produce an image of the kidneys. Biopsies, samples of kidney tissue obtained in a hollow needle, are also useful in diagnosing kidney disease.

Disorders of the urinary tract include urinary tract infections (UTI). An example is cystitis, a disease in which **bacteria** infect the urinary bladder, causing **inflammation**. Most UTIs are treated with **antibiotics**. Sometimes kidney stones, solid salt crystals, form in the urinary tract. Kidney stones can obstruct the urinary passages and cause severe **pain**, and bleeding. If they do not pass out of the body naturally, the physician may use shock wave treatment. In this treatment, a shock wave focused on the stone from outside the body disintegrates it. Physicians also use **surgery** to remove kidney stones. Renal failure is a condition in which the kidneys lose the ability to function. Nitrogenous wastes build up in the blood, the pH drops, and urine production slows down. If left unchecked, this condition can result in death. In chronic renal failure, the urinary system declines, causing permanent loss of kidney function.

Hemodialysis and kidney transplant are two methods of helping chronic renal failure. In hemodialysis, an artificial kidney device cleans the blood of wastes and adjusts the composition of ions. During the procedure, blood is taken out of the radial artery in the patient's arm. It then passes through **dialysis** tubing, which is selectively permeable. The tubing is immersed in a **solution**. As the blood passes through the tubing, wastes pass out of the tubing and into the surrounding solution. The cleansed blood returns to the body. Kidney transplants also help chronic kidney failure. In this procedure, a surgeon replaces a diseased kidney with a closely matched donor kidney. Although about 23,000 people in the United States wait for donor kidneys each year, fewer than 8,000 receive kidney transplants. Current research aims to develop new drugs to help kidney failure better dialysis membranes for the artificial kidney.

See also Transplant, surgical.

Resources

Books

Guyton & Hall. *Textbook of Medical Physiology.* 10th ed. New York: W. B. Saunders Company, 2000.

Marieb, Elaine Nicpon. *Human Anatomy & Physiology.* 5th ed. San Francisco: Benjamin/Cummings, 2000.

Schrier, Robert W., ed. *Renal and Electrolyte Disorders.* Boston: Little, Brown, 1992.

Other

Osmoregulation. Video. Princeton, NJ: Films for the Humanities and Sciences, 1995.

The Urinary Tract, Water! Video and Videodisc. Princeton, NJ: Films for the Humanities and Sciences, 1995.

Bernice Essenfeld

Exercise

By definition, exercise is physical activity that is planned, structured, and repetitive for the purpose of **conditioning** any part of the body. Exercise is utilized to improve health, maintain fitness and is important as a means of physical **rehabilitation**.

Exercise is used in preventing or treating coronary **heart disease**, **osteoporosis**, weakness, diabetes, **obesity**, and **depression**. Range of **motion** is one aspect of exercise important for increasing or maintaining joint function. Strengthening exercises provide appropriate resistance to the muscles to increase endurance and strength. Cardiac rehabilitation exercises are developed and individualized to improve the cardiovascular system for prevention and rehabilitation of cardiac disorders and diseases. A well-balanced exercise program can improve general health, build endurance, and delay many of the effects of aging. The benefits of exercise not only extend into the areas of physical health, but also enhance emotional well-being.

Precautions

Before beginning any exercise program, evaluation by a physician is recommended to rule out any potential health risks. Once health and fitness are determined, and any or all physical restrictions identified, the individual's exercise program should be under the supervision of a health care professional. This is especially the case when exercise is used as a form of rehabilitation. If symptoms of dizziness, nausea, excessive shortness of breath, or chest **pain** are present during any exercise program, the individual should stop the activity and inform the physician about these symptoms before resuming activity. Exercise equipment must be checked to determine if it can bear the weight of people of all sizes and shapes.

Description

Range of motion exercise

Range of motion exercise refers to activity whose goal is improving movement of a specific joint. This motion is influenced by several structures: configuration of bone surfaces within the joint, joint capsule, ligaments, and muscles and tendons acting on the joint. There are three types of range of motion exercises: passive, active, and active assists. Passive range of motion is movement applied to the joint solely by another person or persons or a passive motion machine. When passive range of motion is applied, the joint of the individual receiving exercise is completely relaxed while the outside **force** takes the body part, such as a leg or arm, throughout the available range. Injury, **surgery**, or immobilization of a joint may affect the normal joint range of motion. Active range of motion is movement of the joint provided entirely by the individual performing the exercise. In this case, there is no outside force aiding in the movement. Active assist range of motion is described as the joint receiving partial assistance from an outside force. This range of motion may result from the majority of motion applied by the exerciser or by the person or persons assisting the individual. It may also be a half-and-half effort on the joint from each source.

Strengthening exercise

Strengthening exercise increases muscle strength and **mass**, bone strength, and the body's **metabolism**. It can help attain and maintain proper weight and improve body image and self-esteem. A certain level of muscle strength is needed to do daily activities, such as walking, running and climbing stairs. Strengthening exercises increase this muscle strength by putting more strain on a muscle than it is normally accustomed to receiving. This increased load stimulates the growth of **proteins** inside each muscle **cell** that allow the muscle as a whole to contract. There is evidence indicating that strength training may be better than **aerobic** exercise alone for improving self-esteem and body image. Weight training allows one immediate feedback, through observation of progress in muscle growth and improved muscle tone. Strengthening exercise can take the form of isometric, isotonic and isokinetic strengthening.

Isometric exercise

During isometric exercises muscles contract, however there is no motion in the affected joints. The muscle fibers maintain a constant length throughout the entire contraction. The exercises are usually performed against an immovable surface or object such as pressing the hand against the wall. The muscles of the arm are contracting but the wall is not reacting or moving as a result of the physical effort. Isometric training is effective for developing total strength of a particular muscle or group of muscles. It is often used for rehabilitation since the exact area of muscle weakness can be isolated and strengthening can be administered at the proper joint

angle. This kind of training can provide a relatively quick and convenient method for overloading and strengthening muscles without any special equipment and with little chance of injury.

Isotonic exercise

Isotonic exercise differs from isometric exercise in that there is movement of the joint during the muscle contraction. A classic example of an isotonic exercise is weight training with dumbbells and barbells. As the weight is lifted throughout the range of motion, the muscle shortens and lengthens. Calisthenics are also an example of isotonic exercise. These would include chin-ups, push-ups, and sit-ups, all of which use body weight as the resistance force.

Isokinetic exercise

Isokinetic exercise utilizes machines that control the speed of contraction within the range of motion. Isokinetic exercise attempts to combine the best features of both isometrics and weight training. It provides muscular overload at a constant preset speed while the muscle mobilizes its force through the full range of motion. For example, an isokinetic stationary bicycle set at 90 revolutions per minute means that despite how hard and fast the exerciser works, the isokinetic properties of the bicycle will allow the exerciser to pedal only as fast as 90 revolutions per minute. Machines known as Cybex and Biodex provide isokinetic results; they are generally used by physical therapists and are not readily available to the general population.

Cardiac rehabilitation

Exercise can be very helpful in prevention and rehabilitation of cardiac disorders and disease. With an individually designed exercise program, set at a level considered safe for that individual, heart failure patients can improve their fitness levels substantially. The greatest benefit occurs as the muscles improve the efficiency of their **oxygen** use, which reduces the need for the heart to pump as much **blood**. While such exercise doesn't appear to improve the condition of the heart itself, the increased fitness level reduces the total workload of the heart. The related increase in endurance should also translate into a generally more active lifestyle. Endurance or aerobic routines, such as running, brisk walking, cycling, or swimming, increase the strength and efficiency of the muscles of the heart.

Preparation

A physical examination by a physician is important to determine if strenuous exercise is appropriate or detri-

KEY TERMS

. . . .

Aerobic—Exercise training that is geared to provide a sufficient cardiovascular overload to stimulate increases in cardiac output.

Calisthenics—Exercise involving free movement without the aid of equipment.

Endurance—The time limit of a person's ability to maintain either a specific force or power involving muscular contractions.

Osteoporosis—A disorder characterized by loss of calcium in the bone, leading to thinning of the bones. It occurs most frequently in postmenopausal women.

mental for the individual. Prior to the exercise program, proper stretching is important to prevent the possibility of soft **tissue** injury resulting from tight muscles, tendons, ligaments, and other joint related structures.

Aftercare

Proper cool down after exercise is important in reducing the occurrence of painful muscle spasms. It has been documented that proper cool down may also decrease frequency and intensity of muscle stiffness the day following any exercise program.

Risks

Improper warm up can lead to muscle strains. Overexertion with not enough **time** between exercise sessions to recuperate can also lead to muscle strains, resulting in inactivity due to pain. Stress fractures are also a possibility if activities are strenuous over long periods of time without proper rest. Although exercise is safe for the majority of children and adults, there is still a need for further studies to identify potential risks.

Significant health benefits are obtained by including a moderate amount of physical exercise in the form of an exercise prescription. This is much like a drug prescription in that it also helps enhance the health of those who take it in the proper dosage. Physical activity plays a positive role in preventing disease and improving overall health status. People of all ages, both male and female, benefit from regular physical activity. Regular exercise also provides significant psychological benefits and improves quality of life.

There is a possibility of exercise burnout if the exercise program is not varied and adequate rest periods are

not taken between exercise sessions. Muscle, joint, and cardiac disorders have been noted with people who exercise, however, they often have had preexisting and underlying illnesses.

Resources

Books

McArdie, William D., Frank I. Katch, and Victor L. Katch. *Exercise Physiology: Energy, Nutrition, and Human Performance.* Philadelphia: Lea and Febiger, 1991.
Torg, Joseph S., Joseph J. Vegso, and Elisabeth Torg. *Rehabiliation of Athletic Injuries: An Atlas of Therapeutic Exercise.* Chicago: Year Book Medical Publishers, Inc., 1987.

Periodicals

Colan, Bernard J. "Exercise for Those Who Are Overweight: Just the Start in Fitness Plan." *Advance For Physical Therapy* 8, no. 25 (June 1997).

Jeffrey Peter Larson

Exocrine glands

Glands in the human body are classified as exocrine or endocrine. The secretions of exocrine glands are released through ducts onto an organ's surface, while those of endocrine glands are released directly into the **blood**. The secretions of both types of glands are carefully regulated by the body.

Exocrine gland secretions include saliva, perspiration, oil, earwax, milk, mucus, and digestive enzymes. The pancreas is both an exocrine gland and endocrine gland; it produces digestive enzymes that are released into the intestine via the pancreatic duct, and it produces **hormones**, such as **insulin** and glucagon, which are released from the islets of Langerhans directly into the bloodstream.

Exocrine glands are made up of glandular epithelial **tissue** arranged in single or multilayered sheets. Exocrine gland tissue does not have blood vessels running through it; the cells are nourished by vessels in the **connective tissue** to which the glands are attached. Gland cells communicate with each other and nerves via channels of communication which run through the tissue.

Structural classification

Exocrine glands have two structural classifications, unicellular (one **cell** layer) and multicellular (many cell layers). Goblet cells are unicellular exocrine glands; so named for their shape, these glands secrete mucus and are found in the epithelial lining of the respiratory, urinary, digestive, and reproductive systems. Multicellular exocrine glands are classified by their shape of secretory parts and by the arrangement of their ducts. A gland with one duct is a "simple," whereas a gland with a branched duct is a "compound" gland. The secretory portions of simple glands can be straight tubular, coiled tubular, acinar, or alveolar (flask-like). The secretory portions of compound glands can be tubular, acinar, or a combination: tubulo-acinar.

Functional classification

Exocrine glands can also be classified according to how they secrete their products. There are three categories of functional classification, holocrine glands, merocrine (or eccrine) glands, and apocrine glands. Holocrine glands accumulate their secretions in each cell's cytoplasm and release the whole cell into the duct. This destroys the cell, which is replaced by a new growth cell. Most exocrine glands are merocrine (or eccrine) glands. Here, the gland cells produce their secretions and release it into the duct, causing no damage to the cell. The secretions of apocrine cells accumulate in one part of the cell, called the apical region. This part breaks off from the rest of the cell along with some cytoplasm, releasing its product into the duct. The cells repair themselves quickly and soon repeat the process. An example of apocrine exocrine glands are the apocrine glands in the mammary glands and the arm pits and groin.

Exocrine glands perform a variety of bodily functions. They regulate body **temperature** by producing sweat; nurture young by producing milk; clean, moisten, and lubricate the **eye** by producing tears; and begin digestion and lubricate the mouth by producing saliva. Oil (sebum) from sebaceous glands keeps skin and hair conditioned and protected. Wax (cerumen) from ceruminous glands in the outer **ear** protects ears from foreign **matter**. Exocrine glands in the testes produce seminal fluid, which transports and nourishes sperm. Exocrine gland secretions also aid in the defense against bacterial **infection** by carrying special enzymes, forming protective films, or by washing away microbes.

Humans are not the only living beings that have exocrine glands. Exocrine glands in **plant** life produce **water**, sticky protective fluids, and nectars. The substances necessary for making birds' eggs, caterpillar cocoons, spiders' webs, and beeswax are all produced by exocrine glands. Silk is a product of the silkworm's salivary gland secretion.

See also Endocrine system.

Christine Miner Minderovic

Exothermic reaction *see* **Chemical reactions**
Expansion of universe *see* **Cosmology**

Explosives

Explosives are substances that produce violent chemical or nuclear reactions. These reactions generate large amounts of **heat** and gas in a fraction of a second. Shock waves produced by rapidly expanded gasses are responsible for much of the destruction seen following an explosion.

The power of most chemical explosives comes from the reaction of **oxygen** with other **atoms** such as **nitrogen** and **carbon**. This split-second chemical reaction results in a small amount of material being transformed into a large amount heat and rapidly expanding gas. The heat released in an explosion can incinerate nearby objects. The expanding gas can smash large objects like boulders and buildings to pieces. Chemical explosives can be set off, or detonated, by heat, **electricity**, physical shock, or another explosive.

The power of nuclear explosives comes from **energy** released when the nuclei of particular heavy atoms are split apart, or when the nuclei of certain **light** elements are forced together. These nuclear processes, called fission and fusion, release thousands or even millions of times more energy than chemical explosions. A single nuclear explosive can destroy an entire city and rapidly kill thousands of its inhabitants with lethal **radiation**, intense heat and blast effects.

Chemical explosives are used in peacetime and in wartime. In peacetime they are used to blast rock and stone for **mining** and quarrying, project rockets into **space**, and fireworks into the sky. In wartime, they project missiles carrying warheads toward enemy targets, propel bullets from guns, artillery shells from cannon, and provide the destructive **force** in warheads, mines, artillery shells, torpedoes, bombs, and hand grenades. So far, nuclear explosives have been used only in war.

History

The first chemical explosive was gunpowder, or black powder, a mixture of charcoal, **sulfur**, and **potassium nitrate** (or saltpeter). The Chinese invented it approximately 1,000 years ago. For hundreds of years, gunpowder was used mainly to create fireworks. Remarkably, the Chinese did not use gunpowder as a weapon of war until long after Europeans began using it to shoot stones and spear-like projectiles from tubes and, later, **metal** balls from cannon and guns.

Europeans probably learned about gunpowder from travelers from the Middle East. Clearly by the beginning in the thirteenth century gunpowder was used more often to make war than to make fireworks in the West. The English and the Germans manufactured gunpowder in the early 1300s. It remained the only explosive for 300 hundred years, until 1628, when another explosive called fulminating gold was discovered.

Gunpowder changed the lives of both civilians and soldiers in every Western country that experienced its use. (Eastern nations like China and Japan rejected the widespread use of gunpowder in warfare until the nineteenth century.) Armies and navies who learned to use it first—the rebellious Czech Taborites fighting the Germans in 1420 and the English Navy fighting the Spanish in 1587, for example—scored influential early victories. These victories quickly forced their opponents to learn to use gunpowder as effectively. This changed the way wars were fought, and won, and so changed the relationship between peoples and their rulers. Royalty could no longer hide behind stone walls in castles. Gunpowder blasted the walls away and helped, in part, to end the loyalty and servitude of peasants to local lords and masters. Countries with national armies became more important than local rulers as war became more deadly, due in large part to the use of gunpowder. It was not until the seventeenth century that Europeans began using explosives in peacetime to loosen **rocks** in mines and clear fields of boulders and trees.

Other chemical explosives have been discovered since the invention of gunpowder and fulminating gold. The most common of these are chemical compounds that contain nitrogen such as azides, nitrates, and other nitro-compounds.

In 1846 Italian chemist Ascanio Sobrero (1812-1888) invented the first modern explosive, nitroglycerin, by treating glycerin with nitric and sulfuric acids. Sobrero's discovery was, unfortunately for many early users, too unstable to be used safely. Nitroglycerin readily explodes if bumped or shocked. This inspired Swedish inventor Alfred Nobel (1833-1896) in 1862 to seek a safe way to package nitroglycerin. In the mid-1860s, he succeeded in mixing it with an inert absorbent material. His invention was called dynamite.

Dynamite replaced gunpowder as the most widely used explosive (aside from military uses of gunpowder). But Nobel continued experimenting with explosives and in 1875, invented a gelatinous dynamite, an explosive jelly. It was more powerful and even a little safer than the dynamite he had invented nine years earlier. The addition of ammonium nitrate to dynamite further decreased the chances of accidental explosions. It also made it cheaper to manufacture.

These and other inventions made Nobel very wealthy. Although the explosives he developed and manufactured were used for peaceful purposes, they also greatly increased the destructiveness of warfare. When he died, Nobel used the fortune he made from dynamite and other inventions to establish the Nobel prizes, which were originally awarded for significant accomplishment in the areas of medicine, **chemistry**, **physics**, and peace.

Continued research has produced many more types of chemical explosives than those known in Nobel's time: percholates, chlorates, ammonium nitrate-fuel oil mixtures (ANFO), and liquid oxygen explosives are examples.

Controlling explosives

Explosives are not only useless but dangerous unless the exact time and place they explode can be precisely controlled. Explosives would not have had the influence they have had on world history if two other devices had not been invented. The first device was invented in 1831 by William Bickford, an Englishman. He enclosed gunpowder in a tight fabric wrapping to create the first safety fuse. Lit at one end, the small amount of gunpowder in the core of the fuse burned slowly along the length of the cord that surrounded it. When the thin, burning core of gunpowder reached the end of the cord, it detonated whatever stockpile of explosive was attached. Only when the burning gunpowder in the fuse reached the stockpile did an explosion happen. This enabled users of explosives to set off explosions from a safe **distance** at a fairly predictable time.

In 1865, Nobel invented the blasting cap, a device that increased the ease and safety of handling nitroglycerin. Blasting caps, or detonators, send a shock wave into high explosives causing them to explode. It is itself a low explosive that is easily ignited. Detonators are ignited by primers. Primers burst into flame when heated by a burning fuse or electrical wire, or when mechanically shocked. A blasting cap may contain both a primer and a detonator,

A nuclear explosion at sea. *Photograph by PHOTRI. Stock Market. Reproduced by permission.*

or just a primer. Another technique for setting off explosives is to send an **electric charge** into them, a technique first used before 1900. All these control devices helped increase the use of explosives for peaceful purposes.

Newer explosives

In 1905, nine years after Nobel died, the military found a favorite explosive in TNT (trinitrotoluene). Like nitroglycerin, TNT is highly explosive but unlike nitroglycerin, it does not explode when it is bumped or shocked under normal conditions. It requires a detonator to explode. Many of the wars in this century were fought with TNT as the main explosive and with gunpowder as the main propellant of bullets and artillery shells. Explosives based on ammonium picrate and picric acid were also used by the military.

A completely different type of explosive, a nuclear explosive, was first tested on July 16, 1945, in New Mexico. Instead of generating an explosion from rapid **chemical reactions**, like nitroglycerin or TNT, the atomic bomb releases extraordinary amounts of energy when nuclei of plutonium or **uranium** are split apart in a process called **nuclear fission**. This new type of explo-

sive was so powerful that the first atomic bomb exploded with the force of 20,000 tons of TNT.

Beginning in the early 1950s, atomic bombs were used as detonators for the most powerful explosives of all, thermonuclear **hydrogen** bombs, or H-bombs. Instead of tapping the energy released when atoms are split apart, hydrogen bombs deliver the energy released when types of hydrogen atoms are forced together in a process called **nuclear fusion**. Hydrogen bombs have exploded with as much force as 15 million tons of TNT.

Types of explosives and their sources of power

All chemical explosives, whether solid, liquid, or gas, consist of a fuel, a substance that burns, and an oxidizer, a substance that provides oxygen for the fuel. The burning and the resulting release and expansion of gases during explosions can occur in a few thousandths or a few millionths of a second. The rapid expansion of gases produces a destructive shockwave. The greater the **pressure** of the shockwave, the more powerful the blast.

Fire or **combustion** results when a substance combines with oxygen gas. Many substances that are not explosive by themselves can explode if oxygen is nearby. Turpentine, gasoline, hydrogen, and **alcohol** are not explosives. In the presence of oxygen in the air, however, they can explode if ignited by a flame or spark. This is why drivers are asked to turn off their **automobile** engines, and not smoke, when filling fuel tanks with gasoline. In the automobile engine, the gasoline fuel is mixed with oxygen in the cylinders and ignited by spark plugs. The result is a controlled explosion. The force of the expanding gases drives the piston down and provides power to the wheels.

This type of explosion is not useful for most military and industrial purposes. The amount of oxygen in the air deep in a cannon barrel or a mine shaft may not be enough ensure a dependably powerful blast. For this reason, demolition experts prefer to use explosive chemicals that contain their own supply of concentrated oxygen to sustain the explosion. Potassium nitrate, for example, provides oxygen. Still, if the heat generated by a compound that breaks apart is great enough, the compound can still be an explosive even if it does not contain oxygen. Nitrogen iodide is one of the few examples.

Many chemical explosives contain nitrogen because it does not bind strongly to other atoms. It readily separates from them if heated or shocked. Nitrogen is usually introduced through the action of **nitric acid**, which is often mixed with **sulfuric acid**. Nitrogen is an important component of common chemical explosives like TNT, nitroglycerin, gunpowder, guncotton, nitrocellulose, picric acid, and ammonium nitrate.

Another type of explosion can happen when very fine powders or dust mixes with air in an enclosed space. Anytime a room or building is filled with dust of flammable substances such as **wood**, **coal**, or even flour, a spark can start a fire that will spread so fast through the dust cloud that an explosion will result. Dust explosions such as these have occurred in silos where grain is stored.

Four classifications of chemical explosives

There are four general categories of chemical explosives: blasting agents, primary, low, and high explosives. Blasting agents such as dynamite are relatively safe and inexpensive. Construction workers and miners use them to clear rock and other unwanted objects from work sites. Another blasting agent, a mixture of ammonium nitrate and fuel oil, ANFO, has been used by terrorists around the world because the components are readily available and unregulated. Ammonium nitrate, for instance, is found in **fertilizers**. One thousand pounds of it, packed into a truck or van, can devastate a large building.

Primary explosives are used in detonators, small explosive devices used to set off larger amounts of explo-

sives. Mercury fulminate and **lead** azide are used as primary explosives. They are very sensitive to heat and electricity.

Low, or deflagrating, explosives such as gunpowder do not produce as much pressure as high explosives but they do burn very rapidly. The burning starts at one end of the explosive and burns all the way to the other end in just a few thousandths of a second. This is rapid enough, however, that when it takes place in a sealed cylinder like a rifle cartridge or an artillery shell, the gases released are still powerful enough to propel a bullet or cannon shell from its casing, though the barrel of the rifle or cannon toward a target hundreds or thousands of feet away. In fact this relatively slow burning explosive is preferred in guns and artillery because too rapid an explosion could blow up the weapon itself. The slower explosive has the effects of building up pressure to smoothly force the bullet or shell out of the weapon. Fireworks are also low explosives.

High, or detonating, explosives are much more powerful than primary explosives. When they are detonated, all parts of the explosive explode within a few millionths of a second. Some are also less likely than primary explosives to explode by accident. TNT, PETN (pentaerythritol tetranitrate), and nitroglycerin are all high explosives. They provide the explosive force delivered by hand grenades, bombs, and artillery shells. High explosives that are set off by heat are called primary explosives. High explosives that can only be set off by a detonator are called secondary explosives. When mixed with oil or wax, high explosives become like clay. These plastic explosives can be molded into various shapes to hide them or to direct explosions. In the 1970s and 1980s, plastic explosives became a favorite weapon of terrorists. Plastic explosive can even be pressed flat to fit into an ordinary mailing envelope for use as a "letter bomb."

Nuclear explosives

The power of chemical explosives comes from the rapid release of heat and the formation of gases when atoms in the chemicals break their bonds to other atoms. The power of nuclear explosives comes not from breaking chemical bonds but from the core of the atom itself. When unstable nuclei of heavy elements, such are uranium or plutonium, are split apart, or when the nuclei of light elements, such as the isotopes of hydrogen **deuterium** or **tritium**, are forced together, in nuclear explosives they release tremendous amounts of uncontrolled energy. These nuclear reactions are called fission and fusion. Fission creates the explosive power of the atomic bomb. Fusion creates the power of the thermonuclear or hydrogen bomb. Like chemical explosives, **nuclear weapons** create heat and a shock wave generated by expanding gases. The power of nuclear explosive, however, is far greater than any chemical explosive. A ball of uranium-239 small enough to fit into your hand can explode with the force equal to 20,000 tons of TNT. The heat or thermal radiation released during the explosion travels with the speed of light and the shock wave destroys objects in its path with hurricane-like winds. Nuclear explosives are so much more powerful than chemical explosives that their force is measured in terms of thousands of tons (kilotons) of TNT. Unlike chemical explosives, nuclear explosives also generate **radioactive fallout**.

Current use and development of explosives

Explosives continue to have many important peacetime uses in fields like **engineering**, construction, mining, and quarrying. They propel rockets and space shuttles into **orbit**. Explosives are also used to bond different metals, like those in United States coins, together in a tight sandwich. Explosives carefully applied to carbon produce industrial diamonds for as cutting, grinding and polishing tools.

Today, dynamite is not used as often as it once was. Since 1955 different chemical explosives have been developed. A relatively new type of explosive, "slurry explosives," are liquid and can be poured into place. One popular explosive for industrial use is made from fertilizers like ammonium nitrate or **urea**, fuel oil, and nitric or sulfuric acid. This "ammonium nitrate-fuel oil" or ANFO explosive has replaced dynamite as the explosive of choice for many peacetime uses. An ANFO explosion, although potentially powerful and even devastating, detonates more slowly than an explosion of nitroglycerin or TNT. This creates more of an explosive "push" than a high **velocity** TNT blast. ANFO ingredients are less expensive than other explosives and approximately 25% more powerful than TNT. By 1995, sale of ANFO components were not regulated as TNT and dynamite were. Unfortunately, terrorists also began using bombs made from fertilizer and fuel oil. Two hundred forty-one marines were killed when a truck loaded with such an ANFO mixture exploded in their barracks in Beirut Lebanon in 1983. Six people were killed and more than 1,000 injured by a similar bombing in the World Trade Center in New York in 1993. In 1995, terrorists used the same type of explosive to kill more than 167 people in Oklahoma City.

Other explosives in use today include PETN (pentaerythrite tetranitrate), Cyclonite or RDX, a component of plastic explosives, and Amatol, a mixture of TNT and ammonium nitrite.

Nuclear explosives have evolved too. They are more compact than they were in the mid-part of the century.

Exponent

KEY TERMS

Chemical explosive—A substance that violently and rapidly releases chemical energy creating heat and often a shock wave generated by release of gases.

Dynamite—A explosive made by impregnating an inert, absorbent substance with nitroglycerin or ammonium nitrate mixed with combustible substance, such as wood pulp, and an antacid.

Gunpowder—An explosive mixture of charcoal, potassium nitrate, and sulfur often used to propel bullets from guns and shells from cannons.

Nitroglycerine—An explosive liquid used to make dynamite. Also used as a medicine to dilate blood vessels.

Nuclear explosive—Device which get its explosive force from the release of nuclear energy.

TNT—Trinitrotoluene, a high explosive.

Today they fit into artillery shells and missiles launched from land vehicles. Weapons designers also have created "clean" bombs that generate little radioactive fallout and "dirty" bombs that generate more radioactive fallout than older versions. Explosions of "neutron" bombs have been designed to kill humans with **neutron** radiation but cause little damage to buildings compared to other nuclear explosives.

Resources

Books

Keegan, John. *A History of Warfare.* New York: Alfred A. Knopf, 1994.

Stephenson, Michael, and Roger Hearn. *The Nuclear Casebook.* London: Frederick Muller Limited, 1983.

Periodicals

Treaster, Joseph B. "The Tools of a Terrorist: Everywhere for Anyone." *The New York Times* (April 20, 1995): B8.

Dean Allen Haycock

Exponent

Exponents are numerals that indicate an operation on a number or **variable**. The interpretation of this operation is based upon exponents that symbolize **natural numbers** (also known as positive **integers**). Natural-number exponents are used to indicate that **multiplica-**

tion of a number or variable is to be repeated. For instance, $5 \times 5 \times 5$ is written in exponential notation as 5^3 (read as any of "5 cubed," "5 raised to the exponent 3," or "5 raised to the power 3," or just "5 to the third power"), and $x \times x \times x \times x$ is written x^4. The number that is to be multiplied repeatedly is called the base. The number of times that the base appears in the product is the number represented by the exponent. In the previous examples, 5 and x are the bases, and 3 and 4 are the exponents. The process of repeated multiplication is often referred to as raising a number to a power. Thus the entire expression 5^3 is the power.

Exponents have a number of useful properties:

1) $x^a \cdot x^b = x^{(a+b)}$
2) $x^a \div x^b = x^{(a-b)}$
3) $x^{-a} = 1/x^a$
4) $x^a \cdot y^a = (xy)^a$
5) $(x^a)^b + x^{(ab)}$
6) $x^{a/b} = (x^a)^{1/b} = {}^b\sqrt{x^a} = ({}^b\sqrt{x})^a$

Any of the properties of exponents are easily verified for natural-number exponents by expanding the exponential notation in the form of a product. For example, property number (1) is easily verified for the example $x^3 x^2$ as follows:

$$x^3 \cdot x^2 = (x \cdot x \cdot x) \cdot (x \cdot x) = (x \cdot x \cdot x \cdot x \cdot x) = x^5 = x^{(3+2)}$$

Property (5) is verified for the specific case $x^2 y^2$ in the same fashion:

$$x^2 \cdot y^2 = (x \cdot x) \cdot (y \cdot y) = (x \cdot y \cdot x \cdot y) = (x \cdot y) \cdot (x \cdot y) = (xy)^2$$

Exponents are not limited to the natural numbers. For example, property (3) shows that a base raised to a **negative** exponent is the same as the multiplicative inverse of (1 over) the base raised to the positive value of the same exponent. Thus $2^{-2} = 1/2^2 = 1/4$.

Property (6) shows how the operation of exponentiation is extended to the rational numbers. Note that unit-fraction exponents, such as 1/3 or 1/2, are simply roots; that is, 125 to the 1/3 power is the same as the cube root of 125, while 49 to 1/2 power is the same as the **square root** of 49.

By keeping properties (1) through (6) as central, the operation is extended to all real-number exponents and even to complex-number exponents. For a given base, the real-number exponents **map** into a continuous **curve**.

Extinction

Extinction is the death of all members of a **species** and thus, of the species itself. Extinction may occur as a result of environmental changes (natural or human-caused) or **competition** from other organisms. A species confronted by environmental change or competitors may (1) adapt behaviorally (e.g., shift to a different diet), (2) adapt by evolving new characteristics, or (3) die out. At the present time, human impact on the environment is the leading cause of extinction. **Habitat** destruction by development and resource extraction, hunting, **pollution**, and the introduction of alien species into environments formerly inaccessible to them are causing the greatest burst of extinctions in at least 65 million years.

Extinction as such is a normal event; the great majority of all **plant** and **animal** species that have ever lived are now extinct. Extinction has always occurred at a fluctuating background **rate**. However, the fossil record also reveals a number of exceptional mass-extinction events, each involving the simultaneous or near-simultaneous disappearance of thousands of species of animals, plants, and **microorganisms**: five major **mass** extinctions, and about 20 minor ones, have occurred over the past 540 million years. The extinctions that occurred some 225 million years ago, at the end of the Permian period (the most severe of all mass extinctions, wiping out, for example, over 90% of shallow-water marine species), and some 65 million years ago, at the end of the Cretaceous period, are particularly well known. Almost half of all known species disappeared in the Cretaceous extinction, including all remaining dinosaurs and many marine animals.

The asteroid-impact theory

The primary cause of the Cretaceous **mass extinction** was a mystery for decades, until geologists discovered a thin layer of rock that marks the boundary between the Cretaceous period and following Tertiary period; this layer of sediments is termed the K-T boundary, and gave rise to the asteroid-impact theory of the Cretaceous extinction.

The asteroid-impact theory was first proposed in detail in 1978, by a team led by American geologist Walter Alvarez (1940–) and physicist Luis Alvarez (1911–). The Alvarez team analyzed sediment collected in the 1970s from the K-T layer near the town of Gubbio, Italy. The samples showed a high **concentration** of the element iridium, a substance rare on **Earth** but relatively abundant in meteorites. Other samples of K-T boundary **strata** from around the world were also analyzed; excess iridium was found in these samples as well. Using the average thickness of the sediment as a guide, they calculated that a meteorite about 6 mi (10 km) in diameter would be required to spread that much iridium over the whole Earth.

If a meteorite that size had hit Earth, the dust lofted into the air would have produced an enormous cloud of dust that would have encircled the world and blocked out the sunlight for months, possibly years. This climactic change would have severely depressed **photosynthesis**, resulting in the death of many plants, subsequent deaths of herbivores, and finally the death of their predators as well. (This chain of events would have occurred so rapidly that there would have been no chance for evolutionary **adaptation** to the new environment, which requires thousands of years at minimum.) A major problem with the theory, however, was that a 6-mi (10-km) meteorite would leave a very large crater, 93–124 mi (150–200 km) in diameter—and while Earth has many impact craters on its surface, few are even close to this size, and none of the right age was known.

Because 65 million years had passed since the hypothetical impact, scientists shifted the search underground. A crater that old would almost certainly have been filled in by now. In 1992, an **impact crater** was discovered under the surface near the village of Chicxulub (pronounced CHIX-uh-loob) on Mexico's Yucatan Peninsula. When core samples raised by drilling were analyzed, they showed the crater to be about 112 mi (80 km) in diameter and 65 million years old—the smoking gun that validated the Alvarez asteroid-impact theory.

The asteroid impact theory is now widely accepted as the most probable explanation of the K-T iridium anomaly, but many geologists still debate whether the impact of this large meteorite was the sole cause of the mass extinction of the dinosaurs and other life forms at that time, as the fossil record seems to show an above-average rate of extinctions in the time leading up to the K-T boundary. A number of gradual causes can accelerate extinction: falling **ocean** levels, for example, expose continental shelves, shrinking shallow marine environments and causing drier continental interiors, both changes that encourage extinction. Further, very large volcanic eruptions may stress the global environment. The asteroid that caused the Chicxulub crater may have coincidentally amplified or punctuated an independent extinction process that had already begun. There is no reason why many different causes cannot have acted, independently or in concert, to produce extinction events.

The asteroid-impact theory has been applied to many mass extinctions since the discovery of Chicxulub. Most of the five major mass extinctions of the last 540 million years, and several of the smaller ones, have been shown to coincide in time with large impact craters or iridium spikes (layers of heightened iridium concentration) in the geological column.

The Great Ice Age

The Great Ice Age that occurred during the Pleistocene era (which began about 2 million years ago and ended 10,000 years ago) also caused the extinction of many plants and animal species. This period is of particular interest because it coincides with the **evolution** of the human species. This was a time of diverse animal life, and **mammals** were the dominant large surface forms (i.e., in contrast to the **reptiles** of previous periods; as always, **insects** and **bacteria** dominated the animal world in absolute numbers, numbers of species, and total **biomass**). In the late Pleistocene (50,000–10,000 years ago), several other extinctions occurred. These wiped out several species known collectively as the megafauna ("big animals"), including mammoths, mastodons, ground **sloths**, and giant **beavers**. The late Pleistocene extinctions of megafauna did not occur all at once, nor were they of equal magnitude throughout the world. However, the continents of **Africa**, **Australia**, **Asia**, and **North America** were all affected. Recent work has shown that these extinctions did not, as long thought, single out megafauna as such, but rather all vertebrate species with slow reproduction rates (e.g., one offspring per female per year or less)—megafauna merely happen to be members of this larger class. It has been speculated that human beings caused these late-Pleistocene extinctions, hunting slow-reproducing species to extinction over centuries. Paleontologists continue to debate the pros and cons of different versions of this "overkill" theory.

The causes of the extinction events of the late Pleistocene are still debated, as are those of the larger mass extinctions that have punctuated earlier geological history; most likely several factors were involved, of which the global spread of human beings seems to have been one. Indeed, the past 10,000 years have seen dramatic changes in the **biosphere** caused by human beings. The invention of agriculture and animal husbandry and the eventual spread of these practices throughout the world has allowed humans to utilize a large portion of the available resources of Earth—often in an essentially destructive and nonsustainable way. For example, in pursuit of lumber, farmland, and living room, human beings have reduced Earth's forest area from over one half of total land area less than one third.

The current mass extinction

Eventual extinction awaits all species of plants and animals, but the rate at which extinctions are taking place globally has been greatly accelerated by habitat destruction, pollution, **global climate** change, and overharvesting to 1,000–10,000 times the normal background rate, equaling or exceeding rates of extinction during great mass extinctions of Earth's geological history. The present mass extinction is unique in that it is being caused by a single species—ours—rather than by natural events; furthermore, biologists agree that the effects may be so profound as to threaten the human species itself.

Many countries, including the United States, have passing laws to protect species that are threatened or in danger of extinction, preserving some wild areas from exploitation and some **individual** species from hunting and harassment. Furthermore, several private groups, such as the Nature Conservancy, seek to purchase or obtain easements on large areas of land in order to protect the life they harbor. In the case of species whose numbers have been reduced to a very low level, the role of the zoological park (or "zoo") has changed from one of simple exhibition to a more activist, wildlife-conservation stand.

Despite **conservation** efforts, however, the outlook is decidedly bleak. According to a 1998 poll commissioned by New York Museum of Natural History, 70– of all biologists agree with the statement that one fifth of all living species are likely to become extinct within the next 30 years. According to a statement from the World Conservation Union issued in 2000, over 11,000 species of plants and animals are already close to extinction—and the pace is accelerating. Since only 1.75 million species out of an estimated 14 million have been documented by biologists, even these numbers may understate the magnitude of the developing disaster. Although the Earth has restocked its species diversity after previous mass extinctions, this process invariably takes many millions of years.

See also Biodiversity; Catastrophism; Evolutionary change, rate of; Evolutionary mechanisms; Punctuated equilibrium.

Resources

Books

Gould, Stephen Jay. *The Structure of Evolutionary Theory.* Cambridge, MA: Harvard University Press, 2002.

Periodicals

Cardillo, Marcel and Adrian Lister, "Death in the Slow Lane." *Nature.* (October 3, 2002): 440–441.

Other

American Museum of Natural History. "Scientific Experts Believe We Are in Midst of Fastest Mass Extinction in Earth's History." April 20, 1998 [cited Jan. 6, 2003]. <http://www.amnh.org/museum/press/feature/biofact.html>.

Associated Press. "11,000 Species Said to Face Extinction, With Pace Quickening." *New York Times*, Sep. 29, 2000 [cited Jan. 6, 2003]. <http://www.nytimes.com/2000/09/29/science/29EXTI.html>.

Larry Gilman

Extrasolar planets

Extrasolar planets are planets that **orbit** stars other than the **Sun**; a "planet" is defined as an object too small for gravitational **pressure** at its core to ignite the deuterium-fusion reaction that powers a **star**. (There is no generally-agreed-upon lower limit on the size of a planet.) The existence of extrasolar planets has been suspected since at least the time of Dutch astronomer Christian Huygens (1629–1695). The ancient Greek astronomer Aristarchus (late third century B.C.) may have developed the concept over 2,000 years ago, although this is not known certainly. However, extrasolar planets remained hypothetical until recently because there was no way to detect them. Extrasolar planets are difficult to observe directly because planets shine by reflected **light** and so are only about a billionth as bright as the stars they orbit. Their light is either too dim to see at all with present techniques, or is lost in their stars' glare. Since 1995, thanks to new, indirect observational techniques, over 100 extrasolar planets have been discovered, with masses ranging from that of **Jupiter** to the upper size limit for a **planet** (about 15 Jupiter masses).

The search for extrasolar planets

In 1943, Danish-born United States astronomer K. A. Strand (1907–2000) reported a suspected companion to one of the components of the double star 61 Cygni, based on a slight wobble of the two stars' orbital motions. This seems to be the earliest report of an extrasolar planet; recent data have confirmed the presence of a planet with about eight times the **mass** of Jupiter orbiting the brighter component of 61 Cygni. Strand's estimate of a period of 4.1 years for the planet's orbit has also been confirmed.

In 1963, Dutch-born United States astronomer Peter van de Kamp (1901–1995) reported the detection of a planet orbiting Barnard's Star (Gliese 699) with a 24-year period of revolution. Barnard's Star, which is 5.98 light years away, is the second-closest star system to ours after the Alpha Centauri triple-star system. Van de Kamp suggested that Barnard's Star's wobbling proper **motion** could be explained by two planets in orbits with periods of revolution around Gliese 699 of 12 years and 26 years, respectively.

Independent efforts to confirm Strand's planet (or planets) orbiting Barnard's Star failed, however. By the 1970s, most astronomers had concluded that, instead of discovering extrasolar planets, both Strand's and van de Kamp's studies had only detected a slight systematic change in the characteristics of the **telescope** at the observatory where they had made their observations.

Astronomers kept looking, but for many years the search for extrasolar planets was marked by much-ballyhooed advances followed by summary retreats. No sooner would a group of astronomers announce the discovery of a planet outside of the **solar system**, than an outside group would present evidence refuting the findings discovery. A first breakthrough came in 1991, when it was shown that three approximately Earth-size planets orbit the **pulsar** PSR1257.12. (Subtle timing shifts in the flashes of the pulsar revealed the existence of the two planets; this is the first and, so far, only time that this technique has detected extrasolar planets.) No more discoveries were made for several years; then, in 1995, the dam broke. Using the radial-velocity technique extrasolar planet detection (to be explained below), group after group of astronomers announced extrasolar planet findings, findings quickly confirmed by other independent researchers. Suddenly, extrasolar planets went from rare to commonplace. At present over 100 planets have been detected and confirmed by the astronomical community.

New detection techniques

The recent rush of discoveries has been made possible by new methods of search. Direct visual observation of extrasolar planets remains difficult; all the recent discoveries have been made, therefore, by indirect means, that is, by observing their effects on either the motions or brightness of the stars they orbit.

Apart from the been detected by analyzing the perturbations (disturbances) they cause in their star's motions. A planet does not simply orbit around its star; rather, a star and its planet both orbit around their common center of gravity. Because a star weighs more than a planet, it follows a tighter orbit, but if a planet (or other companion) is massive enough, the orbital motion of a star—its "wobble"—may be detectable from **Earth**. Several techniques have been and are being developed to detect the orbital wobbles caused by planet-size bodies.

All extrasolar planets so far detected (apart from the three orbiting PSR1257.12) have been detected by the radial-velocity technique. This uses **spectroscopy** (analysis of the electromagnetic spectra emitted by stars) to detect perturbations of stars orbited by planets. The mutual orbital motions of a star and a planet around each other manifest in the star's light via the **Doppler effect**, in which spectroscopic lines from a light source such as a star are shifted to longer wavelengths in the case of a source moving away from the observer (red shift), or to shorter wavelengths in the case of a source approaching an observer (blue shift). These shifts are measured relative to the wavelengths of **spectral lines** for a source at rest. Small changes in the wavelengths of spectroscopic

lines in the star's **spectrum** indicate changes in its line-of-sight (radial) **velocity** relative to the observer; periodic (regularly repeated) spectral shifts probably indicate the presence of a planet (or planets) perturbing the star's motion by swinging it toward us and then away from us. (This is true only if the planetary orbits happen to be oriented flat-on to us in space.)

Another new techniques for detecting extrasolar planets involves searching for transits. A transit is the passage of a planet directly between its star and the Earth, and can occur only when the planet's orbit happens to be oriented edge-on to us. When transit does occur, the star's apparent brightness dims for several minutes—perhaps by only a **percent** or so. The amount of dimming and the speed with which dimming occurs as the planet begins to move across the star's disk reveal the planet's diameter, which the "wobble" method cannot do. Furthermore, since a planet with an atmosphere does not block all the light from the star behind it but allows some of that light to filter through its atmosphere, precise measurements of changes in the star's spectrum during transit can supply information about the chemical composition of the transiting planet's atmosphere. In 2001, the **Hubble Space Telescope** detected **sodium** in the atmosphere of a transiting extrasolar planet approximately 150 light years away. This was the first information ever obtained about the composition of an extrasolar planet. This study was limited to the wavelengths in which sodium absorbs light, and were not expected to detect other chemicals; using different wavelengths, astronomers intend to search for potassium, **water** vapor, methane, and other substances in the atmosphere of this and other transiting extrasolar planets.

New discoveries

All of the extrasolar planets detected so far are gas giants with masses on the order of Jupiter's; however, this is probably an artifact of the methods being used, which are only capable of detecting large planets. The orbits observed vary wildly—some planets are closer to their stars than Mercury is to the Sun, with orbits lasting mere days, while others are separated from their stars by many times the Earth-Sun **distance**. Planets with nearly round, Earth-like orbits have been discovered as well as planets whose orbits more closely approximate the eccentric elliptical shape of cometary paths. The stars around which planets have been discovered include dying stars, twin stars, Sunlike stars, and pulsars, with locations ranging from a Barnard's Star to more than 150 light years.

In 1998, astronomers discovered a protoplanet (planet in the process of formation) apparently in the midst of being ejected from its star system. Infrared images from the Hubble Space Telescope showed a pinpoint object with a 130 billion-mile-long filamentary structure trailing behind it toward a pair of binary stars. Although some astronomers speculate that the object could be a **brown dwarf**, others believe that it is a planet flung into deep space by a gravitational "slingshot" effect from its parent stars. This suggests the possibility that rogue planets unattached to any star may also be roving the Universe.

In the spring of 1999, astronomers announced the discovery of a second multiple-planet solar system (not counting our own), detecting three planets circling the star Upsilon Andromedae, some 44 light years away. Though the objects detected are Jupiter-like gas giants, the data does not rule out Earth-type planets, which would not provide sufficient gravitation effect to be detected by the techniques used so far.

Between 2000 and 2003, more than a dozen additional extrasolar planets were discovered.

Although the radial-velocity technique has been responsible for all extrasolar-planet discoveries in recent years, this is expected to change as the transmit method is applied more thoroughly. The advantage of the transit method is that the light from many stars can be monitored for telltale brightness simultaneously; a certain fraction of solar systems are bound to be oriented edge-on to us, allowing for their detection by this means. In 2007, the *Kepler* spacecraft, a space telescope especially designed to scan large areas of the sky for transits by planets as small as Earth, will be launched by the U.S. National Aeronautics and Space Adminstration. By 2011, *Kepler* should have gathered enough data to pinpoint hundreds of extrasolar planets and to determine how typical our own solar system is in the Universe. This is of interest to scientists because estimates of the probability that life exists elsewhere in the Universe depend strongly on the existence of planets not too different from our own. Intelligent life is unlikely to evolve on large gas giants or on bodies of any type that orbit very near to their stars or follow highly eccentric, bake-and-freeze orbits. If solar systems like our own are rare in the Universe, then life (intelligent or otherwise) may be correspondingly rare. Theoretical models of the formation of solar systems have been in a state of rapid change under the pressure of the rush of extrasolar planet discoveries, and revised models indicate that solar systems like our own may be abundant. However, these models supply only educated guesses, and must be checked against observation.

See also Binary star.

Resources

Books

Sagan, Carl. *Cosmos*. New York: Ballantine Books, Inc., 1985.

Periodicals

Kerr, Richard. "Jupiters Like Our Own Await Planet Hunters." *Science*. (January 25, 2002): 605.
Lissauer, Jack J. "Extrasolar Planets." *Nature*. (September 26, 2002): 355–358.
Wilford, John Noble. "New Discoveries Complicate the Meaning of 'Planet'." *New York Times*. January 16, 2001.

Other

Space Telescope Science Institute, Goddard Space Flight Center. "Hubble Makes First Direct Measurements of Atmosphere on World Around Another Star." November 27, 2001 [cited Oct. 22, 2002]. <http://oposite.stsci.edu/pubinfo/PR/2001/38/pr.html>.

Frederick West
Larry Gilman

Eye

The eye is the **organ** of sight in humans and animals which transforms **light** waves into visual images and provides 80% of all information received by the human **brain**. These remarkable organs are almost spherical in shape and are housed in the orbital sockets in the skull. Sight begins when light waves enter the eye through the cornea (the transparent layer at the front of the eye) pass through the pupil (the opening in the center of the iris, the colored portion of the eye) then through a clear **lens** behind the iris. The lens focuses light onto the retina which functions like the film in a camera. Photoreceptor neurons in retinas, called rods and cones, convert light **energy** into electrical impulses, which are then carried to the brain via the optic nerves. At the visual cortex in the occipital lobe of the cerebrum of brain, the electrical impulses are interpreted as images.

Evolution of the eye

Many invertebrate animals have simple light-sensitive eye spots, consisting of a few receptor cells in a cup-

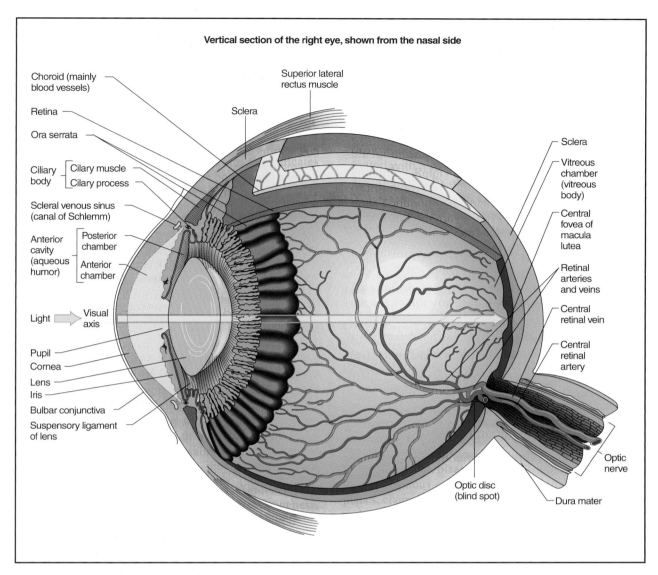

Vertical section of the right eye, shown from the nasal side

Choroid (mainly blood vessels)

Retina

Ora serrata

Ciliary body — Cilary muscle / Cilary process

Scleral venous sinus (canal of Schlemm)

Anterior cavity (aqueous humor) — Posterior chamber / Anterior chamber

Light ⟹ Visual axis

Pupil

Cornea

Lens

Iris

Bulbar conjunctiva

Suspensory ligament of lens

Sclera

Superior lateral rectus muscle

Sclera

Vitreous chamber (vitreous body)

Central fovea of macula lutea

Retinal arteries and veins

Central retinal vein

Central retinal artery

Optic nerve

Optic disc (blind spot)

Dura mater

A cutaway anatomy of the human eye. *Illustration by Hans & Cassidy. Courtesy of Gale Group.*

shaped organ lined with pigmented cells, which detect only changes in light and dark regimes. **Arthropods (insects**, spiders, and **crabs**) have complex compound eyes with thousands of cells which constrict composite pictures of objects. They are very sensitive to detecting movement.

Anatomy and function of the human eye

The human eyeball is about 0.9 in (24 mm) in diameter and is not perfectly round, being slightly flattened in the front and back. The eye consists of three layers: the outer fibrous or sclera, the middle uveal or choroid layer, and the inner nervous layer or retina. Internally the eye is divided into two cavities—the anterior cavity filled with the watery aqueous fluid, and the posterior cavity filled with gel-like vitreous fluid. The internal **pressure** inside the eye (the intraocular pressure) exerted by the aqueous fluid supports the shape of the anterior cavity, while the vitreous fluid holds the shape of the posterior chamber. An irregularly shaped eyeball results in ineffective focusing of light onto the retina and is usually correctable with glasses or contact lenses. An abnormally high intraocular pressure, due to overproduction of aqueous fluid or to the reduction in its outflow through a duct called the **canal** of Schlemm, produces glaucoma, a usually painless and readily treatable condition, which may **lead** to irreversible blindness if left untreated. Elevated intraocular pressure is easily detectable with a simple, sight-saving, pressure test during routine eye examinations. The ophthalmic **arteries** provide the **blood supply** to the eyes, and the movement of the eyeballs is facilitated by six extraocular muscles which run from the bony orbit which insert the sclera, part of the fibrous tunic.

The outer fibrous layer encasing and protecting the eyeball consists of two parts—the cornea and the sclera. The front one-sixth of the fibrous layer is the transparent cornea, which bends incoming light onto the lens inside the eye. A fine mucus **membrane**, the conjunctiva, covers the cornea, and also lines the eyelid. Blinking lubricates the cornea with tears, providing the moisture necessary for its health. The cornea's outside surface is protected by a thin film of tears produced in the lacrimal **glands** located in the lateral part of orbit below the eyebrow. Tears flow through ducts from this gland to the eyelid and eyeball, and drain from the inner corner of the eye into the nasal cavity. A clear watery liquid, the aqueous humor, separates the cornea from the iris and lens. The cornea contains no blood-vessels or pigment and gets its **nutrients** from the aqueous humor. The remaining five-sixths of the fibrous layer of the eye is the sclera, a dense, tough, opaque coat visible as the white of the eye. Its outer layer contains **blood** vessels which produce a "blood-shot eye" when the eye is irritated. The middle or uveal layers of the eye is densely pigmented, well supplied with blood, and includes three major structures—the iris, the ciliary body, and the choroid. The iris is a circular, adjustable diaphragm with a central hole (the pupil), sited in the anterior chamber behind the cornea. The iris gives the eye its **color**, which varies depending on the amount of pigment present. If the pigment is dense, the iris is brown, if there is little pigment the iris is blue, if there is no pigment the iris is pink, as in the eye of a white rabbit. In bright light, muscles in the iris constrict the pupil, reducing the amount of light entering the eye. Conversely, the pupil dilates (enlarges) in dim light, so increasing the amount of incoming light. Extreme fear, head injuries, and certain drugs can also dilate the pupil.

The iris is the anterior extension of the ciliary body, a large, smooth muscle which also connects to the lens via suspensory ligaments. The muscles of the ciliary body continually expand and contract, putting on suspensory ligaments changing the shape of the lens, thereby adjusting the focus of light onto the retina facilitating clear **vision**. The choroid is a thin membrane lying beneath the sclera, and is connected the posterior section of the ciliary body. It is the largest portion of the uveal tract. Along with the sclera the choroid provides a light-tight environment for the inside of the eye, preventing stray light from confusing visual images on the retina. The choroid has a good blood supply and provides **oxygen** and nutrients to the retina.

The front of the eye houses the anterior cavity which is subdivided by the iris into the anterior and posterior chambers. The anterior chamber is the bowl-shaped cavity immediately behind the cornea and in front of the iris which contains aqueous humor. This is a clear watery fluid which facilitates good vision by helping maintain eye shape, regulating the intraocular pressure, providing support for the internal structures, supplying nutrients to the lens and cornea, and disposing of the eye's metabolic waste. The posterior chamber of the anterior cavity lies behind the iris and in front of the lens. The aqueous humor forms in this chamber and flows forward to the anterior chamber through the pupil.

The posterior cavity is lined entirely by the retina, occupies 60% of the human eye, and is filled with a clear gel-like substance called vitreous humor. Light passing through the lens on its way to the retina passes through the vitreous humor. The vitreous humor consists of 99% **water**, contains no cells, and helps to maintain the shape of the eye and support its internal components.

The lens is a crystal-clear, transparent body which is biconvex (curving outward on both surfaces), semi-solid, and flexible, shaped like an **ellipse** or elongated **sphere**. The entire surface of the lens is smooth and shiny, contains no blood vessels, and is encased in an elastic membrane. The lens is sited in the posterior chamber behind the iris and in front of the vitreous humor. The lens is held in place by suspensory ligaments that run from the ciliary muscles to the external circumference of the lens. The continual relaxation and contraction of the ciliary muscles cause the lens to either fatten or became thin, changing its focal length, and allowing it to focus light on the retina. With age, the lens hardens and becomes less flexible, resulting in far-sighted vision that necessitates glasses, bifocals, or contact lenses to restore clear, close-up vision. Clouding of the lens also often occurs with age, creating a cataract that interferes with vision. Clear vision is restored by a relatively simple surgical procedure in which the entire lens is removed and an artificial lens implanted.

Retina

The retina is the innermost layer of the eye. The retina is thin, delicate, extremely complex sensory **tissue** composed of layers of light sensitive nerve cells. The retina begins at the ciliary body and encircles the entire posterior portion of the eye. Photoreceptor cells in the rods and cones, convert light first to chemical energy and then electrical energy. Rods function in dim light, allowing limited nocturnal (night) vision: it is with rods that we see the stars. Rods cannot detect color, but they are the first receptors to detect movement. There are about 126 million rods in each eye and about six million cones. Cones provide acute vision, function best in bright light,

KEY TERMS

. . . .

Aqueous humor—Clear liquid inside the anterior and posterior chambers.

Choroid—Light-impermeable lining behind the sclera.

Cornea—The outer, transparent lens that covers the pupil of the eye and admits light.

Crystalline lens—Focusing mechanism located behind the iris and pupil.

Fovea—Tiny hollow in the retina and area of acute vision.

Iris—Colored portion of the eye.

Ophthalmology—Branch of medicine dealing with the eye.

Pupil—Adjustable opening in the center of the iris.

Retina—An extremely light-sensitive layer of cells at the back part of the eyeball. The image formed by the lens on the retina is carried to the brain by the optic nerve.

Sclera—White of the eye.

Vitreous body—Opaque, gel-like substance inside the vitreous cavity.

Optic nerve

The optic nerve connects the eye to the brain. These fibers of the optic nerve run from the surface of the retina and converge at exit at the optic disc (or blind spot), an area about 0.06 in (1.5 mm) in diameter located at the lower posterior portion of the retina. The fibers of this nerve carry electrical impulses from the retina to the visual cortex in the occipital lobe of the cerebrum. If the optic nerve is severed, vision is lost permanently.

The last two decades have seen an explosion in ophthalmic research. Today, 90% of corneal blindness can be rectified with a corneal transplant, the most frequently performed of all human transplants. Eye banks receive eyes for sight-restoring corneal transplantation just as blood banks receive blood for life-giving transfusions. Many people remain blind, however, because of the lack of eye donors.

See also Vision disorders.

Resources

Books

Davison, Hugh, ed. *The Eye.* San Diego: Academic Press, 1984.

Introduction to Ophthalmology. San Francisco: American Academy of Ophthalmology, 1980.

Moller, Aage R. *Sensory Systems: Anatomy and Physiology.* New York: Academic Press, 2002.

Periodicals

Koretz, Jane F., and George H. Handelman. "How the Human Eye Focuses." *Scientific American* (July 1988): 92-99.

Malhotra, R. "Nasolacrimal Duct Obstruction Following Chicken Pox." *Eye* 16, no. 1 (2002): 88-89.

Marie L. Thompson

and allow color vision. Cones are most heavily concentrated in the central fovea, a tiny hollow in the posterior part of the retina and the point of most acute vision. Dense fields of both rods and cones are found in a circular belt surrounding the fovea, the macula lutea. Continuing outward from this belt, the cone **density** decreases and the **ratio** of rods to cones increases. Both rods and cones disappear completely at the edges of the retina.

Eye diseases *see* **Vision disorders**